STATICS AND DYNAMICS

Engineering Mechanics

STATICS
AND
DYNAMICS

ELEVENTH EDITION

R. C. Hibbeler

PEARSON

Prentice
Hall

Upper Saddle River, NJ 07458

Library of Congress Cataloging-in-Publication Data on File

Vice President and Editorial Director, ECS: *Marcia Horton*
Associate Editor: *Dee Bernhard*
Editorial Assistant: *Dolores Mars*
Executive Managing Editor: *Vince O'Brien*
Managing Editor: *David A. George*
Production Editor: *Rose Kernan*
Director of Creative Services: *Paul Belfanti*
Manager of Electronic Composition & Digital Content: *Allyson Graesser*
Assistant Manager of Electronic Composition & Digital Content: *Clara Bartunek*
Electronic Compositors: *Julita Nazario and Lissette Quiñones*
Creative Director: *Juan Lòpez*
Art Director: *Jonathan Boylan*
Interior and Cover Designer: *Suzanne Behnke*
Art Editor: *Xiaohong Zhu*
Manufacturing Manager: *Alexis Heydt-Long*
Manufacturing Buyer: *Lisa McDowell*
Marketing Manager: *Tim Galligan*

About the Cover: Background image: North America's Longest and Tallest High Level Train Bridge. Litwin Photography / Sutterstock. Foreground image: A passenger train leaves Charing Cross station in central London via a structural steel bridge across the Thames River. Jiri Rezac / Alamy Images. Background image: North America's Longest and Tallest High Level Train Bridge. Litwin Photography / Sutterstock.

© 2007 by R.C. Hibbeler
Published by Pearson Prentice Hall
Pearson Education, Inc.
Upper Saddle River, New Jersey 07458

The author and publisher of this book have used their best efforts in preparing this book. These efforts include the development, research, and testing of the theories and programs to determine their effectiveness. The author and publisher shall not be liable in any event for incidental or consequential damages with, or arising out of, the furnishing, performance, or use of these programs.

Printed in the United States of America

10 9 8 7 6 5 4 3 2

ISBN 0-13-221509-8

Pearson Education Ltd., *London*
Pearson Education Australia Pty. Ltd., *Sydney*
Pearson Education Singapore, Pte. Ltd.
Pearson Education North Asia Ltd., *Hong Kong*
Pearson Education Canada, Inc., *Toronto*
Pearson Educación de Mexico, S.A. de C.V.
Pearson Education—Japan, *Tokyo*
Pearson Education Malaysia, Pte. Ltd.
Pearson Education, Inc., *Upper Saddle River, New Jersey*

To the Student

*With the hope that this work will stimulate
an interest in Engineering Mechanics
and provide an acceptable guide to its understanding.*

Preface

The main purpose of this book is to provide the student with a clear and thorough presentation of the theory and applications of engineering mechanics. To achieve this objective, the author has by no means worked alone; to a large extent, this book, through its 11 editions, has been shaped by the comments and suggestions of hundreds of reviewers in the teaching profession as well as many of the author's students.

New and Enhanced Features

- **Review Material.** New end-of-chapter review sections have been added in response to student requests. These are designed to help students recall and study key chapter concepts.

- **Illustrations.** Instructor and student feedback to the enhanced illustrations in the tenth edition was overwhelmingly positive. In this edition we have enhanced over 150 additional illustrations for a total of over 350 pieces of photo realistic art. These figures provide a strong connection to the 3-D nature of engineering. Particular attention has also been paid to providing a view of any physical object, its dimensions, and the vectors in a manner that can be easily understood.

- **Problems.** The problems sets have been revised so that instructors can select both design and analysis problems having a wide range of difficulty. Apart from the author, three other professionals have checked all the problems for clarity and accuracy of the solutions. At the end of appropriate chapters, design projects are included.

- **Notes.** Closing notes have been added to appropriate Example Problems. These notes add additional context for the example material.

Hallmark Features

Accuracy. You have told us that accuracy is paramount. As with previous editions, each page of the book has been checked by three accuracy checkers in addition to the author.

Organization and Approach. The contents of each chapter are organized into well-defined sections that contain an explanation of specific topics, illustrative example problems, and a set of homework problems. The topics within each section are placed into subgroups defined by boldface titles. The purpose of this is to present a structured method for introducing each new definition or concept and to make the book convenient for later reference and review.

Chapter Contents. Each chapter begins with an illustration demon-strating a broad-range application of the material within the chapter. A bulleted list of the chapter contents is provided to give a general overview of the material that will be covered.

Free-Body Diagrams. The first step to solving most mechanics problems requires drawing a diagram. By doing so, the student forms the habit of tabulating the necessary data while focusing on the physical aspects of the problem and its associated geometry. If this step is performed correctly, applying the relevant equations of mechanics becomes somewhat methodical since the data can be taken directly from the diagram. This step is particularly important when solving equilibrium problems, and for this reason drawing free-body diagrams is strongly emphasized throughout the book. In particular, special sections and examples are devoted to show how to draw free-body diagrams, and specific homework problems in many sections of the book have been added to develop this practice.

Procedures for Analysis. Found after many of the sections of the book, this unique feature provides the student with a logical and orderly method to follow when applying the theory. The example problems are solved using this outlined method in order to clarify its numerical application. It is to be understood, however, that once the relevant principles have been mastered and enough confidence and judgment have been obtained, the student can then develop his or her own procedures for solving problems.

Photographs. Many photographs are used throughout the book to explain how the principles of mechanics apply to real-world situations. In some sections, photographs have been used to show how engineers must first make an idealized model for analysis and then proceed to draw a free-body diagram of this model in order to apply the theory.

Important Points. This feature provides a review or summary of the most important concepts in a section and highlights the most significant points that should be realized when applying the theory to solve problems.

Conceptual Understanding. Through the use of photographs placed throughout the book, theory is applied in a simplified way in order to illustrate some of its more important conceptual features and instill the physical meaning of many of the terms used in the equations. These simplified applications increase interest in the subject matter and better prepare the student to understand the examples and solve problems.

Example Problems. All the example problems are presented in a concise manner and in a style that is easy to understand.

Homework Problems

- **Free-Body Diagram Problems.** Many sections of the book contain introductory problems that only require drawing the free-body diagram for the specific problems within a problem set. These assignments will impress upon the student the importance of mastering this skill as a requirement for a complete solution of any equilibrium problem.

- **General Analysis and Design Problems.** The majority of problems in the book depict realistic situations encountered in engineering practice. Some of these problems come from actual products used in industry and are stated as such. It is hoped that this realism will both stimulate the student's interest in engineering mechanics and provide a means for developing the skill to reduce any such problem from its physical description to a model or symbolic representation to which the principles of mechanics may be applied.

Throughout the book, there is an approximate balance of problems using either SI or FPS units. Furthermore, in any set, an attempt has been made to arrange the problems in order of increasing difficulty. (Review problems at the end of each chapter are presented in random order.) The answers to all but every fourth problem are listed in the back of the book. To alert the user to a problem without a reported answer, an asterisk (∗) is placed before the problem number.

- **Computer Problems.** An effort has been made to include some problems that may be solved using a numerical procedure executed on either a desktop computer or a programmable pocket calculator. Suit-able numerical techniques along with associated computer programs are given in Appendix B. The intent here is to broaden the student's capacity for using other forms of mathematical analysis without sacrificing the time needed to focus on the application of the principles of mechanics. Problems of this type, which either can or must be solved using numerical procedures, are identified by a "square" symbol (■) preceding the problem number.

- **Design Projects.** At the end of some of the chapters, design projects have been included. It is felt that this type of assignment should be given only after the student has developed a basic understanding of the subject matter. These projects focus on solving a problem by specifying the geometry of a structure or mechanical object needed for a specific purpose. A force analysis is required and, in many cases, safety and cost issues must be addressed.

Appendices. The appendices provide a source of mathematical formula and numerical analysis needed to solve the problems in the book. Appendix C provides a set of problems typically found on the Fundamentals of Engineering Examination. By providing a partial solution to all the problems, the student is given a chance to further practice his or her skills.

Contents

Statics

The subject of Statics is covered in the first 11 chapters, in which the principles are applied first to simple, then to more complicated situations. Most often, each principle is applied first to a particle, then to a rigid body subjected to a coplanar system of forces, and finally to a general case of three-dimensional force systems acting on a rigid body.

Chapter 1 begins with an introduction to mechanics and a discussion of units. The notation of a vector and the properties of a concurrent force system are introduced in Chapter 2. This theory is then applied to the equilibrium of a particle in Chapter 3. Chapter 4 contains a general discussion of both concentrated and distributed force systems and the methods used to simplify them. The principles of rigid-body equilibrium are developed in Chapter 5 and then applied to specific problems involving the equilibrium of trusses, frames, and machines in Chapter 6, and to the analysis of internal forces in beams and cables in Chapter 7. Applications to problems involving frictional forces are discussed in Chapter 8, and topics related to the center of gravity and centroid are treated in Chapter 9. If time permits, sections concerning more advanced topics, indicated by stars (★) may be covered. Most of these topics are included in Chapter 10 (Area and Mass Moments of Inertia) and Chapter 11 (Virtual Work and Potential Energy). Note that this material also provides a suitable reference for basic principles when it is discussed in more advanced courses.

Alternative Coverage. At the discretion of the instructor, some of the material may be presented in a different sequence with no loss of continuity. For example, it is possible to introduce the concept of a force and all the necessary methods of vector analysis by first covering Chapter 2 and Section 4.2. Then after covering the rest of Chapter 4 (Force and Moment Systems), the equilibrium methods of Chapters 3 and 5 can be discussed.

Dynamics

The kinematics of a particle is discussed in Chapter 12, followed by a discussion of particle kinetics in Chapter 13 (Equation of Motion), Chapter 14 (Work and Energy), and Chapter 15 (Impulse and Momentum). The concepts of particle dynamics contained in these four chapters are then summarized in a "review" section, and the student is given the chance to identify and solve a variety of problems. A similar sequence of presentation is given for the planar motion of a rigid body: Chapter 16 (Planar Kinematics), Chapter 17 (Equations of Motion), Chapter 18 (Work and Energy), and Chapter 19 (Impulse and Momentum), followed by a summary and review set of problems for these chapters.

If time permits, some of the material involving three-dimensional rigid-body motion may be included in the course. The kinematics and kinetics of this motion are discussed in Chapters 20 and 21, respectively. Chapter 22 (Vibrations) may be included if the student has the necessary mathematical background. Sections of the book that are considered to be beyond the scope of the basic dynamics course are indicated by a star (★) and may be omitted. Note that this material also provides a suitable reference for basic principles when it is discussed in more advanced courses.

Alternative Coverage. At the discretion of the instructor, it is possible to cover Chapters 12 through 19 in the following order with no loss in continuity: Chapters 12 and 16 (Kinematics), Chapters 13 and 17 (Equations of Motion), Chapters 14 and 18 (Work and Energy), and Chapters 15 and 19 (Impulse and Momentum).

Instructor and Student Resources

Hibbeler's robust supplements package supports students and instructors. The eleventh edition features:

OneKey: Custom Homework, Secure Solutions, and Much More

Hibbeler's OneKey course—available at www.prenhall.com/onekey—offers over 3000 *Statics* and *Dynamics* problems that you can personalize and post for your assignments. Editing the values in a problem guarantees a fresh problem for your students. Then, use solutions powered by Mathcad to generate your own personal solution, and if you choose, post the solutions for your students online. OneKey also contains PHGradeAssist—an online assessment tool with approximately 600 algorithmic test bank problems. PHGA generates unique problems for students, grades the answer, and tracks students results automatically.

You'll find Hibbeler's OneKey course contains much more to help you and your students, including

- Student Hints: Each problem contains a student hint that you may choose to provide to your students. This hint is also fully editable should you wish to change it.
- Active Book: A complete, online HTML version of the textbook students can use and refer to while completing homework and assignments.
- An extra bank of student practice problems with solutions.
- Complete bank of .jpg images.
- Complete set of PowerPoint slides.
- Active Learning slides—Perfect for classroom response systems, created by Sudhir Mehta of North Dakota State University and Scott Danielson of Arizona State University.
- Mathcad and MATLAB tutorials.
- Animations and simulations.
- Math review tutorials.
- Mechanics visualization software—Ideal for in-class demonstrations.

Instructors should visit www.prenhall.com/onekey and / or contact your their local sales representative to register and receive more information. You may also send an email requesting information to engineering@prenhall.com.

Instructor's Solutions Manuals.

Statics (0-13-221502-0) and *Dynamics* (0-13-221506-3) These supplements provide complete solutions supported by problem statements and problem figures. The eleventh edition ISMs are fully typeset and have also been triple accuracy checked.

Instructor's Resource CD-ROMs.

Statics (0-13-221503-9) and Dynamics (0-13-221508-X) These supplements offer visual resources in CD-ROM format. These resources are also found on the Hibbeler OneKey course.

Study Packs.

Statics (0-13-221501-2) and *Dynamics* (0-13-221505-5) These supplements contain chapter-by-chapter study materials, a Free-Body Diagram Workbook and access to the Problems Website.

Practice Problems Workbooks.

Statics (0-13-224975-8) and *Dynamics* (0-13-239979-2) These supplements contain additional worked problems. Problems are partially solved and designed to help guide students through difficult topics.

Ordering Options.

The Study Packs and Study Packs with OneKey are available as stand-alone items for student purchase and are also available packaged with the texts. The ISBN for each valuepack option is as follows:

- *Engineering Mechanics: Statics and Dynamics* with Study Pack (0-13-230741-3)
- *Engineering Mechanics: Statics and Dynamics* with Study Pack and OneKey (0-13-229736-1)

Acknowledgments

The author has endeavored to write this book so that it will appeal to both the student and instructor. Through the years, many people have helped in its development, and I will always be grateful for their valued suggestions and comments. Specifically, I wish to personally thank the following individuals who have contributed their comments to the *Statics* and *Dynamics* series:

Zhen Chen, *University of Missouri—Columbia*

Ahmad M. Itani, *University of Nevada, Reno*

John B. Ligon, *Michigan Tech University*

G. N. Jazar, *North Dakota State University*

Peter Sandborn, *University of Maryland*

Thomas H. Miller, *Oregon State University*

Richard Bennett, *The University of Tennessee*

Ali Rostami, *Virginia Commonwealth University*

Robert L. Rennaker, *The University of Oklahoma*

Richard R. Neptune, *The University of Texas at Austin*

Zhikun Hou, *Worcester Polytechnic*

David W. Parish, *North Carolina State University*

Michael H. Santare, *University of Delaware*

A particular note of thanks is also given to Professors Will Liddell, Jr. and Henry Kuhlman for their specific help. A special note of thanks is given to the accuracy checkers:

- Scott Hendricks, *Virginia Polytechnic Institute and State University*
- Karim Nohra, *University of South Florida*
- Kurt Norlin, *Laurel Tech Integrated Publishing Services*

who diligently checked all of the text and problems. I should also like to acknowledge the proofreading assistance of my wife, Conny (Cornelie), during the time it has taken to prepare this manuscript for publication.

Lastly, many thanks are extended to all my students and to members of the teaching profession who have freely taken the time to send me their suggestions and comments. Since this list is too long to mention, it is hoped that those who have given help in this manner will accept this anonymous recognition.

I would greatly appreciate hearing from you if at any time you have any comments, suggestions, or problems related to any matters regarding this edition.

RUSSELL CHARLES HIBBELER
hibbeler@bellsouth.net

Contents

7
Internal
Forces   337

8
Friction   393

9
Center of Gravity
and Centroid   455

10
Moments
of Inertia   519

11
Virtual
Work   573

Appendices

12

Kinematics of a Particle   3

13

Kinetics of a Particle: Force and Acceleration   101

14

Kinetics of a Particle: Work and Energy   165

15

Kinetics of a Particle: Impulse and Momentum   215

Review 1: Kinematics and Kinetics of a Particle

16

Planar Kinematics of a Rigid Body 201

17

Planar Kinetics of a Rigid Body: Force and Acceleration 385

18

Planar Kinetics of a Rigid Body: Work and Energy 445

19

Planar Kinetics of a Rigid Body: Impulse and Momentum 483

20

Three-Dimensional Kinematics of a Rigid Body 535

21

Three-Dimensional Kinetics of a Rigid Body 567

22

Vibrations 619

Appendices

STATICS

The design of this rocket and gantry structure requires a basic knowledge of both statics and dynamics, which form the subject matter of engineering mechanics.

1 General Principles

- To provide an introduction to the basic quantities and idealizations of mechanics.
- To give a statement of Newton's Laws of Motion and Gravitation.
- To review the principles for applying the SI system of units.
- To examine the standard procedures for performing numerical calculations.
- To present a general guide for solving problems.

1.1 Mechanics

Mechanics can be defined as that branch of the physical sciences concerned with the state of rest or motion of bodies that are subjected to the action of forces. In general, this subject is subdivided into three branches: *rigid-body mechanics, deformable-body mechanics*, and *fluid mechanics*. This book treats only rigid-body mechanics since it forms a suitable basis for the design and analysis of many types of structural, mechanical, or electrical devices encountered in engineering. Also, rigid-body mechanics provides part of the necessary background for the study of the mechanics of deformable bodies and the mechanics of fluids.

Rigid-body mechanics is divided into two areas: statics and dynamics. *Statics* deals with the equilibrium of bodies, that is, those that are either at rest or move with a constant velocity; whereas *dynamics* is concerned with the accelerated motion of bodies. Although statics can be considered as a special case of dynamics, in which the acceleration is zero, statics deserves separate treatment in engineering education since many objects are designed with the intention that they remain in equilibrium.

Historical Development.   The subject of statics developed very early in history because the principles involved could be formulated simply from measurements of geometry and force. For example, the writings of Archimedes (287–212 B.C.) deal with the principle of the lever. Studies of the pulley, inclined plane, and wrench are also recorded in ancient writings—at times when the requirements of engineering were limited primarily to building construction.

Since the principles of dynamics depend on an accurate measurement of time, this subject developed much later. Galileo Galilei (1564–1642) was one of the first major contributors to this field. His work consisted of experiments using pendulums and falling bodies. The most significant contributions in dynamics, however, were made by Issac Newton (1642–1727), who is noted for his formulation of the three fundamental laws of motion and the law of universal gravitational attraction. Shortly after these laws were postulated, important techniques for their application were developed by Euler, D'Alembert, Lagrange, and others.

1.2   Fundamental Concepts

Before we begin our study of engineering mechanics, it is important to understand the meaning of certain fundamental concepts and principles.

Basic Quantities.   The following four quantities are used throughout mechanics.

Length.   *Length* is needed to locate the position of a point in space and thereby describe the size of a physical system. Once a standard unit of length is defined, one can then quantitatively define distances and geometric properties of a body as multiples of the unit length.

Time.   *Time* is conceived as a succession of events. Although the principles of statics are time independent, this quantity does play an important role in the study of dynamics.

Mass.   *Mass* is a property of matter by which we can compare the action of one body with that of another. This property manifests itself as a gravitational attraction between two bodies and provides a quantitative measure of the resistance of matter to a change in velocity.

Force.   In general, *force* is considered as a "push" or "pull" exerted by one body on another. This interaction can occur when there is direct contact between the bodies, such as a person pushing on a wall, or it can occur through a distance when the bodies are physically separated. Examples of the latter type include gravitational, electrical, and magnetic forces. In any case, a force is completely characterized by its magnitude, direction, and point of application.

Idealizations.   Models or idealizations are used in mechanics in order to simplify application of the theory. A few of the more important idealizations will now be defined. Others that are noteworthy will be discussed at points where they are needed.

Particle.   A *particle* has a mass, but a size that can be neglected. For example, the size of the earth is insignificant compared to the size of its orbit, and therefore the earth can be modeled as a particle when studying its orbital motion. When a body is idealized as a particle, the principles of mechanics reduce to a rather simplified form since the geometry of the body will not be involved in the analysis of the problem.

Rigid Body.   A *rigid body* can be considered as a combination of a large number of particles in which all the particles remain at a fixed distance from one another both before and after applying a load. As a result, the material properties of any body that is assumed to be rigid will not have to be considered when analyzing the forces acting on the body. In most cases the actual deformations occurring in structures, machines, mechanisms, and the like are relatively small, and the rigid-body assumption is suitable for analysis.

Concentrated Force.   A *concentrated force* represents the effect of a loading which is assumed to act at a point on a body. We can represent a load by a concentrated force, provided the area over which the load is applied is very small compared to the overall size of the body. An example would be the contact force between a wheel and the ground.

Newton's Three Laws of Motion.   The entire subject of rigid-body mechanics is formulated on the basis of Newton's three laws of motion, the validity of which is based on experimental observation. They apply to the motion of a particle as measured from a nonaccelerating reference frame. With reference to Fig. 1–1, they may be briefly stated as follows.

First Law.   A particle originally at rest, or moving in a straight line with constant velocity, will remain in this state provided the particle is *not* subjected to an unbalanced force.

Second Law.   A particle acted upon by an *unbalanced force* **F** experiences an acceleration **a** that has the same direction as the force and a magnitude that is directly proportional to the force.* If **F** is applied to a particle of mass m, this law may be expressed mathematically as

$$\mathbf{F} = m\mathbf{a} \tag{1-1}$$

Third Law.   The mutual forces of action and reaction between two particles are equal, opposite, and collinear.

* Stated another way, the unbalanced force acting on the particle is proportional to the time rate of change of the particle's linear momentum.

Equilibrium

Accelerated motion

Action – reaction

Fig. 1–1

Newton's Law of Gravitational Attraction.   Shortly after formulating his three laws of motion, Newton postulated a law governing the gravitational attraction between any two particles. Stated mathematically,

$$F = G \frac{m_1 m_2}{r^2} \tag{1-2}$$

where

F = force of gravitation between the two particles

G = universal constant of gravitation; according to experimental evidence, $G = 66.73(10^{-12}) \text{ m}^3/(\text{kg} \cdot \text{s}^2)$

m_1, m_2 = mass of each of the two particles

r = distance between the two particles

Weight.   According to Eq. 1–2, any two particles or bodies have a mutual attractive (gravitational) force acting between them. In the case of a particle located at or near the surface of the earth, however, the only gravitational force having any sizable magnitude is that between the earth and the particle. Consequently, this force, termed the *weight*, will be the only gravitational force considered in our study of mechanics.

From Eq. 1–2, we can develop an approximate expression for finding the weight W of a particle having a mass $m_1 = m$. If we assume the earth to be a nonrotating sphere of constant density and having a mass $m_2 = M_e$, then if r is the distance between the earth's center and the particle, we have

$$W = G \frac{m M_e}{r^2}$$

Letting $g = GM_e/r^2$ yields

$$\boxed{W = mg} \tag{1-3}$$

By comparison with $\mathbf{F} = m\mathbf{a}$, we term g the acceleration due to gravity. Since it depends on r, it can be seen that the weight of a body is *not* an absolute quantity. Instead, its magnitude is determined from where the measurement was made. For most engineering calculations, however, g is determined at sea level and at a latitude of 45°, which is considered the "standard location."

1.3 Units of Measurement

The four basic quantities—force, mass, length and time—are not all independent from one another; in fact, they are *related* by Newton's second law of motion, $\mathbf{F} = m\mathbf{a}$. Because of this, the *units* used to measure these quantities cannot *all* be selected arbitrarily. The equality $\mathbf{F} = m\mathbf{a}$ is maintained only if three of the four units, called *base units*, are *arbitrarily defined* and the fourth unit is then *derived* from the equation.

SI Units.   The International System of units, abbreviated SI after the French "Système International d'Unités," is a modern version of the metric system which has received worldwide recognition. As shown in Table 1–1, the SI system specifies length in meters (m), time in seconds (s), and mass in kilograms (kg). The unit of force, called a newton (N), is *derived* from $\mathbf{F} = m\mathbf{a}$. Thus, 1 newton is equal to a force required to give 1 kilogram of mass an acceleration of 1 m/s^2 (N = kg · m/s^2).

If the weight of a body located at the "standard location" is to be determined in newtons, then Eq. 1–3 must be applied. Here $g = 9.806\ 65$ m/s^2; however, for calculations, the value $g = 9.81$ m/s^2 will be used. Thus,

$$W = mg \qquad (g = 9.81 \text{ m/s}^2) \qquad (1\text{–}4)$$

Therefore, a body of mass 1 kg has a weight of 9.81 N, a 2-kg body weighs 19.62 N, and so on, Fig. 1–2a.

U.S. Customary.   In the U.S. Customary system of units (FPS) length is measured in feet (ft), force in pounds (lb), and time in seconds (s), Table 1–1. The unit of mass, called a *slug*, is *derived* from $\mathbf{F} = m\mathbf{a}$. Hence, 1 slug is equal to the amount of matter accelerated at 1 ft/s^2 when acted upon by a force of 1 lb (slug = lb · s^2/ft).

In order to determine the mass of a body having a weight measured in pounds, we must apply Eq. 1–3. If the measurements are made at the "standard location," then $g = 32.2$ ft/s^2 will be used for calculations. Therefore,

$$m = \frac{W}{g} \qquad (g = 32.2 \text{ ft/s}^2) \qquad (1\text{–}5)$$

And so a body weighing 32.2 lb has a mass of 1 slug, a 64.4-lb body has a mass of 2 slugs, and so on, Fig. 1–2b.

(a)

(b)

Fig. 1–2

TABLE 1–1 Systems of Units				
Name	Length	Time	Mass	Force
International System of Units (SI)	meter (m)	second (s)	kilogram (kg)	newton* (N) $\left(\dfrac{\text{kg} \cdot \text{m}}{\text{s}^2}\right)$
U.S. Customary (FPS)	foot (ft)	second (s)	slug* $\left(\dfrac{\text{lb} \cdot \text{s}^2}{\text{ft}}\right)$	pound (lb)

*Derived unit.

Conversion of Units.   Table 1–2 provides a set of direct conversion factors between FPS and SI units for the basic quantities. Also, in the FPS system, recall that 1 ft = 12 in. (inches), 5280 ft = 1 mi (mile), 1000 lb = 1 kip (kilo-pound), and 2000 lb = 1 ton.

TABLE 1–2 Conversion Factors			
Quantity	Unit of Measurement (FPS)	Equals	Unit of Measurement (SI)
Force	lb		4.448 2 N
Mass	slug		14.593 8 kg
Length	ft		0.304 8 m

1.4   The International System of Units

The SI system of units is used extensively in this book since it is intended to become the worldwide standard for measurement. Consequently, the rules for its use and some of its terminology relevant to mechanics will now be presented.

Prefixes.   When a numerical quantity is either very large or very small, the units used to define its size may be modified by using a prefix. Some of the prefixes used in the SI system are shown in Table 1–3. Each represents a multiple or submultiple of a unit which, if applied successively, moves the decimal point of a numerical quantity to every third place.* For example, 4 000 000 N = 4 000 kN (kilo-newton) = 4 MN (mega-newton), or 0.005 m = 5 mm multiple deca (10) or the submultiple centi (0.01), which form part of the metric system. Except for some volume and area measurements, the use of these prefixes is to be avoided in science and engineering.

TABLE 1–3 Prefixes			
	Exponential Form	Prefix	SI Symbol
Multiple			
1 000 000 000	10^9	giga	G
1 000 000	10^6	mega	M
1 000	10^3	kilo	k
Submultiple			
0.001	10^{-3}	milli	m
0.000 001	10^{-6}	micro	μ
0.000 000 001	10^{-9}	nano	n

*The kilogram is the only base unit that is defined with a prefix.

Rules for Use.   The following rules are given for the proper use of the various SI symbols:

1. A symbol is *never* written with a plural "s," since it may be confused with the unit for second (s).

2. Symbols are always written in lowercase letters, with the following exceptions: symbols for the two largest prefixes shown in Table 1–3, giga and mega, are capitalized as G and M, respectively; and symbols named after an individual are also capitalized, e.g., N.

3. Quantities defined by several units which are multiples of one another are separated by a *dot* to avoid confusion with prefix notation, as indicated by $N = kg \cdot m/s^2 = kg \cdot m \cdot s^{-2}$. Also, $m \cdot s$ (meter-second), whereas ms (milli-second).

4. The exponential power represented for a unit having a prefix refers to both the unit *and* its prefix. For example, $\mu N^2 = (\mu N)^2 = \mu N \cdot \mu N$.

   Likewise, mm^2 represents $(mm)^2 = mm \cdot mm$.

5. Physical constants or numbers having several digits on either side of the decimal point should be reported with a *space* between every three digits rather than with a comma; e.g., 73 569.213 427. In the case of four digits on either side of the decimal, the spacing is optional; e.g., 8537 or 8 537. Furthermore, always try to use decimals and avoid fractions; that is, write 15.25 *not* $15\frac{1}{4}$.

6. When performing calculations, represent the numbers in terms of their *base or derived units* by converting all prefixes to powers of 10. The final result should then be expressed using a *single prefix*. Also, after calculation, it is best to keep numerical values between 0.1 and 1000; otherwise, a suitable prefix should be chosen. For example,

$$(50 \text{ kN})(60 \text{ nm}) = [50(10^3) \text{ N}][60(10^{-9}) \text{ m}]$$

$$= 3000(10^{-6}) \text{ N} \cdot m = 3(10^{-3}) \text{ N} \cdot m = 3 \text{ mN} \cdot m$$

7. Compound prefixes should not be used; e.g., $k\mu s$ (kilo-micro-second) should be expressed as ms (milli-second) since $1 \text{ k}\mu s = 1(10^3)(10^{-6}) \text{ s} = 1(10^{-3}) \text{ s} = 1 \text{ ms}$.

8. With the exception of the base unit the kilogram, in general avoid the use of a prefix in the denominator of composite units. For example, do not write N/mm, but rather kN/m; also, m/mg should be written as Mm/kg.

9. Although not expressed in multiples of 10, the minute, hour, etc., are retained for practical purposes as multiples of the second. Furthermore, plane angular measurement is made using radians (rad). In this book, however, degrees will often be used, where $180° = \pi$ rad.

1.5 Numerical Calculations

Numerical work in engineering practice is most often performed by using handheld calculators and computers. It is important, however, that the answers to any problem be reported with both justifiable accuracy and appropriate significant figures. In this section we will discuss these topics together with some other important aspects involved in all engineering calculations.

Dimensional Homogeneity. The terms of any equation used to describe a physical process must be *dimensionally homogeneous;* that is, each term must be expressed in the same units. Provided this is the case, all the terms of an equation can then be combined if numerical values are substituted for the variables. Consider, for example, the equation $s = vt + \frac{1}{2}at^2$, where, in SI units, s is the position in meters, m, t is time in seconds, s, v is velocity in m/s, and a is acceleration in m/s². Regardless of how this equation is evaluated, it maintains its dimensional homogeneity. In the form stated, each of the three terms is expressed in meters [m, (m/s̶) s̶, (m/s²) s̶²,] or solving for a, $a = 2s/t^2 - 2v/t$, the terms are each expressed in units of m/s² [m/s², m/s², (m/s)/s].

Since problems in mechanics involve the solution of dimensionally homogeneous equations, the fact that all terms of an equation are represented by a consistent set of units can be used as a partial check for algebraic manipulations of an equation.

Significant Figures. The accuracy of a number is specified by the number of significant figures it contains. A *significant figure* is any digit, including a zero, provided it is not used to specify the location of the decimal point for the number. For example, the numbers 5604 and 34.52 each have four

Computers are often used in engineering for advanced design and analysis.

significant figures. When numbers begin or end with zeros, however, it is difficult to tell how many significant figures are in the number. Consider the number 400. Does it have one (4), or perhaps two (40), or three (400) significant figures? In order to clarify this situation, the number should be reported using powers of 10. Using *engineering notation*, the exponent is displayed in multiples of three in order to facilitate conversion of SI units to those having an appropriate prefix. Thus, 400 expressed to one significant figure would be $0.4(10^3)$. Likewise, 2500 and 0.00546 expressed to three significant figures would be $2.50(10^3)$ and $5.46(10^{-3})$.

Rounding Off Numbers.   For numerical calculations, the accuracy obtained from the solution of a problem generally can never be better than the accuracy of the problem data. This is what is to be expected, but often handheld calculators or computers involve more figures in the answer than the number of significant figures used for the data. For this reason, a calculated result should always be "rounded off" to an appropriate number of significant figures.

To convey appropriate accuracy, the following rules for rounding off a number to n significant figures apply:

- If the $n + 1$ digit is *less than 5*, the $n + 1$ digit and others following it are dropped. For example, 2.326 and 0.451 rounded off to $n = 2$ significant figures would be 2.3 and 0.45.

- If the $n + 1$ digit is equal to 5 with zeros following it, then round off the nth digit to an *even number*. For example, $1.245(10^3)$ and 0.8655 rounded off to $n = 3$ significant figures become $1.24(10^3)$ and 0.866.

- If the $n + 1$ digit is *greater than 5* or equal to 5 with any nonzero digits following it, then increase the nth digit by 1 and drop the $n + 1$ digit and others following it. For example, 0.723 87 and 565.500 3 rounded off to $n = 3$ significant figures become 0.724 and 566.

Calculations.   As a general rule, to ensure accuracy of a final result when performing calculations on a pocket calculator, always retain a greater number of digits than the problem data. If possible, try to work out the computations so that numbers which are approximately equal are not subtracted since accuracy is often lost from this calculation.

In engineering we generally round off final answers to *three* significant figures since the data for geometry, loads, and other measurements are often reported with this accuracy.* Consequently, in this book the intermediate calculations for the examples are often worked out to four significant figures and the answers are generally reported to *three* significant figures.

---

*Of course, some numbers, such as π, e, or numbers used in derived formulas are exact and are therefore accurate to an infinite number of significant figures.

EXAMPLE 1.1

Convert 2 km/h to m/s. How many ft/s is this?

SOLUTION

Since 1 km = 1000 m and 1 h = 3600 s, the factors of conversion are arranged in the following order, so that a cancellation of the units can be applied:

$$2 \text{ km/h} = \frac{2 \text{ km}}{\text{h}} \left(\frac{1000 \text{ m}}{\text{km}} \right) \left(\frac{1 \text{ h}}{3600 \text{ s}} \right)$$

$$= \frac{2000 \text{ m}}{3600 \text{ s}} = 0.556 \text{ m/s} \qquad Ans.$$

From Table 1–2, 1 ft = 0.3048 m. Thus

$$0.556 \text{ m/s} = \frac{0.556 \text{ m}}{\text{s}} \frac{1 \text{ ft}}{0.3048 \text{ m}}$$

$$= 1.82 \text{ ft/s} \qquad Ans.$$

NOTE: Remember to round off the final answer to three significant figures.

EXAMPLE 1.2

Convert the quantities 300 lb·s and 52 slug/ft^3 to appropriate SI units.

SOLUTION

Using Table 1–2, 1 lb = 4.448 2 N.

$$300 \text{ lb} \cdot \text{s} = 300 \text{ lb} \cdot \text{s} \left(\frac{4.448 \text{ 2 N}}{\text{lb}} \right)$$

$$= 1334.5 \text{ N} \cdot \text{s} = 1.33 \text{ kN} \cdot \text{s} \qquad Ans.$$

Also, 1 slug = 14.593 8 kg and 1 ft = 0.304 8 m.

$$52 \text{ slug/ft}^3 = \frac{52 \text{ slug}}{\text{ft}^3} \left(\frac{14.593 \text{ 8 kg}}{1 \text{ slug}} \right) \left(\frac{1 \text{ ft}}{0.304 \text{ 8 m}} \right)^3$$

$$= 26.8(10^3) \text{ kg/m}^3$$

$$= 26.8 \text{ Mg/m}^3 \qquad Ans.$$

EXAMPLE 1.3

Evaluate each of the following and express with SI units having an appropriate prefix: (a) (50 mN)(6 GN), (b) (400 mm)(0.6 MN)^2, (c) 45 MN^3/900 Gg.

SOLUTION

First convert each number to base units, perform the indicated operations, then choose an appropriate prefix (see Rule 6 on p. 9).

Part (a)

$$(50\text{ mN})(6\text{ GN}) = [50(10^{-3})\text{ N}][6(10^{9})\text{ N}]$$
$$= 300(10^{6})\text{ N}^2$$
$$= 300(10^{6})\text{ N}^2\left(\frac{1\text{ kN}}{10^{3}\text{ N}}\right)\left(\frac{1\text{ kN}}{10^{3}\text{ N}}\right)$$
$$= 300\text{ kN}^2 \qquad\qquad Ans.$$

NOTE: Keep in mind the convention kN^2 = (kN)^2 = 10^6 N^2 (Rule 4 on p. 9).

Part (b)

$$(400\text{ mm})(0.6\text{ MN})^2 = [400(10^{-3})\text{ m}][0.6(10^{6})\text{ N}]^2$$
$$= [400(10^{-3})\text{ m}][0.36(10^{12})\text{ N}^2]$$
$$= 144(10^{9})\text{ m}\cdot\text{N}^2$$
$$= 144\text{ Gm}\cdot\text{N}^2 \qquad\qquad Ans.$$

We can also write

$$144(10^{9})\text{ m}\cdot\text{N}^2 = 144(10^{9})\text{ m}\cdot\text{N}^2\left(\frac{1\text{ MN}}{10^{6}\text{ N}}\right)\left(\frac{1\text{ MN}}{10^{6}\text{ N}}\right)$$
$$= 0.144\text{ m}\cdot\text{MN}^2$$

Part (c)

$$45\text{ MN}^3/900\text{ Gg} = \frac{45(10^{6}\text{ N})^3}{900(10^{6})\text{ kg}}$$
$$= 0.05(10^{12})\text{ N}^3/\text{kg}$$
$$= 0.05(10^{12})\text{ N}^3\left(\frac{1\text{ kN}}{10^{3}\text{ N}}\right)^3\frac{1}{\text{kg}}$$
$$= 0.05(10^{3})\text{ kN}^3/\text{kg}$$
$$= 50\text{ kN}^3/\text{kg} \qquad\qquad Ans.$$

NOTE: Here we have used Rules 4 and 8 on p. 9.

When solving problems, do the work as neatly as possible. Being neat generally stimulates clear and orderly thinking, and vice versa.

1.6   General Procedure for Analysis

The most effective way of learning the principles of engineering mechanics is to *solve problems*. To be successful at this, it is important to always present the work in a *logical* and *orderly manner*, as suggested by the following sequence of steps:

1. Read the problem carefully and try to correlate the actual physical situation with the theory studied.

2. Draw any necessary diagrams and tabulate the problem data.

3. Apply the relevant principles, generally in mathematical form.

4. Solve the necessary equations algebraically as far as practical, then, making sure they are dimensionally homogeneous, use a consistent set of units and complete the solution numerically. Report the answer with no more significant figures than the accuracy of the given data.

5. Study the answer with technical judgment and common sense to determine whether or not it seems reasonable.

Important Points

- Statics is the study of bodies that are at rest or move with constant velocity.

- A particle has a mass but a size that can be neglected.

- A rigid body does not deform under load.

- Concentrated forces are assumed to act at a point on a body.

- Newton's three laws of motion should be memorized.

- Mass is a property of matter that does not change from one location to another.

- Weight refers to the gravitational attraction of the earth on a body or quantity of mass. Its magnitude depends upon the elevation at which the mass is located.

- In the SI system the unit of force, the newton, is a derived unit. The meter, second, and kilogram are base units.

- Prefixes G, M, k, m, μ, n are used to represent large and small numerical quantities. Their exponential size should be known, along with the rules for using the SI units.

- Perform numerical calculations to several significant figures and then report the final answer to three significant figures.

- Algebraic manipulations of an equation can be checked in part by verifying that the equation remains dimensionally homogeneous.

- Know the rules for rounding off numbers.

PROBLEMS

1–1. Represent each of the following combinations of units in the correct SI form using an appropriate prefix: (a) m/ms, (b) μkm, (c) ks/mg, and (d) km $\cdot \mu$N.

1–2. Wood has a density of 4.70 slug/ft^3. What is its density expressed in SI units?

1–3. Represent each of the following combinations of units in the correct SI form using an appropriate prefix: (a) Mg/mm, (b) mN/μs, (c) μm $\cdot$ Mg.

*1–4. Represent each of the following combinations of units in the correct SI form: (a) Mg/ms, (b) N/mm, (c) mN/(kg $\cdot \mu$s).

1–5. Represent each of the following with SI units having an appropriate prefix: (a) 8653 ms, (b) 8368 N, (c) 0.893 kg.

1–6. Represent each of the following to three significant figures and express each answer in SI units using an appropriate prefix: (a) 45 320 kN, (b) 568(10^5)mm, and (c) 0.005 63 mg.

1–7. Evaluate (204 mm)(0.004 57 kg)/(34.6 N) to three significant figures and express the answer in SI units using an appropriate prefix.

*1–8. If a car is traveling at 55 mi/h, determine its speed in kilometers per hour and meters per second.

1–9. Convert: (a) 200 lb $\cdot$ ft to N $\cdot$ m, (b) 350 lb/ft^3 to kN/m^3, (c) 8 ft/h to mm/s. Express the result to three significant figures. Use an appropriate prefix.

1–10. What is the weight in newtons of an object that has a mass of: (a) 10 kg, (b) 0.5 g, (c) 4.50 Mg? Express the result to three significant figures. Use an appropriate prefix.

1–11. If an object has a mass of 40 slugs, determine its mass in kilograms.

*1–12. The specific weight (wt./vol.) of brass is 520 lb/ft^3. Determine its density (mass/vol.) in SI units. Use an appropriate prefix.

1–13. A concrete column has a diameter of 350 mm and a length of 2 m. If the density (mass/volume) of concrete is 2.45 Mg/m^3, determine the weight of the column in pounds.

1–14. The density (mass/volume) of aluminum is 5.26 slug/ft^3. Determine its density in SI units. Use an appropriate prefix.

1–15. Using Table 1–2, determine your own mass in kilograms, your weight in newtons, and your height in meters.

*1–16. Two particles have a mass of 8 kg and 12 kg, respectively. If they are 800 mm apart, determine the force of gravity acting between them. Compare this result with the weight of each particle.

1–17. Using the base units of the SI system, show that Eq. 1–2 is a dimensionally homogeneous equation which gives F in newtons. Compute the gravitational force acting between two identical spheres that are touching each other. The mass of each sphere is 150 kg, and the radius is 275 mm.

1–18. Evaluate each of the following to three significant figures and express each answer in SI units using an appropriate prefix: (a) $(200 \text{ kN})^2$, (b) $(0.005 \text{ mm})^2$, (c) $(400 \text{ m})^3$.

1–19. Evaluate each of the following to three significant figures and express each answer in SI units using an appropriate prefix: (a) $(684 \mu\text{m})/(43 \text{ ms})$, (b) $(28 \text{ ms})(0.0458 \text{ Mm})/(348 \text{ mg})$, (c) $(2.68 \text{ mm})(426 \text{ Mg})$.

*1–20. Evaluate each of the following to three significant figures and express each answer in SI units using an appropriate prefix: (a) $(0.631 \text{ Mm})/(8.60 \text{ kg})^2$, (b) $(35 \text{ mm})^2(48 \text{ kg})^3$.

These communication towers are stabilized by cables that exert forces at the points of connection. In this chapter, we will show how to express these forces as Cartesian vectors.

2

Force Vectors

CHAPTER OBJECTIVES

- To show how to add forces and resolve them into components using the Parallelogram Law.

- To express force and position in Cartesian vector form and explain how to determine the vector's magnitude and direction.

- To introduce the dot product in order to determine the angle between two vectors or the projection of one vector onto another.

2.1 Scalars and Vectors

Most of the physical quantities in mechanics can be expressed mathematically by means of scalars and vectors.

Scalar.   A quantity characterized by a positive or negative number is called a *scalar*. For example, mass, volume, and length are scalar quantities often used in statics. In this book, scalars are indicated by letters in italic type, such as the scalar A.

Vector.   A *vector* is a quantity that has both a magnitude and a direction. In statics the vector quantities frequently encountered are position, force, and moment. For handwritten work, a vector is generally represented by a letter with an arrow written over it, such as $\vec{A}$. The magnitude is designated $|\vec{A}|$ or simply A. In this book vectors will be symbolized in boldface type; for example, **A** is used to designate the vector "A." Its magnitude, which is always a positive quantity, is symbolized in italic type, written as $|A|$, or simply A when it is understood that A is a positive scalar.

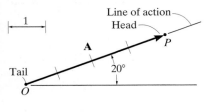

Fig. 2–1

A vector is represented graphically by an arrow, which is used to define its magnitude, direction, and sense. The *magnitude* of the vector is the length of the arrow, the *direction* is defined by the angle between a reference axis and the arrow's line of action, and the *sense* is indicated by the arrowhead. For example, the vector **A** shown in Fig. 2–1 has a magnitude of 4 units, a direction which is 20° measured counterclockwise from the horizontal axis, and a sense which is upward and to the right. The point *O* is called the *tail* of the vector, the point *P* the *tip* or *head*.

2.2 Vector Operations

Multiplication and Division of a Vector by a Scalar.

The product of vector **A** and scalar a, yielding a**A**, is defined as a vector having a magnitude $|aA|$. The *sense* of a**A** is the *same* as **A** provided a is *positive*; it is *opposite* to **A** if a is *negative*. In particular, the negative of a vector is formed by multiplying the vector by the scalar (-1), Fig. 2–2. Division of a vector by a scalar can be defined using the laws of multiplication, since $\mathbf{A}/a = (1/a)\mathbf{A}$, $a \neq 0$. Graphic examples of these operations are shown in Fig. 2–3.

Vector **A** and its negative counterpart

Fig. 2–2

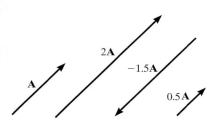

Scalar multiplication and division

Fig. 2–3

Vector Addition.

Two vectors **A** and **B** such as force or position, Fig. 2–4*a*, may be added to form a "resultant" vector $\mathbf{R} = \mathbf{A} + \mathbf{B}$ by using the *parallelogram law*. To do this, **A** and **B** are joined at their tails, Fig. 2–4*b*. Parallel lines drawn from the head of each vector intersect at a common point, thereby forming the adjacent sides of a parallelogram. As shown, the resultant **R** is the diagonal of the parallelogram, which extends from the tails of **A** and **B** to the intersection of the lines.

We can also add **B** to **A** using a *triangle construction*, which is a special case of the parallelogram law, whereby vector **B** is added to vector **A** in a "head-to-tail" fashion, i.e., by connecting the head of **A** to the tail of **B**, Fig. 2–4*c*. The resultant **R** extends from the tail of **A** to the head of **B**. In a similar manner, **R** can also be obtained by adding **A** to **B**, Fig. 2–4*d*. By comparison, it is seen that vector addition is commutative; in other words, the vectors can be added in either order, i.e., $\mathbf{R} = \mathbf{A} + \mathbf{B} = \mathbf{B} + \mathbf{A}$.

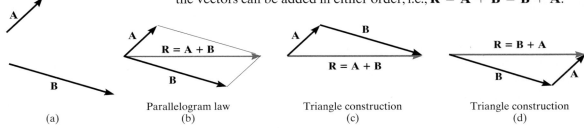

Vector addition

Fig. 2–4

As a special case, if the two vectors **A** and **B** are *collinear*, i.e., both have the same line of action, the parallelogram law reduces to an *algebraic* or *scalar addition* $R = A + B$, as shown in Fig. 2–5.

Vector Subtraction.   The resultant *difference* between two vectors **A** and **B** of the same type may be expressed as

$$\mathbf{R}' = \mathbf{A} - \mathbf{B} = \mathbf{A} + (-\mathbf{B})$$

This vector sum is shown graphically in Fig. 2–6. Subtraction is therefore defined as a special case of addition, so the rules of vector addition also apply to vector subtraction.

Addition of collinear vectors

$R = A + B$

Fig. 2–5

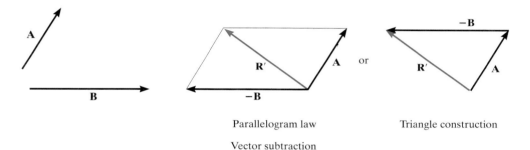

Parallelogram law

Triangle construction

Vector subtraction

Fig. 2–6

Resolution of Vector.   A vector may be resolved into two "components" having known lines of action by using the parallelogram law. For example, if **R** in Fig. 2–7*a* is to be resolved into components acting along the lines *a* and *b*, one starts at the *head* of **R** and extends a line *parallel* to *a* until it intersects *b*. Likewise, a line parallel to *b* is drawn from the *head* of **R** to the point of intersection with *a*, Fig. 2–7*a*. The two components **A** and **B** are then drawn such that they extend from the tail of **R** to the points of intersection, as shown in Fig. 2–7*b*.

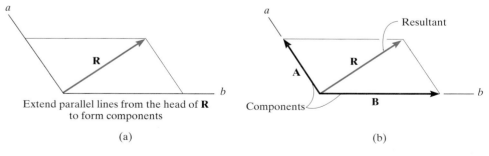

Extend parallel lines from the head of **R** to form components

(a)

Components

(b)

Resolution of a vector

Fig. 2–7

2.3   Vector Addition of Forces

Experimental evidence has shown that a force is a vector quantity since it has a specified magnitude, direction, and sense and it adds according to the parallelogram law. Two common problems in statics involve either finding the resultant force, knowing its components, or resolving a known force into two components. As described in Sec. 2.2, both of these problems require application of the parallelogram law.

If more than two forces are to be added, successive applications of the parallelogram law can be carried out in order to obtain the resultant force. For example, if three forces $\mathbf{F}_1$, $\mathbf{F}_2$, $\mathbf{F}_3$ act at a point O, Fig. 2–8, the resultant of any two of the forces is found—say, $\mathbf{F}_1 + \mathbf{F}_2$—and then this resultant is added to the third force, yielding the resultant of all three forces; i.e., $\mathbf{F}_R = (\mathbf{F}_1 + \mathbf{F}_2) + \mathbf{F}_3$. Using the parallelogram law to add more than two forces, as shown here, often requires extensive geometric and trigonometric calculation to determine the numerical values for the magnitude and direction of the resultant. Instead, problems of this type are easily solved by using the "rectangular-component method," which is explained in Sec. 2.4.

Fig. 2–8

If we know the forces $\mathbf{F}_a$ and $\mathbf{F}_b$ that the two chains a and b exert on the hook, we can find their resultant force $\mathbf{F}_c$ by using the parallelogram law. This requires drawing lines parallel to a and b from the heads of $\mathbf{F}_a$ and $\mathbf{F}_b$ as shown thus forming a parallelogram.

In a similar manner, if the force $\mathbf{F}_c$ along chain c is known, then its two components $\mathbf{F}_a$ and $\mathbf{F}_b$, that act along a and b, can be determined from the parallelogram law. Here we must start at the head of $\mathbf{F}_c$ and construct lines parallel to a and b, thereby forming the parallelogram.

Procedure for Analysis

Problems that involve the addition of two forces can be solved as follows:

Parallelogram Law.

- Make a sketch by placing the vectors together at their tails and then showing the vector addition using the parallelogram law.
- Two "component" forces add according to the parallelogram law, yielding a *resultant* force that forms the diagonal of the parallelogram.
- If a force is to be resolved into *components* along two axes directed from the tail of the force, then start at the head of the force and construct lines parallel to the axes, thereby forming the parallelogram. The sides of the parallelogram represent the components.
- Label all the known and unknown force magnitudes and the angles on the sketch and identify the two unknowns.

Trigonometry.

- Redraw a half portion of the parallelogram to illustrate the triangular head-to-tail addition of the components.
- The magnitude of the resultant force can be determined from the law of cosines, and its direction is determined from the law of sines, Fig. 2–9.
- The magnitudes of two force components are determined from the law of sines, Fig. 2–9.

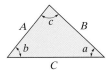

Sine law:
$$\frac{A}{\sin a} = \frac{B}{\sin b} = \frac{C}{\sin c}$$

Cosine law:
$$C = \sqrt{A^2 + B^2 - 2AB \cos c}$$

Fig. 2–9

Important Points

- A scalar is a positive or negative number.
- A vector is a quantity that has a magnitude, direction, and sense.
- Multiplication or division of a vector by a scalar will change the magnitude of the vector. The sense of the vector will change if the scalar is negative.
- As a special case, if the vectors are collinear, the resultant is formed by an algebraic or scalar addition.

EXAMPLE 2.1

(a)

The screw eye in Fig. 2–10a is subjected to two forces, $\mathbf{F}_1$ and $\mathbf{F}_2$. Determine the magnitude and direction of the resultant force.

(b)   (c)

Fig. 2–10

SOLUTION

Parallelogram Law. The parallelogram law of addition is shown in Fig. 2–10b. The two unknowns are the magnitude of $\mathbf{F}_R$ and the angle θ (theta).

Trigonometry. From Fig. 2–10b, the vector triangle, Fig. 2–10c, is constructed. F_R is determined by using the law of cosines:

$$F_R = \sqrt{(100 \text{ N})^2 + (150 \text{ N})^2 - 2(100 \text{ N})(150 \text{ N}) \cos 115°}$$
$$= \sqrt{10\ 000 + 22\ 500 - 30\ 000(-0.4226)} = 212.6 \text{ N}$$
$$= 213 \text{ N} \hspace{4cm} Ans.$$

The angle θ is determined by applying the law of sines, using the computed value of F_R:

$$\frac{150 \text{ N}}{\sin \theta} = \frac{212.6 \text{ N}}{\sin 115°}$$

$$\sin \theta = \frac{150 \text{ N}}{212.6 \text{ N}}(0.9063)$$

$$\theta = 39.8°$$

Thus, the direction ϕ (phi) of $\mathbf{F}_R$, measured from the horizontal, is

$$\phi = 39.8° + 15.0° = 54.8° \ \measuredangle\phi \hspace{2cm} Ans.$$

NOTE: The results seem reasonable, since Fig. 2–10b shows $\mathbf{F}_R$ to have a magnitude larger than its components and a direction that is between them.

EXAMPLE 2.2

Resolve the 200-lb force acting on the pipe, Fig. 2–11a, into components in the (a) x and y directions, and (b) x' and y directions.

(a)

(b)

Fig. 2–11

(c)

(d)

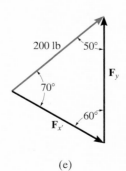

(e)

SOLUTION

In each case the parallelogram law is used to resolve **F** into its two components, and then the vector triangle is constructed to determine the numerical results by trigonometry.

Part (a). The vector addition $\mathbf{F} = \mathbf{F}_x + \mathbf{F}_y$ is shown in Fig. 2–11b. In particular, note that the length of the components is scaled along the x and y axes by first constructing lines from the tip of **F** parallel to the axes in accordance with the parallelogram law. From the vector triangle, Fig. 2–11c,

$$F_x = 200 \text{ lb } \cos 40° = 153 \text{ lb} \qquad \textit{Ans.}$$

$$F_y = 200 \text{ lb } \sin 40° = 129 \text{ lb} \qquad \textit{Ans.}$$

Part (b). The vector addition $\mathbf{F} = \mathbf{F}_{x'} + \mathbf{F}_y$ is shown in Fig. 2–11d. Note carefully how the parallelogram is constructed. Applying the law of sines and using the data listed on the vector triangle, Fig. 2–11e, yields

$$\frac{F_{x'}}{\sin 50°} = \frac{200 \text{ lb}}{\sin 60°}; \qquad F_{x'} = 200 \text{ lb}\left(\frac{\sin 50°}{\sin 60°}\right) = 177 \text{ lb} \qquad \textit{Ans.}$$

$$\frac{F_y}{\sin 70°} = \frac{200 \text{ lb}}{\sin 60°}; \qquad F_y = 200 \text{ lb}\left(\frac{\sin 70°}{\sin 60°}\right) = 217 \text{ lb} \qquad \textit{Ans.}$$

NOTE: A rough sketch drawn to scale will give some idea of the relative magnitude of the components, as calculated here.

EXAMPLE 2.3

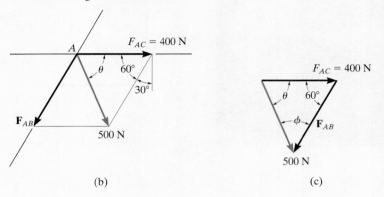

(a)   (b)   (c)

The force **F** acting on the frame shown in Fig. 2–12a has a magnitude of 500 N and is to be resolved into two components acting along members AB and AC. Determine the angle θ, measured *below* the horizontal, so that the component $\mathbf{F}_{AC}$ is directed from A toward C and has a magnitude of 400 N.

SOLUTION

By using the parallelogram law, the vector addition of the two components yielding the resultant is shown in Fig. 2–12b. Note carefully how the resultant force is resolved into the two components $\mathbf{F}_{AB}$ and $\mathbf{F}_{AC}$, which have specified lines of action. The corresponding vector triangle is shown in Fig. 2–12c.

The angle ϕ can be determined by using the law of sines:

$$\frac{400\text{ N}}{\sin \phi} = \frac{500\text{ N}}{\sin 60°}$$

$$\sin \phi = \left(\frac{400\text{ N}}{500\text{ N}}\right) \sin 60° = 0.6928$$

$$\phi = 43.9°$$

Hence,

$$\theta = 180° - 60° - 43.9° = 76.1° \searrow\phi \qquad \text{Ans.}$$

Using this value for θ, apply the law of cosines or the law of sines and show that $\mathbf{F}_{AB}$ has a magnitude of 561 N. This result shows that a component can sometimes have a greater magnitude than the resultant.

NOTE: **F** can also be directed at an angle θ *above* the horizontal, as shown in Fig. 2–12d, and still produce the required component $\mathbf{F}_{AC}$. Show that in this case $\theta = 16.1°$ and $F_{AB} = 161$ N.

(d)

Fig. 2–12

EXAMPLE 2.4

The ring shown in Fig. 2–13a is subjected to two forces, $\mathbf{F}_1$ and $\mathbf{F}_2$. If it is required that the resultant force have a magnitude of 1 kN and be directed vertically downward, determine (a) the magnitudes of $\mathbf{F}_1$ and $\mathbf{F}_2$ provided $\theta = 30°$, and (b) the magnitudes of $\mathbf{F}_1$ and $\mathbf{F}_2$ if F_2 is to have a minimum magnitude.

(a)   (b)   (c)

(d)

SOLUTION

Part (a). A sketch of the vector addition according to the parallelogram law is shown in Fig. 2–13b. From the vector triangle constructed in Fig. 2–13c, the unknown magnitudes F_1 and F_2 are determined by using the law of sines:

$$\frac{F_1}{\sin 30°} = \frac{1000 \text{ N}}{\sin 130°}$$

$$F_1 = 653 \text{ N} \qquad\qquad Ans.$$

$$\frac{F_2}{\sin 20°} = \frac{1000 \text{ N}}{\sin 130°}$$

$$F_2 = 446 \text{ N} \qquad\qquad Ans.$$

Part (b). If θ is not specified, then by the vector triangle, Fig. 2–13d, $\mathbf{F}_2$ may be added to $\mathbf{F}_1$ in various ways to yield the resultant 1000-N force. In particular, the *minimum* length or magnitude of $\mathbf{F}_2$ will occur when its line of action is *perpendicular to* $\mathbf{F}_1$. Any other direction, such as OA or OB, yields a larger value for F_2. Hence, when $\theta = 90° - 20° = 70°$, F_2 is minimum. From the triangle shown in Fig. 2–13e, it is seen that

$$F_1 = 1000 \sin 70° \text{N} = 940 \text{ N} \qquad Ans.$$

$$F_2 = 1000 \cos 70° \text{N} = 342 \text{ N} \qquad Ans.$$

(e)

Fig. 2–13

PROBLEMS

2–1. Determine the magnitude of the resultant force $\mathbf{F}_R = \mathbf{F}_1 + \mathbf{F}_2$ and its direction, measured counterclockwise from the positive x axis.

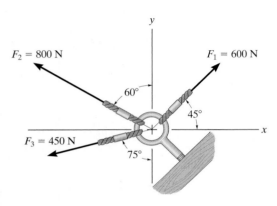

Prob. 2–1

2–2. Determine the magnitude of the resultant force and its direction measured counterclockwise from the positive x axis.

Prob. 2–2

2–3. Determine the magnitude of the resultant force $\mathbf{F}_R = \mathbf{F}_1 + \mathbf{F}_2$ and its direction, measured counterclockwise from the positive x axis.

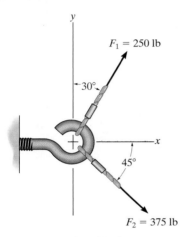

Prob. 2–3

*2–4. Determine the magnitude of the resultant force $\mathbf{F}_R = \mathbf{F}_1 + \mathbf{F}_2$ and its direction, measured clockwise from the positive u axis.

2–5. Resolve the force $\mathbf{F}_1$ into components acting along the u and v axes and determine the magnitudes of the components.

2–6. Resolve the force $\mathbf{F}_2$ into components acting along the u and v axes and determine the magnitudes of the components.

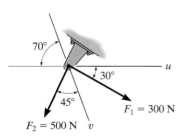

Probs. 2–4/5/6

2–7. Determine the magnitude of the resultant force $\mathbf{F}_R = \mathbf{F}_1 + \mathbf{F}_2$ and its direction measured counterclockwise from the positive u axis.

*2–8. Resolve the force $\mathbf{F}_1$ into components acting along the u and v axes and determine the components.

2–9. Resolve the force $\mathbf{F}_2$ into components acting along the u and v axes and determine the components.

2–11. The force acting on the gear tooth is $F = 20$ lb. Resolve this force into two components acting along the lines aa and bb.

*2–12. The component of force $\mathbf{F}$ acting along line aa is required to be 30 lb. Determine the magnitude of $\mathbf{F}$ and its component along line bb.

Probs. 2–7/8/9

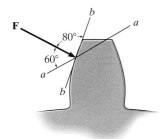

Probs. 2–11/12

2–10. Determine the components of the 250-N force acting along the u and v axes.

2–13. A resultant vertical force of 350 lb is necessary to hold the balloon in place. Resolve this force into components along the tether lines AB and AC, and compute the magnitude of each component.

Prob. 2–10

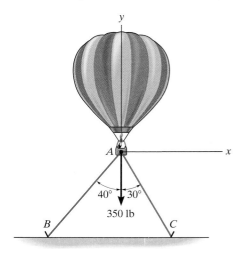

Prob. 2–13

2–14. The post is to be pulled out of the ground using two ropes A and B. Rope A is subjected to a force of 600 lb and is directed at 60° from the horizontal. If the resultant force acting on the post is to be 1200 lb, vertically upward, determine the force T in rope B and the corresponding angle θ.

Prob. 2–14

2–15. Resolve the force $\mathbf{F}_1$ into components acting along the u and v axes and determine the magnitudes of the components.

*2–16. Resolve the force $\mathbf{F}_2$ into components acting along the u and v axes and determine the magnitudes of the components.

Probs. 2–15/16

2–17. Determine the magnitude and direction of the resultant force $\mathbf{F}_R$. Express the result in terms of the magnitudes of the components $\mathbf{F}_1$ and $\mathbf{F}_2$ and the angle ϕ.

Prob. 2–17

2–18. If the tension in the cable is 400 N, determine the magnitude and direction of the resultant force acting on the pulley. This angle defines the same angle θ of line AB on the tailboard block.

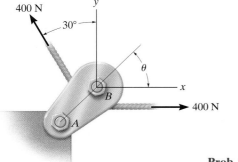

Prob. 2–18

2–19. The riveted bracket supports two forces. Determine the angle θ so that the resultant force is directed along the negative x axis. What is the magnitude of this resultant force?

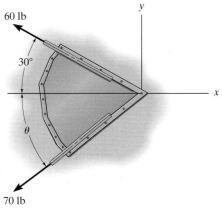

Prob. 2–19

*2–20. The plate is subjected to the forces acting on members A and B as shown. If $\theta = 60°$, determine the magnitude of the resultant of these forces and its direction measured clockwise from the positive x axis.

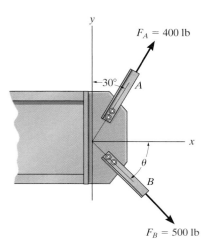

Prob. 2–20

2–21. Determine the angle θ for connecting member B to the plate so that the resultant angle of $\mathbf{F}_A$ and $\mathbf{F}_B$ is directed along the positive x axis. What is the magnitude of the resultant force?

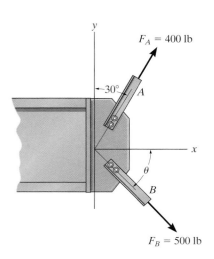

Prob. 2–21

2–22. Determine the magnitude and direction of the resultant $\mathbf{F}_R = \mathbf{F}_1 + \mathbf{F}_2 + \mathbf{F}_3$ of the three forces by first finding the resultant $\mathbf{F}' = \mathbf{F}_1 + \mathbf{F}_2$ and then forming $\mathbf{F}_R = \mathbf{F}' + \mathbf{F}_3$.

2–23. Determine the magnitude and direction of the resultant $\mathbf{F}_R = \mathbf{F}_1 + \mathbf{F}_2 + \mathbf{F}_3$ of the three forces by first finding the resultant $\mathbf{F}' = \mathbf{F}_2 + \mathbf{F}_3$ and then forming $\mathbf{F}_R = \mathbf{F}' + \mathbf{F}_1$.

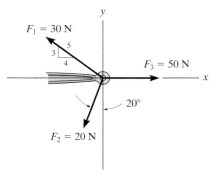

Probs. 2–22/23

*2–24. Resolve the 50-lb force into components acting along (a) the x and y axes, and (b) the x and y' axes.

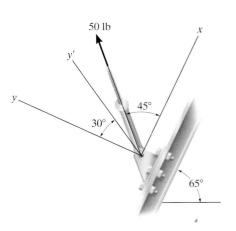

Prob. 2–24

2–25. The boat is to be pulled onto the shore using two ropes. Determine the magnitudes of forces **T** and **P** acting in each rope in order to develop a resultant force of 80 lb, directed along the keel *aa* as shown. Take $\theta = 40°$.

2–26. The boat is to be pulled onto the shore using two ropes. If the resultant force is to be 80 lb, directed along the keel *aa* as shown, determine the magnitudes of forces **T** and **P** acting in each rope and the angle θ of **P** so that the magnitude of **P** is a *minimum*. **T** acts at 30° from the keel as shown.

2–29. Three chains act on the bracket such that they create a resultant force having a magnitude of 500 lb. If two of the chains are subjected to known forces, as shown, determine the orientation θ of the third chain, measured clockwise from the positive *x* axis, so that the magnitude of force **F** in this chain is a *minimum*. All forces lie in the *x–y* plane. What is the magnitude of **F**? *Hint*: First find the resultant of the two known forces. Force **F** acts in this direction.

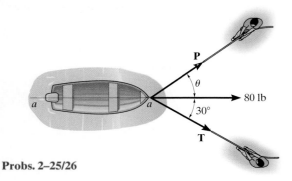

Probs. 2–25/26

2–27. The beam is to be hoisted using two chains. Determine the magnitudes of forces $\mathbf{F}_A$ and $\mathbf{F}_B$ acting on each chain in order to develop a resultant force of 600 N directed along the positive *y* axis. Set $\theta = 45°$.

*2–28. The beam is to be hoisted using two chains. If the resultant force is to be 600 N, directed along the positive *y* axis, determine the magnitudes of forces $\mathbf{F}_A$ and $\mathbf{F}_B$ acting on each chain and the orientation θ of $\mathbf{F}_B$ so that the magnitude of $\mathbf{F}_B$ is a *minimum*. $\mathbf{F}_A$ acts at 30° from the *y* axis as shown.

Prob. 2–29

2–30. Three cables pull on the pipe such that they create a resultant force having a magnitude of 900 lb. If two of the cables are subjected to known forces, as shown in the figure, determine the direction θ of the third cable so that the magnitude of force **F** in this cable is a *minimum*. All forces lie in the *x–y* plane. What is the magnitude of **F**? *Hint*: First find the resultant of the two known forces.

Probs. 2–27/28

Prob. 2–30

2.4 Addition of a System of Coplanar Forces

When the resultant of more than two forces has to be obtained, it is easier to find the components of each force along specified axes, add these components algebraically, and then form the resultant, rather than form the resultant of the forces by successive application of the parallelogram law as discussed in Sec. 2.3.

In this section we will resolve each force into its rectangular components F_x and F_y, which lie along the x and y axes, respectively, Fig. 2–14a. Although the axes are horizontal and vertical, they may in general be directed at any inclination, as long as they remain perpendicular to one another, Fig. 2–14b. In either case, by the parallelogram law, we require

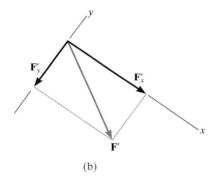

(a)

$$\mathbf{F} = \mathbf{F}_x + \mathbf{F}_y$$

and

$$\mathbf{F}' = \mathbf{F}'_x + \mathbf{F}'_y$$

As shown in Fig. 2–14, the sense of direction of each force component is represented *graphically* by the *arrowhead*. For *analytical* work, however, we must establish a notation for representing the directional sense of the rectangular components. This can be done in one of two ways.

(b)

Fig. 2–14

Scalar Notation.   Since the x and y axes have designated positive and negative directions, the magnitude and directional sense of the rectangular components of a force can be expressed in terms of *algebraic scalars*. For example, the components of $\mathbf{F}$ in Fig. 2–14a can be represented by positive scalars F_x and F_y since their sense of direction is along the *positive x* and *y* axes, respectively. In a similar manner, the components of $\mathbf{F}'$ in Fig. 2–14b are F'_x and $-F'_y$. Here the y component is negative, since $\mathbf{F}'_y$ is directed along the negative y axis.

It is important to keep in mind that this positive and negative scalar notation is to be used only for computational purposes, not for graphical representations in figures. Throughout the book, the *head of a vector arrow* in any figure indicates the sense of the vector *graphically*; algebraic signs are not used for this purpose. Thus, the vectors in Figs. 2–14a and 2–14b are designated by using boldface (vector) notation.* Whenever italic symbols are written near vector arrows in figures, they indicate the *magnitude* of the vector, which is *always* a *positive* quantity.

---

*Negative signs are used only in figures with boldface notation when showing equal but opposite pairs of vectors as in Fig. 2–2.

Cartesian Vector Notation.   It is also possible to represent the components of a force in terms of Cartesian unit vectors. When we do this the methods of vector algebra are easier to apply, and we will see that this becomes particularly advantageous for solving problems in three dimensions.

In two dimensions the *Cartesian unit vectors* **i** and **j** are used to designate the *directions* of the x and y axes, respectively, Fig. 2–15a.*

As shown in Fig. 2–15a, the *magnitude* of each component of **F** is *always a positive quantity*, which is represented by the (positive) scalars F_x and F_y. Therefore, having established notation to represent the magnitude and the direction of each vector component, we can express **F** in Fig. 2–15a as a *Cartesian vector*,

$$\mathbf{F} = F_x\mathbf{i} + F_y\mathbf{j}$$

And in the same way, **F′** in Fig. 2–15b can be expressed as

$$\mathbf{F'} = F'_x\mathbf{i} + F'_y(-\mathbf{j})$$

or simply

$$\mathbf{F'} = F'_x\mathbf{i} - F'_y\mathbf{j}$$

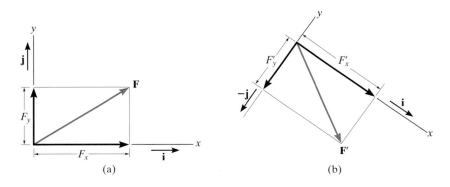

(a)   (b)

Fig. 2–15

*For handwritten work, unit vectors are usually indicated using a circumflex, e.g., $\hat{i}$ and $\hat{j}$. These vectors have a dimensionless magnitude of unity, and their sense (or arrowhead) will be described analytically by a plus or minus sign, depending on whether they are pointing along the positive or negative x or y axis.

Coplanar Force Resultants.   Either of the two methods just described can be used to determine the resultant of several *coplanar forces*. To do this, each force is first resolved into its x and y components, and then the respective components are added using *scalar algebra*. Since they are collinear. The resultant force is then formed by adding the resultants of the x and y components using the parallelogram law. For example, consider the three concurrent forces in Fig. 2–16*a*, which have x and y components as shown in Fig. 2–16*b*. To solve this problem using *Cartesian vector notation*, each force is first represented as a Cartesian vector, i.e.,

$$\mathbf{F}_1 = F_{1x}\mathbf{i} + F_{1y}\mathbf{j}$$
$$\mathbf{F}_2 = -F_{2x}\mathbf{i} + F_{2y}\mathbf{j}$$
$$\mathbf{F}_3 = F_{3x}\mathbf{i} - F_{3y}\mathbf{j}$$

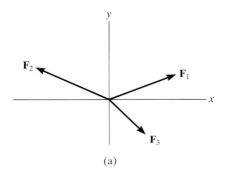

(a)

The vector resultant is therefore

$$\begin{aligned}
\mathbf{F}_R &= \mathbf{F}_1 + \mathbf{F}_2 + \mathbf{F}_3 \\
&= F_{1x}\mathbf{i} + F_{1y}\mathbf{j} - F_{2x}\mathbf{i} + F_{2y}\mathbf{j} + F_{3x}\mathbf{i} - F_{3y}\mathbf{j} \\
&= (F_{1x} - F_{2x} + F_{3x})\mathbf{i} + (F_{1y} + F_{2y} - F_{3y})\mathbf{j} \\
&= (F_{Rx})\mathbf{i} + (F_{Ry})\mathbf{j}
\end{aligned}$$

If *scalar notation* is used, then, from Fig. 2–16*b*, since x is positive to the right and y is positive upward, we have

$(\xrightarrow{+})$ $F_{Rx} = F_{1x} - F_{2x} + F_{3x}$
$(+\uparrow)$ $F_{Ry} = F_{1y} + F_{2y} - F_{3y}$

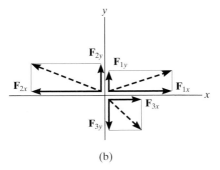

(b)

These results are the *same* as the $\mathbf{i}$ and $\mathbf{j}$ components of $\mathbf{F}_R$ determined above.

In the general case, the x and y components of the resultant of any number of coplanar forces can be represented symbolically by the algebraic sum of the x and y components of all the forces, i.e.,

$$\boxed{\begin{aligned} F_{Rx} &= \Sigma F_x \\ F_{Ry} &= \Sigma F_y \end{aligned}}$$ (2–1)

When applying these equations, it is important to use the *sign convention* established for the components; and that is, components having a directional sense along the positive coordinate axes are considered positive scalars, whereas those having a directional sense along the negative coordinate axes are considered negative scalars. If this convention is followed, then the signs of the resultant components will specify the sense of these components. For example, a positive result indicates that the component has a directional sense which is in the positive coordinate direction.

(c)

Fig. 2–16

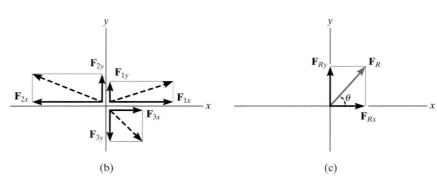

(a)   (b)   (c)

Fig. 2–16

The resultant force of the four cable forces acting on the supporting bracket can be determined by adding algebraically the separate x and y components of each cable force. This resultant $\mathbf{F}_R$ produces the *same pulling effect* on the bracket as all four cables.

Once the resultant components are determined, they may be sketched along the x and y axes in their proper directions, and the resultant force can be determined from vector addition, as shown in Fig. 2–16c. From this sketch, the magnitude of $\mathbf{F}_R$ is then found from the Pythagorean theorem; that is,

$$F_R = \sqrt{F_{Rx}^2 + F_{Ry}^2}$$

Also, the direction angle θ, which specifies the orientation of the force, is determined from trigonometry:

$$\theta = \tan^{-1}\left|\frac{F_{Ry}}{F_{Rx}}\right|$$

The above concepts are illustrated numerically in the examples which follow.

Important Points

- The resultant of several coplanar forces can easily be determined if an x, y coordinate system is established and the forces are resolved along the axes.
- The direction of each force is specified by the angle its line of action makes with one of the axes, or by a sloped triangle.
- The orientation of the x and y axes is arbitrary, and their positive direction can be specified by the Cartesian unit vectors **i** and **j**.
- The x and y components of the *resultant force* are simply the algebraic addition of the components of all the coplanar forces.
- The magnitude of the resultant force is determined from the Pythagorean theorem, and when the components are sketched on the x and y axes, the direction can be determined from trigonometry.

EXAMPLE 2.5

Determine the x and y components of $\mathbf{F}_1$ and $\mathbf{F}_2$ acting on the boom shown in Fig. 2–17a. Express each force as a Cartesian vector.

SOLUTION

Scalar Notation. By the parallelogram law, $\mathbf{F}_1$ is resolved into x and y components, Fig. 2–17b. The magnitude of each component is determined by trigonometry. Since $\mathbf{F}_{1x}$ acts in the $-x$ direction, and $\mathbf{F}_{1y}$ acts in the $+y$ direction, we have

$$F_{1x} = -200 \sin 30° \text{ N} = -100 \text{ N} = 100 \text{ N} \leftarrow \qquad Ans.$$
$$F_{1y} = 200 \cos 30° \text{ N} = 173 \text{ N} = 173 \text{ N} \uparrow \qquad Ans.$$

The force $\mathbf{F}_2$ is resolved into its x and y components as shown in Fig. 2–17c. Here the *slope* of the line of action for the force is indicated. From this "slope triangle" we could obtain the angle θ, e.g., $\theta = \tan^{-1}\left(\frac{5}{12}\right)$, and then proceed to determine the magnitudes of the components in the same manner as for $\mathbf{F}_1$. An easier method, however, consists of using proportional parts of similar triangles, i.e.,

$$\frac{F_{2x}}{260 \text{ N}} = \frac{12}{13} \qquad F_{2x} = 260 \text{ N}\left(\frac{12}{13}\right) = 240 \text{ N}$$

Similarly,

$$F_{2y} = 260 \text{ N}\left(\frac{5}{13}\right) = 100 \text{ N}$$

Hence, using scalar notation,

$$F_{2x} = 240 \text{ N} = 240 \text{ N} \rightarrow \qquad Ans.$$
$$F_{2y} = -100 \text{ N} = 100 \text{ N} \downarrow \qquad Ans.$$

Cartesian Vector Notation. Having determined the magnitudes and directions of the components of each force, we can express each force as a Cartesian vector.

$$\mathbf{F}_1 = \{-100\mathbf{i} + 173\mathbf{j}\} \text{ N} \qquad Ans.$$
$$\mathbf{F}_2 = \{240\mathbf{i} - 100\mathbf{j}\} \text{ N} \qquad Ans.$$

NOTE: The magnitude of the *horizontal component*, F_{2x}, was obtained by multiplying the force magnitude by the ratio of the *horizontal leg* of the slope triangle divided by the hypotenuse; whereas the magnitude of the *vertical component*, F_{2y}, was obtained by multiplying the force magnitude by the ratio of the *vertical leg* divided by the hypotenuse.

(a)

(b)

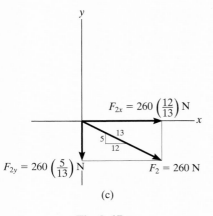

(c)

Fig. 2–17

EXAMPLE 2.6

(a)

(b)

(c)

Fig. 2–18

The link in Fig. 2–18a is subjected to two forces $\mathbf{F}_1$ and $\mathbf{F}_2$. Determine the magnitude and orientation of the resultant force.

SOLUTION I

Scalar Notation. This problem can be solved by using the parallelogram law; however, here we will resolve each force into its x and y components, Fig. 2–18b, and sum these components algebraically. Indicating the "positive" sense of the x and y force components alongside each equation, we have

$$\xrightarrow{+} F_{Rx} = \Sigma F_x; \qquad F_{Rx} = 600 \cos 30° \text{ N} - 400 \sin 45° \text{ N}$$
$$= 236.8 \text{ N} \rightarrow$$
$$+\uparrow F_{Ry} = \Sigma F_y; \qquad F_{Ry} = 600 \sin 30° \text{ N} + 400 \cos 45° \text{ N}$$
$$= 582.8 \text{ N}\uparrow$$

The resultant force, shown in Fig. 2–18c, has a *magnitude* of

$$F_R = \sqrt{(236.8 \text{ N})^2 + (582.8 \text{ N})^2}$$
$$= 629 \text{ N} \qquad\qquad\qquad Ans.$$

From the vector addition, Fig. 2–18c, the direction angle θ is

$$\theta = \tan^{-1}\left(\frac{582.8 \text{ N}}{236.8 \text{ N}}\right) = 67.9° \qquad\qquad Ans.$$

SOLUTION II

Cartesian Vector Notation. From Fig. 2–18b, each force is expressed as a Cartesian vector

$$\mathbf{F}_1 = \{600 \cos 30°\mathbf{i} + 600 \sin 30°\mathbf{j}\} \text{ N}$$
$$\mathbf{F}_2 = \{-400 \sin 45°\mathbf{i} + 400 \cos 45°\mathbf{j}\} \text{ N}$$

Thus,

$$\mathbf{F}_R = \mathbf{F}_1 + \mathbf{F}_2 = (600 \cos 30° \text{ N} - 400 \sin 45° \text{ N})\mathbf{i}$$
$$+ (600 \sin 30° \text{ N} + 400 \cos 45° \text{ N})\mathbf{j}$$
$$= \{236.8\mathbf{i} + 582.8\mathbf{j}\} \text{ N}$$

The magnitude and direction of $\mathbf{F}_R$ are determined in the same manner as shown above.

NOTE: Comparing the two methods of solution, notice that the use of scalar notation is more efficient since the components can be found *directly*, without first having to express each force as a Cartesian vector before adding the components. Later we will show that Cartesian vector analysis is very beneficial for solving three-dimensional problems.

EXAMPLE 2.7

The end of the boom O in Fig. 2–19a is subjected to three concurrent and coplanar forces. Determine the magnitude and orientation of the resultant force.

(a)

SOLUTION

Each force is resolved into its x and y components, Fig. 2–19b. Summing the x components, we have

$$\xrightarrow{+} F_{Rx} = \Sigma F_x; \qquad F_{Rx} = -400 \text{ N} + 250 \sin 45° \text{ N} - 200\left(\tfrac{4}{5}\right)\text{N}$$
$$= -383.2 \text{ N} = 383.2 \text{ N} \leftarrow$$

(b)

The negative sign indicates that F_{Rx} acts to the left, i.e., in the negative x direction as noted by the small arrow. Obviously, this occurs because F_1 and F_3 in Fig. 2–19b contribute a greater pull to the left than F_2, which pulls to the right. Summing the y components yields

$$+\uparrow F_{Ry} = \Sigma F_y; \qquad F_{Ry} = 250 \cos 45° \text{ N} + 200\left(\tfrac{3}{5}\right) \text{ N}$$
$$= 296.8 \text{ N} \uparrow$$

The resultant force, shown in Fig. 2–19c, has a *magnitude* of

$$F_R = \sqrt{(-383.2 \text{ N})^2 + (296.8 \text{ N})^2}$$
$$= 485 \text{ N} \qquad\qquad Ans.$$

From the vector addition in Fig. 2–19c, the direction angle θ is

$$\theta = \tan^{-1}\left(\frac{296.8}{383.2}\right) = 37.8° \qquad\qquad Ans.$$

(c)

Fig. 2–19

NOTE: Application of this method is more convenient, compared to two applications of the parallelogram law.

PROBLEMS

2–31. Determine the x and y components of the 800-lb force.

2–33. Determine the magnitude of the resultant force and its direction measured counterclockwise from the positive x axis.

800 lb

40°

y

60°

x

60°

Prob. 2–31

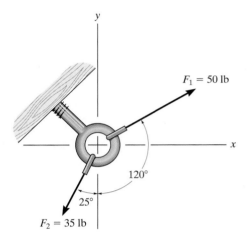

y

$F_1 = 50$ lb

x

120°

25°

$F_2 = 35$ lb

Prob. 2–33

*2–32. Determine the magnitude of the resultant force and its direction, measured clockwise from the positive x axis.

2–34. Determine the magnitude of the resultant force and its direction, measured counterclockwise from the positive x axis.

y

50 N

30°

45°

70 N

x

65 N

Prob. 2–32

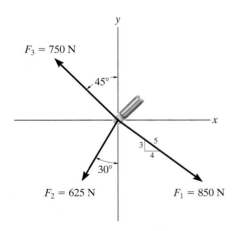

y

$F_3 = 750$ N

45°

x

5
3
4

30°

$F_2 = 625$ N

$F_1 = 850$ N

Prob. 2–34

2–35. Three forces act on the bracket. Determine the magnitude and direction θ of $\mathbf{F}_1$ so that the resultant force is directed along the positive x' axis and has a magnitude of 1 kN.

*2–36. If $F_1 = 300$ N and $\theta = 20°$, determine the magnitude and direction, measured counterclockwise from the x' axis, of the resultant force of the three forces acting on the bracket.

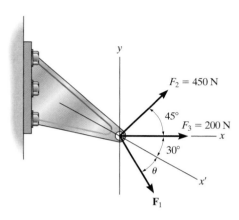

Probs. 2–35/36

2–37. Determine the magnitude of the resultant force and its direction, measured counterclockwise from the positive x axis.

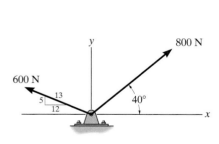

Prob. 2–37

2–38. Determine the magnitude of the resultant force and its direction, measured counterclockwise from the positive x axis.

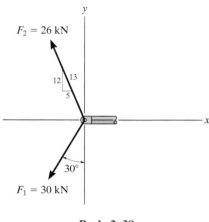

Prob. 2–38

2–39. Determine the magnitude of the resultant force and its direction measured counterclockwise from the positive x axis.

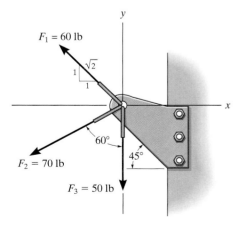

Prob. 2–39

*2–40. Solve Prob. 2–1 by summing the rectangular or x, y components of the forces to obtain the resultant force.

2–41. Solve Prob. 2–22 by summing the rectangular or x, y components of the forces to obtain the resultant force.

2-42. Determine the magnitude and orientation, measured counterclockwise from the positive y axis, of the resultant force acting on the bracket, if $F_B = 600$ N and $\theta = 20°$.

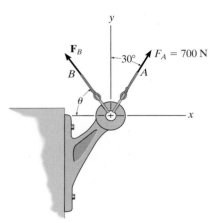

Prob. 2–42

2–43. If $F_1 = 300$ N and $\theta = 10°$, determine the magnitude and direction, measured counterclockwise from the positive x' axis, of the resultant force of the three forces acting on the bracket.

Prob. 2–43

*2–44. Determine the x and y components of $\mathbf{F_1}$ and $\mathbf{F_2}$.

2–45. Determine the magnitude of the resultant force and its direction, measured counterclockwise from the positive x axis.

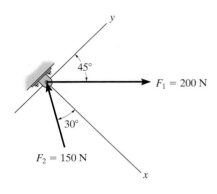

Probs. 2–44/45

2–46. Determine the x and y components of each force acting on the *gusset plate* of the bridge truss. Show that the resultant force is zero.

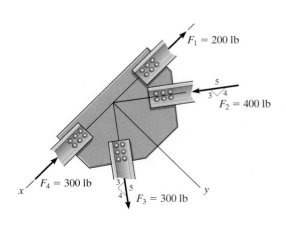

Prob. 2–46

2–47. If $\theta = 60°$ and $F = 20$ kN, determine the magnitude of the resultant force and its direction measured clockwise from the positive x axis.

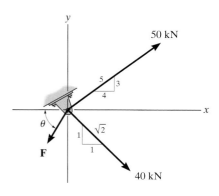

Prob. 2–47

2–50. Express each of the three forces acting on the column in Cartesian vector form and compute the magnitude of the resultant force.

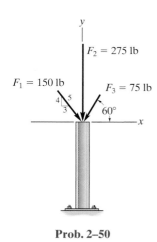

Prob. 2–50

*2–48. Three forces act on the bracket. Determine the magnitude and direction θ of $\mathbf{F}_1$ so that the resultant force is directed along the positive x' axis and has a magnitude of 800 N.

2–49. If $F_1 = 300$ N and $\theta = 10°$, determine the magnitude and direction, measured counterclockwise from the positive x' axis, of the resultant force acting on the bracket.

2–51. Determine the magnitude of force $\mathbf{F}$ so that the resultant $\mathbf{F}_R$ of the three forces is as small as possible. What is the minimum magnitude of $\mathbf{F}_R$?

Probs. 2–48/49

Prob. 2–51

*2–52. Express each of the three forces acting on the bracket in Cartesian vector form with respect to the x and y axes. Determine the magnitude and direction θ of $\mathbf{F}_1$ so that the resultant force is directed along the positive x' axis and has a magnitude of $F_R = 600$ N.

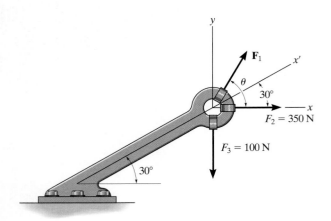

Prob. 2–52

2–53. The three concurrent forces acting on the post produce a resultant force $\mathbf{F}_R = \mathbf{0}$. If $F_2 = \frac{1}{2}F_1$, and $\mathbf{F}_1$ is to be 90° from $\mathbf{F}_2$ as shown, determine the required magnitude F_3 expressed in terms of F_1 and the angle θ.

Prob. 2–53

2–54. Three forces act on the bracket. Determine the magnitude and orientation θ of $\mathbf{F}_2$ so that the resultant force is directed along the positive u axis and has a magnitude of 50 lb.

2–55. If $F_2 = 150$ lb and $\theta = 55°$, determine the magnitude and orientation, measured clockwise from the positive x axis, of the resultant force of the three forces acting on the bracket.

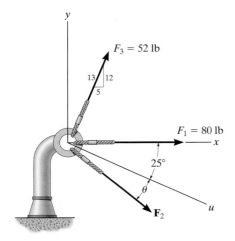

Probs. 2–54/55

*2–56. Three forces act on the ring. Determine the range of values for the magnitude of $\mathbf{P}$ so that the magnitude of the resultant force does not exceed 2500 N. Force $\mathbf{P}$ is always directed to the right.

Prob. 2–56

2.5 Cartesian Vectors

The operations of vector algebra, when applied to solving problems in *three dimensions*, are greatly simplified if the vectors are first represented in Cartesian vector form. In this section we will present a general method for doing this; then in the next section we will apply this method to solving problems involving the addition of forces. Similar applications will be illustrated for the position and moment vectors given in later sections of the book.

Fig. 2–20

Right-Handed Coordinate System.
We will use a right-handed coordinate system to develop the theory of vector algebra that follows. A rectangular or Cartesian coordinate system is said to be *right-handed* provided the thumb of the right hand points in the direction of the positive z axis when the right-hand fingers are curled about this axis and directed from the positive x toward the positive y axis, Fig. 2–20. Furthermore, according to this rule, the z axis for a two-dimensional problem as in Fig. 2–19 would be directed outward, perpendicular to the page.

Rectangular Components of a Vector.
A vector $\mathbf{A}$ may have one, two, or three rectangular components along the x, y, z coordinate axes, depending on how the vector is oriented relative to the axes. In general, though, when $\mathbf{A}$ is directed within an octant of the x, y, z frame, Fig. 2–21, then by two successive applications of the parallelogram law, we may resolve the vector into components as $\mathbf{A} = \mathbf{A}' + \mathbf{A}_z$ and then $\mathbf{A}' = \mathbf{A}_x + \mathbf{A}_y$. Combining these equations, $\mathbf{A}$ is represented by the vector sum of its *three* rectangular components,

$$\mathbf{A} = \mathbf{A}_x + \mathbf{A}_y + \mathbf{A}_z \qquad (2\text{–}2)$$

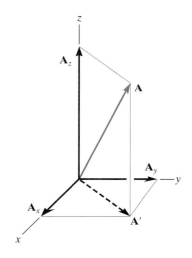

Fig. 2–21

Unit Vector.
The direction of $\mathbf{A}$ can be specified using a unit vector, so named since it has a magnitude of 1. If $\mathbf{A}$ is a vector having a magnitude $A \neq 0$, then the unit vector having the *same direction* as $\mathbf{A}$ is represented by

$$\mathbf{u}_A = \frac{\mathbf{A}}{A} \qquad (2\text{–}3)$$

So that

$$\mathbf{A} = A\mathbf{u}_A \qquad (2\text{–}4)$$

Since $\mathbf{A}$ is of a certain type, e.g., a force vector, it is customary to use the proper set of units for its description. The magnitude A also has this same set of units; hence, from Eq. 2–3, the *unit vector will be dimensionless* since the units will cancel out. Equation 2–4 shows that vector $\mathbf{A}$ can be expressed in terms of both its magnitude and direction *separately*; i.e., A (a positive scalar) defines the *magnitude* of $\mathbf{A}$, and $\mathbf{u}_A$ (a dimensionless vector) defines the *direction* and sense of $\mathbf{A}$, Fig. 2–22.

Fig. 2–22

Fig. 2–23

Fig. 2–24

Fig. 2–25

Cartesian Unit Vectors.   In three dimensions, the set of Cartesian unit vectors, **i**, **j**, **k**, is used to designate the directions of the *x*, *y*, *z* axes respectively. As stated in Sec. 2.4, the *sense* (or arrowhead) of these vectors will be described analytically by a plus or minus sign, depending on whether they are pointing along the positive or negative *x*, *y*, or *z* axes. The positive Cartesian unit vectors are shown in Fig. 2–23.

Cartesian Vector Representation.   Since the three components of **A** in Eq. 2–2 act in the positive **i**, **j**, and **k** directions, Fig. 2–24, we can write **A** in Cartesian vector form as

$$\mathbf{A} = A_x\mathbf{i} + A_y\mathbf{j} + A_z\mathbf{k} \qquad (2\text{–}5)$$

There is a distinct advantage to writing vectors in this manner. Separating the *magnitude* and *direction* of each *component vector* simplifies the operations of vector algebra, particularly in three dimensions.

Magnitude of a Cartesian Vector.   It is always possible to obtain the magnitude of **A** provided it is expressed in Cartesian vector form. As shown in Fig. 2–25, from the blue right triangle, $A = \sqrt{A'^2 + A_z^2}$, and from the gray right triangle, $A' = \sqrt{A_x^2 + A_y^2}$. Combining these equations yields

$$A = \sqrt{A_x^2 + A_y^2 + A_z^2} \qquad (2\text{–}6)$$

*Hence, the magnitude of **A** is equal to the positive square root of the sum of the squares of its components.*

Direction of a Cartesian Vector.   The *orientation* of **A** is defined by the *coordinate direction angles* α (alpha), β (beta), and γ (gamma), measured between the *tail* of **A** and the *positive x, y, z* axes located at the tail of **A**, Fig. 2–26. Note that regardless of where **A** is directed, each of these angles will be between 0° and 180°.

To determine α, β, and γ, consider the projection of **A** onto the *x*, *y*, *z* axes, Fig. 2–27. Referring to the blue colored right triangles shown in each figure, we have

$$\cos\alpha = \frac{A_x}{A} \qquad \cos\beta = \frac{A_y}{A} \qquad \cos\gamma = \frac{A_z}{A} \qquad (2\text{–}7)$$

These numbers are known as the *direction cosines* of **A**. Once they have been obtained, the coordinate direction angles α, β, γ can then be determined from the inverse cosines.

An easy way of obtaining the direction cosines of **A** is to form a unit vector in the direction of **A**, Eq. 2–3. Provided **A** is expressed in Cartesian vector form, $\mathbf{A} = A_x\mathbf{i} + A_y\mathbf{j} + A_z\mathbf{k}$ (Eq. 2–5), then

Fig. 2–26

(a)

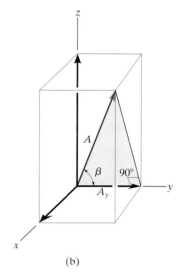

(b)

$$\mathbf{u}_A = \frac{\mathbf{A}}{A} = \frac{A_x}{A}\mathbf{i} + \frac{A_y}{A}\mathbf{j} + \frac{A_z}{A}\mathbf{k} \qquad (2\text{–}8)$$

where $A = \sqrt{A_x^2 + A_y^2 + A_z^2}$ (Eq. 2–6). By comparison with Eqs. 2–7, it is seen that *the* $\mathbf{i}, \mathbf{j}, \mathbf{k}$ *components of* $\mathbf{u}_A$ *represent the direction cosines of* $\mathbf{A}$, i.e.,

$$\mathbf{u}_A = \cos\alpha\,\mathbf{i} + \cos\beta\,\mathbf{j} + \cos\gamma\,\mathbf{k} \qquad (2\text{–}9)$$

Since the magnitude of a vector is equal to the positive square root of the sum of the squares of the magnitudes of its components, and $\mathbf{u}_A$ has a magnitude of 1, then from Eq. 2–9 an important relation between the direction cosines can be formulated as

$$\cos^2\alpha + \cos^2\beta + \cos^2\gamma = 1 \qquad (2\text{–}10)$$

Provided vector $\mathbf{A}$ lies in a known octant, this equation can be used to determine one of the coordinate direction angles if the other two are known.

Finally, if the magnitude and coordinate direction angles of $\mathbf{A}$ are given, $\mathbf{A}$ may be expressed in Cartesian vector form as

$$\begin{aligned} \mathbf{A} &= A\mathbf{u}_A \\ &= A\cos\alpha\,\mathbf{i} + A\cos\beta\,\mathbf{j} + A\cos\gamma\,\mathbf{k} \\ &= A_x\mathbf{i} + A_y\mathbf{j} + A_z\mathbf{k} \end{aligned} \qquad (2\text{–}11)$$

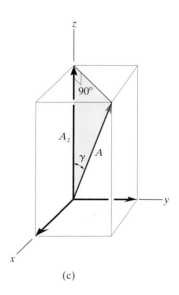

(c)

Fig. 2–27

2.6 Addition and Subtraction of Cartesian Vectors

The vector operations of addition and subtraction of two or more vectors are greatly simplified if the vectors are expressed in terms of their Cartesian components. For example, if $\mathbf{A} = A_x\mathbf{i} + A_y\mathbf{j} + A_z\mathbf{k}$ and $\mathbf{B} = B_x\mathbf{i} + B_y\mathbf{j} + B_z\mathbf{k}$, Fig. 2–28, then the resultant vector, $\mathbf{R}$, has components which represent the scalar sums of the $\mathbf{i}, \mathbf{j}, \mathbf{k}$ components of $\mathbf{A}$ and $\mathbf{B}$, i.e.,

$$\mathbf{R} = \mathbf{A} + \mathbf{B} = (A_x + B_x)\mathbf{i} + (A_y + B_y)\mathbf{j} + (A_z + B_z)\mathbf{k}$$

Vector subtraction, being a special case of vector addition, simply requires a scalar subtraction of the respective $\mathbf{i}, \mathbf{j}, \mathbf{k}$ components of either $\mathbf{A}$ or $\mathbf{B}$. For example,

$$\mathbf{R}' = \mathbf{A} - \mathbf{B} = (A_x - B_x)\mathbf{i} + (A_y - B_y)\mathbf{j} + (A_z - B_z)\mathbf{k}$$

Concurrent Force Systems. If the above concept of vector addition is generalized and applied to a system of several concurrent forces, then the force resultant is the vector sum of all the forces in the system and can be written as

$$\mathbf{F}_R = \Sigma\mathbf{F} = \Sigma F_x\mathbf{i} + \Sigma F_y\mathbf{j} + \Sigma F_z\mathbf{k} \qquad (2\text{–}12)$$

Here ΣF_x, ΣF_y, and ΣF_z represent the algebraic sums of the respective x, y, z or $\mathbf{i}, \mathbf{j}, \mathbf{k}$ components of each force in the system.

The examples which follow illustrate numerically the methods used to apply the above theory to the solution of problems involving force as a vector quantity.

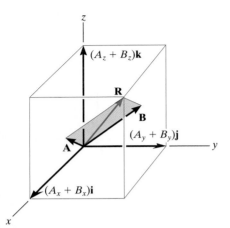

Fig. 2–28

The force $\mathbf{F}$ that the tie-down rope exerts on the ground support at O is directed along the rope. Using the local x, y, z axes, the coordinate direction angles α, β, γ can be measured. The cosines of their values form the components of a unit vector $\mathbf{u}$ which acts in the direction of the rope. If the force has a magnitude F, then the force can be written in Cartesian vector form, as $\mathbf{F} = F\mathbf{u} = F\cos\alpha\mathbf{i} + F\cos\beta\mathbf{j} + F\cos\gamma\mathbf{k}$.

Important Points

- Cartesian vector analysis is often used to solve problems in three dimensions.
- The positive directions of the *x, y, z* axes are defined by the Cartesian unit vectors **i, j, k**, respectively.
- The *magnitude* of a Cartesian vector is $A = \sqrt{A_x^2 + A_y^2 + A_z^2}$.
- The *direction* of a Cartesian vector is specified using coordinate direction angles which the tail of the vector makes with the positive *x, y, z* axes, respectively. The components of the unit vector $\mathbf{u} = \mathbf{A}/A$ represent the direction cosines of α, β, γ. Only two of the angles α, β, γ have to be specified. The third angle is determined from the relationship $\cos^2 \alpha + \cos^2 \beta + \cos^2 \gamma = 1$.
- To find the *resultant* of a concurrent force system, express each force as a Cartesian vector and add the **i, j, k** components of all the forces in the system.

EXAMPLE 2.8

Express the force **F** shown in Fig. 2–29 as a Cartesian vector.

SOLUTION
Since only two coordinate direction angles are specified, the third angle α must be determined from Eq. 2–10; i.e.,

$$\cos^2 \alpha + \cos^2 \beta + \cos^2 \gamma = 1$$
$$\cos^2 \alpha + \cos^2 60° + \cos^2 45° = 1$$
$$\cos \alpha = \sqrt{1 - (0.5)^2 - (0.707)^2} = \pm 0.5$$

Hence, two possibilities exist, namely,

$$\alpha = \cos^{-1}(0.5) = 60° \quad \text{or} \quad \alpha = \cos^{-1}(-0.5) = 120°$$

By inspection of Fig. 2–29, it is necessary that $\alpha = 60°$, since $\mathbf{F}_x$ is in the +*x* direction.

Using Eq. 2–11, with $F = 200$ N, we have

$$\mathbf{F} = F \cos \alpha \mathbf{i} + F \cos \beta \mathbf{j} + F \cos \gamma \mathbf{k}$$
$$= (200 \cos 60° \text{ N})\mathbf{i} + (200 \cos 60° \text{ N})\mathbf{j} + (200 \cos 45° \text{ N})\mathbf{k}$$
$$= \{100.0\mathbf{i} + 100.0\mathbf{j} + 141.4\mathbf{k}\} \text{ N} \qquad \textit{Ans.}$$

NOTE: By applying Eq. 2–6, note that indeed the magnitude of $F = 200$ N.

$$F = \sqrt{F_x^2 + F_y^2 + F_z^2}$$
$$= \sqrt{(100.0)^2 + (100.0)^2 + (141.4)^2} = 200 \text{ N}$$

Fig. 2–29

EXAMPLE 2.9

Determine the magnitude and the coordinate direction angles of the resultant force acting on the ring in Fig. 2–30*a*.

(a)

Fig. 2–30

(b)

SOLUTION

Since each force is represented in Cartesian vector form, the resultant force, shown in Fig. 2–30*b*, is

$$\mathbf{F}_R = \Sigma\mathbf{F} = \mathbf{F}_1 + \mathbf{F}_2 = \{60\mathbf{j} + 80\mathbf{k}\}\ \text{lb} + \{50\mathbf{i} - 100\mathbf{j} + 100\mathbf{k}\}\ \text{lb}$$
$$= \{50\mathbf{i} - 40\mathbf{j} + 180\mathbf{k}\}\ \text{lb}$$

The magnitude of $\mathbf{F}_R$ is found from Eq. 2–6, i.e.,

$$F_R = \sqrt{(50)^2 + (-40)^2 + (180)^2} = 191.0$$
$$= 191\ \text{lb} \qquad\qquad\qquad\qquad Ans.$$

The coordinate direction angles α, β, γ are determined from the components of the unit vector acting in the direction of $\mathbf{F}_R$.

$$\mathbf{u}_{F_R} = \frac{\mathbf{F}_R}{F_R} = \frac{50}{191.0}\mathbf{i} - \frac{40}{191.0}\mathbf{j} + \frac{180}{191.0}\mathbf{k}$$
$$= 0.2617\mathbf{i} - 0.2094\mathbf{j} + 0.9422\mathbf{k}$$

so that

$$\cos\alpha = 0.2617 \qquad \alpha = 74.8° \qquad Ans.$$
$$\cos\beta = -0.2094 \qquad \beta = 102° \qquad Ans.$$
$$\cos\gamma = 0.9422 \qquad \gamma = 19.6° \qquad Ans.$$

These angles are shown in Fig. 2–30*b*.

NOTE: In particular, notice that $\beta > 90°$ since the $\mathbf{j}$ component of $\mathbf{u}_{F_R}$ is negative. This seems reasonable considering how $\mathbf{F}_1$ and $\mathbf{F}_2$ add according to the parallelogram law.

EXAMPLE 2.10

Express the force $\mathbf{F}_1$, shown in Fig. 2–31a as a Cartesian vector.

2-60

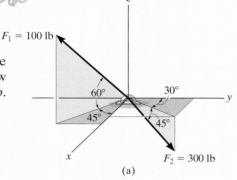

(a)

SOLUTION

The angles of 60° and 45° defining the direction of $\mathbf{F}_1$ are *not* coordinate direction angles. The two successive applications of the parallelogram law needed to resolve $\mathbf{F}_1$ into its x, y, z components are shown in Fig. 2–31b. By trigonometry, the magnitudes of the components are

$$F_{1z} = 100 \sin 60° \text{ lb} = 86.6 \text{ lb}$$
$$F' = 100 \cos 60° \text{ lb} = 50 \text{ lb}$$
$$F_{1x} = 50 \cos 45° \text{ lb} = 35.4 \text{ lb}$$
$$F_{1y} = 50 \sin 45° \text{ lb} = 35.4 \text{ lb}$$

Realizing that $\mathbf{F}_{1y}$ has a direction defined by $-\mathbf{j}$, we have

$$\mathbf{F}_1 = \{35.4\mathbf{i} - 35.4\mathbf{j} + 86.6\mathbf{k}\} \text{ lb} \qquad Ans.$$

To show that the magnitude of this vector is indeed 100 lb, apply Eq. 2–6,

$$F_1 = \sqrt{F_{1x}^2 + F_{1y}^2 + F_{1z}^2}$$
$$= \sqrt{(35.4)^2 + (-35.4)^2 + (86.6)^2} = 100 \text{ lb}$$

(b)

If needed, the coordinate direction angles of $\mathbf{F}_1$ can be determined from the components of the unit vector acting in the direction of $\mathbf{F}_1$. Hence,

$$\mathbf{u}_1 = \frac{\mathbf{F}_1}{F_1} = \frac{F_{1x}}{F_1}\mathbf{i} + \frac{F_{1y}}{F_1}\mathbf{j} + \frac{F_{1z}}{F_1}\mathbf{k}$$

$$= \frac{35.4}{100}\mathbf{i} - \frac{35.4}{100}\mathbf{j} + \frac{86.6}{100}\mathbf{k}$$

$$= 0.354\mathbf{i} - 0.354\mathbf{j} + 0.866\mathbf{k}$$

so that

$$\alpha_1 = \cos^{-1}(0.354) = 69.3°$$
$$\beta_1 = \cos^{-1}(-0.354) = 111°$$
$$\gamma_1 = \cos^{-1}(0.866) = 30.0°$$

These results are shown in Fig. 2–31c.

NOTE: Using this same method, show that $\mathbf{F}_2$ in Fig. 2–31a can be written in Cartesian vector form as

$$\mathbf{F}_2 = \{106\mathbf{i} + 184\mathbf{j} - 212\mathbf{k}\} \text{ N} \qquad Ans.$$

(c)

Fig. 2–31

EXAMPLE 2.11

(a)

(b)

Fig. 2–32

Two forces act on the hook shown in Fig. 2–32a. Specify the coordinate direction angles of $\mathbf{F}_2$ so that the resultant force $\mathbf{F}_R$ acts along the positive y axis and has a magnitude of 800 N.

SOLUTION

To solve this problem, the resultant force $\mathbf{F}_R$ and its two components, $\mathbf{F}_1$ and $\mathbf{F}_2$, will each be expressed in Cartesian vector form. Then, as shown in Fig. 2–32b, it is necessary that $\mathbf{F}_R = \mathbf{F}_1 + \mathbf{F}_2$.

Applying Eq. 2–11,

$$\mathbf{F}_1 = F_1 \cos \alpha_1 \mathbf{i} + F_1 \cos \beta_1 \mathbf{j} + F_1 \cos \gamma_1 \mathbf{k}$$

$$= 300 \cos 45° \, \mathbf{N} \mathbf{i} + 300 \cos 60° \, \mathbf{N} \mathbf{j} + 300 \cos 120° \, \mathbf{N} \mathbf{k}$$

$$= \{212.1\mathbf{i} + 150\mathbf{j} - 150\mathbf{k}\} \, \mathbf{N}$$

$$\mathbf{F}_2 = F_{2x}\mathbf{i} + F_{2y}\mathbf{j} + F_{2z}\mathbf{k}$$

Since the resultant force $\mathbf{F}_R$ has a magnitude of 800 N and acts in the $+\mathbf{j}$ direction.

$$\mathbf{F}_R = (800 \, \mathrm{N})(+\mathbf{j}) = \{800\mathbf{j}\} \, \mathbf{N}$$

We require

$$\mathbf{F}_R = \mathbf{F}_1 + \mathbf{F}_2$$

$$800\mathbf{j} = 212.1\mathbf{i} + 150\mathbf{j} - 150\mathbf{k} + F_{2x}\mathbf{i} + F_{2y}\mathbf{j} + F_{2z}\mathbf{k}$$

$$800\mathbf{j} = (212.1 + F_{2x})\mathbf{i} + (150 + F_{2y})\mathbf{j} + (-150 + F_{2z})\mathbf{k}$$

To satisfy this equation, the corresponding $\mathbf{i}$, $\mathbf{j}$, $\mathbf{k}$ components on the left and right sides must be equal. This is equivalent to stating that the x, y, z components of $\mathbf{F}_R$ must be equal to the corresponding x, y, z components of $(\mathbf{F}_1 + \mathbf{F}_2)$. Hence,

$$0 = 212.1 + F_{2x} \qquad F_{2x} = -212.1 \, \mathrm{N}$$

$$800 = 150 + F_{2y} \qquad F_{2y} = 650 \, \mathrm{N}$$

$$0 = -150 + F_{2z} \qquad F_{2z} = 150 \, \mathrm{N}$$

Since the magnitudes of $\mathbf{F}_2$ and its components are known, we can use Eq. 2–11 to determine α_2, β_2, γ_2.

$$-212.1 = 700 \cos \alpha_2; \qquad \alpha_2 = \cos^{-1}\left(\frac{-212.1}{700}\right) = 108° \qquad Ans.$$

$$650 = 700 \cos \beta_2; \qquad \beta_2 = \cos^{-1}\left(\frac{650}{700}\right) = 21.8° \qquad Ans.$$

$$150 = 700 \cos \gamma_2; \qquad \gamma_2 = \cos^{-1}\left(\frac{150}{700}\right) = 77.6° \qquad Ans.$$

These results are shown in Fig. 2–32b.

PROBLEMS

2–57. Determine the magnitude and coordinate direction angles of $\mathbf{F}_1 = \{60\mathbf{i} - 50\mathbf{j} + 40\mathbf{k}\}$ N and $\mathbf{F}_2 = \{-40\mathbf{i} - 85\mathbf{j} + 30\mathbf{k}\}$ N. Sketch each force on an x, y, z reference.

2–58. Express each force in Cartesian vector form.

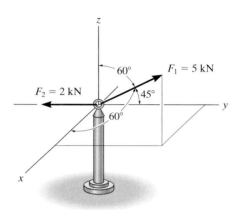

Prob. 2–58

2–59. Determine the magnitude and coordinate direction angles of the force $\mathbf{F}$ acting on the stake.

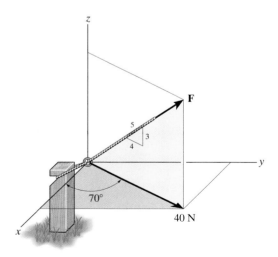

Prob. 2–59

*2–60. Express each force in Cartesian vector form.

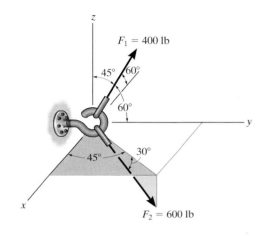

Prob. 2–60

2–61. The stock mounted on the lathe is subjected to a force of 60 N. Determine the coordinate direction angle β and express the force as a Cartesian vector.

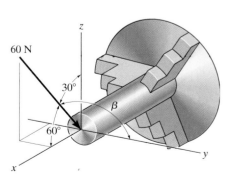

Prob. 2–61

2–62. Determine the magnitude and coordinate direction angles of the resultant force and sketch this vector on the coordinate system.

2–63. Specify the coordinate direction angles of F_1 and F_2 and express each force as a Cartesian vector.

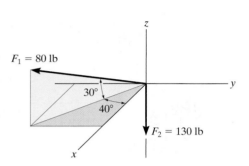

Probs. 2–62/63

2–66. The shaft S exerts three force components on the die D. Find the magnitude and direction of the resultant force. Force F_2 acts within the octant shown.

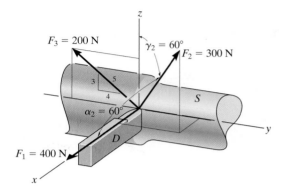

Prob. 2–66

*2–64. The mast is subjected to the three forces shown. Determine the coordinate direction angles $\alpha_1, \beta_1, \gamma_1$ of F_1 so that the resultant force acting on the mast is $F_R = \{350\mathbf{i}\}$ N.

2–65. The mast is subjected to the three forces shown. Determine the coordinate direction angles $\alpha_1, \beta_1, \gamma_1$ of F_1 so that the resultant force acting on the mast is zero.

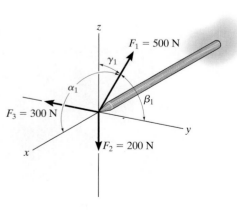

Probs. 2–64/65

2–67. The beam is subjected to the two forces shown. Express each force in Cartesian vector form and determine the magnitude and coordinate direction angles of the resultant force.

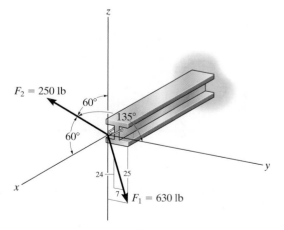

Prob. 2–67

*2–68. Determine the magnitude and coordinate direction angles of the resultant force and sketch this vector on the coordinate system.

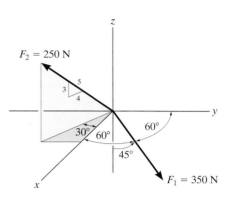

Prob. 2–68

2–69. Determine the magnitude and coordinate direction angles of F_3 so that the resultant of the three forces acts along the positive y axis and has a magnitude of 600 lb.

2–70. Determine the magnitude and coordinate direction angles of F_3 so that the resultant of the three forces is zero.

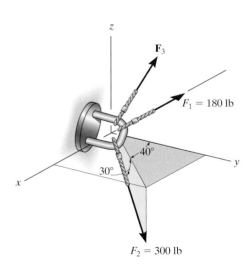

Probs. 2–69/70

2–71. Specify the magnitude F_3 and directions α_3, β_3, and γ_3 of $\mathbf{F}_3$ so that the force of the three forces is $\mathbf{F}_R = \{9\mathbf{j}\}$ kN.

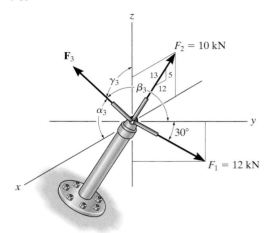

Prob. 2–71

*2–72. The pole is subjected to the force **F**, which has components acting along the x, y, z axes as shown. If the magnitude of **F** is 3 kN, and $\beta = 30°$ and $\gamma = 75°$, determine the magnitudes of its three components.

2–73. The pole is subjected to the force **F** which has components $F_x = 1.5$ kN and $F_z = 1.25$ kN. If $\beta = 75°$, determine the magnitudes of **F** and $\mathbf{F}_y$.

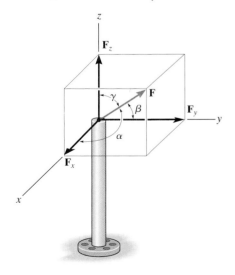

Probs. 2–72/73

2–74. The eye bolt is subjected to the cable force **F** which has a component along the x axis of $F_x = 60$ N, a component along the z axis of $F_z = -80$ N, and a coordinate direction angle of $\beta = 80°$. Determine the magnitude of **F**.

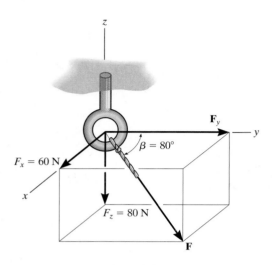

Prob. 2–74

2–75. Three forces act on the hook. If the resultant force $\mathbf{F}_R$ has a magnitude and direction as shown, determine the magnitude and the coordinate direction angles of force $\mathbf{F}_3$.

***2–76.** Determine the coordinate direction angles of $\mathbf{F}_1$ and $\mathbf{F}_R$.

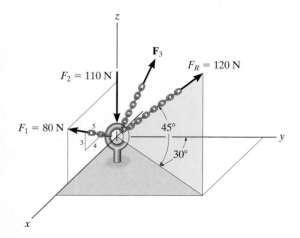

Probs. 2–75/76

2–77. The bolt is subjected to the force **F**, which has components acting along the x, y, z axes as shown. If the magnitude of **F** is 80 N, and $\alpha = 60°$ and $\gamma = 45°$, determine the magnitudes of its components.

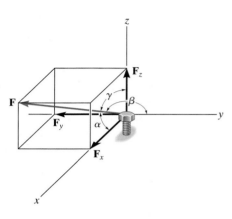

Prob. 2–77

2–78. Two forces $\mathbf{F}_1$ and $\mathbf{F}_2$ act on the bolt. If the resultant force $\mathbf{F}_R$ has a magnitude of 50 lb and coordinate direction angles $\alpha = 110°$ and $\beta = 80°$, as shown, determine the magnitude of $\mathbf{F}_2$ and its coordinate direction angles.

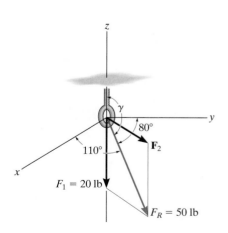

Prob. 2–78

2.7   Position Vectors

In this section we will introduce the concept of a position vector. It will be shown that this vector is of importance in formulating a Cartesian force vector directed between any two points in space. Later, in Chapter 4, we will use it for finding the moment of a force.

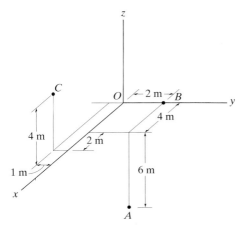

Fig. 2–33

x, y, z Coordinates. Throughout the book we will use a *right-handed* coordinate system to reference the location of points in space. We will also use the convention followed in many technical books, which requires the positive z axis to be directed *upward* (the zenith direction) so that it measures the height of an object or the altitude of a point. The x, y axes then lie in the horizontal plane, Fig. 2–33. Points in space are located relative to the origin of coordinates, O, by successive measurements along the x, y, z axes. For example, in Fig. 2–33 the coordinates of point A are obtained by starting at O and measuring $x_A = +4$ m along the x axis, $y_A = +2$ m along the y axis, and $z_A = -6$ m along the z axis. Thus, $A(4, 2, -6)$. In a similar manner, measurements along the x, y, z axes from O to B yield the coordinates of B, i.e., $B(0, 2, 0)$. Also notice that $C(6, -1, 4)$.

Position Vector. The *position vector* **r** is defined as a fixed vector which locates a point in space relative to another point. For example, if **r** extends from the origin of coordinates, O, to point $P(x, y, z)$, Fig. 2–34*a*, then **r** can be expressed in Cartesian vector form as

$$\mathbf{r} = x\mathbf{i} + y\mathbf{j} + z\mathbf{k}$$

Note how the head-to-tail vector addition of the three components yields vector **r**, Fig. 2–34*b*. Starting at the origin O, one travels x in the $+\mathbf{i}$ direction, then y in the $+\mathbf{j}$ direction, and finally z in the $+\mathbf{k}$ direction to arrive at point $P(x, y, z)$.

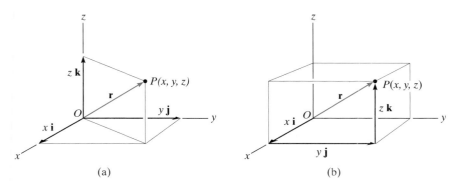

(a)   (b)

Fig. 2–34

In the more general case, the position vector may be directed from point A to point B in space, Fig. 2–35a. This vector is also designated by the symbol **r**. However, as a matter of convention, we will *sometimes* refer to this vector with *two subscripts* to indicate from and to the point where it is directed. Thus, **r** can also be designated as $\mathbf{r}_{AB}$. Also, note that $\mathbf{r}_A$ and $\mathbf{r}_B$ in Fig. 2–35a are referenced with only one subscript since they extend from the origin of coordinates.

From Fig. 2–35a, by the head-to-tail vector addition, we require

$$\mathbf{r}_A + \mathbf{r} = \mathbf{r}_B$$

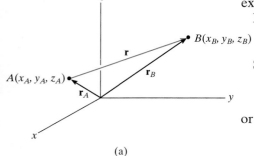

(a)

Solving for **r** and expressing $\mathbf{r}_A$ and $\mathbf{r}_B$ in Cartesian vector form yields

$$\mathbf{r} = \mathbf{r}_B - \mathbf{r}_A = (x_B\mathbf{i} + y_B\mathbf{j} + z_B\mathbf{k}) - (x_A\mathbf{i} + y_A\mathbf{j} + z_A\mathbf{k})$$

or

$$\boxed{\mathbf{r} = (x_B - x_A)\mathbf{i} + (y_B - y_A)\mathbf{j} + (z_B - z_A)\mathbf{k}} \qquad (2\text{–}13)$$

Thus, the **i**, **j**, **k** *components of the position vector* **r** *may be formed by taking the coordinates of the tail of the vector,* $A(x_A, y_A, z_A)$, *and subtracting them from the corresponding coordinates of the head,* $B(x_B, y_B, z_B)$. Again note how the head-to-tail addition of these three components yields **r**, i.e., going from A to B, Fig. 2–35b, one first travels $(x_B - x_A)$ in the $+\mathbf{i}$ direction, then $(y_B - y_A)$ in the $+\mathbf{j}$ direction, and finally $(z_B - z_A)$ in the $+\mathbf{k}$ direction.

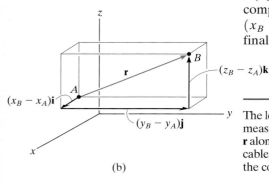

(b)

Fig. 2–35

The length and direction of cable AB used to support the stack can be determined by measuring the coordinates of points A and B using the x, y, z axes. The position vector **r** along the cable can then be established. The magnitude r represents the length of the cable, and the direction of the cable is defined by α, β, γ, which are determined from the components of the unit vector found from the position vector, $\mathbf{u} = \mathbf{r}/r$.

EXAMPLE 2.12

An elastic rubber band is attached to points A and B as shown in Fig. 2–36a. Determine its length and its direction measured from A toward B.

Fig. 2–36

SOLUTION

We first establish a position vector from A to B, Fig. 2–36b. In accordance with Eq. 2–13, the coordinates of the tail $A(1 \text{ m}, 0, -3 \text{ m})$ are subtracted from the coordinates of the head $B(-2 \text{ m}, 2 \text{ m}, 3 \text{ m})$, which yields

$$\mathbf{r} = [-2 \text{ m} - 1 \text{ m}]\mathbf{i} + [2 \text{ m} - 0]\mathbf{j} + [3 \text{ m} - (-3 \text{ m})]\mathbf{k}$$
$$= \{-3\mathbf{i} + 2\mathbf{j} + 6\mathbf{k}\} \text{ m}$$

These components of $\mathbf{r}$ can also be determined *directly* by realizing from Fig. 2–36a that they represent the direction and distance one must go along each axis in order to move from A to B, i.e., along the x axis $\{-3\mathbf{i}\}$ m, along the y axis $\{2\mathbf{j}\}$ m, and finally along the z axis $\{6\mathbf{k}\}$ m.

The magnitude of $\mathbf{r}$ represents the length of the rubber band.

$$r = \sqrt{(-3)^2 + (2)^2 + (6)^2} = 7 \text{ m} \qquad \textit{Ans.}$$

Formulating a unit vector in the direction of $\mathbf{r}$, we have

$$\mathbf{u} = \frac{\mathbf{r}}{r} = \frac{-3}{7}\mathbf{i} + \frac{2}{7}\mathbf{j} + \frac{6}{7}\mathbf{k}$$

The components of this unit vector yield the coordinate direction angles

$$\alpha = \cos^{-1}\left(-\frac{3}{7}\right) = 115° \qquad \textit{Ans.}$$

$$\beta = \cos^{-1}\left(\frac{2}{7}\right) = 73.4° \qquad \textit{Ans.}$$

$$\gamma = \cos^{-1}\left(\frac{6}{7}\right) = 31.0° \qquad \textit{Ans.}$$

NOTE: These angles are measured from the *positive axes* of a localized coordinate system placed at the tail of $\mathbf{r}$, as shown in Fig. 2–36c.

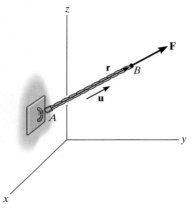

Fig. 2–37

2.8 Force Vector Directed along a Line

Quite often in three-dimensional statics problems, the direction of a force is specified by two points through which its line of action passes. Such a situation is shown in Fig. 2–37, where the force **F** is directed along the cord *AB*. We can formulate **F** as a Cartesian vector by realizing that it has the *same direction* and *sense* as the position vector **r** directed from point *A* to point *B* on the cord. This common direction is specified by the *unit vector* **u** = **r**/*r*. Hence,

$$\mathbf{F} = F\mathbf{u} = F\left(\frac{\mathbf{r}}{r}\right)$$

Although we have represented **F** symbolically in Fig. 2–37, note that it has *units of force*, unlike **r**, which has units of length.

The force F acting along the chain can be represented as a Cartesian vector by first establishing *x, y, z* axes and forming a position vector **r** along the length of the chain, then finding the corresponding unit vector **u** = **r**/*r* that defines the direction of both the chain and the force. Finally, the magnitude of the force is combined with its direction, **F** = *F***u**.

Important Points

- A position vector locates one point in space relative to another point.
- The easiest way to formulate the components of a position vector is to determine the distance and direction that must be traveled along the *x, y, z* directions—going from the tail to the head of the vector.
- A force **F** acting in the direction of a position vector **r** can be represented in Cartesian form if the unit vector **u** of the position vector is determined and this is multiplied by the magnitude of the force, i.e., **F** = *F***u** = *F*(**r**/*r*).

EXAMPLE 2.13

The man shown in Fig. 2–38*a* pulls on the cord with a force of 70 lb. Represent this force, acting on the support *A*, as a Cartesian vector and determine its direction.

SOLUTION

Force **F** is shown in Fig. 2–38*b*. The *direction* of this vector, **u**, is determined from the position vector **r**, which extends from *A* to *B*, Fig. 2–38*b*. The coordinates of the end points of the cord are $A(0, 0, 30 \text{ ft})$ and $B(12 \text{ ft}, -8 \text{ ft}, 6 \text{ ft})$. Forming the position vector by subtracting the corresponding *x*, *y*, and *z* coordinates of *A* from those of *B*, we have

$$\mathbf{r} = (12 \text{ ft} - 0)\mathbf{i} + (-8 \text{ ft} - 0)\mathbf{j} + (6 \text{ ft} - 30 \text{ ft})\mathbf{k}$$

$$= \{12\mathbf{i} - 8\mathbf{j} - 24\mathbf{k}\} \text{ ft}$$

This result can also be determined *directly* by noting in Fig. 2–38*a*, that one must go from *A* $\{-24\mathbf{k}\}$ ft, then $\{-8\mathbf{j}\}$ ft, and finally $\{12\mathbf{i}\}$ ft to get to *B*.

The magnitude of **r**, which represents the *length* of cord *AB*, is

$$r = \sqrt{(12 \text{ ft})^2 + (-8 \text{ ft})^2 + (-24 \text{ ft})^2} = 28 \text{ ft}$$

Forming the unit vector that defines the direction and sense of both **r** and **F** yields

$$\mathbf{u} = \frac{\mathbf{r}}{r} = \frac{12}{28}\mathbf{i} - \frac{8}{28}\mathbf{j} - \frac{24}{28}\mathbf{k}$$

Since **F** has a *magnitude* of 70 lb and a *direction* specified by **u**, then

$$\mathbf{F} = F\mathbf{u} = 70 \text{ lb}\left(\frac{12}{28}\mathbf{i} - \frac{8}{28}\mathbf{j} - \frac{24}{28}\mathbf{k}\right)$$

$$= \{30\mathbf{i} - 20\mathbf{j} - 60\mathbf{k}\} \text{ lb} \qquad \textit{Ans.}$$

The coordinate direction angles are measured between **r** (or **F**) and the *positive axes* of a localized coordinate system with origin placed at *A*, Fig. 2–38*b*. From the components of the unit vector:

$$\alpha = \cos^{-1}\left(\frac{12}{28}\right) = 64.6° \qquad \textit{Ans.}$$

$$\beta = \cos^{-1}\left(\frac{-8}{28}\right) = 107° \qquad \textit{Ans.}$$

$$\gamma = \cos^{-1}\left(\frac{-24}{28}\right) = 149° \qquad \textit{Ans.}$$

NOTE: These results make sense when compared with the angles shown in Fig. 2–38*b*.

(a)

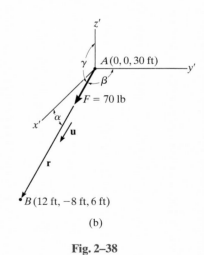

(b)

Fig. 2–38

EXAMPLE 2.14

(a)

(b)

Fig. 2–39

The circular plate in Fig. 2–39a is partially supported by the cable AB. If the force of the cable on the hook at A is $F = 500$ N, express $\mathbf{F}$ as a Cartesian vector.

SOLUTION

As shown in Fig. 2–39b, $\mathbf{F}$ has the same direction and sense as the position vector $\mathbf{r}$, which extends from A to B. The coordinates of the end points of the cable are $A(0, 0, 2 \text{ m})$ and $B(1.707 \text{ m}, 0.707 \text{ m}, 0)$, as indicated in the figure. Thus,

$$\mathbf{r} = (1.707 \text{ m} - 0)\mathbf{i} + (0.707 \text{ m} - 0)\mathbf{j} + (0 - 2 \text{ m})\mathbf{k}$$

$$= \{1.707\mathbf{i} + 0.707\mathbf{j} - 2\mathbf{k}\} \text{ m}$$

Note how one can calculate these components *directly* by going from A, $\{-2\mathbf{k}\}$ m along the z axis, then $\{1.707\mathbf{i}\}$ m along the x axis, and finally $\{0.707\mathbf{j}\}$ m along the y axis to get to B.

The magnitude of $\mathbf{r}$ is

$$r = \sqrt{(1.707)^2 + (0.707)^2 + (-2)^2} = 2.723 \text{ m}$$

Thus,

$$\mathbf{u} = \frac{\mathbf{r}}{r} = \frac{1.707}{2.723}\mathbf{i} + \frac{0.707}{2.723}\mathbf{j} - \frac{2}{2.723}\mathbf{k}$$

$$= 0.6269\mathbf{i} + 0.2597\mathbf{j} - 0.7345\mathbf{k}$$

Since $F = 500$ N and $\mathbf{F}$ has the direction $\mathbf{u}$, we have

$$\mathbf{F} = F\mathbf{u} = 500 \text{ N}(0.6269\mathbf{i} + 0.2597\mathbf{j} - 0.7345\mathbf{k})$$

$$= \{313\mathbf{i} + 130\mathbf{j} - 367\mathbf{k}\} \text{ N} \qquad\qquad\qquad Ans.$$

Using these components, notice that indeed the magnitude of $\mathbf{F}$ is 500 N; i.e.,

$$F = \sqrt{(313)^2 + (130)^2 + (-367)^2} = 500 \text{ N}$$

NOTE: From the unit vector, show that the coordinate direction angle $\gamma = 137°$, and indicate this angle on the figure.

EXAMPLE 2.15

The roof is supported by cables as shown in the photo. If the cables exert forces $F_{AB} = 100$ N and $F_{AC} = 120$ N on the wall hook at A as shown in Fig. 2–40a, determine the magnitude of the resultant force acting at A.

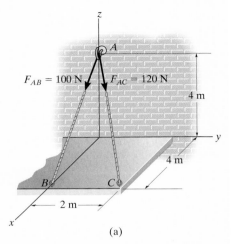

SOLUTION
The resultant force $\mathbf{F}_R$ is shown graphically in Fig. 2–40b. We can express this force as a Cartesian vector by first formulating $\mathbf{F}_{AB}$ and $\mathbf{F}_{AC}$ as Cartesian vectors and then adding their components. The directions of $\mathbf{F}_{AB}$ and $\mathbf{F}_{AC}$ are specified by forming unit vectors $\mathbf{u}_{AB}$ and $\mathbf{u}_{AC}$ along the cables. These unit vectors are obtained from the associated position vectors $\mathbf{r}_{AB}$ and $\mathbf{r}_{AC}$. With reference to Fig. 2–40b, for $\mathbf{F}_{AB}$ we have

$$\mathbf{r}_{AB} = (4\text{ m} - 0)\mathbf{i} + (0 - 0)\mathbf{j} + (0 - 4\text{ m})\mathbf{k}$$
$$= \{4\mathbf{i} - 4\mathbf{k}\}\text{ m}$$
$$r_{AB} = \sqrt{(4)^2 + (-4)^2} = 5.66\text{ m}$$
$$\mathbf{F}_{AB} = 100\text{ N}\left(\frac{\mathbf{r}_{AB}}{r_{AB}}\right) = 100\text{ N}\left(\frac{4}{5.66}\mathbf{i} - \frac{4}{5.66}\mathbf{k}\right)$$
$$\mathbf{F}_{AB} = \{70.7\mathbf{i} - 70.7\mathbf{k}\}\text{ N}$$

For $\mathbf{F}_{AC}$ we have

$$\mathbf{r}_{AC} = (4\text{ m} - 0)\mathbf{i} + (2\text{ m} - 0)\mathbf{j} + (0 - 4\text{ m})\mathbf{k}$$
$$= \{4\mathbf{i} + 2\mathbf{j} - 4\mathbf{k}\}\text{ m}$$
$$r_{AC} = \sqrt{(4)^2 + (2)^2 + (-4)^2} = 6\text{ m}$$
$$\mathbf{F}_{AC} = 120\text{ N}\left(\frac{\mathbf{r}_{AC}}{r_{AC}}\right) = 120\text{ N}\left(\frac{4}{6}\mathbf{i} + \frac{2}{6}\mathbf{j} - \frac{4}{6}\mathbf{k}\right)$$
$$= \{80\mathbf{i} + 40\mathbf{j} - 80\mathbf{k}\}\text{ N}$$

The resultant force is therefore

$$\mathbf{F}_R = \mathbf{F}_{AB} + \mathbf{F}_{AC} = \{70.7\mathbf{i} - 70.7\mathbf{k}\}\text{ N} + \{80\mathbf{i} + 40\mathbf{j} - 80\mathbf{k}\}\text{ N}$$
$$= \{150.7\mathbf{i} + 40\mathbf{j} - 150.7\mathbf{k}\}\text{ N}$$

The magnitude of $\mathbf{F}_R$ is thus

$$F_R = \sqrt{(150.7)^2 + (40)^2 + (-150.7)^2}$$
$$= 217\text{ N} \qquad\qquad Ans.$$

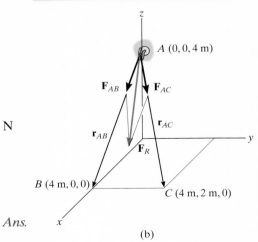

Fig. 2–40

PROBLEMS

2–79. If $\mathbf{r}_1 = \{3\mathbf{i} - 4\mathbf{j} + 3\mathbf{k}\}$ m, $\mathbf{r}_2 = \{4\mathbf{i} - 5\mathbf{k}\}$ m, $\mathbf{r}_3 = \{3\mathbf{i} - 2\mathbf{j} + 5\mathbf{k}\}$ m, determine the magnitude and direction of $\mathbf{r} = 2\mathbf{r}_1 - \mathbf{r}_2 + 3\mathbf{r}_3$.

*2–80. Represent the position vector $\mathbf{r}$ acting from point $A(3$ m, 5 m, 6 m$)$ to point $B(5$ m, -2 m, 1 m$)$ in Cartesian vector form. Determine its coordinate direction angles and find the distance between points A and B.

2–81. A position vector extends from the origin to point $A(2$ m, 3 m, 6 m$)$. Determine the angles α, β, γ which the tail of the vector makes with the x, y, z axes, respectively.

2–82. Express the position vector $\mathbf{r}$ in Cartesian vector form; then determine its magnitude and coordinate direction angles.

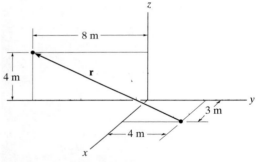

Prob. 2–82

2–83. Express the position vector $\mathbf{r}$ in Cartesian vector form; then determine its magnitude and coordinate direction angles.

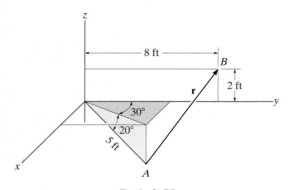

Prob. 2–83

*2–84. Determine the length of the connecting rod AB by first formulating a Cartesian position vector from A to B and then determining its magnitude.

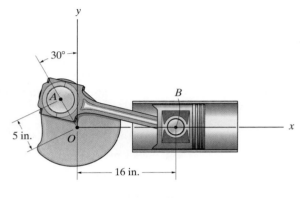

Prob. 2–84

2–85. Determine the length of member AB of the truss by first establishing a Cartesian position vector from A to B and then determining its magnitude.

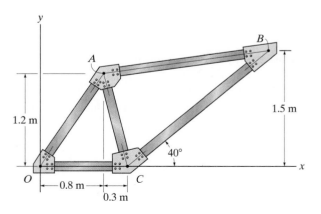

Prob. 2–85

2–86. The positions of point A on the building and point B on the antenna have been measured relative to the electronic distance meter (EDM) at O. Determine the distance between A and B. *Hint:* Formulate a position vector directed from A to B; then determine its magnitude.

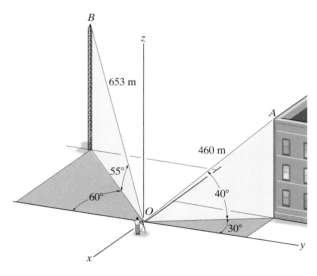

Prob. 2–86

2–87. Determine the lengths of cords ACB and CO. The knot at C is located midway between A and B.

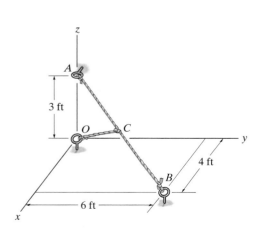

Prob. 2–87

*2–88. At a given instant, the position of a plane at A and a train at B are measured relative to a radar antenna at O. Determine the distance d between A and B at this instant. To solve the problem, formulate a position vector, directed from A to B, and then determine its magnitude.

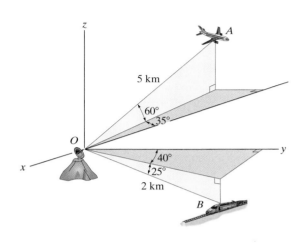

Prob. 2–88

2–89. Determine the lengths of wires AD, BD, and CD. The ring at D is midway between A and B.

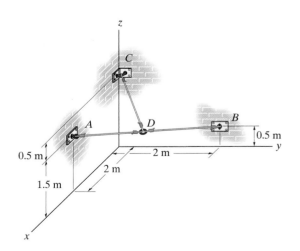

Prob. 2–89

2–90. Determine the magnitude and coordinates on angles of the resultant force.

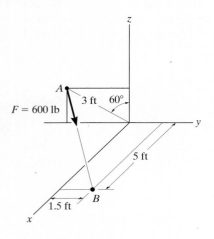

Prob. 2–90

2–91. Express force **F** as a Cartesian vector; then determine its coordinate direction angles.

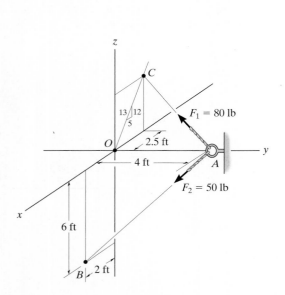

Prob. 2–91

*2–92. Determine the magnitude and coordinate direction angles of the resultant force acting at point A.

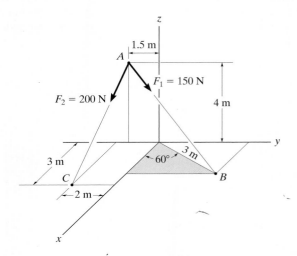

Prob. 2–92

2–93. The plate is suspended using the three cables each exert the forces shown. Express each force as a Cartesian vector.

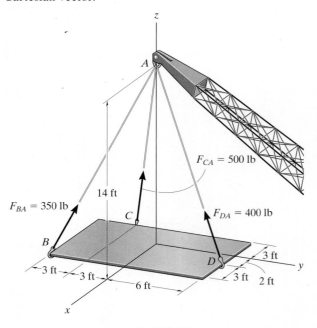

Prob. 2–93

2–94. The engine of the lightweight plane is supported by struts that are connected to the space truss that makes up the structure of the plane. The anticipated loading in two of the struts is shown. Express each of these forces as a Cartesian vector.

Prob. 2–94

2–95. The window is held open by cable AB. Determine the length of the cable and express the 30-N force acting at A along the cable as a Cartesian vector.

Prob. 2–95

*2–96. The force acting on the man, caused by his pulling on the anchor cord, is $\mathbf{F} = \{40\mathbf{i} + 20\mathbf{j} - 50\mathbf{k}\}$ N. If the length of the cord is 25 m, determine the coordinates $A(x, y, -z)$ of the anchor.

Prob. 2–96

2–97. Express each of the forces in Cartesian vector form and determine the magnitude and coordinate direction angles of the resultant force.

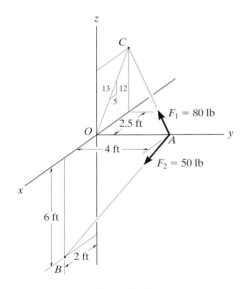

Prob. 2–97

2–98.   The cable attached to the tractor at B exerts a force of 350 lb on the framework. Express this force as a Cartesian vector.

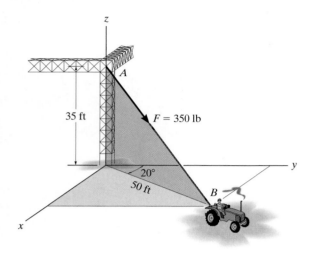

Prob. 2–98

2–99.   The cable OA exerts a force on point O of $F = \{40\mathbf{i} + 60\mathbf{j} + 70\mathbf{k}\}$ N. If the length of the cable is 3 m, what are the coordinates (x, y, z) of point A?

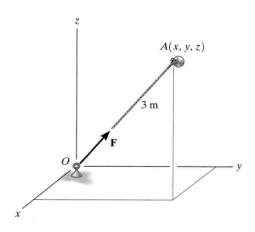

Prob. 2–99

*2–100.   Determine the position $(x, y, 0)$ for fixing cable BA so that the resultant of the forces exerted on the pole is directed along its axis, from B toward O and has a magnitude of 1 kN. Also, what is the magnitude of force $\mathbf{F}_3$?

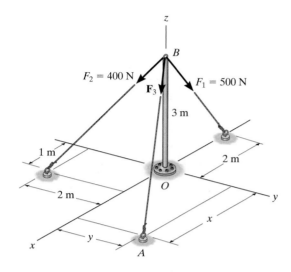

Prob. 2–100

2–101.   The cord exerts a force of $\mathbf{F} = \{12\mathbf{i} + 9\mathbf{j} - 8\mathbf{k}\}$ lb on the hook. If the cord is 8 ft long, determine the location x, y of the point of attachment B, and the height z of the hook.

2–102.   The cord exerts a force of $F = 30$ lb on the hook. If the cord is 8 ft long, $z = 4$ ft, and the x component of the force is $F_x = 25$ lb, determine the location x, y of the point of attachment B of the cord to the ground.

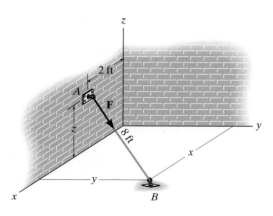

Probs. 2–101/102

2–103. Each of the four forces acting at E has a magnitude of 28 kN. Express each force as a Cartesian vector and determine the resultant force.

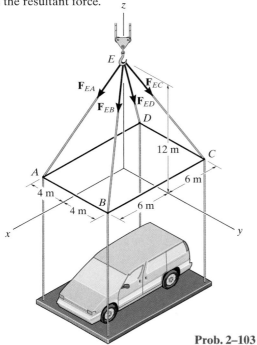

Prob. 2–103

2–105. The chandelier is supported by three chains which are concurrent at point O. If the force in each chain has a magnitude of 60 lb, express each force as a Cartesian vector and determine the magnitude and coordinate direction angles of the resultant force.

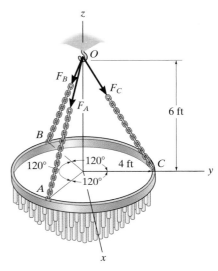

Prob. 2–105

*2–104. The tower is held in place by three cables. If the force of each cable acting on the tower is shown, determine the magnitude and coordinate direction angles α, β, γ of the resultant force. Take $x = 20$ m, $y = 15$ m.

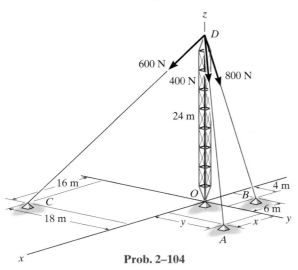

Prob. 2–104

2–106. The chandelier is supported by three chains which are concurrent at point O. If the resultant force at O has a magnitude of 130 lb and is directed along the negative z axis, determine the force in each chain assuming $F_A = F_B = F_C$.

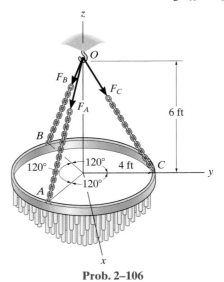

Prob. 2–106

2.9 Dot Product

Fig. 2–41

Occasionally in statics one has to find the angle between two lines or the components of a force parallel and perpendicular to a line. In two dimensions, these problems can readily be solved by trigonometry since the geometry is easy to visualize. In three dimensions, however, this is often difficult, and consequently vector methods should be employed for the solution. The dot product defines a particular method for "multiplying" two vectors and is used to solve the above-mentioned problems.

The *dot product* of vectors **A** and **B**, written **A · B**, and read "**A** dot **B**," is defined as the product of the magnitudes of **A** and **B** and the cosine of the angle θ between their tails, Fig. 2–41. Expressed in equation form,

$$\mathbf{A \cdot B} = AB \cos \theta \qquad (2\text{–}14)$$

where $0° \leq \theta \leq 180°$. The dot product is often referred to as the *scalar product* of vectors since the result is a *scalar* and not a vector.

Laws of Operation.

1. Commutative law:

$$\mathbf{A \cdot B} = \mathbf{B \cdot A}$$

2. Multiplication by a scalar:

$$a(\mathbf{A \cdot B}) = (a\mathbf{A}) \cdot \mathbf{B} = \mathbf{A} \cdot (a\mathbf{B}) = (\mathbf{A \cdot B})a$$

3. Distributive law:

$$\mathbf{A \cdot (B + D)} = (\mathbf{A \cdot B}) + (\mathbf{A \cdot D})$$

It is easy to prove the first and second laws by using Eq. 2–14. The proof of the distributive law is left as an exercise (see Prob. 2–107).

Cartesian Vector Formulation.
Equation 2–14 may be used to find the dot product for each of the Cartesian unit vectors. For example, $\mathbf{i \cdot i} = (1)(1) \cos 0° = 1$ and $\mathbf{i \cdot j} = (1)(1) \cos 90° = 0$. In a similar manner,

$$\mathbf{i \cdot i} = 1 \quad \mathbf{j \cdot j} = 1 \quad \mathbf{k \cdot k} = 1$$
$$\mathbf{i \cdot j} = 0 \quad \mathbf{i \cdot k} = 0 \quad \mathbf{k \cdot j} = 0$$

These results should not be memorized; rather, it should be clearly understood how each is obtained.

Consider now the dot product of two general vectors **A** and **B** which are expressed in Cartesian vector form. We have

$$\mathbf{A \cdot B} = (A_x\mathbf{i} + A_y\mathbf{j} + A_z\mathbf{k}) \cdot (B_x\mathbf{i} + B_y\mathbf{j} + B_z\mathbf{k})$$
$$= A_xB_x(\mathbf{i \cdot i}) + A_xB_y(\mathbf{i \cdot j}) + A_xB_z(\mathbf{i \cdot k})$$
$$+ A_yB_x(\mathbf{j \cdot i}) + A_yB_y(\mathbf{j \cdot j}) + A_yB_z(\mathbf{j \cdot k})$$
$$+ A_zB_x(\mathbf{k \cdot i}) + A_zB_y(\mathbf{k \cdot j}) + A_zB_z(\mathbf{k \cdot k})$$

Carrying out the dot-product operations, the final result becomes

$$\mathbf{A} \cdot \mathbf{B} = A_x B_x + A_y B_y + A_z B_z \qquad (2\text{–}15)$$

Thus, to determine the dot product of two Cartesian vectors, multiply their corresponding x, y, z components and sum their products algebraically. Since the result is a scalar, be careful *not* to include any unit vectors in the final result.

Applications.   The dot product has two important applications in mechanics.

1. *The angle formed between two vectors or intersecting lines.* The angle θ between the tails of vectors **A** and **B** in Fig. 2–41 can be determined from Eq. 2–14 and written as

$$\theta = \cos^{-1}\!\left(\frac{\mathbf{A} \cdot \mathbf{B}}{AB}\right) \qquad 0° \le \theta \le 180°$$

Here $\mathbf{A} \cdot \mathbf{B}$ is found from Eq. 2–15. In particular, notice that if $\mathbf{A} \cdot \mathbf{B} = 0$, $\theta = \cos^{-1} 0 = 90°$, so that **A** will be *perpendicular* to **B**.

2. *The components of a vector parallel and perpendicular to a line.* The component of vector **A** parallel to or collinear with the line aa' in Fig. 2–42 is defined by $\mathbf{A}_\parallel$, where $A_\parallel = A \cos \theta$. This component is sometimes referred to as the *projection* of **A** onto the line, since a right angle is formed in the construction. If the *direction* of the line is specified by the unit vector **u**, then, since $u = 1$, we can determine $A_\parallel$ directly from the dot product (Eq. 2–14); i.e.,

$$A_\parallel = A \cos \theta = \mathbf{A} \cdot \mathbf{u}$$

Fig. 2–42

Hence, the scalar projection of **A** *along a line is determined from the dot product of* **A** *and the unit vector* **u** *which defines the direction of the line.* Notice that if this result is positive, then $\mathbf{A}_\parallel$ has a directional sense which is the same as **u**, whereas if $A_\parallel$ is a negative scalar, then $\mathbf{A}_\parallel$ has the opposite sense of direction to **u**. The component $\mathbf{A}_\parallel$ represented as a *vector* is therefore

$$\mathbf{A}_\parallel = A \cos \theta \, \mathbf{u} = (\mathbf{A} \cdot \mathbf{u})\mathbf{u}$$

The component of **A** that is *perpendicular* to line aa' can also be obtained, Fig. 2–42. Since $\mathbf{A} = \mathbf{A}_\parallel + \mathbf{A}_\perp$, then $\mathbf{A}_\perp = \mathbf{A} - \mathbf{A}_\parallel$. There are two possible ways of obtaining $A_\perp$. One way would be to determine θ from the dot product, $\theta = \cos^{-1}(\mathbf{A} \cdot \mathbf{u}/A)$, then $A_\perp = A \sin \theta$. Alternatively, if $A_\parallel$ is known, then by the Pythagorean theorem we can also write $A_\perp = \sqrt{A^2 - A_\parallel^2}$.

Photo A: The angle θ which is made between the rope and the connecting beam A can be determined by using the dot product. Simply formulate position vectors or unit vectors along the beam, $\mathbf{u}_A = \mathbf{r}_A/r_A$, and along the rope, $\mathbf{u}_r = \mathbf{r}_r/r_r$. Since θ is defined between the tails of these vectors we can solve for θ using $\theta = \cos^{-1}(\mathbf{r}_A \cdot \mathbf{r}_r / r_A r_r) = \cos^{-1} \mathbf{u}_A \cdot \mathbf{u}_r$.

Photo B: If the rope exerts a force $\mathbf{F}$ on the joint, the projection of this force along beam A can be determined by first defining the *direction of the beam* using the unit vector $\mathbf{u}_A = \mathbf{r}_A/r_A$ and then formulating the force as a Cartesian vector $\mathbf{F} = F(\mathbf{r}_r/r_r) = F\mathbf{u}_r$. Applying the dot product, the projection is $F_\parallel = \mathbf{F} \cdot \mathbf{u}_A$.

A   **B**

Important Points

- The dot product is used to determine the angle between two vectors or the projection of a vector in a specified direction.

- If the vectors $\mathbf{A}$ and $\mathbf{B}$ are expressed in Cartesian form, the dot product is determined by multiplying the respective x, y, z scalar components together and algebraically adding the results, i.e., $\mathbf{A} \cdot \mathbf{B} = A_x B_x + A_y B_y + A_z B_z$.

- From the definition of the dot product, the angle formed between the tails of vectors $\mathbf{A}$ and $\mathbf{B}$ is $\theta = \cos^{-1}(\mathbf{A} \cdot \mathbf{B}/AB)$.

- The magnitude of the projection of vector $\mathbf{A}$ along a line whose direction is specified by $\mathbf{u}$ is determined from the dot product $A_\parallel = \mathbf{A} \cdot \mathbf{u}$.

EXAMPLE 2.16

The frame shown in Fig. 2–43a is subjected to a horizontal force $\mathbf{F} = \{300\mathbf{j}\}$ N. Determine the magnitude of the components of this force parallel and perpendicular to member AB.

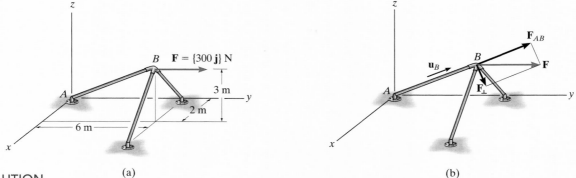

(a)   (b)

Fig. 2–43

SOLUTION
The magnitude of the component of $\mathbf{F}$ along AB is equal to the dot product of $\mathbf{F}$ and the unit vector $\mathbf{u}_B$, which defines the direction of AB, Fig. 2–43b. Since

$$\mathbf{u}_B = \frac{\mathbf{r}_B}{r_B} = \frac{2\mathbf{i} + 6\mathbf{j} + 3\mathbf{k}}{\sqrt{(2)^2 + (6)^2 + (3)^2}} = 0.286\mathbf{i} + 0.857\mathbf{j} + 0.429\mathbf{k}$$

then

$$\begin{aligned}
F_{AB} = F\cos\theta = \mathbf{F}\cdot\mathbf{u}_B &= (300\mathbf{j})\cdot(0.286\mathbf{i} + 0.857\mathbf{j} + 0.429\mathbf{k}) \\
&= (0)(0.286) + (300)(0.857) + (0)(0.429) \\
&= 257.1 \text{ N} \qquad\qquad\qquad\qquad\qquad\qquad Ans.
\end{aligned}$$

Since the result is a positive scalar, $\mathbf{F}_{AB}$ has the same sense of direction as $\mathbf{u}_B$, Fig. 2–43b.

Expressing $\mathbf{F}_{AB}$ in Cartesian vector form, we have

$$\begin{aligned}
\mathbf{F}_{AB} = F_{AB}\mathbf{u}_B &= (257.1 \text{ N})(0.286\mathbf{i} + 0.857\mathbf{j} + 0.429\mathbf{k}) \\
&= \{73.5\mathbf{i} + 220\mathbf{j} + 110\mathbf{k}\} \text{ N} \qquad\qquad Ans.
\end{aligned}$$

The perpendicular component, Fig. 2–43b, is therefore

$$\begin{aligned}
\mathbf{F}_{\perp} = \mathbf{F} - \mathbf{F}_{AB} &= 300\mathbf{j} - (73.5\mathbf{i} + 220\mathbf{j} + 110\mathbf{k}) \\
&= \{-73.5\mathbf{i} + 80\mathbf{j} - 110\mathbf{k}\} \text{ N}
\end{aligned}$$

Its magnitude can be determined either from this vector or from the Pythagorean theorem, Fig. 2–43b:

$$\begin{aligned}
F_{\perp} &= \sqrt{F^2 - F_{AB}^2} \\
&= \sqrt{(300 \text{ N})^2 - (257.1 \text{ N})^2} \\
&= 155 \text{ N} \qquad\qquad\qquad\qquad\qquad Ans.
\end{aligned}$$

EXAMPLE 2.17

(a)

(b)

Fig. 2–44

The pipe in Fig. 2–44a is subjected to the force of $F = 80$ lb. Determine the angle θ between $\mathbf{F}$ and the pipe segment BA, and the magnitudes of the components of $\mathbf{F}$, which are parallel and perpendicular to BA.

SOLUTION

Angle θ. First we will establish position vectors from B to A and B to C. Then we will determine the angle θ between the tails of these two vectors.

$$\mathbf{r}_{BA} = \{-2\mathbf{i} - 2\mathbf{j} + 1\mathbf{k}\} \text{ ft}$$
$$\mathbf{r}_{BC} = \{-3\mathbf{j} + 1\mathbf{k}\} \text{ ft}$$

Thus,

$$\cos \theta = \frac{\mathbf{r}_{BA} \cdot \mathbf{r}_{BC}}{\mathbf{r}_{BA}\mathbf{r}_{BC}} = \frac{(-2)(0) + (-2)(-3) + (1)(1)}{3\sqrt{10}}$$
$$= 0.7379$$
$$\theta = 42.5° \qquad \text{Ans.}$$

Components of F. The force $\mathbf{F}$ is resolved into components as shown in Fig. 2–44b. Since $F_{BA} = \mathbf{F} \cdot \mathbf{u}_{BA}$, we must first formulate the unit vector along BA and force $\mathbf{F}$ as Cartesian vectors.

$$\mathbf{u}_{BA} = \frac{\mathbf{r}_{BA}}{r_{BA}} = \frac{(-2\mathbf{i} - 2\mathbf{j} + 1\mathbf{k})}{3} = -\frac{2}{3}\mathbf{i} - \frac{2}{3}\mathbf{j} + \frac{1}{3}\mathbf{k}$$

$$\mathbf{F} = 80 \text{ lb}\left(\frac{\mathbf{r}_{BC}}{r_{BC}}\right) = 80\left(\frac{-3\mathbf{j} + 1\mathbf{k}}{\sqrt{10}}\right) = -75.89\mathbf{j} + 25.30\mathbf{k}$$

Thus,

$$F_{BA} = \mathbf{F} \cdot \mathbf{u}_{BA} = (-75.89\mathbf{j} + 25.30\mathbf{k}) \cdot \left(-\frac{2}{3}\mathbf{i} - \frac{2}{3}\mathbf{j} + \frac{1}{3}\mathbf{k}\right)$$
$$= 0 + 50.60 + 8.43$$
$$= 59.0 \text{ lb} \qquad \text{Ans.}$$

Since θ was calculated in Fig. 2–44b, this same result can also be obtained directly from trigonometry.

$$F_{BA} = 80 \cos 42.5° \text{ lb} = 59.0 \text{ lb} \qquad \text{Ans.}$$

The perpendicular component can be obtained by trigonometry,

$$F_\perp = F \sin \theta$$
$$= 80 \sin 42.5° \text{ lb}$$
$$= 54.0 \text{ lb} \qquad \text{Ans.}$$

NOTE: We can also obtain this result using the Pythagorean theorem,

$$F_\perp = \sqrt{F^2 - F_{BA}^2} = \sqrt{(80)^2 - (59.0)^2}$$
$$= 54.0 \text{ lb} \qquad \text{Ans.}$$

PROBLEMS

2–107. Given the three vectors **A**, **B**, and **D**, show that $\mathbf{A} \cdot (\mathbf{B} + \mathbf{D}) = (\mathbf{A} \cdot \mathbf{B}) + (\mathbf{A} \cdot \mathbf{D})$.

*2–108. Cable *BC* exerts a force of $F = 28$ N on the top of the flagpole. Determine the projection of this force along the *z* axis of the pole.

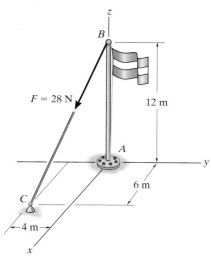

Prob. 2–108

2–109. Determine the angle θ between the tails of the two vectors.

2–110. Determine the magnitude of the projected component of $\mathbf{r}_1$ along $\mathbf{r}_2$, and the projection of $\mathbf{r}_2$ along $\mathbf{r}_1$.

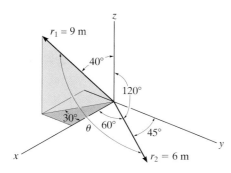

Probs. 2–109/110

2–111. Determine the angles θ and ϕ between the wire segments.

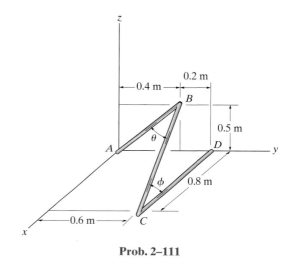

Prob. 2–111

*2–112. Determine the angle θ between the two cords.

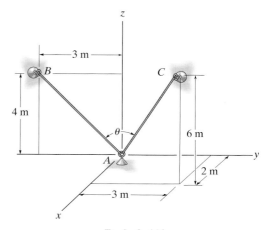

Prob. 2–112

2–113. Determine the angle θ between the two cables.

2–114. Determine the projected component of the force $F = 12$ lb acting in the direction of cable AC. Express the result as a Cartesian vector.

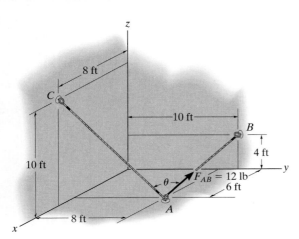

Probs. 2–113/114

2–115. Determine the components of **F** that act along rod AC and perpendicular to it. Point B is located at the midpoint of the rod.

*2–116. Determine the components of **F** that act along rod AC and perpendicular to it. Point B is located 3 m along the rod from end C.

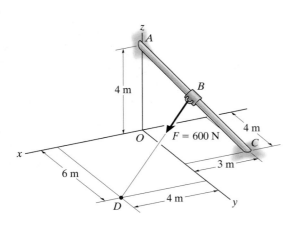

Probs. 2–115/116

2–117. Determine the magnitude of the projected component of the length of cord OA along the Oa axis.

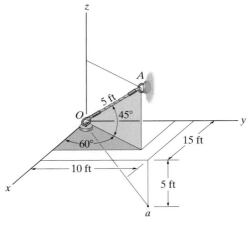

Prob. 2–117

2–118. A force of $\mathbf{F} = \{-40\mathbf{k}\}$ lb acts at the end of the pipe. Determine the magnitudes of the components $\mathbf{F}_1$ and $\mathbf{F}_2$ which are directed along the pipe's axis and perpendicular to it.

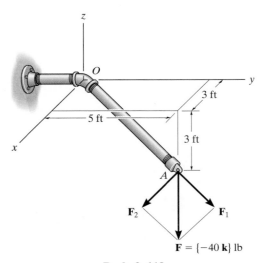

Prob. 2–118

2–119. Determine the projected component of the 80-N force acting along the axis AB of the pipe.

Prob. 2–119

2–122. A force of $F = 80$ N is applied to the handle of the wrench. Determine the angle θ between the tail of the force and the handle AB.

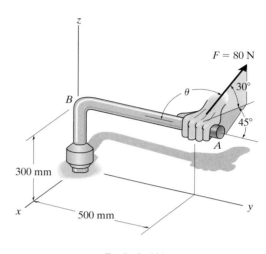

Prob. 2–122

*2–120. Determine the angles θ and ϕ between the axis OA of the pole and each cable, AB and AC.

2–121. The two cables exert the forces shown on the pole. Determine the magnitude of the projected component of each force acting along the axis OA of the pole.

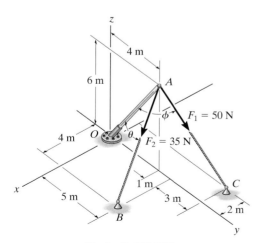

Probs. 2–120/121

2–123. Two cables exert forces on the pipe. Determine the magnitude of the projected component of $\mathbf{F}_1$ along the line of action of $\mathbf{F}_2$.

*2–124. Determine the angle θ between the two cables attached to the pipe.

Probs. 2–123/124

2–125. Determine the angle θ between the two cables.

2–126. Determine the projection of the force $\mathbf{F}_1$ along cable AB. Determine the projection of the force $\mathbf{F}_2$ along cable AC.

*2–128. Determine the magnitude of the projected component of the 100-lb force acting along the axis BC of the pipe.

2–129. Determine the angle θ between pipe segments BA and BC.

Probs. 2–125/126

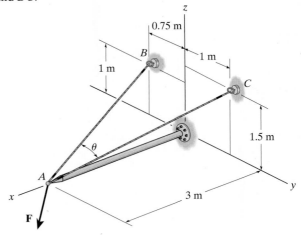

Probs. 2–128/129

2–127. Determine the angle θ between the edges of the sheet-metal bracket.

2–130. Determine the angles θ and ϕ made between the axes OA of the flag pole and AB and AC, respectively, of each cable.

Prob. 2–127

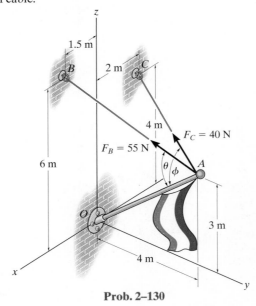

Prob. 2–130

Chapter Review

A scalar is a positive or negative number; e.g., mass and temperature. A vector has a magnitude and an arrowhead sense of direction; e.g., force and position. (*See page 17.*)		
Multiplication or division of a vector by a scalar will change only the magnitude of the vector. If the scalar is negative the sense of the vector will change, so that it acts in the opposite direction. (*See page 18.*)		
If vectors are collinear, the resultant is formed by an algebraic or scalar addition. (*See page 19.*)	$R = A + B$	
Parallelogram Law Two forces add according to the parallelogram law. The *components* form the sides of the parallelogram and the *resultant* is the diagonal. To find the components of a force along any two axes, extend lines from the head of the force, parallel to the axes, to form the components. To obtain the components or the resultant, show how the forces add by a tip-to-tail addition using the triangle rule, and then use the law of sines and the law of cosines to calculate their values. (*See pages 18–21.*)	$$F_R = \sqrt{F_1{}^2 + F_2{}^2 - 2\,F_1F_2\cos\theta_R}$$ $$\frac{F_1}{\sin\theta_1} = \frac{F_2}{\sin\theta_2} = \frac{F_R}{\sin\theta_R}$$	

Rectangular Components: Two Dimensions

Vectors $\mathbf{F}_x$ and $\mathbf{F}_y$ are rectangular components of $\mathbf{F}$.

The resultant force is determined from the algebraic sum of its components.
(*See page 31.*)

$$F_{Rx} = \Sigma F_x$$
$$F_{Ry} = \Sigma F_y$$
$$F_R = \sqrt{(F_{Rx})^2 + (F_{Ry})^2}, \quad \theta = \tan^{-1}\left|\frac{F_{Ry}}{F_{Rx}}\right|$$

Cartesian Vectors

The unit vector $\mathbf{u}_A$ has a length 1, no units, and it points in the direction of the vector $\mathbf{F}$.

$$\mathbf{u} = \frac{\mathbf{F}}{F}$$

A force can be resolved into its Cartesian components along the x, y, z axes so that $\mathbf{F} = F_x\mathbf{i} + F_y\mathbf{j} + F_z\mathbf{k}$.

The magnitude of $\mathbf{F}$ is determined from the positive square root of the sum of the squares of its components.

$$F = \sqrt{F_x^2 + F_y^2 + F_z^2}$$

The coordinate direction angles α, β, γ are determined by formulating a unit vector in the direction of $\mathbf{F}$. The components of $\mathbf{u}$ represent $\cos\alpha$, $\cos\beta$, $\cos\gamma$.

$$\mathbf{u} = \frac{\mathbf{F}}{F} = \frac{F_x}{F}\mathbf{i} + \frac{F_y}{F}\mathbf{j} + \frac{F_z}{F}\mathbf{k}$$

$$\mathbf{u} = \cos\alpha\mathbf{i} + \cos\beta\mathbf{j} + \cos\gamma\mathbf{k}$$

The coordinate direction angles are related so that only two of the three angles are independent of one another.

$$\cos^2 \alpha + \cos^2 \beta + \cos^2 \gamma = 1$$

To find the resultant of a concurrent force system, express each force as a Cartesian vector and add the $\mathbf{i}, \mathbf{j}, \mathbf{k}$ components of all the forces in the system.
(*See pages 43–46.*)

$$\mathbf{F}_R = \Sigma \mathbf{F} = \Sigma F_x \mathbf{i} + \Sigma F_y \mathbf{j} + \Sigma F_z \mathbf{k}$$

Position and Force Vectors

A position vector locates one point in space relative to another. The easiest way to formulate the components of a position vector is to determine the distance and direction that one must travel along the x, y, and z directions—going from the tail to the head of the vector.

$$\mathbf{r} = (x_B - x_A)\mathbf{i}$$
$$+ (y_B - y_A)\mathbf{j}$$
$$+ (z_B - z_A)\mathbf{k}$$

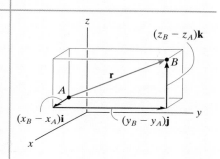

If the line of action of a force passes through points A and B, then the force acts in the same direction $\mathbf{u}$ as the position vector $\mathbf{r}$. The force can then be expressed as a Cartesian vector.
(*See pages 55, 56, and 58.*)

$$\mathbf{F} = F\mathbf{u} = F\left(\frac{\mathbf{r}}{r}\right)$$

Dot Product

The dot product between two vectors $\mathbf{A}$ and $\mathbf{B}$ yields a scalar. If $\mathbf{A}$ and $\mathbf{B}$ are expressed in Cartesian vector form, then the dot product is the sum of the products of their $x, y,$ and z components.

$$\mathbf{A} \cdot \mathbf{B} = AB \cos \theta$$
$$= A_x B_x + A_y B_y + A_z B_z$$

The dot product can be used to determine the angle between $\mathbf{A}$ and $\mathbf{B}$.

$$\theta = \cos^{-1}\left(\frac{\mathbf{A} \cdot \mathbf{B}}{AB}\right)$$

The dot product is also used to determine the projected component of a vector $\mathbf{A}$ onto an axis aa' defined by its unit vector $\mathbf{u}$.
(*See pages 68–69.*)

$$\mathbf{A}_{\parallel} = A \cos \theta \, \mathbf{u} = (\mathbf{A} \cdot \mathbf{u})\mathbf{u}$$

REVIEW PROBLEMS

2–131. Determine the magnitude and coordinate direction angles of $\mathbf{F}_3$ so that the resultant of the three forces acts along the positive y axis and has a magnitude of 600 lb.

***2–132.** Determine the magnitude and coordinate direction angles of $\mathbf{F}_3$ so that the resultant of the three forces is zero.

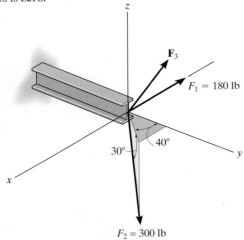

$F_2 = 300$ lb

Probs. 2–131/132

2–133. Resolve the force $\mathbf{F}$ into two components, one acting parallel and the other acting perpendicular to the u axis.

Prob. 2–133

2–134. The force $\mathbf{F}$ has a magnitude of 80 lb and acts at the midpoint C of the thin rod. Express the force as a Cartesian vector.

Prob. 2–134

2–135. Determine the magnitude and direction of the resultant $\mathbf{F}_R = \mathbf{F}_1 + \mathbf{F}_2 + \mathbf{F}_3$ of the three forces by first finding the resultant $\mathbf{F}' = \mathbf{F}_1 + \mathbf{F}_3$ and then forming $\mathbf{F}_R = \mathbf{F}' + \mathbf{F}_2$. Specify its direction measured counterclockwise from the positive x axis.

Prob. 2–135

*2–136. The leg is held in position by the quadriceps AB, which is attached to the pelvis at A. If the force exerted on this muscle by the pelvis is 85 N, in the direction shown, determine the stabilizing force component acting along the positive y axis and the supporting force component acting along the negative x axis.

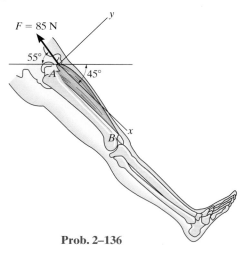

Prob. 2–136

2–137. Determine the magnitudes of the projected components of the force $\mathbf{F} = \{60\mathbf{i} + 12\mathbf{j} - 40\mathbf{k}\}$ N in the direction of the cables AB and AC.

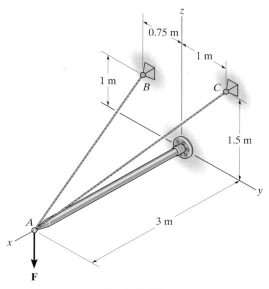

Prob. 2–137

2–138. Determine the magnitude of the projected component of the 100-lb force acting along the axis BC of the pipe.

Prob. 2–138

2–139. If $F_1 = F_2 = 30$ lb, determine the angles θ and ϕ so that the resultant force is directed along the positive x axis and has a magnitude of $F_R = 20$ lb.

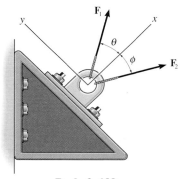

Prob. 2–139

*2–140. Determine the magnitude of the resultant and its direction measured counterclockwise from axis.

Prob. 2–140

Whenever cables are used for hoisting loads, they must be selected so that they do not fail when they are placed at their points of attachment. In this chapter, we will show how to calculate cable loadings for such cases.

3

Equilibrium of a Particle

CHAPTER OBJECTIVES

- To introduce the concept of the free-body diagram for a particle.
- To show how to solve particle equilibrium problems using the equations of equilibrium.

3.1 Condition for the Equilibrium of a Particle

A particle is in *equilibrium* provided it is at rest if originally at rest or has a constant velocity if originally in motion. Most often, however, the term "equilibrium" or, more specifically, "static equilibrium" is used to describe an object at rest. To maintain equilibrium, it is *necessary* to satisfy Newton's first law of motion, which requires the *resultant force* acting on a particle to be equal to *zero*. This condition may be stated mathematically as

$$\Sigma \mathbf{F} = \mathbf{0} \qquad (3–1)$$

where $\Sigma \mathbf{F}$ is the vector *sum of all the forces* acting on the particle.

Not only is Eq. 3–1 a necessary condition for equilibrium, it is also a *sufficient* condition. This follows from Newton's second law of motion, which can be written as $\Sigma \mathbf{F} = m\mathbf{a}$. Since the force system satisfies Eq. 3–1, then $m\mathbf{a} = \mathbf{0}$, and therefore the particle's acceleration $\mathbf{a} = \mathbf{0}$. Consequently the particle indeed moves with constant velocity or remains at rest.

3.2 The Free-Body Diagram

To apply the equation of equilibrium, we must account for *all* the known and unknown forces ($\Sigma \mathbf{F}$) which act *on* the particle. The best way to do this is to draw the particle's *free-body diagram*. This diagram is simply a sketch which shows the particle "free" from its surroundings with *all* the forces that act *on* it.

Before presenting a formal procedure as to how to draw a free-body diagram, we will first consider two types of connections often encountered in particle equilibrium problems.

Springs.   If a *linearly elastic spring* is used for support, the length of the spring will change in direct proportion to the force acting on it. A characteristic that defines the "elasticity" of a spring is the *spring constant* or *stiffness k*. The magnitude of force exerted on a linearly elastic spring which has a stiffness k and is deformed (elongated or compressed) a distance s, measured from its *unloaded* position, is

$$\boxed{F = ks} \tag{3–2}$$

Here s is determined from the difference in the spring's deformed length l and its undeformed length l_0, i.e., $s = l - l_0$. If s is positive, $\mathbf{F}$ "pulls" on the spring; whereas if s is negative, $\mathbf{F}$ must "push" on it. For example, the spring shown in Fig. 3–1 has an undeformed length $l_0 = 0.4$ m and stiffness $k = 500$ N/m. To stretch it so that $l = 0.6$ m, a force $F = ks = (500 \text{ N/m})(0.6 \text{ m} - 0.4 \text{ m}) = 100$ N is needed. Likewise, to compress it to a length $l = 0.2$ m, a force $F = ks = (500 \text{ N/m})(0.2 \text{ m} - 0.4 \text{ m}) = -100$ N is required, Fig. 3–1.

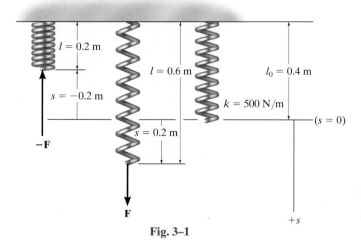

Fig. 3–1

Cables and Pulleys.   Throughout this book, except in Sec. 7.4, all cables (or cords) are assumed to have negligible weight and they cannot stretch. Also, a cable can support *only* a tension or "pulling" force, and this force always acts in the direction of the cable. In Chapter 5, it will be shown that the tension force developed in a *continuous cable* which passes over a frictionless pulley must have a *constant* magnitude to keep the cable in equilibrium. Hence, for any angle θ, shown in Fig. 3–2, the cable is subjected to a constant tension T throughout its length.

Procedure for Drawing a Free-Body Diagram

Since we must account for *all the forces acting on the particle* when applying the equations of equilibrium, the importance of first drawing a free-body diagram cannot be overemphasized. To construct a free-body diagram, the following three steps are necessary.

Draw Outlined Shape.

Imagine the particle to be *isolated* or cut "free" from its surroundings by drawing its outlined shape.

Show All Forces.

Indicate on this sketch *all* the forces that act *on the particle*. These forces can be *active forces*, which tend to set the particle in motion, or they can be *reactive forces* which are the result of the constraints or supports that tend to prevent motion. To account for all these forces, it may help to trace around the particle's boundary, carefully noting each force acting on it.

Identify Each Force.

The forces that are *known* should be labeled with their proper magnitudes and directions. Letters are used to represent the magnitudes and directions of forces that are unknown.

Cable is in tension

Fig. 3–2

Photo A: The bucket is held in equilibrium by the cable, and instinctively we know that the force in the cable must equal the weight of the bucket. By drawing a free-body diagram of the bucket we can understand why this is so. This diagram shows that there are only two forces *acting on the bucket*, namely, its weight **W** and the force **T** of the cable. For equilibrium, the resultant of these forces must be equal to zero, and so $T = W$. The important point is that by *isolating the bucket* the unknown cable force **T** becomes "exposed" and must be considered as a requirement for equilibrium.

Photo B: The spool has a weight W and is suspended from the crane boom. If we wish to obtain the forces in cables AB and AC, then we should consider the free-body diagram of the ring at A. Here the cables AD exert a resultant force of **W** on the ring and the condition of equilibrium is used to obtain $\mathbf{T}_B$ and $\mathbf{T}_C$.

A

B

EXAMPLE 3.1

The sphere in Fig. 3–3a has a mass of 6 kg and is supported as shown. Draw a free-body diagram of the sphere, the cord CE, and the knot at C.

$\mathbf{F}_{CE}$ (Force of cord CE acting on sphere)

58.9N (Weight or gravity acting on sphere)

(b)

SOLUTION

Sphere.   By inspection, there are only two forces acting on the sphere, namely, its weight and the force of cord CE. The sphere has a weight of 6 kg (9.81 m/s^2) = 58.9 N. The free-body diagram is shown in Fig. 3–3b.

Cord CE.   When the cord CE is isolated from its surroundings, its free-body diagram shows only two forces acting on it, namely, the force of the sphere and the force of the knot, Fig. 3–3c. Notice that $\mathbf{F}_{CE}$ shown here is equal but opposite to that shown in Fig. 3–3b, a consequence of Newton's third law. Also, $\mathbf{F}_{CE}$ and $\mathbf{F}_{EC}$ pull on the cord and keep it in tension so that it doesn't collapse. For equilibrium, $F_{CE} = F_{EC}$.

Knot.   The knot at C is subjected to three forces, Fig. 3–3d. They are caused by the cords CBA and CE and the spring CD. As required the free-body diagram shows all these forces labeled with their magnitudes and directions. It is important to recognize that the weight of the sphere does not directly act on the knot. Instead, the cord CE subjects the knot to this force.

$\mathbf{F}_{EC}$ (Force of knot acting on cord CE)

$\mathbf{F}_{CE}$ (Force of sphere acting on cord CE)

(c)

$\mathbf{F}_{CBA}$ (Force of cord CBA acting on knot)

$\mathbf{F}_{CD}$ (Force of spring acting on knot)

$\mathbf{F}_{CE}$ (Force of cord CE acting on knot)

(d)

Fig. 3–3

3.3 Coplanar Force Systems

If a particle is subjected to a system of coplanar forces that lie in the x–y plane, Fig. 3–4, then each force can be resolved into its $\mathbf{i}$ and $\mathbf{j}$ components. For equilibrium, Eq. 3–1 can be written as

$$\Sigma \mathbf{F} = \mathbf{0}$$
$$\Sigma F_x \mathbf{i} + \Sigma F_y \mathbf{j} = \mathbf{0}$$

For this vector equation to be satisfied, both the x and y components must be equal to zero. Hence,

$$\boxed{\begin{aligned} \Sigma F_x &= 0 \\ \Sigma F_y &= 0 \end{aligned}} \qquad (3\text{–}3)$$

These *scalar equations of equilibrium* require that the *algebraic sum* of the x and y components of all the forces acting on the particle be equal to zero. As a result, Eqs. 3–3 can be solved for at most two unknowns, generally represented as angles and magnitudes of forces shown on the particle's free-body diagram.

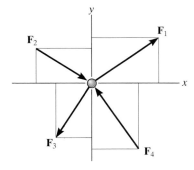

Fig. 3–4

Scalar Notation. Since each of the two equilibrium equations requires the resolution of vector components along a specified x or y axis, we will use scalar notation to represent the components when applying these equations. When doing this, the sense of direction for each component is accounted for by an *algebraic sign* which corresponds to the arrowhead direction of the component along each axis. If a force has an *unknown magnitude*, then the arrowhead sense of the force on the free-body diagram can be *assumed*. Since the magnitude of a force is *always positive*, then if the *solution* yields a *negative scalar*, this indicates that the sense of the force acts in the opposite direction.

The chains exert three forces on the ring at A. The ring will not move, or will move with constant velocity, provided the summation of these forces along the x and along the y axis on the free-body diagram is zero. If one of the three forces is known, the magnitudes of the other two forces can be obtained from the two equations of equilibrium.

Fig. 3–5

For example, consider the free-body diagram of the particle subjected to the two forces shown in Fig. 3–5. Here it is *assumed* that the *unknown force* **F** acts to the right to maintain equilibrium. Applying the equation of equilibrium along the x axis, we have

$$\xrightarrow{+} \Sigma F_x = 0; \qquad\qquad +F + 10\ \text{N} = 0$$

Both terms are "positive" since both forces act in the positive x direction. When this equation is solved, $F = -10\ \text{N}$. Here the *negative sign* indicates that **F** must act to the left to hold the particle in equilibrium, Fig. 3–5. Notice that if the $+x$ axis in Fig. 3–5 was directed to the left, both terms in the above equation would be negative, but again, after solving, $F = -10\ \text{N}$, indicating again **F** would be directed to the left.

Procedure for Analysis

Coplanar force equilibrium problems for a particle can be solved using the following procedure.

Free-Body Diagram.

- Establish the x, y axes in any suitable orientation.
- Label all the known and unknown force magnitudes and directions on the diagram.
- The sense of a force having an unknown magnitude can be assumed.
- All forces on *AFBD* will be concurrent.

Equations of Equilibrium.

- Apply the equations of equilibrium $\Sigma F_x = 0$ and $\Sigma F_y = 0$.
- Components are positive if they are directed along a positive axis, and negative if they are directed along a negative axis.
- If more than two unknowns exist and the problem involves a spring, apply $F = ks$ to relate the spring force to the deformation s of the spring.
- If the solution yields a negative result, this indicates the sense of the force is the reverse of that shown on the free-body diagram.

EXAMPLE 3.2

Determine the tension in cables AB and AD for equilibrium of the 250-kg engine shown in Fig. 3–6a.

SOLUTION

Free-Body Diagram. To solve this problem, we will investigate the equilibrium of the ring at A because this "particle" is subjected to the forces of both cables AB and AD. First, however, note that the engine has a weight $(250 \text{ kg})(9.81 \text{ m/s}^2) = 2.452 \text{ kN}$ which is supported by cable CA. Therefore, as shown in Fig. 3–6b, there are three concurrent forces *acting on the ring*. The forces $\mathbf{T}_B$ and $\mathbf{T}_D$ have unknown magnitudes but known directions, and cable AC exerts a downward force on A equal to 2.452 kN.

(a)

Fig. 3–6

2.452 kN

(b)

Equations of Equilibrium. The two unknown magnitudes T_B and T_D can be obtained from the two scalar equations of equilibrium, $\Sigma F_x = 0$ and $\Sigma F_y = 0$. To apply these equations, the x, y axes are established on the free-body diagram and $\mathbf{T}_B$ must be resolved into its x and y components. Thus,

$$\xrightarrow{+} \Sigma F_x = 0; \qquad T_B \cos 30° - T_D = 0 \qquad (1)$$
$$+\uparrow \Sigma F_y = 0; \qquad T_B \sin 30° - 2.452 \text{ kN} = 0 \qquad (2)$$

Solving Eq. 2 for T_B and substituting into Eq. 1 to obtain T_D yields

$$T_B = 4.90 \text{ kN} \qquad\qquad Ans.$$
$$T_D = 4.25 \text{ kN} \qquad\qquad Ans.$$

NOTE: The accuracy of these results, of course, depends on the accuracy of the data, i.e., measurements of geometry and loads. For most engineering work involving a problem such as this, the data as measured to three significant figures would be sufficient. Also, note that here we have neglected the weights of the cables, a reasonable assumption since they would be small in comparison with the weight of the engine.

EXAMPLE 3.3

(a)

(b)

(c)

(d)

Fig. 3–7

If the sack at A in Fig. 3–7a has a weight of 20 lb, determine the weight of the sack at B and the force in each cord needed to hold the system in the equilibrium position shown.

SOLUTION

Since the weight of A is known, the unknown tension in the two cords EG and EC can be determined by investigating the equilibrium of the ring at E. Why?

Free-Body Diagram.   There are three forces acting on E, as shown in Fig. 3–7b.

Equations of Equilibrium.   Establishing the x, y axes and resolving each force into its x and y components using trigonometry, we have

$$\xrightarrow{+} \Sigma F_x = 0; \qquad\qquad T_{EG} \sin 30° - T_{EC} \cos 45° = 0 \qquad (1)$$

$$+\uparrow \Sigma F_y = 0; \qquad T_{EG} \cos 30° - T_{EC} \sin 45° - 20 \text{ lb} = 0 \qquad (2)$$

Solving Eq. 1 for T_{EG} in terms of T_{EC} and substituting the result into Eq. 2 allows a solution for T_{EC}. One then obtains T_{EG} from Eq. 1. The results are

$$T_{EC} = 38.6 \text{ lb} \qquad\qquad Ans.$$
$$T_{EG} = 54.6 \text{ lb} \qquad\qquad Ans.$$

Using the calculated result for T_{EC}, the equilibrium of the ring at C can now be investigated to determine the tension in CD and the weight of B.

Free-Body Diagram.   As shown in Fig. 3–7c, $T_{EC} = 38.6$ lb "pulls" on C. The reason for this becomes clear when one draws the free-body diagram of cord CE and applies both equilibrium and the principle of action, equal but opposite force reaction (Newton's third law), Fig. 3–7d.

Equations of Equilibrium.   Establishing the x, y axes and noting the components of $\mathbf{T}_{CD}$ are proportional to the slope of the cord as defined by the 3–4–5 triangle, we have

$$\xrightarrow{+} \Sigma F_x = 0; \qquad 38.6 \cos 45° \text{ lb} - \left(\tfrac{4}{5}\right) T_{CD} = 0 \qquad (3)$$

$$\xrightarrow{+} \Sigma F_y = 0; \qquad \left(\tfrac{3}{5}\right) T_{CD} + 38.6 \sin 45° \text{ lb} - W_B = 0 \qquad (4)$$

Solving Eq. 3 and substituting the result into Eq. 4 yields

$$T_{CD} = 34.2 \text{ lb} \qquad\qquad Ans.$$
$$W_B = 47.8 \text{ lb} \qquad\qquad Ans.$$

NOTE: Review this example by constructing the free-body diagrams on a separate sheet and identifying each force.

EXAMPLE 3.4

Determine the required length of cord AC in Fig. 3–8a so that the 8-kg lamp is suspended in the position shown. The *undeformed* length of spring AB is $l'_{AB} = 0.4$ m, and the spring has a stiffness of $k_{AB} = 300$ N/m.

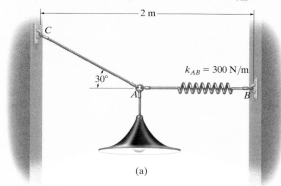

2 m

C

$30°$

A

$k_{AB} = 300$ N/m

B

(a)

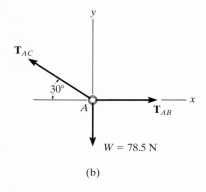

y

$\mathbf{T}_{AC}$

$30°$

A

$\mathbf{T}_{AB}$

x

$W = 78.5$ N

(b)

Fig. 3–8

SOLUTION

If the force in spring AB is known, the stretch of the spring can be found using $F = ks$. From the problem geometry, it is then possible to calculate the required length of AC.

Free-Body Diagram.   The lamp has a weight $W = 8(9.81) = 78.5$ N. The free-body diagram of the ring at A is shown in Fig. 3–8b.

Equations of Equilibrium.   Using the x, y axes,

$\xrightarrow{+} \Sigma F_x = 0;$ $T_{AB} - T_{AC} \cos 30° = 0$

$+\uparrow \Sigma F_y = 0;$ $T_{AC} \sin 30° - 78.5 \text{ N} = 0$

Solving, we obtain

$$T_{AC} = 157.0 \text{ N}$$
$$T_{AB} = 136.0 \text{ N}$$

The stretch of spring AB is therefore

$$T_{AB} = k_{AB} s_{AB}; \qquad 136.0 \text{ N} = 300 \text{ N/m}(s_{AB})$$
$$s_{AB} = 0.453 \text{ m}$$

so the stretched length is

$$l_{AB} = l'_{AB} + s_{AB}$$
$$l_{AB} = 0.4 \text{ m} + 0.453 \text{ m} = 0.853 \text{ m}$$

The horizontal distance from C to B, Fig. 3–8a, requires

$$2 \text{ m} = l_{AC} \cos 30° + 0.853 \text{ m}$$
$$l_{AC} = 1.32 \text{ m} \qquad\qquad Ans.$$

PROBLEMS

3–1. Determine the magnitudes of $\mathbf{F}_1$ and $\mathbf{F}_2$ so that the particle is in equilibrium.

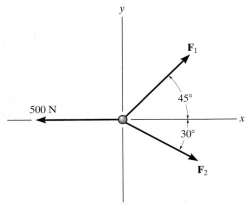

Prob. 3–1

3–3. Determine the magnitude of $\mathbf{F}$ and the orientation θ of the 750-N force so that the particle is in equilibrium.

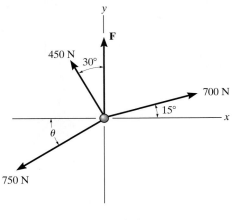

Prob. 3–3

3–2. Determine the magnitude and direction θ of $\mathbf{F}$ so that the particle is in equilibrium.

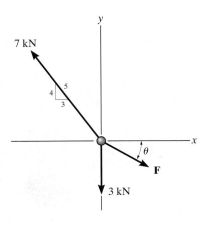

Prob. 3–2

*3–4. Determine the magnitude and angle θ of $\mathbf{F}$ so that the particle is in equilibrium.

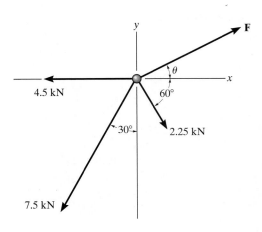

Prob. 3–4

3–5. The members of a truss are connected to the gusset plate. If the forces are concurrent at point O, determine the magnitudes of **F** and **T** for equilibrium. Take $\theta = 30°$.

3–6. The gusset plate is subjected to the forces of four members. Determine the force in member B and its proper orientation θ for equilibrium. The forces are concurrent at point O. Take $F = 12$ kN.

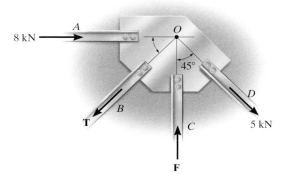

Probs. 3–5/6

3–7. Determine the maximum weight of the engine that can be supported without exceeding a tension of 450 lb in chain AB and 480 lb in chain AC.

Prob. 3–7

*3–8. The 200-kg engine is suspended from a vertical chain at A. A second chain is wrapped around the engine and held in position by the spreader bar BC. Determine the compressive force acting along the axis of the bar and the tension forces in segments BA and CA of the chain. *Hint:* Analyze equilibrium first at A, then at B.

Prob. 3–8

3–9. Cords AB and AC can each sustain a maximum tension of 800 lb. If the drum has a weight of 900 lb, determine the smallest angle θ at which they can be attached to the drum.

Prob. 3–9

3–10. The 500-lb crate is hoisted using the ropes AB and AC. Each rope can withstand a maximum tension of 2500 lb before it breaks. If AB always remains horizontal, determine the smallest angle θ to which the crate can be hoisted.

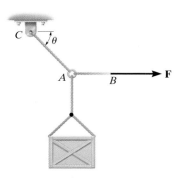

Prob. 3–10

3–11. Two electrically charged pith balls, each having a mass of 0.2 g, are suspended from light threads of equal length. Determine the resultant horizontal force of repulsion, F, acting on each ball if the measured distance between them is $r = 200$ mm.

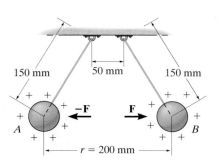

Prob. 3–11

*3–12. The towing pendant AB is subjected to the force of 50 kN which is developed from a tugboat. Determine the force that is in each of the bridles, BC and BD, if the ship is moving forward with constant velocity.

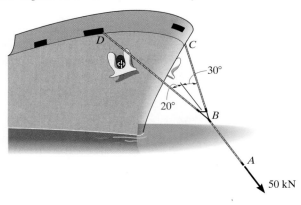

Prob. 3–12

3–13. Determine the stretch in each spring for equilibrium of the 2-kg block. The springs are shown in the equilibrium position.

3–14. The unstretched length of spring AB is 2 m. If the block is held in the equilibrium position shown, determine the mass of the block at D.

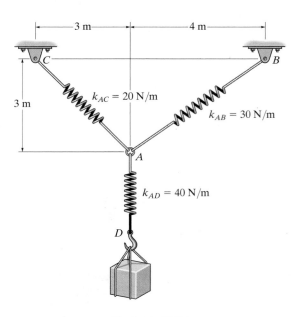

Probs. 3–13/14

3–15. The springs AB and BC have a stiffness of 500 N/m and an unstretched length of 6 m. Determine the horizontal force **F** applied to the cord which is attached to the *small* pulley B so that the displacement of the pulley from the wall is $d = 1.5$ m.

*3–16. The springs AB and BC have a stiffness of 500 N/m and an unstretched length of 6 m. Determine the displacement d of the cord from the wall when a force $F = 175$ N is applied to the cord.

Probs. 3–15/16

3–17. Determine the force in each cable and the force **F** needed to hold the 4-kg lamp in the position shown. *Hint:* First analyze the equilibrium at B; then, using the result for the force in BC, analyze the equilibrium at C.

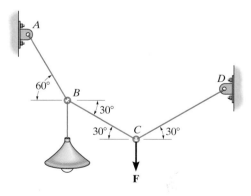

Prob. 3–17

3–18. The motor at B winds up the cord attached to the 65-lb crate with a constant speed. Determine the force in cord CD supporting the pulley and the angle θ for equilibrium. Neglect the size of the pulley at C.

3–19. The cords BCA and CD can each support a maximum load of 100 lb. Determine the maximum weight of the crate that can be hoisted at constant velocity, and the angle θ for equilibrium.

Probs. 3–18/19

*3–20. The sack has a weight of 15 lb and is supported by the six cords tied together as shown. Determine the tension in each cord and the angle θ for equilibrium. Cord BC is horizontal.

3–21. Each cord can sustain a maximum tension of 200 lb. Determine the largest weight of the sack that can be supported. Also, determine θ of cord DC for equilibrium.

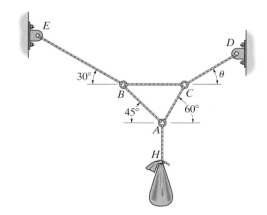

Probs. 3–20/21

3–22.   The block has a weight of 20 lb and is being hoisted at uniform velocity. Determine the angle θ for equilibrium and the required force in each cord.

3–23.   Determine the maximum weight W of the block that can be suspended in the position shown if each cord can support a maximum tension of 80 lb. Also, what is the angle θ for equilibrium?

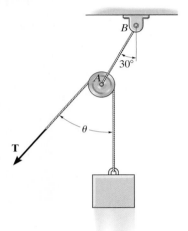

Probs. 3–22/23

3–25.   Blocks D and F weigh 5 lb each and block E weighs 8 lb. Determine the sag s for equilibrium. Neglect the size of the pulleys.

3–26.   If blocks D and F weigh 5 lb each, determine the weight of block E if the sag $s = 3$ ft. Neglect the size of the pulleys.

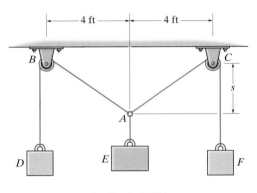

Probs. 3–25/26

*3–24.   Two spheres A and B have an equal mass and are electrostatically charged such that the repulsive force acting between them has a magnitude of 20 mN and is directed along line AB. Determine the angle θ, the tension in cords AC and BC, and the mass m of each sphere.

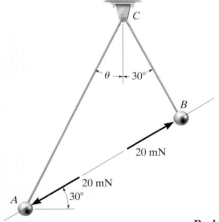

Prob. 3–24

3–27.   The 30-kg block is supported by two springs having the stiffness shown. Determine the unstretched length of each spring after the block is removed.

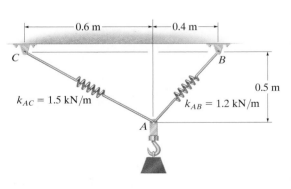

Prob. 3–27

*3–28. Three blocks are supported using the cords and pulleys. If they have weights of $W_A = W$, $W_B = 0.25W$, and $W_C = W$, determine the angle θ for equilibrium.

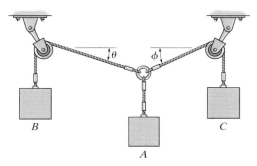

Prob. 3–28

3–30. Prove *Lami's theorem*, which states that if three concurrent forces are in equilibrium, each is proportional to the sine of the angle of the other two; that is, $P/\sin \alpha = Q/\sin \beta = R/\sin \gamma$.

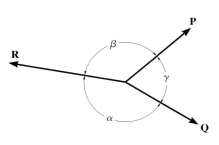

Prob. 3–30

3–29. A continuous cable of total length 4 m is wrapped around the *small* pulleys at A, B, C, and D. If each spring is stretched 300 mm, determine the mass m of each block. Neglect the weight of the pulleys and cords. The springs are unstretched when $d = 2$ m.

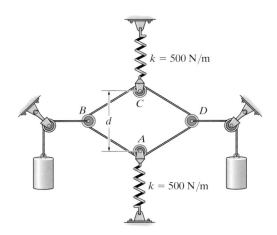

Prob. 3–29

■3–31. A vertical force $P = 10$ lb is applied to the ends of the 2-ft cord AB and spring AC. If the spring has an unstretched length of 2 ft, determine the angle θ for equilibrium. Take $k = 15$ lb/ft.

*3–32. Determine the unstretched length of spring AC if a force $P = 80$ lb causes the angle $\theta = 60°$ for equilibrium. Cord AB is 2 ft long. Take $k = 50$ lb/ft.

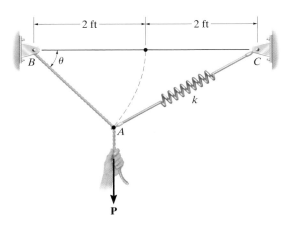

Probs. 3–31/32

■3–33. The 20-kg flowerpot is suspended from three wires and supported by the hooks at B and C. Determine the tension in AB and AC for equilibrium.

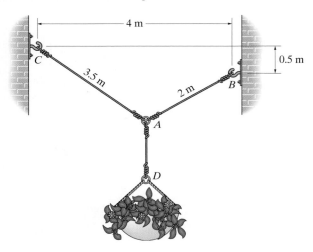

Prob. 3–33

■3–34. A car is to be towed using the rope arrangement shown. The towing force required is 600 lb. Determine the minimum length l of rope AB so that the tension in either rope AB or AC does not exceed 750 lb. *Hint:* Use the equilibrium condition at point A to determine the required angle θ for attachment, then determine l using trigonometry applied to triangle ABC.

Prob. 3–34

3–35. Determine the mass of each of the two cylinders if they cause a sag of $s = 0.5$ m when suspended from the rings at A and B. Note that $s = 0$ when the cylinders are removed.

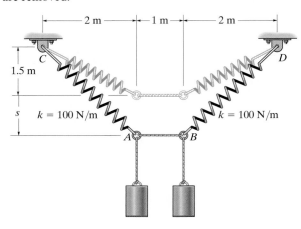

Prob. 3–35

*3–36. The sling BAC is used to lift the 100-lb load with constant velocity. Determine the force in the sling and plot its value T (ordinate) as a function of its orientation θ, where $0 \leq \theta \leq 90°$.

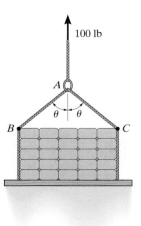

Prob. 3–36

■3–37. The 10-lb lamp fixture is suspended from two springs, each having an unstretched length of 4 ft and stiffness of $k = 5$ lb/ft. Determine the angle θ for equilibrium.

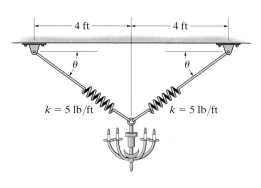

Prob. 3–37

3–38. The 200-lb uniform tank is suspended by means of a 6-ft-long cable, which is attached to the sides of the tank and passes over the small pulley located at O. If the cable can be attached at either points A and B, or C and D, determine which attachment produces the least amount of tension in the cable. What is this tension?

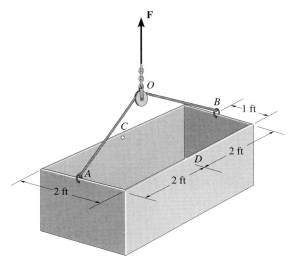

Prob. 3–38

3–39. A 4-kg sphere rests on the smooth parabolic surface. Determine the normal force it exerts on the surface and the mass m_B of block B needed to hold it in the equilibrium position shown.

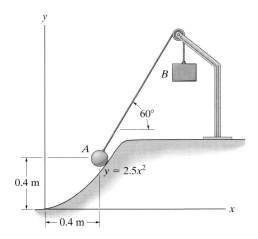

Prob. 3–39

*3–40. The 30-kg pipe is supported at A by a system of five cords. Determine the force in each cord for equilibrium.

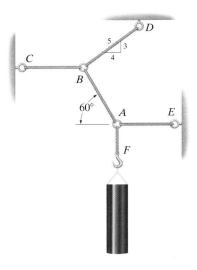

Prob. 3–40

3.4   Three-Dimensional Force Systems

For particle equilibrium we require

$$\Sigma \mathbf{F} = \mathbf{0} \qquad (3\text{--}4)$$

If the forces are resolved into their respective **i**, **j**, **k** components, Fig. 3–9, then we have

$$\Sigma F_x \mathbf{i} + \Sigma F_y \mathbf{j} + \Sigma F_z \mathbf{k} = \mathbf{0}$$

To ensure equilibrium, we must therefore require that the following three scalar component equations be satisfied:

$$
\boxed{
\begin{aligned}
\Sigma F_x &= 0 \\
\Sigma F_y &= 0 \\
\Sigma F_z &= 0
\end{aligned}
} \qquad (3\text{--}5)
$$

These equations represent the *algebraic sums* of the *x, y, z* force components acting on the particle. Using them we can solve for at most three unknowns, generally represented as angles or magnitudes of forces shown on the particle's free-body diagram.

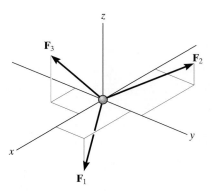

Fig. 3–9

The ring at A is subjected to the force from the hook as well as forces from each of the three chains. If the electromagnet and its load have a weight W, then the force at the hook will be $\mathbf{W}$, and the three scalar equations of equilibrium can be applied to the free-body diagram of the ring in order to determine the chain forces, $\mathbf{F}_B$, $\mathbf{F}_C$ and $\mathbf{F}_D$.

Procedure for Analysis

Three-dimensional force equilibrium problems for a particle can be solved using the following procedure.

Free-Body Diagram.

- Establish the x, y, z axes in any suitable orientation.
- Label all the known and unknown force magnitudes and directions on the diagram.
- The sense of a force having an unknown magnitude can be assumed.

Equations of Equilibrium.

- Use the scalar equations of equilibrium, $\Sigma F_x = 0$, $\Sigma F_y = 0$, $\Sigma F_z = 0$, in cases where it is easy to resolve each force into its x, y, z components.
- If the three-dimensional geometry appears difficult, then first express each force as a Cartesian vector, substitute these vectors into $\Sigma \mathbf{F} = \mathbf{0}$, and then set the $\mathbf{i}, \mathbf{j}, \mathbf{k}$ components equal to zero.
- If the solution yields a negative result, this indicates the sense of the force is the reverse of that shown on the free-body diagram.

EXAMPLE 3.5

(a)

A 90-lb load is suspended from the hook shown in Fig. 3–10a. The load is supported by two cables and a spring having a stiffness $k = 500$ lb/ft. Determine the force in the cables and the stretch of the spring for equilibrium. Cable AD lies in the x–y plane and cable AC lies in the x–z plane.

SOLUTION

The stretch of the spring can be determined once the force in the spring is determined.

Free-Body Diagram. The connection at A is chosen for the equilibrium analysis since the cable forces are concurrent at this point. The free-body diagram is shown in Fig. 3–10b.

Equations of Equilibrium. By inspection, each force can easily be resolved into its x, y, z components, and therefore the three scalar equations of equilibrium can be directly applied. Considering components directed along the positive axes as "positive," we have

$$\Sigma F_x = 0; \qquad F_D \sin 30° - \tfrac{4}{5} F_C = 0 \qquad (1)$$

$$\Sigma F_y = 0; \qquad -F_D \cos 30° + F_B = 0 \qquad (2)$$

$$\Sigma F_z = 0; \qquad \tfrac{3}{5} F_C - 90 \text{ lb} = 0 \qquad (3)$$

Solving Eq. 3 for F_C, then Eq. 1 for F_D, and finally Eq. 2 for F_B, yields

$$F_C = 150 \text{ lb} \qquad \qquad Ans.$$
$$F_D = 240 \text{ lb} \qquad \qquad Ans.$$
$$F_B = 208 \text{ lb} \qquad \qquad Ans.$$

The stretch of the spring is therefore

$$F_B = k s_{AB}$$
$$208 \text{ lb} = 500 \text{ lb/ft}(s_{AB})$$
$$s_{AB} = 0.416 \text{ ft} \qquad \qquad Ans.$$

NOTE: Since the results for all the cable forces are positive, each cable is in tension; that is, it pulls on point A as expected, Fig. 3–10b.

(b)

Fig. 3–10

EXAMPLE 3.6

Determine the magnitude and coordinate direction angles of force $\mathbf{F}$ in Fig. 3–11a that are required for equilibrium of particle O.

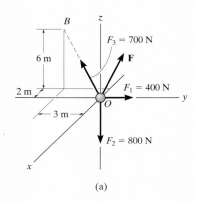

(a)

SOLUTION

Free-Body Diagram. Four forces act on particle O, Fig. 3–11b.

Equations of Equilibrium. Each of the forces can be expressed in Cartesian vector form, and the equations of equilibrium can be applied to determine the x, y, z components of $\mathbf{F}$. Noting that the coordinates of B are $B(-2\text{ m}, -3\text{ m}, 6\text{ m})$, we have

$$\mathbf{F}_1 = \{400\mathbf{j}\}\text{ N}$$

$$\mathbf{F}_2 = \{-800\mathbf{k}\}\text{ N}$$

$$\mathbf{F}_3 = F_3\left(\frac{\mathbf{r}_B}{r_B}\right) = 700\text{ N}\left[\frac{-2\mathbf{i} - 3\mathbf{j} + 6\mathbf{k}}{\sqrt{(-2)^2 + (-3)^2 + (6)^2}}\right]$$

$$= \{-200\mathbf{i} - 300\mathbf{j} + 600\mathbf{k}\}\text{ N}$$

$$\mathbf{F} = F_x\mathbf{i} + F_y\mathbf{j} + F_z\mathbf{k}$$

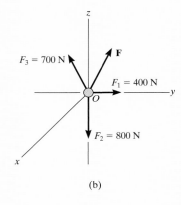

(b)

For equilibrium

$$\Sigma\mathbf{F} = \mathbf{0}; \qquad\qquad \mathbf{F}_1 + \mathbf{F}_2 + \mathbf{F}_3 + \mathbf{F} = \mathbf{0}$$

$$400\mathbf{j} - 800\mathbf{k} - 200\mathbf{i} - 300\mathbf{j} + 600\mathbf{k} + F_x\mathbf{i} + F_y\mathbf{j} + F_z\mathbf{k} = \mathbf{0}$$

Equating the respective $\mathbf{i}, \mathbf{j}, \mathbf{k}$ components to zero, we have

$$\Sigma F_x = 0; \qquad\qquad -200 + F_x = 0 \qquad\qquad F_x = 200\text{ N}$$

$$\Sigma F_y = 0; \qquad\qquad 400 - 300 + F_y = 0 \qquad\qquad F_y = -100\text{ N}$$

$$\Sigma F_z = 0; \qquad\qquad -800 + 600 + F_z = 0 \qquad\qquad F_z = 200\text{ N}$$

Thus,

$$\mathbf{F} = \{200\mathbf{i} - 100\mathbf{j} + 200\mathbf{k}\}\text{ N}$$

$$F = \sqrt{(200)^2 + (-100)^2 + (200)^2} = 300\text{ N} \qquad\qquad Ans.$$

$$\mathbf{u}_F = \frac{\mathbf{F}}{F} = \frac{200}{300}\mathbf{i} - \frac{100}{300}\mathbf{j} + \frac{200}{300}\mathbf{k}$$

(c)

$$\alpha = \cos^{-1}\left(\frac{200}{300}\right) = 48.2° \qquad\qquad Ans.$$

$$\beta = \cos^{-1}\left(\frac{-100}{300}\right) = 109° \qquad\qquad Ans.$$

$$\gamma = \cos^{-1}\left(\frac{200}{300}\right) = 48.2° \qquad\qquad Ans.$$

The magnitude and correct direction of $\mathbf{F}$ are shown in Fig. 3–11c.

Fig. 3–11

EXAMPLE 3.7

(a)

Fig. 3–12

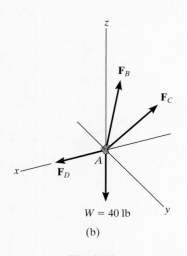

(b)

Determine the force developed in each cable used to support the 40-lb crate shown in Fig. 3–12a.

SOLUTION

Free-Body Diagram.   As shown in Fig. 3–12b, the free-body diagram of point A is considered in order to "expose" the three unknown forces in the cables.

Equations of Equilibrium.   First we will express each force in Cartesian vector form. Since the coordinates of points B and C are $B(-3 \text{ ft}, -4 \text{ ft}, 8 \text{ ft})$ and $C(-3 \text{ ft}, 4 \text{ ft}, 8 \text{ ft})$, we have

$$\mathbf{F}_B = F_B \left[\frac{-3\mathbf{i} - 4\mathbf{j} + 8\mathbf{k}}{\sqrt{(-3)^2 + (-4)^2 + (8)^2}} \right]$$
$$= -0.318F_B\mathbf{i} - 0.424F_B\mathbf{j} + 0.848F_B\mathbf{k}$$
$$\mathbf{F}_C = F_C \left[\frac{-3\mathbf{i} + 4\mathbf{j} + 8\mathbf{k}}{\sqrt{(-3)^2 + (4)^2 + (8)^2}} \right]$$
$$= -0.318F_C\mathbf{i} + 0.424F_C\mathbf{j} + 0.484F_C\mathbf{k}$$
$$\mathbf{F}_D = F_D\mathbf{i}$$
$$\mathbf{W} = \{-40\mathbf{k}\} \text{ lb}$$

Equilibrium requires

$$\Sigma\mathbf{F} = \mathbf{0}; \qquad \mathbf{F}_B + \mathbf{F}_C + \mathbf{F}_D + \mathbf{W} = \mathbf{0}$$
$$-0.318F_B\mathbf{i} - 0.424F_B\mathbf{j} + 0.848F_B\mathbf{k} - 0.318F_C\mathbf{i} + 0.424F_C\mathbf{j}$$
$$+ 0.848F_C\mathbf{k} + F_D\mathbf{i} - 40\mathbf{k} = \mathbf{0}$$

Equating the respective $\mathbf{i}, \mathbf{j}, \mathbf{k}$ components to zero yields

$$\Sigma F_x = 0; \qquad -0.318F_B - 0.318F_C + F_D = 0 \qquad (1)$$
$$\Sigma F_y = 0; \qquad -0.424F_B + 0.424F_C = 0 \qquad (2)$$
$$\Sigma F_z = 0; \qquad 0.848F_B + 0.848F_C - 40 = 0 \qquad (3)$$

Equation 2 states that $F_B = F_C$. Thus, solving Eq. 3 for F_B and F_C and substituting the result into Eq. 1 to obtain F_D, we have

$$F_B = F_C = 23.6 \text{ lb} \qquad \qquad Ans.$$
$$F_D = 15.0 \text{ lb} \qquad \qquad Ans.$$

NOTE: Do you understand why each cable force is in tension? See Fig. 3–7d.

EXAMPLE 3.8

The 100-kg crate shown in Fig. 3–13a is supported by three cords, one of which is connected to a spring. Determine the tension in cords AC and AD and the stretch of the spring.

SOLUTION

Free-Body Diagram. The force in each of the cords can be determined by investigating the equilibrium of point A. The free-body diagram is shown in Fig. 3–13b. The weight of the crate is $W = 100(9.81) = 981$ N.

Equations of Equilibrium. Each vector on the free-body diagram is first expressed in Cartesian vector form. Using Eq. 2–11 for $\mathbf{F}_C$ and noting point $D(-1 \text{ m}, 2 \text{ m}, 2 \text{ m})$ for $\mathbf{F}_D$, we have

$$\mathbf{F}_B = F_B\mathbf{i}$$
$$\mathbf{F}_C = F_C \cos 120°\mathbf{i} + F_C \cos 135°\mathbf{j} + F_C \cos 60°\mathbf{k}$$
$$= -0.5F_C\mathbf{i} - 0.707F_C\mathbf{j} + 0.5F_C\mathbf{k}$$
$$\mathbf{F}_D = F_D\left[\frac{-1\mathbf{i} + 2\mathbf{j} + 2\mathbf{k}}{\sqrt{(-1)^2 + (2)^2 + (2)^2}}\right]$$
$$= -0.333F_D\mathbf{i} + 0.667F_D\mathbf{j} + 0.667F_D\mathbf{k}$$
$$\mathbf{W} = \{-981\mathbf{k}\} \text{ N}$$

(a)

Equilibrium requires

$$\Sigma\mathbf{F} = \mathbf{0}; \qquad\qquad \mathbf{F}_B + \mathbf{F}_C + \mathbf{F}_D + \mathbf{W} = \mathbf{0}$$
$$F_B\mathbf{i} - 0.5F_C\mathbf{i} - 0.707F_C\mathbf{j} + 0.5F_C\mathbf{k} - 0.333F_D\mathbf{i} + 0.667F_D\mathbf{j}$$
$$+ 0.667F_D\mathbf{k} - 981\mathbf{k} = \mathbf{0}$$

Equating the respective $\mathbf{i}, \mathbf{j}, \mathbf{k}$ components to zero,

$$\Sigma F_x = 0; \qquad F_B - 0.5F_C - 0.333F_D = 0 \qquad (1)$$
$$\Sigma F_y = 0; \qquad -0.707F_C + 0.667F_D = 0 \qquad (2)$$
$$\Sigma F_z = 0; \qquad 0.5F_C + 0.667F_D - 981 = 0 \qquad (3)$$

Solving Eq. 2 for F_D in terms of F_C and substituting into Eq. 3 yields F_C. F_D is determined from Eq. 2. Finally, substituting the results into Eq. 1 gives F_B. Hence,

$$F_C = 813 \text{ N} \qquad\qquad Ans.$$
$$F_D = 862 \text{ N} \qquad\qquad Ans.$$
$$F_B = 693.7 \text{ N}$$

The stretch of the spring is therefore

$$F = ks; \qquad\qquad 693.7 = 1500s$$
$$s = 0.462 \text{ m} \qquad\qquad Ans.$$

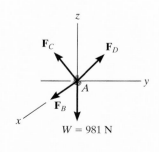

(b)

Fig. 3–13

PROBLEMS

3–41. The joint of a space frame is subjected to four forces. Strut *OA* lies in the *x–y* plane and strut *OB* lies in the *y–z* plane. Determine the forces acting in each of the three struts required for equilibrium.

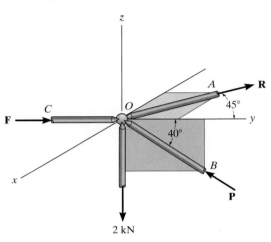

Prob. 3–41

3–43. Determine the magnitudes of F_1, F_2, and F_3 for equilibrium of the particle.

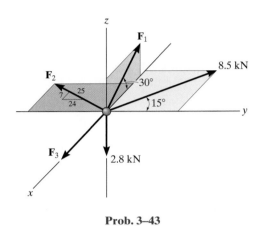

Prob. 3–43

3–42. Determine the magnitudes of F_1, F_2, and F_3 for equilibrium of the particle.

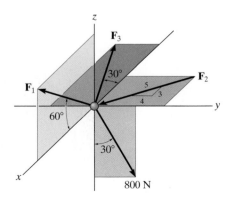

Prob. 3–42

*■**3–44.** Determine the magnitudes of F_1, F_2, and F_3 for equilibrium of the particle. $F = \{-9i - 8j - 5k\}$ kN.

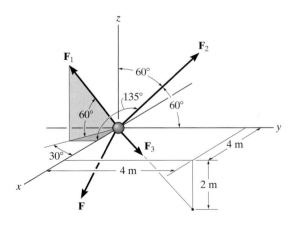

Prob. 3–44

3–45. The three cables are used to support the 800-N lamp. Determine the force developed in each cable for equilibrium.

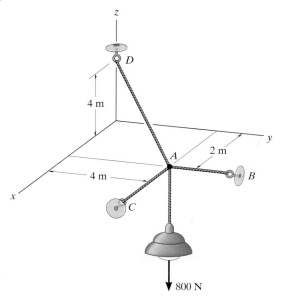

Prob. 3–45

3–46. Determine the force in each cable needed to support the 500-lb load.

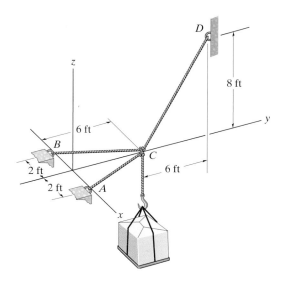

Prob. 3–46

3–47. Determine the stretch in each of the two springs required to hold the 20-kg crate in the equilibrium position shown. Each spring has an unstretched length of 2 m and a stiffness of $k = 300$ N/m.

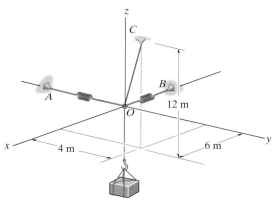

Prob. 3–47

*3–48. If the bucket and its contents have a total weight of 20 lb, determine the force in the supporting cables DA, DB, and DC.

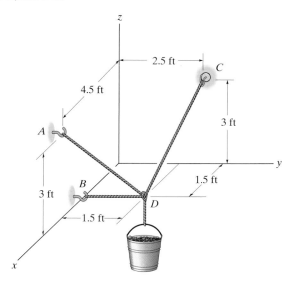

Prob. 3–48

■**3–49.** The 2500-N crate is to be hoisted with constant velocity from the hold of a ship using the cable arrangement shown. Determine the tension in each of the three cables for equilibrium.

Prob. 3–49

■**3–50.** The lamp has a mass of 15 kg and is supported by a pole AO and cables AB and AC. If the force in the pole acts along its axis, determine the forces in AO, AB, and AC for equilibrium.

3–51. Cables AB and AC can sustain a maximum tension of 500 N, and the pole can support a maximum compression of 300 N. Determine the maximum weight of the lamp that can be supported in the position shown. The force in the pole acts along the axis of the pole.

Probs. 3–50/51

*__3–52.__ Determine the tension in cables AB, AC, and AD, required to hold the 60-lb crate in equilibrium.

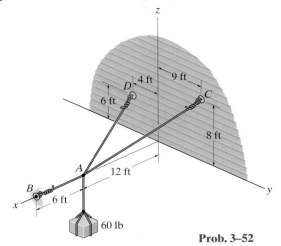

Prob. 3–52

3–53. The bucket has a weight of 20 lb. Determine the tension developed in each cord for equilibrium.

Prob. 3–53

3–54. The mast OA is supported by three cables. If cable AB is subjected to a tension of 500 N, determine the tension in cables AC and AD and the vertical force $\mathbf{F}$ which the mast exerts along its axis on the collar at A.

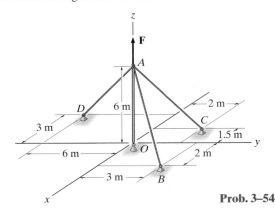

Prob. 3–54

3–55. The ends of the three cables are attached to a ring at A and to the edge of a uniform 150-kg plate. Determine the tension in each of the cables for equilibrium.

*3–56. The ends of the three cables are attached to a ring at A and to the edge of the uniform plate. Determine the largest mass the plate can have if each cable can support a maximum tension of 15 kN.

3–58. The 80-lb chandelier is supported by three wires as shown. Determine the force in each wire for equilibrium.

3–59. If each wire can sustain a maximum tension of 120 lb before it fails, determine the greatest weight of the chandelier the wires will support in the position shown.

Probs. 3–55/56

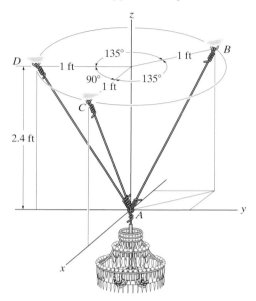

Probs. 3–58/59

3–57. The 500-lb crate is suspended from the cable system shown. Determine the force in each segment of the cable, i.e., AB, AC, CD, CE, and CF. *Hint:* First analyze the equilibrium of point A, then using the result for AC, analyze the equilibrium of point C.

*3–60. Determine the force in each cable used to lift the 9.50-Mg surge arrester at constant velocity.

Prob. 3–57

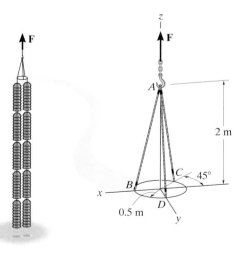

Prob. 3–60

3–61. The 800-lb cylinder is supported by three chains as shown. Determine the force in each chain for equilibrium. Take $d = 1$ ft.

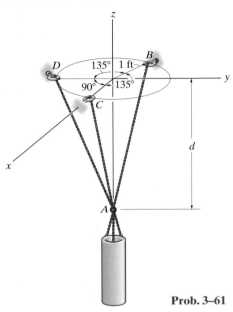

Prob. 3–61

3–62. The triangular frame ABC can be adjusted vertically between the three equal-length cords. If it remains in a horizontal plane, determine the required distance s so that the tension in each of the cords, OA, OB, and OC, equals 20 N. The lamp has a mass of 5 kg.

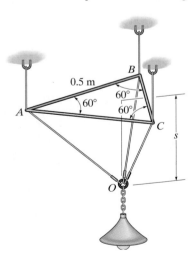

Prob. 3–62

3–63. Determine the force in each cable needed to support the 3500-lb platform. Set $d = 4$ ft.

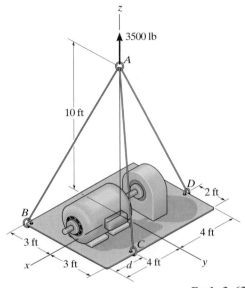

Prob. 3–63

*3–64. The 25-kg flowerpot is supported at A by the three cords. Determine the force acting in each cord for equilibrium.

3–65. If each cord can sustain a maximum tension of 50 N before it fails, determine the greatest weight of the flowerpot the cords can support.

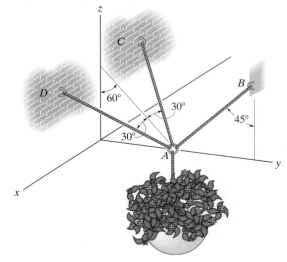

Probs. 3–64/65

Chapter Review

Particle Equilibrium		
When a particle is at rest or moves with constant velocity, it is in equilibrium. This requires that all the forces acting on the particle form a zero force resultant.		
In order to account for all the forces that act on a particle, it is necessary to draw its free-body diagram. This diagram is an outlined shape of the particle that shows all the forces listed with their known or unknown magnitudes and directions. (*See page 83.*)	$$\mathbf{F}_R = \Sigma \mathbf{F} = \mathbf{0}$$	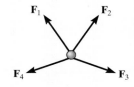

Two Dimensions		
The two scalar equations of force equilibrium can be applied when referenced from an established x, y coordinate system.	$$\Sigma F_x = 0$$ $$\Sigma F_y = 0$$	
If the solution for a force magnitude yields a negative scalar, then the force acts in the opposite direction to that shown on the free-body diagram.	$$\Sigma F = 0;$$ $$F + 10\,\text{N} = 0, F = -10\,\text{N}$$	
The tension force developed in a *continuous cable* that passes over a frictionless pulley must have a *constant* magnitude to keep the cable in equilibrium. (*See pages 84–88.*)		Cable is in tension

If the problem involves a linearly elastic spring, then the stretch or compression s of the spring can be related to the force applied to it. (*See page 84.*)	$F = ks$	

Three Dimensions

Since three-dimensional geometry can be difficult to visualize, the equilibrium equation should be applied using a Cartesian vector analysis. This requires first expressing each force in the free-body diagram as a Cartesian vector. When the forces are summed and set equal to zero, then the **i**, **j**, and **k** components are also zero. (*See page 100.*)

$$\Sigma \mathbf{F} = \mathbf{0}$$

$$\Sigma F_x = 0$$
$$\Sigma F_y = 0$$
$$\Sigma F_z = 0$$

REVIEW PROBLEMS

3–66. The pipe is held in place by the vice. If the bolt exerts a force of 50 lb on the pipe in the direction shown, determine the forces F_A and F_B that the smooth contacts at A and B exert on the pipe.

Prob. 3–66

3–67. When y is zero, the springs sustain a force of 60 lb. Determine the magnitude of the applied vertical forces $\mathbf{F}$ and $-\mathbf{F}$ required to pull point A away from point B a distance of $y = 2$ ft. The ends of cords CAD and CBD are attached to rings at C and D.

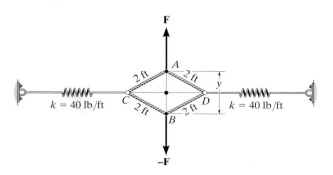

Prob. 3–67

***3–68.** When y is zero, the springs are each stretched 1.5 ft. Determine the distance y if a force of $F = 60$ lb is applied to points A and B as shown. The ends of cords CAD and CBD are attached to rings at C and D.

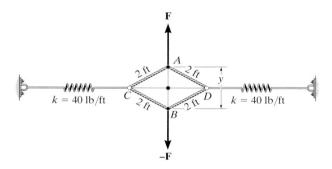

Prob. 3–68

■3–69. The 5-ft-long cord AB is attached to the end B of a spring having an unstretched length of 5 ft. The other end of the spring is attached to a roller C so that the spring remains horizontal as it stretches. If a 10-lb weight is suspended from B, determine the angle θ of cord AB for equilibrium.

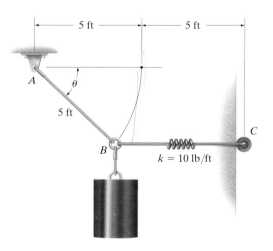

Prob. 3–69

3–70. The uniform 50-kg crate is suspended by using a 2-m-long cord that is attached to the sides of the crate and passes over the small pulley at O. If the cord can be attached at either points A and B, or C and D, determine which attachment produces the least amount of tension in the cord and specify the cord tension in this case.

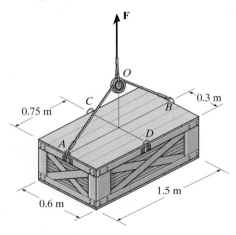

Prob. 3–70

3–71. The man attempts to pull the log at C by using the three ropes. Determine the direction θ in which he should pull on his rope with a force of 80 lb, so that he exerts a maximum force on the log. What is the force on the log for this case? Also, determine the direction in which he should pull in order to maximize the force in the rope attached to B. What is this maximum force?

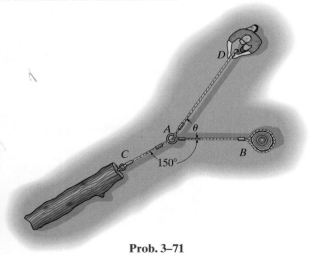

Prob. 3–71

*3–72. The "scale" consists of a known weight W which is suspended at A from a cord of total length L. Determine the weight w at B if A is at a distance y for equilibrium. Neglect the sizes and weights of the pulleys.

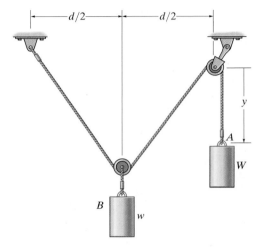

Prob. 3–72

3–73. Determine the maximum weight W that can be supported in the position shown if each cable AC and AB can support a maximum tension of 600 lb before it fails.

Prob. 3–73

3–74. If the spring on rope OB has been stretched 2 in. and fixed in place as shown, determine the tension developed in each of the other three ropes in order to hold the 225-lb weight in equilibrium. Rope OD lies in the x–y plane.

3–75. The joint of a space frame is subjected to four member forces. Member OA lies in the x–y plane and member OB lies in the y–z plane. Determine the forces acting in each of the members required for equilibrium of the joint.

Prob. 3–75

Prob. 3–74

Application of forces to the handle of this wrench will produce a tendency to rotate the wrench about its end. It is important to calculate this effect and, in some cases, to be able to simplify this system to a single resultant force and specify where this resultant acts on the wrench.

4

Force System Resultants

CHAPTER OBJECTIVES

- To discuss the concept of the moment of a force and show how to calculate it in two and three dimensions.

- To provide a method for finding the moment of a force about a specified axis.

- To define the moment of a couple.

- To present methods for determining the resultants of nonconcurrent force systems.

- To indicate how to reduce a simple distributed loading to a resultant force having a specified location.

4.1 Moment of a Force— Scalar Formulation

The *moment* of a force about a point or axis provides a measure of the tendency of the force to cause a body to rotate about the point or axis. For example, consider the horizontal force $\mathbf{F}_x$, which acts perpendicular to the handle of the wrench and is located a distance d_y from point O, Fig. 4–1a. It is seen that this force tends to cause the pipe to turn about the z axis. The larger the force or the distance d_y, the greater the turning effect. This tendency for rotation caused by $\mathbf{F}_x$ is sometimes called a *torque*, but most often it is called the *moment of a force* or simply the *moment* $(\mathbf{M}_O)_z$. Note that the *moment axis* (z) is perpendicular to the shaded plane $(x$–$y)$ which contains both $\mathbf{F}_x$ and d_y and that this axis intersects the plane at point O.

(a)

Fig. 4–1

(b)

(c)

Fig. 4–1 (cont.)

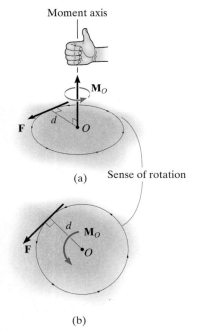

(a)

(b)

Fig. 4–2

Now consider applying the force $\mathbf{F}_z$ to the wrench, Fig. 4–1b. This force will *not* rotate the pipe about the z axis. Instead, it tends to rotate it about the x axis. Keep in mind that although it may not be possible to actually "rotate" or turn the pipe in this manner, $\mathbf{F}_z$ still creates the *tendency* for rotation and so the moment $(\mathbf{M}_O)_x$ is produced. As before, the force and distance d_y lie in the shaded plane (y–z) which is perpendicular to the moment axis (x). Lastly, if a force $\mathbf{F}_y$ is applied to the wrench, Fig. 4–1c, no moment is produced about point O. This results in a lack of turning since the line of action of the force passes through O and therefore no tendency for rotation is possible.

We will now generalize the above discussion and consider the force $\mathbf{F}$ and point O which lie in a shaded plane as shown in Fig. 4–2a. The moment $\mathbf{M}_O$ about point O, or about an axis passing through O and perpendicular to the plane, is a *vector quantity* since it has a specified magnitude and direction.

Magnitude.   The magnitude of M_O is

$$\boxed{M_O = Fd} \qquad (4\text{–}1)$$

where d is referred to as the *moment arm* or perpendicular distance from the axis at point O to the line of action of the force. Units of moment magnitude consist of force times distance, e.g., N · m or lb · ft.

Direction.   The direction of $\mathbf{M}_O$ will be specified by using the "right-hand rule." To do this, the fingers of the right hand are curled such that they follow the sense of rotation, which would occur if the force could rotate about point O, Fig. 4–2a. The *thumb* then *points* along the *moment axis* so that it gives the direction and sense of the moment vector, which is upward and *perpendicular* to the shaded plane containing $\mathbf{F}$ and d.

In three dimensions, $\mathbf{M}_O$ is illustrated by a vector arrow with a curl on it to *distinguish* it from a force vector, Fig. 4–2a. Many problems in mechanics, however, involve coplanar force systems that may be conveniently viewed in two dimensions. For example, a two-dimensional view of Fig. 4–2a is given in Fig. 4–2b. Here $\mathbf{M}_O$ is simply represented by the (counterclockwise) curl, which indicates the action of $\mathbf{F}$. The arrowhead on this curl is used to show the *sense of rotation* caused by $\mathbf{F}$. Using the right-hand rule, however, realize that the direction and sense of the moment vector in Fig. 4–2b are specified by the thumb, which points *out* of the page since the fingers follow the curl. In particular, notice that *this curl or sense of rotation can always be determined by observing in which direction the force would "orbit" about point O* (counterclockwise in Fig. 4–2b). In two dimensions we will often refer to finding the moment of a force "about a point" (O). Keep in mind, however, that the moment *always acts about an axis* which is perpendicular to the plane containing $\mathbf{F}$ and d, and this axis intersects the plane at the point (O), Fig. 4–2a.

Resultant Moment of a System of Coplanar Forces.   If a system of forces lies in an x–y plane, then the moment produced by each force about point O will be directed along the z axis, Fig. 4–3. Consequently, the resultant moment $\mathbf{M}_{R_O}$ of the system can be determined by simply adding the moments of all forces *algebraically* since all the moment vectors are collinear. We can write this vector sum symbolically as

$$\curvearrowleft + M_{R_O} = \Sigma Fd \qquad (4\text{–}2)$$

Here the counterclockwise curl written alongside the equation indicates that, by the scalar sign convention, the moment of any force will be positive if it is directed along the $+z$ axis, whereas a negative moment is directed along the $-z$ axis.

The following examples illustrate numerical application of Eqs. 4–1 and 4–2.

Fig. 4–3

By pushing down on the pry bar the load on the ground at A can be lifted. The turning effect, caused by the applied force, is due to the moment about A. To produce this moment with minimum effort we instinctively know that the force should be applied to the *end* of the bar; however, the *direction* in which this force is applied is also important. This is because a moment is the product of the force and the moment arm. Notice that when the force is at an angle $\theta < 90°$, then the moment arm distance is *shorter* than when the force is applied perpendicular to the bar, $\theta = 90°$, i.e., $d' < d$. Hence the greatest moment is produced when the force is farthest from point A and applied perpendicular to the axis of the bar so as to maximize the moment arm.

In the photo to the right, the moment of a force does not always cause a rotation. For example, the force $\mathbf{F}$ tends to rotate the beam clockwise about its support at A with a moment $M_A = Fd_A$. The actual rotation would occur if the support at B were removed. In the same manner, $\mathbf{F}$ creates a tendency to rotate the beam counterclockwise about B with a moment $M_B = Fd_B$. Here the support at A prevents the rotation.

EXAMPLE 4.1

For each case illustrated in Fig. 4–4, determine the moment of the force about point O.

SOLUTION (SCALAR ANALYSIS)

The line of action of each force is extended as a dashed line in order to establish the moment arm d. Also illustrated is the tendency of rotation of the member as caused by the force. Furthermore, the orbit of the force is shown as a colored curl. Thus,

Fig. 4–4a	$M_O = (100\text{ N})(2\text{ m}) = 200\text{ N}\cdot\text{m} \downarrow$	*Ans.*
Fig. 4–4b	$M_O = (50\text{ N})(0.75\text{ m}) = 37.5\text{ N}\cdot\text{m} \downarrow$	*Ans.*
Fig. 4–4c	$M_O = (40\text{ lb})(4\text{ ft} + 2\cos 30°\text{ ft}) = 229\text{ lb}\cdot\text{ft} \downarrow$	*Ans.*
Fig. 4–4d	$M_O = (60\text{ lb})(1\sin 45°\text{ ft}) = 42.4\text{ lb}\cdot\text{ft} \uparrow$	*Ans.*
Fig. 4–4e	$M_O = (7\text{ kN})(4\text{ m} - 1\text{ m}) = 21.0\text{ kN}\cdot\text{m} \uparrow$	*Ans.*

(a)

(b)

(c)

(d)

(e)

Fig. 4–4

EXAMPLE 4.2

Determine the moments of the 800-N force acting on the frame in Fig. 4–5 about points A, B, C, and D.

SOLUTION (SCALAR ANALYSIS)
In general, $M = Fd$, where d is the moment arm or *perpendicular distance* from the *point* on the moment axis to the *line of action* of the force. Hence,

$M_A = 800 \text{ N}(2.5 \text{ m}) = 2000 \text{ N} \cdot \text{m}$ ↲   *Ans.*

$M_B = 800 \text{ N}(1.5 \text{ m}) = 1200 \text{ N} \cdot \text{m}$ ↲   *Ans.*

$M_C = 800 \text{ N}(0) = 0$ (line of action of **F** passes through C)   *Ans.*

$M_D = 800 \text{ N}(0.5 \text{ m}) = 400 \text{ N} \cdot \text{m}$ ↰   *Ans.*

The curls indicate the sense of rotation of the moment, which is defined by the direction the force orbits about each point.

Fig. 4–5

EXAMPLE 4.3

Determine the resultant moment of the four forces acting on the rod shown in Fig. 4–6 about point O.

SOLUTION
Assuming that positive moments act in the $+\mathbf{k}$ direction, i.e., counterclockwise, we have

$\zeta + M_{R_O} = \Sigma Fd;$

$\quad M_{R_O} = -50 \text{ N}(2 \text{ m}) + 60 \text{ N}(0) + 20 \text{ N}(3 \sin 30° \text{ m})$

$\quad\quad\quad -40 \text{ N}(4 \text{ m} + 3 \cos 30° \text{ m})$

$\quad M_{R_O} = -334 \text{ N} \cdot \text{m} = 334 \text{ N} \cdot \text{m}$ ↲   *Ans.*

For this calculation, note how the moment-arm distances for the 20-N and 40-N forces are established from the extended (dashed) lines of action of each of these forces.

Fig. 4–6

4.2   Cross Product

The moment of a force will be formulated using Cartesian vectors in the next section. Before doing this, however, it is first necessary to expand our knowledge of vector algebra and introduce the cross-product method of vector multiplication.

The *cross product* of two vectors **A** and **B** yields the vector **C**, which is written

$$\mathbf{C} = \mathbf{A} \times \mathbf{B}$$

and is read "**C** equals **A** cross **B**."

Magnitude.   The *magnitude* of **C** is defined as the product of the magnitudes of **A** and **B** and the sine of the angle θ between their tails ($0° \leq \theta \leq 180°$). Thus, $C = AB \sin \theta$.

Direction.   Vector **C** has a *direction* that is perpendicular to the plane containing **A** and **B** such that **C** is specified by the right-hand rule; i.e., curling the fingers of the right hand from vector **A** (cross) to vector **B**, the thumb then points in the direction of **C**, as shown in Fig. 4–7.

Knowing both the magnitude and direction of **C**, we can write

$$\mathbf{C} = \mathbf{A} \times \mathbf{B} = (AB \sin \theta)\mathbf{u}_C \qquad (4\text{–}3)$$

where the scalar $AB \sin \theta$ defines the *magnitude* of **C** and the unit vector $\mathbf{u}_C$ defines the *direction* of **C**. The terms of Eq. 4–3 are illustrated graphically in Fig. 4–8.

$\mathbf{C} = \mathbf{A} \times \mathbf{B}$

Fig. 4–7

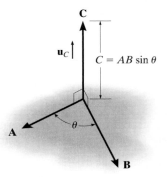

Fig. 4–8

Laws of Operation.

1. The commutative law is *not* valid; i.e.,

$$\mathbf{A} \times \mathbf{B} \neq \mathbf{B} \times \mathbf{A}$$

Rather,

$$\mathbf{A} \times \mathbf{B} = -\mathbf{B} \times \mathbf{A}$$

This is shown in Fig. 4–9 by using the right-hand rule. The cross product **B** × **A** yields a vector that acts in the opposite direction to **C**; i.e., **B** × **A** = −**C**.

2. Multiplication by a scalar:

$$a(\mathbf{A} \times \mathbf{B}) = (a\mathbf{A}) \times \mathbf{B} = \mathbf{A} \times (a\mathbf{B}) = (\mathbf{A} \times \mathbf{B})a$$

This property is easily shown since the magnitude of the resultant vector ($|a| AB \sin \theta$) and its direction are the same in each case.

3. The distributive law:

$$\mathbf{A} \times (\mathbf{B} + \mathbf{D}) = (\mathbf{A} \times \mathbf{B}) + (\mathbf{A} \times \mathbf{D})$$

The proof of this identity is left as an exercise (see Prob. 4–1). It is important to note that *proper order* of the cross products must be maintained, since they are not commutative.

Cartesian Vector Formulation.

Equation 4–3 may be used to find the cross product of a pair of Cartesian unit vectors. For example, to find **i** × **j**, the magnitude of the resultant vector is $(i)(j)(\sin 90°) = (1)(1)(1) = 1$, and its direction is determined using the right-hand rule. As shown in Fig. 4–10, the resultant vector points in the +**k** direction. Thus, **i** × **j** = (1)**k**. In a similar manner,

$$\mathbf{i} \times \mathbf{j} = \mathbf{k} \quad \mathbf{i} \times \mathbf{k} = -\mathbf{j} \quad \mathbf{i} \times \mathbf{i} = 0$$
$$\mathbf{j} \times \mathbf{k} = \mathbf{i} \quad \mathbf{j} \times \mathbf{i} = -\mathbf{k} \quad \mathbf{j} \times \mathbf{j} = 0$$
$$\mathbf{k} \times \mathbf{i} = \mathbf{j} \quad \mathbf{k} \times \mathbf{j} = -\mathbf{i} \quad \mathbf{k} \times \mathbf{k} = 0$$

These results should *not* be memorized; rather, it should be clearly understood how each is obtained by using the right-hand rule and the definition of the cross product. A simple scheme shown in Fig. 4–11 is helpful for obtaining the same results when the need arises. If the circle is constructed as shown, then "crossing" two unit vectors in a *counterclockwise* fashion around the circle yields the *positive* third unit vector; e.g., **k** × **i** = **j**. Moving *clockwise*, a *negative* unit vector is obtained; e.g., **i** × **k** = −**j**.

$$\mathbf{C} = \mathbf{A} \times \mathbf{B}$$
$$-\mathbf{C} = \mathbf{B} \times \mathbf{A}$$

Fig. 4–9

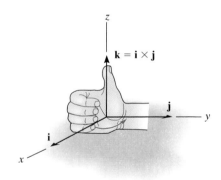

$$\mathbf{k} = \mathbf{i} \times \mathbf{j}$$

Fig. 4–10

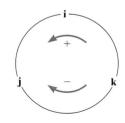

Fig. 4–11

Consider now the cross product of two general vectors **A** and **B** which are expressed in Cartesian vector form. We have

$$\mathbf{A} \times \mathbf{B} = (A_x\mathbf{i} + A_y\mathbf{j} + A_z\mathbf{k}) \times (B_x\mathbf{i} + B_y\mathbf{j} + B_z\mathbf{k})$$
$$= A_xB_x(\mathbf{i} \times \mathbf{i}) + A_xB_y(\mathbf{i} \times \mathbf{j}) + A_xB_z(\mathbf{i} \times \mathbf{k})$$
$$+ A_yB_x(\mathbf{j} \times \mathbf{i}) + A_yB_y(\mathbf{j} \times \mathbf{j}) + A_yB_z(\mathbf{j} \times \mathbf{k})$$
$$+ A_zB_x(\mathbf{k} \times \mathbf{i}) + A_zB_y(\mathbf{k} \times \mathbf{j}) + A_zB_z(\mathbf{k} \times \mathbf{k})$$

Carrying out the cross-product operations and combining terms yields

$$\mathbf{A} \times \mathbf{B} = (A_yB_z - A_zB_y)\mathbf{i} - (A_xB_z - A_zB_x)\mathbf{j} + (A_xB_y - A_yB_x)\mathbf{k} \quad (4\text{--}4)$$

This equation may also be written in a more compact determinant form as

$$\mathbf{A} \times \mathbf{B} = \begin{vmatrix} \mathbf{i} & \mathbf{j} & \mathbf{k} \\ A_x & A_y & A_z \\ B_x & B_y & B_z \end{vmatrix} \quad (4\text{--}5)$$

Thus, to find the cross product of any two Cartesian vectors **A** and **B**, it is necessary to expand a determinant whose first row of elements consists of the unit vectors **i**, **j**, and **k** and whose second and third rows represent the *x, y, z* components of the two vectors **A** and **B**, respectively.*

*A determinant having three rows and three columns can be expanded using three minors, each of which is multiplied by one of the three terms in the first row. There are four elements in each minor, e.g.,

By *definition*, this notation represents the terms $(A_{11}A_{22} - A_{12}A_{21})$, which is simply the product of the two elements of the arrow slanting downward to the right $(A_{11}A_{22})$ *minus* the product of the two elements intersected by the arrow slanting downward to the left $(A_{12}A_{21})$. For a 3 × 3 determinant, such as Eq. 4–5, the three minors can be generated in accordance with the following scheme:

For element **i**: $\begin{vmatrix} \oplus & \mathbf{j} & \mathbf{k} \\ A_x & A_y & A_z \\ B_x & B_y & B_z \end{vmatrix} = \mathbf{i}(A_yB_z - A_zB_y)$

For element **j**: $\begin{vmatrix} \mathbf{i} & \oplus & \mathbf{k} \\ A_x & A_y & A_z \\ B_x & B_y & B_z \end{vmatrix} = -\mathbf{j}(A_xB_z - A_zB_x)$

For element **k**: $\begin{vmatrix} \mathbf{i} & \mathbf{j} & \oplus \\ A_x & A_y & A_z \\ B_x & B_y & B_z \end{vmatrix} = \mathbf{k}(A_xB_y - A_yB_x)$

Adding the results and noting that the **j** element *must include the minus sign* yields the expanded form of **A** × **B** given by Eq. 4–4.

4.3   Moment of a Force—Vector Formulation

The moment of a force $\mathbf{F}$ about point O, or actually about the moment axis passing through O and perpendicular to the plane containing O and $\mathbf{F}$, Fig. 4–12a, can be expressed using the vector cross product, namely,

$$\mathbf{M}_O = \mathbf{r} \times \mathbf{F} \tag{4-6}$$

Here $\mathbf{r}$ represents a position vector drawn *from O* to *any point* lying on the line of action of $\mathbf{F}$. We will now show that indeed the moment $\mathbf{M}_O$, when determined by this cross product, has the proper magnitude and direction.

Magnitude.   The magnitude of the cross product is defined from Eq. 4–3 as $M_O = rF \sin \theta$, where the angle θ is measured between the *tails* of $\mathbf{r}$ and $\mathbf{F}$. To establish this angle, $\mathbf{r}$ must be treated as a sliding vector so that θ can be constructed properly, Fig. 4–12b. Since the moment arm $d = r \sin \theta$, then

$$M_O = rF \sin \theta = F(r \sin \theta) = Fd$$

which agrees with Eq. 4–1.

Direction.   The direction and sense of $\mathbf{M}_O$ in Eq. 4–6 are determined by the right-hand rule as it applies to the cross product. Thus, extending $\mathbf{r}$ to the dashed position and curling the right-hand fingers from $\mathbf{r}$ toward $\mathbf{F}$, "$\mathbf{r}$ cross $\mathbf{F}$," the thumb is directed upward or perpendicular to the plane containing $\mathbf{r}$ and $\mathbf{F}$ and this is in the *same direction* as $\mathbf{M}_O$, the moment of the force about point O, Fig. 4–12b. Note that the "curl" of the fingers, like the curl around the moment vector, indicates the sense of rotation caused by the force. Since the cross product is not commutative, it is important that the *proper order* of $\mathbf{r}$ and $\mathbf{F}$ be maintained in Eq. 4–6.

(a)   (b)

Fig. 4–12

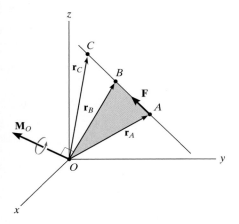

Fig. 4–13

Principle of Transmissibility.

Consider the force **F** applied at point A in Fig. 4–13. The moment created by **F** about O is $\mathbf{M}_O = \mathbf{r}_A \times \mathbf{F}$; however, it was shown that "**r**" can extend from O to *any point* on the line of action of **F**. Consequently, **F** may be applied at point B or C, and the same moment $\mathbf{M}_O = \mathbf{r}_B \times \mathbf{F} = \mathbf{r}_C \times \mathbf{F}$ will be computed. As a result, **F** has the properties of a *sliding vector* and can therefore act at *any point along its line of action* and still create the same moment about point O. We refer to this as the *principle of transmissibility*, and we will discuss this property further in Sec. 4.7.

Cartesian Vector Formulation.

If we establish x, y, z coordinate axes, then the position vector **r** and force **F** can be expressed as Cartesian vectors, Fig. 4–14. Applying Eq. 4–5 we have

$$\mathbf{M}_O = \mathbf{r} \times \mathbf{F} = \begin{vmatrix} \mathbf{i} & \mathbf{j} & \mathbf{k} \\ r_x & r_y & r_z \\ F_x & F_y & F_z \end{vmatrix} \qquad (4\text{–}7)$$

where

r_x, r_y, r_z represent the x, y, z components of the position vector drawn from point O to *any point* on the line of action of the force

F_x, F_y, F_z represent the x, y, z components of the force vector

If the determinant is expanded, then like Eq. 4–4 we have

$$\mathbf{M}_O = (r_y F_z - r_z F_y)\mathbf{i} - (r_x F_z - r_z F_x)\mathbf{j} + (r_x F_y - r_y F_x)\mathbf{k} \qquad (4\text{–}8)$$

The physical meaning of these three moment components becomes evident by studying Fig. 4–14a. For example, the **i** component of $\mathbf{M}_O$ is determined from the moments of $\mathbf{F}_x$, $\mathbf{F}_y$, and $\mathbf{F}_z$ about the x axis. In particular, note that $\mathbf{F}_x$ does *not* create a moment or tendency to cause

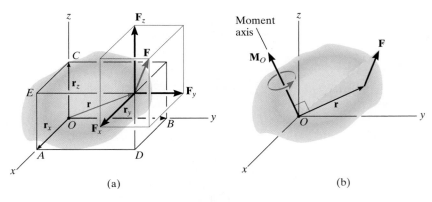

(a)   (b)

Fig. 4–14

turning about the x axis since this force is *parallel* to the x axis. The line of action of $\mathbf{F}_y$ passes through point E, and so the magnitude of the moment of $\mathbf{F}_y$ about point A on the x axis is r_zF_y. By the right-hand rule this component acts in the negative $\mathbf{i}$ direction. Likewise, $\mathbf{F}_z$ contributes a moment component of $r_yF_z\mathbf{i}$. Thus, $(M_O)_x = (r_yF_z - r_zF_y)$ as shown in Eq. 4–8. As an exercise, establish the $\mathbf{j}$ and $\mathbf{k}$ components of $\mathbf{M}_O$ in this manner and show that indeed the expanded form of the determinant, Eq. 4–8, represents the moment of $\mathbf{F}$ about point O. Once $\mathbf{M}_O$ is determined, realize that it will always be *perpendicular* to the shaded plane containing vectors $\mathbf{r}$ and $\mathbf{F}$, Fig. 4–14b.

It will be shown in Example 4.4 that the computation of the moment using the cross product has a distinct advantage over the scalar formulation when solving problems in *three dimensions*. This is because it is generally easier to establish the position vector $\mathbf{r}$ to the force, rather than determining the moment-arm distance d that must be directed *perpendicular* to the line of action of the force.

Resultant Moment of a System of Forces.   If a body is acted upon by a system of forces, Fig. 4–15, the resultant moment of the forces about point O can be determined by vector addition resulting from successive applications of Eq. 4–6. This resultant can be written symbolically as

$$\mathbf{M}_{R_O} = \Sigma(\mathbf{r} \times \mathbf{F}) \qquad (4\text{–}9)$$

and is shown in Fig. 4–15.

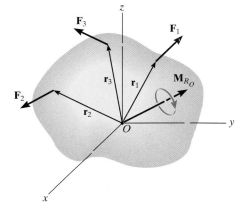

Fig. 4–15

If we pull on cable BC with a force $\mathbf{F}$ at *any point along the cable*, the moment of this force about the base of the utility pole at A will always be the *same*. This is a consequence of the principle of transmissibility. Note that the moment arm, or perpendicular distance from A to the cable, is r_d, and so $M_A = r_dF$. In three dimensions this distance is often difficult to determine, and so we can use the vector cross product to obtain the moment in a more direct manner. For example, $\mathbf{M}_A = \mathbf{r}_{AB} \times \mathbf{F} = \mathbf{r}_{AC} \times \mathbf{F}$. As required, both of these vectors are directed from point A to a point on the line of action of the force.

EXAMPLE 4.4

(a)

(b)

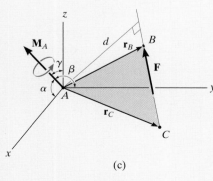

(c)

Fig. 4–16

The pole in Fig. 4–16a is subjected to a 60-N force that is directed from C to B. Determine the magnitude of the moment created by this force about the support at A.

SOLUTION (VECTOR ANALYSIS)

As shown in Fig. 4–16b, either one of two position vectors can be used for the solution, since $\mathbf{M}_A = \mathbf{r}_B \times \mathbf{F}$ or $\mathbf{M}_A = \mathbf{r}_C \times \mathbf{F}$. The position vectors are represented as

$$\mathbf{r}_B = \{1\mathbf{i} + 3\mathbf{j} + 2\mathbf{k}\} \text{ m} \quad \text{and} \quad \mathbf{r}_C = \{3\mathbf{i} + 4\mathbf{j}\} \text{ m}$$

The force has a magnitude of 60 N and a direction specified by the unit vector $\mathbf{u}_F$, directed from C to B. Thus,

$$\mathbf{F} = (60 \text{ N})\mathbf{u}_F = (60 \text{ N})\left[\frac{(1-3)\mathbf{i} + (3-4)\mathbf{j} + (2-0)\mathbf{k}}{\sqrt{(-2)^2 + (-1)^2 + (2)^2}}\right]$$

$$= \{-40\mathbf{i} - 20\mathbf{j} + 40\mathbf{k}\} \text{ N}$$

Substituting into the determinant formulation, Eq. 4–7, and following the scheme for determinant expansion as stated in the footnote on page 124, we have

$$\mathbf{M}_A = \mathbf{r}_B \times \mathbf{F} = \begin{vmatrix} \mathbf{i} & \mathbf{j} & \mathbf{k} \\ 1 & 3 & 2 \\ -40 & -20 & 40 \end{vmatrix}$$

$$= [3(40) - 2(-20)]\mathbf{i} - [1(40) - 2(-40)]\mathbf{j} + [1(-20) - 3(-40)]\mathbf{k}$$

or

$$\mathbf{M}_A = \mathbf{r}_C \times \mathbf{F} = \begin{vmatrix} \mathbf{i} & \mathbf{j} & \mathbf{k} \\ 3 & 4 & 0 \\ -40 & -20 & 40 \end{vmatrix}$$

$$= [4(40) - 0(-20)]\mathbf{i} - [3(40) - 0(-40)]\mathbf{j} + [3(-20) - 4(-40)]\mathbf{k}$$

In both cases,

$$\mathbf{M}_A = \{160\mathbf{i} - 120\mathbf{j} + 100\mathbf{k}] \text{ N} \cdot \text{m}$$

The *magnitude* of $\mathbf{M}_A$ is therefore

$$M_A = \sqrt{(160)^2 + (-120)^2 + (100)^2} = 224 \text{ N} \cdot \text{m} \qquad \textit{Ans.}$$

NOTE: As expected, $\mathbf{M}_A$ acts perpendicular to the shaded plane containing vectors $\mathbf{F}$, $\mathbf{r}_B$, and $\mathbf{r}_C$, Fig. 4–16c. (How would you find its coordinate direction angles $\alpha = 44.3°$, $\beta = 122°$, $\gamma = 63.4°$?) Had this problem been worked using a scalar approach, where $M_A = Fd$, notice the difficulty that can arise in obtaining the moment arm d.

EXAMPLE 4.5

Three forces act on the rod shown in Fig. 4–17a. Determine the resultant moment they create about the flange at O and determine the coordinate direction angles of the moment axis.

SOLUTION

Position vectors are directed from point O to each force as shown in Fig. 4–17b. These vectors are

$$\mathbf{r}_A = \{5\mathbf{j}\} \text{ ft}$$
$$\mathbf{r}_B = \{4\mathbf{i} + 5\mathbf{j} - 2\mathbf{k}\} \text{ ft}$$

The resultant moment about O is therefore

(a)

$$
\begin{aligned}
\mathbf{M}_{R_O} &= \Sigma(\mathbf{r} \times \mathbf{F}) \\
&= \mathbf{r}_A \times \mathbf{F}_1 + \mathbf{r}_A \times \mathbf{F}_2 + \mathbf{r}_B \times \mathbf{F}_3 \\
&= \begin{vmatrix} \mathbf{i} & \mathbf{j} & \mathbf{k} \\ 0 & 5 & 0 \\ -60 & 40 & 20 \end{vmatrix} + \begin{vmatrix} \mathbf{i} & \mathbf{j} & \mathbf{k} \\ 0 & 5 & 0 \\ 0 & 50 & 0 \end{vmatrix} + \begin{vmatrix} \mathbf{i} & \mathbf{j} & \mathbf{k} \\ 4 & 5 & -2 \\ 80 & 40 & -30 \end{vmatrix} \\
&= [5(20) - 40(0)]\mathbf{i} - [0\mathbf{j}] + [0(40) - (-60)(5)]\mathbf{k} + [0\mathbf{i} - 0\mathbf{j} + 0\mathbf{k}] \\
&\quad + [5(-30) - (40)(-2)]\mathbf{i} - [4(-30) - 80(-2)]\mathbf{j} + [4(40) - 80(5)]\mathbf{k} \\
&= \{30\mathbf{i} - 40\mathbf{j} + 60\mathbf{k}\} \text{ lb} \cdot \text{ft} \qquad\qquad Ans.
\end{aligned}
$$

The moment axis is directed along the line of action of $\mathbf{M}_{R_O}$. Since the magnitude of this moment is

$$M_{R_O} = \sqrt{(30)^2 + (-40)^2 + (60)^2} = 78.10 \text{ lb} \cdot \text{ft}$$

the unit vector which defines the direction of the moment axis is

$$\mathbf{u} = \frac{\mathbf{M}_{R_O}}{M_{R_O}} = \frac{30\mathbf{i} - 40\mathbf{j} + 60\mathbf{k}}{78.10} = 0.3841\mathbf{i} - 0.5121\mathbf{j} + 0.7682\mathbf{k}$$

(b)

Therefore, the coordinate direction angles of the moment axis are

$$
\begin{aligned}
\cos\alpha &= 0.3841; & \alpha &= 67.4° & Ans. \\
\cos\beta &= -0.5121; & \beta &= 121° & Ans. \\
\cos\gamma &= 0.7682; & \gamma &= 39.8° & Ans.
\end{aligned}
$$

(c)

NOTE: These results are shown in Fig. 4–17c. Realize that the three forces tend to cause the rod to rotate about this axis in the manner shown by the curl indicated on the moment vector.

Fig. 4–17

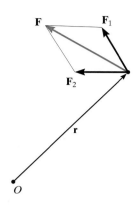

Fig. 4-18

4.4   Principle of Moments

A concept often used in mechanics is the *principle of moments*, which is sometimes referred to as *Varignon's theorem* since it was originally developed by the French mathematician Varignon (1654–1722). It states that *the moment of a force about a point is equal to the sum of the moments of the force's components about the point.* The proof follows directly from the distributive law of the vector cross product. To show this, consider the force **F** and two of its rectangular components, where $\mathbf{F} = \mathbf{F}_1 + \mathbf{F}_2$, Fig. 4–18. We have

$$\mathbf{M}_O = \mathbf{r} \times \mathbf{F}_1 + \mathbf{r} \times \mathbf{F}_2 = \mathbf{r} \times (\mathbf{F}_1 + \mathbf{F}_2) = \mathbf{r} \times \mathbf{F}$$

This concept has important applications to the solution of problems and proofs of theorems that follow, since it is often easier to determine the moments of a force's components rather than the moment of the force itself.

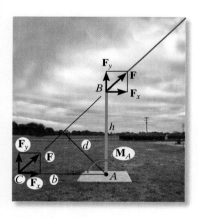

The guy cable exerts a force **F** on the pole and this creates a moment about the base at A of $M_A = Fd$. If the force is replaced by its two components $\mathbf{F}_x$ and $\mathbf{F}_y$ at point B where the cable acts on the pole, then the sum of the moments of these two components about A will yield the *same* resultant moment. For the calculation $\mathbf{F}_y$ will create zero moment about A and so $M_A = F_x h$. This is an application of the *principle of moments*. In addition we can apply the *principle of transmissibility* and slide the force to where its line of action intersects the ground at C. In this case $\mathbf{F}_x$ will create zero moment about A, and so $M_A = F_y b$.

Important Points

- The moment of a force indicates the tendency of a body to turn about an axis passing through a specific point O.

- Using the right-hand rule, the sense of rotation is indicated by the fingers, and the thumb is directed along the moment axis, or line of action of the moment.

- The magnitude of the moment is determined from $M_O = Fd$, where d is the perpendicular or shortest distance from point O to the line of action of the force **F**.

- In three dimensions use the vector cross product to determine the moment, i.e., $\mathbf{M}_O = \mathbf{r} \times \mathbf{F}$. Remember that **r** is directed *from* point O *to any point* on the line of action of **F**.

- The principle of moments states that the moment of a force about a point is equal to the sum of the moments of the force's components about the point. This is a very convenient method to use in two dimensions.

EXAMPLE 4.6

A 200-N force acts on the bracket shown in Fig. 4–19a. Determine the moment of the force about point A.

(a)

SOLUTION I

The moment arm d can be found by trigonometry, using the construction shown in Fig. 4–19b. From the right triangle BCD,

$$CB = d = 100 \cos 45° = 70.71 \text{ mm} = 0.070\ 71 \text{ m}$$

Thus,

$$M_A = Fd = 200 \text{ N}(0.070\ 71 \text{ m}) = 14.1 \text{ N} \cdot \text{m} \curvearrowleft$$

According to the right-hand rule, $\mathbf{M}_A$ is directed in the $+\mathbf{k}$ direction since the force tends to rotate or orbit *counterclockwise* about point A. Hence, reporting the moment as a Cartesian vector, we have

$$\mathbf{M}_A = \{14.1\mathbf{k}\} \text{ N} \cdot \text{m} \qquad\qquad Ans.$$

(b)

SOLUTION II

The 200-N force may be resolved into x and y components, as shown in Fig. 4–19c. In accordance with the principle of moments, the moment of $\mathbf{F}$ computed about point A is equivalent to the sum of the moments produced by the two force components. Assuming counterclockwise rotation as positive, i.e., in the $+\mathbf{k}$ direction, we can apply Eq. 4–2 ($M_A = \Sigma Fd$), in which case

$$\zeta + M_A = (200 \sin 45° \text{ N})(0.20 \text{ m}) - (200 \cos 45° \text{ N})(0.10 \text{ m})$$
$$= 14.1 \text{ N} \cdot \text{m} \curvearrowleft$$

Thus

$$\mathbf{M}_A = \{14.1\mathbf{k}\} \text{ N} \cdot \text{m} \qquad\qquad Ans.$$

NOTE: By comparison, it is seen that Solution II provides a more *convenient method* for analysis than Solution I since the moment arm for each component force is easier to establish.

(c)

Fig. 4–19

EXAMPLE 4.7

(a)

(b)

(c)

Fig. 4–20

The force **F** acts at the end of the angle bracket shown in Fig. 4–20a. Determine the moment of the force about point O.

SOLUTION I *(SCALAR ANALYSIS)*

The force is resolved into its x and y components as shown in Fig. 4–20b, and the moments of the components are computed about point O. Taking positive moments as counterclockwise, i.e., in the +**k** direction, we have

$$\zeta +M_O = 400 \sin 30° \text{ N}(0.2 \text{ m}) - 400 \cos 30° \text{ N}(0.4 \text{ m})$$
$$= -98.6 \text{ N} \cdot \text{m} = 98.6 \text{ N} \cdot \text{m} \downarrow$$

or

$$\mathbf{M}_O = \{-98.6\mathbf{k}\} \text{ N} \cdot \text{m} \qquad\qquad Ans.$$

SOLUTION II *(VECTOR ANALYSIS)*

Using a Cartesian vector approach, the force and position vectors shown in Fig. 4–20c can be represented as

$$\mathbf{r} = \{0.4\mathbf{i} - 0.2\mathbf{j}\} \text{ m}$$
$$\mathbf{F} = \{400 \sin 30°\mathbf{i} - 400 \cos 30°\mathbf{j}\} \text{ N}$$
$$= \{200.0\mathbf{i} - 346.4\mathbf{j}\} \text{ N}$$

The moment is therefore

$$\mathbf{M}_O = \mathbf{r} \times \mathbf{F} = \begin{vmatrix} \mathbf{i} & \mathbf{j} & \mathbf{k} \\ 0.4 & -0.2 & 0 \\ 200.0 & -346.4 & 0 \end{vmatrix}$$
$$= 0\mathbf{i} - 0\mathbf{j} + [0.4(-346.4) - (-0.2)(200.0)]\mathbf{k}$$
$$= \{-98.6\mathbf{k}\} \text{ N} \cdot \text{m} \qquad\qquad Ans.$$

NOTE: By comparison, it is seen that the scalar analysis (Solution I) provides a more *convenient method* for analysis than Solution II since the direction of the moment and the moment arm for each component force are easy to establish. Hence, this method is generally recommended for solving problems displayed in two dimensions. On the other hand, Cartesian vector analysis is generally recommended only for solving three-dimensional problems, where the moment arms and force components are often more difficult to determine.

PROBLEMS

4–1. If **A**, **B**, and **D** are given vectors, prove the distributive law for the vector cross product, i.e., $\mathbf{A} \times (\mathbf{B} + \mathbf{D}) = (\mathbf{A} \times \mathbf{B}) + (\mathbf{A} \times \mathbf{D})$.

4–2. Prove the triple scalar product identity $\mathbf{A} \cdot (\mathbf{B} \times \mathbf{C}) = (\mathbf{A} \times \mathbf{B}) \cdot \mathbf{C}$.

4–3. Given the three nonzero vectors **A**, **B**, and **C**, show that if $\mathbf{A} \cdot (\mathbf{B} \times \mathbf{C}) = 0$, the three vectors *must* lie in the same plane.

*4–4. Determine the magnitude and directional sense of the resultant moment of the forces at A and B about point O.

4–5. Determine the magnitude and directional sense of the resultant moment of the forces at A and B about point P.

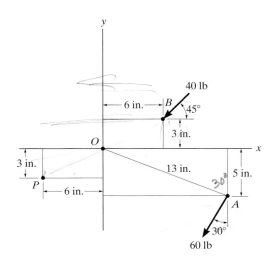

Probs. 4–4/5

4–6. Determine the magnitude of the force **F** that should be applied at the end of the lever such that this force creates a clockwise moment of 15 N·m about point O when $\theta = 30°$.

■4–7. If the force $F = 100$ N, determine the angle θ $(0 \le \theta \le 90°)$ so that the force develops a clockwise moment about point O of 20 N·m.

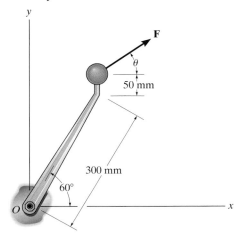

Probs. 4–6/7

*4–8. Determine the magnitude and directional sense of the resultant moment of the forces about point O.

4–9. Determine the magnitude and directional sense of the resultant moment of the forces about point P.

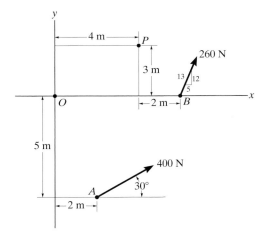

Probs. 4–8/9

4–10.   A force of 40 N is applied to the wrench. Determine the moment of this force about point O. Solve the problem using both a scalar analysis and a vector analysis.

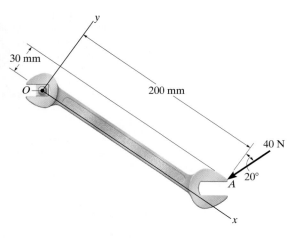

Prob. 4–10

4–11.   Determine the magnitude and directional sense of the resultant moment of the forces about point O.

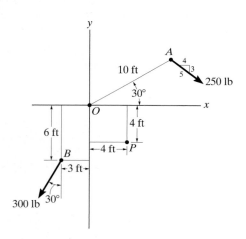

Prob. 4–11

*4–12.   To correct a birth defect, the tibia of the leg is straightened using three wires that are attached through holes made in the bone and then to an external brace that is worn by the patient. Determine the moment of each wire force about joint A.

4–13.   To correct a birth defect, the tibia of the leg is straightened using three wires that are attached through holes made in the bone and then to an external brace that is worn by the patient. Determine the moment of each wire force about joint B.

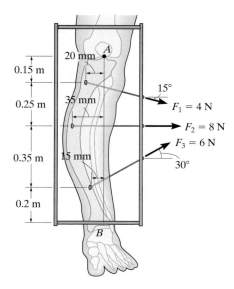

Probs. 4–12/13

4–14.   Determine the moment of each force about the bolt located at A. Take $F_B = 40$ lb, $F_C = 50$ lb.

4–15.   If $F_B = 30$ lb and $F_C = 45$ lb, determine the resultant moment about the bolt located at A.

Probs. 4–14/15

*4–16. The elbow joint is flexed using the biceps brachii muscle, which remains essentially vertical as the arm moves in the vertical plane. If this muscle is located a distance of 16 mm from the pivot point A on the humerus, determine the variation of the moment capacity about A if the constant force developed by the muscle is 2.30 kN. Plot these results of M vs. θ for $-60° \leq \theta \leq 80°$.

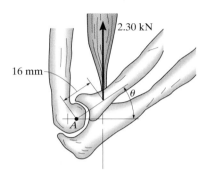

Prob. 4–16

4–17. The Snorkel Co. produces the articulating boom platform that can support a weight of 550 lb. If the boom is in the position shown, determine the moment of this force about points A, B, and C.

Prob. 4–17

4–18. Determine the direction θ $(0° \leq \theta \leq 180°)$ of the force $F = 40$ lb so that it produces (a) the maximum moment about point A and (b) the minimum moment about point A. Compute the moment in each case.

Prob. 4–18

4–19. The rod on the power control mechanism for a business jet is subjected to a force of 80 N. Determine the moment of this force about the bearing at A.

Prob. 4–19

*4–20. The boom has a length of 30 ft, a weight of 800 lb, and mass center at G. If the maximum moment that can be developed by the motor at A is $M = 20(10^3)$ lb·ft, determine the maximum load W, having a mass center at G', that can be lifted. Take $\theta = 30°$.

Prob. 4–20

4–21. The tool at A is used to hold a power lawnmower blade stationary while the nut is being loosened with the wrench. If a force of 50 N is applied to the wrench at B in the direction shown, determine the moment it creates about the nut at C. What is the magnitude of force $\mathbf{F}$ at A so that it creates the opposite moment about C?

Prob. 4–21

4–22. Determine the clockwise direction θ $(0° \le \theta \le 180°)$ of the force $F = 80$ lb so that it produces (a) the maximum moment about point A and (b) no moment about point A. Compute the moment in each case.

Prob. 4–22

4–23. The Y-type structure is used to support the high voltage transmission cables. If the supporting cables each exert a force of 275 lb on the structure at B, determine the moment of each force about point A. Also, by the principle of transmissibility, locate the forces at points C and D and determine the moments.

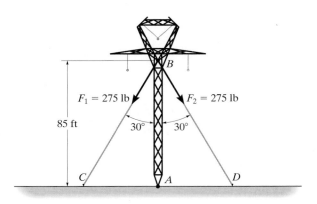

Prob. 4–23

***4–24.** The 70-N force acts on the end of the pipe at B. Determine (a) the moment of this force about point A, and (b) the magnitude and direction of a horizontal force, applied at C, which produces the same moment. Take $\theta = 60°$.

4–25. The 70-N force acts on the end of the pipe at B. Determine the angles θ $(0° \le \theta \le 180°)$ of the force that will produce maximum and minimum moments about point A. What are the magnitudes of these moments?

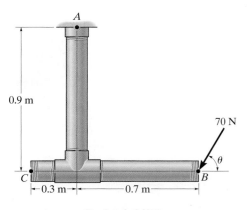

Probs. 4–24/25

4–26. The towline exerts a force of $P = 4$ kN at the end of the 20-m-long crane boom. If $\theta = 30°$, determine the placement x of the hook at A so that this force creates a maximum moment about point O. What is this moment?

4–27. The towline exerts a force of $P = 4$ kN at the end of the 20-m-long crane boom. If $x = 25$ m, determine the position θ of the boom so that this force creates a maximum moment about point O. What is this moment?

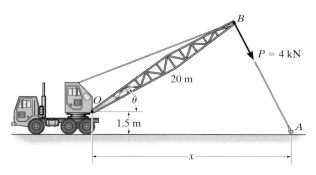

Probs. 4–26/27

*4–28. Determine the resultant moment of the forces about point A. Solve the problem first by considering each force as a whole, and then by using the principle of moments. Take $F_1 = 250$ N, $F_2 = 300$ N, $F_3 = 500$ N.

4–29. If the resultant moment about point A is 4800 N · m clockwise, determine the magnitude of $\mathbf{F}_3$ if $F_1 = 300$ N and $F_2 = 400$ N.

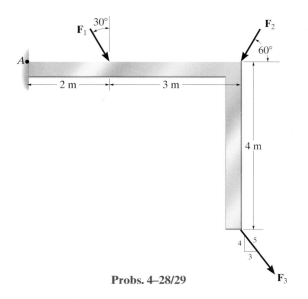

Probs. 4–28/29

4–30. The flat-belt tensioner is manufactured by the Daton Co. and is used with V-belt drives on poultry and livestock fans. If the tension in the belt is 52 lb, when this pulley is not turning, determine the moment of each of these forces about the pin at A.

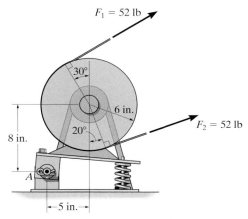

Prob. 4–30

4–31. The worker is using the bar to pull two pipes together in order to complete the connection. If he applies a horizontal force of 80 lb to the handle of the lever, determine the moment of this force about the end A. What would be the tension T in the cable needed to cause the opposite moment about point A?

Prob. 4–31

*4–32.   If it takes a force of $F = 125$ lb to pull the nail out, determine the smallest vertical force **P** that must be applied to the handle of the crowbar. *Hint:* This requires the moment of **F** about point A to be equal to the moment of **P** about A. Why?

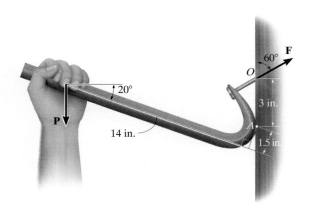

Prob. 4–32

4–34.   Determine the moment of the force **F** at A about point O. Express the result as a Cartesian vector.

4–35.   Determine the moment of the force **F** at A about point P. Express the result as a Cartesian vector.

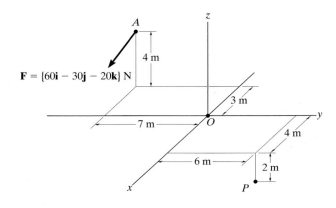

Probs. 4–34/35

■4–33.   The pipe wrench is activated by pulling on the cable segment with a horizontal force of 500 N. Determine the moment M_A produced by the wrench on the pipe when $\theta = 20°$. Neglect the size of the pulley.

Prob. 4–33

*4–36.   Determine the moment of the force **F** at A about point O. Express the result as a Cartesian vector.

4–37.   Determine the moment of the force **F** at A about point P. Express the result as a Cartesian vector.

Probs. 4–36/37

4–38. The curved rod lies in the *x–y* plane and has a radius of 3 m. If a force of $F = 80$ N acts at its end as shown, determine the moment of this force about point *O*.

4–39. The curved rod lies in the *x–y* plane and has a radius of 3 m. If a force of $F = 80$ N acts at its end as shown, determine the moment of this force about point *B*.

4–41. The pole supports a 22-lb traffic light. Using Cartesian vectors, determine the moment of the weight of the traffic light about the base of the pole at *A*.

Probs. 4–38/39

Prob. 4–41

*4–40. The force $\mathbf{F} = \{600\mathbf{i} + 300\mathbf{j} - 600\mathbf{k}\}$ N acts at the end of the beam. Determine the moment of the force about point *A*.

4–42. The man pulls on the rope with a force of $F = 20$ N. Determine the moment that this force exerts about the base of the pole at *O*. Solve the problem two ways, i.e., by using a position vector from *O* to *A*, then *O* to *B*.

Prob. 4–40

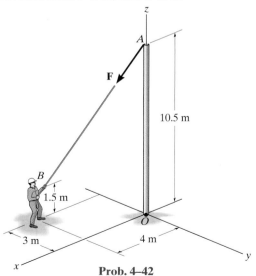

Prob. 4–42

4–43. Determine the smallest force F that must be applied along the rope in order to cause the curved rod, which has a radius of 5 ft, to fail at the support C. This requires a moment of $M = 80$ lb · ft to be developed at C.

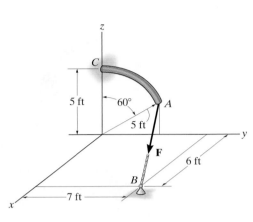

Prob. 4–43

4–45. The pipe assembly is subjected to the 80-N force. Determine the moment of this force about point B.

Prob. 4–45

*4–44. The pipe assembly is subjected to the 80-N force. Determine the moment of this force about point A.

Prob. 4–44

4–46. The x-ray machine is used for medical diagnosis. If the camera and housing at C have a mass of 150 kg and a mass center at G, determine the moment of its weight about point O when it is in the position shown.

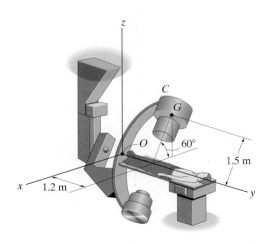

Prob. 4–46

4–47. Using Cartesian vector analysis, determine the resultant moment of the three forces about the base of the column at A. Take $\mathbf{F}_1 = \{400\mathbf{i} + 300\mathbf{j} + 120\mathbf{k}\}$ N.

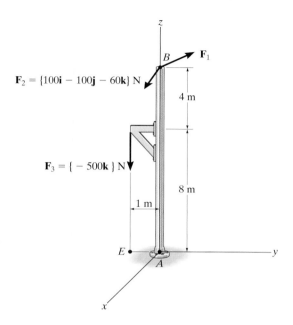

Prob. 4–47

4–49. The force $\mathbf{F} = \{6\mathbf{i} + 8\mathbf{j} + 10\mathbf{k}\}$ N creates a moment about point O of $\mathbf{M}_O = \{-14\mathbf{i} + 8\mathbf{j} + 2\mathbf{k}\}$ N·m. If the force passes through a point having an x coordinate of 1 m, determine the y and z coordinates of the point. Also, realizing that $M_O = Fd$, determine the perpendicular distance d from point O to the line of action of $\mathbf{F}$.

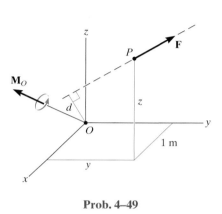

Prob. 4–49

*4–48. A force of $\mathbf{F} = \{6\mathbf{i} - 2\mathbf{j} + 1\mathbf{k}\}$ kN produces a moment of $\mathbf{M}_O = \{4\mathbf{i} + 5\mathbf{j} - 14\mathbf{k}\}$ kN·m about the origin of coordinates, point O. If the force acts at a point having an x coordinate of $x = 1$ m, determine the y and z coordinates.

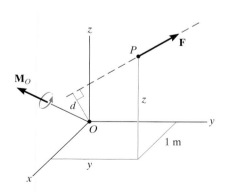

Prob. 4–48

4–50. A force of $\mathbf{F} = \{6\mathbf{i} - 2\mathbf{j} + 1\mathbf{k}\}$ kN produces a moment of $\mathbf{M}_O = \{4\mathbf{i} + 5\mathbf{j} - 14\mathbf{k}\}$ kN·m about the origin of coordinates, point O. If the force acts at a point having an x coordinate of $x = 1$ m, determine the y and z coordinates.

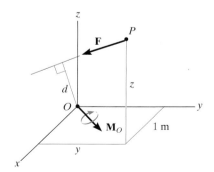

Prob. 4–50

4.5   Moment of a Force about a Specified Axis

Recall that when the moment of a force is computed about a point, the moment and its axis are *always* perpendicular to the plane containing the force and the moment arm. In some problems it is important to find the *component* of this moment along a *specified axis* that passes through the point. To solve this problem either a scalar or vector analysis can be used.

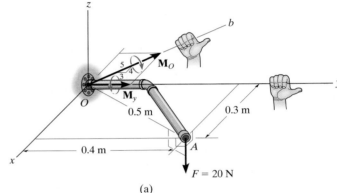

Fig. 4–21   (a)

Scalar Analysis.   As a numerical example of this problem, consider the pipe assembly shown in Fig. 4–21*a*, which lies in the horizontal plane and is subjected to the vertical force of $F = 20$ N applied at point A. The moment of this force about point O has a *magnitude* of $M_O = (20\text{ N})(0.5\text{ m}) = 10\text{ N} \cdot \text{m}$, and a *direction* defined by the right-hand rule, as shown in Fig. 4–21*a*. This moment tends to turn the pipe about the Ob axis. For practical reasons, however, it may be necessary to determine the *component* of M_O about the y axis, M_y, since this component tends to unscrew the pipe from the flange at O. From Fig. 4–21*a*, M_y has a magnitude of $M_y = \frac{3}{5}(10\text{ N} \cdot \text{m}) = 6\text{ N} \cdot \text{m}$ and a sense of direction shown by the vector resolution. Rather than performing this *two-step* process of first finding the moment of the force about point O and then resolving the moment along the y axis, it is also possible to solve this problem *directly*. To do so, it is necessary to determine the perpendicular or moment-arm distance from the line of action of F to the y axis. From Fig. 4–21*a* this distance is 0.3 m. Thus the *magnitude* of the moment of the force about the y axis is again $M_y = 0.3(20\text{ N}) = 6\text{ N} \cdot \text{m}$, and the *direction* is determined by the right-hand rule as shown.

In general, then, *if the line of action of a force* **F** *is perpendicular to any specified axis aa*, the magnitude of the moment of **F** about the axis can be determined from the equation

$$M_a = Fd_a$$ (4–10)

Here d_a is the *perpendicular or shortest distance* from the force line of action to the axis. The direction is determined from the thumb of the right hand when the fingers are curled in accordance with the direction of rotation as produced by the force. In particular, realize that a *force will not contribute a moment about a specified axis if the force line of action is parallel to the axis or its line of action passes through the axis.*

If a horizontal force **F** is applied to the handle of the flex-headed wrench, it tends to turn the socket at A about the z axis. This effect is caused by the moment of **F** about the z axis. The *maximum moment* is determined when the wrench is in the horizontal plane so that full leverage from the handle can be achieved, i.e., $(M_z)_{max} = Fd$. If the handle is not in the horizontal position, then the moment about the z axis is determined from $M_z = Fd'$, where d' is the perpendicular distance from the force line of action to the axis. We can also determine this moment by first finding the moment of **F** about A, $M_A = Fd$, then finding the projection or component of this moment along z, i.e., $M_z = M_A \cos \theta$.

Vector Analysis. The previous two-step solution of first finding the moment of the force about a point on the axis and then finding the projected component of the moment about the axis can also be performed using a vector analysis, Fig. 4–21b. Here the moment about point O is first determined from $\mathbf{M}_O = \mathbf{r}_A \times \mathbf{F} = (0.3\mathbf{i} + 0.4\mathbf{j}) \times (-20\mathbf{k}) = \{-8\mathbf{i} + 6\mathbf{j}\}$ N · m. The component or projection of this moment along the y axis is then determined from the dot product (Sec. 2.9). Since the unit vector for this axis (or line) is $\mathbf{u}_a = \mathbf{j}$, then $M_y = \mathbf{M}_O \cdot \mathbf{u}_a = (-8\mathbf{i} + 6\mathbf{j}) \cdot \mathbf{j} = 6$ N · m. This result, of course, is to be expected, since it represents the **j** component of $\mathbf{M}_O$.

(b)

Fig. 4–21 (cont.)

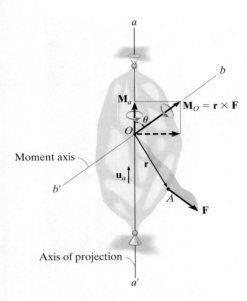

Moment axis

Axis of projection

Fig. 4–22

A vector analysis such as this is particularly advantageous for finding the moment of a force about an axis when the force components or the appropriate moment arms are difficult to determine. For this reason, the above two-step process will now be generalized and applied to a body of arbitrary shape. To do so, consider the body in Fig. 4–22, which is subjected to the force **F** acting at point A. Here we wish to determine the effect of **F** in tending to rotate the body about the aa' axis. This tendency for rotation is measured by the moment component $\mathbf{M}_a$. To determine $\mathbf{M}_a$ we first compute the moment of **F** about any *arbitrary point O* that lies on the aa' axis. In this case, $\mathbf{M}_O$ is expressed by the cross product $\mathbf{M}_O = \mathbf{r} \times \mathbf{F}$, where **r** is directed from O to A. Here $\mathbf{M}_O$ acts along the moment axis bb', and so the component or projection of $\mathbf{M}_O$ onto the aa' axis is then $\mathbf{M}_a$. The *magnitude* of $\mathbf{M}_a$ is determined by the dot product, $M_a = M_O \cos \theta = \mathbf{M}_O \cdot \mathbf{u}_a$ where $\mathbf{u}_a$ is a unit vector that defines the direction of the aa' axis. Combining these two steps as a general expression, we have $M_a = (\mathbf{r} \times \mathbf{F}) \cdot \mathbf{u}_a$. Since the dot product is commutative, we can also write

$$M_a = \mathbf{u}_a \cdot (\mathbf{r} \times \mathbf{F})$$

In vector algebra, this combination of dot and cross product yielding the scalar M_a is called the *triple scalar product*. Provided x, y, z axes are established and the Cartesian components of each of the vectors can be determined, then the triple scalar product may be written in determinant form as

$$M_a = (u_{a_x}\mathbf{i} + u_{a_y}\mathbf{j} + u_{a_z}\mathbf{k}) \cdot \begin{vmatrix} \mathbf{i} & \mathbf{j} & \mathbf{k} \\ r_x & r_y & r_z \\ F_x & F_y & F_z \end{vmatrix}$$

or simply

$$M_a = \mathbf{u}_a \cdot (\mathbf{r} \times \mathbf{F}) = \begin{vmatrix} u_{a_x} & u_{a_y} & u_{a_z} \\ r_x & r_y & r_z \\ F_x & F_y & F_z \end{vmatrix} \qquad (4\text{–}11)$$

where

$u_{a_x}, u_{a_y}, u_{a_z}$ represent the x, y, z components of the unit vector defining the direction of the aa' axis

r_x, r_y, r_z represent the x, y, z components of the position vector drawn from *any point O* on the aa' axis to *any point A* on the line of action of the force

F_x, F_y, F_z represent the x, y, z components of the force vector.

When M_a is evaluated from Eq. 4–11, it will yield a positive or negative scalar. The sign of this scalar indicates the sense of direction of $\mathbf{M}_a$ along the aa' axis. If it is positive, then $\mathbf{M}_a$ will have the same sense as $\mathbf{u}_a$, whereas if it is negative, then $\mathbf{M}_a$ will act opposite to $\mathbf{u}_a$.

Once M_a is determined, we can then express $\mathbf{M}_a$ as a Cartesian vector, namely,

$$\mathbf{M}_a = M_a\mathbf{u}_a = [\mathbf{u}_a \cdot (\mathbf{r} \times \mathbf{F})]\mathbf{u}_a \qquad (4\text{–}12)$$

Finally, if the resultant moment of a series of forces is to be computed about the aa' axis, then the moment components of each force are added together *algebraically*, since each component lies along the same axis. Thus the magnitude of $\mathbf{M}_a$ is

$$M_a = \Sigma[\mathbf{u}_a \cdot (\mathbf{r} \times \mathbf{F})] = \mathbf{u}_a \cdot \Sigma(\mathbf{r} \times \mathbf{F})$$

The examples which follow illustrate a numerical application of the above concepts.

Wind blowing on the face of this traffic sign creates a resultant force $\mathbf{F}$ that tends to tip the sign over due to the moment $\mathbf{M}_A$ created about the a–a axis. The moment of $\mathbf{F}$ about a point A that lies on the axis is $\mathbf{M}_A = \mathbf{r} \times \mathbf{F}$. The projection of this moment along the axis, whose direction is defined by the unit vector $\mathbf{u}_a$, is $M_a = \mathbf{u}_a \cdot (\mathbf{r} \times \mathbf{F})$. Had this moment been calculated using scalar methods, then the perpendicular distance from the force line of action to the a–a axis would have to be determined, which in this case would be a more difficult task.

Important Points

- The moment of a force about a specified axis can be determined provided the perpendicular distance d_a from *both* the force line of action and the axis can be determined. $M_a = Fd_a$.

- If vector analysis is used, $M_a = \mathbf{u}_a \cdot (\mathbf{r} \times \mathbf{F})$, where $\mathbf{u}_a$ defines the direction of the axis and $\mathbf{r}$ is directed from *any point* on the axis to *any point* on the line of action of the force.

- If M_a is calculated as a negative scalar, then the sense of direction of $\mathbf{M}_a$ is opposite to $\mathbf{u}_a$.

- The moment $\mathbf{M}_a$ expressed as a Cartesian vector is determined from $\mathbf{M}_a = M_a\mathbf{u}_a$.

EXAMPLE 4.8

(a)

(b)

(c)

Fig. 4–23

The force $\mathbf{F} = \{-40\mathbf{i} + 20\mathbf{j} + 10\mathbf{k}\}$ N acts at point A shown in Fig. 4–23a. Determine the moments of this force about the x and a axes.

SOLUTION I *(VECTOR ANALYSIS)*

We can solve this problem by using the position vector $\mathbf{r}_A$. Why? Since $\mathbf{r}_A = \{-3\mathbf{i} + 4\mathbf{j} + 6\mathbf{k}\}$ m and $\mathbf{u}_x = \mathbf{i}$, then applying Eq. 4–11,

$$M_x = \mathbf{i} \cdot (\mathbf{r}_A \times \mathbf{F}) = \begin{vmatrix} 1 & 0 & 0 \\ -3 & 4 & 6 \\ -40 & 20 & 10 \end{vmatrix}$$

$$= 1[4(10)-6(20)]-0[(-3)(10)-6(-40)]+0[(-3)(20)-4(-40)]$$

$$= -80 \text{ N} \cdot \text{m} \qquad\qquad\qquad\qquad\qquad\qquad Ans.$$

The negative sign indicates that the sense of $\mathbf{M}_x$ is opposite to $\mathbf{i}$.

We can compute M_a also using $\mathbf{r}_A$ because $\mathbf{r}_A$ extends from a point on the a axis to the force. Also, $\mathbf{u}_a = -\frac{3}{5}\mathbf{i} + \frac{4}{5}\mathbf{j}$. Thus,

$$M_a = \mathbf{u}_a \cdot (\mathbf{r}_A \times \mathbf{F}) = \begin{vmatrix} -\frac{3}{5} & \frac{4}{5} & 0 \\ -3 & 4 & 6 \\ -40 & 20 & 10 \end{vmatrix}$$

$$= -\tfrac{3}{5}[4(10) - 6(20)]-\tfrac{4}{5}[(-3)(10)-6(-40)]+0[(-3)(20)-4(-40)]$$

$$= -120 \text{ N} \cdot \text{m} \qquad\qquad\qquad\qquad\qquad\qquad Ans.$$

What does the negative sign indicate?

The moment components are shown in Fig. 4–23b.

SOLUTION II *(SCALAR ANALYSIS)*

Since the force components and moment arms are easy to determine for computing M_x, a scalar analysis can be used to solve this problem. Referring to Fig. 4–23c, only the 10-N and 20-N forces contribute moments about the x axis. (The line of action of the 40-N force is parallel to this axis and hence its moment about the x axis is zero.) Using the right-hand rule, the algebraic sum of the moment components about the x axis is therefore

$$M_x = (10 \text{ N})(4 \text{ m}) - (20 \text{ N})(6 \text{ m}) = -80 \text{ N} \cdot \text{m} \qquad Ans.$$

Although not required here, note also that

$$M_y = (10 \text{ N})(3 \text{ m}) - (40 \text{ N})(6 \text{ m}) = -210 \text{ N} \cdot \text{m}$$
$$M_z = (40 \text{ N})(4 \text{ m}) - (20 \text{ N})(3 \text{ m}) = 100 \text{ N} \cdot \text{m}$$

If we were to determine M_a by this scalar method, it would require much more effort since the force components of 40 N and 20 N are *not perpendicular* to the direction of the a axis. The vector analysis yields a more direct solution.

EXAMPLE 4.9

The rod shown in Fig. 4–24a is supported by two brackets at A and B. Determine the moment $\mathbf{M}_{AB}$ produced by $\mathbf{F} = \{-600\mathbf{i} + 200\mathbf{j} - 300\mathbf{k}\}$ N, which tends to rotate the rod about the AB axis.

SOLUTION

A vector analysis using $M_{AB} = \mathbf{u}_B \cdot (\mathbf{r} \times \mathbf{F})$ will be considered for the solution since the moment arm or perpendicular distance from the line of action of $\mathbf{F}$ to the AB axis is difficult to determine. Each of the terms in the equation will now be identified.

Unit vector $\mathbf{u}_B$ defines the direction of the AB axis of the rod, Fig. 4–24b, where

$$\mathbf{u}_B = \frac{\mathbf{r}_B}{r_B} = \frac{0.4\mathbf{i} + 0.2\mathbf{j}}{\sqrt{(0.4)^2 + (0.2)^2}} = 0.894\mathbf{i} + 0.447\mathbf{j}$$

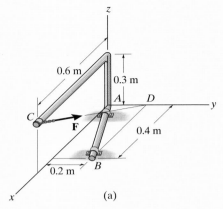

(a)

Vector $\mathbf{r}$ is directed from *any point* on the AB axis to *any point* on the line of action of the force. For example, position vectors $\mathbf{r}_C$ and $\mathbf{r}_D$ are suitable, Fig. 4–24b. (Although not shown, $\mathbf{r}_{BC}$ or $\mathbf{r}_{BD}$ can also be used.) For simplicity, we choose $\mathbf{r}_D$, where

$$\mathbf{r}_D = \{0.2\mathbf{j}\} \text{ m}$$

The force is

$$\mathbf{F} = \{-600\mathbf{i} + 200\mathbf{j} - 300\mathbf{k}\} \text{ N}$$

Substituting these vectors into the determinant form and expanding, we have

$$M_{AB} = \mathbf{u}_B \cdot (\mathbf{r}_D \times \mathbf{F}) = \begin{vmatrix} 0.894 & 0.447 & 0 \\ 0 & 0.2 & 0 \\ -600 & 200 & -300 \end{vmatrix}$$

$$= 0.894[0.2(-300) - 0(200)] - 0.447[0(-300) - 0(-600)]$$
$$+ 0[0(200) - 0.2(-600)]$$

$$= -53.67 \text{ N} \cdot \text{m}$$

(b)

Fig. 4–24

The negative sign indicates that the sense of $\mathbf{M}_{AB}$ is opposite to that of $\mathbf{u}_B$.
Expressing $\mathbf{M}_{AB}$ as a Cartesian vector yields

$$\mathbf{M}_{AB} = M_{AB}\mathbf{u}_B = (-53.67 \text{ N} \cdot \text{m})(0.894\mathbf{i} + 0.447\mathbf{j})$$
$$= \{-48.0\mathbf{i} - 24.0\mathbf{j}\} \text{ N} \cdot \text{m} \qquad\qquad Ans.$$

The result is shown in Fig. 4–24b.

NOTE: If axis AB is defined using a unit vector directed from B toward A, then in the above formulation $-\mathbf{u}_B$ would have to be used. This would lead to $M_{AB} = +53.67 \text{ N} \cdot \text{m}$. Consequently, $\mathbf{M}_{AB} = M_{AB}(-\mathbf{u}_B)$, and the above result would again be determined.

PROBLEMS

4–51. Determine the moment of the force **F** about the *Oa* axis. Express the result as a Cartesian vector.

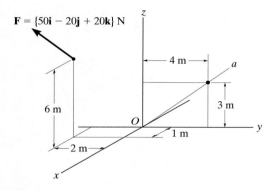

Prob. 4–51

4–53. Determine the resultant moment of the two forces about the *Oa* axis. Express the result as a Cartesian vector.

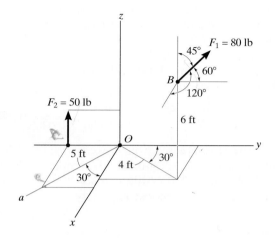

Prob. 4–53

*4–52. Determine the moment of the force **F** about the *aa* axis. Express the result as a Cartesian vector.

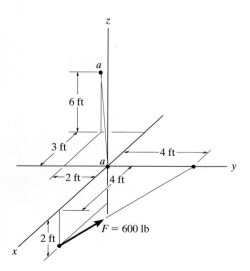

Prob. 4–52

4–54. A force of $\mathbf{F} = \{8\mathbf{i} - 1\mathbf{j} + 1\mathbf{k}\}$ lb is applied to the handle of the box wrench. Determine the component of the moment of this force about the z axis which is effective in loosening the bolt.

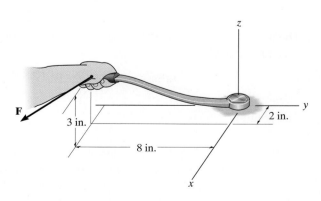

Prob. 4–54

4–55. The 50-lb force acts on the gear in the direction shown. Determine the moment of this force about the y axis.

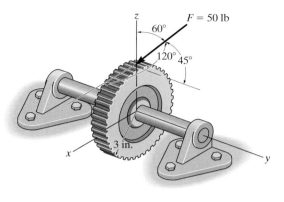

Prob. 4–55

4–57. The cutting tool on the lathe exerts a force **F** on the shaft in the direction shown. Determine the moment of this force about the y axis of the shaft.

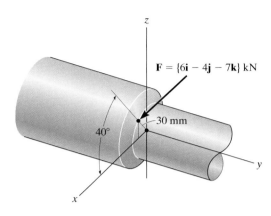

Prob. 4–57

*4–56. The Rollerball skate is an in-line tandem skate that uses two large spherical wheels on each skate, rather than traditional wafer-shape wheels. During skating the two forces acting on the wheel of one skate consist of a 78-lb normal force and a 13-lb friction force. Determine the moment of both of these forces about the axle AB of the wheel.

Prob. 4–56

4–58. The hood of the automobile is supported by the strut AB, which exerts a force of $F = 24$ lb on the hood. Determine the moment of this force about the hinged axis y.

Prob. 4–58

4–59. The lug nut on the wheel of the automobile is to be removed using the wrench and applying the vertical force of $F = 30$ N at A. Determine if this force is adequate, provided 14 N · m of torque about the x axis is initially required to turn the nut. If the 30-N force can be applied at A in any other direction, will it be possible to turn the nut?

*4–60. Solve Prob. 4–59 if the cheater pipe AB is slipped over the handle of the wrench and the 30-N force can be applied at any point and in any direction on the assembly.

Probs. 4–59/60

4–61. The bevel gear is subjected to the force $\mathbf{F}$ which is caused from contact with another gear. Determine the moment of this force about the y axis of the gear shaft.

Prob. 4–61

4–62. The wooden shaft is held in a lathe. The cutting tool exerts a force $\mathbf{F}$ on the shaft in the direction shown. Determine the moment of this force about the x axis of the shaft. Express the result as a Cartesian vector. The distance OA is 25 mm.

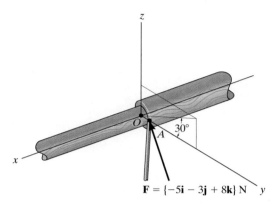

Prob. 4–62

4–63. Determine the magnitude of the moment of the force $\mathbf{F} = \{50\mathbf{i} - 20\mathbf{j} - 80\mathbf{k}\}$ N about the base line CA of the tripod.

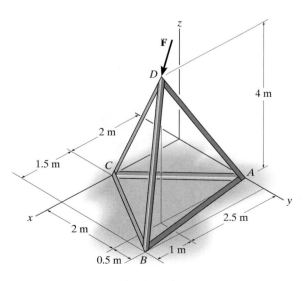

Prob. 4–63

*4–64. The flex-headed ratchet wrench is subjected to a force of $P = 16$ lb, applied perpendicular to the handle as shown. Determine the moment or torque this imparts along the vertical axis of the bolt at A.

4–65. If a torque or moment of 80 lb·in. is required to loosen the bolt at A, determine the force P that must be applied perpendicular to the handle of the flex-headed ratchet wrench.

4–67. Determine the moment of each force acting on the handle of the wrench about the a axis. Take $\mathbf{F}_1 = \{-2\mathbf{i} + 4\mathbf{j} - 8\mathbf{k}\}$ lb, $\mathbf{F}_2 = \{3\mathbf{i} + 2\mathbf{j} - 6\mathbf{k}\}$ lb.

Prob. 4–67

Probs. 4–64/65

4–66. The A-frame is being hoisted into an upright position by the vertical force of $F = 80$ lb. Determine the moment of this force about the y axis when the frame is in the position shown.

*4–68. Determine the moment of each force acting on the handle of the wrench about the z axis. Take $\mathbf{F}_1 = \{-2\mathbf{i} + 4\mathbf{j} - 8\mathbf{k}\}$ lb, $\mathbf{F}_2 = \{3\mathbf{i} + 2\mathbf{j} - 6\mathbf{k}\}$ lb.

Prob. 4–66

Prob. 4–68

Fig. 4–25

Fig. 4–26

Fig. 4–27

4.6   Moment of a Couple

A *couple* is defined as two parallel forces that have the same magnitude, have opposite directions, and are separated by a perpendicular distance d, Fig. 4–25. Since the resultant force is zero, the only effect of a couple is to produce a rotation or tendency of rotation in a specified direction.

The moment produced by a couple is called a *couple moment*. We can determine its value by finding the sum of the moments of both couple forces about *any* arbitrary point. For example, in Fig. 4–26, position vectors $\mathbf{r}_A$ and $\mathbf{r}_B$ are directed from point O to points A and B lying on the line of action of $-\mathbf{F}$ and $\mathbf{F}$. The couple moment computed about O is therefore

$$\mathbf{M} = \mathbf{r}_A \times (-\mathbf{F}) + \mathbf{r}_B \times \mathbf{F}$$

Rather than sum the moments of both forces to determine the couple moment, it is simpler to take moments about a point lying on the line of action of one of the forces. If point A is chosen, then the moment of $-\mathbf{F}$ about A is zero, and we have

$$\mathbf{M} = \mathbf{r} \times \mathbf{F} \qquad (4\text{–}13)$$

The fact that we obtain the *same result* in both cases can be demonstrated by noting that in the first case we can write $\mathbf{M} = (\mathbf{r}_B - \mathbf{r}_A) \times \mathbf{F}$; and by the triangle rule of vector addition, $\mathbf{r}_A + \mathbf{r} = \mathbf{r}_B$ or $\mathbf{r} = \mathbf{r}_B - \mathbf{r}_A$, so that upon substitution we obtain Eq. 4–13. This result indicates that a couple moment is a *free vector*, i.e., it can act at *any point* since $\mathbf{M}$ depends *only* upon the position vector $\mathbf{r}$ directed *between* the forces and *not* the position vectors $\mathbf{r}_A$ and $\mathbf{r}_B$, directed from the arbitrary point O to the forces. This concept is therefore unlike the moment of a force, which requires a definite point (or axis) about which moments are determined.

Scalar Formulation.   The moment of a couple, $\mathbf{M}$, Fig. 4–27, is defined as having a *magnitude* of

$$\boxed{M = Fd} \qquad (4\text{–}14)$$

where F is the magnitude of one of the forces and d is the perpendicular distance or moment arm between the forces. The *direction* and sense of the couple moment are determined by the right-hand rule, where the thumb indicates the direction when the fingers are curled with the sense of rotation caused by the two forces. In all cases, $\mathbf{M}$ acts perpendicular to the plane containing these forces.

Vector Formulation.   The moment of a couple can also be expressed by the vector cross product using Eq. 4–13, i.e.,

$$\boxed{\mathbf{M} = \mathbf{r} \times \mathbf{F}} \qquad (4\text{–}15)$$

Application of this equation is easily remembered if one thinks of taking the moments of both forces about a point lying on the line of action of one of the forces. For example, if moments are taken about point A in

Fig. 4–26, the moment of $-\mathbf{F}$ is *zero* about this point, and the moment or $\mathbf{F}$ is defined from Eq. 4–15. Therefore, in the formulation $\mathbf{r}$ is crossed with the force $\mathbf{F}$ to which it is directed.

Equivalent Couples.   Two couples are said to be equivalent if they produce the same moment. Since the moment produced by a couple is always perpendicular to the plane containing the couple forces, it is therefore necessary that the forces of equal couples lie either in the same plane or in planes that are *parallel* to one another. In this way, the direction of each couple moment will be the same, that is, perpendicular to the parallel planes.

Resultant Couple Moment.   Since couple moments are free vectors, they may be applied at any point P on a body and added vectorially. For example, the two couples acting on different planes of the body in Fig. 4–28a may be replaced by their corresponding couple moments $\mathbf{M}_1$ and $\mathbf{M}_2$, Fig. 4–28b, and then these free vectors may be moved to the *arbitrary point P* and added to obtain the resultant couple moment $\mathbf{M}_R = \mathbf{M}_1 + \mathbf{M}_2$, shown in Fig. 4–28c.

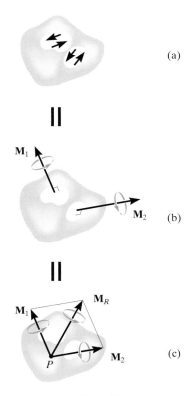

(a)

(b)

(c)

Fig. 4–28

If more than two couple moments act on the body, we may generalize this concept and write the vector resultant as

$$\mathbf{M}_R = \Sigma(\mathbf{r} \times \mathbf{F}) \qquad (4\text{–}16)$$

These concepts are illustrated numerically in the examples which follow. In general, problems projected in two dimensions should be solved using a scalar analysis since the moment arms and force components are easy to compute.

A moment of 12 N · m is needed to turn the shaft connected to the center of the wheel. To do this it is efficient to apply a couple since this effect produces a pure rotation. The couple forces can be made as small as possible by placing the hands on the *rim* of the wheel, where the spacing is 0.4 m. In this case 12 N · m = $F(0.4 \text{ m})$, $F = 30$ N. An equivalent couple moment of 12 N · m can be produced if one grips the wheel within the inner hub, although here much larger forces are needed. If the distance between the hands becomes 0.3 m, then 12 N · m = $F'(0.3)$, $F' = 40$ N. Also, realize that if the wheel was connected to the shaft at a point other than at its center, the wheel would still turn when the forces are applied since the 12-N · m couple moment is a *free vector*.

The frictional forces of the floor on the blades of the concrete finishing machine create a couple moment $\mathbf{M}_c$ on the machine that tends to turn it. An equal but opposite couple moment must be applied by the hands of the operator to prevent the turning. Here the couple moment, $M_c = Fd$, is applied on the handle, although it could be applied at any other point on the machine.

Important Points

- A couple moment is produced by two noncollinear forces that are equal but opposite. Its effect is to produce pure rotation, or tendency for rotation in a specified direction.

- A couple moment is a free vector, and as a result it causes the same effect of rotation on a body regardless of where the couple moment is applied to the body.

- The moment of the two couple forces can be computed about *any point*. For convenience, this point is often chosen on the line of action of one of the forces in order to eliminate the moment of this force about the point.

- In three dimensions the couple moment is often determined using the vector formulation, $\mathbf{M} = \mathbf{r} \times \mathbf{F}$, where $\mathbf{r}$ is directed from *any point* on the line of action of one of the forces to *any point* on the line of action of the other force $\mathbf{F}$.

- A resultant couple moment is simply the vector sum of all the couple moments of the system.

EXAMPLE 4.10

A couple acts on the gear teeth as shown in Fig. 4–29a. Replace it by an equivalent couple having a pair of forces that act through points A and B.

(a)

(b)

SOLUTION (*SCALAR ANALYSIS*)

The couple has a magnitude of $M = Fd = 40(0.6) = 24 \text{ N} \cdot \text{m}$ and a direction that is out of the page since the forces tend to rotate counterclockwise. $\mathbf{M}$ is a free vector, and so it can be placed at any point on the gear, Fig. 4–29b. To preserve the counterclockwise rotation of $\mathbf{M}$, *vertical* forces acting through points A and B must be directed as shown in Fig. 4–29c. The magnitude of each force is

$$M = Fd \qquad 24 \text{ N} \cdot \text{m} = F(0.2 \text{ m})$$
$$F = 120 \text{ N} \qquad\qquad\qquad\qquad Ans.$$

(c)

Fig. 4–29

EXAMPLE 4.11

Determine the moment of the couple acting on the member shown in Fig. 4–30a.

(a)

(b)

$M = 390$ lb · ft

(c)

Fig. 4–30

SOLUTION (SCALAR ANALYSIS)

Here it is somewhat difficult to determine the perpendicular distance between the forces and compute the couple moment as $M = Fd$. Instead, we can resolve each force into its horizontal and vertical components, $F_x = \frac{4}{5}(150 \text{ lb}) = 120 \text{ lb}$ and $F_y = \frac{3}{5}(150 \text{ lb}) = 90 \text{ lb}$, Fig. 4–30b, and then use the principle of moments. The couple moment can be determined about *any point*. For example, if point D is chosen, we have for all four forces,

$$\zeta + M = 120 \text{ lb } (0 \text{ ft}) - 90 \text{ lb } (2 \text{ ft}) + 90 \text{ lb } (5 \text{ ft}) + 120 \text{ lb } (1 \text{ ft})$$
$$= 390 \text{ lb} \cdot \text{ft} \, \uparrow \qquad\qquad\qquad\qquad\qquad\qquad Ans.$$

It is easier, however, to determine the moments about point A or B in order to *eliminate* the moment of the forces acting at the moment point. For point A, Fig. 4–30b, we have

$$\zeta + M = 90 \text{ lb } (3 \text{ ft}) + 120 \text{ lb } (1 \text{ ft})$$
$$= 390 \text{ lb} \cdot \text{ft} \, \uparrow \qquad\qquad\qquad\qquad\qquad\qquad Ans.$$

NOTE: Show that one obtains the same result if moments are summed about point B. Notice also that the couple in Fig. 4–30a can be replaced by *two* couples in Fig. 4–30b. Using $M = Fd$, one couple has a moment of $M_1 = 90 \text{ lb } (3 \text{ ft}) = 270 \text{ lb} \cdot \text{ft}$ and the other has a moment of $M_2 = 120 \text{ lb } (1 \text{ ft}) = 120 \text{ lb} \cdot \text{ft}$. By the right-hand rule, both couple moments are counterclockwise and are therefore directed out of the page. Since these couples are free vectors, they can be moved to any point and added, which yields $M = 270 \text{ lb} \cdot \text{ft} + 120 \text{ lb} \cdot \text{ft} = 390 \text{ lb} \cdot \text{ft} \, \uparrow$, the same result determined above. **M** is a free vector and can therefore act at any point on the member, Fig. 4–30c. Also, realize that the external effects, such as the support reactions on the member, will be the *same* if the member supports the couple, Fig. 4–30a, or the couple moment, Fig. 4–30c.

EXAMPLE 4.12

Determine the couple moment acting on the pipe shown in Fig. 4–31a. Segment AB is directed 30° below the x–y plane.

(a)

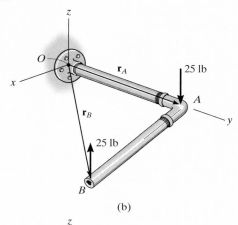

(b)

SOLUTION I (VECTOR ANALYSIS)

The moment of the two couple forces can be found about *any point*. If point O is considered, Fig. 4–31b, we have

$$\mathbf{M} = \mathbf{r}_A \times (-25\mathbf{k}) + \mathbf{r}_B \times (25\mathbf{k})$$
$$= (8\mathbf{j}) \times (-25\mathbf{k}) + (6\cos 30°\mathbf{i} + 8\mathbf{j} - 6\sin 30°\mathbf{k}) \times (25\mathbf{k})$$
$$= -200\mathbf{i} - 129.9\mathbf{j} + 200\mathbf{i}$$
$$= \{-130\mathbf{j}\} \text{ lb} \cdot \text{in.} \qquad\qquad Ans.$$

It is *easier* to take moments of the couple forces about a point lying on the line of action of one of the forces, e.g., point A, Fig. 4–31c. In this case the moment of the force A is zero, so that

$$\mathbf{M} = \mathbf{r}_{AB} \times (25\mathbf{k})$$
$$= (6\cos 30°\mathbf{i} - 6\sin 30°\mathbf{k}) \times (25\mathbf{k})$$
$$= \{-130\mathbf{j}\} \text{ lb} \cdot \text{in.} \qquad\qquad Ans.$$

(c)

SOLUTION II (SCALAR ANALYSIS)

Although this problem is shown in three dimensions, the geometry is simple enough to use the scalar equation $M = Fd$. The perpendicular distance between the lines of action of the forces is $d = 6\cos 30° = 5.20$ in., Fig. 4–31d. Hence, taking moments of the forces about either point A or B yields

$$M = Fd = 25 \text{ lb } (5.20 \text{ in.}) = 129.9 \text{ lb} \cdot \text{in.}$$

Applying the right-hand rule, $\mathbf{M}$ acts in the $-\mathbf{j}$ direction. Thus,

$$\mathbf{M} = \{-130\mathbf{j}\} \text{ lb} \cdot \text{in.} \qquad\qquad Ans.$$

(d)

Fig. 4–31

EXAMPLE 4.13

Replace the two couples acting on the pipe column in Fig. 4–32a by a resultant couple moment.

(a)

(b)

(c)

Fig. 4–32

SOLUTION (VECTOR ANALYSIS)

The couple moment $\mathbf{M}_1$, developed by the forces at A and B, can easily be determined from a scalar formulation.

$$M_1 = Fd = 150 \text{ N}(0.4 \text{ m}) = 60 \text{ N} \cdot \text{m}$$

By the right-hand rule, $\mathbf{M}_1$ acts in the $+\mathbf{i}$ direction, Fig. 4–32b. Hence,

$$\mathbf{M}_1 = \{60\mathbf{i}\} \text{ N} \cdot \text{m}$$

Vector analysis will be used to determine $\mathbf{M}_2$, caused by forces at C and D. If moments are computed about point D, Fig. 4–32a, $\mathbf{M}_2 = \mathbf{r}_{DC} \times \mathbf{F}_C$, then

$$\begin{aligned} \mathbf{M}_2 = \mathbf{r}_{DC} \times \mathbf{F}_C &= (0.3\mathbf{i}) \times \left[125\left(\tfrac{4}{5}\right)\mathbf{j} - 125\left(\tfrac{3}{5}\right)\mathbf{k} \right] \\ &= (0.3\mathbf{i}) \times [100\mathbf{j} - 75\mathbf{k}] = 30(\mathbf{i} \times \mathbf{j}) - 22.5(\mathbf{i} \times \mathbf{k}) \\ &= \{22.5\mathbf{j} + 30\mathbf{k}\} \text{ N} \cdot \text{m} \end{aligned}$$

Since $\mathbf{M}_1$ and $\mathbf{M}_2$ are free vectors, they may be moved to some arbitrary point P and added vectorially, Fig. 4–32c. The resultant couple moment becomes

$$\mathbf{M}_R = \mathbf{M}_1 + \mathbf{M}_2 = \{60\mathbf{i} + 22.5\mathbf{j} + 30\mathbf{k}\} \text{ N} \cdot \text{m} \qquad Ans.$$

NOTE: Try to establish $\mathbf{M}_2$ by using a scalar formulation, Fig. 4–32b.

PROBLEMS

4–69.   Determine the magnitude and sense of the couple moment.

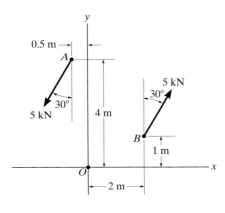

Prob. 4–69

4–71.   Determine the magnitude and sense of the couple moment.

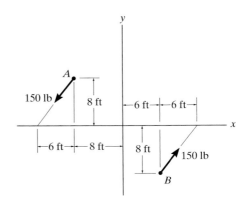

Prob. 4–71

4–70.   Determine the magnitude and sense of the couple moment. Each force has a magnitude of $F = 65$ lb..

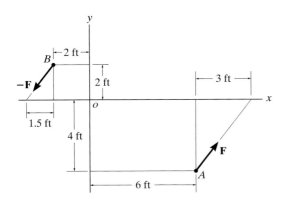

Prob. 4–70

*4–72.   If the couple moment has a magnitude of 300 lb · ft, determine the magnitude F of the couple forces.

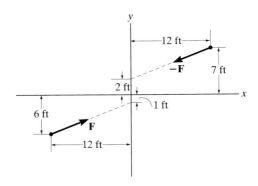

Prob. 4–72

4–73. A clockwise couple $M = 5\,N\cdot m$ is resisted by the shaft of the electric motor. Determine the magnitude of the reactive forces $-R$ and R which act at supports A and B so that the resultant of the two couples is zero.

Prob. 4–73

4–74. The resultant couple moment created by the two couples acting on the disk is $M_R = \{10k\}\,kip\cdot in.$ Determine the magnitude of force T.

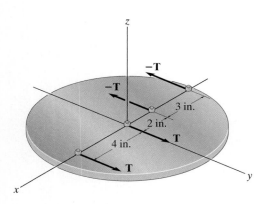

Prob. 4–74

4–75. Three couple moments act on the pipe assembly. Determine the magnitude of M_3 and the bend angle θ so that the resultant couple moment is zero.

Prob. 4–75

***4–76.** The floor causes a couple moment of $M_A = 40\,N\cdot m$ and $M_B = 30\,N\cdot m$ on the brushes of the polishing machine. Determine the magnitude of the couple forces that must be developed by the operator on the handles so that the resultant couple moment on the polisher is zero. What is the magnitude of these forces if the brush at B suddenly stops so that $M_B = 0$?

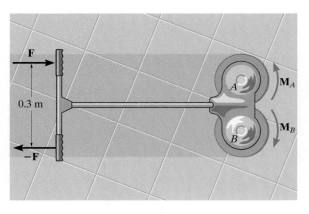

Prob. 4–76

4–77. The ends of the triangular plate are subjected to three couples. Determine the magnitude of the force **F** so that the resultant couple moment is 400 N · m clockwise.

4–79. Express the moment of the couple acting on the pipe assembly in Cartesian vector form. Solve the problem (a) using Eq. 4–13, and (b) summing the moment of each force about point O. Take **F** = {25**k**} N.

***4–80.** If the couple moment acting on the pipe has a magnitude of 400 N · m, determine the magnitude F of the vertical force applied to each wrench.

Prob. 4–77

Probs. 4–79/80

4–78. Two couples act on the beam. Determine the magnitude of **F** so that the resultant couple moment is 450 lb · ft, counterclockwise. Where on the beam does the resultant couple moment act?

4–81. Determine the resultant couple moment acting on the beam. Solve the problem two ways: (a) sum moments about point O; and (b) sum moments about point A.

Prob. 4–78

Prob. 4–81

4–82. Two couples act on the beam as shown. Determine the magnitude of **F** so that the resultant couple moment is 300 lb·ft counterclockwise. Where on the beam does the resultant couple act?

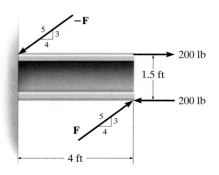

Prob. 4–82

4–83. Two couples act on the frame. If the resultant couple moment is to be zero, determine the distance d between the 80-lb couple forces.

*4–84. Two couples act on the frame. If $d = 4$ ft, determine the resultant couple moment. Compute the result by resolving each force into x and y components and (a) finding the moment of each couple (Eq. 4–13) and (b) summing the moments of all the force components about point A.

4–85. Two couples act on the frame. If $d = 4$ ft, determine the resultant couple moment. Compute the result by resolving each force into x and y components and (a) finding the moment of each couple (Eq. 4–13) and (b) summing the moments of all the force components about point B.

Probs. 4–83/84/85

4–86. Determine the couple moment. Express the result as a Cartesian vector.

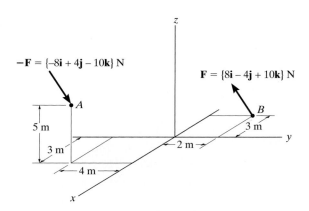

Prob. 4–86

4–87. Determine the couple moment. Express the result as a Cartesian vector.

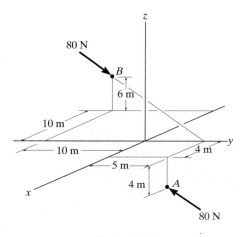

Prob. 4–87

*4–88. If the resultant couple of the two couples acting on the fire hydrant is $\mathbf{M}_R = \{-15\mathbf{i} + 30\mathbf{j}\}$ N·m, determine the force magnitude P.

4–90. If $\mathbf{F} = \{100\mathbf{k}\}$ N, determine the couple moment that acts on the assembly. Express the result as a Cartesian vector. Member BA lies in the x–y plane.

4–91. If the magnitude of the resultant couple moment is 15 N·m, determine the magnitude F of the forces applied to the wrenches.

Prob. 4–88

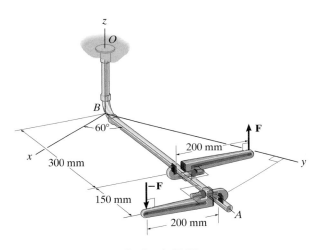

Probs. 4–90/91

4–89. If the resultant couple of the three couples acting on the triangular block is to be zero, determine the magnitude of forces $\mathbf{F}$ and $\mathbf{P}$.

*4–92. The gears are subjected to the couple moments shown. Determine the magnitude and coordinate direction angles of the resultant couple moment.

Prob. 4–89

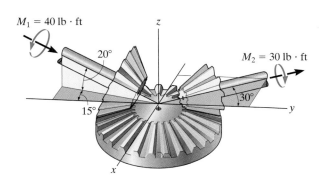

Prob. 4–92

4–93. Express the moment of the couple acting on the rod in Cartesian vector form. What is the magnitude of the couple moment?

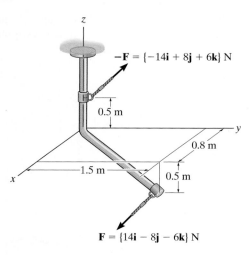

$$-\mathbf{F} = \{-14\mathbf{i} + 8\mathbf{j} + 6\mathbf{k}\}\ N$$

0.5 m

0.8 m

1.5 m

0.5 m

$$\mathbf{F} = \{14\mathbf{i} - 8\mathbf{j} - 6\mathbf{k}\}\ N$$

Prob. 4–93

4–95. A couple acts on each of the handles of the dual valve. Determine the magnitude and coordinate direction angles of the resultant couple moment.

35 N

25 N

60°

175 mm

175 mm

25 N

35 N

Prob. 4–95

*4–96.** Express the moment of the couple acting on the pipe in Cartesian vector form. What is the magnitude of the couple moment? Take $F = 125$ N.

4–97. If the couple moment acting on the pipe has a magnitude of 300 N · m, determine the magnitude F of the forces applied to the wrenches.

4–94. Express the moment of the couple acting on the pipe assembly in Cartesian vector form. Solve the problem (a) using Eq. 4–13, and (b) summing the moment of each force about point O.

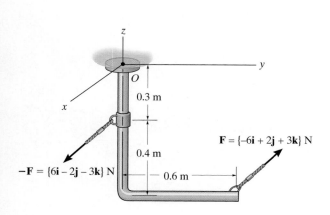

O

0.3 m

$$\mathbf{F} = \{-6\mathbf{i} + 2\mathbf{j} + 3\mathbf{k}\}\ N$$

0.4 m

$$-\mathbf{F} = \{6\mathbf{i} - 2\mathbf{j} - 3\mathbf{k}\}\ N$$

0.6 m

Prob. 4–94

−F

150 mm

600 mm

A

B

200 mm

150 mm

F

Probs. 4–96/97

4.7   Equivalent System

A force has the effect of both translating and rotating a body, and the amount by which it does so depends upon where and how the force is applied. In the next section we will discuss the method used to *simplify* a system of forces and couple moments acting on a body to a single resultant force and couple moment acting at a specified point O. To do this, however, it is necessary that the force and couple moment system produce the *same* "external" effects of translation and rotation of the body as their resultants. When this occurs these two sets of loadings are said to be *equivalent*.

In this section we wish to show how to maintain this equivalency when a single force is applied to a specific point on a body and when it is located at another point O. Two cases for the location of point O will now be considered.

Point O Is On the Line of Action of the Force.   Consider the body shown in Fig. 4–33a, which is subjected to the force $\mathbf{F}$ applied to point A. In order to apply the force to point O without altering the external effects on the body, we will first apply equal but opposite forces $\mathbf{F}$ and $-\mathbf{F}$ at O, as shown in Fig. 4–33b. The two forces indicated by the slash across them can be canceled, leaving the force at point O as required, Fig. 4–33c. By using this construction procedure, an *equivalent system* has been maintained between each of the diagrams, as shown by the equal signs. Note, however, that the force has simply been "transmitted" along its line of action, from point A, Fig. 4–33a, to point O, Fig. 4–33c. In other words, the force can be considered as a *sliding vector* since it can act at any point O along its line of action. In Sec. 4.3 we referred to this concept as the *principle of transmissibility*. It is important to realize that only the *external effects*, such as the body's motion or the forces needed to support the body if it is stationary, remain *unchanged* after $\mathbf{F}$ is moved. Certainly the *internal effects* depend on where $\mathbf{F}$ is located. For example, when $\mathbf{F}$ acts at A, the internal forces in the body have a high intensity around A; whereas movement of $\mathbf{F}$ away from this point will cause these internal forces to decrease.

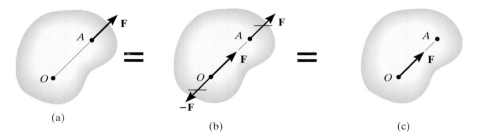

(a)   (b)   (c)

Fig. 4–33

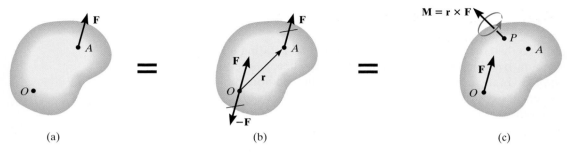

Fig. 4–34

Point *O* Is Not On the Line of Action of the Force.

This case is shown in Fig. 4–34*a*, where **F** is to be moved to point *O* without altering the external effects on the body. Following the same procedure as before, we first apply equal but opposite forces **F** and −**F** at point *O*, Fig. 4–34*b*. Here the two forces indicated by a slash across them form a couple which has a moment that is perpendicular to **F** and is defined by the cross product **M** = **r** × **F**. Since the couple moment is a *free vector*, it may be applied at *any point P* on the body as shown in Fig. 4–34*c*. In addition to this couple moment, **F** now acts at point *O* as required.

To summarize these concepts, when the point on the body is *on the line of action of the force*, simply transmit or slide the force along its line of action to the point. When the point is not on the line of action of the force, then move the force to the point and add a couple moment anywhere to the body. This couple moment is found by taking the moment of the force about the point. When these rules are carried out, equivalent external effects will be produced.

Photo A: Consider the effects on the hand when a stick of negligible weight supports a force **F** at its end. When the force is applied horizontally, the same force is felt at the grip, regardless of where it is applied along its line of action. This is a consequence of the principle of transmissibility.

Photo B: When the force is applied vertically it causes both a downward force **F** to be felt at the grip and a clockwise couple moment or twist of $M = Fd$. These same effects are felt if **F** is applied at the grip and **M** is applied anywhere on the stick. In both cases the systems are equivalent.

4.8 Resultants of a Force and Couple System

When a rigid body is subjected to a *system* of forces and couple moments, it is often simpler to study the external effects on the body by *replacing* the system by an equivalent single resultant force acting at a specified point O and a resultant couple moment. To show how to determine these resultants we will consider the rigid body in Fig. 4–35a and use the concepts discussed in the previous section. Since point O is not on the line of action of the forces, an equivalent effect is produced if the forces are moved to point O *and* the corresponding couple moments $\mathbf{M}_1 = \mathbf{r}_1 \times \mathbf{F}_1$ and $\mathbf{M}_2 = \mathbf{r}_2 \times \mathbf{F}_2$ are applied to the body. Furthermore, the couple moment $\mathbf{M}_c$ is simply moved to point O since it is a free vector. These results are shown in Fig. 4–35b. By vector addition, the resultant force is $\mathbf{F}_R = \mathbf{F}_1 + \mathbf{F}_2$, and the resultant couple moment is $\mathbf{M}_{R_O} = \mathbf{M}_c + \mathbf{M}_1 + \mathbf{M}_2$, Fig. 4–35$c$. Since equivalency is maintained between the diagrams in Fig. 4–35, each force and couple system will cause the *same external effects*, i.e., the same translation and rotation of the body. Note that both the magnitude and direction of $\mathbf{F}_R$ are independent of the location of point O; however, $\mathbf{M}_{R_O}$ depends upon this location since the moments $\mathbf{M}_1$ and $\mathbf{M}_2$ are determined using the position vectors $\mathbf{r}_1$ and $\mathbf{r}_2$. Also note that $\mathbf{M}_{R_O}$ is a free vector and can act at *any point* on the body, although point O is generally chosen as its point of application.

The above method of simplifying any force and couple moment system to a resultant force acting at point O and a resultant couple moment can be generalized and represented by application of the following two equations.

$$
\begin{aligned}
\mathbf{F}_R &= \Sigma \mathbf{F} \\
\mathbf{M}_{R_O} &= \Sigma \mathbf{M}_c + \Sigma \mathbf{M}_O
\end{aligned}
\tag{4–17}
$$

The first equation states that the resultant force of the system is equivalent to the sum of all the forces; and the second equation states that the resultant couple moment of the system is equivalent to the sum of all the couple moments $\Sigma \mathbf{M}_c$, plus the moments about point O of all the forces $\Sigma \mathbf{M}_O$. If the force system lies in the x–y plane and any couple moments are perpendicular to this plane, that is along the z axis, then the above equations reduce to the following three scalar equations.

$$
\begin{aligned}
F_{R_x} &= \Sigma F_x \\
F_{R_y} &= \Sigma F_y \\
M_{R_o} &= \Sigma M_c + \Sigma M_O
\end{aligned}
\tag{4–18}
$$

Note that the resultant force $\mathbf{F}_R$ is equivalent to the vector sum of its two components $\mathbf{F}_{R_x}$ and $\mathbf{F}_{R_y}$.

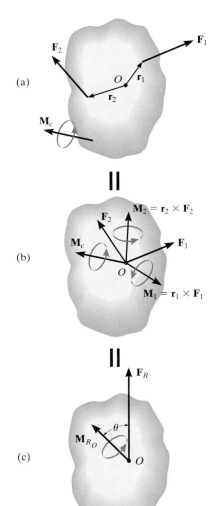

(a)

(b)

(c)

Fig. 4–35

If the two forces acting on the stick are replaced by an equivalent resultant force and couple momen t at point A, or by the equivalent resultant force and couple moment at point B, then in each case the hand must provide the same resistance to translation and rotation in order to keep the stick in the horizontal position. In other words, the external effects on the stick are the *same* in each case.

Procedure for Analysis

The following points should be kept in mind when applying Eqs. 4–17 or 4–18.

- Establish the coordinate axes with the origin located at point O and the axes having a selected orientation.

Force Summation.

- If the force system is *coplanar*, resolve each force into its x and y components. If a component is directed along the positive x or y axis, it represents a positive scalar; whereas if it is directed along the negative x or y axis, it is a negative scalar.

- In three dimensions, represent each force as a Cartesian vector before summing the forces.

Moment Summation.

- When determining the moments of a *coplanar* force system about point O, it is generally advantageous to use the principle of moments, i.e., determine the moments of the components of each force rather than the moment of the force itself.

- In three dimensions use the vector cross product to determine the moment of each force about the point. Here the position vectors extend from point O to any point on the line of action of each force.

EXAMPLE 4.14

Replace the forces acting on the brace shown in Fig. 4–36a by an equivalent resultant force and couple moment acting at point A.

Fig. 4–36

SOLUTION (SCALAR ANALYSIS)

The principle of moments will be applied to the 400-N force, whereby the moments of its two rectangular components will be considered.

Force Summation. The resultant force has x and y components of

$$\xrightarrow{+} F_{R_x} = \Sigma F_x; \quad F_{R_x} = -100\ \text{N} - 400\cos 45°\ \text{N} = -382.8\ \text{N} = 382.8\ \text{N} \leftarrow$$

$$+\uparrow F_{R_y} = \Sigma F_y; \quad F_{R_y} = -600\ \text{N} - 400\sin 45°\ \text{N} = -882.8\ \text{N} = 882.8\ \text{N} \downarrow$$

As shown in Fig. 4–36b, $\mathbf{F}_R$ has a magnitude of

$$F_R = \sqrt{(F_{R_x})^2 + (F_{R_y})^2} = \sqrt{(382.8)^2 + (882.8)^2} = 962\ \text{N} \qquad Ans.$$

and a direction of

$$\theta = \tan^{-1}\!\left(\frac{F_{R_y}}{F_{R_x}}\right) = \tan^{-1}\!\left(\frac{882.8}{382.8}\right) = 66.6° \quad \theta\nearrow \qquad Ans.$$

Moment Summation. The resultant couple moment $\mathbf{M}_{R_A}$ is determined by summing the moments of the forces about point A. Assuming that positive moments act counterclockwise, i.e., in the $+\mathbf{k}$ direction, we have

$$\curvearrowleft +M_{R_A} = \Sigma M_A;$$

$$M_{R_A} = 100\ \text{N}(0) - 600\ \text{N}(0.4\ \text{m}) - (400\sin 45°\ \text{N})(0.8\ \text{m})$$

$$- (400\cos 45°\ \text{N})(0.3\ \text{m})$$

$$= -551\ \text{N} \cdot \text{m} = 551\ \text{N} \cdot \text{m} \downarrow \qquad Ans.$$

NOTE: Realize that, when $\mathbf{M}_{R_A}$ and $\mathbf{F}_R$ act on the brace at point A, Fig. 4–36b, they will produce the *same* external effects or reactions at the supports as those produced by the force system in Fig. 4–36a.

EXAMPLE 4.15

(a)

Fig. 4–37

(b)

A structural member is subjected to a couple moment $\mathbf{M}$ and forces $\mathbf{F}_1$ and $\mathbf{F}_2$ as shown in Fig. 4–37a. Replace this system by an equivalent resultant force and couple moment acting at its base, point O.

SOLUTION *(VECTOR ANALYSIS)*
The three-dimensional aspects of the problem can be simplified by using a Cartesian vector analysis. Expressing the forces and couple moment as Cartesian vectors, we have

$$\mathbf{F}_1 = \{-800\mathbf{k}\} \text{ N}$$

$$\mathbf{F}_2 = (300 \text{ N})\mathbf{u}_{CB} = (300 \text{ N})\left(\frac{\mathbf{r}_{CB}}{r_{CB}}\right)$$

$$= 300\left[\frac{-0.15\mathbf{i} + 0.1\mathbf{j}}{\sqrt{(-0.15)^2 + (0.1)^2}}\right] = \{-249.6\mathbf{i} + 166.4\mathbf{j}\} \text{ N}$$

$$\mathbf{M} = -500\left(\tfrac{4}{5}\right)\mathbf{j} + 500\left(\tfrac{3}{5}\right)\mathbf{k} = \{-400\mathbf{j} + 300\mathbf{k}\} \text{ N} \cdot \text{m}$$

Force Summation.

$$\mathbf{F}_R = \Sigma\mathbf{F}; \qquad \mathbf{F}_R = \mathbf{F}_1 + \mathbf{F}_2 = -800\mathbf{k} - 249.6\mathbf{i} + 166.4\mathbf{j}$$

$$= \{-249.6\mathbf{i} + 166.4\mathbf{j} - 800\mathbf{k}\} \text{ N} \qquad \textit{Ans.}$$

Moment Summation.

$$\mathbf{M}_{R_O} = \Sigma\mathbf{M}_C + \Sigma\mathbf{M}_O$$

$$\mathbf{M}_{R_O} = \mathbf{M} + \mathbf{r}_C \times \mathbf{F}_1 + \mathbf{r}_B \times \mathbf{F}_2$$

$$\mathbf{M}_{R_O} = (-400\mathbf{j} + 300\mathbf{k}) + (1\mathbf{k}) \times (-800\mathbf{k}) + \begin{vmatrix} \mathbf{i} & \mathbf{j} & \mathbf{k} \\ -0.15 & 0.1 & 1 \\ -249.6 & 166.4 & 0 \end{vmatrix}$$

$$= (-400\mathbf{j} + 300\mathbf{k}) + (\mathbf{0}) + (-166.4\mathbf{i} - 249.6\mathbf{j})$$

$$= \{-166\mathbf{i} - 650\mathbf{j} + 300\mathbf{k}\} \text{ N} \cdot \text{m} \qquad \textit{Ans.}$$

The results are shown in Fig. 4–37b.

4.9   Further Reduction of a Force and Couple System

Simplification to a Single Resultant Force.   Consider now a special case for which the system of forces and couple moments acting on a rigid body, Fig. 4–38a, reduces at point O to a resultant force $\mathbf{F}_R = \Sigma\mathbf{F}$ and resultant couple moment $\mathbf{M}_{R_O} = \Sigma\mathbf{M}_O$, which are *perpendicular* to one another, Fig. 4–38b. Whenever this occurs, we can further simplify the force and couple moment system by moving $\mathbf{F}_R$ to another point P, located either on or off the body so that no resultant couple moment has to be applied to the body, Fig. 4–38c. In other words, if the force and couple moment system in Fig. 4–38a is reduced to a resultant system at point P, only the force resultant will have to be applied to the body, Fig. 4–38c.

The location of point P, measured from point O, can always be determined provided $\mathbf{F}_R$ and $\mathbf{M}_{R_O}$ are known, Fig. 4–38b. As shown in Fig. 4–38c, P must lie on the bb axis, which is perpendicular to both the line of action of $\mathbf{F}_R$ and the aa axis. This point is chosen such that the distance d satisfies the scalar equation $M_{R_O} = F_R d$ or $d = M_{R_O}/F_R$. With $\mathbf{F}_R$ so located, it will produce the same external effects on the body as the force and couple moment system in Fig. 4–38a, or the force and couple moment resultants in Fig. 4–38b.

If a system of forces is either concurrent, coplanar, or parallel, it can always be reduced, as in the above case, to a single resultant force $\mathbf{F}_R$ acting through a specific point. This is because in each of these cases $\mathbf{F}_R$ and $\mathbf{M}_{R_O}$ will always be perpendicular to each other when the force system is simplified at *any* point O.

Fig. 4–38

Fig. 4–39

Concurrent Force Systems.

A concurrent force system has been treated in detail in Chapter 2. Obviously, all the forces act at a point for which there is no resultant couple moment, so the point P is automatically specified, Fig. 4–39.

Coplanar Force Systems.

Coplanar force systems, which may include couple moments directed perpendicular to the plane of the forces as shown in Fig. 4–40a, can be reduced to a single resultant force, because when each force in the system is moved to any point O in the x–y plane, it produces a couple moment that is *perpendicular* to the plane, i.e., in the $\pm\mathbf{k}$ direction. The resultant moment $\mathbf{M}_{R_O} = \Sigma\mathbf{M} + \Sigma(\mathbf{r} \times \mathbf{F})$ is thus perpendicular to the resultant force $\mathbf{F}_R$, Fig. 4–40b; and so $\mathbf{F}_R$ can be positioned a distance d from O so as to create the same moment $\mathbf{M}_{R_O}$ about O, Fig. 4–40c.

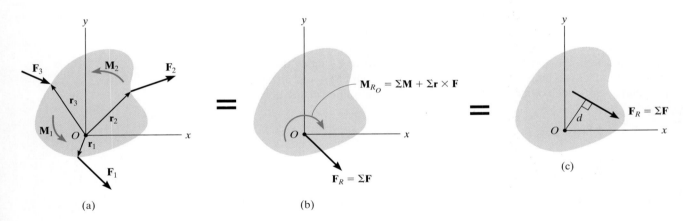

Fig. 4–40

Parallel Force Systems.

Parallel force systems, which can include couple moments that are perpendicular to the forces, as shown in Fig. 4–41a, can be reduced to a single resultant force because when each force is moved to any point O in the x–y plane, it produces a couple moment that has components only about the x and y axes. The resultant moment $\mathbf{M}_{R_O} = \Sigma\mathbf{M}_O + \Sigma(\mathbf{r} \times \mathbf{F})$ is thus perpendicular to the resultant force $\mathbf{F}_R$, Fig. 4–41b; and so $\mathbf{F}_R$ can be moved to a point a distance d away so that it produces the same moment about O.

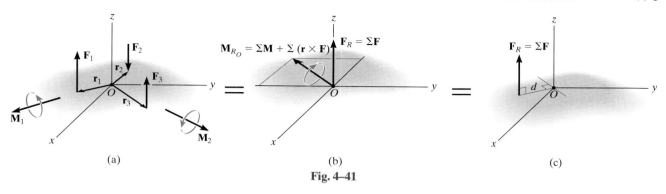

Fig. 4–41

The three parallel forces acting on the stick can be replaced by a single resultant force F_R acting at a distance d from the grip. To be equivalent we require the resultant force to equal the sum of the forces, $F_R = F_1 + F_2 + F_3$, and to find the distance d the moment of the resultant force about the grip must be equal to the moment of all the forces about the grip, $F_R d = F_1 d_1 + F_2 d_2 + F_3 d_3$.

Procedure for Analysis

The technique used to reduce a coplanar or parallel force system to a single resultant force follows a similar procedure outlined in the previous section.

- Establish the x, y, z, axes and locate the resultant force $\mathbf{F}_R$ an arbitrary distance away from the origin of the coordinates.

Force Summation.

- The resultant force is equal to the sum of all the forces in the system.
- For a coplanar force system, resolve each force into its x and y components. Positive components are directed along the positive x and y axes, and negative components are directed along the negative x and y axes.

Moment Summation.

- The moment of the resultant force about point O is equal to the sum of all the couple moments in the system plus the moments about point O of all the forces in the system.
- This moment condition is used to find the location of the resultant force from point O.

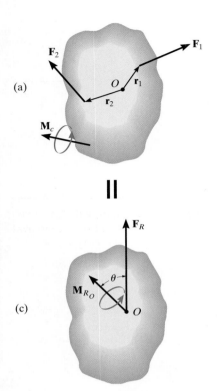

(a)

(c)

Fig. 4–35 (Repeated)

Reduction to a Wrench.

In the general case, the force and couple moment system acting on a body, Fig. 4–35a, will reduce to a single resultant force $\mathbf{F}_R$ and couple moment $\mathbf{M}_{R_O}$ at O which are *not* perpendicular. Instead, $\mathbf{F}_R$ will act at an angle θ from $\mathbf{M}_{R_O}$, Fig. 4–35c. As shown in Fig. 4–42a, however, $\mathbf{M}_{R_O}$ may be resolved into two components: one perpendicular, $\mathbf{M}_\perp$, and the other parallel $\mathbf{M}_\|$, to the line of action of $\mathbf{F}_R$. As in the previous discussion, the perpendicular component $\mathbf{M}_\perp$ may be *eliminated* by moving $\mathbf{F}_R$ to point P, as shown in Fig. 4–42b. This point lies on axis bb, which is perpendicular to both $\mathbf{M}_{R_O}$ and $\mathbf{F}_R$. In order to maintain an equivalency of loading, the distance from O to P is $d = M_\perp/F_R$. Furthermore, when $\mathbf{F}_R$ is applied at P, the moment of $\mathbf{F}_R$ tending to cause rotation of the body *about O* is in the *same direction* as $\mathbf{M}_\perp$, Fig. 4–42a. Finally, since $\mathbf{M}_\|$ is a free vector, it may be moved to P so that it is collinear with $\mathbf{F}_R$, Fig. 4–42c. This combination of a collinear force and couple moment is called a *wrench* or *screw*. The *axis of the wrench* has the same line of action as the force. Hence, the wrench tends to cause both a translation along and a rotation about this axis. Comparing Fig. 4–42a to Fig. 4–42c, it is seen that a general force and couple moment system acting on a body can be reduced to a wrench. The axis of the wrench and the point through which this axis passes can always be determined.

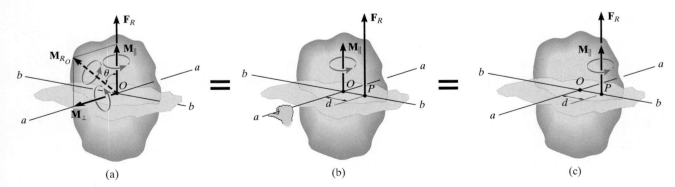

(a)   (b)   (c)

Fig. 4–42

EXAMPLE 4.16

The beam AE in Fig. 4–43a is subjected to a system of coplanar forces. Determine the magnitude, direction, and location on the beam of a resultant force which is equivalent to the given system of forces measured from E.

(a)   Fig. 4–43   (b)

SOLUTION

The origin of coordinates is located at point E as shown in Fig. 4–43a.

Force Summation. Resolving the 500-N force into x and y components and summing the force components yields

$$\xrightarrow{+} F_{R_x} = \Sigma F_x; \quad F_{R_x} = 500 \cos 60° \text{ N} + 100 \text{ N} = 350.0 \text{ N} \rightarrow$$
$$+\uparrow F_{R_y} = \Sigma F_y; \quad F_{R_y} = -500 \sin 60° \text{ N} + 200 \text{ N} = -233.0 \text{ N}$$
$$= 233.0 \text{ N} \downarrow$$

The magnitude and direction of the resultant force are established from the vector addition shown in Fig. 4–43b. We have

$$F_R = \sqrt{(350.0)^2 + (233.0)^2} = 420.5 \text{ N} \qquad Ans.$$

$$\theta = \tan^{-1}\left(\frac{233.0}{350.0}\right) = 33.7° \searrow\theta \qquad Ans.$$

Moment Summation. Moments will be summed about point E. Hence, from Figs. 4–43a and 4–43b, we require the moments of the components of $\mathbf{F}_R$ (or the moment of $\mathbf{F}_R$) about point E to equal the moments of the force system about E. Assuming positive moments are counterclockwise, we have

$$\zeta + M_{R_E} = \Sigma M_E;$$

$$233.0 \text{ N}(d) + 350.0 \text{ N}(0) = (500 \sin 60° \text{ N})(4 \text{ m}) + (500 \cos 60° \text{ N})(0)$$
$$- (100 \text{ N})(0.5 \text{ m}) - (200 \text{ N})(2.5 \text{ m})$$

$$d = \frac{1182.1}{233.0} = 5.07 \text{ m} \qquad Ans.$$

NOTE: Using a clockwise sign convention would yield this same result. Since d is *positive*, $\mathbf{F}_R$ acts to the left of E as shown. Try to solve this problem by summing moments about point A and show $d' = 0.927$ m, measured to the right of A.

EXAMPLE 4.17

(a)

(b)

Fig. 4–44

The jib crane shown in Fig. 4–44a is subjected to three coplanar forces. Replace this loading by an equivalent resultant force and specify where the resultant's line of action intersects the column AB and boom BC.

SOLUTION

Force Summation.   Resolving the 250-lb force into x and y components and summing the force components yields

$$\xrightarrow{+} F_{R_x} = \Sigma F_x; \quad F_{R_x} = -250 \text{ lb}\left(\tfrac{3}{5}\right) - 175 \text{ lb} = -325 \text{ lb} = 325 \text{ lb} \leftarrow$$
$$+\uparrow F_{R_y} = \Sigma F_y; \quad F_{R_y} = -250 \text{ lb}\left(\tfrac{4}{5}\right) - 60 \text{ lb} = -260 \text{ lb} = 260 \text{ lb}\downarrow$$

As shown by the vector addition in Fig. 4–44b,

$$F_R = \sqrt{(325)^2 + (260)^2} = 416 \text{ lb} \qquad \qquad Ans.$$

$$\theta = \tan^{-1}\left(\frac{260}{325}\right) = 38.7° \, \theta \angle \qquad \qquad Ans.$$

Moment Summation.   Moments will be summed about the arbitrary point A. Assuming the line of action of $\mathbf{F}_R$ intersects AB, Fig. 4–44b, we require the moment of the components of $\mathbf{F}_R$ in Fig. 4–44b about A to equal the moments of the force system in Fig. 4–44a about A; i.e.,

$$\zeta + M_{R_A} = \Sigma M_A; \qquad 325 \text{ lb } (y) + 260 \text{ lb } (0)$$
$$= 175 \text{ lb } (5 \text{ ft}) - 60 \text{ lb } (3 \text{ ft}) + 250 \text{ lb}\left(\tfrac{3}{5}\right)(11 \text{ ft}) - 250 \text{ lb}\left(\tfrac{4}{5}\right)(8 \text{ ft})$$
$$y = 2.29 \text{ ft} \qquad \qquad Ans.$$

By the principle of transmissibility, $\mathbf{F}_R$ can also be treated as intersecting BC, Fig. 4–44b, in which case we have

$$\zeta + M_{R_A} = \Sigma M_A; \qquad 325 \text{ lb } (11 \text{ ft}) - 260 \text{ lb } (x)$$
$$= 175 \text{ lb } (5 \text{ ft}) - 60 \text{ lb } (3 \text{ ft}) + 250 \text{ lb}\left(\tfrac{3}{5}\right)(11 \text{ ft}) - 250 \text{ lb}\left(\tfrac{4}{5}\right)(8 \text{ ft})$$
$$x = 10.9 \text{ ft} \qquad \qquad Ans.$$

NOTE: We can also solve for these positions by assuming $\mathbf{F}_R$ acts at the arbitrary point (x, y) on its line of action, Fig. 4–44b. Summing moments about point A yields

$$\zeta + M_{R_A} = \Sigma M_A; 325 \text{ lb } (y) - 260 \text{ lb } (x)$$
$$= 175 \text{ lb } (5 \text{ ft}) - 60 \text{ lb } (3 \text{ ft}) + 250 \text{ lb}\left(\tfrac{3}{5}\right)(11 \text{ ft}) - 250 \text{ lb}\left(\tfrac{4}{5}\right)(8 \text{ ft})$$
$$325y - 260x = 745$$

which is the equation of the colored dashed line in Fig. 4–44b. To find the points of intersection with the crane along AB, set $x = 0$, then $y = 2.29$ ft, and along BC set $y = 11$ ft, then $x = 10.9$ ft.

EXAMPLE 4.18

The slab in Fig. 4–45a is subjected to four parallel forces. Determine the magnitude and direction of a resultant force equivalent to the given force system and locate its point of application on the slab.

(a)   (b)

SOLUTION (SCALAR ANALYSIS)

Fig. 4–45

Force Summation.   From Fig. 4–45a, the resultant force is

$$+\uparrow F_R = \Sigma F; \qquad F_R = -600 \text{ N} + 100 \text{ N} - 400 \text{ N} - 500 \text{ N}$$
$$= -1400 \text{ N} = 1400 \text{ N}\downarrow \qquad\qquad Ans.$$

Moment Summation.   We require the moment about the x axis of the resultant force, Fig. 4–45b, to be equal to the sum of the moments about the x axis of all the forces in the system, Fig. 4–45a. The moment arms are determined from the y coordinates since these coordinates represent the *perpendicular distances* from the x axis to the lines of action of the forces. Using the right-hand rule, where positive moments act in the $+\mathbf{i}$ direction, we have

$$M_{R_x} = \Sigma M_x;$$
$$-(1400 \text{ N})y = 600 \text{ N}(0) + 100 \text{ N}(5 \text{ m}) - 400 \text{ N}(10 \text{ m}) + 500 \text{ N}(0)$$
$$-1400y = -3500 \qquad y = 2.50 \text{ m} \qquad Ans.$$

In a similar manner, assuming that positive moments act in the $+\mathbf{j}$ direction, a moment equation can be written about the y axis using moment arms defined by the x coordinates of each force.

$$M_{R_y} = \Sigma M_y;$$
$$(1400 \text{ N})x = 600 \text{ N}(8 \text{ m}) - 100 \text{ N}(6 \text{ m}) + 400 \text{ N}(0) + 500 \text{ N}(0)$$
$$1400x = 4200 \qquad x = 3.00 \text{ m} \qquad Ans.$$

NOTE: A force of $F_R = 1400$ N placed at point $P(3.00 \text{ m}, 2.50 \text{ m})$ on the slab, Fig. 4–45b, is therefore equivalent to the parallel force system acting on the slab in Fig. 4–45a.

EXAMPLE 4.19

(a)

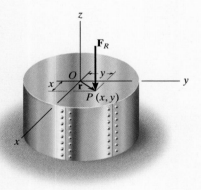

(b)

Fig. 4–46

Three parallel bolting forces act on the rim of the circular cover plate in Fig. 4–46a. Determine the magnitude and direction of a resultant force equivalent to the given force system and locate its point of application, P, on the cover plate.

SOLUTION (*VECTOR ANALYSIS*)

Force Summation.   From Fig. 4–46a, the force resultant $\mathbf{F}_R$ is

$$\mathbf{F}_R = \Sigma\mathbf{F}; \qquad \mathbf{F}_R = -300\mathbf{k} - 200\mathbf{k} - 150\mathbf{k}$$
$$= \{-650\mathbf{k}\}\ \text{lb} \qquad\qquad Ans.$$

Moment Summation.   Choosing point O as a reference for computing moments and assuming that $\mathbf{F}_R$ acts at a point $P(x, y)$, Fig. 4–46b, we require

$$\mathbf{M}_{R_O} = \Sigma\mathbf{M}_O;$$
$$\mathbf{r} \times \mathbf{F}_R = \mathbf{r}_A \times (-300\mathbf{k}) + \mathbf{r}_B \times (-200\mathbf{k}) + \mathbf{r}_C \times (-150\mathbf{k})$$
$$(x\mathbf{i} + y\mathbf{j}) \times (-650\mathbf{k}) = (8\mathbf{i}) \times (-300\mathbf{k}) + (-8\mathbf{j}) \times (-200\mathbf{k})$$
$$+ (-8 \sin 45°\mathbf{i} + 8 \cos 45°\mathbf{j}) \times (-150\mathbf{k})$$
$$650x\mathbf{j} - 650y\mathbf{i} = 2400\mathbf{j} + 1600\mathbf{i} - 848.5\mathbf{j} - 848.5\mathbf{i}$$

Equating the corresponding $\mathbf{j}$ and $\mathbf{i}$ components yields

$$650x = 2400 - 848.5 \qquad\qquad (1)$$
$$-650y = 1600 - 848.5 \qquad\qquad (2)$$

Solving these equations, we obtain the coordinates of point P,

$$x = 2.39\ \text{ft} \qquad y = -1.16\ \text{ft} \qquad\qquad Ans.$$

The negative sign indicates that it was wrong to have assumed a $+y$ position for $\mathbf{F}_R$ as shown in Fig. 4–46b.

NOTE: It is also possible to establish Eqs. 1 and 2 directly by summing moments about the y and x axes. Using the right-hand rule we have

$$M_{R_y} = \Sigma M_y; \qquad 650x = 300\ \text{lb}\ (8\ \text{ft}) - 150\ \text{lb}\ (8 \sin 45°\ \text{ft})$$
$$M_{R_x} = \Sigma M_x; \qquad -650y = 200\ \text{lb}\ (8\ \text{ft}) - 150\ \text{lb}\ (8 \cos 45°\ \text{ft})$$

PROBLEMS

4–98. Replace the force at *A* by an equivalent force and couple moment at point *O*.

4–99. Replace the force at *A* by an equivalent force and couple moment at point *P*.

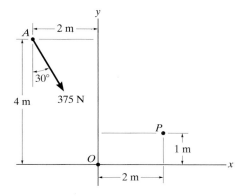

Probs. 4–98/99

4–102. Replace the force system by an equivalent force and couple moment at point *O*.

4–103. Replace the force system by an equivalent force and couple moment at point *P*.

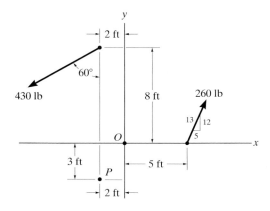

Probs. 4–102/103

*4–100. Replace the force system by an equivalent resultant force and couple moment at point *O*.

4–101. Replace the force system by an equivalent resultant force and couple moment at point *P*.

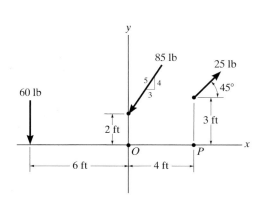

Probs. 4–100/101

*4–104. Replace the loading system acting on the post by an equivalent resultant force and couple moment at point *O*.

4–105. Replace the loading system acting on the post by an equivalent resultant force and couple moment at point *P*.

Probs. 4–104/105

4–106. Replace the force and couple system by an equivalent force and couple moment at point O.

4–107. Replace the force and couple system by an equivalent force and couple moment at point P.

4–110. The forces and couple moments which are exerted on the toe and heel plates of a snow ski are $\mathbf{F}_t = \{-50\mathbf{i} + 80\mathbf{j} - 158\mathbf{k}\}\,\text{N}, \mathbf{M}_t = \{-6\mathbf{i} + 4\mathbf{j} + 2\mathbf{k}\}\,\text{N}\cdot\text{m}$, and $\mathbf{F}_h = \{-20\mathbf{i} + 60\mathbf{j} - 250\mathbf{k}\}\,\text{N}, \mathbf{M}_h = \{-20\mathbf{i} + 8\mathbf{j} + 3\mathbf{k}\}\,\text{N}\cdot\text{m}$, respectively. Replace this system by an equivalent force and couple moment acting at point O. Express the results in Cartesian vector form.

4–111. The forces and couple moments that are exerted on the toe and heel plates of a snow ski are $\mathbf{F}_t = \{-50\mathbf{i} + 80\mathbf{j} - 158\mathbf{k}\}\,\text{N}, \mathbf{M}_t = \{-6\mathbf{i} + 4\mathbf{j} + 2\mathbf{k}\}\,\text{N}\cdot\text{m}$, and $\mathbf{F}_h = \{-20\mathbf{i} + 60\mathbf{j} - 250\mathbf{k}\}\,\text{N}, \mathbf{M}_h = \{-20\mathbf{i} + 8\mathbf{j} + 3\mathbf{k}\}\,\text{N}\cdot\text{m}$, respectively. Replace this system by an equivalent force and couple moment acting at point P. Express the results in Cartesian vector form.

Probs. 4–106/107

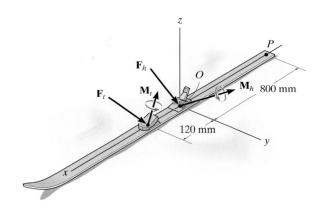

Probs. 4–110/111

*4–108. Replace the force system by a single force resultant and specify its point of application, measured along the x axis from point O.

4–109. Replace the force system by a single force resultant and specify its point of application, measured along the x axis from point P.

*4–112. Replace the three forces acting on the shaft by a single resultant force. Specify where the force acts, measured from end A.

4–113. Replace the three forces acting on the shaft by a single resultant force. Specify where the force acts, measured from end B.

Probs. 4–108/109

Probs. 4–112/113

4–114. Replace the loading on the frame by a single resultant force. Specify where its line of action intersects member AB, measured from A.

Prob. 4–114

4–115. Replace the loading acting on the beam by a single resultant force. Specify where the force acts, measured from end A.

*4–116. Replace the loading acting on the beam by a single resultant force. Specify where the force acts, measured from B.

Probs. 4–115/116

4–117. Replace the loading system acting on the beam by an equivalent resultant force and couple moment at point O.

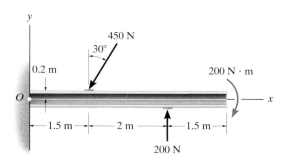

Prob. 4–117

4–118. Determine the magnitude and direction θ of force **F** and its placement d on the beam so that the loading system is equivalent to a resultant force of 12 kN acting vertically downward at point A and a clockwise couple moment of 50 kN · m.

4–119. Determine the magnitude and direction θ of force **F** and its placement d on the beam so that the loading system is equivalent to a resultant force of 10 kN acting vertically downward at point A and a clockwise couple moment of 45 kN · m.

Probs. 4–118/119

*4–120. Replace the loading on the frame by a single resultant force. Specify where its line of action intersects member AB, measured from A.

4–121. Replace the loading on the frame by a single resultant force. Specify where its line of action intersects member CD, measured from end C.

Probs. 4–120/121

4–122. Replace the force system acting on the frame by an equivalent resultant force and specify where the resultant's line of action intersects member AB, measured from point A.

4–123. Replace the force system acting on the frame by an equivalent resultant force and specify where the resultant's line of action intersects member BC, measured from point B.

*4–124. Replace the force system acting on the frame by an equivalent resultant force and couple moment acting at point A.

Probs. 4–122/123/124

4–125. Replace the force and couple-moment system by an equivalent resultant force and couple moment at point O. Express the results in Cartesian vector form.

4–126. Replace the force and couple-moment system by an equivalent resultant force and couple moment at point P. Express the results in Cartesian vector form.

4–127. Replace the force and couple-moment system by an equivalent resultant force and couple moment at point Q. Express the results in Cartesian vector form.

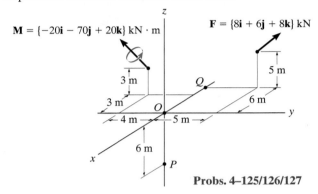

Probs. 4–125/126/127

*4–128. The belt passing over the pulley is subjected to forces $\mathbf{F}_1$ and $\mathbf{F}_2$, each having a magnitude of 40 N. $\mathbf{F}_1$ acts in the $-\mathbf{k}$ direction. Replace these forces by an equivalent force and couple moment at point A. Express the result in Cartesian vector form. Set $\theta = 0°$ so that $\mathbf{F}_2$ acts in the $-\mathbf{j}$ direction.

4–129. The belt passing over the pulley is subjected to two forces $\mathbf{F}_1$ and $\mathbf{F}_2$, each having a magnitude of 40 N. $\mathbf{F}_1$ acts in the $-\mathbf{k}$ direction. Replace these forces by an equivalent force and couple moment at point A. Express the result in Cartesian vector form. Take $\theta = 45°$.

Probs. 4–128/129

4–130. A force and couple act on the pipe assembly. If $F_1 = 50$ N and $F_2 = 80$ N, replace this system by an equivalent resultant force and couple moment acting at O. Express the results in Cartesian vector form.

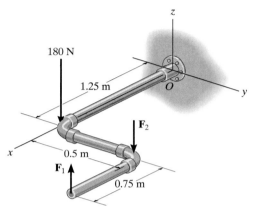

Prob. 4–130

4–131. Handle forces $\mathbf{F}_1$ and $\mathbf{F}_2$ are applied to the electric drill. Replace this system by an equivalent resultant force and couple moment acting at point O. Express the results in Cartesian vector form.

Prob. 4–131

*4–132. A biomechanical model of the lumbar region of the human trunk is shown. The forces acting in the four muscle groups consist of $F_R = 35$ N for the rectus, $F_O = 45$ N for the oblique, $F_L = 23$ N for the lumbar latissimus dorsi, and $F_E = 32$ N for the erector spinae. These loadings are symmetric with respect to the y–z plane. Replace this system of parallel forces by an equivalent force and couple moment acting at the spine, point O. Express the results in Cartesian vector form.

Prob. 4–132

4–133. The building slab is subjected to four parallel column loadings. Determine the equivalent resultant force and specify its location (x, y) on the slab. Take $F_1 = 30$ kN, $F_2 = 40$ kN.

4–134. The building slab is subjected to four parallel column loadings. Determine the equivalent resultant force and specify its location (x, y) on the slab. Take $F_1 = 20$ kN, $F_2 = 50$ kN.

Probs. 4–133/134

4–135. The pipe assembly is subjected to the action of a wrench at B and a couple at A. Determine the magnitude F of the couple forces so that the system can be simplified to a wrench acting at point C.

Prob. 4–135

4–137. Replace the three forces acting on the plate by a wrench. Specify the magnitude of the force and couple moment for the wrench and the point $P(x, y)$ where its line of action intersects the plate.

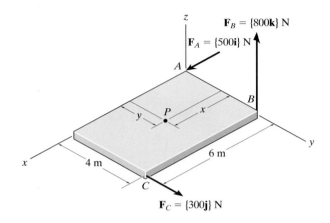

Prob. 4–137

*4–136. The three forces acting on the block each have a magnitude of 10 lb. Replace this system by a wrench and specify the point where the wrench intersects the z axis, measured from point O.

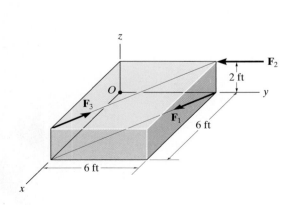

Prob. 4–136

4–138. Replace the three forces acting on the plate by a wrench. Specify the magnitude of the force and couple moment for the wrench and the point $P(y, z)$ where its line of action intersects the plate.

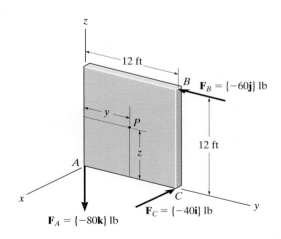

Prob. 4–138

4.10   Reduction of a Simple Distributed Loading

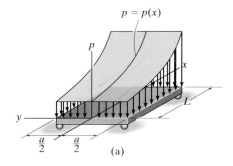

In many situations a very large surface area of a body may be subjected to *distributed loadings* such as those caused by wind, fluids, or simply the weight of material supported over the body's surface. The *intensity* of these loadings at each point on the surface is defined as the *pressure p* (force per unit area), which can be measured in units of lb/ft^2 or pascals (Pa), where $1\ Pa = 1\ N/m^2$.

In this section we will consider the most common case of a distributed pressure loading, which is *uniform* along one axis of a flat rectangular body upon which the loading is applied.* An example of such a loading is shown in Fig. 4–47a. The direction of the intensity of the pressure load is indicated by arrows shown on the *load-intensity diagram*. The entire loading on the plate is therefore a system of parallel forces, infinite in number and each acting on a separate differential area of the plate. Here the *loading function*, $p = p(x)$ Pa, is only a function of x since the pressure is uniform along the y axis. If we multiply $p = p(x)$ by the *width a* m of the plate, we obtain $w = [p(x)\ N/m^2]a\ m = w(x)\ N/m$. This loading function, shown in Fig. 4–47b, is a measure of load distribution along the line $y = 0$ which is in the plane of symmetry of the loading, Fig. 4–47a. As noted, it is measured as a force per unit length, rather than a force per unit area. Consequently, the load-intensity diagram for $w = w(x)$ can be represented by a system of *coplanar* parallel forces, shown in two dimensions in Fig. 4–47b. Using the methods of Sec. 4.9, this system of forces can be simplified to a single resultant force $\mathbf{F}_R$ and its location $\bar{x}$ can be specified, Fig. 4–47c.

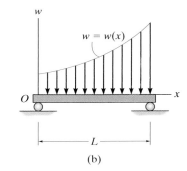

Magnitude of Resultant Force.   From Eq. 4–17 ($F_R = \Sigma F$), the magnitude of $\mathbf{F}_R$ is equivalent to the sum of all the forces in the system. In this case integration must be used since there is an infinite number of parallel forces $d\mathbf{F}$ acting along the plate, Fig. 4–47b. Since $d\mathbf{F}$ is acting on an element of length dx and $w(x)$ is a force per unit length, then at the location x, $dF = w(x)\ dx = dA$. In other words, the magnitude of $d\mathbf{F}$ is determined from the colored differential *area dA* under the loading curve. For the entire plate length,

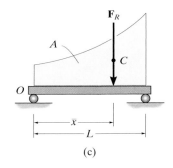

$$+\downarrow F_R = \Sigma F; \qquad \boxed{F_R = \int_L w(x)\ dx = \int_A dA = A} \qquad (4\text{--}19)$$

Hence, the magnitude of the resultant force is equal to the total area A under the loading diagram $w = w(x)$, Fig. 4–47c.

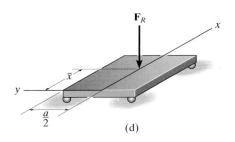

*The more general case of a nonuniform surface loading acting on a body is considered in Sec. 9.5.

Fig. 4–47

Location of Resultant Force. Applying Eq. 4–17 ($M_{R_O} = \Sigma M_O$), the location $\bar{x}$ of the line of action of $\mathbf{F}_R$ can be determined by equating the moments of the force resultant and the force distribution about point O (the y axis). Since $d\mathbf{F}$ produces a moment of $x\,dF = xw(x)\,dx$ about O, Fig. 4–47b, then for the entire plate, Fig. 4–47c,

$$\curvearrowleft + M_{R_O} = \Sigma M_O; \qquad \bar{x}F_R = \int_L xw(x)\,dx$$

Solving for $\bar{x}$, using Eq. 4–19, we can write

$$\bar{x} = \frac{\displaystyle\int_L xw(x)\,dx}{\displaystyle\int_L w(x)\,dx} = \frac{\displaystyle\int_A x\,dA}{\displaystyle\int_A dA} \qquad (4\text{–}20)$$

This equation represents the x coordinate for the geometric center or *centroid* of the *area* under the distributed-loading diagram $w(x)$. *Therefore, the resultant force has a line of action which passes through the centroid C (geometric center) of the area defined by the distributed-loading diagram $w(x)$,* Fig. 4–47c.

Once $\bar{x}$ is determined, $\mathbf{F}_R$ by symmetry passes through point $(\bar{x}, 0)$ on the surface of the plate, Fig. 4–47d. If we now consider the three-dimensional pressure loading $p(x)$, Fig. 4–47a, we can therefore conclude that *the resultant force has a magnitude equal to the volume under the distributed-loading curve $p = p(x)$ and a line of action which passes through the centroid (geometric center) of this volume.* Detailed treatment of the integration techniques for computing the centroids of volumes or areas is given in Chapter 9. In many cases, however, the distributed-loading diagram is in the shape of a rectangle, triangle, or some other simple geometric form. The centroids for such common shapes do not have to be determined from Eq. 4–20; rather, they can be obtained directly from the tabulation given on the inside back cover.

The beam supporting this stack of lumber is subjected to a *uniform* distributed loading, and so the load-intensity diagram has a rectangular shape. If the load intensity is w_0, then the resultant force is determined from the area of the rectangle, $F_R = w_0 b$. The line of action of this force passes through the centroid or center of this area, $\bar{x} = a + b/2$. This resultant is equivalent to the distributed load, and so both loadings produce the same "external" effects or support reactions on the beam.

Important Points

- Distributed loadings are defined by using a loading function $w = w(x)$ that indicates the intensity of the loading along the length of the member. This intensity is measured in N/m or lb/ft.
- The external effects caused by a coplanar distributed load acting on a body can be represented by a single resultant force.
- The resultant force is equivalent to the *area* under the distributed loading diagram, and has a line of action that passes through the *centroid* or geometric center of this area.

EXAMPLE 4.20

Determine the magnitude and location of the equivalent resultant force acting on the shaft in Fig. 4–48a.

SOLUTION

Since $w = w(x)$ is given, this problem will be solved by integration. The colored differential area element $dA = w\, dx = 60x^2\, dx$. Applying Eq. 4–19, by summing these elements from $x = 0$ to $x = 2$ m, we obtain the resultant force $\mathbf{F}_R$.

$$F_R = \Sigma F;$$

$$F_R = \int_A dA = \int_0^2 60x^2\, dx = 60\left[\frac{x^3}{3}\right]_0^2 = 60\left[\frac{2^3}{3} - \frac{0^3}{3}\right]$$

$$= 160 \text{ N} \qquad\qquad\qquad Ans.$$

Since the element of area dA is located an arbitrary distance x from O, the location $\bar{x}$ of $\mathbf{F}_R$ *measured from* O, Fig. 4–48b, is determined from Eq. 4–20.

$$\bar{x} = \frac{\displaystyle\int_A x\, dA}{\displaystyle\int_A dA} = \frac{\displaystyle\int_0^2 x(60x^2)\, dx}{160} = \frac{60\left[\dfrac{x^4}{4}\right]_0^2}{160} = \frac{60\left[\dfrac{2^4}{4} - \dfrac{0^4}{4}\right]}{160}$$

$$= 1.5 \text{ m} \qquad\qquad\qquad Ans.$$

NOTE: These results may be checked by using the table on the inside back cover, where it is shown that for an exparabolic area of length a, height b, and shape shown in Fig. 4–48a,

$$A = \frac{ab}{3} = \frac{2 \text{ m}(240 \text{ N/m})}{3} = 160 \text{ N} \text{ and } \bar{x} = \frac{3}{4}a = \frac{3}{4}(2 \text{ m}) = 1.5 \text{ m}$$

Fig. 4–48

EXAMPLE 4.21

A distributed loading of $p = 800x$ Pa acts over the top surface of the beam shown in Fig. 4–49a. Determine the magnitude and location of the equivalent resultant force.

(a)

SOLUTION

The loading function $p = 800x$ Pa indicates that the load intensity varies uniformly from $p = 0$ at $x = 0$ to $p = 7200$ Pa at $x = 9$ m. Since the intensity is uniform along the width of the beam (the y axis), the loading may be viewed in two dimensions as shown in Fig. 4–49b. Here

$$w = (800x \text{ N/m}^2)(0.2 \text{ m})$$
$$= (160x) \text{ N/m}$$

(b)

At $x = 9$ m, note that $w = 1440$ N/m. Although we may again apply Eqs. 4–19 and 4–20 as in Example 4.20, it is simpler to use the table on the inside back cover.

The magnitude of the resultant force is equivalent to the area under the triangle.

$$F_R = \tfrac{1}{2}(9 \text{ m})(1440 \text{ N/m}) = 6480 \text{ N} = 6.48 \text{ kN} \qquad Ans.$$

The line of action of $\mathbf{F}_R$ passes through the *centroid C* of the triangle. Hence,

$$\bar{x} = 9 \text{ m} - \tfrac{1}{3}(9 \text{ m}) = 6 \text{ m} \qquad Ans.$$

The results are shown in Fig. 4–49c.

(c)

Fig. 4–49

NOTE: We may also view the resultant $\mathbf{F}_R$ as *acting* through the *centroid* of the *volume* of the loading diagram $p = p(x)$ in Fig. 4–49a. Hence $\mathbf{F}_R$ intersects the x–y plane at the point (6 m, 0). Furthermore, the magnitude of $\mathbf{F}_R$ is equal to the volume under the loading diagram; i.e.,

$$F_R = V = \tfrac{1}{2}(7200 \text{ N/m}^2)(9 \text{ m})(0.2 \text{ m}) = 6.48 \text{ kN} \qquad Ans.$$

EXAMPLE 4.22

The granular material exerts the distributed loading on the beam as shown in Fig. 4–50a. Determine the magnitude and location of the equivalent resultant of this load.

(a)

SOLUTION

The area of the loading diagram is a *trapezoid*, and therefore the solution can be obtained directly from the area and centroid formulas for a trapezoid listed on the inside back cover. Since these formulas are not easily remembered, instead we will solve this problem by using "composite" areas. In this regard, we can divide the trapezoidal loading into a rectangular and triangular loading as shown in Fig. 4–50b. The magnitude of the force represented by each of these loadings is equal to its associated *area*,

$$F_1 = \tfrac{1}{2}(9 \text{ ft})(50 \text{ lb/ft}) = 225 \text{ lb}$$
$$F_2 = (9 \text{ ft})(50 \text{ lb/ft}) = 450 \text{ lb}$$

The lines of action of these parallel forces act through the *centroid* of their associated areas and therefore intersect the beam at

$$\overline{x}_1 = \tfrac{1}{3}(9 \text{ ft}) = 3 \text{ ft}$$
$$\overline{x}_2 = \tfrac{1}{2}(9 \text{ ft}) = 4.5 \text{ ft}$$

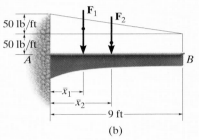

(b)

The two parallel forces $\mathbf{F}_1$ and $\mathbf{F}_2$ can be reduced to a single resultant $\mathbf{F}_R$. The magnitude of $\mathbf{F}_R$ is

$$+\downarrow F_R = \Sigma F; \qquad F_R = 225 + 450 = 675 \text{ lb} \qquad\qquad Ans.$$

With reference to point A, Fig. 4–50b and 4–50c, we can find the location of $\mathbf{F}_R$. We require

$$\curvearrowright + M_{R_A} = \Sigma M_A; \quad \overline{x}(675) = 3(225) + 4.5(450)$$
$$\overline{x} = 4 \text{ ft} \qquad\qquad Ans.$$

(c)

NOTE: The trapezoidal area in Fig. 4–50a can also be divided into two triangular areas as shown in Fig. 4–50d. In this case

$$F_1 = \tfrac{1}{2}(9 \text{ ft})(100 \text{ lb/ft}) = 450 \text{ lb}$$
$$F_2 = \tfrac{1}{2}(9 \text{ ft})(50 \text{ lb/ft}) = 225 \text{ lb}$$

and

$$\overline{x}_1 = \tfrac{1}{3}(9 \text{ ft}) = 3 \text{ ft}$$
$$\overline{x}_2 = \tfrac{1}{3}(9 \text{ ft}) = 3 \text{ ft}$$

NOTE: Using these results, show that again $F_R = 675$ lb and $\overline{x} = 4$ ft.

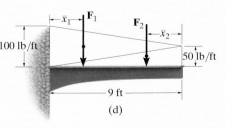

(d)

Fig. 4–50

PROBLEMS

4–139. The loading on the bookshelf is distributed as shown. Determine the magnitude of the equivalent resultant location, measured from point O.

4–141. Replace the loading by an equivalent force and couple moment acting at point O.

4–142. Replace the loading by a single resultant force, and specify the location of the force on the beam measured from point O.

Prob. 4–139

Probs. 4–141/142

*4–140. Replace the loading by an equivalent resultant force and couple moment acting at point A.

4–143. The column is used to support the floor which exerts a force of 3000 lb on the top of the column. The effect of soil pressure along its side is distributed as shown. Replace this loading by an equivalent resultant force and specify where it acts along the column, measured from its base A.

Prob. 4–140

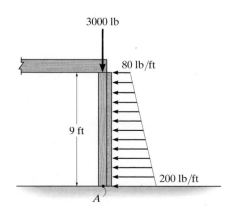

Prob. 4–143

*4–144.   Replace the loading by an equivalent force and couple moment acting at point O.

Prob. 4–144

4–146.   The beam supports the distributed load caused by the sandbags. Determine the resultant force on the beam and specify its location measured from point A.

Prob. 4–146

4–145.   Replace the distributed loading by an equivalent resultant force, and specify its location on the beam, measured from the pin at C.

Prob. 4–145

4–147.   Determine the length b of the triangular load and its position a on the beam such that the equivalent resultant force is zero and the resultant couple moment is 8 kN · m clockwise.

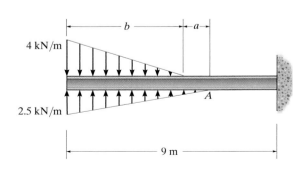

Prob. 4–147

*4–148.   Replace the distributed loading by an equivalent resultant force and specify its location, measured from point A.

4–150.   The beam is subjected to the distributed loading. Determine the length b of the uniform load and its position a on the beam such that the resultant force and couple moment acting on the beam are zero.

Prob. 4–148

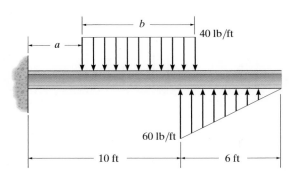

Prob. 4–150

4–149.   The distribution of soil loading on the bottom of a building slab is shown. Replace this loading by an equivalent resultant force and specify its location, measured from point O.

4–151.   Replace the loading by an equivalent resultant force and specify its location on the beam, measured from point B.

Prob. 4–149

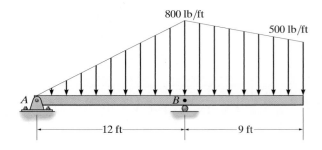

Prob. 4–151

*4–152. Replace the distributed loading by an equivalent resultant force and specify where its line of action intersects member AB, measured from A.

4–153. Replace the distributed loading by an equivalent resultant force and specify where its line of action intersects member BC, measured from C.

4–155. Determine the equivalent resultant force and couple moment at point O.

Prob. 4–155

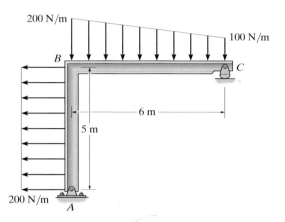

Probs. 4–152/153

4–154. Replace the loading by an equivalent resultant force and couple moment acting at point O.

*4–156. Wind has blown sand over a platform such that the intensity of the load can be approximated by the function $w = (0.5x^3)$ N/m. Simplify this distributed loading to an equivalent resultant force and specify the magnitude and location of the force, measured from A.

Prob. 4–154

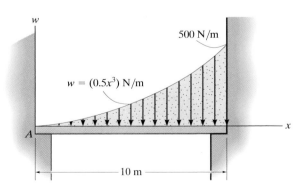

Prob. 4–156

4–157. Determine the equivalent resultant force and its location, measured from point O.

4–159. Currently eighty-five percent of all neck injuries are caused by rear-end car collisions. To alleviate this problem, an automobile seat restraint has been developed that provides additional pressure contact with the cranium. During dynamic tests the distribution of load on the cranium has been plotted and shown to be parabolic. Determine the equivalent resultant force and its location, measured from point A.

Prob. 4–157

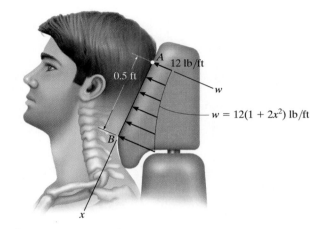

Prob. 4–159

4–158. Determine the equivalent resultant force acting on the bottom of the wing due to air pressure and specify where it acts, measured from point A.

***4–160.** Determine the equivalent resultant force of the distributed loading and its location, measured from point A. Evaluate the integral using Simpson's rule.

Prob. 4–158

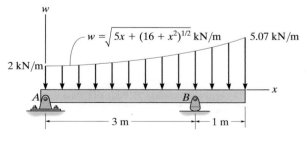

Prob. 4–160

Chapter Review

Moment of Force — Scalar Definition

A force produces a turning effect about a point O that does not lie on its line of action. In scalar form, the moment *magnitude* is the product of the force and the moment arm or perpendicular distance from point O to the line of action of the force.

$$M_O = Fd$$

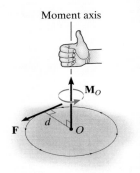

The *direction* of the moment is defined using the right-hand rule. $\mathbf{M}_O$ always acts along an axis perpendicular to the plane containing $\mathbf{F}$ and d, and passes through the point O. If the force acts through the point O, the moment $\mathbf{M}_O$ will be zero.

Rather than finding d, it is normally easier to resolve the force into its x and y components, determine the moment of each component about the point, and then sum the results. This is called the principle of moments.
(*See pages 118–119.*)

$$M_O = Fd = F_x y - F_y x$$

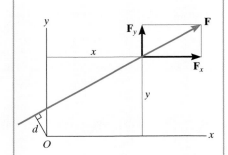

Moment of a Force — Vector Definition

Since three-dimensional geometry is generally more difficult to visualize, the vector cross product can be used to determine the moment.

$$\mathbf{M}_O = \mathbf{r}_A \times \mathbf{F} = \mathbf{r}_B \times \mathbf{F} = \mathbf{r}_C \times \mathbf{F}$$

$\mathbf{r}$ is a position vector that extends from point O to any point A, B, C on the line of action of $\mathbf{F}$.

If the position vector $\mathbf{r}$ and force $\mathbf{F}$ can be expressed as Cartesian vectors, then the cross product can be expressed by the expansion of a determinant.
(*See pages 125–127.*)

$$\mathbf{M}_O = \mathbf{r} \times \mathbf{F} = \begin{vmatrix} \mathbf{i} & \mathbf{j} & \mathbf{k} \\ r_x & r_y & r_z \\ F_x & F_y & F_z \end{vmatrix}$$

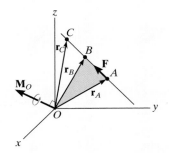

Moment about an Axis

If the moment of a force $\mathbf{F}$ is to be determined about an arbitrary axis, then the projection of the moment onto the axis must be obtained. Provided the distance d_a that is perpendicular to *both* the line of action of the force and the axis can be found, then the moment of the force about the axis can be determined from a scalar equation.

It's a scalar equation when the line of action of $\mathbf{F}$ intersects an axis and the moment of $\mathbf{F}$ about the axis is zero. Also, when the line of action of $\mathbf{F}$ is parallel to the axis, the moment of $\mathbf{F}$ about the axis is zero.

In three dimensions, the scalar triple product should be used. Here $\mathbf{u}_a$ is the unit vector that specifies the direction of the axis, and $\mathbf{r}$ is a position vector that is directed from any point on the axis to any point on the line of action of the force. If M_a is calculated as a negative scalar, then the sense of direction of $\mathbf{M}_a$ is opposite to $\mathbf{u}_a$. (*See pages 142–145.*)

$$M_a = d_a F$$

$$M_a = \mathbf{u}_a \cdot (\mathbf{r} \times \mathbf{F}) = \begin{vmatrix} u_{a_x} & u_{a_y} & u_z \\ r_x & r_y & r_z \\ F_x & F_y & F_z \end{vmatrix}$$

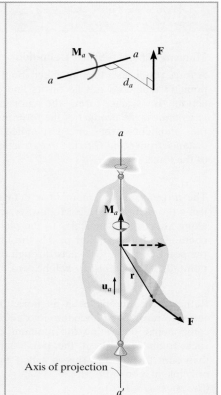

Axis of projection

Couple Moment

A couple consists of two equal but opposite forces that act a perpendicular distance d apart. Couples tend to produce a rotation without translation.

The moment of the couple has a direction that is established using the right-hand rule.

If the vector cross product is used to determine the moment of a couple, then $\mathbf{r}$ extends from any point on the line of action of one of the forces to any point on the line of action of the other force $\mathbf{F}$ that is used in the cross product. (*See pages 152 and 153.*)

$$M_c = F d$$

$$\mathbf{M}_c = \mathbf{r} \times \mathbf{F}$$

Simplification of a Force and Couple System

Any system of forces and couples can be reduced to a single resultant force and resultant couple moment acting at a point. The resultant force is the sum of all the forces in the system, $F_R = \Sigma F$, and the resultant couple moment is equal to the sum of all the moments and couple moments about the point, $M_{R_O} = \Sigma M_c + \Sigma M_O$

Further simplification to a single resultant force is possible provided the force system is *concurrent*, *coplanar*, or *parallel*. For this case, to find the location of the resultant force from a point, it is necessary to equate the moment of the resultant force about the point to the moment of the forces and couples in the system about the same point.

Equating the moment of a resultant force about a point to the moment of the forces and couples in the system about the same point, for any type of force system that is not concurrent, coplanar, or parallel, would yield a wrench, which consists of the resultant force and a resultant collinear couple moment.
(*See pages 167, 171–172, and 174.*)

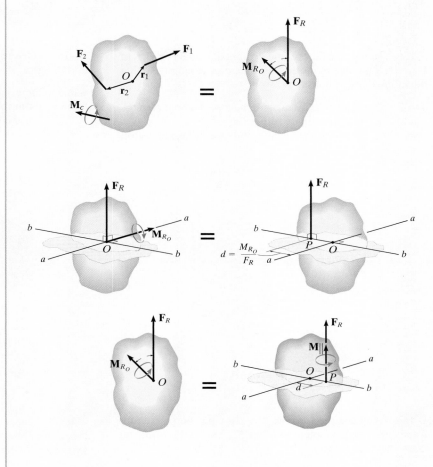

Coplanar Distributed Loading

A resultant force can replace a simple distributed loading, which is equivalent to the *area* under the loading curve. This resultant has a line of action that passes through the *centroid* or geometric center of the area or volume under the loading diagram.
(*See pages 185 and 186.*)

REVIEW PROBLEMS

4–161.   Determine the coordinate direction angles α, β, γ of **F**, which is applied to the end A of the pipe assembly, so that the moment of **F** about O is zero.

4–162.   Determine the moment of the force **F** about point O. The force has coordinate direction angles of $\alpha = 60°$, $\beta = 120°$, $\gamma = 45°$. Express the result as a Cartesian vector.

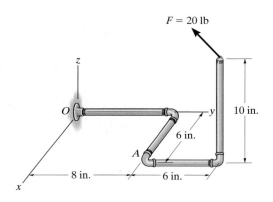

Probs. 4–161/162

*4–164.   Determine the moment of the force $\mathbf{F}_c$ about the door hinge at A. Express the result as a Cartesian vector.

4–165.   Determine the magnitude of the moment of the force $\mathbf{F}_c$ about the hinged axis aa of the door.

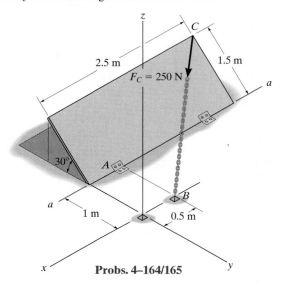

Probs. 4–164/165

4–163.   Replace the force at A by an equivalent resultant force and couple moment at point P. Express the results in Cartesian vector form.

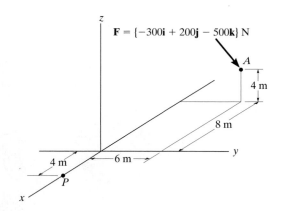

Prob. 4–163

4–166.   A force of $F = 80$ N acts vertically downward on the Z-bracket. Determine the moment of this force about the bolt axis (z axis), which is directed at $15°$ from the vertical.

Prob. 4–166

4–167. Replace the force **F** having a magnitude of $F = 50$ lb and acting at point A by an equivalent force and couple moment at point C.

4–169. The horizontal 30-N force acts on the handle of the wrench. Determine the moment of this force about point O. Specify the coordinate direction angles α, β, γ of the moment axis.

Prob. 4–169

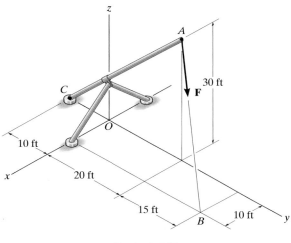

Prob. 4–167

*4–168. The horizontal 30-N force acts on the handle of the wrench. What is the magnitude of the moment of this force about the z axis?

4–170. If the resultant couple moment of the three couples acting on the triangular block is to be zero, determine the magnitudes of forces **F** and **P**.

Prob. 4–168

Prob. 4–170

The tower crane is subjected to its weight and the load it supports. In order to calculate the support reactions for the crane, it is necessary to apply the principles of equilibrium.

5

Equilibrium of a Rigid Body

CHAPTER OBJECTIVES

- To develop the equations of equilibrium for a rigid body.
- To introduce the concept of the free-body diagram for a rigid body.
- To show how to solve rigid-body equilibrium problems using the equations of equilibrium.

5.1 Conditions for Rigid-Body Equilibrium

In this section we will develop both the necessary and sufficient conditions required for equilibrium of a rigid body. To do this, consider the rigid body in Fig. 5–1a, which is fixed in the x, y, z reference and is either at rest or moves with the reference at constant velocity. A free-body diagram of the arbitrary ith particle of the body is shown in Fig. 5–1b. There are two types of forces which act on it. The resultant *internal force*, $\mathbf{f}_i$, is caused by interactions with adjacent particles. The resultant *external force* $\mathbf{F}_i$ represents, for example, the effects of gravitational, electrical, magnetic, or contact forces between the ith particle and adjacent bodies or particles *not* included within the body. If the particle is in equilibrium, then applying Newton's first law we have

$$\mathbf{F}_i + \mathbf{f}_i = \mathbf{0}$$

When the equation of equilibrium is applied to each of the other particles of the body, similar equations will result. If all these equations are added together *vectorially*, we obtain

$$\Sigma \mathbf{F}_i + \Sigma \mathbf{f}_i = \mathbf{0}$$

(a)

(b)

Fig. 5–1

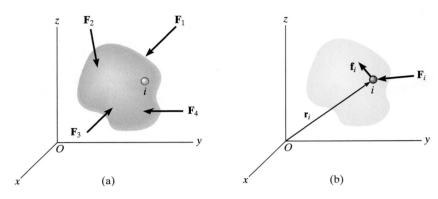

Fig. 5–1

The summation of the internal forces will equal zero since the internal forces between particles within the body will occur in equal but opposite collinear pairs, Newton's third law. Consequently, only the sum of the *external forces* will remain; and therefore, letting $\Sigma \mathbf{F}_i = \Sigma \mathbf{F}$, the above equation can be written as

$$\Sigma \mathbf{F} = \mathbf{0}$$

Let us now consider the moments of the forces acting on the ith particle about the arbitrary point O, Fig. 5–1b. Using the above particle equilibrium equation and the distributive law of the vector cross product we have

$$\mathbf{r}_i \times (\mathbf{F}_i + \mathbf{f}_i) = \mathbf{r}_i \times \mathbf{F}_i + \mathbf{r}_i \times \mathbf{f}_i = \mathbf{0}$$

Similar equations can be written for the other particles of the body, and adding them together vectorially, we obtain

$$\Sigma \mathbf{r}_i \times \mathbf{F}_i + \Sigma \mathbf{r}_i \times \mathbf{f}_i = \mathbf{0}$$

The second term is zero since, as stated above, the internal forces occur in equal but opposite collinear pairs, and therefore the resultant moment of each pair of forces about point O is zero. Using the notation $\Sigma \mathbf{M}_O = \Sigma \mathbf{r}_i \times \mathbf{F}_i$, we have

$$\Sigma \mathbf{M}_O = \mathbf{0}$$

Hence the two *equations of equilibrium* for a rigid body can be summarized as follows:

$$\boxed{\begin{aligned} \Sigma \mathbf{F} &= \mathbf{0} \\ \Sigma \mathbf{M}_o &= \mathbf{0} \end{aligned}}$$

(5–1)

These equations require that a rigid body will remain in equilibrium provided the sum of all the *external forces* acting on the body is equal to zero and the sum of the moments of the external forces about a point is equal to zero. The fact that these conditions are *necessary* for equilibrium has now been proven. They are also *sufficient* for maintaining equilibrium. To show this, let us assume that the body is in equilibrium and the force system acting on the body satisfies Eqs. 5–1. Suppose that an *additional force* $\mathbf{F}'$ is applied to the body. As a result, the equilibrium equations become

$$\Sigma\mathbf{F} + \mathbf{F}' = \mathbf{0}$$

$$\Sigma\mathbf{M}_O + \mathbf{M}'_O = \mathbf{0}$$

where $\mathbf{M}'_O$ is the moment of $\mathbf{F}'$ about O. Since $\Sigma\mathbf{F} = \mathbf{0}$ and $\Sigma\mathbf{M}_O = \mathbf{0}$, then we require $\mathbf{F}' = \mathbf{0}$ (also $\mathbf{M}'_O = \mathbf{0}$). Consequently, the additional force $\mathbf{F}'$ is not required, and indeed Eqs. 5–1 are also sufficient conditions for maintaining equilibrium.

Many types of engineering problems involve symmetric loadings and can be solved by projecting all the forces acting on a body onto a single plane. And so in the next section, the equilibrium of a body subjected to a *coplanar* or *two-dimensional force system* will be considered. Ordinarily the geometry of such problems is not very complex, so a scalar solution is suitable for analysis. The more general discussion of rigid bodies subjected to *three-dimensional force systems* is given in the latter part of this chapter. It will be seen that many of these types of problems are best solved using vector analysis.

EQUILIBRIUM IN TWO DIMENSIONS

5.2 Free-Body Diagrams

Successful application of the equations of equilibrium requires a complete specification of *all* the known and unknown external forces that act *on* the body. The best way to account for these forces is to draw the body's free-body diagram. This diagram is a sketch of the outlined shape of the body, which represents it as being *isolated* or "free" from its surroundings, i.e., a "free body." On this sketch it is necessary to show *all* the forces and couple moments that the surroundings exert *on the body* so that these effects can be accounted for when the equations of equilibrium are applied. For this reason, *a thorough understanding of how to draw a free-body diagram is of primary importance for solving problems in mechanics.*

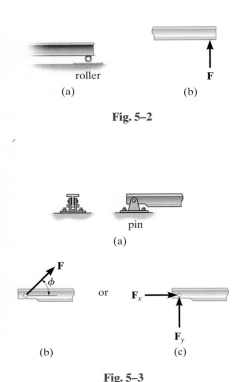

roller

(a)   (b)

F

Fig. 5–2

pin

(a)

F

ϕ

or F_x

(b)   (c)

F_y

Fig. 5–3

Support Reactions.   Before presenting a formal procedure as to how to draw a free-body diagram, we will first consider the various types of reactions that occur at supports and points of support between bodies subjected to coplanar force systems. *As a general rule, if a support prevents the translation of a body in a given direction, then a force is developed on the body in that direction. Likewise, if rotation is prevented, a couple moment is exerted on the body.*

For example, let us consider three ways in which a horizontal member, such as a beam, is supported at its end. One method consists of a *roller* or cylinder, Fig. 5–2a. Since this support only prevents the beam from *translating* in the vertical direction, the roller can only exert a *force* on the beam in this direction, Fig. 5–2b.

The beam can be supported in a more restrictive manner by using a *pin* as shown in Fig. 5–3a. The pin passes through a hole in the beam and two leaves which are fixed to the ground. Here the pin can prevent *translation* of the beam in *any direction* ϕ, Fig. 5–3b, and so the pin must exert a *force* **F** on the beam in this direction. For purposes of analysis, it is generally easier to represent this resultant force **F** by its two components $\mathbf{F}_x$ and $\mathbf{F}_y$, Fig. 5–3c. If F_x and F_y are known, then F and ϕ can be calculated.

The most restrictive way to support the beam would be to use a *fixed support* as shown in Fig. 5–4a. This support will prevent both *translation and rotation* of the beam, and so to do this a *force and couple moment* must be developed on the beam at its point of connection, Fig. 5–4b. As in the case of the pin, the force is usually represented by its components $\mathbf{F}_x$ and $\mathbf{F}_y$.

Table 5–1 lists other common types of supports for bodies subjected to coplanar force systems. (In all cases the angle θ is assumed to be known.) Carefully study each of the symbols used to represent these supports and the types of reactions they exert on their contacting members. Although concentrated forces and couple moments are shown in this table, they actually represent the *resultants* of small *distributed surface loads* that exist between each support and its contacting member. It is these *resultants* which will be determined from the equations of equilibrium.

M

F_x

fixed support

(a) F_y (b)

Fig. 5–4

TABLE 5–1 Supports for Rigid Bodies Subjected to Two-Dimensional Force Systems

Types of Connection	Reaction	Number of Unknowns
(1) cable		One unknown. The reaction is a tension force which acts away from the member in the direction of the cable.
(2) weightless link	or	One unknown. The reaction is a force which acts along the axis of the link.
(3) roller		One unknown. The reaction is a force which acts perpendicular to the surface at the point of contact.
(4) roller or pin in confined smooth slot	or	One unknown. The reaction is a force which acts perpendicular to the slot.
(5) rocker		One unknown. The reaction is a force which acts perpendicular to the surface at the point of contact.
(6) smooth contacting surface		One unknown. The reaction is a force which acts perpendicular to the surface at the point of contact.
(7) member pin connected to collar on smooth rod	or	One unknown. The reaction is a force which acts perpendicular to the rod.

continued

TABLE 5–1   Continued

Types of Connection	Reaction	Number of Unknowns
(8) smooth pin or hinge	F_y, F_x or F, ϕ	Two unknowns. The reactions are two components of force, or the magnitude and direction ϕ of the resultant force. Note that ϕ and θ are not necessarily equal [usually not, unless the rod shown is a link as in (2)].
(9) member fixed connected to collar on smooth rod	F, M	Two unknowns. The reactions are the couple moment and the force which acts perpendicular to the rod.
(10) fixed support	F_y, F_x, M or F, ϕ, M	Three unknowns. The reactions are the couple moment and the two force components, or the couple moment and the magnitude and direction ϕ of the resultant force.

Typical examples of actual supports are shown in the following sequence of photos. The numbers refer to the connection types in Table 5–1.

The cable exerts a force on the bracket in the direction of the cable. (1)

The rocker support for this bridge girder allows horizontal movement so the bridge is free to expand and contract due to temperature. (5)

This concrete girder rests on the ledge that is assumed to act as a smooth contacting surface. (6)

This utility building is pin supported at the top of the column. (8)

The floor beams of this building are welded together and thus form fixed connections. (10)

External and Internal Forces. Since a rigid body is a composition of particles, both *external* and *internal* loadings may act on it. It is important to realize, however, that if the free-body diagram for the body is drawn, the forces that are *internal* to the body are *not represented* on the free-body diagram. As discussed in Sec. 5.1, these forces always occur in equal but opposite collinear pairs, and therefore their *net effect* on the body is zero.

In some problems, a free-body diagram for a "system" of connected bodies may be used for an analysis. An example would be the free-body diagram of an entire automobile (system) composed of its many parts. Obviously, the connecting forces between its parts would represent *internal forces* which would *not* be included on the free-body diagram of the automobile. To summarize, internal forces act between particles which are contained within the boundary of the free-body diagram. Particles or bodies outside this boundary exert external forces on the system, and these alone must be shown on the free-body diagram.

Weight and the Center of Gravity. When a body is subjected to a gravitational field, then each of its particles has a specified weight. For the entire body it is appropriate to consider these gravitational forces to be represented as a *system of parallel forces* acting on all the particles contained within the boundary of the body. It was shown in Sec. 4.9 that such a system can be reduced to a single resultant force acting through a specified point. We refer to this force resultant as the *weight* **W** of the body and to the location of its point of application as the *center of gravity*. The methods used for its calculation will be developed in Chapter 9.

In the examples and problems that follow, if the weight of the body is important for the analysis, this force will then be reported in the problem statement. Also, when the body is *uniform* or made of homogeneous material, the center of gravity will be located at the body's *geometric center* or *centroid*; however, if the body is nonhomogeneous or has an unusual shape, then the location of its center of gravity will be given.

Idealized Models. When an engineer performs a force analysis of any object, he or she considers a corresponding analytical or idealized model that gives results that approximate as closely as possible the actual situation. To do this, careful choices have to be made so that selection of the type of supports, the material behavior, and the object's dimensions can be justified. This way the engineer can feel confident that any design or analysis will yield results which can be trusted. In complex cases this process may require developing several different models of the object that must be analyzed, but in any case, this selection process requires both skill and experience.

(a)

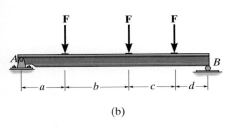

(b)

Fig. 5–5

The cases that follow illustrate what is required to develop a proper model. In Fig. 5–5a, the steel beam is to be used to support the roof joists of a building. For a force analysis it is reasonable to assume the material is rigid since only very small deflections will occur when the beam is loaded. A bolted connection at A will allow for any slight rotation that occurs when the load is applied, and so a *pin* can be considered for this support. At B a *roller* can be considered since the support offers no resistance to horizontal movement here. Building code requirements are used to specify the roof loading which results in a calculation of the joist loads **F**. These forces will be larger than any actual loading on the beam since they account for extreme loading cases and for dynamic or vibrational effects. The weight of the beam is generally neglected when it is small compared to the load the beam supports. The idealized model of the beam is shown with average dimensions a, b, c, and d in Fig. 5–5b.

As a second case, consider the lift boom in Fig. 5–6a. By inspection, it is supported by a pin at A and by the hydraulic cylinder BC, which can be approximated as a weightless link. The material can be assumed rigid, and with its density known, the weight of the boom and the location of its center of gravity G are determined. When a design loading **P** is specified, the idealized model shown in Fig. 5–6b can be used for a force analysis. Average dimensions (not shown) are used to specify the location of the loads and the supports.

Idealized models of specific objects will be given in some of the examples throughout the text. It should be realized, however, that each case represents the reduction of a practical situation using simplifying assumptions like the ones illustrated here.

(a)

(b)

Fig. 5–6

Procedure for Drawing a Free-Body Diagram

To construct a free-body diagram for a rigid body or group of bodies considered as a single system, the following steps should be performed:

Draw Outlined Shape.

Imagine the body to be *isolated* or cut "free" from its constraints and connections and draw (sketch) its outlined shape.

Show All Forces and Couple Moments.

Identify all the external forces and couple moments that act on the body. Those generally encountered are due to (1) applied loadings, (2) reactions occurring at the supports or at points of contact with other bodies (see Table 5–1), and (3) the weight of the body. To account for all these effects, it may help to trace over the boundary, carefully noting each force or couple moment acting on it.

Identify Each Loading and Give Dimensions.

The forces and couple moments that are known should be labeled with their proper magnitudes and directions. Letters are used to represent the magnitudes and direction angles of forces and couple moments that are *unknown*. Establish an *x, y* coordinate system so that these unknowns, A_x, B_y, etc., can be identified. Indicate the dimensions of the body necessary for calculating the moments of forces.

Important Points

- No equilibrium problem should be solved without *first* drawing the free-body diagram, so as to account for all the forces and couple moments that act on the body.

- If a support *prevents translation* of a body in a particular direction, then the support exerts a *force* on the body in that direction.

- If *rotation is prevented*, then the support exerts a *couple moment* on the body.

- Study Table 5–1.

- Internal forces are never shown on the free-body diagram since they occur in equal but opposite collinear pairs and therefore cancel out.

- The weight of a body is an external force, and its effect is shown as a single resultant force acting through the body's center of gravity *G*.

- *Couple moments* can be placed anywhere on the free-body diagram since they are *free vectors. Forces* can act at any point along their lines of action since they are *sliding vectors*.

EXAMPLE 5.1

Draw the free-body diagram of the uniform beam shown in Fig. 5–7a. The beam has a mass of 100 kg.

(a)

SOLUTION

The free-body diagram of the beam is shown in Fig. 5–7b. Since the support at A is a fixed wall, there are three reactions acting *on the beam* at A, denoted as $\mathbf{A}_x$, $\mathbf{A}_y$, and $\mathbf{M}_A$ drawn in an arbitrary direction. The magnitudes of these vectors are *unknown*, and their sense has been *assumed*. The weight of the beam, $W = 100(9.81) = 981$ N, acts through the beam's center of gravity G, which is 3 m from A since the beam is uniform.

(b)

Fig. 5–7

EXAMPLE 5.2

Draw the free-body diagram of the foot lever shown in Fig. 5–8a. The operator applies a vertical force to the pedal so that the spring is stretched 1.5 in. and the force in the short link at B is 20 lb.

(a)

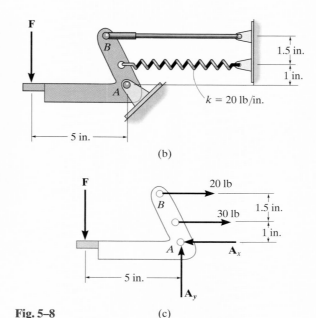

(b)

(c)

Fig. 5–8

SOLUTION

By inspection, the lever is loosely bolted to the frame at A. The rod at B is pinned at its ends and acts as a "short link." After making the proper measurements, the idealized model of the lever is shown in Fig. 5–8b. From this the free-body diagram must be drawn. As shown in Fig. 5–8c, the pin support at A exerts force components $\mathbf{A}_x$ and $\mathbf{A}_y$ on the lever, each force has a known line of action but unknown magnitude. The link at B exerts a force of 20 lb, acting in the direction of the link. In addition the spring also exerts a horizontal force on the lever. If the stiffness is measured and found to be $k = 20$ lb/in., then since the stretch $s = 1.5$ in., using Eq. 3–2, $F_s = ks = 20$ lb/in. $(1.5$ in.$) = 30$ lb. Finally, the operator's shoe applies a vertical force of $\mathbf{F}$ on the pedal. The dimensions of the lever are also shown on the free-body diagram, since this information will be useful when computing the moments of the forces. As usual, the senses of the unknown forces at A have been assumed. The correct senses will become apparent after solving the equilibrium equations.

EXAMPLE 5.3

Two smooth pipes, each having a mass of 300 kg, are supported by the forks of the tractor in Fig. 5–9a. Draw the free-body diagrams for each pipe and both pipes together.

(a)

(b)

(c)

Effect of B acting on A

Effect of sloped blade acting on A

Effect of gravity (weight) acting on A

Effect of sloped fork acting on A

Fig. 5–9

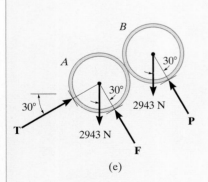

(d)

(e)

SOLUTION

The idealized model from which we must draw the free-body diagrams is shown in Fig. 5–9b. Here the pipes are identified, the dimensions have been added, and the physical situation reduced to its simplest form.

The free-body diagram for pipe A is shown in Fig. 5–9c. Its weight is $W = 300(9.81) = 2943$ N. Assuming all contacting surfaces are *smooth*, the reactive forces **T**, **F**, **R** act in a direction *normal* to the tangent at their surfaces of contact.

The free-body diagram of pipe B is shown in Fig. 5–9d. Can you identify each of the three forces acting *on this pipe*? In particular, note that **R**, representing the force of A on B, Fig. 5–9d, is equal and opposite to **R** representing the force of B on A, Fig. 5–9c. This is a consequence of Newton's third law of motion.

The free-body diagram of both pipes combined ("system") is shown in Fig. 5–9e. Here the contact force **R**, which acts between A and B, is considered as an *internal* force and hence is not shown on the free-body diagram. That is, it represents a pair of equal but opposite collinear forces which cancel each other.

EXAMPLE 5.4

Draw the free-body diagram of the unloaded platform that is suspended off the edge of the oil rig shown in Fig. 5–10a. The platform has a mass of 200 kg.

(a)

Fig. 5–10

(b)

SOLUTION

The idealized model of the platform will be considered in two dimensions because by observation the loading and the dimensions are all symmetrical about a vertical plane passing through its center, Fig. 5–10b. Here the connection at A is assumed to be a pin, and the cable supports the platform at B. The direction of the cable and average dimensions of the platform are listed, and the center of gravity G has been determined. It is from this model that we must proceed to draw the free-body diagram, which is shown in Fig. 5–10c. The platform's weight is $200(9.81) = 1962$ N. The force components A_x and A_y along with the cable force T represent the reactions that both pins and both cables exert on the platform, Fig. 5–10a. Consequently, after the solution for these reactions, half their magnitude is developed at A and half is developed at B.

(c)

EXAMPLE 5.5

The free-body diagram of each object in Fig. 5–11 is drawn. Carefully study each solution and identify what each loading represents, as was done in Fig. 5–7b.

SOLUTION

(a)

(c)

(b)

Note: Internal forces of one member on another are equal but opposite collinear forces which are not to be included here since they cancel out.

(d)

Fig. 5–11

PROBLEMS

5–1. Draw the free-body diagram of the 10-lb sphere resting between the smooth inclined planes. Explain the significance of each force on the diagram. (See Fig. 5–7b.)

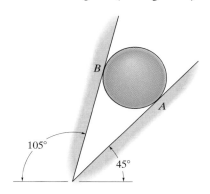

Prob. 5–1

5–2. Draw the free-body diagram of the hand punch, which is pinned at A and bears down on the smooth surface at B.

Prob. 5–2

5–3. Draw the free-body diagram of the beam supported at A by a fixed support and at B by a roller. Explain the significance of each force on the diagram. (See Fig. 5–7b.)

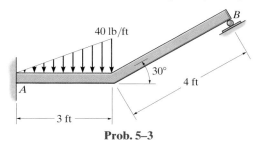

Prob. 5–3

*5–4. Draw the free-body diagram of the jib crane AB, which is pin-connected at A and supported by member (link) BC.

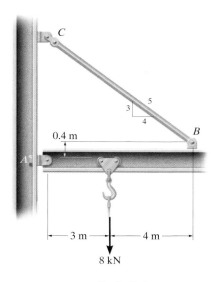

Prob. 5–4

5–5. Draw the free-body diagram of the C-bracket supported at A, B, and C by rollers. Explain the significance of each force on the diagram. (See Fig. 5–7b.)

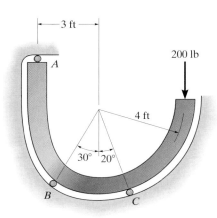

Prob. 5–5

5–6. Draw the free-body diagram of the smooth 20-g rod which rests inside the glass. Explain the significance of each force on the diagram. (See Fig. 5–7b.)

Prob. 5–6

5–7. Draw the free-body diagram of the "spanner wrench" subjected to the 20-lb force. The support at *A* can be considered a pin, and the surface of contact at *B* is smooth. Explain the significance of each force on the diagram. (See Fig. 5–7b.)

Prob. 5–7

*5–8. Draw the free-body diagram of the automobile, which is being towed at constant velocity up the incline using the cable at *C*. The automobile has a mass of 5 Mg and center of mass at *G*. The tires are free to roll. Explain the significance of each force on the diagram. (See Fig. 5–7b.)

Prob. 5–8

5–9. Draw the free-body diagram of the uniform bar, which has a mass of 100 kg and a center of mass at *G*. The supports *A*, *B*, and *C* are smooth.

Prob. 5–9

5–10. Draw the free-body diagram of the beam, which is pin-connected at *A* and rocker-supported at *B*.

Prob. 5–10

5.3 Equations of Equilibrium

In Sec. 5.1 we developed the two equations which are both necessary and sufficient for the equilibrium of a rigid body, namely, $\Sigma \mathbf{F} = \mathbf{0}$ and $\Sigma \mathbf{M}_O = \mathbf{0}$. When the body is subjected to a system of forces, which all lie in the x–y plane, then the forces can be resolved into their x and y components. Consequently, the conditions for equilibrium in two dimensions are

$$\begin{aligned} \Sigma F_x &= 0 \\ \Sigma F_y &= 0 \\ \Sigma M_o &= 0 \end{aligned} \tag{5–2}$$

Here ΣF_x and ΣF_y represent, respectively, the algebraic sums of the x and y components of all the forces acting on the body, and ΣM_O represents the algebraic sum of the couple moments and the moments of all the force components about an axis perpendicular to the x–y plane and passing through the arbitrary point O, which may lie either on or off the body.

Alternative Sets of Equilibrium Equations.

Although Eqs. 5–2 are *most often* used for solving coplanar equilibrium problems, two *alternative* sets of three independent equilibrium equations may also be used. One such set is

$$\begin{aligned} \Sigma F_a &= 0 \\ \Sigma M_A &= 0 \\ \Sigma M_B &= 0 \end{aligned} \tag{5–3}$$

When using these equations it is required that a line passing through points A and B is *not perpendicular* to the a axis. To prove that Eqs. 5–3 provide the *conditions* for equilibrium, consider the free-body diagram of an arbitrarily shaped body shown in Fig. 5–12a. Using the methods of Sec. 4.8, all the forces on the free-body diagram may be replaced by an equivalent resultant force $\mathbf{F}_R = \Sigma \mathbf{F}$, acting at point A, and a resultant couple moment $\mathbf{M}_{R_A} = \Sigma \mathbf{M}_A$, Fig. 5–12$b$. If $\Sigma M_A = 0$ is satisfied, it is necessary that $\mathbf{M}_{R_A} = \mathbf{0}$. Furthermore, in order that $\mathbf{F}_R$ satisfy $\Sigma F_a = 0$, it must have *no component* along the a axis, and therefore its line of action must be perpendicular to the a axis, Fig. 5–12c. Finally, if it is required that $\Sigma M_B = 0$, where B does not lie on the line of action of $\mathbf{F}_R$, then $\mathbf{F}_R = \mathbf{0}$. Since $\Sigma \mathbf{F} = \mathbf{0}$ and $\Sigma \mathbf{M}_A = \mathbf{0}$, indeed the body in Fig. 5–12a must be in equilibrium.

(a)

(b)

(c)

Fig. 5–12

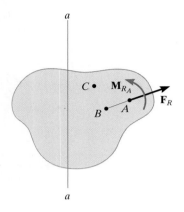

a

a

Fig. 5–13

A second alternative set of equilibrium equations is

$$\Sigma M_A = 0$$
$$\Sigma M_B = 0 \qquad\qquad (5\text{–}4)$$
$$\Sigma M_C = 0$$

Here it is necessary that points A, B, and C do not lie on the same line. To prove that these equations, when satisfied, ensure equilibrium, consider the free-body diagram in Fig. 5–13. If $\Sigma M_A = 0$ is to be satisfied, then $\mathbf{M}_{R_A} = \mathbf{0}$. $\Sigma M_B = 0$ is satisfied if the line of action of $\mathbf{F}_R$ passes through point B as shown. Finally, if we require $\Sigma M_C = 0$, where C does not lie on line AB, it is necessary that $\mathbf{F}_R = \mathbf{0}$, and the body in Fig. 5–12*a* must then be in equilibrium.

Procedure for Analysis

Coplanar force equilibrium problems for a rigid body can be solved using the following procedure.

Free-Body Diagram.

- Establish the x, y coordinate axes in any suitable orientation.
- Draw an outlined shape of the body.
- Show all the forces and couple moments acting on the body.
- Label all the loadings and specify their directions relative to the x, y axes. The sense of a force or couple moment having an *unknown* magnitude but known line of action can be *assumed*.
- Indicate the dimensions of the body necessary for computing the moments of forces.

Equations of Equilibrium.

- Apply the moment equation of equilibrium, $\Sigma M_O = 0$, about a point (O) that lies at the intersection of the lines of action of two unknown forces. In this way, the moments of these unknowns are zero about O, and a *direct solution* for the third unknown can be determined.
- When applying the force equilibrium equations, $\Sigma F_x = 0$ and $\Sigma F_y = 0$, orient the x and y axes along lines that will provide the simplest resolution of the forces into their x and y components.
- If the solution of the equilibrium equations yields a negative scalar for a force or couple moment magnitude, this indicates that the sense is opposite to that which was assumed on the free-body diagram.

EXAMPLE 5.6

Determine the horizontal and vertical components of reaction on the beam caused by the pin at B and the rocker at A as shown in Fig. 5–14a. Neglect the weight of the beam.

Fig. 5–14

SOLUTION

Free-Body Diagram. Identify each of the forces shown on the free-body diagram of the beam, Fig. 5–14b. (See Example 5.1.) For simplicity, the 600-N force is represented by its x and y components as shown. Also, note that a 200-N force acts on the beam at B and is independent of the force components $\mathbf{B}_x$ and $\mathbf{B}_y$, which represent the effect of the pin on the beam.

Equations of Equilibrium. Summing forces in the x direction yields

$$\xrightarrow{+} \Sigma F_x = 0; \qquad 600 \cos 45° \text{ N} - B_x = 0$$
$$B_x = 424 \text{ N} \qquad\qquad Ans.$$

A direct solution for $\mathbf{A}_y$ can be obtained by applying the moment equation $\Sigma M_B = 0$ about point B. For the calculation, it should be apparent that forces 200 N, $\mathbf{B}_x$, and $\mathbf{B}_y$ all create zero moment about B. Assuming counterclockwise rotation about B to be positive (in the $+\mathbf{k}$ direction), Fig. 5–14b, we have

$$\curvearrowleft + \Sigma M_B = 0; \qquad 100 \text{ N}(2 \text{ m}) + (600 \sin 45° \text{ N})(5 \text{ m})$$
$$- (600 \cos 45° \text{ N})(0.2 \text{ m}) - A_y(7 \text{ m}) = 0$$
$$A_y = 319 \text{ N} \qquad\qquad Ans.$$

Summing forces in the y direction, using this result, gives

$$+\uparrow \Sigma F_y = 0; \quad 319 \text{ N} - 600 \sin 45° \text{ N} - 100 \text{ N} - 200 \text{ N} + B_y = 0$$
$$B_y = 405 \text{ N} \qquad\qquad Ans.$$

NOTE: We can check this result by summing moments about point A.

$$\curvearrowleft + \Sigma M_A = 0; \; -(600 \sin 45° \text{ N})(2 \text{ m}) - (600 \cos 45° \text{ N})(0.2 \text{ m})$$
$$-(100 \text{ N})(5 \text{ m}) - (200 \text{ N})(7 \text{ m}) + B_y(7 \text{ m}) = 0$$
$$B_y = 405 \text{ N} \qquad\qquad Ans.$$

EXAMPLE 5.7

The cord shown in Fig. 5–15a supports a force of 100 lb and wraps over the frictionless pulley. Determine the tension in the cord at C and the horizontal and vertical components of reaction at pin A.

0.5 ft

A

$\theta = 30°$

C

100 lb

(a)

Fig. 5–15

SOLUTION

Free-Body Diagrams. The free-body diagrams of the cord and pulley are shown in Fig. 5–15b. Note that the principle of action, equal but opposite reaction must be carefully observed when drawing each of these diagrams: the cord exerts an unknown load distribution p along part of the pulley's surface, whereas the pulley exerts an equal but opposite effect on the cord. For the solution, however, it is simpler to *combine* the free-body diagrams of the pulley and the contacting portion of the cord, so that the distributed load becomes *internal* to the system and is therefore eliminated from the analysis, Fig. 5–15c.

Equations of Equilibrium. Summing moments about point A to eliminate $\mathbf{A}_x$ and $\mathbf{A}_y$, Fig. 5–15c, we have

$$\curvearrowleft + \Sigma M_A = 0; \qquad 100 \text{ lb } (0.5 \text{ ft}) - T(0.5 \text{ ft}) = 0$$

$$T = 100 \text{ lb} \qquad\qquad Ans.$$

Using the result for T, a force summation is applied to determine the components of reaction at pin A.

$$\xrightarrow{+} \Sigma F_x = 0; \qquad -A_x + 100 \sin 30° \text{ lb} = 0$$

$$A_x = 50.0 \text{ lb} \qquad\qquad Ans.$$

$$+\uparrow \Sigma F_y = 0; \qquad A_y - 100 \text{ lb} - 100 \cos 30° \text{ lb} = 0$$

$$A_y = 187 \text{ lb} \qquad\qquad Ans.$$

NOTE: It is seen that the tension remains *constant* as the cord passes over the pulley. (This of course is true for *any angle* θ at which the cord is directed and for *any radius* r of the pulley.)

p

p

30°

100 lb

A

$\mathbf{A}_x$

$\mathbf{A}_y$

T

(b)

0.5 ft

A

$\mathbf{A}_x$

$\mathbf{A}_y$

100 lb

$\theta = 30°$

T

y

x

(c)

EXAMPLE 5.8

The link shown in Fig. 5–16a is pin-connected at A and rests against a smooth support at B. Compute the horizontal and vertical components of reaction at the pin A.

(a)

(b)

Fig. 5–16

SOLUTION

Free-Body Diagram.   As shown in Fig. 5–16b, the reaction $\mathbf{N}_B$ is perpendicular to the link at B. Also, horizontal and vertical components of reaction are represented at A.

Equations of Equilibrium.   Summing moments about A, we obtain a direct solution for N_B,

$$\zeta + \Sigma M_A = 0; \quad -90 \text{ N} \cdot \text{m} - 60 \text{ N}(1 \text{ m}) + N_B(0.75 \text{ m}) = 0$$
$$N_B = 200 \text{ N}$$

Using this result,

$$\xrightarrow{+} \Sigma F_x = 0; \qquad A_x - 200 \sin 30° \text{ N} = 0$$
$$A_x = 100 \text{ N} \qquad\qquad Ans.$$
$$+\uparrow \Sigma F_y = 0; \qquad A_y - 200 \cos 30° \text{ N} - 60 \text{ N} = 0$$
$$A_y = 233 \text{ N} \qquad\qquad Ans.$$

EXAMPLE 5.9

The box wrench in Fig. 5–17a is used to tighten the bolt at A. If the wrench does not turn when the load is applied to the handle, determine the torque or moment applied to the bolt and the force of the wrench on the bolt.

SOLUTION

Free-Body Diagram. The free-body diagram for the wrench is shown in Fig. 5–17b. Since the bolt acts as a "fixed support," it exerts force components A_x and A_y and a torque M_A on the wrench at A.

Equations of Equilibrium.

$$\xrightarrow{+} \Sigma F_x = 0; \qquad A_x - 52\left(\tfrac{5}{13}\right)\text{N} + 30\cos 60° \text{ N} = 0$$
$$A_x = 5.00 \text{ N} \qquad\qquad Ans.$$

$$+\uparrow \Sigma F_y = 0; \qquad A_y - 52\left(\tfrac{12}{13}\right)\text{N} - 30\sin 60° \text{ N} = 0$$
$$A_y = 74.0 \text{ N} \qquad\qquad Ans.$$

$$\downarrow + \Sigma M_A = 0; \quad M_A - 52\left(\tfrac{12}{13}\right)\text{N}(0.3 \text{ m}) - (30\sin 60° \text{ N})(0.7 \text{ m}) = 0$$
$$M_A = 32.6 \text{ N} \cdot \text{m} \qquad\qquad Ans.$$

Point A was chosen for summing moments because the lines of action of the *unknown* forces A_x and A_y pass through this point, and therefore these forces were not included in the moment summation. Realize, however, that M_A must be *included* in this moment summation. This couple moment is a free vector and represents the twisting resistance of the bolt on the wrench. By Newton's third law, the wrench exerts an equal but opposite moment or torque on the bolt. Furthermore, the resultant force on the wrench is

$$F_A = \sqrt{(5.00)^2 + (74.0)^2} = 74.1 \text{ N} \qquad\qquad Ans.$$

Because the force components A_x and A_y were calculated as positive quantities, their directional sense is shown correctly on the free-body diagram in Fig. 5–17b. Hence

$$\theta = \tan^{-1}\frac{74.0 \text{ N}}{5.00 \text{ N}} = 86.1° \angle \theta$$

Realize that $\mathbf{F}_A$ acts in the opposite direction on the bolt. Why?

NOTE: Although only *three* independent equilibrium equations can be written for a rigid body, it is a good practice to *check* the calculations using a fourth equilibrium equation. For example, the above computations may be verified in part by summing moments about point C:

$$\downarrow + \Sigma M_C = 0; \quad 52\left(\tfrac{12}{13}\right)\text{N}(0.4 \text{ m}) + 32.6 \text{ N} \cdot \text{m} - 74.0 \text{ N}(0.7 \text{ m}) = 0$$
$$19.2 \text{ N} \cdot \text{m} + 32.6 \text{ N} \cdot \text{m} - 51.8 \text{ N} \cdot \text{m} = 0$$

(a)

(b)

Fig. 5–17

EXAMPLE 5.10

Placement of concrete from the truck is accomplished using the chute shown in the photos, Fig. 5–18a. Determine the force that the hydraulic cylinder and the truck frame exert on the chute to hold it in the position shown. The chute and wet concrete contained along its length have a uniform weight of 35 lb/ft.

SOLUTION

The idealized model of the chute is shown in Fig. 5–18b. Here the dimensions are given, and it is assumed the chute is pin connected to the frame at A and the hydraulic cylinder BC acts as a short link.

(b)

(c)

(a)

Fig. 5–18

Free-Body Diagram.   Since the chute has a length of 16 ft, the total supported weight is $(35 \text{ lb/ft})(16 \text{ ft}) = 560 \text{ lb}$, which is assumed to act at its midpoint, G. The hydraulic cylinder exerts a horizontal force $\mathbf{F}_{BC}$ on the chute, Fig. 5–18c.

Equations of Equilibrium.   A direct solution for $\mathbf{F}_{BC}$ is possible by summing moments about the pin at A. To do this we will use the principle of moments and resolve the weight into components parallel and perpendicular to the chute. We have,

$\zeta + \Sigma M_A = 0;$
$$-F_{BC}(2 \text{ ft}) + 560 \cos 30° \text{ lb } (8 \text{ ft}) + 560 \sin 30° \text{ lb } (0.25 \text{ ft}) = 0$$
$$F_{BC} = 1975 \text{ lb} \qquad \qquad \textit{Ans.}$$

Summing forces to obtain A_x and A_y, we obtain

$\xrightarrow{+} \Sigma F_x = 0;$
$$-A_x + 1975 \text{ lb} = 0$$
$$A_x = 1975 \text{ lb} \qquad \qquad \textit{Ans.}$$

$+\uparrow \Sigma F_y = 0;$
$$A_y - 560 \text{ lb} = 0$$
$$A_y = 560 \text{ lb} \qquad \qquad \textit{Ans.}$$

NOTE: To verify this solution we can sum moments about point B.

$\zeta + \Sigma M_B = 0; \qquad -1975 \text{ lb } (2 \text{ ft}) + 560 \text{ lb } (4 \cos 30° \text{ ft}) +$
$$560 \cos 30° \text{ lb } (4 \text{ ft}) + 560 \sin 30° \text{ lb } (0.25 \text{ ft}) = 0$$

EXAMPLE 5.11

The uniform smooth rod shown in Fig. 5–19a is subjected to a force and couple moment. If the rod is supported at A by a smooth wall and at B and C either at the top or bottom by rollers, determine the reactions at these supports. Neglect the weight of the rod.

(a)

SOLUTION

Free-Body Diagram.   As shown in Fig. 5–19b, all the support reactions act normal to the surface of contact since the contacting surfaces are smooth. The reactions at B and C are shown acting in the positive y' direction. This assumes that only the rollers located on the bottom of the rod are used for support.

Equations of Equilibrium.   Using the x, y coordinate system in Fig. 5–19b, we have

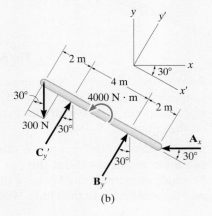

$$\xrightarrow{+}\ \Sigma F_x = 0; \quad C_{y'} \sin 30° + B_{y'} \sin 30° - A_x = 0 \tag{1}$$

$$+\uparrow \Sigma F_y = 0; \quad -300\ \text{N} + C_{y'} \cos 30° + B_{y'} \cos 30° = 0 \tag{2}$$

$$\curvearrowright{+}\Sigma M_A = 0; \quad -B_{y'}(2\ \text{m}) + 4000\ \text{N} \cdot \text{m} - C_{y'}(6\ \text{m})$$
$$+ (300 \cos 30°\ \text{N})(8\ \text{m}) = 0 \tag{3}$$

When writing the moment equation, it should be noticed that the line of action of the force component $300 \sin 30°$ N passes through point A, and therefore this force is not included in the moment equation.
    Solving Eqs. 2 and 3 simultaneously, we obtain

$$B_{y'} = -1000.0\ \text{N} = -1\ \text{kN} \qquad \qquad \textit{Ans.}$$

$$C_{y'} = 1346.4\ \text{N} = 1.35\ \text{kN} \qquad \qquad \textit{Ans.}$$

Since $B_{y'}$ is a negative scalar, the sense of $\mathbf{B}_{y'}$ is opposite to that shown on the free-body diagram in Fig. 5–19b. Therefore, the top roller at B serves as the support rather than the bottom one. Retaining the negative sign for $B_{y'}$ (Why?) and substituting the results into Eq. 1, we obtain

$$1346.4 \sin 30°\ \text{N} - 1000.0 \sin 30°\ \text{N} - A_x = 0$$

$$A_x = 173\ \text{N} \qquad \qquad \textit{Ans.}$$

Fig. 5–19

EXAMPLE 5.12

(a)

The uniform truck ramp shown in Fig. 5–20a has a weight of 400 lb and is pinned to the body of the truck at each end and held in the position shown by the two side cables. Determine the tension in the cables.

SOLUTION

The idealized model of the ramp, which indicates all necessary dimensions and supports, is shown in Fig. 5–20b. Here the center of gravity is located at the midpoint since the ramp is approximately uniform.

Free-Body Diagram. Working from the idealized model, the ramp's free-body diagram is shown in Fig. 5–20c.

Equations of Equilibrium. Summing moments about point A will yield a direct solution for the cable tension. Using the principle of moments, there are several ways of determining the moment of $\mathbf{T}$ about A. If we use x and y components, with $\mathbf{T}$ applied at B, we have

(b)

$$\zeta + \Sigma M_A = 0; \quad -T\cos 20°(7\sin 30°\ \text{ft}) + T\sin 20°(7\cos 30°\ \text{ft})$$
$$+ 400\ \text{lb}\,(5\cos 30°\ \text{ft}) = 0$$
$$T = 1425\ \text{lb}$$

By the principle of transmissibility, we can locate $\mathbf{T}$ at C, even though this point is not on the ramp, Fig. 5–20c. In this case the horizontal component of $\mathbf{T}$ does not create a moment about A. First we must determine d using the sine law.

$$\frac{d}{\sin 10°} = \frac{7\ \text{ft}}{\sin 20°}; \qquad d = 3.554\ \text{ft}$$

$$\zeta + \Sigma M_A = 0; \quad -T\sin 20°(3.554\ \text{ft}) + 400\ \text{lb}\,(5\cos 30°\ \text{ft}) = 0$$

$$T = 1425\ \text{lb}$$

The simplest way to compute the moment of $\mathbf{T}$ about A is to resolve it into components parallel and perpendicular to the ramp at B. Then the moment of the parallel component is zero about A, so that

$$\zeta + \Sigma M_A = 0; \quad -T\sin 10°(7\ \text{ft}) + 400\ \text{lb}\,(5\cos 30°\ \text{ft}) = 0$$

$$T = 1425\ \text{lb}$$

Since there are two cables supporting the ramp,

$$T' = \frac{T}{2} = 712\ \text{lb} \qquad\qquad Ans.$$

Fig. 5–20   (c)

NOTE: As an exercise, show that $A_x = 1339$ lb and $A_y = 887.4$ lb.

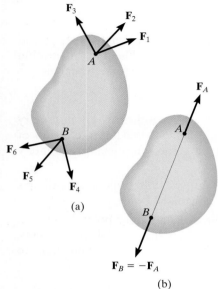

(a)

$F_B = -F_A$

(b)

Two-force member

Fig. 5–21

5.4   Two- and Three-Force Members

The solution to some equilibrium problems can be simplified by recognizing members that are subjected to only two or three forces.

Two-Force Members.   When a member is subjected to *no couple moments* and forces are applied at only two points on the member, the member is called a *two-force member*. An example is shown in Fig. 5–21a. The forces at A and B are summed to obtain their respective *resultants* F_A and F_B, Fig. 5–21b. These two forces will maintain *translational or force equilibrium* ($\Sigma F = 0$) provided F_A is of equal magnitude and opposite direction to F_B. Furthermore, *rotational or moment equilibrium* ($\Sigma M_O = 0$) is satisfied if F_A is *collinear* with F_B. As a result, the line of action of both forces is known since it always passes through A and B. Hence, only the force magnitude must be determined or stated. Other examples of two-force members held in equilibrium are shown in Fig. 5–22.

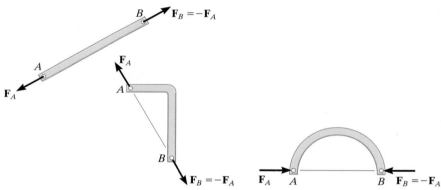

Two-force members

Fig. 5–22

Three-Force Members.   If a member is subjected to only three forces, then it is necessary that the forces be either *concurrent* or *parallel* for the member to be in equilibrium. To show the concurrency requirement, consider the body in Fig. 5–23a and suppose that any two of the three forces acting on the body have lines of action that intersect at point O. To satisfy moment equilibrium about O, i.e., $\Sigma M_O = 0$, the third force must also pass through O, which then makes the force system *concurrent*. If two of the three forces are parallel, Fig. 5–23b, the point of concurrency, O, is considered to be at "infinity," and the third force must be parallel to the other two forces to intersect at this "point."

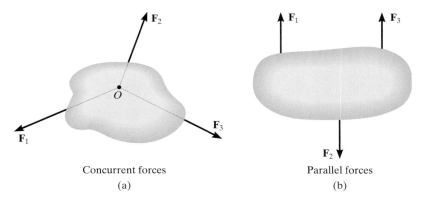

Concurrent forces
(a)

Parallel forces
(b)

Three-force members

Fig. 5–23

Many mechanical elements act as two- or three-force members, and the ability to recognize them in a problem will considerably simplify an equilibrium analysis.

- The bucket link AB on the back-hoe is a typical example of a two-force member since it is pin connected at its ends and, provided its weight is neglected, no other force acts on this member.

- The hydraulic cylinder BC is pin connected at its ends. It is a two-force member. The boom ABD is subjected to the weight of the suspended motor at D, the force of the hydraulic cylinder at B, and the force of the pin at A. If the boom's weight is neglected, it is a three-force member.

- The dump bed of the truck operates by extending the telescopic hydraulic cylinder AB. If the weight of AB is neglected, we can classify it as a two-force member since it is pin connected at its end points.

EXAMPLE 5.13

(a)

(b)

(c)

Fig. 5–24

The lever ABC is pin-supported at A and connected to a short link BD as shown in Fig. 5–24a. If the weight of the members is negligible, determine the force of the pin on the lever at A.

SOLUTION

Free-Body Diagrams.   As shown by the free-body diagram, Fig. 5–24b, the short link BD is a *two-force member*, so the *resultant forces* at pins D and B must be equal, opposite, and collinear. Although the magnitude of the force is unknown, the line of action is known since it passes through B and D.

Lever ABC is a *three-force member*, and therefore, in order to satisfy moment equilibrium, the three nonparallel forces acting on it must be concurrent at O, Fig. 5–24c. In particular, note that the force F on the lever at B is equal but opposite to the force F acting at B on the link. Why? The distance CO must be 0.5 m since the lines of action of **F** and the 400-N force are known.

Equations of Equilibrium.   By requiring the force system to be concurrent at O, since $\Sigma M_O = 0$, the angle θ which defines the line of action of $\mathbf{F}_A$ can be determined from trigonometry,

$$\theta = \tan^{-1}\left(\frac{0.7}{0.4}\right) = 60.3° \quad \angle\theta \qquad Ans.$$

Using the x, y axes and applying the force equilibrium equations, we can obtain F_A and F.

$$\xrightarrow{+} \Sigma F_x = 0; \qquad F_A \cos 60.3° - F \cos 45° + 400 \text{ N} = 0$$

$$+\uparrow \Sigma F_y = 0; \qquad F_A \sin 60.3° - F \sin 45° = 0$$

Solving, we get

$$F_A = 1.07 \text{ kN} \qquad\qquad Ans.$$
$$F = 1.32 \text{ kN}$$

NOTE: We can also solve this problem by representing the force at A by its two components $\mathbf{A}_x$ and $\mathbf{A}_y$ and applying $\Sigma M_A = 0$, $\Sigma F_x = 0$, $\Sigma F_y = 0$ to the lever. Once A_x and A_y are determined, how would you find F_A and θ?

PROBLEMS

5–11. Determine the reactions at the supports in Prob. 5–1.

*5–12. Determine the magnitude of the resultant force acting at pin A of the handpunch in Prob. 5–2.

5–13. Determine the reactions at the supports at C in Prob. 5–5.

5–14. Determine the reactions on the rod in Prob. 5–6.

5–15. Determine the support reactions on the spanner wrench in Prob. 5–7.

*5–16. Determine the reactions on the car in Prob. 5–8.

5–17. Determine the reactions at the points of contact at A, B, and C of the bar in Prob. 5–9.

5–18. Determine the reactions at the pin A and at the roller at B of the beam in Prob. 5–10.

5–19. Determine the magnitude of the reactions on the beam at A and B. Neglect the thickness of the beam.

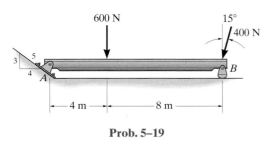

Prob. 5–19

*5–20. Determine the reactions at the supports.

Prob. 5–20

5–21. When holding the 5-lb stone in equilibrium, the humerus H, assumed to be smooth, exerts normal forces $\mathbf{F}_C$ and $\mathbf{F}_A$ on the radius C and ulna A as shown. Determine these forces and the force $\mathbf{F}_B$ that the biceps B exerts on the radius for equilibrium. The stone has a center of mass at G. Neglect the weight of the arm.

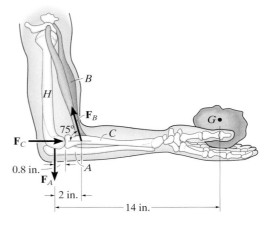

Prob. 5–21

5–22. The uniform door has a weight of 100 lb and a center of gravity at G. Determine the reactions at the hinges if the hinge at A supports only a horizontal reaction on the door, whereas the hinge at B exerts both horizontal and vertical reactions.

Prob. 5–22

5–23. The ramp of a ship has a weight of 200 lb and a center of gravity at G. Determine the cable force in CD needed to just start lifting the ramp, (i.e., so the reaction at B becomes zero). Also, determine the horizontal and vertical components of force at the hinge (pin) at A.

5–25. While *slowly* walking, a man having a total mass of 80 kg places all his weight on *one foot*. Assuming that the normal force $\mathbf{N}_C$ of the ground acts on his foot at C, determine the resultant vertical compressive force $\mathbf{F}_B$ which the tibia T exerts on the astragalus B, and the vertical tension $\mathbf{F}_A$ in the achilles tendon A at the instant shown.

Prob. 5–23

Prob. 5–25

*■5–24. The 1.4-Mg drainpipe is held in the tines of the fork lift. Determine the normal forces at A and B as functions of the blade angle θ and plot the results of force (ordinate) versus θ (abscissa) for $0 \le \theta \le 90°$.

5–26. Determine the reactions at the roller A and pin B.

Prob. 5–24

Prob. 5–26

5–27. The platform assembly has a weight of 250 lb and center of gravity at G_1. If it is intended to support a maximum load of 400 lb placed at point G_2, determine the smallest counterweight W that should be placed at B in order to prevent the platform from tipping over.

Prob. 5–27

5–29. The device is used to hold an elevator door open. If the spring has a stiffness of $k = 40$ N/m and it is compressed 0.2 m, determine the horizontal and vertical components of reaction at the pin A and the resultant force at the wheel bearing B.

Prob. 5–29

*5–28. The articulated crane boom has a weight of 125 lb and mass center at G. If it supports a load of 600 lb, determine the force acting at the pin A and the compression in the hydraulic cylinder BC when the boom is in the position shown.

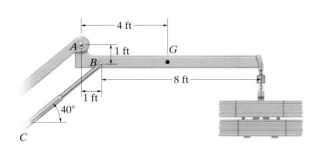

Prob. 5–28

5–30. Determine the reactions on the bent rod which is supported by a smooth surface at B and by a collar at A, which is fixed to the rod and is free to slide over the fixed inclined rod.

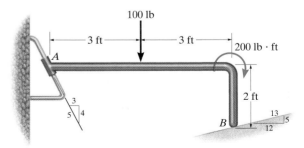

Prob. 5–30

5–31. The cantilevered jib crane is used to support the load of 780 lb. If the trolley T can be placed anywhere between $1.5 \text{ ft} \leq x \leq 7.5 \text{ ft}$, determine the maximum magnitude of reaction at the supports A and B. Note that the supports are collars that allow the crane to rotate freely about the vertical axis. The collar at B supports a force in the vertical direction, whereas the one at A does not.

5–33. The power pole supports the three lines, each line exerting a vertical force on the pole due to its weight as shown. Determine the reactions at the fixed support D. If it is possible for wind or ice to snap the lines, determine which line(s) when removed create(s) a condition for the greatest moment reaction at D.

Prob. 5–31

Prob. 5–33

*5–32. The uniform rod AB has a weight of 15 lb. Determine the force in the cable when the rod is in the position shown.

5–34. The picnic table has a weight of 50 lb and a center of gravity at G_T. If a man weighing 225 lb has a center of gravity at G_M and sits down in the centered position shown, determine the vertical reaction at each of the two legs at B. Neglect the thickness of the legs. What can you conclude from the results?

Prob. 5–32

Prob. 5–34

5–35. If the wheelbarrow and its contents have a mass of 60 kg and center of mass at G, determine the magnitude of the resultant force which the man must exert on *each* of the two handles in order to hold the wheelbarrow in equilibrium.

Prob. 5–35

5–37. When no force is applied to the brake pedal of the lightweight truck, the retainer spring AB keeps the pedal in contact with the smooth brake light switch at C. If the force on the switch is 3 N, determine the unstretched length of the spring if the stiffness of the spring is $k = 80$ N/m.

Prob. 5–37

*5–36.** The man has a weight W and stands at the center of the plank. If the planes at A and B are smooth, determine the tension in the cord in terms of W and θ.

Prob. 5–36

5–38. The telephone pole of negligible thickness is subjected to the force of 80 lb directed as shown. It is supported by the cable BCD and can be assumed pinned at its base A. In order to provide clearance for a sidewalk right of way, where D is located, the strut CE is attached at C, as shown by the dashed lines (cable segment CD is removed). If the tension in CD' is to be twice the tension in BCD, determine the height h for placement of the strut CE.

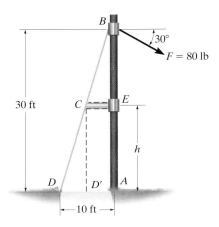

Prob. 5–38

5–39. The worker uses the hand truck to move material down the ramp. If the truck and its contents are held in the position shown and have a weight of 100 lb with center of gravity at G, determine the resultant normal force of both wheels on the ground A and the magnitude of the force required at the grip B.

Prob. 5–39

*5–40. Soil pressure acting on the concrete retaining wall is represented as a loading per foot length of wall. If concrete has a specific weight of 150 lb/ft³, determine the magnitudes of the soil distribution, w_1 and w_2, and the frictional force **F** for equilibrium.

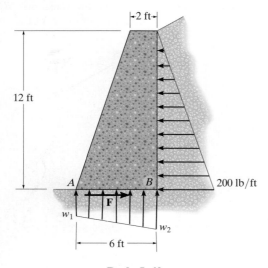

Prob. 5–40

5–41. The shelf supports the electric motor which has a mass of 15 kg and mass center at G_m. The platform upon which it rests has a mass of 4 kg and mass center at G_p. Assuming that a single bolt B holds the shelf up and the bracket bears against the smooth wall at A, determine this normal force at A and the horizontal and vertical components of reaction of the bolt on the bracket.

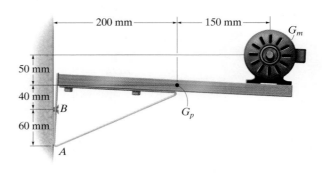

Prob. 5–41

5–42. A cantilever beam, having an extended length of 3 m, is subjected to a vertical force of 500 N. Assuming that the wall resists this load with linearly varying distributed loads over the 0.15-m length of the beam portion inside the wall, determine the intensities w_1 and w_2 for equilibrium.

Prob. 5–42

5–43. The upper portion of the crane boom consists of the jib AB, which is supported by the pin at A, the guy line BC, and the backstay CD, each cable being separately attached to the mast at C. If the 5-kN load is supported by the hoist line, which passes over the pulley at B, determine the magnitude of the resultant force the pin exerts on the jib at A for equilibrium, the tension in the guy line BC, and the tension T in the hoist line. Neglect the weight of the jib. The pulley at B has a radius of 0.1 m.

Prob. 5–43

5–46. The man attempts to support the load of boards having a weight W and a center of gravity at G. If he is standing on a smooth floor, determine the smallest angle θ at which he can hold them up in the position shown. Neglect his weight.

Prob. 5–46

*5–44. The mobile crane has a weight of 120,000 lb and center of gravity at G_1; the boom has a weight of 30,000 lb and center of gravity at G_2. Determine the smallest angle of tilt θ of the boom, without causing the crane to overturn if the suspended load is $W = 40,000$ lb. Neglect the thickness of the tracks at A and B.

5–45. The mobile crane has a weight of 120,000 lb and center of gravity at G_1; the boom has a weight of 30,000 lb and center of gravity at G_2. If the suspended load has a weight of $W = 16,000$ lb, determine the normal reactions at the tracks A and B. For the calculation, neglect the thickness of the tracks and take $\theta = 30°$.

Probs. 5–44/45

5–47. The motor has a weight of 850 lb. Determine the force that each of the chains exerts on the supporting hooks at A, B, and C. Neglect the size of the hooks and the thickness of the beam.

Prob. 5–47

*5–48. The boom supports the two vertical loads. Neglect the size of the collars at D and B and the thickness of the boom, and compute the horizontal and vertical components of force at the pin A and the force in cable CB. Set $F_1 = 800$ N and $F_2 = 350$ N.

5–49. The boom is intended to support two vertical loads, F_1 and F_2. If the cable CB can sustain a maximum load of 1500 N before it fails, determine the critical loads if $F_1 = 2F_2$. Also, what is the magnitude of the maximum reaction at pin A?

5–51. The toggle switch consists of a cocking lever that is pinned to a fixed frame at A and held in place by the spring which has an unstretched length of 200 mm. Determine the magnitude of the resultant force at A and the normal force on the peg at B when the lever is in the position shown.

Prob. 5–51

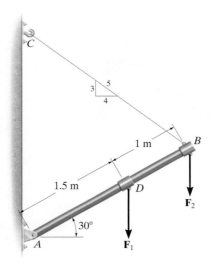

Probs. 5–48/49

*5–52. The rigid beam of negligible weight is supported horizontally by two springs and a pin. If the springs are uncompressed when the load is removed, determine the force in each spring when the load P is applied. Also, compute the vertical deflection of end C. Assume the spring stiffness k is large enough so that only small deflections occur. *Hint:* The beam rotates about A so the deflections in the springs can be related.

5–50. The uniform rod of length L and weight W is supported on the smooth planes. Determine its position θ for equilibrium. Neglect the thickness of the rod.

Prob. 5–50

Prob. 5–52

5–53. The rod supports a weight of 200 lb and is pinned at its end A. If it is also subjected to a couple moment of 100 lb · ft, determine the angle θ for equilibrium. The spring has an unstretched length of 2 ft and a stiffness of $k = 50$ lb/ft.

5–55. The rigid metal strip of negligible weight is used as part of an electromagnetic switch. If the stiffness of the springs at A and B is $k = 5$ N/m, and the strip is originally horizontal when the springs are unstretched, determine the smallest force needed to close the contact gap at C.

*5–56. The rigid metal strip of negligible weight is used as part of an electromagnetic switch. Determine the maximum stiffness k of the springs at A and B so that the contact at C closes when the vertical force developed there is 0.5 N. Originally the strip is horizontal as shown.

Prob. 5–53

Probs. 5–55/56

5–54. The smooth pipe rests against the wall at the points of contact A, B, and C. Determine the reactions at these points needed to support the vertical force of 45 lb. Neglect the pipe's thickness in the calculation.

5–57. Determine the distance d for placement of the load **P** for equilibrium of the smooth bar in the position θ as shown. Neglect the weight of the bar.

Prob. 5–54

Prob. 5–57

5–58. The wheelbarrow and its contents have a mass m and center of mass at G. Determine the greatest angle of tilt θ without causing the wheelbarrow to tip over.

Prob. 5–58

5–59. Determine the force P needed to pull the 50-kg roller over the smooth step. Take $\theta = 60°$.

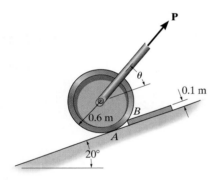

Prob. 5–59

*5–60. Determine the magnitude and direction θ of the minimum force P needed to pull the 50-kg roller over the smooth step.

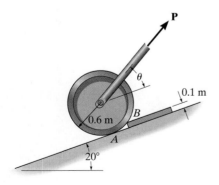

Prob. 5–60

5–61. A uniform glass rod having a length L is placed in the smooth hemispherical bowl having a radius r. Determine the angle of inclination θ for equilibrium.

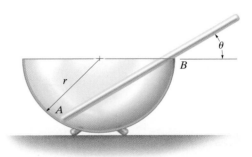

Prob. 5–61

5–62. The disk B has a mass of 20 kg and is supported on the smooth cylindrical surface by a spring having a stiffness of $k = 400$ N/m and unstretched length of $l_0 = 1$ m. The spring remains in the horizontal position since its end A is attached to the small roller guide which has negligible weight. Determine the angle θ to the nearest degree for equilibrium of the roller.

Prob. 5–62

EQUILIBRIUM IN THREE DIMENSIONS

5.5 Free-Body Diagrams

The first step in solving three-dimensional equilibrium problems, as in the case of two dimensions, is to draw a free-body diagram of the body (or group of bodies considered as a system). Before we show this, however, it is necessary to discuss the types of reactions that can occur at the supports.

Support Reactions.   The reactive forces and couple moments acting at various types of supports and connections, when the members are viewed in three dimensions, are listed in Table 5–2. It is important to recognize the symbols used to represent each of these supports and to understand clearly how the forces and couple moments are developed by each support. As in the two-dimensional case, *a force is developed by a support that restricts the translation of the attached member, whereas a couple moment is developed when rotation of the attached member is prevented*. For example, in Table 5–2, item (4), the ball-and-socket joint prevents any translation of the connecting member; therefore, a force must act on the member at the point of connection. This force has three components having unknown magnitudes, F_x, F_y, F_z. Provided these components are known, one can obtain the magnitude of force. $F = \sqrt{F_x^2 + F_y^2 + F_z^2}$, and the force's orientation defined by the coordinate direction angles α, β, γ, Eqs. 2–7.* Since the connecting member is allowed to rotate freely about *any* axis, no couple moment is resisted by a ball-and-socket joint.

It should be noted that the *single* bearing supports in items (5) and (7), the *single* pin (8), and the *single* hinge (9) are shown to support both force and couple-moment components. If, however, these supports are used in conjunction with *other* bearings, pins, or hinges to hold a rigid body in equilibrium and the supports are *properly aligned* when connected to the body, then the *force reactions* at these supports *alone* may be adequate for supporting the body. In other words, the couple moments become redundant and are not shown on the free-body diagram. The reason for this should become clear after studying the examples which follow.

---

*The three unknowns may also be represented as an unknown force magnitude F and two unknown coordinate direction angles. The third direction angle is obtained using the identity $\cos^2 \alpha + \cos^2 \beta + \cos^2 \gamma = 1$, Eq. 2–10.

TABLE 5–2 Supports for Rigid Bodies Subjected to Three-Dimensional Force Systems

Types of Connection	Reaction	Number of Unknowns
(1) cable	**F**	One unknown. The reaction is a force which acts away from the member in the known direction of the cable.
(2) smooth surface support	**F**	One unknown. The reaction is a force which acts perpendicular to the surface at the point of contact.
(3) roller	**F**	One unknown. The reaction is a force which acts perpendicular to the surface at the point of contact.
(4) ball and socket	$\mathbf{F}_z$ $\mathbf{F}_x$ $\mathbf{F}_y$	Three unknowns. The reactions are three rectangular force components.
(5) single journal bearing	$\mathbf{M}_z$ $\mathbf{F}_z$ $\mathbf{M}_x$ $\mathbf{F}_x$	Four unknowns. The reactions are two force and two couple-moment components which act perpendicular to the shaft.

continued

TABLE 5–2 Continued

Types of Connection	Reaction	Number of Unknowns
(6) single journal bearing with square shaft	M_z F_z M_x M_y F_x	Five unknowns. The reactions are two force and three couple-moment components.
(7) single thrust bearing	M_z F_y F_z M_x F_x	Five unknowns. The reactions are three force and two couple-moment components.
(8) single smooth pin	M_z F_z F_y M_y F_x	Five unknowns. The reactions are three force and two couple-moment components.
(9) single hinge	M_z F_z F_y F_x M_x	Five unknowns. The reactions are three force and two couple-moment components.
(10) fixed support	M_z F_z F_x F_y M_y M_x	Six unknowns. The reactions are three force and three couple-moment components.

Typical examples of actual supports that are referenced to Table 5–2 are shown in the following sequence of photos.

This ball-and-socket joint provides a connection for the housing of an earth grader to its frame. (4)

This journal bearing supports the end of the shaft. (5)

This thrust bearing is used to support the drive shaft on a machine. (7)

This pin is used to support the end of the strut used on a tractor. (8)

Free-Body Diagrams.   The general procedure for establishing the free-body diagram of a rigid body has been outlined in Sec. 5.2. Essentially it requires first "isolating" the body by drawing its outlined shape. This is followed by a careful *labeling* of *all* the forces and couple moments in reference to an established *x, y, z* coordinate system. As a general rule, *components of reaction* having an *unknown magnitude* are shown acting on the free-body diagram in the *positive sense*. In this way, if any negative values are obtained, they will indicate that the components act in the negative coordinate directions.

(a)

(b)

It is a mistake to support a door using a single hinge (a) since the hinge must develop a force $\mathbf{C}_y$ to support the weight $\mathbf{W}$ of the door and a couple moment $\mathbf{M}$ to support the moment of $\mathbf{W}$, i.e., $M = Wd$. If instead two properly aligned hinges are used, (b) then the weight is carried by both hinges, $A_y + B_y = W$, and the moment of the door is resisted by the two hinge forces $\mathbf{F}_x$ and $-\mathbf{F}_x$. These forces form a couple, such that $F_x d' = Wd$. In other words, no couple moments are produced by the hinges on the door provided they are in *proper alignment*. Instead, the forces $\mathbf{F}_x$ and $-\mathbf{F}_x$ resist the rotation caused by $\mathbf{W}$.

EXAMPLE 5.14

Several examples of objects along with their associated free-body diagrams are shown in Fig. 5–25. In all cases, the x, y, z axes are established and the unknown reaction components are indicated in the positive sense. The weight of the objects is neglected.

SOLUTION

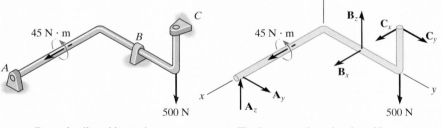

Properly aligned journal bearings at $A, B, C.$

The force reactions developed by the bearings are sufficient for equilibrium since they prevent the shaft from rotating about each of the coordinate axes.

(a)

Fig. 5–25

continued

Pin at A and cable BC.

Moment components are developed
by the pin on the rod to prevent
rotation about the x and z axes.

Properly aligned journal bearing
at A and hinge at C. Roller at B.

Only force reactions are developed by
the bearing and hinge on the plate to
prevent rotation about each coordinate axis.
No moments at the hinge are developed.

Thrust bearing at A and
cable BC

Moment components are developed
by the bearing on the rod in order to
prevent rotation about the y and z axes.

Fig. 5–25 (cont.)

5.6   Equations of Equilibrium

As stated in Sec. 5.1, the conditions for equilibrium of a rigid body subjected to a three-dimensional force system require that both the *resultant* force and *resultant* couple moment acting on the body be equal to *zero*.

Vector Equations of Equilibrium.   The two conditions for equilibrium of a rigid body may be expressed mathematically in vector form as

$$\Sigma \mathbf{F} = \mathbf{0}$$
$$\Sigma \mathbf{M}_o = \mathbf{0}$$

(5–5)

where $\Sigma \mathbf{F}$ is the vector sum of all the external forces acting on the body and $\Sigma \mathbf{M}_O$ is the sum of the couple moments and the moments of all the forces about any point O located either on or off the body.

Scalar Equations of Equilibrium.   If all the applied external forces and couple moments are expressed in Cartesian vector form and substituted into Eqs. 5–5, we have

$$\Sigma \mathbf{F} = \Sigma F_x \mathbf{i} + \Sigma F_y \mathbf{j} + \Sigma F_z \mathbf{k} = \mathbf{0}$$
$$\Sigma \mathbf{M}_O = \Sigma M_x \mathbf{i} + \Sigma M_y \mathbf{j} + \Sigma M_z \mathbf{k} = \mathbf{0}$$

Since the $\mathbf{i}, \mathbf{j}$, and $\mathbf{k}$ components are independent from one another, the above equations are satisfied provided

$$\Sigma F_x = 0$$
$$\Sigma F_y = 0$$
$$\Sigma F_z = 0$$

(5–6*a*)

and

$$\Sigma M_x = 0$$
$$\Sigma M_y = 0$$
$$\Sigma M_z = 0$$

(5–6*b*)

These *six scalar equilibrium equations* may be used to solve for at most six unknowns shown on the free-body diagram. Equations 5–6*a* express the fact that the sum of the external force components acting in the *x*, *y*, and *z* directions must be zero, and Eqs. 5–6*b* require the sum of the moment components about the *x*, *y*, and *z* axes to be zero.

5.7   Constraints for a Rigid Body

To ensure the equilibrium of a rigid body, it is not only necessary to satisfy the equations of equilibrium, but the body must also be properly held or constrained by its supports. Some bodies may have more supports than are necessary for equilibrium, whereas others may not have enough or the supports may be arranged in a particular manner that could cause the body to collapse. Each of these cases will now be discussed.

Redundant Constraints.   When a body has redundant supports, that is, more supports than are necessary to hold it in equilibrium, it becomes statically indeterminate. *Statically indeterminate* means that there will be more unknown loadings on the body than equations of equilibrium available for their solution. For example, the two-dimensional problem, Fig. 5–26a, and the three-dimensional problem, Fig. 5–26b, shown together with their free-body diagrams, are both statically indeterminate because of additional support reactions. In the two-dimensional case, there are five unknowns, that is, M_A, A_x, A_y, B_y, and C_y, for which only three equilibrium equations can be written ($\Sigma F_x = 0$, $\Sigma F_y = 0$, and $\Sigma M_O = 0$, Eqs. 5–2). The three-dimensional problem has eight unknowns, for which only six equilibrium equations can be written, Eqs. 5–6. The additional equations needed to solve indeterminate problems of the type shown in Fig. 5–26 are generally obtained from the deformation conditions at the points of support. These equations involve the physical properties of the body which are studied in subjects dealing with the mechanics of deformation, such as "mechanics of materials."*

Fig. 5–26

(a)

(b)

*See R. C. Hibbeler, *Mechanics of Materials*, 6th edition (Pearson Education/Prentice Hall, Inc., 2005).

Improper Constraints.   In some cases, there may be as many unknown forces on the body as there are equations of equilibrium; however, *instability* of the body can develop because of *improper constraining* by the supports. In the case of three-dimensional problems, the body is improperly constrained if the support reactions *all intersect a common axis*. For two-dimensional problems, this axis is *perpendicular* to the plane of the forces and therefore appears as a point. Hence, when all the reactive forces are *concurrent* at this point, the body is improperly constrained. Examples of both cases are given in Fig. 5–27. From the free-body diagrams it is seen that the summation of moments about the *x* axis, Fig. 5–27*a*, or point *O*, Fig. 5–27*b*, will *not* be equal to zero; thus rotation about the *x* axis or point *O* will take place.* Furthermore, in both cases, it becomes *impossible* to solve *completely* for all the unknowns since one can write a moment equation that *does not* involve any of the unknown support reactions, and as a result, this reduces the number of available equilibrium equations by one.

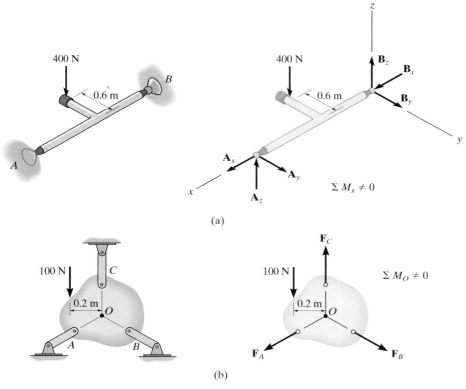

(a)

(b)

Fig. 5–27

*For the three-dimensional problem, $\Sigma M_x = (400\ \text{N})(0.6\ \text{m}) \neq 0$, and for the two-dimensional problem, $\Sigma M_O = (100\ \text{N})(0.2\ \text{m}) \neq 0$.

(a)

(b)

Fig. 5–28

(a)

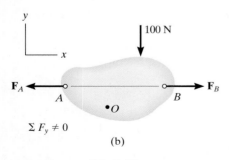

$\Sigma F_y \neq 0$

(b)

Fig. 5–29

Another way in which improper constraining leads to instability occurs when the *reactive forces* are all *parallel*. Three- and two-dimensional examples of this are shown in Fig. 5–28. In both cases, the summation of forces along the x axis will not equal zero.

In some cases, a body may have *fewer* reactive forces than equations of equilibrium that must be satisfied. The body then becomes only *partially constrained*. For example, consider the body shown in Fig. 5–29a with its corresponding free-body diagram in Fig. 5–29b. If O is a point not located on the line AB, the equation $\Sigma F_x = 0$ gives $F_A = F_B$, however, $\Sigma F_y = 0$ will not be satisfied for the loading conditions and therefore equilibrium will not be maintained.

Proper constraining therefore requires that:

1. The lines of action of the reactive forces do not intersect points on a common axis.

2. The reactive forces must not all be parallel to one another.

When the minimum number of reactive forces is needed to properly constrain the body in question, the problem will be statically determinate, and therefore the equations of equilibrium can be used to determine *all* the reactive forces.

Important Points

- Always draw the free-body diagram first.
- If a support *prevents translation* of a body in a specific direction, then it exerts a *force* on the body in that direction.
- If *rotation about an axis is prevented*, then the support exerts a *couple moment* on the body about the axis.
- If a body is subjected to more unknown reactions than available equations of equilibrium, then the problem is *statically indeterminate*.
- To avoid instability of a body require that the lines of action of the reactive forces do not intersect a common axis and are not parallel to one another.

Procedure for Analysis

Three-dimensional equilibrium problems for a rigid body can be solved using the following procedure.

Free-Body Diagram.

- Draw an outlined shape of the body.
- Show all the forces and couple moments acting on the body.
- Establish the origin of the x, y, z axes at a convenient point and orient the axes so that they are parallel to as many of the external forces and moments as possible.
- Label all the loadings and specify their directions relative to the x, y, z axes. In general, show all the unknown components having a positive sense along the x, y, z axes if the sense cannot be determined.
- Indicate the dimensions of the body necessary for computing the moments of forces.

Equations of Equilibrium.

- If the x, y, z force and moment components seem easy to determine, then apply the six scalar equations of equilibrium; otherwise use the vector equations.
- It is not necessary that the set of axes chosen for force summation coincide with the set of axes chosen for moment summation. Also, any set of nonorthogonal axes may be chosen for this purpose.
- Choose the direction of an axis for moment summation such that it intersects the lines of action of as many unknown forces as possible. The moments of forces passing through points on this axis and forces which are parallel to the axis will then be zero.
- If the solution of the equilibrium equations yields a negative scalar for a force or couple moment magnitude, it indicates that the sense is opposite to that assumed on the free-body diagram.

EXAMPLE 5.15

(a)

(b)

Fig. 5–30

The homogeneous plate shown in Fig. 5–30a has a mass of 100 kg and is subjected to a force and couple moment along its edges. If it is supported in the horizontal plane by means of a roller at A, a ball-and-socket joint at B, and a cord at C, determine the components of reaction at the supports.

SOLUTION *(SCALAR ANALYSIS)*

Free-Body Diagram.   There are five unknown reactions acting on the plate, as shown in Fig. 5–30b. Each of these reactions is assumed to act in a positive coordinate direction.

Equations of Equilibrium.   Since the three-dimensional geometry is rather simple, a *scalar analysis* provides a *direct solution* to this problem. A force summation along each axis yields

$$\Sigma F_x = 0; \qquad B_x = 0 \qquad\qquad\qquad\qquad Ans.$$
$$\Sigma F_y = 0; \qquad B_y = 0 \qquad\qquad\qquad\qquad Ans.$$
$$\Sigma F_z = 0; \qquad A_z + B_z + T_C - 300\ \text{N} - 981\ \text{N} = 0 \qquad (1)$$

Recall that the moment of a force about an axis is equal to the product of the force magnitude and the perpendicular distance (moment arm) from the line of action of the force to the axis. The sense of the moment is determined by the right-hand rule. Also, forces that are parallel to an axis or pass through it create no moment about the axis. Hence, summing moments of the forces on the free-body diagram, with positive moments acting along the positive x or y axis, we have

$$\Sigma M_x = 0; \qquad T_C(2\ \text{m}) - 981\ \text{N}(1\ \text{m}) + B_z(2\ \text{m}) = 0 \qquad (2)$$
$$\Sigma M_y = 0;$$
$$300\ \text{N}(1.5\ \text{m}) + 981\ \text{N}(1.5\ \text{m}) - B_z(3\ \text{m}) - A_z(3\ \text{m}) - 200\ \text{N}\cdot\text{m} = 0 \qquad (3)$$

The components of force at B can be eliminated if the x', y', z' axes are used. We obtain

$$\Sigma M_{x'} = 0; \qquad 981\ \text{N}(1\ \text{m}) + 300\ \text{N}(2\ \text{m}) - A_z(2\ \text{m}) = 0 \qquad (4)$$
$$\Sigma M_{y'} = 0;$$
$$-300\ \text{N}(1.5\ \text{m}) - 981\ \text{N}(1.5\ \text{m}) - 200\ \text{N}\cdot\text{m} + T_C(3\ \text{m}) = 0 \qquad (5)$$

Solving Eqs. 1 through 3 or the more convenient Eqs. 1, 4, and 5 yields

$$A_z = 790\ \text{N} \qquad B_z = -217\ \text{N} \qquad T_C = 707\ \text{N} \qquad Ans.$$

The negative sign indicates that $\mathbf{B}_z$ acts downward.

NOTE: The solution of this problem does not require the use of a summation of moments about the z axis. The plate is partially constrained since the supports cannot prevent it from turning about the z axis if a force is applied to it in the x–y plane.

EXAMPLE 5.16

The windlass shown in Fig. 5–31a is supported by a thrust bearing at A and a smooth journal bearing at B, which are properly aligned on the shaft. Determine the magnitude of the vertical force $\mathbf{P}$ that must be applied to the handle to maintain equilibrium of the 100-kg bucket. Also calculate the reactions at the bearings.

(a)   **Fig. 5–31**   (b)

SOLUTION (SCALAR ANALYSIS)

Free-Body Diagram.   Since the bearings at A and B are aligned correctly, *only* force reactions occur at these supports, Fig. 5–31b. Why are there no moment reactions?

Equations of Equilibrium.   Summing moments about the x axis yields a direct solution for $\mathbf{P}$. Why? For a scalar moment summation, it is necessary to determine the moment of each force as the product of the force magnitude and the *perpendicular distance* from the x axis to the line of action of the force. Using the right-hand rule and assuming positive moments act in the $+\mathbf{i}$ direction, we have

$$\Sigma M_x = 0; \quad 981 \text{ N}(0.1 \text{ m}) - P(0.3 \cos 30° \text{ m}) = 0$$
$$P = 377.6 \text{ N} \qquad\qquad Ans.$$

Using this result for P and summing moments about the y and z axes yields

$$\Sigma M_y = 0; \quad -981 \text{ N}(0.5 \text{ m}) + A_z(0.8 \text{ m}) + (377.6 \text{ N})(0.4 \text{ m}) = 0$$
$$A_z = 424.3 \text{ N} \qquad\qquad Ans.$$
$$\Sigma M_z = 0; \quad -A_y(0.8 \text{ m}) = 0 \quad A_y = 0$$

The reactions at B are determined by a force summation using these results.

$$\Sigma F_x = 0; \quad A_x = 0$$
$$\Sigma F_y = 0; \quad 0 + B_y = 0 \quad B_y = 0$$
$$\Sigma F_z = 0; \quad 424.3 - 981 + B_z - 377.6 = 0 \quad B_z = 934 \text{ N} \quad Ans.$$

NOTE: As expected, the force on the handle is much smaller than the weight of the bucket.

EXAMPLE 5.17

(a)

(b)

$F_B = -F_A$

(c)

Fig. 5–32

Determine the tension in cables BC and BD and the reactions at the ball-and-socket joint A for the mast shown in Fig. 5–32a.

SOLUTION (VECTOR ANALYSIS)

Free-Body Diagram. There are five unknown force magnitudes shown on the free-body diagram, Fig. 5–32b.

Equations of Equilibrium. Expressing each force in Cartesian vector form, we have

$$\mathbf{F} = \{-1000\mathbf{j}\}\ \text{N}$$
$$\mathbf{F}_A = A_x\mathbf{i} + A_y\mathbf{j} + A_z\mathbf{k}$$
$$\mathbf{T}_C = 0.707T_C\mathbf{i} - 0.707T_C\mathbf{k}$$
$$\mathbf{T}_D = T_D\left(\frac{\mathbf{r}_{BD}}{r_{BD}}\right) = -\frac{3}{9}T_D\mathbf{i} + \frac{6}{9}T_D\mathbf{j} - \frac{6}{9}T_D\mathbf{k}$$

Applying the force equation of equilibrium gives

$$\Sigma\mathbf{F} = \mathbf{0}; \qquad \mathbf{F} + \mathbf{F}_A + \mathbf{T}_C + \mathbf{T}_D = \mathbf{0}$$

$$\left(A_x + 0.707T_C - \frac{3}{9}T_D\right)\mathbf{i} + \left(-1000 + A_y + \frac{6}{9}T_D\right)\mathbf{j}$$
$$+ \left(A_z - 0.707T_C - \frac{6}{9}T_D\right)\mathbf{k} = \mathbf{0}$$

$$\Sigma F_x = 0; \qquad A_x + 0.707T_C - \frac{3}{9}T_D = 0 \qquad (1)$$

$$\Sigma F_y = 0; \qquad A_y + \frac{6}{9}T_D - 1000 = 0 \qquad (2)$$

$$\Sigma F_z = 0; \qquad A_z - 0.707T_C - \frac{6}{9}T_D = 0 \qquad (3)$$

Summing moments about point A, we have

$$\Sigma\mathbf{M}_A = \mathbf{0}; \qquad \mathbf{r}_B \times (\mathbf{F} + \mathbf{T}_C + \mathbf{T}_D) = \mathbf{0}$$

$$6\mathbf{k} \times \left(-1000\mathbf{j} + 0.707T_C\mathbf{i} - 0.707T_C\mathbf{k} - \frac{3}{9}T_D\mathbf{i} + \frac{6}{9}T_D\mathbf{j} - \frac{6}{9}T_D\mathbf{k}\right) = \mathbf{0}$$

Evaluating the cross product and combining terms yields

$$(-4T_D + 6000)\mathbf{i} + (4.24T_C - 2T_D)\mathbf{j} = \mathbf{0}$$

$$\Sigma M_x = 0; \qquad -4T_D + 6000 = 0 \qquad (4)$$
$$\Sigma M_y = 0; \qquad 4.24T_C - 2T_D = 0 \qquad (5)$$

The moment equation about the z axis, $\Sigma M_z = 0$, is automatically satisfied. Why? Solving Eqs. 1 through 5 we have

$$T_C = 707\ \text{N} \qquad T_D = 1500\ \text{N} \qquad\qquad Ans.$$
$$A_x = 0\ \text{N} \qquad A_y = 0\ \text{N} \qquad A_z = 1500\ \text{N} \qquad Ans.$$

NOTE: Since the mast is a two-force member, Fig. 5–32c, the value $A_x = A_y = 0$ could have been determined *by inspection*.

EXAMPLE 5.18

Rod AB shown in Fig. 5–33a is subjected to the 200-N force. Determine the reactions at the ball-and-socket joint A and the tension in cables BD and BE.

SOLUTION (VECTOR ANALYSIS)

Free-Body Diagram.   Fig. 5–33b.

Equations of Equilibrium.   Representing each force on the free-body diagram in Cartesian vector form, we have

$$\mathbf{F}_A = A_x\mathbf{i} + A_y\mathbf{j} + A_z\mathbf{k}$$
$$\mathbf{T}_E = T_E\mathbf{i}$$
$$\mathbf{T}_D = T_D\mathbf{j}$$
$$\mathbf{F} = \{-200\mathbf{k}\}\ \text{N}$$

Applying the force equation of equilibrium.

$$\Sigma \mathbf{F} = \mathbf{0}; \qquad \mathbf{F}_A + \mathbf{T}_E + \mathbf{T}_D + \mathbf{F} = \mathbf{0}$$
$$(A_x + T_E)\mathbf{i} + (A_y + T_D)\mathbf{j} + (A_z - 200)\mathbf{k} = \mathbf{0}$$

$\Sigma F_x = 0;$	$A_x + T_E = 0$	(1)
$\Sigma F_y = 0;$	$A_y + T_D = 0$	(2)
$\Sigma F_z = 0;$	$A_z - 200 = 0$	(3)

Summing moments about point A yields

$$\Sigma \mathbf{M}_A = \mathbf{0}; \qquad \mathbf{r}_C \times \mathbf{F} + \mathbf{r}_B \times (\mathbf{T}_E + \mathbf{T}_D) = \mathbf{0}$$

Since $\mathbf{r}_C = \frac{1}{2}\mathbf{r}_B$, then

$$(0.5\mathbf{i} + 1\mathbf{j} - 1\mathbf{k}) \times (-200\mathbf{k}) + (1\mathbf{i} + 2\mathbf{j} - 2\mathbf{k}) \times (T_E\mathbf{i} + T_D\mathbf{j}) = \mathbf{0}$$

Expanding and rearranging terms gives

$$(2T_D - 200)\mathbf{i} + (-2T_E + 100)\mathbf{j} + (T_D - 2T_E)\mathbf{k} = \mathbf{0}$$

$\Sigma M_x = 0;$	$2T_D - 200 = 0$	(4)
$\Sigma M_y = 0;$	$-2T_E + 100 = 0$	(5)
$\Sigma M_z = 0;$	$T_D - 2T_E = 0$	(6)

Solving Eqs. 1 through 6, we get

$T_D = 100\ \text{N}$		*Ans.*
$T_E = 50\ \text{N}$		*Ans.*
$A_x = -50\ \text{N}$		*Ans.*
$A_y = -100\ \text{N}$		*Ans.*
$A_z = 200\ \text{N}$		*Ans.*

NOTE: The negative sign indicates that $\mathbf{A}_x$ and $\mathbf{A}_y$ have a sense which is opposite to that shown on the free-body diagram, Fig. 5–33b.

(a)

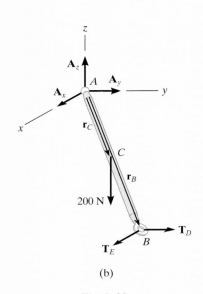

(b)

Fig. 5–33

EXAMPLE 5.19

(a)

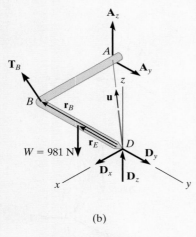

(b)

Fig. 5–34

The bent rod in Fig. 5–34a is supported at A by a journal bearing, at D by a ball-and-socket joint, and at B by means of cable BC. Using only *one equilibrium equation*, obtain a direct solution for the tension in cable BC. The bearing at A is capable of exerting force components only in the z and y directions since it is properly aligned on the shaft.

SOLUTION (VECTOR ANALYSIS)

Free-Body Diagram. As shown in Fig. 5–34b, there are six unknowns: three force components caused by the ball-and-socket joint, two caused by the bearing, and one caused by the cable.

Equations of Equilibrium. The cable tension $\mathbf{T}_B$ may be obtained *directly* by summing moments about an axis passing through points D and A. Why? The direction of the axis is defined by the unit vector $\mathbf{u}$, where

$$\mathbf{u} = \frac{\mathbf{r}_{DA}}{r_{DA}} = -\frac{1}{\sqrt{2}}\mathbf{i} - \frac{1}{\sqrt{2}}\mathbf{j}$$
$$= -0.707\mathbf{i} - 0.707\mathbf{j}$$

Hence, the sum of the moments about this axis is zero provided

$$\Sigma M_{DA} = \mathbf{u} \cdot \Sigma(\mathbf{r} \times \mathbf{F}) = 0$$

Here $\mathbf{r}$ represents a position vector drawn from *any point* on the axis DA to any point on the line of action of force $\mathbf{F}$ (see Eq. 4–11). With reference to Fig. 5–34b, we can therefore write

$$\mathbf{u} \cdot (\mathbf{r}_B \times \mathbf{T}_B + \mathbf{r}_E \times \mathbf{W}) = \mathbf{0}$$
$$(-0.707\mathbf{i} - 0.707\mathbf{j}) \cdot \left[(-1\mathbf{j}) \times \left(\tfrac{0.2}{0.7}T_B\mathbf{i} - \tfrac{0.3}{0.7}T_B\mathbf{j} + \tfrac{0.6}{0.7}T_B\mathbf{k}\right)\right.$$
$$\left. + (-0.5\mathbf{j}) \times (-981\mathbf{k})\right] = \mathbf{0}$$
$$(-0.707\mathbf{i} - 0.707\mathbf{j}) \cdot [(-0.857T_B + 490.5)\mathbf{i} + 0.286T_B\mathbf{k}] = \mathbf{0}$$
$$- 0.707(-0.857T_B + 490.5) + 0 + 0 = 0$$

$$T_B = \frac{490.5}{0.857} = 572 \text{ N} \qquad\qquad Ans.$$

The advantage of using Cartesian vectors for this solution should be noted. It would be especially tedious to determine the perpendicular distance from the DA axis to the line of action of $\mathbf{T}_B$ using scalar methods.

NOTE: In a similar manner, we can obtain D_z ($= 490.5$ N) by summing moments about an axis passing through AB. Also, A_z ($= 0$) is obtained by summing moments about the y axis.

PROBLEMS

5–63. Determine the x, y, z components of reaction at the fixed wall A. The 150-N force is parallel to the z axis and the 200-N force is parallel to the y axis.

Prob. 5–63

5–65. The uniform concrete slab has a weight of 5500 lb. Determine the tension in each of the three parallel supporting cables when the slab is held in the horizontal plane as shown.

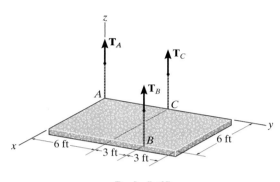

Prob. 5–65

*5–64. The wing of the jet aircraft is subjected to a thrust of $T = 8$ kN from its engine and the resultant lift force $L = 45$ kN. If the mass of the wing is 2.1 Mg and the mass center is at G, determine the x, y, z components of reaction where the wing is fixed to the fuselage at A.

Prob. 5–64

5–66. The air-conditioning unit is hoisted to the roof of a building using the three cables. If the tensions in the cables are $T_A = 250$ lb, $T_B = 300$ lb, and $T_C = 200$ lb, determine the weight of the unit and the location (x, y) of its center of gravity G.

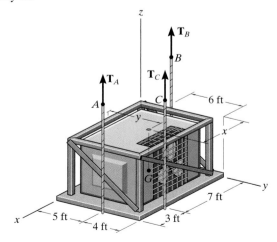

Prob. 5–66

5–67. The platform truck supports the three loadings shown. Determine the normal reactions on each of its three wheels.

Prob. 5–67

5–69. If the cable can be subjected to a maximum tension of 300 lb, determine the maximum force F which may be applied to the plate. Compute the x, y, z components of reaction at the hinge A for this loading.

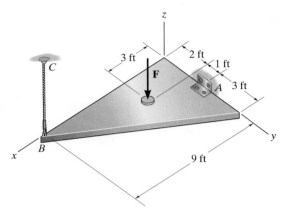

Prob. 5–69

***5–68.** Due to an unequal distribution of fuel in the wing tanks, the centers of gravity for the airplane fuselage A and wings B and C are located as shown. If these components have weights $W_A = 45\,000$ lb, $W_B = 8000$ lb, and $W_C = 6000$ lb, determine the normal reactions of the wheels D, E, and F on the ground.

5–70. The boom AB is held in equilibrium by a ball-and-socket joint A and a pulley and cord system as shown. Determine the x, y, z components of reaction at A and the tension in cable DEC if $\mathbf{F} = \{-1500\mathbf{k}\}$ lb.

5–71. The cable CED can sustain a maximum tension of 800 lb before it fails. Determine the greatest vertical force F that can be applied to the boom. Also, what are the x, y, z components of reaction at the ball-and-socket joint A?

Prob. 5–68

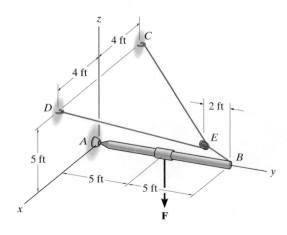

Probs. 5–70/71

*5–72. The uniform table has a weight of 20 lb and is supported by the framework shown. Determine the smallest vertical force **P** that can be applied to its surface that will cause it to tip over. Where should this force be applied?

Prob. 5–72

5–73. The windlass is subjected to a load of 150 lb. Determine the horizontal force **P** needed to hold the handle in the position shown, and the components of reaction at the ball-and-socket joint A and the smooth journal bearing B. The bearing at B is in proper alignment and exerts only force reactions on the windlass.

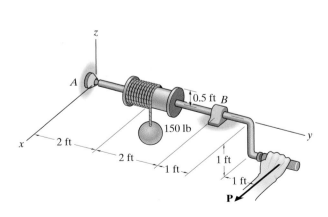

Prob. 5–73

5–74. The 2-kg ball rests between the 45° grooves A and B of the 10° incline and against a vertical wall at C. If all three surfaces of contact are smooth, determine the reactions of the surfaces on the ball. *Hint:* Use the x, y, z axes, with origin at the center of the ball, and the z axis inclined as shown.

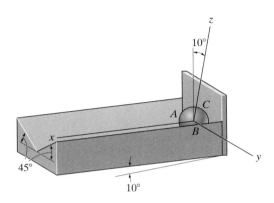

Prob. 5–74

5–75. Member AB is supported by a cable BC and at A by a *square* rod which fits loosely through the square hole at the end joint of the member as shown. Determine the components of reaction at A and the tension in the cable needed to hold the 800-lb cylinder in equilibrium.

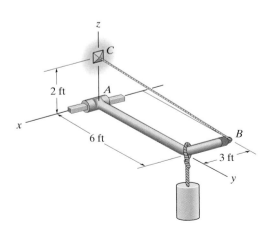

Prob. 5–75

*5–76. The pipe assembly supports the vertical loads shown. Determine the components of reaction at the ball-and-socket joint A and the tension in the supporting cables BC and BD.

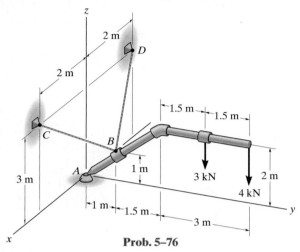

Prob. 5–76

5–77. The hatch door has a weight of 80 lb and center of gravity at G. If the force $\mathbf{F}$ applied to the handle at C has coordinate direction angles of $\alpha = 60°$, $\beta = 45°$, and $\gamma = 60°$, determine the magnitude of $\mathbf{F}$ needed to hold the door slightly open as shown. The hinges are in proper alignment and exert only force reactions on the door. Determine the components of these reactions if A exerts only x and z components of force and B exerts x, y, z force components.

5–78. The hatch door has a weight of 80 lb and center of gravity at G. If the force $\mathbf{F}$ applied to the handle at C has coordinate direction angles of $\alpha = 60°$, $\beta = 45°$, and $\gamma = 60°$, determine the magnitude of $\mathbf{F}$ needed to hold the door slightly open as shown. If the hinge at A becomes loose from its attachment and is ineffective, what are the x, y, z components of reaction at hinge B?

Probs. 5–77/78

5–79. The bent rod is supported at A, B, and C by smooth journal bearings. Compute the x, y, z components of reaction at the bearings if the rod is subjected to forces $F_1 = 300$ lb and $F_2 = 250$ lb. F_1 lies in the y–z plane. The bearings are in proper alignment and exert only force reactions on the rod.

*5–80. The bent rod is supported at A, B, and C by smooth journal bearings. Determine the magnitude of $\mathbf{F}_2$ which will cause the reaction $\mathbf{C}_y$ at the bearing C to be equal to zero. The bearings are in proper alignment and exert only force reactions on the rod. Set $F_1 = 300$ lb.

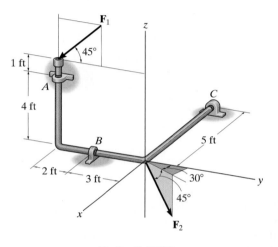

Probs. 5–79/80

5–81. Determine the tension in cables BD and CD and the x, y, z components of reaction at the ball-and-socket joint at A.

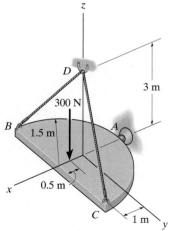

Prob. 5–81

5–82. Determine the tensions in the cables and the components of reaction acting on the smooth collar at A necessary to hold the 50-lb sign in equilibrium. The center of gravity for the sign is at G.

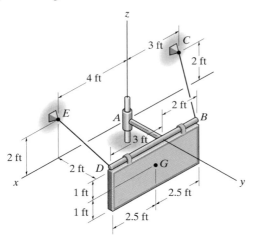

Prob. 5–82

5–83. The member is supported by a pin at A and a cable BC. If the load at D is 300 lb, determine the x, y, z components of reaction at these supports.

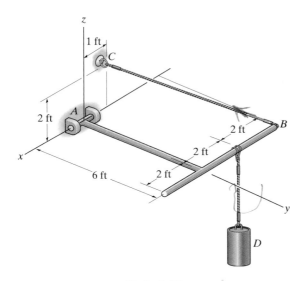

Prob. 5–83

*5–84. Determine the x, y, z components of reaction at the pin A and the tension in the cable BC necessary for equilibrium of the rod.

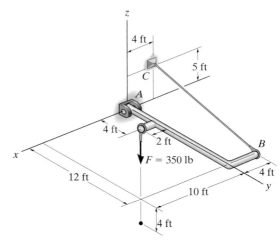

Prob. 5–84

5–85. Rod AB is supported by a ball-and-socket joint at A and a cable at the smooth wall at B. Determine the x, y, z components of reaction at these supports if the rod is subjected to a 50-lb vertical force as shown.

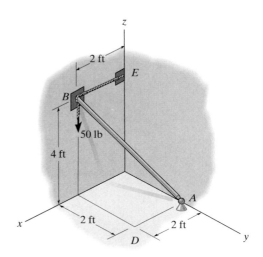

Prob. 5–85

5–86. The member is supported by a square rod which fits loosely through a smooth square hole of the attached collar at *A* and by a roller at *B*. Determine the *x, y, z* components of reaction at these supports when the member is subjected to the loading shown.

$$F = \{20i - 40j - 30k\} \text{ lb}$$

Prob. 5–86

5–87. The platform has a mass of 3 Mg and center of mass located at *G*. If it is lifted with constant velocity using the three cables, determine the force in each of the cables.

*****5–88.** The platform has a mass of 2 Mg and center of mass located at *G*. If it is lifted using the three cables, determine the force in each of the cables. Solve for each force by using a single moment equation of equilibrium.

Probs. 5–87/88

5–89. The cables exert the forces shown on the pole. Assuming the pole is supported by a ball-and-socket joint at its base, determine the components of reaction at *A*. The forces of 140 lb and 75 lb lie in a horizontal plane.

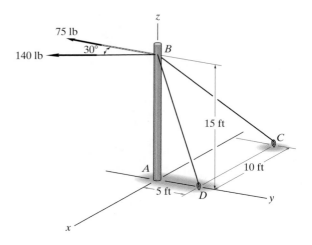

Prob. 5–89

5–90. The silo has a weight of 3500 lb and a center of gravity at *G*. Determine the vertical component of force that each of the three struts at *A*, *B*, and *C* exerts on the silo if it is subjected to a resultant wind loading of 250 lb which acts in the direction shown.

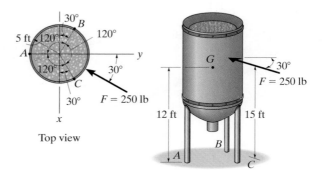

Prob. 5–90

Chapter Review

Equilibrium

A body in equilibrium does not accelerate, so it either moves at constant velocity, or does not move at all or is at rest.
(*See page 201.*)

$$\Sigma \mathbf{F} = 0$$
$$\Sigma \mathbf{M} = 0$$

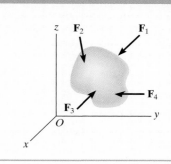

Two Dimensions

Before analyzing the equilibrium of a body, it is first necessary to draw its free-body diagram. This is an outlined shape of the body, which shows all the forces and couple moments that act on it.

Couple moments can be placed anywhere on a free-body diagram since they are free vectors. Forces can act at any point along their line of action since they are sliding vectors.

Angles used to resolve forces, and dimensions used to take moments of the forces, should also be shown on the free-body diagram.

Some common types of supports and their reactions are shown below in two dimensions.

Remember that a support will exert a force on the body in a particular direction if it prevents translation of the body in that direction, and it will exert a couple moments on the body if it prevents rotation.
(*See pages 203–204.*)

roller

smooth pin or hinge

fixed support

The three scalar equations of equilibrium can be applied when solving problems in two dimensions, since the geometry is easy to visualize.

$$\Sigma F_x = 0$$
$$\Sigma F_y = 0$$
$$\Sigma M_O = 0$$

For the most direct solution, try to sum forces along an axis that will eliminate as many unknown forces as possible. Sum moments about a point A that passes through the line of action of as many unknown forces as possible. (*See pages 217 and 218.*)

$\Sigma F_x = 0;$
$\quad A_x - P_2 = 0 \qquad A_x = P_2$

$\Sigma M_A = 0;$
$\quad P_2 d_2 + B_y d_B - P_1 d_1 = 0$

$$B_y = \frac{P_1 d_1 - P_2 d_2}{d_B}$$

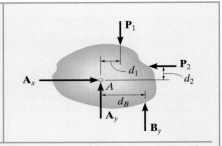

Three Dimensions

Some common types of supports and their reactions are shown in three dimensions.

roller   ball and socket   fixed support

In three dimensions, it is often advantageous to use a Cartesian vector analysis when applying the equations of equilibrium. To do this, first express each known and unknown force and couple moment shown on the free-body diagram as a Cartesian vector. Then set the force summation equal to zero. Take moments about a point O that lies on the line of action of as many unknown force components as possible. From point O direct position vectors to each force, and then use the cross product to determine the moment of each force. The six scalar equations of equilibrium are established by setting the respective $\mathbf{i}, \mathbf{j}$, and $\mathbf{k}$ components of these force and moment summations equal to zero. (*See page 245.*)

$$\Sigma \mathbf{F} = \mathbf{0}$$
$$\Sigma \mathbf{M}_O = \mathbf{0}$$

$$\Sigma F_x = 0$$
$$\Sigma F_y = 0$$
$$\Sigma F_z = 0$$

$$\Sigma M_x = 0$$
$$\Sigma M_y = 0$$
$$\Sigma M_z = 0$$

Determinacy and Stability

If a body is supported by a minimum number of constraints to ensure equilibrium, then it is statically determinate. If it has more constraints than required, then it is statically indeterminate.

To properly constrain the body, the reactions must not all be parallel to one another or concurrent. (*See pages 246–248.*)

Statically indeterminate,
five reactions, three
equilibrium equations

Proper constraint, statically determinant

REVIEW PROBLEMS

5–91. The shaft assembly is supported by two smooth journal bearings A and B and a short link DC. If a couple moment is applied to the shaft as shown, determine the components of force reaction at the bearings and the force in the link. The link lies in a plane parallel to the y–z plane and the bearings are properly aligned on the shaft.

5–93. If the maximum intensity of the distributed load acting on the beam is $w = 4$ kN/m, determine the reactions at the pin A and roller B.

Prob. 5–91

Prob. 5–93

*5–92. If neither the pin at A nor the roller at B can support a load no greater than 6 kN, determine the maximum intensity of the distributed load w, measured in kN/m, so that failure of a support does not occur.

5–94. Determine the normal reaction at the roller A and horizontal and vertical components at pin B for equilibrium of the member.

Prob. 5–92

Prob. 5–94

5–95. The symmetrical shelf is subjected to a uniform load of 4 kPa. Support is provided by a bolt (or pin) located at each end A and A' and by the symmetrical brace arms, which bear against the smooth wall on both sides at B and B'. Determine the force resisted by each bolt at the wall and the normal force at B for equilibrium.

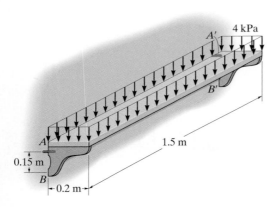

Prob. 5–95

5–97. The uniform ladder rests along the wall of a building at A and on the roof at B. If the ladder has a weight of 25 lb and the surfaces at A and B are assumed smooth, determine the angle θ for equilibrium.

Prob. 5–97

*5–96. A uniform beam having a weight of 200 lb supports a vertical load of 800 lb. If the ground pressure varies linearly as shown, determine the load intensities w_1 and w_2, measured in lb/ft, necessary for equilibrium.

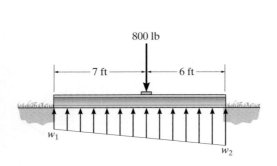

Prob. 5–96

5–98. Determine the x, y, z components of reaction at the ball supports B and C and the ball-and-socket A (not shown) for the uniformly loaded plate.

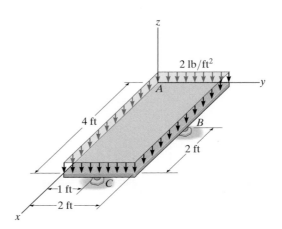

Prob. 5–98

5–99. A vertical force of 80 lb acts on the crankshaft. Determine the horizontal equilibrium force P that must be applied to the handle and the x, y, z components of force at the smooth journal bearing A and the thrust bearing B. The bearings are properly aligned and exert only force reactions on the shaft.

Prob. 5–99

*5–100.** The horizontal beam is supported by springs at its ends. If the stiffness of the spring at A is $k_A = 5 \text{ kN/m}$, determine the required stiffness of the spring at B so that if the beam is loaded with the 800-N force, it remains in the horizontal position both before and after loading.

Prob. 5–100

The forces within the members of these truss bridges must be determined if they are to be properly designed.

6

Structural Analysis

CHAPTER OBJECTIVES

- To show how to determine the forces in the members of a truss using the method of joints and the method of sections.
- To analyze the forces acting on the members of frames and machines composed of pin-connected members.

6.1 Simple Trusses

A *truss* is a structure composed of slender members joined together at their end points. The members commonly used in construction consist of wooden struts or metal bars. The joint connections are usually formed by bolting or welding the ends of the members to a common plate, called a *gusset plate,* as shown in Fig. 6–1a, or by simply passing a large bolt or pin through each of the members, Fig. 6–1b. Gusset plates are used for the member connections of the bridge trusses shown on the chapter opening photo.

(a)

(b)

Fig. 6–1

(a)

Purlin

(b)

Roof truss

Fig. 6–2

Planar Trusses.   *Planar* trusses lie in a single plane and are often used to support roofs and bridges. The truss *ABCDE*, shown in Fig. 6–2a, is an example of a typical roof-supporting truss. In this figure, the roof load is transmitted to the truss *at the joints* by means of a series of *purlins,* such as *DD′*. Since the imposed loading acts in the same plane as the truss, Fig. 6–2b, the analysis of the forces developed in the truss members is two-dimensional.

In the case of a bridge, such as shown in Fig. 6–3a, the load on the deck is first transmitted to *stringers,* then to *floor beams,* and finally to the *joints B, C,* and *D* of the two supporting side trusses. Like the roof truss, the bridge truss loading is also coplanar, Fig. 6–3b.

When bridge or roof trusses extend over large distances, a rocker or roller is commonly used for supporting one end, e.g., joint *E* in Figs. 6–2a and 6–3a. This type of support allows freedom for expansion or contraction of the members due to temperature or application of loads.

(a)

Bridge truss

(b)

Fig. 6–3

Assumptions for Design. To design both the members and the connections of a truss, it is first necessary to determine the *force* developed in each member when the truss is subjected to a given loading. In this regard, two important assumptions will be made:

1. *All loadings are applied at the joints.* In most situations, such as for bridge and roof trusses, this assumption is true. Frequently in the force analysis the weight of the members is neglected since the forces supported by the members are usually large in comparison with their weight. If the member's weight is to be included in the analysis, it is generally satisfactory to apply it as a vertical force, with half of its magnitude applied at each end of the member.

2. *The members are joined together by smooth pins.* In cases where bolted or welded joint connections are used, this assumption is satisfactory provided the center lines of the joining members are *concurrent,* as in Fig. 6–1a.

Because of these two assumptions, *each truss member acts as a two-force member,* and therefore the forces at the ends of the member must be directed along the axis of the member. If the force tends to *elongate* the member, it is a *tensile force* (T), Fig. 6–4a; whereas if it tends to *shorten* the member, it is a *compressive force* (C), Fig. 6–4b. In the actual design of a truss it is important to state whether the nature of the force is tensile or compressive. Often, compression members must be made *thicker* than tension members because of the buckling or column effect that occurs when a member is in compression.

Simple Truss. To prevent collapse, the form of a truss must be rigid. Obviously, the four-bar shape *ABCD* in Fig. 6–5 will collapse unless a diagonal member, such as *AC*, is added for support. The simplest form that is rigid or stable is a *triangle*. Consequently, a *simple truss* is constructed by *starting* with a basic triangular element, such as *ABC* in Fig. 6–6, and connecting two members (*AD* and *BD*) to form an additional element. As each additional element consisting of two members and a joint is placed on the truss, it is possible to construct a simple truss.

Tension
(a)

Compression
(b)

Fig. 6–4

Fig. 6–5

Fig. 6–6

These Howe trusses are used to support the roof of the metal building. This form was named after William Howe, who patented the design in 1840. Note how the members come together at a common point on the gusset plate and how the roof purlins transmit the load to the joints.

6.2 The Method of Joints

In order to analyze or design a truss, we must obtain the force in each of its members. If we were to consider a free-body diagram of the entire truss, then the forces in the members would be *internal forces,* and they could not be obtained from an equilibrium analysis. Instead, if we consider the equilibrium of a joint of the truss then a member force becomes an *external force* on the joint's free-body diagram, and the equations of equilibrium can be applied to obtain its magnitude. This forms the basis for the *method of joints.*

Because the truss members are all straight two-force members lying in the same plane, the force system acting at each joint is *coplanar and concurrent.* Consequently, rotational or moment equilibrium is automatically satisfied at the joint (or pin), and it is only necessary to satisfy $\Sigma F_x = 0$ and $\Sigma F_y = 0$ to ensure equilibrium.

When using the method of joints, it is *first* necessary to draw a free-body diagram of a joint before applying the equilibrium equations. To do this, recall that the *line of action* of each member force acting on the joint is *specified* from the geometry of the truss since the force in a member passes along the axis of the member. As an example, consider the pin at joint B of the truss in Fig. 6–7a. Three forces act on the pin, namely, the 500-N force and the forces exerted by members BA and BC. The free-body diagram is shown in Fig. 6–7b. As shown, $\mathbf{F}_{BA}$ is "pulling" on the pin, which means that member BA is in *tension;* whereas $\mathbf{F}_{BC}$ is "pushing" on the pin, and consequently member BC is in *compression.* These effects are clearly demonstrated by isolating the joint with small segments of the member connected to the pin, Fig. 6–7c. The pushing or pulling on these small segments indicates the effect of the member being either in compression or tension.

The analysis always should start at a joint having at least one known force and at most two unknown forces, as in Fig. 6–7b. In this way, application of $\Sigma F_x = 0$ and $\Sigma F_y = 0$ yields two algebraic equations which can be solved for the two unknowns. When applying these equations, the correct sense of an unknown member force can be determined using one of two possible methods:

(a)

(b)

(c)

Fig. 6–7

- *Always assume* the *unknown member forces* acting on the joint's free-body diagram to be in *tension,* i.e., "pulling" on the pin. If this is done, then numerical solution of the equilibrium equations will yield *positive scalars for members in tension and negative scalars for members in compression.* Once an unknown member force is found, use its *correct* magnitude and sense (T or C) on subsequent joint free-body diagrams.

- The *correct* sense of direction of an unknown member force can, in many cases, be determined "by inspection." For example, $\mathbf{F}_{BC}$ in Fig. 6–7b must push on the pin (compression) since its horizontal component, $F_{BC} \sin 45°$, must balance the 500-N force ($\Sigma F_x = 0$). Likewise, $\mathbf{F}_{BA}$ is a tensile force since it balances the vertical component, $F_{BC} \cos 45°$ ($\Sigma F_y = 0$). In more complicated cases, the sense of an unknown member force can be *assumed;* then, after applying the equilibrium equations, the assumed sense can be verified from the numerical results. A *positive* answer indicates that the sense is *correct,* whereas a *negative* answer indicates that the sense shown on the free-body diagram must be *reversed.* This is the method we will use in the example problems which follow.

Procedure For Analysis

The following procedure provides a typical means for analyzing a truss using the method of joints.

- Draw the free-body diagram of a joint having at least one known force and at most two unknown forces. (If this joint is at one of the supports, then it may be necessary to know the external reactions at the truss support.)

- Use one of the two methods described above for establishing the sense of an unknown force.

- Orient the x and y axes such that the forces on the free-body diagram can be easily resolved into their x and y components and then apply the two force equilibrium equations $\Sigma F_x = 0$ and $\Sigma F_y = 0$. Solve for the two unknown member forces and verify their correct sense.

- Continue to analyze each of the other joints. Again, it is necessary to choose a joint having at most two unknowns and at least one known force.

- Once the force in a member is found from the analysis of a joint at one of its ends, the result can be used to analyze the forces acting on the joint at its other end. Remember that a member in *compression* "pushes" on the joint and a member in *tension* "pulls" on the joint.

EXAMPLE 6.1

(a)

(b)

(c)

(d)

(e)

Fig. 6–8

Determine the force in each member of the truss shown in Fig. 6–8a and indicate whether the members are in tension or compression.

SOLUTION

By inspection of Fig. 6–8a, there are two unknown member forces at joint B, two unknown member forces and an unknown reaction force at joint C, and two unknown member forces and two unknown reaction forces at joint A. Since we should have no more than two unknowns at the joint and at least one known force acting there, we will begin the analysis at joint B.

Joint B.   The free-body diagram of the pin at B is shown in Fig. 6–8b. Applying the equations of joint equilibrium, we have

$$\xrightarrow{+} \Sigma F_x = 0; \qquad 500 \text{ N} - F_{BC} \sin 45° = 0 \qquad F_{BC} = 707.1 \text{ N (C)} \textit{ Ans.}$$
$$+\uparrow \Sigma F_y = 0; \qquad F_{BC} \cos 45° - F_{BA} = 0 \qquad F_{BA} = 500 \text{ N (T)} \quad \textit{Ans.}$$

Since the force in member BC has been calculated, we can proceed to analyze joint C in order to determine the force in member CA and the support reaction at the rocker.

Joint C.   From the free-body diagram of joint C, Fig. 6–8c, we have

$$\xrightarrow{+} \Sigma F_x = 0; \quad -F_{CA} + 707.1 \cos 45° \text{ N} = 0 \quad F_{CA} = 500 \text{ N (T)} \textit{ Ans.}$$
$$+\uparrow \Sigma F_y = 0; \qquad C_y - 707.1 \sin 45° \text{ N} = 0 \qquad C_y = 500 \text{ N} \qquad \textit{Ans.}$$

Joint A.   Although it is not necessary, we can determine the support reactions at joint A using the results of $F_{CA} = 500$ N and $F_{BA} = 500$ N. From the free-body diagram, Fig. 6–8d, we have

$$\xrightarrow{+} \Sigma F_x = 0; \qquad 500 \text{ N} - A_x = 0 \quad A_x = 500 \text{ N}$$
$$+\uparrow \Sigma F_y = 0; \qquad 500 \text{ N} - A_y = 0 \quad A_y = 500 \text{ N}$$

NOTE: The results of the analysis are summarized in Fig. 6–8e. Note that the free-body diagram of each pin shows the effects of all the connected members and external forces applied to the pin, whereas the free-body diagram of each member shows only the effects of the end pins on the member.

EXAMPLE 6.2

Determine the forces acting in all the members of the truss shown in Fig. 6–9a.

(a)

SOLUTION

By inspection, there are more than two unknowns at each joint. Consequently, the support reactions on the truss must first be determined. Show that they have been correctly calculated on the free-body diagram in Fig. 6–9b. We can now begin the analysis at joint C. Why?

Joint C.   From the free-body diagram, Fig. 6–9c,

$$\xrightarrow{+} \Sigma F_x = 0; \qquad\qquad -F_{CD} \cos 30° + F_{CB} \sin 45° = 0$$

$$+\uparrow \Sigma F_y = 0; \qquad 1.5 \text{ kN} + F_{CD} \sin 30° - F_{CB} \cos 45° = 0$$

(b)

These two equations must be solved *simultaneously* for each of the two unknowns. Note, however, that a *direct solution* for one of the unknown forces may be obtained by applying a force summation along an axis that is *perpendicular* to the direction of the other unknown force. For example, summing forces along the y' axis, which is perpendicular to the direction of $\mathbf{F}_{CD}$, Fig. 6–9d, yields a direct solution for F_{CB}.

$$+\nearrow \Sigma F_{y'} = 0;$$

$$1.5 \cos 30° \text{ kN} - F_{CB} \sin 15° = 0 \quad F_{CB} = 5.02 \text{ kN} \quad (\text{C}) \quad Ans.$$

In a similar fashion, summing forces along the y'' axis, Fig. 6–9e, yields a direct solution for F_{CD}.

$$+\nearrow \Sigma F_{y''} = 0;$$

$$1.5 \cos 45° \text{ kN} - F_{CD} \sin 15° = 0 \qquad F_{CD} = 4.10 \text{ kN} \quad (\text{T}) \quad Ans.$$

(c)   (d)

Joint D.   We can now proceed to analyze joint D. The free-body diagram is shown in Fig. 6–9f.

$$\xrightarrow{+} \Sigma F_x = 0; \quad -F_{DA} \cos 30° + 4.10 \cos 30° \text{ kN} = 0$$

$$F_{DA} = 4.10 \text{ kN} \quad (\text{T}) \qquad\qquad Ans.$$

$$+\uparrow \Sigma F_y = 0; \qquad F_{DB} - 2(4.10 \sin 30° \text{ kN}) = 0$$

$$F_{DB} = 4.10 \text{ kN} \quad (\text{T}) \qquad\qquad Ans.$$

NOTE: The force in the last member, BA, can be obtained from joint B or joint A. As an exercise, draw the free-body diagram of joint B, sum the forces in the horizontal direction, and show that $F_{BA} = 0.776 \text{ kN}$ (C).

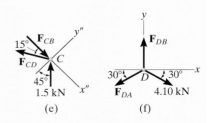

(e)   (f)

Fig. 6–9

EXAMPLE 6.3

Determine the force in each member of the truss shown in Fig. 6–10a. Indicate whether the members are in tension or compression.

(a)   (b)

Fig. 6–10

SOLUTION

Support Reactions. No joint can be analyzed until the support reactions are determined. Why? A free-body diagram of the entire truss is given in Fig. 6–10b. Applying the equations of equilibrium, we have

$$\xrightarrow{+} \Sigma F_x = 0; \qquad 600\ \text{N} - C_x = 0 \qquad C_x = 600\ \text{N}$$
$$\downarrow + \Sigma M_C = 0; \qquad -A_y(6\ \text{m}) + 400\ \text{N}(3\ \text{m}) + 600\ \text{N}(4\ \text{m}) = 0$$
$$A_y = 600\ \text{N}$$
$$+\uparrow \Sigma F_y = 0; \qquad 600\ \text{N} - 400\ \text{N} - C_y = 0 \qquad C_y = 200\ \text{N}$$

The analysis can now start at either joint A or C. The choice is arbitrary since there are one known and two unknown member forces acting on the pin at each of these joints.

Joint A (Fig. 6–10c). As shown on the free-body diagram, there are three forces that act on the pin at joint A. The inclination of $\mathbf{F}_{AB}$ is determined from the geometry of the truss. By inspection, can you see why this force is assumed to be compressive and $\mathbf{F}_{AD}$ tensile? Applying the equations of equilibrium, we have

(c)

$$+\uparrow \Sigma F_y = 0; \qquad 600\ \text{N} - \tfrac{4}{5}F_{AB} = 0 \qquad F_{AB} = 750\ \text{N} \quad (C) \qquad Ans.$$
$$\xrightarrow{+} \Sigma F_x = 0; \qquad F_{AD} - \tfrac{3}{5}(750\ \text{N}) = 0 \qquad F_{AD} = 450\ \text{N} \quad (T) \qquad Ans.$$

Joint D   (Fig. 6–10d). The pin at this joint is chosen next since, by inspection of Fig. 6–10a, the force in AD is known and the unknown forces in DB and DC can be determined. Summing forces in the horizontal direction, Fig. 6–10d, we have

(d)

$$\xrightarrow{+} \Sigma F_x = 0; \qquad -450\text{ N} + \tfrac{3}{5} F_{DB} + 600\text{ N} = 0 \quad F_{DB} = -250\text{ N}$$

The negative sign indicates that $\mathbf{F}_{DB}$ acts in the *opposite sense* to that shown in Fig. 6–10d.* Hence,

$$F_{DB} = 250\text{ N} \qquad (\text{T}) \qquad\qquad Ans.$$

To determine $\mathbf{F}_{DC}$, we can either correct the sense of $\mathbf{F}_{DB}$ and then apply $\Sigma F_y = 0$, or apply this equation and retain the negative sign for F_{DB}, i.e.,

$$+\uparrow \Sigma F_y = 0; \quad -F_{DC} - \tfrac{4}{5}(-250\text{ N}) = 0 \qquad F_{DC} = 200\text{ N} \quad (\text{C}) \;\; Ans.$$

Joint C   (Fig. 6–10e).

$$\xrightarrow{+} \Sigma F_x = 0; \qquad F_{CB} - 600\text{ N} = 0 \qquad F_{CB} = 600\text{ N} \quad (\text{C}) \qquad\qquad Ans.$$

$$+\uparrow \Sigma F_y = 0; \qquad 200\text{ N} - 200\text{ N} \equiv 0 \quad (\text{check})$$

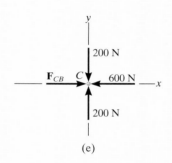

(e)

NOTE: The analysis is summarized in Fig. 6–10f, which shows the correct free-body diagram for each pin and member.

(f)

*The proper sense could have been determined by inspection, prior to applying $\Sigma F_x = 0$.

6.3 Zero-Force Members

Truss analysis using the method of joints is greatly simplified if we can first determine those members which support *no loading*. These *zero-force members* are used to increase the stability of the truss during construction and to provide support if the applied loading is changed.

The zero-force members of a truss can generally be determined *by inspection* of each of its joints. For example, consider the truss shown in Fig. 6–11*a*. If a free-body diagram of the pin at joint *A* is drawn, Fig. 6–11*b*, it is seen that members *AB* and *AF* are zero-force members. On the other hand, notice that we could not have come to this conclusion if we had considered the free-body diagrams of joints *F* or *B* simply because there are five unknowns at each of these joints. In a similar manner, consider the free-body diagram of joint *D*, Fig. 6–11*c*. Here again it is seen that *DC* and *DE* are zero-force members. As a general rule, *if only two members form a truss joint and no external load or support reaction is applied to the joint, the members must be zero-force members*. The load on the truss in Fig. 6–11*a* is therefore supported by only five members as shown in Fig. 6–11*d*.

(a)

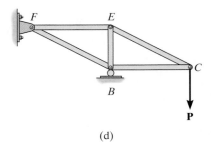

(b)

$$\xrightarrow{+} \Sigma F_x = 0; \quad F_{AB} = 0$$
$$+\uparrow \Sigma F_y = 0; \quad F_{AF} = 0$$

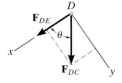

$$+\searrow \Sigma F_y = 0; F_{DC} \sin \theta = 0; \quad F_{DC} = 0 \text{ since } \sin \theta \neq$$
$$+\swarrow \Sigma F_x = 0; F_{DE} + 0 = 0; \quad F_{DE} = 0$$

(c)

(d)

Fig. 6–11

Now consider the truss shown in Fig. 6–12a. The free-body diagram of the pin at joint D is shown in Fig. 6–12b. By orienting the y axis along members DC and DE and the x axis along member DA, it is seen that DA is a zero-force member. Note that this is also the case for member CA, Fig. 6–12c. In general, *if three members form a truss joint for which two of the members are collinear, the third member is a zero-force member provided no external force or support reaction is applied to the joint.* The truss shown in Fig. 6–12d is therefore suitable for supporting the load **P**.

(a)

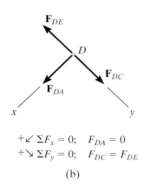

$$+\swarrow \Sigma F_x = 0; \quad F_{DA} = 0$$
$$+\searrow \Sigma F_y = 0; \quad F_{DC} = F_{DE}$$

(b)

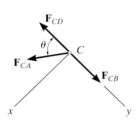

$$+\swarrow \Sigma F_x = 0; \quad F_{CA} \sin\theta = 0; \quad F_{CA} = 0 \text{ since } \sin\theta \neq 0;$$
$$+\searrow \Sigma F_y = 0; \quad F_{CB} = F_{CD}$$

(c)

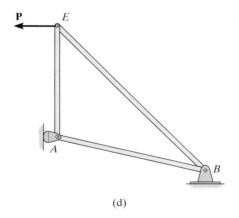

(d)

Fig. 6–12

EXAMPLE 6.4

(b)

(c)

(d)

(e)

(f)

Using the method of joints, determine all the zero-force members of the *Fink roof truss* shown in Fig. 6–13a. Assume all joints are pin connected.

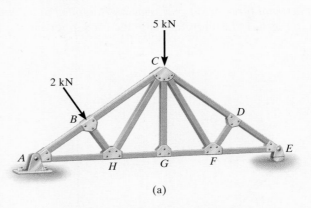

(a)

Fig. 6–13

SOLUTION

Look for joint geometries that have three members for which two are collinear. We have

Joint G   (Fig. 6–13b).

$$+\uparrow \Sigma F_y = 0; \qquad F_{GC} = 0 \qquad\qquad\qquad Ans.$$

Realize that we could not conclude that GC is a zero-force member by considering joint C, where there are five unknowns. The fact that GC is a zero-force member means that the 5-kN load at C must be supported by members CB, CH, CF, and CD.

Joint D   (Fig. 6–13c).

$$+\swarrow \Sigma F_x = 0; \qquad F_{DF} = 0 \qquad\qquad\qquad Ans.$$

Joint F   (Fig. 6–13d).

$$+\uparrow \Sigma F_y = 0; \quad F_{FC}\cos\theta = 0 \quad \text{Since } \theta \neq 90°, \quad F_{FC} = 0 \quad Ans.$$

NOTE: If joint B is analyzed, Fig. 6–13e,

$$+\searrow \Sigma F_x = 0; \qquad\qquad 2 \text{ kN} - F_{BH} = 0 \quad F_{BH} = 2 \text{ kN} \qquad \text{(C)}$$

Also, F_{HC} must satisfy $\Sigma F_y = 0$, Fig. 6–13f, and therefore HC is *not* a zero-force member.

PROBLEMS

6–1. Determine the force in each member of the truss and state if the members are in tension or compression. Set $P_1 = 7$ kN, $P_2 = 7$ kN.

6–2. Determine the force in each member of the truss and state if the members are in tension or compression. Set $P_1 = 8$ kN, $P_2 = 10$ kN.

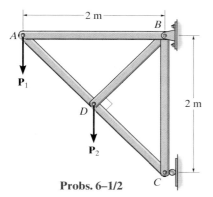

Probs. 6–1/2

6–3. The truss, used to support a balcony, is subjected to the loading shown. Approximate each joint as a pin and determine the force in each member. State whether the members are in tension or compression. Set $P_1 = 600$ lb, $P_2 = 400$ lb.

*6–4. The truss, used to support a balcony, is subjected to the loading shown. Approximate each joint as a pin and determine the force in each member. State whether the members are in tension or compression. Set $P_1 = 800$ lb, $P_2 = 0$.

Probs. 6–3/4

6–5. Determine the force in each member of the truss and state if the members are in tension or compression. Set $P_1 = 20$ kN, $P_2 = 10$ kN.

6–6. Determine the force in each member of the truss and state if the members are in tension or compression. Set $P_1 = 40$ kN, $P_2 = 20$ kN.

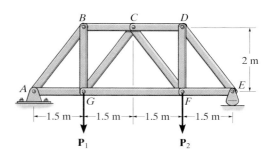

Probs. 6–5/6

6–7. Determine the force in each member of the truss and state if the members are in tension or compression.

Prob. 6–7

*6–8.   Determine the force in each member of the truss in terms of the external loading and state if the members are in tension or compression.

6–9.   The maximum allowable tensile force in the members of the truss is $(F_t)_{max} = 1500$ lb, and the maximum allowable compressive force is $(F_c)_{max} = 800$ lb. Determine the maximum magnitude P of the two loads that can be applied to the truss. Take $a = 8$ ft.

*6–12.   Determine the force in each member of the truss and state if the members are in tension or compression. Set $P_1 = 10$ kN, $P_2 = 15$ kN.

6–13.   Determine the force in each member of the truss and state if the members are in tension or compression. Set $P_1 = 0$, $P_2 = 20$ kN.

Probs. 6–8/9

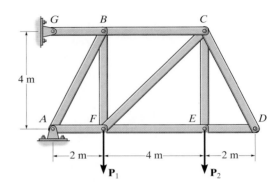

Probs. 6–12/13

6–10.   Determine the force in each member of the truss and state if the members are in tension or compression. Set $P_1 = 0$, $P_2 = 1000$ lb.

6–11.   Determine the force in each member of the truss and state if the members are in tension or compression. Set $P_1 = 500$ lb, $P_2 = 1500$ lb.

6–14.   Determine the force in each member of the truss and state if the members are in tension or compression. Set $P_1 = 100$ lb, $P_2 = 200$ lb, $P_3 = 300$ lb.

6–15.   Determine the force in each member of the truss and state if the members are in tension or compression. Set $P_1 = 400$ lb, $P_2 = 400$ lb, $P_3 = 0$.

Probs. 6–10/11

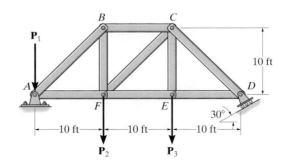

Probs. 6–14/15

*6–16.   Determine the force in each member of the truss in terms of the load P and state if the members are in tension or compression.

6–17.   The maximum allowable tensile force in the members of the truss is $(F_t)_{max} = 5$ kN, and the maximum allowable compressive force is $(F_c)_{max} = 3$ kN. Determine the maximum magnitude of the load **P** that can be applied to the truss. Take $d = 2$ m.

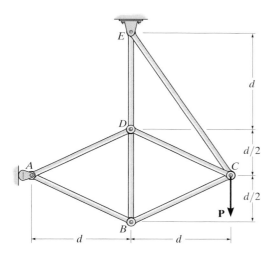

Probs. 6–16/17

6–18.   Determine the force in each member of the truss and state if the members are in tension or compression. *Hint:* The horizontal force component at A must be zero. Why?

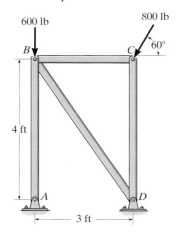

Prob. 6–18

6–19.   Determine the force in each member of the truss and state if the members are in tension or compression. *Hint:* The resultant force at the pin E acts along member ED. Why?

*6–20.   Each member of the truss is uniform and has a mass of 8 kg/m. Remove the external loads of 3 kN and 2 kN and determine the approximate force in each member due to the weight of the truss. State if the members are in tension or compression. Solve the problem by *assuming* the weight of each member can be represented as a vertical force, half of which is applied at each end of the member.

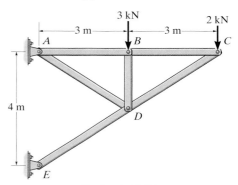

Probs. 6–19/20

6–21.   Determine the force in each member of the truss in terms of the external loading and state if the members are in tension or compression.

6–22.   The maximum allowable tensile force in the members of the truss is $(F_t)_{max} = 2$ kN, and the maximum allowable compressive force is $(F_c)_{max} = 1.2$ kN. Determine the maximum magnitude P of the two loads that can be applied to the truss. Take $L = 2$ m and $\theta = 30°$.

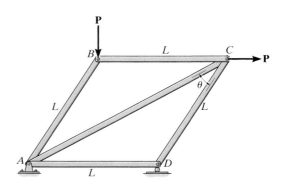

Probs. 6–21/22

6–23.   The *Fink truss* supports the loads shown. Determine the force in each member and state if the members are in tension or compression. Approximate each joint as a pin.

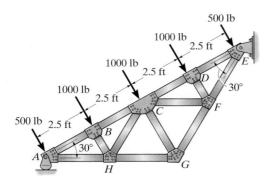

Prob. 6–23

*6–24.   Determine the force in each member of the double scissors truss in terms of the load *P* and state if the members are in tension or compression.

Prob. 6–24

6–25.   Determine the force in each member of the truss and state if the members are in tension or compression. *Hint:* The vertical component of force at *C* must equal zero. Why?

6–26.   Each member of the truss is uniform and has a mass of 8 kg/m. Remove the external loads of 6 kN and 8 kN and determine the approximate force in each member due to the weight of the truss. State if the members are in tension or compression. Solve the problem by *assuming* the weight of each member can be represented as a vertical force, half of which is applied at each end of the member.

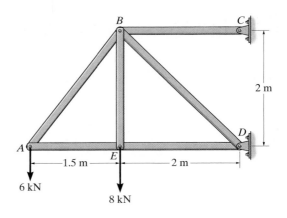

Probs. 6–25/26

6–27.   Determine the force in each member of the truss and state if the members are in tension or compression. Set $P_1 = 4$ kN, $P_2 = 0$.

*6–28.   Determine the force in each member of the truss and state if the members are in tension or compression. Set $P_1 = 2$ kN, $P_2 = 4$ kN.

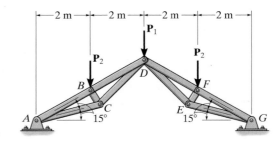

Probs. 6–27/28

6–29.   Determine the force in each member of the truss and state if the members are in tension or compression.

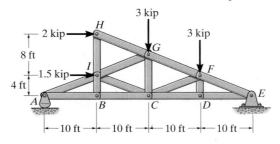

Prob. 6–29

6.4 The Method of Sections

When we need to find the force in only a few members of a truss, we can analyze the truss using the *method of sections*. It is based on the principle that if a body is in equilibrium then any part of the body is also in equilibrium. For example, consider the two truss members shown on the left in Fig. 6–14. If the forces within the members are to be determined, then an imaginary section indicated by the blue line, can be used to cut each member into two parts and thereby "expose" each internal force as "external" to the free-body diagrams shown on the right. Clearly, it can be seen that equilibrium requires that the member in tension (T) be subjected to a "pull," whereas the member in compression (C) is subjected to a "push."

The method of sections can also be used to "cut" or section the members of an entire truss. If the section passes through the truss and the free-body diagram of either of its two parts is drawn, we can then apply the equations of equilibrium to that part to determine the member forces at the "cut section." Since only *three* independent equilibrium equations ($\Sigma F_x = 0$, $\Sigma F_y = 0$, $\Sigma M_O = 0$) can be applied to the isolated part of the truss, try to select a section that, in general, passes through not more than *three* members in which the forces are unknown. For example, consider the truss in Fig. 6–15a. If the force in member *GC* is to be determined, section *aa* would be appropriate. The free-body diagrams of the two parts are shown in Figs. 6–15b and 6–15c. In particular, note that the line of action of each member force is specified from the *geometry* of the truss, since the force in a member passes along its axis. Also, the member forces acting on one part of the truss are equal but opposite to those acting on the other part—Newton's third law. As noted above, the members assumed to be in *tension* (*BC* and *GC*) are subjected to a "pull," whereas the member in *compression* (*GF*) is subjected to a "push."

Tension

Compression

Fig. 6–14

(a)

(b)

(c)

Fig. 6–15

(a)

(c)

Fig. 6–15

Two Pratt trusses are used to construct this pedestrian bridge.

The three unknown member forces $\mathbf{F}_{BC}$, $\mathbf{F}_{GC}$, and $\mathbf{F}_{GF}$ can be obtained by applying the three equilibrium equations to the free-body diagram in Fig. 6–15b. If, however, the free-body diagram in Fig. 6–15c is considered, the three support reactions $\mathbf{D}_x$, $\mathbf{D}_y$ and $\mathbf{E}_x$ will have to be known, because only three equations of equilibrium are available. (This, of course, is done in the usual manner by considering a free-body diagram of the *entire truss.*)

When applying the equilibrium equations, one should consider ways of writing the equations so as to yield a *direct solution* for each of the unknowns, rather than having to solve simultaneous equations. For example, summing moments about C in Fig. 6–15b would yield a direct solution for $\mathbf{F}_{GF}$ since $\mathbf{F}_{BC}$ and $\mathbf{F}_{GC}$ create zero moment about C. Likewise, $\mathbf{F}_{BC}$ can be directly obtained by summing moments about G. Finally, $\mathbf{F}_{GC}$ can be found directly from a force summation in the vertical direction since $\mathbf{F}_{GF}$ and $\mathbf{F}_{BC}$ have no vertical components. This ability to *determine directly* the force in a particular truss member is one of the main advantages of using the method of sections.*

*By comparison, if the method of joints were used to determine, say, the force in member GC, it would be necessary to analyze joints A, B, and G in sequence.

As in the method of joints, there are two ways in which one can determine the correct sense of an unknown member force:

- *Always assume* that the unknown member forces at the cut section are in *tension*, i.e., "pulling" on the member. By doing this, the numerical solution of the equilibrium equations will yield *positive scalars for members in tension and negative scalars for members in compression.*

- The correct sense of an unknown member force can in many cases be determined "by inspection." For example, $\mathbf{F}_{BC}$ is a tensile force as represented in Fig. 6–15*b* since moment equilibrium about *G* requires that $\mathbf{F}_{BC}$ create a moment opposite to that of the 1000-N force. Also, $\mathbf{F}_{GC}$ is tensile since its vertical component must balance the 1000-N force which acts downward. In more complicated cases, the sense of an unknown member force may be *assumed.* If the solution yields a *negative* scalar, it indicates that the force's sense is *opposite* to that shown on the free-body diagram. This is the method we will use in the example problems which follow.

Procedure for Analysis

The forces in the members of a truss may be determined by the method of sections using the following procedure.

Free-Body Diagram.

- Make a decision as to how to "cut" or section the truss through the members where forces are to be determined.

- Before isolating the appropriate section, it may first be necessary to determine the truss's *external* reactions. Then three equilibrium equations are available to solve for member forces at the cut section.

- Draw the free-body diagram of that part of the sectioned truss which has the least number of forces acting on it.

- Use one of the two methods described above for establishing the sense of an unknown member force.

Equations of Equilibrium.

- Moments should be summed about a point that lies at the intersection of the lines of action of two unknown forces, so that the third unknown force is determined directly from the moment equation.

- If two of the unknown forces are *parallel*, forces may be summed *perpendicular* to the direction of these unknowns to determine *directly* the third unknown force.

EXAMPLE 6.5

(a)

(b)

(c)

Fig. 6–16

Determine the force in members *GE*, *GC*, and *BC* of the truss shown in Fig. 6–16*a*. Indicate whether the members are in tension or compression.

SOLUTION

Section *aa* in Fig. 6–16*a* has been chosen since it cuts through the *three* members whose forces are to be determined. In order to use the method of sections, however, it is *first* necessary to determine the external reactions at *A* or *D*. Why? A free-body diagram of the entire truss is shown in Fig. 6–16*b*. Applying the equations of equilibrium, we have

$$\xrightarrow{+}\ \Sigma F_x = 0; \qquad 400\ \text{N} - A_x = 0 \qquad A_x = 400\ \text{N}$$

$$\zeta+\Sigma M_A = 0; \qquad -1200\ \text{N}(8\ \text{m}) - 400\ \text{N}(3\ \text{m}) + D_y(12\ \text{m}) = 0$$

$$D_y = 900\ \text{N}$$

$$+\uparrow\Sigma F_y = 0; \qquad A_y - 1200\ \text{N} + 900\ \text{N} = 0 \qquad A_y = 300\ \text{N}$$

Free-Body Diagram.   The free-body diagram of the left portion of the sectioned truss is shown in Fig. 6–16*c*. For the analysis this diagram will be used since it involves the least number of forces.

Equations of Equilibrium.   Summing moments about point *G* eliminates $\mathbf{F}_{GE}$ and $\mathbf{F}_{GC}$ and yields a direct soluion for F_{BC}.

$$\zeta+\Sigma M_G = 0; \qquad -300\ \text{N}(4\ \text{m}) - 400\ \text{N}(3\ \text{m}) + F_{BC}(3\ \text{m}) = 0$$

$$F_{BC} = 800\ \text{N}\quad(\text{T}) \qquad\qquad Ans.$$

In the same manner, by summing moments about point *C* we obtain a direct solution for F_{GE}.

$$\zeta+\Sigma M_C = 0; \qquad -300\ \text{N}(8\ \text{m}) + F_{GE}(3\ \text{m}) = 0$$

$$F_{GE} = 800\ \text{N}\quad(\text{C}) \qquad\qquad Ans.$$

Since $\mathbf{F}_{BC}$ and $\mathbf{F}_{GE}$ have no vertical components, summing forces in the *y* direction directly yields F_{GC}, i.e.,

$$+\uparrow\Sigma F_y = 0; \qquad 300\ \text{N} - \tfrac{3}{5}F_{GC} = 0$$

$$F_{GC} = 500\ \text{N}\quad(\text{T}) \qquad\qquad Ans.$$

NOTE: As an exercise, obtain these results by applying the equations of equilibrium to the free-body diagram of the right portion of the sectioned truss. You should be able to tell, by inspection, the proper direction for each member force. For example, $\Sigma M_C = 0$ requires $\mathbf{F}_{GE}$ to be compressive to balance the moment of the 300-N force about *C*.

EXAMPLE 6.6

Determine the force in member CF of the bridge truss shown in Fig. 6–17a. Indicate whether the member is in tension or compression. Assume each member is pin connected.

(a)

(b)

Fig. 6–17

SOLUTION

Free-Body Diagram. Section aa in Fig. 6–17a will be used since this section will "expose" the internal force in member CF as "external" on the free-body diagram of either the right or left portion of the truss. It is first necessary, however, to determine the external reactions on either the left or right side. Verify the results shown on the free-body diagram in Fig. 6–17b.

The free-body diagram of the right portion of the truss, which is the easiest to analyze, is shown in Fig. 6–17c. There are three unknowns, F_{FG}, F_{CF}, and F_{CD}.

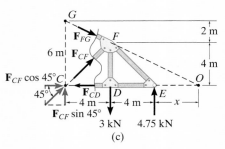

(c)

Equations of Equilibrium. The most direct method for solving this problem requires application of the moment equation about a point that eliminates two of the unknown forces. Hence, to obtain $\mathbf{F}_{CF}$, we will eliminate $\mathbf{F}_{FG}$ and $\mathbf{F}_{CD}$ by summing moments about point O, Fig. 6–17c. Note that the location of point O measured from E is determined from proportional triangles, i.e., $4/(4 + x) = 6/(8 + x)$, $x = 4$ m. Or, stated in another manner, the slope of member GF has a drop of 2 m to a horizontal distance of 4 m. Since FD is 4 m, Fig. 6–17c, then from D to O the distance must be 8 m.

An easy way to determine the moment of $\mathbf{F}_{CF}$ about point O is to use the principle of transmissibility and move $\mathbf{F}_{CF}$ to point C, and then resolve $\mathbf{F}_{CF}$ into its two rectangular components. We have

$$\zeta + \Sigma M_O = 0;$$
$$-F_{CF} \sin 45°(12 \text{ m}) + (3 \text{ kN})(8 \text{ m}) - (4.75 \text{ kN})(4 \text{ m}) = 0$$
$$F_{CF} = 0.589 \text{ kN} \quad \text{(C)} \qquad\qquad Ans.$$

EXAMPLE 6.7

(a)

Fig. 6–18

Determine the force in member EB of the roof truss shown in Fig. 6–18a. Indicate whether the member is in tension or compression.

SOLUTION

Free-Body Diagrams.   By the method of sections, any imaginary vertical section that cuts through EB, Fig. 6–18a, will also have to cut through three other members for which the forces are unknown. For example, section aa cuts through ED, EB, FB, and AB. If the components of reaction at A are calculated first ($A_x = 0$, $A_y = 4000$ N) and a free-body diagram of the left side of this section is considered, Fig. 6–18b, it is possible to obtain $\mathbf{F}_{ED}$ by summing moments about B to eliminate the other three unknowns; however, $\mathbf{F}_{EB}$ cannot be determined from the remaining two equilibrium equations. One possible way of obtaining $\mathbf{F}_{EB}$ is first to determine $\mathbf{F}_{ED}$ from section aa, then use this result on section bb, Fig. 6–18a, which is shown in Fig. 6–18c. Here the force system is concurrent and our sectioned free-body diagram is the same as the free-body diagram for the pin at E (method of joints).

(c)

Equations of Equilibrium.   In order to determine the moment of $\mathbf{F}_{ED}$ about point B, Fig. 6–18b, we will resolve the force into its rectangular components and, by the principle of transmissibility, extend it to point C as shown. The moments of 1000 N, F_{AB}, F_{FB}, F_{EB}, and $F_{ED} \cos 30°$ are all zero about B. Therefore,

$$\zeta + \Sigma M_B = 0; \quad 1000 \text{ N}(4 \text{ m}) + 3000 \text{ N}(2 \text{ m}) - 4000 \text{ N}(4 \text{ m})$$
$$+ F_{ED} \sin 30°(4) = 0$$
$$F_{ED} = 3000 \text{ N} \quad (C)$$

Considering now the free-body diagram of section bb, Fig. 6–18c, we have

$$\xrightarrow{+} \Sigma F_x = 0; \quad F_{EF} \cos 30° - 3000 \cos 30° \text{ N} = 0$$
$$F_{EF} = 3000 \text{ N} \quad (C)$$
$$+\uparrow \Sigma F_y = 0; \quad 2(3000 \sin 30° \text{ N}) - 1000 \text{ N} - F_{EB} = 0$$
$$F_{EB} = 2000 \text{ N} \quad (T) \qquad Ans.$$

PROBLEMS

6–30. The *Howe bridge truss* is subjected to the loading shown. Determine the force in members *DE, EH*, and *HG*, and state if the members are in tension or compression.

6–33. The roof truss supports the vertical loading shown. Determine the force in members *BC, CK*, and *KJ* and state if these members are in tension or compression.

Prob. 6–30

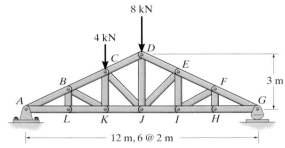

Prob. 6–33

6–31. The *Pratt bridge truss* is subjected to the loading shown. Determine the force in members *LD, LK, CD*, and *KD*, and state if the members are in tension or compression.

*6–32. The *Pratt bridge truss* is subjected to the loading shown. Determine the force in members *JI, JE*, and *DE*, and state if the members are in tension or compression.

6–34. Determine the force in members *CD, CJ, KJ*, and *DJ* of the truss which serves to support the deck of a bridge. State if these members are in tension or compression.

6–35. Determine the force in members *EI* and *JI* of the truss which serves to support the deck of a bridge. State if these members are in tension or compression.

Probs. 6–31/32

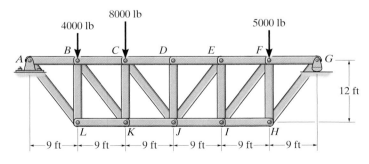

Probs. 6–34/35

*6–36. Determine the force in members *BE, EF,* and *CB,* and state if the members are in tension or compression.

6–37. Determine the force in members *BF, BG,* and *AB,* and state if the members are in tension or compression.

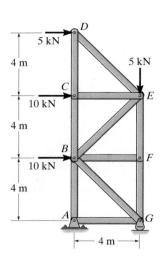

Probs. 6–36/37

6–38. Determine the force developed in members *GB* and *GF* of the bridge truss and state if these members are in tension or compression.

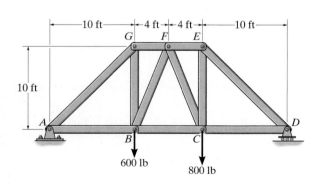

Prob. 6–38

6–39. Determine the force members *BC, FC,* and *FE,* and state if the members are in tension or compression.

Prob. 6–39

*6–40. Determine the force in members *IC* and *CG* of the truss and state if these members are in tension or compression. Also, indicate all zero-force members.

6–41. Determine the force in members *JE* and *GF* of the truss and state if these members are in tension or compression. Also, indicate all zero-force members.

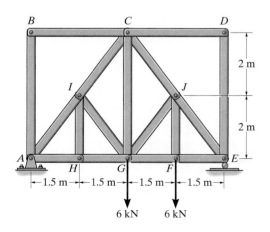

Probs. 6–40/41

6–42. Determine the force in members *BC, HC*, and *HG*. After the truss is sectioned use a single equation of equilibrium for the calculation of each force. State if these members are in tension or compression.

6–43. Determine the force in members *CD, CF*, and *CG* and state if these members are in tension or compression.

6–45. Determine the force in member *GJ* of the truss and state if this member is in tension or compression.

6–46. Determine the force in member *GC* of the truss and state if this member is in tension or compression.

Probs. 6–42/43

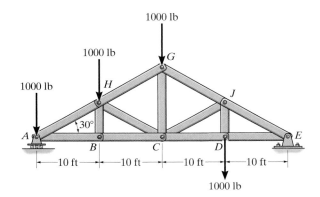

Probs. 6–45/46

*6–44. Determine the force in members *OE, LE*, and *LK* of the *Baltimore truss* and state if the members are in tension or compression.

6–47. Determine the force in members *KJ, JN*, and *CD*, and state if the members are in tension or compression. Also indicate all zero-force members.

Prob. 6–44

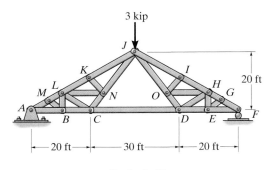

Prob. 6–47

*6–48. Determine the force in members *BG, HG,* and *BC* of the truss and state if the members are in tension or compression.

Prob. 6–48

6–49. The skewed truss carries the load shown. Determine the force in members *CB, BE,* and *EF* and state if these members are in tension or compression. Assume that all joints are pinned.

6–50. The skewed truss carries the load shown. Determine the force in members *AB, BF,* and *EF* and state if these members are in tension or compression. Assume that all joints are pinned.

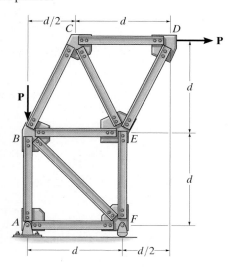

Probs. 6–49/50

6–51. Determine the force developed in members *BC* and *CH* of the roof truss and state if the members are in tension or compression.

*6–52. Determine the force in members *CD* and *GF* of the truss and state if the members are in tension or compression. Also indicate all zero-force members.

Probs. 6–51/52

6–53. Determine the force in members *DE, DL,* and *ML* of the roof truss and state if the members are in tension or compression.

6–54. Determine the force in members *EF* and *EL* of the roof truss and state if the members are in tension or compression.

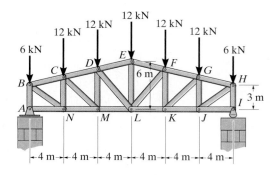

Probs. 6–53/54

*6.5 Space Trusses

A *space truss* consists of members joined together at their ends to form a stable three-dimensional structure. The simplest element of a space truss is a *tetrahedron*, formed by connecting six members together, as shown in Fig. 6–19. Any additional members added to this basic element would be redundant in supporting the force **P**. A *simple space truss* can be built from this basic tetrahedral element by adding three additional members and a joint, forming a system of multiconnected tetrahedrons.

Assumptions for Design. The members of a space truss may be treated as two-force members provided the external loading is applied at the joints and the joints consist of ball-and-socket connections. These assumptions are justified if the welded or bolted connections of the joined members intersect at a common point and the weight of the members can be neglected. In cases where the weight of a member is to be included in the analysis, it is generally satisfactory to apply it as a vertical force, half of its magnitude applied at each end of the member.

Fig. 6–19

Procedure for Analysis

Either the method of joints or the method of sections can be used to determine the forces developed in the members of a simple space truss.

Method of Joints.

Generally, if the forces in *all* the members of the truss must be determined, the method of joints is most suitable for the analysis. When using the method of joints, it is necessary to solve the three scalar equilibrium equations $\Sigma F_x = 0$, $\Sigma F_y = 0$, $\Sigma F_z = 0$ at each joint. The solution of many simultaneous equations can be avoided if the force analysis begins at a joint having at least one known force and at most three unknown forces. If the three-dimensional geometry of the force system at the joint is hard to visualize, it is recommended that a Cartesian vector analysis be used for the solution.

Method of Sections.

If only a *few* member forces are to be determined, the method of sections may be used. When an imaginary section is passed through a truss and the truss is separated into two parts, the force system acting on one of the parts must satisfy the *six* scalar equilibrium equations: $\Sigma F_x = 0$, $\Sigma F_y = 0$, $\Sigma F_z = 0$, $\Sigma M_x = 0$, $\Sigma M_y = 0$, $\Sigma M_z = 0$ (Eqs. 5–6). By proper choice of the section and axes for summing forces and moments, many of the unknown member forces in a space truss can be computed *directly*, using a single equilibrium equation.

Typical roof-supporting space truss. Notice the use of ball-and-socket joints for the connections.

EXAMPLE 6.8

(a)

(b)

(c)

Fig. 6–20

Determine the forces acting in the members of the space truss shown in Fig. 6–20a. Indicate whether the members are in tension or compression.

SOLUTION
Since there are one known force and three unknown forces acting at joint A, the force analysis of the truss will begin at this joint.

Joint A   (Fig. 6–20b). Expressing each force that acts on the free-body diagram of joint A in vector notation, we have

$$\mathbf{P} = \{-4\mathbf{j}\} \text{ kN}, \quad \mathbf{F}_{AB} = F_{AB}\mathbf{j}, \quad \mathbf{F}_{AC} = -F_{AC}\mathbf{k},$$

$$\mathbf{F}_{AE} = F_{AE}\left(\frac{\mathbf{r}_{AE}}{r_{AE}}\right) = F_{AE}(0.577\mathbf{i} + 0.577\mathbf{j} - 0.577\mathbf{k})$$

For equilibrium,

$$\Sigma \mathbf{F} = \mathbf{0}; \qquad\qquad \mathbf{P} + \mathbf{F}_{AB} + \mathbf{F}_{AC} + \mathbf{F}_{AE} = \mathbf{0}$$

$$-4\mathbf{j} + F_{AB}\mathbf{j} - F_{AC}\mathbf{k} + 0.577F_{AE}\mathbf{i} + 0.577F_{AE}\mathbf{j} - 0.577F_{AE}\mathbf{k} = \mathbf{0}$$

$$\Sigma F_x = 0; \qquad\qquad 0.577F_{AE} = 0$$

$$\Sigma F_y = 0; \qquad -4 + F_{AB} + 0.577F_{AE} = 0$$

$$\Sigma F_z = 0; \qquad\quad -F_{AC} - 0.577F_{AE} = 0$$

$$F_{AC} = F_{AE} = 0 \qquad\qquad Ans.$$

$$F_{AB} = 4 \text{ kN} \quad (\text{T}) \qquad Ans.$$

Since F_{AB} is known, joint B may be analyzed next.

Joint B   (Fig. 6–20c).

$$\Sigma F_x = 0; \qquad -R_B \cos 45° + 0.707F_{BE} = 0$$

$$\Sigma F_y = 0; \qquad\qquad -4 + R_B \sin 45° = 0$$

$$\Sigma F_z = 0; \qquad\qquad 2 + F_{BD} - 0.707F_{BE} = 0$$

$$R_B = F_{BE} = 5.66 \text{ kN} \quad (\text{T}), \quad F_{BD} = 2 \text{ kN} \quad (\text{C}) \qquad Ans.$$

The *scalar* equations of equilibrium may also be applied directly to the force systems on the free-body diagrams of joints D and C since the force components are easily determined. Show that

$$F_{DE} = F_{DC} = F_{CE} = 0 \qquad\qquad Ans.$$

PROBLEMS

6–55. Two space trusses are used to equally support the uniform 50-kg sign. Determine the force developed in members AB, AC, and BC of truss $ABCD$ and state if the members are in tension or compression. Horizontal short links support the truss at joints B and D and there is a ball-and-socket joint at C.

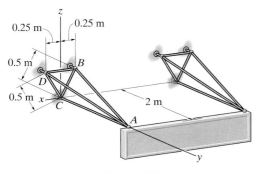

Prob. 6–55

*6–56. Determine the force in each member of the space truss and state if the members are in tension or compression. The truss is supported by short links at B, C, and D.

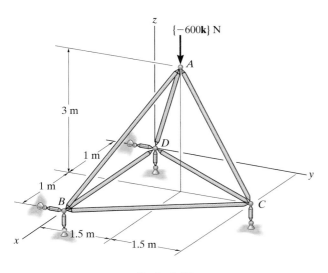

Prob. 6–56

6–57. Determine the force in each member of the space truss and state if the members are in tension or compression. The truss is supported by short links at A, B, and C.

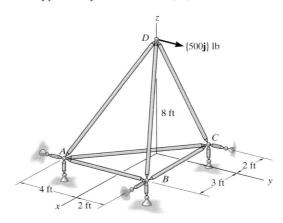

Prob. 6–57

6–58. The space truss is supported by a ball-and-socket joint at D and short links at C and E. Determine the force in each member and state if the members are in tension or compression. Take $\mathbf{F}_1 = \{-500\mathbf{k}\}$ lb and $\mathbf{F}_2 = \{400\mathbf{j}\}$ lb.

6–59. The space truss is supported by a ball-and-socket joint at D and short links at C and E. Determine the force in each member and state if the members are in tension or compression. Take $\mathbf{F}_1 = \{200\mathbf{i} + 300\mathbf{j} - 500\mathbf{k}\}$ lb and $\mathbf{F}_2 = \{400\mathbf{j}\}$ lb.

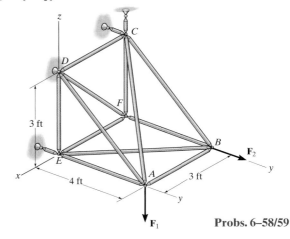

Probs. 6–58/59

*6–60. Determine the force in each member of the space truss and state if the members are in tension or compression. The truss is supported by a ball-and-socket joints at A, B, and E. Set **F** = {−200**i** + 400**j**} N. *Hint*: The support reaction at E acts along member E C. Why?

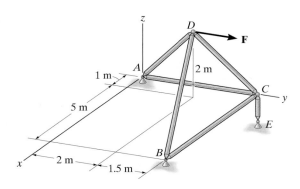

Prob. 6–60

6–61. Determine the force in each member of the space truss and state if the members are in tension or compression. The truss is supported by ball-and-socket joints at C, D, E, and G.

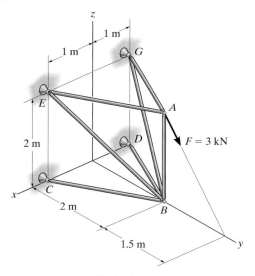

Prob. 6–61

6–62. Determine the force in members BD, AD, and AF of the space truss and state if the members are in tension or compression. The truss is supported by short links at A, B, D, and F.

6–63. Determine the force in members CF, EF, and DF of the space truss and state if the members are in tension or compression. The truss is supported by short links at A, B, D, and F.

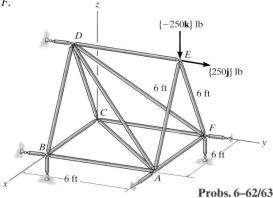

Probs. 6–62/63

*6–64. Determine the force developed in each member of the space truss and state if the members are in tension or compression. The crate has a weight of 150 lb.

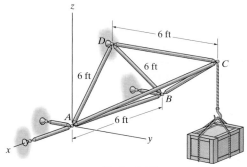

Prob. 6–64

6–65. The space truss is used to support vertical forces at joints B, C, and D. Determine the force in each member and state if the members are in tension or compression.

Prob. 6–65

6.6   Frames and Machines

Frames and machines are two common types of structures which are often composed of pin-connected *multiforce members*, i.e., members that are subjected to more than two forces. *Frames* are generally stationary and are used to support loads, whereas *machines* contain moving parts and are designed to transmit and alter the effect of forces. Provided a frame or machine is properly constrained and contains no more supports or members than are necessary to prevent collapse, the forces acting at the joints and supports can be determined by applying the equations of equilibrium to each member. Once the forces at the joints are obtained, it is then possible to *design* the size of the members, connections, and supports using the theory of mechanics of materials and an appropriate engineering design code.

Free-Body Diagrams.   In order to determine the forces acting at the joints and supports of a frame or machine, the structure must be disassembled and the free-body diagrams of its parts must be drawn. The following important points *must* be observed:

* Isolate each part by drawing its *outlined shape*. Then show all the forces and/or couple moments that act on the part. Make sure to *label* or *identify* each known and unknown force and couple moment with reference to an established *x, y* coordinate system. Also, indicate any dimensions used for taking moments. Most often the equations of equilibrium are easier to apply if the forces are represented by their rectangular components. As usual, the sense of an unknown force or couple moment can be assumed.

* Identify all the two-force members in the structure and represent their free-body diagrams as having two equal but opposite collinear forces acting at their points of application. (See Sec. 5.4.) By recognizing the two-force members, we can avoid solving an unnecessary number of equilibrium equations.

* Forces common to any two *contacting* members act with equal magnitudes but opposite sense on the respective members. If the two members are treated as a *"system" of connected members*, then these forces are *"internal"* and are *not shown* on the *free-body diagram of the system*; however, if the free-body diagram of *each member* is drawn, the forces are *"external"* and *must* be shown on each of the free-body diagrams.

The following examples graphically illustrate application of these points in drawing the free-body diagrams of a dismembered frame or machine. In all cases, the weight of the members is neglected.

EXAMPLE 6.9

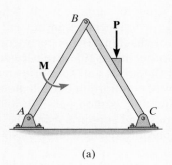

(a)

(b)

For the frame shown in Fig. 6–21a, draw the free-body diagram of (a) each member, (b) the pin at B, and (c) the two members connected together.

SOLUTION

Part (a). By inspection, members BA and BC are *not* two-force members. Instead, as shown on the free-body diagrams, Fig. 6–21b, BC is subjected to *not* five but *three forces*, namely, the resultant forces from the pins at B and C and the external force $\mathbf{P}$. Likewise, AB is subjected to the *resultant* forces from the pins at A and B and the external couple moment $\mathbf{M}$.

Part (b). It can be seen in Fig. 6–21a that the pin at B is subjected to only *two forces*, i.e., the force of member BC on the pin and the force of member AB on the pin. For *equilibrium* these forces and therefore their respective components must be equal but opposite, Fig. 6–21c. Notice carefully how Newton's third law is applied between the pin and its contacting members, i.e., the effect of the pin on the two members, Fig. 6–21b, and the equal but opposite effect of the two members on the pin, Fig. 6–21c. Also note that $\mathbf{B}_x$ and $\mathbf{B}_y$ shown equal but opposite in Fig. 6–21b on members AB and BC, is *not* the effect of Newton's third law; instead, this results from the *equilibrium* analysis of the pin, Fig. 6–21c.

Part (c). The free-body diagram of both members connected together, yet removed from the supporting pins at A and C, is shown in Fig. 6–21d. The force components $\mathbf{B}_x$ and $\mathbf{B}_y$ are *not shown* on this diagram since they form equal but opposite collinear pairs of *internal* forces (Fig. 6–21b) and therefore cancel out. Also, to be consistent when later applying the equilibrium equations, the unknown force components at A and C must act in the *same sense* as those shown in Fig. 6–21b. Here the couple moment $\mathbf{M}$ can be applied at any point on the frame in order to determine the reactions at A and C. Note, however, that it must act on member AB in Fig. 6–21b and *not* on member BC.

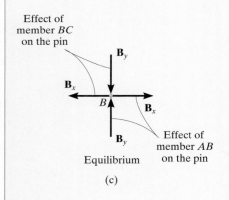

Effect of
member BC
on the pin

Effect of
member AB
on the pin

Equilibrium

(c)

(d)

Fig. 6–21

EXAMPLE 6.10

A constant tension in the conveyor belt is maintained by using the device shown in Fig. 6–22a. Draw the free-body diagrams of the frame and the cylinder that the belt surrounds. The suspended block has a weight of W.

(b)

(a)

Fig. 6–22

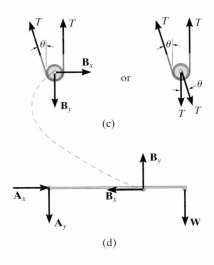

(c)

or

(d)

SOLUTION

The idealized model of the device is shown in Fig. 6–22b. Here the angle θ is assumed to be known. Notice that the tension in the belt is the same on each side of the cylinder, since the cylinder is free to turn. From this model, the free-body diagrams of the cylinder and frame are shown in Figs. 6–22c and 6–22d, respectively. Note that the force that the pin at B exerts on the cylinder can be represented by either its horizontal and vertical components $\mathbf{B}_x$ and $\mathbf{B}_y$, which can be determined by using the force equations of equilibrium applied to the cylinder, or by the two components T, which provide equal but opposite couple moments on the cylinder and thus keep it from turning. Also, realize that once the pin reactions at A have been determined, half of their values act on each side of the frame since pin connections occur on each side, Fig. 6–22a.

EXAMPLE 6.11

Draw the free-body diagram of each part of the smooth piston and link mechanism used to crush recycled cans, which is shown in Fig. 6–23a.

(a)

(b)

Fig. 6–23

(c)

SOLUTION

By inspection, member AB is a two-force member. The free-body diagrams of the parts are shown in Fig. 6–23b. Since the pins at B and D *connect only two parts together,* the forces there are shown as equal but opposite on the separate free-body diagrams of their connected members. In particular, four components of force act on the piston: $\mathbf{D}_x$ and $\mathbf{D}_y$ represent the effect of the pin (or lever EBD), $\mathbf{N}_w$ is the *resultant force* of the floor, and $\mathbf{P}$ is the resultant compressive force caused by the can C.

NOTE: A free-body diagram of the entire assembly is shown in Fig. 6–23c. Here the forces between the components are internal and are not shown on the free-body diagram.

EXAMPLE 6.12

For the frame shown in Fig. 6–24*a*, draw the free-body diagrams of (a) the entire frame including the pulleys and cords, (b) the frame without the pulleys and cords, and (c) each of the pulleys.

(a)

SOLUTION

Part (a). When the entire frame including the pulleys and cords is considered, the interactions at the points where the pulleys and cords are connected to the frame become pairs of *internal forces* which cancel each other and therefore are not shown on the free-body diagram, Fig. 6–24*b*.

Part (b). When the cords and pulleys are removed, their effect *on the frame* must be shown, Fig. 6–24*c*.

Part (c). The force components $\mathbf{B}_x$, $\mathbf{B}_y$, $\mathbf{C}_x$, $\mathbf{C}_y$ of the pins on the pulleys, Fig. 6–24*d*, are equal but opposite to the force components exerted by the pins on the frame, Fig. 6–24*c*. Why?

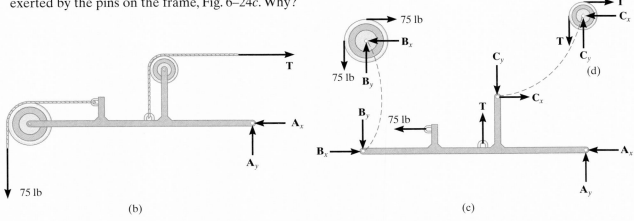

(b)

(c)

(d)

Fig. 6–24

EXAMPLE 6.13

(a)

Fig. 6–25

Draw the free-body diagrams of the bucket and the vertical boom of the back hoe shown in the photo, Fig. 6–25a. The bucket and its contents have a weight W. Neglect the weight of the members.

SOLUTION

The idealized model of the assembly is shown in Fig. 6–25b. Not shown are the required dimensions and angles that must be obtained, along with the location of the center of gravity G of the load. By inspection, members AB, BC, BE, and HI are all two-force members since they are pin connected at their end points and no other forces act on them. The free-body diagrams of the bucket and the boom are shown in Fig. 6–25c. Note that pin C is subjected to only two forces, the force of link BC and the force of the boom. For equilibrium, these forces must be equal in magnitude but opposite in direction, Fig. 6–25d. The pin at B is subjected to three forces, Fig. 6–25e. The force $\mathbf{F}_{BE}$ is caused by the hydraulic cylinder, and the forces $\mathbf{F}_{BA}$ and $\mathbf{F}_{BC}$ are caused by the links. These three forces are related by the two equations of force equilibrium applied to the pin.

NOTE: The free-body diagram of the entire assembly is shown in Fig. 6–25f.

(b)

(c)

(d)

(e)

(f)

Before proceeding, it is recommended to cover the solutions to the previous examples and attempt to draw the requested free-body diagrams. When doing so, make sure the work is neat and that all the forces and couple moments are properly labeled.

Equations of Equilibrium. Provided the structure (frame or machine) is properly supported and contains no more supports or members than are necessary to prevent its collapse, then the unknown forces at the supports and connections can be determined from the equations of equilibrium. If the structure lies in the x–y plane, then for *each* free-body diagram drawn the loading must satisfy $\Sigma F_x = 0$, $\Sigma F_y = 0$, and $\Sigma M_O = 0$. The selection of the free-body diagrams used for the analysis is *completely arbitrary*. They may represent each of the members of the structure, a portion of the structure, or its entirety. For example, consider finding the six components of the pin reactions at A, B, and C for the frame shown in Fig. 6–26a. If the frame is dismembered, as it is in Fig. 6–26b, these unknowns can be determined by applying the three equations of equilibrium to each of the two members (total of six equations). The free-body diagram of the *entire frame* can also be used for part of the analysis, Fig. 6–26c. Hence, if so desired, all six unknowns can be determined by applying the three equilibrium equations to the entire frame, Fig. 6–26c, and also to either one of its members. Furthermore, the answers can be checked in part by applying the three equations of equilibrium to the remaining "second" member. In general, then, this problem can be solved by writing *at most* six equilibrium equations using free-body diagrams of the members and/or the combination of connected members. Any more than six equations written would *not* be unique from the original six and would only serve to check the results.

(a)

(b)

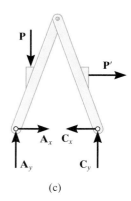

(c)

Fig. 6–26

Procedure For Analysis

The joint reactions on frames or machines (structures) composed of multiforce members can be determined using the following procedure.

Free-Body Diagram.

- Draw the free-body diagram of the entire structure, a portion of the structure, or each of its members. The choice should be made so that it leads to the most direct solution of the problem.

- When the free-body diagram of a group of members of a structure is drawn, the forces at the connected parts of this group are internal forces and are not shown on the free-body diagram of the group.

- Forces common to two members which are in contact act with equal magnitude but opposite sense on the respective free-body diagrams of the members.

- Two-force members, regardless of their shape, have equal but opposite collinear forces acting at the ends of the member.

- In many cases it is possible to tell by inspection the proper sense of the unknown forces acting on a member; however, if this seems difficult, the sense can be assumed.

- A couple moment is a free vector and can act at any point on the free-body diagram. Also, a force is a sliding vector and can act at any point along its line of action.

Equations of Equilibrium.

- Count the number of unknowns and compare it to the total number of equilibrium equations that are available. In two dimensions, there are three equilibrium equations that can be written for each member.

- Sum moments about a point that lies at the intersection of the lines of action of as many unknown forces as possible.

- If the solution of a force or couple moment magnitude is found to be negative, it means the sense of the force is the reverse of that shown on the free-body diagram.

EXAMPLE 6.14

Determine the horizontal and vertical components of force which the pin at *C* exerts on member *BC* of the frame in Fig. 6–27a.

SOLUTION I

Free-Body Diagrams. By inspection it can be seen that *AB* is a two-force member. The free-body diagrams are shown in Fig. 6–27b.

Equations of Equilibrium. The *three unknowns*, C_x, C_y, and F_{AB}, can be determined by applying the three equations of equilibrium to member *CB*.

$\zeta+\Sigma M_C = 0; \; 2000 \text{ N}(2 \text{ m}) - (F_{AB} \sin 60°)(4 \text{ m}) = 0 \; F_{AB} = 1154.7 \text{ N}$

$\xrightarrow{+} \Sigma F_x = 0; \; 1154.7 \cos 60° \text{ N} - C_x = 0 \quad C_x = 577 \text{ N} \qquad Ans.$

$+\uparrow \Sigma F_y = 0; \; 1154.7 \sin 60° \text{ N} - 2000 \text{ N} + C_y = 0 \; C_y = 1000 \text{ N} \quad Ans.$

SOLUTION II

Free-Body Diagrams. If one does not recognize that *AB* is a two-force member, then more work is involved in solving this problem. The free-body diagrams are shown in Fig. 6–27c.

Equations of Equilibrium. The *six unknowns*, A_x, A_y, B_x, B_y, C_x, and C_y, are determined by applying the three equations of equilibrium to each member.

Member AB

$\zeta+\Sigma M_A = 0; \quad B_x(3 \sin 60° \text{ m}) - B_y(3 \cos 60° \text{ m}) = 0 \qquad (1)$

$\xrightarrow{+} \Sigma F_x = 0; \quad A_x - B_x = 0 \qquad (2)$

$+\uparrow \Sigma F_y = 0; \quad A_y - B_y = 0 \qquad (3)$

Member BC

$\zeta+\Sigma M_C = 0; \quad 2000 \text{ N}(2 \text{ m}) - B_y(4 \text{ m}) = 0 \qquad (4)$

$\xrightarrow{+} \Sigma F_x = 0; \quad B_x - C_x = 0 \qquad (5)$

$+\uparrow \Sigma F_y = 0; \quad B_y - 2000 \text{ N} + C_y = 0 \qquad (6)$

The results for C_x and C_y can be determined by solving these equations in the following sequence: 4, 1, 5, then 6. The results are

$$B_y = 1000 \text{ N}$$
$$B_x = 577 \text{ N}$$
$$C_x = 577 \text{ N} \qquad Ans.$$
$$C_y = 1000 \text{ N} \qquad Ans.$$

By comparison, Solution I is simpler since the requirement that F_{AB} in Fig. 6–27b be equal, opposite, and collinear at the ends of member *AB* automatically satisfies Eqs. 1, 2, and 3 above and therefore eliminates the need to write these equations. *As a result, always identify the two-force members before starting the analysis!*

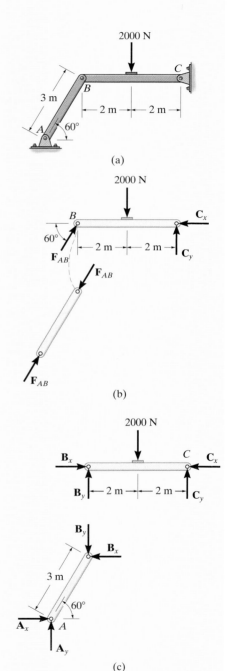

Fig. 6–27

EXAMPLE 6.15

The compound beam shown in Fig. 6–28*a* is pin connected at *B*. Determine the reactions at its supports. Neglect its weight and thickness.

(a)   (b)

Fig. 6–28

SOLUTION

Free-Body Diagrams. By inspection, if we consider a free-body diagram of the entire beam ABC, there will be three unknown reactions at A and one at C. These four unknowns cannot all be obtained from the three equations of equilibrium, and so it will become necessary to dismember the beam into its two segments as shown in Fig. 6–28*b*.

Equations of Equilibrium.   The six unknowns are determined as follows:

Segment BC

$$\xrightarrow{+} \Sigma F_x = 0; \qquad\qquad\qquad B_x = 0$$

$$\zeta + \Sigma M_B = 0; \qquad -8 \text{ kN}(1 \text{ m}) + C_y(2 \text{ m}) = 0$$

$$+\uparrow \Sigma F_y = 0; \qquad\qquad B_y - 8 \text{ kN} + C_y = 0$$

Segment AB

$$\xrightarrow{+} \Sigma F_x = 0; \qquad\qquad A_x - (10 \text{ kN})\left(\tfrac{3}{5}\right) + B_x = 0$$

$$\zeta + \Sigma M_A = 0; \quad M_A - (10 \text{ kN})\left(\tfrac{4}{5}\right)(2 \text{ m}) - B_y(4 \text{ m}) = 0$$

$$+\uparrow \Sigma F_y = 0; \qquad\qquad A_y - (10 \text{ kN})\left(\tfrac{4}{5}\right) - B_y = 0$$

Solving each of these equations successively, using previously calculated results, we obtain

$$A_x = 6 \text{ kN} \qquad\quad A_y = 12 \text{ kN} \qquad M_A = 32 \text{ kN} \cdot \text{m} \qquad \textit{Ans.}$$

$$B_x = 0 \qquad\qquad B_y = 4 \text{ kN}$$

$$C_y = 4 \text{ kN} \qquad\qquad\qquad\qquad\qquad\qquad\qquad\qquad\quad \textit{Ans.}$$

EXAMPLE 6.16

Determine the horizontal and vertical components of force which the pin at C exerts on member $ABCD$ of the frame shown in Fig. 6–29a.

SOLUTION

Free-Body Diagrams.   By inspection, the three components of reaction that the supports exert on $ABCD$ can be determined from a free-body diagram of the entire frame, Fig. 6–29b. Also, the free-body diagram of each frame member is shown in Fig. 6–29c. Notice that member BE is a two-force member. As shown by the colored dashed lines, the forces at B, C, and E have equal magnitudes but opposite directions on the separate free-body diagrams.

Equations of Equilibrium.   The six unknowns A_x, A_y, F_B, C_x, C_y, and D_x will be determined from the equations of equilibrium applied to the entire frame and then to member CEF. We have

(a)

Entire Frame

$$\curvearrowleft + \Sigma M_A = 0; \quad -981 \text{ N}(2 \text{ m}) + D_x(2.8 \text{ m}) = 0 \qquad D_x = 700.7 \text{ N}$$
$$\overset{+}{\rightarrow} \Sigma F_x = 0; \quad A_x - 700.7 \text{ N} = 0 \qquad\qquad A_x = 700.7 \text{ N}$$
$$+\uparrow \Sigma F_y = 0; \quad A_y - 981 \text{ N} = 0 \qquad\qquad A_y = 981 \text{ N}$$

Member CEF

$$\curvearrowleft + \Sigma M_C = 0; \quad -981 \text{ N}(2 \text{ m}) - (F_B \sin 45°)(1.6 \text{ m}) = 0$$
$$F_B = -1734.2 \text{ N}$$
$$\overset{+}{\rightarrow} \Sigma F_x = 0; \quad -Cx - (-1734.2 \cos 45° \text{ N}) = 0$$
$$C_x = 1226 \text{ N} \qquad\qquad\qquad Ans.$$
$$+\uparrow \Sigma F_y = 0; \quad C_y - (-1734.2 \sin 45° \text{ N}) - 981 \text{ N} = 0$$
$$C_y = -245 \text{ N} \qquad\qquad\qquad Ans.$$

(b)

Since the magnitudes of $\mathbf{F}_B$ and $\mathbf{C}_y$ were calculated as negative quantities, they were assumed to be acting in the wrong sense on the free-body diagrams, Fig. 6–29c. The correct sense of these forces might have been determined "by inspection" *before* applying the equations of equilibrium to member CEF. As shown in Fig. 6–29c, moment equilibrium about point E on member CEF indicates that $\mathbf{C}_y$ must actually act *downward* to counteract the moment created by the 981-N force about E. Similarly, summing moments about C, it is seen that the vertical component of $\mathbf{F}_B$ must actually act *upward*, and so $\mathbf{F}_B$ must act upward and to the right.

NOTE: The above calculations can be checked by applying the three equilibrium equations to member $ABCD$, Fig. 6–29c.

(c)

Fig. 6–29

EXAMPLE 6.17

The smooth disk shown in Fig. 6–30a is pinned at D and has a weight of 20 lb. Neglecting the weights of the other members, determine the horizontal and vertical components of reaction at pins B and D.

(a)

SOLUTION

Free-Body Diagrams. By inspection, the three components of reaction at the supports can be determined from a free-body diagram of the entire frame, Fig. 6–30b. Also, free-body diagrams of the members are shown in Fig. 6–30c.

Equations of Equilibrium. The eight unknowns can of course be obtained by applying the eight equilibrium equations to each member— three to member AB, three to member BCD, and two to the disk. (Moment equilibrium is automatically satisfied for the disk.) If this is done, however, all the results can be obtained only from a simultaneous solution of some of the equations. (Try it and find out.) To avoid this situation, it is best to first determine the three support reactions on the *entire* frame; then, using these results, the remaining five equilibrium equations can be applied to two other parts in order to solve successively for the other unknowns.

(b)

(c)

Fig. 6–30

Entire Frame

$$\zeta + \Sigma M_A = 0; \qquad -20\ \text{lb}\,(3\ \text{ft}) + C_x(3.5\ \text{ft}) = 0 \qquad C_x = 17.1\ \text{lb}$$

$$\xrightarrow{+} \Sigma F_x = 0; \qquad A_x - 17.1\ \text{lb} = 0 \qquad A_x = 17.1\ \text{lb}$$

$$+\uparrow \Sigma F_y = 0; \qquad A_y - 20\ \text{lb} = 0 \qquad A_y = 20\ \text{lb}$$

Member AB

$$\xrightarrow{+} \Sigma F_x = 0; \qquad 17.1\ \text{lb} - B_x = 0 \qquad B_x = 17.1\ \text{lb} \quad Ans.$$

$$\zeta + \Sigma M_B = 0; \qquad -20\ \text{lb}\,(6\ \text{ft}) + N_D(3\ \text{ft}) = 0 \quad N_D = 40\ \text{lb}$$

$$+\uparrow \Sigma F_y = 0; \qquad 20\ \text{lb} - 40\ \text{lb} + B_y = 0 \qquad B_y = 20\ \text{lb} \quad Ans.$$

Disk

$$\xrightarrow{+} \Sigma F_x = 0; \qquad D_x = 0 \qquad\qquad\qquad\qquad\qquad Ans.$$

$$+\uparrow \Sigma F_y = 0; \qquad 40\ \text{lb} - 20\ \text{lb} - D_y = 0 \quad D_y = 20\ \text{lb} \qquad Ans.$$

EXAMPLE 6.18

Determine the tension in the cables and also the force **P** required to support the 600-N force using the frictionless pulley system shown in Fig. 6–31*a*.

(a)   (b)

Fig. 6–31

SOLUTION

Free-Body Diagram.   A free-body diagram of each pulley *including* its pin and a portion of the contacting cable is shown in Fig. 6–31*b*. Since the cable is *continuous* and the pulleys are frictionless, the cable has a *constant tension P* acting throughout its length (see Example 5.7). The link connection between pulleys *B* and *C* is a two-force member, and therefore it has an unknown tension *T* acting on it. Notice that the *principle of action, equal but opposite reaction* must be carefully observed for forces **P** and **T** when the *separate* free-body diagrams are drawn.

Equations of Equilibrium.   The three unknowns are obtained as follows:

Pulley A

$+\uparrow \Sigma F_y = 0;$ $3P - 600 \text{ N} = 0$ $P = 200 \text{ N}$ *Ans.*

Pulley B

$+\uparrow \Sigma F_y = 0;$ $T - 2P = 0$ $T = 400 \text{ N}$ *Ans.*

Pulley C

$+\uparrow \Sigma F_y = 0;$ $R - 2P - T = 0$ $R = 800 \text{ N}$ *Ans.*

EXAMPLE 6.19

(a)

A man having a weight of 150 lb supports himself by means of the cable and pulley system shown in Fig. 6–32a. If the seat has a weight of 15 lb, determine the force that he must exert on the cable at A and the force he exerts on the seat. Neglect the weight of the cables and pulleys.

SOLUTION I

Free-Body Diagrams. The free-body diagrams of the man, seat, and pulley C are shown in Fig. 6–32b. The *two* cables are subjected to tensions T_A and T_E, respectively. The man is subjected to three forces: his weight, the tension T_A of cable AC, and the reaction N_s of the seat.

Equations of Equilibrium. The three unknowns are obtained as follows:

Man

$$+\uparrow \Sigma F_y = 0; \qquad T_A + N_s - 150 \text{ lb} = 0 \qquad (1)$$

Seat

$$+\uparrow \Sigma F_y = 0; \qquad T_E - N_s - 15 \text{ lb} = 0 \qquad (2)$$

Pulley C

$$+\uparrow \Sigma F_y = 0; \qquad 2T_E - T_A = 0 \qquad (3)$$

Here T_E can be determined by adding Eqs. 1 and 2 to eliminate N_s and then using Eq. 3. The other unknowns are then obtained by resubstitution of T_E.

$$T_A = 110 \text{ lb} \qquad\qquad Ans.$$
$$T_E = 55 \text{ lb}$$
$$N_s = 40 \text{ lb} \qquad\qquad Ans.$$

(b)

SOLUTION II

Free-Body Diagrams. By using the blue section shown in Fig. 6–32a, the man, pulley, and seat can be considered as a *single system*, Fig. 6–32c. Here N_s and T_A are *internal* forces and hence are not included on this "combined" free-body diagram.

Equations of Equilibrium. Applying $\Sigma F_y = 0$ yields a *direct* solution for T_E.

$$+\uparrow \Sigma F_y = 0; \quad 3T_E - 15 \text{ lb} - 150 \text{ lb} = 0 \qquad T_E = 55 \text{ lb}$$

The other unknowns can be obtained from Eqs. 2 and 3.

(c)

Fig. 6–32

EXAMPLE 6.20

The hand exerts a force of 8 lb on the grip of the spring compressor shown in Fig. 6–33a. Determine the force in the spring needed to maintain equilibrium of the mechanism.

Fig. 6–33

SOLUTION

Free-Body Diagrams. By inspection, members EA, ED, and EF are all two-force members. The free-body diagrams for parts DC and ABG are shown in Fig. 6–33b. The pin at E has also been included here since *three* force interactions occur on this pin. They represent the effects of members ED, EA, and EF. Note carefully how equal and opposite force reactions occur between each of the parts.

Equations of Equilibrium. By studying the free-body diagrams, the most direct way to obtain the spring force is to apply the equations of equilibrium in the following sequence:

Lever ABG

$\curvearrowleft + \Sigma M_B = 0;$ $\quad F_{EA}(1 \text{ in.}) - 8 \text{ lb} (4 \text{ in.}) = 0$ $\quad F_{EA} = 32 \text{ lb}$

Pin E

$+\uparrow \Sigma F_y = 0;$ $\quad F_{ED} \sin 60° - F_{EF} \sin 60° = 0$ $\quad F_{ED} = F_{EF} = F$

$\xrightarrow{+} \Sigma F_x = 0;$ $\quad 2F \cos 60° - 32 \text{ lb} = 0$ $\quad\quad F = 32 \text{ lb}$

Arm DC

$\curvearrowleft + \Sigma M_C = 0;$ $\quad -F_s(6 \text{ in.}) + 32 \cos 30° \text{ lb} (3 \text{ in.}) = 0$

$$F_s = 13.9 \text{ lb} \qquad\qquad Ans.$$

EXAMPLE 6.21

0.6 m

C

0.8 m

A

B

45°

D

(a)

The 100-kg block is held in equilibrium by means of the pulley and continuous cable system shown in Fig. 6–34a. If the cable is attached to the pin at B, compute the forces which this pin exerts on each of its connecting members.

SOLUTION

Free-Body Diagrams. A free-body diagram of each member of the frame is shown in Fig. 6–34b. By inspection, members AB and CB are two-force members. Furthermore, the cable must be subjected to a force of 490.5 N in order to hold pulley D and the block in equilibrium. A free-body diagram of the pin at B is needed since *four interactions* occur at this pin. These are caused by the attached cable (490.5 N), member AB ($\mathbf{F}_{AB}$), member CB ($\mathbf{F}_{CB}$), and pulley B ($\mathbf{B}_x$ and $\mathbf{B}_y$).

Equations of Equilibrium. Applying the equations of force equilibrium to pulley B, we have

$\xrightarrow{+} \Sigma F_x = 0; \quad B_x - 490.5 \cos 45° \text{ N} = 0 \quad B_x = 346.8 \text{ N}$ \hfill *Ans.*

$+\uparrow \Sigma F_y = 0; \quad B_y - 490.5 \sin 45° \text{ N} - 490.5 \text{ N} = 0$

$$B_y = 837.3 \text{ N} \hspace{4cm} \textit{Ans.}$$

Using these results, equilibrium of the pin requires that

$+\uparrow \Sigma F_y = 0; \quad \frac{4}{5} F_{CB} - 837.3 \text{ N} - 490.5 \text{ N} \quad F_{CB} = 1660 \text{ N} \quad \textit{Ans.}$

$\xrightarrow{+} \Sigma F_x = 0; \quad F_{AB} - \frac{3}{5}(1660 \text{ N}) - 346.8 \text{ N} = 0 \; F_{AB} = 1343 \text{ N} \quad \textit{Ans.}$

NOTE: It may be noted that the two-force member CB is subjected to bending as caused by the force $\mathbf{F}_{CB}$. From the standpoint of design, it would be better to make this member *straight* (from C to B) so that the force $\mathbf{F}_{CB}$ would create only tension in the member.

$\mathbf{F}_{CB}$

4   5

3

$\mathbf{F}_{CB}$

$\mathbf{F}_{AB}$ $\mathbf{F}_{AB}$

490.5 N   490.5 N

D

(b)

981 N

F_{CB}

4   5

3

B_y

F_{AB} B_x

490.5 N

B

45°

B_x

490.5 N

B_y

490.5 N

(c)

Fig. 6–34

PROBLEMS

Before solving the following problems, it is suggested that a brief review be made of all the previous examples. This may be done by covering each solution, trying to locate the two-force members, drawing the free-body diagrams, and conceiving ways of applying the equations of equilibrium to obtain the solution.

6–66. A force of $P = 8$ lb is applied to the handles of the pliers. Determine the force developed on the smooth bolt B and the reaction that pin A exerts on its attached members.

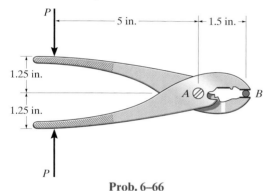

Prob. 6–66

6–67. The eye hook has a positive locking latch when it supports the load because its two parts are pin-connected at A and they bear against one another along the smooth surface at B. Determine the resultant force at the pin and the normal force at B when the eye hook supports a load of 800 lb.

Prob. 6–67

*6–68. Determine the force P needed to hold the 20-lb block in equilibrium.

Prob. 6–68

6–69. The link is used to hold the rod in place. Determine the required axial force on the screw at E if the largest force to be exerted on the rod at B, C, or D is to be 100 N. Also, find the magnitude of the force reaction at pin A. Assume all surfaces of contact are smooth.

Prob. 6–69

6–70. The 150-lb man attempts to lift himself and the 10-lb seat using the rope and pulley system shown. Determine the force at A needed to do so, and also find his reaction on the seat.

Prob. 6–70

6–71. Determine the horizontal and vertical components of force that pins A and C exert on the frame.

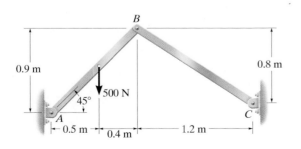

Prob. 6–71

*6–72. Determine the horizontal and vertical components of force that pins A and C exert on the frame.

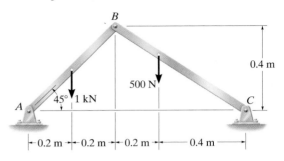

Prob. 6–72

6–73. The truck exerts the three forces shown on the girders of the bridge. Determine the reactions at the supports when the truck is in the position shown. The girders are connected together by a short vertical link DC.

Prob. 6–73

6–74. Determine the greatest force P that can be applied to the frame if the largest force resultant acting at A can have a magnitude of 2 kN.

Prob. 6–74

6–75. The compound beam is pin supported at B and supported by rockers at A and C. There is a hinge (pin) at D. Determine the reactions at the supports.

Prob. 6–75

*6–76. The compound beam is fixed supported at A and supported by rockers at B and C. If there are hinges (pins) at D and E, determine the reactions at the supports A, B, and C.

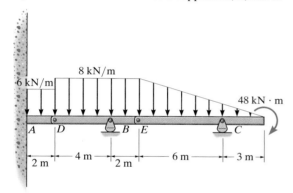

Prob. 6–76

6–77. Determine the reactions at supports A and B.

Prob. 6–77

6–78. Determine the horizontal and vertical components of force at C which member ABC exerts on member CEF.

Prob. 6–78

6–79. Determine the horizontal and vertical components of force that the pins at A, B, and C exert on their connecting members.

Prob. 6–79

*6–80. Operation of exhaust and intake valves in an automobile engine consists of the cam C, push rod DE, rocker arm EFG which is pinned at F, and a spring and valve, V. If the compression in the spring is 20 mm when the valve is open as shown, determine the normal force acting on the cam lobe at C. Assume the cam and bearings at H, I, and J are smooth. The spring has a stiffness of 300 N/m.

Prob. 6–80

6–81. Determine the force P on the cord, and the angle θ that the pulley-supporting link AB makes with the vertical. Neglect the mass of the pulleys and the link. The block has a weight of 200 lb and the cord is attached to the pin at B. The pulleys have radii of $r_1 = 2$ in. and $r_2 = 1$ in.

Prob. 6–81

6–82. The nail cutter consists of the handle and the two cutting blades. Assuming the blades are pin connected at B and the surface at D is smooth, determine the normal force on the fingernail when a force of 1 lb is applied to the handles as shown. The pin AC slides through a smooth hole at A and is attached to the bottom member at C.

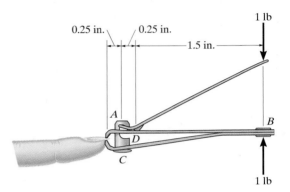

Prob. 6–82

6–83. The wall crane supports a load of 700 lb. Determine the horizontal and vertical components of reaction at the pins A and D. Also, what is the force in the cable at the winch W?

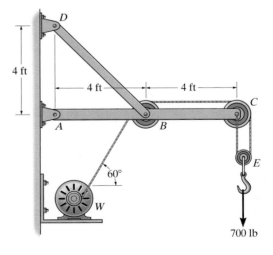

Prob. 6–83

*6–84. Determine the force that the smooth roller C exerts on beam AB. Also, what are the horizontal and vertical components of reaction at pin A? Neglect the weight of the frame and roller.

Prob. 6–84

6–85. Determine the horizontal and vertical components of force which the pins exert on member ABC.

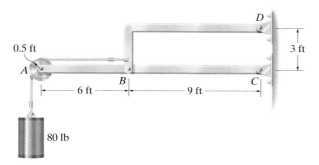

Prob. 6–85

6–86. The floor beams AB and BC are stiffened using the two tie rods CD and AD. Determine the force along each rod when the floor beams are subjected to a uniform load of 80 lb/ft. Assume the three contacting members at B are smooth and the joints at A, C, and D are pins. *Hint:* Members AD, CD, and BD are two-force members.

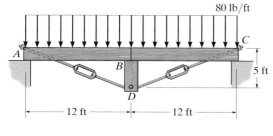

Prob. 6–86

6–87. Determine the horizontal and vertical components of force at pins B and C.

Prob. 6–87

*6–88. The skid steer loader has a mass of 1.18 Mg, and in the position shown the center of mass is at G_1. If there is a 300-kg stone in the bucket, with center of mass at G_2, determine the reactions of each pair of wheels A and B on the ground and the force in the hydraulic cylinder CD and at the pin E. There is a similar linkage on each side of the loader.

Prob. 6–88

6–89. Determine the horizontal and vertical components of force at each pin. The suspended cylinder has a weight of 80 lb.

Prob. 6–89

6–91. Determine the horizontal and vertical components of force which the pins at A, B, and C exert on member ABC of the frame.

Prob. 6–91

6–90. The two-member frame is pin connected at C, D, and E. The cable is attached to A, passes over the smooth peg at B, and is attached to a 100-lb load. Determine the horizontal and vertical reactions at each pin.

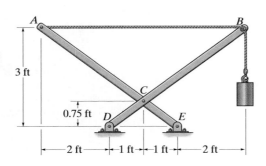

Prob. 6–90

*6–92. The derrick is pin-connected to the pivot at A. Determine the largest mass that can be supported by the derrick if the maximum force that can be sustained by the pin at A is 18 kN.

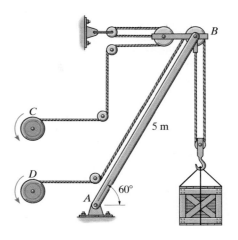

Prob. 6–92

6–93. Determine the required mass of the suspended cylinder if the tension in the chain wrapped around the freely turning gear is to be 2 kN. Also, what is the magnitude of the resultant force on pin A?

Prob. 6–93

6–94. The tongs consist of two jaws pinned to links at A, B, C, and D. Determine the horizontal and vertical components of force exerted on the 500-lb stone at F and G in order to lift it.

Prob. 6–94

6–95. Determine the force P on the cable if the spring is compressed 0.5 in. when the mechanism is in the position shown. The spring has a stiffness of $k = 800$ lb/ft.

Prob. 6–95

***6–96.** The scale consists of five pin-connected members. Determine the load W on the pan EG if a weight of 3 lb is suspended from the hook at A.

Prob. 6–96

6–97.   The machine shown is used for forming metal plates. It consists of two toggles ABC and DEF, which are operated by the hydraulic cylinder H. The toggles push the movable bar G forward, pressing the plate p into the cavity. If the force which the plate exerts on the head is $P = 12$ kN, determine the force F in the hydraulic cylinder when $\theta = 30°$.

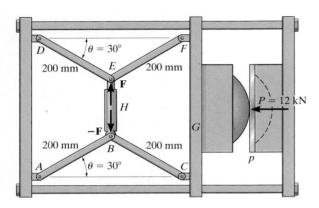

Prob. 6–97

6–98.   Determine the horizontal and vertical components of force at pins A and C of the two-member frame.

Prob. 6–98

6–99.   The truck rests on the scale, which consists of a series of compound levers. If a mass of 15 kg is placed on the pan P and it is required that $x = 0.480$ m to balance the "beam" ABC, determine the mass of the truck. There are pins at all lettered points. Take $EF = 0.2$ m, $FD = 3$ m, $KG = GH = 2.5$ m, and $HI = KJ = 0.1$ m. Is it necessary for the truck to be symmetrically placed on the scale? Explain.

Prob. 6–99

*6–100.   By squeezing on the hand brake of the bicycle, the rider subjects the brake cable to a tension of 50 lb. If the caliper mechanism is pin-connected to the bicycle frame at B, determine the normal force each brake pad exerts on the rim of the wheel. Is this the force that stops the wheel from turning? Explain.

Prob. 6–100

6–101. If a force of $P = 6$ lb is applied perpendicular to the handle of the mechanism, determine the magnitude of force **F** for equilibrium. The members are pin-connected at A, B, C, and D.

Prob. 6–101

6–103. The tower truss has a weight of 575 lb and a center of gravity at G. The rope system is used to hoist it into the vertical position. If rope CB is attached to the top of the shear leg AC and a second rope CD is attached to the truss, determine the required tension in BC to hold the truss in the position shown. The base of the truss and the shear leg bears against the stake at A, which can be considered as a pin. Also, compute the compressive force acting along the shear leg.

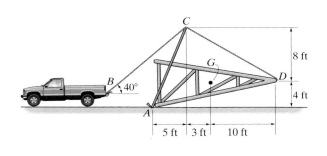

Prob. 6–103

6–102. The pillar crane is subjected to the load having a mass of 500 kg. Determine the force developed in the tie rod AB and the horizontal and vertical reactions at the pin support C when the boom is tied in the position shown.

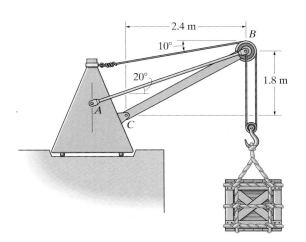

Prob. 6–102

*■**6–104.** The constant moment of 50 N · m is applied to the crank shaft. Determine the compressive force P that is exerted on the piston for equilibrium as a function of θ. Plot the results of P (ordinate) versus θ (abscissa) for $0° \leq \theta \leq 90°$.

Prob. 6–104

6–105. Five coins are stacked in the smooth plastic container shown. If each coin weighs 0.0235 lb, determine the normal reactions of the bottom coin on the container at points A and B.

6–107. A 5-lb force is applied to the handles of the vise grip. Determine the compressive force developed on the smooth bolt shank A at the jaws.

Prob. 6–107

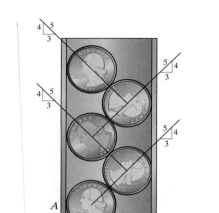

Prob. 6–105

6–106. Determine the horizontal and vertical components of force at pin B and the normal force the pin at C exerts on the smooth slot. Also, determine the moment and horizontal and vertical reactions of force at A. There is a pulley at E.

*6–108. If a force of 10 lb is applied to the grip of the clamp, determine the compressive force F that the wood block exerts on the clamp.

Prob. 6–106

Prob. 6–108

6–109. The hoist supports the 125-kg engine. Determine the force in member *DB* and in the hydraulic cylinder *H* of member *FB*.

6–111. Determine the force created in the hydraulic cylinders *EF* and *AD* in order to hold the shovel in equilibrium. The shovel load has a mass of 1.25 Mg and a center of gravity at *G*. All joints are pin connected.

Prob. 6–109

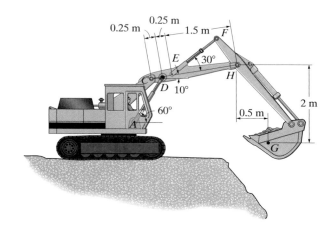

Prob. 6–111

6–110. The flat-bed trailer has a weight of 7000 lb and center of gravity at G_T. It is pin-connected to the cab at *D*. The cab has a weight of 6000 lb and center of gravity at G_C. Determine the range of values *x* for the position of the 2000-lb load *L* so that when it is placed over the rear axle, no axle is subjected to more than 5500 lb. The load has a center of gravity at G_L.

*6–112. The aircraft-hangar door opens and closes slowly by means of a motor which draws in the cable *AB*. If the door is made in two sections (bifold) and each section has a uniform weight *W* and length *L*, determine the force in the cable as a function of the door's position θ. The sections are pin-connected at *C* and *D* and the bottom is attached to a roller that travels along the vertical track.

Prob. 6–110

Prob. 6–112

6–113. A man having a weight of 175 lb attempts to lift himself using one of the two methods shown. Determine the total force he must exert on bar AB in each case and the normal reaction he exerts on the platform at C. Neglect the weight of the platform.

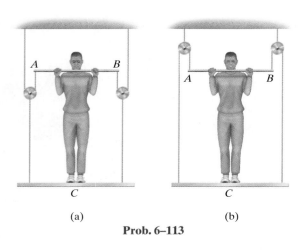

(a)   (b)

Prob. 6–113

6–114. A man having a weight of 175 lb attempts to lift himself using one of the two methods shown. Determine the total force he must exert on bar AB in each case and the normal reaction he exerts on the platform at C. The platform has a weight of 30 lb.

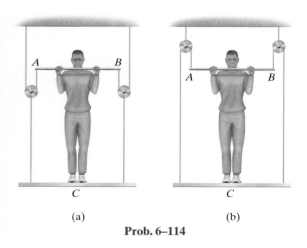

(a)   (b)

Prob. 6–114

6–115. The piston C moves vertically between the two smooth walls. If the spring has a stiffness of $k = 15$ lb/in., and is unstretched when $\theta = 0°$, determine the couple $\mathbf{M}$ that must be applied to AB to hold the mechanism in equilibrium when $\theta = 30°$.

Prob. 6–115

*6–116.** The compound shears are used to cut metal parts. Determine the vertical cutting force exerted on the rod R if a force of $F = 20$ lb is applied at the grip G. The lobe CDE is in smooth contact with the head of the shear blade at E.

Prob. 6–116

6–117. The handle of the sector press is fixed to gear G, which in turn is in mesh with the sector gear C. Note that AB is pinned at its ends to gear C and the underside of the table EF, which is allowed to move vertically due to the smooth guides at E and F. If the gears exert tangential forces between them, determine the compressive force developed on the cylinder S when a vertical force of 40 N is applied to the handle of the press.

6–119. If each of the three links of the mechanism has a weight of 25 lb, determine the angle θ for equilibrium. The spring, which always remains horizontal, is unstretched when $\theta = 0°$.

Prob. 6–119

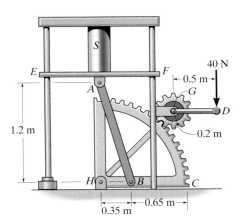

Prob. 6–117

6–118. The mechanism is used to hide kitchen appliances under a cabinet by allowing the shelf to rotate downward. If the mixer weighs 10 lb, is centered on the shelf, and has a mass center at G, determine the stretch in the spring necessary to hold the shelf in the equilibrium position shown. There is a similar mechanism on each side of the shelf, so that each mechanism supports 5 lb of the load. The springs each have a stiffness of $k = 4$ lb/in. spring.

*6–120. Determine the required force P that must be applied at the blade of the pruning shears so that the blade exerts a normal force of 20 lb on the twig at E.

Prob. 6–118

Prob. 6–120

6–121. The three power lines exert the forces shown on the truss joints, which in turn are pin-connected to the poles *AH* and *EG*. Determine the force in the guy cable *AI* and the pin reaction at the support *H*.

6–123. The kinetic sculpture requires that each of the three pinned beams be in perfect balance at all times during its slow motion. If each member has a uniform weight of 2 lb/ft and length of 3 ft, determine the necessary counterweights W_1, W_2, and W_3 which must be added to the ends of each member to keep the system in balance for any position. Neglect the size of the counterweights.

Prob. 6–121

Prob. 6–123

6–122. The hydraulic crane is used to lift the 1400-lb load. Determine the force in the hydraulic cylinder *AB* and the force in links *AC* and *AD* when the load is held in the position shown.

*6–124. The three-member frame is connected at its ends using ball-and-socket joints. Determine the *x, y, z* components of reaction at *B* and the tension in member *ED*. The force acting at *D* is $\mathbf{F} = \{135\mathbf{i} + 200\mathbf{j} - 180\mathbf{k}\}$ lb.

Prob. 6–122

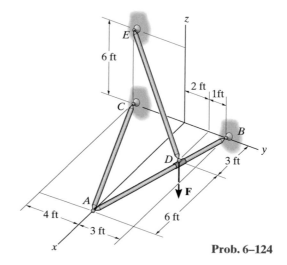

Prob. 6–124

6–125. The four-member "A" frame is supported at A and E by smooth collars and at G by a pin. All the other joints are ball-and-sockets. If the pin at G will fail when the resultant force there is 800 N, determine the largest vertical force P that can be supported by the frame. Also, what are the x, y, z force components which member BD exerts on members EDC and ABC? The collars at A and E and the pin at G only exert force components on the frame.

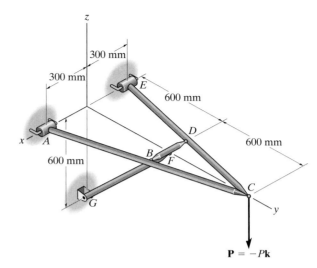

Prob. 6–125

6–126. The structure is subjected to the loading shown. Member AD is supported by a cable AB and roller at C and fits through a smooth circular hole at D. Member ED is supported by a roller at D and a pole that fits in a smooth snug circular hole at E. Determine the x, y, z components of reaction at E and the tension in cable AB.

6–127. The structure is subjected to the loadings shown. Member AB is supported by a ball-and-socket at A and smooth collar at B. Member CD is supported by a pin at C. Determine the x, y, z components of reaction at A and C.

Prob. 6–126

Prob. 6–127

Chapter Review

Simple Truss

A simple truss consists of triangular elements connected together by pin joints. The forces within its members can be determined by assuming the members are all two-force members, connected concurrently at each joint. They are either in tension or compression or carry no force.
(*See pages 267–269.*)

Roof truss

Method of Joints

The method of joints states that if a truss is in equilibrium, then each of its joints is also in equilibrium. For a coplanar truss, the concurrent force system at each joint must satisfy force equilibrium.

$$\Sigma F_x = 0$$
$$\Sigma F_y = 0$$

To obtain a numerical solution for the forces in the members, select a joint that has a free-body diagram with at most two unknown forces and one known force. (This may require first finding the reactions at the supports.)

Once a member force is determined, use its value and apply it to an adjacent joint.

Remember that forces that are found to *pull* on the joint are *tension forces*, and those that *push* on the joint are *compression forces*.

To avoid a simultaneous solution of two equations, try to sum forces in a direction that is perpendicular to one of the unknowns. This will allow a direct solution for the other unknown.

$$\Sigma F_{y'} = 0$$

$$1.5 \text{ kN} \cos 30° - F_{CB} \sin 15° = 0$$

$$F_{CB} = 5.02 \text{ kN (C)}$$

To further simplify the analysis, first identify all the zero-force members.
(*See pages 270–271.*)

Method of Sections

The method of sections states that if a truss is in equilibrium, then each section of the truss is also in equilibrium. Pass a section through the member whose force is to be determined. Then draw the free-body diagram of the sectioned part having the least number of forces on it.

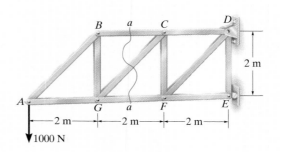

Sectioned members subjected to *pulling* are in *tension*, and those that are subjected to *pushing* are in *compression*.

Three equations of equilibrium are available to determine the unknowns.

$$\Sigma F_x = 0$$
$$\Sigma F_y = 0$$
$$\Sigma M_O = 0$$

If possible, sum forces in a direction that is perpendicular to two of the three unknown forces. This will yield a direct solution for the third force.

$$+\uparrow \Sigma F_y = 0$$
$$-1000 \text{ N} + F_{GC} \sin 45° = 0$$
$$F_{GC} = 1.41 \text{ kN (T)}$$

Sum moments about a point that passes through the line of action of two of the three unknown forces, so that the third unknown force can be determined directly.
(*See pages 283–285.*)

$$\downarrow + \Sigma M_C = 0$$
$$1000 \text{ N}(4 \text{ m}) - F_{GF} (2 \text{ m}) = 0$$
$$F_{GF} = 2 \text{ kN (C)}$$

Space Truss

A space truss is a three-dimensional truss built from tetrahedral elements, and is analyzed using the same methods as for planar trusses. The joints are assumed to be ball and socket connections.
(*See page 293.*)

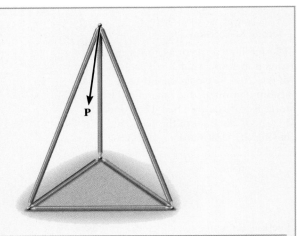

Frames and Machines

Frames and machines are structures that contain one or more multiforce members, that is, members with three or more forces or couples acting on them. Frames are designed to support loads, and machines transmit and alter the effect of forces.

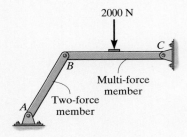

The forces acting at the joints of a frame or machine can be determined by drawing the free-body diagrams of each of its members or parts. The principle of action–reaction should be carefully observed when drawing these forces on each adjacent member or pin. For a coplanar force system, there are three equilibrium equations available for each member.

To simplify the analysis, be sure to recognize all two-force members. They have equal but opposite collinear forces at their ends.
(*See pages 297, 303–304.*)

REVIEW PROBLEMS

*6–128. Determine the resultant forces at pins B and C on member ABC of the four-member frame.

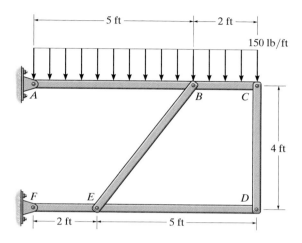

Prob. 6–128

6–129. The mechanism consists of identical meshed gears A and B and arms which are fixed to the gears. The spring attached to the ends of the arms has an unstretched length of 100 mm and a stiffness of $k = 250$ N/m. If a torque of $M = 6$ N·m is applied to gear A, determine the angle θ through which each arm rotates. The gears are each pinned to fixed supports at their centers.

Prob. 6–129

6–130. Determine the force in each member of the truss and state if the members are in tension or compression.

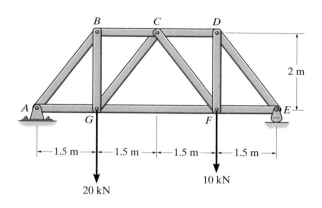

Prob. 6–130

6–131. The spring has an unstretched length of 0.3 m. Determine the angle θ for equilibrium if the uniform links each have a mass of 5 kg.

*6–132. The spring has an unstretched length of 0.3 m. Determine the mass m of each uniform link if the angle $\theta = 20°$ for equilibrium.

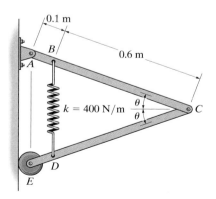

Probs. 6–131/132

6–133. Determine the horizontal and vertical components of force that the pins A and B exert on the two-member frame. Set $F = 0$.

6–134. Determine the horizontal and vertical components of force that pins A and B exert on the two-member frame. Set $F = 500$ N.

Probs. 6–133/134

*6–136. Determine the force in each member of the truss and state if the members are in tension or compression.

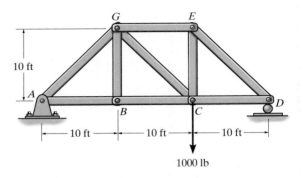

Prob. 6–136

6–135. Determine the force in each member of the truss and indicate whether the members are in tension or compression.

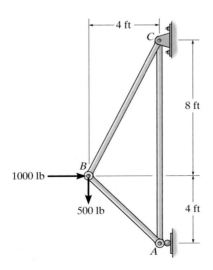

Prob. 6–135

6–137. Determine the force in members AB, AD, and AC of the space truss and state if the members are in tension or compression.

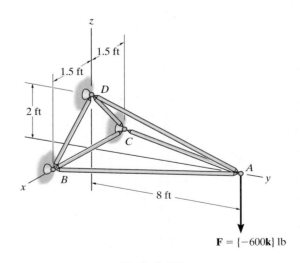

Prob. 6–137

Design Projects

6–1D Design of a Bridge Truss

A bridge having a horizontal top cord is to span between the two piers A and B having an arbitrary height. It is required that a pin-connected truss be used, consisting of steel members bolted together to steel gusset plates, such as the one shown in the figure. The end supports are assumed to be a pin at A and a roller at B. A vertical loading of 5 kN is to be supported within the middle 3 m of the span. This load can be applied in part to several joints on the top cord within this region, or to a single joint at the middle of the top cord. The force of the wind and the weight of the members are to be neglected.

Prob. 6–1D

Assume the maximum tensile force in each member cannot exceed 4.25 kN; and regardless of the length of the member, the maximum compressive force cannot exceed 3.5 kN. Design the most economical truss that will support the loading. The members cost $7.50/m, and the gusset plates cost $8.00 each. Submit your cost analysis for the materials, along with a scaled drawing of the truss, identifying on this drawing the tensile and compressive force in each member. Also, include your calculations of the complete force analysis.

6–2D Design of a Cart Lift

A hand cart is used to move a load from one loading dock to another. Any dock will have a different elevation relative to the bed of a truck that backs up to it. It is necessary that the loading platform on the hand cart will bring the load resting on it up to the elevation of each truck bed as shown. The maximum elevation difference between the frame of the hand cart and a truck bed is 1 ft. Design a hand-operated mechanical system that will allow the load to be lifted this distance from the frame of the hand cart. Assume the operator can exert a (comfortable) force of 20 lb to make the lift, and that the maximum load, centered on the loading platform, is 400 lb. Submit a scaled drawing of your design, and explain how it works based on a force analysis.

Prob. 6–2D

6–3D Design of a Pulley System

The steel beam AB, having a length of 5 m and a mass of 700 kg, is to be hoisted in its horizontal position to a height of 4 m. Design a pulley-and-rope system, which can be suspended from the overhead beam CD, that will allow a single worker to hoist the beam. Assume that the maximum (comfortable) force that he can apply to the rope is 180 N. Submit a drawing of your design, specify its approximate material cost, and discuss the safety aspects of its operation. Rope costs $1.25/m and each pulley costs $8.00.

Prob. 6–3D

6–4D Design of a Tool used to Position a Suspended Load

Heavy loads are suspended from an overhead pulley and each load must be positioned over a depository. Design a tool that can be used to shorten or lengthen the pulley cord AB a small amount in order to make the location adjustment. Assume the worker can apply a maximum (comfortable) force of 25 lb to the tool, and the maximum force allowed in cord AB is 500 lb. Submit a scale drawing of the tool, and a brief paragraph to explain how it works using a force analysis. Include a discussion on the safety aspects of its use.

Prob. 6–4D

6–5D Design of a Fence-Post Remover

A farmer wishes to remove several fence posts. Each post is buried 18 in. in the ground and will require a maximum vertical pulling force of 175 lb to remove it. He can use his truck to develop the force, but he needs to devise a method for their removal without breaking the posts. Design a method that can be used, considering that the only materials available are a strong rope and several pieces of wood having various sizes and lengths. Submit a sketch of your design and discuss the safety and reliability of its use. Also, provide a force analysis to show how it works and why it will cause minimal damage to a post when it is removed.

175 lb maximum

Prob. 6–5D

The design and analysis of any structural member requires knowledge of the internal loadings acting within it. In this chapter, we will discuss how engineers determine these loadings.

7

Internal Forces

CHAPTER OBJECTIVES

- To show how to use the method of sections for determining the internal loadings in a member.

- To generalize this procedure by formulating equations that can be plotted so that they describe the internal shear and moment throughout a member.

- To analyze the forces and study the geometry of cables supporting a load.

7.1 Internal Forces Developed in Structural Members

The design of any structural or mechanical member requires an investigation of the loading acting within the member in order to be sure the material can resist this loading. These internal loadings can be determined by using the *method of sections*. To illustrate this method, consider the "simply supported" beam shown in Fig. 7–1a, which is subjected to the forces $\mathbf{F}_1$ and $\mathbf{F}_2$ and the *support reactions* $\mathbf{A}_x$, $\mathbf{A}_y$, and $\mathbf{B}_y$, Fig. 7–1b. If the *internal loadings* acting on the cross section at C are to be determined, then an

(a)

Fig. 7–1

337

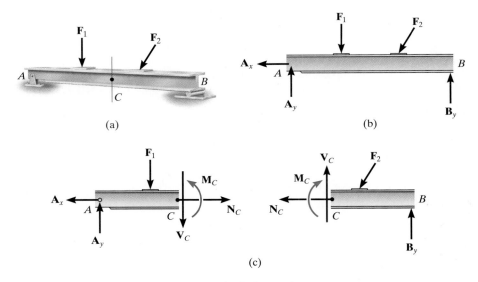

(a)   (b)

(c)

Fig. 7–1 (cont.)

The designer of this shop crane realized the need for additional reinforcement around the joint in order to prevent severe internal bending of the joint when a large load is suspended from the chain hoist.

imaginary section is passed through the beam, cutting it into two segments. By doing this the internal loadings at the section become *external* on the free-body diagram of each segment, Fig. 7–1c. Since both segments (*AC* and *CB*) were in equilibrium *before* the beam was sectioned, equilibrium of each segment is maintained provided rectangular force components $\mathbf{N}_C$ and $\mathbf{V}_C$ and a resultant couple moment $\mathbf{M}_C$ are developed at the section. Note that these loadings must be equal in magnitude and opposite in direction on each of the segments (Newton's third law). The magnitude of each of these loadings can now be determined by applying the three equations of equilibrium to either segment *AC* or *CB*. A *direct solution* for $\mathbf{N}_C$ is obtained by applying $\Sigma F_x = 0$; $\mathbf{V}_C$ is obtained directly from $\Sigma F_y = 0$; and $\mathbf{M}_C$ is determined by summing moments about point *C*, $\Sigma M_C = 0$, in order to eliminate the moments of the unknowns $\mathbf{N}_C$ and $\mathbf{V}_C$.

In mechanics, the force components $\mathbf{N}$, acting normal to the beam at the cut section, and $\mathbf{V}$, acting tangent to the section, are termed the *normal or axial force* and the *shear force*, respectively. The couple moment $\mathbf{M}$ is referred to as the *bending moment*, Fig. 7–2a. In three dimensions, a general internal force and couple moment resultant will act at the section. The *x, y, z* components of these loadings are shown in Fig. 7–2b. Here $\mathbf{N}_y$ is the *normal force*, and $\mathbf{V}_x$ and $\mathbf{V}_z$ are *shear force components*. $\mathbf{M}_y$ is a *torsional or twisting moment*, and $\mathbf{M}_x$ and $\mathbf{M}_z$ are *bending moment components*. For most applications, these *resultant loadings* will act at the geometric center or centroid (*C*) of the section's cross-sectional area. Although the magnitude for each loading generally will be different at various points along the axis of the member, the method of sections can always be used to determine their values.

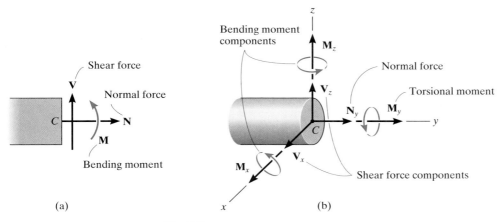

Fig. 7–2

Free-Body Diagrams.   Trusses are composed of two-force members that only support normal loads. On the other hand, frames and machines are composed of *multiforce members*, and so each of these members will generally be subjected to internal normal, shear, and bending loadings. For example, consider the frame shown in Fig. 7–3*a*. If the blue section is passed through the frame to determine the internal loadings at points *H*, *G*, and *F*, the resulting free-body diagram of the top portion of this section is shown in Fig. 7–3*b*. At each point where a member is sectioned there is an unknown normal force, shear force, and bending moment. As a result, we cannot apply the *three* equations of equilibrium to this section in order to obtain these *nine unknowns*. Instead, to solve this problem we must *first dismember* the frame and determine the reactions at the connections of the members using the techniques of Sec. 6.6. Once this is done, *each member* may then be sectioned at its appropriate point, and the three equations of equilibrium can be applied to determine **N**, **V**, and **M**. For example, the free-body diagram of segment *DG*, Fig. 7–3*c*, can be used to determine the internal loadings at *G* provided the reactions of the pin, $\mathbf{D}_x$ and $\mathbf{D}_y$, are known.

(a)

(b)

(c)

Fig. 7–3

In each case, the link on the backhoe is a two-force member. In the top photo it is subjected to both bending and axial load at its center. By making the member straight, as in the bottom photo, then only an axial force acts within the member.

Procedure for Analysis

The method of sections can be used to determine the internal loadings at a specific location in a member using the following procedure.

Support Reactions.

- Before the member is "cut" or sectioned, it may first be necessary to determine its support reactions, so that the equilibrium equations are used only to solve for the internal loadings when the member is sectioned.

- If the member is part of a frame or machine, the reactions at its connections are determined using the methods of Sec. 6.6.

Free-Body Diagram.

- Keep all distributed loadings, couple moments, and forces acting on the member in their *exact locations*, then pass an imaginary section through the member, perpendicular to its axis at the point where the internal loading is to be determined.

- After the section is made, draw a free-body diagram of the segment that has the least number of loads on it, and indicate the x, y, z components of the force and couple moment resultants at the section.

- If the member is subjected to a *coplanar* system of forces, only **N**, **V**, and **M** act at the section.

- In many cases it may be possible to tell by inspection the proper sense of the unknown loadings; however, if this seems difficult, the sense can be assumed.

Equations of Equilibrium.

- Moments should be summed at the section about axes passing through the *centroid* or geometric center of the member's cross-sectional area in order to eliminate the unknown normal and shear forces and thereby obtain direct solutions for the moment components.

- If the solution of the equilibrium equations yields a negative scalar, the assumed sense of the quantity is opposite to that shown on the free-body diagram.

EXAMPLE 7.1

The bar is fixed at its end and is loaded as shown in Fig. 7–4a. Determine the internal normal force at points B and C.

SOLUTION

Fig. 7–4

Support Reactions. A free-body diagram of the entire bar is shown in Fig. 7–4b. By inspection, only a normal force $\mathbf{A}_y$ acts at the fixed support since the loads are applied symmetrically along the bar's axis. ($A_x = 0$, $M_A = 0$.)

$+\uparrow \Sigma F_y = 0;$ $A_y - 16 \text{ kN} + 12 \text{ kN} - 4 \text{ kN} = 0$ $A_y = 8 \text{ kN}$

Free-Body Diagrams. The internal forces at B and C will be found using the free-body diagrams of the sectioned bar shown in Fig. 7–4c. No shear or moment act on the sections since they are not required for equilibrium. In particular, segments AB and DC will be chosen here, since they contain the *least* number of forces.

Equations of Equilibrium.

  Segment AB

$+\uparrow \Sigma F_y = 0;$ $8 \text{ kN} - N_B = 0$ $N_B = 8 \text{ kN}$ *Ans.*

  Segment DC

$+\uparrow \Sigma F_y = 0;$ $N_C - 4 \text{ kN} = 0$ $N_C = 4 \text{ kN}$ *Ans.*

NOTE: Try working this problem in the following manner: Determine N_B from segment BD. (Note that this approach *does not require* solution for the support reaction at A.) Using the result for N_B, isolate segment BC to determine N_C.

EXAMPLE 7.2

The circular shaft is subjected to three concentrated torques as shown in Fig. 7–5a. Determine the internal torques at points B and C.

SOLUTION

(a)

Support Reactions. Since the shaft is subjected only to collinear torques, a torque reaction occurs at the support, Fig. 7–5b. Using the right-hand rule to define the positive directions of the torques, we require

$$\Sigma M_x = 0; \qquad -10 \text{ N} \cdot \text{m} + 15 \text{ N} \cdot \text{m} + 20 \text{ N} \cdot \text{m} - T_D = 0$$
$$T_D = 25 \text{ N} \cdot \text{m}$$

(b)

Fig. 7–5

(c)

Free-Body Diagrams. The internal torques at B and C will be found using the free-body diagrams of the shaft segments AB and CD shown in Fig. 7–5c.

Equations of Equilibrium. Applying the equation of moment equilibrium along the shaft's axis, we have

Segment AB

$$\Sigma M_x = 0; \qquad -10 \text{ N} \cdot \text{m} + 15 \text{ N} \cdot \text{m} - T_B = 0 \qquad T_B = 5 \text{ N} \cdot \text{m} \quad \textit{Ans.}$$

Segment CD

$$\Sigma M_x = 0; \qquad T_C - 25 \text{ N} \cdot \text{m} = 0 \qquad T_C = 25 \text{ N} \cdot \text{m} \qquad \textit{Ans.}$$

NOTE: Try to solve for T_C by using segment CA. Note that this approach *does not require* a solution for the support reaction at D.

EXAMPLE 7.3

The beam supports the loading shown in Fig. 7–6a. Determine the internal normal force, shear force, and bending moment acting just to the left, point B, and just to the right, point C, of the 6-kN force.

(a)

SOLUTION

Support Reactions.   The free-body diagram of the beam is shown in Fig. 7–6b. When determining the *external reactions*, realize that the 9-kN · m couple moment is a free vector and therefore it can be placed *anywhere* on the free-body diagram of the entire beam. Here we will only determine $\mathbf{A}_y$, since segments AB and AC will be used for the analysis.

$$\zeta + \Sigma M_D = 0; \quad 9 \text{ kN} \cdot \text{m} + (6 \text{ kN})(6 \text{ m}) - A_y(9 \text{ m}) = 0$$
$$A_y = 5 \text{ kN}$$

(b)

Free-Body Diagrams.   The free-body diagrams of the left segments AB and AC of the beam are shown in Figs. 7–6c and 7–6d. In this case the 9-kN · m couple moment is *not included* on these diagrams since it must be kept in its *original position* until *after* the section is made and the appropriate body is isolated. In other words, the free-body diagrams of the left segments of the beam do not show the couple moment since this moment does not actually act on these segments.

(c)

Equations of Equilibrium.

Segment AB

$$\xrightarrow{+} \Sigma F_x = 0; \qquad\qquad N_B = 0 \qquad\qquad Ans.$$
$$+\uparrow \Sigma F_y = 0; \qquad\quad 5 \text{ kN} - V_B = 0 \quad V_B = 5 \text{ kN} \qquad Ans.$$
$$\zeta + \Sigma M_B = 0; \quad -(5 \text{ kN})(3 \text{ m}) + M_B = 0 \quad M_B = 15 \text{ kN} \cdot \text{m} \quad Ans.$$

Segment AC

$$\xrightarrow{+} \Sigma F_x = 0; \qquad\qquad N_C = 0 \qquad\qquad Ans.$$
$$+\uparrow \Sigma F_y = 0; \quad 5 \text{ kN} - 6 \text{ kN} + V_C = 0 \quad V_C = 1 \text{ kN} \qquad Ans.$$
$$\zeta + \Sigma M_C = 0; \quad -(5 \text{ kN})(3 \text{ m}) + M_C = 0 \quad M_C = 15 \text{ kN} \cdot \text{m} \quad Ans.$$

Here the moment arm for the 5-kN force in both cases is approximately 3 m since B and C are "almost" coincident.

(d)

Fig. 7–6

EXAMPLE 7.4

(a)

Determine the internal normal force, shear force, and bending moment acting at point B of the two-member frame shown in Fig. 7–7a.

SOLUTION

Support Reactions. A free-body diagram of each member is shown in Fig. 7–7b. Since CD is a two-force member, the equations of equilibrium need to be applied only to member AC.

$$\zeta + \Sigma M_A = 0; \quad -400 \text{ lb } (4 \text{ ft}) + \left(\tfrac{3}{5}\right)F_{DC}(8 \text{ ft}) = 0 \qquad F_{DC} = 333.3 \text{ lb}$$

$$\xrightarrow{+} \Sigma F_x = 0; \qquad -A_x + \left(\tfrac{4}{5}\right)(333.3 \text{ lb}) = 0 \qquad A_x = 266.7 \text{ lb}$$

$$+\uparrow \Sigma F_y = 0; \qquad A_y - 400 \text{ lb} + \tfrac{3}{5}(333.3 \text{ lb}) = 0 \qquad A_y = 200 \text{ lb}$$

Free-Body Diagrams. Passing an imaginary section perpendicular to the axis of member AC through point B yields the free-body diagrams of segments AB and BC shown in Fig. 7–7c. When constructing these diagrams it is important to keep the distributed loading exactly as it is until *after* the section is made. Only then can it be replaced by a single resultant force. Why? Also, notice that $\mathbf{N}_B$, $\mathbf{V}_B$, and $\mathbf{M}_B$ act with equal magnitude but opposite direction on each segment—Newton's third law.

Equations of Equilibrium. Applying the equations of equilibrium to segment AB, we have

$$\xrightarrow{+} \Sigma F_x = 0; \qquad N_B - 266.7 \text{ lb} = 0 \qquad N_B = 267 \text{ lb} \qquad Ans.$$

$$+\uparrow \Sigma F_y = 0; \qquad 200 \text{ lb} - 200 \text{ lb} - V_B = 0 \qquad V_B = 0 \qquad Ans.$$

$$\zeta + \Sigma M_B = 0; \qquad M_B - 200 \text{ lb } (4 \text{ ft}) + 200 \text{ lb } (2 \text{ ft}) = 0$$

$$M_B = 400 \text{ lb} \cdot \text{ft} \qquad Ans.$$

(b)

NOTE: As an exercise, try to obtain these same results using segment BC.

(c)

Fig. 7–7

EXAMPLE 7.5

Determine the normal force, shear force, and bending moment acting at point E of the frame loaded as shown in Fig. 7–8a.

(a)   (b)

SOLUTION

Support Reactions.   By inspection, members AC and CD are two-force members, Fig. 7–8b. In order to determine the internal loadings at E, we must first determine the force $\mathbf{R}$ at the end of member AC. To do this we must analyze the equilibrium of the pin at C. Why? Summing forces in the vertical direction on the pin, Fig. 7–8b, we have

$$+\uparrow \Sigma F_y = 0; \quad R \sin 45° - 600 \text{ N} = 0 \quad R = 848.5 \text{ N}$$

Free-Body Diagram.   The free-body diagram of segment CE is shown in Fig. 7–8c.

Equations of Equilibrium.

$$\xrightarrow{+} \Sigma F_x = 0; \quad 848.5 \cos 45° \text{ N} - V_E = 0 \qquad V_E = 600 \text{ N} \qquad Ans.$$

$$+\uparrow \Sigma F_y = 0; \quad -848.5 \sin 45° \text{ N} + N_E = 0 \qquad N_E = 600 \text{ N} \qquad Ans.$$

$$\zeta + \Sigma M_E = 0; \quad 848.5 \cos 45° \text{ N}(0.5 \text{ m}) - M_E = 0 \quad M_E = 300 \text{ N} \cdot \text{m} \quad Ans.$$

(c)

Fig. 7–8

NOTE: These results indicate a poor design. Member AC should be *straight* (from A to C) so that bending within the member is *eliminated*. If AC is straight then the internal force would only create tension in the member. See Example 6.21.

EXAMPLE 7.6

The uniform sign shown in Fig. 7–9a has a mass of 650 kg and is supported on the fixed column. Design codes indicate that the expected maximum uniform wind loading that will occur in the area where it is located is 900 Pa. Determine the internal loadings at A.

(a)   (b)

SOLUTION

The idealized model for the sign is shown in Fig. 7–9b. Here the necessary dimensions are indicated. We can consider the free-body diagram of a section above point A since it does not involve the support reactions.

Free-Body Diagram.   The sign has a weight of $W = 650(9.81) = 6.376$ kN, and the wind creates a resultant force of $F_w = 900$ N/m^2(6 m)(2.5 m) $= 13.5$ kN perpendicular to the face of the sign. These loadings are shown on the free-body diagram, Fig. 7–9c.

Equations of Equilibrium.   Since the problem is three dimensional, a vector analysis will be used.

$$\Sigma F = 0; \qquad F_A - 13.5i - 6.376k = 0$$
$$F_A = \{13.5i + 6.38k\} \text{ kN} \qquad\qquad Ans.$$

$$\Sigma M_A = 0; \qquad M_A + r \times (F_w + W) = 0$$

$$M_A + \begin{vmatrix} i & j & k \\ 0 & 3 & 5.25 \\ -13.5 & 0 & 6.376 \end{vmatrix} = 0$$

$$M_A = \{-19.1i + 70.9j + 40.5k\} \text{ kN} \cdot \text{m} \qquad Ans.$$

NOTE: Here $F_{A_z} = \{6.38k\}$ kN represents the normal force N, whereas $F_{A_x} = \{13.5i\}$ kN is the shear force. Also, the torsional moment is $M_{A_z} = \{40.5k\}$ kN $\cdot$ m, and the bending moment is determined from its components $M_{A_x} = \{-19.1i\}$ kN $\cdot$ m and $M_{A_y} = \{-70.9j\}$ kN $\cdot$ m; i.e., $M_b = \sqrt{M_x^2 + M_y^2}$.

(c)

Fig. 7–9

PROBLEMS

7–1. The column is fixed to the floor and is subjected to the loads shown. Determine the internal normal force, shear force, and moment at points A and B.

6 kN   6 kN
150 mm ——— 150 mm

A

8 kN

150 mm

B

Prob. 7–1

7–2. The axial forces act on the shaft as shown. Determine the internal normal force at points A and B.

50 lb

20 lb   20 lb

A

—10 lb

B

Prob. 7–2

7–3. The shaft is supported by smooth bearings at A and B and subjected to the torques shown. Determine the internal torque at points C, D, and E.

A

400 N · m

150 N · m

550 N · m

C

D

E

B

Prob. 7–3

*7–4. Three torques act on the shaft. Determine the internal torque at points A, B, C, and D.

400 N · m

A B C D

300 N · m   200 N · m

Prob. 7–4

7–5. The shaft is supported by a journal bearing at A and a thrust bearing at B. Determine the normal force, shear force, and moment at a section passing through (a) point C, which is just to the right of the bearing at A, and (b) point D, which is just to the left of the 3000-lb force.

2500 lb

75 lb/ft

3000 lb

C

A

D B

6 ft

12 ft

2 ft

Prob. 7–5

7–6. Determine the internal normal force, shear force, and moment at point C.

Prob. 7–6

7–7. Determine the internal normal force, shear force, and moment at point C.

Prob. 7–7

***7–8.** Determine the normal force, shear force, and moment at a section passing through point C. Assume the support at A can be approximated by a pin and B as a roller.

Prob. 7–8

7–9. The beam AB will fail if the maximum internal moment at D reaches $800 \text{ N} \cdot \text{m}$ or the normal force in member BC becomes 1500 N. Determine the largest load w it can support.

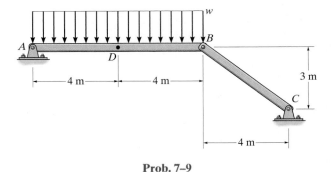

Prob. 7–9

7–10. Determine the shear force and moment acting at a section passing through point C in the beam.

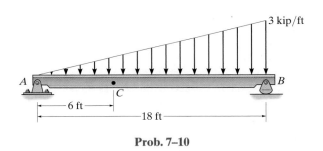

Prob. 7–10

7–11. Determine the internal normal force, shear force, and moment at points E and D of the compound beam.

Prob. 7–11

*7–12. The boom DF of the jib crane and the column DE have a uniform weight of 50 lb/ft. If the hoist and load weigh 300 lb, determine the normal force, shear force, and moment in the crane at sections passing through points A, B, and C. *Hint*: Treat the boom tip, beyond the hoist, as weightless.

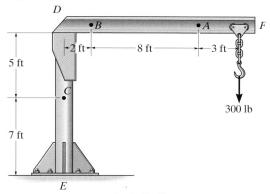

Prob. 7–12

7–13. Determine the internal normal force, shear force, and moment at point C.

Prob. 7–13

7–14. Determine the normal force, shear force, and moment at a section passing through point D of the two-member frame.

Prob. 7–14

7–15. The beam weighs 280 lb/ft. Determine the internal normal force, shear force, and moment at point C.

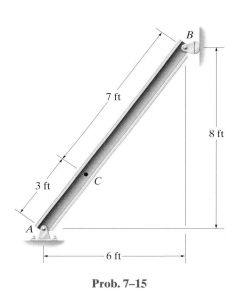

Prob. 7–15

*7–16. Determine the internal normal force, shear force, and moment at points C and D of the beam.

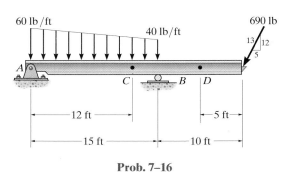

Prob. 7–16

7–17. Determine the normal force, shear force, and moment acting at a section passing through point C.

7–18. Determine the normal force, shear force, and moment acting at a section passing through point D.

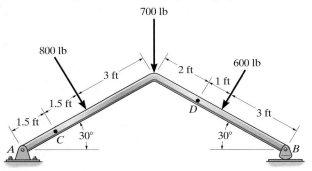

Probs. 7–17/18

7–19. Determine the normal force, shear force, and moment at a section passing through point C. Take $P = 8$ kN.

*7–20. The cable will fail when subjected to a tension of 2 kN. Determine the largest vertical load P the frame will support and calculate the internal normal force, shear force, and moment at a section passing through point C for this loading.

Probs. 7–19/20

7–21. Determine the internal shear force and moment acting at point C of the beam.

Prob. 7–21

7–22. Determine the internal shear force and moment acting at point D of the beam.

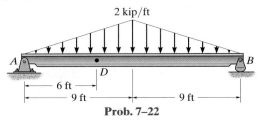

Prob. 7–22

7–23. The shaft is supported by a journal bearing at A and a thrust bearing at B. Determine the internal normal force, shear force, and moment at (a) point C, which is just to the right of the bearing at A, and (b) point D, which is just to the left of the 3000-lb force.

Prob. 7–23

*7–24. The jack AB is used to straighten the bent beam DE using the arrangement shown. If the axial compressive force in the jack is 5000 lb, determine the internal moment developed at point C of the top beam. Neglect the weight of the beams.

7–25. Solve Prob. 7–24 assuming that each beam has a uniform weight of 150 lb/ft.

Probs. 7–24/25

7–26. Determine the normal force, shear force, and moment in the beam at sections passing through points *D* and *E*. Point *E* is just to the right of the 3-kip load.

Prob. 7–26

7–27. Determine the normal force, shear force, and moment at a section passing through point *D* of the two-member frame.

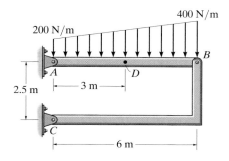

Prob. 7–27

*7–28. Determine the normal force, shear force, and moment at sections passing through points *E* and *F*. Member *BC* is pinned at *B* and there is a smooth slot in it at *C*. The pin at *C* is fixed to member *CD*.

Prob. 7–28

7–29. The bolt shank is subjected to a tension of 80 lb. Determine the internal normal force, shear force, and moment at point *C*.

Prob. 7–29

7–30. Determine the normal force, shear force, and moment acting at sections passing through points *B* and *C* on the curved rod.

Prob. 7–30

7–31. The cantilevered rack is used to support each end of a smooth pipe that has a total weight of 300 lb. Determine the normal force, shear force, and moment that act in the arm at its fixed support *A* along a vertical section.

Prob. 7–31

*7–32. Determine the normal force, shear force, and moment at a section passing through point D of the two-member frame.

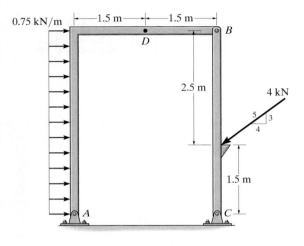

Prob. 7–32

7–34. Determine the internal normal force, shear force, and moment acting at points B and C on the curved rod.

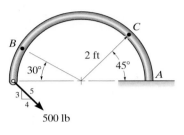

Prob. 7–34

7–33. Determine the internal normal force, shear force, and moment acting at point A of the smooth hook.

Prob. 7–33

7–35. Determine the ratio of a/b for which the shear force will be zero at the midpoint C of the beam.

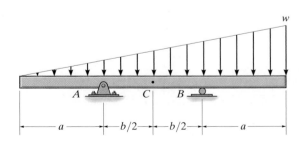

Prob. 7–35

*7–36.   The semicircular arch is subjected to a uniform distributed load along its axis of w_0 per unit length. Determine the internal normal force, shear force, and moment in the arch at $\theta = 45°$.

7–37.   Solve Prob. 7–36 for $\theta = 120°$.

Probs. 7–36/37

7–38.   Determine the x, y, z components of internal loading at a section passing through point C in the pipe assembly. Neglect the weight of the pipe. Take $\mathbf{F}_1 = \{350\mathbf{j} - 400\mathbf{k}\}$ lb and $\mathbf{F}_2 = \{150\mathbf{i} - 300\mathbf{k}\}$ lb.

7–39.   Determine the x, y, z components of internal loading at a section passing through point C in the pipe assembly. Neglect the weight of the pipe. Take $\mathbf{F}_1 = \{-80\mathbf{i} + 200\mathbf{j} - 300\mathbf{k}\}$ lb and $\mathbf{F}_2 = \{250\mathbf{i} - 150\mathbf{j} - 200\mathbf{k}\}$ lb.

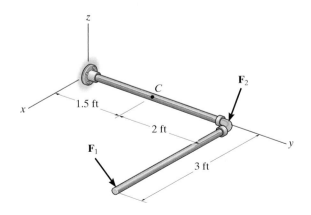

Probs. 7–38/39

*7–40.   Determine the x, y, z components of internal loading in the rod at point D. $\mathbf{F} = \{7\mathbf{i} - 12\mathbf{j} - 5\mathbf{k}\}$ kN.

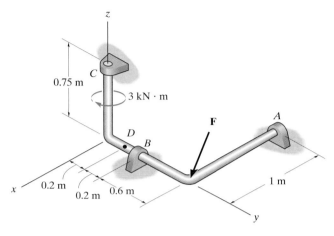

Prob. 7–40

7–41.   Determine the x, y, z components of internal loading in the rod at point E. $\mathbf{F} = \{7\mathbf{i} - 12\mathbf{j} - 5\mathbf{k}\}$ kN.

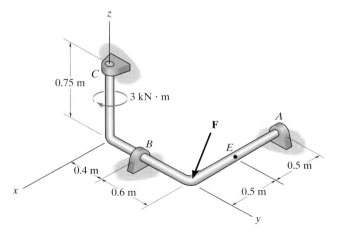

Prob. 7–41

*7.2   Shear and Moment Equations and Diagrams

To save on material the beams used to support the roof of this shelter were tapered since the roof loading will produce a larger internal moment at the beams' centers than at their ends.

Beams are structural members designed to support loadings applied perpendicular to their axes. In general, beams are long, straight bars having a constant cross-sectional area. Often they are classified as to how they are supported. For example, a *simply supported beam* is pinned at one end and roller-supported at the other, Fig. 7–10*a*, whereas a *cantilevered beam* is fixed at one end and free at the other. The actual design of a beam requires a detailed knowledge of the *variation* of the internal shear force V and bending moment M acting at *each point* along the axis of the beam. After this force and bending-moment analysis is complete, the theory of mechanics of materials and an appropriate engineering design code then can be used to determine the beam's required cross-sectional area.

The *variations* of V and M as functions of the position x along the beam's axis can be obtained by using the method of sections discussed in Sec. 7.1. Here, however, it is necessary to section the beam at an arbitrary distance x from one end rather than at a specified point. If the results are plotted, the graphical variations of V and M as functions of x are termed the *shear diagram* and *bending-moment diagram*, respectively, Figs. 7–10*b* and 7–10*c*.

In general, the internal shear and bending-moment functions generally will be discontinuous, or their slopes will be discontinuous at points where a distributed load changes or where concentrated forces or couple moments are applied. Because of this, these functions must be determined for *each segment* of the beam located between any two discontinuities of loading. For example, sections located at x_1, x_2, and x_3 will have to be used to describe the variation of V and M along the length of the beam in Fig. 7–10. These functions will be valid *only* within regions from O to a for x_1, from a to b for x_2, and from b to L for x_3.

The internal normal force will not be considered in the following discussion for two reasons. In most cases, the loads applied to a beam act perpendicular to the beam's axis and hence produce only an internal shear force and bending moment. For design purposes, the beam's resistance to shear, and particularly to bending, is more important than its ability to resist a normal force.

(a)

(b)

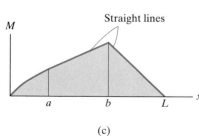

(c)

Fig. 7–10

Sign Convention.   Before presenting a method for determining the shear and bending moment as functions of x and later plotting these functions (shear and bending-moment diagrams), it is first necessary to establish a *sign convention* so as to define a "positive" and "negative" shear force and bending moment acting in the beam. [This is analogous to assigning coordinate directions x positive to the right and y positive upward when plotting a function $y = f(x)$.] Although the choice of a sign convention is arbitrary, here we will choose the one used for the majority of engineering applications. It is illustrated in Fig. 7–11. Here the positive directions are denoted by an internal *shear force* that causes *clockwise rotation* of the member on which it acts, and by an internal *moment* that causes *compression or pushing on the upper part* of the member. Also, positive moment would tend to bend the member if it were elastic, concave upward. Loadings that are opposite to these are considered negative.

Positive shear

Positive moment

Beam sign convention

Fig. 7–11

Procedure for Analysis

The shear and bending-moment diagrams for a beam can be constructed using the following procedure.

Support Reactions.

- Determine all the reactive forces and couple moments acting on the beam and resolve all the forces into components acting perpendicular and parallel to the beam's axis.

Shear and Moment Functions.

- Specify separate coordinates x having an origin at the beam's *left end* and extending to regions of the beam *between* concentrated forces and/or couple moments, or where there is no discontinuity of distributed loading.

- Section the beam perpendicular to its axis at each distance x and draw the free-body diagram of one of the segments. Be sure **V** and **M** are shown acting in their *positive sense*, in accordance with the sign convention given in Fig. 7–11.

- The shear V is obtained by summing forces perpendicular to the beam's axis.

- The moment M is obtained by summing moments about the sectioned end of the segment.

Shear and Moment Diagrams.

- Plot the shear diagram (V versus x) and the moment diagram (M versus x). If computed values of the functions describing V and M are *positive*, the values are plotted above the x axis, whereas *negative* values are plotted below the x axis.

- Generally, it is convenient to plot the shear and bending-moment diagrams directly below the free-body diagram of the beam.

EXAMPLE 7.7

(a)

(b)

(c)

(d)

Fig. 7–12

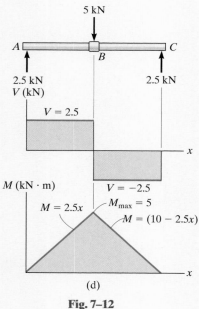

Draw the shear and bending-moment diagrams for the shaft shown in Fig. 7–12a. The support at A is a thrust bearing and the support at C is a journal bearing.

SOLUTION

Support Reactions.   The support reactions have been computed, as shown on the shaft's free-body diagram, Fig. 7–12d.

Shear and Moment Functions.   The shaft is sectioned at an arbitrary distance x from point A, extending within the region AB, and the free-body diagram of the left segment is shown in Fig. 7–12b. The unknowns $\mathbf{V}$ and $\mathbf{M}$ are assumed to act in the *positive sense* on the right-hand face of the segment according to the established sign convention. Why? Applying the equilibrium equations yields

$$+\uparrow \Sigma F_y = 0; \qquad\qquad\qquad V = 2.5 \text{ kN} \qquad\qquad (1)$$

$$\zeta + \Sigma M = 0; \qquad\qquad\qquad M = 2.5x \text{ kN} \cdot \text{m} \qquad\qquad (2)$$

A free-body diagram for a left segment of the shaft extending a distance x within the region BC is shown in Fig. 7–12c. As always, $\mathbf{V}$ and $\mathbf{M}$ are shown acting in the positive sense. Hence,

$$+\uparrow \Sigma F_y = 0; \qquad\quad 2.5 \text{ kN} - 5 \text{ kN} - V = 0$$

$$V = -2.5 \text{ kN} \qquad\qquad (3)$$

$$\zeta + \Sigma M = 0; \quad M + 5 \text{ kN}(x - 2 \text{ m}) - 2.5 \text{ kN}(x) = 0$$

$$M = (10 - 2.5x) \text{ kN} \cdot \text{m} \qquad\qquad (4)$$

Shear and Moment Diagrams.   When Eqs. 1 through 4 are plotted within the regions in which they are valid, the shear and bending-moment diagrams shown in Fig. 7–12d are obtained. The shear diagram indicates that the internal shear force is always 2.5 kN (positive) within shaft segment AB. Just to the right of point B, the shear force changes sign and remains at a constant value of -2.5 kN for segment BC. The moment diagram starts at zero, increases linearly to point B at $x = 2$ m, where $M_{max} = 2.5 \text{ kN}(2 \text{ m}) = 5 \text{ kN} \cdot \text{m}$, and thereafter decreases back to zero.

NOTE: It is seen in Fig. 7–12d that the graph of the shear and moment diagrams is discontinuous at points of concentrated force, i.e., points A, B, and C. For this reason, as stated earlier, it is necessary to express both the shear and bending-moment functions separately for regions between concentrated loads. It should be realized, however, that all loading discontinuities are mathematical, arising from the *idealization of a concentrated force and couple moment*. Physically, loads are always applied over a finite area, and if the load variation could actually be accounted for, the shear and bending-moment diagrams would then be continuous over the shaft's entire length.

EXAMPLE 7.8

Draw the shear and bending-moment diagrams for the beam shown in Fig. 7–13a.

(a)

SOLUTION

Support Reactions. The support reactions have been computed as shown on the beam's free-body diagram, Fig. 7–13c.

Shear and Moment Functions. A free-body diagram for a left segment of the beam having a length x is shown in Fig. 7–13b. The distributed loading acting on this segment has an intensity of $\frac{2}{3}x$ at its end and is replaced by a resultant force *after* the segment is isolated as a free-body diagram. The *magnitude* of the resultant force is equal to $\frac{1}{2}(x)(\frac{2}{3}x) = \frac{1}{3}x^2$. This force *act through the centroid* of the distributed loading area, a distance $\frac{1}{3}x$ from the right end. Applying the two equations of equilibrium yields

(b)

$$+\uparrow \Sigma F_y = 0; \qquad 9 - \frac{1}{3}x^2 - V = 0$$

$$V = \left(9 - \frac{x^2}{3}\right) \text{kN} \qquad (1)$$

$$\zeta + \Sigma M = 0; \qquad M + \frac{1}{3}x^2\left(\frac{x}{3}\right) - 9x = 0$$

$$M = \left(9x - \frac{x^3}{9}\right) \text{kN} \cdot \text{m} \qquad (2)$$

Shear and Moment Diagrams. The shear and bending-moment diagrams shown in Fig. 7–13c are obtained by plotting Eqs. 1 and 2. The point of *zero shear* can be found using Eq. 1:

$$V = 9 - \frac{x^2}{3} = 0$$

$$x = 5.20 \text{ m}$$

NOTE: It will be shown in Sec. 7.3 that this value of x happens to represent the point on the beam where the *maximum moment* occurs. Using Eq. 2, we have

$$M_{\text{max}} = \left(9(5.20) - \frac{(5.20)^3}{9}\right) \text{kN} \cdot \text{m}$$

$$= 31.2 \text{ kN} \cdot \text{m}$$

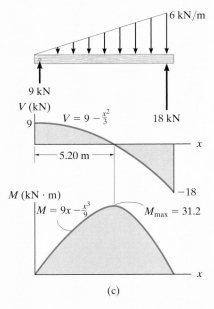

(c)

Fig. 7–13

PROBLEMS

For each of the following problems, establish the x axis with the origin at the left side of the beam, and obtain the internal shear and moment as a function of x. Use these results to plot the shear and moment diagrams.

7–42. Draw the shear and moment diagrams for the shaft (a) in terms of the parameters shown; (b) set $P = 9$ kN, $a = 2$ m, $L = 6$ m. There is a thrust bearing at A and a journal bearing at B.

Prob. 7–42

7–43. Draw the shear and moment diagrams for the beam (a) in terms of the parameters shown; (b) set $P = 800$ lb, $a = 5$ ft, $L = 12$ ft.

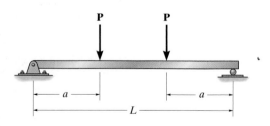

Prob. 7–43

*7–44. Draw the shear and moment diagrams for the beam (a) in terms of the parameters shown; (b) set $M_0 = 500$ N $\cdot$ m, $L = 8$ m.

7–45. If $L = 9$ m, the beam will fail when the maximum shear force is $V_{max} = 5$ kN or the maximum bending moment is $M_{max} = 2$ kN $\cdot$ m. Determine the magnitude M_0 of the largest couple moments it will support.

Probs. 7–44/45

7–46. The shaft is supported by a thrust bearing at A and a journal bearing at B. Draw the shear and moment diagrams for the shaft (a) in terms of the parameters shown; (b) set $w = 500$ lb/ft, $L = 10$ ft.

Prob. 7–46

7–47. The shaft is supported by a thrust bearing at A and a journal bearing at B. If $L = 10$ ft, the shaft will fail when the maximum moment is $M_{max} = 5$ kip $\cdot$ ft. Determine the largest uniform distributed load w the shaft will support.

Prob. 7–47

*7–48. Draw the shear and moment diagrams for the beam.

Prob. 7–48

7–49. Draw the shear and moment diagrams for the beam.

Prob. 7–49

7–50. Draw the shear and moment diagrams for the beam.

Prob. 7–50

7–51. Draw the shear and moment diagrams for the beam.

Prob. 7–51

*7–52. Draw the shear and moment diagrams for the beam.

Prob. 7–52

7–53. Draw the shear and moment diagrams for the beam.

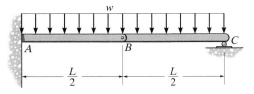

Prob. 7–53

7–54. Draw the shear and bending-moment diagrams for beam ABC. Note that there is a pin at B.

Prob. 7–54

7–55. The beam has a depth of 2 ft and is subjected to a uniform distributed loading of 50 lb/ft which acts at an angle of 30° from the vertical as shown. Determine the internal normal force, shear force, and moment in the beam as a function of x. *Hint:* The moment loading is to be determined from a point along the centerline of the beam (x axis).

Prob. 7–55

*7–56. Draw the shear and moment diagrams for the beam (a) in terms of the parameters shown; (b) set $w = 250$ lb/ft, $L = 12$ ft.

Prob. 7–56

7–57. If $L = 18$ ft, the beam will fail when the maximum shear force is $V_{max} = 800$ lb, or the maximum moment is $M_{max} = 1200$ lb · ft. Determine the largest intensity w of the distributed loading it will support.

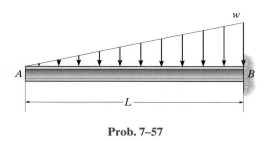

Prob. 7–57

7–58. The beam will fail when the maximum internal moment is M_{max}. Determine the position x of the concentrated force **P** and its smallest magnitude that will cause failure.

Prob. 7–58

7–59. Draw the shear and moment diagrams for the beam.

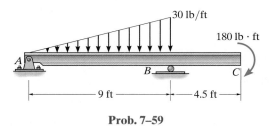

Prob. 7–59

*7–60. The cantilevered beam is made of material having a specific weight γ. Determine the shear and moment in the beam as a function of x.

Prob. 7–60

7–61. Draw the shear and moment diagrams for the beam.

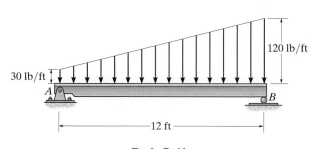

Prob. 7–61

7–62. Draw the shear and moment diagrams for the beam.

Prob. 7–62

7–63. Express the x, y, z components of internal loading in the rod as a function of y, where $0 \leq y \leq 4$ ft.

Prob. 7–63

*7–64. Determine the normal force, shear force, and moment in the curved rod as a function of θ.

Prob. 7–64

7–65. The quarter circular rod lies in the horizontal plane and supports a vertical force **P** at its end. Determine the magnitudes of the components of the internal shear force, moment, and torque acting in the rod as a function of the angle θ.

Prob. 7–65

(a)

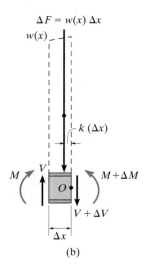

(b)

Fig. 7–14

*7.3   Relations between Distributed Load, Shear, and Moment

In cases where a beam is subjected to several concentrated forces, couple moments, and distributed loads, the method of constructing the shear and bending-moment diagrams discussed in Sec. 7.2 may become quite tedious. In this section a simpler method for constructing these diagrams is discussed—a method based on differential relations that exist between the load, shear, and bending moment.

Distributed Load.   Consider the beam AD shown in Fig. 7–14a, which is subjected to an arbitrary load $w = w(x)$ and a series of concentrated forces and couple moments. In the following discussion, the *distributed load* will be considered *positive* when the *loading acts downward* as shown. A free-body diagram for a small segment of the beam having a length Δx is chosen at a point x along the beam which is *not* subjected to a concentrated force or couple moment, Fig. 7–14b. Hence any results obtained will not apply at points of concentrated loading. The internal shear force and bending moment shown on the free-body diagram are assumed to act in the *positive sense* according to the established sign convention. Note that both the shear force and moment acting on the right-hand face must be increased by a small, finite amount in order to keep the segment in equilibrium. The distributed loading has been replaced by a resultant force $\Delta F = w(x) \, \Delta x$ that acts at a fractional distance $k(\Delta x)$ from the right end, where $0 < k < 1$ [for example, if $w(x)$ is *uniform*, $k = \frac{1}{2}$]. Applying the equations of equilibrium, we have

$$+\uparrow \Sigma F_y = 0; \qquad V - w(x) \, \Delta x - (V + \Delta V) = 0$$
$$\Delta V = -w(x) \, \Delta x$$

$$\zeta + \Sigma M_O = 0; \qquad -V \, \Delta x - M + w(x) \, \Delta x [k(\Delta x)] + (M + \Delta M) = 0$$
$$\Delta M = V \, \Delta x - w(x) k (\Delta x)^2$$

Dividing by Δx and taking the limit as $\Delta x \rightarrow 0$, these two equations become

$$\frac{dV}{dx} = -w(x)$$

$$\frac{\text{Slope of}}{\text{shear diagram}} = \frac{\text{Negative of distributed}}{\text{load intensity}} \qquad (7\text{–}1)$$

$$\frac{dM}{dx} = V$$

$$\frac{\text{Slope of}}{\text{moment diagram}} = \text{Shear} \qquad (7\text{–}2)$$

These two equations provide a convenient means for plotting the shear and moment diagrams for a beam. At a specific point in a beam, Eq. 7–1 states that the *slope of the shear diagram is equal to the negative of the intensity of the distributed load*, while Eq. 7–2 states that the *slope of the moment diagram is equal to the shear*. In particular, if the shear is equal to zero, $dM/dx = 0$, and therefore *a point of zero shear corresponds to a point of maximum (or possibly minimum) moment.*

Equations 7–1 and 7–2 may also be rewritten in the form $dV = -w(x)\,dx$ and $dM = V\,dx$. Noting that $w(x)\,dx$ and $V\,dx$ represent differential areas under the distributed-loading and shear diagrams, respectively, we can integrate these areas between two points B and C along the beam, Fig. 7–14a, and write

This concrete beam is used to support the roof. Its size and the placement of steel reinforcement within it can be determined once the shear and moment diagrams have been established.

$$\Delta V_{BC} = -\int w(x)\,dx$$

$$\begin{array}{ccc} \text{Change} & & \text{Negative of area under} \\ \text{in shear} & = & \text{loading curve} \end{array}$$

(7–3)

and

$$\Delta M_{BC} = \int V\,dx$$

$$\begin{array}{ccc} \text{Change} & & \text{Area under} \\ \text{in moment} & = & \text{shear diagram} \end{array}$$

(7–4)

Equation 7–3 states that the *change in shear between points B and C is equal to the negative of the area under the distributed-loading curve between these points*. Similarly, from Eq. 7–4, the *change in moment between B and C is equal to the area under the shear diagram within region BC*. Because two integrations are involved, first to determine the change in shear, Eq. 7–3, then to determine the change in moment, Eq. 7–4, we can state that if the loading curve $w = w(x)$ is a polynomial of degree n, then $V = V(x)$ will be a curve of degree $n + 1$, and $M = M(x)$ will be a curve of degree $n + 2$.

As stated previously, the above equations do not apply at points where a *concentrated* force or couple moment acts. These two special cases create *discontinuities* in the shear and moment diagrams, and as a result, each deserves separate treatment.

(a)

(b)

Fig. 7–15

Each outrigger such as AB supporting this crane acts as a beam which is fixed to the frame of the crane at one end and subjected to a force $\mathbf{F}$ on the footing at its other end. A proper design requires that the outrigger is able to resist its maximum internal shear and moment. The shear and moment diagrams indicate that the shear will be constant throughout its length and the maximum moment occurs at the support A.

Force. A free-body diagram of a small segment of the beam in Fig. 7–14a, taken from under one of the forces, is shown in Fig. 7–15a. Here it can be seen that force equilibrium requires

$$+\uparrow \Sigma F_y = 0; \qquad\qquad \Delta V = -F \qquad\qquad (7\text{–}5)$$

Thus, the *change in shear is negative*, so that on the shear diagram the shear will "jump" *downward when* $\mathbf{F}$ *acts downward* on the beam. Likewise, the jump in shear (ΔV) is upward when $\mathbf{F}$ acts upward.

Couple Moment. If we remove a segment of the beam in Fig. 7–14a that is located at the couple moment, the free-body diagram shown in Fig. 7–15b results. In this case letting $\Delta x \rightarrow 0$, moment equilibrium requires

$$\zeta + \Sigma M = 0; \qquad\qquad \Delta M = M_0 \qquad\qquad (7\text{–}6)$$

Thus, the *change in moment is positive*, or the moment diagram will "jump" *upward if* $\mathbf{M}_0$ *is clockwise*. Likewise, the jump ΔM is downward when $\mathbf{M}_0$ is counterclockwise.

The examples which follow illustrate application of the above equations for the construction of the shear and moment diagrams. After working through these examples, it is recommended that Examples 7.7 and 7.8 be solved using this method.

Important Points

- The slope of the shear diagram is equal to the negative of the intensity of the distributed loading, where positive distributed loading is downward, i.e., $dV/dx = -w(x)$.

- If a concentrated force acts downward on the beam, the shear will jump downward by the amount of the force.

- The change in the shear ΔV between two points is equal to *the negative of the area* under the distributed-loading curve between the points.

- The slope of the moment diagram is equal to the shear (i.e., $dM/dx = V$).

- The change in the moment ΔM between two points is equal to the *area* under the shear diagram between the two points.

- If a *clockwise* couple moment acts on the beam, the shear will not be affected, however, the moment diagram will jump *upward* by the amount of the moment.

- Points of *zero shear* represent points of *maximum or minimum moment* since $dM/dx = 0$.

EXAMPLE 7.9

Draw the shear and moment diagrams for the beam shown in Fig. 7–16a.

(a)   (b)

SOLUTION

Support Reactions.   The reactions at the fixed support have been calculated and are shown on the free-body diagram of the beam, Fig. 7–16b.

Shear Diagram.   The shear at the end points is plotted first, Fig. 7–16c. From the sign convention, Fig. 7–11, $V = +400$ at $x = 0$ and $V = 0$ at $x = 8$. Since $dV/dx = -w = -50$, a straight, *negative* sloping line connects the end points.

Moment Diagram.   From our sign convention, Fig. 7–11, the moments at the beam's end points, $M = -1600$ at $x = 0$ and $M = 0$ at $x = 8$, are plotted first, Fig. 7–16d. Successive values of shear taken from the shear diagram, Fig. 7–16c, indicate that the *slope $dM/dx = V$* of the moment diagram, Fig. 7–16d, is always positive yet *linearly decreasing* from $dM/dx = 400$ at $x = 0$ to $dM/dx = 0$ at $x = 8$.

NOTE: Due to the integrations, w a constant yields V a sloping line (first-degree curve) and M a parabola (second-degree curve).

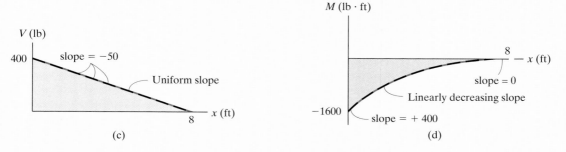

(c)   (d)

Fig. 7–16

EXAMPLE 7.10

(a)

(b)

(c)

(d)

(e)

Fig. 7–17

Draw the shear and moment diagrams for the cantilevered beam shown in Fig. 7–17a.

SOLUTION

Support Reactions. The reactions at the fixed support have been calculated and are shown on the free-body diagram of the beam, Fig. 7–17b.

Shear Diagram. Using the established sign convention, Fig. 7–11, the shear at the ends of the beam is plotted first; i.e., $x = 0$, $V = +1080$; $x = 20$, $V = +600$, Fig. 7–17c.

Since the uniform distributed load is downward and *constant*, the slope of the shear diagram is $dV/dx = -w = -40$ for $0 \le x < 12$ as indicated.

The magnitude of shear at $x = 12$ is $V = +600$. This can be determined by first finding the area under the load diagram between $x = 0$ and $x = 12$. This represents the change in shear. That is, $\Delta V = -\int w(x)\,dx = -40(12) = -480$. Thus $V|_{x=12} = V|_{x=0} + (-480) = 1080 - 480 = 600$. Also, we can obtain this value by using the method of sections, Fig. 7–17e, where for equilibrium $V = +600$.

Since the load between $12 < x \le 20$ is $w = 0$, the slope $dV/dx = 0$ as indicated. This brings the shear to the required value of $V = +600$ at $x = 20$.

Moment Diagram. Again, using the established sign convention, the moments at the ends of the beam are plotted first; i.e., $x = 0$, $M = -15\,880$; $x = 20$, $M = -1000$, Fig. 7–17d.

Each value of shear gives the slope of the moment diagram since $dM/dx = V$. As indicated, at $x = 0$, $dM/dx = +1080$; and at $x = 12$, $dM/dx = +600$. For $0 \le x < 12$, specific values of the shear diagram are positive but linearly decreasing. Hence, the moment diagram is parabolic with a linearly decreasing positive slope.

The magnitude of moment at $x = 12$ is -5800. This can be found by first determining the trapezoidal area under the shear diagram, which represents the change in moment, $\Delta M = \int V\,dx = 600(12) + \frac{1}{2}(1080 - 600)(12) = +10\,080$. Thus, $M|_{x=12} = M|_{x=0} + 10\,080 = -15\,880 + 10\,080 = -5800$. The more "basic" method of sections can also be used, where equilibrium at $x = 12$ requires $M = -5800$, Fig. 7–17e.

The moment diagram has a constant slope for $12 < x \le 20$ since, from the shear diagram, $dM/dx = V = +600$. This brings the value of $M = -1000$ at $x = 20$, as required.

EXAMPLE 7.11

Draw the shear and moment diagrams for the shaft in Fig. 7–18a. The support at A is a thrust bearing and the support at B is a journal bearing.

(a)

SOLUTION

Support Reactions.   The reactions at the supports are shown on the free-body diagram in Fig. 7–18b.

(b)

Shear Diagram.   The end points $x = 0$, $V = +3.5$ and $x = 8$, $V = -3.5$ are plotted first, as shown in Fig. 7–18c.

Since there is no distributed load on the shaft, the slope of the shear diagram throughout the shaft's length is zero; i.e., $dV/dx = -w = 0$. There is a discontinuity or "jump" of the shear diagram, however, at each concentrated force. From Eq. 7–5, $\Delta V = -F$, the change in shear is negative when the force acts downward and positive when the force acts upward. Stated another way, the "jump" follows the force, i.e., a downward force causes a downward jump, and vice versa. Thus, the 2-kN force at $x = 2$ m changes the shear from 3.5 kN to 1.5 kN; the 3-kN force at $x = 4$ m changes the shear from 1.5 kN to −1.5 kN, etc. We can *also* obtain numerical values for the shear at a specified point in the shaft by using the method of sections, as for example, $x = 2^+$ m, $V = 1.5$ kN in Fig. 7–18e.

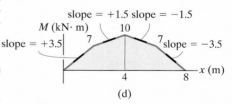

(c)

(d)

Moment Diagram.   The end points $x = 0$, $M = 0$ and $x = 8$, $M = 0$ are plotted first, as shown in Fig. 7–18d.

Since the shear is constant in each region of the shaft, the moment diagram has a corresponding constant positive or negative slope as indicated on the diagram. Numerical values for the change in moment at any point can be computed from the *area* under the shear diagram. For example, at $x = 2$ m, $\Delta M = \int V \, dx = 3.5(2) = 7$. Thus, $M|_{x=2} = M|_{x=0} + 7 = 0 + 7 = 7$. Also, by the method of sections, we can determine the moment at a specified point, as for example, $x = 2^+$ m, $M = 7$ kN·m, Fig. 7–18e.

(e)

Fig. 7–18

EXAMPLE 7.12

(b)

(c)

(d)

(e)

(a)

Sketch the shear and moment diagrams for the beam shown in Fig. 7–19a.

SOLUTION

Support Reactions.   The reactions are calculated and indicated on the free-body diagram, Fig. 7–19b.

Shear Diagram.   As in Example 7.11, the shear diagram can be constructed by "following the load" on the free-body diagram. In this regard, beginning at A, the reaction is up so $V_A = +100$ lb, Fig. 7–19c. No load acts between A and C, so the shear remains constant; i.e., $dV/dx = -w(x) = 0$. At C the 600-lb force acts downward, so the shear jumps down 600 lb, from 100 lb to -500 lb. Again the shear is constant (no load) and ends at -500 lb, point B. Notice that no jump or discontinuity in shear occurs at D, the point where the 4000-lb · ft couple moment is applied, Fig. 7–19b. This is because, for force equilibrium, $\Delta V = 0$ in Fig. 7–15b.

Moment Diagram.   The moment at each end of the beam is zero. These two points are plotted first, Fig. 7–19d. The slope of the moment diagram from A to C is constant since $dM/dx = V = +100$. The value of the moment at C can be determined by the method of sections, Fig. 7–19e, where $M_C = +1000$ lb · ft; or by first computing the rectangular area under the shear diagram between A and C to obtain the change in moment $\Delta M_{AC} = (100 \text{ lb})(10 \text{ ft}) = 1000$ lb · ft. Since $M_A = 0$, then $M_C = 0 + 1000$ lb · ft $= 1000$ lb · ft. From C to D the slope of the moment diagram is $dM/dx = V = -500$, Fig. 7–19c. The area under the shear diagram between points C and D is $\Delta M_{CD} = (-500 \text{ lb})(5 \text{ ft}) = -2500$ lb · ft, so that $M_D = M_C + \Delta M_{CD}$ $\Delta M_{CD} = 1000 - 2500 = -1500$ lb · ft. A jump in the moment diagram occurs at point D, which is caused by the concentrated couple moment of 4000 lb · ft. From Eq. 7–6, the jump is *positive* since the couple moment is *clockwise*. Thus, at $x = 15^+$ ft, the moment is $M_D = -1500 + 4000 = 2500$ lb · ft. This value can *also* be determined by the method of sections, Fig. 7–19f. From point D the slope of $dM/dx = -500$ is maintained until the diagram closes to zero at B, Fig. 7–19d.

Fig. 7–19

PROBLEMS

7–66. Draw the shear and moment diagrams for the beam.

Prob. 7–66

7–67. Draw the shear and moment diagrams for the beam.

Prob. 7–67

*7–68. Draw the shear and moment diagrams for the beam.

Prob. 7–68

7–69. Draw the shear and moment diagrams for the beam.

Prob. 7–69

7–70. Draw the shear and moment diagrams for the beam.

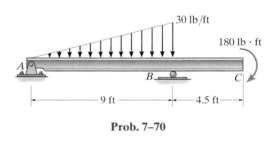

Prob. 7–70

7–71. Draw the shear and moment diagrams for the beam.

Prob. 7–71

*7–72.　Draw the shear and moment diagrams for the shaft. The support at A is a journal bearing and at B it is a thrust bearing.

Prob. 7–72

7–73.　Draw the shear and moment diagrams for the beam.

Prob. 7–73

7–74.　Draw the shear and moment diagrams for the shaft. The support at A is a journal bearing and at B it is a thrust bearing.

Prob. 7–74

7–75.　Draw the shear and moment diagrams for the beam.

Prob. 7–75

*7–76.　Draw the shear and moment diagrams for the shaft. The support at A is a thrust bearing and at B it is a journal bearing.

Prob. 7–76

7–77.　Draw the shear and moment diagrams for the beam.

Prob. 7–77

7–78. The beam will fail when the maximum moment is $M_{max} = 30 \text{ kip} \cdot \text{ft}$ or the maximum shear is $V_{max} = 8 \text{ kip}$. Determine the largest distributed load w the beam will support.

Prob. 7–78

7–79. The beam consists of two segments pin connected at B. Draw the shear and moment diagrams for the beam.

Prob. 7–79

*7–80. The beam consists of three segments pin connected at B and E. Draw the shear and moment diagrams for the beam.

Prob. 7–80

7–81. Draw the shear and moment diagrams for the beam.

Prob. 7–81

7–82. Draw the shear and moment diagrams for the beam.

Prob. 7–82

7–83. Draw the shear and moment diagrams for the beam.

Prob. 7–83

*7–84. Draw the shear and moment diagrams for the beam.

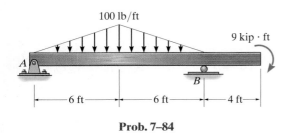

Prob. 7–84

7–85. Draw the shear and moment diagrams for the beam.

Prob. 7–85

7–86. Draw the shear and moment diagrams for the beam.

Prob. 7–86

7–87. Draw the shear and moment diagrams for the beam.

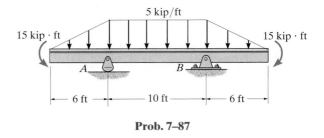

Prob. 7–87

*7–88. Draw the shear and moment diagrams for the beam.

Prob. 7–88

*7.4 Cables

Flexible cables and chains combine strength with lightness and often are used in engineering structures for support and to transmit loads from one member to another. When used to support suspension bridges and trolley wheels, cables form the main load-carrying element of the structure. In the force analysis of such systems, the weight of the cable itself may be neglected because it is often small compared to the load it carries. On the other hand, when cables are used as transmission lines and guys for radio antennas and derricks, the cable weight may become important and must be included in the structural analysis. Three cases are considered in the analysis that follows: (1) a cable subjected to concentrated loads; (2) a cable subjected to a distributed load; and (3) a cable subjected to its own weight. In each case, provided the loading is coplanar with the cable, the requirements for equilibrium are formulated in an identical manner.

When deriving the necessary relations between the force in the cable and its slope, we will make the assumption that the cable is *perfectly flexible* and *inextensible*. Due to its flexibility, the cable offers no resistance to bending, and therefore, the tensile force acting in the cable is always tangent to the cable at points along its length. Being inextensible, the cable has a constant length both before and after the load is applied. As a result, once the load is applied, the geometry of the cable remains fixed, and the cable or a segment of it can be treated as a rigid body.

Each of the cable segments remains approximately straight as they support the weight of these traffic lights.

Cable Subjected to Concentrated Loads.   When a cable of negligible weight supports several concentrated loads, the cable takes the form of several straight-line segments, each of which is subjected to a constant tensile force. Consider, for example, the cable shown in Fig. 7–20, where the distances h, L_1, L_2, and L_3 and the loads $\mathbf{P}_1$ and $\mathbf{P}_2$ are known. The problem here is to determine the *nine unknowns* consisting of the tension in each of the *three* segments, the *four* components of reaction at A and B, and the sags y_C and y_D at the *two* points C and D. For the solution we can write *two* equations of force equilibrium at each of points A, B, C, and D. This results in a total of *eight equations.** To complete the solution, we need to know something about the geometry of the cable in order to obtain the necessary ninth equation. For example, if the cable's total *length* L is specified, then the Pythagorean theorem can be used to relate each of the three segmental lengths, written in terms of h, y_C, y_D, L_1, L_2, and L_3, to the total length L. Unfortunately, this type of problem cannot be solved easily by hand. Another possibility, however, is to specify one of the sags, either y_C or y_D, instead of the cable length. By doing this, the equilibrium equations are then sufficient for obtaining the unknown forces and the remaining sag. Once the sag at each point of loading is obtained, the length of the cable can be determined by trigonometry. The following example illustrates a procedure for performing the equilibrium analysis for a problem of this type.

Fig. 7–20

*As will be shown in the following example, the eight equilibrium equations *also* can be written for the entire cable, or any part thereof. But *no more* than *eight* equations are available.

EXAMPLE 7.13

Determine the tension in each segment of the cable shown in Fig. 7–21a.

(a)

SOLUTION

By inspection, there are four unknown external reactions (A_x, A_y, E_x, and E_y) and four unknown cable tensions, one in each cable segment. These eight unknowns along with the two unknown sags y_B and y_D can be determined from *ten* available equilibrium equations. One method is to apply these equations as force equilibrium ($\Sigma F_x = 0$, $\Sigma F_y = 0$) to each of the five points A through E. Here, however, we will take a more direct approach.

Consider the free-body diagram for the entire cable, Fig. 7–21b. Thus,

$$\xrightarrow{+} \Sigma F_x = 0; \qquad\qquad -A_x + E_x = 0$$

$$\downarrow + \Sigma M_E = 0;$$

$$-A_y(18\text{ m}) + 4\text{ kN }(15\text{ m}) + 15\text{ kN }(10\text{ m}) + 3\text{ kN }(2\text{ m}) = 0$$

$$A_y = 12\text{ kN}$$

$$+\uparrow \Sigma F_y = 0; \qquad 12\text{ kN} - 4\text{ kN} - 15\text{ kN} - 3\text{ kN} + E_y = 0$$

$$E_y = 10\text{ kN}$$

Since the sag $y_C = 12$ m is known, we will now consider the leftmost section, which cuts cable BC, Fig. 7–21c.

$$\downarrow + \Sigma M_C = 0; \quad A_x(12\text{ m}) - 12\text{ kN }(8\text{ m}) + 4\text{ kN }(5\text{ m}) = 0$$

$$A_x = E_x = 6.33\text{ kN}$$

$$\xrightarrow{+} \Sigma F_x = 0; \qquad T_{BC}\cos\theta_{BC} - 6.33\text{ kN} = 0$$

$$+\uparrow \Sigma F_y = 0; \qquad 12\text{ kN} - 4\text{ kN} - T_{BC}\sin\theta_{BC} = 0$$

Thus,

$$\theta_{BC} = 51.6°$$

$$T_{BC} = 10.2\text{ kN} \qquad\qquad\qquad Ans.$$

(b)

(c)

Fig. 7–21

Fig. 7–21 (cont.)

Proceeding now to analyze the equilibrium of points A, C, and E in sequence, we have

Point A   (Fig. 7–21d).

$\xrightarrow{+} \Sigma F_x = 0;$ $T_{AB} \cos \theta_{AB} - 6.33 \text{ kN} = 0$

$+\uparrow \Sigma F_y = 0;$ $-T_{AB} \sin \theta_{AB} + 12 \text{ kN} = 0$

$$\theta_{AB} = 62.2°$$
$$T_{AB} = 13.6 \text{ kN} \qquad\qquad Ans.$$

Point C   (Fig. 7–21e).

$\xrightarrow{+} \Sigma F_x = 0;$ $T_{CD} \cos \theta_{CD} - 10.2 \cos 51.6° \text{ kN} = 0$

$+\uparrow \Sigma F_y = 0;$ $T_{CD} \sin \theta_{CD} + 10.2 \sin 51.6° \text{ kN} - 15 \text{ kN} = 0$

$$\theta_{CD} = 47.9°$$
$$T_{CD} = 9.44 \text{ kN} \qquad\qquad Ans.$$

Point E   (Fig. 7–21f).

$\xrightarrow{+} \Sigma F_x = 0;$ $6.33 \text{ kN} - T_{ED} \cos \theta_{ED} = 0$

$+\uparrow \Sigma F_y = 0;$ $10 \text{ kN} - T_{ED} \sin \theta_{ED} = 0$

$$\theta_{ED} = 57.7°$$
$$T_{ED} = 11.8 \text{ kN} \qquad\qquad Ans.$$

NOTE: By comparison, the maximum cable tension is in segment AB since this segment has the greatest slope (θ) and it is required that for any cable segment the horizontal component $T \cos \theta = A_x = E_x$ (a constant). Also, since the slope angles that the cable segments make with the horizontal have now been determined, it is possible to determine the sags y_B and y_D, Fig. 7–21a, using trigonometry.

Cable Subjected to a Distributed Load. Consider the weightless cable shown in Fig. 7–22a, which is subjected to a loading function $w = w(x)$ *as measured in the x direction.* The free-body diagram of a small segment of the cable having a length Δs is shown in Fig. 7–22b. Since the tensile force in the cable changes continuously in both magnitude and direction along the cable's length, this change is denoted on the free-body diagram by ΔT. The distributed load is represented by its resultant force $w(x)(\Delta x)$, which acts at a fractional distance $k(\Delta x)$ from point O, where $0 < k < 1$. Applying the equations of equilibrium yields

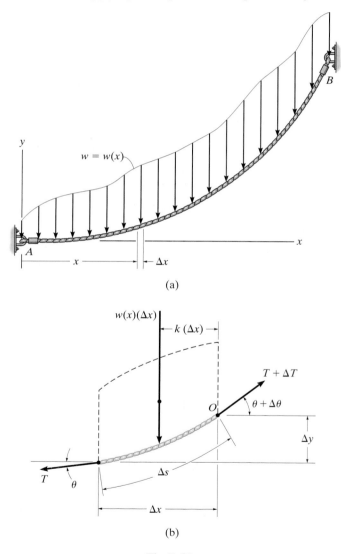

(a)

(b)

Fig. 7–22

$$\xrightarrow{+} \Sigma F_x = 0; \qquad -T \cos \theta + (T + \Delta T) \cos(\theta + \Delta \theta) = 0$$

$$+\uparrow \Sigma F_y = 0; \qquad -T \sin \theta - w(x)(\Delta x) + (T + \Delta T) \sin(\theta + \Delta \theta) = 0$$

$$\zeta + \Sigma M_O = 0; \qquad w(x)(\Delta x)k(\Delta x) - T \cos \theta\, \Delta y + T \sin \theta\, \Delta x = 0$$

Dividing each of these equations by Δx and taking the limit as $\Delta x \to 0$, and hence $\Delta y \to 0$, $\Delta \theta \to 0$, and $\Delta T \to 0$, we obtain

$$\frac{d(T \cos \theta)}{dx} = 0 \qquad (7\text{--}7)$$

$$\frac{d(T \sin \theta)}{dx} - w(x) = 0 \qquad (7\text{--}8)$$

$$\frac{dy}{dx} = \tan \theta \qquad (7\text{--}9)$$

Integrating Eq. 7–7, we have

$$T \cos \theta = \text{constant} = F_H \qquad (7\text{--}10)$$

Here F_H represents the horizontal component of tensile force at *any point* along the cable.

Integrating Eq. 7–8 gives

$$T \sin \theta = \int w(x)\, dx \qquad (7\text{--}11)$$

Dividing Eq. 7–11 by Eq. 7–10 eliminates T. Then, using Eq. 7–9, we can obtain the slope

$$\tan \theta = \frac{dy}{dx} = \frac{1}{F_H} \int w(x)\, dx$$

Performing a second integration yields

$$\boxed{y = \frac{1}{F_H} \int \left(\int w(x)\, dx \right) dx} \qquad (7\text{--}12)$$

This equation is used to determine the curve for the cable, $y = f(x)$. The horizontal force component F_H and the two constants, say C_1 and C_2, resulting from the integration are determined by applying the boundary conditions for the cable.

The cable and suspenders are used to support the uniform load of a gas pipe which crosses the river.

EXAMPLE 7.14

The cable of a suspension bridge supports half of the uniform road surface between the two columns at A and B, as shown in Fig. 7–23a. If this distributed loading is w_0, determine the maximum force developed in the cable and the cable's required length. The span length L and sag h are known.

(a)

Fig. 7–23

SOLUTION

We can determine the unknowns in the problem by first finding the curve that defines the shape of the cable by using Eq. 7–12. For reasons of symmetry, the origin of coordinates has been placed at the cable's center. Noting that $w(x) = w_0$, we have

$$y = \frac{1}{F_H} \int \left(\int w_0 \, dx \right) dx$$

Performing the two integrations gives

$$y = \frac{1}{F_H} \left(\frac{w_0 x^2}{2} + C_1 x + C_2 \right) \tag{1}$$

The constants of integration may be determined by using the boundary conditions $y = 0$ at $x = 0$ and $dy/dx = 0$ at $x = 0$. Substituting into Eq. 1 yields $C_1 = C_2 = 0$. The curve then becomes

$$y = \frac{w_0}{2F_H} x^2 \tag{2}$$

This is the equation of a *parabola*. The constant F_H may be obtained by using the boundary condition $y = h$ at $x = L/2$. Thus,

$$F_H = \frac{w_0 L^2}{8h} \qquad (3)$$

Therefore, Eq. 2 becomes

$$y = \frac{4h}{L^2} x^2 \qquad (4)$$

Since F_H is known, the tension in the cable may be determined using Eq. 7–10, written as $T = F_H/\cos\theta$. For $0 \le \theta < \pi/2$, the maximum tension will occur when θ is *maximum*, i.e., at point B, Fig. 7–23a. From Eq. 2, the slope at this point is

$$\left.\frac{dy}{dx}\right|_{x=L/2} = \tan\theta_{\max} = \left.\frac{w_0}{F_H} x\right|_{x=L/2}$$

or

$$\theta_{\max} = \tan^{-1}\left(\frac{w_0 L}{2 F_H}\right) \qquad (5)$$

Therefore,

$$T_{\max} = \frac{F_H}{\cos(\theta_{\max})} \qquad (6)$$

Using the triangular relationship shown in Fig. 7–23b, which is based on Eq. 5, Eq. 6 may be written as

$$T_{\max} = \frac{\sqrt{4 F_H^2 + w_0^2 L^2}}{2}$$

Substituting Eq. 3 into the above equation yields

$$T_{\max} = \frac{w_0 L}{2} \sqrt{1 + \left(\frac{L}{4h}\right)^2} \qquad Ans.$$

For a differential segment of cable length ds, we can write

$$ds = \sqrt{(dx)^2 + (dy)^2} = \sqrt{1 + \left(\frac{dy}{dx}\right)^2}\, dx$$

Hence, the total length of the cable, $\mathscr{L}$, can be determined by integration. Using Eq. 4, we have

$$\mathscr{L} = \int ds = 2 \int_0^{L/2} \sqrt{1 + \left(\frac{8h}{L^2} x\right)^2}\, dx \qquad (7)$$

Integrating yields

$$\mathscr{L} = \frac{L}{2}\left[\sqrt{1 + \left(\frac{4h}{L}\right)^2} + \frac{L}{4h}\sinh^{-1}\left(\frac{4h}{L}\right)\right] \qquad Ans.$$

(b)

Fig. 7–23 (cont.)

Over time the forces the cables exert on this telephone pole have caused it to tilt. Proper bracing of the pole should be required.

Cable Subjected to Its Own Weight. When the weight of the cable becomes important in the force analysis, the loading function along the cable becomes a function of the arc length s rather than the projected length x. A generalized loading function $w = w(s)$ acting along the cable is shown in Fig. 7–24a. The free-body diagram for a segment of the cable is shown in Fig. 7–24b. Applying the equilibrium equations to the force system on this diagram, one obtains relationships identical to those given by Eqs. 7–7 through 7–9, but with ds replacing dx. Therefore, it may be shown that

$$T \cos \theta = F_H$$

$$T \sin \theta = \int w(s)\, ds \tag{7–13}$$

$$\frac{dy}{dx} = \frac{1}{F_H} \int w(s)\, ds \tag{7–14}$$

To perform a direct integration of Eq. 7–14, it is necessary to replace dy/dx by ds/dx. Since

$$ds = \sqrt{dx^2 + dy^2}$$

then

$$\frac{dy}{dx} = \sqrt{\left(\frac{ds}{dx}\right)^2 - 1}$$

Therefore,

$$\frac{ds}{dx} = \left\{1 + \frac{1}{F_H^2}\left(\int w(s)\, ds\right)^2\right\}^{1/2}$$

Separating the variables and integrating yields

$$x = \int \frac{ds}{\left\{1 + \frac{1}{F_H^2}\left(\int w(s)\, ds\right)^2\right\}^{1/2}} \tag{7–15}$$

The two constants of integration, say C_1 and C_2, are found using the boundary conditions for the cable.

(a)

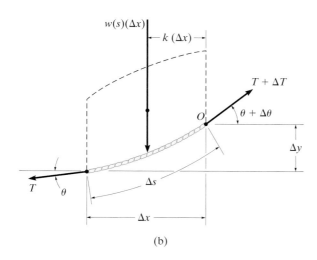

(b)

Fig. 7–24

EXAMPLE 7.15

Determine the deflection curve, the length, and the maximum tension in the uniform cable shown in Fig. 7–25. The cable weighs $w_0 = 5$ N/m.

Fig. 7–25

SOLUTION

For reasons of symmetry, the origin of coordinates is located at the center of the cable. The deflection curve is expressed as $y = f(x)$. We can determine it by first applying Eq. 7–15, where $w(s) = w_0$.

$$x = \int \frac{ds}{\left[1 + (1/F_H^2)\left(\int w_0 \, ds \right)^2 \right]^{1/2}}$$

Integrating the term under the integral sign in the denominator, we have

$$x = \int \frac{ds}{[1 + (1/F_H^2)(w_0 s + C_1)^2]^{1/2}}$$

Substituting $u = (1/F_H)(w_0 s + C_1)$ so that $du = (w_0/F_H) \, ds$, a second integration yields

$$x = \frac{F_H}{w_0}(\sinh^{-1} u + C_2)$$

or

$$x = \frac{F_H}{w_0}\left\{ \sinh^{-1}\left[\frac{1}{F_H}(w_0 s + C_1) \right] + C_2 \right\} \tag{1}$$

To evaluate the constants note that, from Eq. 7–14,

$$\frac{dy}{dx} = \frac{1}{F_H} \int w_0 \, ds \quad \text{or} \quad \frac{dy}{dx} = \frac{1}{F_H}(w_0 s + C_1)$$

Since $dy/dx = 0$ at $s = 0$, then $C_1 = 0$. Thus,

$$\frac{dy}{dx} = \frac{w_0 s}{F_H} \tag{2}$$

The constant C_2 may be evaluated by using the condition $s = 0$ at $x = 0$ in Eq. 1, in which case $C_2 = 0$. To obtain the deflection curve, solve for s in Eq. 1, which yields

$$s = \frac{F_H}{w_0}\sinh\left(\frac{w_0}{F_H}x \right) \tag{3}$$

Now substitute into Eq. 2, in which case

$$\frac{dy}{dx} = \sinh\left(\frac{w_0}{F_H}x \right)$$

Hence,

$$y = \frac{F_H}{w_0}\cosh\left(\frac{w_0}{F_H}x\right) + C_3 \qquad (4)$$

If the boundary condition $y = 0$ at $x = 0$ is applied, the constant $C_3 = -F_H/w_0$, and therefore the deflection curve becomes

$$y = \frac{F_H}{w_0}\left[\cosh\left(\frac{w_0}{F_H}x\right) - 1\right]$$

This equation defines the shape of a *catenary curve*. The constant F_H is obtained by using the boundary condition that $y = h$ at $x = L/2$, in which case

$$h = \frac{F_H}{w_0}\left[\cosh\left(\frac{w_0 L}{2F_H}\right) - 1\right] \qquad (5)$$

Since $w_0 = 5\text{ N/m}$, $h = 6\text{ m}$, and $L = 20\text{ m}$, Eqs. 4 and 5 become

$$y = \frac{F_H}{5\text{ N/m}}\left[\cosh\left(\frac{5\text{ N/m}}{F_H}x\right) - 1\right] \qquad (6)$$

$$6\text{ m} = \frac{F_H}{5\text{ N/m}}\left[\cosh\left(\frac{50\text{ N}}{F_H}\right) - 1\right] \qquad (7)$$

Equation 7 can be solved for F_H by using a trial-and-error procedure. The result is

$$F_H = 45.9\text{ N}$$

and therefore the deflection curve, Eq. 6, becomes

$$y = 9.19[\cosh(0.109x) - 1]\text{ m} \qquad \textit{Ans.}$$

Using Eq. 3, with $x = 10\text{ m}$, the half-length of the cable is

$$\frac{\mathscr{L}}{2} = \frac{45.9\text{ N}}{5\text{ N/m}}\sinh\left[\frac{5\text{ N/m}}{45.9\text{ N}}(10\text{ m})\right] = 12.1\text{ m}$$

Hence,

$$\mathscr{L} = 24.2\text{ m} \qquad \textit{Ans.}$$

Since $T = F_H/\cos\theta$, Eq. 7–13, the maximum tension occurs when θ is maximum, i.e., at $s = \mathscr{L}/2 = 12.1\text{ m}$. Using Eq. 2 yields

$$\left.\frac{dy}{dx}\right|_{s=12.1\text{ m}} = \tan\theta_{max} = \frac{5\text{ N/m}(12.1\text{ m})}{45.9\text{ N}} = 1.32$$

$$\theta_{max} = 52.8°$$

Thus,

$$T_{max} = \frac{F_H}{\cos\theta_{max}} = \frac{45.9\text{ N}}{\cos 52.8°} = 75.9\text{ N} \qquad \textit{Ans.}$$

PROBLEMS

Neglect the weight of the cable in the following problems, unless specified.

7–89. Determine the force P needed to hold the cable in the position shown, i.e., so segment BC remains horizontal. Also, compute the sag y_B and the maximum tension in the cable.

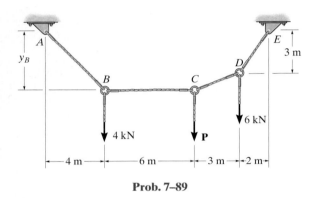

Prob. 7–89

7–90. Cable $ABCD$ supports the 10-kg lamp E and the 15-kg lamp F. Determine the maximum tension in the cable and the sag of point B.

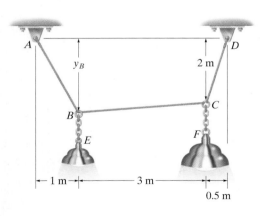

Prob. 7–90

7–91. The cable supports the three loads shown. Determine the sags y_B and y_D of points B and D. Take $P_1 = 400$ lb, $P_2 = 250$ lb.

*7–92. The cable supports the three loads shown. Determine the magnitude of $\mathbf{P}_1$ if $P_2 = 300$ lb and $y_B = 8$ ft. Also find the sag y_D.

Probs. 7–91/92

7–93. The cable supports the loading shown. Determine the distance x_B the force at point B acts from A. Set $P = 40$ lb.

7–94. The cable supports the loading shown. Determine the magnitude of the horizontal force $\mathbf{P}$ so that $x_B = 6$ ft.

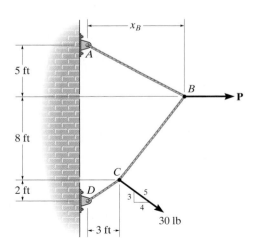

Probs. 7–93/94

7–95. Determine the forces P_1 and P_2 needed to hold the cable in the position shown, i.e., so segment CD remains horizontal. Also, compute the sag y_D and the maximum tension in the cable.

7–97. Determine the maximum uniform loading w, measured in lb/ft, that the cable can support if it is capable of sustaining a maximum tension of 3000 lb before it will break.

Prob. 7–95

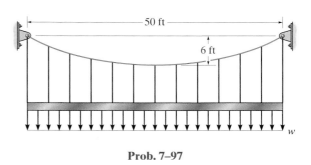

Prob. 7–97

*7–96.** The cable supports the loading shown. Determine the distance x_B from the wall to point B.

7–98. The cable is subjected to a uniform loading of $w = 250$ lb/ft. Determine the maximum and minimum tension in the cable.

Prob. 7–96

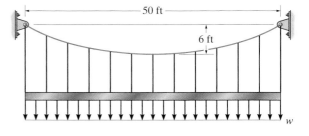

Prob. 7–98

7–99. The cable is subjected to the triangular loading. If the slope of the cable at A is zero, determine the equation of the curve $y = f(x)$ which defines the cable shape AB, and the maximum tension developed in the cable.

7–101. The cable is subjected to the triangular loading. If the slope of the cable at point O is zero, determine the equation of the curve $y = f(x)$ which defines the cable shape OB, and the maximum tension developed in the cable.

Prob. 7–99

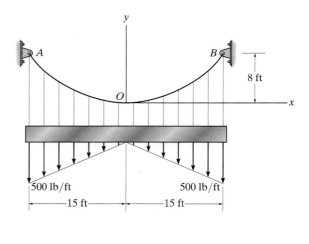

Prob. 7–101

***7–100.** The cable supports a girder which weighs 850 lb/ft. Determine the tension in the cable at points A, B, and C.

7–102. The cable is subjected to the parabolic loading $w = 150(1 - (x/50)^2)$ lb/ft, where x is in ft. Determine the equation $y = f(x)$ which defines the cable shape AB and the maximum tension in the cable.

Prob. 7–100

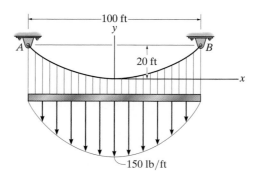

Prob. 7–102

7–103. The cable will break when the maximum tension reaches $T_{max} = 10$ kN. Determine the minimum sag h if it supports the uniform distributed load of $w = 600$ N/m.

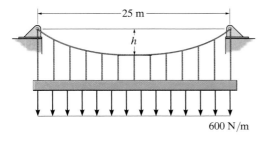

Prob. 7–103

■7–105. A cable has a weight of 3 lb/ft and is supported at points that are 500 ft apart and at the same elevation. If it has a length of 600 ft, determine the sag.

7–106. Show that the deflection curve of the cable discussed in Example 7.15 reduces to Eq. (4) in Example 7.14 when the *hyperbolic cosine function* is expanded in terms of a series and only the first two terms are retained. (The answer indicates that the *catenary* may be replaced by a *parabola* in the analysis of problems in which the sag is small. In this case, the cable weight is assumed to be uniformly distributed along the horizontal.)

7–107. A uniform cord is suspended between two points having the same elevation. Determine the sag-to-span ratio so that the maximum tension in the cord equals the cord's total weight.

*■7–108. A cable has a weight of 2 lb/ft. If it can span 100 ft and has a sag of 12 ft, determine the length of the cable. The ends of the cable are supported from the same elevation.

*7–104. A fiber optic cable is suspended over the poles so that the angle at the supports is $\theta = 22°$. Determine the minimum tension in the cable and the sag. The cable has a mass of 0.9 kg/m and the supports are at the same elevation.

7–109. The transmission cable having a weight of 20 lb/ft is strung across the river as shown. Determine the required force that must be applied to the cable at its points of attachment to the towers at B and C.

Prob. 7–104

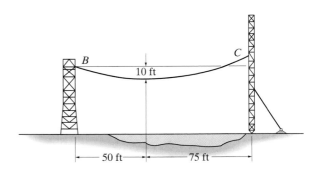

Prob. 7–109

7–110.   Determine the maximum tension developed in the cable if it is subjected to a uniform load of 600 N/m.

Prob. 7–110

■7–111.   A 40-m-long chain has a total mass of 100 kg and is suspended between two points 10 m apart at the same elevation. Determine the maximum tension and the sag in the chain.

*■**7–112.**   The cable has a mass of 0.5 kg/m, and is 25 m long. Determine the vertical and horizontal components of force it exerts on the top of the tower.

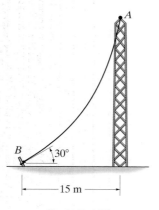

Prob. 7–112

■7–113.   A 50-ft cable is suspended between two points a distance of 15 ft apart and at the same elevation. If the minimum tension in the cable is 200 lb, determine the total weight of the cable and the maximum tension developed in the cable.

7–114.   The 80-ft-long chain is fixed at its ends and hoisted at its midpoint *B* using a crane. If the chain has a weight of 0.5 lb/ft, determine the minimum height *h* of the hook in order to lift the chain *completely* off the ground. What is the horizontal force at pin *A* or *C* when the chain is in this position? *Hint:* When *h* is a minimum, the slope at *A* and *C* is zero.

Prob. 7–114

■7–115.   A steel tape used for measurement in surveying has a length of 100.00 ft and a total weight of 2 lb. How much horizontal tension must be applied to the tape so that the distance marked on the ground is 99.90 ft? In practice the calculation should also include the effects of elastic stretching and temperature changes on the tape's length.

Prob. 7–115

Chapter Review

Internal Loadings

If a coplanar force system acts on a member, then in general a resultant internal *normal force N, shear force V,* and *bending moment M* will act at any cross section along the member.

Shear force

Normal force

Bending moment

The resultant internal normal force, shear force, and bending moment are determined using the method of sections. To find them, the member is sectioned at the point where the internal loadings are to be determined. A free-body diagram of one of the sectioned parts is then drawn.

The resultant normal force is determined by summing forces normal to the cross section. The resultant shear force is found by summing forces tangent to the cross section, and the resultant bending moment is found by summing moments about the centroid of the cross-sectional area.

$$\Sigma F_x = 0$$
$$\Sigma F_y = 0$$
$$\Sigma M_C = 0$$

If the member is subjected to a three-dimensional loading, then, in general, a *torsional loading* will also act on the cross section. It can be determined by summing moments about an axis that is perpendicular to the cross section and passes through its centroid.
(*See pages 337–339.*)

Bending moment components

Normal force

Torsional moment

Shear force components

Shear and Moment Diagrams

To construct the shear and moment diagrams for a member, it is necessary to section the member at an arbitrary point, located a distance x from one end.

If the external loading consists of changes in the distributed load, or a series of concentrated forces and couple moments act on the member, then different expressions for V and M must be determined within regions between these different loadings.

The unknown shear and moment are indicated on the cross section in the positive direction according to the established sign convention.

Application of the equilibrium equations will give the shear and moment as a function of x, which can then be plotted. (*See pages 354 and 355.*)

Positive moment   Positive shear

Relations between Shear and Moment

It is possible to plot the shear and moment diagrams quickly by using differential relationships that exist between the distributed loading w and V and M.

The slope of the shear diagram is equal to the negative of the distributed loading at any point.

$$\frac{dV}{dx} = -w$$

The slope of the moment diagram is equal to the shear at any point.

$$\frac{dM}{dx} = V$$

The change in shear between any two points is equal to the area under the distributed loading between the points.

$$\Delta V = -\int w\, dx$$

The change in the moment is equal to the area under the shear diagram between the points.
(*See pages 362–364.*)

$$\Delta M = \int V\, dx$$

Cables

When a flexible and inextensible cable is subjected to a series of concentrated forces, then the analysis of the cable can be performed by using the equations of equilibrium applied to free-body diagrams of either segments or points of application of the loading.

$$y = \frac{1}{F_H} \int \left(\int w(x)\, dx \right) dx$$

Distributed load

If external distributed loads or the weight of the cable are to be considered, then the forces and shape of the cable must be determined by first analyzing the forces on a differential segment of the cable and then integrating this result. The two constants, say C_1 and C_2, resulting from the integration are determined by applying the boundary conditions for the cable. (*See pages 376–377 and 380.*)

$$x = \int \frac{ds}{\left\{ 1 + \frac{1}{F_H^2} \left(\int w(s)\, ds \right)^2 \right\}^{1/2}}$$

Cable weight

REVIEW PROBLEMS

*▪7–116. A 100-lb cable is attached between two points at a distance 50 ft apart having equal elevations. If the maximum tension developed in the cable is 75 lb, determine the length of the cable and the sag.

7–117. Determine the distance a between the supports in terms of the beam's length L so that the moment in the *symmetric* beam is zero at the beam's center.

Prob. 7–117

7–118. Determine the internal normal force, shear force, and moment at point D.

Prob. 7–118

7–119. The beam is supported by a pin at C and a rod AB. Determine the internal normal force, shear force, and moment at point D.

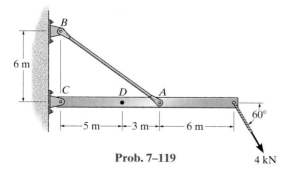

Prob. 7–119

*7–120. Express the shear and moment acting in the pipe as a function of y, where $0 \leq y \leq 4$ ft.

Prob. 7–120

7–121. Determine the normal force, shear force, and moment at points B and C of the beam.

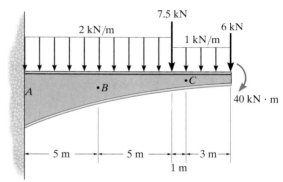

Prob. 7–121

7–122. A chain is suspended between points at the same elevation and spaced a distance of 60 ft apart. If it has a weight of 0.5 lb/ft and the sag is 3 ft, determine the maximum tension in the chain.

7–123. Draw the shear and moment diagrams for the beam.

Prob. 7–123

The effective design of a brake system, such as the one for this bicycle, requires efficient capacity for the mechanism to resist frictional forces. In this chapter, we will study the nature of friction and show how these forces are considered in engineering analysis.

8 Friction

1, 3, 16, 40, 62, 65

CHAPTER OBJECTIVES

- To introduce the concept of dry friction and show how to analyze the equilibrium of rigid bodies subjected to this force.
- To present specific applications of frictional force analysis on wedges, screws, belts, and bearings.
- To investigate the concept of rolling resistance.

8.1 Characteristics of Dry Friction

Friction can be defined as a force of resistance acting on a body that prevents or retards slipping of the body relative to a second body or surface with which it is in contact. This force always acts *tangent* to the surface at points of contact with other bodies and is directed so as to oppose the possible or existing motion of the body relative to these points.

In general, two types of friction can occur between surfaces. *Fluid friction* exists when the contacting surfaces are separated by a film of fluid (gas or liquid). The nature of fluid friction is studied in fluid mechanics since it depends upon knowledge of the velocity of the fluid and the fluid's ability to resist shear force. In this book, we will focus only on the effects of *dry friction*. This type of friction is often called *Coulomb friction* since its characteristics were studied extensively by C. A. Coulomb in 1781. Dry friction occurs between the contacting surfaces of bodies when there is no lubricating fluid.

The heat generated by the abrasive action of friction can be noticed when using this grinder to sharpen a metal blade.

Rough surface

(a)

(b)

(c)

Resultant Normal
and Frictional Forces

(d)

Fig. 8–1

Theory of Dry Friction.   The theory of dry friction is best explained by considering the effects caused by pulling horizontally on a block of uniform weight **W** which is resting on a rough horizontal surface, Fig. 8–1*a*. To properly develop a full understanding of the nature of friction, it is necessary to consider the surfaces of contact to be *nonrigid or deformable*. The upper portion of the block, however, can be considered rigid. As shown on the free-body diagram of the block, Fig. 8–1*b*, the floor exerts a *distribution* of both *normal force* $\Delta \mathbf{N}_n$ and *frictional force* $\Delta \mathbf{F}_n$ along the contacting surface. For equilibrium, the normal forces must act *upward* to balance the block's weight **W**, and the frictional forces act to the left to prevent the applied force **P** from moving the block to the right. Close examination of the contacting surfaces between the floor and block reveals how these frictional and normal forces develop, Fig. 8–1*c*. It can be seen that many microscopic irregularities exist between the two surfaces and, as a result, reactive forces $\Delta \mathbf{R}_n$ are developed at each of the protuberances.* These forces act at all points of contact, and, as shown, each reactive force contributes both a frictional component $\Delta \mathbf{F}_n$ and a normal component $\Delta \mathbf{N}_n$.

Equilibrium.   For simplicity in the following analysis, the effect of the *distributed* normal and frictional loadings will be indicated by their *resultants* **N** and **F**, which are represented on the free-body diagram as shown in Fig. 8-1*d*. Clearly, the distribution of $\Delta \mathbf{F}_n$ in Fig. 8–1*b* indicates that **F** always acts *tangent to the contacting surface, opposite* to the direction of **P**. On the other hand, the normal force **N** is determined from the distribution of $\Delta \mathbf{N}_n$ in Fig. 8–1*b* and is directed upward to balance the block's weight **W**. Notice that **N** acts a distance x to the right of the line of action of **W**, Fig. 8–1*d*. This location, which coincides with the centroid or geometric center of the loading diagram in Fig. 8–1*b*, is necessary in order to balance the "tipping effect" caused by **P**. For example, if **P** is applied at a height h from the surface, Fig. 8–1*d*, then moment equilibrium about point O is satisfied if $Wx = Ph$ or $x = Ph/W$. In particular, the block will be on the verge of *tipping* if N acts at the right corner of the block, $x = a/2$.

*Besides mechanical interactions as explained here, which is referred to as a classical approach, a detailed treatment of the nature of frictional forces must also include the effects of temperature, density, cleanliness, and atomic or molecular attraction between the contacting surfaces. See J. Krim, *Scientific American*, October, 1996.

Equilibrium

(e)

Impending Motion.

In cases where h is small or the surfaces of contact are rather "slippery," the frictional force $\mathbf{F}$ may *not* be great enough to balance $\mathbf{P}$, and consequently the block will tend to slip *before* it can tip. In other words, as P is slowly increased, F correspondingly increases until it attains a certain *maximum value F_s*, called the *limiting static frictional force*, Fig. 8–1e. When this value is reached, the block is in *unstable equilibrium* since any further increase in P will cause deformations and fractures at the points of surface contact, and consequently the block will begin to move. Experimentally, it has been determined that the limiting static frictional force F_s is *directly proportional* to the resultant normal force N. This may be expressed mathematically as

$$F_s = \mu_s N \qquad (8\text{–}1)$$

where the constant of proportionality, μ_s (mu "sub" s), is called the *coefficient of static friction*.

Thus, when the block is on the *verge of sliding*, the normal force $\mathbf{N}$ and frictional force $\mathbf{F}_s$ combine to create a resultant $\mathbf{R}_s$, Fig. 8–1e. The angle ϕ_s that $\mathbf{R}_s$ makes with $\mathbf{N}$ is called the *angle of static friction*. From the figure,

$$\phi_s = \tan^{-1}\left(\frac{F_s}{N}\right) = \tan^{-1}\left(\frac{\mu_s N}{N}\right) = \tan^{-1} \mu_s$$

Tabular Values of μ_s.

Typical values for μ_s, found in many engineering handbooks, are given in Table 8–1. Although this coefficient is generally less than 1, be aware that in some cases it is possible, as in the case of aluminum on aluminum, for μ_s to be greater than 1. Physically this means, of course, that in this case the frictional force is greater than the corresponding normal force. Furthermore, it should be noted that μ_s is dimensionless and depends only on the characteristics of the two surfaces in contact. A wide range of values is given for each value of μ_s since experimental testing was done under variable conditions of roughness and cleanliness of the contacting surfaces. For applications, therefore, it is important that both caution and judgment be exercised when selecting a coefficient of friction for a given set of conditions. When a more accurate calculation of F_s is required, the coefficient of friction should be determined directly by an experiment that involves the two materials to be used.

TABLE 8–1
Typical Values for μ_s

Contact Materials	Coefficient of Static Friction (μ_s)
Metal on ice	0.03–0.05
Wood on wood	0.30–0.70
Leather on wood	0.20–0.50
Leather on metal	0.30–0.60
Aluminum on aluminum	1.10–1.70

(a)

(b)

Fig. 8–2

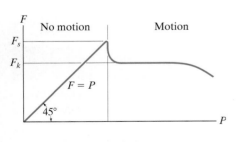

Fig. 8–3

Motion. If the magnitude of **P** acting on the block is increased so that it becomes greater than F_s, the frictional force at the contacting surfaces drops slightly to a smaller value F_k, called the *kinetic frictional force*. The block will *not* be held in equilibrium ($P > F_k$); instead, it will begin to slide with increasing speed, Fig. 8–2a. The drop made in the frictional force magnitude, from F_s (static) to F_k (kinetic), can be explained by again examining the surfaces of contact, Fig. 8–2b. Here it is seen that when $P > F_s$, then P has the capacity to shear off the peaks at the contact surfaces and cause the block to "lift" somewhat out of its settled position and "ride" on top of these peaks. Once the block begins to slide, high local temperatures at the points of contact cause momentary adhesion (welding) of these points. The continued shearing of these welds is the dominant mechanism creating kinetic friction. Since the resultant contact forces $\Delta \mathbf{R}_n$ are aligned slightly more in the vertical direction than before, they thereby contribute *smaller* frictional components, $\Delta \mathbf{F}_n$, than when the irregularities are meshed.

Experiments with sliding blocks indicate that the magnitude of the resultant frictional force $\mathbf{F}_k$ is directly proportional to the magnitude of the resultant normal force $\mathbf{N}$. This may be expressed mathematically as

$$F_k = \mu_k N \qquad (8-2)$$

Here the constant of proportionality, μ_k, is called the *coefficient of kinetic friction*. Typical values for μ_k are approximately 25 percent *smaller* than those listed in Table 8–1 for μ_s.

As shown in Fig. 8–2a, in this case, the resultant $\mathbf{R}_k$ has a line of action defined by ϕ_k. This angle is referred to as the *angle of kinetic friction*, where

$$\phi_k = \tan^{-1}\left(\frac{F_k}{N}\right) = \tan^{-1}\left(\frac{\mu_k N}{N}\right) = \tan^{-1}\mu_k$$

By comparison, $\phi_s \geq \phi_k$.

The above effects regarding friction can be summarized by reference to the graph in Fig. 8–3, which shows the variation of the frictional force F versus the applied load P. Here the frictional force is categorized in three different ways: namely, F is a *static frictional force* if equilibrium is maintained; F is a *limiting static frictional force* F_s when it reaches a maximum value needed to maintain equilibrium; and finally, F is termed a *kinetic-frictional force* F_k when sliding occurs at the contacting surface. Notice also from the graph that for very large values of P or for high speeds, because of aerodynamic effects, F_k and likewise μ_k begin to decrease.

Characteristics of Dry Friction.   As a result of *experiments* that pertain to the foregoing discussion, we can state the following rules which apply to bodies subjected to dry friction.

- The frictional force acts *tangent* to the contacting surfaces in a direction *opposed* to the *relative motion* or tendency for motion of one surface against another.

- The maximum static frictional force F_s that can be developed is independent of the area of contact, provided the normal pressure is not very low nor great enough to severely deform or crush the contacting surfaces of the bodies.

- The maximum static frictional force is generally greater than the kinetic frictional force for any two surfaces of contact. However, if one of the bodies is moving with a *very low velocity* over the surface of another, F_k becomes approximately equal to F_s, i.e., $\mu_s \approx \mu_k$.

- When *slipping* at the surface of contact is *about to occur*, the maximum static frictional force is proportional to the normal force, such that $F_s = \mu_s N$.

- When *slipping* at the surface of contact is *occurring*, the kinetic frictional force is proportional to the normal force, such that $F_k = \mu_k N$.

8.2   Problems Involving Dry Friction

If a rigid body is in equilibrium when it is subjected to a system of forces that includes the effect of friction, the force system must satisfy not only the equations of equilibrium but *also* the laws that govern the frictional forces.

Types of Friction Problems.   In general, there are three types of mechanics problems involving dry friction. They can easily be classified once free-body diagrams are drawn and the total number of unknowns are identified and compared with the total number of available equilibrium equations. Each type of problem will now be explained and illustrated graphically by examples. In all these cases the geometry and dimensions for the problem are assumed to be known.

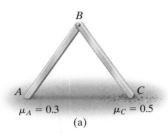

$\mu_A = 0.3$ $\mu_C = 0.5$

(a)

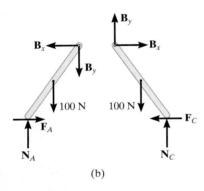

(b)

Fig. 8–4

Equilibrium. Problems in this category are strictly equilibrium problems which require *the total number of unknowns to be equal to the total number of available equilibrium equations.* Once the frictional forces are determined from the solution, however, their numerical values must be checked to be sure they satisfy the inequality $F \leq \mu_s N$; otherwise, slipping will occur and the body will not remain in equilibrium. A problem of this type is shown in Fig. 8–4a. Here we must determine the frictional forces at A and C to check if the equilibrium position of the two-member frame can be maintained. If the bars are uniform and have known weights of 100 N each, then the free-body diagrams are as shown in Fig. 8–4b. There are six unknown force components which can be determined *strictly* from the six equilibrium equations (three for each member). Once F_A, N_A, F_C, and N_C are determined, then the bars will remain in equilibrium provided $F_A \leq 0.3N_A$ and $F_C \leq 0.5N_C$ are satisfied.

Impending Motion at All Points. In this case *the total number of unknowns will equal the total number of available equilibrium equations plus the total number of available frictional equations, $F = \mu N$.* In particular, if *motion is impending* at the points of contact, then $F_s = \mu_s N$; whereas if the body is *slipping*, then $F_k = \mu_k N$. For example, consider the problem of finding the smallest angle θ at which the 100-N bar in Fig. 8–5a can be placed against the wall without slipping. The free-body diagram is shown in Fig. 8–5b. Here there are *five* unknowns: F_A, N_A, F_B, N_B, θ. For the solution there are *three* equilibrium equations and *two* static frictional equations which apply at *both* points of contact, so that $F_A = 0.3N_A$ and $F_B = 0.4N_B$.

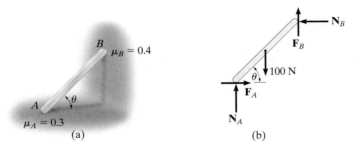

Fig. 8–5

Impending Motion at Some Points.

Here *the total number of unknowns will be less than the number of available equilibrium equations plus the total number of frictional equations or conditional equations for tipping.* As a result, several possibilities for motion or impending motion will exist and the problem will involve a determination of the kind of motion which actually occurs. For example, consider the two-member frame shown in Fig. 8–6a. In this problem we wish to determine the horizontal force P needed to cause movement. If each member has a weight of 100 N, then the free-body diagrams are as shown in Fig. 8–6b. There are *seven* unknowns: N_A, F_A, N_C, F_C, B_x, B_y, P. For a unique solution we must satisfy the *six* equilibrium equations (three for each member) and only *one* of two possible static frictional equations. This means that as P increases it will either cause slipping at A and no slipping at C, so that $F_A = 0.3N_A$ and $F_C \leq 0.5N_C$; or slipping occurs at C and no slipping at A, in which case $F_C = 0.5N_C$ and $F_A \leq 0.3N_A$. The actual situation can be determined by calculating P for each case and then choosing the case for which P is *smaller*. If in both cases the *same value* for P is calculated, which in practice would be highly improbable, then slipping at both points occurs simultaneously; i.e., the *seven unknowns* will satisfy *eight equations*.

(a)

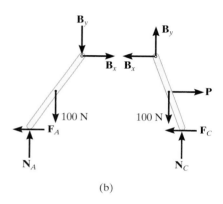

(b)

Fig. 8–6

Consider pushing on the uniform crate that has a weight W and sits on the rough surface. As shown on the first free-body diagram, if the magnitude of **P** is small, the crate will remain in equilibrium. As P increases the crate will either be on the verge of slipping on the surface ($F = \mu_s W$), or if the surface is very rough (large μ_s) then the resultant normal force will shift to the corner, $x = b/2$, as shown on the second free-body diagram, and the crate will tip over. The crate has a greater chance of tipping if **P** is applied at a greater height h above the surface, or if the crate's width b is smaller.

Equilibrium Versus Frictional Equations.   It was stated earlier that the frictional force *always* acts so as to either oppose the relative motion or impede the motion of a body over its contacting surface. Realize, however, that we can *assume* the sense of the frictional force in problems which require F to be an "equilibrium force" and satisfy the inequality $F < \mu_s N$. The correct sense is made known *after* solving the equations of equilibrium for F. For example, if F is a negative scalar the sense of **F** is the reverse of that which was assumed. This convenience of *assuming* the sense of **F** is possible because the equilibrium equations equate to zero the *components of vectors* acting in the *same direction*. In cases where the frictional equation $F = \mu N$ is used in the solution of a problem, however, the convenience of *assuming* the sense of **F** is *lost*, since the frictional equation relates only the *magnitudes* of two perpendicular vectors. Consequently, **F** *must always* be shown acting with its *correct sense* on the free-body diagram whenever the frictional equation is used for the solution of a problem.

Procedure for Analysis

Equilibrium problems involving dry friction can be solved using the following procedure.

Free-Body Diagrams.

- Draw the necessary free-body diagrams, and unless it is stated in the problem that impending motion or slipping occurs, *always* show the frictional forces as unknowns (i.e., *do not assume* $F = \mu N$).

- Determine the number of unknowns and compare this with the number of available equilibrium equations.

- If there are more unknowns than equations of equilibrium, it will be necessary to apply the frictional equation at some, if not all, points of contact to obtain the extra equations needed for a complete solution.

- If the equation $F = \mu N$ is to be used, it will be necessary to show **F** acting in the proper direction on the free-body diagram.

Equations of Equilibrium and Friction.

- Apply the equations of equilibrium and the necessary frictional equations (or conditional equations if tipping is possible) and solve for the unknowns.

- If the problem involves a three-dimensional force system such that it becomes difficult to obtain the force components or the necessary moment arms, apply the equations of equilibrium using Cartesian vectors.

EXAMPLE 8.1

The uniform crate shown in Fig. 8–7a has a mass of 20 kg. If a force $P = 80$ N is applied to the crate, determine if it remains in equilibrium. The coefficient of static friction is $\mu = 0.3$.

(a)

Fig. 8–7

SOLUTION

Free-Body Diagram.   As shown in Fig. 8–7b, the *resultant* normal force $\mathbf{N}_C$ must act a distance x from the crate's center line in order to counteract the tipping effect caused by **P**. There are *three unknowns, F,* N_C, and x, which can be determined strictly from the *three* equations of equilibrium.

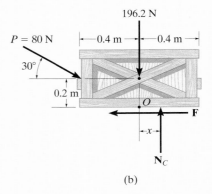

(b)

Equations of Equilibrium.

$$\xrightarrow{+} \Sigma F_x = 0; \qquad\qquad 80 \cos 30^\circ \text{ N} - F = 0$$

$$+\uparrow \Sigma F_y = 0; \qquad\qquad -80 \sin 30^\circ \text{ N} + N_C - 196.2 \text{ N} = 0$$

$$\llcorner + \Sigma M_O = 0; \; 80 \sin 30^\circ \text{ N}(0.4 \text{ m}) - 80 \cos 30^\circ \text{ N}(0.2 \text{ m}) + N_C(x) = 0$$

Solving,

$$F = 69.3 \text{ N}$$

$$N_C = 236 \text{ N}$$

$$x = -0.00908 \text{ m} = -9.08 \text{ mm}$$

Since x is negative it indicates the *resultant* normal force acts (slightly) to the *left* of the crate's center line. No tipping will occur since $x \leq 0.4$ m. Also, the *maximum* frictional force which can be developed at the surface of contact is $F_{\max} = \mu_s N_C = 0.3(236 \text{ N}) = 70.8$ N.   Since $F = 69.3 \text{ N} < 70.8$ N, the crate will *not slip*, although it is very close to doing so.

EXAMPLE 8.2

(a)

(b)

(c)

Fig. 8–8

It is observed that when the bed of the dump truck is raised to an angle of $\theta = 25°$ the vending machines begin to slide off the bed, Fig. 8–8a. Determine the static coefficient of friction between a vending machine and the surface of the truckbed.

SOLUTION

An idealized model of a vending machine resting on the truckbed is shown in Fig. 8–8b. The dimensions have been measured and the center of gravity has been located. We will assume that the machine weighs W.

Free-Body Diagram.   As shown in Fig. 8–8c, the dimension x is used to locate the position of the resultant normal force $\mathbf{N}$. There are four unknowns, N, F, μ_s, and x.

Equations of Equilibrium.

$$+\searrow \Sigma F_x = 0; \qquad\qquad W \sin 25° - F = 0 \qquad\qquad (1)$$

$$+\nearrow \Sigma F_y = 0; \qquad\qquad N - W \cos 25° = 0 \qquad\qquad (2)$$

$$\zeta + \Sigma M_O = 0; \quad -W \sin \theta (2.5 \text{ ft}) + W \cos \theta(x) = 0 \qquad (3)$$

Since slipping impends at $\theta = 25°$, using the first two equations, we have

$$F_s = \mu_s N; \qquad W \sin 25° = \mu_s(W \cos 25°)$$

$$\mu_s = \tan 25° = 0.466 \qquad\qquad Ans.$$

The angle of $\theta = 25°$ is referred to as the *angle of repose*, and by comparison, it is equal to the angle of static friction, $\theta = \phi_s$. Notice from the calculation that θ is independent of the weight of the vending machine, and so knowing θ provides a convenient method for determining the coefficient of static friction.

NOTE: From Eq. 3, with $\theta = 25°$, we find $x = 1.17$ ft. Since 1.17 ft < 1.5 ft, indeed the vending machine will slip before it can tip as observed in Fig. 8–8a.

EXAMPLE 8.3

The uniform rod having a weight W and length l is supported at its ends against the surface at A and B in Fig. 8–9a. If the rod is on the verge of slipping when $\theta = 30°$, determine the coefficient of static friction μ_s at A and B. Neglect the thickness of the rod for the calculation.

(a)

SOLUTION

Free-Body Diagram.   As shown in Fig. 8–9b, there are *five* unknowns: F_A, N_A, F_B, N_B, and μ_s. These can be determined from the *three* equilibrium equations and *two* frictional equations applied at points A and B. The frictional forces must be drawn with their correct sense so that they oppose the tendency for motion of the rod. Why? (Refer to p. 400.)

Equations of Friction and Equilibrium.   Writing the frictional equations,

$$F = \mu_s N; \qquad\qquad F_A = \mu_s N_A$$
$$F_B = \mu_s N_B$$

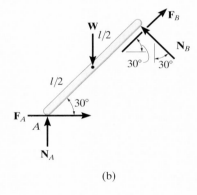

Using these results and applying the equations of equilibrium yields

$$\xrightarrow{+} \Sigma F_x = 0; \qquad \mu_s N_A + \mu_s N_B \cos 30° - N_B \sin 30° = 0 \qquad (1)$$

$$+\uparrow \Sigma F_y = 0; \qquad N_A - W + N_B \cos 30° + \mu_s N_B \sin 30° = 0 \qquad (2)$$

$$\zeta + \Sigma M_A = 0; \qquad N_B l - W\left(\frac{l}{2}\right)\cos 30° = 0 \qquad (3)$$

$$N_B = 0.4330W$$

(b)

Fig. 8–9

From Eqs. 1 and 2,

$$\mu_s N_A = 0.2165W - (0.3750W)\mu_s$$
$$N_A = 0.6250W - (0.2165W)\mu_s$$

By division,

$$0.6250\mu_s - 0.2165\mu_s^2 = 0.2165 - 0.375\mu_s$$

or,

$$\mu_s^2 - 4.619\mu_s + 1 = 0$$

Solving for the smallest root,

$$\mu_s = 0.228 \qquad\qquad\qquad Ans.$$

EXAMPLE 8.4

(a)

(b)

Fig. 8–10

The concrete pipes are stacked in the yard as shown in Fig. 8–10a. Determine the minimum coefficient of static friction at each point of contact so that the pile does not collapse.

SOLUTION

Free-body Diagrams.   Recognize that the coefficient of static friction between two pipes, at A and B, and between a pipe and the ground, at C, will be different since the contacting surfaces are different. We will assume each pipe has an outer radius r and weight W. The free-body diagrams for two of the pipes are shown in Fig. 8–10b. There are six unknowns, N_A, F_A, N_B, F_B, N_C, F_C. (Note that when collapse is about to occur the normal force at D is zero.) Since only the six equations of equilibrium are necessary to obtain the unknowns, the sense of direction of the frictional forces can be verified from the solution.

Equations of Equilibrium.   For the top pipe, we have

$$\curvearrowleft + \Sigma M_O = 0; \quad -F_A(r) + F_B(r) = 0; \quad F_A = F_B = F$$
$$\xrightarrow{+} \Sigma F_x = 0; \quad N_A \sin 30° - F \cos 30° - N_B \sin 30° + F \cos 30° = 0$$
$$N_A = N_B = N$$
$$+\uparrow \Sigma F_y = 0; \quad 2N \cos 30° + 2F \sin 30° - W = 0 \qquad (1)$$

For the bottom pipe, using $F_A = F$ and $N_A = N$, we have

$$\curvearrowleft + \Sigma M_{O'} = 0; \quad F_C(r) - F(r) = 0; \qquad\qquad F_C = F$$
$$\xrightarrow{+} \Sigma F_x = 0; \quad -N \sin 30° + F \cos 30° + F = 0 \qquad (2)$$
$$+\uparrow \Sigma F_y = 0; \quad N_C - W - N \cos 30° - F \sin 30° = 0 \qquad (3)$$

From Eq. 2, $F = 0.2679N$, so that between the pipes

$$(\mu_s)_{min} = \frac{F}{N} = 0.268 \qquad\qquad Ans.$$

Using this result in Eq. 1,

$$N = 0.5W$$

From Eq. 3,

$$N_C - W - (0.5W) \cos 30° - 0.2679(0.5W) \sin 30° = 0$$
$$N_C = 1.5W$$

At the ground, the smallest required coefficient of static friction would be

$$(\mu'_s)_{min} = \frac{F}{N_C} = \frac{0.2679(0.5W)}{1.5W} = 0.0893 \qquad Ans.$$

NOTE: A greater coefficient of static friction is required between the pipes than that required at the ground; and so it is likely that if slipping would occur between the pipes the bottom two pipes would roll away from one another without slipping as the top pipe falls downward.

EXAMPLE 8.5

Beam AB is subjected to a uniform load of 200 N/m and is supported at B by post BC, Fig. 8–11a. If the coefficients of static friction at B and C are $\mu_B = 0.2$ and $\mu_C = 0.5$, determine the force $\mathbf{P}$ needed to pull the post out from under the beam. Neglect the weight of the members and the thickness of the post.

SOLUTION

Free-Body Diagrams.   The free-body diagram of beam AB is shown in Fig. 8–11b. Applying $\Sigma M_A = 0$, we obtain $N_B = 400$ N. This result is shown on the free-body diagram of the post, Fig. 8–11c. Referring to this member, the *four* unknowns F_B, P, F_C, and N_C are determined from the *three* equations of equilibrium and *one* frictional equation applied either at B or C.

Equations of Equilibrium and Friction.

$$\xrightarrow{+} \Sigma F_x = 0; \qquad P - F_B - F_C = 0 \qquad\qquad (1)$$

$$+\uparrow \Sigma F_y = 0; \qquad N_C - 400\ \text{N} = 0 \qquad\qquad (2)$$

$$\zeta + \Sigma M_C = 0; \qquad -P(0.25\ \text{m}) + F_B(1\ \text{m}) = 0 \qquad (3)$$

(Post Slips Only at B.)   This requires $F_C \le \mu_C N_C$ and

$$F_B = \mu_B N_B; \qquad F_B = 0.2(400\ \text{N}) = 80\ \text{N}$$

Using this result and solving Eqs. 1 through 3, we obtain

$$P = 320\ \text{N}$$
$$F_C = 240\ \text{N}$$
$$N_C = 400\ \text{N}$$

Since $F_C = 240\ \text{N} > \mu_C N_C = 0.5(400\ \text{N}) = 200\ \text{N}$, the other case of movement must be investigated.

(Post Slips Only at C.)   Here $F_B \le \mu_B N_B$ and

$$F_C = \mu_C N_C; \qquad F_C = 0.5 N_C \qquad\qquad (4)$$

Solving Eqs. 1 through 4 yields

$$P = 267\ \text{N}$$
$$N_C = 400\ \text{N} \qquad\qquad Ans.$$
$$F_C = 200\ \text{N}$$
$$F_B = 66.7\ \text{N}$$

Obviously, this case occurs first since it requires a *smaller* value for P.

(a)

(b)

(c)

Fig. 8–11

EXAMPLE 8.6

(a)

(b)

(c)

Fig. 8–12

Determine the normal force P that must be exerted on the rack to begin pushing the 100-kg pipe shown in Fig. 8–12a up the 20° incline. The coefficients of static friction at the points of contact are $(\mu_s)_A = 0.15$ and $(\mu_s)_B = 0.4$.

SOLUTION

Free-Body Diagram. As shown in Fig. 8–12b, the rack must exert a force P on the pipe due to force equilibrium in the x direction. There are four unknowns P, F_A, N_A, and F_B acting on the pipe Fig. 8–12c. These can be determined from the *three* equations of equilibrium and *one* frictional equation, which apply either at A or B. If slipping begins to occur only at B, the pipe will begin to roll up the incline; whereas if slipping occurs only at A, the pipe will begin to *slide* up the incline. Here we must find N_B.

Equations of Equilibrium and Friction. (Fig. 8–12c)

$$+\nearrow \Sigma F_x = 0; \qquad -F_A + P - 981 \sin 20° \text{ N} = 0 \qquad (1)$$

$$+\nwarrow \Sigma F_y = 0; \qquad N_A - F_B - 981 \cos 20° \text{ N} = 0 \qquad (2)$$

$$\downarrow + \Sigma M_O = 0; \qquad F_B(400 \text{ mm}) - F_A(400 \text{ mm}) = 0 \qquad (3)$$

(Pipe Rolls up Incline.) In this case, $F_A \le 0.15 N_A$ and

$$(F_s)_B = (\mu_s)_B N_B; \qquad F_B = 0.4P \qquad (4)$$

The direction of the frictional force at B must be specified correctly. Why? Since the spool is being forced up the incline, $\mathbf{F}_B$ acts downward to prevent any clockwise rolling motion of the pipe, Fig. 8–12c. Solving Eqs. 1 through 4, we have

$$N_A = 1146 \text{ N} \quad F_A = 224 \text{ N} \quad F_B = 224 \text{ N} \quad P = 559 \text{ N}$$

The assumption regarding no slipping at A should be checked.

$$F_A \le (\mu_s)_A N_A; \qquad 224 \text{ N} \overset{?}{\le} 0.15(1146 \text{ N}) = 172 \text{ N}$$

The inequality does *not apply*, and therefore slipping occurs at A and not at B. Hence, the other case of motion will occur.

(Pipe Slides up Incline.) In this case, $P \le 0.4 N_B$ and

$$(F_s)_A = (\mu_s)_A N_A; \qquad F_A = 0.15 N_A \qquad (5)$$

Solving Eqs. 1 through 3 and 5 yields

$$N_A = 1085 \text{ N} \quad F_A = 163 \text{ N} \quad F_B = 163 \text{ N} \quad P = 498 \text{ N} \qquad \textit{Ans.}$$

NOTE: The validity of the solution ($P = 498$ N) can be checked by testing the assumption that indeed no slipping occurs at B.

$$F_B \le (\mu_s)_B P; \qquad 163 \text{ N} < 0.4(498 \text{ N}) = 199 \text{ N} \qquad \text{(check)}$$

PROBLEMS

8–1. If the horizontal force $P = 80$ lb, determine the normal and frictional forces acting on the 300-lb crate. Take $\mu_s = 0.3$, $\mu_k = 0.2$.

Prob. 8–1

8–2. Determine the magnitude of force **P** needed to start towing the 40-kg crate. Also determine the location of the resultant normal force acting on the crate, measured from point A. Take $\mu_s = 0.3$.

8–3. Determine the friction force on the 40-kg crate, and the resultant normal force and its position x, measured from point A, if the force $P = 300$ N. Take $\mu_s = 0.5$ and $\mu_k = 0.2$.

Probs. 8–2/3

*8–4. The loose-fitting collar is supported by the pipe for which the coefficient of static friction at the points of contact A and B is $\mu_s = 0.2$. Determine the largest dimension d so the rod will not slip when the load **P** is applied.

Prob. 8–4

8–5. The spool of wire having a mass of 150 kg rests on the ground at A and against the wall at B. Determine the force P required to begin pulling the wire horizontally off the spool. The coefficient of static friction between the spool and its points of contact is $\mu_s = 0.25$.

8–6. The spool of wire having a mass of 150 kg rests on the ground at A and against the wall at B. Determine the forces acting on the spool at A and B if $P = 800$ N. The coefficient of static friction between the spool and the ground at point A is $\mu_s = 0.35$. The wall at B is smooth.

Probs. 8–5/6

8–7. The crate has a mass of 350 kg and is subjected to a towing force **P** acting at a 20° angle with the horizontal. If the coefficient of static friction is $\mu_s = 0.5$, determine the magnitude of **P** to just start the crate moving down the plane.

Prob. 8–7

*8–8.   The winch on the truck is used to hoist the garbage bin onto the bed of the truck. If the loaded bin has a weight of 8500 lb and center of gravity at G, determine the force in the cable needed to begin the lift. The coefficients of static friction at A and B are $\mu_A = 0.3$ and $\mu_B = 0.2$, respectively. Neglect the height of the support at A.

Prob. 8–8

8–9.   The motorcyclist travels with constant velocity along a straight, horizontal, banked road. If he aligns his bike so that the tires are perpendicular to the road at A, determine the frictional force at A. The man has a mass of 60 kg and a mass center at G_C, and the motorcycle has a mass of 120 kg and a mass center at G_m. If the coefficient of static friction at A is $\mu_A = 0.4$, will the bike slip?

Prob. 8–9

8–10.   The block brake is used to stop the wheel from rotating when the wheel is subjected to a couple moment $\mathbf{M}_0$. If the coefficient of static friction between the wheel and the block is μ_s, determine the smallest force P that should be applied.

8–11.   Show that the brake in Prob. 8–10 is self locking, i.e., $P \le 0$, provided $b/c \le \mu_s$.

*8–12.   Solve Prob. 8–10 if the couple moment $\mathbf{M}_0$ is applied counterclockwise.

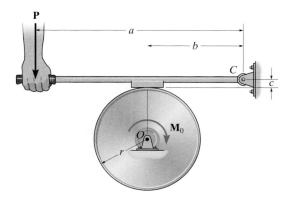

Probs. 8–10/11/12

8–13.   The block brake consists of a pin-connected lever and friction block at B. The coefficient of static friction between the wheel and the lever is $\mu_s = 0.3$, and a torque of 5 N · m is applied to the wheel. Determine if the brake can hold the wheel stationary when the force applied to the lever is (a) $P = 30$ N, (b) $P = 70$ N.

8–14.   Solve Prob. 8–13 if the 5-N · m torque is applied counter-clockwise.

Probs. 8–13/14

8–15. The doorstop of negligible weight is pin connected at A and the coefficient of static friction at B is $\mu_s = 0.3$. Determine the required distance s from A to the floor so that the stop will resist opening of the door for any force P applied to the handle.

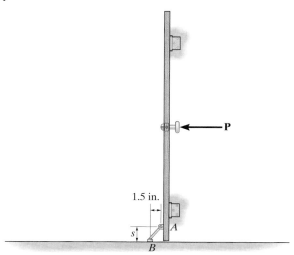

Prob. 8–15

*8–16. The chair has a weight of 10 lb and center of gravity at G. It is propped against the door as shown. If the coefficient of static friction at A is $\mu_A = 0.3$, determine the smallest force P that must be applied to the handle to open the door.

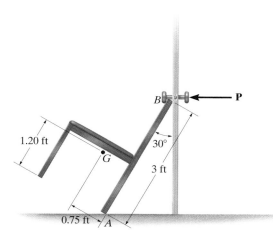

Prob. 8–16

8–17. The uniform hoop of weight W is suspended from the peg at A and a horizontal force $\mathbf{P}$ is slowly applied at B. If the hoop begins to slip at A when $\theta = 30°$, determine the coefficient of static friction between the hoop and the peg.

8–18. The uniform hoop of weight W is suspended from the peg at A and a horizontal force $\mathbf{P}$ is slowly applied at B. If the coefficient of static friction between the hoop and peg is $\mu_s = 0.2$, determine if it is possible for the angle $\theta = 30°$ before the hoop begins to slip.

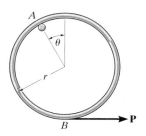

Prob. 8–17/18

8–19. The coefficient of static friction between the shoes at A and B of the tongs and the pallet is $\mu_s' = 0.5$, and between the pallet and the floor $\mu_s = 0.4$. If a horizontal towing force of $P = 300$ N is applied to the tongs, determine the largest mass that can be towed.

Prob. 8–19

*8–20. The pipe is hoisted using the tongs. If the coefficient of static friction at A and B is μ_s, determine the smallest dimension b so that any pipe of inner diameter d can be lifted.

Prob. 8–20

8–21. A very thin bookmark having a width of 1 in. is in the middle of a 10-lb dictionary. If the pages are 8 in. by 10 in., determine the force P needed to start to pull the bookmark out. The coefficient of static friction between the bookmark and the paper is $\mu_s = 0.7$. Assume the pressure on each page and the bookmark is uniform.

Prob. 8–21

8–22. The uniform dresser has a weight of 90 lb and rests on a tile floor for which $\mu_s = 0.25$. If the man pushes on it in the horizontal direction $\theta = 0°$, determine the smallest magnitude of force $\mathbf{F}$ needed to move the dresser. Also, if the man has a weight of 150 lb, determine the smallest coefficient of static friction between his shoes and the floor so that he does not slip.

8–23. The uniform dresser has a weight of 90 lb and rests on a tile floor for which $\mu_s = 0.25$. If the man pushes on it in the direction $\theta = 30°$, determine the smallest magnitude of force $\mathbf{F}$ needed to move the dresser. Also, if the man has a weight of 150 lb, determine the smallest coefficient of static friction between his shoes and the floor so that he does not slip.

Probs. 8–22/23

*8–24. The cam is subjected to a couple moment of $5 \text{ N} \cdot \text{m}$. Determine the minimum force P that should be applied to the follower in order to hold the cam in the position shown. The coefficient of static friction between the cam and the follower is $\mu_s = 0.4$. The guide at A is smooth.

Prob. 8–24

8–25. The board can be adjusted vertically by tilting it up and sliding the smooth pin A along the vertical guide G. When placed horizontally, the bottom C then bears along the edge of the guide, where $\mu_s = 0.4$. Determine the largest dimension d which will support any applied force **F** without causing the board to slip downward.

Prob. 8–25

8–26. The homogeneous semicylinder has a mass m and mass center at G. Determine the largest angle θ of the inclined plane upon which it rests so that it does not slip down the plane. The coefficient of static friction between the plane and the cylinder is $\mu_s = 0.3$. Also, what is the angle ϕ for this case?

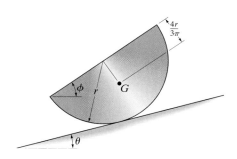

Prob. 8–26

8–27. A chain having a length of 20 ft and weight of 8 lb/ft rests on a street for which the coefficient of static friction is $\mu_s = 0.2$. If a crane is used to hoist the chain, determine the force P it applies to the chain if the length of chain remaining on the ground begins to slip when the horizontal component of P becomes 10 lb. What length of chain remains on the ground?

Prob. 8–27

***8–28.** The fork lift has a weight of 2400 lb and a center of gravity at G. If the rear wheels are powered, whereas the front wheels are free to roll, determine the maximum number of 300-lb crates the fork lift can push forward. The coefficient of static friction between the wheels and the ground is $\mu_s = 0.4$, and between each crate and the ground $\mu_s' = 0.35$.

Prob. 8–28

8–29. The brake is to be designed to be self locking, that is, it will not rotate when no load **P** is applied to it when the disk is subjected to a clockwise couple moment **M**$_0$. Determine the distance d of the lever that will allow this to happen. The coefficient of static friction at B is $\mu_s = 0.5$.

Prob. 8–29

8–30. The 800-lb concrete pipe is being lowered from the truck bed when it is in the position shown. If the coefficient of static friction at the points of support A and B is $\mu_s = 0.4$, determine where it begins to slip first: at A or B, or both at A and B.

Prob. 8–30

8–31. A 5-kg wedge is placed in the grooved slot of an inclined plane. Determine the maximum angle θ for the incline without causing the wedge to slip. The coefficient of static friction between the wedge and the surfaces of contact is $\mu_s = 0.2$.

Prob. 8–31

*8–32.** A roll of paper has a uniform weight of 0.75 lb and is suspended from the wire hanger so that it rests against the wall. If the hanger has a negligible weight and the bearing at O can be considered frictionless, determine the force P needed to start turning the roll if $\theta = 30°$. The coefficient of static friction between the wall and the paper is $\mu_s = 0.25$.

8–33. A roll of paper has a uniform weight of 0.75 lb and is suspended from the wire hanger so that it rests against the wall. If the hanger has a negligible weight and the bearing at O can be considered frictionless, determine the minimum force P and the associated angle θ needed to start turning the roll. The coefficient of static friction between the wall and the paper is $\mu_s = 0.25$.

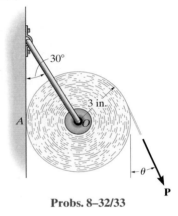

Probs. 8–32/33

8–34. The door brace AB is to be designed to prevent opening the door. If the brace forms a pin connection under the doorknob and the coefficient of static friction with the floor is $\mu_s = 0.5$, determine the largest length L the brace can have to prevent the door from being opened. Neglect the weight of the brace.

Prob. 8–34

8–35. The man has a weight of 200 lb, and the coefficient of static friction between his shoes and the floor is $\mu_s = 0.5$. Determine where he should position his center of gravity G at d in order to exert the maximum horizontal force on the door. What is this force?

Prob. 8–35

*8–36.** In an effort to move the two 100-lb crates, which are stacked on top of one another, the man pushes horizontally on them at the bottom of crate A as shown. Determine the smallest force P that must be applied in order to cause impending motion. Explain what happens. The coefficient of static friction between the crates is $\mu_s = 0.8$, and between the bottom crate and the floor $\mu_s' = 0.3$.

8–37. The man having a weight of 150 lb pushes horizontally on the bottom of crate A, which is stacked on top of crate B. Each crate has a weight of 100 lb. If the coefficient of static friction between each crate is $\mu_s = 0.8$ and between the bottom crate, his shoes, and the floor is $\mu_s' = 0.3$, determine if he can cause impending motion.

Probs. 8–36/37

8–38. The crate has a weight of 200 lb and a center of gravity at G. Determine the horizontal force P required to tow it. Also, determine the location of the resultant normal force measured from A. Take $h = 4$ ft and $\mu_s = 0.4$.

8–39. The crate has a weight of 200 lb and a center of gravity at G. Determine the height h of the tow rope so that the crate slips and tips at the same time. What horizontal force P is required to do this? Take $\mu_s = 0.4$.

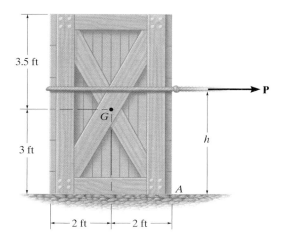

Probs. 8–38/39

*8–40.** Determine the smallest force the man must exert on the rope in order to move the 80-kg crate. Also, what is the angle θ at this moment? The coefficient of static friction between the crate and the floor is $\mu_s = 0.3$.

Prob. 8–40

8–41. The symmetrical crab hook is used to lift packages by means of friction developed between the shoes A and B and a package. Determine the smallest coefficient of static friction at the shoes so that the package of weight W can be lifted.

8–43. The crate has a weight of 300 lb and a center of gravity at G. If the coefficient of static friction between the crate and the floor is $\mu_s = 0.2$, determine if the 200-lb man can *push* the crate to the left. The coefficient of static friction between his shoes and the floor is $\mu'_s = 0.4$. Assume the man exerts only a horizontal force on the crate.

*8–44. The crate has a weight of 300 lb and a center of gravity at G. If the coefficient of static friction between the crate and the floor is $\mu_s = 0.2$, determine the smallest weight of the man so he can push the crate to the left. The coefficient of static friction between his shoes and the floor is $\mu'_s = 0.35$. Assume the man exerts only a horizontal force on the crate.

Prob. 8–41

Probs. 8–43/44

8–42. The friction hook is made from a fixed frame which is shown colored and a cylinder of negligible weight. A piece of paper is placed between the smooth wall and the cylinder. If $\theta = 20°$, determine the smallest coefficient of static friction μ at all points of contact so that any weight W of paper p can be held.

8–45. The wheel weighs 20 lb and rests on a surface for which $\mu_B = 0.2$. A cord wrapped around it is attached to the top of the 30-lb homogeneous block. If the coefficient of static friction at D is $\mu_D = 0.3$, determine the smallest vertical force that can be applied tangentially to the wheel which will cause motion to impend.

Prob. 8–42

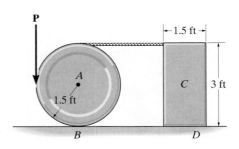

Prob. 8–45

8–46. Determine the smallest couple moment which can be applied to the 20-lb wheel that will cause impending motion. The cord is attached to the 30-lb block, and the coefficients of static friction are $\mu_B = 0.2$ and $\mu_D = 0.3$.

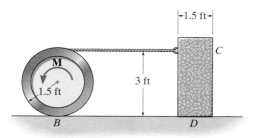

Prob. 8–46

8–47. The beam AB has a negligible mass and thickness and is subjected to a triangular distributed loading. It is supported at one end by a pin and at the other end by a post having a mass of 50 kg and negligible thickness. Determine the minimum force P needed to move the post. The coefficients of static friction at B and C are $\mu_B = 0.4$ and $\mu_C = 0.2$, respectively.

*8–48. The beam AB has a negligible mass and thickness and is subjected to a triangular distributed loading. It is supported at one end by a pin and at the other end by a post having a mass of 50 kg and negligible thickness. Determine the two coefficients of static friction at B and at C so that when the magnitude of the applied force is increased to $P = 150$ N, the post slips at both B and C simultaneously.

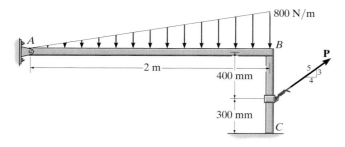

Probs. 8–47/48

8–49. The block of weight W is being pulled up the inclined plane of slope α using a force $\mathbf{P}$. If $\mathbf{P}$ acts at the angle ϕ as shown, show that for slipping to occur, $P = W \sin(\alpha + \theta)/\cos(\phi - \theta)$, where θ is the angle of friction; $\theta = \tan^{-1} \mu$.

8–50. Determine the angle ϕ at which $\mathbf{P}$ should act on the block so that the magnitude of $\mathbf{P}$ is as small as possible to begin pushing the block up the incline. What is the corresponding value of P? The block weighs W and the slope α is known.

Probs. 8–49/50

8–51. Two blocks A and B, each having a mass of 6 kg, are connected by the linkage shown. If the coefficient of static friction at the contacting surfaces is $\mu_s = 0.5$, determine the largest vertical force P that may be applied to pin C of the linkage without causing the blocks to move. Neglect the weight of the links.

Prob. 8–51

*8–52. Block C has a mass of 50 kg and is confined between two walls by smooth rollers. If the block rests on top of the 40-kg spool, determine the minimum cable force P needed to move the spool. The cable is wrapped around the spool's inner core. The coefficients of static friction at A and B are $\mu_A = 0.3$ and $\mu_B = 0.6$.

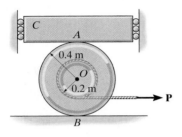

Prob. 8–52

8–54. Determine the minimum force P needed to push the tube E up the incline. The tube has a mass of 75 kg and the roller D has a mass of 100 kg. The force acts parallel to the plane, and the coefficients of static friction at the contacting surfaces are $\mu_A = 0.3$, $\mu_B = 0.25$, and $\mu_C = 0.4$. Each cylinder has a radius of 150 mm.

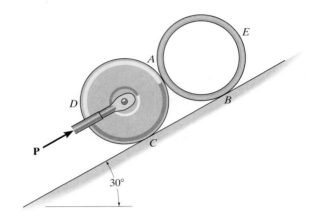

Prob. 8–54

8–53. The 50-lb board is placed across the channel and a 100-lb boy attempts to walk across. If the coefficient of static friction at A and B is $\mu_s = 0.4$, determine if he can make the crossing; and if not, how far will he get from A before the board slips?

Prob. 8–53

8–55. The concrete pipe at A rests on top of B and C. If the coefficient of static friction between the pipes is μ_s and at the ground μ'_s, determine their smallest values so that the pipes will not slip. Each pipe has a radius r and weight W, and the angle between the centers as indicated is θ.

Prob. 8–55

*8–56. The uniform pole has a weight W and length L. Its end B is tied to a supporting cord, and end A is placed against the wall, for which the coefficient of static friction is μ_s. Determine the largest angle θ at which the pole can be placed without slipping.

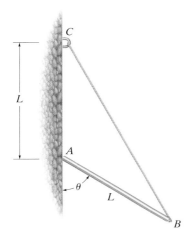

Prob. 8–56

8–57. The carpenter slowly pushes the uniform board horizontally over the top of the saw horse. The board has a uniform weight of 3 lb/ft, and the saw horse has a weight of 15 lb and a center of gravity at G. Determine if the saw horse will stay in position, slip, or tip if the board is pushed forward when $d = 10$ ft. The coefficients of static friction are shown in the figure.

8–58. The carpenter slowly pushes the uniform board horizontally over the top of the saw horse. The board has a uniform weight of 3 lb/ft, and the saw horse has a weight of 15 lb and a center of gravity at G. Determine if the saw horse will stay in position, slip, or tip if the board is pushed forward when $d = 14$ ft. The coefficients of static friction are shown in the figure.

Probs. 8–57/58

8–59. The 45-kg disk rests on the surface for which the coefficient of static friction is $\mu_A = 0.2$. Determine the largest couple moment M that can be applied to the bar without causing motion.

*8–60. The 45-kg disk rests on the surface for which the coefficient of static friction is $\mu_A = 0.15$. If $M = 50$ N · m, determine the friction force at A.

Probs. 8–59/60

■8–61. The 6-lb block is attached to a light rod AD that pivots at pin A. If the coefficient of static friction between the plane and the block is $\mu_s = 0.4$, determine the minimum angle θ at which the block may be placed on the plane without slipping. Neglect the *size* of the block in the calculation.

Prob. 8–61

Wedges are often used to adjust the elevation of structural or mechanical parts. Also, they provide stability for objects such as this tank.

8.3 Wedges

A *wedge* is a simple machine that is often used to transform an applied force into much larger forces, directed at approximately right angles to the applied force. Wedges also can be used to slightly move or adjust heavy loads.

Consider, for example, the wedge shown in Fig. 8–13a, which is used to *lift* a block of weight **W** by applying a force **P** to the wedge. Free-body diagrams of the block and wedge are shown in Fig. 8–13b. Here we have excluded the weight of the wedge since it is usually *small* compared to the weight of the block. Also, note that the frictional forces F_1 and F_2 must oppose the motion of the wedge. Likewise, the frictional force F_3 of the wall on the block must act downward so as to oppose the block's upward motion. The locations of the resultant normal forces are not important in the force analysis since neither the block nor wedge will "tip." Hence the moment equilibrium equations will not be considered. There are seven unknowns, consisting of the applied force **P**, needed to cause motion of the wedge, and six normal and frictional forces. The seven available equations consist of two force equilibrium equations ($\Sigma F_x = 0$, $\Sigma F_y = 0$) applied to the wedge and block (four equations total) and the frictional equation $F = \mu N$ applied at each surface of contact (three equations total).

If the block is to be *lowered*, the frictional forces will all act in a sense opposite to that shown in Fig. 8–13b. The applied force **P** must act to the right to hold the block as shown, provided the coefficient of friction is very *small* or the wedge angle θ is *large*. Otherwise, **P** may have a reverse sense of direction in order to *pull* on the wedge to remove it. If **P** is *not applied*, or **P** = **0**, and friction forces hold the block in place, then the wedge is referred to as *self-locking*.

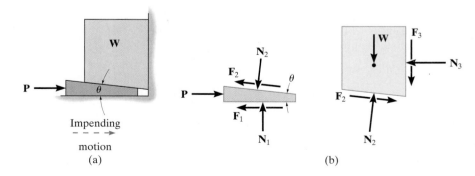

Fig. 8–13

EXAMPLE 8.7

The uniform stone in Fig. 8–14a has a mass of 500 kg and is held in the horizontal position using a wedge at B. If the coefficient of static friction is $\mu_s = 0.3$ at, the surfaces of contact, determine the minimum force **P** needed to remove the wedge. Is the wedge self-locking? Assume that the stone does not slip at A.

(a)

(b)

Fig. 8–14

SOLUTION

The minimum force P requires $F = \mu_s N$ at the surfaces of contact with the wedge. The free-body diagrams of the stone and wedge are shown in Fig. 8–14b. On the wedge the friction force opposes the motion, and on the stone at A, $F_A \leq \mu_s N_A$, since slipping does not occur there. There are five unknowns F_A, N_A, N_B, N_C, and P. Three equilibrium equations for the stone and two for the wedge are available for solution. From the free-body diagram of the stone,

$$\zeta + \Sigma M_A = 0; \qquad -4905 \text{ N}(0.5 \text{ m}) + (N_B \cos 7° \text{ N})(1 \text{ m})$$
$$+ (0.3 N_B \sin 7° \text{ N})(1 \text{ m}) = 0$$
$$N_B = 2383.1 \text{ N}$$

Using this result for the wedge, we have

$$\xrightarrow{+} \Sigma F_x = 0;$$
$$2383.1 \sin 7° \text{ N} - 0.3(2383.1 \cos 7° \text{ N}) + P - 0.3 N_C = 0$$
$$+\uparrow \Sigma F_y = 0;$$
$$N_C - 2383.1 \cos 7° \text{ N} - 0.3(2383.1 \sin 7° \text{ N}) = 0$$
$$N_C = 2452.5 \text{ N}$$
$$P = 1154.9 \text{ N} = 1.15 \text{ kN} \qquad\qquad \textit{Ans.}$$

NOTE: Since P is positive, indeed the wedge must be pulled out. If P was zero, the wedge would remain in place (self-locking) and the frictional forces developed at B and C would satisfy $F_B < \mu_s N_B$ and $F_C < \mu_s N_C$.

Square-threaded screws find applications on valves, jacks, and vises, where particularly large forces must be developed along the axis of the screw.

8.4   Frictional Forces on Screws

In most cases screws are used as fasteners; however, in many types of machines they are incorporated to transmit power or motion from one part of the machine to another. A *square-threaded screw* is commonly used for the latter purpose, especially when large forces are applied along its axis. In this section we will analyze the forces acting on square-threaded screws. The analysis of other types of screws, such as the V-thread, is based on these same principles.

A *screw* may be thought of simply as an inclined plane or wedge wrapped around a cylinder. A nut initially at position A on the screw shown in Fig. 8–15*a* will move up to B when rotated 360° around the screw. This rotation is equivalent to translating the nut up an inclined plane of height l and length $2\pi r$, where r is the mean radius of the thread, Fig. 8–15*b*. The rise l for a single revolution is referred to as the *lead* of the screw, where the *lead angle* is given by $\theta = \tan^{-1}(l/2\pi r)$.

Frictional Analysis.   When a screw is subjected to large axial loads, the frictional forces developed on the thread become important if we are to determine the moment **M*** needed to turn the screw. Consider, for example, the square-threaded jack screw shown in Fig. 8–16, which supports the vertical load **W**. The reactive forces of the jack to this load are actually distributed over the circumference of the screw thread in contact with the screw hole in the jack, that is, within region h shown in Fig. 8–16. For simplicity, this portion of thread can be imagined as being unwound from the screw and represented as a simple block resting on an inclined plane having the screw's lead angle θ, Fig. 8–17*a*. Here the inclined plane represents the inside *supporting thread* of the jack base. Three forces act on the block or screw. The force **W** is the total axial load applied to the screw. The horizontal force **S** is caused by the applied moment **M**, such that by summing moments about the axis of the screw, $M = Sr$, where r is the screw's mean radius. As a result of **W** and **S**, the inclined plane exerts a resultant force **R** on the block, which is shown to have components acting normal, **N**, and tangent, **F**, to the contacting surfaces.

(a)

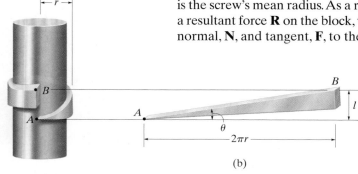

(b)

Fig. 8–15

*For applications, **M** is developed by applying a horizontal force **P** at a right angle to the end of a lever that would be fixed to the screw.

Fig. 8–16

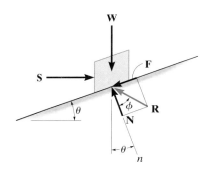

Upward screw motion
(a)

Upward Screw Motion.

Provided M is great enough, the screw (and hence the block) can either be brought to the verge of upward impending motion or motion can be occurring. Under these conditions, **R** acts at an angle $(\theta + \phi)$ from the vertical as shown in Fig. 8–17a, where $\phi = \tan^{-1}(F/N) = \tan^{-1}(\mu N/N) = \tan^{-1}\mu$. Applying the two force equations of equilibrium to the block, we obtain

$$\xrightarrow{+} \Sigma F_x = 0; \qquad S - R\sin(\theta + \phi) = 0$$
$$+\uparrow \Sigma F_y = 0; \qquad R\cos(\theta + \phi) - W = 0$$

Eliminating R and solving for S, then substituting this value into the equation $M = Sr$, yields

$$\boxed{M = Wr\tan(\theta + \phi)} \qquad (8\text{–}3)$$

As indicated, M is the moment necessary to cause upward impending motion of the screw, provided $\phi = \phi_s = \tan^{-1}\mu_s$ (the angle of static friction). If ϕ is replaced by $\phi_k = \tan^{-1}\mu_k$ (the angle of kinetic friction), Eq. 8–3 will give a smaller value M necessary to maintain uniform upward motion of the screw.

Downward Screw Motion $(\theta > \phi)$.

If the surface of the screw is very *slippery*, it may be possible for the screw to rotate downward if the magnitude of the moment is reduced to, say, $M' < M$. As shown in Fig. 8–17b, this causes the effect of **M'** to become **S'**, and it requires the angle ϕ (ϕ_s or ϕ_k) to lie on the opposite side of the normal n to the plane supporting the block, such that $\theta > \phi$. For this case, Eq. 8–3 becomes

$$\boxed{M' = Wr\tan(\theta - \phi)} \qquad (8\text{–}4)$$

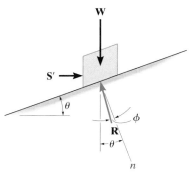

Downward screw motion $(\theta > \phi)$
(b)

Fig. 8–17

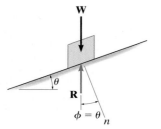

Self-locking screw ($\theta = \phi$)
(on the verge of rotating downward)

(c)

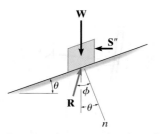

Downward screw motion ($\theta < \phi$)

(d)

Self-Locking Screw.   If the moment **M** (or its effect **S**) is *removed*, the screw will remain *self-locking;* i.e., it will support the load **W** by *friction forces alone* provided $\phi \geq \theta$. To show this, consider the necessary limiting case when $\phi = \theta$, Fig. 8–17c. Here vertical equilibrium is maintained since **R** is vertical and thus balances **W**.

Downward Screw Motion ($\theta < \phi$).   When the surface of the screw is very rough, the screw will not rotate downward as stated above. Instead, the direction of the applied moment must be reversed in order to cause the motion. The free-body diagram shown in Fig. 8–17d is representative of this case. Here **S″** is caused by the applied (reverse) moment **M″**. Hence Eq. 8–3 becomes

$$\boxed{M'' = Wr \tan(\phi - \theta)} \tag{8–5}$$

Each of the above cases should be thoroughly understood before proceeding to solve problems.

Fig. 8–17 (cont.)

EXAMPLE 8.8

Fig. 8–18

The turnbuckle shown in Fig. 8–18 has a square thread with a mean radius of 5 mm and a lead of 2 mm. If the coefficient of static friction between the screw and the turnbuckle is $\mu_s = 0.25$, determine the moment **M** that must be applied to draw the end screws closer together.

SOLUTION
The moment may be obtained by using Eq. 8–3. Why? Since friction at *two screws* must be overcome, this requires

$$M = 2[Wr \tan(\theta + \phi)] \tag{1}$$

Here $W = 2000$ N, $r = 5$ mm, $\phi_s = \tan^{-1} \mu_s = \tan^{-1}(0.25) = 14.04°$, and $\theta = \tan^{-1}(l/2\pi r) = \tan^{-1}(2 \text{ mm}/[2\pi(5 \text{ mm})]) = 3.64°$. Substituting these values into Eq. 1 and solving gives

$$M = 2[(2000 \text{ N})(5 \text{ mm}) \tan(14.04° + 3.64°)]$$
$$= 6374.7 \text{ N} \cdot \text{mm} = 6.37 \text{ N} \cdot \text{m} \qquad \textit{Ans.}$$

NOTE: When the moment is *removed*, the turnbuckle will be self-locking; i.e., it will not unscrew since $\phi_s > \theta$.

PROBLEMS

8–62. Determine the force *P* needed to lift the 100-lb load. Smooth rollers are placed between the wedges. The coefficient of static friction between *A* and *C* and between *B* and *D* is $\mu_s = 0.3$. Neglect the weight of each wedge.

Prob. 8–62

8–63. The wedge is used to level the floor of a building. For the floor loading shown, determine the horizontal force *P* that must be applied to move the wedge forward. The coefficient of static friction between the wedge and the two surfaces of contact is $\mu_s = 0.25$. Neglect the size and weight of the wedge and the thickness of the beam.

Prob. 8–63

*8–64. The three stone blocks have weights of $W_A = 600$ lb, $W_B = 150$ lb, and $W_C = 500$ lb. Determine the smallest horizontal force *P* that must be applied to block *C* in order to move this block. The coefficient of static friction between the blocks is $\mu_s = 0.3$, and between the floor and each block $\mu_s' = 0.5$.

Prob. 8–64

8–65. If the spring is compressed 60 mm and the coefficient of static friction between the tapered stub *S* and the slider *A* is $\mu_{SA} = 0.5$, determine the horizontal force **P** needed to move the slider forward. The stub is free to move without friction within the fixed collar *C*. The coefficient of static friction between *A* and surface *B* is $\mu_{AB} = 0.4$. Neglect the weights of the slider and stub.

Prob. 8–65

8–66. The coefficient of static friction between wedges B and C is $\mu_s = 0.6$ and between the surfaces of contact B and A and C and D, $\mu_s' = 0.4$. If the spring is compressed 200 mm when in the position shown, determine the smallest force P needed to move wedge C to the left. Neglect the weight of the wedges.

8–67. The coefficient of static friction between the wedges B and C is $\mu_s = 0.6$ and between the surfaces of contact B and A and C and D, $\mu_s' = 0.4$. If $P = 50$ N, determine the largest allowable compression of the spring without causing wedge C to move to the left. Neglect the weight of the wedges.

Probs. 8–66/67

8–69. The wedge is used to level the member. Determine the reversed horizontal force $-P$ that must be applied to pull the wedge out to the left. The coefficient of static friction between the wedge and the two surfaces of contact is $\mu_s = 0.15$. Neglect the weight of the wedge.

Prob. 8–69

*8–68.** The wedge blocks are used to hold the specimen in a tension testing machine. Determine the design angle θ of the wedges so that the specimen will not slip regardless of the applied load. The coefficients of static friction are $\mu_A = 0.1$ at A and $\mu_B = 0.6$ at B. Neglect the weight of the blocks.

Prob. 8–68

8–70. If the coefficient of static friction between all the surfaces of contact is μ_s, determine the force P that must be applied to the wedge in order to lift the brace that supports the load F.

Prob. 8–70

8–71. The column is used to support the upper floor. If a force $F = 80$ N is applied perpendicular to the handle to tighten the screw, determine the compressive force in the column. The square-threaded screw on the jack has a coefficient of static friction of $\mu_s = 0.4$, mean diameter of 25 mm, and a lead of 3 mm.

*8–72. If the force **F** is removed from the handle of the jack in Prob. 8–71, determine if the screw is self-locking.

Probs. 8–71/72

8–73. The vise is used to grip the pipe. If a horizontal force of 25 lb is applied perpendicular to the end of the 10-in. handle, determine the compressive force F developed in the pipe. The square threads have a mean diameter of 1.5 in. and a lead of 0.2 in. How much force must be applied perpendicular to the handle to loosen the vise? Take $\mu_s = 0.3$.

Prob. 8–73

8–74. Determine the couple forces F that must be applied to the handle of the machinist's vise in order to create a compressive force of 400 N in the block. Neglect friction at the bearing A. The guide at B is smooth so that the axial force on the screw is 400 N. The single square-threaded screw has a mean radius of 6 mm and a lead of 8 mm, and the coefficient of static friction is $\mu_s = 0.27$.

Prob. 8–74

8–75. If couple forces of $F = 35$ N are applied to the handle of the machinist's vise, determine the compressive force developed in the block. Neglect friction at the bearing A. The guide at B is smooth. The single square-threaded screw has a mean radius of 6 mm and a lead of 8 mm, and the coefficient of static friction is $\mu_s = 0.27$.

Prob. 8–75

*8–76. The machine part is held in place using the double-end clamp. The bolt at B has square threads with a mean radius of 4 mm and a lead of 2 mm, and the coefficient of static friction with the nut is $\mu_s = 0.5$. If a torque of $M = 0.4\,\text{N} \cdot \text{m}$ is applied to the nut to tighten it, determine the normal force of the clamp at the smooth contacts A and C.

Prob. 8–76

8–77. Determine the clamping force on the board A if the screw of the "C" clamp is tightened with a twist of $M = 8\,\text{N} \cdot \text{m}$. The single square-threaded screw has a mean radius of 10 mm, a lead of 3 mm, and the coefficient of static friction is $\mu_s = 0.35$.

8–78. If the required clamping force at the board A is to be 50 N, determine the torque M that must be applied to the handle of the "C" clamp to tighten it down. The single square-threaded screw has a mean radius of 10 mm, a lead of 3 mm, and the coefficient of static friction is $\mu_s = 0.35$.

Probs. 8–77/78

8–79. Determine the clamping force on the board at A if the screw of the hold-down clamp is tightened with a twist of $M = 0.2\,\text{N} \cdot \text{m}$. The single square-threaded screw has a mean radius of 8 mm and a lead of 2 mm, and the coefficient of static friction is $\mu_s = 0.38$.

Prob. 8–79

*8–80. If the required clamping force at the board A is to be 70 N, determine the torque M that must be applied to the handle of the hold-down clamp to tighten it down. The single square-threaded screw has a mean radius of 8 mm and a lead of 2 mm, and the coefficient of static friction is $\mu_s = 0.38$.

Prob. 8–80

8–81. The fixture clamp consist of a square-threaded screw having a coefficient of static friction of $\mu_s = 0.3$, mean diameter of 3 mm, and a lead of 1 mm. The five points indicated are pin connections. Determine the clamping force at the smooth blocks D and E when a torque of $M = 0.08$ N $\cdot$ m is applied to the handle of the screw.

Prob. 8–81

8–82. The clamp provides pressure from several directions on the edges of the board. If the square-threaded screw has a lead of 3 mm, radius of 10 mm, and the coefficient of static friction is $\mu_s = 0.4$, determine the horizontal force developed on the board at A and the vertical forces developed at B and C if a torque of $M = 1.5$ N $\cdot$ m is applied to the handle to tighten it further. The blocks at B and C are pin-connected to the board.

Prob. 8–82

8–83. The two blocks under the double wedge are brought together using a left and right square-threaded screw. If the mean diameter is 20 mm, the lead is 5 mm, and the coefficient of static friction is $\mu_s = 0.4$, determine the torque needed to draw the blocks together. The coefficient of static friction between each block and its surfaces of contact is $\mu_s' = 0.4$.

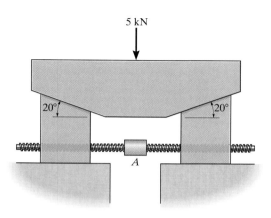

Prob. 8–83

*8–84. The two blocks under the double wedge are brought together using a left and right square-threaded screw. If the mean diameter is 20 mm, the lead is 5 mm, and the coefficient of static friction is $\mu_s = 0.4$, determine the torque needed to spread the blocks apart. The coefficient of static friction between each block and its surfaces of contact is $\mu_s' = 0.4$.

Prob. 8–84

8.5   Frictional Forces on Flat Belts

Whenever belt drives or band brakes are designed, it is necessary to determine the frictional forces developed between the belt and its contacting surface. In this section we will analyze the frictional forces acting on a flat belt, although the analysis of other types of belts, such as the V-belt, is based on similar principles.

Here we will consider the flat belt shown in Fig. 8–19a, which passes over a fixed curved surface, such that the total angle of belt to surface contact in radians is β and the coefficient of friction between the two surfaces is μ. We will determine the tension T_2 in the belt which is needed to pull the belt counterclockwise over the surface and thereby overcome both the frictional forces at the surface of contact and the known tension T_1. Obviously, $T_2 > T_1$.

Frictional Analysis.   A free-body diagram of the belt segment in contact with the surface is shown in Fig. 8–19b. Here the normal force **N** and the frictional force **F**, acting at different points along the belt, will vary both in magnitude and direction. Due to this *unknown* force distribution, the analysis of the problem will proceed on the basis of initially studying the forces acting on a differential element of the belt.

A free-body diagram of an element having a length ds is shown in Fig. 8–19c. Assuming either impending motion or motion of the belt, the magnitude of the frictional force $dF = \mu \, dN$. This force opposes the sliding motion of the belt and thereby increases the magnitude of the tensile force acting in the belt by dT. Applying the two force equations of equilibrium, we have

$$\searrow + \Sigma F_x = 0; \qquad T \cos\left(\frac{d\theta}{2}\right) + \mu \, dN - (T + dT)\cos\left(\frac{d\theta}{2}\right) = 0$$

$$+ \nearrow \Sigma F_y = 0; \qquad dN - (T + dT)\sin\left(\frac{d\theta}{2}\right) - T \sin\left(\frac{d\theta}{2}\right) = 0$$

Flat or V-belts are often used to transmit the torque developed by a motor to a fan or blower.

Motion or impending motion of belt relative to surface

(a)

(b)

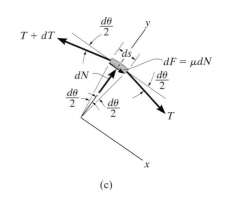

(c)

Fig. 8–19

Since $d\theta$ is of *infinitesimal size*, $\sin(d\theta/2)$ and $\cos(d\theta/2)$ can be replaced by $d\theta/2$ and 1, respectively. Also, the *product* of the two infinitesimals dT and $d\theta/2$ may be neglected when compared to infinitesimals of the first order. The above two equations therefore reduce to

$$\mu \, dN = dT$$

and

$$dN = T \, d\theta$$

Eliminating dN yields

$$\frac{dT}{T} = \mu \, d\theta$$

Integrating this equation between all the points of contact that the belt makes with the drum, and noting that $T = T_1$ at $\theta = 0$ and $T = T_2$ at $\theta = \beta$, yields

$$\int_{T_1}^{T_2} \frac{dT}{T} = \mu \int_0^{\beta} d\theta$$

$$\ln \frac{T_2}{T_1} = \mu\beta$$

Solving for T_2, we obtain

$$\boxed{T_2 = T_1 e^{\mu\beta}} \qquad (8\text{--}6)$$

where

T_2, T_1 = belt tensions; T_1 opposes the direction of motion (or impending motion) of the belt measured relative to the surface, while T_2 acts in the direction of the relative belt motion (or impending motion); because of friction, $T_2 > T_1$

μ = coefficient of static or kinetic friction between the belt and the surface of contact

β = angle of belt to surface contact, measured in radians

e = 2.718..., base of the natural logarithm

Note that T_2 is *independent* of the *radius* of the drum, and instead it is a function of the angle of belt to surface contact, β. Furthermore, as indicated by the integration, this equation is valid for flat belts placed on *any shape* of contacting surface. For application, Eq. 8–6 is valid only when *impending motion* or *motion* occurs.

EXAMPLE 8.9

The maximum tension that can be developed in the cord shown in Fig. 8–20a is 500 N. If the pulley at A is free to rotate and the coefficient of static friction at the fixed drums B and C is $\mu_s = 0.25$, determine the largest mass of the cylinder that can be lifted by the cord. Assume that the force **F** applied at the end of the cord is directed vertically downward, as shown.

(a)

SOLUTION

Lifting the cylinder, which has a weight $W = mg$, causes the cord to move counterclockwise over the drums at B and C; hence, the maximum tension T_2 in the cord occurs at D. Thus, $F = T_2 = 500$ N. A section of the cord passing over the drum at B is shown in Fig. 8–20b. Since $180° = \pi$ rad, the angle of contact between the drum and the cord is $\beta = (135°/180°)\pi = 3\pi/4$ rad. Using Eq. 8–6, we have

$$T_2 = T_1 e^{\mu_s \beta}; \qquad\qquad 500\text{ N} = T_1 e^{0.25[(3/4)\pi]}$$

Hence,

$$T_1 = \frac{500\text{ N}}{e^{0.25[(3/4)\pi]}} = \frac{500\text{ N}}{1.80} = 277.4\text{ N}$$

Since the pulley at A is free to rotate, equilibrium requires that the tension in the cord remains the *same* on both sides of the pulley.

The section of the cord passing over the drum at C is shown in Fig. 8–20c. The weight $W < 277.4$ N. Why? Applying Eq. 8–6, we obtain

$$T_2 = T_1 e^{\mu_s \beta}; \qquad\qquad 277.4\text{ N} = W e^{0.25[(3/4)\pi]}$$
$$W = 153.9\text{ N}$$

so that

$$m = \frac{W}{g} = \frac{153.9\text{ N}}{9.81\text{ m/s}^2}$$
$$= 15.7\text{ kg} \qquad\qquad\qquad Ans.$$

(b)

(c)

Fig. 8–20

PROBLEMS

8–85. The cord supporting the 6-kg cylinder passes around three pegs, A, B, C, where $\mu_s = 0.2$. Determine the range of values for the magnitude of the horizontal force **P** for which the cylinder will not move up or down.

Prob. 8–85

8–86. The truck, which has a mass of 3.4 Mg, is to be lowered down the slope by a rope that is wrapped around a tree. If the wheels are free to roll and the man at A can resist a pull of 300 N, determine the minimum number of turns the rope should be wrapped around the tree to lower the truck at a constant speed. The coefficient of kinetic friction between the tree and rope is $\mu_k = 0.3$.

Prob. 8–86

8–87. The wheel is subjected to a torque of $M = 50\,\text{N}\cdot\text{m}$. If the coefficient of kinetic friction between the band brake and the rim of the wheel is $\mu_k = 0.3$, determine the smallest horizontal force P that must be applied to the lever to stop the wheel.

Prob. 8–87

*__8–88.__ The cylinder A has a mass of 75 kg. Determine the smallest force P applied to the handle of the lever required for equilibrium. The coefficient of static friction between the belt and the wheel is $\mu_s = 0.3$. The drum is pin connected at its center, B.

8–89. Determine the largest mass of cylinder A that can be supported from the drum if a force $P = 20\,\text{N}$ is applied to the handle of the lever. The coefficient of static friction between the belt and the wheel is $\mu_s = 0.3$. The drum is pin supported at its center, B.

Probs. 8–88/89

8–90. The uniform bar AB is supported by a rope that passes over a frictionless pulley at C and a fixed peg at D. If the coefficient of static friction between the rope and the peg is $\mu_D = 0.3$, determine the smallest distance x from the end of the bar at which a 20-N force may be placed and not cause the bar to move.

20 N

x

A B

1 m

Prob. 8–90

8–91. Determine the smallest lever force P needed to prevent the wheel from rotating if it is subjected to a torque of $M = 250\ \text{N} \cdot \text{m}$. The coefficient of static friction between the belt and the wheel is $\mu_s = 0.3$. The wheel is pin connected at its center, B.

400 mm

B

M

A 200 mm

750 mm

P

Prob. 8–91

*8–92. Determine the torque M that can be resisted by the band brake if a force of $P = 30\ \text{N}$ is applied to the handle of the lever. The coefficient of static friction between the belt and the wheel is $\mu_s = 0.3$. The wheel is pin connected at its center, B.

400 mm

B

M

A 200 mm

750 mm

P

Prob. 8–92

8–93. Blocks A and B weigh 50 lb and 30 lb, respectively. Using the coefficients of static friction indicated, determine the greatest weight of block D without causing motion.

$\mu = 0.5$

B

$20°$ $\mu_{BA} = 0.6$

A

C $\mu_{AC} = 0.4$

D

Prob. 8–93

8–94. Blocks A and B weigh 75 lb each, and D weighs 30 lb. Using the coefficients of static friction indicated, determine the frictional force between blocks A and B and between block A and the floor C.

$\mu = 0.5$

B

$20°$ $\mu_{BA} = 0.6$

A

C $\mu_{AC} = 0.4$

D

Prob. 8–94

8–95. Show that the frictional relationship between the belt tensions, the coefficient of friction μ, and the angular contacts α and β for the V-belt is $T_2 = T_1 e^{\mu\beta/\sin(\alpha/2)}$.

8–97. A cable is attached to the 20-kg plate B, passes over a fixed peg at C, and is attached to the block at A. Using the coefficients of static friction shown, determine the smallest mass of block A so that it will prevent sliding motion of B down the plane.

Prob. 8–95

Prob. 8–97

*8–96.** The 60° V-fan-belt of an automobile engine passes around the hub H of a generator G and over the housing F to a fan. If the generator locks, and the maximum tension the belt can sustain is $T_{max} = 175$ lb, determine the maximum possible torque M resisted by the axle as the belt slips over the hub. Assume that slipping of the belt occurs only at H and that the coefficient of static friction for the hub is $\mu_s = 0.25$. Use the result of Prob. 8–95.

8–98. The simple band brake is constructed so that the ends of the friction strap are connected to the pin at A and the lever arm at B. If the wheel is subjected to a torque of $M = 80$ lb·ft, determine the smallest force P applied to the lever that is required to hold the wheel stationary. The coefficient of static friction between the strap and wheel is $\mu_s = 0.5$.

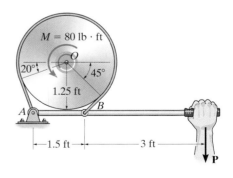

Prob. 8–96

Prob. 8–98

8–99. The uniform 50-lb beam is supported by the rope which is attached to the end of the beam, wraps over the rough peg, and is then connected to the 100-lb block. If the coefficient of static friction between the beam and the block, and between the rope and the peg, is $\mu_s = 0.4$, determine the maximum distance that the block can be placed from A and still remain in equilibrium. Assume the block will not tip.

Prob. 8–99

8–101. A cord having a weight of 0.5 lb/ft and a total length of 10 ft is suspended over a peg P as shown. If the coefficient of static friction between the peg and cord is $\mu_s = 0.5$, determine the longest length h which one side of the suspended cord can have without causing motion. Neglect the size of the peg and the length of cord draped over it.

Prob. 8–101

*8–100. The uniform concrete pipe has a weight of 800 lb and is unloaded slowly from the truck bed using the rope and skids shown. If the coefficient of kinetic friction between the rope and pipe is $\mu_k = 0.3$, determine the force the worker must exert on the rope to lower the pipe at constant speed. There is a pulley at B, and the pipe does not slip on the skids. The lower portion of the rope is parallel to the skids.

8–102. Granular material, having a density of 1.5 Mg/m³, is transported on a conveyor belt that slides over the fixed surface, having a coefficient of kinetic friction of $\mu_k = 0.3$. Operation of the belt is provided by a motor that supplies a torque **M** to wheel A. The wheel at B is free to turn, and the coefficient of static friction between the wheel at A and the belt is $\mu_A = 0.4$. If the belt is subjected to a pretension of 300 N when no load is on the belt, determine the greatest volume V of material that is permitted on the belt at any time without allowing the belt to stop. What is the torque **M** required to drive the belt when it is subjected to this maximum load?

Prob. 8–100

Prob. 8–102

8–103. Blocks A and B have a mass of 100 kg and 150 kg, respectively. If the coefficient of static friction between A and B and between B and C is $\mu_s = 0.25$, and between the ropes and the pegs D and E $\mu'_s = 0.5$, determine the smallest force F needed to cause motion of block B if $P = 30$ N.

Prob. 8–103

8–105. Block A has a mass of 50 kg and rests on surface for which $\mu_s = 0.25$. If the coefficient of static friction between the cord and the fixed peg at C is $\mu'_s = 0.3$, determine the greatest mass of the suspended cylinder D without causing motion.

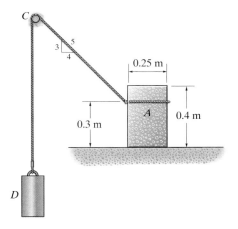

Prob. 8–105

*__8–104.__ Blocks A and B weigh 50 lb and 30 lb, respectively. Using the coefficients of static friction indicated, determine the greatest weight of block E without causing motion.

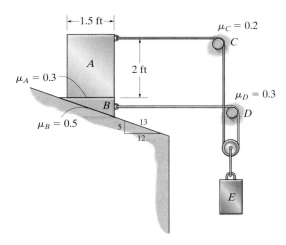

Prob. 8–104

8–106. Block A rests on the surface for which $\mu_s = 0.25$. If the mass of the suspended cylinder D is 4 kg, determine the smallest mass of block A so that it does not slip or tip. The coefficient of static friction between the cord and the fixed peg at C is $\mu'_s = 0.3$.

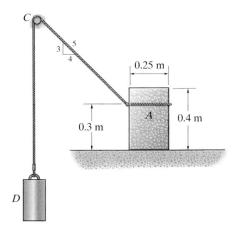

Prob. 8–106

*8.6   Frictional Forces on Collar Bearings, Pivot Bearings, and Disks

Pivot and *collar bearings* are commonly used in machines to support an *axial load* on a rotating shaft. These two types of support are shown in Fig. 8–21. Provided the bearings are not lubricated, or are only partially lubricated, the laws of dry friction may be applied to determine the moment **M** needed to turn the shaft when it supports an axial force **P**.

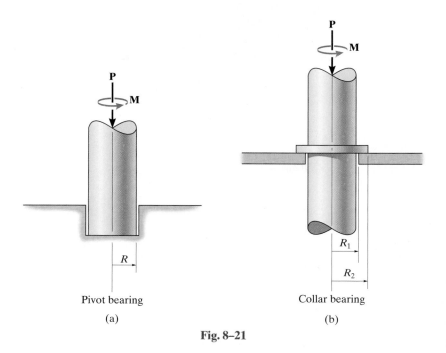

Pivot bearing

(a)

Collar bearing

(b)

Fig. 8–21

Frictional Analysis.   The collar bearing on the shaft shown in Fig. 8–22 is subjected to an axial force **P** and has a total bearing or contact area $\pi(R_2^2 - R_1^2)$. In the following analysis, the normal pressure p is considered to be *uniformly distributed* over this area—a reasonable assumption provided the bearing is new and evenly supported. Since $\Sigma F_z = 0$, then p, measured as a force per unit area, is $p = P/\pi(R_2^2 - R_1^2)$.

The moment needed to cause impending rotation of the shaft can be determined from moment equilibrium about the z axis. A small area element $dA = (r\,d\theta)(dr)$, shown in Fig. 8–22, is subjected to both a normal force $dN = p\,dA$ and an associated frictional force,

$$dF = \mu_s\,dN = \mu_s p\,dA = \frac{\mu_s P}{\pi(R_2^2 - R_1^2)}\,dA$$

Fig. 8–22

The normal force does not create a moment about the z axis of the shaft; however, the frictional force does; namely, $dM = r\,dF$. Integration is needed to compute the total moment created by all the frictional forces acting on differential areas dA. Therefore, for impending rotational motion,

$$\Sigma M_z = 0; \qquad M - \int_A r\,dF = 0$$

Substituting for dF and dA and integrating over the entire bearing area yields

$$M = \int_{R_1}^{R_2} \int_0^{2\pi} r \left[\frac{\mu_s P}{\pi(R_2^2 - R_1^2)} \right] (r\,d\theta\,dr) = \frac{\mu_s P}{\pi(R_2^2 - R_1^2)} \int_{R_1}^{R_2} r^2\,dr \int_0^{2\pi} d\theta$$

or

$$M = \frac{2}{3}\mu_s P \left(\frac{R_2^3 - R_1^3}{R_2^2 - R_1^2} \right) \qquad (8\text{–}7)$$

This equation gives the magnitude of moment required for impending rotation of the shaft. The frictional moment developed at the end of the shaft, when it is *rotating* at constant speed, can be found by substituting μ_k for μ_s in Eq. 8–7.

When $R_2 = R$ and $R_1 = 0$, as in the case of a pivot bearing, Fig. 8–21a, Eq. 8–7 reduces to

$$M = \frac{2}{3}\mu_s P R \qquad (8\text{–}8)$$

Recall from the initial assumption that both Eqs. 8–7 and 8–8 apply only for bearing surfaces subjected to *constant pressure*. If the pressure is not uniform, a variation of the pressure as a function of the bearing area must be determined before integrating to obtain the moment. The following example illustrates this concept.

The motor that turns the disk of this sanding machine develops a torque that must overcome the frictional forces acting on the disk.

EXAMPLE 8.10

(a)

The uniform bar shown in Fig. 8–23a has a total mass m. If it is assumed that the normal pressure acting at the contacting surface varies linearly along the length of the bar as shown, determine the couple moment **M** required to rotate the bar. Assume that the bar's width a is negligible in comparison to its length l. The coefficient of static friction is equal to μ_s.

SOLUTION

A free-body diagram of the bar is shown in Fig. 8–23b. Since the bar has a total weight of $W = mg$, the intensity w_0 of the distributed lead at the center ($x = 0$) is determined from vertical force equilibrium, Fig. 8–23a.

$$+\uparrow \Sigma F_z = 0; \quad -mg + 2\left[\frac{1}{2}\left(\frac{l}{2}\right)w_0\right] = 0 \qquad w_0 = \frac{2mg}{l}$$

Since $w = 0$ at $x = l/2$, the distributed load expressed as a function of x is

$$w = w_0\left(1 - \frac{2x}{l}\right) = \frac{2mg}{l}\left(1 - \frac{2x}{l}\right)$$

The magnitude of the normal force acting on a segment of area having a length dx is therefore

$$dN = w\, dx = \frac{2mg}{l}\left(1 - \frac{2x}{l}\right) dx$$

The magnitude of the frictional force acting on the same element of area is

$$dF = \mu_s\, dN = \frac{2\mu_s mg}{l}\left(1 - \frac{2x}{l}\right) dx$$

Hence, the moment created by this force about the z axis is

$$dM = x\, dF = \frac{2\mu_s mg}{l} x\left(1 - \frac{2x}{l}\right) dx$$

The summation of moments about the z axis of the bar is determined by integration, which yields

$$\Sigma M_z = 0; \qquad M - 2\int_0^{l/2} \frac{2\mu_s mg}{l} x\left(1 - \frac{2x}{l}\right) dx = 0$$

$$M = \frac{4\mu_s mg}{l}\left(\frac{x^2}{2} - \frac{2x^3}{3l}\right)\Bigg|_0^{l/2}$$

$$M = \frac{\mu_s mgl}{6} \qquad\qquad Ans.$$

(b)

Fig. 8–23

8.7   Frictional Forces on Journal Bearings

When a shaft or axle is subjected to lateral loads, a *journal bearing* is commonly used for support. Well-lubricated journal bearings are subjected to the laws of fluid mechanics, in which the viscosity of the lubricant, the speed of rotation, and the amount of clearance between the shaft and bearing are needed to determine the frictional resistance of the bearing. When the bearing is not lubricated or is only partially lubricated, however, a reasonable analysis of the frictional resistance can be based on the laws of dry friction.

Frictional Analysis.   A typical journal-bearing support is shown in Fig. 8–24a. As the shaft rotates in the direction shown in the figure, it rolls up against the wall of the bearing to some point A where slipping occurs. If the lateral load acting at the end of the shaft is $\mathbf{P}$, it is necessary that the bearing reactive force $\mathbf{R}$ acting at A be equal and opposite to $\mathbf{P}$, Fig. 8–24b. The moment needed to maintain constant rotation of the shaft can be found by summing moments about the z axis of the shaft; i.e.,

$$\Sigma M_z = 0; \qquad M - (R \sin \phi_k)r = 0$$

or

$$M = Rr \sin \phi_k \qquad (8\text{–}9)$$

where ϕ_k is the angle of kinetic friction defined by $\tan \phi_k = F/N = \mu_k N/N = \mu_k$. In Fig. 8–24c, it is seen that $r \sin \phi_k = r_f$. The dashed circle with radius r_f is called the *friction circle*, and as the shaft rotates, the reaction $\mathbf{R}$ will always be tangent to it. If the bearing is partially lubricated, μ_k is small, and therefore $\mu_k = \tan \phi_k \approx \sin \phi_k \approx \phi_k$. Under these conditions, a reasonable *approximation* to the moment needed to overcome the frictional resistance becomes

$$M \approx Rr\mu_k \qquad (8\text{–}10)$$

The following example illustrates a common application of this analysis.

Unwinding the cable from this spool requires overcoming friction from the supporting shaft.

(a)

Fig. 8–24

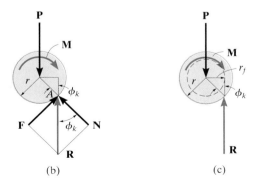

(b)   (c)

EXAMPLE 8.11

The 100-mm-diameter pulley shown in Fig. 8–25a fits loosely on a 10-mm-diameter shaft for which the coefficient of static friction is $\mu_s = 0.4$. Determine the minimum tension T in the belt needed to (a) raise the 100-kg block and (b) lower the block. Assume that no slipping occurs between the belt and pulley and neglect the weight of the pulley.

(a)

SOLUTION

Part (a).   A free-body diagram of the pulley is shown in Fig. 8–25b. When the pulley is subjected to belt tensions of 981 N each, it makes contact with the shaft at point P_1. As the tension T is *increased*, the pulley will roll around the shaft to point P_2 before motion impends. From the figure, the friction circle has a radius $r_f = r \sin \phi_s$. Using the simplification that $\sin \phi_s \approx (\tan \phi_s \approx \phi_s)$, then $r_f \approx r\mu_s = (5 \text{ mm})(0.4) = 2$ mm, so that summing moments about P_2 gives

$$\zeta + \Sigma M_{P_2} = 0; \quad 981 \text{ N}(52 \text{ mm}) - T(48 \text{ mm}) = 0$$
$$T = 1063 \text{ N} = 1.06 \text{ kN} \qquad \qquad Ans.$$

If a more exact analysis is used, then $\phi_s = \tan^{-1} 0.4 = 21.8°$. Thus, the radius of the friction circle would be $r_f = r \sin \phi_s = 5 \sin 21.8° = 1.86$ mm. Therefore,

$$\zeta + \Sigma M_{P_2} = 0;$$
$$981 \text{ N}(50 \text{ mm} + 1.86 \text{ mm}) - T(50 \text{ mm} - 1.86 \text{ mm}) = 0$$
$$T = 1057 \text{ N} = 1.06 \text{ kN} \qquad \qquad Ans.$$

Part (b).   When the block is lowered, the resultant force **R** acting on the shaft passes through point P_3, as shown in Fig. 8–25c. Summing moments about this point yields

$$\zeta + \Sigma M_{P_3} = 0; \quad 981 \text{ N}(48 \text{ mm}) - T(52 \text{ mm}) = 0$$
$$T = 906 \text{ N} \qquad \qquad Ans.$$

NOTE: The difference between raising and lowering the block is thus 54 N.

(b)

(c)

Fig. 8–25

*8.8   Rolling Resistance

If a *rigid* cylinder of weight **W** rolls at constant velocity along a *rigid* surface, the normal force exerted by the surface on the cylinder acts at the tangent point of contact, as shown in Fig. 8–26a. Under these conditions, provided the cylinder does not encounter frictional resistance from the air, motion will continue indefinitely. Actually, however, no materials are perfectly rigid, and therefore the reaction of the surface on the cylinder consists of a distribution of normal pressure. For example, consider the cylinder to be made of a very hard material, and the surface on which it rolls to be relatively soft. Due to its weight, the cylinder compresses the surface underneath it, Fig. 8–26b. As the cylinder rolls, the surface material in front of the cylinder *retards* the motion since it is being *deformed*, whereas the material in the rear is *restored* from the deformed state and therefore tends to *push* the cylinder forward. The normal pressures acting on the cylinder in this manner are represented in Fig. 8–26b by their resultant forces $\mathbf{N}_d$ and $\mathbf{N}_r$. Unfortunately, the magnitude of the force of *deformation*, $\mathbf{N}_d$, and its horizontal component is *always greater* than that of *restoration*, $\mathbf{N}_r$, and consequently a horizontal driving force **P** must be applied to the cylinder to maintain the motion. Fig. 8–26b.*

Rolling resistance is caused primarily by this effect, although it is also, to a lesser degree, the result of surface adhesion and relative microsliding between the surfaces of contact. Because the actual force **P** needed to overcome these effects is difficult to determine, a simplified method will be developed here to explain one way engineers have analyzed this phenomenon. To do this, we will consider the resultant of the *entire* normal pressure, $\mathbf{N} = \mathbf{N}_d + \mathbf{N}_r$, acting on the cylinder, Fig. 8–26c. As shown in Fig. 8–26d, this force acts at an angle θ with the vertical. To keep the cylinder in equilibrium, i.e., rolling at a constant rate, it is necessary that **N** be *concurrent* with the driving force **P** and the weight **W**. Summing moments about point A gives $Wa = P(r \cos \theta)$. Since the deformations are generally very small in relation to the cylinder's radius, $\cos \theta \approx 1$; hence,

$$Wa \approx Pr$$

or

$$\boxed{P \approx \frac{Wa}{r}} \qquad (8\text{–}11)$$

*Actually, the deformation force $\mathbf{N}_d$ causes *energy* to be stored in the material as its magnitude is increased, whereas the restoration force $\mathbf{N}_r$, as its magnitude is decreased, allows some of this energy to be released. The remaining energy is *lost* since it is used to heat up the surface, and if the cylinder's weight is very large, it accounts for permanent deformation of the surface. Work must be done by the horizontal force **P** to make up for this loss.

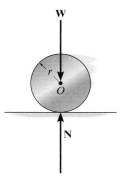

Rigid surface of contact

(a)

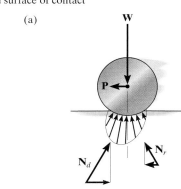

Soft surface of contact

(b)

(c)

(d)

Fig. 8–26

The distance a is termed the *coefficient of rolling resistance,* which has the dimension of length. For instance, $a \approx 0.5$ mm for a wheel rolling on a rail, both of which are made of mild steel. For hardened steel ball bearings on steel, $a \approx 0.1$ mm. Experimentally, though, this factor is difficult to measure, since it depends on such parameters as the rate of rotation of the cylinder, the elastic properties of the contacting surfaces, and the surface finish. For this reason, little reliance is placed on the data for determining a. The analysis presented here does, however, indicate why a heavy load (W) offers greater resistance to motion (P) than a light load under the same conditions. Furthermore, since the ratio Wa/r is generally very small compared to $\mu_k W$, the force needed to *roll* the cylinder over the surface will be much less than that needed to *slide* the cylinder across the surface. Hence, the analysis indicates why roller or ball bearings are often used to minimize the frictional resistance between moving parts.

Rolling resistance of railroad wheels on the rails is small since steel is very stiff. By comparison, the rolling resistance of the wheels of a tractor in a wet field is very large.

EXAMPLE 8.12

A 10-kg steel wheel shown in Fig. 8–27a has a radius of 100 mm and rests on an inclined plans made of wood. If θ is increased so that the wheel begins to roll-down the incline with constant velocity when $\theta = 1.2°$, determine the coefficient of rolling resistance.

(a)

(b)

Fig. 8–27

SOLUTION

As shown on the free-body diagram, Fig. 8–27b, when the wheel has impending motion, the normal reaction **N** acts at point A defined by the dimension a. Resolving the weight into components parallel and perpendicular to the incline, and summing moments about point A, yields (approximately)

$$\zeta + \Sigma M_A = 0; \quad 98.1 \cos 1.2° \, N(a) - 98.1 \sin 1.2° \, N(100 \text{ mm}) = 0$$

Solving, we obtain

$$a = 2.09 \text{ mm} \qquad\qquad\qquad Ans.$$

PROBLEMS

8–107. The collar bearing uniformly supports an axial force of $P = 800$ lb. If the coefficient of static friction is $\mu_s = 0.3$, determine the torque M required to overcome friction.

*8–108. The collar bearing uniformly supports an axial force of $P = 500$ lb. If a torque of $M = 3$ lb $\cdot$ ft is applied to the shaft and causes it to rotate at constant velocity, determine the coefficient of kinetic friction at the surface of contact.

Probs. 8–107/108

8–109. The *double-collar bearing* is subjected to an axial force $P = 4$ kN. Assuming that collar A supports $0.75P$ and collar B supports $0.25P$, both with a uniform distribution of pressure, determine the maximum frictional moment M that may be resisted by the bearing. Take $\mu_s = 0.2$ for both collars.

Prob. 8–109

8–110. The annular ring bearing is subjected to a thrust of 800 lb. If $\mu_s = 0.35$, determine the torque M that must be applied to overcome friction.

Prob. 8–110

8–111. The floor-polishing machine rotates at a constant angular velocity. If it has a weight of 80 lb. determine the couple forces F the operator must apply to the handles to hold the machine stationary. The coefficient of kinetic friction between the floor and brush is $\mu_k = 0.3$. Assume the brush exerts a uniform pressure on the floor.

Prob. 8–111

*8–112. The plate clutch consists of a flat plate A that slides over the rotating shaft S. The shaft is fixed to the driving plate gear B. If the gear C, which is in mesh with B, is subjected to a torque of $M = 0.8 \text{ N} \cdot \text{m}$, determine the smallest force P, that must be applied via the control arm, to stop the rotation. The coefficient of static friction between the plates A and D is $\mu_s = 0.4$. Assume the bearing pressure between A and D to be uniform.

Prob. 8–112

8–113. The 4-in.-diameter shaft is held in the hole such that the normal pressure acting around the shaft varies linearly with its depth as shown. Determine the frictional torque that must be overcome to rotate the shaft. Take $\mu_s = 0.2$.

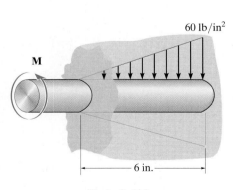

Prob. 8–113

8–114. Because of wearing at the edges, the pivot bearing is subjected to a conical pressure distribution at its surface of contact. Determine the torque M required to overcome friction and turn the shaft, which supports an axial force $\mathbf{P}$. The coefficient of static friction is μ_s. For the solution, it is necessary to determine the peak pressure p_0 in terms of P and the bearing radius R.

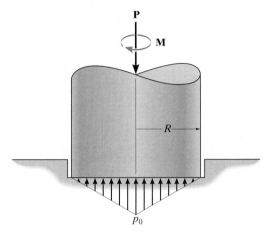

Prob. 8–114

8–115. The conical bearing is subjected to a constant pressure distribution at its surface of contact. If the coefficient of static friction is μ_s, determine the torque $\mathbf{M}$ required to overcome friction if the shaft supports an axial force $\mathbf{P}$.

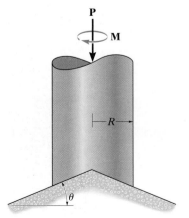

Prob. 8–115

*8–116. The tractor is used to push the 1500-lb pipe. To do this it must overcome the frictional forces at the ground, caused by sand. Assuming that the sand exerts a pressure on the bottom of the pipe as shown, and the coefficient of static friction between the pipe and the sand is $\mu_s = 0.3$, determine the force required to push the pipe forward. Also, determine the peak pressure p_0.

15 in.

θ

p_0 $p = p_0 \cos \theta$

12 ft

Prob. 8–116

8–117. Assuming that the variation of pressure at the bottom of the pivot bearing is defined as $p = p_0(R_2/r)$, determine the torque M needed to overcome friction if the shaft is subjected to an axial force **P**. The coefficient of static friction is μ_s. For the solution, it is necessary to determine p_0 in terms of P and the bearing dimensions R_1 and R_2.

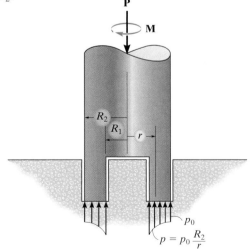

P

M

R_2

R_1 r

p_0

$p = p_0 \dfrac{R_2}{r}$

Prob. 8–117

8–118. A disk having an outer diameter of 8 in. fits loosely over a fixed shaft having a diameter of 3 in. If the coefficient of static friction between the disk and the shaft is $\mu_s = 0.15$, determine the smallest vertical force P, acting on the rim, which must be applied to the disk to cause it to slip over the shaft. The disk weighs 10 lb.

8 in.

3 in.

P

Prob. 8–118

8–119. The pulley has a radius of 3 in. and fits loosely on the 0.5-in.-diameter shaft. If the loadings acting on the belt cause the pulley to rotate with constant angular velocity, determine the frictional force between the shaft and the pulley and compute the coefficient of kinetic friction. The pulley weighs 18 lb.

*8–120. The pulley has a radius of 3 in. and fits loosely on the 0.5-in.-diameter shaft. If the loadings acting on the belt cause the pulley to rotate with constant angular velocity, determine the frictional force between the shaft and the pulley and compute the coefficient of kinetic friction. Neglect the weight of the pulley.

3 in.

5 lb   5.5 lb

Probs. 8–119/120

8–121. The 5-kg pulley has a diameter of 240 mm and the axle has a diameter of 40 mm. If the coefficient of kinetic friction between the axle and the pulley is $\mu_k = 0.15$, determine the vertical force P on the rope required to lift the 80-kg block at constant velocity.

8–122. Solve Prob. 8–121 if the force **P** is applied horizontally to the right.

*8–124.** The trailer has a total weight of 850 lb and center of gravity at G which is directly over its axle. If the axle has a diameter of 1 in., the radius of the wheel is $r = 1.5$ ft, and the coefficient of kinetic friction at the bearing is $\mu_k = 0.08$, determine the horizontal force P needed to pull the trailer.

Prob. 8–124

Probs. 8–121/122

8–123. A wheel on a freight car carries a load of 20 kip. If the axle of the car has a diameter of 2 in., determine the horizontal force P that must be applied to the axle to rotate the wheel. The coefficient of kinetic friction is $\mu_k = 0.05$.

8–125. The collar fits *loosely* around a fixed shaft that has a radius of 2 in. If the coefficient of kinetic friction between the shaft and the collar is $\mu_k = 0.3$, determine the force P on the horizontal segment of the belt so that the collar rotates *counterclockwise* with a constant angular velocity. Assume that the belt does not slip on the collar; rather, the collar slips on the shaft. Neglect the weight and thickness of the belt and collar. The radius, measured from the center of the collar to the mean thickness of the belt, is 2.25 in.

Prob. 8–123

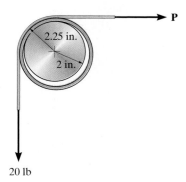

Prob. 8–125

8–126. The collar fits *loosely* around a fixed shaft that has a radius of 2 in. If the coefficient of kinetic friction between the shaft and the collar is $\mu_k = 0.3$, determine the force P on the horizontal segment of the belt so that the collar rotates *clockwise* with a constant angular velocity. Assume that the belt does not slip on the collar; rather, the collar slips on the shaft. Neglect the weight and thickness of the belt and collar. The radius, measured from the center of the collar to the mean thickness of the belt, is 2.25 in.

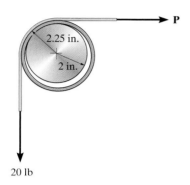

2.25 in.

2 in.

20 lb

Prob. 8–126

8–127. The connecting rod is attached to the piston by a 0.75-in.-diameter pin at B and to the crank shaft by a 2-in.-diameter bearing A. If the piston is moving downwards, and the coefficient of static friction at these points is $\mu_s = 0.2$, determine the radius of the friction circle at each connection.

B

A

Prob. 8–127

*8–128. The connecting rod is attached to the piston by a 20-mm-diameter pin at B and to the crank shaft by a 50-mm-diameter bearing A. If the piston is moving upwards, and the coefficient of static friction at these points is $\mu_s = 0.3$, determine the radius of the friction circle at each connection.

B

A

Prob. 8–128

8–129. The lawn roller has a mass of 80 kg. If the arm BA is held at an angle of 30° from the horizontal and the coefficient of rolling resistance for the roller is 25 mm, determine the force P needed to push the roller at constant speed. Neglect friction developed at the axle, A, and assume that the resultant force $\mathbf{P}$ acting on the handle is applied along arm BA.

P

B

250 mm

A

30°

Prob. 8–129

8–130. The handcart has wheels with a diameter of 6 in. If a crate having a weight of 1500 lb is placed on the cart, determine the force P that must be applied to the handle to overcome the rolling resistance. The coefficient of rolling resistance is 0.04 in. Neglect the weight of the cart.

*8–132. The 1.2-Mg steel beam is moved over a level surface using a series of 30-mm-diameter rollers for which the coefficient of rolling resistance is 0.4 mm at the ground and 0.2 mm at the bottom surface of the beam. Determine the horizontal force P needed to push the beam forward at a constant speed. *Hint:* Use the result of Prob. 8–131.

Prob. 8–130

Prob. 8–132

8–131. The cylinder is subjected to a load that has a weight W. If the coefficients of rolling resistance for the cylinder's top and bottom surfaces are a_A and a_B, respectively, show that a force having a magnitude of $P = [W(a_A + a_B)]/2r$ is required to move the load and thereby roll the cylinder forward. Neglect the weight of the cylinder.

8–133. The 1.4-Mg machine is to be moved over a level surface using a series of rollers for which the coefficient of rolling resistance is 0.5 mm at the ground and 0.2 mm at the bottom surface of the machine. Determine the appropriate diameter of the rollers so that the machine can be pushed forward with a horizontal force of $P = 250$ N. *Hint:* Use the result of Prob. 8–131.

Prob. 8–131

Prob. 8–133

Chapter Review

Dry Friction

Frictional forces exist at rough surfaces of contact. They act on a body so as to oppose its motion or tendency of motion.

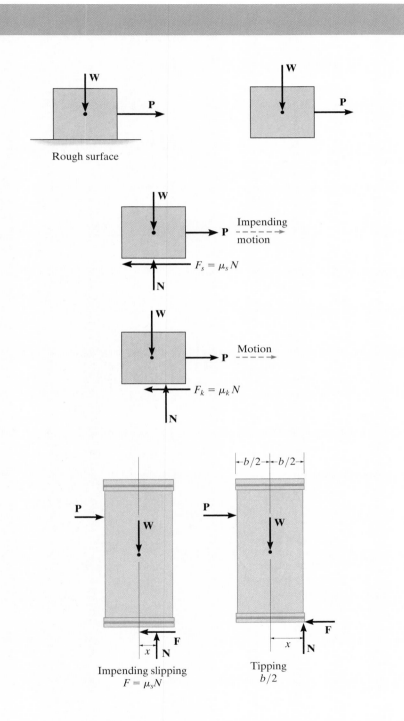

Rough surface

A static friction force approaches a maximum value of $F_s = \mu_s N$, where μ_s is the *coefficient of static friction*. In this case, motion between the contacting surfaces is *about to impend*.

Impending motion

$F_s = \mu_s N$

If slipping occurs, then the friction force remains essentially constant and equal to a value of $F_k = \mu_k N$. Here μ_k is the *coefficient of kinetic friction*.

Motion

$F_k = \mu_k N$

The solution of a problem involving friction requires first drawing the free-body diagram of the body. If the unknowns cannot be determined strictly from the equations of equilibrium, and the possibility of slipping can occur, then the friction equation should be applied at the appropriate points of contact in order to complete the solution.

It may also be possible for slender objects, like crates, to tip over, and this situation should also be investigated. (*See pages 393–400.*)

Impending slipping
$F = \mu_s N$

Tipping
$b/2$

Wedges

Wedges are inclined planes used to increase the application of a force. The two-force equilibrium equations are used to relate the forces acting on the wedge.

An applied force **P** must push on the wedge to move it to the right.

If the coefficients of friction between the surfaces are large enough, then **P** can be removed, and the wedge will be self-locking and remain in place. (*See page 418.*)

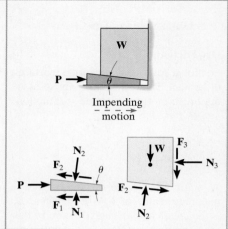

Screws

Square-threaded screws are used to move heavy loads. They represent an inclined plane, wrapped around a cylinder.

The moment needed to turn a screw depends upon the coefficient of friction and the screw's lead angle θ.

If the coefficient of friction between the surfaces is large enough, then the screw will support the load without tending to turn, i.e., it will be self-locking. (*See pages 420–422.*)

$$M = Wr \tan(\theta + \phi)$$

Upward Screw Motion

$$M' = Wr \tan(\theta - \phi)$$

Downward Screw Motion

$$\theta > \phi$$
$$M'' = Wr \tan(\phi - \theta)$$

Downward Screw Motion

$$\phi > \theta$$

Flat Belts

The force needed to move a flat belt over a rough curved surface depends only on the angle of belt contact, β, and the coefficient of friction. (*See pages 428–429.*)

$$T_2 = T_1 e^{\mu\beta}$$

Collar Bearings and Disks

The frictional analysis of a collar bearing or disk requires looking at a differential segment of the contact area. The force acting on the differential element is determined from force equilibrium along the shaft, and the moment needed to turn the shaft at a constant rate is determined from moment equilibrium about the shaft's axis.

If the pressure on the surface of a collar bearing is uniform, then integration gives the result.
(*See pages 436–437.*)

$$M = \frac{2}{3}\mu_s P\left(\frac{R_2^3 - R_1^3}{R_2^2 - R_1^2}\right)$$

Journal Bearings

When a moment is applied to a shaft in a nonlubricated journal bearing, the shaft will tend to roll up the side of the bearing until slipping occurs. This defines the radius of a friction circle, and from it the moment needed to turn the shaft can be determined.
(*See page 439.*)

$$M = Rr \sin \phi_k$$

Rolling Resistance

The resistance of a wheel to rolling over a surface is caused by localized *deformation* of the two materials in contact. This causes the resultant normal force acting on the rolling body to be inclined so that it provides a component that acts in the opposite direction of the force causing the motion. This effect is characterized using the *coefficient of rolling resistance*, which is determined from experiment.
(*See pages 441–442.*)

$$P \approx \frac{Wa}{r}$$

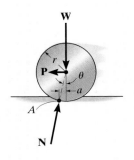

REVIEW PROBLEMS

8–134. A single force **P** is applied to the handle of the drawer. If friction is neglected at the bottom side and the coefficient of static friction along the sides is $\mu_s = 0.4$, determine the largest spacing s between the symmetrically placed handles so that the drawer does not bind at the corners A and B when the force **P** is applied to one of the handles.

Prob. 8–134

8–135. The truck has a mass of 1.25 Mg and a center of mass at G. Determine the greatest load it can pull if (a) the truck has rear-wheel drive while the front wheels are free to roll, and (b) the truck has four-wheel drive. The coefficient of static friction between the wheels and the ground is $\mu_s = 0.5$, and between the crate and the ground, it is $\mu'_s = 0.4$.

*8–136. Solve Prob. 8–135 if the truck and crate are traveling up a 10° incline.

Probs. 8–135/136

8–137. A roofer, having a mass of 70 kg, walks slowly in an upright position down along the surface of a dome that has a radius of curvature of $r = 20$ m. If the coefficient of static friction between his shoes and the dome is $\mu_s = 0.7$, determine the angle θ at which he first begins to slip.

Prob. 8–137

8–138. A man attempts to lift the uniform 40-lb ladder to an upright position by applying a force **P** perpendicular to the ladder at rung R. Determine the coefficient of static friction between the ladder and the ground at A if the ladder begins to slip on the ground when his hands reach a height of 6 ft.

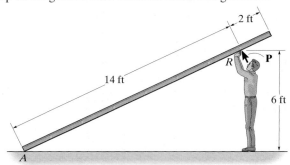

Prob. 8–138

8–139. Column D is subjected to a vertical load of 8000 lb. It is supported on two identical wedges A and B for which the coefficient of static friction at the contacting surfaces between A and B and between B and C is $\mu_s = 0.4$. Determine the force P needed to raise the column and the equilibrium force P' needed to hold wedge A stationary. The contacting surface between A and D is smooth.

*8–140. Column D is subjected to a vertical load of 8000 lb. It is supported on two identical wedges A and B for which the coefficient of static friction at the contacting surfaces between A and B and between B and C is $\mu_s = 0.4$. If the forces **P** and **P'** are removed, are the wedges self-locking? The contacting surface between A and D is smooth.

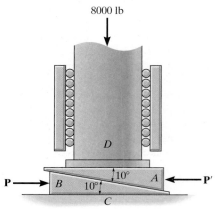

Probs. 8–139/140

Design Projects

8–1D Design of a Rope-and-Pulley System for Pulling a Crate up an Incline

A large 300-kg packing crate is to be hoisted up the 25° incline. The coefficient of static friction between the incline and the crate is $\mu_s = 0.5$, and the coefficient of kinetic friction is $\mu_k = 0.4$. Using a system of ropes and pulleys, design a method that will allow a single worker to pull the crate up the ramp. Pulleys can be attached to any point on the wall AB. Assume the worker can exert a maximum (comfortable) pull of 200 N on a rope. Submit a drawing of your design and a force analysis to show how it operates. Estimate the material cost required for its construction. Assume rope costs $1.25/m and a pulley costs $3.50.

8–2D Design of a Device for Lifting Stainless-Steel Pipes

Stainless-steel pipes are stacked vertically in a manufacturing plant and are to be moved by an overhead crane from one point to another. The pipes have inner diameters ranging from 100 mm $\leq d \leq$ 250 mm and the maximum mass of any pipe is 500 kg. Design a device that can be connected to the hook and used to lift each pipe. The device should be made of structural steel and should be able to grip the pipe only at its inside surface, since the outside surface is required not to be scratched or damaged. Assume the smallest coefficient of static friction between the two steels is $\mu_s = 0.25$. Submit a scaled drawing of your device, along with a brief explanation of how it works based on a force analysis.

8–3D Design of a Tool Used to Turn Plastic Pipe

PVC plastic is often used for sewer pipe. If the outer diameter of any pipe ranges from 4 in. $\leq d \leq$ 8 in., design a tool that can be used by a worker to turn the pipe when it is subjected to a maximum anticipated ground resistance of 80 lb · ft. The device is to be made of steel and should be designed so that it does not cut into the pipe and leave any significant marks on its surface. Assume a worker can apply a maximum (comfortable) force of 40 lb, and take the minimum coefficient of static friction between the PVC and the steel to be $\mu_s = 0.35$. Submit a scaled drawing of the device, and a brief paragraph to explain how it works based on a force analysis.

Prob. 8–1D

Prob. 8–2D

Prob. 8–3D

When a pressure vessel is designed, it is important to be able to determine the center of gravity of its component parts, calculate its volume and surface area, and reduce three-dimensional distributed loadings to their resultants. These topics are discussed in this chapter.

9 Center of Gravity and Centroid

CHAPTER OBJECTIVES

- To discuss the concept of the center of gravity, center of mass, and the centroid.

- To show how to determine the location of the center of gravity and centroid for a system of discrete particles and a body of arbitrary shape.

- To use the theorems of Pappus and Guldinus for finding the area and volume for a surface of revolution.

- To present a method for finding the resultant of a general distributed loading and show how it applies to finding the resultant of a fluid.

9.1 Center of Gravity and Center of Mass for a System of Particles

Center of Gravity. The *center of gravity* G is a point which locates the resultant weight of a system of particles. To show how to determine this point consider the system of n particles fixed within a region of space as shown in Fig. 9–1a. The weights of the particles comprise a system of parallel forces* which can be replaced by a single (equivalent) resultant weight having the defined point G of application. To find the $\bar{x}, \bar{y}, \bar{z}$ coordinates of G, we must use the principles outlined in Sec. 4.9.

---

*This is not true in the exact sense, since the weights are not parallel to each other; rather they are all *concurrent* at the earth's center. Furthermore, the acceleration of gravity g is actually different for each particle since it depends on the distance from the earth's center to the particle. For all practical purposes, however, both of these effects can generally be neglected.

(a)

(b)

Fig. 9–1

This requires that the resultant weight be equal to the total weight of all n particles; that is,

$$W_R = \Sigma W$$

The sum of the moments of the weights of all the particles about the x, y, and z axes is then equal to the moment of the resultant weight about these axes. Thus, to determine the $\bar{x}$ coordinate of G, we can sum moments about the y axis. This yields

$$\bar{x}W_R = \tilde{x}_1W_1 + \tilde{x}_2W_2 + \cdots + \tilde{x}_nW_n$$

Likewise, summing moments about the x axis, we can obtain the $\bar{y}$ coordinate; i.e.,

$$\bar{y}W_R = \tilde{y}_1W_1 + \tilde{y}_2W_2 + \cdots + \tilde{y}_nW_n$$

Although the weights do not produce a moment about the z axis, we can obtain the $\bar{z}$ coordinate of G by imagining the coordinate system, with the particles fixed in it, as being rotated 90° about the x (or y) axis, Fig. 9–1b. Summing moments about the x axis, we have

$$\bar{z}W_R = \tilde{z}_1W_1 + \tilde{z}_2W_2 + \cdots + \tilde{z}_nW_n$$

We can generalize these formulas, and write them symbolically in the form

$$\bar{x} = \frac{\Sigma \tilde{x}W}{\Sigma W} \qquad \bar{y} = \frac{\Sigma \tilde{y}W}{\Sigma W} \qquad \bar{z} = \frac{\Sigma \tilde{z}W}{\Sigma W} \qquad (9\text{–}1)$$

Here

$\bar{x}, \bar{y}, \bar{z}$ represent the coordinates of the center of gravity G of the system of particles.

$\tilde{x}, \tilde{y}, \tilde{z}$ represent the coordinates of each particle in the system.

ΣW is the resultant sum of the weights of all the particles in the system.

These equations are easily remembered if it is kept in mind that they simply represent a balance between the sum of the moments of the weights of each particle of the system and the moment of the *resultant* weight for the system.

Center of Mass.   To study problems concerning the motion of *matter* under the influence of force, i.e., dynamics, it is necessary to locate a point called the *center of mass*. Provided the acceleration due to gravity g for every particle is constant, then $W = mg$. Substituting into Eqs. 9–1 and canceling g from both the numerator and denominator yields

$$\bar{x} = \frac{\Sigma \tilde{x}m}{\Sigma m} \qquad \bar{y} = \frac{\Sigma \tilde{y}m}{\Sigma m} \qquad \bar{z} = \frac{\Sigma \tilde{z}m}{\Sigma m} \qquad (9\text{–}2)$$

By comparison, then, the location of the center of gravity *coincides* with that of the center of mass.* Recall, however, that particles have "weight" only when under the influence of a gravitational attraction, whereas the center of mass is independent of gravity. For example, it would be meaningless to define the center of gravity of a system of particles representing the planets of our solar system, while the center of mass of this system is important.

9.2 Center of Gravity, Center of Mass, and Centroid for a Body

Center of Gravity. A rigid body is composed of an infinite number of particles, and so if the principles used to determine Eqs. 9–1 are applied to the system of particles composing a rigid body, it becomes necessary to use integration rather than a discrete summation of the terms. Considering the arbitrary particle located at $(\tilde{x}, \tilde{y}, \tilde{z})$ and having a weight dW, Fig. 9–2, the resulting equations are

$$\bar{x} = \frac{\int \tilde{x}\, dW}{\int dW} \qquad \bar{y} = \frac{\int \tilde{y}\, dW}{\int dW} \qquad \bar{z} = \frac{\int \tilde{z}\, dW}{\int dW} \qquad (9\text{–}3)$$

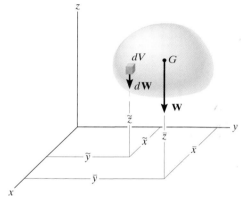

Fig. 9–2

In order to apply these equations, the differential weight dW must be expressed in terms of its associated volume dV. If γ represents the *specific weight* of the body, measured as a weight per unit volume, then $dW = \gamma\, dV$ and therefore

$$\bar{x} = \frac{\int_V \tilde{x}\gamma\, dV}{\int_V \gamma\, dV} \qquad \bar{y} = \frac{\int_V \tilde{y}\gamma\, dV}{\int_V \gamma\, dV} \qquad \bar{z} = \frac{\int_V \tilde{z}\gamma\, dV}{\int_V \gamma\, dV} \qquad (9\text{–}4)$$

Here integration must be performed throughout the entire volume of the body.

Center of Mass. The *density* ρ, or mass per unit volume, is related to γ by the equation $\gamma = \rho g$, where g is the acceleration due to gravity. Substituting this relationship into Eqs. 9–4 and canceling g from both the numerators and denominators yields similar equations (with ρ replacing γ) that can be used to determine the body's *center of mass.*

*This is true as long as the gravity field is assumed to have the same magnitude and direction everywhere. That assumption is appropriate for most engineering applications, since gravity does not vary appreciably between, for instance, the bottom and the top of a building.

Volume centroid

Fig. 9–3

Area centroid

Fig. 9–4

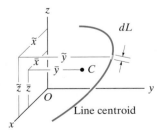

Line centroid

Fig. 9–5

Centroid.   The *centroid C* is a point which defines the *geometric center* of an object. Its location can be determined from formulas similar to those used to determine the body's center of gravity or center of mass. In particular, if the material composing a body is uniform or *homogeneous*, the *density or specific weight* will be *constant* throughout the body, and therefore this term will factor out of the integrals and *cancel* from both the numerators and denominators of Eqs. 9–4. The resulting formulas define the centroid of the body since they are independent of the body's weight and instead depend only on the body's geometry. Three specific cases will be considered.

Volume.   If an object is subdivided into volume elements dV, Fig. 9–3, the location of the centroid $C(\bar{x}, \bar{y}, \bar{z})$ for the volume of the object can be determined by computing the "moments" of the elements about each of the coordinate axes. The resulting formulas are

$$\bar{x} = \frac{\displaystyle\int_V \tilde{x}\, dV}{\displaystyle\int_V dV} \qquad \bar{y} = \frac{\displaystyle\int_V \tilde{y}\, dV}{\displaystyle\int_V dV} \qquad \bar{z} = \frac{\displaystyle\int_V \tilde{z}\, dV}{\displaystyle\int_V dV} \qquad (9\text{–}5)$$

Area.   In a similar manner, the centroid for the surface area of an object, such as a plate or shell, Fig. 9–4, can be found by subdividing the area into differential elements dA and computing the "moments" of these area elements about each of the coordinate axes, namely,

$$\bar{x} = \frac{\displaystyle\int_A \tilde{x}\, dA}{\displaystyle\int_A dA} \qquad \bar{y} = \frac{\displaystyle\int_A \tilde{y}\, dA}{\displaystyle\int_A dA} \qquad \bar{z} = \frac{\displaystyle\int_A \tilde{z}\, dA}{\displaystyle\int_A dA} \qquad (9\text{–}6)$$

Line.   If the geometry of the object, such as a thin rod or wire, takes the form of a line, Fig. 9–5, the balance of moments of the differential elements dL about each of the coordinate axes yields

$$\bar{x} = \frac{\displaystyle\int_L \tilde{x}\, dL}{\displaystyle\int_L dL} \qquad \bar{y} = \frac{\displaystyle\int_L \tilde{y}\, dL}{\displaystyle\int_L dL} \qquad \bar{z} = \frac{\displaystyle\int_L \tilde{z}\, dL}{\displaystyle\int_L dL} \qquad (9\text{–}7)$$

Remember that when applying Eqs. 9–4 through 9–7 it is best to choose a coordinate system that simplifies as much as possible the equation used to describe the object's boundary. For example, polar coordinates are generally appropriate for areas having circular boundaries. Also, the terms $\tilde{x}$, $\tilde{y}$, $\tilde{z}$ in the equations refer to the "moment arms" or coordinates of the *center of gravity or centroid for the differential element* used. If possible, this differential element should be chosen such that it has a differential size or thickness in only *one direction*. When this is done, only a single integration is required to cover the entire region.

Symmetry. The *centroids* of some shapes may be partially or completely specified by using conditions of *symmetry*. In cases where the shape has an axis of symmetry, the centroid of the shape will lie along that axis. For example, the centroid C for the curve shown in Fig. 9–6 must lie along the y axis since for every elemental length dL at a distance $+\tilde{x}$ to the right of the y axis there is an identical element at a distance $-\tilde{x}$ to the left. The total moment for all the elements about the axis of symmetry will therefore cancel; i.e., $\int \tilde{x}\, dL = 0$ (Eq. 9–7), so that $\bar{x} = 0$. In cases where a shape has two or three axes of symmetry, it follows that the centroid lies at the intersection of these axes, Fig. 9–7 and Fig. 9–8.

Integration must be used to determine the location of the center of gravity of this goal post due to the curvature of the supporting member.

Fig. 9–6

Fig. 9–7

Fig. 9–8

Important Points

- The centroid represents the geometric center of a body. This point coincides with the center of mass or the center of gravity only if the material composing the body is uniform or homogeneous.

- Formulas used to locate the center of gravity or the centroid simply represent a balance between the sum of moments of all the parts of the system and the moment of the "resultant" for the system.

- In some cases the centroid is located at a point that is not on the object, as in the case of a ring, where the centroid is at its center. Also, this point will lie on any axis of symmetry for the body.

Procedure for Analysis

The center of gravity or centroid of an object or shape can be determined by single integrations using the following procedure.

Differential Element.

- Select an appropriate coordinate system, specify the coordinate axes, and then choose a differential element for integration.
- For lines the element dL is represented as a differential line segment.
- For areas the element dA is generally a rectangle having a finite length and differential width.
- For volumes the element dV is either a circular disk having a finite radius and differential thickness, or a shell having a finite length and radius and a differential thickness.
- Locate the element at an arbitrary point (x, y, z) on the curve that defines the shape.

Size and Moment Arms.

- Express the length dL, area dA, or volume dV of the element in terms of the coordinates of the curve used to define the geometric shape.
- Determine the coordinates or moment arms $\tilde{x}, \tilde{y}, \tilde{z}$ for the centroid or center of gravity of the element.

Integrations.

- Substitute the formulations for $\tilde{x}, \tilde{y}, \tilde{z}$ and $dL, dA,$ or dV into the appropriate equations (Eqs. 9–4 through 9–7) and perform the integrations.*
- Express the function in the integrand in terms of the *same variable as the differential thickness of the element* in order to perform the integration.
- The limits of the integral are defined from the two extreme locations of the element's differential thickness, so that when the elements are "summed" or the integration performed, the entire region is covered.

*Formulas for integration are given in Appendix A.

EXAMPLE 9.1

Locate the centroid of the rod bent into the shape of a parabolic arc, shown in Fig. 9–9.

Fig. 9–9

SOLUTION

Differential Element.   The differential element is shown in Fig. 9–9. It is located on the curve at the *arbitrary point* (x, y).

Area and Moment Arms.   The differential length of the element dL can be expressed in terms of the differentials dx and dy by using the Pythagorean theorem.

$$dL = \sqrt{(dx)^2 + (dy)^2} = \sqrt{\left(\frac{dx}{dy}\right)^2 + 1}\, dy$$

Since $x = y^2$, then $dx/dy = 2y$. Therefore, expressing dL in terms of y and dy, we have

$$dL = \sqrt{(2y)^2 + 1}\, dy$$

The centroid is located at $\widetilde{x} = x$, $\widetilde{y} = y$.

Integrations.   Applying Eqs. 9–7 and integrating with respect to y using the formulas in Appendix A, we have

$$\bar{x} = \frac{\displaystyle\int_L \widetilde{x}\, dL}{\displaystyle\int_L dL} = \frac{\displaystyle\int_0^1 x\sqrt{4y^2 + 1}\, dy}{\displaystyle\int_0^1 \sqrt{4y^2 + 1}\, dy} = \frac{\displaystyle\int_0^1 y^2\sqrt{4y^2 + 1}\, dy}{\displaystyle\int_0^1 \sqrt{4y^2 + 1}\, dy}$$

$$= \frac{0.6063}{1.479} = 0.410 \text{ m} \qquad\qquad Ans.$$

$$\bar{y} = \frac{\displaystyle\int_L \widetilde{y}\, dL}{\displaystyle\int_L dL} = \frac{\displaystyle\int_0^1 y\sqrt{4y^2 + 1}\, dy}{\displaystyle\int_0^1 \sqrt{4y^2 + 1}\, dy} = \frac{0.8484}{1.479} = 0.574 \text{ m} \qquad Ans.$$

NOTE: These results seem reasonable when they are plotted on Fig. 9–9.

EXAMPLE 9.2

Locate the centroid of the circular wire segment shown in Fig. 9–10.

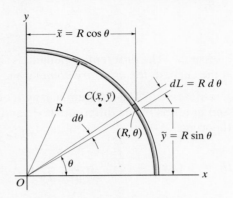

Fig. 9–10

SOLUTION

Polar coordinates will be used to solve this problem since the arc is circular.

Differential Element.   A differential circular arc is selected as shown in the figure. This element intersects the curve at (R, θ).

Length and Moment Arm.   The differential length of the element is $dL = R\,d\theta$, and its centroid is located at $\tilde{x} = R\cos\theta$ and $\tilde{y} = R\sin\theta$.

Integrations.   Applying Eqs. 9–7 and integrating with respect to θ, we obtain

$$\bar{x} = \frac{\displaystyle\int_L \tilde{x}\,dL}{\displaystyle\int_L dL} = \frac{\displaystyle\int_0^{\pi/2} (R\cos\theta)R\,d\theta}{\displaystyle\int_0^{\pi/2} R\,d\theta} = \frac{R^2\displaystyle\int_0^{\pi/2}\cos\theta\,d\theta}{R\displaystyle\int_0^{\pi/2} d\theta} = \frac{2R}{\pi} \quad Ans.$$

$$\bar{y} = \frac{\displaystyle\int_L \tilde{y}\,dL}{\displaystyle\int_L dL} = \frac{\displaystyle\int_0^{\pi/2} (R\sin\theta)R\,d\theta}{\displaystyle\int_0^{\pi/2} R\,d\theta} = \frac{R^2\displaystyle\int_0^{\pi/2}\sin\theta\,d\theta}{R\displaystyle\int_0^{\pi/2} d\theta} = \frac{2R}{\pi} \quad Ans.$$

NOTE: As expected, the two coordinates are numericallly the same due to the symmetry of the wire.

EXAMPLE 9.3

Determine the distance $\bar{y}$ from the x axis to the centroid of the area of the triangle shown in Fig. 9–11.

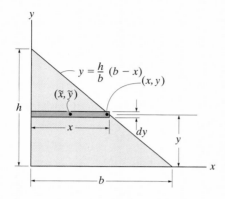

Fig. 9–11

SOLUTION

Differential Element. Consider a rectangular element having a thickness dy, which intersects the boundary at (x, y), Fig. 9–11.

Area and Moment Arms. The area of the element is $dA =$
$= x\,dy = \dfrac{b}{h}(h - y)\,dy$, and its centroid is located a distance $\tilde{y} = y$ from the x axis.

Integrations. Applying the second of Eqs. 9–6 and integrating with respect to y yields

$$\bar{y} = \frac{\displaystyle\int_A \tilde{y}\,dA}{\displaystyle\int_A dA} = \frac{\displaystyle\int_0^h y\frac{b}{h}(h - y)\,dy}{\displaystyle\int_0^h \frac{b}{h}(h - y)\,dy} = \frac{\frac{1}{6}bh^2}{\frac{1}{2}bh}$$

$$= \frac{h}{3} \qquad\qquad Ans.$$

EXAMPLE 9.4

Locate the centroid for the area of a quarter circle shown in Fig. 9–12a.

SOLUTION I

Differential Element.   Polar coordinates will be used, since the boundary is circular. We choose the element in the shape of a *triangle*, Fig. 9–12a. (Actually the shape is a circular sector; however, neglecting higher-order differentials, the element becomes triangular.) The element intersects the curve at point (R, θ).

Area and Moment Arms.   The area of the element is

$$dA = \tfrac{1}{2}(R)(R\,d\theta) = \frac{R^2}{2}d\theta$$

and using the results of Example 9.3, the centroid of the (triangular) element is located at $\widetilde{x} = \tfrac{2}{3}R\cos\theta$, $\widetilde{y} = \tfrac{2}{3}R\sin\theta$.

Integrations.   Applying Eqs. 9–6 and integrating with respect to θ, we obtain

$$\bar{x} = \frac{\displaystyle\int_A \widetilde{x}\,dA}{\displaystyle\int_A dA} = \frac{\displaystyle\int_0^{\pi/2}\left(\frac{2}{3}R\cos\theta\right)\frac{R^2}{2}d\theta}{\displaystyle\int_0^{\pi/2}\frac{R^2}{2}d\theta}$$

$$= \frac{\displaystyle\left(\frac{2}{3}R\right)\int_0^{\pi/2}\cos\theta\,d\theta}{\displaystyle\int_0^{\pi/2}d\theta} = \frac{4R}{3\pi} \qquad\qquad \textit{Ans.}$$

$$\bar{y} = \frac{\displaystyle\int_A \widetilde{y}\,dA}{\displaystyle\int_A dA} = \frac{\displaystyle\int_0^{\pi/2}\left(\frac{2}{3}R\sin\theta\right)\frac{R^2}{2}d\theta}{\displaystyle\int_0^{\pi/2}\frac{R^2}{2}d\theta}$$

$$= \frac{\displaystyle\left(\frac{2}{3}R\right)\int_0^{\pi/2}\sin\theta\,d\theta}{\displaystyle\int_0^{\pi/2}d\theta} = \frac{4R}{3\pi} \qquad\qquad \textit{Ans.}$$

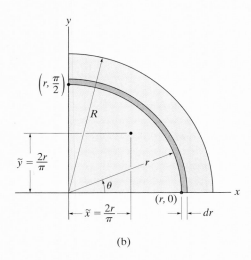

(a)   (b)

Fig. 9–12

SOLUTION II

Differential Element.   The differential element may be chosen in the form of a *circular arc* having a thickness dr as shown in Fig. 9–12b. The element intersects the axes at points $(r, 0)$ and $(r, \pi/2)$.

Area and Moment Arms.   The area of the element is $dA = (2\pi r/4)\, dr$. Since the centroid of a 90° circular arc was determined in Example 9.2, then for the element $\tilde{x} = 2r/\pi$, $\tilde{y} = 2r/\pi$.

Integrations.   Using Eqs. 9–6 and integrating with respect to r, we obtain

$$\overline{x} = \frac{\displaystyle\int_A \tilde{x}\, dA}{\displaystyle\int_A dA} = \frac{\displaystyle\int_0^R \frac{2r}{\pi}\left(\frac{2\pi r}{4}\right) dr}{\displaystyle\int_0^R \frac{2\pi r}{4} dr} = \frac{\displaystyle\int_0^R r^2\, dr}{\dfrac{\pi}{2}\displaystyle\int_0^R r\, dr} = \frac{4}{3}\frac{R}{\pi} \qquad Ans.$$

$$\overline{y} = \frac{\displaystyle\int_A \tilde{y}\, dA}{\displaystyle\int_A dA} = \frac{\displaystyle\int_0^R \frac{2r}{\pi}\left(\frac{2\pi r}{4}\right) dr}{\displaystyle\int_0^R \frac{2\pi r}{4} dr} = \frac{\displaystyle\int_0^R r^2\, dr}{\dfrac{\pi}{2}\displaystyle\int_0^R r\, dr} = \frac{4}{3}\frac{R}{\pi} \qquad Ans.$$

NOTE: The two coordinates are numerically the same because of the symmetry of the area.

EXAMPLE 9.5

(a)

(b)

Fig. 9–13

Locate the centroid of the area shown in Fig. 9–13a.

SOLUTION I

Differential Element.   A differential element of thickness dx is shown in Fig. 9–13a. The element intersects the curve at the *arbitrary point* (x, y), and so it has a height y.

Area and Moment Arms.   The area of the element is $dA = y\,dx$, and its centroid is located at $\tilde{x} = x$, $\tilde{y} = y/2$.

Integrations.   Applying Eqs. 9–6 and integrating with respect to x yields

$$\bar{x} = \frac{\displaystyle\int_A \tilde{x}\,dA}{\displaystyle\int_A dA} = \frac{\displaystyle\int_0^1 xy\,dx}{\displaystyle\int_0^1 y\,dx} = \frac{\displaystyle\int_0^1 x^3\,dx}{\displaystyle\int_0^1 x^2\,dx} = \frac{0.250}{0.333} = 0.75\text{ m} \qquad Ans.$$

$$\bar{y} = \frac{\displaystyle\int_A \tilde{y}\,dA}{\displaystyle\int_A dA} = \frac{\displaystyle\int_0^1 (y/2)y\,dx}{\displaystyle\int_0^1 y\,dx} = \frac{\displaystyle\int_0^1 (x^2/2)x^2\,dx}{\displaystyle\int_0^1 x^2\,dx} = \frac{0.100}{0.333} = 0.3\text{ m }Ans.$$

SOLUTION II

Differential Element.   The differential element of thickness dy is shown in Fig. 9–13b. The element intersects the curve at the *arbitrary point* (x, y), and so it has a length $(1 - x)$.

Area and Moment Arms.   The area of the element is $dA = (1 - x)\,dy$, and its centroid is located at

$$\tilde{x} = x + \left(\frac{1 - x}{2}\right) = \frac{1 + x}{2}, \quad \tilde{y} = y$$

Integrations.   Applying Eqs. 9–6 and integrating with respect to y, we obtain

$$\tilde{x} = \frac{\displaystyle\int_A \tilde{x}\,dA}{\displaystyle\int_A dA} = \frac{\displaystyle\int_0^1 [(1 + x)/2](1 - x)\,dy}{\displaystyle\int_0^1 (1 - x)\,dy} = \frac{\dfrac{1}{2}\displaystyle\int_0^1 (1 - y)\,dy}{\displaystyle\int_0^1 (1 - \sqrt{y})\,dy} = \frac{0.250}{0.333} = 0.75\text{ m }\ Ans.$$

$$\bar{y} = \frac{\displaystyle\int_A \tilde{y}\,dA}{\displaystyle\int_A dA} = \frac{\displaystyle\int_0^1 y(1 - x)\,dy}{\displaystyle\int_0^1 (1 - x)\,dy} = \frac{\displaystyle\int_0^1 (y - y^{3/2})\,dy}{\displaystyle\int_0^1 (1 - \sqrt{y})\,dy} = \frac{0.100}{0.333} = 0.3\text{ m }\ Ans.$$

NOTE: Plot these results and notice that they seem reasonable. Also, by comparison, elements of thickness dx offer a simpler solution.

EXAMPLE 9.6

Locate the $\bar{x}$ centroid of the shaded area bounded by the two curves $y = x$ and $y = x^2$, Fig. 9–14.

SOLUTION I

Differential Element.   A differential element of thickness dx is shown in Fig. 9–14a. The element intersects the curves at *arbitrary points* (x, y_1) and (x, y_2), and so it has a height $(y_2 - y_1)$.

Area and Moment Arm.   The area of the element is $dA = (y_2 - y_1)\,dx$, and its centroid is located at $\widetilde{x} = x$.

Integration.   Applying Eq. 9–6, we have

$$\bar{x} = \frac{\displaystyle\int_A \widetilde{x}\,dA}{\displaystyle\int_A dA} = \frac{\displaystyle\int_0^1 x(y_2 - y_1)\,dx}{\displaystyle\int_0^1 (y_2 - y_1)\,dx} = \frac{\displaystyle\int_0^1 x(x - x^2)\,dx}{\displaystyle\int_0^1 (x - x^2)\,dx} = \frac{\frac{1}{12}}{\frac{1}{6}} = 0.5 \text{ ft } \textit{Ans.}$$

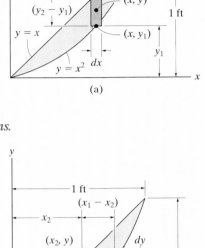

(a)

SOLUTION II

Differential Element.   A differential element having a thickness dy is shown in Fig. 9–14b. The element intersects the curves at the *arbitrary points* (x_2, y) and (x_1, y), and so it has a length $(x_1 - x_2)$.

Area and Moment Arm.   The area of the element is $dA = (x_1 - x_2)\,dy$, and its centroid is located at

$$\widetilde{x} = x_2 + \frac{x_1 - x_2}{2} = \frac{x_1 + x_2}{2}$$

Integration.   Applying Eq. 9–6, we have

(b)

Fig. 9–14

$$\bar{x} = \frac{\displaystyle\int_A \widetilde{x}\,dA}{\displaystyle\int_A dA} = \frac{\displaystyle\int_0^1 [(x_1 + x_2)/2](x_1 - x_2)\,dy}{\displaystyle\int_0^1 (x_1 - x_2)\,dy} = \frac{\displaystyle\int_0^1 [(\sqrt{y} + y)/2](\sqrt{y} - y)\,dy}{\displaystyle\int_0^1 (\sqrt{y} - y)\,dy}$$

$$= \frac{\dfrac{1}{2}\displaystyle\int_0^1 (y - y^2)\,dy}{\displaystyle\int_0^1 (\sqrt{y} - y)\,dy} = \frac{\frac{1}{12}}{\frac{1}{6}} = 0.5 \text{ ft} \qquad\qquad \textit{Ans.}$$

EXAMPLE 9.7

Locate the $\bar{y}$ centroid for the paraboloid of revolution, which is generated by revolving the shaded area shown in Fig. 9–15a about the y axis.

(a)

Fig. 9–15

SOLUTION I

Differential Element. An element having the shape of a *thin disk* is chosen, Fig. 9–15a. This element has a thickness dy. In this "disk" method of analysis, the element of planar area, dA, is always taken *perpendicular* to the axis of revolution. Here the element intersects the generating curve at the *arbitrary point* $(0, y, z)$, and so its radius is $r = z$.

Area and Moment Arm. The volume of the element is $dV = (\pi z^2)\, dy$, and its centroid is located at $\tilde{y} = y$.

Integration. Applying the second of Eqs. 9–5 and integrating with respect to y yields

$$\bar{y} = \frac{\displaystyle\int_V \tilde{y}\, dV}{\displaystyle\int_V dV} = \frac{\displaystyle\int_0^{100} y(\pi z^2)\, dy}{\displaystyle\int_0^{100} (\pi z^2)\, dy} = \frac{100\pi \displaystyle\int_0^{100} y^2\, dy}{100\pi \displaystyle\int_0^{100} y\, dy} = 66.7\ \text{mm}\quad Ans.$$

(b)

SOLUTION II

Differential Element.   As shown in Fig. 9–15b, the volume element can be chosen in the form of a *thin cylindrical shell*, where the shell's thickness is dz. In this "shell" method of analysis, the element of planar area, dA, is always taken *parallel* to the axis of revolution. Here the element intersects the generating curve at point $(0, y, z)$, and so the radius of the shell is $r = z$.

Area and Moment Arm.   The volume of the element is $dV = 2\pi r\, dA = 2\pi z(100 - y)\, dz$, and its centroid is located at $\tilde{y} = y + (100 - y)/2 = (100 + y)/2$.

Integrations.   Applying the second of Eqs. 9–5 and integrating with respect to z yields

$$\bar{y} = \frac{\displaystyle\int_V \tilde{y}\, dV}{\displaystyle\int_V dV} = \frac{\displaystyle\int_0^{100} [(100 + y)/2]\, 2\pi z(100 - y)\, dz}{\displaystyle\int_0^{100} 2\pi z(100 - y)\, dz}$$

$$= \frac{\pi \displaystyle\int_0^{100} z(10^4 - 10^{-4}z^4)\, dz}{2\pi \displaystyle\int_0^{100} z(100 - 10^{-2}z^2)\, dz} = 66.7 \text{ mm} \qquad Ans.$$

NOTE: By comparison, choosing a disk element offers a simpler solution.

EXAMPLE 9.8

Determine the location of the center of mass of the cylinder shown in Fig. 9–16a if its density varies directly with the distance from its base, i.e., $\rho = 200z$ kg/m^3.

(a)

SOLUTION

For reasons of material symmetry,

$$\overline{x} = \overline{y} = 0 \qquad \textit{Ans.}$$

Differential Element. A disk element of radius 0.5 m and thickness dz is chosen for integration, Fig. 9–16a, since the *density of the entire element is constant* for a given value of z. The element is located along the z axis at the *arbitrary point* $(0, 0, z)$.

Volume and Moment Arm. The volume of the element is $dV = \pi(0.5)^2 \, dz$, and its centroid is located at $\widetilde{z} = z$.

Integrations. Using an equation similar to the third of Eqs. 9–4 and integrating with respect to z, noting that $\rho = 200z$, we have

$$\overline{z} = \frac{\displaystyle\int_v \widetilde{z}\rho \, dV}{\displaystyle\int_v \rho \, dV} = \frac{\displaystyle\int_0^1 z(200z)\pi(0.5)^2 \, dz}{\displaystyle\int_0^1 (200z)\pi(0.5)^2 \, dz}$$

$$= \frac{\displaystyle\int_0^1 z^2 \, dz}{\displaystyle\int_0^1 z \, dz} = 0.667 \text{ m} \qquad \textit{Ans.}$$

(b)

Fig. 9–16

NOTE: It is not possible to use a shell element for integration such as shown in Fig. 9–16b since the density of the material composing the shell would *vary* along the shell's height and hence the location of $\widetilde{z}$ for the element cannot be specified.

PROBLEMS

9–1. Locate the center of mass of the homogeneous rod bent in the form of a parabola.

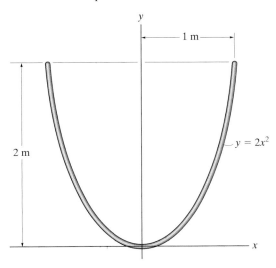

Prob. 9–1

9–2. Locate the center of gravity $\bar{x}$ of the homogeneous rod. If the rod has a weight per unit length of 0.5 lb/ft, determine the vertical reaction at A and the x and y components of reaction at the pin B.

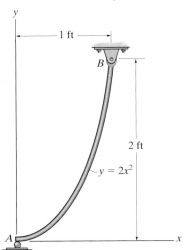

Prob. 9–2

9–3. Locate the center of mass of the homogeneous rod bent into the shape of a circular arc.

Prob. 9–3

*9–4. Locate the center of gravity $\bar{x}$ of the homogeneous rod bent in the form of a semicircular arc. The rod has a weight per unit length of 0.5 lb/ft. Also, determine the horizontal reaction at the smooth support B and the x and y components of reaction at the pin A.

Prob. 9–4

9–5. Determine the distance $\bar{x}$ to the center of gravity of the homogeneous rod bent into the parabolic shape. If the rod has a weight per unit length of 0.5 lb/ft, determine the reactions at the fixed support O.

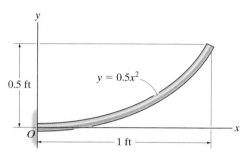

$y = 0.5x^2$

0.5 ft

1 ft

Prob. 9–5

9–6. Determine the distance $\bar{y}$ to the center of gravity of the homogeneous rod bent into the parabolic shape.

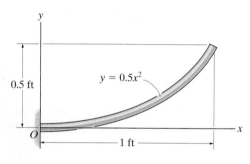

$y = 0.5x^2$

0.5 ft

1 ft

Prob. 9–6

9–7. Locate the centroid of the parabolic area.

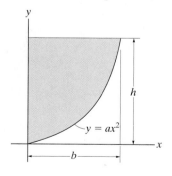

h

$y = ax^2$

b

Prob. 9–7

*9–8. Locate the centroid $\bar{y}$ of the shaded area.

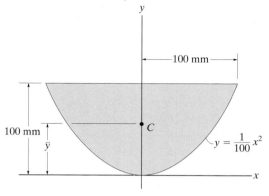

100 mm

100 mm

C

$y = \dfrac{1}{100}x^2$

$\bar{y}$

Prob. 9–8

9–9. Locate the centroid of the shaded area.

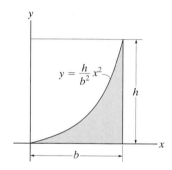

$y = \dfrac{h}{b^2}x^2$

h

b

Prob. 9–9

9–10. Determine the location $(\bar{x}, \bar{y})$ of the centroid of the triangular area.

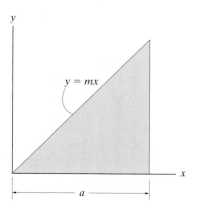

$y = mx$

a

Prob. 9–10

9–11. Determine the location $(\bar{x}, \bar{y})$ of the center of gravity of the quartercircular plate. Also determine the force in each of the supporting wires. The plate has a weight per unit area of 5 lb/ft^2.

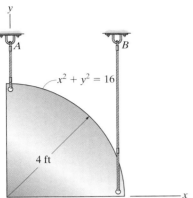

Prob. 9–11

9–14. Locate the centroid $(\bar{x}, \bar{y})$ of the exparabolic segment of area.

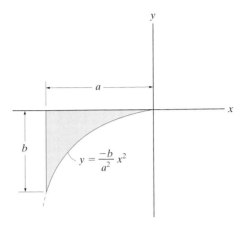

Prob. 9–14

*9–12. Locate the centroid of the shaded area.

Prob. 9–12

9–15. Locate the centroid of the shaded area.

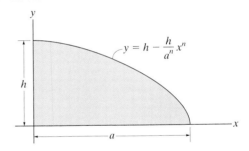

Prob. 9–15

9–13. Locate the center of gravity of the homogeneous cantilever beam and determine the reactions at the fixed support. The material has a density of 8 Mg/m^3.

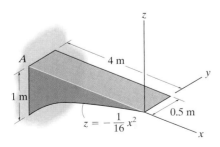

Prob. 9–13

*9–16. Locate the centroid of the shaded area bounded by the parabola and the line $y = a$.

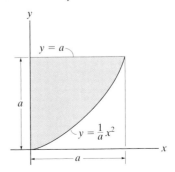

Prob. 9–16

9–17.   Locate the centroid of the quarter elliptical area.

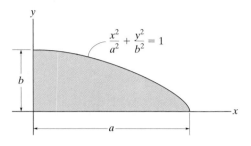

$$\frac{x^2}{a^2} + \frac{y^2}{b^2} = 1$$

Prob. 9–17

9–18.   Locate the centroid $\bar{x}$ of the triangular area.

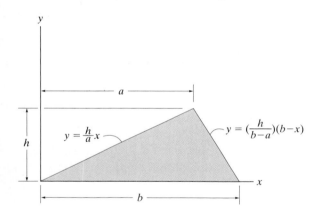

$$y = \frac{h}{a}x$$

$$y = \left(\frac{h}{b-a}\right)(b-x)$$

Prob. 9–18

9–19.   Locate the centroid of the shaded area.

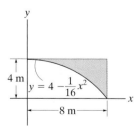

$$y = 4 - \frac{1}{16}x^2$$

Prob. 9–19

*■**9–20.**   Locate the centroid $\bar{x}$ of the shaded area. Solve the problem by evaluating the integrals using Simpson's rule.

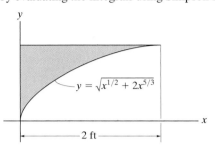

$$y = \sqrt{x^{1/2} + 2x^{5/3}}$$

Prob. 9–20

■**9–21.**   Locate the centroid $\bar{y}$ of the shaded area. Solve the problem by evaluating the integrals using Simpson's rule.

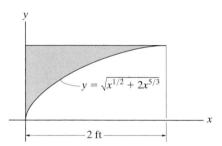

$$y = \sqrt{x^{1/2} + 2x^{5/3}}$$

Prob. 9–21

9–22.   The steel plate is 0.3 m thick and has a density of 7850 kg/m³. Determine the location of its center of mass. Also compute the reactions at the pin and roller support.

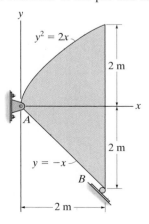

$$y^2 = 2x$$

$$y = -x$$

Prob. 9–22

9–23.   Locate the centroid $\bar{x}$ of the shaded area.

*9–24.   Locate the centroid $\bar{y}$ of the shaded area.

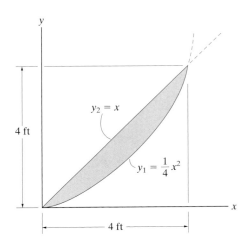

Probs. 9–23/24

9–25.   Locate the centroid $\bar{x}$ of the shaded area.

9–26.   Locate the centroid $\bar{y}$ of the shaded area.

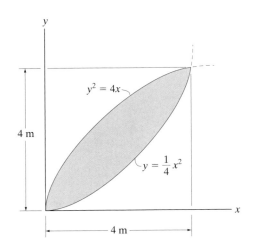

Probs. 9–25/26

9–27.   Locate the centroid $\bar{x}$ of the shaded area.

*9–28.   Locate the centroid $\bar{y}$ of the shaded area.

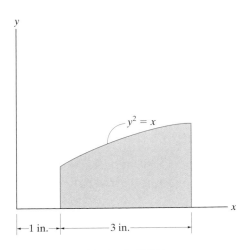

Probs. 9–27/28

9–29.   Locate the centroid $\bar{x}$ of the shaded area.

9–30.   Locate the centroid $\bar{y}$ of the shaded area.

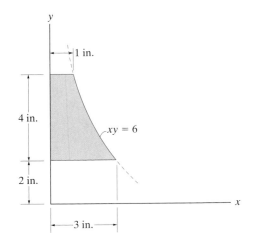

Probs. 9–29/30

9–31. Determine the location $\bar{r}$ of the centroid C of the cardioid, $r = a(1 - \cos\theta)$.

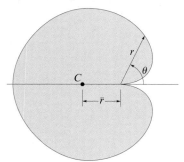

Prob. 9–31

*9–32. Locate the centroid of the ellipsoid of revolution.

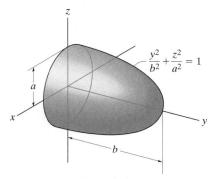

Prob. 9–32

9–33. Locate the centroid $\bar{z}$ of the very thin conical shell. *Hint:* Use thin ring elements having a center at $(0, 0, z)$, radius y, and width $dL = \sqrt{(dy)^2 + (dz)^2}$.

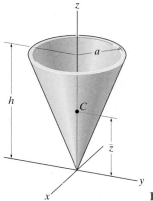

Prob. 9–33

9–34. Locate the centroid $\bar{z}$ of the volume segment.

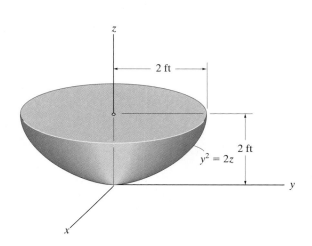

Prob. 9–34

9–35. Locate the centroid of the solid.

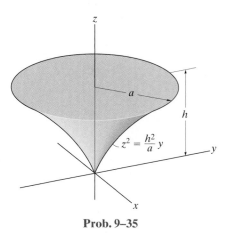

Prob. 9–35

*9–36.   Locate the centroid of the quarter-cone.

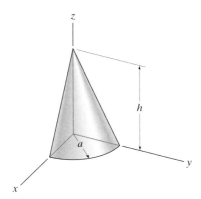

Prob. 9–36

9–37.   Locate the center of mass $\bar{x}$ of the hemisphere. The density of the material varies linearly from zero at the origin O to ρ_0 at the surface. *Hint:* Choose a hemispherical shell element for integration.

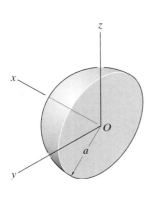

Prob. 9–37

9–38.   Locate the centroid $\bar{z}$ of the right-elliptical cone.

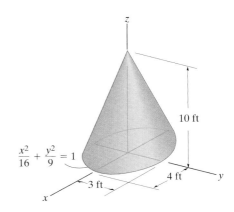

$$\frac{x^2}{16} + \frac{y^2}{9} = 1$$

Prob. 9–38

9–39.   Locate the center of gravity $\bar{z}$ of the frustum of the paraboloid. The material is homogeneous.

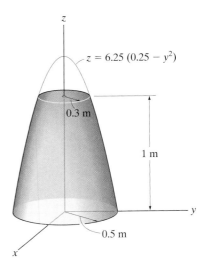

$z = 6.25\,(0.25 - y^2)$

Prob. 9–39

*9–40. Locate the center of gravity $\bar{y}$ of the volume. The material is homogeneous.

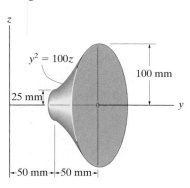

Prob. 9–40

9–41. Locate the center of gravity for the homogeneous half-cone.

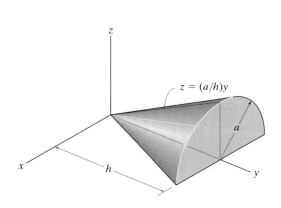

Prob. 9–41

9–42. Locate the centroid $\bar{z}$ of the spherical segment.

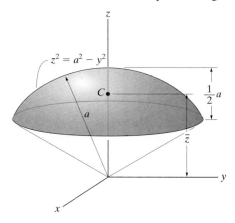

Prob. 9–42

9–43. Determine the location $\bar{z}$ of the centroid for the tetrahedron. *Hint:* Use a triangular "plate" element parallel to the x–y plane and of thickness dz.

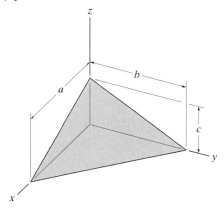

Prob. 9–43

9.3 Composite Bodies

A *composite body* consists of a series of connected "simpler" shaped bodies, which may be rectangular, triangular, semicircular, etc. Such a body can often be sectioned or divided into its composite parts and, provided the *weight* and location of the center of gravity of each of these parts are known, we can eliminate the need for integration to determine the center of gravity for the entire body. The method for doing this requires treating each composite part like a particle and following the procedure outlined in Sec. 9.1. Formulas analogous to Eqs. 9–1 result since we must account for a finite number of weights. Rewriting these formulas, we have

$$\bar{x} = \frac{\Sigma \tilde{x} W}{\Sigma W} \qquad \bar{y} = \frac{\Sigma \tilde{y} W}{\Sigma W} \qquad \bar{z} = \frac{\Sigma \tilde{z} W}{\Sigma W} \qquad (9\text{--}8)$$

Here

$\bar{x}, \bar{y}, \bar{z}$ represent the coordinates of the center of gravity G of the composite body.

$\tilde{x}, \tilde{y}, \tilde{z}$ represent the coordinates of the center of gravity of each composite part of the body.

ΣW is the sum of the weights of all the composite parts of the body, or simply the total weight of the body.

When the body has a *constant density or specific weight*, the center of gravity *coincides* with the centroid of the body. The centroid for composite lines, areas, and volumes can be found using relations analogous to Eqs. 9–8; however, the W's are replaced by L's, A's, and V's, respectively. Centroids for common shapes of lines, areas, shells, and volumes that often make up a composite body are given in the table on the inside back cover.

In order to determine the force required to tip over this concrete barrier it is first necessary to determine the location of its center of gravity G. Due to symmetry, G will lie on the axis of symmetry.

Procedure for Analysis

The location of the center of gravity of a body or the centroid of a composite geometrical object represented by a line, area, or volume can be determined using the following procedure.

Composite Parts.
- Using a sketch, divide the body or object into a finite number of composite parts that have simpler shapes.
- If a composite part has a *hole*, or a geometric region having no material, then consider the composite part without the hole and consider the hole as an *additional* composite part having *negative* weight or size.

Moment Arms.
- Establish the coordinate axes on the sketch and determine the coordinates $\tilde{x}, \tilde{y}, \tilde{z}$ of the center of gravity or centroid of each part.

Summations.
- Determine $\bar{x}, \bar{y}, \bar{z}$ by applying the center of gravity equations, Eqs. 9–8, or the analogous centroid equations.
- If an object is *symmetrical* about an axis, the centroid of the object lies on this axis.

If desired, the calculations can be arranged in tabular form, as indicated in the following three examples.

EXAMPLE 9.9

Locate the centroid of the wire shown in Fig. 9–17a.

SOLUTION

Composite Parts.   The wire is divided into three segments as shown in Fig. 9–17b.

Moment Arms.   The location of the centroid for each piece is determined and indicated in the figure. In particular, the centroid of segment ① is determined either by integration or by using the table on the inside back cover.

Summations.   The calculations are tabulated as follows:

Segment	L (mm)	$\tilde{x}$ (mm)	$\tilde{y}$ (mm)	$\tilde{z}$ (mm)	$\tilde{x}L$ (mm^2)	$\tilde{y}L$ (mm^2)	$\tilde{z}L$ (mm^2)
1	$\pi(60) = 188.5$	60	-38.2	0	11 310	-7200	0
2	40	0	20	0	0	800	0
3	20	0	40	-10	0	800	-200
	$\Sigma L = 248.5$				$\Sigma \tilde{x}L = 11\,310$	$\Sigma \tilde{y}L = -5600$	$\Sigma \tilde{z}L = -200$

Thus,

$$\bar{x} = \frac{\Sigma \tilde{x}L}{\Sigma L} = \frac{11310}{248.5} = 45.5 \text{ mm} \qquad \textit{Ans.}$$

$$\bar{y} = \frac{\Sigma \tilde{y}L}{\Sigma L} = \frac{-5600}{248.5} = -22.5 \text{ mm} \qquad \textit{Ans.}$$

$$\bar{z} = \frac{\Sigma \tilde{z}L}{\Sigma L} = \frac{-200}{248.5} = -0.805 \text{ mm} \qquad \textit{Ans.}$$

(a)

(b)

Fig. 9–17

EXAMPLE 9.10

Locate the centroid of the plate area shown in Fig. 9–18a.

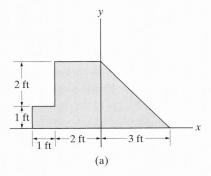

(a)

Fig. 9–18

SOLUTION

Composite Parts.   The plate is divided into three segments as shown in Fig. 9–18b. Here the area of the small rectangle ③ is considered "negative" since it must be subtracted from the larger one ②.

Moment Arms.   The centroid of each segment is located as indicated in the figure. Note that the $\tilde{x}$ coordinates of ② and ③ are *negative*.

Summations.   Taking the data from Fig. 9–18b, the calculations are tabulated as follows:

Segment	A (ft^2)	$\tilde{x}$ (ft)	$\tilde{y}$ (ft)	$\tilde{x}A$ (ft^3)	$\tilde{y}A$ (ft^3)
1	$\frac{1}{2}(3)(3) = 4.5$	1	1	4.5	4.5
2	$(3)(3) = 9$	−1.5	1.5	−13.5	13.5
3	$-(2)(1) = -2$	−2.5	2	5	−4
	$\Sigma A = 11.5$			$\Sigma\tilde{x}A = -4$	$\Sigma\tilde{y}A = 14$

(b)

Thus,

$$\bar{x} = \frac{\Sigma\tilde{x}A}{\Sigma A} = \frac{-4}{11.5} = -0.348 \text{ ft} \qquad Ans.$$

$$\bar{y} = \frac{\Sigma\tilde{y}A}{\Sigma A} = \frac{14}{11.5} = 1.22 \text{ ft} \qquad Ans.$$

NOTE: If these results are plotted in Fig. 9–18, the location of point C seems reasonable.

EXAMPLE 9.11

25 mm

100 mm

50 mm

50 mm

(a)

Fig. 9–19

Locate the center of mass of the composite assembly shown in Fig. 9–19a. The conical frustum has a density of $\rho_c = 8 \text{ Mg/m}^3$, and the hemisphere has a density of $\rho_h = 4 \text{ Mg/m}^3$. There is a 25-mm-radius cylindrical hole in the center.

SOLUTION

Composite Parts. The assembly can be thought of as consisting of four segments as shown in Fig. 9–19b. For the calculations, ③ and ④ must be considered as "negative" volumes in order that the four segments, when added together, yield the total composite shape shown in Fig. 9–19a.

Moment Arm. Using the table on the inside back cover, the computations for the centroid $\tilde{z}$ of each piece are shown in the figure.

Summations. Because of *symmetry*, note that

$$\overline{x} = \overline{y} = 0 \qquad\qquad Ans.$$

Since $W = mg$ and g is constant, the third of Eqs. 9–8 becomes $\overline{z} = \Sigma\tilde{z}m/\Sigma m$. The mass of each piece can be computed from $m = \rho V$ and used for the calculations. Also, $1 \text{ Mg/m}^3 = 10^{-6} \text{ kg/mm}^3$, so that

Segment	m (kg)	$\tilde{z}$ (mm)	$\tilde{z}m$ (kg · mm)
1	$8(10^{-6})\left(\frac{1}{3}\right)\pi(50)^2(200) = 4.189$	50	209.440
2	$4(10^{-6})\left(\frac{2}{3}\right)\pi(50)^3 = 1.047$	-18.75	-19.635
3	$-8(10^{-6})\left(\frac{1}{3}\right)\pi(25)^2(100) = -0.524$	$100 + 25 = 125$	-65.450
4	$-8(10^{-6})\pi(25)^2(100) = -1.571$	50	-78.540
	$\Sigma m = 3.141$		$\Sigma\tilde{z}m = 45.815$

Thus, $\tilde{z} = \dfrac{\Sigma\tilde{z}m}{\Sigma m} = \dfrac{45.815}{3.141} = 14.6 \text{ mm}$ *Ans.*

200 mm

$\dfrac{200 \text{ mm}}{4} = 50 \text{ mm}$

50 mm

①

50 mm

$-\dfrac{3}{8}(50) = -18.75 \text{ mm}$

②

$\dfrac{100 \text{ mm}}{4} = 25 \text{ mm}$

③

100 mm

25 mm

100 mm

25 mm

④

50 mm

50 mm

(b)

PROBLEMS

*9–44. Determine the location (x, y) of the 7-kg particle so that the three particles, which lie in the x–y plane, have a center of mass located at the origin O.

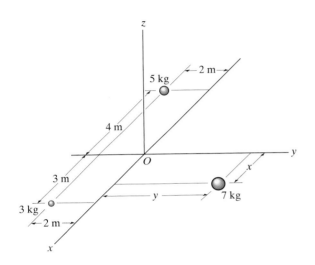

Prob. 9–44

9–45. Locate the center of gravity $(\bar{x}, \bar{y}, \bar{z})$ of the four particles.

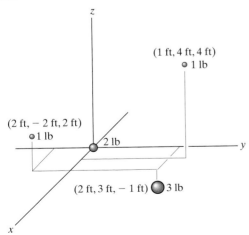

Prob. 9–45

9–46. A rack is made from roll-formed sheet steel and has the cross section shown. Determine the location $(\bar{x}, \bar{y})$ of the centroid of the cross section. The dimensions are indicated at the center thickness of each segment.

Prob. 9–46

9–47. The steel and aluminum plate assembly is bolted together and fastened to the wall. Each plate has a constant width in the z direction of 200 mm and thickness of 20 mm. If the density of A and B is $\rho_s = 7.85$ Mg/m^3, and for C, $\rho_{al} = 2.71$ Mg/m^3, determine the location $\bar{x}$ of the center of mass. Neglect the size of the bolts.

Prob. 9–47

*9–48. The truss is made from five members, each having a length of 4 m and a mass of 7 kg/m. If the mass of the gusset plates at the joints and the thickness of the members can be neglected, determine the distance d to where the hoisting cable must be attached, so that the truss does not tip (rotate) when it is lifted.

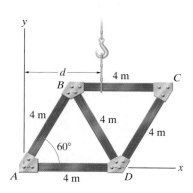

Prob. 9–48

9–49. Locate the center of gravity $(\bar{x}, \bar{y}, \bar{z})$ of the homogeneous wire.

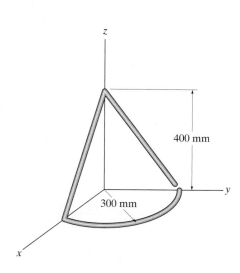

Prob. 9–49

9–50. Determine the location $(\bar{x}, \bar{y})$ of the center of gravity of the homogeneous wire bent in the form of a triangle. Neglect any slight bends at the corners. If the wire is suspended using a thread T attached to it at C, determine the angle of tilt AB makes with the horizontal when the wire is in equilibrium.

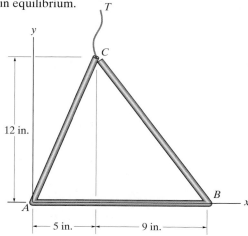

Prob. 9–50

9–51. The three members of the frame each have a weight per unit length of 4 kg/m. Locate the position $(\bar{x}, \bar{y})$ of the center of gravity. Neglect the size of the pins at the joints and the thickness of the members. Also, calculate the reactions at the fixed support A.

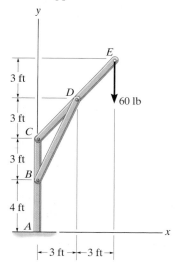

Prob. 9–51

*9–52. Locate the center of gravity $G(\bar{x}, \bar{y})$ of the streetlight. Neglect the thickness of each segment. The mass per unit length of each segment is as follows: $\rho_{AB} = 12$ kg/m, $\rho_{BC} = 8$ kg/m, $\rho_{CD} = 5$ kg/m, and $\rho_{DE} = 2$ kg/m.

Prob. 9–52

9–53. Determine the location $\bar{y}$ of the centroid of the beam's cross-sectional area. Neglect the size of the corner welds at A and B for the calculation.

Prob. 9–53

9–54. The gravity wall is made of concrete. Determine the location $(\bar{x}, \bar{y})$ of the center of gravity G for the wall.

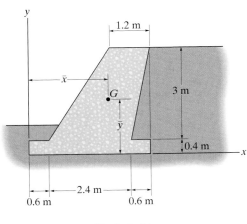

Prob. 9–54

9–55. Locate the centroid $(\bar{x}, \bar{y})$ of the shaded area.

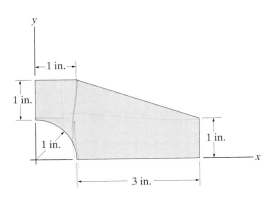

Prob. 9–55

*9–56.   Locate the centroid $(\bar{x}, \bar{y})$ of the shaded area.

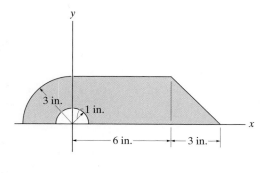

Prob. 9–56

9–58.   Determine the location $(\bar{x}, \bar{y})$ of the centroid C of the area.

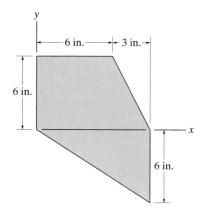

Prob. 9–58

9–57.   Determine the location $\bar{y}$ of the centroidal axis $\bar{x}\,\bar{x}$ of the beam's cross-sectional area. Neglect the size of the corner welds at A and B for the calculation.

9–59.   Determine the location $\bar{y}$ of the centroid C for a beam having the cross-sectional area shown. The beam is symmetric with respect to the y axis.

Prob. 9–57

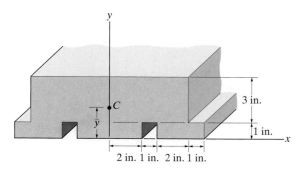

Prob. 9–59

*9–60. The wooden table is made from a square board having a weight of 15 lb. Each of the legs weighs 2 lb and is 3 ft long. Determine how high its center of gravity is from the floor. Also, what is the angle, measured from the horizontal, through which its top surface can be tilted on two of its legs before it begins to overturn? Neglect the thickness of each leg.

Prob. 9–60

9–61. Locate the centroid $\bar{y}$ for the beam's cross-sectional area.

Prob. 9–61

9–62. Determine the location $\bar{x}$ of the centroid C of the shaded area which is part of a circle having a radius r.

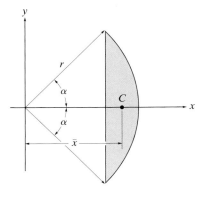

Prob. 9–62

9–63. Locate the centroid $\bar{y}$ for the strut's cross-sectional area.

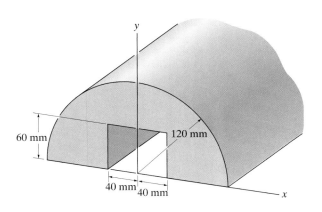

Prob. 9–63

*9–64. The "New Jersey" concrete barrier is commonly used during highway construction. Determine the location $\bar{y}$ of its centroid.

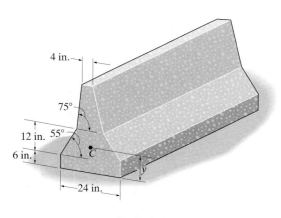

Prob. 9–64

9–65. The composite plate is made from both steel (A) and brass (B) segments. Determine the mass and location $(\bar{x}, \bar{y}, \bar{z})$ of its mass center G. Take $\rho_{st} = 7.85 \text{ Mg/m}^3$ and $\rho_{br} = 8.74 \text{ Mg/m}^3$.

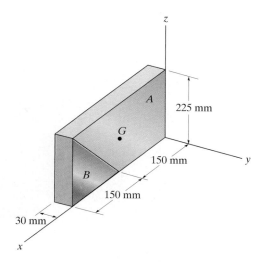

Prob. 9–65

9–66. Locate the centroid $\bar{y}$ of the concrete beam having the tapered cross section shown.

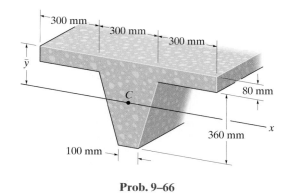

Prob. 9–66

9–67. The anatomical center of gravity G of a person can be determined by using a scale and a rigid board having a uniform weight W_1 and length l. With the person's weight W known, the person lies down on the board and the scale reading P is recorded. From this show how to calculate the location $\bar{x}$. Discuss the best place l_1 for the smooth support at B in order to improve the accuracy of this experiment.

Prob. 9–67

*9–68. The tank and compressor have a mass of 15 kg and mass center at G_T, and the motor has a mass of 70 kg and a mass center at G_M. Determine the angle of tilt, θ, of the tank so that the unit will be on the verge of tipping over.

Prob. 9–68

9–69. Determine the distance h to which a 100-mm-diameter hole must be bored into the base of the cone so that the center of mass of the resulting shape is located at $\bar{z} = 115$ mm. The material has a density of 8 Mg/m³.

9–70. Determine the distance $\bar{z}$ to the centroid of the shape which consists of a cone with a hole of height $h = 50$ mm bored into its base.

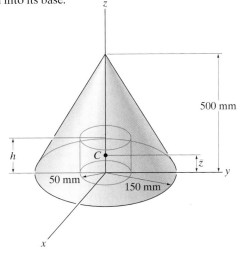

Probs. 9–69/70

9–71. The sheet metal part has the dimensions shown. Determine the location $(\bar{x}, \bar{y}, \bar{z})$ of its centroid.

*9–72. The sheet metal part has a weight per unit area of 2 lb/ft² and is supported by the smooth rod and at C. If the cord is cut, the part will rotate about the y axis until it reaches equilibrium. Determine the equilibrium angle of tilt, measured downward from the negative x axis, that AD makes with the $-x$ axis.

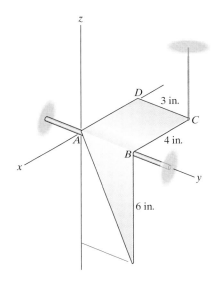

Probs. 9–71/72

9–73. A toy skyrocket consists of a solid conical top, $\rho_t = 600$ kg/m³, a hollow cylinder, $\rho_c = 400$ kg/m³, and a stick having a circular cross section, $\rho_s = 300$ kg/m³. Determine the length of the stick, x, so that the center of gravity G of the skyrocket is located along line aa.

Prob. 9–73

9–74. Determine the location $(\bar{x}, \bar{y})$ of the center of mass of the turbine and compressor assembly. The mass and the center of mass of each of the various components are indicated below.

1. Intake housing   25 kg
2. Turbine   80 kg
3. Exhaust housing   30 kg
4. Compressor   105 kg

Prob. 9–74

9–75. The solid is formed by boring a conical hole into the hemisphere. Determine the distance $\bar{z}$ to the center of gravity.

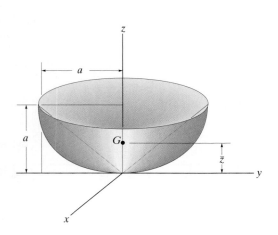

Prob. 9–75

*9–76.** Determine the location $\bar{x}$ of the centroid of the solid made from a hemisphere, cylinder, and cone.

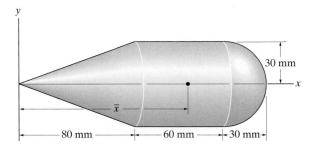

Prob. 9–76

9–77. The buoy is made from two homogeneous cones each having a radius of 1.5 ft. If $h = 1.2$ ft, find the distance $\bar{z}$ to the buoy's center of gravity G.

9–78. The buoy is made from two homogeneous cones each having a radius of 1.5 ft. If it is required that the buoy's center of gravity G be located at $\bar{z} = 0.5$ ft, determine the height h of the top cone.

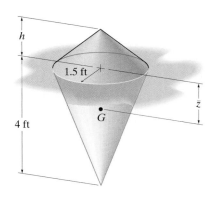

Probs. 9–77/78

9–79. Locate the center of mass $\bar{z}$ of the forked lever, which is made from a homogeneous material and has the dimensions shown.

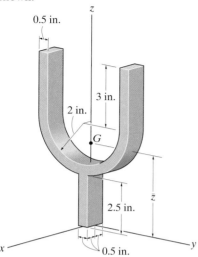

0.5 in.

3 in.

2 in.

G

2.5 in.

$\bar{z}$

x

0.5 in.

y

Prob. 9–79

***9–80.** A triangular plate made of homogeneous material has a constant thickness which is very small. If it is folded over as shown, determine the location $\bar{y}$ of the plate's center of gravity G.

9–81. A triangular plate made of homogeneous material has a constant thickness which is very small. If it is folded over as shown, determine the location $\bar{z}$ of the plate's center of gravity G.

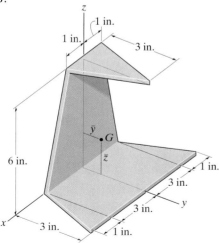

1 in.

1 in.

3 in.

$\bar{y}$
G

6 in.
$\bar{z}$

1 in.

3 in.
3 in.
y

x

3 in.

1 in.

Probs. 9–80/81

9–82. Each of the three homogeneous plates welded to the rod has a density of 6 Mg/m³ and a thickness of 10 mm. Determine the length l of plate C and the angle of placement, θ, so that the center of mass of the assembly lies on the y axis. Plates A and B lie in the x–y and z–y planes, respectively.

z

100 mm

10 mm

150 mm

l

A

C

θ

x 10 mm

200 mm

B

150 mm

y

250 mm

10 mm

Prob. 9–82

9–83. The assembly consists of a 20-in. wooden dowel rod and a tight-fitting steel collar. Determine the distance $\bar{x}$ to its center of gravity if the specific weights of the materials are $\gamma_w = 150\ \text{lb/ft}^3$ and $\gamma_{st} = 490\ \text{lb/ft}^3$. The radii of the dowel and collar are shown.

5 in.

5 in.

10 in.

G

$\bar{x}$

x

2 in.

1 in.

Prob. 9–83

Fig. 9–20

Fig. 9–21

Fig. 9–22

*9.4   Theorems of Pappus and Guldinus

The two *theorems of Pappus and Guldinus* are used to find the surface area and volume of any object of revolution. They were first developed by Pappus of Alexandria during the third century A.D. and then restated at a later time by the Swiss mathematician Paul Guldin or Guldinus (1577–1643).

A *surface area of revolution* is generated by revolving a *plane curve* about a nonintersecting fixed axis in the plane of the curve; whereas a *volume of revolution* is generated by revolving a *plane area* about a nonintersecting fixed axis in the plane of the area. For example, if the *line AB* shown in Fig. 9–20 is rotated about a fixed axis, it generates the *surface area* of a cone (less the area of the base); if the triangular *area ABC* shown in Fig. 9–21 is rotated about the axis, it generates the *volume* of a cone.

The statements and proofs of the theorems of Pappus and Guldinus follow. The proofs require that the generating curves and areas do *not* cross the axis about which they are rotated. If they did, the two sections on either side of the axis would generate areas or volumes having opposite signs and hence cancel each other.

Surface Area.   *The area of a surface of revolution equals the product of the length of the generating curve and the distance traveled by the centroid of the curve in generating the surface area.*

Proof.   When a differential length dL of the curve shown in Fig. 9–22 is revolved about an axis through a distance $2\pi r$, it generates a ring having a surface area $dA = 2\pi r \, dL$. The entire surface area, generated by revolving the entire curve about the axis, is therefore $A = 2\pi \int_L r \, dL$. This equation may be simplified, however, by noting that the location r of the centroid for the line of total length L can be determined from an equation having the form of Eqs. 9–7, namely, $\int_L r \, dL = \bar{r} L$. Thus, the total surface area becomes $A = 2\pi \bar{r} L$. In general, though, if the line does not undergo a complete revolution, then,

$$A = \theta \bar{r} L \qquad (9\text{–}9)$$

where

A = surface area of revolution

θ = angle of revolution measured in radians, $\theta \le 2\pi$

$\bar{r}$ = perpendicular distance from the axis of revolution to the centroid of the generating curve

L = length of the generating curve

Volume.   *The volume of a body of revolution equals the product of the generating area and the distance traveled by the centroid of the area in generating the volume.*

Proof.   When the differential area dA shown in Fig. 9–23 is revolved about an axis through a distance $2\pi r$, it generates a ring having a volume $dV = 2\pi r\, dA$. The entire volume, generated by revolving A about the axis, is therefore $V = 2\pi \int_V r\, dA$. Here the integral can be eliminated by using an equation analogous to Eqs. 9–6, $\int_V r\, dA = \bar{r}A$, where $\bar{r}$ locates the centroid C of the generating area A. The volume becomes $V = 2\pi\bar{r}A$. In general, though,

Fig. 9–23

$$\boxed{V = \theta\bar{r}A} \tag{9–10}$$

where

$V =$ volume of revolution

$\theta =$ angle of revolution measured in radians, $\theta \le 2\pi$

$\bar{r} =$ perpendicular distance from the axis of revolution to the centroid of the generating area

$A =$ generating area

Composite Shapes.   We may also apply the above two theorems to lines or areas that may be composed of a series of composite parts. In this case the total surface area or volume generated is the addition of the surface areas or volumes generated by each of the composite parts. Since each part undergoes the *same* angle of revolution, θ, and the distance from the axis of revolution to the centroid of each composite part is $\tilde{r}$, then

The surface area and the amount of water that can be stored in this water tank can be determined by using the theorems of Pappus and Guldinus.

$$A = \theta\Sigma(\tilde{r}L) \tag{9–11}$$

and

$$V = \theta\Sigma(\tilde{r}A) \tag{9–12}$$

Application of the above theorems is illustrated numerically in the following example.

The amount of roofing material used on this storage building can be estimated by using the first theorem of Pappus and Guldinus to determine its surface area.

EXAMPLE 9.12

Show that the surface area of a sphere is $A = 4\pi R^2$ and its volume is $V = \frac{4}{3}\pi R^3$.

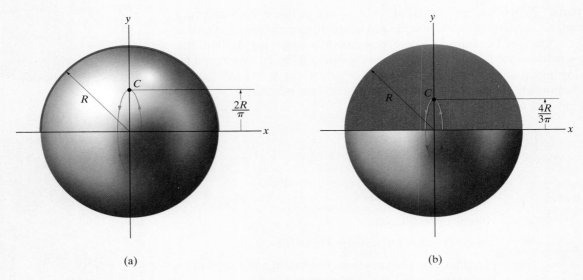

(a)   (b)

Fig. 9–24

SOLUTION

Surface Area. The surface area of the sphere in Fig. 9–24a is generated by rotating a semicircular *arc* about the x axis. Using the table on the inside back cover, it is seen that the centroid of this arc is located at a distance $\bar{r} = 2R/\pi$ from the x axis of rotation. Since the centroid moves through an angle of $\theta = 2\pi$ rad in generating the sphere, then applying Eq. 9–9 we have

$$A = \theta \bar{r} L; \qquad A = 2\pi \left(\frac{2R}{\pi}\right)\pi R = 4\pi R^2 \qquad \textit{Ans.}$$

Volume. The volume of the sphere is generated by rotating the semicircular *area* in Fig. 9–24b about the x axis. Using the table on the inside back cover to locate the centroid of the area, i.e., $\bar{r} = 4R/3\pi$, and applying Eq. 9–10, we have

$$V = \theta \bar{r} A; \qquad V = 2\pi \left(\frac{4R}{3\pi}\right)\left(\frac{1}{2}\pi R^2\right) = \frac{4}{3}\pi R^3 \qquad \textit{Ans.}$$

PROBLEMS

*9–84. Determine the surface area and the volume of the ring formed by rotating the square about the vertical axis.

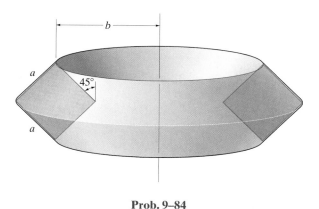

Prob. 9–84

9–85. The anchor ring is made of steel having a specific weight of $\gamma_{st} = 490 \text{ lb/ft}^3$. Determine the surface area of the ring. The cross section is circular as shown.

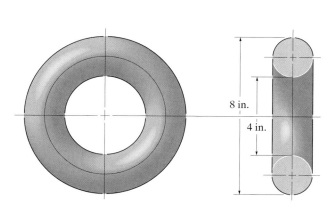

Prob. 9–85

9–86. Using integration, determine both the area and the distance $\bar{y}$ to the centroid of the shaded area. Then using the second theorem of Pappus–Guldinus, determine the volume of the solid generated by revolving the shaded area about the x axis.

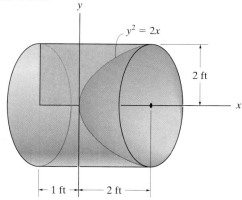

Prob. 9–86

9–87. The grain bin of the type shown is manufactured by Grain Systems, Inc. Determine the required square footage of the sheet metal needed to form it, and also the maximum storage capacity (volume) within it.

Prob. 9–87

*9–88. Determine the surface area and the volume of the conical solid.

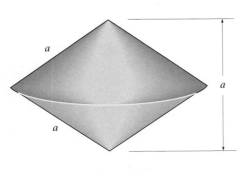

Prob. 9–88

9–90. The *rim* of a flywheel has the cross section *A–A* shown. Determine the volume of material needed for its construction.

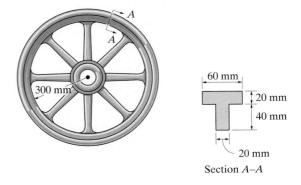

Section *A–A*

Prob. 9–90

9–89. Sand is piled between two walls as shown. Assume the pile to be a quarter section of a cone and that 26 percent of this volume is voids (air space). Use the second theorem of Pappus–Guldinus to determine the volume of sand.

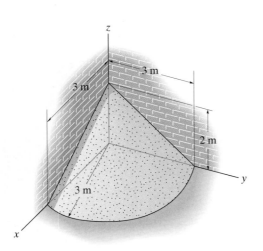

Prob. 9–89

9–91. The Gates Manufacturing Co. produces pulley wheels such as the one shown. Determine the weight of the wheel if it is made from steel having a specific weight of 490 lb/ft³.

*9–92. The Gates Manufacturing Co. produces pulley wheels such as the one shown. Determine the total surface area of the wheel in order to estimate the amount of paint needed to protect its surface from rust.

Probs. 9–91/92

9–93. Determine the volume of material needed to make the casting.

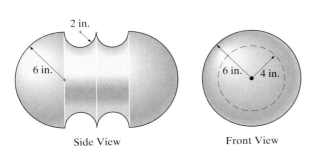

2 in.

6 in.

Side View

6 in.   4 in.

Front View

Prob. 9–93

9–95. Determine the surface area of the tank, which consists of a cylinder and hemispherical cap.

*9–96.** Determine the volume of the tank, which consists of a cylinder and hemispherical cap.

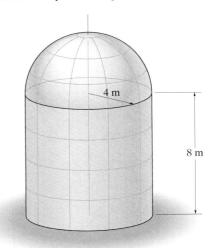

4 m

8 m

Probs. 9–95/96

9–94. A circular sea wall is made of concrete. Determine the total weight of the wall if the concrete has a specific weight of $\gamma_c = 150 \text{ lb/ft}^3$.

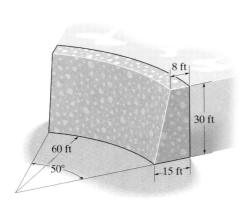

8 ft

30 ft

60 ft

50°

15 ft

Prob. 9–94

9–97. Determine the surface area of the silo which consists of a cylinder and hemispherical cap. Neglect the thickness of the plates.

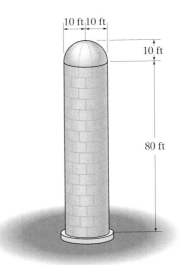

10 ft 10 ft

10 ft

80 ft

Prob. 9–97

9–98. Determine the volume of the silo which consists of a cylinder and hemispherical cap. Neglect the thickness of the plates.

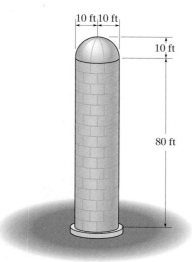

10 ft 10 ft

10 ft

80 ft

Prob. 9–98

9–99. The process tank is used to store liquids during manufacturing. Estimate both the volume of the tank and its surface area. The tank has a flat top and the plates from which the tank is made have negligible thickness.

3 m   3 m

6 m

4 m

Prob. 9–99

*9–100. Determine the height h to which liquid should be poured into the cup so that it contacts half the surface area on the inside of the cup. Neglect the cup's thickness for the calculation.

30 mm

50 mm

h

10 mm

Prob. 9–100

9–101. Using integration, compute both the area and the centroidal distance $\bar{x}$ of the shaded region. Then, using the second theorem of Pappus–Guldinus, compute the volume of the solid generated by revolving the shaded area about the aa axis.

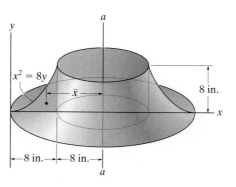

a

y

$x^2 = 8y$

$\bar{x}$

8 in.

x

8 in.   8 in.

a

Prob. 9–101

9–102. Using integration, determine the area and the centroidal distance $\bar{y}$ of the shaded area. Then, using the second theorem of Pappus–Guldinus, determine the volume of a solid formed by revolving the area about the x axis.

Prob. 9–102

9–103. Determine the surface area of the roof of the structure if it is formed by rotating the parabola about the y axis.

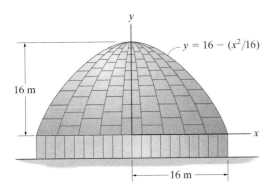

Prob. 9–103

*9–104. The suspension bunker is made from plates which are curved to the natural shape which a completely flexible membrane would take if subjected to a full load of coal. This curve may be approximated by a parabola, $y = 0.2x^2$. Determine the weight of coal which the bunker would contain when completely filled. Coal has a specific weight of $\gamma = 50 \text{ lb/ft}^3$, and assume there is a 20% loss in volume due to air voids. Solve the problem by integration to determine the cross-sectional area of ABC; then use the second theorem of Pappus–Guldinus to find the volume.

Prob. 9–104

9–105. Determine the interior surface area of the brake piston. It consists of a full circular part. Its cross section is shown in the figure.

Prob. 9–105

(a)

(b)

Fig. 9–25

*9.5 Resultant of a General Distributed Loading

In Sec. 4.10, we discussed the method used to simplify a two-dimensional distributed loading to a single resultant force acting at a specific point. In this section we will generalize this method to include surfaces that have an arbitrary shape and are subjected to a variable load distribution. As a specific application, in the next section, we will find the resultant loading acting on the surface of a body that is submerged in a fluid.

Pressure Distribution over a Surface.   Consider the flat plate shown in Fig. 9–25a, which is subjected to the loading function $p = p(x, y)$ Pa, where Pa (pascal) = 1 N/m^2. Knowing this function, we can determine the force $d\mathbf{F}$ acting on the differential area dA m^2 of the plate, located at the arbitrary point (x, y). This force magnitude is simply $dF = [p(x, y) \text{ N/m}^2](dA \text{ m}^2) = [p(x, y) \, dA]$ N. The entire loading on the plate is therefore represented as a system of *parallel forces* infinite in number and each acting on a separate differential area dA. This system will now be simplified to a single resultant force $\mathbf{F}_R$ acting through a unique point $(\bar{x}, \bar{y})$ on the plate, Fig. 9–25b.

Magnitude of Resultant Force.   To determine the *magnitude* of $\mathbf{F}_R$, it is necessary to sum each of the differential forces $d\mathbf{F}$ acting over the plate's *entire surface area A*. This sum may be expressed mathematically as an integral:

$$F_R = \Sigma F; \qquad \boxed{F_R = \int_A p(x, y) \, dA = \int_V dV} \qquad (9\text{–}13)$$

Here $p(x, y) \, dA = dV$, the colored differential *volume element* shown in Fig. 9–25a. Therefore, the result indicates that the *magnitude of the resultant force is equal to the total volume under the distributed-loading diagram.*

Location of Resultant Force.   The location $(\bar{x}, \bar{y})$ of $\mathbf{F}_R$ is determined by setting the moments of $\mathbf{F}_R$ equal to the moments of all the forces $d\mathbf{F}$ about the respective y and x axes: From Figs. 9–25a and 9–25b, using Eq. 9–13, this results in

$$\boxed{\bar{x} = \frac{\int_A x p(x, y) \, dA}{\int_A p(x, y) \, dA} = \frac{\int_V x \, dV}{\int_V dV} \qquad \bar{y} = \frac{\int_A y p(x, y) \, dA}{\int_A p(x, y) \, dA} = \frac{\int_V y \, dV}{\int_V dV}} \qquad (9\text{–}14)$$

Hence, it can be seen that the *line of action of the resultant force passes through the geometric center or centroid of the volume under the distributed-loading diagram.*

*9.6   Fluid Pressure

According to Pascal's law, a fluid at rest creates a pressure p at a point that is the *same* in *all* directions. The magnitude of p, measured as a force per unit area, depends on the specific weight γ or mass density ρ of the fluid and the depth z of the point from the fluid surface.* The relationship can be expressed mathematically as

$$p = \gamma z = \rho g z \qquad (9\text{–}15)$$

where g is the acceleration due to gravity. Equation 9–15 is valid only for fluids that are assumed *incompressible*, as in the case of most liquids. Gases are compressible fluids, and since their density changes significantly with both pressure and temperature, Eq. 9–15 cannot be used.

To illustrate how Eq. 9–15 is applied, consider the submerged plate shown in Fig. 9–26. Three points on the plate have been specified. Since point B is at depth z_1 from the liquid surface, the *pressure* at this point has a magnitude $p_1 = \gamma z_1$. Likewise, points C and D are both at depth z_2; hence, $p_2 = \gamma z_2$. In all cases, the pressure acts *normal* to the surface area dA located at the specified point. Using Eq. 9–15 and the results of Sec. 9.5, it is possible to determine the resultant force caused by a liquid pressure distribution and specify its location on the surface of a submerged plate. Three different shapes of plates will now be considered.

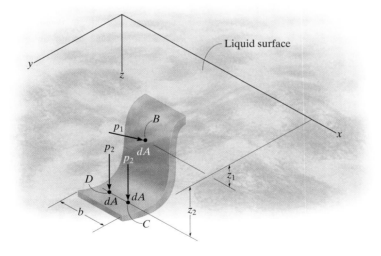

Fig. 9–26

*In particular, for water $\gamma = 62.4 \text{ lb/ft}^3$, or $\gamma = \rho g = 9810 \text{ N/m}^3$ since $\rho = 1000 \text{ kg/m}^3$ and $g = 9.81 \text{ m/s}^2$.

Flat Plate of Constant Width.   A flat rectangular plate of constant width, which is submerged in a liquid having a specific weight γ, is shown in Fig. 9–27a. The plane of the plate makes an angle with the horizontal, such that its top edge is located at a depth z_1 from the liquid surface and its bottom edge is located at a depth z_2. Since pressure varies linearly with depth, Eq. 9–15, the distribution of pressure over the plate's surface is represented by a trapezoidal volume having an intensity of $p_1 = \gamma z_1$ at depth z_1 and $p_2 = \gamma z_2$ at depth z_2. As noted in Sec. 9.5, the magnitude of the *resultant force* $\mathbf{F}_R$ is equal to the *volume* of this loading diagram and $\mathbf{F}_R$ has a *line of action* that passes through the volume's centroid C. Hence, $\mathbf{F}_R$ does *not* act at the centroid of the plate; rather, it acts at point P, called the *center of pressure*.

Since the plate has a *constant width*, the loading distribution may also be viewed in two dimensions, Fig. 9–27b. Here the loading intensity is measured as force/length and varies linearly from $w_1 = bp_1 = b\gamma z_1$ to $w_2 = bp_2 = b\gamma z_2$. The magnitude of $\mathbf{F}_R$ in this case equals the trapezoidal *area*, and $\mathbf{F}_R$ has a *line of action* that passes through the area's *centroid* C. For numerical applications, the area and location of the centroid for a trapezoid are tabulated on the inside back cover.

(a)

(b)

Fig. 9–27

(a)

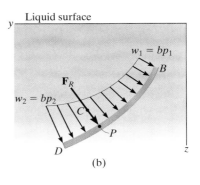

(b)

Fig. 9–28

Curved Plate of Constant Width.   When the submerged plate is curved, the pressure acting normal to the plate continually changes direction, and therefore calculation of the magnitude of $\mathbf{F}_R$ and its location P is more difficult than for a flat plate. Three- and two-dimensional views of the loading distribution are shown in Figs. 9–28a and 9–28b, respectively. Here integration can be used to determine both F_R and the location of the centroid C or center of pressure P.

A simpler method exists, however, for calculating the magnitude of $\mathbf{F}_R$ and its location along a curved (or flat) plate having a *constant width*. This method requires separate calculations for the horizontal and vertical *components* of $\mathbf{F}_R$. For example, the distributed loading acting on the curved plate DB in Fig. 9–28b can be represented by the *equivalent loading* shown in Fig. 9–29. Here the plate supports the weight of liquid W_f contained within the block BDA. This force has a magnitude $W_f = (\gamma b)(\text{area}_{BDA})$ and acts through the centroid of BDA. In addition, there are the pressure distributions caused by the liquid acting along the vertical and horizontal sides of the block. Along the vertical side AD, the force $\mathbf{F}_{AD}$ has a magnitude that equals the area under the trapezoid and acts through the centroid C_{AD} of this area. The distributed loading along the horizontal side AB is constant since all points lying in this plane are at the same depth from the surface of the liquid. The magnitude of $\mathbf{F}_{AB}$ is simply the area of the rectangle. This force acts through the area's centroid C_{AB} or the midpoint of AB. Summing the three forces in Fig. 9–29 yields $\mathbf{F}_R = \Sigma \mathbf{F} = \mathbf{F}_{AD} + \mathbf{F}_{AB} + \mathbf{W}_f$, which is shown in Fig. 9–28. Finally, the location of the center of pressure P on the plate is determined by applying the equation $M_{R_O} = \Sigma M_O$, which states that the moment of the resultant force about a convenient reference point O, such as D or B, in Fig. 9–28, is equal to the sum of the moments of the three forces in Fig. 9–29 about the same point.

Fig. 9–29

The resultant force of the water and its location on the elliptical back plate of this tank truck must be determined by integration.

Flat Plate of Variable Width.   The pressure distribution acting on the surface of a submerged plate having a variable width is shown in Fig. 9–30. The resultant force of this loading equals the volume described by the plate area as its base and linearly varying pressure distribution as its altitude. The shaded element shown in Fig. 9–30 may be used if integration is chosen to determine this volume. The element consists of a rectangular strip of area $dA = x\, dy'$ located at a depth z below the liquid surface. Since a uniform pressure $p = \gamma z$ (force/area) acts on dA, the magnitude of the differential force $d\mathbf{F}$ is equal to $dF = dV = p\, dA = \gamma z(x\, dy')$. Integrating over the entire volume yields Eq. 9–13; i.e.,

$$F_R = \int_A p\, dA = \int_V dV = V$$

From Eq. 9–14, the centroid of V defines the point through which $\mathbf{F}_R$ acts. The center of pressure, which lies on the surface of the plate just below C, has coordinates $P(\bar{x}, \bar{y}')$ defined by the equations

$$\bar{x} = \frac{\int_V \tilde{x}\, dV}{\int_V dV} \qquad \bar{y}' = \frac{\int_V \tilde{y}'\, dV}{\int_V dV}$$

This point should *not* be mistaken for the centroid of the plate's *area*.

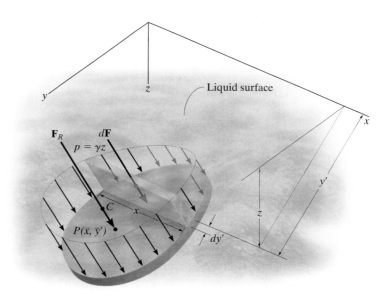

Fig. 9–30

EXAMPLE 9.13

Determine the magnitude and location of the resultant hydrostatic force acting on the submerged rectangular plate AB shown in Fig. 9–31a. The plate has a width of 1.5 m; $\rho_w = 1000$ kg/m^3.

(a)

SOLUTION

The water pressures at depths A and B are

$$p_A = \rho_w g z_A = (1000 \text{ kg/m}^3)(9.81 \text{ m/s}^2)(2 \text{ m}) = 19.62 \text{ kPa}$$
$$p_B = \rho_w g z_B = (1000 \text{ kg/m}^3)(9.81 \text{ m/s}^2)(5 \text{ m}) = 49.05 \text{ kPa}$$

Since the plate has a constant width, the distributed loading can be viewed in two dimensions as shown in Fig. 9–31b. The intensities of the load at A and B are

$$w_A = b p_A = (1.5 \text{ m})(19.62 \text{ kPa}) = 29.43 \text{ kN/m}$$
$$w_B = b p_B = (1.5 \text{ m})(49.05 \text{ kPa}) = 73.58 \text{ kN/m}$$

From the table on the inside back cover, the magnitude of the resultant force $\mathbf{F}_R$ created by the distributed load is

$$
\begin{aligned}
F_R &= \text{area of trapezoid} \\
&= \tfrac{1}{2}(3)(29.4 + 73.6) = 154.5 \text{ kN} \qquad \textit{Ans.}
\end{aligned}
$$

This force acts through the centroid of the area,

$$h = \frac{1}{3}\left(\frac{2(29.43) + 73.58}{29.43 + 73.58}\right)(3) = 1.29 \text{ m} \qquad \textit{Ans.}$$

(b)

measured upward from B, Fig. 9–31b.

The same results can be obtained by considering two components of $\mathbf{F}_R$ defined by the triangle and rectangle shown in Fig. 9–31c. Each force acts through its associated centroid and has a magnitude of

$$F_{Re} = (29.43 \text{ kN/m})(3 \text{ m}) = 88.3 \text{ kN}$$
$$F_t = \tfrac{1}{2}(44.15 \text{ kN/m})(3 \text{ m}) = 66.2 \text{ kN}$$

Hence,

$$F_R = F_{Re} + F_t = 88.3 + 66.2 = 154.5 \text{ kN} \qquad \textit{Ans.}$$

The location of $\mathbf{F}_R$ is determined by summing moments about B, Fig. 9–31b and c, i.e.,

$$\zeta + (M_R)_B = \Sigma M_B; \qquad (154.5)h = 88.3(1.5) + 66.2(1)$$
$$h = 1.29 \text{ m} \qquad \textit{Ans.}$$

(c)

NOTE: The result of $\mathbf{F}_R$ should give you some idea as to how large hydrostatic forces can be.

Fig. 9–31

EXAMPLE 9.14

Determine the magnitude of the resultant hydrostatic force acting on the surface of a seawall shaped in the form of a parabola as shown in Fig. 9–32a. The wall is 5 m long; $\rho_w = 1020$ kg/m^3.

(a)

(b)

Fig. 9–32

SOLUTION

The horizontal and vertical components of the resultant force will be calculated, Fig. 9–32b. Since

$$p_B = \rho_w g z_B = (1020 \text{ kg/m}^3)(9.81 \text{ m/s}^2)(3 \text{ m}) = 30.02 \text{ kPa}$$

then

$$w_B = b p_B = 5 \text{ m}(30.02 \text{ kPa}) = 150.1 \text{ kN/m}$$

Thus,

$$F_x = \tfrac{1}{2}(3 \text{ m})(150.1 \text{ kN/m}) = 225.1 \text{ kN}$$

The area of the parabolic sector ABC can be determined using the table on the inside back cover. Hence, the weight of water within this region is

$$F_y = (\rho_w g b)(\text{area}_{ABC})$$
$$= (1020 \text{ kg/m}^3)(9.81 \text{ m/s}^2)(5 \text{ m})\left[\tfrac{1}{3}(1 \text{ m})(3 \text{ m})\right] = 50.0 \text{ kN}$$

The resultant force is therefore

$$F_R = \sqrt{F_x^2 + F_y^2} = \sqrt{(225.1)^2 + (50.0)^2}$$
$$= 231 \text{ kN} \qquad\qquad Ans.$$

EXAMPLE 9.15

Determine the magnitude and location of the resultant force acting on the triangular end plates of the water trough shown in Fig. 9–33a; $\rho_w = 1000 \text{ kg/m}^3$.

(a)

SOLUTION

The pressure distribution acting on the end plate E is shown in Fig. 9–33b. The magnitude of the resultant force $\mathbf{F}$ is equal to the volume of this loading distribution. We will solve the problem by integration. Choosing the differential volume element shown in the figure, we have

$$dF = dV = p\, dA = \rho_w g z (2x\, dz) = 19\,620 z x\, dz$$

The equation of line AB is

$$x = 0.5(1 - z)$$

Hence, substituting and integrating with respect to z from $z = 0$ to $z = 1$ m yields

$$F = V = \int_V dV = \int_0^1 (19\,620)z[0.5(1 - z)]\, dz$$

$$= 9810 \int_0^1 (z - z^2)\, dz = 1635 \text{ N} = 1.64 \text{ kN} \qquad Ans.$$

This resultant passes through the centroid of the volume. Because of symmetry,

$$\bar{x} = 0 \qquad Ans.$$

Since $\tilde{z} = z$ for the volume element in Fig. 9–33b, then

$$\bar{z} = \frac{\int_V \tilde{z}\, dV}{\int_V dV} = \frac{\int_0^1 z(19\,620)z[0.5(1 - z)]\, dz}{1635} = \frac{9810 \int_0^1 (z^2 - z^3)\, dz}{1635}$$

$$= 0.5 \text{ m} \qquad Ans.$$

(b)

Fig. 9–33

PROBLEMS

9–106. Determine the magnitude of the resultant hydrostatic force acting on the dam and its location, measured from the top surface of the water. The width of the dam is 8 m; $\rho_w = 1.0 \text{ Mg/m}^3$.

Prob. 9–106

9–107. The tank is filled with water to a depth of $d = 4$ m. Determine the resultant force the water exerts on side A and side B of the tank. If oil instead of water is placed in the tank, to what depth d should it reach so that it creates the same resultant forces? $\rho_o = 900 \text{ kg/m}^3$ and $\rho_w = 1000 \text{ kg/m}^3$.

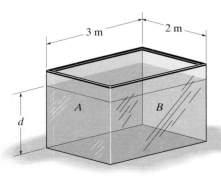

Prob. 9–107

*9–108. The factor of safety for tipping of the concrete dam is defined as the ratio of the stabilizing moment about O due to the dam's weight divided by the overturning moment about O due to the water pressure. Determine this factor if the concrete has a specific weight of $\gamma_{conc} = 150 \text{ lb/ft}^3$ and for water $\gamma_w = 62.4 \text{ lb/ft}^3$.

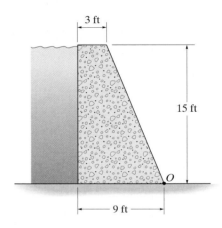

Prob. 9–108

9–109. The concrete "gravity" dam is held in place by its own weight. If the density of concrete is $\rho_c = 2.5 \text{ Mg/m}^3$, and water has a density of $\rho_w = 1.0 \text{ Mg/m}^3$, determine the smallest dimension d that will prevent the dam from overturning about its end A.

Prob. 9–109

9–110. The concrete dam is designed so that its face AB has a gradual slope into the water as shown. Because of this, the frictional force at the base BD of the dam is increased due to the hydrostatic force of the water acting on the dam. Calculate the hydrostatic force acting on the face AB of the dam. The dam is 60 ft wide. $\gamma_w = 62.4\ \text{lb/ft}^3$.

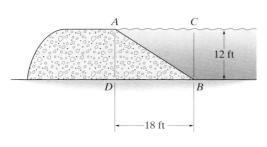

Prob. 9–110

*9–112.** The tank is used to store a liquid having a specific weight of 80 lb/ft³. If it is filled to the top, determine the magnitude of force the liquid exerts on each of its two sides $ABDC$ and $BDFE$.

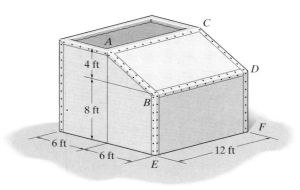

Prob. 9–112

9–111. The symmetric concrete "gravity" dam is held in place by its own weight. If the density of concrete is $\rho_c = 2.5\ \text{Mg/m}^3$, and water has a density of $\rho_w = 1.0\ \text{Mg/m}^3$, determine the smallest distance d at its base that will prevent the dam from overturning about its end A. The dam has a width of 8 m.

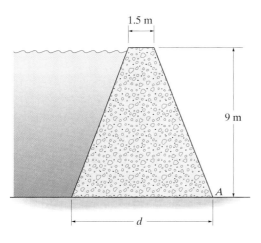

Prob. 9–111

9–113. The 2-m-wide rectangular gate is pinned at its center A and is prevented from rotating by the block at B. Determine the reactions at these supports due to hydrostatic pressure. $\rho_w = 1.0\ \text{Mg/m}^3$.

Prob. 9–113

9–114. The gate AB is 8 m wide. Determine the horizontal and vertical components of force acting on the pin at B and the vertical reaction at the smooth support A. $\rho_w = 1.0 \text{ Mg/m}^3$.

Prob. 9–114

*9–116. The arched surface AB is shaped in the form of a quarter circle. If it is 8 m long, determine the horizontal and vertical components of the resultant force caused by the water acting on the surface. $\rho_w = 1.0 \text{ Mg/m}^3$.

Prob. 9–116

9–115. The storage tank contains oil having a specific weight of $\gamma_o = 56 \text{ lb/ft}^3$. If the tank is 6 ft wide, calculate the resultant force acting on the inclined side BC of the tank, caused by the oil, and specify its location along BC, measured from B. Also compute the total resultant force acting on the bottom of the tank.

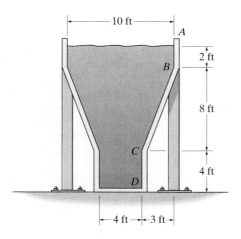

Prob. 9–115

9–117. The rectangular bin is filled with coal, which creates a pressure distribution along wall A that varies as shown, i.e., $p = 4z^3 \text{ lb/ft}^2$, where z is measured in feet. Determine the resultant force created by the coal, and specify its location measured from the top surface of the coal.

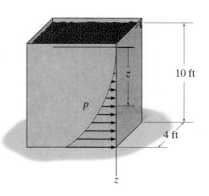

Prob. 9–117

9–118. The semicircular drainage pipe is filled with water. Determine the resultant horizontal and vertical force components that the water exerts on the side AB of the pipe per foot of pipe length; $\gamma_w = 62.4 \text{ lb/ft}^3$.

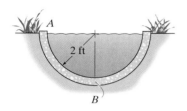

Prob. 9–118

*9–120. The drum is filled to its top ($y = 1.5$ ft) with oil. having a density of $\rho_o = 55 \text{ lb/ft}^3$. Determine the resultant force of the oil pressure acting on the flat end A of the drum and specify its location measured from the top of the drum.

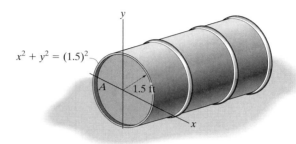

Prob. 9–120

9–119. The load over the plate varies linearly along the sides of the plate such that $p = 10[y(2 - x)] \text{ lb/ft}^2$. Determine the magnitude of the resultant force and the coordinates $(\bar{x}, \bar{y})$ of the point where the line of action of the force intersects the plate.

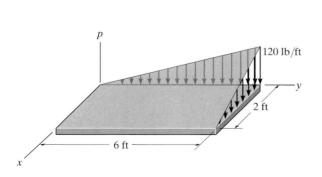

Prob. 9–119

9–121. The gasoline tank is constructed with elliptical ends on each side of the tank. Determine the resultant force and its location on these ends if the tank is half full. Take $\gamma = 41 \text{ lb/ft}^3$.

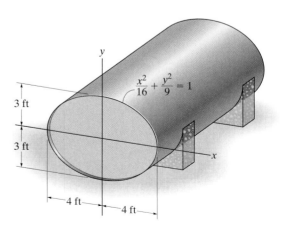

Prob. 9–121

9–122. The loading acting on a square plate is represented by a parabolic pressure distribution. Determine the magnitude of the resultant force and the coordinates $(\overline{x}, \overline{y})$ of the point where the line of action of the force intersects the plate. Also, what are the reactions at the rollers B and C and the ball-and-socket joint A? Neglect the weight of the plate.

9–123. The tank is filled with a liquid which has a density of 900 kg/m³. Determine the resultant force that it exerts on the elliptical end plate, and the location of the center of pressure, measured from the x axis.

Prob. 9–122

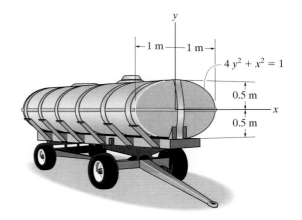

Prob. 9–123

Chapter Review

Center of Gravity and Centroid

The *center of gravity* represents a point where the weight of the body can be considered concentrated. The distance s to this point can be determined from a balance of moments. This requires that the moment of the weight of all the particles of the body about some point must equal the moment of the entire body about the point.

$$\bar{x} = \frac{\int \tilde{x}\, dW}{\int dW}$$

$$\bar{y} = \frac{\int \tilde{y}\, dW}{\int dW}$$

$$\bar{z} = \frac{\int \tilde{z}\, dW}{\int dW}$$

The center of mass will coincide with the center of gravity provided the acceleration of gravity is constant.

The *centroid* is the location of the geometric center for the body. It is determined in a similar manner, using a moment balance of geometric elements such as line, area, or volume segments. For bodies having a continuous shape, moments are summed (integrated) using differential elements.

$$\bar{x} = \frac{\int_L \tilde{x}\, dL}{\int_L dL} \qquad \bar{y} = \frac{\int_L \tilde{y}\, dL}{\int_L dL} \qquad \bar{z} = \frac{\int_L \tilde{z}\, dL}{\int_L dL}$$

$$\bar{x} = \frac{\int_A \tilde{x}\, dA}{\int_A dA} \qquad \bar{y} = \frac{\int_A \tilde{y}\, dA}{\int_A dA} \qquad \bar{z} = \frac{\int_A \tilde{z}\, dA}{\int_A dA}$$

$$\bar{x} = \frac{\int_V \tilde{x}\, dV}{\int_V dV} \qquad \bar{y} = \frac{\int_V \tilde{y}\, dV}{\int_V dV} \qquad \bar{z} = \frac{\int_V \tilde{z}\, dV}{\int_V dV}$$

The center of gravity will coincide with the centroid provided the material is homogeneous, i.e., the density of the material is the same throughout. The centroid will lie on an axis of symmetry. (*See pages 455–460.*)

Composite Body

If the body is a composite of several shapes, each having a known location for its center of gravity or centroid, then the location is determined from a discrete summation using its composite parts. (*See pages 478–479.*)

$$\bar{x} = \frac{\Sigma \tilde{x} W}{\Sigma W} \qquad \bar{y} = \frac{\Sigma \tilde{y} W}{\Sigma W} \qquad \bar{z} = \frac{\Sigma \tilde{z} W}{\Sigma W}$$

Theorems of Pappus and Guldinus

The theorems of Pappus and Guldinus can be used to determine the surface area and volume of a body of revolution.

The *surface area* equals the product of the length of the generating curve and the distance traveled by the centroid of the curve needed to generate the area, $A = \theta \bar{r} L$.

The *volume* of the body equals the product of the generating area and the distance traveled by the centroid of this area needed to generate the volume, $V = \theta \bar{r} A$. (*See pages 492–493.*)

General Distributed Loading

The magnitude of the resutant force is equal to the total volume under the distributed-loading diagram. The line of action of the resultant force passes through the geometric center or centroid of the volume under the distributed-loading diagram.
(*See page 500.*)

$$F_R = \int_A p(x, y)\, dA = \int_V dV$$

$$\overline{x} = \frac{\displaystyle\int_V x\, dV}{\displaystyle\int_V dV}$$

$$\overline{y} = \frac{\displaystyle\int_V y\, dV}{\displaystyle\int_V dV}$$

Fluid Pressure

The pressure developed by a liquid at a point on a submerged surface depends upon the depth of the point and the density of the liquid in accordance with Pascal's law, $p = \rho g h = \gamma h$. This pressure will create a *linear distribution* of loading on a flat vertical or inclined surface.

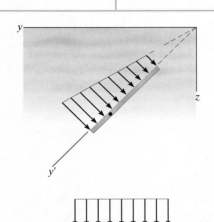

If the surface is horizontal, then the loading will be *uniform*.

In any case, the resultants of these loadings can be determined by finding the volume under the loading curve. The line of action of the resultant force passes through the centroid of the loading diagram.
(*See pages 501–504.*)

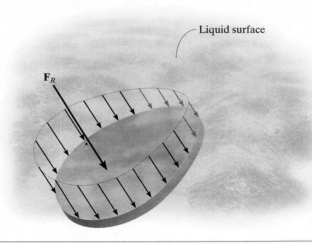

REVIEW PROBLEMS

*9–124.   A circular V-belt has an inner radius of 600 mm and a cross-sectional area as shown. Determine the volume of material required to make the belt.

9–125.   A circular V-belt has an inner radius of 600 mm and a cross-sectional area as shown. Determine the surface area of the belt.

Probs. 9–124/125

9–126.   Locate the center of mass of the homogeneous rod.

Prob. 9–126

9–127.   Locate the centroid of the solid.

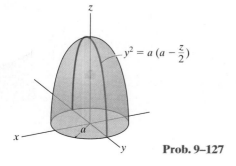

Prob. 9–127

*9–128.   Locate the centroid $(\bar{x}, \bar{y})$ of the thin plate.

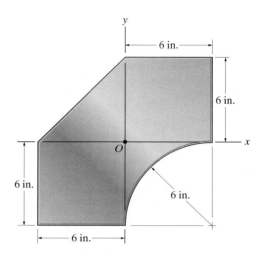

Prob. 9–128

9–129.   Determine the weight and location $(\bar{x}, \bar{y})$ of the center of gravity G of the concrete retaining wall. The wall has a length of 10 ft, and concrete has a specific gravity of $\gamma = 150 \text{ lb/ft}^3$.

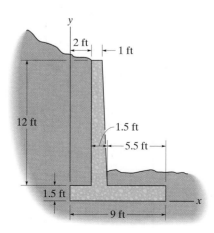

Prob. 9–129

9–130. The hopper is filled to its top with coal. Determine the volume of coal if the voids (air space) are 35 percent of the volume of the hopper.

Prob. 9–130

9–131. Locate the centroid $(\bar{x}, \bar{y})$ of the shaded area.

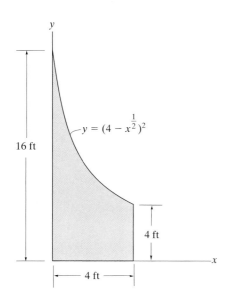

$$y = (4 - x^{\frac{1}{2}})^2$$

Prob. 9–131

*9–132. The rectangular bin is filled with coal, which creates a pressure distribution along wall A that varies as shown, i.e., $p = 4z^{1/3}$ lb/ft², where z is in feet. Compute the resultant force created by the coal, and its location, measured from the top surface of the coal.

$$p = 4z^{1/3}$$

Prob. 9–132

9–133. The load over the plate varies linearly along the sides of the plate such that $p = \frac{2}{3}[x(4 - y)]$ kPa. Determine the resultant force and its position $(\bar{x}, \bar{y})$ on the plate.

Prob. 9–133

9–134. The pressure loading on the plate is described by the function $p = \{-240/(x + 1) + 340\}$ Pa. Determine the magnitude of the resultant force and coordinates of the point where the line of action of the force intersects the plate.

Prob. 9–134

The design of a structural member, such as a beam or column, requires calculation of its cross-sectional moment of inertia. In this chapter, we will discuss how this is done.

10 Moments of Inertia

CHAPTER OBJECTIVES

- To develop a method for determining the moment of inertia for an area.
- To introduce the product of inertia and show how to determine the maximum and minimum moments of inertia for an area.
- To discuss the mass moment of inertia.

10.1 Definition of Moments of Inertia for Areas

In the last chapter, we determined the centroid for an area by considering the first moment of the area about an axis; that is, for the computation we had to evaluate an integral of the form $\int x \, dA$. An integral of the second moment of an area, such as $\int x^2 \, dA$, is referred to as the *moment of inertia* for the area. The terminology "moment of inertia" as used here is actually a misnomer; however, it has been adopted because of the similarity with integrals of the same form related to mass.

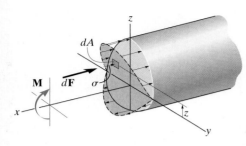

Fig. 10–1

The moment of inertia for an area is a quantity that relates the normal stress σ (sigma), or force per unit area, acting on the transverse cross section of an elastic beam, to the applied external moment $\mathbf{M}$, which causes bending of the beam. The theory of mechanics of materials shows that the stress within the beam varies linearly with its distance from an axis passing through the centroid C of the beam's cross-sectional area; i.e., $\sigma = kz$, Fig. 10–1. The magnitude of force acting on the area element dA, shown in the figure, is therefore $dF = \sigma \, dA = kz \, dA$. Since this force is located a distance z from the y axis, the moment of $d\mathbf{F}$ about the y axis is $dM = dFz = kz^2 \, dA$. The resulting moment of the entire stress distribution is equal to the applied moment $\mathbf{M}$; hence, $M = k \int z^2 \, dA$. Here the integral represents the moment of inertia of the area about the y axis. Integrals of this form often arise in formulas used in mechanics of materials, structural mechanics, fluid mechanics, and machine design, and the engineer needs to be familiar with the methods used for their computation.

Moment of Inertia.   Consider the area A, shown in Fig. 10–2, which lies in the x–y plane. By definition, the moments of inertia for the differential planar area dA about the x and y axes are $dI_x = y^2 \, dA$ and $dI_y = x^2 \, dA$, respectively. For the entire area the *moments of inertia* are determined by integration; i.e.,

$$
\begin{aligned}
I_x &= \int_A y^2 \, dA \\
I_y &= \int_A x^2 \, dA
\end{aligned}
\qquad (10\text{–}1)
$$

We can also formulate the second moment of dA about the "pole" O or z axis, Fig. 10–2. This is referred to as the polar moment of inertia. It is used to determine the torsional stress in a shaft. It is defined as $dJ_O = r^2 \, dA$, where r is the perpendicular distance from the pole (z axis) to the element dA. For the entire area the *polar moment of inertia* is

$$
J_O = \int_A r^2 \, dA = I_x + I_y
\qquad (10\text{–}2)
$$

The relationship between J_O and I_x, I_y is possible since $r^2 = x^2 + y^2$, Fig. 10–2.

From the above formulations it is seen that I_x, I_y, and J_O will *always* be *positive* since they involve the product of distance squared and area. Furthermore, the units for moment of inertia involve length raised to the fourth power, e.g., m^4, mm^4, or ft^4, in^4.

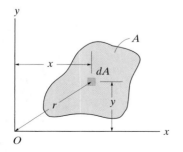

Fig. 10–2

10.2   Parallel-Axis Theorem for an Area

If the moment of inertia for an area is known about an axis passing through its centroid, which is often the case, it is convenient to determine the moment of inertia of the area about a corresponding parallel axis using the *parallel-axis theorem*. To derive this theorem, consider finding the moment of inertia of the shaded area shown in Fig. 10–3 about the x axis. In this case, a differential element dA is located at an arbitrary distance y' from the *centroidal* x' axis, whereas the *fixed distance* between the parallel x and x' axes is defined as d_y. Since the moment of inertia of dA about the x axis is $dI_x = (y' + d_y)^2 \, dA$, then for the entire area,

$$I_x = \int_A (y' + d_y)^2 \, dA$$

$$= \int_A y'^2 \, dA + 2d_y \int_A y' \, dA + d_y^2 \int_A dA$$

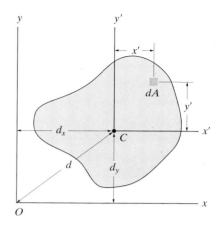

Fig. 10–3

The first integral represents the moment of inertia of the area about the centroidal axis, $\bar{I}_{x'}$. The second integral is zero since the x' axis passes through the area's centroid C; i.e., $\int y' \, dA = \bar{y} \int dA = 0$ since $\bar{y} = 0$. Realizing that the third integral represents the total area A, the final result is therefore

$$\boxed{I_x = \bar{I}_{x'} + Ad_y^2} \qquad (10–3)$$

A similar expression can be written for I_y; i.e.,

$$\boxed{I_y = \bar{I}_{y'} + Ad_x^2} \qquad (10–4)$$

And finally, for the polar moment of inertia about an axis perpendicular to the x–y plane and passing through the "pole" O (z axis), Fig. 10–3, we have

$$\boxed{J_O = \bar{J}_C + Ad^2} \qquad (10–5)$$

The form of each of these three equations states that *the moment of inertia for an area about an axis is equal to the moment of inertia for the area about a parallel axis passing through the area's centroid plus the product of the area and the square of the perpendicular distance between the axes.*

10.3   Radius of Gyration of an Area

The *radius of gyration* of a planar area has units of length and is a quantity that is often used for the design of columns in structural mechanics. Provided the areas and moments of inertia are *known*, the radii of gyration are determined from the formulas

$$k_x = \sqrt{\frac{I_x}{A}} \qquad k_y = \sqrt{\frac{I_y}{A}} \qquad k_O = \sqrt{\frac{J_O}{A}} \qquad (10–6)$$

The form of these equations is easily remembered since it is similar to that for finding the moment of inertia for a differential area about an axis. For example, $I_x = k_x^2 A$; whereas for a differential area, $dI_x = y^2\, dA$.

10.4   Moments of Inertia for an Area by Integration

When the boundaries for a planar area are expressed by mathematical functions, Eqs. 10–1 may be integrated to determine the moments of inertia for the area. If the element of area chosen for integration has a differential size in two directions as shown in Fig. 10–2, a double integration must be performed to evaluate the moment of inertia. Most often, however, it is easier to perform only a single integration by choosing an element having a differential size or thickness in only one direction.

Fig. 10–4

Procedure for Analysis

- If a single integration is performed to determine the moment of inertia for an area about an axis, it will be necessary to specify the differential element dA.
- Most often this element will be rectangular, such that it will have a finite length and differential width.
- The element should be located so that it intersects the boundary of the area at the *arbitrary point* (x, y). There are two possible ways to orient the element with respect to the axis about which the moment of inertia is to be determined.

Case 1

- The length of the element can be oriented *parallel* to the axis. This situation occurs when the rectangular element shown in Fig. 10–4 is used to determine I_y for the area. Direct application of Eq. 10–1, i.e., $I_y = \int x^2\, dA$, can be made in this case since the element has an infinitesimal thickness dx and therefore *all parts* of the element lie at the *same* moment-arm distance x from the y axis.*

Case 2

- The length of the element can be oriented *perpendicular* to the axis. Here Eq. 10–1 *does not apply* since all parts of the element will *not* lie at the same moment-arm distance from the axis. For example, if the rectangular element in Fig. 10–4 is used for determining I_x for the area, it will first be necessary to calculate the moment of inertia of the *element* about a horizontal axis passing through the element's centroid and then determine the moment of inertia of the *element* about the x axis by using the parallel-axis theorem. Integration of this result will yield I_x.

*In the case of the element $dA = dx\, dy$, Fig. 10–2, the moment arms y and x are appropriate for the formulation of I_x and I_y (Eq. 10–1) since the *entire* element, because of its infinitesimal size, lies at the specified y and x perpendicular distances from the x and y axes.

EXAMPLE 10.1

Determine the moment of inertia for the rectangular area shown in Fig. 10–5 with respect to (a) the centroidal x' axis, (b) the axis x_b passing through the base of the rectangle, and (c) the pole or z' axis perpendicular to the $x'-y'$ plane and passing through the centroid C.

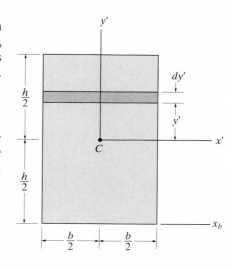

Fig. 10–5

SOLUTION (Case 1)

Part (a).   The differential element shown in Fig. 10–5 is chosen for integration. Because of its location and orientation, the *entire element* is at a distance y' from the x' axis. Here it is necessary to integrate from $y' = -h/2$ to $y' = h/2$. Since $dA = b \, dy'$, then

$$\bar{I}_{x'} = \int_A y'^2 \, dA = \int_{-h/2}^{h/2} y'^2 (b \, dy') = b \int_{-h/2}^{h/2} y'^2 \, dy$$

$$= \frac{1}{12} bh^3 \qquad\qquad\qquad Ans.$$

Part (b).   The moment of inertia about an axis passing through the base of the rectangle can be obtained by using the result of part (a) and applying the parallel-axis theorem, Eq. 10–3.

$$I_{x_b} = \bar{I}_{x'} + A d_y^2$$

$$= \frac{1}{12} bh^3 + bh \left(\frac{h}{2}\right)^2 = \frac{1}{3} bh^3 \qquad\qquad Ans.$$

Part (c).   To obtain the polar moment of inertia about point C, we must first obtain $\bar{I}_{y'}$, which may be found by interchanging the dimensions b and h in the result of part (a), i.e.,

$$\bar{I}_{y'} = \frac{1}{12} hb^3$$

Using Eq. 10–2, the polar moment of inertia about C is therefore

$$\bar{J}_C = \bar{I}_{x'} + \bar{I}_{y'} = \frac{1}{12} bh(h^2 + b^2) \qquad\qquad Ans.$$

EXAMPLE 10.2

(a)

Fig. 10–6

Determine the moment of inertia for the shaded area shown in Fig. 10–6a about the x axis.

SOLUTION I (Case 1)
A differential element of area that is *parallel* to the x axis, as shown in Fig. 10–6a, is chosen for integration. Since the element has a thickness dy and intersects the curve at the *arbitrary point* (x, y), the area is $dA = (100 - x)\, dy$. Furthermore, all parts of the element lie at the same distance y from the x axis. Hence, integrating with respect to y, from $y = 0$ to $y = 200$ mm, yields

$$I_x = \int_A y^2\, dA = \int_A y^2(100 - x)\, dy$$

$$= \int_0^{200} y^2\left(100 - \frac{y^2}{400}\right) dy = 100 \int_0^{200} y^2\, dy - \frac{1}{400} \int_0^{200} y^4\, dy$$

$$= 107(10^6)\ \text{mm}^4 \qquad\qquad\qquad Ans.$$

SOLUTION II (Case 2)
A differential element *parallel* to the y axis, as shown in Fig. 10–6b, is chosen for integration. It intersects the curve at the *arbitrary point* (x, y). In this case, all parts of the element do *not* lie at the same distance from the x axis, and therefore the parallel-axis theorem must be used to determine the *moment of inertia of the element* with respect to this axis. For a rectangle having a base b and height h, the moment of inertia about its centroidal axis has been determined in part (a) of Example 10.1. There it was found that $\bar{I}_{x'} = \frac{1}{12}bh^3$. For the differential element shown in Fig. 10–6b, $b = dx$ and $h = y$, and thus $d\bar{I}_{x'} = \frac{1}{12}dx\, y^3$. Since the centroid of the element is at $\tilde{y} = y/2$ from the x axis, the moment of inertia of the element about this axis is

$$dI_x = d\bar{I}_{x'} + dA\,\tilde{y}^2 = \frac{1}{12}dx\, y^3 + y\, dx\left(\frac{y}{2}\right)^2 = \frac{1}{3}y^3\, dx$$

This result can also be concluded from part (b) of Example 10.1. Integrating with respect to x, from $x = 0$ to $x = 100$ mm, yields

$$I_x = \int dI_x = \int_A \frac{1}{3}y^3\, dx = \int_0^{100} \frac{1}{3}(400x)^{3/2}\, dx$$

$$= 107(10^6)\ \text{mm}^4 \qquad\qquad\qquad Ans.$$

EXAMPLE 10.3

Determine the moment of inertia with respect to the x axis for the circular area shown in Fig. 10–7a.

(a)

SOLUTION I (Case 1)

Using the differential element shown in Fig. 10–7a, since $dA = 2x\,dy$, we have

$$I_x = \int_A y^2\,dA = \int_A y^2(2x)\,dy$$

$$= \int_{-a}^{a} y^2\left(2\sqrt{a^2 - y^2}\right)dy = \frac{\pi a^4}{4} \qquad \textit{Ans.}$$

SOLUTION II (Case 2)

When the differential element is chosen as shown in Fig. 10–7b, the centroid for the element happens to lie on the x axis, and so, applying Eq. 10–3, noting that $d_y = 0$ and for a rectangle $\bar{I}_{x'} = \frac{1}{12}bh^3$, we have

$$dI_x = \frac{1}{12}dx(2y)^3$$

$$= \frac{2}{3}y^3\,dx$$

Integrating with respect to x yields

$$I_x = \int_{-a}^{a} \frac{2}{3}(a^2 - x^2)^{3/2}\,dx = \frac{\pi a^4}{4} \qquad \textit{Ans.}$$

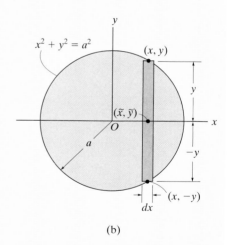

(b)

Fig. 10–7

EXAMPLE 10.4

(a)

Fig. 10–8

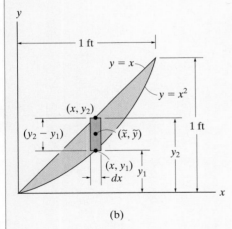

(b)

Determine the moment of inertia for the shaded area shown in Fig. 10–8a about the x axis.

SOLUTION I (Case 1)

The differential element parallel to the x axis is chosen for integration, Fig. 10–8a. The element intersects the curve at the *arbitrary points* (x_2, y) and (x_1, y). Consequently, its area is $dA = (x_1 - x_2)\,dy$. Since all parts of the element lie at the same distance y from the x axis, we have

$$I_x = \int_A y^2\,dA = \int_0^1 y^2(x_1 - x_2)\,dy = \int_0^1 y^2(\sqrt{y} - y)\,dy$$

$$I_x = \frac{2}{7}y^{7/2} - \frac{1}{4}y^4 \Big|_0^1 = 0.0357 \text{ ft}^4 \qquad\qquad Ans.$$

SOLUTION II (Case 2)

The differential element parallel to the y axis is shown in Fig. 10–8b. It intersects the curves at the *arbitrary points* (x, y_2) and (x, y_1). Since all parts of its entirety do *not* lie at the same distance from the x axis, we must first use the parallel-axis theorem to find the *element's* moment of inertia about the x axis, using $\overline{I}_{x'} = \frac{1}{12}bh^3$, then integrate this result to determine I_x. Thus,

$$dI_x = d\overline{I}_{x'} + dA\,\tilde{y}^2 = \frac{1}{12}dx(y_2 - y_1)^3$$

$$+ (y_2 - y_1)\,dx\left(y_1 + \frac{y_2 - y_1}{2}\right)^2$$

$$= \frac{1}{3}(y_2^3 - y_1^3)\,dx = \frac{1}{3}(x^3 - x^6)\,dx$$

$$I_x = \frac{1}{3}\int_0^1 (x^3 - x^6)\,dx = \frac{1}{12}x^4 - \frac{1}{21}x^7 \Big|_0^1 = 0.0357 \text{ ft}^4 \qquad Ans.$$

NOTE: By comparison, Solution I requires much less computation. Therefore, if an integral using a particular element appears difficult to evaluate, try solving the problem using an element oriented in the other direction.

PROBLEMS

10–1. Determine the moment of inertia for the shaded area about the x axis.

10–2. Determine the moment of inertia for the shaded area about the y axis.

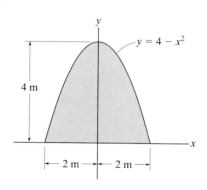

Probs. 10–1/2

10–3. Determine the moment of inertia for the thin strip of area about the x axis. The strip is oriented at an angle θ from the x axis. Assume that $t \ll l$.

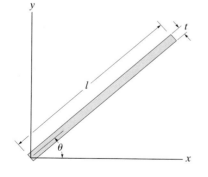

Prob. 10–3

*10–4. Determine the moment of inertia for the shaded area about the x axis.

10–5. Determine the moment of inertia for the shaded area about the y axis.

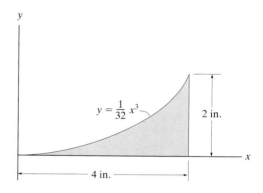

Probs. 10–4/5

10–6. Determine the moment of inertia for the shaded area about the x axis.

Prob. 10–6

10–7. Determine the moment of inertia for the shaded area about the x axis.

*10–8. Determine the moment of inertia for the shaded area about the y axis.

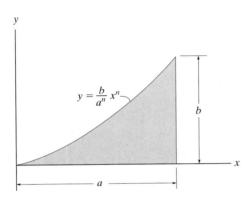

$$y = \frac{b}{a^n} x^n$$

Probs. 10–7/8

10–9. Determine the moment of inertia for the shaded area about the x axis.

10–10. Determine the moment of inertia for the shaded area about the y axis.

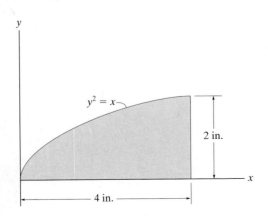

$y^2 = x$

2 in.

4 in.

Probs. 10–9/10

10–11. Determine the moment of inertia for the shaded area about the x axis.

$y^3 = x$

2 in.

8 in.

Prob. 10–11

*10–12. Determine the moment of inertia for the shaded area about the x axis.

10–13. Determine the moment of inertia for the shaded area about the y axis.

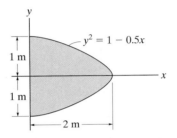

$y^2 = 1 - 0.5x$

1 m

1 m

2 m

Probs. 10–12/13

10–14. Determine the moment of inertia for the shaded area about the x axis.

10–15. Determine the moment of inertia for the shaded area about the y axis.

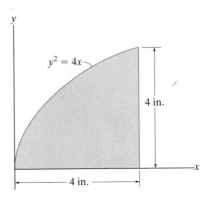

$y^2 = 4x$

4 in.

4 in.

Probs. 10–14/15

*10–16. Determine the moment of inertia for the shaded area about the x axis.

10–17. Determine the moment of inertia for the shaded area about the y axis.

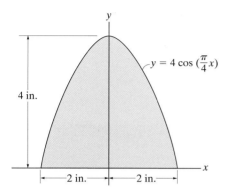

$y = 4 \cos \left(\frac{\pi}{4} x\right)$

4 in.

2 in.   2 in.

Probs. 10–16/17

10–18. Determine the moment of inertia for the shaded area about the x axis.

10–19. Determine the moment of inertia for the shaded area about the y axis.

$y = 2 \cos \left(\frac{\pi}{8} x\right)$

2 in.

4 in.   4 in.

Probs. 10–18/19

*10–20. Determine the moment of inertia for the shaded area about the x axis.

10–21. Determine the moment of inertia for the shaded area about the y axis.

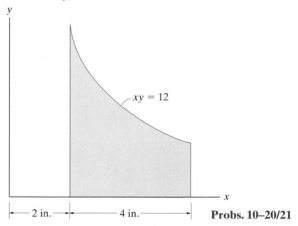

$xy = 12$

2 in.   4 in.

Probs. 10–20/21

10–22. Determine the moment of inertia for the shaded area about the x axis.

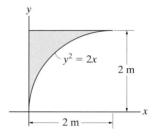

$y^2 = 2x$

2 m

2 m

Prob. 10–22

■10–23. Determine the moment of inertia for the shaded area about the y axis. Use Simpson's rule to evaluate the integral.

*■10–24. Determine the moment of inertia for the shaded area about the x axis. Use Simpson's rule to evaluate the integral.

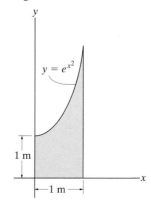

$y = e^{x^2}$

1 m

1 m

Probs. 10–23/24

10.5 Moments of Inertia for Composite Areas

A composite area consists of a series of connected "simpler" parts or shapes, such as semicircles, rectangles, and triangles. Provided the moment of inertia of each of these parts is known or can be determined about a common axis, then the moment of inertia for the composite area equals the *algebraic sum* of the moments of inertia of all its parts.

Structural members have various cross-sectional shapes, and it is necessary to calculate their moments of inertia in order to determine the stress in these members.

Procedure for Analysis

The moment of inertia for a composite area about a reference axis can be determined using the following procedure.

Composite Parts.

- Using a sketch, divide the area into its composite parts and indicate the perpendicular distance from the centroid of each part to the reference axis.

Parallel-Axis Theorem.

- The moment of inertia of each part should be determined about its centroidal axis, which is parallel to the reference axis. For the calculation use the table given on the inside back cover.

- If the centroidal axis does not coincide with the reference axis, the parallel-axis theorem, $I = \bar{I} + Ad^2$, should be used to determine the moment of inertia of the part about the reference axis.

Summation.

- The moment of inertia of the entire area about the reference axis is determined by summing the results of its composite parts.

- If a composite part has a "hole," its moment of inertia is found by "subtracting" the moment of inertia for the hole from the moment of inertia of the entire part including the hole.

EXAMPLE 10.5

Compute the moment of inertia for the composite area shown in Fig. 10–9a about the x axis.

Fig. 10–9

SOLUTION

Composite Parts.   The composite area is obtained by *subtracting* the circle from the rectangle as shown in Fig. 10–9b. The centroid of each area is located in the figure.

Parallel-Axis Theorem.   The moments of inertia about the x axis are determined using the parallel-axis theorem and the data in the table on the inside back cover.

Circle

$$I_x = \overline{I}_{x'} + Ad_y^2$$

$$= \frac{1}{4}\pi(25)^4 + \pi(25)^2(75)^2 = 11.4(10^6) \text{ mm}^4$$

Rectangle

$$I_x = \overline{I}_{x'} + Ad_y^2$$

$$= \frac{1}{12}(100)(150)^3 + (100)(150)(75)^2 = 112.5(10^6) \text{ mm}^4$$

Summation.   The moment of inertia for the composite area is thus

$$I_x = -11.4(10^6) + 112.5(10^6)$$

$$= 101(10^6) \text{ mm}^4 \qquad\qquad \textit{Ans.}$$

EXAMPLE 10.6

(a)

(b)

Fig. 10–10

Determine the moments of inertia for the beam's cross-sectional area shown in Fig. 10–10a about the x and y centroidal axes.

SOLUTION

Composite Parts.   The cross section can be considered as three composite rectangular areas A, B, and D shown in Fig. 10–10b. For the calculation, the centroid of each of these rectangles is located in the figure.

Parallel-Axis Theorem.   From the table on the inside back cover, or Example 10.1, the moment of inertia of a rectangle about its centroidal axis is $\bar{I} = \frac{1}{12}bh^3$. Hence, using the parallel-axis theorem for rectangles A and D, the calculations are as follows:

Rectangle A

$$I_x = \bar{I}_{x'} + Ad_y^2 = \frac{1}{12}(100)(300)^3 + (100)(300)(200)^2$$

$$= 1.425(10^9)\ \text{mm}^4$$

$$I_y = \bar{I}_{y'} + Ad_x^2 = \frac{1}{12}(300)(100)^3 + (100)(300)(250)^2$$

$$= 1.90(10^9)\ \text{mm}^4$$

Rectangle B

$$I_x = \frac{1}{12}(600)(100)^3 = 0.05(10^9)\ \text{mm}^4$$

$$I_y = \frac{1}{12}(100)(600)^3 = 1.80(10^9)\ \text{mm}^4$$

Rectangle D

$$I_x = \bar{I}_{x'} + Ad_y^2 = \frac{1}{12}(100)(300)^3 + (100)(300)(200)^2$$

$$= 1.425(10^9)\ \text{mm}^4$$

$$I_y = \bar{I}_{y'} + Ad_x^2 = \frac{1}{12}(300)(100)^3 + (100)(300)(250)^2$$

$$= 1.90(10^9)\ \text{mm}^4$$

Summation.   The moments of inertia for the entire cross section are thus

$$I_x = 1.425(10^9) + 0.05(10^9) + 1.425(10^9)$$

$$= 2.90(10^9)\ \text{mm}^4 \qquad\qquad \textit{Ans.}$$

$$I_y = 1.90(10^9) + 1.80(10^9) + 1.90(10^9)$$

$$= 5.60(10^9)\ \text{mm}^4 \qquad\qquad \textit{Ans.}$$

PROBLEMS

10–25. The polar moment of inertia for the area is $\bar{J}_C = 28$ in⁴ about the z axis passing through the centroid C. The moment of inertia about the x axis is 17 in⁴, and the moment of inertia about the y' axis is 56 in⁴. Determine the area A.

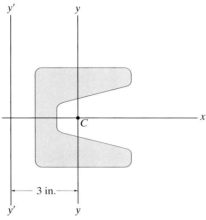

Prob. 10–25

10–26. The polar moment of inertia for the area is $\bar{J}_C = 548(10^6)$ mm⁴, about the z' axis passing through the centroid C. The moment of inertia about the y' axis is $383(10^6)$ mm⁴, and the moment of inertia about the x axis is $856(10^6)$ mm⁴. Determine the area A.

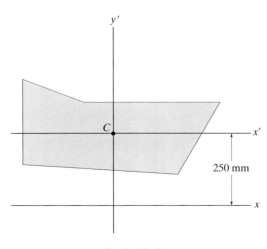

Prob. 10–26

10–27. Determine the radius of gyration k_x of the column's cross-sectional area.

*10–28. Determine the radius of gyration k_y of the column's cross-sectional area.

Probs. 10–27/28

10–29. Determine the moment of inertia for the beam's cross-sectional area with respect to the x' centroidal axis. Neglect the size of all the rivet heads. R, for the calculation. Handbook values for the area, moment of inertia, and location of the centroid C of one of the angles are listed in the figure.

$(Ia)x_a' = 1.32(10^6)$ mm⁴
$A_a = 1.36(10^3)$ mm²

Prob. 10–29

10–30.   Locate the centroid $\bar{y}$ of the cross-sectional area for the angle. Then find the moment of inertia $\bar{I}_{x'}$ about the x' centroidal axis.

10–31.   Locate the centroid $\bar{x}$ of the cross-sectional area for the angle. Then find the moment of inertia $\bar{I}_{y'}$ about the y' centroidal axis.

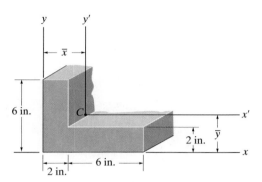

Probs. 10–30/31

10–34.   Determine the moments of inertia for the shaded area about the x and y axes.

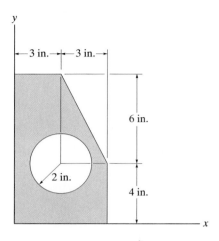

Prob. 10–34

*10–32.   Determine the distance $\bar{x}$ to the centroid of the beam's cross-sectional area; then find the moment of inertia about the y' axis.

10–33.   Determine the moment of inertia for the beam's cross-sectional area about the x' axis.

Probs. 10–32/33

10–35.   Determine the location of the centroid $\bar{y}$ of the beam constructed from the two channels and the cover plate. If each channel has a cross-sectional area of $A_c = 11.8 \text{ in}^2$ and a moment of inertia about a horizontal axis passing through its own centroid, C_c, of $\bar{I}_{x_c'} = 349 \text{ in}^4$, determine the moment of inertia for the beam's cross-sectional area about the x' axis.

Prob. 10–35

*10–36.   Compute the moments of inertia I_x and I_y for the beam's cross-sectional area about the x and y axes.

Prob. 10–36

10–37.   Determine the distance $\bar{y}$ to the centroid C of the beam's cross-sectional area and then compute the moment of inertia $\bar{I}_{x'}$ about the x' axis.

10–38.   Determine the distance $\bar{x}$ to the centroid C of the beam's cross-sectional area and then compute the moment of inertia $\bar{I}_{y'}$ about the y' axis.

Probs. 10–37/38

10–39.   Determine the location $\bar{y}$ of the centroid C of the beam's cross-sectional area. Then compute the moment of inertia for the area about the x' axis.

Prob. 10–39

*10–40.   Determine $\bar{y}$, which locates the centroidal axis x' for the cross-sectional area of the T-beam, and then find the moments of inertia $\bar{I}_{x'}$ and $\bar{I}_{y'}$.

Prob. 10–40

10–41. Determine the centroid $\bar{y}$ for the beam's cross-sectional area; then find $\bar{I}_{x'}$.

10–42. Determine the moment of inertia for the beam's cross-sectional area about the y axis.

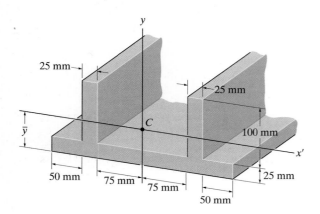

Probs. 10–41/42

10–43. Determine the moment of inertia I_x for the shaded area about the x axis.

*10–44. Determine the moment of inertia I_y for the shaded area about the y axis.

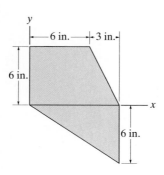

Probs. 10–43/44

10–45. Locate the centroid $\bar{y}$ of the channel's cross-sectional area, and then determine the moment of inertia with respect to the x' axis passing through the centroid.

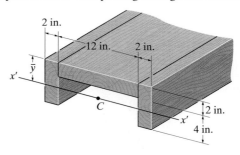

Prob. 10–45

10–46. Determine the moments of inertia I_x and I_y for the shaded area.

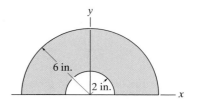

Prob. 10–46

10–47. Determine the moment of inertia for the parallelogram about the x' axis, which passes through the centroid C of the area.

*10–48. Determine the moment of inertia for the parallelogram about the y' axis, which passes through the centroid C of the area.

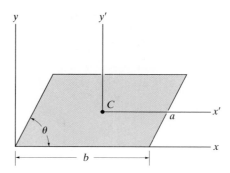

Probs. 10–47/48

10–49. Determine the moments of inertia for the triangular area about the x' and y' axes, which pass through the centroid C of the area.

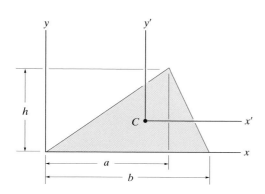

Prob. 10–49

10–50. Determine the moment of inertia for the beam's cross-sectional area about the x' axis passing through the centroid C of the cross section.

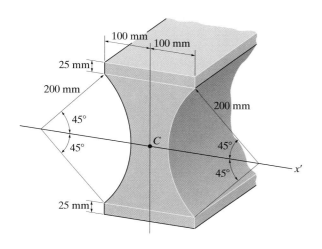

Prob. 10–50

10–51. Determine the moment of inertia for the composite area about the x axis.

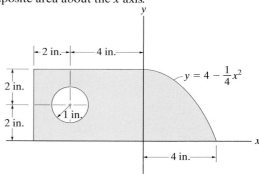

Prob. 10–51

*10–52. Determine the moment of inertia for the composite area about the y axis.

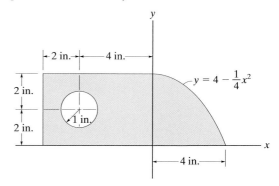

Prob. 10–52

10–53. Determine the radius of gyration k_x for the column's cross-sectional area.

Prob. 10–53

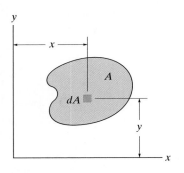

Fig. 10–11

*10.6   Product of Inertia for an Area

In general, the moment of inertia for an area is different for every axis about which it is computed. In some applications of structural or mechanical design it is necessary to know the orientation of those axes which give, respectively, the maximum and minimum moments of inertia for the area. The method for determining this is discussed in the next section. To use this method, however, we first need to compute the product of inertia for the area as well as its moments of inertia for given x, y axes.

The product of inertia for an element of area dA located at point (x, y), Fig. 10–11, is defined as $dI_{xy} = xy\, dA$. Thus, for the entire area A, the *product of inertia* is

$$I_{xy} = \int_A xy\, dA \qquad (10\text{–}7)$$

If the element of area chosen has a differential size in two directions, as shown in Fig. 10–11, a double integration must be performed to evaluate I_{xy}. Most often, however, it is easier to choose an element having a differential size or thickness in only one direction in which case the evaluation requires only a single integration (see Example 10.7).

Like the moment of inertia, the product of inertia has units of length raised to the fourth power, e.g., m^4, mm^4 or ft^4, in^4. However, since x or y may be a negative quantity, while the element of area is always positive, the product of inertia may be positive, negative, or zero, depending on the location and orientation of the coordinate axes. For example, the product of inertia I_{xy} for an area will be *zero* if either the x or y axis is an axis of *symmetry* for the area. To show this, consider the shaded area in Fig. 10–12, where for every element dA located at point (x, y) there is a corresponding element dA located at $(x, -y)$. Since the products of inertia for these elements are, respectively, $xy\, dA$ and $-xy\, dA$, the algebraic sum or integration of all the elements that are chosen in this way will cancel each other. Consequently, the product of inertia for the

Fig. 10–12

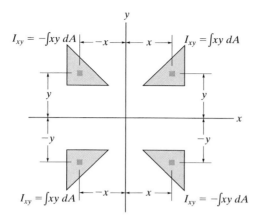

$I_{xy} = -\int xy\, dA$ $I_{xy} = \int xy\, dA$

$I_{xy} = \int xy\, dA$ $I_{xy} = -\int xy\, dA$

Fig. 10–13

total area becomes zero. It also follows from the definition of I_{xy} that the "sign" of this quantity depends on the quadrant where the area is located. As shown in Fig. 10–13, if the area is rotated from one quadrant to another, the sign of I_{xy} will change.

Parallel-Axis Theorem.   Consider the shaded area shown in Fig. 10–14, where x' and y' represent a set of axes passing through the *centroid* of the area, and x and y represent a corresponding set of parallel axes. Since the product of inertia of dA with respect to the x and y axes is $dI_{xy} = (x' + d_x)(y' + d_y)\, dA$, then for the entire area,

$$I_{xy} = \int_A (x' + d_x)(y' + d_y)\, dA$$

$$= \int_A x'y'\, dA + d_x \int_A y'\, dA + d_y \int_A x'\, dA + d_x d_y \int_A dA$$

The first term on the right represents the product of inertia for the area with respect to the centroidal axis, $\bar{I}_{x'y'}$. The integrals in the second and third terms are zero since the moments of the area are taken about the centroidal axis. Realizing that the fourth integral represents the total area A, the final result is therefore

$$\boxed{I_{xy} = \bar{I}_{x'y'} + A d_x d_y} \qquad (10\text{–}8)$$

Note the similarity between this equation and the parallel-axis theorem for moments of inertia. In particular, it is important that the *algebraic signs* for d_x and d_y be maintained when applying Eq. 10–8. As illustrated in Example 10.8, the parallel-axis theorem finds important application in determining the product of inertia for a *composite area* with respect to a set of x, y axes.

Fig. 10–14

EXAMPLE 10.7

(a)

(b)

(c)

Fig. 10–15

Determine the product of inertia I_{xy} for the triangle shown in Fig. 10–15a.

SOLUTION I
A differential element that has a thickness dx, Fig. 10–15b, has an area $dA = y\, dx$. The product of inertia of the element about the x, y axes is determined using the parallel-axis theorem.

$$dI_{xy} = d\overline{I}_{x'y'} + dA\, \widetilde{x}\widetilde{y}$$

where $(\widetilde{x}, \widetilde{y})$ locates the *centroid* of the element or the origin of the x', y' axes. Since $d\overline{I}_{x'y'} = 0$, due to symmetry, and $\widetilde{x} = x$, $\widetilde{y} = y/2$, then

$$dI_{xy} = 0 + (y\,dx)x\left(\frac{y}{2}\right) = \left(\frac{h}{b}x\,dx\right)x\left(\frac{h}{2b}x\right)$$

$$= \frac{h^2}{2b^2}x^3\,dx$$

Integrating with respect to x from $x = 0$ to $x = b$ yields

$$I_{xy} = \frac{h^2}{2b^2}\int_0^b x^3\,dx = \frac{b^2 h^2}{8} \qquad\qquad Ans.$$

SOLUTION II
The differential element that has a thickness dy, Fig. 10–15c, and area $dA = (b - x)\,dy$ can also be used. The *centroid* is located at point $\widetilde{x} = x + (b - x)/2 = (b + x)/2$, $\widetilde{y} = y$, so the product of inertia of the element becomes

$$dI_{xy} = d\overline{I}_{x'y'} + dA\, \widetilde{x}\widetilde{y}$$

$$= 0 + (b - x)\,dy\left(\frac{b + x}{2}\right)y$$

$$= \left(b - \frac{b}{h}y\right)dy\left[\frac{b + (b/h)y}{2}\right]y = \frac{1}{2}y\left(b^2 - \frac{b^2}{h^2}y^2\right)dy$$

Integrating with respect to y from $y = 0$ to $y = h$ yields

$$I_{xy} = \frac{1}{2}\int_0^h y\left(b^2 - \frac{b^2}{h^2}y^2\right)dy = \frac{b^2 h^2}{8} \qquad\qquad Ans.$$

EXAMPLE 10.8

Determine the product of inertia for the beam's cross-sectional area, shown in Fig. 10–16a, about the x and y centroidal axes.

(a)   **Fig. 10–16**   (b)

SOLUTION
As in Example 10.6, the cross section can be considered as three composite rectangular areas A, B, and D, Fig. 10–16b. The coordinates for the centroid of each of these rectangles are shown in the figure. Due to symmetry, the product of inertia of *each rectangle* is *zero* about each set of x', y' axes that passes through the rectangle's centroid. Hence, application of the parallel-axis theorem to each of the rectangles yields

Rectangle A

$$I_{xy} = \bar{I}_{x'y'} + Ad_xd_y$$
$$= 0 + (300)(100)(-250)(200) = -1.50(10^9) \text{ mm}^4$$

Rectangle B

$$I_{xy} = \bar{I}_{x'y'} + Ad_xd_y$$
$$= 0 + 0 = 0$$

Rectangle D

$$I_{xy} = \bar{I}_{x'y'} + Ad_xd_y$$
$$= 0 + (300)(100)(250)(-200) = -1.50(10^9) \text{ mm}^4$$

The product of inertia for the entire cross section is therefore

$$I_{xy} = -1.50(10^9) + 0 - 1.50(10^9) = -3.00(10^9) \text{ mm}^4 \qquad Ans.$$

NOTE: This negative result is due to the fact that rectangles A and D have centroids located with negative x and negative y coordinates, respectively.

*10.7 Moments of Inertia for an Area about Inclined Axes

In structural and mechanical design, it is sometimes necessary to calculate the moments and product of inertia I_u, I_v, and I_{uv} for an area with respect to a set of inclined u and v axes when the values for θ, I_x, I_y, and I_{xy} are *known*. To do this we will use *transformation equations* which relate the x, y and u, v coordinates. From Fig. 10–17, these equations are

$$u = x \cos \theta + y \sin \theta$$
$$v = y \cos \theta - x \sin \theta$$

Using these equations, the moments and product of inertia of dA about the u and v axes become

$$dI_u = v^2\, dA = (y \cos \theta - x \sin \theta)^2\, dA$$
$$dI_v = u^2\, dA = (x \cos \theta + y \sin \theta)^2\, dA$$
$$dI_{uv} = uv\, dA = (x \cos \theta + y \sin \theta)(y \cos \theta - x \sin \theta)\, dA$$

Expanding each expression and integrating, realizing that $I_x = \int y^2\, dA$, $I_y = \int x^2\, dA$, and $I_{xy} = \int xy\, dA$, we obtain

$$I_u = I_x \cos^2 \theta + I_y \sin^2 \theta - 2I_{xy} \sin \theta \cos \theta$$
$$I_v = I_x \sin^2 \theta + I_y \cos^2 \theta + 2I_{xy} \sin \theta \cos \theta$$
$$I_{uv} = I_x \sin \theta \cos \theta - I_y \sin \theta \cos \theta + I_{xy}(\cos^2 \theta - \sin^2 \theta)$$

These equations may be simplified by using the trigonometric identities $\sin 2\theta = 2 \sin \theta \cos \theta$ and $\cos 2\theta = \cos^2 \theta - \sin^2 \theta$, in which case

$$I_u = \frac{I_x + I_y}{2} + \frac{I_x - I_y}{2} \cos 2\theta - I_{xy} \sin 2\theta$$

$$I_v = \frac{I_x + I_y}{2} - \frac{I_x - I_y}{2} \cos 2\theta + I_{xy} \sin 2\theta$$

$$I_{uv} = \frac{I_x - I_y}{2} \sin 2\theta + I_{xy} \cos 2\theta \tag{10–9}$$

If the first and second equations are added together, we can show that the polar moment of inertia about the z axis passing through point O is *independent* of the orientation of the u and v axes; i.e.,

$$J_O = I_u + I_v = I_x + I_y$$

Principal Moments of Inertia.

Equations 10–9 show that I_u, I_v, and I_{uv} depend on the angle of inclination, θ, of the u, v axes. We will now determine the orientation of these axes about which the moments of inertia for the area, I_u and I_v, are maximum and minimum. This particular set of axes is called the *principal axes* of the area, and the corresponding

Fig. 10–17

moments of inertia with respect to these axes are called the *principal moments of inertia*. In general, there is a set of principal axes for every chosen origin O. For the structural and mechanical design of a member, the origin O is generally located at the cross-sectional area's centroid.

The angle $\theta = \theta_p$, which defines the orientation of the principal axes for the area, may be found by differentiating the first of Eqs. 10–9 with respect to θ and setting the result equal to zero. Thus,

$$\frac{dI_u}{d\theta} = -2\left(\frac{I_x - I_y}{2}\right)\sin 2\theta - 2I_{xy}\cos 2\theta = 0$$

Therefore, at $\theta = \theta_p$,

$$\tan 2\theta_p = \frac{-I_{xy}}{(I_x - I_y)/2} \qquad (10\text{–}10)$$

This equation has two roots, θ_{p_1} and θ_{p_2}, which are 90° apart and so specify the inclination of the principal axes. In order to substitute them into Eq. 10–9, we must first find the sine and cosine of $2\theta_{p_1}$ and $2\theta_{p_2}$. This can be done using the triangles shown in Fig. 10–18, which are based on Eq. 10–10.

For θ_{p_1},

$$\sin 2\theta_{p_1} = -I_{xy}\Big/ \sqrt{\left(\frac{I_x - I_y}{2}\right)^2 + I_{xy}^2}$$

$$\cos 2\theta_{p_1} = \left(\frac{I_x - I_y}{2}\right)\Big/ \sqrt{\left(\frac{I_x - I_y}{2}\right)^2 + I_{xy}^2}$$

For θ_{p_2},

$$\sin 2\theta_{p_2} = I_{xy}\Big/ \sqrt{\left(\frac{I_x - I_y}{2}\right)^2 + I_{xy}^2}$$

$$\cos 2\theta_{p_2} = -\left(\frac{I_x - I_y}{2}\right)\Big/ \sqrt{\left(\frac{I_x - I_y}{2}\right)^2 + I_{xy}^2}$$

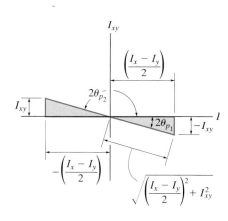

Fig. 10–18

Substituting these two sets of trigonometric relations into the first or second of Eqs. 10–9 and simplifying, we obtain

$$I^{max}_{min} = \frac{I_x + I_y}{2} \pm \sqrt{\left(\frac{I_x - I_y}{2}\right)^2 + I_{xy}^2} \qquad (10\text{–}11)$$

Depending on the sign chosen, this result gives the maximum or minimum moment of inertia for the area. Furthermore, if the above trigonometric relations for θ_{p_1} and θ_{p_2} are substituted into the third of Eqs. 10–9, it can be shown that $I_{uv} = 0$; that is, the *product of inertia with respect to the principal axes is zero*. Since it was indicated in Sec. 10.6 that the product of inertia is zero with respect to any symmetrical axis, it therefore follows that *any symmetrical axis represents a principal axis of inertia for the area.*

EXAMPLE 10.9

(a)

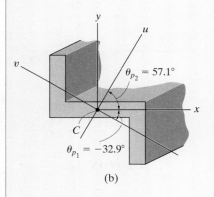

(b)

Fig. 10–19

Determine the principal moments of inertia for the beam's cross-sectional area shown in Fig. 10–19a with respect to an axis passing through the centroid.

SOLUTION
The moments and product of inertia of the cross section with respect to the x, y axes have been computed in Examples 10.6 and 10.8. The results are

$$I_x = 2.90(10^9) \text{ mm}^4 \quad I_y = 5.60(10^9) \text{ mm}^4 \quad I_{xy} = -3.00(10^9) \text{ mm}^4$$

Using Eq. 10–10, the angles of inclination of the principal axes u and v are

$$\tan 2\theta_p = \frac{-I_{xy}}{(I_x - I_y)/2} = \frac{3.00(10^9)}{[2.90(10^9) - 5.60(10^9)]/2} = -2.22$$

$$2\theta_{p_1} = -65.8° \quad \text{and} \quad 2\theta_{p_2} = 114.2°$$

Thus, as shown in Fig. 10–19b,

$$\theta_{p_1} = -32.9° \quad \text{and} \quad \theta_{p_2} = 57.1°$$

The principal moments of inertia with respect to the u and v axes are determined from Eq. 10–11. Hence,

$$I_{\substack{max \\ min}} = \frac{I_x + I_y}{2} \pm \sqrt{\left(\frac{I_x - I_y}{2}\right)^2 + I_{xy}^2}$$

$$= \frac{2.90(10^9) + 5.60(10^9)}{2}$$

$$\pm \sqrt{\left[\frac{2.90(10^9) - 5.60(10^9)}{2}\right]^2 + [-3.00(10^9)]^2}$$

$$I_{\substack{max \\ min}} = 4.25(10^9) \pm 3.29(10^9)$$

or

$$I_{max} = 7.54(10^9) \text{ mm}^4 \quad I_{min} = 0.960(10^9) \text{ mm}^4 \qquad Ans.$$

NOTE: The maximum moment of inertia, $I_{max} = 7.54(10^9) \text{ mm}^4$, occurs with respect to the selected u axis since *by inspection* most of the cross-sectional area is farthest away from this axis. Or, stated in another manner, I_{max} occurs about the u axis since it is located within $\pm45°$ of the y axis, which has the largest value of $I(I_y > I_x)$. Also, this may be concluded by substituting the data with $\theta = 57.1°$ into the first of Eqs. 10–9.

*10.8   Mohr's Circle for Moments of Inertia

Equations 10–9 to 10–11 have a graphical solution that is convenient to use and generally easy to remember. Squaring the first and third of Eqs. 10–9 and adding, it is found that

$$\left(I_u - \frac{I_x + I_y}{2}\right)^2 + I_{uv}^2 = \left(\frac{I_x - I_y}{2}\right)^2 + I_{xy}^2 \qquad (10\text{–}12)$$

In a given problem, I_u and I_{uv} are *variables*, and I_x, I_y, and I_{xy} are *known constants*. Thus, Eq. 10–12 may be written in compact form as

$$(I_u - a)^2 + I_{uv}^2 = R^2$$

When this equation is plotted on a set of axes that represent the respective moment of inertia and the product of inertia, Fig. 10–20, the resulting graph represents a *circle* of radius

$$R = \sqrt{\left(\frac{I_x - I_y}{2}\right)^2 + I_{xy}^2}$$

having its center located at point $(a, 0)$, where $a = (I_x + I_y)/2$. The circle so constructed is called *Mohr's circle*, named after the German engineer Otto Mohr (1835–1918).

(a)

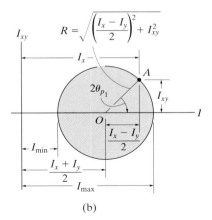

(b)

Fig. 10–20

Procedure for Analysis

The main purpose in using Mohr's circle here is to have a convenient means for transforming I_x, I_y, and I_{xy} into the principal moments of inertia. The following procedure provides a method for doing this.

Determine I_x, I_y, and I_{xy}.

- Establish the x, y axes for the area, with the origin located at the point P of interest, and determine I_x, I_y, and I_{xy}, Fig. 10–20a.

Construct the Circle.

- Construct a rectangular coordinate system such that the abscissa represents the moment of inertia I, and the ordinate represents the product of inertia I_{xy}, Fig. 10–20b.

- Determine the center of the circle, O, which is located at a distance $(I_x + I_y)/2$ from the origin, and plot the reference point A having coordinates (I_x, I_{xy}). By definition, I_x is always positive, whereas I_{xy} will be either positive or negative.

- Connect the reference point A with the center of the circle and determine the distance OA by trigonometry. This distance represents the radius of the circle, Fig. 10–20b. Finally, draw the circle.

Principal Moments of Inertia.

- The points where the circle intersects the abscissa give the values of the principal moments of inertia I_{min} and I_{max}. Notice that the *product of inertia will be zero at these points*, Fig. 10–20b.

Principal Axes.

- To find the direction of the major principal axis, determine by trigonometry the angle $2\theta_{p_1}$, *measured from the radius OA to the positive I axis*, Fig. 10–20b. This angle represents *twice* the angle from the x axis of the area in question to the axis of maximum moment of inertia I_{max}, Fig. 10–20a. Both the angle on the circle, $2\theta_{p_1}$, and the angle to the axis on the area, θ_{p_1}, *must be measured in the same sense*, as shown in Fig. 10–20. The axis for minimum moment of inertia I_{min} is perpendicular to the axis for I_{max}.

Using trigonometry, the above procedure may be verified to be in accordance with the equations developed in Sec. 10.7.

EXAMPLE 10.10

Using Mohr's circle, determine the principal moments of inertia for the beam's cross-sectional area shown in Fig. 10–21a, with respect to an axis passing through the centroid.

(a)

(b)

SOLUTION

Determine I_x, I_y, I_{xy}.   The moment of inertia and the product of inertia have been determined in Examples 10.6 and 10.8 with respect to the x, y axes shown in Fig. 10–21a. The results are $I_x = 2.90(10^9)$ mm^4, $I_y = 5.60(10^9)$ mm^4, and $I_{xy} = -3.00(10^9)$ mm^4.

Construct the Circle.   The I and I_{xy} axes are shown in Fig. 10–21b. The center of the circle, O, lies at a distance $(I_x+I_y)/2 = (2.90+5.60)/2 = 4.25$ from the origin. When the reference point $A(2.90, -3.00)$ is connected to point O, the radius OA is determined from the triangle OBA using the Pythagorean theorem.

$$OA = \sqrt{(1.35)^2 + (-3.00)^2} = 3.29$$

The circle is constructed in Fig. 10–21c.

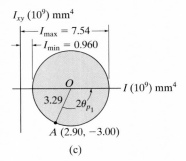

(c)

Principal Moments of Inertia.   The circle intersects the I axis at points $(7.54, 0)$ and $(0.960, 0)$. Hence,

$$I_{max} = 7.54(10^9) \text{ mm}^4 \qquad\qquad Ans.$$
$$I_{min} = 0.960(10^9) \text{ mm}^4 \qquad\qquad Ans.$$

Principal Axes.   As shown in Fig. 10–21c, the angle $2\theta_{p_1}$ is determined from the circle by measuring counterclockwise from OA to the direction of the *positive I* axis. Hence,

$$2\theta_{p_1} = 180° - \sin^{-1}\left(\frac{|BA|}{|OA|}\right) = 180° - \sin^{-1}\left(\frac{3.00}{3.29}\right) = 114.2°$$

The principal axis for $I_{max} = 7.54(10^9)$ mm^4 is therefore oriented at an angle $\theta_{p_1} = 57.1°$, measured *counterclockwise*, from the *positive x* axis to the *positive u* axis. The v axis is perpendicular to this axis. The results are shown in Fig. 10–21d.

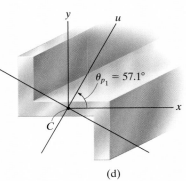

(d)

Fig. 10–21

PROBLEMS

10–54. Determine the product of inertia for the shaded portion of the parabola with respect to the x and y axes.

Prob. 10–54

10–55. Determine the product of inertia for the shaded area with respect to the x and y axes.

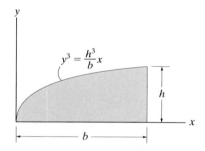

Prob. 10–55

*10–56. Determine the product of inertia for the shaded area of the ellipse with respect to the x and y axes.

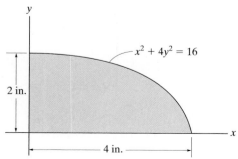

Prob. 10–56

10–57. Determine the product of inertia for the parabolic area with respect to the x and y axes.

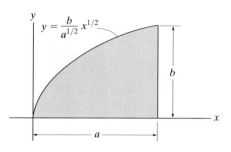

Prob. 10–57

10–58. Determine the product of inertia for the shaded area with respect to the x and y axes.

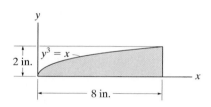

Prob. 10–58

10–59. Determine the product of inertia for the shaded parabolic area with respect to the x and y axes.

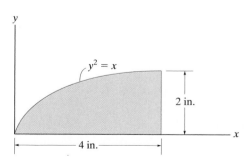

Prob. 10–59

*10–60. Determine the product of inertia for the shaded area with respect to the x and y axes.

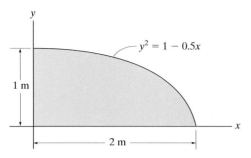

Prob. 10–60

10–61. Determine the product of inertia for the shaded area with respect to the x and y axes.

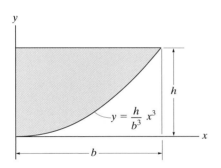

Prob. 10–61

10–62. Determine the product of inertia for the shaded area with respect to the x and y axes.

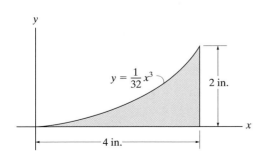

Prob. 10–62

10–63. Determine the product of inertia for the shaded area with respect to the x and y axes.

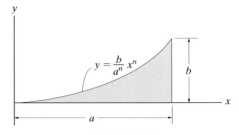

Prob. 10–63

*10–64. Determine the product of inertia for the shaded area with respect to the x and y axes.

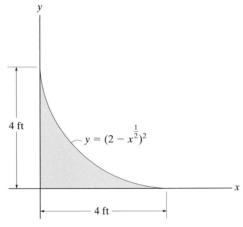

Prob. 10–64

■10–65. Determine the product of inertia for the shaded area with respect to the x and y axes. Use Simpson's rule to evaluate the integral.

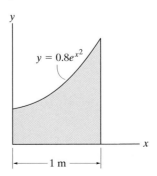

Prob. 10–65

10–66.   Determine the product of inertia for the parabolic area with respect to the x and y axes.

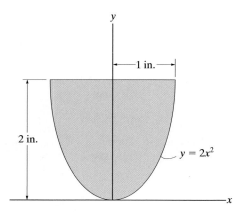

Prob. 10–66

10–67.   Determine the product of inertia for the cross-sectional area with respect to the x and y axes that have their origin located at the centroid C.

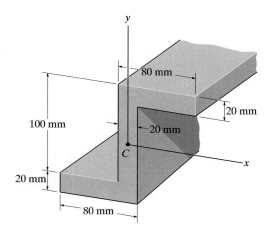

Prob. 10–67

***10–68.**   Determine the product of inertia for the beam's cross-sectional area with respect to the x and y axes.

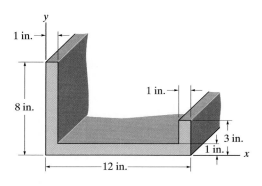

Prob. 10–68

10–69.   Determine the location $(\bar{x}, \bar{y})$ of the centroid C of the angle's cross-sectional area, and then compute the product of inertia with respect to the x' and y' axes.

Prob. 10–69

10–70. Determine the product of inertia for the beam's cross-sectional area with respect to the x and y axes that have their origin located at the centroid C.

Prob. 10–70

10–71. Determine the product of inertia for the shaded area with respect to the x and y axes.

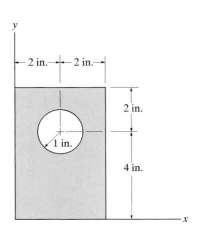

Prob. 10–71

*10–72. Determine the product of inertia for the beam's cross-sectional area with respect to the x and y axes that have their origin located at the centroid C.

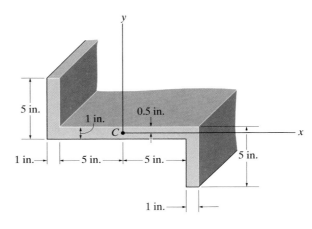

Prob. 10–72

10–73. Determine the product of inertia for the cross-sectional area with respect to the x and y axes.

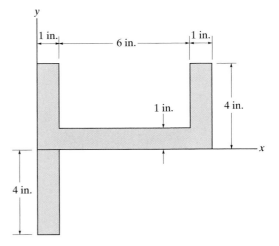

Prob. 10–73

10–74. Determine the product of inertia for the beam's cross-sectional area with respect to the u and v axes.

Prob. 10–74

10–75. Determine the moments of inertia I_u and I_v and the product of inertia I_{uv} for the rectangular area. The u and v axes pass through the centroid C.

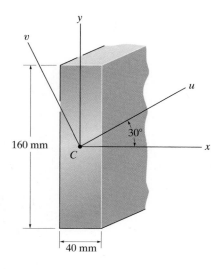

Prob. 10–75

*10–76. Determine the distance $\bar{y}$ to the centroid of the area and then calculate the moments of inertia I_u and I_v for the channel's cross-sectional area. The u and v axes have their origin at the centroid C. For the calculation, assume all corners to be square.

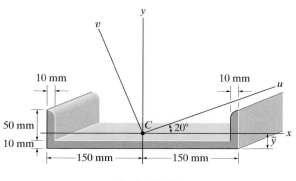

Prob. 10–76

10–77. Determine the moments of inertia for the shaded area with respect to the u and v axes.

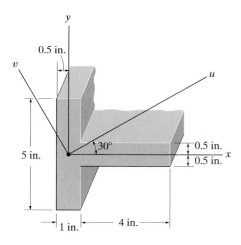

Prob. 10–77

10–78. Determine the directions of the principal axes with origin located at point O, and the principal moments of inertia for the rectangular area about these axes.

Prob. 10–78

10–79. Determine the moments of inertia I_u, I_v and the product of inertia I_{uv} for the beam's cross-sectional area. Take $\theta = 45°$.

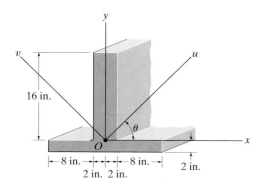

Prob. 10–79

*10–80. Determine the directions of the principal axes with origin located at point O, and the principal moments of inertia for the area about these axes.

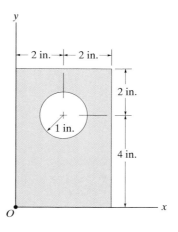

Prob. 10–80

10–81. Determine the principal moments of inertia for the beam's cross-sectional area about the principal axes that have their origin located at the centroid C. Use the equations developed in Section 10.7. For the calculation, assume all corners to be square.

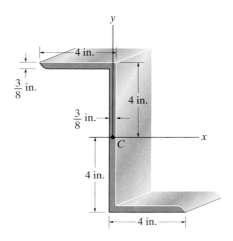

Prob. 10–81

10–82. Determine the principal moments of inertia for the angle's cross-sectional area with respect to a set of principal axes that have their origin located at the centroid C. Use the equation developed in Section 10.7. For the calculation, assume all corners to be square.

Prob. 10–82

10–83. The area of the cross section of an airplane wing has the following properties about the x and y axes passing through the centroid C: $\bar{I}_x = 450 \text{ in}^4$, $\bar{I}_y = 1730 \text{ in}^4$, $\bar{I}_{xy} = 138 \text{ in}^4$. Determine the orientation of the principal axes and the principal moments of inertia.

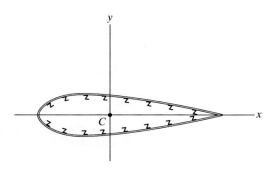

Prob. 10–83

*10–84. Using Mohr's circle, determine the principal moments of inertia for the triangular area and the orientation of the principal axes of inertia having an origin at point O.

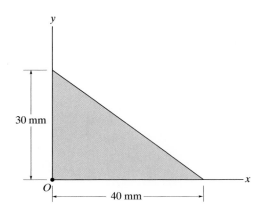

Prob. 10–84

10–85. Solve Prob. 10–78 using Mohr's circle.

10–86. Solve Prob. 10–81 using Mohr's circle.

10–87. Solve Prob. 10–82 using Mohr's circle.

*10–88. Solve Prob. 10–80 using Mohr's circle.

10–89. Solve Prob. 10–83 using Mohr's circle.

10.9   Mass Moment of Inertia

The mass moment of inertia of a body is a property that measures the resistance of the body to angular acceleration. Since it is used in dynamics to study rotational motion, methods for its calculation will now be discussed.

We define the *mass moment of inertia* as the integral of the "second moment" about an axis of all the elements of mass dm which compose the body.* For example, consider the rigid body shown in Fig. 10–22. The body's moment of inertia about the z axis is

$$I = \int_m r^2 \, dm \qquad (10\text{--}13)$$

Here the "moment arm" r is the perpendicular distance from the axis to the arbitrary element dm. Since the formulation involves r, the value of I is *unique* for each axis z about which it is computed. However, the axis which is generally chosen for analysis passes through the body's mass center G. The moment of inertia computed about this axis will be defined as I_G. Realize that because r is squared in Eq. 10–13, the mass moment of inertia is always a *positive quantity*. Common units used for its measurement are $\text{kg} \cdot \text{m}^2$ or $\text{slug} \cdot \text{ft}^2$.

If the body consists of material having a variable density, $\rho = \rho(x, y, z)$, the elemental mass dm of the body may be expressed in terms of its density and volume as $dm = \rho \, dV$. Substituting dm into Eq. 10–13, the body's moment of inertia is then computed using *volume elements* for integration; i.e.

$$I = \int_V r^2 \rho \, dV \qquad (10\text{--}14)$$

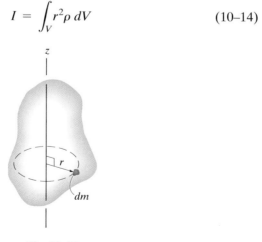

Fig. 10–22

*Another property of the body which measures the symmetry of the body's mass with respect to a coordinate system is the mass product of inertia. This property most often applies to the three-dimensional motion of a body and is discussed in *Engineering Mechanics: Dynamics* (Chapter 21).

(a)

(b)

(c)

Fig. 10–23

In the special case of ρ being a *constant*, this term may be factored out of the integral, and the integration is then purely a function of geometry:

$$I = \rho \int_V r^2 \, dV \qquad (10\text{–}15)$$

When the elemental volume chosen for integration has differential sizes in all three directions, e.g., $dV = dx\,dy\,dz$, Fig. 10–23a, the moment of inertia of the body must be determined using "triple integration." The integration process can, however, be simplified to a *single integration* provided the chosen elemental volume has a differential size or thickness in only *one direction*. Shell or disk elements are often used for this purpose.

Procedure For Analysis

For integration, we will consider only symmetric bodies having surfaces which are generated by revolving a curve about an axis. An example of such a body which is generated about the z axis is shown in Fig. 10–23.

Shell Element.

- If a *shell element* having a height z, radius y, and thickness dy is chosen for integration, Fig. 10–23b, then the volume $dV = (2\pi y)(z)\,dy$.
- This element may be used in Eq. 10–14 or 10–15 for determining the moment of inertia I_z of the body about the z axis since the *entire element*, due to its "thinness," lies at the *same* perpendicular distance $r = y$ from the z axis (see Example 10.11).

Disk Element.

- If a disk element having a radius y and a thickness dz is chosen for integration, Fig. 10–23c, then the volume $dV = (\pi y^2)\,dz$.
- In this case the element is *finite* in the radial direction, and consequently its parts *do not* all lie at the *same radial distance r* from the z axis. As a result, Eqs. 10–14 or 10–15 *cannot* be used to determine I_z. Instead, to perform the integration using this element, it is first necessary to determine the moment of inertia *of the element* about the z axis and then integrate this result (see Example 10.12).

EXAMPLE 10.11

Determine the mass moment of inertia of the cylinder shown in Fig. 10–24a about the z axis. The density of the material, ρ, is constant.

Fig. 10–24

SOLUTION

Shell Element. This problem may be solved using the *shell element* in Fig. 10–24b and single integration. The volume of the element is $dV = (2\pi r)(h)\,dr$, so that its mass is $dm = \rho\,dV = \rho(2\pi hr\,dr)$. Since the *entire element* lies at the same distance r from the z axis, the moment of inertia *of the element* is

$$dI_z = r^2\,dm = \rho 2\pi h r^3\,dr$$

Integrating over the entire region of the cylinder yields

$$I_z = \int_m r^2\,dm = \rho 2\pi h \int_0^R r^3\,dr = \frac{\rho\pi}{2}R^4 h$$

The mass of the cylinder is

$$m = \int_m dm = \rho 2\pi h \int_0^R r\,dr = \rho\pi h R^2$$

so that

$$I_z = \frac{1}{2}mR^2 \qquad\qquad Ans.$$

EXAMPLE 10.12

A solid is formed by revolving the shaded area shown in Fig. 10–25a about the y axis. If the density of the material is 5 slug/ft³, determine the mass moment of inertia about the y axis.

Fig. 10–25

SOLUTION

Disk Element.   The moment of inertia will be determined using a *disk element*, as shown in Fig. 10–25b. Here the element intersects the curve at the arbitrary point (x, y) and has a mass

$$dm = \rho \, dV = \rho(\pi x^2) \, dy$$

Although all portions of the element are *not* located at the same distance from the y axis, it is still possible to determine the moment of inertia dI_y *of the element* about the y axis. In Example 10.11 it was shown that the moment of inertia of a cylinder about its longitudinal axis is $I = \frac{1}{2}mR^2$, where m and R are the mass and radius of the cylinder. Since the height of the cylinder is not involved in this formula, we can also use it for a disk. Thus, for the disk element in Fig. 10–25b, we have

$$dI_y = \frac{1}{2}(dm)x^2 = \frac{1}{2}[\rho(\pi x^2) \, dy]x^2$$

Substituting $x = y^2$, $\rho = 5$ slug/ft³, and integrating with respect to y, from $y = 0$ to $y = 1$ ft, yields the moment of inertia for the entire solid:

$$I_y = \frac{5\pi}{2} \int_0^1 x^4 \, dy = \frac{5\pi}{2} \int_0^1 y^8 \, dy = 0.873 \text{ slug} \cdot \text{ft}^2 \qquad \textit{Ans.}$$

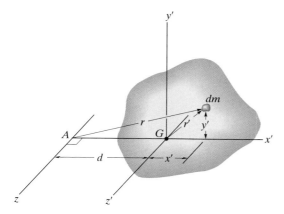

Fig. 10–26

Parallel-Axis Theorem.   If the moment of inertia of the body about an axis passing through the body's mass center is known, then the moment of inertia about any other *parallel axis* may be determined by using the *parallel-axis theorem.* This theorem can be derived by considering the body shown in Fig. 10–26. The z' axis passes through the mass center G, whereas the corresponding *parallel z axis* lies at a constant distance d away. Selecting the differential element of mass dm which is located at point (x', y') and using the Pythagorean theorem, $r^2 = (d + x')^2 + y'^2$, we can express the moment of inertia of the body about the z axis as

$$I = \int_m r^2 \, dm = \int_m [(d + x')^2 + y'^2] \, dm$$

$$= \int_m (x'^2 + y'^2) \, dm + 2d \int_m x' \, dm + d^2 \int_m dm$$

Since $r'^2 = x'^2 + y'^2$, the first integral represents I_G. The second integral equals *zero*, since the z' axis passes through the body's mass center, i.e., $\int x' \, dm = \bar{x} \int dm = 0$ since $\bar{x} = 0$. Finally, the third integral represents the total mass m of the body. Hence, the moment of inertia about the z axis can be written as

$$\boxed{I = I_G + md^2} \qquad (10\text{–}16)$$

where

I_G = moment of inertia about the z' axis passing through the mass center G

m = mass of the body

d = distance between the parallel axes

Radius of Gyration.

Occasionally, the moment of inertia of a body about a specified axis is reported in handbooks using the *radius of gyration, k*. This value has units of length, and when it and the body's mass m are known, the moment of inertia is determined from the equation

$$I = mk^2 \quad \text{or} \quad k = \sqrt{\frac{I}{m}} \qquad (10\text{–}17)$$

Note the *similarity* between the definition of k in this formula and r in the equation $dI = r^2\,dm$, which defines the moment of inertia of an elemental mass dm of the body about an axis.

Composite Bodies.

If a body is constructed from a number of simple shapes such as disks, spheres, and rods, the moment of inertia of the body about any axis z can be determined by adding algebraically the moments of inertia of all the composite shapes computed about the z axis. Algebraic addition is necessary since a composite part must be considered as a negative quantity if it has already been included within another part— for example, a "hole" subtracted from a solid plate. The parallel-axis theorem is needed for the calculations if the center of mass of each composite part does not lie on the z axis. For the calculation, then, $I = \Sigma(I_G + md^2)$, where I_G for each of the composite parts is computed by integration or can be determined from a table, such as the one given on the inside back cover of this book.

This flywheel, which operates a metal cutter, has a large moment of inertia about its center. Once it begins rotating it is difficult to stop it and therefore a uniform motion can be effectively transferred to the cutting blade.

EXAMPLE 10.13

If the plate shown in Fig. 10–27a has a density of 8000 kg/m³ and a thickness of 10 mm, determine its mass moment of inertia about an axis directed perpendicular to the page and passing through point O.

(a)   (b)

Fig. 10–27

SOLUTION

The plate consists of two composite parts, the 250-mm-radius disk *minus* a 125-mm-radius disk, Fig. 10–27b. The moment of inertia about O can be determined by computing the moment of inertia of each of these parts about O and then *algebraically* adding the results. The computations are performed by using the parallel-axis theorem in conjunction with the data listed in the table on the inside back cover.

Disk.   The moment of inertia of a disk about an axis perpendicular to the plane of the disk is $I_G = \frac{1}{2}mr^2$. The mass center of the disk is located at a distance of 0.25 m from point O. Thus,

$$m_d = \rho_d V_d = 8000 \text{ kg/m}^3 \left[\pi(0.25 \text{ m})^2(0.01 \text{ m})\right] = 15.71 \text{ kg}$$

$$
\begin{aligned}
(I_O)_d &= \tfrac{1}{2}m_d r_d^2 + m_d d^2 \\
&= \tfrac{1}{2}(15.71 \text{ kg})(0.25 \text{ m})^2 + (15.71 \text{ kg})(0.25 \text{ m})^2 \\
&= 1.473 \text{ kg} \cdot \text{m}^2
\end{aligned}
$$

Hole.   For the 125-mm-radius disk (hole), we have

$$m_h = \rho_h V_h = 8000 \text{ kg/m}^3 \left[\pi(0.125 \text{ m})^2(0.01 \text{ m})\right] = 3.93 \text{ kg}$$

$$
\begin{aligned}
(I_O)_h &= \tfrac{1}{2}m_h r_h^2 + m_h d^2 \\
&= \tfrac{1}{2}(3.93 \text{ kg})(0.125 \text{ m})^2 + (3.93 \text{ kg})(0.25 \text{ m})^2 \\
&= 0.276 \text{ kg} \cdot \text{m}^2
\end{aligned}
$$

The moment of inertia of the plate about point O is therefore

$$
\begin{aligned}
I_O &= (I_O)_d - (I_O)_h \\
&= 1.473 \text{ kg} \cdot \text{m}^2 - 0.276 \text{ kg} \cdot \text{m}^2 \\
&= 1.20 \text{ kg} \cdot \text{m}^2 \qquad\qquad\qquad\qquad \textit{Ans.}
\end{aligned}
$$

EXAMPLE 10.14

Fig. 10–28

The pendulum in Fig. 10–28 consists of two thin rods each having a weight of 10 lb. Determine the pendulum's mass moment of inertia about an axis passing through (a) the pin at O, and (b) the mass center G of the pendulum.

SOLUTION

Part (a).   Using the table on the inside back cover, the moment of inertia of rod OA about an axis perpendicular to the page and passing through the end point O of the rod is $I_O = \frac{1}{3}ml^2$. Hence,

$$(I_{OA})_O = \frac{1}{3}ml^2 = \frac{1}{3}\left(\frac{10\text{ lb}}{32.2\text{ ft/s}^2}\right)(2\text{ ft})^2 = 0.414\text{ slug}\cdot\text{ft}^2$$

The same value may be computed using $I_G = \frac{1}{12}ml^2$ and the parallel-axis theorem; i.e.,

$$(I_{OA})_O = \frac{1}{12}ml^2 + md^2 = \frac{1}{12}\left(\frac{10\text{ lb}}{32.2\text{ ft/s}^2}\right)(2\text{ ft})^2 + \frac{10\text{ lb}}{32.2\text{ ft/s}^2}(1\text{ ft})^2$$

$$= 0.414\text{ slug}\cdot\text{ft}^2$$

For rod BC we have

$$(I_{BC})_O = \frac{1}{12}ml^2 + md^2 = \frac{1}{12}\left(\frac{10\text{ lb}}{32.2\text{ ft/s}^2}\right)(2\text{ ft})^2 + \frac{10\text{ lb}}{32.2\text{ ft/s}^2}(2\text{ ft})^2$$

$$= 1.346\text{ slug}\cdot\text{ft}^2$$

The moment of inertia of the pendulum about O is therefore

$$I_O = 0.414 + 1.346 = 1.76\text{ slug}\cdot\text{ft}^2 \qquad \textit{Ans.}$$

Part (b).   The mass center G will be located relative to the pin at O. Assuming this distance to be $\bar{y}$, Fig. 10–28, and using the formula for determining the mass center, we have

$$\bar{y} = \frac{\Sigma\tilde{y}m}{\Sigma m} = \frac{1(10/32.2) + 2(10/32.2)}{(10/32.2) + (10/32.2)} = 1.50\text{ ft}$$

NOTE: The moment of inertia I_G may be computed in the same manner as I_O, which requires successive applications of the parallel-axis theorem in order to transfer the moments of inertia of rods OA and BC to G. A more direct solution, however, involves applying the parallel-axis theorem using the result for I_O determined above; i.e.,

$$I_O = I_G + md^2; \qquad 1.76\text{ slug}\cdot\text{ft}^2 = I_G + \left(\frac{20\text{ lb}}{32.2\text{ ft/s}^2}\right)(1.50\text{ ft})^2$$

$$I_G = 0.362\text{ slug}\cdot\text{ft}^2 \qquad \textit{Ans.}$$

PROBLEMS

10–90. The right circular cone is formed by revolving the shaded area around the x axis. Determine the moment of inertia I_x and express the result in terms of the total mass m of the cone. The cone has a constant density ρ.

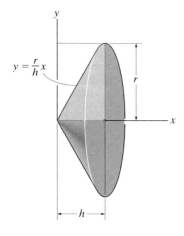

$$y = \frac{r}{h}x$$

Prob. 10–90

10–91. Determine the moment of inertia of the thin ring about the z axis. The ring has a mass m.

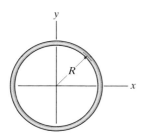

Prob. 10–91

*10–92. The solid is formed by revolving the shaded area around the y axis. Determine the radius of gyration k_y. The specific weight of the material is $\gamma = 380 \text{ lb/ft}^3$.

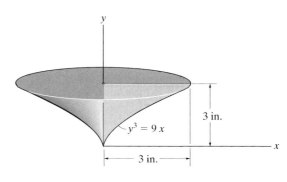

3 in.

$y^3 = 9x$

3 in.

Prob. 10–92

10–93. Determine the moment of inertia I_x of the sphere and express the result in terms of the total mass m of the sphere. The sphere has a constant density ρ.

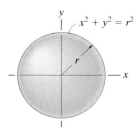

$$x^2 + y^2 = r^2$$

Prob. 10–93

10–94. Determine the radius of gyration k_x of the paraboloid. The density of the material is $\rho = 5\ \mathrm{Mg/m^3}$.

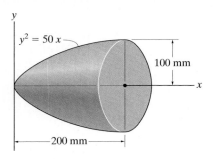

$y^2 = 50\,x$

100 mm

x

200 mm

Prob. 10–94

10–95. Determine the moment of inertia of the semi-ellipsoid with respect to the x axis and express the result in terms of the mass m of the semiellipsoid. The material has a constant density ρ.

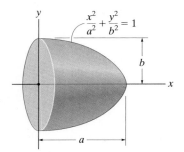

$\dfrac{x^2}{a^2} + \dfrac{y^2}{b^2} = 1$

b

x

a

Prob. 10–95

*10–96. Determine the radius of gyration k_x of the body. The specific weight of the material is $\gamma = 380\ \mathrm{lb/ft^3}$.

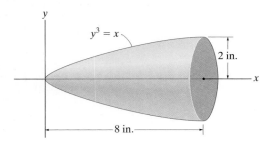

$y^3 = x$

2 in.

x

8 in.

Prob. 10–96

10–97. Determine the moment of inertia of the ellipsoid with respect to the x axis and express the result in terms of the mass m of the ellipsoid. The material has a constant density ρ.

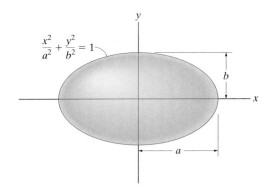

$\dfrac{x^2}{a^2} + \dfrac{y^2}{b^2} = 1$

b

x

a

Prob. 10–97

10–98. Determine the moment of inertia of the homogeneous pyramid of mass m with respect to the z axis. The density of the material is ρ. *Hint:* Use a rectangular plate element having a volume of $dV = (2x)(2y)\,dz$.

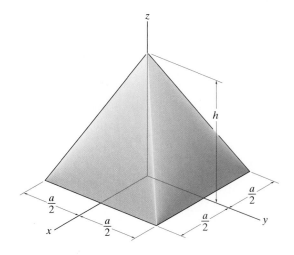

z

h

$\dfrac{a}{2}$

$\dfrac{a}{2}$

$\dfrac{a}{2}$

$\dfrac{a}{2}$

y

x

Prob. 10–98

10–99. The concrete shape is formed by rotating the shaded area about the y axis. Determine the moment of inertia I_y. The specific weight of concrete is $\gamma = 150 \text{ lb/ft}^3$.

Prob. 10–99

10–101. Determine the moment of inertia I_z of the frustum of the cone which has a conical depression. The material has a density of 200 kg/m³.

Prob. 10–101

*10–100.** Determine the moment of inertia of the thin plate about an axis perpendicular to the page and passing through the pin at O. The plate has a hole in its center. Its thickness is 50 mm, and the material has a density of $\rho = 50 \text{ kg/m}^3$.

Prob. 10–100

10–102. Determine the moment of inertia of the assembly about an axis which is perpendicular to the page and passes through the center of mass G. The material has a specific weight of $\gamma = 90 \text{ lb/ft}^3$.

10–103. Determine the moment of inertia of the assembly about an axis which is perpendicular to the page and passes through point O. The material has a specific weight of $\gamma = 90 \text{ lb/ft}^3$.

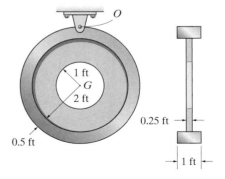

Probs. 10–102/103

*10–104. The wheel consists of a thin ring having a mass of 10 kg and four spokes made from slender rods, each having a mass of 2 kg. Determine the wheel's moment of inertia about an axis perpendicular to the page and passing through point A.

Prob. 10–104

10–105. The slender rods have a weight of 3 lb/ft. Determine the moment of inertia of the assembly about an axis perpendicular to the page and passing through point A.

Prob. 10–105

10–106. Each of the three rods has a mass m. Determine the moment of inertia of the assembly about an axis which is perpendicular to the page and passes through the center point O.

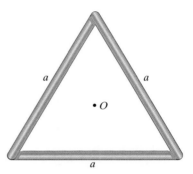

Prob. 10–106

10–107. The slender rods have a weight of 3 lb/ft. Determine the moment of inertia of the assembly about an axis perpendicular to the page and passing through point A.

Prob. 10–107

*10–108. The pendulum consists of a plate having a weight of 12 lb and a slender rod having a weight of 4 lb. Determine the radius of gyration of the pendulum about an axis perpendicular to the page and passing through point O.

Prob. 10–108

10–109. Determine the moment of inertia of the overhung crank about the x axis. The material is steel having a density of $\rho = 7.85 \ \text{Mg/m}^3$.

Prob. 10–109

10–111. Determine the moment of inertia of the solid steel assembly about the x axis. Steel has a specific weight of $\gamma_{st} = 490 \ \text{lb/ft}^3$.

Prob. 10–111

10–110. Determine the moment of inertia of the overhung crank about the x' axis. The material is steel having a density of $\rho = 7.85 \ \text{Mg/m}^3$.

*10–112. The pendulum consists of two slender rods AB and OC which have a mass of 3 kg/m. The thin plate has a mass of $12 \ \text{kg/m}^2$. Determine the location $\bar{y}$ of the center of mass G of the pendulum, then calculate the moment of inertia of the pendulum about an axis perpendicular to the page and passing through G.

Prob. 10–110

Prob. 10–112

Chapter Review

Area Moment of Inertia

The *area moment of inertia* represents the second moment of the area about an axis. It is frequently used in formulas related to strength and stability of structural members or mechanical elements.

$$I_x = \int_A y^2 \, dA$$

If the area shape is irregular, then a differential element must be selected and integration over the entire area must be performed.

$$I_y = \int_A x^2 \, dA$$

To determine the moment of inertia of common shapes about some *other axis*, the parallel-axis theorem must be used.

$$I_x = \overline{I}_{x'} + Ad_y^2$$

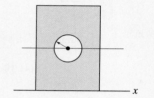

If an area is a composite of common shapes, then its moment of inertia is equal to the sum of the moments of inertia of each of its parts.
(*See pages 519–522.*)

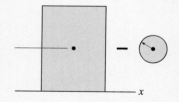

Product of Inertia

The *product of inertia* of an area is used to determine the location of an axis about which the moment of inertia for the area is a maximum or minimum.

$$I_{xy} = \int_A xy \, dA$$

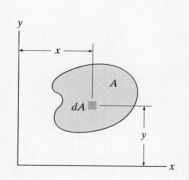

If the product of inertia for an area is known about its centroidal x', y' axes, then its value can be determined about any x, y axes using the parallel-axis theorem for the product of inertia.
(*See pages 538–539.*)

$$I_{xy} = \overline{I}_{x'y'} + Ad_x d_y$$

Principal Moments of Inertia

Provided the moments of inertia, I_x and I_y, and the product of inertia, I_{xy}, are known, then the transformation formulas, or Mohr's circle, can be used to determine the maximum and minimum or *principal moments of inertia* for the area, as well as finding the orientation of the principal axes of inertia.

(*See pages 542–543.*)

$$I_{\substack{max \\ min}} = \frac{I_x + I_y}{2} \pm \sqrt{\left(\frac{I_x - I_y}{2}\right)^2 + I_{xy}^2}$$

$$\tan 2\theta_p = \frac{-I_{xy}}{(I_x - I_y)/2}$$

Mass Moment of Inertia

The *mass moment of inertia* is a property of a body that measures its resistance to a change in its rotation. It is defined as the second moment of the mass elements of the body about an axis.

For bodies having axial symmetry, the mass moment of inertia can be determined by integration, using a disk element.

$$I = \int_m r^2 \, dm$$

The mass moment of inertia of a composite body is determined by using tabular values of its composite shapes found on the inside back cover, along with the parallel-axis theorem.

(*See pages 555–556.*)

$$I = I_G + md^2$$

REVIEW PROBLEMS

10–113. Determine the moment of inertia for the shaded area about the x axis.

10–114. Determine the moment of inertia for the shaded area about the y axis.

*10–116.** Determine the product of inertia for the shaded area with respect to the x and y axes.

Prob. 10–116

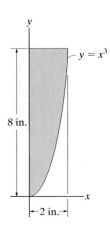

Probs. 10–113/114

10–115. Determine the mass moment of inertia I_x for the body and express the result in terms of the total mass m of the body. The density is constant.

10–117. Determine the area moments of inertia I_u and I_v and the product of inertia I_{uv} for the semicircular area.

Prob. 10–115

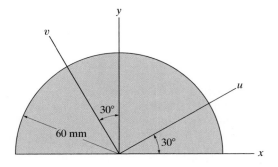

Prob. 10–117

10–118.   Determine the moment of inertia for the shaded area about the x axis.

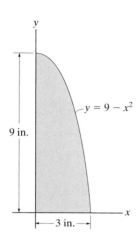

$y = 9 - x^2$

9 in.

3 in.

Prob. 10–118

10–119.   Determine the moment of inertia for the shaded area about the y axis.

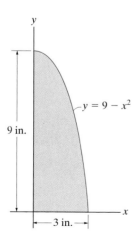

$y = 9 - x^2$

9 in.

3 in.

Prob. 10–119

*10–120.   Determine the area moment of inertia for the area about the x axis. Then, using the parallel-axis theorem, find the area moment of inertia about the x' axis that passes through the centroid C of the area. $\bar{y} = 120$ mm.

200 mm

200 mm

C

$\bar{y}$

x'

$y = \dfrac{1}{200} x^2$

x

Prob. 10–120

10–121.   Determine the area moment of inertia for the triangular area about (a) the x axis, and (b) the centroidal x' axis.

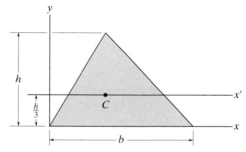

h

$\dfrac{h}{3}$

C

x'

b

x

Prob. 10–121

10–122.   Determine the product of inertia for the shaded area with respect to the x and y axes.

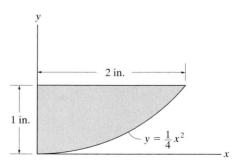

2 in.

1 in.

$y = \dfrac{1}{4} x^2$

x

Prob. 10–122

Equilibrium and stability of this crane as a function of the boom position can be analyzed using methods based on work and energy, which are explained in this chapter.

11 Virtual Work

CHAPTER OBJECTIVES

- To introduce the principle of virtual work and show how it applies to determining the equilibrium configuration of a series of pin-connected members.

- To establish the potential energy function and use the potential-energy method to investigate the type of equilibrium or stability of a rigid body or configuration.

11.1   Definition of Work and Virtual Work

Work of a Force.   In mechanics a force $\mathbf{F}$ does work on a particle or body only when it has a displacement in the direction of the force. For example, consider the force $\mathbf{F}$ in Fig. 11–1, which is located on the path s specified by the position vector $\mathbf{r}$. If the force moves the particle along the path to a new position $\mathbf{r}' = \mathbf{r} + d\mathbf{r}$, the displacement is $d\mathbf{r}$, and therefore the work dU is a *scalar quantity* defined by the dot product

$$dU = \mathbf{F} \cdot d\mathbf{r}$$

Fig. 11–1

573

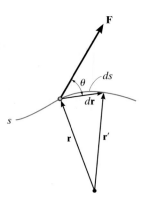

Fig. 11–1

Because $d\mathbf{r}$ is infinitesimal, the magnitude of $d\mathbf{r}$ can be represented by ds, the differential arc segment along the path. If the angle between the tails of $d\mathbf{r}$ and $\mathbf{F}$ is θ, Fig. 11–1, then by definition of the dot product, the above equation may also be written as

$$dU = F\, ds \cos \theta$$

Work expressed by this equation may be interpreted in one of two ways: either as the product of F and the component of displacement in the direction of the force, i.e., $ds \cos \theta$; or as the product of ds and the component of force in the direction of displacement, i.e., $F \cos \theta$. When the force component and the displacement have the *same sense*, $0° \le \theta < 90°$, the work is *positive*. If $90° < \theta \le 180°$, these vectors have an *opposite sense*, and therefore the work is *negative*. Also, $dU = 0$ if the force is *perpendicular* to displacement, since $\cos 90° = 0$, or if the force is applied at a *fixed point*, in which case the displacement $ds = 0$.

The basic unit for work combines the units of force and displacement. In the SI system a *joule* (J) is equivalent to the work done by a force of 1 newton which moves 1 meter in the direction of the force $(1\text{ J} = 1\text{ N} \cdot \text{m})$. In the FPS system, work is defined in units of ft $\cdot$ lb. The moment of a force has the same combination of units; however, the concepts of moment and work are in no way related. A moment is a vector quantity, whereas work is a scalar.

(a)

Translation
(b)

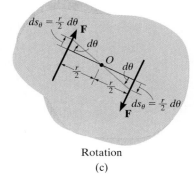

Rotation
(c)

Fig. 11–2

Work of a Couple. The two forces of a couple do work when the couple *rotates* about an axis perpendicular to the plane of the couple. To show this, consider the body in Fig. 11–2a, which is subjected to a couple whose moment has a magnitude $M = Fr$. Any general differential displacement of the body can be considered as a combination of a translation and rotation. When the body *translates* such that the *component of displacement* along the line of action of each force is ds_t, clearly the "positive" work of one force ($F\, ds_t$) cancels the "negative" work of the other ($-F\, ds_t$), Fig. 11–2b. Consider now a differential *rotation* $d\theta$ of the body about an axis perpendicular to the plane of the couple, which intersects the plane at point O, Fig. 11–2c. (For the derivation, any other point in the plane may also be considered.) As shown, each force has a displacement $ds_\theta = (r/2)\, d\theta$ in the direction of the force; hence, the work of both forces is

$$dU = F\left(\frac{r}{2}d\theta\right) + F\left(\frac{r}{2}d\theta\right) = (Fr)\, d\theta$$

or

$$dU = M\, d\theta$$

The resultant work is *positive* when the sense of $\mathbf{M}$ is the *same* as that of $d\boldsymbol{\theta}$, and negative when they have an opposite sense. As in the case of the moment vector, the *direction and sense* of $d\boldsymbol{\theta}$ are defined by the right-hand rule, where the fingers of the right hand follow the rotation or "curl" and the thumb indicates the direction of $d\boldsymbol{\theta}$. Hence, the line of action of $d\boldsymbol{\theta}$ will be *parallel* to the line of action of $\mathbf{M}$ if movement of the body occurs in the *same plane*. If the body rotates in space, however, the *component* of $d\boldsymbol{\theta}$ in the direction of $\mathbf{M}$ is required. Thus, in general, the work done by a couple is defined by the dot product, $dU = \mathbf{M} \cdot d\boldsymbol{\theta}$.

Virtual Work. The definitions of the work of a force and a couple have been presented in terms of *actual movements* expressed by differential displacements having magnitudes of ds and $d\theta$. Consider now an *imaginary* or *virtual movement* of a body in static equilibrium, which indicates a displacement or rotation that is *assumed* and *does not actually exist*. These movements are first-order differential quantities and will be denoted by the symbols δs and $\delta \theta$ (delta s and delta θ), respectively. The *virtual work* done by a force having a virtual displacement δs is

$$\boxed{\delta U = F \cos \theta\, \delta s} \qquad (11\text{–}1)$$

Similarly, when a couple undergoes a virtual rotation $\delta \theta$ in the plane of the couple forces, the *virtual work* is

$$\boxed{\delta U = M\, \delta \theta} \qquad (11\text{–}2)$$

Fig. 11–3

Fig. 11–4

(a)

(b)

Fig. 11–5

11.2 Principle of Virtual Work for a Particle and a Rigid Body

Particle. If the particle in Fig. 11–3 undergoes an imaginary or virtual displacement $\delta\mathbf{r}$, then the virtual work (δU) done by the force system becomes

$$\delta U = \Sigma\mathbf{F}\cdot\delta\mathbf{r}$$
$$= (\Sigma F_x\mathbf{i} + \Sigma F_y\mathbf{j} + \Sigma F_z\mathbf{k})\cdot(\delta x\mathbf{i} + \delta y\mathbf{j} + \delta z\mathbf{k})$$
$$= \Sigma F_x\,\delta x + \Sigma F_y\,\delta y + \Sigma F_z\,\delta z$$

For equilibrium $\Sigma F_x = 0$, $\Sigma F_y = 0$, $\Sigma F_z = 0$, and so the virtual work must also be zero, i.e.,

$$\delta U = 0$$

In other words, we can write three independent virtual work equations corresponding to the three equations of equilibrium.

For example, consider the free-body diagram of the ball which rests on the floor, Fig. 11–4. If we "imagine" the ball to be displaced downwards a virtual amount δy, then the weight does positive virtual work, $W\,\delta y$, and the normal force does negative virtual work, $-N\,\delta y$. For equilibrium the total virtual work must be zero, so that $\delta U = W\,\delta y - N\,\delta y = (W - N)\,\delta y = 0$. Since $\delta y \neq 0$, then $N = W$ as required.

Rigid Body. In a similar manner, we can also write a set of three virtual work equations ($\delta U = 0$) for a rigid body subjected to a coplanar force system. If these equations involve separate virtual translations in the x and y directions and a virtual rotation about an axis perpendicular to the x–y plane and passing through an arbitrary point O, then it can be shown that they will correspond to the three equilibrium equations, $\Sigma F_x = 0$, $\Sigma F_y = 0$, and $\Sigma M_O = 0$. When writing these equations, it is *not necessary* to include the work done by the *internal forces* acting within the body since a rigid body *does not deform* when subjected to an external loading, and furthermore, when the body moves through a virtual displacement, the internal forces occur in equal but opposite collinear pairs, so that the corresponding work done by each pair of forces *cancels*.

To demonstrate an application, consider the simply supported beam in Fig. 11–5a. When the beam is given a virtual rotation $\delta\theta$ about point B, Fig. 11–5b, the only forces that do work are $\mathbf{P}$ and $\mathbf{A}_y$. Since $\delta y = l\,\delta\theta$ and $\delta y' = (l/2)\,\delta\theta$, the virtual work equation for this case is $\delta U = A_y(l\,\delta\theta) - P(l/2)\,\delta\theta = (A_y - P/2)l\,\delta\theta = 0$. Since $\delta\theta \neq 0$, then $A_y = P/2$. Excluding $\delta\theta$, notice that the terms in parentheses actually represent moment equilibrium about point B.

As in the case of a particle, no added advantage is gained by solving rigid-body equilibrium problems using the principle of virtual work. This is because for each application of the virtual-work equation the virtual displacement, common to every term, factors out, leaving an equation that could have been obtained in a more *direct manner* by simply applying the equations of equilibrium.

11.3   Principle of Virtual Work for a System of Connected Rigid Bodies

The method of virtual work is most suitable for solving equilibrium problems that involve a system of several *connected* rigid bodies such as the ones shown in Fig. 11–6. Before we can apply the principle of virtual work to these systems, however, we must first specify the number of degrees of freedom for a system and establish coordinates that define the position of the system.

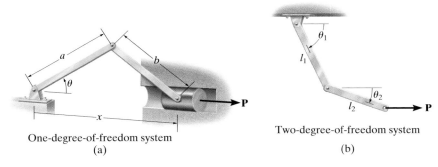

One-degree-of-freedom system
(a)

Two-degree-of-freedom system
(b)

Fig. 11–6

Degrees of Freedom.   A system of connected bodies takes on a unique shape that can be specified provided we know the position of a number of specific points on the system. These positions are defined using *independent coordinates q*, which are measured from fixed reference points. For every coordinate established, the system will have a *degree of freedom* for displacement along the coordinate axis such that it is consistent with the constraining action of the supports. Thus, an *n*-degree-of-freedom system requires *n* independent coordinates q_n to specify the location of all its members. For example, the link and sliding-block arrangement shown in Fig. 11–6a is an example of a one-degree-of-freedom system. The independent coordinate $q = \theta$ may be used to specify the location of the two connecting links and the block. The coordinate *x* could also be used as the independent coordinate. However, since the block is constrained to move within the slot, *x* is not independent of θ; rather, it can be related to θ using the cosine law, $b^2 = a^2 + x^2 - 2ax \cos \theta$. The double-link arrangement, shown in Fig. 11–6b, is an example of a two-degrees-of-freedom system. To specify the location of each link, the coordinate angles θ_1 and θ_2 must be known since a rotation of one link is independent of a rotation of the other.

During operation the scissors lift has one degree of freedom. Without the need for dismembering the mechanism, the hydraulic force required to provide the lift can be determined *directly* by using the principle of virtual work.

Principle of Virtual Work.   The principle of virtual work for a system of rigid bodies whose connections are *frictionless* may be stated as follows: *A system of connected rigid bodies is in equilibrium provided the virtual work done by all the external forces and couples acting on the system is zero for each independent virtual displacement of the system.* Mathematically, this may be expressed as

$$\delta U = 0 \qquad\qquad (11\text{--}3)$$

where δU represents the virtual work of all the external forces (and couples) acting on the system during any independent virtual displacement.

As stated above, if a system has n degrees of freedom it takes n independent coordinates q_n to completely specify the location of the system. Hence, for the system it is possible to write n independent virtual-work equations, one for every virtual displacement taken along each of the independent coordinate axes, while the remaining $n - 1$ independent coordinates are held *fixed.**

Important Points

- A force does work when it moves through a displacement in the direction of the force. A couple moment does work when it moves through a collinear rotation. Specifically, positive work is done when the force or couple moment and its displacement have the same sense of direction.

- The principle of virtual work is generally used to determine the equilibrium configuration for a series of multiply-connected members.

- A virtual displacement is imaginary, i.e., does not really happen. It is a differential that is given in the positive direction of the position coordinate.

- Forces or couple moments that do not virtually displace do no virtual work.

*This method of applying the principle of virtual work is sometimes called the *method of virtual displacements* since a virtual displacement is applied, resulting in the calculation of a real force. Although it is not to be used here, realize that we can also apply the principle of virtual work as a method of virtual forces. This method is often used to determine the displacements of points on deformable bodies. See R. C. Hibbeler, *Mechanics of Materials*, 6th edition, Prentice Hall, Inc., 2005.

Procedure for Analysis

The equation of virtual work can be used to solve problems involving a system of frictionless connected rigid bodies having a single degree of freedom by using the following procedure.

Free-Body Diagram.

- Draw the free-body diagram of the entire system of connected bodies and define the *independent coordinate q*.
- Sketch the "deflected position" of the system on the free-body diagram when the system undergoes a positive virtual displacement δq.

Virtual Displacements.

- Indicate *position coordinates* s_i, measured from a *fixed point* on the free-body diagram to each of the i number of "active" forces and couples, i.e., those that do work.
- Each coordinate axis should be parallel to the line of action of the "active" force to which it is directed, so that the virtual work along the coordinate axis can be calculated.
- Relate each of the position coordinates s_i to the independent coordinate q; then *differentiate* these expressions in order to express the virtual displacements δs_i in terms of δq.

Virtual-Work Equation.

- Write the *virtual-work equation* for the system assuming that, whether possible or not, all the position coordinates s_i undergo *positive* virtual displacements δs_i.
- Using the relations for δs_i, express the work of *each* "active" force and couple in the equation in terms of the *single* independent virtual displacement δq.
- Factor out this common displacement from all the terms and solve for the unknown force, couple, or equilibrium position, q.
- If the system contains n degrees of freedom, n independent coordinates q_n must be specified. Follow the above procedure and let *only one* of the independent coordinates have a virtual displacement, while the remaining $n - 1$ coordinates are *held fixed*. In this way, n virtual-work equations can be written, one for each independent coordinate.

EXAMPLE 11.1

(a)

Fig. 11–7

(b)

Determine the angle θ for equilibrium of the two-member linkage shown in Fig. 11–7a. Each member has a mass of 10 kg.

SOLUTION

Free-Body Diagram. The system has only one degree of freedom since the location of both links may be specified by the single independent coordinate $(q =)\ \theta$. As shown on the free-body diagram in Fig. 11–7b, when θ has a *positive* (clockwise) virtual rotation $\delta\theta$, only the active forces, **F** and the two 98.1-N weights, do work. (The reactive forces $\mathbf{D}_x$ and $\mathbf{D}_y$ are fixed, and $\mathbf{B}_y$ does not move along its line of action.)

Virtual Displacements. If the origin of coordinates is established at the *fixed* pin support D, the location of **F** and **W** may be specified by the *position coordinates* x_B and y_w, as shown in the figure. In order to determine the work, note that these coordinates are parallel to the lines of action of their associated forces.

Expressing the position coordinates in terms of the independent coordinate θ and taking the derivatives yields

$$x_B = 2(1 \cos \theta) \text{ m} \qquad \delta x_B = -2 \sin \theta\ \delta\theta \text{ m} \qquad (1)$$
$$y_w = \tfrac{1}{2}(1 \sin \theta) \text{ m} \qquad \delta y_w = 0.5 \cos \theta\ \delta\theta \text{ m} \qquad (2)$$

It is seen by the *signs* of these equations, and indicated in Fig. 11–7b, that an *increase* in θ (i.e., $\delta\theta$) causes a *decrease* in x_B and an *increase* in y_w.

Virtual-Work Equation. If the virtual displacements δx_B and δy_w were *both positive*, then the forces **W** and **F** would do positive work since the forces and their corresponding displacements would have the same sense. Hence, the virtual-work equation for the displacement $\delta\theta$ is

$$\delta U = 0; \qquad W\ \delta y_w + W\ \delta y_w + F\ \delta x_B = 0 \qquad (3)$$

Substituting Eqs. 1 and 2 into Eq. 3 in order to relate the virtual displacements to the common virtual displacement $\delta\theta$ yields

$$98.1(0.5 \cos \theta\ \delta\theta) + 98.1(0.5 \cos \theta\ \delta\theta) + 25(-2 \sin \theta\ \delta\theta) = 0$$

Notice that the "negative work" done by **F** (force in the opposite sense to displacement) has been *accounted for* in the above equation by the "negative sign" of Eq. 1. Factoring out the *common displacement* $\delta\theta$ and solving for θ, noting that $\delta\theta \neq 0$, yields

$$(98.1 \cos \theta - 50 \sin \theta)\ \delta\theta = 0$$

$$\theta = \tan^{-1} \frac{98.1}{50} = 63.0° \qquad \qquad Ans.$$

NOTE: If this problem had been solved using the equations of equilibrium, it would have been necessary to dismember the links and apply three scalar equations to *each* link. The principle of virtual work, by means of calculus, has eliminated this task so that the answer is obtained directly.

EXAMPLE 11.2

Determine the angle θ required to maintain equilibrium of the mechanism in Fig. 11–8a. Neglect the weight of the links. The spring is unstretched when $\theta = 0°$, and it maintains a horizontal position due to the roller.

(a)

SOLUTION

Free-Body Diagram.   The mechanism has one degree of freedom, and therefore the location of each member may be specified using the independent coordinate θ. When θ undergoes a *positive* virtual displacement $\delta\theta$, as shown on the free-body diagram in Fig. 11–8b, links *AB* and *EC* rotate by the same amount since they have the same length, and link *BC* only translates. Since a couple moment does work *only* when it rotates, the work done by $\mathbf{M}_2$ is zero. The reactive forces at *A* and *E* do no work. Why?

Virtual Displacements.   The position coordinates x_B and x_D are *parallel* to the lines of action of $\mathbf{P}$ and $\mathbf{F}_s$, and these coordinates locate the forces with respect to the *fixed points A* and *E*. From Fig. 11–8b,

$$x_B = 0.4 \sin\theta \text{ m}$$
$$x_D = 0.2 \sin\theta \text{ m}$$

Thus,

$$\delta x_B = 0.4 \cos\theta \, \delta\theta \text{ m}$$
$$\delta x_D = 0.2 \cos\theta \, \delta\theta \text{ m}$$

(b)

Fig. 11–8

Virtual-Work Equation.   For *positive* virtual displacements, $\mathbf{F}_s$ is opposite to δx_D and hence does negative work. Thus,

$$\delta U = 0; \quad M_1 \, \delta\theta + P \, \delta x_B - F_s \, \delta x_D = 0$$

Relating each of the virtual displacements to the *common* virtual displacement $\delta\theta$ yields

$$0.5\delta\theta + 2(0.4 \cos\theta \, \delta\theta) - F_s(0.2 \cos\theta \, \delta\theta) = 0$$
$$(0.5 + 0.8 \cos\theta - 0.2F_s \cos\theta) \, \delta\theta = 0 \qquad (1)$$

For the arbitrary angle θ, the spring is stretched a distance of $x_D = (0.2 \sin\theta)$ m; and therefore, $F_s = (60 \text{ N/m})(0.2 \sin\theta)$ m = $(12 \sin\theta)$ N. Substituting into Eq. 1 and noting that $\delta\theta \neq 0$, we have

$$0.5 + 0.8 \cos\theta - 0.2(12 \sin\theta) \cos\theta = 0$$

Since $\sin 2\theta = 2 \sin\theta \cos\theta$, then

$$1 = 2.4 \sin 2\theta - 1.6 \cos\theta$$

Solving for θ by trial and error yields

$$\theta = 36.3° \qquad\qquad Ans.$$

EXAMPLE 11.3

(a)

(b)

Fig. 11–9

Determine the horizontal force C_x that the pin at C must exert on BC in order to hold the mechanism shown in Fig. 11–9a in equilibrium when $\theta = 45°$. Neglect the weight of the members.

SOLUTION

Free-Body Diagram. The reaction $\mathbf{C}_x$ can be obtained by *releasing* the pin constraint at C in the x direction and allowing the frame to be displaced in this direction. The system then has only one degree of freedom, defined by the independent coordinate θ, Fig. 11–9b. When θ has a *positive* virtual displacement $\delta\theta$, only $\mathbf{C}_x$ and the 200-N force do work.

Virtual Displacements. Forces $\mathbf{C}_x$ and 200 N are located from the fixed origin A using position coordinates y_B and x_C. From Fig. 11–9b, x_C can be related to θ by the "law of cosines." Hence,

$$(0.7)^2 = (0.6)^2 + x_C^2 - 2(0.6)x_C \cos\theta \qquad (1)$$

$$0 = 0 + 2x_C\,\delta x_C - 1.2\delta x_C \cos\theta + 1.2x_C \sin\theta\,\delta\theta$$

$$\delta x_C = \frac{1.2x_C \sin\theta}{1.2\cos\theta - 2x_C}\delta\theta \qquad (2)$$

Also,

$$y_B = 0.6 \sin\theta$$

$$\delta y_B = 0.6 \cos\theta\,\delta\theta \qquad (3)$$

Virtual-Work Equation. When y_B and x_C have *positive* virtual displacements δy_B and δx_C, $\mathbf{C}_x$ and 200 N do *negative work* since they both act in the opposite sense to $\delta \mathbf{y}_B$ and $\delta \mathbf{x}_C$. Hence,

$$\delta U = 0; \qquad -200\delta y_B - C_x\,\delta x_C = 0$$

Substituting Eqs. 2 and 3 into this equation, factoring out $\delta\theta$, and solving for C_x yields

$$-200(0.6\cos\theta\,\delta\theta) - C_x\frac{1.2x_C\sin\theta}{1.2\cos\theta - 2x_C}\delta\theta = 0$$

$$C_x = \frac{-120\cos\theta(1.2\cos\theta - 2x_C)}{1.2x_C\sin\theta} \qquad (4)$$

At the required equilibrium position $\theta = 45°$, the corresponding value of x_C can be found by using Eq. 1, in which case

$$x_C^2 - 1.2\cos 45°\,x_C - 0.13 = 0$$

Solving for the positive root yields

$$x_C = 0.981 \text{ m}$$

Thus, from Eq. 4,

$$C_x = 114 \text{ N} \qquad\qquad\qquad Ans.$$

EXAMPLE 11.4

Determine the equilibrium position of the two-bar linkage shown in Fig. 11–10a. Neglect the weight of the links.

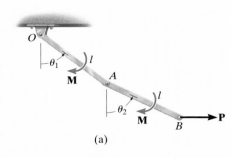

(a)

SOLUTION

The system has two degrees of freedom since the *independent coordinates* θ_1 and θ_2 must be known to locate the position of both links. The position coordinate x_B, measured from the fixed point O, is used to specify the location of P, Fig. 11–10b and c.

If θ_1 is held *fixed* and θ_2 varies by an amount $\delta\theta_2$, as shown in Fig. 11–10b, the virtual-work equation becomes

$$[\delta U = 0]_{\theta_2}; \qquad P(\delta x_B)_{\theta_2} - M\,\delta\theta_2 = 0 \qquad (1)$$

Here P and M represent the magnitudes of the applied force and couple moment acting on link AB.

When θ_2 is held *fixed* and θ_1 varies by an amount $\delta\theta_1$, as shown in Fig. 11–10c, then AB translates and the virtual-work equation becomes

$$[\delta U = 0]_{\theta_1}; \qquad P(\delta x_B)_{\theta_1} - M\,\delta\theta_1 = 0 \qquad (2)$$

The *position coordinate* x_B may be related to the independent coordinates θ_1 and θ_2 by the equation

$$x_B = l\sin\theta_1 + l\sin\theta_2 \qquad (3)$$

To obtain the variation δx_B in terms of $\delta\theta_2$, it is necessary to take the *partial derivative* of x_B with respect to θ_2 since x_B is a function of both θ_1 and θ_2. Hence,

$$\frac{\partial x_B}{\partial \theta_2} = l\cos\theta_2 \qquad (\delta x_B)_{\theta_2} = l\cos\theta_2\,\delta\theta_2$$

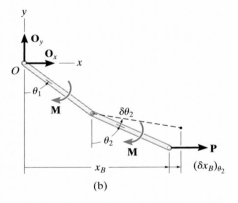

(b)

Substituting into Eq. 1, we have

$$(Pl\cos\theta_2 - M)\,\delta\theta_2 = 0$$

Since $\delta\theta_2 \neq 0$, then

$$\theta_2 = \cos^{-1}\!\left(\frac{M}{Pl}\right) \qquad Ans.$$

Using Eq. 3 to obtain the variation of x_B with θ_1 yields

$$\frac{\partial x_B}{\partial \theta_1} = l\cos\theta_1 \qquad (\delta x_B)_{\theta_1} = l\cos\theta_1\,\delta\theta_1$$

Substituting into Eq. 2, we have

$$(Pl\cos\theta_1 - M)\,\delta\theta_1 = 0$$

Since $\delta\theta_1 \neq 0$, then

$$\theta_1 = \cos^{-1}\!\left(\frac{M}{Pl}\right) \qquad Ans.$$

(c)

Fig. 11–10

PROBLEMS

11–1.   The thin rod of weight W rests against the smooth wall and floor. Determine the magnitude of force **P** needed to hold it in equilibrium.

Prob. 11–1

11–2.   The disk has a weight of 10 lb and is subjected to a vertical force $P = 8$ lb and a couple moment $M = 8$ lb $\cdot$ ft. Determine the disk's rotation θ if the end of the spring wraps around the periphery of the disk as the disk turns. The spring is originally unstretched.

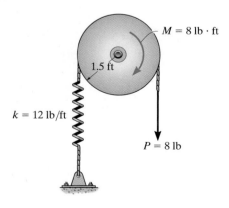

Prob. 11–2

11–3.   The platform supports a load W. Determine the horizontal force P that must be supplied by the screw in order to support the platform when the links are at the arbitrary angle θ.

Prob. 11–3

*■11–4.   Each member of the pin-connected mechanism has a mass of 8 kg. If the spring is unstretched when $\theta = 0°$, determine the angle θ for equilibrium. Set $k = 2500$ N/m and $M = 50$ N $\cdot$ m.

11–5.   Each member of the pin-connected mechanism has a mass of 8 kg. If the spring is unstretched when $\theta = 0°$, determine the required stiffness k so that the mechanism is in equilibrium when $\theta = 30°$. Set **M** $= 0$.

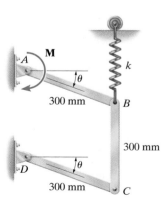

Probs. 11–4/5

11–6. The crankshaft is subjected to a torque of $M = 50$ N·m. Determine the horizontal compressive force F applied to the piston for equilibrium when $\theta = 60°$.

Prob. 11–6

11–7. The crankshaft is subjected to a torque of $M = 50$ N·m. Determine the horizontal compressive force F and plot the result of F (ordinate) versus θ (abscissa) for $0° \leq \theta \leq 90°$.

Prob. 11–7

*11–8. If a force of $P = 30$ lb is applied perpendicular to the handle of the toggle press, determine the compressive force developed at C; $\theta = 30°$.

Prob. 11–8

11–9. A force **P** is applied to the end of the lever. Determine the horizontal force F on the piston for equilibrium.

Prob. 11–9

11–10.   The mechanism consists of the four pin-connected bars and three springs, each having a stiffness k and an unstretched length l_0. Determine the horizontal forces P that must be applied to the pins in order to hold the mechanism in the horizontal position for equilibrium.

*11–12.   The spring is unstretched when $\theta = 0°$. If $P = 8$ lb, determine the angle θ for equilibrium. Due to the roller guide, the spring always remains vertical. Neglect the weight of the links.

Prob. 11–10

Prob. 11–12

11–11.   When $\theta = 20°$, the 50-lb uniform block compresses the two vertical springs 4 in. If the uniform links AB and CD each weigh 10 lb, determine the magnitude of the applied couple moments $\mathbf{M}$ needed to maintain equilibrium when $\theta = 20°$.

11–13.   Determine the force P required to lift the 15-kg block using the differential hoist. The lever arm is fixed to the upper pulley and turns with it.

Prob. 11–11

Prob. 11–13

11–14. Determine the magnitude of the applied couple moments **M** needed to maintain equilibrium when $\theta = 20°$. The plate E has a weight of 50 lb. Neglect the weight of the links AB and CD.

*11–16. Determine the force F needed to lift the block having a weight of 100 lb. *Hint:* Note that the coordinates s_A and s_B can be related to the *constant* vertical length l of the cord.

Prob. 11–14

Prob. 11–16

11–15. The members of the mechanism are pin connected. If a horizontal force of $400 \cdot$ N acts at A, determine the angle θ for equilibrium. The spring is unstretched when $\theta = 90°$.

■11–17. Each member of the pin-connected mechanism has a mass of 8 kg. If the spring is unstretched when $\theta = 0°$, determine the angle θ for equilibrium.

Prob. 11–15

Prob. 11–17

11–18. The bar is supported by the spring and smooth collar that allows the spring to be always perpendicular to the bar for any angle θ. If the unstretched length of the spring is l_0, determine the force P needed to hold the bar in the equilibrium position θ. Neglect the weight of the bar.

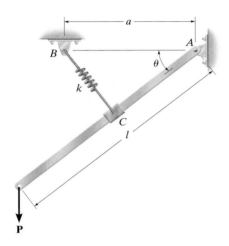

Prob. 11–18

11–19. The scissors jack supports a load **P**. Determine the axial force in the screw necessary for equilibrium when the jack is in the position θ. Each of the four links has a length L and is pin-connected at its center. Points B and D can move horizontally.

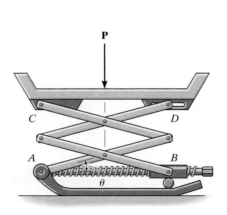

Prob. 11–19

***11–20.** Determine the mass of A and B required to hold the 400-g desk lamp in balance for any angles θ and ϕ. Neglect the weight of the mechanism and the size of the lamp.

Prob. 11–20

11–21. The *Roberval balance* is in equilibrium when no weights are placed on the pans A and B. If two masses m_A and m_B are placed at *any* location a and b on the pans, show that equilibrium is maintained if $m_A d_A = m_B d_B$.

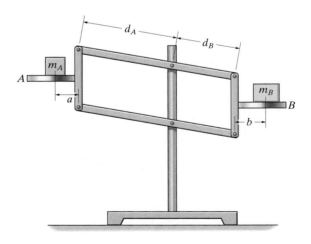

Prob. 11–21

11–22. The chain puller is used to draw two ends of a chain together in order to attach the "master link." The device is operated by turning the screw S, which pushes the bar AB downward, thereby drawing the tips C and D towards one another. If the sliding contacts at A and B are smooth, determine the force F maintained by the screw at E which is required to develop a drawing tension of 120 lb in the chains.

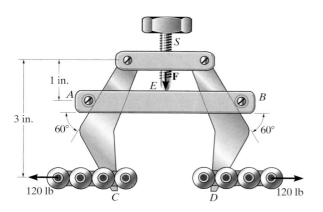

Prob. 11–22

*11–24.** A horizontal force acts on the end of the link as shown. Determine the angles θ_1 and θ_2 for equilibrium of the two links. Each link is uniform and has a mass m.

Prob. 11–24

11–23. The service window at a fast-food restaurant consists of glass doors that open and close automatically using a motor which supplies a torque **M** to each door. The far ends, A and B, move along the horizontal guides. If a food tray becomes stuck between the doors as shown, determine the horizontal force the doors exert on the tray at the position θ.

Prob. 11–23

11–25. Rods AB and BC have centers of mass located at their midpoints. If all contacting surfaces are smooth and BC has a mass of 100 kg, determine the appropriate mass of AB required for equilibrium.

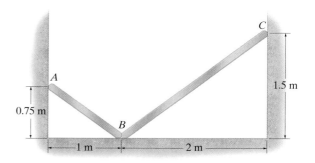

Prob. 11–25

*11.4   Conservative Forces

The work done by a force when it undergoes a *differential displacement* has been defined as $dU = F \cos \theta \, ds$, Fig. 11–1. If the force is displaced over a path that has a *finite length s*, the work is determined by integrating over the path: i.e.,

$$U = \int_s F \cos \theta \, ds$$

To evaluate the integral, it is necessary to obtain a relationship between F and the component of displacement $ds \cos \theta$. In some instances, however, the work done by a force will be *independent* of its path and, instead, will depend only on the initial and final locations of the force along the path. A force that has this property is called a *conservative force*.

Weight.   Consider the body in Fig. 11–11, which is initially at P'. If the body is moved *down* along the *arbitrary path A* to the second position, then, for a given displacement ds along the path, the displacement component in the direction of $\mathbf{W}$ has a magnitude of $dy = ds \cos \theta$, as shown. Since both the force and displacement are in the same direction, the work is positive; hence,

$$U = \int_s W \cos \theta \, ds = \int_0^y W \, dy$$

or

$$U = Wy$$

In a similar manner, the work done by the weight when the body moves up a distance y back to P', along the arbitrary path A', is

$$U = -Wy$$

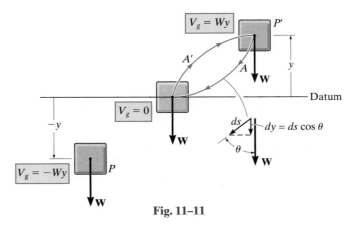

Fig. 11–11

Why is the work negative?

The weight of a body is therefore a conservative force since the work done by the weight depends *only* on the body's *vertical displacement* and is independent of the path along which the body moves.

Elastic Spring.   The force developed by an elastic spring ($F_s = ks$) is also a conservative force. If the spring is attached to a body and the body is displaced along *any path*, such that it causes the spring to elongate or compress from a position s_1 to a further position s_2, the work will be negative since the spring exerts a force $\mathbf{F}_s$ *on the body* that is opposite to the body's displacement ds, Fig. 11–12. For either extension or compression, the work is independent of the path and is simply

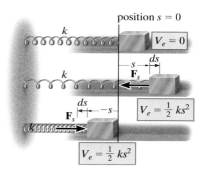

Fig. 11–12

$$U = \int_{s_1}^{s_2} F_s \, ds = \int_{s_1}^{s_2} (-ks) \, ds$$

$$= -\left(\tfrac{1}{2} k s_2^2 - \tfrac{1}{2} k s_1^2\right)$$

Friction.   In contrast to a conservative force, consider the force of *friction* exerted on a sliding body by a fixed surface. The work done by the frictional force depends on the path; the longer the path, the greater the work. Consequently, frictional forces are *nonconservative*, and the work done is dissipated from the body in the form of heat.

*11.5   Potential Energy

When a conservative force acts on a body, it gives the body the capacity to do work. This capacity, measured as *potential energy*, depends on the location of the body.

Gravitational Potential Energy.   If a body is located a distance *y above* a fixed horizontal reference or datum, Fig. 11–11, the weight of the body has *positive* gravitational potential energy V_g since $\mathbf{W}$ has the capacity of doing positive work when the body is moved back down to the datum. Likewise, if the body is located a distance *y below* the datum, V_g is *negative* since the weight does negative work when the body is moved back up to the datum. At the datum, $V_g = 0$.

Measuring *y* as *positive upward*, the gravitational potential energy of the body's weight $\mathbf{W}$ is thus

$$\boxed{V_g = Wy}$$ (11–4)

Datum

(a)

(b)

Fig. 11–13

Elastic Potential Energy.   The elastic potential energy V_e that a spring produces on an attached body, when the spring is elongated or compressed from an undeformed position ($s = 0$) to a final position s, is

$$V_e = \tfrac{1}{2}ks^2 \qquad (11\text{--}5)$$

Here V_e is *always positive* since in the deformed position the spring has the capacity of doing *positive work* in *returning* the body back to the spring's undeformed position, Fig. 11–12.

Potential Function.   In the general case, if a body is subjected to *both* gravitational and elastic forces, the *potential energy or potential function V* of the body can be expressed as the algebraic sum

$$V = V_g + V_e \qquad (11\text{--}6)$$

where measurement of V depends on the location of the body with respect to a selected datum in accordance with Eqs. 11–4 and 11–5.

In general, if a system of frictionless connected rigid bodies has a *single degree of freedom* such that its position from the datum is defined by the independent coordinate q, then the potential function for the system can be expressed as $V = V(q)$. The work done by all the conservative forces acting on the system in moving it from q_1 to q_2 is measured by the *difference* in V; i.e.,

$$U_{1-2} = V(q_1) - V(q_2) \qquad (11\text{--}7)$$

For example, the potential function for a system consisting of a block of weight **W** supported by a spring, Fig. 11–13a, can be expressed in terms of its independent coordinate $(q =) y$, measured from a fixed datum located at the unstretched length of the spring; we have

$$V = V_g + V_e$$
$$= -Wy + \tfrac{1}{2}ky^2 \qquad (11\text{--}8)$$

If the block moves from y_1 to a farther downward position y_2, then the work of **W** and $\mathbf{F}_s$ is

$$U_{1-2} = V(y_1) - V(y_2) = -W[y_1 - y_2] + \tfrac{1}{2}ky_1^2 - \tfrac{1}{2}ky_2^2$$

*11.6   Potential-Energy Criterion for Equilibrium

System Having One Degree of Freedom.   When the displacement of a frictionless connected system is *infinitesimal*, i.e., from q to $q + dq$, Eq. 11–7 becomes

$$dU = V(q) - V(q + dq)$$

or

$$dU = -dV$$

Furthermore, if the system undergoes a *virtual displacement* δq, rather than an actual displacement dq, then $\delta U = -\delta V$. For equilibrium, the principle of virtual work requires that $\delta U = 0$, and therefore, provided the potential function for the system is known, this also requires that $\delta V = 0$. We can also express this requirement as

$$\boxed{\frac{dV}{dq} = 0} \qquad (11\text{--}9)$$

Hence, *when a frictionless connected system of rigid bodies is in equilibrium, the first variation or change in V is zero.* This change is determined by taking the *first derivative* of the potential function and setting it equal to zero. For example, using Eq. 11–8 to determine the equilibrium position for the spring and block in Fig. 11–13a, we have

$$\frac{dV}{dy} = -W + ky = 0$$

Hence, the equilibrium position $y = y_{eq}$ is

$$y_{eq} = \frac{W}{k}$$

Of course, the *same result* is obtained by applying $\Sigma F_y = 0$ to the forces acting on the free-body diagram of the block, Fig. 11–13b.

System Having *n* Degrees of Freedom.

When the system of connected bodies has *n* degrees of freedom, the total potential energy stored in the system will be a function of *n* independent coordinates q_n, i.e., $V = V(q_1, q_2, \ldots, q_n)$. In order to apply the equilibrium criterion $\delta V = 0$, it is necessary to determine the change in potential energy δV by using the "chain rule" of differential calculus; i.e.,

$$\delta V = \frac{\partial V}{\partial q_1}\delta q_1 + \frac{\partial V}{\partial q_2}\delta q_2 + \cdots + \frac{\partial V}{\partial q_n}\delta q_n = 0$$

Since the virtual displacements $\delta q_1, \delta q_2, \ldots, \delta q_n$ are independent of one another, the equation is satisfied provided

$$\frac{\partial V}{\partial q_1} = 0, \quad \frac{\partial V}{\partial q_2} = 0, \quad \ldots, \quad \frac{\partial V}{\partial q_n} = 0$$

Hence *it is possible to write n independent equations for a system having n degrees of freedom.*

A

B

Stable equilibrium   Neutral equilibrium

C

Unstable equilibrium

Fig. 11–14

During high winds and when going around a curve, these sugar-cane trucks can become unstable and tip over since their center of gravity is high off the road when they are fully loaded.

*11.7   Stability of Equilibrium

Once the equilibrium configuration for a body or system of connected bodies is defined, it is sometimes important to investigate the "type" of equilibrium or the stability of the configuration. For example, consider the position of a ball resting at a point on each of the three paths shown in Fig. 11–14. Each situation represents an equilibrium state for the ball. When the ball is at *A*, it is said to be in *stable equilibrium* because if it is given a small displacement up the hill, it will always *return* to its original, lowest, position. At *A*, its total potential energy is a *minimum*. When the ball is at *B*, it is in *neutral equilibrium*. A small displacement to either the left or right of *B* will not alter this condition. The ball *remains* in equilibrium in the displaced position, and therefore its potential energy is *constant*. When the ball is at *C*, it is in *unstable equilibrium*. Here a small displacement will cause the ball's potential energy to be *decreased*, and so it will roll farther *away* from its original, highest position. At *C*, the potential energy of the ball is a *maximum*.

Types of Equilibrium.   The example just presented illustrates that one of three types of equilibrium positions can be specified for a body or system of connected bodies.

1. *Stable equilibrium* occurs when a small displacement of the system causes the system to return to its original position. In this case the original potential energy of the system is a minimum.

2. *Neutral equilibrium* occurs when a small displacement of the system causes the system to remain in its displaced state. In this case the potential energy of the system remains constant.

3. *Unstable equilibrium* occurs when a small displacement of the system causes the system to move farther away from its original position. In this case the original potential energy of the system is a maximum.

System Having One Degree of Freedom.   For *equilibrium* of a system having a single degree of freedom, defined by the independent coordinate *q*, it has been shown that the first derivative of the potential function for the system must be equal to zero; i.e., $dV/dq = 0$. If the potential function $V = V(q)$ is plotted, Fig. 11–15, the first derivative (equilibrium position) is represented as the slope dV/dq, which is zero when the function is maximum, minimum, or an inflection point.

If the *stability* of a body is to be investigated, it is necessary to determine the *second derivative* of *V* and evaluate it at the equilibrium position $q = q_{eq}$. As shown in Fig. 11–15a, if $V = V(q)$ is a *minimum*, then

$$\frac{dV}{dq} = 0, \quad \frac{d^2V}{dq^2} > 0 \quad \text{stable equilibrium} \qquad (11\text{–}10)$$

If $V = V(q)$ is a *maximum*, Fig. 11–15b, then

$$\frac{dV}{dq} = 0, \quad \frac{d^2V}{dq^2} < 0 \qquad \text{unstable equilibrium} \qquad (11\text{–}11)$$

If the second derivative is zero, it will be necessary to investigate *higher-order* derivatives to determine the stability. In particular, stable equilibrium will occur if the order of the lowest remaining nonzero derivative is *even* and the sign of this nonzero derivative is positive when it is evaluated at $q = q_{eq}$; otherwise, it is unstable.

If the system is in neutral equilibrium, Fig. 11–15c, it is required that

$$\frac{dV}{dq} = \frac{d^2V}{dq^2} = \frac{d^3V}{dq^3} = \cdots = 0 \qquad \text{neutral equilibrium} \qquad (11\text{–}12)$$

since then V must be constant at and around the "neighborhood" of q_{eq}.

System Having Two Degrees of Freedom. A criterion for investigating stability becomes increasingly complex as the number of degrees of freedom for the system increases. For a system having two degrees of freedom, defined by independent coordinates (q_1, q_2), it may be verified (using the calculus of functions of two variables) that equilibrium and stability occur at a point (q_{1eq}, q_{2eq}) when

$$\frac{\partial V}{\partial q_1} = \frac{\partial V}{\partial q_2} = 0$$

$$\left[\left(\frac{\partial^2 V}{\partial q_1 \, \partial q_2} \right)^2 - \left(\frac{\partial^2 V}{\partial q_1^2} \right)\left(\frac{\partial^2 V}{\partial q_2^2} \right) \right] < 0$$

$$\frac{\partial^2 V}{\partial q_1^2} > 0 \quad \text{or} \quad \frac{\partial^2 V}{\partial q_2^2} > 0$$

Both equilibrium and instability occur when

$$\frac{\partial V}{\partial q_1} = \frac{\partial V}{\partial q_2} = 0$$

$$\left[\left(\frac{\partial^2 V}{\partial q_1 \, \partial q_2} \right)^2 - \left(\frac{\partial^2 V}{\partial q_1^2} \right)\left(\frac{\partial^2 V}{\partial q_2^2} \right) \right] < 0$$

$$\frac{\partial^2 V}{\partial q_1^2} < 0 \quad \text{or} \quad \frac{\partial^2 V}{\partial q_2^2} < 0$$

Stable equilibrium
(a)

Unstable equilibrium
(b)

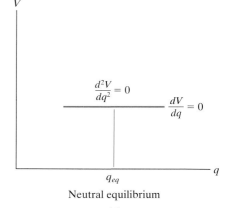

Neutral equilibrium
(c)

Fig. 11–15

Procedure for Analysis

Using potential-energy methods, the equilibrium positions and the stability of a body or a system of connected bodies having a single degree of freedom can be obtained by applying the following procedure.

Potential Function.

- Sketch the system so that it is located at some *arbitrary position* specified by the independent coordinate q.
- Establish a horizontal *datum* through a *fixed point** and express the *gravitational potential energy* V_g in terms of the weight W of each member and its vertical distance y from the datum, $V_g = Wy$.
- Express the elastic potential energy V_e of the system in terms of the stretch or compression, s, of any connecting spring and the spring's stiffness k, $V_e = \frac{1}{2}ks^2$.
- Formulate the potential function $V = V_g + V_e$ and express the *position coordinates* y and s in terms of the independent coordinate q.

Equilibrium Position.

- The equilibrium position is determined by taking the first derivative of V and setting it equal to zero, $\delta V = 0$.

Stability.

- Stability at the equilibrium position is determined by evaluating the second or higher-order derivatives of V.
- If the second derivative is greater than zero, the body is stable, if all derivatives are equal to zero the body is in neutral equilibrium, and if the second derivative is less than zero, the body is unstable.

*The location of the datum is *arbitrary* since only the *changes* or differentials of V are required for investigation of the equilibrium position and its stability.

EXAMPLE 11.5

The uniform link shown in Fig. 11–16a has a mass of 10 kg. The spring is unstretched when $\theta = 0°$. Determine the angle θ for equilibrium and investigate the stability at the equilibrium position.

SOLUTION

Potential Function.   The datum is established at the top of the link when the *spring is unstretched*, Fig. 11–16b. When the link is located at the arbitrary position θ, the spring increases its potential energy by stretching and the weight decreases its potential energy. Hence,

$$V = V_e + V_g = \frac{1}{2}ks^2 - W\left(s + \frac{l}{2}\cos\theta\right)$$

Since $l = s + l\cos\theta$ or $s = l(1 - \cos\theta)$, then

$$V = \frac{1}{2}kl^2(1 - \cos\theta)^2 - \frac{Wl}{2}(2 - \cos\theta)$$

Equilibrium Position.   The first derivative of V gives

$$\frac{dV}{d\theta} = kl^2(1 - \cos\theta)\sin\theta - \frac{Wl}{2}\sin\theta = 0$$

or

$$l\left[kl(1 - \cos\theta) - \frac{W}{2}\right]\sin\theta = 0$$

This equation is satisfied provided

$$\sin\theta = 0 \qquad \theta = 0° \qquad\qquad Ans.$$

$$\theta = \cos^{-1}\left(1 - \frac{W}{2kl}\right) = \cos^{-1}\left[1 - \frac{10(9.81)}{2(200)(0.6)}\right] = 53.8° \quad Ans.$$

Stability.   Determining the second derivative of V gives

$$\frac{d^2V}{d\theta^2} = kl^2(1 - \cos\theta)\cos\theta + kl^2\sin\theta\sin\theta - \frac{Wl}{2}\cos\theta$$

$$= kl^2(\cos\theta - \cos 2\theta) - \frac{Wl}{2}\cos\theta$$

Substituting values for the constants, with $\theta = 0°$ and $\theta = 53.8°$, yields

$$\left.\frac{d^2V}{d\theta^2}\right|_{\theta=0°} = 200(0.6)^2(\cos 0° - \cos 0°) - \frac{10(9.81)(0.6)}{2}\cos 0°$$

$$= -29.4 < 0 \qquad \text{(unstable equilibrium at } \theta = 0°) \qquad Ans.$$

$$\left.\frac{d^2V}{d\theta^2}\right|_{\theta=53.8°} = 200(0.6)^2(\cos 53.8° - \cos 107.6°) - \frac{10(9.81)(0.6)}{2}\cos 53.8°$$

$$= 46.9 > 0 \qquad \text{(stable equilibrium at } \theta = 53.8°) \qquad Ans.$$

(a)

(b)

Fig. 11–16

EXAMPLE 11.6

(a)

$(1.2 - 1.5 \sin \theta)$ m

$(1.5 \sin \theta)$ m

Datum B'

98.1 N

$\theta = 20$

$(1.5 \cos \theta)$ m

$(y_W)_1$

$(y_W)_2$

Δy

$m (9.81)$

(b)

Fig. 11–17

Determine the mass m of the block required for equilibrium of the uniform 10-kg rod shown in Fig. 11–17a when $\theta = 20°$. Investigate the stability at the equilibrium position.

SOLUTION

Potential Function.   The datum is established through point A, Fig. 11–17b. When $\theta = 0°$, the block is assumed to be suspended $(y_W)_1$ below the datum. Hence, in the position θ,

$$V = V_e + V_g = 98.1\left(\frac{1.5 \sin \theta}{2}\right) - m(9.81)(\Delta y) \qquad (1)$$

The distance $\Delta y = (y_W)_2 - (y_W)_1$ may be related to the independent coordinate θ by measuring the difference in cord lengths $B'C$ and BC. Since

$$B'C = \sqrt{(1.5)^2 + (1.2)^2} = 1.92$$
$$BC = \sqrt{(1.5 \cos \theta)^2 + (1.2 - 1.5 \sin \theta)^2} = \sqrt{3.69 - 3.60 \sin \theta}$$

then

$$\Delta y = B'C - BC = 1.92 - \sqrt{3.69 - 3.60 \sin \theta}$$

Substituting the above result into Eq. 1 yields

$$V = 98.1\left(\frac{1.5 \sin \theta}{2}\right) - m(9.81)(1.92 - \sqrt{3.69 - 3.60 \sin \theta}) \quad (2)$$

Equilibrium Position.

$$\frac{dV}{d\theta} = 73.6 \cos \theta - \left[\frac{m(9.81)}{2}\right]\left(\frac{3.60 \cos \theta}{\sqrt{3.69 - 3.60 \sin \theta}}\right) = 0$$

$$\left.\frac{dV}{d\theta}\right|_{\theta = 20°} = 69.14 - 10.58m = 0$$

$$m = \frac{69.14}{10.58} = 6.53 \text{ kg} \qquad\qquad Ans.$$

Stability.   Taking the second derivative of Eq. 2, we obtain

$$\frac{d^2V}{d\theta^2} = -73.6 \sin \theta - \left[\frac{m(9.81)}{2}\right]\left(\frac{-1}{2}\right)\frac{-(3.60 \cos \theta)^2}{(3.69 - 3.60 \sin \theta)^{3/2}}$$
$$- \left[\frac{m(9.81)}{2}\right]\left(\frac{-3.60 \sin \theta}{\sqrt{3.69 - 3.60 \sin \theta}}\right)$$

For the equilibrium position $\theta = 20°$, with $m = 6.53$ kg, then

$$\frac{d^2V}{d\theta^2} = -47.6 < 0 \qquad \text{(unstable equilibrium at } \theta = 20°\text{)} \quad Ans.$$

EXAMPLE 11.7

The homogeneous block having a mass m rests on the top surface of the cylinder, Fig. 11–18a. Show that this is a condition of unstable equilibrium if $h > 2R$.

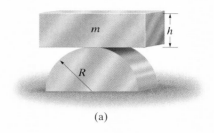

(a)

SOLUTION

Potential Function.   The datum is established at the base of the cylinder, Fig. 11–18b. If the block is displaced by an amount θ from the equilibrium position, the potential function may be written in the form

$$V = V_e + V_g$$
$$= 0 + mgy$$

From Fig. 11–18b,

$$y = \left(R + \frac{h}{2}\right)\cos\theta + R\theta\sin\theta$$

Thus,

$$V = mg\left[\left(R + \frac{h}{2}\right)\cos\theta + R\theta\sin\theta\right]$$

(b)

Fig. 11–18

Equilibrium Position.

$$\frac{dV}{d\theta} = mg\left[-\left(R + \frac{h}{2}\right)\sin\theta + R\sin\theta + R\theta\cos\theta\right] = 0$$

$$= mg\left(-\frac{h}{2}\sin\theta + R\theta\cos\theta\right) = 0$$

Obviously, $\theta = 0°$ is the equilibrium position that satisfies this equation.

Stability.   Taking the second derivative of V yields

$$\frac{d^2V}{d\theta^2} = mg\left(-\frac{h}{2}\cos\theta + R\cos\theta - R\theta\sin\theta\right)$$

At $\theta = 0°$,

$$\frac{d^2V}{d\theta^2}\bigg|_{\theta=0°} = -mg\left(\frac{h}{2} - R\right)$$

Since all the constants are positive, the block is in unstable equilibrium if $h > 2R$, because then $d^2V/d\theta^2 < 0$.

PROBLEMS

11–26. If the potential energy for a conservative two-degree-of-freedom system is expressed by the relation $V = (3y^2 + 2x^2)$ J, where y and x are given in meters, determine the equilibrium positions and investigate the stability at each position.

11–27. If the potential energy for a conservative one-degree-of-freedom system is expressed by the relation $V = (4x^3 - x^2 - 3x + 10)$ ft · lb, where x is given in feet, determine the equilibrium positions and investigate the stability at each position.

*11–28. If the potential energy for a conservative one-degree-of-freedom system is expressed by the relation $V = (24 \sin \theta + 10 \cos 2\theta)$ ft · lb, $0° \le \theta \le 180°$, determine the equilibrium positions and investigate the stability at each position.

11–29. If the potential energy for a conservative two-degree-of-freedom system is expressed by the relation $V = (6y^2 + 2x^2)$ J, where x and y are given in meters, determine the equilibrium position and investigate the stability at this position.

11–30. The spring of the scale has an unstretched length of a. Determine the angle θ for equilibrium when a weight W is supported on the platform. Neglect the weight of the members. What value W would be required to keep the scale in neutral equilibrium when $\theta = 0°$?

11–31. The two bars each have a weight of 8 lb. Determine the required stiffness k of the spring so that the two bars are in equilibrium when $\theta = 30°$. The spring has an unstretched length of 1 ft.

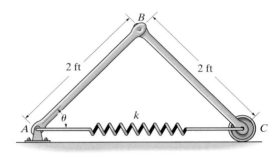

Prob. 11–31

*11–32. Each of the two springs has an unstretched length of 500 mm. Determine the mass m of the cylinder when it is held in the equilibrium position shown, i.e., $y = 1$ m.

Prob. 11–30

Prob. 11–32

■11–33. The uniform beam has a mass of 200 kg. If the contacting surfaces are smooth, determine the angle θ for equilibrium and investigate the stability of the beam when it is in this position. The spring has an unstretched length of 0.5 m.

Prob. 11–33

11–35. The uniform rod AB has a mass of 80 kg. If spring DC is unstretched when $\theta = 90°$, determine the angle θ for equilibrium and investigate the stability at the equilibrium position. The spring always acts in the horizontal position due to the roller guide at D.

Prob. 11–35

11–34. The bar supports a weight of $W = 500$ lb at its end. If the springs are originally unstretched when the bar is vertical, determine the required stiffness $k_1 = k_2 = k$ of the springs so that the bar is in neutral equilibrium when it is vertical.

Prob. 11–34

*11–36. Determine the angle θ for equilibrium and investigate the stability at this position. The bars each have a mass of 3 kg and the suspended block D has a mass of 7 kg. Cord DC has a total length of 1 m.

Prob. 11–36

11–37. The bar supports a weight of 500 lb at its end. If the springs are originally unstretched when the bar is vertical, and $k_1 = 300$ lb/ft, $k_2 = 500$ lb/ft, investigate the stability of the bar when it is in the vertical position.

3 ft

k_2

k_1

3 ft

3 ft

A

Prob. 11–37

11–38. If each of the three links of the mechanism has a weight W, determine the angle θ for equilibrium. The spring, which always remains vertical, is unstretched when $\theta = 0°$.

k

θ

a

a

$2\,a$

$2\,a$

Prob. 11–38

11–39. The small postal scale consists of a counterweight W_1, connected to the members having negligible weight. Determine the weight W_2 that is on the pan in terms of the angles θ and ϕ and the dimensions shown. All members are pin connected.

W_2

a

ϕ

θ

b

a

ϕ

W_1

Prob. 11–39

*****11–40.** The uniform right circular cone having a mass m is suspended from the cord as shown. Determine the angle θ at which it hangs from the wall for equilibrium. Is the cone in stable equilibrium?

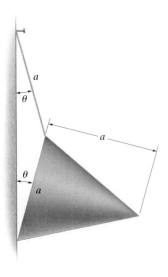

a

θ

a

θ

a

Prob. 11–40

11–41. The homogeneous cylinder has a conical cavity cut into its base as shown. Determine the depth d of the cavity so that the cylinder balances on the pivot and remains in neutral equilibrium.

150 mm

d

50 mm

Prob. 11–41

11–42. The conical manhole cap is made of concrete and has the dimensions shown. Determine the critical location $h = h_{cr}$ of the pick-up connectors at A and B so that when hoisted with constant velocity the cap is in neutral equilibrium. Explain what would happen if the connectors were placed at a point $h > h_{cr}$.

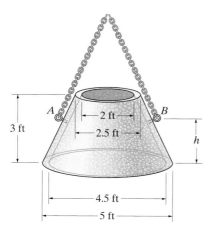

3 ft

A 2 ft B

2.5 ft

h

4.5 ft

5 ft

Prob. 11–42

11–43. Each bar has a mass per length of m_0. Determine the angles θ and ϕ at which they are suspended in equilibrium. The contact at A is smooth, and both are pin connected at B.

B

$\frac{3}{2}l$

l

A

$\frac{l}{2}$

Prob. 11–43

*11–44. The triangular block of weight W rests on the smooth corners which are a distance a apart. If the block has three equal sides of length d, determine the angle θ for equilibrium.

60° d

•G 60°

θ

a

Prob. 11–44

11–45. A homogeneous cone rests on top of the cylindrical surface. Derive a relationship between the radius r of the cylinder and the height h of the cone for neutral equilibrium. *Hint:* Establish the potential function for a *small* angle θ of tilt of the cone, i.e., approximate $\sin\theta \approx \theta$ and $\cos\theta \approx 1 - \theta^2/2$.

h

a a

r

Prob. 11–45

11–46.   The door has a uniform weight of 50 lb. It is hinged at A and is held open by the 30-lb weight and the pulley. Determine the angle θ for equilibrium.

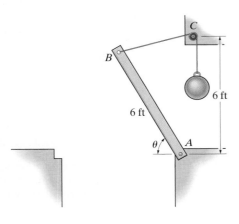

Prob. 11–46

11–47.   The 60-lb hemisphere supports a cylinder having a specific weight of $\gamma = 311$ lb/ft^3. If the radii of the cylinder and hemisphere are both 5 in., determine the height h of the cylinder which will produce neutral equilibrium in the position shown.

Prob. 11–47

*11–48.   Compute the force developed in the spring required to keep the 6-kg rod in equilibrium when $\theta = 30°$. The spring remains horizontal due to the roller guide.

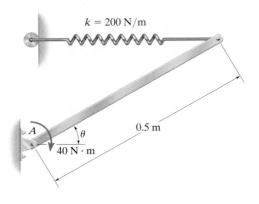

Prob. 11–48

11–49.   Determine the force **P** acting on the cord which is required to maintain equilibrium of the horizontal 20-kg bar *CB*. *Hint:* First show that the coordinates s_A and s_B are related to the *constant* vertical length l of the cord by the equation $5s_B - s_A = l$.

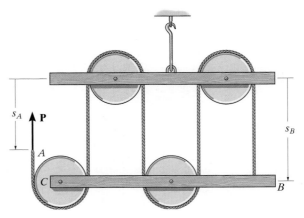

Prob. 11–49

Chapter Review

Principle of Virtual Work

The forces on a body will do *virtual work* when the body undergoes an *imaginary* differential displacement or rotation.

$\delta y, \delta y'$ –virtual displacement

$\delta \theta$ –virtual rotation

For equilibrium, the sum of the virtual work done by all the forces acting on the body must be equal to zero for any virtual displacement. This is referred to as the *principle of virtual work*, and it is useful for finding the equilibrium configuration for a mechanism or a reactive force acting on a series of connected members.

$$\delta U = 0$$

If this system has one degree of freedom, then its position can be specified by one independent coordinate $q = x$ or $q = \theta$.

One-degree-of-freedom system
(a)

To apply the principle of virtual work, it is first necessary to use *position coordinates* to locate all the forces and moments on the mechanism that will do work when the mechanism undergoes a virtual movement δq.

$$q = \theta$$

θ, x –position coordinates

The coordinates are related to the independent coordinate q and then these expressions are differentiated in order to relate the *virtual* coordinate displacements to the virtual displacement δq.

$$b = \sqrt{x^2 + a^2 - 2ax \cos \theta}$$

Finally, the equation of virtual work is written for the mechanism in terms of the common displacement δq, and then it is set equal to zero. By factoring δq out of the equation, it is then possible to determine either the unknown force or couple moment, or the equilibrium position q.
(*See pages 573–579*)

Potential-Energy Criterion for Equilibrium

When a system is subjected only to conservative forces, such as weight or spring forces, then the equilibrium configuration can be determined using the *potential-energy function V* for the system.

$$q = \theta$$

The potential-energy function is established by expressing the weight and spring potential energy for the system in terms of the independent coordinate q.

Once the potential-energy function is formulated, its first derivative is set equal to zero. The solution yields the equilibrium position q_{eq} for the system.

$$\frac{dV}{dq} = 0$$

The stability of the system can be investigated by taking the second derivative of V. (*See pages 591–596.*)

$$\frac{dV}{dq} = 0, \quad \frac{d^2V}{dq^2} > 0 \qquad \text{stable equilibrium}$$

$$\frac{dV}{dq} = 0, \quad \frac{d^2V}{dq^2} < 0 \qquad \text{unstable equilibrium}$$

$$\frac{dV}{dq} = \frac{d^2V}{dq^2} = \frac{d^3V}{dq^3} = \cdots = 0 \qquad \text{neutral equilibrium}$$

REVIEW PROBLEMS

11–50. The uniform bar AB weighs 10 lb. If the attached spring is unstretched when $\theta = 90°$, use the method of virtual work and determine the angle θ for equilibrium. Note that the spring always remains in the vertical position due to the roller guide.

11–51. Solve Prob. 11–50 using the principle of potential energy. Investigate the stability of the bar when it is in the equilibrium position.

Probs. 11–50/51

11–53 The truck has a mass of 20 Mg and a mass center at G. Determine the steepest grade θ along which it can park without overturning and investigate the stability in this position.

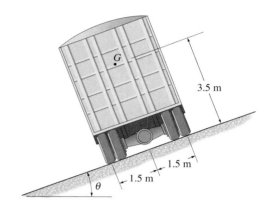

Prob. 11–53

***11–52.** The punch press consists of the ram R, connecting rod AB, and a flywheel. If a torque of $M = 50 \text{ N} \cdot \text{m}$ is applied to the flywheel, determine the force **F** applied at the ram to hold the rod in the position $\theta = 60°$.

11–54. Use the method of virtual work to determine the tensions in cable AC. The lamp weighs 10 lb.

Prob. 11–52

Prob. 11–54

Mathematical Review and Expressions

Geometry and Trigonometry Review

The angles θ in Fig. A–1 are equal between the transverse and two parallel lines.

Fig. A–1

For a line and its normal, the angles θ in Fig. A–2 are equal.

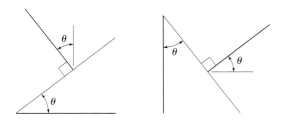

Fig. A–2

For the circle in Fig. A–3 $s = \theta r$, so that when $\theta = 360° = 2\pi$ rad then the circumference is $s = 2\pi r$. Also, since $180° = \pi$ rad, then θ (rad) $= (\pi/180°)\theta°$. The area of the circle is $A = \pi r^2$.

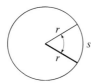

Fig. A–3

The sides of a similar triangle can be obtained by proportion as in Fig. A–4, where $\dfrac{a}{A} = \dfrac{b}{B} = \dfrac{c}{C}$.

For the right triangle in Fig. A-5, the Pythagorean theorem is

$$h = \sqrt{(o)^2 + (a)^2}$$

The trigonometric functions are

$$\sin \theta = \frac{o}{h}$$

$$\cos \theta = \frac{a}{h}$$

$$\tan \theta = \frac{o}{a}$$

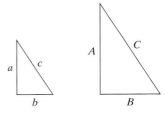

Fig. A–4

This is easily remembered as "soh, cah, toa", i.e., the sine is the opposite over the hypotenuse, etc. The other trigonometric functions follow from this.

$$\csc \theta = \frac{1}{\sin \theta} = \frac{h}{o}$$

$$\sec \theta = \frac{1}{\cos \theta} = \frac{h}{a}$$

$$\cot \theta = \frac{1}{\tan \theta} = \frac{a}{o}$$

Fig. A–5

Trigonometric Identities

$$\sin^2\theta + \cos^2\theta = 1$$

$$\sin(\theta \pm \phi) = \sin\theta\cos\phi \pm \cos\theta\sin\phi$$

$$\sin 2\theta = 2\sin\theta\cos\theta$$

$$\cos(\theta \pm \phi) = \cos\theta\cos\phi \mp \sin\theta\sin\phi$$

$$\cos 2\theta = \cos^2\theta - \sin^2\theta$$

$$\cos\theta = \pm\sqrt{\frac{1+\cos 2\theta}{2}}, \ \sin\theta = \pm\sqrt{\frac{1-\cos 2\theta}{2}}$$

$$\tan\theta = \frac{\sin\theta}{\cos\theta}$$

$$1 + \tan^2\theta = \sec^2\theta \qquad 1 + \cot^2\theta = \csc^2\theta$$

Quadratic Formula

If $ax^2 + bx + c = 0$, then $x = \dfrac{-b \pm \sqrt{b^2 - 4ac}}{2a}$

Hyperbolic Functions

$$\sinh x = \frac{e^x - e^{-x}}{2},$$

$$\cosh x = \frac{e^x + e^{-x}}{2},$$

$$\tanh x = \frac{\sinh x}{\cosh x}$$

Power-Series Expansions

$$\sin x = x - \frac{x^3}{3!} + \cdots, \ \cos x = 1 - \frac{x^2}{2!} + \cdots$$

$$\sinh x = x + \frac{x^3}{3!} + \cdots, \ \cosh x = 1 + \frac{x^2}{2!} + \cdots$$

Derivatives

$$\frac{d}{dx}(u^n) = nu^{n-1}\frac{du}{dx} \qquad \frac{d}{dx}(\sin u) = \cos u\frac{du}{dx}$$

$$\frac{d}{dx}(uv) = u\frac{dv}{dx} + v\frac{du}{dx} \qquad \frac{d}{dx}(\cos u) = -\sin u\frac{du}{dx}$$

$$\frac{d}{dx}\left(\frac{u}{v}\right) = \frac{v\frac{du}{dx} - u\frac{dv}{dx}}{v^2} \qquad \frac{d}{dx}(\tan u) = \sec^2 u\frac{du}{dx}$$

$$\frac{d}{dx}(\cot u) = -\csc^2 u\frac{du}{dx} \qquad \frac{d}{dx}(\sinh u) = \cosh u\frac{du}{dx}$$

$$\frac{d}{dx}(\sec u) = \tan u\sec u\frac{du}{dx} \quad \frac{d}{dx}(\cosh u) = \sinh u\frac{du}{dx}$$

$$\frac{d}{dx}(\csc u) = -\csc u\cot u\frac{du}{dx}$$

Integrals

$$\int x^n \, dx = \frac{x^{n+1}}{n+1} + C, \, n \neq -1$$

$$\int \frac{dx}{a + bx} = \frac{1}{b} \ln(a + bx) + C$$

$$\int \frac{dx}{a + bx^2} = \frac{1}{2\sqrt{-ba}} \ln\left[\frac{a + x\sqrt{-ab}}{a - x\sqrt{-ab}}\right] + C,$$
$$ab < 0$$

$$\int \frac{x \, dx}{a + bx^2} = \frac{1}{2b} \ln(bx^2 + a) + C$$

$$\int \frac{x^2 \, dx}{a + bx^2} = \frac{x}{b} - \frac{a}{b\sqrt{ab}} \tan^{-1} \frac{x\sqrt{ab}}{a} + C, \, ab > 0$$

$$\int \sqrt{a + bx} \, dx = \frac{2}{3b} \sqrt{(a + bx)^3} + C$$

$$\int x\sqrt{a + bx} \, dx = \frac{-2(2a - 3bx)\sqrt{(a + bx)^3}}{15b^2} + C$$

$$\int x^2 \sqrt{a + bx} \, dx =$$
$$\frac{2(8a^2 - 12abx + 15b^2x^2)\sqrt{(a + bx)^3}}{105b^3} + C$$

$$\int \sqrt{a^2 - x^2} \, dx = \frac{1}{2}\left[x\sqrt{a^2 - x^2} + a^2 \sin^{-1}\frac{x}{a}\right] + C,$$
$$a > 0$$

$$\int x\sqrt{a^2 - x^2} \, dx = -\frac{1}{3}\sqrt{(a^2 - x^2)^3} + C$$

$$\int x^2\sqrt{a^2 - x^2} \, dx = -\frac{x}{4}\sqrt{(a^2 - x^2)^3}$$
$$+ \frac{a^2}{8}\left(x\sqrt{a^2 - x^2} + a^2 \sin^{-1}\frac{x}{a}\right) + C, \, a > 0$$

$$\int \sqrt{x^2 \pm a^2} \, dx =$$
$$\frac{1}{2}\left[x\sqrt{x^2 \pm a^2} \pm a^2 \ln\left(x + \sqrt{x^2 \pm a^2}\right)\right] + C$$

$$\int x\sqrt{x^2 \pm a^2} \, dx = \frac{1}{3}\sqrt{(x^2 \pm a^2)^3} + C$$

$$\int x^2\sqrt{x^2 \pm a^2} \, dx = \frac{x}{4}\sqrt{(x^2 \pm a^2)^3}$$
$$\mp \frac{a^2}{8}x\sqrt{x^2 \pm a^2} - \frac{a^4}{8}\ln\left(x + \sqrt{x^2 \pm a^2}\right) + C$$

$$\int \frac{dx}{\sqrt{a + bx}} = \frac{2\sqrt{a + bx}}{b} + C$$

$$\int \frac{x \, dx}{\sqrt{x^2 \pm a^2}} = \sqrt{x^2 \pm a^2} + C$$

$$\int \frac{dx}{\sqrt{a + bx + cx^2}} = \frac{1}{\sqrt{c}} \ln\left[\sqrt{a + bx + cx^2} + \right.$$
$$\left. x\sqrt{c} + \frac{b}{2\sqrt{c}}\right] + C, \, c > 0$$

$$= \frac{1}{\sqrt{-c}} \sin^{-1}\left(\frac{-2cx - b}{\sqrt{b^2 - 4ac}}\right) + C, \, c < 0$$

$$\int \sin x \, dx = -\cos x + C$$

$$\int \cos x \, dx = \sin x + C$$

$$\int x \cos(ax) \, dx = \frac{1}{a^2}\cos(ax) + \frac{x}{a}\sin(ax) + C$$

$$\int x^2 \cos(ax) \, dx = \frac{2x}{a^2}\cos(ax) + \frac{a^2x^2 - 2}{a^3}\sin(ax) + C$$

$$\int e^{ax} \, dx = \frac{1}{a}e^{ax} + C$$

$$\int xe^{ax} \, dx = \frac{e^{ax}}{a^2}(ax - 1) + C$$

$$\int \sinh x \, dx = \cosh x + C$$

$$\int \cosh x \, dx = \sinh x + C$$

B

Numerical and Computer Analysis

Occasionally the application of the laws of mechanics will lead to a system of equations for which a closed-form solution is difficult or impossible to obtain. When confronted with this situation, engineers will often use a numerical method which in most cases can be programmed on a microcomputer or programmable pocket calculator. Here we will briefly present a computer program for solving a set of linear algebraic equations and three numerical methods which can be used to solve an algebraic or transcendental equation, evaluate a definite integral, and solve an ordinary differential equation. Application of each method will be explained by example, and an associated computer program written in Microsoft BASIC, which is designed to run on most personal computers, is provided.* A text on numerical analysis should be consulted for further discussion regarding a check of the accuracy of each method and the inherent errors that can develop from the methods.

B.1 Linear Algebraic Equations

Application of the equations of static equilibrium or the equations of motion sometimes requires solving a set of linear algebraic equations. The computer program listed in Fig. B–1 can be used for this purpose. It is based on the method of a Gaussian elimination and can solve at most 10 equations with 10 unknowns. To do so, the equations should first be written in the following general format:

$$A_{11}x_1 + A_{12}x_2 + \cdots + A_{1n}x_n = B_1$$
$$A_{21}x_1 + A_{22}x_2 + \cdots + A_{2n}x_n = B_2$$
$$\vdots$$
$$A_{n1}x_1 + A_{n2}x_2 + \cdots + A_{nn}x_n = B_n$$

*Similar types of programs can be written or purchased for programmable pocket calculators.

The A and B coefficients are called for when running the program. The output presents the unknowns $x_1, \ldots, x_n$.

```
1 PRINT"Linear system of equations":PRINT
2 DIM A(10,11)
3 INPUT"Input number of equations : ",N
4 PRINT
5 PRINT"A  coefficients"
6 FOR I = 1 TO N
7 FOR J = 1 TO N
8 PRINT "A(";I;",";J;
9 INPUT")=",A(I,J)
10 NEXT J
11 NEXT I
12 PRINT
13 PRINT"B  coefficients"
14 FOR I = 1 TO N
15 PRINT "B(";I;
16 INPUT")=",A(I,N+1)
17 NEXT I
18 GOSUB 25
19 PRINT
```

```
20 PRINT"Unknowns"
21 FOR I = 1 TO N
22 PRINT "X(";I;")=";A(I,N+1)
23 NEXT I
24 END
25 REM Subroutine Guassian
26 FOR M=1 TO N
27 NP=M
28 BG=ABS(A(M,M))
29 FOR I = M TO N
30 IF ABS(A(I,M))<=BG THEN 33
31 BG=ABS(A(I,M))
32 NP=I
33 NEXT I
34 IF NP=M THEN 40
35 FOR I = M TO N+1
36 TE=A(M,I)
37 A(M,I)=A(NP,I)
38 A(NP,I)=TE
```

```
39 NEXT I
40 FOR I = M+1 TO N
41 FC=A(I,M)/A(M,M)
42 FOR J = M+1 TO N+1
43 A(I,J)=A(I,J)-FC*A(M,J)
44 NEXT J
45 NEXT I
46 NEXT M
47 A(N,N+1)=A(N,N+1)/A(N,N)
48 FOR I = N-1 TO 1 STEP -1
49 SM=0
50 FOR J=I+1 TO N
51 SM=SM+A(I,J)*A(J,N+1)
52 NEXT J
53 A(I,N+1)=(A(I,N+1)-SM)/A(I,I)
54 NEXT I
55 RETURN
```

Fig. B–1

EXAMPLE B.1

Solve the two equations

$$3x_1 + x_2 = 4$$
$$2x_1 - x_2 = 10$$

SOLUTION
When the program begins to run, it first calls for the number of equations (2); then the A coefficients in the sequence $A_{11} = 3$, $A_{12} = 1$, $A_{21} = 2$, $A_{22} = -1$; and finally the B coefficients $B_1 = 4$, $B_2 = 10$. The output appears as

Unknowns

$$X(1) = 2.8 \qquad\qquad\qquad Ans.$$
$$X(2) = -4.4 \qquad\qquad\qquad Ans.$$

B.2 Simpson's Rule

Simpson's rule is a numerical method that can be used to determine the area under a curve given as a graph or as an explicit function $y = f(x)$. Likewise, it can be used to compute the value of a definite integral which involves the function $y = f(x)$. To do so, the area must be subdivided into an *even number* of strips or intervals having a width h. The curve between three consecutive ordinates is approximated by a parabola, and the entire area or definite integral is then determined from the formula

$$\int_{x_0}^{x_n} f(x)\, dx \simeq \frac{h}{3}[y_0 + 4(y_1 + y_3 + \cdots + y_{n-1})$$

$$+ 2(y_2 + y_4 + \cdots + y_{n-2}) + y_n] \quad (B\text{--}1)$$

The computer program for this equation is given in Fig. B–2. For its use, we must first specify the function (on line 6 of the program). The upper and lower limits of the integral and the number of intervals are called for when the program is executed. The value of the integral is then given as the output.

```
1  PRINT"Simpson's rule":PRINT
2  PRINT" To execute this program :":PRINT
3  PRINT"    1- Modify right-hand side of the equation given below,
4  PRINT"       then press RETURN key"
5  PRINT"    2- Type   RUN 6":PRINT:EDIT 6
6  DEF FNF(X)=LOG(X)
7  PRINT:INPUT" Enter Lower Limit = ",A
8  INPUT" Enter Upper Limit = ",B
9  INPUT" Enter Number (even) of Intervals = ",N%
10 H=(B-A)/N%:AR=FNF(A):X=A+H
11 FOR J%=2 TO N%
12 K=2*(2-J%+2*INT(J%/2))
13 AR=AR+K*FNF(X)
14 X=X+H:NEXT J%
15 AR=H*(AR+FNF(B))/3
16 PRINT" Integral = ",AR
17 END
```

Fig. B–2

EXAMPLE B.2

Evaluate the definite integral

$$\int_2^5 \ln x \, dx$$

SOLUTION

The interval $x_0 = 2$ to $x_6 = 5$ will be divided into six equal parts ($n = 6$), each having a width $h = (5 - 2)/6 = 0.5$. We then compute $y = f(x) = \ln x$ at each point of subdivision.

n	x_n	y_n
0	2	0.693
1	2.5	0.916
2	3	1.099
3	3.5	1.253
4	4	1.386
5	4.5	1.504
6	5	1.609

Thus, Eq. B–1 becomes

$$\int_2^5 \ln x \, dx \simeq \frac{0.5}{3}[0.693 + 4(0.916 + 1.253 + 1.504)$$

$$+ 2(1.099 + 1.386) + 1.609]$$

$$\simeq 3.66 \qquad\qquad Ans.$$

This answer is equivalent to the exact answer to three significant figures. Obviously, accuracy to a greater number of significant figures can be improved by selecting a smaller interval h (or larger n).

Using the computer program, we first specify the function $\ln x$, line 6 in Fig. B–2. During execution, the program input requires the upper and lower limits 2 and 5, and the number of intervals $n = 6$. The output appears as

$$\text{Integral} = 3.66082 \qquad\qquad Ans.$$

B.3 The Secant Method

The secant method is used to find the real roots of an algebraic or transcendental equation $f(x) = 0$. The method derives its name from the fact that the formula used is established from the slope of the secant line to the graph $y = f(x)$. This slope is $[f(x_n) - f(x_{n-1})]/(x_n - x_{n-1})$, and the secant formula is

$$x_{n+1} = x_n - f(x_n)\left[\frac{x_n - x_{n-1}}{f(x_n) - f(x_{n-1})}\right] \qquad (B-2)$$

For application it is necessary to provide two initial guesses, x_0 and x_1, and thereby evaluate x_2 from Eq. B–2 ($n = 1$). One then proceeds to reapply Eq. B–2 with x_1 and the calculated value of x_2 and obtain x_3 ($n = 2$), etc., until the value $x_{n+1} \simeq x_n$. One can see this will occur if x_n is approaching the root of the function $f(x) = 0$, since the correction term on the right of Eq. B–2 will tend toward zero. In particular, the larger the slope, the smaller the correction to x_n, and the faster the root will be found. On the other hand, if the slope is very small in the neighborhood of the root, the method leads to large corrections for x_n, and convergence to the root is slow and may even lead to a failure to find it. In such cases other numerical techniques must be used for solution.

A computer program based on Eq. B–2 is listed in Fig. B–3. We must first specify the function on line 7 of the program. When the program is executed, two initial guesses, x_0 and x_1, must be entered in order to approximate the solution. The output specifies the value of the root. If it cannot be determined, this is so stated.

```
 1 PRINT"Secant method":PRINT
 2 PRINT" To execute this program :":PRINT
 3 PRINT"    1) Modify right hand side of the equation given below,"
 4 PRINT"       then press RETURN key."
 5 PRINT"    2) Type  RUN 7"
 6 PRINT:EDIT 7
 7 DEF FNF(X)=.5*SIN(X)-2*COS(X)+1.3
 8 INPUT"Enter point #1 =",X
 9 INPUT"Enter point #2 =",X1
10 IF X=X1 THEN 14
11 EP=.00001:TL=2E-20
12 FP=(FNF(X1)-FNF(X))/(X1-X)
13 IF ABS(FP)>TL THEN 15
14 PRINT"Root can not be found.":END
15 DX=FNF(X1)/FP
16 IF ABS(DX)>EP THEN 19
17 PRINT "Root = ";X1;"      Function evaluated at this root = ";FNF(X1)
18 END
19 X=X1:X1=X1-DX
20 GOTO 12
```

Fig. B–3

EXAMPLE B.3

Determine the root of the equation

$$f(x) = 0.5 \sin x - 2 \cos x + 1.30 = 0$$

SOLUTION
Guesses of the initial roots will be $x_0 = 45°$ and $x_1 = 30°$. Applying Eq. B–2,

$$x_2 = 30° - (-0.1821)\frac{(30° - 45°)}{(-0.1821 - 0.2393)} = 36.48°$$

Using this value in Eq. B–2, along with $x_1 = 30°$, we have

$$x_3 = 36.48° - (-0.0108)\frac{36.48° - 30°}{(-0.0108 + 0.1821)} = 36.89°$$

Repeating the process with this value and $x_2 = 36.48°$ yields

$$x_4 = 36.89° - (0.0005)\left[\frac{36.89° - 36.48°}{(0.0005 + 0.0108)}\right] = 36.87°$$

Thus $x = 36.9°$ is appropriate to three significant figures.

If the problem is solved using the computer program, first we specify the function, line 7 in Fig. B–3. During execution, the first and second guesses must be entered in radians. Choosing these to be 0.8 rad and 0.5 rad, the result appears as

Root = 0.6435022
Function evaluated at this root = 1.66893E − 06

This result converted from radians to degrees is therefore

$$x = 36.9° \qquad\qquad Ans.$$

Review for the Fundamentals of Engineering Examination

The Fundamentals of Engineering (FE) exam is given semiannually by the National Council of Engineering Examiners (NCEE) and is one of the requirements for obtaining a Professional Engineering License. A portion of this exam contains problems in statics, and this appendix provides a review of the subject matter most often asked on this exam. Before solving any of the problems, you should review the sections indicated in each chapter in order to become familiar with the boldfaced definitions and the procedures used to solve the various types of problems. Also, review the example problems in these sections.

The following problems are arranged in the same sequence as the topics in each chapter. Besides helping as a preparation for the FE exam, these problems also provide additional examples for general practice of the subject matter. Solutions to *all the problems* are given at the back of this appendix.

Chapter 2—Review All Sections

C-1. Two forces act on the hook. Determine the magnitude of the resultant force.

C-2. The force $F = 450$ lb acts on the frame. Resolve this force into components acting along members AB and AC, and determine the magnitude of each component.

Prob. C–1

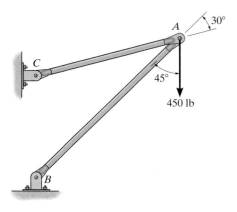

Prob. C–2

618

C-3. Determine the magnitude and direction of the resultant force.

C-5. The force has a component of 20 N directed along the $-y$ axis as shown. Represent the force **F** as a Cartesian vector.

Prob. C–3

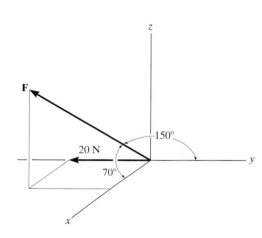

Prob. C–5

C-4. If **F** $= \{30\mathbf{i} + 50\mathbf{j} - 45\mathbf{k}\}$ N, determine the magnitude and coordinate direction angles of the force.

C-6. The force acts on the beam as shown. Determine its coordinate direction angles.

Prob. C–4

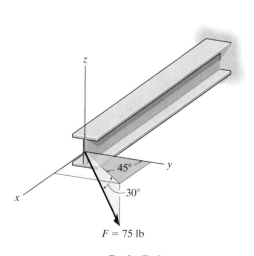

Prob. C–6

C-7. The cables supporting the antenna are subjected to the forces shown. Represent each force as a Cartesian vector.

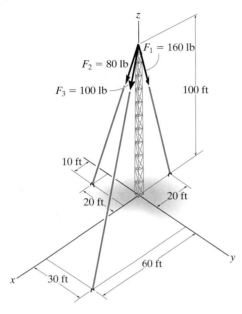

Prob. C–7

C-8. Determine the angle θ between the two cords.

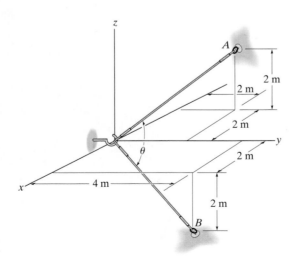

Prob. C–8

C-9. Determine the component of projection of the force **F** along the pipe AB.

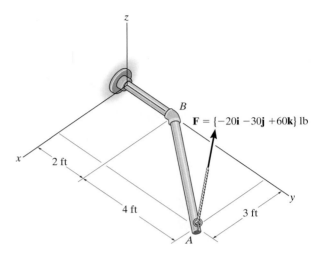

Prob. C–9

Chapter 3—Review Sections 3.1–3.3

C-10. The crate at D has a weight of 550 lb. Determine the force in each supporting cable.

Prob. C–10

C-11. The beam has a weight of 700 lb. Determine the shortest cable ABC that can be used to lift it if the maximum force the cable can sustain is 1500 lb.

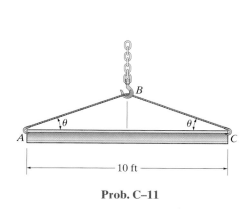

Prob. C–11

C-12. The block has a mass of 5 kg and rests on the smooth plane. Determine the unstretched length of the spring.

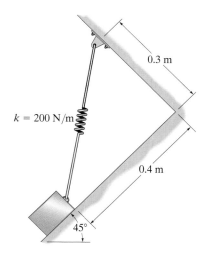

Prob. C–12

C-13. The post can be removed by a vertical force of 400 lb. Determine the force P that must be applied to the cord in order to pull the post out of the ground.

Prob. C–13

Chapter 4—Review All Sections

C-14. Determine the moment of the force about point O.

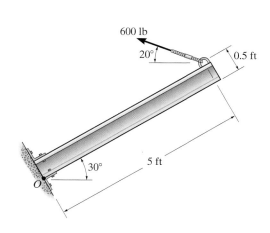

Prob. C–14

C-15. Determine the moment of the force about point O. Neglect the thickness of the member.

Prob. C–15

C-17. Determine the moment of the force about point A. Express the result as a Cartesian vector.

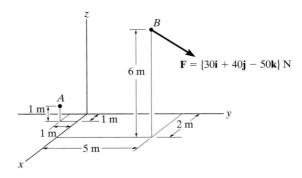

$F = \{30\mathbf{i} + 40\mathbf{j} - 50\mathbf{k}\}$ N

Prob. C–17

C-16. Determine the moment of the force about point O.

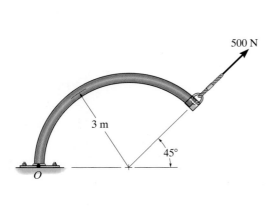

Prob. C–16

C-18. Determine the moment of the force about point A. Express the result as a Cartesian vector.

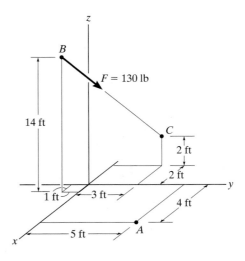

$F = 130$ lb

Prob. C–18

C-19. Determine the resultant couple moment acting on the beam.

Prob. C–19

C-21. Replace the loading shown by an equivalent resultant force and couple-moment system at point A.

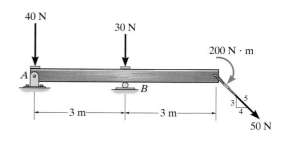

Prob. C–21

C-20. Determine the resultant couple moment acting on the triangular plate.

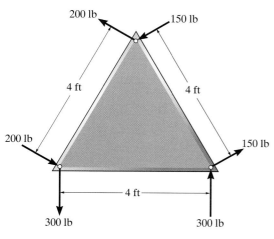

Prob. C–20

C-22. Replace the loading shown by an equivalent resultant force and couple-moment system at point A.

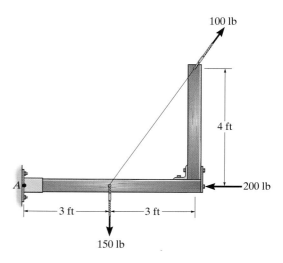

Prob. C–22

C-23. Replace the loading shown by an equivalent single resultant force and specify where the force acts, measured from point O.

Prob. C–23

C-24. Replace the loading shown by an equivalent single resultant force and specify the x and y coordinates of its line of action.

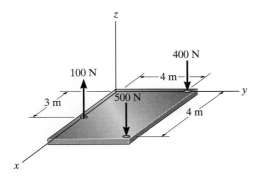

Prob. C–24

C-25. Replace the loading shown by an equivalent single resultant force and specify the x and y coordinates of its line of action.

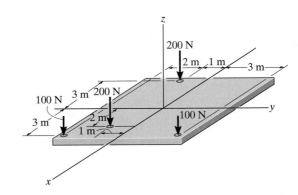

Prob. C–25

C-26. Determine the resultant force and specify where it acts on the beam measured from A.

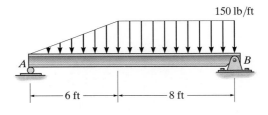

Prob. C–26

C-27. Determine the resultant force and specify where it acts on the beam measured from A.

C-29. Determine the horizontal and vertical components of reaction at the supports. Neglect the thickness of the beam.

Prob. C–27

Prob. C–29

C-28. Determine the resultant force and specify where it acts on the beam measured from A.

C-30. Determine the horizontal and vertical components of reaction at the supports.

Prob. C–28

Prob. C–30

C-31. Determine the components of reaction at the fixed support A. Neglect the thickness of the beam.

Prob. C–31

C-32. Determine the tension in the cable and the horizontal and vertical components of reaction at the pin A. Neglect the size of the pulley.

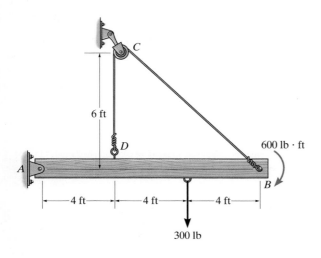

Prob. C–32

C-33. The uniform plate has a weight of 500 lb. Determine the tension in each of the supporting cables.

Prob. C–33

Chapter 6—Review Sections 6.1–6.4, 6.6

C-34. Determine the force in each member of the truss. State if the members are in tension or compression.

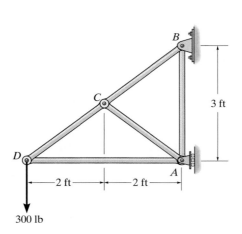

Prob. C–34

C-35. Determine the force in members *AE* and *DC*. State if the members are in tension or compression.

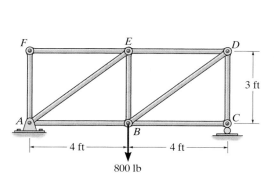

Prob. C–35

C-36. Determine the force in members *BC*, *CF*, and *FE*. State if the members are in tension or compression.

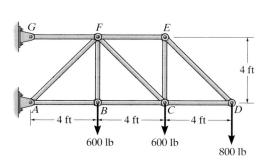

Prob. C–36

C-37. Determine the force in members *GF*, *FC*, and *CD*. State if the members are in tension or compression.

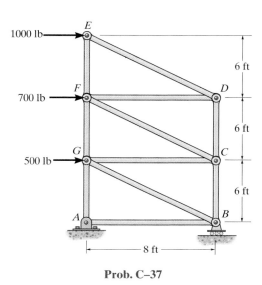

Prob. C–37

C-38. Determine the force *P* needed to hold the 60-lb weight in equilibrium.

Prob. C–38

C-39. Determine the horizontal and vertical components of reaction at pin C.

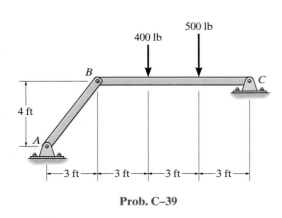

Prob. C–39

C-41. Determine the normal force that the 100-lb plate A exerts on the 30-lb plate B.

Prob. C–41

C-40. Determine the horizontal and vertical components of reaction at pin C.

Prob. C–40

C-42. Determine the force P needed to lift the load. Also, determine the proper placement x of the hook for equilibrium. Neglect the weight of the beam.

Prob. C–42

Chapter 7—Review Section 7.1

C-43. Determine the internal normal force, shear force, and moment acting in the beam at point B.

Prob. C–43

C-44. Determine the internal normal force, shear force, and moment acting in the beam at point B, which is located just to the left of the 800-lb force.

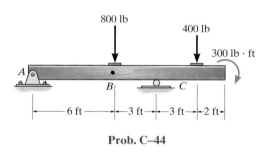

Prob. C–44

C-45. Determine the internal normal force, shear force, and moment acting in the beam at point B.

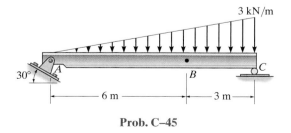

Prob. C–45

Chapter 8—Review Sections 8.1–8.2

C-46. Determine the force P needed to move the 100-lb block. The coefficient of static friction is $\mu_s = 0.3$, and the coefficient of kinetic friction is $\mu_k = 0.25$. Neglect tipping.

Prob. C–46

C-47. Determine the vertical force P needed to rotate the 200-lb spool. The coefficient of static friction at all contacting surfaces is $\mu_s = 0.4$.

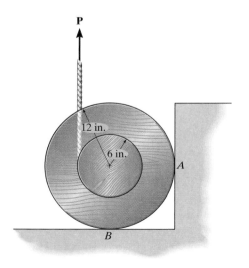

Prob. C–47

C-48. Block *A* has a weight of 30 lb and block *B* weighs 50 lb. If the coefficient of static friction is $\mu_s = 0.4$ between all contacting surfaces, determine the frictional force at each surface.

C-50. The filing cabinet *A* has a mass of 60 kg and center of mass at *G*. It rests on a 10-kg plank. Determine the smallest force *P* needed to move it. The coefficient of static friction between the cabinet *A* and the plank *B* is $\mu_s = 0.4$, and between the plank and the floor $\mu_s = 0.3$.

Prob. C–48

Prob. C–50

C-49. Determine the force *P* necessary to move the 250-lb crate which has a center of gravity at *G*. The coefficient of static friction at the floor is $\mu_s = 0.4$.

Chapter 9—Review Sections 9.1–9.3

(Integration is covered in the mathematics portion of the exam.)

C-51. Determine the location $(\bar{x}, \bar{y})$ of the centroid of the area.

Prob. C–49

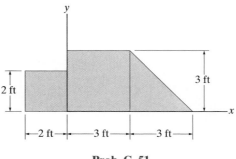

Prob. C–51

C-52. Determine the location $(\bar{x}, \bar{y})$ of the centroid of the area.

Prob. C–52

C-54. Determine the moment of inertia of the area with respect to the x axis.

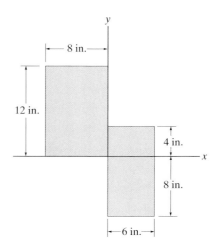

Prob. C–54

Chapter 10—Review Sections 10.1–10.5

(Integration is covered in the mathematics portion of the exam.)

C-53. Determine the moment of inertia of the cross-sectional area of the channel with respect to the y axis.

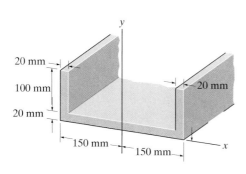

Prob. C–53

C-55. Determine the moment of inertia of the cross-sectional area of the T-beam with respect to the x' axis passing through the centroid of the cross section.

Prob. C–55

Partial Solutions and Answers

C–1. $F_R = \sqrt{200^2 + 500^2 - 2(200)(500)\cos 140°}$
    $= 666 \text{ N}$ *Ans.*

C–2. $\dfrac{F_{AB}}{\sin 105°} = \dfrac{450}{\sin 30°}$
        $= 869 \text{ lb}$ *Ans.*

$\dfrac{F_{AC}}{\sin 45°} = \dfrac{450}{\sin 30°}$
        $F_{AC} = 636 \text{ lb}$ *Ans.*

C–3. $F_{Rx} = 300 + 400\cos 30° - 250\left(\dfrac{4}{5}\right) = 446.4 \text{ N}$

$F_{Ry} = 400\sin 30° + 250\left(\dfrac{3}{5}\right) = 350 \text{ N}$

$F_R = \sqrt{(446.4)^2 + 350^2} = 567 \text{ N}$ *Ans.*

$\theta = \tan^{-1}\dfrac{350}{446.4} = 38.1° \nearrow$ *Ans.*

C–4. $F = \sqrt{30^2 + 50^2 + (-45)^2} = 73.7 \text{ N}$ *Ans.*

$\alpha = \cos^{-1}\left(\dfrac{30}{73.7}\right) = 66.0°$ *Ans.*

$\beta = \cos^{-1}\left(\dfrac{50}{73.7}\right) = 47.2°$ *Ans.*

$\gamma = \cos^{-1}\left(\dfrac{-45}{73.7}\right) = 128°$ *Ans.*

C–5. $F_y = -20$

$\dfrac{F_y}{|F|} = \cos\beta$

$|F| = \left|\dfrac{-20}{\cos 150°}\right| = 23.09 \text{ N}$

$\cos\gamma = \sqrt{1 - \cos^2 70° - \cos^2 150°}$
    $\gamma = 68.61° \text{ (From Fig. } \gamma < 90°)$
    $\mathbf{F} = 23.09\cos 70°\mathbf{i} + 23.09\cos 150°\mathbf{j}$
    $\quad + 23.09\cos 68.61°\mathbf{k}$
    $= \{7.90\mathbf{i} - 20\mathbf{j} + 8.42\mathbf{k}\} \text{ N}$ *Ans.*

C–6. $F_x = 75\cos 30°\sin 45° = 45.93$
    $F_y = 75\cos 30°\cos 45° = 45.93$
    $F_z = -75\sin 30° = -37.5$

$\alpha = \cos^{-1}\left(\dfrac{45.93}{75}\right) = 52.2°$ *Ans.*

$\beta = \cos^{-1}\left(\dfrac{45.93}{75}\right) = 52.2°$ *Ans.*

$\gamma = \cos^{-1}\left(\dfrac{-37.5}{75}\right) = 120°$ *Ans.*

C–7. $\mathbf{F}_1 = 160 \text{ lb}\left(-\dfrac{20}{102.0}\mathbf{i} - \dfrac{100}{102.0}\mathbf{k}\right)$
    $= \{-31.4\mathbf{i} - 157\mathbf{k}\} \text{ lb}$ *Ans.*

$\mathbf{F}_2 = 80 \text{ lb}\left(\dfrac{10}{102.5}\mathbf{i} - \dfrac{20}{102.5}\mathbf{j} - \dfrac{100}{102.5}\mathbf{k}\right)$
    $= \{7.81\mathbf{i} - 15.6\mathbf{j} - 78.1\mathbf{k}\} \text{ lb}$ *Ans.*

$\mathbf{F}_3 = 100 \text{ lb}\left(\dfrac{60}{120.4}\mathbf{i} + \dfrac{30}{120.4}\mathbf{j} - \dfrac{100}{120.4}\mathbf{k}\right)$
    $= \{49.8\mathbf{i} + 24.9\mathbf{j} - 83.0\mathbf{k}\} \text{ lb}$ *Ans.*

C–8. $\mathbf{r}_{OA} = \{-2\mathbf{i} + 2\mathbf{j} + 2\mathbf{k}\} \text{ m}$
    $\mathbf{r}_{OB} = \{2\mathbf{i} + 4\mathbf{j} - 2\mathbf{k}\} \text{ m}$

$\cos\theta = \dfrac{\mathbf{r}_{OA}\cdot\mathbf{r}_{OB}}{|r_{OA}||r_{OB}|}$

$\dfrac{(-2\mathbf{i} + 2\mathbf{j} + 2\mathbf{k})\cdot(2\mathbf{i} + 4\mathbf{j} - 2\mathbf{k})}{\sqrt{12}\sqrt{24}} = 0$

$\theta = 90°$ *Ans.*

C–9. $|F_{AB}| = \mathbf{F}\cdot\mathbf{u}_{AB}$

$= (-20\mathbf{i} - 30\mathbf{j} + 60\mathbf{k})\cdot\left(-\dfrac{3}{5}\mathbf{i} - \dfrac{4}{5}\mathbf{j}\right) = 36 \text{ lb}$ *Ans.*

C–10. $\xrightarrow{+}\Sigma F_x = 0; \quad \dfrac{4}{5}F_{AC} - F_{AB}\cos 30° = 0$

$+\uparrow\Sigma F_y = 0; \quad \dfrac{3}{5}F_{AC} + F_{AB}\sin 30° - 550 = 0$

$F_{AB} = 478 \text{ lb}$ *Ans.*
$F_{AC} = 518 \text{ lb}$ *Ans.*

C–11. $+\uparrow\Sigma F_y = 0; \quad -2(1500)\sin\theta + 700 = 0$
    $\theta = 13.5°$

$L_{ABC} = 2\left(\dfrac{5 \text{ ft}}{\cos 13.5°}\right) = 10.3 \text{ ft}$

C–12. $+\nearrow \Sigma F_x = 0;\ \dfrac{4}{5}(F_{sp}) - 5(9.81)\sin 45° = 0$

$F_{sp} = 43.35\ \text{N}$

$F_{sp} = k(l - l_0);\ 43.35 = 200(0.5 - l_0)$

$l_0 = 0.283\ \text{m}$ *Ans.*

C–13. At A:

$\xleftarrow{+}\Sigma F_x = 0;\ \dfrac{3}{5}P - T_{AC}\cos 30° = 0$

$+\uparrow \Sigma F_y = 0;\ \dfrac{4}{5}P + T_{AC}\sin 30° - 400 = 0$

$P = 349\ \text{lb}$ *Ans.*

$T_{AC} = 242\ \text{lb}$ *Ans.*

C–14. $\downarrow+M_O = 600\sin 50°\,(5) + 600\cos 50°\,(0.5)$

$= 2.49\ \text{kip}\cdot\text{ft}$ *Ans.*

C–15. $\uparrow+M_O = 50\sin 60°\,(0.1 + 0.2\cos 45° + 0.1)$

$- 50\cos 60°\,(0.2\sin 45°)$

$= 11.2\ \text{N}\cdot\text{m}$ *Ans.*

C–16. $\downarrow+M_O = 500\sin 45°\,(3 + 3\cos 45°)$

$- 500\cos 45°\,(3\sin 45°)$

$= 1.06\ \text{kN}\cdot\text{m}$ *Ans.*

C–17. $\mathbf{M}_A = \mathbf{r}_{AB}\times \mathbf{F} = \begin{vmatrix} \mathbf{i} & \mathbf{j} & \mathbf{k} \\ 1 & 6 & 5 \\ 30 & 40 & -50 \end{vmatrix}$

$= \{-500\mathbf{i} + 200\mathbf{j} - 140\mathbf{k}\}\ \text{N}\cdot\text{m}$ *Ans.*

C–18. $\mathbf{F} = 130\ \text{lb}\left(-\dfrac{3}{13}\mathbf{i} + \dfrac{4}{13}\mathbf{j} - \dfrac{12}{13}\mathbf{k}\right)$

$= \{-30\mathbf{i} + 40\mathbf{j} - 120\mathbf{k}\}\ \text{lb}$

$\mathbf{M}_A = \mathbf{r}_{AB}\times \mathbf{F} = \begin{vmatrix} \mathbf{i} & \mathbf{j} & \mathbf{k} \\ -3 & -6 & 14 \\ -30 & 40 & -120 \end{vmatrix}$

$= \{160\mathbf{i} - 780\mathbf{j} - 300\mathbf{k}\}\ \text{lb}\cdot\text{ft}$ *Ans.*

C–19. $\uparrow+M_{C_R} = \Sigma M_A = 400(3) - 400(5) + 300(5)$

$+\ 200(0.2) = 740\ \text{N}\cdot\text{m}$ *Ans.*

Also,

$\uparrow+M_{C_R} = 300(5) - 400(2) + 200(0.2)$

$= 740\ \text{N}\cdot\text{m}$ *Ans.*

C–20. $\downarrow+M_{C_R} = 300(4) + 200(4) + 150(4)$

$= 2600\ \text{lb}\cdot\text{ft}$ *Ans.*

C–21. $\xrightarrow{+} F_{Rx} = \Sigma F_x;\ F_{Rx} = \dfrac{4}{5}(50) = 40\ \text{N}$

$+\downarrow F_{Ry} = \Sigma F_y;\ F_{Ry} = 40 + 30 + \dfrac{3}{5}(50)$

$= 100\ \text{N}$

$F_R = \sqrt{(40)^2 + (100)^2} = 108\ \text{N}$ *Ans.*

$\theta = \tan^{-1}\left(\dfrac{100}{40}\right) = 68.2°$ *Ans.*

$+\downarrow M_{A_R} = \Sigma M_A;$

$M_{A_R} = 30(3) + \dfrac{3}{5}(50)(6) + 200$

$= 470\ \text{N}\cdot\text{m}$ *Ans.*

C–22. $\xleftarrow{+} F_{Rx} = \Sigma F_x;\ F_{Rx} = 200 - \dfrac{3}{5}(100) = 140\ \text{lb}$

$+\downarrow F_{Ry} = \Sigma F_y;\ F_{Ry} = 150 - \dfrac{4}{5}(100) = 70\ \text{lb}$

$F_R = \sqrt{140^2 + 70^2} = 157\ \text{lb}$ *Ans.*

$\theta = \tan^{-1}\left(\dfrac{70}{140}\right) = 26.6°$ *Ans.*

$+\downarrow M_{A_R} = \Sigma M_A;$

$M_{A_R} = \dfrac{3}{5}(100)(4) - \dfrac{4}{5}(100)(6) + 150(3)$

$M_{R_A} = 210\ \text{lb}\cdot\text{ft}$ *Ans.*

C–23. $+\downarrow F_R = \Sigma F_y;\ F_R = 500 + 250 + 500$

$= 1250\ \text{lb}$ *Ans.*

$+\downarrow F_R x = \Sigma M_O;$

$1250(x) = 500(3) + 250(6) + 500(9)$

$x = 6\ \text{ft}$ *Ans.*

C–24. $+\downarrow F_R = \Sigma F_z;\ F_R = 400 + 500 - 100$

$= 800\ \text{N}$ *Ans.*

$M_{Rx} = \Sigma M_x;\ -800y = -400(4) - 500(4)$

$y = 4.50\ \text{m}$ *Ans.*

$M_{Ry} = \Sigma M_y;\ 800x = 500(4) - 100(3)$

$x = 2.125\ \text{m}$ *Ans.*

C–25. $\quad +\downarrow F_R = \Sigma F_y; \quad F_R = 200 + 200 + 100 + 100$
$$= 600 \text{ N} \qquad Ans.$$
$$M_{Rx} = \Sigma M_x; \quad -600y = 200(1) + 200(1)$$
$$+ 100(3) - 100(3)$$
$$y = -0.667 \text{ m} \qquad Ans.$$
$$M_{Ry} = \Sigma M_y; \quad 600x = 100(3) + 100(3)$$
$$+ 200(2) - 200(3)$$
$$x = 0.667 \text{ m} \qquad Ans.$$

C–26. $\qquad F_R = \dfrac{1}{2}(6)(150) + 8(150) = 1650 \text{ lb} \qquad Ans.$
$$+\downarrow M_{A_R} = \Sigma M_A;$$
$$1650d = \left[\dfrac{1}{2}(6)(150)\right](4) + [8(150)](10)$$
$$d = 8.36 \text{ ft} \qquad Ans.$$

C–27. $\qquad F_R = \displaystyle\int w(x)\, dx = \int_0^4 2.5x^3\, dx = 160 \text{ N} \qquad Ans.$
$$+\downarrow M_{A_R} = \Sigma M_A;$$
$$x = \dfrac{\displaystyle\int xw(x)\, dx}{\displaystyle\int w(x)\, dx} = \dfrac{\displaystyle\int_0^4 2.5x^4\, dx}{160} = 3.20 \text{ m} \qquad Ans.$$

C–28. $\quad +\downarrow F_R = \Sigma F_y; \quad F_R = \dfrac{1}{2}(50)(6) + 150(6) + 500$
$$= 1550 \text{ lb} \quad Ans.$$
$$+\downarrow M_{A_R} = \Sigma M_A;$$
$$1550d = \left[\dfrac{1}{2}(50)(6)\right](4) + [150(6)](3) + 500(9)$$
$$d = 5.03 \text{ ft} \qquad Ans.$$

C–29. $\quad \xrightarrow{+} \Sigma F_x = 0; \quad -A_x + 500\left(\dfrac{3}{5}\right) = 0$
$$A_x = 300 \text{ lb} \quad Ans.$$
$$+\uparrow\Sigma M_A = 0; \quad B_y(10) - 500\left(\dfrac{4}{5}\right)(5) - 600 = 0$$
$$B_y = 260 \text{ lb} \quad Ans.$$
$$+\uparrow\Sigma F_y = 0; \quad A_y + 260 - 500\left(\dfrac{4}{5}\right) = 0$$
$$A_y = 140 \text{ lb} \quad Ans.$$

C–30. $\quad \xrightarrow{+} \Sigma F_x = 0; \quad -A_x + 400 = 0; \quad A_x = 400 \text{ N} \quad Ans.$
$$\downarrow+\Sigma M_A = 0; \quad B_y(4) - 400(0.5) - 500(2) = 0$$
$$B_y = 300 \text{ N} \quad Ans.$$
$$+\uparrow\Sigma F_y = 0; \quad A_y + 300 - 500 = 0$$
$$A_y = 200 \text{ N} \quad Ans.$$

C–31. $\quad \xrightarrow{+} \Sigma F_x = 0; \quad -A_x + 400\cos 30° = 0$
$$A_x = 346 \text{ N} \quad Ans.$$
$$+\uparrow\Sigma F_y = 0;$$
$$A_y - 200 - 200 - 200 - 400\sin 30° = 0$$
$$A_y = 800 \text{ N} \quad Ans.$$
$$\downarrow+\Sigma M_A = 0;$$
$$M_A - 200(2.5) - 200(3.5) - 200(4.5)$$
$$-400\sin 30°(4.5) - 400\cos 30°(3\sin 60°) = 0$$
$$M_A = 3.90 \text{ kN}\cdot\text{m} \quad Ans.$$

C–32. $\quad +\!\uparrow\!\Sigma M_A = 0; \quad T(4) + \dfrac{3}{5}T(12) - 300(8) - 600 = 0$
$$T = 267.9 = 268 \text{ lb} \quad Ans.$$
$$\xrightarrow{+} \Sigma F_x = 0; \quad A_x - \left(\dfrac{4}{5}\right)(267.9) = 0$$
$$A_x = 214 \text{ lb} \quad Ans.$$
$$+\uparrow\Sigma F_y = 0; \quad A_y + 267.9 + \left(\dfrac{3}{5}\right)(267.9) - 300 = 0$$
$$A_y = -129 \text{ lb} \quad Ans.$$

C–33. $\quad \Sigma F_z = 0; \quad T_A + T_B + T_C - 200 - 500 = 0$
$$\Sigma M_x = 0; \quad T_A(3) + T_C(3) - 500(1.5) - 200(3) = 0$$
$$\Sigma M_y = 0; \quad -T_B(4) - T_C(4) + 500(2) + 200(2) = 0$$
$$T_A = 350 \text{ lb}, \ T_B = 250 \text{ lb}, \ T_C = 100 \text{ lb} \qquad Ans.$$

C–34. Joint D:
$$+\uparrow\Sigma F_y = 0; \quad \dfrac{3}{5}F_{CD} - 300 = 0;$$
$$F_{CD} = 500 \text{ lb (T)} \qquad Ans.$$
$$\xrightarrow{+} \Sigma F_x = 0; \quad -F_{AD} + \dfrac{4}{5}(500) = 0;$$
$$F_{AD} = 400 \text{ lb (C)} \qquad Ans.$$

Joint C:

$+\searrow \Sigma F_y = 0; F_{CA} = 0$ *Ans.*

$+\nearrow \Sigma F_x = 0; F_{CB} - 500 = 0;$

$F_{CB} = 500$ lb (T) *Ans.*

Joint A:

$+\uparrow \Sigma F_y = 0; F_{AB} = 0$ *Ans.*

C–35. $Ax = 0, \ Ay = Cy = 400$ lb

Joint A:

$+\uparrow \Sigma F_y = 0; \ -\dfrac{3}{5}F_{AE} + 400 = 0;$

$F_{AE} = 667$ lb (C)   *Ans.*

Joint C:

$+\uparrow \Sigma F_y = 0; -F_{DC} + 400 = 0;$

$F_{DC} = 400$ lb (C) *Ans.*

C–36.   Section truss through *FE, FC, BC*. Use the right segment.

$+\uparrow \Sigma F_y = 0; \ F_{CF} \sin 45° - 600 - 800 = 0$

$\qquad F_{CF} = 1980$ lb (T) *Ans.*

$+\uparrow \Sigma M_C = 0; \ F_{FE}(4) - 800(4) = 0$

$\qquad F_{FE} = 800$ lb (T) *Ans.*

$\downarrow + \Sigma M_F = 0; \ F_{BC}(4) - 600(4) - 800(8) = 0$

$\qquad F_{BC} = 2200$ lb (C) *Ans.*

C–37.   Section truss through *GF, FC, DC*. Use the top segment.

$+\uparrow \Sigma M_C = 0; \ F_{GF}(8) - 700(6) - 1000(12) = 0$

$\qquad F_{GF} = 2025$ lb (T) *Ans.*

$\xrightarrow{+} \Sigma F_x = 0; \ -\dfrac{4}{5}F_{FC} + 700 + 1000 = 0$

$\qquad F_{FC} = 2125$ lb (C) *Ans.*

$\downarrow + \Sigma M_F = 0; \ F_{CD}(8) - 1000(6) = 0$

$\qquad F_{CD} = 750$ lb (C) *Ans.*

C–38. $+\uparrow \Sigma F_y = 0; \ 3P - 60 = 0$

$\qquad P = 20$ lb *Ans.*

C–39. $+\uparrow \Sigma M_C = 0;$

$-\left(\dfrac{4}{5}\right)(F_{AB})(9) + 400(6) + 500(3) = 0$

$F_{AB} = 541.67$ lb

$\xrightarrow{+} \Sigma F_x = 0; -C_x + \dfrac{3}{5}(541.67) = 0$

$\qquad C_x = 325$ lb *Ans.*

$+\uparrow \Sigma F_y = 0; C_y + \dfrac{4}{5}(541.67) - 400 - 500 = 0$

$\qquad C_y = 467$ lb *Ans.*

C–40. $+\uparrow \Sigma M_C = 0;$

$F_{AB} \cos 45°(1) - F_{AB} \sin 45°(3) + 800$

$+ 400(2) = 0$

$F_{AB} = 1131.37$ N

$\xrightarrow{+} \Sigma F_x = 0; \ -C_x + 1131.37 \cos 45° = 0$

$\qquad C_x = 800$ N *Ans.*

$+\uparrow \Sigma F_y = 0; \ -C_y + 1131.37 \sin 45° - 400 = 0$

$\qquad C_y = 400$ N *Ans.*

C–41.   Plate A:

$+\uparrow \Sigma F_y = 0; \ 2T + N_{AB} - 100 = 0$

Plate B:

$+\uparrow \Sigma F_y = 0; \ 2T - N_{AB} - 30 = 0$

$T = 32.5$ lb, $N_{AB} = 35$ lb   *Ans.*

C–42.   Pulley C:

$+\uparrow \Sigma F_y = 0; \ T - 2P = 0; T = 2P$

Beam:

$+\uparrow \Sigma F_y = 0; \ 2P + P - 6 = 0$

$\qquad P = 2$ kN *Ans.*

$+\uparrow \Sigma M_A = 0; \ 2(1) - 6(x) = 0$

$\qquad x = 0.333$ m *Ans.*

C–43. $A_y = 8.75$ kN. Use segment AB.

$\xrightarrow{+} \Sigma F_x = 0; \ N_B = 0$ *Ans.*

$+\uparrow \Sigma F_y = 0; \ 8.75 - 3(1.5) - V_B = 0$

$V_B = 4.25$ kN *Ans.*

$+\curvearrowleft \Sigma M_B = 0; \ M_B + 3(1.5)(0.75) - 8.75(1.5) = 0$

$M_B = 9.75$ kN $\cdot$ m *Ans.*

C–44. $A_x = 0, \ A_y = 100$ lb. Use segment AB.

$\xrightarrow{+} \Sigma F_x = 0; \ N_B = 0$ *Ans.*

$+\uparrow \Sigma F_y = 0; \ 100 - V_B = 0$

$V_B = 100$ lb *Ans.*

$+\curvearrowleft \Sigma M_B = 0; \ M_B - 100(6) = 0$

$M_B = 600$ lb $\cdot$ ft *Ans.*

C–45. $A_x = 0, \ A_y = 4.5$ kN, $w_B = 2$ kN/m.
Use segment AB.

$\xrightarrow{+} \Sigma F_x = 0; \ N_B = 0$ *Ans.*

$+\uparrow \Sigma F_y = 0; \ 4.5 - \dfrac{1}{2}(6)(2) + V_B = 0$

$V_B = 1.5$ kN *Ans.*

$+\curvearrowleft \Sigma M_B = 0; \ M_B + \left[\dfrac{1}{2}(6)(2)\right](2) - 4.5(6) = 0$

$M_B = 15$ kN $\cdot$ m *Ans.*

C–46. $+\uparrow \Sigma F_y = 0; \ N_b - P \sin 30° - 100 = 0$

$\xrightarrow{+} \Sigma F_x = 0; \ -P \cos 30° + 0.3N_b = 0$

$P = 41.9$ lb *Ans.*

C–47. $\xrightarrow{+} \Sigma F_x = 0; \ 0.4N_B - N_A = 0$

$+\curvearrowleft \Sigma M_B = 0; \ 0.4N_A(12) + N_A(12) - P(6) = 0$

$+\curvearrowleft \Sigma F_y = 0; \ P + 0.4N_A + N_B - 200 = 0$

$P = 98.2$ lb *Ans.*

C–48. Block B:

$+\uparrow \Sigma F_y = 0; \ N_B - 20 \sin 30° - 50 = 0$

$N_B = 60$ lb

$\xrightarrow{+} \Sigma F_x = 0; \ F_B - 20 \cos 30° = 0$

$F_B = 17.3$ lb $(<0.4(60 \text{ lb}))$ *Ans.*

Blocks A and B:

$+\uparrow \Sigma F_y = 0; \ N_A - 30 - 50 - 20 \sin 30° = 0$

$N_A = 90$ lb

$\xrightarrow{+} \Sigma F_x = 0; \ F_A - 20 \cos 30° - 10 = 0$

$F_A = 27.3$ lb $(<0.4(90 \text{ lb}))$ *Ans.*

C–49. If slipping occurs:

$+\uparrow \Sigma F_y = 0; \ N_C - 250 \text{ lb} = 0$

$N_C = 250$ lb

$\xrightarrow{+} \Sigma F_x = 0; \ P - 0.4(250) = 0$

$P = 100$ lb

If tipping occurs:

$\curvearrowleft +\Sigma M_A = 0; \ -P(4.5) + 250(1.5) = 0$

$P = 83.3$ lb *Ans.*

C–50. P for A to slip on B:

$+\uparrow \Sigma F_y = 0; \ N_A - 60(9.81) = 0$

$N_A = 588.6$ N

$\xrightarrow{+} \Sigma F_x = 0; \ 0.4(588.6) - P = 0$

$P = 235$ N

P for B to slip:

$+\uparrow \Sigma F_y = 0; \ N_B - 60(9.81) - 10(9.81) = 0$

$N_B = 686.7$ N

$\xrightarrow{+} \Sigma F_x = 0; \ 0.3(686.7) - P = 0$

$P = 206$ N

P to tip A:

$\curvearrowleft +\Sigma M_C = 0; \ P(1.3) - 60(9.81)(0.2) = 0$

$P = 90.6$ N *Ans.*

C–51. $\bar{x} = \dfrac{\Sigma \tilde{x} A}{\Sigma A} =$

$$\dfrac{(-1)(2)(2) + 1.5(3)(3) + 4\left(\dfrac{1}{2}\right)(3)(3)}{2(2) + 3(3) + \dfrac{1}{2}(3)(3)} = 1.57 \text{ ft}$$

Ans.

$\bar{y} = \dfrac{\Sigma \tilde{y} A}{\Sigma A} =$

$$\dfrac{1(2)(2) + 1.5(3)(3) + 1\left(\dfrac{1}{2}\right)(3)(3)}{2(2) + 3(3) + \dfrac{1}{2}(3)(3)} = 1.26 \text{ ft}$$

Ans.

C–52. $\bar{x} = 0$ (symmetry)   *Ans.*

$\bar{y} = \dfrac{\Sigma \tilde{y} A}{\Sigma A} = \dfrac{4(1(8)) + 9(6)(2)}{1(8) + 6(2)} = 7 \text{ in.}$ *Ans.*

C–53. $I_y = \dfrac{1}{12}(120)(300)^3 - \dfrac{1}{12}(100)(260)^3$

$= 124\,(10^6)\text{ mm}^4$ *Ans.*

C–54. $I = \Sigma(\bar{I} + Ad^2) = \left[\dfrac{1}{12}(8)(12)^3 + (8)(12)(6)^2\right]$

$+ \left[\dfrac{1}{12}(6)(12)^3 + (6)(12)(-2)^2\right] = 5760 \text{ in}^4$ *Ans.*

C–55. $\bar{x} = \dfrac{\Sigma \tilde{x} A}{\Sigma A} = \dfrac{4(8)(2) + 9(2)(8)}{8(2) + 2(8)} = 6.5 \text{ in.}$

$\bar{I}_{x'} = \Sigma(\bar{I} + Ad^2) = \left[\dfrac{1}{12}(2)(8)^3 + (8)(2)(6.5 - 4)^2\right]$

$+ \left[\dfrac{1}{12}(8)(2)^3 + 2(8)(9 - 6.5)^2\right] = 291 \text{ in}^4$ *Ans.*

Answers to Selected Problems

Chapter 1

1–1. (a) km/s, **(b)** mm, **(c)** Gs/kg, **(d)** mm · N

1–2. 2.42 Mg/m^3

1–3. (a) Mg/mm = Gg/m, **(b)** mN/μs = kN/s,
(c) μm · Mg = mm · kg

1–5. (a) 8.653 s, **(b)** 8.368 kN, **(c)** 893 g

1–6. (a) 45.3 MN, **(b)** 56.8 km, **(c)** 5.63 μg

1–7. 26.9 μm · kg/N

1–9. (a) 271 N · m, **(b)** 55.0 kN/m^3, **(c)** 0.677 mm/s

1–10. (a) 98.1 N, **(b)** 4.90 mN, **(c)** 44.1 kN

1–11. 584 kg

1–13. 1.04 kip

1–14. 2.71 Mg/m^3

1–17. 4.96 μN

1–18. (a) 0.04 MN^2, **(b)** 25 μm^2, **(c)** 0.064 km^3

1–19. (a) 15.9 mm/s, **(b)** 3.69 Mm · s/kg, **(c)** 1.14 km · kg

Chapter 2

2–1. $F_R = 867$ N, $\phi = 108°$

2–2. $F_R = 72.1$ lb, $\theta = 73.9°$

2–3. $F_R = 393$ lb, $\phi = 353°$

2–5. $F_{1u} = 205$ N, $F_{1v} = 160$ N

2–6. $F_{2u} = 376$ N, $F_{2v} = 482$ N

2–7. $F_R = 61.4$ lb, $\theta = 6.85°$

2–9. $F_u = 86.6$ lb, $F_v = -50$ lb

2–10. $F_u = 320$ N, $F_v = 332$ N

2–11. $F_a = 30.6$ lb, $F_b = 26.9$ lb

2–13. $F_{AB} = 186$ lb, $F_{AC} = 239$ lb

2–14. $T = 744$ lb, $\theta = 23.8°$

2–15. $F_{1v} = 129$ N, $F_{1u} = 183$ N

2–17. $F_R = \sqrt{F_1^2 + F_2^2 + 2F_1F_2 \cos \phi}$,

$$\tan \theta = \frac{F_1 \sin \phi}{F_2 + F_1 \cos \phi}$$

2–18. $F_R = 400$ N, $\theta = 60°$

2–19. $\theta = 25.4°$, $F_R = 115$ lb

2–21. $\theta = 43.9°$, $F_R = 561$ lb

2–22. $F_R = 19.2$ N, $\theta = 2.37°$ ⦨θ

2–23. $F_R = 19.2$ N, $\theta = 2.37°$ ⦨θ

2–25. $T = 54.7$ lb, $P = 42.6$ lb

2–26. $\theta = 60°$, $P = 40$ lb, $T = 69.3$ lb

2–27. $F_A = 439$ N, $F_B = 311$ N

2–29. $\theta = 10.9°$, $F_{min} = 235$ lb

2–30. $F_{min} = 97.4$ lb, $\theta = 16.2°$

2–31. $F_x = 514$ lb, $F_y = -613$ lb

2–33. $F_R = 29.3$ lb, $\theta = 347°$

2–34. $F_R = 546$ N, $\theta = 253°$

2–35. $\theta = 37.0°$, $F_1 = 889$ N

2–37. $F_R = 747$ N, $\theta = 85.5°$

2–38. $F_R = 25.1$ kN, $\theta = 185°$

2–39. $F_R = 111$ lb, $\theta = 202°$

2–41. $F_R = 19.2$ N, $\theta = 2.37°$ ⦨θ

2–42. $F_R = 839$ N, $\theta = 14.8°$

2–43. $F_R = 389$ N, $42.7°$

2–45. $F_R = 217$ N, $\theta = 87.0°$

2–46. $F_{1x} = -200$ lb, $F_{1y} = 0$, $F_{2x} = 320$ lb,
$F_{2y} = -240$ lb, $F_{3x} = 180$ lb, $F_{3y} = 240$ lb,
$F_{4x} = -300$ lb, $F_{4y} = 0$

2–47. $F_R = 60.3$ kN, $\theta = 15.0°$

2–49. $F_R = 389$ N, $\phi' = 42.7°$

2–50. $\mathbf{F}_1 = \{90\mathbf{i} - 120\mathbf{j}\}$ lb, $\mathbf{F}_2 = \{-275\mathbf{j}\}$ lb,
$\mathbf{F}_3 = \{-37.5\mathbf{i} - 65.0\mathbf{j}\}$ lb, $F_R = 463$ lb

2–51. $F = 5.96$ kN, $F_R = 2.33$ kN

2–53. $\theta = 117°$, $F_3 = 1.12F_1$

2–54. $\theta = 103°$, $F_2 = 88.1$ lb

2–55. $F_R = 161$ lb, $\theta = 38.3°$

2–57. $F_1 = 87.7$ N, $\alpha_1 = 46.9°$, $\beta_1 = 125°$, $\gamma_1 = 62.9°$,
$F_2 = 98.6$ N, $\alpha_2 = 114°$, $\beta_2 = 150°$, $\gamma_2 = 72.3°$

2–58. $\mathbf{F}_1 = \{2.50\mathbf{i} + 3.54\mathbf{j} + 2.50\mathbf{k}\}$ kN,
$\mathbf{F}_2 = \{-2\mathbf{j}\}$ kN

2–59. $F = 50$ N, $\alpha = 74.1°$, $\beta = 41.3°$, $\gamma = 53.1°$

2–61. $\beta = 90°$, $\mathbf{F} = \{-30\mathbf{i} - 52.0\mathbf{k}\}$ N

2–62. $F_R = 114$ lb, $\alpha = 62.1°$, $\beta = 113°$, $\gamma = 142°$

2–63. $\mathbf{F}_1 = \{53.1\mathbf{i} - 44.5\mathbf{j} + 40\mathbf{k}\}$ lb, $\alpha_1 = 48.4°$,
$\beta_1 = 124°$, $\gamma_1 = 60°$, $\mathbf{F}_2 = \{-130\mathbf{k}\}$ lb,
$\alpha_2 = 90°$, $\beta_2 = 90°$, $\gamma_2 = 180°$

2–65. $\alpha_1 = 90°$, $\beta_1 = 53.1°$, $\gamma_1 = 66.4°$

2–66. $F_R = 615$ N, $\alpha = 26.6°$, $\beta = 85.1°$, $\gamma = 64.0°$

2–67. $\mathbf{F}_1 = \{176\mathbf{j} - 605\mathbf{k}\}$ lb,
$\mathbf{F}_2 = \{125\mathbf{i} - 177\mathbf{j} + 125\mathbf{k}\}$ lb,
$F_R = 496$ lb, $\alpha = 75.4°$, $\beta = 90.0°$, $\gamma = 165°$

2–69. $F_3 = 428$ lb, $\alpha = 88.3°$, $\beta = 20.6°$, $\gamma = 69.5°$

2–70. $F_3 = 250$ lb, $\alpha = 87.0°$, $\beta = 143°$, $\gamma = 53.1°$
2–71. $F_3 = 9.58$ kN, $\alpha_3 = 15.5°$, $\beta_3 = 98.4°$, $\gamma_3 = 77.0°$
2–73. $F = 2.02$ kN, $F_y = 0.523$ kN
2–74. $F = 102$ N
2–75. $\alpha = 97.5°$, $\beta = 63.7°$, $\gamma = 27.5°$
2–77. $F_x = 40$ N, $F_y = 40$ N, $F_z = 56.6$ N
2–78. $F_2 = 32.4$ lb, $\alpha_2 = 122°$, $\beta_2 = 74.5°$, $\gamma_2 = 144°$
2–79. $\alpha = 69.6°$, $\beta = 116°$, $\gamma = 34.4°$
2–81. $\alpha = 73.4°$, $\beta = 64.6°$, $\gamma = 31.0°$
2–82. $\mathbf{r} = \{-3\mathbf{i} - 12\mathbf{j} + 4\mathbf{k}\}$ m, $r = 13$ m, $\alpha = 103°$, $\beta = 157°$, $\gamma = 72.1°$
2–83. $\mathbf{r} = \{-2.35\mathbf{i} + 3.93\mathbf{j} + 3.71\mathbf{k}\}$ ft, $r = 5.89$ ft, $\alpha = 113°$, $\beta = 48.2°$, $\gamma = 51.0°$
2–85. $r_{AB} = 2.11$ m
2–86. $r_{AB} = 567$ m
2–87. $r_{AB} = 7.81$ ft, $r_{OC} = 3.91$ ft
2–89. $r_{AD} = 1.50$ m, $r_{BD} = 1.50$ m, $r_{CD} = 1.73$ m
2–90. $F_R = 52.2$ lb, $\alpha = 87.8°$, $\beta = 63.7°$, $\gamma = 154°$
2–91. $\mathbf{F} = \{452\mathbf{i} + 370\mathbf{j} - 136\mathbf{k}\}$ lb, $\alpha = 41.1°$, $\beta = 51.9°$, $\gamma = 103°$
2–93. $\mathbf{F}_{BA} = \{-109\mathbf{i} + 131\mathbf{j} + 306\mathbf{k}\}$ lb, $\mathbf{F}_{CA} = \{103\mathbf{i} + 103\mathbf{j} + 479\mathbf{k}\}$ lb, $\mathbf{F}_{DA} = \{-52.1\mathbf{i} - 156\mathbf{j} + 365\mathbf{k}\}$ lb
2–94. $\mathbf{F}_1 = \{389\mathbf{i} - 64.9\mathbf{j} + 64.9\mathbf{k}\}$ lb, $\mathbf{F}_2 = \{-584\mathbf{i} + 97.3\mathbf{j} - 97.3\mathbf{k}\}$ lb
2–95. $r_{AB} = 592$ mm, $\mathbf{F} = \{-13.2\mathbf{i} - 17.7\mathbf{j} + 20.3\mathbf{k}\}$ N
2–97. $\mathbf{F}_1 = \{-26.2\mathbf{i} - 41.9\mathbf{j} + 62.9\mathbf{k}\}$ lb, $\mathbf{F}_2 = \{13.4\mathbf{i} - 26.7\mathbf{j} - 40.1\mathbf{k}\}$ lb, $F_R = 73.5$ lb, $\alpha = 100°$, $\beta = 159°$, $\gamma = 71.9°$
2–98. $\mathbf{F} = \{98.1\mathbf{i} + 269\mathbf{j} - 201\mathbf{k}\}$ lb
2–99. $x = 1.19$ m, $y = 1.79$ m, $z = 2.09$ m
2–101. $x = 7.65$ ft, $y = 4.24$ ft, $z = 3.76$ ft
2–102. $x = 8.67$ ft, $y = 1.89$ ft
2–103. $\mathbf{F}_{EA} = \{12\mathbf{i} - 8\mathbf{j} - 24\mathbf{k}\}$ kN, $\mathbf{F}_{EB} = \{12\mathbf{i} + 8\mathbf{j} - 24\mathbf{k}\}$ kN, $\mathbf{F}_{EC} = \{-12\mathbf{i} + 8\mathbf{j} - 24\mathbf{k}\}$ kN, $\mathbf{F}_{ED} = \{-12\mathbf{i} - 8\mathbf{j} - 24\mathbf{k}\}$ kN, $\mathbf{F}_R = \{-96\mathbf{k}\}$ kN
2–105. $\mathbf{F}_A = \{28.8\mathbf{i} - 16.6\mathbf{j} - 49.9\mathbf{k}\}$ lb, $\mathbf{F}_B = \{-28.8\mathbf{i} - 16.6\mathbf{j} - 49.9\mathbf{k}\}$ lb, $\mathbf{F}_C = \{33.3\mathbf{j} - 49.9\mathbf{k}\}$ lb, $F_R = 150$ lb, $\alpha = 90°$, $\beta = 90°$, $\gamma = 180°$
2–106. $F = 52.1$ N
2–109. $\theta = 109°$
2–110. Proj. $r_1 = |2.99$ m$|$, Proj. $r_2 = |1.99$ m$|$

2–111. $\theta = 74.0°$, $\phi = 33.9°$
2–113. $\theta = 82.9°$
2–114. Proj. $\mathbf{F}_{AB} = \{0.229\mathbf{i} - 0.916\mathbf{j} + 1.15\mathbf{k}\}$ lb
2–115. $F_\parallel = 99.1$ N, $F_\perp = 592$ N
2–117. Proj. $r_{OA} = 2.11$ ft
2–118. $F_1 = 18.3$ lb, $F_2 = 35.6$ lb
2–119. Proj. $F = 31.1$ N
2–121. Proj. $F_1 = 18.5$ N, Proj. $F_2 = 21.3$ N
2–122. $\theta = 128°$
2–123. 5.44 lb
2–125. $\theta = 59.2°$
2–126. Proj. $F_1 = 30.8$ lb, Proj. $F_2 = 15.4$ lb
2–127. $\theta = 82.0°$
2–129. Proj. $F = 10.5$ lb, $\theta = 143°$
2–130. $\theta = 74.4°$, $\phi = 55.4°$
2–131. $F_3 = 428$ lb, $\alpha = 88.3°$, $\beta = 20.6°$, $\gamma = 69.5°$
2–133. $F_\perp = 460$ lb, $F_\parallel = 386$ lb
2–134. $F = \{-34.3\mathbf{i} + 22.9\mathbf{j} - 68.6\mathbf{k}\}$ lb
2–135. $F_R = 178$ N, $\theta = 85.2°$
2–137. Proj. $F_{AB} = 70.5$ N, Proj. $F_{AC} = 65.1$ N
2–138. $F_{CB} = 10.5$ lb
2–139. $70.5°$

Chapter 3

3–1. $F_1 = 259$ N, $F_2 = 366$ N
3–2. $\theta = 31.8°$, $F = 4.94$ kN
3–3. $\theta = 53.0°$, $F = 28.3$ N
3–5. $T = 13.3$ kN, $F = 10.2$ kN
3–6. $T = 14.3$ kN, $\theta = 36.3°$
3–7. $W = 240$ lb
3–9. $\theta = 34.2°$
3–10. $\theta = 11.5°$
3–11. $F = 1.13$ mN
3–13. $x_{AC} = 0.793$ m, $x_{AB} = 0.467$ m
3–14. $m = 12.8$ kg
3–15. $F = 158$ N
3–17. At B: $T_{BC} = 39.2$ N, $T_{BA} = 68.0$ N; At C: $T_{CD} = 39.2$ N, $F = 39.2$ N
3–18. $\theta = 78.7°$, $F_{CD} = 127$ lb
3–19. $\theta = 78.7°$, $W = 51.0$ lb
3–21. $\theta = 45°$, $W = 200$ lb
3–22. $\theta = 60°$, $T_{AB} = 34.6$ lb
3–23. $\theta = 60°$, $W = 46.2$ lb
3–25. $s = 5.33$ ft
3–26. $W = 6$ lb
3–27. $l'_{AB} = 0.452$ m, $l'_{AC} = 0.658$ m

3–29. $m = 15.6$ kg
3–31. $\theta = 35.0°$
3–33. $T_{AB} = 157$ N, $T_{AC} = 118$ N
3–34. $l = 2.65$ ft
3–35. $m = 2.37$ kg
3–37. $\theta = 43.0°$
3–38. C and D, $T = 106$ lb
3–39. $m_B = 3.58$ kg, $N = 19.7$ N
3–41. $R = 0$, $P = 3.11$ kN, $F = 2.38$ kN
3–42. $F_1 = 800$ N, $F_2 = 147$ N, $F_3 = 564$ N
3–43. $F_1 = 5.60$ kN, $F_2 = 8.55$ kN, $F_3 = 9.44$ kN
3–45. $F_{AD} = 1.20$ kN, $F_{AC} = 0.40$ kN, $F_{AB} = 0.80$ kN
3–46. $F_{CD} = 625$ lb, $F_{CA} = F_{CB} = 198$ lb
3–47. $s_{OB} = 327$ mm, $s_{OA} = 218$ mm
3–49. $F_{AB} = 0.980$ kN, $F_{AC} = 0.463$ kN,
$\quad\quad F_{AD} = 1.55$ kN
3–50. $F_{AO} = 319$ N, $F_{AB} = 110$ N, $F_{AC} = 85.8$ N
3–51. $W = 138$ N
3–53. $F_{DA} = 21.5$ lb, $F_{DB} = 14.0$ lb, $F_{DC} = 17.6$ lb
3–54. $F_{AC} = 92.9$ N, $F_{AD} = 364$ N, $F = 757$ N
3–55. $F_B = 858$ N, $F_C = 0$, $F_D = 858$ N
3–57. At A: $F_{AC} = 574$ lb, $F_{AB} = 500$ lb
$\quad\quad$ At C: $F_{CD} = 1.22$ kip, $F_{CE} = F_{CF} = 416$ lb
3–58. $F_{AB} = 35.9$ lb, $F_{AC} = F_{AD} = 25.4$ lb
3–59. $W = 267$ lb
3–61. $F_{AB} = 469$ lb, $F_{AC} = F_{AD} = 331$ lb
3–62. $s = 410$ mm
3–63. $F_{AD} = 1.42$ kip, $F_{AC} = 0.914$ kip, $F_{AB} = 1.47$ kip
3–65. $W = 55.8$ N
3–66. $F_A = 34.6$ lb, $F_B = 57.3$ lb
3–67. $F = 40.8$ lb
3–69. $\theta = 40.2°$
3–70. $T = 257$ N
3–71. $\theta = 90°$, $(F_{AC})_{max} = 160$ lb, $\theta = 120°$,
$\quad\quad F_{AB} = 160$ lb
3–73. $W = 954$ lb
3–74. $F_{OA} = 202$ lb, $F_{OC} = 216$ lb, $F_{OD} = 58.6$ lb
3–75. $F_1 = 0$, $F_2 = 311$ lb, $F_3 = 238$ lb

Chapter 4

4–5. $M_{RP} = 1.17$ kip · in. $\downarrow$
4–6. $F = 77.6$ N
4–7. $\theta = 28.6°$
4–9. $M_P = 3.15$ kN · m $\uparrow$
4–10. $\mathbf{M}_O = \{-7.11\mathbf{k}\}$ N · m
4–11. $M_O = 2.42$ kip · ft $\downarrow$

4–13. $(M_B)_1 = 3.07$ N · m $\downarrow$, $(M_B)_2 = 4.40$ N · m $\downarrow$,
$\quad\quad (M_B)_3 = 1.08$ N · m $\downarrow$
4–14. $M_B = 90.6$ lb · ft $\uparrow$, $M_C = 141$ lb · ft $\uparrow$
4–15. $M_A = 195$ lb · ft $\uparrow$
4–17. $M_A = 1.65$ kip · ft $\uparrow$, $M_B = 9.27$ kip · ft $\uparrow$,
$\quad\quad M_C = 6.45$ kip · ft $\uparrow$
4–18. (a) $(M_A)_{max} = 330$ lb · ft, $\theta = 76.0°$
$\quad\quad$ (b) $(M_A)_{min} = 0$, $\theta = 166°$
4–19. $M_A = 7.71$ N · m $\uparrow$
4–21. (a) $M_A = 13.0$ N · m, (b) $F = 35.2$ N
4–22. (a) $(M_A)_{max} = 330$ lb · ft, $\theta = 104°$
$\quad\quad$ (b) For $M_A = 0$, $\theta = 14.04°$
4–23. $(M_A)_1 = 11.7$ kip · ft $\uparrow$, $(M_A)_2 = 11.7$ kip · ft $\downarrow$
4–25. Maximum moment: $\theta = 37.9°$,
$\quad\quad (M_A)_{max} = 79.8$ N · m $\downarrow$
$\quad\quad$ Minimum moment: $\theta = 128°$, $(M_A)_{min} = 0$
4–26. $(M_O)_{max} = 80$ kN · m $\downarrow$, $x = 24.0$ m
4–27. Maximum moment: $\uparrow+(M_O)_{max} = 80.0$ kN · m,
$\quad\quad \theta = 33.6°$
4–29. $F_3 = 1.59$ kN
4–30. $(M_A)_1 = 542$ lb · in. $\downarrow$, $(M_A)_2 = 10.0$ lb · in. $\uparrow$
4–31. $M_A = 306$ lb · ft $\uparrow$, $T = 652$ lb
4–33. $M_A = 199$ N · m
4–34. $\mathbf{M}_O = \{260\mathbf{i} + 180\mathbf{j} + 510\mathbf{k}\}$ N · m
4–35. $\mathbf{M}_P = \{440\mathbf{i} + 220\mathbf{j} - 990\mathbf{k}\}$ N · m
4–37. $\mathbf{M}_P = \{-116\mathbf{i} + 16\mathbf{j} - 135\mathbf{k}\}$ kN · m
4–38. $\mathbf{M}_O = \{-128\mathbf{i} + 128\mathbf{j} - 257\mathbf{k}\}$ N · m
4–39. $\mathbf{M}_B = \{-37.6\mathbf{i} + 90.7\mathbf{j} - 155\mathbf{k}\}$ N · m
4–41. $\mathbf{M}_A = \{-229\mathbf{i} + 132\mathbf{j}\}$ lb · ft
4–42. $\mathbf{M}_O = \{61.2\mathbf{i} + 81.6\mathbf{j}\}$ N · m
4–43. $F_{AB} = 18.6$ lb
4–45. $\mathbf{M}_B = \{10.6\mathbf{i} + 13.1\mathbf{j} + 29.2\mathbf{k}\}$ N · m
4–46. $\mathbf{M}_O = \{-1.77\mathbf{i} - 1.10\mathbf{j}\}$ kN · m
4–47. $\mathbf{M}_A = \{-1.90\mathbf{i} + 6.00\mathbf{j}\}$ kN · m
4–49. $y = 1$ m, $z = 3$ m, $d = 1.15$ m
4–50. $y = 2$ m, $z = 1$ m
4–51. $(\mathbf{M}_{Oa})_F = \{218\mathbf{j} + 163\mathbf{k}\}$ N · m
4–53. $(\mathbf{M}_R)_{Oa} = \{26.1\mathbf{i} - 15.1\mathbf{j}\}$ lb · ft
4–54. $M_z = 62$ lb · in.
4–55. $M_y = 75$ lb · in.
4–57. $M_y = 0.277$ N · m
4–58. $\mathbf{M}_y = \{-78.4\mathbf{j}\}$ lb · ft
4–59. $M_x = 12$ N · m < 14 N · m, No
$\quad\quad (M_x)_{max} = 15$ N · m > 14 N · m, Yes
4–61. $M_y = 0$
4–62. $\mathbf{M}_x = \{0.211\mathbf{i}\}$ N · m

4–63. $|M_{CA}| = 226$ N·m
4–65. $P = 8.50$ lb
4–66. $M_y = 282$ lb·ft
4–67. $(M_a)_1 = 30$ lb·in., $(M_a)_2 = 8$ lb·in.
4–69. $M_C = 18.3$ kN·m ↖
4–70. $M_C = 650$ lb·ft ↖
4–71. $M_C = 3.12$ kip·ft ↖
4–73. $R = 28.9$ N
4–74. $T = 0.909$ kip
4–75. $\theta = 32.9°$, $M_3 = 651$ N·m
4–77. $F = 830$ N
4–78. $F = 139$ lb. The resultant couple moment is a free vector. It can act at any point on the beam.
4–79. $\mathbf{M}_C = \{-5\mathbf{i} + 8.75\mathbf{j}\}$ N·m
4–81. $M_R = 9.69$ kN·m ↙
4–82. $F = 167$ lb. Resultant couple can act anywhere.
4–83. $d = 2.03$ ft
4–85. $M_C = 126$ lb·ft ↖
4–86. $\mathbf{M}_C = \{40\mathbf{i} + 20\mathbf{j} - 24\mathbf{k}\}$ N·m
4–87. $\mathbf{M}_C = \{-253\mathbf{i} + 67.4\mathbf{j} - 253\mathbf{k}\}$ N·m
4–89. $F = 75$ N, $P = 100$ N
4–90. $\mathbf{M}_C = \{7.01\mathbf{i} + 42.1\mathbf{j}\}$ N·m
4–91. $F = 35.1$ N
4–93. $\mathbf{M}_C = \{-17.0\mathbf{i} - 9.20\mathbf{j} - 27.4\mathbf{k}\}$ N·m, $M_C = 33.5$ N·m
4–94. $\mathbf{M}_C = \{2.60\mathbf{i} + 2.40\mathbf{j} + 3.60\mathbf{k}\}$ N·m
4–95. 18.3 N·m, $\alpha = 155°$, $\beta = 115°$, $\gamma = 90°$
4–97. 832 N
4–98. $F = 375$ N, $M_O = 100$ N·m ↙
4–99. $F_P = 375$ N, $M_P = 737$ N·m ↖
4–101. $F_R = 115$ lb, $\theta = 73.2°$ ↗, $M_{RP} = 921$ lb·ft ↖
4–102. $F_R = 274$ lb, $\theta = 5.24°$ ↖, $M_O = 4.61$ kip·ft ↖
4–103. $F_R = 274$ lb, $\theta = 5.24°$ ↖, $M_P = 5.48$ kip·ft ↖
4–105. $F_R = 78.1$ lb, $\theta = 87.1°$ ↙, $M_{R_P} = 124$ lb·ft ↖
4–106. $F_R = 2.10$ kN, $\theta = 81.6°$ ∡θ, $M_O = 10.6$ kN·m ↙
4–107. $F_R = 2.10$ kN, $\theta = 81.6°$ ∡θ, $M_P = 16.8$ kN·m ↙
4–109. $F_R = 375$ lb ↑, $x = 2.47$ ft
4–110. $\mathbf{F}_R = \{-70\mathbf{i} + 140\mathbf{j} - 408\mathbf{k}\}$ N, $\mathbf{M}_{RO} = \{-26\mathbf{i} + 31.0\mathbf{j} + 14.6\mathbf{k}\}$ N·m
4–111. $\mathbf{M}_{RP} = \{-26\mathbf{i} + 357\mathbf{j} + 127\mathbf{k}\}$ N·m
4–113. $F = 798$ lb, $\theta = 67.9°$ ↗, $x = 6.57$ ft
4–114. $F = 922$ lb, $\theta = 77.5°$ ↗, $x = 3.56$ ft

4–115. $F = 1.30$ kN, $\theta = 84.5°$ ↗, $x = 7.36$ m
4–117. $F_R = 294$ N, $\theta = 40.1°$ ↗, $M_{RO} = 39.6$ N·m ↙
4–118. $F_R = 4.43$ kN, $\theta = 71.6°$ ↙, $d = 3.52$ m
4–119. $\theta = 57.5°$, $F = 2.61$ kN, $d = 2.64$ m
4–121. $F_R = 991$ N, $\theta = 63.0°$ ↗, $x = 2.64$ m
4–122. $F_R = 65.9$ lb, $\theta = 49.8°$ ↙, $d = 2.10$ ft
4–123. $F_R = 65.9$ lb, $\theta = 49.8°$ ↙, $d = 4.62$ ft
4–125. $\mathbf{F}_R = \{8\mathbf{i} + 6\mathbf{j} + 8\mathbf{k}\}$ kN, $\mathbf{M}_{RO} = \{-10\mathbf{i} + 18\mathbf{j} - 56\mathbf{k}\}$ kN·m
4–126. $\mathbf{F}_R = \{8\mathbf{i} + 6\mathbf{j} + 8\mathbf{k}\}$ kN, $\mathbf{M}_{RP} = \{-46\mathbf{i} + 66\mathbf{j} - 56\mathbf{k}\}$ kN·m
4–127. $\mathbf{F}_R = \{8\mathbf{i} + 6\mathbf{j} + 8\mathbf{k}\}$ kN, $\mathbf{M}_{RQ} = \{-10\mathbf{i} - 30\mathbf{j} - 20\mathbf{k}\}$ kN·m
4–129. $\mathbf{F}_R = \{-28.3\mathbf{j} - 68.3\mathbf{k}\}$ N, $\mathbf{M}_{RA} = \{-20.5\mathbf{j} + 8.49\mathbf{k}\}$ N·m
4–130. $\mathbf{F}_R = \{-210\mathbf{k}\}$ N, $\mathbf{M}_{RO} = \{-15\mathbf{i} + 225\mathbf{j}\}$ N·m
4–131. $\mathbf{F}_R = \{6\mathbf{i} - 1\mathbf{j} - 14\mathbf{k}\}$ N, $\mathbf{M}_{RO} = \{1.30\mathbf{i} + 3.30\mathbf{j} - 0.450\mathbf{k}\}$ N·m
4–133. $F_R = 140$ kN ↓, $y = 7.14$ m, $x = 5.71$ m
4–134. $F_R = 140$ kN, $x = 6.43$ m, $y = 7.29$ m
4–135. $F = 53.3$ N, $\mathbf{F}_{RW} = \{-40\mathbf{i}\}$ N, $\mathbf{M}_{RW} = \{-30\mathbf{i}\}$ N·m
4–137. $F_R = 990$ N, $M_R = 3.07$ kN·m, $x = 1.16$ m, $y = 2.06$ m
4–138. $F_R = 108$ lb, $M_R = -624$ lb·ft, $z = 8.69$ ft, $y = 0.414$ ft
4–139. $F_{RO} = 13.2$ lb ↓, $x = 0.340$ ft
4–141. $F_R = 51.0$ kN ↓, $M_{R_O} = 914$ kN·m ↙
4–142. $F_R = 51.0$ kN ↓, $d = 17.9$ m
4–143. $F_R = 3.25$ kip, $\theta = 67.2°$ ↗, $x = 3.86$ ft
4–145. $F_R = 18.0$ kip ↓, $x = 11.7$ ft
4–146. $F_R = 11.25$ kN ↓, $d = 4.05$ m
4–147. $b = 5.625$ m, $a = 1.54$ m
4–149. $F_R = 3.90$ kip ↑, $d = 11.3$ ft
4–150. $b = 4.50$ ft, $a = 9.75$ ft
4–151. $F_R = 10.6$ kip ↓, $x = 0.479$ ft
4–153. $F_R = 1.35$ kN, $\theta = 42.0°$ ↗, $x = 0.556$ m
4–154. $F_R = 95.6$ kN →, $M_{R_O} = 349$ kN·m ↙
4–155. $F_R = 3$ kN ↓, $M_{RO} = 2.25$ kN·m ↖
4–157. $F_R = \dfrac{2w_0 L}{\pi}$ ↓, $\bar{x} = \dfrac{L}{2}$
4–158. $F_R = 774$ lb ↑, $\bar{x} = 2.25$ ft
4–159. $F_R = 7$ lb, $\bar{x} = 0.268$ ft
4–161. $\alpha = 70.8°$, $\beta = 39.8°$, $\gamma = 56.7°$, $\alpha = 109°$, $\beta = 140°$, $\gamma = 123°$

4–162. $\mathbf{M}_O = \{298\mathbf{i} + 15.1\mathbf{j} - 200\mathbf{k}\}$ lb $\cdot$ in

4–163. $\mathbf{F}_R = \{-300\mathbf{i} + 200\mathbf{j} - 500\mathbf{k}\}$ N,
$\mathbf{M}_{RO} = \{-3.80\mathbf{i} - 7.20\mathbf{j} - 0.600\mathbf{k}\}$ kN $\cdot$ m

4–165. $M_{a-a} = 59.7$ N $\cdot$ m

4–166. $\mathbf{M}_z = \{-6.21\mathbf{k}\}$ N $\cdot$ m

4–167. $\mathbf{F}_R = \{14.3\mathbf{i} + 21.4\mathbf{j} - 42.9\mathbf{k}\}$ lb,
$\mathbf{M}_A = \{-1.93\mathbf{i} + 0.429\mathbf{j} - 0.429\mathbf{k}\}$ kip $\cdot$ ft

4–169. $\mathbf{M}_O = \{1.06\mathbf{i} + 1.06\mathbf{j} - 4.03\mathbf{k}\}$ N $\cdot$ m,
$\alpha = 75.7°, \beta = 75.7°, \gamma = 160°$

4–170. $F = 3$ lb, $P = 4$ lb

Chapter 5

5–1. N_A, N_B force of plane on sphere,
10 lb force of gravity on sphere

5–3. A_x, A_y, M_A effect of wall on beam,
N_B force of roller on beam, 60 lb resultant force
of distributed load on beam

5–5. N_A, N_B, N_C force of rollers on beam

5–6. A_x, A_y, N_B force of glass on rod,
0.02(9.81) N force of gravity on rod

5–7. A_x, A_y, N_B force of cylinder on wrench

5–11. $N_A = 19.3$ lb, $N_B = 14.1$ lb

5–13. $N_A = 267$ lb, $N_B = 208$ lb, $N_C = 305$ lb

5–14. $N_B = 0.103$ N, $A_x = 0.0664$ N, $A_y = 0.117$ N

5–15. $N_B = 140$ lb, $A_x = 140$ lb, $A_y = 20$ lb

5–17. $N_C = 493$ N, $N_B = 554$ N, $N_A = 247$ N

5–18. $B_y = 642$ N, $A_x = 192$ N, $A_y = 180$ N

5–19. $B_y = 586$ N, $F_A = 413$ N

5–21. $F_A = 30$ lb, $F_B = 36.2$ lb, $F_C = 9.38$ lb

5–22. $A_x = 33.3$ lb, $B_x = 33.3$ lb, $B_y = 100$ lb

5–23. $F_{CD} = 195$ lb, $A_x = 97.4$ lb, $A_y = 31.2$ lb

5–25. $N_C = 785$ N, $F_B = 4.71$ kN, $F_A = 3.92$ kN

5–26. $R_A = 105$ lb, $B_x = 97.4$ lb, $B_y = 269$ lb

5–27. $W_B = 78.6$ lb

5–29. $F_B = 6.38$ N, $A_x = 3.19$ N, $A_y = 2.48$ N

5–30. $N_A = 39.7$ lb, $N_B = 82.5$ lb, $M_A = 106$ lb $\cdot$ ft

5–31. $A_x = 1462$ lb, $F_B = 1.66$ kip

5–33. $D_x = 0$, $D_y = 1.65$ kip, $M_D = 1.40$ kip $\cdot$ ft,
$(M_D)_{max} = 3.00$ kip $\cdot$ ft

5–34. The table will tip over.

5–35. $F_B = 105$ N

5–37. $l_0 = 70.3$ mm

5–38. $h = 15.8$ ft

5–39. $N_A = 81.6$ lb, $F_B = 50.2$ lb

5–41. $B_x = 989$ N, $A_x = 989$ N, $B_y = 186$ N

5–42. $w_1 = 413$ kN/m, $w_2 = 407$ kN/m

5–43. $T = 5$ kN, $T_{BC} = 16.4$ kN, $F_A = 20.6$ kN

5–45. $R_A = 40.9$ kip, $R_B = 125$ kip

5–46. $\theta = 41.4°$

5–47. $F_A = 432$ lb, $F_B = 0$, $F_C = 432$ lb

5–49. $F_2 = 724$ lb, $F_1 = 1.45$ kip, $F_A = 1.75$ kip

5–50. $\theta = \tan^{-1}\left(\dfrac{1}{2}\cot\psi - \dfrac{1}{2}\cot\phi\right)$

5–51. $N_B = 2.11$ N, $F_A = 2.81$ N

5–53. $\theta = 23.2°, 85.2°$

5–54. $R_A = 26.0$ lb, $R_B = 11.9$ lb, $R_C = 63.9$ lb

5–55. $F_C = 10$ mN

5–57. $d = \dfrac{a}{\cos^3\theta}$

5–58. $\theta = \tan^{-1}\dfrac{b}{a}$

5–59. $P = 441$ N

5–61. $\theta = \cos^{-1}\left(\dfrac{L + \sqrt{L^2 + 128r^2}}{16r}\right)$

5–62. $\theta = 27.1°$ or $\theta = 50.2°$

5–63. $A_x = 0$, $A_y = -200$ N, $A_z = 150$ N,
$(M_A)_x = 100$ N $\cdot$ m, $(M_A)_y = 0$, $(M_A)_z = 500$ N $\cdot$ m

5–65. $T_B = 2.75$ kip, $T_C = 1.375$ kip, $T_A = 1.375$ kip

5–66. $W = 750$ lb, $x = 5.20$ ft, $y = 5.27$ ft

5–67. $F_A = 663$ lb, $F_C = 569$ lb, $F_B = 449$ lb

5–69. $F = 900$ lb, $A_x = 0$, $A_y = 0$,
$A_z = 600$ lb, $M_{Ax} = 0$, $M_{Az} = 0$

5–70. $A_x = 0$, $A_y = 1.50$ kip, $A_z = 750$ lb, $T = 919$ lb

5–71. $F = 1.31$ kip, $A_x = 0$, $A_y = 1.31$ kip, $A_z = 653$ lb

5–73. $P = 75$ lb, $A_y = 0$, $B_z = 75$ lb,
$A_z = 75$ lb, $B_x = 112$ lb, $A_x = 37.5$ lb

5–74. $F_C = 3.46$ N, $N_A = N_B = 14.1$ N

5–75. $F_{BC} = 0$, $A_y = 0$, $A_z = 800$ lb,
$(M_A)_x = 4.80$ kip $\cdot$ ft, $(M_A)_y = 0$, $(M_A)_z = 0$

5–77. $F = 96.0$ lb, $B_z = 45.7$ lb, $B_y = -67.9$ lb,
$B_x = 48.5$ lb, $A_x = -96.5$ lb, $A_z = -13.7$ lb

5–78. $F = 96.0$ lb, $B_x = -48.0$ lb, $B_y = -67.9$ lb,
$B_z = 32.0$ lb, $M_{Bx} = 96.0$ lb $\cdot$ ft, $M_{Bz} = -675$ lb $\cdot$ ft

5–79. $A_x = 633$ lb, $A_y = -141$ lb, $B_x = -721$ lb,
$B_z = 895$ lb, $C_y = 200$ lb, $C_z = -506$ lb

5–81. $T_{BD} = T_{CD} = 117$ N, $A_x = 66.7$ N,
$A_y = 0$, $A_z = 100$ N

5–82. $T_{DE} = 32.1$ lb, $T_{BC} = 42.9$ lb, $A_x = 3.57$ lb,
$A_y = 50$ lb, $(M_A)_x = 0$, $(M_A)_y = -17.9$ lb $\cdot$ ft

5–83. $T_{BC} = 1.05$ kip, $A_x = -450$ lb, $A_y = 900$ lb,
$A_z = 0$, $M_{Ay} = -600$ lb $\cdot$ ft, $M_{Az} = -900$ lb $\cdot$ ft

5–85. $T_B = 25$ lb, $A_x = 25$ lb, $A_y = -25$ lb,
$A_z = 50$ lb, $B_y = 25$ lb

5–86. $A_x = -20$ lb, $A_y = 40$ lb, $B_z = 30$ lb,
$M_{Ax} = 110$ lb $\cdot$ ft, $M_{Ay} = 40$ lb $\cdot$ ft, $M_{Az} = 110$ lb $\cdot$ ft

5–87. $F_{AC} = F_{BC} = 6.13$ kN, $F_{DE} = 19.6$ kN

5–89. $T_{BC} = 131$ lb, $T_{BD} = 510$ lb, $A_x = 0$,
$A_y = 0$, $A_z = 589$ lb

5–90. $B_z = 1167$ lb, $C_z = 734$ lb, $A_z = 1600$ lb

5–91. $F_{CD} = 1.02$ kN, $A_z = -208$ N, $B_z = -139$ N,
$A_y = 573$ N, $B_y = 382$ N

5–93. $A_x = 0$, $B_y = 7$ kN, $A_y = 11$ kN

5–94. $N_A = 8.00$ kN, $B_x = 5.20$ kN, $B_y = 5.00$ kN

5–95. $N_B = 400$ N, $F_A = 721$ N

5–97. $\theta = 30.8°$

5–98. $A_x = 0$, $A_y = 0$, $A_z = B_z = C_z = 5.33$ lb

5–99. $P = 100$ lb, $B_z = 40$ lb, $B_x = -35.7$ lb,
$A_x = 136$ lb, $B_y = 0$, $A_z = 40$ lb

Chapter 6

6–1. Joint A: $F_{AD} = 9.90$ kN (C), $F_{AB} = 7$ kN (T)
Joint D: $F_{DB} = 4.95$ kN (T), $F_{DC} = 14.8$ kN (C)
Joint C: $F_{CB} = 10.5$ kN (T)

6–2. Joint A: $F_{AD} = 11.3$ kN (C), $F_{AB} = 8$ kN (T)
Joint D: $F_{DB} = 7.07$ kN (T), $F_{DC} = 18.4$ kN (C)
Joint C: $F_{CB} = 13.0$ kN (T)

6–3. Joint A: $F_{AD} = 849$ lb (C), $F_{AB} = 600$ lb (T)
Joint B: $F_{BD} = 400$ lb (C), $F_{BC} = 600$ lb (T)
Joint D: $F_{DC} = 1.41$ kip (T), $F_{DE} = 1.60$ kip (C)

6–5. Joint A: $F_{AB} = 21.9$ kN (C), $F_{AG} = 13.1$ kN (T)
Joint B: $F_{BC} = 13.1$ kN (C), $F_{BG} = 17.5$ kN (T)
Joint G: $F_{CG} = 3.12$ kN (T), $F_{FG} = 11.2$ kN (T)
Joint C: $F_{CF} = 3.12$ kN (C), $F_{CD} = 9.38$ kN (C)
Joint D: $F_{DE} = 15.6$ kN (C), $F_{DF} = 12.5$ kN (T)
Joint F: $F_{EF} = 9.38$ kN (T)

6–6. Joint A: $F_{AB} = 43.8$ kN (C), $F_{AG} = 26.2$ kN (T)
Joint B: $F_{BC} = 26.2$ kN (C), $F_{BG} = 35.0$ kN (T)
Joint G: $F_{GC} = 6.25$ kN (T), $F_{GF} = 22.5$ kN (T)
Joint E: $F_{ED} = 31.2$ kN (C), $F_{EF} = 18.8$ kN (T)
Joint D: $F_{DC} = 18.8$ kN (C), $F_{DF} = 25.0$ kN (T)
Joint F: $F_{FC} = 6.25$ kN (C)

6–7. Joint B: $F_{BC} = 3$ kN (C), $F_{BA} = 8$ kN (C)
Joint A: $F_{AC} = 1.46$ kN (C), $F_{AF} = 4.17$ kN (T)
Joint C: $F_{CD} = 4.17$ kN (C), $F_{CF} = 3.12$ kN (C)
Joint E: $F_{EF} = 0$, $F_{ED} = 13.1$ kN (C)
Joint D: $F_{DF} = 5.21$ kN (T)

6–9. $P = 424$ lb

6–10. Joint A: $F_{AG} = 471$ lb (C), $F_{AB} = 333$ lb (T)
Joint B: $F_{BG} = 0$, $F_{BC} = 333$ lb (T)
Joint D: $F_{DE} = 943$ lb (C), $F_{DC} = 667$ lb (T)
Joint E: $F_{EC} = 667$ lb (T), $F_{EG} = 667$ lb (C)
Joint C: $F_{CG} = 471$ lb (T)

6–11. Joint A: $F_{AG} = 1179$ lb (C), $F_{AB} = 833$ lb (T)
Joint B: $F_{BC} = 833$ lb (T), $F_{BG} = 500$ lb (T)
Joint D: $F_{DE} = 1650$ lb (C), $F_{DC} = 1167$ lb (T)
Joint E: $F_{EC} = 1167$ lb (T), $F_{EG} = 1167$ lb (C)
Joint C: $F_{CG} = 471$ lb (T)

6–13. Joint G: $F_{GB} = 30$ kN (T)
Joint A: $F_{AF} = 20$ kN (C), $F_{AB} = 22.4$ kN (C)
Joint B: $F_{BF} = 20$ kN (T), $F_{BC} = 20$ kN (T)
Joint F: $F_{FC} = 28.3$ kN (C), $F_{FE} = 0$
Joint E: $F_{ED} = 0$, $F_{EC} = 20.0$ kN (T)
Joint C: $F_{DC} = 0$

6–14. Joint A: $F_{AB} = 330$ lb (C), $F_{AF} = 79.4$ lb (T)
Joint B: $F_{BF} = 233$ lb (T), $F_{BC} = 233$ lb (C)
Joint F: $F_{FC} = 47.1$ lb (C), $F_{FE} = 113$ lb (T)
Joint E: $F_{EC} = 300$ lb (T), $F_{ED} = 113$ lb (T)
Joint C: $F_{CD} = 377$ lb (C)

6–15. Joint A: $F_{AB} = 377$ lb (C), $F_{AF} = 190$ lb (T)
Joint B: $F_{BF} = 267$ lb (T), $F_{BC} = 267$ lb (C)
Joint F: $F_{FC} = 189$ lb (T), $F_{FE} = 56.7$ lb (T)
Joint E: $F_{ED} = 56.7$ lb (T), $F_{EC} = 0$
Joint C: $F_{CD} = 189$ lb (C)

6–17. $P = 4.16$ kN

6–18. Joint C: $F_{CB} = 400$ lb (C), $F_{CD} = 693$ lb (C)
Joint B: $F_{BD} = 667$ lb (T),
Member AB is a two-force member and exerts only a vertical force along AB at A.

6–19. Joint C: $F_{CD} = 3.61$ kN (C), $F_{CB} = 3$ kN (T)
Joint B: $F_{BA} = 3$ kN (T), $F_{BD} = 3$ kN (C)
Joint D: $F_{DA} = 2.70$ kN (T), $F_{DE} = 6.31$ kN (C)

6–21. Joint B: $F_{BA} = P \csc 2\theta$ (C), $F_{BC} = P \cot 2\theta$ (C)
Joint C: $F_{CA} = (\cot \theta \cos \theta - \sin \theta + 2 \cos \theta)P$ (T),
$F_{CD} = (\cot 2\theta + 1)P$ (C)
Joint D: $F_{DA} = (\cot 2\theta + 1)(\cos 2\theta)P$ (C)

6–22. $P_{\max} = 732$ N

6–23. Joint A: $F_{AB} = 3.75$ kip (C), $F_{AH} = 3$ kip (T)
Joint B: $F_{BC} = 3.75$ kip (C), $F_{BH} = 1$ kip (C)
Joint H: $F_{HC} = 1$ kip (T), $F_{GH} = 2$ kip (T)
Joint E: $F_{EF} = 3$ kip (T), $F_{ED} = 3.75$ kip (C)
Joint D: $F_{DC} = 3.75$ kip (C), $F_{DF} = 1$ kip (C)
Joint C: $F_{CF} = 1$ kip (T), $F_{CG} = 2$ kip (C)
Joint F: $F_{FG} = 2$ kip (T)

6–25. Joint A: $F_{AB} = 7.5$ kN (T), $F_{AE} = 4.5$ kN (C)
Joint E: $F_{ED} = 4.5$ kN (C), $F_{EB} = 8$ kN (T)
Joint B: $F_{BD} = 19.8$ kN (C), $F_{BC} = 18.5$ kN (T)

6–26. Joint A: $F_{AB} = 196$ N (T), $F_{AE} = 118$ N (C)
Joint E: $F_{ED} = 118$ N (C), $F_{EB} = 216$ N (T)
Joint B: $F_{BD} = 1.04$ kN (C), $F_{BC} = 857$ N (T)

6–27. Joint B: $F_{BC} = 0$
Joint C: $F_{CA} = 0$, $F_{CD} = 0$
By symmetry, $F_{EF} = F_{ED} = F_{EG} = 0$
Joint D: $F_{DF} = 4$ kN (C)
$F_{DB} = F_{BA} = F_{FG} = 4$ kN (C)

6–29. Joint A: $F_{AI} = 4.04$ kip (C), $F_{AB} = 3.75$ kip (T)
Joint E: $F_{EF} = 12.1$ kip (C), $F_{ED} = 7.75$ kip (T)
Joint B: $F_{BI} = 0$, $F_{BC} = 3.75$ kip (T)
Joint D: $F_{DF} = 0$, $F_{DC} = 7.75$ kip (T)
Joint F: $F_{FC} = 4.04$ kip (C), $F_{FG} = 8.08$ kip (C)
Joint H: $F_{HG} = 2.15$ kip (C), $F_{HI} = 0.8$ kip (T)
Joint C: $F_{CI} = 0.269$ kip (T), $F_{CG} = 1.40$ kip (T)
Joint G: $F_{GI} = 5.92$ kip (C)

6–30. $F_{DE} = 45$ kN (C), $F_{EH} = 5$ kN (T),
$F_{HG} = 45$ kN (T)

6–31. $F_{LD} = 0$, $F_{LK} = 112.5$ kN (C),
$F_{CD} = 112.5$ kN (T), $F_{KD} = 50$ kN (C)

6–33. $F_{KJ} = 13.3$ kN (T), $F_{BC} = 14.9$ kN (C),
$F_{CK} = 0$

6–34. $F_{KJ} = 11.2$ kip (T), $F_{CD} = 9.38$ kip (C),
$F_{CJ} = 3.12$ kip (C), $F_{DJ} = 0$

6–35. $F_{JI} = 7.50$ kip (T), $F_{EI} = 2.50$ kip (C)

6–37. $F_{BF} = 0$, $F_{BG} = 35.4$ kN (C),
$F_{AB} = 45$ kN (T)

6–38. $F_{GF} = 671$ lb (C), $F_{GB} = 671$ lb (T)

6–39. $F_{FE} = 8.49$ kN (T), $F_{BC} = 8.49$ kN (C), $F_{FC} = 0$

6–41. $F_{JE} = 9.38$ kN (C), $F_{GF} = 5.625$ kN (T)
AB, BC, CD, DE, HI, and GI are zero-force
members.

6–42. $F_{BC} = 10.4$ kN (C), $F_{HG} = 9.16$ kN (T),
$F_{HC} = 2.24$ kN (T)

6–43. $F_{CD} = 11.2$ kN (C), $F_{CF} = 3.21$ kN (T),
$F_{CG} = 6.80$ kN (C)

6–45. $F_{GJ} = 2.00$ kip (C)

6–46. $F_{GC} = 1.00$ kip (T)

6–47. $F_{CD} = 2.62$ kip (T), $F_{KJ} = 3.02$ kip (C), $F_{JN} = 0$

6–49. $F_{EF} = P$ (C), $F_{CB} = 1.12P$ (T), $F_{BE} = 0.5P$ (T)

6–50. $F_{AB} = P$ (T), $F_{EF} = P$ (C), $F_{BF} = 1.41P$ (C)

6–51. $F_{BC} = 3.25$ kN (C), $F_{CH} = 1.92$ kN (T)

6–53. $F_{ML} = 38.4$ kN (T), $F_{DE} = 37.1$ kN (C),
$F_{DL} = 3.84$ kN (C)

6–54. $F_{EF} = 37.1$ kN (C), $F_{EL} = 6$ kN (T)

6–55. $F_{AB} = 584$ N (T), $F_{AC} = 1.13$ kN (C),
$F_{BC} = 142$ N (C)

6–57. Joint B: $F_{BD} = 731$ lb (C), $F_{BC} = 250$ lb (T),
$F_{BA} = 167$ lb (T)
Joint A: $F_{AD} = 786$ lb (T), $F_{AC} = 391$ lb (C)
Joint C: $F_{CD} = 0$

6–58. Joint F: $F_{BF} = 0$
Joint B: $F_{BC} = 0$, $F_{BE} = 500$ lb (T),
$F_{AB} = 300$ lb (C)
Joint A: $F_{AC} = 583$ lb (T), $F_{AD} = 333$ lb (T),
$F_{AE} = 667$ lb (C)
Joint E: $F_{DE} = 0$, $F_{EF} = 300$ lb (C)
Joint C: $F_{CD} = 300$ lb (C), $F_{CF} = 300$ lb (C)
Joint F: $F_{DF} = 424$ lb (T)

6–59. Joint F: $F_{BF} = 0$
Joint B: $F_{BC} = 0$, $F_{BE} = 500$ lb (T),
$F_{AB} = 300$ lb (C)
Joint A: $F_{AC} = 972$ lb (T), $F_{AD} = 0$,
$F_{AE} = 367$ lb (C)
Joint E: $F_{DE} = 0$, $F_{EF} = 300$ lb (C)
Joint C: $F_{CD} = 500$ lb (C), $F_{CF} = 300$ lb (C)
Joint F: $F_{DF} = 424$ lb (T)

6–61. $F_{BC} = F_{BD} = 1.34$ kN (C)
Joint A: $F_{AB} = 2.4$ kN (C)
$F_{AG} = F_{AE} = 1.01$ kN (T)
Joint B: $F_{BG} = 1.80$ kN (T), $F_{BE} = 1.80$ kN (T)

6–62. $F_{BD} = 144$ lb (C), $F_{AD} = 204$ lb (T),
$F_{AF} = 72.2$ lb (T)

6–63. $F_{EF} = 144$ lb (C), $F_{DF} = 0$, $F_{CF} = 72.2$ lb (T)

6–65. Joint C: $F_{BC} = 0$, $F_{CD} = 0$, $F_{CF} = 8$ kN (C)
Joint B: $F_{BD} = 0$, $F_{BA} = 6$ kN (C)
Joint D: $F_{AD} = 0$, $F_{DF} = 0$, $F_{DE} = 9$ kN (C)
Joint E: $F_{EF} = 0$, $F_{EA} = 0$,
Joint A: $F_{AF} = 0$

6–66. $R_B = 26.7$ lb, $A_x = 0$, $A_y = 34.7$ lb

6–67. $F_B = 61.9$ lb, $F_A = 854$ lb

6–69. $R_E = 177$ N, $R_A = 128$ N

6–70. $T = 16$ lb, $N = 134$ lb

6–71. $C_x = 167$ N, $C_y = 111$ N,
$A_x = 167$ N, $A_y = 389$ N

6–73. $E_y = 80$ lb, $B_y = 10.8$ kip, $A_y = 96.7$ lb

6–74. $P = 743$ N

6–75. $C_y = 1.52$ kN, $B_y = 23.5$ kN,
$A_y = 3.09$ kN, $B_x = 3.5$ kN

6–77. $B_x = 2.80$ kip, $B_y = 1.05$ kip, $A_x = 2.80$ kip,
$A_y = 5.10$ kip, $M_A = 43.2$ kip $\cdot$ ft

6–78. $C_x = 75$ lb, $C_y = 100$ lb

6–79. $B_x = 4.20$ kN, $A_y = 4.00$ kN, $B_y = 3.20$ kN, $C_x = 3.40$ kN, $C_y = 4.00$ kN

6–81. $T = 100$ lb, $\theta = 14.6°$

6–82. $F_N = 5.25$ lb

6–83. $T = 350$ lb, $A_y = 700$ lb, $A_z = 1.88$ kip, $D_x = 1.70$ kip, $D_y = 1.70$ kip

6–85. $A_x = 80$ lb, $A_y = 80$ lb, $B_y = 133$ lb, $B_x = 333$ lb, $C_x = 413$ lb, $C_y = 53.3$ lb

6–86. $T = 1.25$ kip

6–87. $C_y = 34.4$ lb, $C_x = 16.7$ lb, $B_x = 66.7$ lb, $B_y = 15.6$ lb

6–89. $C_x = D_x = 160$ lb, $C_y = D_y = 107$ lb, $B_y = 26.7$ lb, $B_x = 80.0$ lb, $E_x = 0$, $E_y = 26.7$ lb, $A_x = 160$ lb

6–90. $C_y = 200$ lb, $C_x = 133$ lb, $D_x = 33.3$ lb, $D_y = 100$ lb, $E_x = 33.3$ lb, $E_y = 200$ lb

6–91. $A_y = 657$ N, $C_y = 229$ N, $C_x = 0$, $B_x = 0$, $B_y = 429$ N

6–93. $m = 366$ kg, $F_A = 2.93$ kN

6–94. $F_x = 333$ lb, $F_y = 250$ lb, $G_x = 333$ lb, $G_y = 250$ lb

6–95. $P = 46.9$ lb

6–97. $F = 6.93$ kN

6–98. $A_y = 250$ N, $A_x = 1.40$ kN, $C_x = 500$ N, $C_y = 1.70$ kN

6–99. $m = 15.0$ Mg; No ($K_y + H_y$ is constant.)

6–101. $F = 9.42$ lb

6–102. $F_{AB} = 9.70$ kN, $C_x = 11.5$ kN, $C_y = 8.65$ kN

6–103. $T_{BC} = 769$ lb, $F_{CA} = 739$ lb

6–105. $N_B = 0.1175$ lb, $N_A = 0.0705$ lb

6–106. $N_C = 20$ lb, $B_x = 34$ lb, $B_y = 62$ lb, $A_x = 34$ lb, $A_y = 12$ lb, $M_A = 336$ lb · ft

6–107. $N_A = 36.0$ lb

6–109. $F_{FB} = 1.94$ kN, $F_{DB} = 2.60$ kN

6–110. 1.75 ft $\leq x \leq 17.4$ ft

6–111. $F_{EF} = 8.18$ kN (T), $F_{AD} = 158$ kN (C)

6–113. **(a)** $F = 175$ lb, $N_C = 350$ lb
(b) $F = 87.5$ lb, $N_C = 87.5$ lb

6–114. **(a)** $F = 205$ lb, $N_C = 380$ lb
(b) $F = 102$ lb, $N_C = 72.5$ lb

6–115. $M = 14.2$ lb · ft

6–117. $F_S = 286$ N

6–118. $x = 4.38$ in.

6–119. $\theta = 33.6°$

6–121. $T_{AI} = T_{EF} = 2.88$ kip, $F_H = F_G = 3.99$ kip

6–122. $F_{CA} = 12.9$ kip, $F_{AB} = 11.9$ kip, $F_{AD} = 2.39$ kip

6–123. $W_1 = 3$ lb, $W_2 = 21$ lb, $W_3 = 75$ lb

6–125. $P = 471$ N, $B_x = D_x = 283$ N, $B_y = D_y = 283$ N, $B_z = D_z = 0$

6–126. $F_{AB} = 1.56$ kN, $M_{Ex} = 0.5$ kN · m, $M_{Ey} = 0$, $E_y = 0$, $E_x = 0$

6–127. $A_z = 0$, $A_x = 172$ N, $A_y = 115$ N, $C_x = 47.3$ N, $C_y = 61.9$ N, $C_z = 125$ N, $M_{Cy} = -429$ N · m, $M_{Cz} = 0$

6–129. $\theta = 16.1°$

6–130. Joint D: $F_{CD} = 0$, $F_{ED} = 0$
Joint E: $F_{FE} = 0$, $F_{EC} = 15$ kN (T)
Joint C: $F_{CF} = 21.2$ kN (C), $F_{CB} = 15$ kN (T)
Joint F: $F_{AF} = 15$ kN (C), $F_{FB} = 25$ kN (T)
Joint B: $F_{AB} = 28.0$ kN (C), $F_{BG} = 27.5$ kN (T)

6–131. $\theta = 21.7°$

6–133. $B_x = B_y = 220$ N, $A_x = 300$ N, $A_y = 80.4$ N

6–134. $A_x = 117$ N, $A_y = 397$ N, $B_x = 97.4$ N, $B_y = 97.4$ N

6–135. Joint B: $F_{BC} = 373$ lb (C), $F_{BA} = 1.18$ kip (C)
Joint A: $F_{AC} = 833$ lb (T)

6–137. $F_{AD} = 2.47$ kip (T), $F_{AC} = F_{AB} = 1.22$ kip (C)

Chapter 7

7–1. $V_A = 0$, $N_A = 12.0$ kN, $M_A = 0$, $V_B = 0$, $N_B = 20.0$ kN, $M_B = 1.20$ kN · m

7–2. $N_A = 10$ lb, $N_B = 0$

7–3. $T_C = 0$, $T_D = 400$ N · m, $T_E = 550$ N · m

7–5. $M_C = -15.0$ kip · ft, $N_C = 0$, $V_C = 2.01$ kip, $M_D = 3.77$ kip · ft, $N_D = 0$, $V_D = 1.11$ kip

7–6. $N_C = 0$, $V_C = 25$ lb, $M_C = 100$ lb · ft

7–7. $N_C = 21.7$ kN, $V_C = 62.5$ kN, $M_C = -225$ kN · m

7–9. $w = 100$ N/m

7–10. $M_C = 48$ kip · ft, $V_C = 6$ kip

7–11. $N_E = 0$, $V_E = -50$ N, $M_E = -100$ N · m, $N_D = 0$, $V_D = 750$ N, $M_D = -1.30$ kN · m

7–13. $N_C = -4.32$ kip, $V_C = 1.35$ kip, $M_C = 4.72$ kip · ft

7–14. $N_D = 1.92$ kN, $V_D = 100$ N, $M_D = 900$ N · m

7–15. $N_C = 2.20$ kip, $V_C = 0.336$ kip, $M_C = 1.76$ kip · ft

7–17. $N_C = -406$ lb, $V_C = 903$ lb, $M_C = 1.35$ kip · ft

7–18. $N_D = -464$ lb, $V_D = -203$ lb, $M_D = 2.61$ kip · ft

7–19. $N_C = -30$ kN, $V_C = -8$ kN, $M_C = 6$ kN · m

7–21. $N_C = 0$, $V_C = 0$, $M_C = 54$ kip · ft

7–22. $N_D = 0$, $V_D = 5$ kip, $M_D = 46$ kip · ft

7–23. $M_C = -15.0$ kip · ft, $N_C = 0$, $V_C = 2.01$ kip, $M_D = 3.77$ kip · ft, $N_D = 0$, $V_D = 1.11$ kip

7–25. $M_C = -17.8$ kip · ft

7–26. $N_D = 0$, $V_D = 0.75$ kip, $M_D = 13.5$ kip · ft, $N_E = 0$, $V_E = -9$ kip, $M_E = -24.0$ kip · ft

7–27. $N_D = 2.40$ kN, $V_D = 50$ N, $M_D = 1.35$ kN $\cdot$ m

7–29. $N_C = -80$ lb, $V_C = 0$, $M_C = -480$ lb $\cdot$ in.

7–30. $N_B = 59.8$ lb, $V_B = -496$ lb, $M_B = -480$ lb $\cdot$ ft,
$N_C = -495$ lb, $V_C = 70.7$ lb, $M_C = -1.59$ kip $\cdot$ ft

7–31. $N_A = 86.6$ lb, $V_A = 150$ lb, $M_A = 1.80$ kip $\cdot$ in.

7–33. $N_A = 14.1$ lb, $V_A = 14.1$ lb, $M_A = 28.3$ lb $\cdot$ in.

7–34. $N_B = 59.8$ lb, $V_B = -496$ lb, $M_B = -480$ lb $\cdot$ ft,
$N_C = -495$ lb, $V_C = 70.7$ lb, $M_C = -1.59$ kip $\cdot$ ft

7–35. $\dfrac{a}{b} = \dfrac{1}{4}$

7–37. $N = -0.866rw_0$, $V = -1.5rw_0$, $M = 1.23r^2w_0$

7–38. $C_x = -150$ lb, $C_y = -350$ lb, $C_z = 700$ lb,
$M_{Cx} = 1.40$ kip $\cdot$ ft, $M_{Cy} = -1.20$ kip $\cdot$ ft,
$M_{Cz} = -750$ lb $\cdot$ ft

7–39. $C_x = -170$ lb, $C_y = -50$ lb, $C_z = 500$ lb,
$M_{Cx} = 1$ kip $\cdot$ ft, $M_{Cy} = -900$ lb $\cdot$ ft,
$M_{Cz} = -260$ lb $\cdot$ ft

7–41. $(N_E)_x = 0$, $(V_E)_y = 53.6$ kN, $(V_E)_z = -87.0$ kN,
$(M_E)_x = 0$, $(M_E)_y = -43.5$ kN $\cdot$ m,
$(M_E)_z = -26.8$ kN $\cdot$ m

7–42. $0 \le x \le a$: $V = \left(1 - \dfrac{a}{L}\right)P$, $M = \left(1 - \dfrac{a}{L}\right)Px$,

$a < x < L$: $V = -\left(\dfrac{a}{L}\right)P$, $M = P\left(a - \dfrac{a}{L}x\right)$

7–43. **(a)** For $0 \le x < a$: $V = P$, $M = Px$
For $a < x < L - a$: $V = 0$, $M = Pa$
For $L - a < x \le L$: $V = -P$, $M = P(L - x)$
(b) For $0 \le x < 5$ ft: $V = 800$ lb, $M = 800x$ lb $\cdot$ ft
For 5 ft $< x < 7$ ft: $V = 0$, $M = 4000$ lb $\cdot$ ft
For 7 ft $< x \le 12$ ft: $V = -800$ lb,
$M = (9600 - 800x)$ lb $\cdot$ ft

7–45. $0 \le x < \dfrac{L}{3}$: $V = 0$, $M = 0$,

$\dfrac{L}{3} < x < \dfrac{2L}{3}$: $V = 0$, $M = M_0$,

$\dfrac{2L}{3} < x \le L$: $V = 0$, $M = 0$, $M_{\max} = 2$ kN $\cdot$ m

7–46. **(a)** $V = \dfrac{w}{2}(L - 2x)$, $M = \dfrac{w}{2}(Lx - x^2)$
(b) $V = (2500 - 500x)$ lb,
$M = (2500x - 250x^2)$ lb $\cdot$ ft

7–47. $w = 400$ lb/ft

7–49. $V = 28 - 3x$, $M = 28x - 1.5x^2 - 114$

7–50. For $0 \le x \le 2$ m: $V = 0.75$ kN, $M = 0.75x$ kN $\cdot$ m
For 2 m $< x < 4$ m: $V = 3.75 - 1.5x$ kN,
$M = -0.75x^2 + 3.75x - 3$ kN $\cdot$ m

7–51. $V = 250(10 - x)$, $M = 25(100x - 5x^2 - 6)$

7–53. For $0 \le x < a$: $V = -wx$, $M = -\dfrac{w}{2}x^2$

For $a < x \le 2a$: $V = w(2a - x)$,
$M = 2wax - 2wa^2 - \dfrac{w}{2}x^2$

7–54. $V = \dfrac{w}{4}(3L - 4x)$, $M = \dfrac{w}{4}(3Lx - 2x^2 - L^2)$

7–55. $N = -25x$, $V = -43.3x$, $M = 25x - 21.7x^2$

7–57. $w = 22.2$ lb/ft

7–58. $x = \dfrac{L}{2}$, $P = \dfrac{4M_{\max}}{L}$

7–59. For $0 \le x < 9$ ft: $V = 25 - 1.667x^2$,
$M = 25x - 0.5556x^3$,
For 9 ft $< x < 13.5$ ft: $V = 0$, $M = -180$

7–61. $V = 0$, $M = -1.25x^3 - 15x^2 + 360x$,
$M_{\max} = 1.36$ kip $\cdot$ ft

7–62. $x = 15^-$, $V = -20$, $M = -300$,
$x = 30^+$, $V = 0$, $M = 150$,
$x = 45^-$, $V = -60$, $M = -300$

7–63. $V_x = 1.5$ kip, $V_y = 0$, $V_z = 800(4 - y)$ lb,
$M_x = 400(4 - y)^2$ lb $\cdot$ ft, $M_y = -3$ kip $\cdot$ ft,
$M_z = -1500(4 - y)$ lb $\cdot$ ft

7–65. $V = |P|$, $M = |Pr \cos \theta|$, $T = |Pr(1 - \sin \theta)|$

7–73. $x = 2^-$, $V = 8.5$, $M = 17$,
$x = 6^+$, $V = -11.5$, $M = 23$

7–74. $x = 1^+$, $V = 175$, $M = 2.75$,
$x = 6$, $V = 0$, $M = -300$

7–75. $x = 2^+$, $V = -14.3$, $M = -8.6$

7–77. $x = 20^+$, $V = -104$, $M = 2080$

7–78. $w = 2$ kip/ft

7–79. $x = 8^+$, $V = 317$, $M = -1267$

7–81. $x = 8^-$, $V = 1.10$, $M = -1.60$

7–82. $x = 9^+$, $V = -1.375$, $M = 25.9$

7–83. $x = 3$, $V = 0$, $M = 9$

7–85. $x = \dfrac{L}{2}$, $V = \dfrac{wL}{4}$, $M = -\dfrac{wL^2}{24}$

7–86. $x = 3$, $V = 3$, $M = -6$

7–87. $x = 11$, $V = 0$, $M = 17.5$

7–89. $y_B = 3.53$ m, $P = 0.8$ kN, $T_{max} = 8.17$ kN
7–90. $y_B = 2.43$ m, $T_{max} = 157$ N
7–91. $y_B = 8.67$ ft, $y_D = 7.04$ ft
7–93. $x_B = 4.36$ ft
7–94. $P = 71.4$ lb
7–95. $P_1 = 2.50$ kN, $P_2 = 6.25$ kN, $F_{max} = 12.5$ kN
7–97. $w = 51.9$ lb/ft
7–98. $T_{max} = 14.4$ kip, $T_{min} = 13.0$ kip
7–99. $y = 0.741(10^{-3})x^3$, $T_{max} = 4.19$ kip
7–101. $y = 2.37(10^{-3})x^3$, $T_{max} = 4.42$ kip
7–102. $y = \dfrac{x^2}{7813}\left(75 - \dfrac{x^2}{200}\right)$, $T_{max} = 9.28$ kip
7–103. $h = 7.09$ m
7–105. $h = 146$ ft
7–107. $\dfrac{h}{L} = 0.141$
7–109. $(T_{max})_B = 2.73$ kip, $(T_{max})_C = 2.99$ kip
7–110. $T_{max} = 1.30$ MN
7–111. $h = 18.5$ m, $T_{max} = 492$ N
7–113. 4.00 kip, $T_{max} = 2.01$ kip
7–114. $F_H = 11.1$ lb, $h = 23.5$ ft
7–115. $F_H = 12.9$ lb
7–117. $a = 0.366L$
7–118. $N_D = -800$ N, $V_D = 0$, $M_D = 1.20$ kN·m
7–119. $N_D = -6.08$ kN, $V_D = -2.60$ kN, $M_D = -13.0$ kN·m
7–121. $N_C = 0$, $V_C = 9.00$ kN, $M_C = -62.5$ kN·m, $N_B = 0$, $V_B = 27.5$ kN, $M_B = -184.5$ kN·m
7–122. $T_{max} = 76.7$ lb

Chapter 8

8–1. $F_C = 27.4$ lb, $N_C = 309$ lb
8–2. $P = 140$ N, distance from A is 500 mm
8–3. $F_C = 30.5$ N, $N_C = 152$ N, $x = 0.794$ m
8–5. $P = 1.14$ kN
8–6. $F_A = 444$ N, $N_A = 1.47$ kN, $N_B = 1.24$ kN, no slipping at A
8–7. $P = 981$ N
8–9. $F_A = 604$ N, no slipping
8–10. $P = \dfrac{M_0}{\mu_s r a}(b - \mu_s c)$
8–13. (a) $P = 30$ N < 39.8 N, No
(b) $P = 70$ N > 39.8 N, Yes

8–14. (a) $P = 30$ N < 34.26 N, No
(b) $P = 70$ N > 34.26 N, Yes
8–15. $s = 5$ in.
8–17. $\mu = 0.268$
8–18. No, hoop slips first
8–19. $m = 54.9$ kg
8–21. $P = 0.875$ lb
8–22. $F = 22.5$ lb, $\mu_m = 0.15$
8–23. $F = 30.4$ lb, $\mu_m = 0.195$
8–25. $d = 2.70$ in.
8–26. $\theta = 16.7°$, $\phi = 42.6°$
8–27. $P = 110$ lb, $L = 6.25$ ft
8–29. $d = 3$ ft
8–30. Slipping occurs at A
8–31. $\theta = 21.8°$
8–33. $\theta = 120°$, $P = 0.0946$ lb
8–34. $L = 3.35$ ft
8–35. $P = 100$ lb, $d = 1.50$ ft
8–37. Man cannot cause motion
8–38. $P = 80$ lb, the distance from A is 0.4 ft
8–39. $P = 80$ lb, $h = 5$ ft
8–41. $\mu = 0.556$
8–42. $\mu = 0.176$
8–43. Yes, the man can push the crate
8–45. $P = 13.3$ lb
8–46. $M = 12$ lb·ft
8–47. $P = 355$ N
8–50. $\phi = \theta$, $P = W\sin(\alpha + \phi)$
8–51. $P = 23.9$ N
8–53. Board will slip, $d = 6.47$ ft < 10 ft
8–54. $P = 1.17$ kN
8–55. For pipe A: $\mu_s \geq \dfrac{\sin\theta}{(\cos\theta + 1)}$

For pipe C: $\mu_s{}' \geq \dfrac{1}{3}\dfrac{\sin\theta}{(\cos\theta + 1)}$
8–57. The saw horse will start to slip
8–58. The saw horse will start to slip
8–59. $M = 77.3$ N·m
8–61. $\theta = 20.4°$
8–62. $P = 50.3$ lb
8–63. $P = 4.83$ kN
8–65. $P = 34.5$ N
8–66. $P = 304$ N
8–67. $x = 32.9$ mm

8–69. $P = 3.29$ kN

8–70. $P = F\left(\dfrac{(1 - \mu_s^2)\tan\alpha + 2\mu_s}{1 - 2\mu_s\tan\alpha - \mu_s^2}\right)$

8–71. $W = 7.19$ kN

8–73. $F = 961$ lb, $P = 18.3$ lb

8–74. $F = 4.91$ N

8–75. $P = 2.85$ kN

8–77. $P = 1.98$ kN

8–78. $M = 0.202$ N $\cdot$ m

8–79. $F = 58.7$ N

8–81. 72.7 N

8–82. $A_x = 329$ N, $B_y = C_y = 164$ N

8–83. $M = 32.0$ N $\cdot$ m

8–85. 22.9 N $\le P \le 151$ N

8–86. Approx. 2 turns (695°)

8–87. $P = 53.6$ N

8–89. $m = 156$ kg

8–90. $x = 0.384$ m

8–91. $P = 42.3$ N

8–93. $W_D = 12.7$ lb

8–94. $F_C = 13.7$ lb, $F_B = 38.5$ lb

8–97. $m_A = 2.22$ kg

8–98. $P = 17.1$ lb

8–99. $d = 4.60$ ft

8–101. $h = 8.28$ ft

8–102. $M = 75.4$ N $\cdot$ m, $V = 0.171$ m^3

8–103. $F = 2.49$ kN

8–105. $m = 25.6$ kg

8–106. $m = 7.82$ kg

8–107. $M = 304$ lb $\cdot$ in.

8–109. $M = 16.1$ N $\cdot$ m

8–110. $M = 36.3$ lb $\cdot$ ft

8–111. $F = 10.7$ lb

8–113. $T = 905$ lb $\cdot$ in.

8–114. $M = \dfrac{1}{2}\mu PR$

8–115. $M = \dfrac{2\mu_s PR}{3\cos\theta}$

8–117. $M = \dfrac{1}{2}\mu_s P(R_2 + R_1)$

8–118. $P = 0.589$ lb, by approximation, $P = 0.596$ lb

8–119. $\mu = 0.215$, by approximation, $\mu = 0.211$, $F = 6$ lb

8–121. $P = 826$ N

8–122. $P = 814$ N

8–123. $P = 62.4$ lb, by approximation, $P = 62.5$ lb

8–125. 13.8 lb

8–126. 29.0 lb

8–127. $(r_f)_A = 0.2$ in., $(r_f)_B = 0.075$ in.

8–129. $P = 96.7$ N

8–130. $P = 25.3$ lb

8–133. $d = 38.5$ mm

8–134. $s = 0.750$ m

8–135. (a) $W = 6.97$ kN, (b) $W = 15.3$ kN

8–137. $\theta = 35.0°$

8–138. $\mu_A = 0.415$

8–139. $P' = 4.96$ kip, $P = 8.16$ kip

Chapter 9

9–1. $\bar{x} = 0$, $\bar{y} = 0.912$ m

9–2. $\bar{x} = 0.620$ ft, $A_y = 0.442$ lb, $B_x = 0$, $B_y = 0.720$ lb

9–3. $\bar{x} = 124$ mm, $\bar{y} = 0$

9–5. $\bar{x} = 0.531$ ft, $O_x = 0$, $O_y = 0.574$ lb, $M_o = 0.305$ lb $\cdot$ ft

9–6. $\bar{y} = 0.183$ ft

9–7. $\bar{x} = \dfrac{3}{8}b$, $\bar{y} = \dfrac{3}{5}h$

9–9. $\bar{x} = \dfrac{3}{4}b$, $\bar{y} = \dfrac{3}{10}h$

9–10. $\bar{x} = \dfrac{2}{3}a$, $\bar{y} = \dfrac{m}{3}a$

9–11. $\bar{x} = \bar{y} = 1.70$ ft, $T_B = 26.7$ lb, $T_A = 36.2$ lb

9–13. $\bar{x} = -3$ m, $\bar{y} = 0.25$ m, $\bar{z} = -0.3$ m, $A_x = A_y = 0$, $A_z = 52.3$ kN, $M_A = 52.3$ kN $\cdot$ m

9–14. $\bar{x} = -\dfrac{3}{4}a$, $\bar{y} = -\dfrac{3}{10}b$

9–15. $\bar{x} = \dfrac{n + 1}{2(n + 2)}a$, $\bar{y} = \dfrac{n}{2n + 1}h$

9–17. $\bar{y} = \dfrac{4b}{3\pi}$, $\bar{x} = \dfrac{4a}{3\pi}$

9–18. $\bar{x} = \dfrac{1}{3}(b + a)$

9–19. $\bar{y} = 2.80$ m, $\bar{x} = 3.00$ m

9–21. $\bar{y} = 2.04$ ft

9–22. $\bar{x} = 1.26$ m, $\bar{y} = 0.143$ m, $N_B = 47.9$ kN, $A_x = 33.9$ kN, $A_y = 73.9$ kN

9–23. $\bar{x} = 2$ ft

9–25. $\bar{x} = 1.80$ m

9–26. $\bar{y} = 1.80$ m

9–27. $\bar{x} = 2.66$ in.

9–29. $\bar{x} = 0.910$ in.

9–30. $\bar{y} = 3.64$ in.

9–31. $\bar{r} = 0.833a$

9–33. $\bar{z} = \dfrac{2}{3}h$

9–34. $\bar{z} = 1.33$ ft

9–35. $\bar{x} = \bar{y} = 0, \bar{z} = \dfrac{5}{6}h$

9–37. $\bar{x} = 0.4a$

9–38. $\bar{z} = 2.50$ ft

9–39. $\bar{z} = 0.422$ m

9–41. $\bar{x} = 0, \bar{y} = \dfrac{3}{4}h, \bar{z} = \dfrac{a}{\pi}$

9–42. $\bar{z} = 0.675a$

9–43. $\bar{z} = \dfrac{c}{4}$

9–45. $\bar{x} = 1.29$ ft, $\bar{y} = 1.57$ ft, $\bar{z} = 0.429$ ft

9–46. $\bar{x} = 24.4$ mm, $\bar{y} = 40.6$ mm

9–47. $\bar{x} = 179$ mm

9–49. $\bar{x} = 112$ mm, $\bar{y} = 112$ mm, $\bar{z} = 136$ mm

9–50. $\bar{x} = 6.5$ in., $\bar{y} = 4$ in., $\theta = 10.6°$

9–51. $\bar{x} = 1.60$ ft, $\bar{y} = 7.04$ ft, $A_x = 0$,
$A_y = 149$ lb, $M_A = 502$ lb·ft

9–53. $\bar{y} = 85.9$ mm

9–54. $\bar{x} = 2.22$ m, $\bar{y} = 1.41$ m

9–55. $\bar{x} = 1.95$ in., $\bar{y} = 0.904$ in.

9–57. $\bar{y} = 154$ mm

9–58. $\bar{x} = 4.62$ in., $\bar{y} = 1.00$ in.

9–59. $\bar{y} = 2$ in.

9–61. $\bar{y} = 229$ mm

9–62. $\bar{x} = \dfrac{\frac{2}{3}r\sin^3\alpha}{\alpha - \frac{\sin 2\alpha}{2}}$

9–63. $\bar{y} = 56.6$ mm

9–65. 16.4 kg, $\bar{x} = 153$ mm, $\bar{y} = -15$ mm, $\bar{z} = 111$ mm

9–66. $\bar{y} = 135$ mm

9–67. $\bar{x} = \dfrac{Pl_1 - W_1\left(l_1 - \frac{l}{2}\right)}{W}$

9–69. $h = 323$ mm

9–70. $\bar{z} = 128$ mm

9–71. $\bar{x} = -1.14$ in., $\bar{y} = 1.71$ in., $\bar{z} = -0.857$ in.

9–73. $x = 490$ mm

9–74. $\bar{x} = 1.59$ m, $\bar{y} = 0.940$ m

9–75. $\bar{z} = \dfrac{1}{2}a$

9–77. $\bar{z} = 0.70$ ft

9–78. $h = 2.00$ ft

9–79. $\bar{z} = 4.32$ in.

9–81. $\bar{z} = 1.625$ in.

9–82. $\theta = 70.4°, l = 265$ mm

9–83. $\bar{x} = 8.22$ in.

9–85. $A = 118$ in^2

9–86. $A = 3.33$ ft^2, $\bar{y} = 1.20$ ft, $V = 25.1$ ft^3

9–87. $A = 11.9(10^3)$ ft^2, $V = 146(10^3)$ ft^3

9–89. $V = 3.49$ m^3

9–90. $V = 4.25(10^6)$ mm^3

9–91. $W = 3.01$ lb

9–93. $V = 1.40(10^3)$ in^3

9–94. $W = 3.12(10^6)$ lb

9–95. $A = 302$ m^2

9–97. $A = 5.65(10^3)$ ft^2

9–98. $V = 27.2(10^3)$ ft^3

9–99. $V = 207$ m^3, $A = 188$ m^2

9–101. $A = 21.3$ in^2, $\bar{x} = 6.0$ in., $V = 1.34(10^3)$ in^3

9–102. $A = 1.39$ ft^2, $\bar{y} = 0.541$ ft, $V = 4.71$ ft^3

9–103. 1365 m^3

9–105. $A = 119(10^3)$ mm^2

9–106. $F = 1.41$ MN, $h = 4$ m

9–107. For water: $F_{R_A} = 157$ kN, $F_{R_B} = 235$ kN
For oil: $d = 4.22$ m

9–109. $d = 2.68$ m

9–110. $F_{AB} = 486$ kip

9–111. $d = 3.65$ m

9–113. $F_B = 29.4$ kN, $F_A = 235$ kN

9–114. $A_y = 2.51$ MN, $B_x = 2.20$ MN, $B_y = 859$ kN

9–115. $F_R = 17.2$ kip, $d = 5.22$ ft
At bottom: $F_R = 18.8$ kip

9–117. $F_R = 40.0$ kip, $\bar{z} = 8.00$ ft

9–118. $F_{R_v} = 196$ lb, $F_{R_h} = 125$ lb

9–119. $F_R = 360$ lb, $\bar{x} = 0.667$ ft, $\bar{y} = 4$ ft

9–121. $F_R = 984$ lb, $\bar{y} = -1.77$ ft, $\bar{x} = 0$

9–122. $\bar{x} = 0, \bar{y} = 2.40$ m, $F_R = 42.7$ kN,
$B_y = C_y = 12.8$ kN, $A_y = 17.1$ kN

9–123. $F_R = 6.93$ kN, $\bar{y} = -0.125$ m

9–125. $A = 1.25$ m^2

9–126. $\bar{x} = 154$ mm, $\bar{y} = 172$ mm, $\bar{z} = 50$ mm

9–127. $\bar{x} = \bar{y} = 0, \bar{z} = \dfrac{2}{3}a$

9–129. $\bar{x} = 3.52$ ft, $\bar{y} = 4.09$ ft, $W = 42.8$ kip

9–130. $V = 20.5$ m^3

9–131. $\bar{x} = 1.60$ ft, $\bar{y} = 4.15$ ft

9–133. $F_R = 24.0$ kN, $\bar{x} = 2.00$ m, $\bar{y} = 1.33$ m

9–134. $F_R = 7.62$ kN, $\bar{x} = 2.74$ m, $\bar{y} = 3.00$ m

Chapter 10

10–1. $I_x = 39.0 \text{ m}^4$

10–2. $I_y = 8.53 \text{ m}^4$

10–3. $I_x = \dfrac{1}{3}(t)\, l^3 \sin^2 \theta$

10–5. $I_y = 21.3 \text{ in}^4$

10–6. $I_x = \dfrac{2}{15} bh^3$

10–7. $I_x = \dfrac{ab^3}{3(1 + 3n)}$

10–9. $I_x = 4.27 \text{ in}^4$

10–10. $I_y = 36.6 \text{ in}^4$

10–11. $I_x = 10.7 \text{ in}^4$

10–13. $I_y = 2.44 \text{ m}^4$

10–14. $I_x = 34.1 \text{ in}^4$

10–15. $I_y = 73.1 \text{ in}^4$

10–17. $I_y = 7.72 \text{ in}^4$

10–18. $I_x = 9.05 \text{ in}^4$

10–19. $I_y = 30.9 \text{ in}^4$

10–21. $I_y = 192 \text{ in}^4$

10–22. $I_x = 3.20 \text{ m}^4$

10–23. $I_y = 0.628 \text{ m}^4$

10–25. $A = 5.00 \text{ in}^2$

10–26. $A = 11.1(10^3) \text{ mm}^2$

10–27. $k_x = 74.7 \text{ mm}$

10–29. $162(10^6) \text{ mm}^4$

10–30. $\bar{y} = 2.00 \text{ in.}, \bar{I}_{x'} = 64.0 \text{ in}^4$

10–31. $\bar{x} = 3.00 \text{ in.}, \bar{I}_{y'} = 136 \text{ in}^4$

10–33. $49.5(10^6) \text{ mm}^4$

10–34. $I_x = 1.192(10^3), I_y = 364.8 \text{ in}^4$

10–35. $\bar{y} = 15.7 \text{ in.}, \bar{I}_{x'} = 2.16(10^3) \text{ in}^4$

10–37. $\bar{y} = 80.7 \text{ mm}, \bar{I}_{x'} = 67.6(10^6) \text{ mm}^4$

10–38. $\bar{x} = 61.6 \text{ mm}, \bar{I}_{y'} = 41.2(10^6) \text{ mm}^4$

10–39. $\bar{y} = 48.25 \text{ mm}, \bar{I}_{x'} = 15.1(10^6) \text{ mm}^4$

10–41. $\bar{y} = 37.5 \text{ mm}, I_{x'} = 16.3(10)^6 \text{ mm}^4$

10–42. $I_y = 94.8(10^6) \text{ mm}^4$

10–43. $I_x = 648 \text{ in}^4$

10–45. $\bar{y} = 2 \text{ in.}, \bar{I}_{x'} = 128 \text{ in}^4$

10–46. $I_x = I_y = 503 \text{ in}^4$

10–47. $\bar{I}_{x'} = \dfrac{1}{12} a^3 b \sin^3 \theta$

10–49. $\bar{I}_{x'} = \dfrac{1}{36} bh^3, \bar{I}_{y'} = \dfrac{1}{36} hb(b^2 - ab + a^2)$

10–50. $\bar{I}_{x'} = 520(10^6) \text{ mm}^4$

10–51. $I_x = 154 \text{ in}^4$

10–53. $k_x = 109 \text{ mm}$

10–54. $I_{xy} = 0$

10–55. $I_{xy} = \dfrac{3}{16} b^2 h^2$

10–57. $I_{xy} = \dfrac{1}{6} a^2 b^2$

10–58. $I_{xy} = 48 \text{ in}^4$

10–59. $I_{xy} = 10.7 \text{ in}^4$

10–61. $I_{xy} = \dfrac{3}{16} b^2 h^2$

10–62. $I_{xy} = 4 \text{ in}^4$

10–63. $I_{xy} = \dfrac{a^2 b^2}{4(n + 1)}$

10–65. $I_{xy} = 0.511 \text{ m}^4$

10–66. $I_{xy} = 0$

10–67. $I_{xy} = 4.80(10^6) \text{ mm}^4$

10–69. $\bar{x} = \bar{y} = 44.1 \text{ mm}, I_{x'y'} = -6.26(10^6) \text{ mm}^4$

10–70. $I_{xy} = -28.1(10^3) \text{ mm}^4$

10–71. $I_{xy} = 119 \text{ in}^4$

10–73. $I_{xy} = 72 \text{ in}^4$

10–74. $I_{uv} = 135(10)^6 \text{ mm}^4$

10–75. $I_u = 10.5(10^6) \text{ mm}^4, I_v = 4.05(10^6) \text{ mm}^4,$
$I_{uv} = 5.54(10^6) \text{ mm}^4$

10–77. $I_u = 15.75 \text{ in}^4, I_u = 25.75 \text{ in}^4$

10–78. $\theta = -22.5°, I_{max} = 250 \text{ in}^4, I_{min} = 20.4 \text{ in}^4$

10–79. $I_u = 3.47(10^3) \text{ in}^4, I_v = 3.47(10^3) \text{ in}^4,$
$I_{uv} = 2.05(10^3) \text{ in}^4$

10–81. $I_{max} = 64.1 \text{ in}^4, I_{min} = 5.33 \text{ in}^4$

10–82. $I_{max} = 4.92(10^6) \text{ mm}^4, I_{min} = 1.36(10^6) \text{ mm}^4$

10–83. $\theta = 6.08°, I_{max} = 1.74(10^3) \text{ in}^4,$
$I_{min} = 435 \text{ in}^4$

10–85. $I_{max} = 250 \text{ in}^4, I_{min} = 20.4 \text{ in}^4$

10–86. $I_{max} = 64.1 \text{ in}^4, I_{min} = 5.33 \text{ in}^4$

10–87. $I_{max} = 4.92(10^6) \text{ mm}^4, I_{min} = 1.36(10^6) \text{ mm}^4$

10–89. $I_{max} = 1.74(10^3) \text{ in}^4, I_{min} = 435 \text{ in}^4$

10–90. $I_x = \dfrac{3}{10} mr^2$

10–91. $I_z = mR^2$

10–93. $I_x = \frac{2}{5}mr^2$

10–94. $k_x = 57.7$ mm

10–95. $I_x = \frac{2}{5}mb^2$

10–97. $I_x = \frac{2}{5}mb^2$

10–98. $I_z = \frac{m}{10}a^2$

10–99. $I_y = 2.25$ slug·ft²

10–101. $I_z = 1.53$ kg·m²

10–102. $I_G = 118$ slug·ft²

10–103. $I_O = 282$ slug·ft²

10–105. $I_A = 2.17$ slug·ft²

10–106. $I_O = \frac{1}{2}ma^2$

10–107. $I_A = 1.58$ slug·ft²

10–109. $I_x = 3.25(10^{-3})$ kg·m²

10–110. $I_{x'} = 7.20(10^{-3})$ kg·m²

10–111. $I_x = 5.64$ slug·ft²

10–113. $I_x = 307$ in⁴

10–114. $I_y = 10.7$ in⁴

10–115. $I_x = \frac{93}{70}mb^2$

10–117. $I_u = 5.09(10^6)$ mm⁴, $I_v = 5.09(10^6)$ mm⁴, $I_{uv} = 0$

10–118. $I_x = 333$ in⁴

10–119. $I_y = 32.4$ in⁴

10–121. $I_x = \frac{bh^3}{12}, \bar{I}_{x'} = \frac{bh^3}{36}$

10–122. $I_{xy} = 0.667$ in⁴

Chapter 11

11–1. $P = \frac{W}{2} \operatorname{ctn}\theta$

11–2. $\theta = 42.4°$

11–3. $P = 2W \operatorname{ctn}\theta$

11–5. $k = 1.05$ kN/m

11–6. $F = 512$ N

11–7. $F = \dfrac{500\sqrt{0.04\cos^2\theta + 0.6}}{(0.2\cos\theta + \sqrt{0.04\cos^2\theta + 0.6})\sin\theta}$

11–9. $F = P \csc\theta$

11–10. $P = 3k \operatorname{ctn}\theta\,(2l\sin\theta - l_0)$

11–11. $M = 52.0$ lb·ft

11–13. $P = 13.8$ N

11–14. $M = 17.1$ lb·ft

11–15. $\theta = 78.5°$

11–17. $\theta = 10.7°, \theta = 89.3°$

11–18. $P = \dfrac{ka(a\sin\theta - l_0)}{l}$

11–19. $F = 2P \operatorname{ctn}\theta$

11–22. $F = 312$ lb

11–23. $F = \dfrac{M}{2a\sin\theta}$

11–25. $m = 100$ kg

11–26. **Stable** at $(0, 0)$

11–27. At $x = 0.590$ ft: **Stable** At $x = -0.424$ ft: **Unstable**

11–29. $x = 0, y = 0$: **Stable** at $(0, 0)$

11–30. $\theta = 0°$ or $\theta = \cos^{-1}\left(\dfrac{W}{2kL}\right), W = 2kL$

11–31. $k = 2.81$ lb/ft

11–33. At $\theta = 36.4°$: **Unstable**

11–34. $k = 100$ lb/ft

11–35. At $\theta = 90°$: **Stable** At $\theta = 36.1°$: **Unstable**

11–37. At $\theta = 0°$: **Stable**

11–38. $\theta = \sin^{-1}\left(\dfrac{4W}{ka}\right), \theta = 90°$

11–39. $W_2 = W_1\left(\dfrac{b}{a}\right)\dfrac{\sin\theta}{\cos\phi}$

11–41. $d = 87.9$ mm

11–42. $h_{cr} = 1.32$ ft If $h > h_{cr}$ then stable

11–43. $\phi = 17.4°, \theta = 9.18°$

11–45. $r = \dfrac{h}{4}$

11–46. $\theta = 16.3°$

11–47. $h = 3.99$ in.

11–49. $P = 39.2$ N

11–50. $\theta = 90°, \theta = 30°$

11–51. $\theta = 90°$: **Unstable**, $\theta = 30°$: **Stable**

Index

DYNAMICS

Although each of these planes is rather large, from a distance their motion can be modeled as if each plane were a particle.

12 Kinematics of a Particle

CHAPTER OBJECTIVES

- To introduce the concepts of position, displacement, velocity, and acceleration.
- To study particle motion along a straight line and represent this motion graphically.
- To investigate particle motion along a curved path using different coordinate systems.
- To present an analysis of dependent motion of two particles.
- To examine the principles of relative motion of two particles using translating axes.

12.1 Introduction

Mechanics is a branch of the physical sciences that is concerned with the state of rest or motion of bodies subjected to the action of forces. The mechanics of rigid bodies is divided into two areas: statics and dynamics. *Statics* is concerned with the equilibrium of a body that is either at rest or moves with constant velocity. Here we are concerned with *dynamics*, which deal with the accelerated motion of a body. The subject of dynamics will be presented in two parts: *kinematics*, which treats only the geometric aspects of the motion, and *kinetics*, which is the analysis of the forces causing the motion. To develop these principles, the dynamics of a particle will be discussed first, followed by topics in rigid-body dynamics in two and then three dimensions.

Historically, the principles of dynamics developed when it was possible to make an accurate measurement of time. Galileo Galilei (1564–1642) was one of the first major contributors to this field. His work consisted of experiments using pendulums and falling bodies. The most significant contributions in dynamics, however, were made by Isaac Newton (1642–1727), who is noted for his formulation of the three fundamental laws of motion and the law of universal gravitational attraction. Shortly after these laws were postulated, important techniques for their application were developed by Euler, D'Alembert, Lagrange, and others.

There are many problems in engineering whose solutions require application of the principles of dynamics. Typically the structural design of any vehicle, such as an automobile or airplane, requires consideration of the motion to which it is subjected. This is also true for many mechanical devices, such as motors, pumps, movable tools, industrial manipulators, and machinery. Furthermore, predictions of the motions of artificial satellites, projectiles, and spacecraft are based on the theory of dynamics. With further advances in technology, there will be an even greater need for knowing how to apply the principles of this subject.

Problem Solving.   Dynamics is considered to be more involved than statics since both the forces applied to a body and its motion must be taken into account. Also, many applications require using calculus, rather than just algebra and trigonometry. In any case, the most effective way of learning the principles of dynamics is *to solve problems*. To be successful at this, it is necessary to present the work in a logical and orderly manner as suggested by the following sequence of steps:

1. Read the problem carefully and try to correlate the actual physical situation with the theory you have studied.

2. Draw any necessary diagrams and tabulate the problem data.

3. Establish a coordinate system and apply the relevant principles, generally in mathematical form.

4. Solve the necessary equations algebraically as far as practical; then, use a consistent set of units and complete the solution numerically. Report the answer with no more significant figures than the accuracy of the given data.

5. Study the answer using technical judgment and common sense to determine whether or not it seems reasonable.

6. Once the solution has been completed, review the problem. Try to think of other ways of obtaining the same solution.

In applying this general procedure, do the work as neatly as possible. Being neat generally stimulates clear and orderly thinking, and vice versa.

12.2   Rectilinear Kinematics: Continuous Motion

We will begin our study of dynamics by discussing the kinematics of a particle that moves along a rectilinear or straight line path. Recall that a *particle* has a mass but negligible size and shape. Therefore we must limit application to those objects that have dimensions that are of no consequence in the analysis of the motion. In most problems, we will be interested in bodies of finite size, such as rockets, projectiles, or vehicles. Such objects can be considered as particles, as long as the motion of the body is characterized by motion of its mass center and any rotation of the body is neglected.

Rectilinear Kinematics.   The kinematics of a particle is characterized by specifying, at any given instant, the particle's position, velocity, and acceleration.

Position.   The straight-line path of a particle will be defined using a single coordinate axis s, Fig. 12–1a. The origin O on the path is a fixed point, and from this point the *position vector* $\mathbf{r}$ is used to specify the location of the particle P at any given instant. Notice that $\mathbf{r}$ is *always* along the s axis, and so its direction never changes. What will change is its magnitude and its sense or arrowhead direction. For analytical work it is therefore convenient to represent $\mathbf{r}$ by an *algebraic scalar s*, representing the *position coordinate* of the particle, Fig. 12–1a. The magnitude of s (and $\mathbf{r}$) is the distance from O to P, usually measured in meters (m) or feet (ft), and the sense (or arrowhead direction of $\mathbf{r}$) is defined by the algebraic sign on s. Although the choice is arbitrary, in this case s is positive since the coordinate axis is positive to the right of the origin. Likewise, it is negative if the particle is located to the left of O.

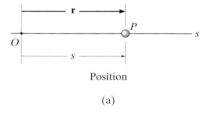

Position

(a)

Displacement.   The *displacement* of the particle is defined as the *change* in its *position*. For example, if the particle moves from P to P', Fig. 12–1b, the displacement is $\Delta\mathbf{r} = \mathbf{r}' - \mathbf{r}$. Using algebraic scalars to represent $\Delta\mathbf{r}$, we also have

$$\Delta s = s' - s$$

Here Δs is *positive* since the particle's final position is to the *right* of its initial position, i.e., $s' > s$. Likewise, if the final position were to the *left* of its initial position, Δs would be *negative*.

Since the displacement of a particle is a *vector quantity*, it should be distinguished from the distance the particle travels. Specifically, the *distance traveled* is a *positive scalar* which represents the total length of path over which the particle travels.

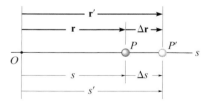

Displacement

(b)

Fig. 12–1

Velocity.   If the particle moves through a displacement $\Delta \mathbf{r}$ from P to P' during the time interval Δt, Fig. 12–1b, the *average velocity* of the particle during this time interval is

$$\mathbf{v}_{\text{avg}} = \frac{\Delta \mathbf{r}}{\Delta t}$$

If we take smaller and smaller values of Δt, the magnitude of $\Delta \mathbf{r}$ becomes smaller and smaller. Consequently, the *instantaneous velocity* is defined as $\mathbf{v} = \lim_{\Delta t \to 0}(\Delta \mathbf{r}/\Delta t)$, or

$$\mathbf{v} = \frac{d\mathbf{r}}{dt}$$

Representing $\mathbf{v}$ as an algebraic scalar, Fig. 12–1c, we can also write

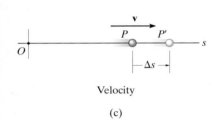

Velocity

(c)

$$(\xrightarrow{+}) \qquad \boxed{v = \frac{ds}{dt}} \qquad (12\text{–}1)$$

Since Δt or dt is always positive, the sign used to define the *sense* of the velocity is the same as that of Δs or ds. For example, if the particle is moving to the *right*, Fig. 12–1c, the velocity is *positive;* whereas if it is moving to the *left*, the velocity is *negative*. (This is emphasized here by the arrow written at the left of Eq. 12–1.) The *magnitude* of the velocity is known as the *speed*, and it is generally expressed in units of m/s or ft/s.

Occasionally, the term "average speed" is used. The *average speed* is always a positive scalar and is defined as the total distance traveled by a particle, s_T, divided by the elapsed time Δt; i.e.,

$$(v_{\text{sp}})_{\text{avg}} = \frac{s_T}{\Delta t}$$

For example, the particle in Fig. 12–1d travels along the path of length s_T in time Δt, so its average speed is $(v_{\text{sp}})_{\text{avg}} = s_T/\Delta t$, but its average velocity is $v_{\text{avg}} = -\Delta s/\Delta t$.

Average velocity and
Average speed

(d)

Fig. 12–1 (cont.)

Acceleration.   Provided the velocity of the particle is known at the two points P and P', the *average acceleration* of the particle during the time interval Δt is defined as

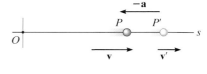

Acceleration

(e)

$$a_{\text{avg}} = \frac{\Delta v}{\Delta t}$$

Here $\Delta \mathbf{v}$ represents the difference in the velocity during the time interval Δt, i.e., $\Delta \mathbf{v} = \mathbf{v}' - \mathbf{v}$, Fig. 12–1e.

The *instantaneous acceleration* at time t is found by taking smaller and smaller values of Δt and corresponding smaller and smaller values of $\Delta \mathbf{v}$, so that $\boldsymbol{a} = \lim_{\Delta t \to 0} (\Delta \mathbf{v}/\Delta t)$ or, using algebraic scalars,

($\xrightarrow{+}$)
$$\boxed{a = \frac{dv}{dt}}$$
(12–2)

Substituting Eq. 12–1 into this result, we can also write

($\xrightarrow{+}$)
$$a = \frac{d^2 s}{dt^2}$$

Both the average and instantaneous acceleration can be either positive or negative. In particular, when the particle is *slowing down*, or its speed is decreasing, it is said to be *decelerating*. In this case, v' in Fig. 12–1f is *less* than v, and so $\Delta v = v' - v$ will be negative. Consequently, a will also be negative, and therefore it will act to the *left*, in the opposite *sense* to v. Also, note that when the *velocity* is *constant*, the *acceleration is zero* since $\Delta v = v - v = 0$. Units commonly used to express the magnitude of acceleration are m/s² or ft/s².

A differential relation involving the displacement, velocity, and acceleration along the path may be obtained by eliminating the time differential dt between Eqs. 12–1 and 12–2, which gives

Deceleration

(f)

($\xrightarrow{+}$)
$$\boxed{a \, ds = v \, dv}$$
(12–3)

Although we have now produced three important kinematic equations, realize that Eq. 12–3 is not independent of Eqs. 12–1 and 12–2.

Constant Acceleration, $a = a_c$. When the acceleration is constant, each of the three kinematic equations $a_c = dv/dt$, $v = ds/dt$, and $a_c\,ds = v\,dv$ may be integrated to obtain formulas that relate a_c, v, s, and t.

Velocity as a Function of Time. Integrate $a_c = dv/dt$, assuming that initially $v = v_0$ when $t = 0$.

$$\int_{v_0}^{v} dv = \int_{0}^{t} a_c\,dt$$

($\xrightarrow{+}$)
$$\boxed{v = v_0 + a_c t}$$
Constant Acceleration
(12–4)

Position as a Function of Time. Integrate $v = ds/dt = v_0 + a_c t$, assuming that initially $s = s_0$ when $t = 0$.

$$\int_{s_0}^{s} ds = \int_{0}^{t} (v_0 + a_c t)\,dt$$

($\xrightarrow{+}$)
$$\boxed{s = s_0 + v_0 t + \tfrac{1}{2} a_c t^2}$$
Constant Acceleration
(12–5)

Velocity as a Function of Position. Either solve for t in Eq. 12–4 and substitute into Eq. 12–5, or integrate $v\,dv = a_c\,ds$, assuming that initially $v = v_0$ at $s = s_0$.

$$\int_{v_0}^{v} v\,dv = \int_{s_0}^{s} a_c\,ds$$

($\xrightarrow{+}$)
$$\boxed{v^2 = v_0^2 + 2a_c(s - s_0)}$$
Constant Acceleration
(12–6)

This equation is not independent of Eqs. 12–4 and 12–5 since it can be obtained by eliminating t between these equations.

The magnitudes and signs of s_0, v_0, and a_c, used in the above three equations are determined from the chosen origin and positive direction of the s axis as indicated by the arrow written at the left of each equation. Also, it is important to remember that these equations are useful *only when the acceleration is constant and when $t = 0$, $s = s_0$, $v = v_0$.* A common example of constant accelerated motion occurs when a body falls freely toward the earth. If air resistance is neglected and the distance of fall is short, then the *downward* acceleration of the body when it is close to the earth is constant and approximately 9.81 m/s^2 or 32.2 ft/s^2. The proof of this is given in Example 13.2.

Important Points

- Dynamics is concerned with bodies that have accelerated motion.
- Kinematics is a study of the geometry of the motion.
- Kinetics is a study of the forces that cause the motion.
- Rectilinear kinematics refers to straight-line motion.
- Speed refers to the magnitude of velocity.
- Average speed is the total distance traveled divided by the total time. This is different from the average velocity which is the displacement divided by the time.
- The acceleration, $a = dv/dt$, is negative when the particle is slowing down or decelerating.
- A particle can have an acceleration and yet have zero velocity.
- The relationship $a\,ds = v\,dv$ is derived from $a = dv/dt$ and $v = ds/dt$, by eliminating dt.

During the time this rocket undergoes rectilinear motion, its altitude as a function of time can be measured and expressed as $s = s(t)$. Its velocity can then be found using $v = ds/dt$, and its acceleration can be determined from $a = dv/dt$.

Procedure for Analysis

The equations of rectilinear kinematics should be applied using the following procedure.

Coordinate System.

- Establish a position coordinate s along the path and specify its *fixed origin* and positive direction.
- Since motion is along a straight line, the particle's position, velocity, and acceleration can be represented as algebraic scalars. For analytical work the sense of s, v, and a is then determined from their *algebraic signs*.
- The positive sense for each scalar can be indicated by an arrow shown alongside each kinematic equation as it is applied.

Kinematic Equations.

- If a relationship is known between any *two* of the four variables a, v, s and t, then a third variable can be obtained by using one of the kinematic equations, $a = dv/dt$, $v = ds/dt$ or $a\,ds = v\,dv$, which relates all three variables.*
- Whenever integration is performed, it is important that the position and velocity be known at a given instant in order to evaluate either the constant of integration if an indefinite integral is used, or the limits of integration if a definite integral is used.
- Remember that Eqs. 12–4 through 12–6 have only a limited use. Never apply these equations unless it is absolutely certain that the *acceleration is constant*.

*Some standard differentiation and integration formulas are given in Appendix A.

EXAMPLE 12.1

The car in Fig. 12–2 moves in a straight line such that for a short time its velocity is defined by $v = (3t^2 + 2t)$ ft/s, where t is in seconds. Determine its position and acceleration when $t = 3$ s. When $t = 0$, $s = 0$.

Fig. 12–2

SOLUTION

Coordinate System.   The position coordinate extends from the fixed origin O to the car, positive to the right.

Position.   Since $v = f(t)$, the car's position can be determined from $v = ds/dt$, since this equation relates v, s, and t. Noting that $s = 0$ when $t = 0$, we have*

$(\xrightarrow{+})$

$$v = \frac{ds}{dt} = (3t^2 + 2t)$$

$$\int_0^s ds = \int_0^t (3t^2 + 2t)\, dt$$

$$s\Big|_0^s = t^3 + t^2\Big|_0^t$$

$$s = t^3 + t^2$$

When $t = 3$ s,

$$s = (3)^3 + (3)^2 = 36 \text{ ft} \qquad\qquad Ans.$$

Acceleration.   Knowing $v = f(t)$, the acceleration is determined from $a = dv/dt$, since this equation relates a, v, and t.

$(\xrightarrow{+})$

$$a = \frac{dv}{dt} = \frac{d}{dt}(3t^2 + 2t)$$

$$= 6t + 2$$

When $t = 3$ s,

$$a = 6(3) + 2 = 20 \text{ ft/s}^2 \rightarrow \qquad\qquad Ans.$$

NOTE: The formulas for constant acceleration *cannot* be used to solve this problem, because the acceleration is a function of time.

---

*The *same result* can be obtained by evaluating a constant of integration C rather than using definite limits on the integral. For example, integrating $ds = (3t^2 + 2t)\, dt$ yields $s = t^3 + t^2 + C$. Using the condition that at $t = 0$, $s = 0$, then $C = 0$.

EXAMPLE 12.2

A small projectile is fired vertically *downward* into a fluid medium with an initial velocity of 60 m/s. Due to the resistance of the fluid the projectile experiences a deceleration equal to $a = (-0.4v^3)$ m/s^2, where v is in m/s. Determine the projectile's velocity and position 4 s after it is fired.

SOLUTION

Coordinate System.   Since the motion is downward, the position coordinate is positive downward, with origin located at O, Fig. 12–3.

Velocity.   Here $a = f(v)$ and so we must determine the velocity as a function of time using $a = dv/dt$, since this equation relates v, a, and t. (Why not use $v = v_0 + a_c t$?) Separating the variables and integrating, with $v_0 = 60$ m/s when $t = 0$, yields

$(+\downarrow)$
$$a = \frac{dv}{dt} = -0.4v^3$$

$$\int_{60 \text{ m/s}}^{v} \frac{dv}{-0.4v^3} = \int_0^t dt$$

$$\frac{1}{-0.4}\left(\frac{1}{-2}\right)\frac{1}{v^2}\bigg|_{60}^{v} = t - 0$$

$$\frac{1}{0.8}\left[\frac{1}{v^2} - \frac{1}{(60)^2}\right] = t$$

$$v = \left\{\left[\frac{1}{(60)^2} + 0.8t\right]^{-1/2}\right\} \text{ m/s}$$

Fig. 12–3

Here the positive root is taken, since the projectile is moving downward. When $t = 4$ s,

$$v = 0.559 \text{ m/s} \downarrow \qquad\qquad Ans.$$

Position.   Knowing $v = f(t)$, we can obtain the projectile's position from $v = ds/dt$, since this equation relates s, v, and t. Using the initial condition $s = 0$, when $t = 0$, we have

$(+\downarrow)$
$$v = \frac{ds}{dt} = \left[\frac{1}{(60)^2} + 0.8t\right]^{-1/2}$$

$$\int_0^s ds = \int_0^t \left[\frac{1}{(60)^2} + 0.8t\right]^{-1/2} dt$$

$$s = \frac{2}{0.8}\left[\frac{1}{(60)^2} + 0.8t\right]^{1/2}\bigg|_0^t$$

$$s = \frac{1}{0.4}\left\{\left[\frac{1}{(60)^2} + 0.8t\right]^{1/2} - \frac{1}{60}\right\} \text{ m}$$

When $t = 4$ s,

$$s = 4.43 \text{ m} \qquad\qquad Ans.$$

EXAMPLE 12.3

Fig. 12–4

During a test a rocket is traveling upward at 75 m/s, and when it is 40 m from the ground its engine fails. Determine the maximum height s_B reached by the rocket and its speed just before it hits the ground. While in motion the rocket is subjected to a constant downward acceleration of 9.81 m/s² due to gravity. Neglect the effect of air resistance.

SOLUTION

Coordinate System.   The origin O for the position coordinate s is taken at ground level with positive upward, Fig. 12–4.

Maximum Height.   Since the rocket is traveling *upward*, $v_A = +75$ m/s when $t = 0$. At the maximum height $s = s_B$ the velocity $v_B = 0$. For the entire motion, the acceleration is $a_c = -9.81$ m/s² (negative since it acts in the *opposite* sense to positive velocity or positive displacement). Since a_c is *constant* the rocket's position may be related to its velocity at the two points A and B on the path by using Eq. 12–6, namely,

$$(+\uparrow) \qquad v_B^2 = v_A^2 + 2a_c(s_B - s_A)$$

$$0 = (75 \text{ m/s})^2 + 2(-9.81 \text{ m/s}^2)(s_B - 40 \text{ m})$$

$$s_B = 327 \text{ m} \qquad\qquad\qquad\qquad Ans.$$

Velocity.   To obtain the velocity of the rocket just before it hits the ground, we can apply Eq. 12–6 between points B and C, Fig. 12–4.

$$(+\uparrow) \qquad v_C^2 = v_B^2 + 2a_c(s_C - s_B)$$

$$= 0 + 2(-9.81 \text{ m/s}^2)(0 - 327 \text{ m})$$

$$v_C = -80.1 \text{ m/s} = 80.1 \text{ m/s} \downarrow \qquad\qquad Ans.$$

The negative root was chosen since the rocket is moving downward. Similarly, Eq. 12–6 may also be applied between points A and C, i.e.,

$$(+\uparrow) \qquad v_C^2 = v_A^2 + 2a_c(s_C - s_A)$$

$$= (75 \text{ m/s})^2 + 2(-9.81 \text{ m/s}^2)(0 - 40 \text{ m})$$

$$v_C = -80.1 \text{ m/s} = 80.1 \text{ m/s} \downarrow$$

NOTE: It should be realized that the rocket is subjected to a *deceleration* from A to B of 9.81 m/s², and then from B to C it is *accelerated* at this rate. Furthermore, even though the rocket momentarily comes to *rest* at B ($v_B = 0$) the acceleration at B is 9.81 m/s² downward!

EXAMPLE 12.4

A metallic particle is subjected to the influence of a magnetic field as it travels downward through a fluid that extends from plate A to plate B, Fig. 12–5. If the particle is released from rest at the midpoint C, $s = 100$ mm, and the acceleration is $a = (4s)$ m/s^2, where s is in meters, determine the velocity of the particle when it reaches plate B, $s = 200$ mm, and the time it needs to travel from C to B.

SOLUTION

Coordinate System.   As shown in Fig. 12–5, s is taken positive downward, measured from plate A.

Velocity.   Since $a = f(s)$, the velocity as a function of position can be obtained by using $v\, dv = a\, ds$. Realizing that $v = 0$ at $s = 0.1$ m, we have

$(+\downarrow)$

$$v\, dv = a\, ds$$

$$\int_0^v v\, dv = \int_{0.1}^s 4s\, ds$$

$$\tfrac{1}{2}v^2 \Big|_0^v = \frac{4}{2}s^2 \Big|_{0.1}^s$$

$$v = 2(s^2 - 0.01)^{1/2} \qquad\qquad (1)$$

Fig. 12–5

At $s = 200$ mm $= 0.2$ m,

$$v_B = 0.346 \text{ m/s} = 346 \text{ mm/s} \downarrow \qquad\qquad Ans.$$

The positive root is chosen since the particle is traveling downward, i.e., in the $+s$ direction.

Time.   The time for the particle to travel from C to B can be obtained using $v = ds/dt$ and Eq. 1, where $s = 0.1$ m when $t = 0$. From Appendix A,

$(+\downarrow)$

$$ds = v\, dt$$

$$= 2(s^2 - 0.01)^{1/2}\, dt$$

$$\int_{0.1}^s \frac{ds}{(s^2 - 0.01)^{1/2}} = \int_0^t 2\, dt$$

$$\ln\!\left(\sqrt{s^2 - 0.01} + s\right)\Big|_{0.1}^s = 2t \Big|_0^t$$

$$\ln\!\left(\sqrt{s^2 - 0.01} + s\right) + 2.303 = 2t$$

At $s = 0.2$ m,

$$t = \frac{\ln\!\left(\sqrt{(0.2)^2 - 0.01} + 0.2\right) + 2.303}{2} = 0.658 \text{ s} \qquad Ans.$$

NOTE: The formulas for constant acceleration cannot be used here, because the acceleration changes with position, i.e., $a = 4s$.

EXAMPLE 12.5

A particle moves along a horizontal path with a velocity of $v = (3t^2 - 6t)$ m/s, where t is the time in seconds. If it is initially located at the origin O, determine the distance traveled in 3.5 s, and the particle's average velocity and average speed during the time interval.

SOLUTION

Coordinate System.   Here we will assume positive motion to the right, measured from the origin O, Fig. 12–6a.

Distance Traveled.   Since $v = f(t)$, the position as a function of time may be found by integrating $v = ds/dt$ with $t = 0$, $s = 0$.

(a)

$$(\overset{+}{\rightarrow}) \qquad\qquad ds = v\, dt$$

$$= (3t^2 - 6t)\, dt$$

$$\int_0^s ds = 3 \int_0^t t^2\, dt - 6 \int_0^t t\, dt$$

$$s = (t^3 - 3t^2)\ \text{m} \qquad\qquad (1)$$

In order to determine the distance traveled in 3.5 s, it is necessary to investigate the path of motion. The graph of the velocity function, Fig. 12–6b, reveals that for $0 \le t < 2$ s the velocity is *negative*, which means the particle is traveling to the *left*, and for $t > 2$ s the velocity is *positive*, and hence the particle is traveling to the *right*. Also, $v = 0$ at $t = 2$ s. The particle's position when $t = 0$, $t = 2$ s, and $t = 3.5$ s can be determined from Eq. 1. This yields

$$s|_{t=0} = 0 \qquad s|_{t=2\,\text{s}} = -4.0\ \text{m} \qquad s|_{t=3.5\,\text{s}} = 6.125\ \text{m}$$

(b)

Fig. 12–6

The path is shown in Fig. 12–6a. Hence, the distance traveled in 3.5 s is

$$s_T = 4.0 + 4.0 + 6.125 = 14.125\ \text{m} = 14.1\ \text{m} \qquad\qquad Ans.$$

Velocity.   The *displacement* from $t = 0$ to $t = 3.5$ s is

$$\Delta s = s|_{t=3.5\,\text{s}} - s|_{t=0} = 6.12 - 0 = 6.12\ \text{m}$$

and so the average velocity is

$$v_{\text{avg}} = \frac{\Delta s}{\Delta t} = \frac{6.12}{3.5 - 0} = 1.75\ \text{m/s} \rightarrow \qquad\qquad Ans.$$

The average speed is defined in terms of the *distance traveled* s_T. This positive scalar is

$$(v_{\text{sp}})_{\text{avg}} = \frac{s_T}{\Delta t} = \frac{14.125}{3.5 - 0} = 4.04\ \text{m/s} \qquad\qquad Ans.$$

NOTE: In this problem, the acceleration is $a = dv/dt = (6t - 6)$ m/s², which is not constant.

PROBLEMS

12–1. A truck, traveling along a straight road at 20 km/h, increases its speed to 120 km/h in 15 s. If its acceleration is constant, determine the distance traveled.

12–2. A car starts from rest and reaches a speed of 80 ft/s after traveling 500 ft along a straight road. Determine its constant acceleration and the time of travel.

12–3. A baseball is thrown downward from a 50-ft tower with an initial speed of 18 ft/s. Determine the speed at which it hits the ground and the time of travel.

*12–4. Starting from rest, a particle moving in a straight line has an acceleration of $a = (2t - 6)$ m/s^2, where t is in seconds. What is the particle's velocity when $t = 6$ s, and what is its position when $t = 11$ s?

12–5. Traveling with an initial speed of 70 km/h, a car accelerates at 6000 km/h^2 along a straight road. How long will it take to reach a speed of 120 km/h? Also, through what distance does the car travel during this time?

12–6. A freight train travels at $v = 60(1 - e^{-t})$ ft/s, where t is the elapsed time in seconds. Determine the distance traveled in three seconds, and the acceleration at this time.

Prob. 12–6

12–7. The position of a particle along a straight line is given by $s = (t^3 - 9t^2 + 15t)$ ft, where t is in seconds. Determine its maximum acceleration and maximum velocity during the time interval $0 \le t \le 10$ s.

*12–8. From approximately what floor of a building must a car be dropped from an at-rest position so that it reaches a speed of 80.7 ft/s (55 mi/h) when it hits the ground? Each floor is 12 ft higher than the one below it. (*Note:* You may want to remember this when traveling 55 mi/h.)

12–9. A particle moves along a straight line such that its position is defined by $s = (t^3 - 3t^2 + 2)$ m. Determine the average velocity, the average speed, and the acceleration of the particle when $t = 4$ s.

12–10. A particle is moving along a straight line such that its acceleration is defined as $a = (-2v)$ m/s^2, where v is in meters per second. If $v = 20$ m/s when $s = 0$ and $t = 0$, determine the particle's velocity as a function of position and the distance the particle moves before it stops.

12–11. The acceleration of a particle as it moves along a straight line is given by $a = (2t - 1)$ m/s^2, where t is in seconds. If $s = 1$ m and $v = 2$ m/s when $t = 0$, determine the particle's velocity and position when $t = 6$ s. Also, determine the total distance the particle travels during this time period.

*12–12. A particle, initially at the origin, moves along a straight line through a fluid medium such that its velocity is defined as $v = 1.8(1 - e^{-0.3t})$ m/s, where t is in seconds. Determine the displacement of the particle during the first 3 s.

12–13. The velocity of a particle traveling in a straight line is given by $v = (6t - 3t^2)$ m/s, where t is in seconds. If $s = 0$ when $t = 0$, determine the particle's deceleration and position when $t = 3$ s. How far has the particle traveled during the 3-s time interval, and what is its average speed?

12–14. A particle moves along a straight line such that its position is defined by $s = (t^2 - 6t + 5)$ m. Determine the average velocity, the average speed, and the acceleration of the particle when $t = 6$ s.

12–15. A particle is moving along a straight line such that when it is at the origin it has a velocity of 4 m/s. If it begins to decelerate at the rate of $a = (-1.5v^{1/2})$ m/s^2, where v is in m/s, determine the particle's position and velocity when $t = 2$ s.

*12–16. A particle travels to the right along a straight line with a velocity $v = [5/(4 + s)]$ m/s, where s is in meters. Determine its deceleration when $s = 2$ m.

12–17. Two particles A and B start from rest at the origin $s = 0$ and move along a straight line such that $a_A = (6t - 3)$ ft/s^2 and $a_B = (12t^2 - 8)$ ft/s^2, where t is in seconds. Determine the distance between them when $t = 4$ s and the total distance each has traveled in $t = 4$ s.

12–18. A car is to be hoisted by elevator to the fourth floor of a parking garage, which is 48 ft above the ground. If the elevator can accelerate at 0.6 ft/s^2, decelerate at 0.3 ft/s^2, and reach a maximum speed of 8 ft/s, determine the shortest time to make the lift, starting from rest and ending at rest.

12–19. A stone A is dropped from rest down a well, and in 1 s another stone B is dropped from rest. Determine the distance between the stones another second later.

*12–20. A stone A is dropped from rest down a well, and in 1 s another stone B is dropped from rest. Determine the time interval between the instant A strikes the water and the instant B strikes the water. Also, at what speed do they strike the water?

80 ft

Probs. 12–19/20

12–21. A particle has an initial speed of 27 m/s. If it experiences a deceleration of $a = (-6t)$ m/s^2, where t is in seconds, determine the distance traveled before it stops.

12–22. The acceleration of a rocket traveling upward is given by $a = (6 + 0.02s)$ m/s^2, where s is in meters. Determine the rocket's velocity when $s = 2$ km and the time needed to reach this altitude. Initially, $v = 0$ and $s = 0$ when $t = 0$.

s

Prob. 12–22

12–23. The acceleration of a rocket traveling upward is given by $a = (6 + 0.02s)$ m/s^2, where s is in meters. Determine the time needed for the rocket to reach an altitude of $s = 100$ m. Initially, $v = 0$ and $s = 0$ when $t = 0$.

s

Prob. 12–23

*12–24.   A particle is moving with a velocity of v_0 when $s = 0$ and $t = 0$. If it is subjected to a deceleration of $a = -kv^3$, where k is a constant, determine its velocity and position as functions of time.

■12–25.   A particle has an initial speed of 27 m/s. If it experiences a deceleration of $a = (-6t)$ m/s², where t is in seconds, determine its velocity when it travels 10 m. How much time does this take?

12–26.   Ball A is released from rest at a height of 40 ft at the same time that a second ball B is thrown upward 5 ft from the ground. If the balls pass one another at a height of 20 ft, determine the speed at which ball B was thrown upward.

A

40 ft

B

5 ft

Prob. 12–26

12–27.   A car starts from rest and moves along a straight line with an acceleration of $a = (3s^{-1/3})$ m/s², where s is in meters. Determine the car's velocity and position when $t = 6$ s.

*12–28.   The acceleration of a particle along a straight line is defined by $a = (2t - 9)$ m/s², where t is in seconds. At $t = 0$, $s = 1$ m and $v = 10$ m/s. When $t = 9$ s, determine (a) the particle's position, (b) the total distance traveled, and (c) the velocity.

12–29.   A particle is moving along a straight line such that its acceleration is defined as $a = (4s^2)$ m/s², where s is in meters. If $v = -100$ m/s when $s = 10$ m and $t = 0$, determine the particle's velocity as a function of position.

12–30.   A car can have an acceleration and a deceleration of 5 m/s². If it starts from rest, and can have a maximum speed of 60 m/s, determine the shortest time it can travel a distance of 1200 m when it stops.

12–31.   Determine the time required for a car to travel 1 km along a road if the car starts from rest, reaches a maximum speed at some intermediate point, and then stops at the end of the road. The car can accelerate at 1.5 m/s² and decelerate at 2 m/s².

*12–32.   When two cars A and B are next to one another, they are traveling in the same direction with speeds v_A and v_B, respectively. If B maintains its constant speed, while A begins to decelerate at a_A, determine the distance d between the cars at the instant A stops.

A   B

d

Prob. 12–32

12–33. If the effects of atmospheric resistance are accounted for, a freely falling body has an acceleration defined by the equation $a = 9.81[1 - v^2(10^{-4})]$ m/s^2, where v is in m/s and the positive direction is downward. If the body is released from rest at a *very high altitude*, determine (a) the velocity when $t = 5$ s, and (b) the body's terminal or maximum attainable velocity (as $t \to \infty$).

12–34. As a body is projected to a high altitude above the earth's *surface*, the variation of the acceleration of gravity with respect to altitude y must be taken into account. Neglecting air resistance, this acceleration is determined from the formula $a = -g_0[R^2/(R + y)^2]$, where g_0 is the constant gravitational acceleration at sea level, R is the radius of the earth, and the positive direction is measured upward. If $g_0 = 9.81$ m/s^2 and $R = 6356$ km, determine the minimum initial velocity (escape velocity) at which a projectile should be shot vertically from the earth's surface so that it does not fall back to the earth. *Hint:* This requires that $v = 0$ as $y \to \infty$.

12–35. Accounting for the variation of gravitational acceleration a with respect to altitude y (see Prob. 12–34), derive an equation that relates the velocity of a freely falling particle to its altitude. Assume that the particle is released from rest at an altitude y_0 from the earth's surface. With what velocity does the particle strike the earth if it is released from rest at an altitude $y_0 = 500$ km? Use the numerical data in Prob. 12–34.

*****12–36.** When a particle falls through the air, its initial acceleration $a = g$ diminishes until it is zero, and thereafter it falls at a constant or terminal velocity v_f. If this variation of the acceleration can be expressed as $a = (g/v^2_f)(v^2_f - v^2)$, determine the time needed for the velocity to become $v = v_f$. Initially the particle falls from rest.

12.3   Rectilinear Kinematics: Erratic Motion

When a particle's motion during a time period is erratic or there is a discontinuity in the motion, then it may be difficult to obtain a continuous mathematical function to describe its position, velocity, or acceleration. Instead, the motion may best be described graphically using a series of curves that can be generated experimentally. If the resulting graph describes the relationship between any two of the variables, a, v, s, t, a graph describing the relationship between the other variables can be established by using the kinematic equations $a = dv/dt$, $v = ds/dt$, $a\,ds = v\,dv$. Several situations occur frequently.

Given the s–t Graph, Construct the v–t Graph.   If the position of a particle can be *determined experimentally* during a time period t, the s–t graph for the particle can be plotted, Fig. 12–7a. To determine the particle's velocity as a function of time, i.e., the v–t graph, we must use $v = ds/dt$ since this equation relates v, s, and t. Therefore, the velocity at any instant is determined by measuring the *slope* of the s–t graph, i.e.,

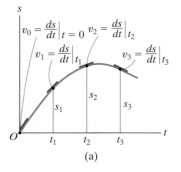

(a)

Fig. 12–7

$$\boxed{\begin{array}{c} \dfrac{ds}{dt} = v \\[4pt] \text{slope of} \\ s\text{–}t \text{ graph} = \text{velocity} \end{array}}$$

$(12{-}36)$

For example, measurement of the slopes v_0, v_1, v_2, v_3 at the intermediate points $(0, 0)$, (t_1, s_1), (t_2, s_2), (t_3, s_3) on the s–t graph, Fig. 12–7a, gives the corresponding points on the v–t graph shown in Fig. 12–7b.

It may also be possible to establish the v–t graph *mathematically*, provided the segments of the s–t graph can be expressed in the form of equations $s = f(t)$. Corresponding equations describing the segments of the v–t graph are then determined by time *differentiation*, since $v = ds/dt$.

Given the v–t Graph, Construct the a–t Graph.   When the particle's v–t graph is known, as in Fig. 12–8a, the acceleration as a function of time, i.e., the a–t graph, can be determined using $a = dv/dt$. Hence, the acceleration at any instant is determined by measuring the slope of the v–t graph, i.e.,

$$\frac{dv}{dt} = a$$

slope of
v–t graph $=$ acceleration

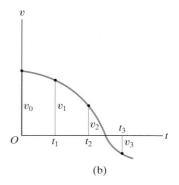

(b)

For example, measurement of the slopes a_0, a_1, a_2, a_3 at the intermediate points $(0, v_0)$, (t_1, v_1), (t_2, v_2), (t_3, v_3) on the v–t graph, Fig. 12–8a, yields the corresponding points on the a–t graph shown in Fig. 12–8b.

Any segments of the a–t graph can also be determined *mathematically*, provided the equations of the corresponding segments of the v–t graph are known, $v = g(t)$. This is done by simply taking the time *derivative* of $v = g(t)$, since $a = dv/dt$.

Since differentiation reduces a polynomial of degree n to that of degree $n - 1$, then if the s–t graph is parabolic (a second-degree curve), the v–t graph will be a sloping line (a first-degree curve), and the a–t graph will be a constant or a horizontal line (a zero-degree curve).

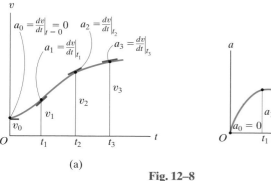

(a)

(b)

Fig. 12–8

EXAMPLE 12.6

A bicycle moves along a straight road such that its position is described by the graph shown in Fig. 12–9a. Construct the v–t and a–t graphs for $0 \leq t \leq 30$ s.

(a)

SOLUTION

v–t Graph.   Since $v = ds/dt$, the v–t graph can be determined by differentiating the equations defining the s–t graph, Fig. 12–9a. We have

$$0 \leq t < 10 \text{ s}; \qquad s = t^2 \qquad\qquad v = \frac{ds}{dt} = 2t$$

$$10 \text{ s} < t \leq 30 \text{ s}; \qquad s = 20t - 100 \qquad v = \frac{ds}{dt} = 20$$

The results are plotted in Fig. 12–9b. We can also obtain specific values of v by measuring the *slope* of the s–t graph at a given instant. For example, at $t = 20$ s, the slope of the s–t graph is determined from the straight line from 10 s to 30 s, i.e.,

$$t = 20 \text{ s}; \qquad v = \frac{\Delta s}{\Delta t} = \frac{500 - 100}{30 - 10} = 20 \text{ ft/s}$$

a–t Graph.   Since $a = dv/dt$, the a–t graph can be determined by differentiating the equations defining the lines of the v–t graph. This yields

$$0 \leq t < 10 \text{ s}; \qquad v = 2t \qquad a = \frac{dv}{dt} = 2$$

$$10 < t \leq 30 \text{ s}; \qquad v = 20 \qquad a = \frac{dv}{dt} = 0$$

The results are plotted in Fig. 12–9c.

NOTE: Show that $a = 2$ ft/s^2 when $t = 5$ s by measuring the slope of the v–t graph.

(b)

(c)

Fig. 12–9

Given the *a–t* Graph, Construct the *v–t* Graph.   If the *a–t* graph is given, Fig. 12–10a, the *v–t* graph may be constructed using $a = dv/dt$, written in integrated form as

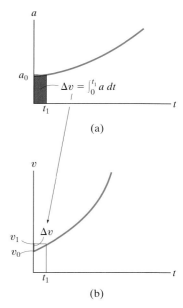

$$\Delta v = \int a\, dt$$
$$\text{change in} \atop \text{velocity} \quad = \quad {\text{area under} \atop a\text{–}t \text{ graph}}$$

Hence, to construct the *v–t* graph, we begin with the particle's initial velocity v_0 and then add to this small increments of area (Δv) determined from the *a–t* graph. In this manner successive points, $v_1 = v_0 + \Delta v$, etc., for the *v–t* graph are determined, Fig. 12–10b. Notice that an algebraic addition of the area increments is necessary, since areas lying above the *t* axis correspond to an increase in *v* ("positive" area), whereas those lying below the axis indicate a decrease in *v* ("negative" area).

   If segments of the *a–t* graph can be described by a series of equations, then each of these equations may be *integrated* to yield equations describing the corresponding segments of the *v–t* graph. For example, if the *a–t* graph is linear (a first-degree curve), integration will yield a *v–t* graph that is parabolic (a second-degree curve), etc.

Given the *v–t* Graph, Construct the *s–t* Graph.   When the *v–t* graph is given, Fig. 12–11a, it is possible to determine the *s–t* graph using $v = ds/dt$, written in integrated form

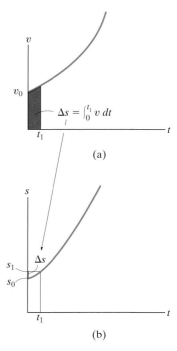

$$\Delta s = \int v\, dt$$
$$\text{displacement} \quad = \quad {\text{area under} \atop v\text{–}t \text{ graph}}$$

In the same manner as stated above, we begin with the particle's initial position s_0 and add (algebraically) to this small area increments Δs determined from the *v–t* graph, Fig. 12–11b.

   If it is possible to describe segments of the *v–t* graph by a series of equations, then each of these equations may be *integrated* to yield equations that describe corresponding segments of the *s–t* graph.

Fig. 12–10

Fig. 12–11

EXAMPLE 12.7

(a)

(b)

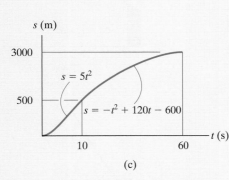

(c)

Fig. 12–12

The test car in Fig. 12–12a starts from rest and travels along a straight track such that it accelerates at a constant rate for 10 s and then decelerates at a constant rate. Draw the v–t and s–t graphs and determine the time t' needed to stop the car. How far has the car traveled?

SOLUTION

v–t Graph.   Since $dv = a\, dt$, the v–t graph is determined by integrating the straight-line segments of the a–t graph. Using the *initial condition* $v = 0$ when $t = 0$, we have

$$0 \le t < 10 \text{ s}; \qquad a = 10; \qquad \int_0^v dv = \int_0^t 10\, dt, \qquad v = 10t$$

When $t = 10$ s, $v = 10(10) = 100$ m/s. Using this as the *initial condition* for the next time period, we have

$$10 \text{ s} < t \le t'; \qquad a = -2; \qquad \int_{100}^v dv = \int_{10}^t -2\, dt, \qquad v = -2t + 120$$

When $t = t'$ we require $v = 0$. This yields, Fig. 12–12b,

$$t' = 60 \text{ s} \qquad\qquad Ans.$$

A more direct solution for t' is possible by realizing that the area under the a–t graph is equal to the change in the car's velocity. We require $\Delta v = 0 = A_1 + A_2$, Fig. 12–12a. Thus

$$0 = 10 \text{ m/s}^2(10 \text{ s}) + (-2 \text{ m/s}^2)(t' - 10 \text{ s})$$

$$t' = 60 \text{ s} \qquad\qquad Ans.$$

s–t Graph.   Since $ds = v\, dt$, integrating the equations of the v–t graph yields the corresponding equations of the s–t graph. Using the *initial condition* $s = 0$ when $t = 0$, we have

$$0 \le t \le 10 \text{ s}; \qquad v = 10t; \qquad \int_0^s ds = \int_0^t 10t\, dt, \qquad s = 5t^2$$

When $t = 10$ s, $s = 5(10)^2 = 500$ m. Using this *initial condition*,

$$10 \text{ s} \le t \le 60 \text{ s}; \qquad v = -2t + 120; \qquad \int_{500}^s ds = \int_{10}^t (-2t + 120)\, dt$$

$$s - 500 = -t^2 + 120t - [-(10)^2 + 120(10)]$$

$$s = -t^2 + 120t - 600$$

When $t' = 60$ s, the position is

$$s = -(60)^2 + 120(60) - 600 = 3000 \text{ m} \qquad\qquad Ans.$$

The s–t graph is shown in Fig. 12–12c.

NOTE: A direct solution for s is possible when $t' = 60$ s, since the *triangular area* under the v–t graph would yield the displacement $\Delta s = s - 0$ from $t = 0$ to $t' = 60$ s. Hence,

$$\Delta s = \tfrac{1}{2}(60)(100) = 3000 \text{ m} \qquad\qquad Ans.$$

Given the a–s Graph, Construct the v–s Graph.   In some cases an a–s graph for the particle can be constructed, so that points on the v–s graph can be determined by using $v\,dv = a\,ds$. Integrating this equation between the limits $v = v_0$ at $s = s_0$ and $v = v_1$ at $s = s_1$, we have,

(a)

$$\tfrac{1}{2}(v_1^2 - v_0^2) = \int_{s_0}^{s_1} a\,ds$$
area under
a–s graph

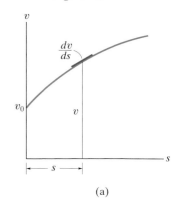

(b)

Fig. 12–13

Thus, the initial small segment of area under the a–s graph, $\int_{s_0}^{s_1} a\,ds$, shown colored in Fig. 12–13a, equals one-half the difference in the squares of the speed, $\tfrac{1}{2}(v_1^2 - v_0^2)$. Therefore, if the area is determined and the initial value of v_0 at $s_0 = 0$ is known, then $v_1 = \left(2\int_{s_0}^{s_1} a\,ds + v_0^2\right)^{1/2}$, Fig. 12–13b. Successive points on the v–s graph can be constructed in this manner starting from the initial velocity v_0.

Another way to construct the v–s graph is to first determine the equations which define the segments of the a–s graph. Then the corresponding equations defining the segments of the v–s graph can be obtained directly from integration, using $v\,dv = a\,ds$.

Given the v–s Graph, Construct the a–s Graph.   If the v–s graph is known, the acceleration a at any position s can be determined using $a\,ds = v\,dv$, written as

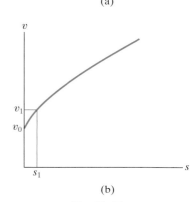

(a)

$$a = v\left(\frac{dv}{ds}\right)$$
acceleration = velocity times
slope of
v–s graph

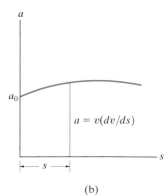

(b)

Fig. 12–14

Thus, at any point (s, v) in Fig. 12–14a, the slope dv/ds of the v–s graph is measured. Then since v and dv/ds are known, the value of a can be calculated, Fig. 12–14b.

We can also determine the segments describing the a–s graph analytically, provided the equations of the corresponding segments of the v–s graph are known. As above, this requires integration using $a\,ds = v\,dv$.

EXAMPLE 12.8

(a)

(b)

Fig. 12–15

The v–s graph describing the motion of a motorcycle is shown in Fig. 12–15a. Construct the a–s graph of the motion and determine the time needed for the motorcycle to reach the position $s = 400$ ft.

SOLUTION

a–s Graph. Since the equations for segments of the v–s graph are given, the a–s graph can be determined using $a\,ds = v\,dv$.

$0 \le s < 200$ ft; $v = 0.2s + 10$

$$a = v\frac{dv}{ds} = (0.2s + 10)\frac{d}{ds}(0.2s + 10) = 0.04s + 2$$

200 ft $< s \le 400$ ft; $v = 50$;

$$a = v\frac{dv}{ds} = (50)\frac{d}{ds}(50) = 0$$

The results are plotted in Fig. 12–15b.

Time. The time can be obtained using the v–s graph and $v = ds/dt$, because this equation relates v, s, and t. For the first segment of motion, $s = 0$ at $t = 0$, so

$0 \le s < 200$ ft; $v = 0.2s + 10$; $dt = \dfrac{ds}{v} = \dfrac{ds}{0.2s + 10}$

$$\int_0^t dt = \int_0^s \frac{ds}{0.2s + 10}$$

$$t = 5\ln(0.2s + 10) - 5\ln 10$$

At $s = 200$ ft, $t = 5\ln[0.2(200) + 10] - 5\ln 10 = 8.05$ s. Therefore, for the second segment of motion,

200 ft $< s \le 400$ ft; $v = 50$; $dt = \dfrac{ds}{v} = \dfrac{ds}{50}$

$$\int_{8.05}^t dt = \int_{200}^s \frac{ds}{50}; \quad t - 8.05 = \frac{s}{50} - 4$$

$$t = \frac{s}{50} + 4.05$$

Therefore, at $s = 400$ ft,

$$t = \frac{400}{50} + 4.05 = 12.0 \text{ s} \qquad\qquad Ans.$$

NOTE: The graphical results can be checked in part by calculating slopes. For example, at $s = 0$, $a = v(dv/ds) = 10(50 - 10)/200 = 2$ m/s^2. Also, the results can be checked in part by inspection. The v–s graph indicates the initial increase in velocity (acceleration) followed by constant velocity ($a = 0$).

PROBLEMS

12–37. An airplane starts from rest, travels 5000 ft down a runway, and after uniform acceleration, takes off with a speed of 162 mi/h. It then climbs in a straight line with a uniform acceleration of 3 ft/s^2 until it reaches a constant speed of 220 mi/h. Draw the s–t, v–t, and a–t graphs that describe the motion.

12–38. The elevator starts from rest at the first floor of the building. It can accelerate at 5 ft/s^2 and then decelerate at 2 ft/s^2. Determine the shortest time it takes to reach a floor 40 ft above the ground. The elevator starts from rest and then stops. Draw the a–t, v–t, and s–t graphs for the motion.

Prob. 12–38

12–39. If the position of a particle is defined as $s = (5t - 3t^2)$ ft, where t is in seconds, construct the s–t, v–t, and a–t graphs for $0 \le t \le 10$ s.

*12–40. If the position of a particle is defined by $s = [2\sin(\pi/5)t + 4]$ m, where t is in seconds, construct the s–t, v–t, and a–t graphs for $0 \le t \le 10$ s.

12–41. The v–t graph for a particle moving through an electric field from one plate to another has the shape shown in the figure. The acceleration and deceleration that occur are constant and both have a magnitude of 4 m/s^2. If the plates are spaced 200 mm apart, determine the maximum velocity v_{max} and the time t' for the particle to travel from one plate to the other. Also draw the s–t graph. When $t = t'/2$ the particle is at $s = 100$ mm.

12–42. The v–t graph for a particle moving through an electric field from one plate to another has the shape shown in the figure, where $t' = 0.2$ s and $v_{max} = 10$ m/s. Draw the s–t and a–t graphs for the particle. When $t = t'/2$ the particle is at $s = 0.5$ m.

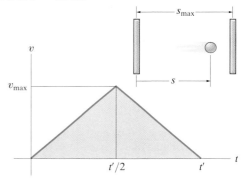

Probs. 12–41/42

12–43. A car starting from rest moves along a straight track with an acceleration as shown. Determine the time –t for the car to reach a speed of 50 m/s and construct the v–t graph that describes the motion until the time t.

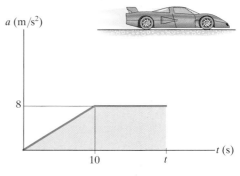

Prob. 12–43

*12–44.   A motorcycle starts from rest at $s = 0$ and travels along a straight road with the speed shown by the v–t graph. Determine the motorcycle's acceleration and position when $t = 8$ s and $t = 12$ s.

Prob. 12–44

12–46.   A car travels along a straight road with the speed shown by the v–t graph. Determine the total distance the car travels until it stops when $t = 48$ s. Also plot the s–t and a–t graphs.

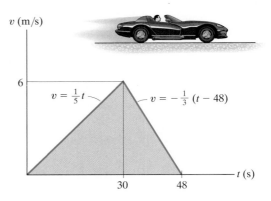

Prob. 12–46

12–45.   From experimental data, the motion of a jet plane while traveling along a runway is defined by the v–t graph shown. Construct the s–t and a–t graphs for the motion.

Prob. 12–45

12–47.   The v–t graph for the motion of a train as it moves from station A to station B is shown. Draw the a–t graph and determine the average speed and the distance between the stations.

Prob. 12–47

***12–48.** The s–t graph for a train has been experimentally determined. From the data, construct the v–t and a–t graphs for the motion; $0 \le t \le 40$ s. For $0 \le t \le 30$ s, the curve is $s = (0.4t^2)$ m, and then it becomes straight for $t \ge 30$ s.

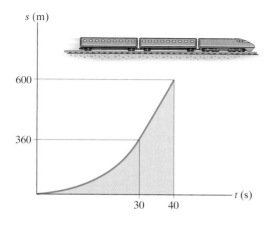

s (m)

600

360

30   40 t (s)

Prob. 12–48

12–49. The v–t graph for the motion of a car as it moves along a straight road is shown. Draw the a–t graph and determine the maximum acceleration during the 30-s time interval. The car starts from rest at $s = 0$.

12–50. The v–t graph for the motion of a car as it moves along a straight road is shown. Draw the s–t graph and determine the average speed and the distance traveled for the 30-s time interval. The car starts from rest at $s = 0$.

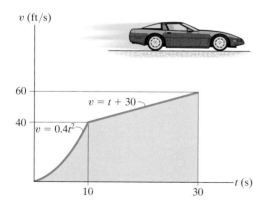

v (ft/s)

60

40

$v = t + 30$

$v = 0.4t^2$

10   30 t (s)

Probs. 12–49/50

■**12–51.** The a–s graph for a boat moving along a straight path is given. If the boat starts at $s = 0$ when $v = 0$, determine its speed when it is at $s = 75$ ft, and 125 ft, respectively. Use Simpson's rule with $n = 100$ to evaluate v at $s = 125$ ft.

a (ft/s^2)

$a = 5 + 6(\sqrt{s} - 10)^{5/3}$

5

100   s (ft)

Prob. 12–51

***12–52.** A man riding upward in a freight elevator accidentally drops a package off the elevator when it is 100 ft from the ground. If the elevator maintains a constant upward speed of 4 ft/s, determine how high the elevator is from the ground the instant the package hits the ground. Draw the v–t curve for the package during the time it is in motion. Assume that the package was released with the same upward speed as the elevator.

12–53. Two cars start from rest side by side and travel along a straight road. Car A accelerates at 4 m/s^2 for 10 s and then maintains a constant speed. Car B accelerates at 5 m/s^2 until reaching a constant speed of 25 m/s and then maintains this speed. Construct the a–t, v–t, and s–t graphs for each car until $t = 15$ s. What is the distance between the two cars when $t = 15$ s?

12–54. A two-stage rocket is fired vertically from rest at $s = 0$ with an acceleration as shown. After 30 s the first stage A burns out and the second stage B ignites. Plot the v–t and s–t graphs which describe the motion of the second stage for $0 \leq t \leq 60$ s.

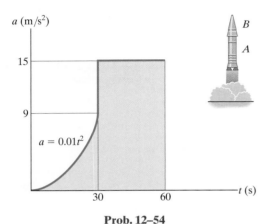

Prob. 12–54

*12–55.** The a–t graph for a motorcycle traveling along a straight road has been estimated as shown. Determine the time needed for the motorcycle to reach a maximum speed of 100 ft/s and the distance traveled in this time. Draw the v–t and s–t graphs. The motorcycle starts from rest at $s = 0$.

Prob. 12–55

*12–56.** The jet plane starts from rest at $s = 0$ and is subjected to the acceleration shown. Determine the speed of the plane when it has traveled 200 ft. Also, how much time is required for it to travel 200 ft?

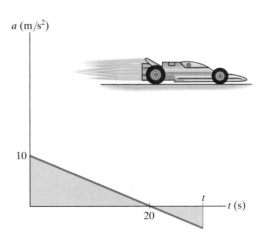

Prob. 12–56

12–57. The jet car is originally traveling at a speed of 20 m/s when it is subjected to the acceleration shown in the graph. Determine the car's maximum speed and the time t when it stops.

Prob. 12–57

12–58. A motorcyclist at A is traveling at 60 ft/s when he wishes to pass the truck T which is traveling at a constant speed of 60 ft/s. To do so the motorcyclist accelerates at 6 ft/s² until reaching a maximum speed of 85 ft/s. If he then maintains this speed, determine the time needed for him to reach a point located 100 ft in front of the truck. Draw the v–t and s–t graphs for the motorcycle during this time.

$(v_m)_1 = 60$ ft/s $(v_m)_2 = 85$ ft/s

$v_t = 60$ ft/s

40 ft — 55 ft — 100 ft

Prob. 12–58

12–59. The v–s graph for a go-cart traveling on a straight road is shown. Determine the acceleration of the go-cart at $s = 50$ m and $s = 150$ m. Draw the a–s graph.

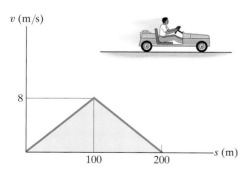

v (m/s)

8

100   200 s (m)

Prob. 12–59

*12–60. The a–t graph for a car is shown. Construct the v–t and s–t graphs if the car starts from rest at $t = 0$. At what time t' does the car stop?

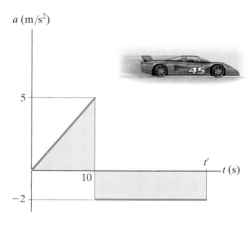

a (m/s²)

5

10   t' t (s)

-2

Prob. 12–60

12–61. The a–s graph for a train traveling along a straight track is given for the first 400 m of its motion. Plot the v–s graph. $v = 0$ at $s = 0$.

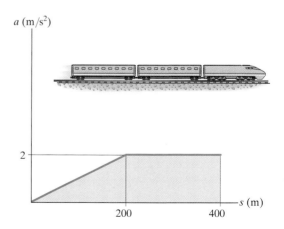

a (m/s²)

2

200   400 s (m)

Prob. 12–61

12–62. The *v–s* graph for an airplane traveling on a straight runway is shown. Determine the acceleration of the plane at $s = 100$ m and $s = 150$ m. Draw the *a–s* graph.

$v = 0.1s + 30$

$v = 0.4s$

Prob. 12–62

*12–64. The *v–s* graph for a test vehicle is shown. Determine its acceleration when $s = 100$ m and when $s = 175$ m.

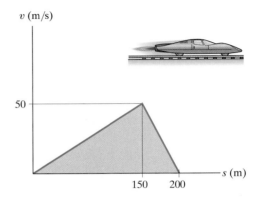

Prob. 12–64

12–63. Starting from rest at $s = 0$, a boat travels in a straight line with an acceleration as shown by the *a–s* graph. Determine the boat's speed when $s = 40$, 90, and 200 ft.

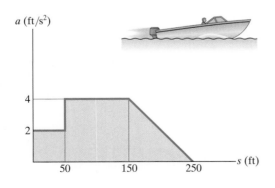

Prob. 12–63

12–65. The *v–s* graph was determined experimentally to describe the straight-line motion of a rocket sled. Determine the acceleration of the sled when $s = 100$ m, and when $s = 200$ m.

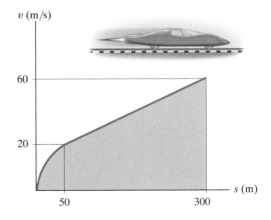

Prob. 12–65

12.4   General Curvilinear Motion

Curvilinear motion occurs when a particle moves along a curved path. Since this path is often described in three dimensions, vector analysis will be used to formulate the particle's position, velocity, and acceleration.* In this section the general aspects of curvilinear motion are discussed, and in subsequent sections we will consider three types of coordinate systems often used to analyze this motion.

Position.   Consider a particle located at point P on a space curve defined by the path function s, Fig. 12–16a. The position of the particle, measured from a fixed point O, will be designated by the *position vector* $\mathbf{r} = \mathbf{r}(t)$. This vector is a function of time since, in general, both its magnitude and direction change as the particle moves along the curve.

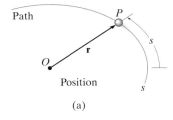

Position
(a)

Displacement.   Suppose that during a small time interval Δt the particle moves a distance Δs along the curve to a new position P', defined by $\mathbf{r}' = \mathbf{r} + \Delta \mathbf{r}$, Fig. 12–16b. The *displacement* $\Delta \mathbf{r}$ represents the change in the particle's position and is determined by vector subtraction; i.e., $\Delta \mathbf{r} = \mathbf{r}' - \mathbf{r}$.

Velocity.   During the time Δt, the *average velocity* of the particle is defined as

$$\mathbf{v}_{\text{avg}} = \frac{\Delta \mathbf{r}}{\Delta t}$$

The *instantaneous velocity* is determined from this equation by letting $\Delta t \to 0$, and consequently the direction of $\Delta \mathbf{r}$ *approaches* the *tangent* to the curve at point P. Hence, $\mathbf{v} = \lim_{\Delta t \to 0}(\Delta \mathbf{r}/\Delta t)$ or

$$\mathbf{v} = \frac{d\mathbf{r}}{dt} \tag{12–7}$$

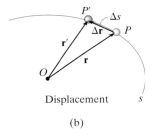

Displacement
(b)

Since $d\mathbf{r}$ will be tangent to the curve at P, the *direction* of $\mathbf{v}$ is also *tangent to the curve*, Fig. 12–16c. The *magnitude* of $\mathbf{v}$, which is called the *speed*, is obtained by noting that the magnitude of the displacement $\Delta \mathbf{r}$ is the length of the straight line segment from P to P', Fig. 12–16b. Realizing that this length, Δr, approaches the arc length Δs as $\Delta t \to 0$, we have $v = \lim_{\Delta t \to 0}(\Delta r/\Delta t) = \lim_{\Delta t \to 0}(\Delta s/\Delta t)$, or

$$v = \frac{ds}{dt} \tag{12–8}$$

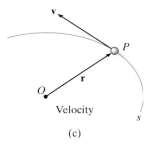

Velocity
(c)

Fig. 12–16

Thus, the *speed* can be obtained by differentiating the path function s with respect to time.

*A summary of some of the important concepts of vector analysis is given in Appendix C.

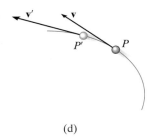

(d)

Acceleration.   If the particle has a velocity **v** at time *t* and a velocity **v**′ = **v** + Δ**v** at *t* + Δ*t*, Fig. 12–16*d*, then the *average acceleration* of the particle during the time interval Δ*t* is

$$\mathbf{a}_{\text{avg}} = \frac{\Delta \mathbf{v}}{\Delta t}$$

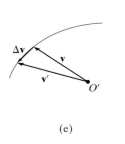

(e)

where Δ**v** = **v**′ − **v**. To study this time rate of change, the two velocity vectors in Fig. 12–16*d* are plotted in Fig. 12–16*e* such that their tails are located at the fixed point *O*′ and their arrowheads touch points on a curve. This curve is called a *hodograph*, and when constructed, it describes the locus of points for the arrowhead of the velocity vector in the same manner as the *path s* describes the locus of points for the arrowhead of the position vector, Fig. 12–16*a*.

To obtain the *instantaneous acceleration*, let Δ*t* → 0 in the above equation. In the limit Δ**v** will approach the *tangent to the hodograph*, and so **a** = lim_{Δt→0} (Δ**v**/Δ*t*), or

$$\mathbf{a} = \frac{d\mathbf{v}}{dt} \tag{12–9}$$

Substituting Eq. 12–7 into this result, we can also write

$$\mathbf{a} = \frac{d^2\mathbf{r}}{dt^2}$$

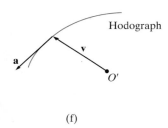

(f)

By definition of the derivative, **a** acts *tangent to the hodograph*, Fig. 12–16*f*, and therefore, *in general*, **a** *is not tangent to the path of motion*, Fig. 12–16*g*. To clarify this point, realize that Δ**v** and consequently **a** must account for the change made in *both* the magnitude *and* direction of the velocity **v** as the particle moves from *P* to *P*′, Fig. 12–16*d*. Just a magnitude change increases (or decreases) the "length" of **v**, and this in itself would allow **a** to remain tangent to the path. However, in order for the particle to follow the path, the directional change always "swings" the velocity vector toward the "inside" or "concave side" of the path, and therefore **a** *cannot* remain tangent to the path. In summary, **v** is always tangent to the *path* and **a** is always tangent to the *hodograph*.

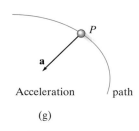

(g)

Fig. 12–16 (cont.)

12.5   Curvilinear Motion: Rectangular Components

Occasionally the motion of a particle can best be described along a path that is represented using a fixed x, y, z frame of reference.

Position. If at a given instant the particle P is at point (x, y, z) on the curved path s, Fig. 12–17a, its location is then defined by the *position vector*

$$\boxed{\mathbf{r} = x\mathbf{i} + y\mathbf{j} + z\mathbf{k}} \tag{12–10}$$

Because of the particle motion and the shape of the path, the x, y, z components of $\mathbf{r}$ are generally all functions of time; i.e., $x = x(t)$, $y = y(t)$, $z = z(t)$, so that $\mathbf{r} = \mathbf{r}(t)$.

In accordance with the discussion in Appendix C, the *magnitude* of $\mathbf{r}$ is *always positive* and defined from Eq. C–3 as

$$r = \sqrt{x^2 + y^2 + z^2}$$

The *direction* of $\mathbf{r}$ is specified by the components of the unit vector $\mathbf{u}_r = \mathbf{r}/r$.

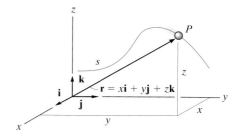

Position

(a)

Velocity. The first time derivative of $\mathbf{r}$ yields the velocity $\mathbf{v}$ of the particle. Hence,

$$\mathbf{v} = \frac{d\mathbf{r}}{dt} = \frac{d}{dt}(x\mathbf{i}) + \frac{d}{dt}(y\mathbf{j}) + \frac{d}{dt}(z\mathbf{k})$$

When taking this derivative, it is necessary to account for changes in *both* the magnitude and direction of each of the vector's components. The derivative of the $\mathbf{i}$ component of $\mathbf{v}$ is therefore

$$\frac{d}{dt}(x\mathbf{i}) = \frac{dx}{dt}\mathbf{i} + x\frac{d\mathbf{i}}{dt}$$

The second term on the right side is zero, since the x, y, z reference frame is *fixed*, and therefore the *direction* (and the *magnitude*) of $\mathbf{i}$ does not change with time. Differentiation of the $\mathbf{j}$ and $\mathbf{k}$ components may be carried out in a similar manner, which yields the final result,

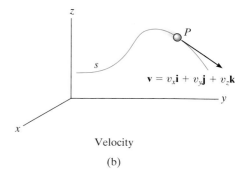

Velocity

(b)

Fig. 12–17

$$\boxed{\mathbf{v} = \frac{d\mathbf{r}}{dt} = v_x\mathbf{i} + v_y\mathbf{j} + v_z\mathbf{k}} \tag{12–11}$$

where

$$\boxed{v_x = \dot{x} \quad v_y = \dot{y} \quad v_z = \dot{z}} \tag{12–12}$$

The "dot" notation $\dot{x}$, $\dot{y}$, $\dot{z}$ represents the first time derivatives of the parametric equations $x = x(t)$, $y = y(t)$, $z = z(t)$, respectively.

The velocity has a *magnitude* defined as the positive value of

$$v = \sqrt{v_x^2 + v_y^2 + v_z^2}$$

and a *direction* that is specified by the components of the unit vector $\mathbf{u}_v = \mathbf{v}/v$. This direction is *always tangent to the path*, as shown in Fig. 12–17b.

Acceleration.

The acceleration of the particle is obtained by taking the first time derivative of Eq. 12–11 (or the second time derivative of Eq. 12–10). Using dots to represent the derivatives of the components, we have

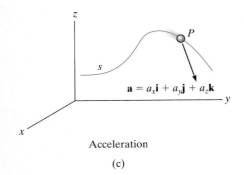

$$\mathbf{a} = \frac{d\mathbf{v}}{dt} = a_x\mathbf{i} + a_y\mathbf{j} + a_z\mathbf{k} \qquad (12\text{–}13)$$

Acceleration

(c)

Fig. 12–17 (cont.)

where

$$\begin{aligned} a_x &= \dot{v}_x = \ddot{x} \\ a_y &= \dot{v}_y = \ddot{y} \\ a_z &= \dot{v}_z = \ddot{z} \end{aligned} \qquad (12\text{–}14)$$

Here a_x, a_y, a_z represent, respectively, the first time derivatives of the functions $v_x = v_x(t)$, $v_y = v_y(t)$, $v_z = v_z(t)$, or the second time derivatives of the functions $x = x(t)$, $y = y(t)$, $z = z(t)$.

The acceleration has a *magnitude* defined by the positive value of

$$a = \sqrt{a_x^2 + a_y^2 + a_z^2}$$

and a *direction* specified by the components of the unit vector $\mathbf{u}_a = \mathbf{a}/a$. Since $\mathbf{a}$ represents the time rate of *change* in velocity, in general $\mathbf{a}$ will *not* be tangent to the path, Fig. 12–17c.

Important Points

- Curvilinear motion can cause changes in *both* the magnitude and direction of the position, velocity, and acceleration vectors.

- The velocity vector is always directed *tangent* to the path.

- In general, the acceleration vector is *not* tangent to the path, but rather, it is tangent to the hodograph.

- If the motion is described using rectangular coordinates, then the components along each of the axes do not change direction, only their magnitude and sense (algebraic sign) will change.

- By considering the component motions, the direction of motion of the particle is automatically taken into account.

Procedure for Analysis

Coordinate System

- A rectangular coordinate system can be used to solve problems for which the motion can conveniently be expressed in terms of its x, y, z components.

Kinematic Quantities

- Since *rectilinear motion* occurs along *each* coordinate axis, the motion of each component is found using $v = ds/dt$ and $a = dv/dt$; or in cases where the motion is not expressed as a function of time, the equation $a\,ds = v\,dv$ can be used.

- Once the x, y, z components of **v** and **a** have been determined, the magnitudes of these vectors are found from the Pythagorean theorem, Eq. C–3, and their directions from the components of their unit vectors, Eqs. C–4 and C–5.

As the airplane takes off, its path of motion can be established by knowing its horizontal position $x = x(t)$, and its vertical position or altitude $y = y(t)$, both of which can be found from navigation equipment. By plotting the results from these equations the path can be shown, and by taking the time derivatives, the velocity and acceleration of the plane at any instant can be determined.

EXAMPLE 12.9

(a)

(b)

(c)

Fig. 12–18

At any instant the horizontal position of the weather balloon in Fig. 12–18a is defined by $x = (8t)$ ft, where t is in seconds. If the equation of the path is $y = x^2/10$, determine (a) the distance of the balloon from the station at A when $t = 2$ s, (b) the magnitude and direction of the velocity when $t = 2$ s, and (c) the magnitude and direction of the acceleration when $t = 2$ s.

SOLUTION

Position.   When $t = 2$ s, $x = 8(2)$ ft $= 16$ ft, and so

$$y = (16)^2/10 = 25.6 \text{ ft}$$

The straight-line distance from A to B is therefore

$$r = \sqrt{(16)^2 + (25.6)^2} = 30.2 \text{ ft} \qquad \textit{Ans.}$$

Velocity.   Using Eqs. 12–12 and application of the chain rule of calculus the components of velocity when $t = 2$ s are

$$v_x = \dot{x} = \frac{d}{dt}(8t) = 8 \text{ ft/s} \rightarrow$$

$$v_y = \dot{y} = \frac{d}{dt}(x^2/10) = 2x\dot{x}/10 = 2(16)(8)/10 = 25.6 \text{ ft/s} \uparrow$$

When $t = 2$ s, the magnitude of velocity is therefore

$$v = \sqrt{(8)^2 + (25.6)^2} = 26.8 \text{ ft/s} \qquad \textit{Ans.}$$

The direction is tangent to the path, Fig. 12–18b, where

$$\theta_v = \tan^{-1}\frac{v_y}{v_x} = \tan^{-1}\frac{25.6}{8} = 72.6° \qquad \textit{Ans.}$$

Acceleration.   The components of acceleration are determined from Eqs. 12–14 and application of the chain rule, noting that $\ddot{x} = d^2(8t)/dt^2 = 0$. We have

$$a_x = \dot{v}_x = 0$$

$$a_y = \dot{v}_y = \frac{d}{dt}(2x\dot{x}/10) = 2(\dot{x})\dot{x}/10 + 2x(\ddot{x})/10$$

$$= 2(8)^2/10 + 2(16)(0)/10 = 12.8 \text{ ft/s}^2 \uparrow$$

Thus,

$$a = \sqrt{(0)^2 + (12.8)^2} = 12.8 \text{ ft/s}^2 \qquad \textit{Ans.}$$

The direction of **a**, as shown in Fig. 12–18c, is

$$\theta_a = \tan^{-1}\frac{12.8}{0} = 90° \qquad \textit{Ans.}$$

NOTE: It is also possible to obtain v_y and a_y by first expressing $y = f(t) = (8t)^2/10 = 6.4t^2$ and then taking successive time derivatives.

EXAMPLE 12.10

The motion of a box B moving along the spiral conveyor shown in Fig. 12–19 is defined by the position vector $\mathbf{r} = \{0.5 \sin(2t)\mathbf{i} + 0.5 \cos(2t)\mathbf{j} - 0.2t\mathbf{k}\}$ m, where t is in seconds and the arguments for sine and cosine are in radians (π rad = 180°). Determine the location of the box when $t = 0.75$ s and the magnitudes of its velocity and acceleration at this instant.

SOLUTION

Position. Evaluating $\mathbf{r}$ when $t = 0.75$ s yields

$$\mathbf{r}|_{t=0.75 \text{ s}} = \{0.5 \sin(1.5 \text{ rad})\mathbf{i} + 0.5 \cos(1.5 \text{ rad})\mathbf{j} - 0.2(0.75)\mathbf{k}\} \text{ m}$$
$$= \{0.499\mathbf{i} + 0.0354\mathbf{j} - 0.150\mathbf{k}\} \text{ m} \qquad Ans.$$

The distance of the box from the origin O is

$$r = \sqrt{(0.499)^2 + (0.0354)^2 + (-0.150)^2} = 0.522 \text{ m} \qquad Ans.$$

The direction of $\mathbf{r}$ is obtained from the components of the unit vector,

$$\mathbf{u}_r = \frac{\mathbf{r}}{r} = \frac{0.499}{0.522}\mathbf{i} + \frac{0.0354}{0.522}\mathbf{j} - \frac{0.150}{0.522}\mathbf{k}$$
$$= 0.955\mathbf{i} + 0.0678\mathbf{j} - 0.287\mathbf{k}$$

Fig. 12–19

Hence, the coordinate direction angles α, β, γ, Fig. 12–19, are

$$\alpha = \cos^{-1}(0.955) = 17.2° \qquad Ans.$$
$$\beta = \cos^{-1}(0.0678) = 86.1° \qquad Ans.$$
$$\gamma = \cos^{-1}(-0.287) = 107° \qquad Ans.$$

Velocity. The velocity is defined by

$$\mathbf{v} = \frac{d\mathbf{r}}{dt} = \frac{d}{dt}[0.5 \sin(2t)\mathbf{i} + 0.5 \cos(2t)\mathbf{j} - 0.2t\mathbf{k}]$$
$$= \{1 \cos(2t)\mathbf{i} - 1 \sin(2t)\mathbf{j} - 0.2\mathbf{k}\} \text{ m/s}$$

Hence, when $t = 0.75$ s the magnitude of velocity, or the speed, is

$$v = \sqrt{v_x^2 + v_y^2 + v_z^2}$$
$$= \sqrt{[1 \cos(1.5 \text{ rad})]^2 + [-1 \sin(1.5 \text{ rad})]^2 + (-0.2)^2}$$
$$= 1.02 \text{ m/s} \qquad Ans.$$

Acceleration. The acceleration $\mathbf{a}$ of the box is

$$\mathbf{a} = \frac{d\mathbf{v}}{dt} = \{-2 \sin(2t)\mathbf{i} - 2 \cos(2t)\mathbf{j}\} \text{ m/s}^2$$

At $t = 0.75$ s, $a = 2$ m/s^2 \qquad Ans.

NOTE: The velocity will be tangent to the path and its coordinate angles can be determined from $\mathbf{u}_v = \mathbf{v}/v$. The acceleration will *not* be tangent to the path, Fig. 12–19.

12.6   Motion of a Projectile

The free-flight motion of a projectile is often studied in terms of its rectangular components, since the projectile's acceleration *always* acts in the vertical direction. To illustrate the kinematic analysis, consider a projectile launched at point (x_0, y_0), as shown in Fig. 12–20. The path is defined in the *x–y* plane such that the initial velocity is $\mathbf{v}_0$, having components $(\mathbf{v}_0)_x$ and $(\mathbf{v}_0)_y$. When air resistance is neglected, the only force acting on the projectile is its weight, which causes the projectile to have a *constant downward acceleration* of approximately $a_c = g = 9.81$ m/s^2 or $g = 32.2$ ft/s^2.*

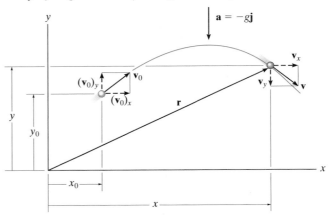

Fig. 12–20

Horizontal Motion.   Since $a_x = 0$, application of the constant acceleration equations, 12–4 to 12–6, yields

$$(\xrightarrow{+}) \quad v = v_0 + a_c t; \qquad\qquad v_x = (v_0)_x$$

$$(\xrightarrow{+}) \quad x = x_0 + v_0 t + \tfrac{1}{2}a_c t^2; \qquad x = x_0 + (v_0)_x t$$

$$(\xrightarrow{+}) \quad v^2 = v_0^2 + 2a_c(s - s_0); \qquad v_x = (v_0)_x$$

The first and last equations indicate that *the horizontal component of velocity always remains constant during the motion.*

Vertical Motion.   Since the positive *y* axis is directed upward, then $a_y = -g$. Applying Eqs. 12–4 to 12–6, we get

$$(+\uparrow) \quad v = v_0 + a_c t; \qquad\qquad v_y = (v_0)_y - gt$$

$$(+\uparrow) \quad y = y_0 + v_0 t + \tfrac{1}{2}a_c t^2; \qquad y = y_0 + (v_0)_y t - \tfrac{1}{2}gt^2$$

$$(+\uparrow) \quad v^2 = v_0^2 + 2a_c(s - s_0); \qquad v_y^2 = (v_0)_y^2 - 2g(y - y_0)$$

Recall that the last equation can be formulated on the basis of eliminating the time *t* between the first two equations, and therefore *only two of the above three equations are independent of one another.*

*This assumes that the earth's gravitational field does not vary with altitude.

Each picture in this sequence is taken after the same time interval. The red ball falls from rest, whereas the yellow ball is given a horizontal velocity when released. Both balls accelerate downward at the same rate, and so they remain at the same elevation at any instant. This acceleration causes the difference in elevation to increase between successive photos. Also, note the horizontal distance between successive photos of the yellow ball is constant since the velocity in the horizontal direction remains constant.

To summarize, problems involving the motion of a projectile can have at most three unknowns since only three independent equations can be written; that is, *one* equation in the *horizontal direction* and *two* in the *vertical direction*. Once $\mathbf{v}_x$ and $\mathbf{v}_y$ are obtained, the resultant velocity $\mathbf{v}$, which is *always tangent* to the path, is defined by the *vector sum* as shown in Fig. 12–20.

Procedure for Analysis

Free-flight projectile motion problems can be solved using the following procedure.

Coordinate System.

- Establish the fixed x, y coordinate axes and sketch the trajectory of the particle. Between any *two points* on the path specify the given problem data and the *three unknowns*. In all cases the acceleration of gravity acts downward. The particle's initial and final velocities should be represented in terms of their x and y components.

- Remember that positive and negative position, velocity, and acceleration components always act in accordance with their associated coordinate directions.

Kinematic Equations.

- Depending upon the known data and what is to be determined, a choice should be made as to which three of the following four equations should be applied between the two points on the path to obtain the most direct solution to the problem.

Horizontal Motion.

- The *velocity* in the horizontal or x direction is *constant*, i.e., $(v_x) = (v_0)_x$, and

$$x = x_0 + (v_0)_x t$$

Vertical Motion.

- In the vertical or y direction *only two* of the following three equations can be used for solution.

$$v_y = (v_0)_y + a_c t$$
$$y = y_0 + (v_0)_y t + \tfrac{1}{2} a_c t^2$$
$$v_y^2 = (v_0)_y^2 + 2a_c(y - y_0)$$

- For example, if the particle's final velocity v_y is not needed, then the first and third of these equations (for y) will not be useful.

Gravel falling off the end of this conveyor belt follows a path that can be predicted using the equations of constant acceleration. In this way the location of the accumulated pile can be determined. Rectilinear coordinates are used for the analysis since the acceleration is only in the vertical direction.

EXAMPLE 12.11

A sack slides off the ramp, shown in Fig. 12–21, with a horizontal velocity of 12 m/s. If the height of the ramp is 6 m from the floor, determine the time needed for the sack to strike the floor and the range R where sacks begin to pile up.

Fig. 12–21

SOLUTION

Coordinate System.   The origin of coordinates is established at the beginning of the path, point A, Fig. 12–21. The initial velocity of a sack has components $(v_A)_x = 12$ m/s and $(v_A)_y = 0$. Also, between points A and B the acceleration is $a_y = -9.81$ m/s^2. Since $(v_B)_x = (v_A)_x = 12$ m/s, the three unknowns are $(v_B)_y$, R, and the time of flight t_{AB}. Here we do not need to determine $(v_B)_y$.

Vertical Motion.   The vertical distance from A to B is known, and therefore we can obtain a direct solution for t_{AB} by using the equation

$(+\uparrow)$ $$y = y_0 + (v_0)_y t_{AB} + \tfrac{1}{2} a_c t_{AB}^2$$

$$-6 \text{ m} = 0 + 0 + \tfrac{1}{2}(-9.81 \text{ m/s}^2) t_{AB}^2$$

$$t_{AB} = 1.11 \text{ s} \qquad\qquad Ans.$$

Horizontal Motion.   Since t has been calculated, R is determined as follows:

$(\xrightarrow{+})$ $$x = x_0 + (v_0)_x t_{AB}$$

$$R = 0 + 12 \text{ m/s } (1.11 \text{ s})$$

$$R = 13.3 \text{ m} \qquad\qquad Ans.$$

NOTE: The calculation for t_{AB} also indicates that if a sack were released *from rest* at A, it would take the same amount of time to strike the floor at C, Fig. 12–21.

EXAMPLE 12.12

The chipping machine is designed to eject wood chips at $v_O = 25$ ft/s as shown in Fig. 12–22. If the tube is oriented at 30° from the horizontal, determine how high, h, the chips strike the pile if they land on the pile 20 ft from the tube.

$v_O = 25$ ft/s
30°
O
A
x
4 ft
h
20 ft

Fig. 12–22

SOLUTION

Coordinate System.   When the motion is analyzed between points O and A, the three unknowns are represented as the height h, time of flight t_{OA}, and vertical component of velocity $(v_A)_y$. (Note that $(v_A)_x = (v_O)_x$.) With the origin of coordinates at O, Fig. 12–22, the initial velocity of a chip has components of

$$(v_O)_x = (25 \cos 30°) \text{ ft/s} = 21.65 \text{ ft/s} \rightarrow$$

$$(v_O)_y = (25 \sin 30°) \text{ ft/s} = 12.5 \text{ ft/s} \uparrow$$

Also, $(v_A)_x = (v_O)_x = 21.65$ ft/s and $a_y = -32.2$ ft/s². Since we do not need to determine $(v_A)_y$, we have

Horizontal Motion.

$(\xrightarrow{+})$
$$x_A = x_O + (v_O)_x t_{OA}$$
$$20 \text{ ft} = 0 + (21.65 \text{ ft/s}) t_{OA}$$
$$t_{OA} = 0.9238 \text{ s}$$

Vertical Motion.   Relating t_{OA} to the initial and final elevations of a chip, we have

$(+\uparrow)$
$$y_A = y_O + (v_O)_y t_{OA} + \tfrac{1}{2} a_c t_{OA}^2$$
$$(h - 4 \text{ ft}) = 0 + (12.5 \text{ ft/s})(0.9238 \text{ s}) + \tfrac{1}{2}(-32.2 \text{ ft/s}^2)(0.9238 \text{ s})^2$$
$$h = 1.81 \text{ ft} \qquad\qquad\qquad\qquad\qquad\qquad Ans.$$

NOTE: We can determine $(v_A)_y$ by using $(v_A)_y = (v_O)_y + a_c t_{OA}$.

EXAMPLE 12.13

The track for this racing event was designed so that riders jump off the slope at 30°, from a height of 1 m. During a race it was observed that the rider shown in Fig. 12–23a remained in mid air for 1.5 s. Determine the speed at which he was traveling off the slope, the horizontal distance he travels before striking the ground, and the maximum height he attains. Neglect the size of the bike and rider.

(a)

(b)

Fig. 12–23

SOLUTION

Coordinate System.   As shown in Fig. 12–23b, the origin of the coordinates is established at A. Between the end points of the path AB the three unknowns are the initial speed v_A, range R, and the vertical component of velocity v_B.

Vertical Motion.   Since the time of flight and the vertical distance between the ends of the path are known, we can determine v_A.

$$(+\uparrow) \qquad (s_B)_y = (s_A)_y + (v_A)_y t_{AB} + \tfrac{1}{2} a_c t_{AB}^2$$

$$-1 = 0 + v_A \sin 30°(1.5) + \tfrac{1}{2}(-9.81)(1.5)^2$$

$$v_A = 13.38 \text{ m/s} = 13.4 \text{ m/s} \qquad Ans.$$

Horizontal Motion.   The range R can now be determined.

$$(\overset{+}{\rightarrow}) \qquad (s_B)_x = (s_A)_x + (v_A)_x t_{AB}$$

$$R = 0 + 13.38 \cos 30°(1.5)$$

$$= 17.4 \text{ m} \qquad Ans.$$

In order to find the maximum height h we will consider the path AC, Fig. 12–23b. Here the three unknowns become the time of flight t_{AC}, the horizontal distance from A to C, and the height h. At the maximum height $(v_C)_y = 0$, and since v_A is known, we can determine h *directly* without considering t_{AC} using the following equation.

$$(v_C)_y^2 = (v_A)_y^2 + 2a_c[(s_C)_y - (s_A)_y]$$

$$(0)^2 = (13.38 \sin 30°)^2 + 2(-9.81)[(h - 1) - 0]$$

$$h = 3.28 \text{ m} \qquad Ans.$$

NOTE: Show that the bike will strike the ground at B with a velocity having components of

$$(v_B)_x = 11.6 \text{ m/s} \rightarrow, \quad (v_B)_y = 8.02 \text{ m/s} \downarrow$$

PROBLEMS

12–66. A particle, originally at rest and located at point (3 ft, 2 ft, 5 ft), is subjected to an acceleration of $\mathbf{a} = \{6t\mathbf{i} + 12t^2\mathbf{k}\}$ ft/s². Determine the particle's position (x, y, z) at $t = 1$ s.

12–67. The velocity of a particle is given by $v = \{16t^2\mathbf{i} + 4t^3\mathbf{j} + (5t + 2)\mathbf{k}\}$ m/s, where t is in seconds. If the particle is at the origin when $t = 0$, determine the magnitude of the particle's acceleration when $t = 2$ s. Also, what is the x, y, z coordinate position of the particle at this instant?

*■12–68. A particle is traveling with a velocity of $\mathbf{v} = \{3\sqrt{t}e^{-0.2t}\mathbf{i} + 4e^{-0.8t^2}\mathbf{j}\}$ m/s, where t is in seconds. Determine the magnitude of the particle's displacement from $t = 0$ to $t = 3$ s. Use Simpson's rule with $n = 100$ to evaluate the integrals. What is the magnitude of the particle's acceleration when $t = 2$ s?

12–69. The position of a particle is defined by $r = \{5(\cos 2t)\mathbf{i} + 4(\sin 2t)\mathbf{j}\}$ m, where t is in seconds and the arguments for the sine and cosine are given in radians. Determine the magnitudes of the velocity and acceleration of the particle when $t = 1$ s. Also, prove that the path of the particle is elliptical.

12–70. A particle travels along the curve from A to B in 1 s. If it takes 3 s for it to go from A to C, determine its *average velocity* when it goes from B to C.

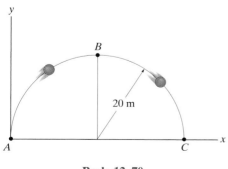

Prob. 12–70

12–71. A particle travels along the curve from A to B in 2 s. It takes 4 s for it to go from B to C and then 3 s to go from C to D. Determine its average speed when it goes from A to D.

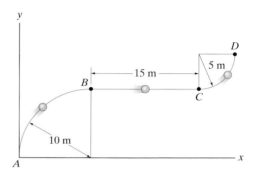

Prob. 12–71

*12–72. A car travels east 2 km for 5 minutes, then north 3 km for 8 minutes, and then west 4 km for 10 minutes. Determine the total distance traveled and the magnitude of displacement of the car. Also, what is the magnitude of the average velocity and the average speed?

12–73. A car traveling along the straight portions of the road has the velocities indicated in the figure when it arrives at points A, B, and C. If it takes 3 s to go from A to B, and then 5 s to go from B to C, determine the average acceleration between points A and B and between points A and C.

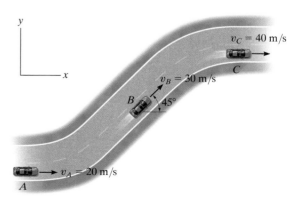

Prob. 12–73

12–74. A particle moves along the curve $y = e^{2x}$ such that its velocity has a constant magnitude of $v = 4$ ft/s. Determine the x and y components of velocity when the particle is at $y = 5$ ft.

12–75. The path of a particle is defined by $y^2 = 4kx$, and the component of velocity along the y axis is $v_y = ct$, where both k and c are constants. Determine the x and y components of acceleration.

*12–76. A particle is moving along the curve $y = x - (x^2/400)$, where x and y are in ft. If the velocity component in the x direction is $v_x = 2$ ft/s and remains *constant*, determine the magnitudes of the velocity and acceleration when $x = 20$ ft.

12–77. The flight path of the helicopter as it takes off from A is defined by the parametric equations $x = (2t^2)$ m and $y = (0.04t^3)$ m, where t is the time in seconds. Determine the distance the helicopter is from point A and the magnitudes of its velocity and acceleration when $t = 10$ s.

Prob. 12–77

12–78. At the instant shown particle A is traveling to the right at 10 ft/s and has an acceleration of 2 ft/s². Determine the initial speed v_0 of particle B so that when it is fired at the same instant from the angle shown it strikes A. Also, at what speed does it strike A?

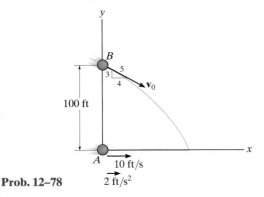

Prob. 12–78

12–79. When a rocket reaches an altitude of 40 m it begins to travel along the parabolic path $(y - 40)^2 = 160x$, where the coordinates are measured in meters. If the component of velocity in the vertical direction is constant at $v_y = 180$ m/s, determine the magnitudes of the rocket's velocity and acceleration when it reaches an altitude of 80 m.

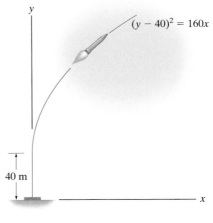

Prob. 12–79

*12–80. Determine the minimum speed of the stunt rider, so that when he leaves the ramp at A he passes through the center of the hoop at B. Also, how far h should the landing ramp be from the hoop so that he lands on it safely at C? Neglect the size of the motorcycle and rider.

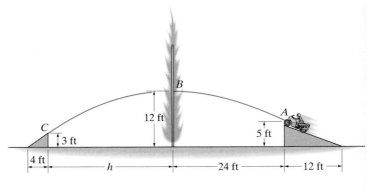

Prob. 12–80

12–81. Show that if a projectile is fired at an angle θ from the horizontal with an initial velocity $\mathbf{v}_0$, the *maximum* range the projectile can travel is given by $R_{max} = v_0^2/g$, where g is the acceleration of gravity. What is the angle θ for this condition?

12–82. The balloon A is ascending at the rate $v_A = 12$ km/h and is being carried horizontally by the wind at $v_w = 20$ km/h. If a ballast bag is dropped from the balloon at the instant $h = 50$ m, determine the time needed for it to strike the ground. Assume that the bag was released from the balloon with the same velocity as the balloon. Also, with what speed does the bag strike the ground?

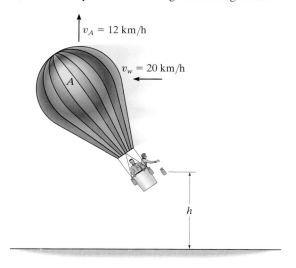

Prob. 12–82

12–83. Determine the height on the wall to which the firefighter can project water from the hose, if $\theta = 40°$ and the speed of the water at the nozzle is $v_C = 48$ ft/s.

*12–84. Determine the smallest angle θ, measured above the horizontal, that the hose should be directed so that the water stream strikes the bottom of the wall at B. The speed of the water at the nozzle is $v_C = 48$ ft/s.

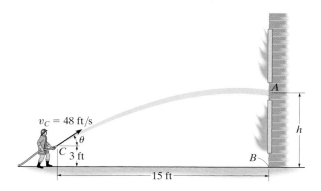

Probs. 12–83/84

12–85. The catapult is used to launch a ball such that it strikes the wall of the building at the maximum height of its trajectory. If it takes 1.5 s to travel from A to B, determine the velocity $\mathbf{v}_A$ at which it was launched, the angle of release θ, and the height h.

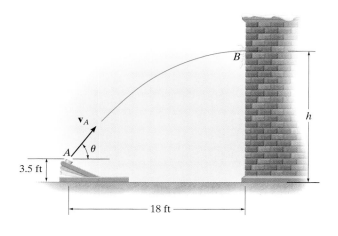

Prob. 12–85

12–86. The buckets on the conveyor travel with a speed of 15 ft/s. Each bucket contains a block which falls out of the bucket when $\theta = 120°$. Determine the distance s to where the block strikes the conveyor. Neglect the size of the block.

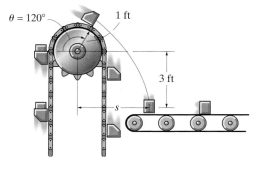

Prob. 12–86

12–87. Measurements of a shot recorded on a videotape during a basketball game are shown. The ball passed through the hoop even though it barely cleared the hands of the player B who attempted to block it. Neglecting the size of the ball, determine the magnitude v_A of its initial velocity and the height h of the ball when it passes over player B.

Prob. 12–87

12–89. The projectile is launched with a velocity $\mathbf{v}_0$. Determine the range R, the maximum height h attained, and the time of flight. Express the results in terms of the angle θ and v_0. The acceleration due to gravity is g.

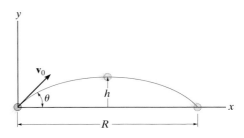

Prob. 12–89

*12–88. The snowmobile is traveling at 10 m/s when it leaves the embankment at A. Determine the time of flight from A to B and the range R of the trajectory.

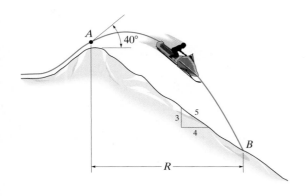

Prob. 12–88

12–90. The fireman standing on the ladder directs the flow of water from his hose to the fire at B. Determine the velocity of the water at A if it is observed that the hose is held at $\theta = 20°$.

Prob. 12–90

12–91. A ball bounces on the 30° inclined plane such that it rebounds perpendicular to the incline with a velocity of $v_A = 40$ ft/s. Determine the distance R to where it strikes the plane at B.

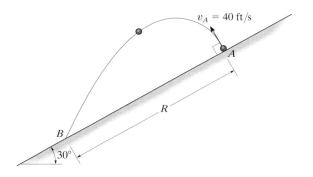

Prob. 12–91

*12–92.** The man stands 60 ft from the wall and throws a ball at it with a speed $v_0 = 50$ ft/s. Determine the angle θ at which he should release the ball so that it strikes the wall at the highest point possible. What is this height? The room has a ceiling height of 20 ft.

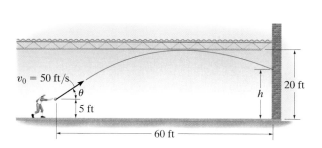

Prob. 12–92

12–93. The stones are thrown off the conveyor with a horizontal velocity of 10 ft/s as shown. Determine the distance d down the slope to where the stones hit the ground at B.

12–94. The stones are thrown off the conveyor with a horizontal velocity of 10 ft/s as shown. Determine the speed at which the stones hit the ground at B.

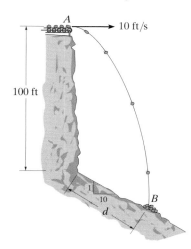

Probs. 12–93/94

12–95. The drinking fountain is designed such that the nozzle is located from the edge of the basin as shown. Determine the maximum and minimum speed at which water can be ejected from the nozzle so that it does not splash over the sides of the basin at B and C.

Prob. 12–95

*12–96.   A boy at O throws a ball in the air with a speed v_O at an angle θ_1. If he then throws another ball at the same speed v_O at an angle $\theta_2 < \theta_1$, determine the time between the throws so the balls collide in mid air at B.

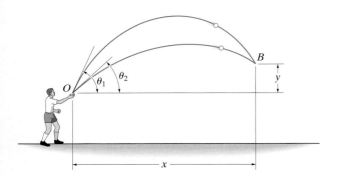

Prob. 12–96

12–97.   The man at A wishes to throw two darts at the target at B so that they arrive at the *same time*. If each dart is thrown with a speed of 10 m/s, determine the angles θ_C and θ_D at which they should be thrown and the time between each throw. Note that the first dart must be thrown at θ_C ($>\theta_D$), then the second dart is thrown at θ_D.

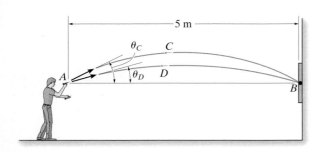

Prob. 12–97

12–98.   The water sprinkler, positioned at the base of a hill, releases a stream of water with a velocity of 15 ft/s as shown. Determine the point $B(x, y)$ where the water strikes the ground on the hill. Assume that the hill is defined by the equation $y = (0.05x^2)$ ft and neglect the size of the sprinkler.

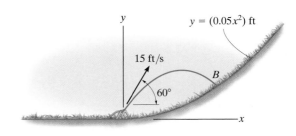

Prob. 12–98

12–99.   The projectile is launched from a height h with a velocity $\mathbf{v}_0$. Determine the range R.

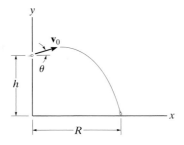

Prob. 12–99

12.7   Curvilinear Motion: Normal and Tangential Components

When the path along which a particle is moving is *known*, then it is often convenient to describe the motion using n and t coordinates which act normal and tangent to the path, respectively, and at the instant considered have their *origin located at the particle*.

Planar Motion.   Consider the particle P shown in Fig. 12–24a, which is moving in a plane along a fixed curve, such that at a given instant it is at position s, measured from point O. We will now consider a coordinate system that has its origin at a *fixed point* on the curve, and at the instant considered this origin happens to *coincide* with the location of the particle. The t axis is *tangent* to the curve at P and is positive in the direction of *increasing* s. We will designate this positive direction with the unit vector $\mathbf{u}_t$. A unique choice for the *normal axis* can be made by noting that geometrically the curve is constructed from a series of differential arc segments ds, Fig. 12–24b. Each segment ds is formed from the arc of an associated circle having a *radius of curvature* ρ (rho) and *center of curvature* O'. The normal axis n is perpendicular to the t axis and is directed from P *toward* the center of curvature O', Fig. 12–24a. This positive direction, which is *always* on the concave side of the curve, will be designated by the unit vector $\mathbf{u}_n$. The plane which contains the n and t axes is referred to as the *osculating plane*, and in this case it is fixed in the plane of motion.*

Velocity.   Since the particle is moving, s is a function of time. As indicated in Sec. 12.4, the particle's velocity $\mathbf{v}$ has a *direction* that is *always tangent to the path*, Fig. 12–24c, and a *magnitude* that is determined by taking the time derivative of the path function $s = s(t)$, i.e., $v = ds/dt$ (Eq. 12–8). Hence

$$\boxed{\mathbf{v} = v\mathbf{u}_t} \qquad (12\text{–}15)$$

where

$$\boxed{v = \dot{s}} \qquad (12\text{–}16)$$

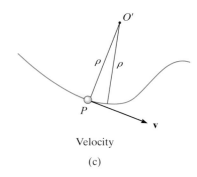

Position

(a)

Radius of curvature

(b)

Velocity

(c)

Fig. 12–24

---

*The osculating plane may also be defined as that plane which has the greatest contact with the curve at a point. It is the limiting position of a plane contacting both the point and the arc segment ds. As noted above, the osculating plane is always coincident with a plane curve; however, each point on a three-dimensional curve has a unique osculating plane.

(d)

(e)

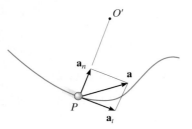

Acceleration

(f)

Fig. 12–24

Acceleration.   The acceleration of the particle is the time rate of change of the velocity. Thus,

$$\mathbf{a} = \dot{\mathbf{v}} = \dot{v}\mathbf{u}_t + v\dot{\mathbf{u}}_t \tag{12–17}$$

In order to determine the time derivative $\dot{\mathbf{u}}_t$, note that as the particle moves along the arc ds in time dt, $\mathbf{u}_t$ preserves its magnitude of unity; however, its *direction* changes, and becomes $\mathbf{u}_t'$, Fig. 12–24d. As shown in Fig. 12–24e, we require $\mathbf{u}_t' = \mathbf{u}_t + d\mathbf{u}_t$. Here $d\mathbf{u}_t$ stretches between the arrowheads of $\mathbf{u}_t$ and $\mathbf{u}_t'$, which lie on an infinitesimal arc of radius $u_t = 1$. Hence, $d\mathbf{u}_t$ has a *magnitude* of $du_t = (1)\, d\theta$, and its *direction* is defined by $\mathbf{u}_n$. Consequently, $d\mathbf{u}_t = d\theta \mathbf{u}_n$, and therefore the time derivative becomes $\dot{\mathbf{u}}_t = \dot{\theta}\mathbf{u}_n$. Since $ds = \rho\, d\theta$, Fig. 12–24d, then $\dot{\theta} = \dot{s}/\rho$, and therefore

$$\dot{\mathbf{u}}_t = \dot{\theta}\mathbf{u}_n = \frac{\dot{s}}{\rho}\mathbf{u}_n = \frac{v}{\rho}\mathbf{u}_n$$

Substituting into Eq. 12–17, **a** can be written as the sum of its two components,

$$\boxed{\mathbf{a} = a_t\mathbf{u}_t + a_n\mathbf{u}_n} \tag{12–18}$$

where

$$\boxed{a_t = \dot{v}} \qquad \text{or} \qquad \boxed{a_t\, ds = v\, dv} \tag{12–19}$$

and

$$\boxed{a_n = \frac{v^2}{\rho}} \tag{12–20}$$

These two mutually perpendicular components are shown in Fig. 12–24f, in which case the *magnitude* of acceleration is the positive value of

$$a = \sqrt{a_t^2 + a_n^2} \tag{12–21}$$

To summarize these concepts, consider the following two special cases of motion.

1. If the particle moves along a straight line, then $\rho \to \infty$ and from Eq. 12–20, $a_n = 0$. Thus $a = a_t = \dot{v}$, and we can conclude that the *tangential component of acceleration represents the time rate of change in the magnitude of the velocity.*

2. If the particle moves along a curve with a constant speed, then $a_t = \dot{v} = 0$ and $a = a_n = v^2/\rho$. Therefore, the *normal component of acceleration represents the time rate of change in the direction of the velocity.* Since $\mathbf{a}_n$ *always* acts towards the center of curvature, this component is sometimes referred to as the *centripetal acceleration.*

As a result of these interpretations, a particle moving along the curved path in Fig. 12–25 will have accelerations directed as shown.

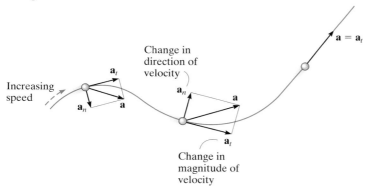

Fig. 12–25

Three-Dimensional Motion.   If the particle is moving along a space curve, Fig. 12–26, then at a given instant the t axis is uniquely specified; however, an infinite number of straight lines can be constructed normal to the tangent axis at P. As in the case of planar motion, we will choose the positive n axis directed from P toward the path's center of curvature O'. This axis is referred to as the *principal normal* to the curve at P. With the n and t axes so defined, Eqs. 12–15 to 12–21 can be used to determine $\mathbf{v}$ and $\mathbf{a}$. Since $\mathbf{u}_t$ and $\mathbf{u}_n$ are always perpendicular to one another and lie in the osculating plane, for spatial motion a third unit vector, $\mathbf{u}_b$, defines a *binormal axis* b which is perpendicular to $\mathbf{u}_t$ and $\mathbf{u}_n$, Fig. 12–26.

Since the three unit vectors are related to one another by the vector cross product, e.g., $\mathbf{u}_b = \mathbf{u}_t \times \mathbf{u}_n$, Fig. 12–26, it may be possible to use this relation to establish the direction of one of the axes, if the directions of the other two are known. For example, no motion occurs in the $\mathbf{u}_b$ direction, and so if this direction and $\mathbf{u}_t$ are known, then $\mathbf{u}_n$ can be determined, where in this case $\mathbf{u}_n = \mathbf{u}_b \times \mathbf{u}_t$, Fig. 12–26. Remember, though, that $\mathbf{u}_n$ is always on the concave side of the curve.

Fig. 12–26

Motorists traveling along this clover-leaf interchange experience a normal acceleration due to the change in direction of their velocity. A tangential component of acceleration occurs when the cars' speed is increased or decreased.

Procedure for Analysis

Coordinate System.

- Provided the *path* of the particle is *known*, we can establish a set of *n* and *t* coordinates having a *fixed origin,* which is coincident with the particle at the instant considered.
- The positive tangent axis acts in the direction of motion and the positive normal axis is directed toward the path's center of curvature.
- The *n* and *t* axes are particularly advantageous for studying the velocity and acceleration of the particle, because the *t* and *n* components of **a** are expressed by Eqs. 12–19 and 12–20, respectively.

Velocity.

- The particle's *velocity* is always tangent to the path.
- The magnitude of velocity is found from the time derivative of the path function.

$$v = \dot{s}$$

Tangential Acceleration.

- The tangential component of acceleration is the result of the time rate of change in the magnitude of velocity. This component acts in the positive *s* direction if the particle's speed is increasing or in the opposite direction if the speed is decreasing.
- The relations between a_t, v, t and s are the same as for rectilinear motion, namely,

$$a_t = \dot{v} \qquad a_t \, ds = v \, dv$$

- If a_t is constant, $a_t = (a_t)_c$, the above equations, when integrated, yield

$$s = s_0 + v_0 t + \tfrac{1}{2}(a_t)_c t^2$$
$$v = v_0 + (a_t)_c t$$
$$v^2 = v_0^2 + 2(a_t)_c(s - s_0)$$

Normal Acceleration.

- The normal component of acceleration is the result of the time rate of change in the direction of the particle's velocity. This component is *always* directed toward the center of curvature of the path, i.e., along the positive *n* axis.
- The magnitude of this component is determined from

$$a_n = \frac{v^2}{\rho}$$

- If the path is expressed as $y = f(x)$, the radius of curvature ρ at any point on the path is determined from the equation

$$\rho = \frac{[1 + (dy/dx)^2]^{3/2}}{|d^2 y/dx^2|}$$

The derivation of this result is given in any standard calculus text.

EXAMPLE 12.14

When the skier reaches point A along the parabolic path in Fig. 12–27a, he has a speed of 6 m/s which is increasing at 2 m/s^2. Determine the direction of his velocity and the direction and magnitude of his acceleration at this instant. Neglect the size of the skier in the calculation.

SOLUTION

Coordinate System.   Although the path has been expressed in terms of its x and y coordinates, we can still establish the origin of the n, t axes at the fixed point A on the path and determine the components of $\mathbf{v}$ and $\mathbf{a}$ along these axes, Fig. 12–27a.

Velocity.   By definition, the velocity is always directed tangent to the path. Since $y = \frac{1}{20}x^2$, $dy/dx = \frac{1}{10}x$, then $dy/dx|_{x=10} = 1$. Hence, at A, $\mathbf{v}$ makes an angle of $\theta = \tan^{-1} 1 = 45°$ with the x axis, Fig. 12–27. Therefore,

$$v_A = 6 \text{ m/s} \qquad 45° \nearrow \mathbf{v}_A \qquad\qquad Ans.$$

(a)

Acceleration.   The acceleration is determined from $\mathbf{a} = \dot{v}\mathbf{u}_t + (v^2/\rho)\mathbf{u}_n$. However, it is first necessary to determine the radius of curvature of the path at A (10 m, 5 m). Since $d^2y/dx^2 = \frac{1}{10}$, then

$$\rho = \frac{[1 + (dy/dx)^2]^{3/2}}{|d^2y/dx^2|} = \frac{[1 + \left(\frac{1}{10}x\right)^2]^{3/2}}{\left|\frac{1}{10}\right|}\bigg|_{x=10 \text{ m}} = 28.28 \text{ m}$$

The acceleration becomes

$$\mathbf{a}_A = \dot{v}\mathbf{u}_t + \frac{v^2}{\rho}\mathbf{u}_n$$

$$= 2\mathbf{u}_t + \frac{(6 \text{ m/s})^2}{28.28 \text{ m}}\mathbf{u}_n$$

$$= \{2\mathbf{u}_t + 1.273\mathbf{u}_n\} \text{ m/s}^2$$

As shown in Fig. 12–27b,

$$a = \sqrt{(2)^2 + (1.273)^2} = 2.37 \text{ m/s}^2$$

$$\phi = \tan^{-1}\frac{2}{1.273} = 57.5°$$

(b)

Fig. 12–27

Thus, $57.5° - 45° = 12.5°$ so that,

$$a = 2.37 \text{ m/s}^2 \qquad 12.5° \nwarrow \mathbf{a}_A \qquad\qquad Ans.$$

NOTE: By using n, t coordinates, we were able to readily solve this problem since the n and t components account for the *separate* changes in the magnitude and direction of $\mathbf{v}$.

EXAMPLE 12.15

A race car C travels around the horizontal circular track that has a radius of 300 ft, Fig. 12–28. If the car increases its speed at a constant rate of 7 ft/s^2, starting from rest, determine the time needed for it to reach an acceleration of 8 ft/s^2. What is its speed at this instant?

Fig. 12–28

SOLUTION

Coordinate System. The origin of the n and t axes is coincident with the car at the instant considered. The t axis is in the direction of motion, and the positive n axis is directed toward the center of the circle. This coordinate system is selected since the path is known.

Acceleration. The magnitude of acceleration can be related to its components using $a = \sqrt{a_t^2 + a_n^2}$. Here $a_t = 7$ ft/s^2. Since $a_n = v^2/\rho$, the velocity as a function of time is

$$v = v_0 + (a_t)_c t$$
$$v = 0 + 7t$$

Thus

$$a_n = \frac{v^2}{\rho} = \frac{(7t)^2}{300} = 0.163t^2 \text{ ft/s}^2$$

The time needed for the acceleration to reach 8 ft/s^2 is therefore

$$a = \sqrt{a_t^2 + a_n^2}$$
$$8 = \sqrt{(7)^2 + (0.163t^2)^2}$$

Solving for the positive value of t yields

$$0.163t^2 = \sqrt{(8)^2 - (7)^2}$$
$$t = 4.87 \text{ s} \qquad\qquad Ans.$$

Velocity. The speed at time $t = 4.87$ s is

$$v = 7t = 7(4.87) = 34.1 \text{ ft/s} \qquad\qquad Ans.$$

NOTE: Remember the velocity will be tangent to the path, whereas the acceleration will be directed inwards, towards the $+n$ axis.

EXAMPLE 12.16

The boxes in Fig. 12–29a travel along the industrial conveyor. If a box as in Fig. 12–29b starts from rest at A and increases its speed such that $a_t = (0.2t)$ m/s^2, where t is in seconds, determine the magnitude of its acceleration when it arrives at point B.

SOLUTION

Coordinate System.   The position of the box at any instant is defined from the fixed point A using the position or path coordinate s, Fig. 12–29b. The acceleration is to be determined at B, so the origin of the n, t axes is at this point.

Acceleration.   To determine the acceleration components $a_t = \dot{v}$ and $a_n = v^2/\rho$, it is first necessary to formulate v and $\dot{v}$ so that they may be evaluated at B. Since $v_A = 0$ when $t = 0$, then

$$a_t = \dot{v} = 0.2t \qquad (1)$$

$$\int_0^v dv = \int_0^t 0.2t\, dt$$

$$v = 0.1t^2 \qquad (2)$$

(a)

The time needed for the box to reach point B can be determined by realizing that the position of B is $s_B = 3 + 2\pi(2)/4 = 6.142$ m, Fig. 12–29b, and since $s_A = 0$ when $t = 0$ we have

$$v = \frac{ds}{dt} = 0.1t^2$$

$$\int_0^{6.142} ds = \int_0^{t_B} 0.1t^2\, dt$$

$$6.142 = 0.0333t_B^3$$

$$t_B = 5.690 \text{ s}$$

Substituting into Eqs. 1 and 2 yields

$$(a_B)_t = \dot{v}_B = 0.2(5.690) = 1.138 \text{ m/s}^2$$

$$v_B = 0.1(5.69)^2 = 3.238 \text{ m/s}$$

At B, $\rho_B = 2$ m, so that

$$(a_B)_n = \frac{v_B^2}{\rho_B} = \frac{(3.238 \text{ m/s})^2}{2 \text{ m}} = 5.242 \text{ m/s}^2$$

The magnitude of $\mathbf{a}_B$, Fig. 12–29c, is therefore

$$a_B = \sqrt{(1.138)^2 + (5.242)^2} = 5.36 \text{ m/s}^2 \qquad Ans.$$

(b)

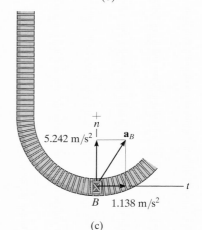

(c)

Fig. 12–29

PROBLEMS

*12–100. A car is traveling along a circular curve that has a radius of 50 m. If its speed is 16 m/s and is increasing uniformly at 8 m/s^2, determine the magnitude of its acceleration at this instant.

12–101. A car moves along a circular track of radius 250 ft such that its speed for a short period of time, $0 \le t \le 4$ s, is $v = 3(t + t^2)$ ft/s, where t is in seconds. Determine the magnitude of its acceleration when $t = 3$ s. How far has it traveled in $t = 3$ s?

12–102. At a given instant the jet plane has a speed of 400 ft/s and an acceleration of 70 ft/s^2 acting in the direction shown. Determine the rate of increase in the plane's speed and the radius of curvature ρ of the path.

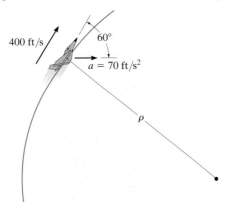

Prob. 12–102

12–103. A particle is moving along a curved path at a constant speed of 60 ft/s. The radii of curvature of the path at points P and P' are 20 and 50 ft, respectively. If it takes the particle 20 s to go from P to P', determine the acceleration of the particle at P and P'.

*12–104. A boat is traveling along a circular path having a radius of 20 m. Determine the magnitude of the boat's acceleration when the speed is $v = 5$ m/s and the rate of increase in the speed is $\dot{v} = 2$ m/s^2.

■12–105. Starting from rest, a bicyclist travels around a horizontal circular path, $\rho = 10$ m, at a speed of $v = (0.09t^2 + 0.1t)$ m/s, where t is in seconds. Determine the magnitudes of his velocity and acceleration when he has traveled $s = 3$ m.

12–106. The jet plane travels along the vertical parabolic path. When it is at point A it has a speed of 200 m/s, which is increasing at the rate of 0.8 m/s^2. Determine the magnitude of acceleration of the plane when it is at point A.

Prob. 12–106

12–107. The car travels along the curve having a radius of 300 m. If its speed is uniformly increased from 15 m/s to 27 m/s in 3 s, determine the magnitude of its acceleration at the instant its speed is 20 m/s.

Prob. 12–107

*12–108. The satellite S travels around the earth in a circular path with a constant speed of 20 Mm/h. If the acceleration is 2.5 m/s², determine the altitude h. Assume the earth's diameter to be 12 713 km.

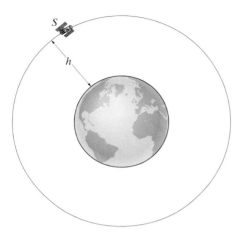

Prob. 12–108

12–109. A particle P moves along the curve $y = (x^2 - 4)$ m with a constant speed of 5 m/s. Determine the point on the curve where the maximum magnitude of acceleration occurs and compute its value.

12–110. The Ferris wheel turns such that the speed of the passengers is increased by $\dot{v} = (4t)$ ft/s², where t is in seconds. If the wheel starts from rest when $\theta = 0°$, determine the magnitudes of the velocity and acceleration of the passengers when the wheel turns $\theta = 30°$.

Prob. 12–110

12–111. At a given instant the train engine at E has a speed of 20 m/s and an acceleration of 14 m/s² acting in the direction shown. Determine the rate of increase in the train's speed and the radius of curvature ρ of the path.

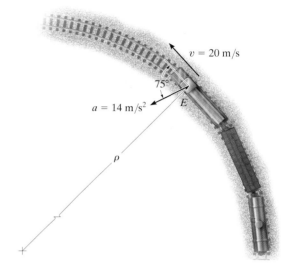

Prob. 12–111

*12–112. A package is dropped from the plane which is flying with a constant horizontal velocity of $v_A = 150$ ft/s. Determine the normal and tangential components of acceleration and the radius of curvature of the path of motion (a) at the moment the package is released at A, where it has a horizontal velocity of $v_A = 150$ ft/s, and (b) *just before* it strikes the ground at B.

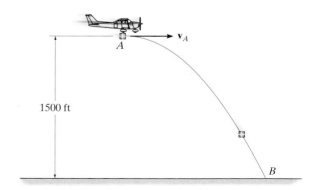

Prob. 12–112

12–113. The automobile is originally at rest at $s = 0$. If its speed is increased by $\dot{v} = (0.05t^2)$ ft/s^2, where t is in seconds, determine the magnitudes of its velocity and acceleration when $t = 18$ s.

Prob. 12–113

12–114. The automobile is originally at rest $s = 0$. If it then starts to increase its speed at $\dot{v} = (0.05t^2)$ ft/s^2, where t is in seconds, determine the magnitudes of its velocity and acceleration at $s = 550$ ft.

Prob. 12–114

12–115. The truck travels in a circular path havng a radius of 50 m at a speed of $4 = $ m/s. For a short distance from $s = 0$, its speed is increased by $\dot{v} = (0.05s)$ m/s^2, where s is in meters. Determine its speed and the magnitude of its acceleration when it has moved $s = 10$ m.

Prob. 12–115

*__12–116.__ The particle travels with a constant speed of 300 mm/s along the curve. Determine the particle's acceleration when it is located at point (200 mm, 100 mm) and sketch this vector on the curve.

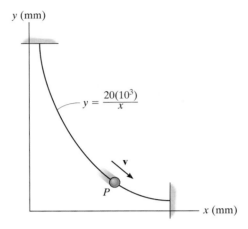

Prob. 12–116

12–117. Cars move around the "traffic circle" which is in the shape of an ellipse. If the speed limit is posted at 60 km/h, determine the maximum acceleration experienced by the passengers.

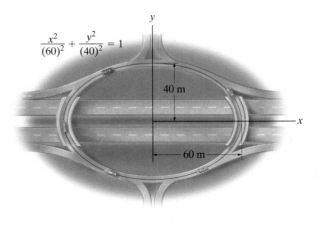

Prob. 12–117

12–118. Cars move around the "traffic circle" which is in the shape of an ellipse. If the speed limit is posted at 60 km/h, determine the minimum acceleration experienced by the passengers.

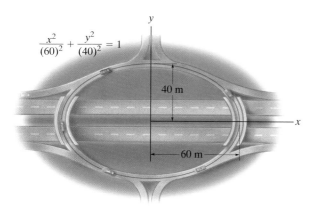

$$\frac{x^2}{(60)^2} + \frac{y^2}{(40)^2} = 1$$

40 m

60 m

Prob. 12–118

*12–120.** The car B turns such that its speed is increased by $\dot{v}_B = (0.5e^t)$ m/s², where t is in seconds. If the car starts from rest when $\theta = 0°$, determine the magnitudes of its velocity and acceleration when $t = 2$ s. Neglect the size of the car. Also, through what angle θ has it traveled?

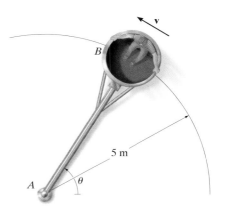

B

5 m

A θ

Prob. 12–120

12–119. The car B turns such that its speed is increased by $\dot{v}_B = (0.5e^t)$ m/s², where t is in seconds. If the car starts from rest when $\theta = 0°$, determine the magnitudes of its velocity and acceleration when the arm AB rotates $\theta = 30°$. Neglect the size of the car.

B

5 m

A θ

Prob. 12–119

12–121. The motorcycle is traveling at 1 m/s when it is at A. If the speed is then increased at $\dot{v} = 0.1$ m/s², determine its speed and acceleration at the instant $t = 5$ s.

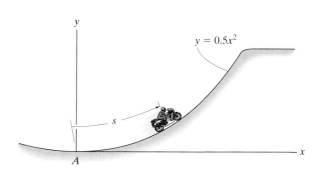

y

$y = 0.5x^2$

s

A   x

Prob. 12–121

12–122. The ball is ejected horizontally from the tube with a speed of 8 m/s. Find the equation of the path, $y = f(x)$, and then find the ball's velocity and the normal and tangential components of acceleration when $t = 0.25$ s.

Prob. 12–122

■12–123. The car travels around the circular track having a radius of $r = 300$ m such that when it is at point A it has a velocity of 5 m/s, which is increasing at the rate of $\dot{v} = (0.06t)$ m/s^2, where t is in seconds. Determine the magnitudes of its velocity and acceleration when it has traveled one-third the way around the track.

*12–124. The car travels around the portion of a circular track having a radius of $r = 500$ ft such that when it is at point A it has a velocity of 2 ft/s, which is increasing at the rate of $\dot{v} = (0.002s)$ ft/s^2, where t is in seconds. Determine the magnitudes of its velocity and acceleration when it has traveled three-fourths the way around the track.

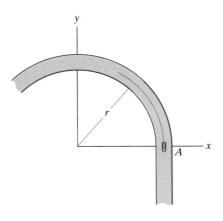

Prob. 12–124

12–125. The two particles A and B start at the origin O and travel in opposite directions along the circular path at constant speeds $v_A = 0.7$ m/s and $v_B = 1.5$ m/s, respectively. Determine in $t = 2$ s, (a) the displacement along the path of each particle, (b) the position vector to each particle, and (c) the shortest distance between the particles.

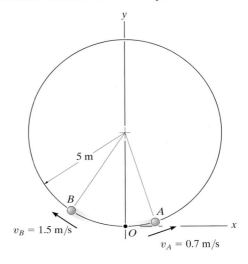

Prob. 12–125

12–126. The two particles A and B start at the origin O and travel in opposite directions along the circular path at constant speeds $v_A = 0.7$ m/s and $v_B = 1.5$ m/s, respectively. Determine the time when they collide and the magnitude of the acceleration of B just before this happens.

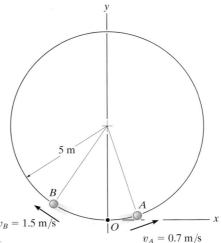

Prob. 12–126

12–127. The race car has an initial speed $v_A = 15$ m/s at A. If it increases its speed along the circular track at the rate $a_t = (0.4s)$ m/s², where s is in meters, determine the time needed for the car to travel 20 m. Take $\rho = 150$ m.

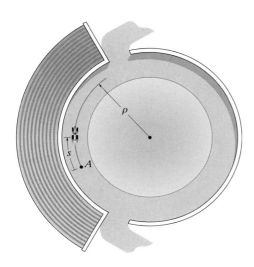

Prob. 12–127

*12–128. A boy sits on a merry-go-round so that he is always located at $r = 8$ ft from the center of rotation. The merry-go-round is originally at rest, and then due to rotation the boy's speed is increased at 2 ft/s². Determine the time needed for his acceleration to become 4 ft/s².

12–129. A particle moves along the curve $y = \sin x$ with a constant speed $v = 2$ m/s. Determine the normal and tangential components of its velocity and acceleration at any instant.

12–130. The motion of a particle along a fixed path is defined by the parametric equations $r = 8$ ft, $\theta = (4t)$ rad, and $z = (6t^2)$ ft, where t is in seconds. Determine the unit vector that specifies the direction of the binormal axis to the osculating plane with respect to a set of fixed x, y, z coordinate axes when $t = 2$ s. *Hint:* Formulate the particle's velocity $\mathbf{v}_P$ and acceleration $\mathbf{a}_P$ in terms of their $\mathbf{i}, \mathbf{j}, \mathbf{k}$ components. Note that $x = r \cos \theta$ and $y = r \sin \theta$. The binormal is parallel to $\mathbf{v}_P \times \mathbf{a}_P$. Why?

■**12–131.** Particles A and B are traveling counterclockwise around a circular track at a constant speed of 8 m/s. If at the instant shown the speed of A is increased by $\dot{v}_A = (4s_A)$ m/s², where s_A is in meters, determine the distance measured counterclockwise along the track from B to A when $t = 1$ s. What is the magnitude of the acceleration of each particle at this instant?

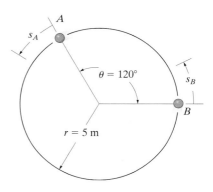

Prob. 12–131

*12–132. Particles A and B are traveling around a circular track at a speed of 8 m/s at the instant shown. If the speed of B is increased by $\dot{v}_B = 4$ m/s², and at the same instant A has an increase in speed $\dot{v}_A = 0.8t$ m/s², determine how long it takes for a collision to occur. What is the magnitude of the acceleration of each particle just before the collision occurs?

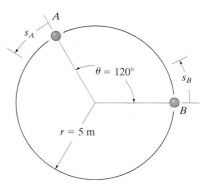

Prob. 12–132

12–133. The truck travels at a speed of 4 m/s along a circular road that has a radius of 50 m. For a short distance from $s = 0$, its speed is then increased by $\dot{v} = (0.05s)$ m/s^2, where s is in meters. Determine its speed and the magnitude of its acceleration when it has moved $s = 10$ m.

50 m

Prob. 12–133

■**12–134.** A go-cart moves along a circular track of radius 100 ft such that its speed for a short period of time, $0 \leq t \leq 4$ s, is $v = 60(1 - e^{-t^2})$ ft/s. Determine the magnitude of its acceleration when $t = 2$ s. How far has it traveled in $t = 2$ s? Use Simpson's rule with $n = 50$ to evaluate the integral.

12–135. A particle P travels along an elliptical spiral path such that its position vector $\mathbf{r}$ is defined by $\mathbf{r} = \{2\cos(0.1t)\mathbf{i} + 1.5\sin(0.1t)\mathbf{j} + (2t)\mathbf{k}\}$ m, where t is in seconds and the arguments for the sine and cosine are given in radians. When $t = 8$ s, determine the coordinate direction angles α, β, and γ, which the binormal axis to the osculating plane makes with the x, y, and z axes. *Hint:* Solve for the velocity $\mathbf{v}_P$ and acceleration $\mathbf{a}_P$ of the particle in terms of their $\mathbf{i}$, $\mathbf{j}$, $\mathbf{k}$ components. The binormal is parallel to $\mathbf{v}_P \times \mathbf{a}_P$. Why?

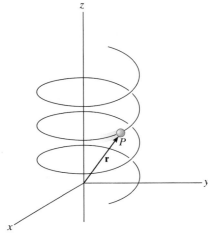

Prob. 12–135

12.8 Curvilinear Motion: Cylindrical Components

Some engineering problems involve angular position and a radial distance. For these cases, it is often convenient to express the path of motion in terms of cylindrical coordinates, r, θ, z. If motion is restricted to the plane, the polar coordinates r and θ are used.

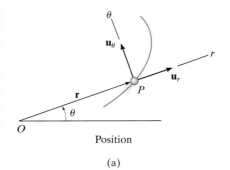

Position

(a)

Fig. 12–30

Polar Coordinates.   We can specify the location of particle P shown in Fig. 12–30a using both the *radial coordinate r*, which extends outward from the fixed origin O to the particle, and a *transverse coordinate* θ, which is the counterclockwise angle between a fixed reference line and the r axis. The angle is generally measured in degrees or radians, where 1 rad $= 180°/\pi$. The positive directions of the r and θ coordinates are defined by the unit vectors $\mathbf{u}_r$ and $\mathbf{u}_\theta$, respectively. Here $\mathbf{u}_r$ or the radial direction $+r$ extends from P along increasing r, when θ is held fixed, and $\mathbf{u}_\theta$ or $+\theta$ extends from P in a direction that occurs when r is held fixed and θ is increased. Note that these directions are perpendicular to one another.

Position.   At any instant the position of the particle, Fig. 12–30a, is defined by the position vector

$$\mathbf{r} = r\mathbf{u}_r \qquad (12\text{--}22)$$

Velocity.   The instantaneous velocity $\mathbf{v}$ is obtained by taking the time derivative of $\mathbf{r}$. Using a dot to represent time differentiation, we have

$$\mathbf{v} = \dot{\mathbf{r}} = \dot{r}\mathbf{u}_r + r\dot{\mathbf{u}}_r$$

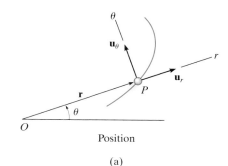

Position

(a)

To evaluate $\dot{\mathbf{u}}_r$, notice that $\mathbf{u}_r$ changes only its direction with respect to time, since by definition the magnitude of this vector is always one unit. Hence, during the time Δt, a change Δr will not cause a change in the direction of $\mathbf{u}_r$; however, a change $\Delta \theta$ will cause $\mathbf{u}_r$ to become $\mathbf{u}'_r$, where $\mathbf{u}'_r = \mathbf{u}_r + \Delta\mathbf{u}_r$, Fig. 12–30b. The time change in $\mathbf{u}_r$ is then $\Delta\mathbf{u}_r$. For small angles $\Delta\theta$ this vector has a magnitude $\Delta u_r \approx 1(\Delta\theta)$ and acts in the $\mathbf{u}_\theta$ direction. Therefore, $\Delta\mathbf{u}_r = \Delta\theta\,\mathbf{u}_\theta$, and so

(b)

$$\dot{\mathbf{u}}_r = \lim_{\Delta t \to 0} \frac{\Delta\mathbf{u}_r}{\Delta t} = \left(\lim_{\Delta t \to 0} \frac{\Delta\theta}{\Delta t} \right)\mathbf{u}_\theta$$

$$\dot{\mathbf{u}}_r = \dot{\theta}\mathbf{u}_\theta \qquad (12\text{--}23)$$

Substituting into the above equation for $\mathbf{v}$, the velocity can be written in component form as

$$\boxed{\mathbf{v} = v_r\mathbf{u}_r + v_\theta\mathbf{u}_\theta} \qquad (12\text{--}24)$$

where

$$\boxed{\begin{aligned} v_r &= \dot{r} \\ v_\theta &= r\dot{\theta} \end{aligned}} \qquad (12\text{--}25)$$

These components are shown graphically in Fig. 12–30c. The *radial component* $\mathbf{v}_r$ is a measure of the rate of increase or decrease in the length of the radial coordinate, i.e., $\dot{r}$; whereas the *transverse component* $\mathbf{v}_\theta$ can be interpreted as the rate of motion along the circumference of a circle having a radius r. In particular, the term $\dot{\theta} = d\theta/dt$ is called the *angular velocity*, since it indicates the time rate of change of the angle θ. Common units used for this measurement are rad/s.

Since $\mathbf{v}_r$ and $\mathbf{v}_\theta$ are mutually perpendicular, the *magnitude* of velocity or speed is simply the positive value of

$$v = \sqrt{(\dot{r})^2 + (r\dot{\theta})^2} \qquad (12\text{--}26)$$

and the *direction* of $\mathbf{v}$ is, of course, tangent to the path at P, Fig. 12–30c.

Velocity

(c)

Fig. 12–30

(d)

Acceleration. Taking the time derivatives of Eq. 12–24, using Eqs. 12–25, we obtain the particle's instantaneous acceleration,

$$\mathbf{a} = \dot{\mathbf{v}} = \ddot{r}\mathbf{u}_r + \dot{r}\dot{\mathbf{u}}_r + \dot{r}\dot{\theta}\mathbf{u}_\theta + r\ddot{\theta}\mathbf{u}_\theta + r\dot{\theta}\dot{\mathbf{u}}_\theta$$

To evaluate the term involving $\dot{\mathbf{u}}_\theta$, it is necessary only to find the change made in the direction of $\mathbf{u}_\theta$ since its magnitude is always unity. During the time Δt, a change Δr will not change the direction of $\mathbf{u}_\theta$, although a change $\Delta\theta$ will cause $\mathbf{u}_\theta$ to become $\mathbf{u}_\theta'$, where $\mathbf{u}_\theta' = \mathbf{u}_\theta + \Delta\mathbf{u}_\theta$, Fig. 12–30d. The time change in $\mathbf{u}_\theta$ is thus $\Delta\mathbf{u}_\theta$. For small angles this vector has a magnitude $\Delta u_\theta \approx 1(\Delta\theta)$ and acts in the $-\mathbf{u}_r$, direction; i.e., $\Delta\mathbf{u}_\theta = -\Delta\theta\mathbf{u}_r$. Thus,

$$\dot{\mathbf{u}}_\theta = \lim_{\Delta t \to 0}\frac{\Delta\mathbf{u}_\theta}{\Delta t} = -\left(\lim_{\Delta t \to 0}\frac{\Delta\theta}{\Delta t}\right)\mathbf{u}_r$$

$$\dot{\mathbf{u}}_\theta = -\dot{\theta}\mathbf{u}_r \tag{12–27}$$

Substituting this result and Eq. 12–23 into the above equation for $\mathbf{a}$, we can write the acceleration in component form as

$$\boxed{\mathbf{a} = a_r\mathbf{u}_r + a_\theta\mathbf{u}_\theta} \tag{12–28}$$

where

$$\boxed{\begin{aligned} a_r &= \ddot{r} - r\dot{\theta}^2 \\ a_\theta &= r\ddot{\theta} + 2\dot{r}\dot{\theta} \end{aligned}} \tag{12–29}$$

Acceleration

(e)

Fig. 12–30 (cont.)

The term $\ddot{\theta} = d^2\theta/dt^2 = d/dt(d\theta/dt)$ is called the *angular acceleration* since it measures the change made in the angular velocity during an instant of time. Units for this measurement are rad/s^2.

Since a_r, and a_θ are always perpendicular, the *magnitude* of acceleration is simply the positive value of

$$a = \sqrt{(\ddot{r} - r\dot{\theta}^2)^2 + (r\ddot{\theta} + 2\dot{r}\dot{\theta})^2} \tag{12–30}$$

The *direction* is determined from the vector addition of its two components. In general, $\mathbf{a}$ will *not* be tangent to the path, Fig. 12–30e.

Cylindrical Coordinates.

If the particle P moves along a space curve as shown in Fig. 12–31, then its location may be specified by the three *cylindrical coordinates, r, θ, z*. The z coordinate is identical to that used for rectangular coordinates. Since the unit vector defining its direction, $\mathbf{u}_z$, is constant, the time derivatives of this vector are zero, and therefore the position, velocity, and acceleration of the particle can be written in terms of its cylindrical coordinates as follows:

$$\mathbf{r}_p = r\mathbf{u}_r + z\mathbf{u}_z$$

$$\mathbf{v} = \dot{r}\mathbf{u}_r + r\dot{\theta}\mathbf{u}_\theta + \dot{z}\mathbf{u}_z$$

$$\mathbf{a} = (\ddot{r} - r\dot{\theta}^2)\mathbf{u}_r + (r\ddot{\theta} + 2\dot{r}\dot{\theta})\mathbf{u}_\theta + \ddot{z}\mathbf{u}_z$$

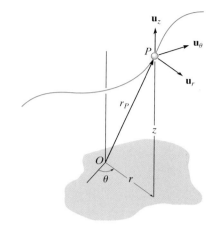

Fig. 12–31

Time Derivatives.

The equations of kinematics require that we obtain the time derivatives $\dot{r}$, $\ddot{r}$, $\dot{\theta}$, and $\ddot{\theta}$ in order to evaluate the r and θ components of $\mathbf{v}$ and $\mathbf{a}$. Two types of problems generally occur:

1. If the coordinates are specified as time parametric equations, $r = r(t)$ and $\theta = \theta(t)$, then the time derivatives can be found directly. For example, consider

$$r = 4t^2 \qquad \theta = (8t^3 + 6)$$

$$\dot{r} = 8t \qquad \dot{\theta} = 24t^2$$

$$\ddot{r} = 8 \qquad \ddot{\theta} = 48t$$

2. If the time-parametric equations are not given, then we have to specify the path $r = f(\theta)$ and find the *relationship* between the time derivatives using the chain rule of calculus. Consider the following examples.

$$r = 5\theta^2$$

$$\dot{r} = 10\theta\dot{\theta}$$

$$\ddot{r} = 10[(\dot{\theta})\dot{\theta} + \theta(\ddot{\theta})]$$

$$= 10\dot{\theta}^2 + 10\theta\ddot{\theta}$$

The spiral motion of this boy can be followed by using cylindrical components. Here the radial coordinate r is constant, the transverse coordinate θ will increase with time as the boy rotates about the vertical, and his altitude z will decrease with time.

or

$$r^2 = 6\theta^3$$

$$2r\dot{r} = 18\theta^2\dot{\theta}$$

$$2[(\dot{r})\dot{r} + r(\ddot{r})] = 18[(2\theta\dot{\theta})\dot{\theta} + \theta^2(\ddot{\theta})]$$

$$\dot{r}^2 + r\ddot{r} = 9(2\theta\dot{\theta}^2 + \theta^2\ddot{\theta})$$

If two of the *four* time derivatives $\dot{r}, \ddot{r}, \dot{\theta},$ and $\ddot{\theta}$ are *known*, then the other two can be obtained from the equations for first and second time derivatives of $r = f(\theta)$. See Example 12.19. In some problems, however, two of these time derivatives may *not* be known; instead the magnitude of the particle's velocity or acceleration may be specified. If this is the case, Eqs. 12–26 and 12–30 $[v^2 = \dot{r}^2 + (r\dot{\theta})^2$ and $a^2 = (\ddot{r} - r\dot{\theta}^2)^2 + (r\ddot{\theta} + 2\dot{r}\dot{\theta})^2]$ may be used to obtain the necessary relationships involving $\dot{r}, \ddot{r}, \dot{\theta},$ and $\ddot{\theta}$. See Example 12.20.

Procedure for Analysis

Coordinate System.

- Polar coordinates are a suitable choice for solving problems when data regarding the angular motion of the radial coordinate r is given to describe the particle's motion. Also, some paths of motion can conveniently be described in terms of these coordinates.

- To use polar coordinates, the origin is established at a fixed point, and the radial line r is directed to the particle.

- The transverse coordinate θ is measured from a fixed reference line to the radial line.

Velocity and Acceleration.

- Once r and the four time derivatives $\dot{r}, \ddot{r}, \dot{\theta},$ and $\ddot{\theta}$ have been evaluated at the instant considered, their values can be substituted into Eqs. 12–25 and 12–29 to obtain the radial and transverse components of **v** and **a**.

- If it is necessary to take the time derivatives of $r = f(\theta)$, it is very important to use the chain rule of calculus.

- Motion in three dimensions requires a simple extension of the above procedure to include $\dot{z}$ and $\ddot{z}$.

Besides the examples which follow, further examples involving the calculation of a_r and a_θ can be found in the "kinematics" sections of Examples 13.10 through 13.12.

EXAMPLE 12.17

The amusement park ride shown in Fig. 12–32a consists of a chair that is rotating in a horizontal circular path of radius r such that the arm OB has an angular velocity $\dot{\theta}$ and angular acceleration $\ddot{\theta}$. Determine the radial and transverse components of velocity and acceleration of the passenger. Neglect his size in the calculation.

(a)   (b)

Fig. 12–32

SOLUTION

Coordinate System.   Since the angular motion of the arm is reported, polar coordinates are chosen for the solution, Fig. 12–32a. Here θ is not related to r, since the radius is constant for all θ.

Velocity and Acceleration.   Equations 12–25 and 12–29 will be used for the solution, and so it is first necessary to specify the first and second time derivatives of r and θ. Since r is *constant*, we have

$$r = r \qquad \dot{r} = 0 \qquad \ddot{r} = 0$$

Thus,

$$v_r = \dot{r} = 0 \qquad\qquad\qquad Ans.$$
$$v_\theta = r\dot{\theta} \qquad\qquad\qquad Ans.$$
$$a_r = \ddot{r} - r\dot{\theta}^2 = -r\dot{\theta}^2 \qquad\qquad Ans.$$
$$a_\theta = r\ddot{\theta} + 2\dot{r}\dot{\theta} = r\ddot{\theta} \qquad\qquad Ans.$$

These results are shown in Fig. 12–32b.

NOTE: Also shown are the n, t axes, which in this special case of circular motion happen to be *collinear* with the r and θ axes, respectively. In particular note that $v = v_\theta = v_t = r\dot{\theta}$. Also,

$$-a_r = a_n = \frac{v^2}{\rho} = \frac{(r\dot{\theta})^2}{r} = r\dot{\theta}^2$$

$$a_\theta = a_t = \frac{dv}{dt} = \frac{d}{dt}(r\dot{\theta}) = \frac{dr}{dt}\dot{\theta} + r\frac{d\dot{\theta}}{dt} = 0 + r\ddot{\theta}$$

EXAMPLE 12.18

(a)

(b)

The rod *OA* in Fig. 12–33*a* is rotating in the horizontal plane such that $\theta = (t^3)$ rad. At the same time, the collar *B* is sliding outward along *OA* so that $r = (100t^2)$ mm. If in both cases *t* is in seconds, determine the velocity and acceleration of the collar when $t = 1$ s.

SOLUTION

Coordinate System. Since time-parametric equations of the path are given, it is not necessary to relate *r* to θ.

Velocity and Acceleration. Determining the time derivatives and evaluating when $t = 1$ s, we have

$$r = 100t^2 \Big|_{t=1\text{ s}} = 100 \text{ mm} \quad \theta = t^3 \Big|_{t=1\text{ s}} = 1 \text{ rad} = 57.3°$$

$$\dot{r} = 200t \Big|_{t=1\text{ s}} = 200 \text{ mm/s} \quad \dot{\theta} = 3t^2 \Big|_{t=1\text{ s}} = 3 \text{ rad/s}$$

$$\ddot{r} = 200 \Big|_{t=1\text{ s}} = 200 \text{ mm/s}^2 \quad \ddot{\theta} = 6t \Big|_{t=1\text{ s}} = 6 \text{ rad/s}^2.$$

As shown in Fig. 12–33*b*,

$$\mathbf{v} = \dot{r}\mathbf{u}_r + r\dot{\theta}\mathbf{u}_\theta$$

$$= 200\mathbf{u}_r + 100(3)\mathbf{u}_\theta = \{200\mathbf{u}_r + 300\mathbf{u}_\theta\} \text{ mm/s}$$

The magnitude of **v** is

$$v = \sqrt{(200)^2 + (300)^2} = 361 \text{ mm/s} \qquad \textit{Ans.}$$

$$\delta = \tan^{-1}\left(\frac{300}{200}\right) = 56.3° \qquad \delta + 57.3° = 114° \qquad \textit{Ans.}$$

As shown in Fig. 12–33*c*,

$$\mathbf{a} = (\ddot{r} - r\dot{\theta}^2)\mathbf{u}_r + (r\ddot{\theta} + 2\dot{r}\dot{\theta})\mathbf{u}_\theta$$

$$= [200 - 100(3)^2]\mathbf{u}_r + [100(6) + 2(200)3]\mathbf{u}_\theta$$

$$= \{-700\mathbf{u}_r + 1800\mathbf{u}_\theta\} \text{ mm/s}^2$$

The magnitude of **a** is

$$a = \sqrt{(700)^2 + (1800)^2} = 1930 \text{ mm/s}^2 \qquad \textit{Ans.}$$

$$\phi = \tan^{-1}\left(\frac{1800}{700}\right) = 68.7° \qquad (180° - \phi) + 57.3° = 169° \qquad \textit{Ans.}$$

NOTE: The velocity is tangent to the path; however, the acceleration is directed towards the center of curvature, as expected.

(c)

Fig. 12–33

EXAMPLE 12.19

The searchlight in Fig. 12–34a casts a spot of light along the face of a wall that is located 100 m from the searchlight. Determine the magnitudes of the velocity and acceleration at which the spot appears to travel across the wall at the instant $\theta = 45°$. The searchlight is rotating at a constant rate of $\dot{\theta} = 4$ rad/s.

(a)

SOLUTION

Coordinate System.   Polar coordinates will be used to solve this problem since the angular rate of the searchlight is given. To find the necessary time derivatives it is first necessary to relate r to θ. From Fig. 12–34a, this relation is

$$r = 100/\cos \theta = 100 \sec \theta$$

Velocity and Acceleration.   Using the chain rule of calculus, noting that $d(\sec \theta) = \sec \theta \tan \theta \, d\theta$, and $d(\tan \theta) = \sec^2 \theta \, d\theta$, we have

$$\dot{r} = 100(\sec \theta \tan \theta)\dot{\theta}$$

$$\ddot{r} = 100(\sec \theta \tan \theta)\dot{\theta}(\tan \theta)\dot{\theta} + 100 \sec \theta(\sec^2 \theta)\dot{\theta}(\dot{\theta})$$
$$+ 100 \sec \theta \tan \theta(\ddot{\theta})$$
$$= 100 \sec \theta \tan^2 \theta(\dot{\theta})^2 + 100 \sec^3 \theta(\dot{\theta})^2 + 100(\sec \theta \tan \theta)\ddot{\theta}$$

Since $\dot{\theta} = 4$ rad/s = constant, then $\ddot{\theta} = 0$, and the above equations, when $\theta = 45°$, become

$$r = 100 \sec 45° = 141.4$$
$$\dot{r} = 400 \sec 45° \tan 45° = 565.7$$
$$\ddot{r} = 1600(\sec 45° \tan^2 45° + \sec^3 45°) = 6788.2$$

As shown in Fig. 12–34b,

$$\mathbf{v} = \dot{r}\mathbf{u}_r + r\dot{\theta}\mathbf{u}_\theta$$
$$= 565.7\mathbf{u}_r + 141.4(4)\mathbf{u}_\theta$$
$$= \{565.7\mathbf{u}_r + 565.7\mathbf{u}_\theta\} \text{ m/s}$$
$$v = \sqrt{v_r^2 + v_\theta^2} = \sqrt{(565.7)^2 + (565.7)^2}$$
$$= 800 \text{ m/s} \qquad\qquad Ans.$$

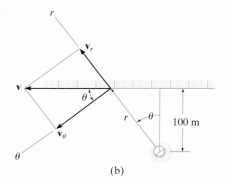

(b)

As shown in Fig. 12–34c,

$$\mathbf{a} = (\ddot{r} - r\dot{\theta}^2)\mathbf{u}_r + (r\ddot{\theta} + 2\dot{r}\dot{\theta})\mathbf{u}_\theta$$
$$= [6788.2 - 141.4(4)^2]\mathbf{u}_r + [141.4(0) + 2(565.7)4]\mathbf{u}_\theta$$
$$= \{4525.5\mathbf{u}_r + 4525.5\mathbf{u}_\theta\} \text{ m/s}^2$$
$$a = \sqrt{a_r^2 + a_\theta^2} = \sqrt{(4525.5)^2 + (4525.5)^2}$$
$$= 6400 \text{ m/s}^2 \qquad\qquad Ans.$$

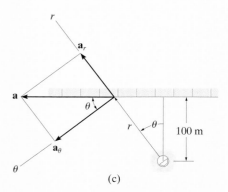

(c)

NOTE: It is also possible to find a without having to calculate $\ddot{r}$ (or a_r). As shown in Fig. 12–34d, since $a_\theta = 4525.5$ m/s^2, then by vector resolution, $a = 4525.5/\cos 45° = 6400$ m/s^2.

(d)

Fig. 12–34

EXAMPLE 12.20

$r = 0.5 (1 - \cos\theta)$ ft

(a)

Due to the rotation of the forked rod, the ball A in Fig. 12–35a travels around the slotted path, a portion of which is in the shape of a cardioid, $r = 0.5(1 - \cos\theta)$ ft, where θ is in radians. If the ball's velocity is $v = 4$ ft/s and its acceleration is $a = 30$ ft/s^2 at the instant $\theta = 180°$, determine the angular velocity $\dot{\theta}$ and angular acceleration $\ddot{\theta}$ of the fork.

SOLUTION
Coordinate System.   This path is most unusual, and mathematically it is best expressed using polar coordinates, as done here, rather than rectangular coordinates. Also, $\dot{\theta}$ and $\ddot{\theta}$ must be determined so r, θ coordinates are an obvious choice.

Velocity and Acceleration.   Determining the time derivatives of r using the chain rule of calculus yields

$$r = 0.5(1 - \cos\theta)$$
$$\dot{r} = 0.5(\sin\theta)\dot{\theta}$$
$$\ddot{r} = 0.5(\cos\theta)\dot{\theta}(\dot{\theta}) + 0.5(\sin\theta)\ddot{\theta}$$

Evaluating these results at $\theta = 180°$, we have

$$r = 1 \text{ ft} \qquad \dot{r} = 0 \qquad \ddot{r} = -0.5\dot{\theta}^2$$

Since $v = 4$ ft/s, using Eq. 12–26 to determine $\dot{\theta}$ yields

$$v = \sqrt{(\dot{r})^2 + (r\dot{\theta})^2}$$
$$4 = \sqrt{(0)^2 + (1\dot{\theta})^2}$$
$$\dot{\theta} = 4 \text{ rad/s} \qquad\qquad\qquad Ans.$$

In a similar manner, $\ddot{\theta}$ can be found using Eq. 12–30.

$$a = \sqrt{(\ddot{r} - r\dot{\theta}^2)^2 + (r\ddot{\theta} + 2\dot{r}\dot{\theta})^2}$$
$$30 = \sqrt{[-0.5(4)^2 - 1(4)^2]^2 + [1\ddot{\theta} + 2(0)(4)]^2}$$
$$(30)^2 = (-24)^2 + \ddot{\theta}^2$$
$$\ddot{\theta} = 18 \text{ rad/s}^2 \qquad\qquad\qquad Ans.$$

r

$v = 4$ ft/s

$a = 30$ ft/s^2

θ

(b)

Fig. 12–35

Vectors **a** and **v** are shown in Fig. 12–35b.

NOTE: At this location A, the θ and t (tangential) axes will coincide. The $+n$ (normal) axis is directed to the right, opposite to $+r$.

PROBLEMS

*12–136. The time rate of change of acceleration is referred to as the *jerk*, which is often used as a means of measuring passenger discomfort. Calculate this vector, $\dot{\mathbf{a}}$, in terms of its cylindrical components, using Eq. 12–32.

12–137. If a particle's position is described by the polar coordinates $r = 4(1 + \sin t)$ m and $\theta = (2e^{-t})$ rad, where t is in seconds and the argument for the sine is in radians, determine the radial and transverse components of the particle's velocity and acceleration when $t = 2$ s.

12–138. The slotted fork is rotating about O at a constant rate of $\dot{\theta} = 3$ rad/s. Determine the radial and transverse components of the velocity and acceleration of the pin A at the instant $\theta = 360°$. The path is defined by the spiral groove $r = (5 + \theta/\pi)$ in., where θ is in radians.

12–139. The slotted fork is rotating about O at $\dot{\theta} = 3$ rad/s, which is increasing at $\ddot{\theta} = 2$ rad/s² when $\theta = 360°$. Determine the radial and transverse components of the velocity and acceleration of the pin A at this instant. The path is defined by the spiral groove $r = (5 + \theta/\pi)$ in., where θ is in radians.

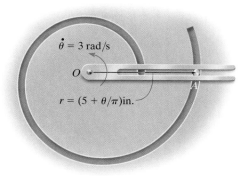

Probs. 12–138/139

*12–140. If a particle moves along a path such that $r = (2 \cos t)$ ft and $\theta = (t/2)$ rad, where t is in seconds, plot the path $r = f(\theta)$ and determine the particle's radial and transverse components of velocity and acceleration.

12–141. If a particle's position is described by the polar coordinates $r = (2 \sin 2\theta)$ m and $\theta = (4t)$ rad, where t is in seconds, determine the radial and transverse components of its velocity and acceleration when $t = 1$ s.

12–142. A particle is moving along a circular path having a 400-mm radius. Its position as a function of time is given by $\theta = (2t^2)$ rad, where t is in seconds. Determine the magnitude of the particle's acceleration when $\theta = 30°$. The particle starts from rest when $\theta = 0°$.

12–143. A particle moves in the x–y plane such that its position is defined by $\mathbf{r} = \{2t\mathbf{i} + 4t^2\mathbf{j}\}$ ft, where t is in seconds. Determine the radial and transverse components of the particle's velocity and acceleration when $t = 2$ s.

*12–144. A truck is traveling along the horizontal circular curve of radius $r = 60$ m with a constant speed $v = 20$ m/s. Determine the angular rate of rotation $\dot{\theta}$ of the radial line r and the magnitude of the truck's acceleration.

12–145. A truck is traveling along the horizontal circular curve of radius $r = 60$ m with a speed of 20 m/s which is increasing at 3 m/s². Determine the truck's radial and transverse components of acceleration.

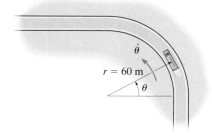

Probs. 12–144/145

12–146. A particle is moving along a circular path having a radius of 6 in. such that its position as a function of time is given by $\theta = \sin 3t$, where θ is in radians, the argument for the sine is in degrees, and t is in seconds. Determine the acceleration of the particle at $\theta = 30°$. The particle starts from rest at $\theta = 0°$.

12–147. The slotted link is pinned at O, and as a result of the constant angular velocity $\dot{\theta} = 3$ rad/s it drives the peg P for a short distance along the spiral guide $r = (0.4\theta)$ m, where θ is in radians. Determine the radial and transverse components of the velocity and acceleration of P at the instant $\theta = \pi/3$ rad.

*12–148. Solve Prob. 12–147 if the slotted link has an angular acceleration $\ddot{\theta} = 8$ rad/s² when $\dot{\theta} = 3$ rad/s at $\theta = \pi/3$ rad.

12–149. The slotted link is pinned at O, and as a result of the constant angular velocity $\dot{\theta} = 3$ rad/s it drives the peg P for a short distance along the spiral guide $r = (0.4\theta)$ m, where θ is in radians. Determine the velocity and acceleration of the particle at the instant it leaves the slot in the link, i.e., when $r = 0.5$ m.

Probs. 12–147/148/149

12–150. A train is traveling along the circular curve of radius $r = 600$ ft. At the instant shown, its angular rate of rotation is $\dot{\theta} = 0.02$ rad/s, which is decreasing at $\ddot{\theta} = -0.001$ rad/s². Determine the magnitudes of the train's velocity and acceleration at this instant.

Prob. 12–150

12–151. A particle travels along a portion of the "four-leaf rose" defined by the equation $r = (5 \cos 2\theta)$ m. If the angular velocity of the radial coordinate line is $\dot{\theta} = (3t^2)$ rad/s, where t is in seconds, determine the radial and transverse components of the particle's velocity and acceleration at the instant $\theta = 30°$. When $t = 0, \theta = 0°$.

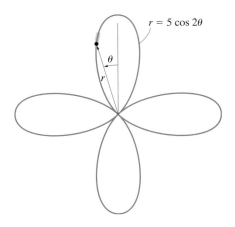

Prob. 12–151

*12–152. At the instant shown, the water sprinkler is rotating with an angular speed $\dot{\theta} = 2$ rad/s and an angular acceleration $\ddot{\theta} = 3$ rad/s². If the nozzle lies in the vertical plane and water is flowing through it at a constant rate of 3 m/s, determine the magnitudes of the velocity and acceleration of a water particle as it exits the open end, $r = 0.2$ m.

Prob. 12–152

12–153. The boy slides down the slide at a constant speed of 2 m/s. If the slide is in the form of a helix, defined by the equations $r = 1.5$ m and $z = -\theta/\pi$, determine the boy's angular velocity about the z axis, $\dot{\theta}$, and the magnitude of his acceleration.

Prob. 12–153

12–154. A cameraman standing at A is following the movement of a race car, B, which is traveling along a straight track at a constant speed of 80 ft/s. Determine the angular rate at which he must turn in order to keep the camera directed on the car at the instant $\theta = 60°$.

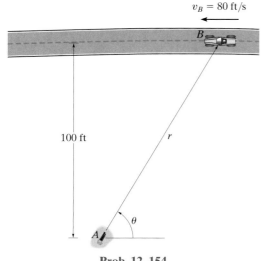

Prob. 12–154

12–155. For a short distance the train travels along a track having the shape of a spiral, $r = (1000/\theta)$ m, where θ is in radians. If it maintains a constant speed $v = 20$ m/s, determine the radial and transverse components of its velocity when $\theta = (9\pi/4)$ rad.

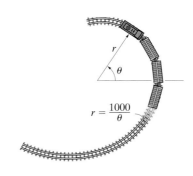

Prob. 12–155

*****12–156.** For a short distance the train travels along a track having the shape of a spiral, $r = (1000/\theta)$ m, where θ is in radians. If the angular rate is constant, $\dot{\theta} = 0.2$ rad/s, determine the radial and transverse components of its velocity and acceleration when $\theta = (9\pi/4)$ rad.

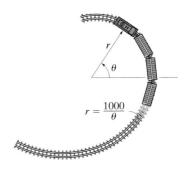

Prob. 12–156

12–157. The arm of the robot has a fixed length so that $r = 3$ ft and its grip A moves along the path $z = (3 \sin 4\theta)$ ft, where θ is in radians. If $\theta = (0.5t)$ rad, where t is in seconds, determine the magnitudes of the grip's velocity and acceleration when $t = 3$ s.

12–158. For a short time the arm of the robot is extending at a constant rate such that $\dot{r} = 1.5$ ft/s when $r = 3$ ft, $z = (4t^2)$ ft, and $\theta = 0.5t$ rad, where t is in seconds. Determine the magnitudes of the velocity and acceleration of the grip A when $t = 3$ s.

Probs. 12–157/158

12–159. The rod OA rotates counterclockwise with a constant angular velocity of $\dot{\theta} = 5$ rad/s. Two pin-connected slider blocks, located at B, move freely on OA and the curved rod whose shape is a limaçon described by the equation $r = 100(2 - \cos \theta)$ mm. Determine the speed of the slider blocks at the instant $\theta = 120°$.

*12–160. Determine the magnitude of the acceleration of the slider blocks in Prob. 12–159 when $\theta = 120°$.

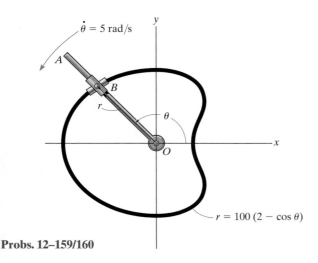

Probs. 12–159/160

12–161. The searchlight on the boat anchored 2000 ft from shore is turned on the automobile, which is traveling along the straight road at a constant speed of 80 ft/s. Determine the angular rate of rotation of the light when the automobile is $r = 3000$ ft from the boat.

12–162. If the car in Prob. 12–161 is accelerating at 15 ft/s² at the instant $r = 3000$ ft, determine the required angular acceleration $\ddot{\theta}$ of the light at this instant.

Probs. 12–161/162

12–163. For a short time the bucket of the backhoe traces the path of the cardioid $r = 25(1 - \cos \theta)$ ft. Determine the magnitudes of the velocity and acceleration of the bucket when $\theta = 120°$ if the boom is rotating with an angular velocity of $\dot{\theta} = 2$ rad/s and an angular acceleration of $\ddot{\theta} = 0.2$ rad/s² at the instant shown.

Prob. 12–163

*12–164. A car is traveling along the circular curve having a radius $r = 400$ ft. At the instant shown, its angular rate of rotation is $\dot{\theta} = 0.025$ rad/s, which is decreasing at the rate $\ddot{\theta} = -0.008$ rad/s². Determine the radial and transverse components of the car's velocity and acceleration at this instant and sketch these components on the curve.

$r = 400$ ft

$\dot{\theta}$

Prob. 12–164

12–165. The mechanism of a machine is constructed so that for a short time the roller at A follows the surface of the cam described by the equation $r = (0.3 + 0.2 \cos \theta)$ m. If $\dot{\theta} = 0.5$ rad/s and $\ddot{\theta} = 0$, determine the magnitudes of the roller's velocity and acceleration at the instant, $\theta = 30°$. Neglect the size of the roller. Also determine the velocity components $(\mathbf{v}_A)_x$ and $(\mathbf{v}_A)_y$ of the roller at this instant. The rod to which the roller is attached remains vertical and can slide up or down along the guides while the guides move horizontally to the left.

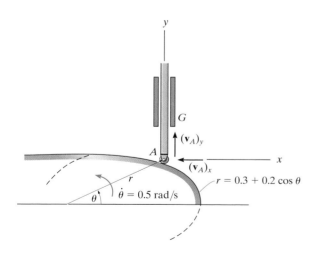

$(\mathbf{v}_A)_y$

G

A

$(\mathbf{v}_A)_x$

x

$r = 0.3 + 0.2 \cos \theta$

r

θ $\dot{\theta} = 0.5$ rad/s

y

Prob. 12–165

12–166. The roller coaster is traveling down along the spiral ramp with a constant speed $v = 6$ m/s. If the track descends a distance of 10 m for every full revolution, $\theta = 2\pi$ rad, determine the magnitude of the roller coaster's acceleration as it moves along the track, $r = 5$ m. *Hint:* For part of the solution, note that the tangent to the ramp at any point is at an angle $\phi = \tan^{-1}[10/2\pi(5)] = 17.66°$ from the horizontal. Use this to determine the velocity components v_θ and v_z, which in turn are used to determine $\dot{\theta}$ and $\dot{z}$.

$r = 5$ m

10 m

Prob. 12–166

12–167. A cameraman standing at A is following the movement of a race car, B, which is traveling around a curved track at a constant speed of 30 m/s. Determine the angular rate $\dot{\theta}$ at which the man must turn in order to keep the camera directed on the car at the instant $\theta = 30°$.

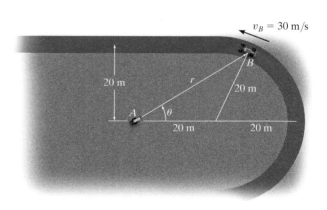

$v_B = 30$ m/s

B

20 m

r

20 m

A θ

20 m   20 m

Prob. 12–167

*12–168. The pin follows the path described by the equation $r = (0.2 + 0.15 \cos \theta)$ m. At the instant $\theta = 30°$, $\dot{\theta} = 0.7$ rad/s and $\ddot{\theta} = 0.5$ rad/s². Determine the magnitudes of the pin's velocity and acceleration at this instant. Neglect the size of the pin.

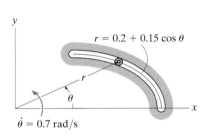

$r = 0.2 + 0.15 \cos \theta$

$\dot{\theta} = 0.7$ rad/s

Prob. 12–168

12–169. For a short time the position of the roller-coaster car along its path is defined by the equations $r = 25$ m, $\theta = (0.3t)$ rad, and $z = (-8 \cos \theta)$ m, where t is in seconds. Determine the magnitude of the car's velocity and acceleration when $t = 4$ s.

Prob. 12–169

12–170. The small washer is sliding down the cord OA. When it is at the midpoint, its speed is 200 mm/s and its acceleration is 10 mm/s². Express the velocity and acceleration of the washer at this point in terms of its cylindrical components .

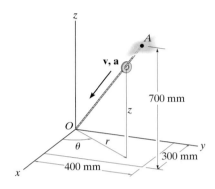

v, a

700 mm

300 mm

400 mm

Prob. 12–170

12–171. A double collar C is pin connected together such that one collar slides over a fixed rod and the other slides over a rotating rod. If the geometry of the fixed rod for a short distance can be defined by a lemniscate, $r^2 = (4 \cos 2\theta)$ ft², determine the collar's radial and transverse components of velocity and acceleration at the instant $\theta = 0°$ as shown. Rod OA is rotating at a constant rate of $\dot{\theta} = 6$ rad/s.

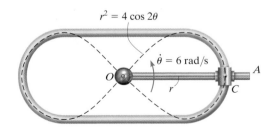

$r^2 = 4 \cos 2\theta$

$\dot{\theta} = 6$ rad/s

Prob. 12–171

12.9   Absolute Dependent Motion Analysis of Two Particles

In some types of problems the motion of one particle will *depend* on the corresponding motion of another particle. This dependency commonly occurs if the particles are interconnected by inextensible cords which are wrapped around pulleys. For example, the movement of block A downward along the inclined plane in Fig. 12–36 will cause a corresponding movement of block B up the other incline. We can show this mathematically by first specifying the location of the blocks using *position coordinates* s_A and s_B. Note that each of the coordinate axes is (1) referenced from a *fixed* point (O) or *fixed* datum line, (2) measured along each inclined plane in the direction of motion of block A and block B, and (3) has a positive sense from C to A and D to B. If the total cord length is l_T, the position coordinates are related by the equation

$$s_A + l_{CD} + s_B = l_T$$

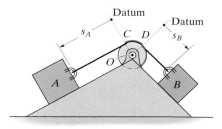

Fig. 12–36

Here l_{CD} is the length of the cord passing over arc CD. Taking the time derivative of this expression, realizing that l_{CD} and l_T *remain constant*, while s_A and s_B measure the lengths of the changing segments of the cord, we have

$$\frac{ds_A}{dt} + \frac{ds_B}{dt} = 0 \qquad \text{or} \qquad v_B = -v_A$$

The negative sign indicates that when block A has a velocity downward, i.e., in the direction of positive s_A, it causes a corresponding upward velocity of block B; i.e., B moves in the negative s_B direction.

In a similar manner, time differentiation of the velocities yields the relation between the accelerations, i.e.,

$$a_B = -a_A$$

A more complicated example involving dependent motion of two blocks is shown in Fig. 12–37a. In this case, the position of block A is specified by s_A, and the position of the *end* of the cord from which block B is suspended is defined by s_B. Here we have chosen coordinate axes which are (1) referenced from fixed points or datums, (2) measured in the direction of motion of each block, and (3) positive to the right (s_A) and positive downward (s_B). During the motion, the length of the red colored segments of the cord in Fig. 12–37a *remains constant*. If l represents the total length of cord minus these segments, then the position coordinates can be related by the equation

$$2s_B + h + s_A = l$$

Since l and h are constant during the motion, the two time derivatives yield

$$2v_B = -v_A \qquad 2a_B = -a_A$$

Hence, when B moves downward ($+s_B$), A moves to the left ($-s_A$) with two times the motion.

(a)

Fig. 12–37

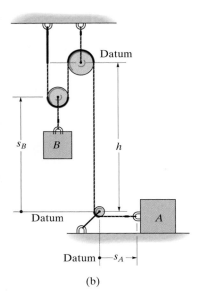

Datum

s_B B

h

Datum

A

Datum s_A

(b)

Fig. 12–37 (cont.)

The motion of the traveling block on this oil rig depends upon the motion of the cable connected to the winch which operates it. It is important to be able to relate these motions in order to determine the power requirements of the winch and the force in the cable caused by accelerated motion.

This example can also be worked by defining the position of block B from the center of the bottom pulley (a fixed point), Fig. 12–37b. In this case

$$2(h - s_B) + h + s_A = l$$

Time differentiation yields

$$2v_B = v_A \quad 2a_B = a_A$$

Here the signs are the same. Why?

Procedure for Analysis

The above method of relating the dependent motion of one particle to that of another can be performed using algebraic scalars or position coordinates provided each particle moves along a rectilinear path. When this is the case, only the magnitudes of the velocity and acceleration of the particles will change, not their line of direction. The following procedure is required.

Position-Coordinate Equation.

- Establish position coordinates which have their origin located at a *fixed* point or datum.
- The coordinates are directed along the path of motion and extend to a point having the same motion as each of the particles.
- It is *not necessary* that the *origin* be the *same* for each of the coordinates; however, it is *important* that each coordinate axis selected be directed along the *path of motion* of the particle.
- Using geometry or trigonometry, relate the coordinates to the total length of the cord, l_T, or to that portion of cord, l, which *excludes* the segments that do not change length as the particles move—such as arc segments wrapped over pulleys.
- If a problem involves a *system* of two or more cords wrapped around pulleys, then the position of a point on one cord must be related to the position of a point on another cord using the above procedure. Separate equations are written for a fixed length of each cord of the system and the positions of the two particles are then related by these equations (see Examples 12.22 and 12.23).

Time Derivatives.

- Two successive time derivatives of the position-coordinate equations yield the required velocity and acceleration equations which relate the motions of the particles.
- The signs of the terms in these equations will be consistent with those that specify the positive and negative sense of the position coordinates.

EXAMPLE 12.21

Determine the speed of block A in Fig. 12–38 if block B has an upward speed of 6 ft/s.

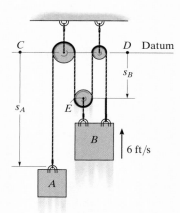

Fig. 12–38

SOLUTION

Position-Coordinate Equation.   There is *one cord* in this system having segments which are changing length. Position coordinates s_A and s_B will be used since each is measured from a fixed point (C or D) and extends along each block's *path of motion*. In particular, s_B is directed to point E since motion of B and E is the *same*.

The red colored segments of the cord in Fig. 12–38 remain at a constant length and do not have to be considered as the blocks move. The remaining length of cord, l, is also constant and is related to the changing position coordinates s_A and s_B by the equation

$$s_A + 3s_B = l$$

Time Derivative.   Taking the time derivative yields

$$v_A + 3v_B = 0$$

so that when $v_B = -6$ ft/s (upward),

$$v_A = 18 \text{ ft/s } \downarrow \qquad\qquad\qquad \textit{Ans.}$$

EXAMPLE 12.22

Determine the speed of block A in Fig. 12–39 if block B has an upward speed of 6 ft/s.

Fig. 12–39

SOLUTION

Position-Coordinate Equation.   As shown, the positions of blocks A and B are defined using coordinates s_A and s_B. Since the system has *two* cords which change length, it will be necessary to use a third coordinate, s_C, in order to relate s_A to s_B. In other words, the length of one of the cords can be expressed in terms of s_A and s_C, and the length of the other cord can be expressed in terms of s_B and s_C.

The red colored segments of the cords in Fig. 12–39 do not have to be considered in the analysis. Why? For the remaining cord lengths, say l_1 and l_2, we have

$$s_A + 2s_C = l_1 \qquad s_B + (s_B - s_C) = l_2$$

Eliminating s_C yields an equation defining the positions of both blocks, i.e.,

$$s_A + 4s_B = 2l_2 + l_1$$

Time Derivative.   The time derivative gives

$$v_A + 4v_B = 0$$

so that when $v_B = -6$ ft/s (upward),

$$v_A = +24 \text{ ft/s} = 24 \text{ ft/s} \downarrow \qquad\qquad Ans.$$

EXAMPLE 12.23

Determine the speed with which block B rises in Fig. 12–40 if the end of the cord at A is pulled down with a speed of 2 m/s.

Fig. 12–40

SOLUTION

Position-Coordinate Equation.   The position of point A is defined by s_A, and the position of block B is specified by s_B since point E on the pulley will have the *same motion* as the block. Both coordinates are measured from a horizontal datum passing through the *fixed* pin at pulley D. Since the system consists of *two* cords, the coordinates s_A and s_B cannot be related directly. Instead, by establishing a third position coordinate, s_C, we can now express the length of one of the cords in terms of s_B and s_C, and the length of the other cord in terms of s_A, s_B, and s_C.

Excluding the red colored segments of the cords in Fig. 12–40, the remaining constant cord lengths l_1 and l_2 (along with the hook and link dimensions) can be expressed as

$$s_C + s_B = l_1$$

$$(s_A - s_C) + (s_B - s_C) + s_B = l_2$$

Eliminating s_C yields

$$s_A + 4s_B = l_2 + 2l_1$$

As required, this equation relates the position s_B of block B to the position s_A of point A.

Time Derivative.   The time derivative gives

$$v_A + 4v_B = 0$$

so that when $v_A = 2$ m/s (downward),

$$v_B = -0.5 \text{ m/s} = 0.5 \text{ m/s} \uparrow \qquad\qquad \textit{Ans.}$$

EXAMPLE 12.24

A man at A is hoisting a safe S as shown in Fig. 12–41 by walking to the right with a constant velocity $v_A = 0.5$ m/s. Determine the velocity and acceleration of the safe when it reaches the elevation at E. The rope is 30 m long and passes over a small pulley at D.

SOLUTION

Position-Coordinate Equation. This problem is unlike the previous examples since rope segment DA changes both direction and magnitude. However, the ends of the rope, which define the positions of S and A, are specified by means of the x and y coordinates measured from a fixed point and *directed along the paths of motion* of the ends of the rope.

The x and y coordinates may be related since the rope has a fixed length $l = 30$ m, which at all times is equal to the length of segment DA plus CD. Using the Pythagorean theorem to determine l_{DA}, we have $l_{DA} = \sqrt{(15)^2 + x^2}$; also, $l_{CD} = 15 - y$. Hence,

$$l = l_{DA} + l_{CD}$$
$$30 = \sqrt{(15)^2 + x^2} + (15 - y)$$
$$y = \sqrt{225 + x^2} - 15 \qquad (1)$$

Time Derivatives. Taking the time derivative, using the chain rule, where $v_S = dy/dt$ and $v_A = dx/dt$, yields

$$v_S = \frac{dy}{dt} = \left[\frac{1}{2}\frac{2x}{\sqrt{225 + x^2}}\right]\frac{dx}{dt}$$
$$= \frac{x}{\sqrt{225 + x^2}}v_A \qquad (2)$$

Fig. 12–41

At $y = 10$ m, x is determined from Eq. 1, i.e., $x = 20$ m. Hence, from Eq. 2 with $v_A = 0.5$ m/s,

$$v_S = \frac{20}{\sqrt{225 + (20)^2}}(0.5) = 0.4 \text{ m/s} = 400 \text{ mm/s} \uparrow \qquad Ans.$$

The acceleration is determined by taking the time derivative of Eq. 2. Since v_A is constant, then $a_A = dv_A/dt = 0$, and we have

$$a_S = \frac{d^2y}{dt^2} = \left[\frac{-x(dx/dt)}{(225 + x^2)^{3/2}}\right]xv_A + \left[\frac{1}{\sqrt{225 + x^2}}\right]\left(\frac{dx}{dt}\right)v_A + \left[\frac{1}{\sqrt{225 + x^2}}\right]x\frac{dv_A}{dt} = \frac{225v_A^2}{(225 + x^2)^{3/2}}$$

At $x = 20$ m, with $v_A = 0.5$ m/s, the acceleration becomes

$$a_S = \frac{225(0.5 \text{ m/s})^2}{[225 + (20 \text{ m})^2]^{3/2}} = 0.00360 \text{ m/s}^2 = 3.60 \text{ mm/s}^2 \uparrow \qquad Ans.$$

NOTE: The constant velocity at A causes the other end C of the rope to have an acceleration since $\mathbf{v}_A$ causes segment DA to change its direction as well as its length.

12.10   Relative-Motion Analysis of Two Particles Using Translating Axes

Throughout this chapter the absolute motion of a particle has been determined using a single fixed reference frame for measurement. There are many cases, however, where the path of motion for a particle is complicated, so that it may be feasible to analyze the motion in parts by using two or more frames of reference. For example, the motion of a particle located at the tip of an airplane propeller, while the plane is in flight, is more easily described if one observes first the motion of the airplane from a fixed reference and then superimposes (vectorially) the circular motion of the particle measured from a reference attached to the airplane. Any type of coordinates—rectangular, cylindrical, etc.—can be chosen to describe these two different motions.

In this section *translating frames of reference* will be considered for the analysis. Relative-motion analysis of particles using rotating frames of reference will be treated in Secs. 16.8 and 20.4, since such an analysis depends on prior knowledge of the kinematics of line segments.

Position.   Consider particles A and B, which move along the arbitrary paths aa and bb, respectively, as shown in Fig. 12–42a. The *absolute position* of each particle, $\mathbf{r}_A$ and $\mathbf{r}_B$, is measured from the common origin O of the fixed x, y, z reference frame. The origin of a second frame of reference x', y', z' is attached to and moves with particle A. The axes of this frame are *only permitted to translate* relative to the fixed frame. The *relative position* of "B with respect to A" is designated by a *relative-position vector* $\mathbf{r}_{B/A}$. Using vector addition, the three vectors shown in Fig. 12–42a can be related by the equation*

$$\mathbf{r}_B = \mathbf{r}_A + \mathbf{r}_{B/A} \qquad (12\text{--}33)$$

Velocity.   An equation that relates the velocities of the particles can be determined by taking the time derivative of Eq. 12–33, i.e.,

$$\mathbf{v}_B = \mathbf{v}_A + \mathbf{v}_{B/A} \qquad (12\text{--}34)$$

Here $\mathbf{v}_B = d\mathbf{r}_B/dt$ and $\mathbf{v}_A = d\mathbf{r}_A/dt$ refer to *absolute velocities*, since they are observed from the fixed frame; whereas the *relative velocity* $\mathbf{v}_{B/A} = d\mathbf{r}_{B/A}/dt$ is observed from the translating frame. It is important to note that since the x', y', z' axes translate, the *components* of $\mathbf{r}_{B/A}$ will *not* change direction and therefore the time derivative of this vector's components will only have to account for the change in the vector's magnitude. Equation 12–34 therefore states that the velocity of B is

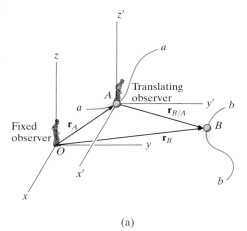

(a)

Fig. 12–42

*An easy way to remember the setup of this equation, and others like it, is to note the "cancellation" of the subscript A between the two terms, i.e., $\mathbf{r}_B = \mathbf{r}_A + \mathbf{r}_{B/A}$.

(b)

(c)

Fig. 12–42 (cont.)

equal to the velocity of *A* plus (vectorially) the relative velocity of "*B* with respect to *A*," as measured by the *translating observer* fixed in the *x'*, *y'*, *z'* reference, Fig. 12–42*b*.

Acceleration.   The time derivative of Eq. 12–34 yields a similar vector relationship between the *absolute* and *relative accelerations* of particles *A* and *B*.

$$\mathbf{a}_B = \mathbf{a}_A + \mathbf{a}_{B/A} \qquad (12\text{–}35)$$

Here $\mathbf{a}_{B/A}$ is the acceleration of *B* as seen by the observer located at *A* and translating with the *x'*, *y'*, *z'* reference frame. The vector addition is shown in Fig. 12–42*c*.

Procedure For Analysis

- When applying the relative-position equation, $\mathbf{r}_B = \mathbf{r}_A + \mathbf{r}_{B/A}$, it is first necessary to specify the locations of the fixed *x, y, z,* and translating *x', y', z'* axes.
- Usually, the origin *A* of the translating axes is located at a point having a *known position*, $\mathbf{r}_A$, Fig. 12–42*a*.
- A graphical representation of the vector addition $\mathbf{r}_B = \mathbf{r}_A + \mathbf{r}_{B/A}$ can be shown, and both the known and unknown quantities labeled on this sketch.
- Since vector addition forms a triangle, there can be at most *two unknowns*, represented by the magnitudes and/or directions of the vector quantities.
- These unknowns can be solved for either graphically, using trigonometry (law of sines, law of cosines), or by resolving each of the three vectors $\mathbf{r}_B$, $\mathbf{r}_A$, and $\mathbf{r}_{B/A}$ into rectangular or Cartesian components, thereby generating a set of scalar equations.
- The relative-motion equations $\mathbf{v}_B = \mathbf{v}_A + \mathbf{v}_{B/A}$ and $\mathbf{a}_B = \mathbf{a}_A + \mathbf{a}_{B/A}$ are applied in the same manner as explained above, except in this case the origin *O* of the fixed *x, y, z* axes does not have to be specified, Figs. 12–42*b* and 12–42*c*.

The pilots of these jet planes flying close to one another must be aware of their relative positions and velocities at all times in order to avoid a collision.

EXAMPLE 12.25

A train, traveling at a constant speed of 60 mi/h, crosses over a road as shown in Fig. 12–43a. If automobile A is traveling at 45 mi/h along the road, determine the magnitude and direction of the relative velocity of the train with respect to the automobile.

(a)

SOLUTION I

Vector Analysis.   The relative velocity $\mathbf{v}_{T/A}$ is measured from the translating x', y' axes attached to the automobile, Fig. 12–43a. It is determined from $\mathbf{v}_T = \mathbf{v}_A + \mathbf{v}_{T/A}$. Since $\mathbf{v}_T$ and $\mathbf{v}_A$ are known in *both* magnitude and direction, the unknowns become the x and y components of $\mathbf{v}_{T/A}$. Using the x, y axes in Fig. 12–43a and a Cartesian vector analysis, we have

$$\mathbf{v}_T = \mathbf{v}_A + \mathbf{v}_{T/A}$$

$$60\mathbf{i} = (45 \cos 45°\mathbf{i} + 45 \sin 45°\mathbf{j}) + \mathbf{v}_{T/A}$$

$$\mathbf{v}_{T/A} = \{28.2\mathbf{i} - 31.8\mathbf{j}\} \text{ mi/h} \qquad \textit{Ans.}$$

The magnitude of $\mathbf{v}_{T/A}$ is thus

$$v_{T/A} = \sqrt{(28.2)^2 + (-31.8)^2} = 42.5 \text{ mi/h} \qquad \textit{Ans.}$$

From the direction of each component, Fig. 12–43b, the direction of $\mathbf{v}_{T/A}$ defined from the x axis is

$$\tan \theta = \frac{(v_{T/A})_y}{(v_{T/A})_x} = \frac{31.8}{28.2}$$

$$\theta = 48.5° \quad \searrow^{\theta} \qquad \textit{Ans.}$$

Note that the vector addition shown in Fig. 12–43b indicates the correct sense for $\mathbf{v}_{T/A}$. This figure anticipates the answer and can be used to check it.

(b)

SOLUTION II

Scalar Analysis.   The unknown components of $\mathbf{v}_{T/A}$ can also be determined by applying a scalar analysis. We will assume these components act in the *positive* x and y directions. Thus,

$$\mathbf{v}_T = \mathbf{v}_A + \mathbf{v}_{T/A}$$

$$\begin{bmatrix} 60 \text{ mi/h} \\ \rightarrow \end{bmatrix} = \begin{bmatrix} 45 \text{ mi/h} \\ \nearrow^{45°} \end{bmatrix} + \begin{bmatrix} (v_{T/A})_x \\ \rightarrow \end{bmatrix} + \begin{bmatrix} (v_{T/A})_y \\ \uparrow \end{bmatrix}$$

Resolving each vector into its x and y components yields

$(\xrightarrow{+})$ $60 = 45 \cos 45° + (v_{T/A})_x + 0$

$(+\uparrow)$ $0 = 45 \sin 45° + 0 + (v_{T/A})_y$

Solving, we obtain the previous results,

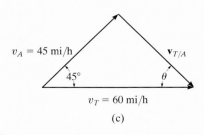

$v_A = 45$ mi/h

$\mathbf{v}_{T/A}$

$45°$

θ

$v_T = 60$ mi/h

(c)

$$(v_{T/A})_x = 28.2 \text{ mi/h} = 28.2 \text{ mi/h} \rightarrow$$

$$(v_{T/A})_y = -31.8 \text{ mi/h} = 31.8 \text{ mi/h} \downarrow \qquad \textit{Ans.}$$

Fig. 12–43

EXAMPLE 12.26

(a)

(b)

(c)

Fig. 12–44

Plane A in Fig. 12–44a is flying along a straight-line path, whereas plane B is flying along a circular path having a radius of curvature of $\rho_B = 400$ km. Determine the velocity and acceleration of B as measured by the pilot of A.

SOLUTION

Velocity.   The x, y axes are located at an arbitrary fixed point. Since the motion relative to plane A is to be determined, the *translating frame of reference* x', y' is attached to it, Fig. 12–44a. Applying the relative-velocity equation in scalar form since the velocity vectors of both planes are parallel at the instant shown, we have

$$(+\uparrow) \qquad\qquad v_B = v_A + v_{B/A}$$

$$600 = 700 + v_{B/A}$$

$$v_{B/A} = -100 \text{ km/h} = 100 \text{ km/h} \downarrow \qquad\qquad Ans.$$

The vector addition is shown in Fig. 12–44b.

Acceleration.   Plane B has both tangential and normal components of acceleration, since it is flying along a *curved path*. From Eq. 12–20, the magnitude of the normal component is

$$(a_B)_n = \frac{v_B^2}{\rho} = \frac{(600 \text{ km/h})^2}{400 \text{ km}} = 900 \text{ km/h}^2$$

Applying the relative-acceleration equation, we have

$$\mathbf{a}_B = \mathbf{a}_A + \mathbf{a}_{B/A}$$

$$900\mathbf{i} - 100\mathbf{j} = 50\mathbf{j} + \mathbf{a}_{B/A}$$

Thus,

$$\mathbf{a}_{B/A} = \{900\mathbf{i} - 150\mathbf{j}\} \text{ km/h}^2$$

From Fig. 12–44c, the magnitude and direction of $\mathbf{a}_{B/A}$ are therefore

$$a_{B/A} = 912 \text{ km/h}^2 \quad \theta = \tan^{-1}\frac{150}{900} = 9.46° \quad\text{↘} \qquad Ans.$$

NOTE: The solution to this problem was possible using a translating frame of reference, since the pilot in plane A is "translating." Observation of plane A with respect to the pilot of plane B, however, must be obtained using a *rotating* set of axes attached to plane B. (This assumes, of course, that the pilot of B is fixed in the rotating frame, so he does not turn his eyes to follow the motion of A.) The analysis for this case is given in Example 16.21.

EXAMPLE 12.27

At the instant shown in Fig. 12–45 cars A and B are traveling with speeds of 18 m/s and 12 m/s, respectively. Also at this instant, A has a decrease in speed of 2 m/s^2, and B has an increase in speed of 3 m/s^2. Determine the velocity and acceleration of B with respect to A.

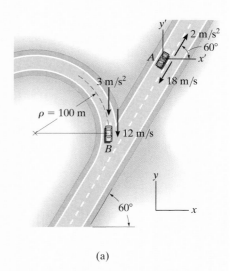

(a)

SOLUTION

Velocity.   The fixed x, y axes are established at a point on the ground and the translating x', y' axes are attached to car A, Fig. 12–45a. Why? The relative velocity is determined from $\mathbf{v}_B = \mathbf{v}_A + \mathbf{v}_{B/A}$. What are the two unknowns? Using a Cartesian vector analysis, we have

$$\mathbf{v}_B = \mathbf{v}_A + \mathbf{v}_{B/A}$$

$$-12\mathbf{j} = (-18 \cos 60°\mathbf{i} - 18 \sin 60°\mathbf{j}) + \mathbf{v}_{B/A}$$

$$\mathbf{v}_{B/A} = \{9\mathbf{i} + 3.588\mathbf{j}\} \text{ m/s}$$

Thus,

$$v_{B/A} = \sqrt{(9)^2 + (3.588)^2} = 9.69 \text{ m/s} \qquad Ans.$$

Noting that $\mathbf{v}_{B/A}$ has $+\mathbf{i}$ and $+\mathbf{j}$ components, Fig. 12–45b, its direction is

$$\tan \theta = \frac{(v_{B/A})_y}{(v_{B/A})_x} = \frac{3.588}{9}$$

$$\theta = 21.7° \quad \swarrow \qquad Ans.$$

(b)

Acceleration.   Car B has both tangential and normal components of acceleration. Why? The magnitude of the normal component is

$$(a_B)_n = \frac{v_B^2}{\rho} = \frac{(12 \text{ m/s})^2}{100 \text{ m}} = 1.440 \text{ m/s}^2$$

Applying the equation for relative acceleration yields

$$\mathbf{a}_B = \mathbf{a}_A + \mathbf{a}_{B/A}$$

$$(-1.440\mathbf{i} - 3\mathbf{j}) = (2 \cos 60°\mathbf{i} + 2 \sin 60°\mathbf{j}) + \mathbf{a}_{B/A}$$

$$\mathbf{a}_{B/A} = \{-2.440\mathbf{i} - 4.732\mathbf{j}\} \text{ m/s}^2$$

Here $\mathbf{a}_{B/A}$ has $-\mathbf{i}$ and $-\mathbf{j}$ components. Thus, from Fig. 12–45c,

$$a_{B/A} = \sqrt{(2.440)^2 + (4.732)^2} = 5.32 \text{ m/s}^2 \qquad Ans.$$

$$\tan \phi = \frac{(a_{B/A})_y}{(a_{B/A})_x} = \frac{4.732}{2.440}$$

$$\phi = 62.7° \quad \searrow \qquad Ans.$$

(c)

Fig. 12–45

NOTE: Is it possible to obtain the relative acceleration of $\mathbf{a}_{A/B}$ using this method? Refer to the comment made at the end of Example 12.26.

PROBLEMS

*12–172.   If the end of the cable at A is pulled down with a speed of 2 m/s, determine the speed at which block B rises.

Prob. 12–172

12–173.   If the end of the cable at A is pulled down with a speed of 2 m/s, determine the speed at which block B rises.

Prob. 12–173

12–174.   Determine the constant speed at which the cable at A must be drawn in by the motor in order to hoist the load at B 15 ft in 5 s.

Prob. 12–174

12–175.   Determine the time needed for the load at B to attain a speed of 8 m/s, starting from rest, if the cable is drawn into the motor with an acceleration of 0.2 m/s^2.

Prob. 12–175

*12–176.   If the hydraulic cylinder at H draws rod BC in by 8 in., determine how far the slider at A moves.

Prob. 12–176

12–177.   The crate is being lifted up the inclined plane using the motor M and the rope and pulley arrangement shown. Determine the speed at which the cable must be taken up by the motor in order to move the crate up the plane with a constant speed of 4 ft/s.

Prob. 12–177

12–178. Determine the displacement of the block at B if A is pulled down 4 ft.

Prob. 12–178

12–179. The hoist is used to lift the load at D. If the end A of the chain is traveling downward at $v_A = 5$ ft/s and the end B is traveling upward at $v_B = 2$ ft/s, determine the velocity of the load at D.

Prob. 12–179

*12–180. The pulley arrangement shown is designed for hoisting materials. If *BC remains fixed* while the plunger P is pushed downward with a speed of 4 ft/s, determine the speed of the load at A.

Prob. 12–180

12–181. If block A is moving downward with a speed of 4 ft/s while C is moving up at 2 ft/s, determine the speed of block B.

12–182. If block A is moving downward at 6 ft/s while block C is moving down at 18 ft/s, determine the relative velocity of block B with respect to C.

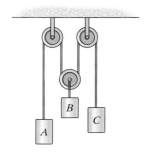

Prob. 12–182

12–183. The motor draws in the cable at C with a constant velocity of $v_C = 4$ m/s. The motor draws in the cable at D with a constant acceleration of $a_D = 8$ m/s². If $v_D = 0$ when $t = 0$, determine (a) the time needed for block A to rise 3 m, and (b) the relative velocity of block A with respect to block B when this occurs.

Prob. 12–183

*12–184. If block *A* of the pulley system is moving downward with a speed of 4 ft/s while block *C* is moving up at 2 ft/s, determine the speed of block *B*.

Prob. 12–184

12–185. If the end *A* of the cable is moving upwards at $v_A = 14$ m/s, determine the speed of block *B*.

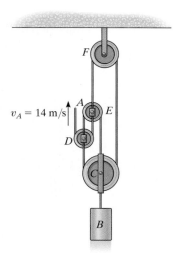

Prob. 12–185

12–186. The cylinder *C* is being lifted using the cable and pulley system shown. If point *A* on the cable is being drawn toward the drum with a speed of 2 m/s, determine the speed of the cylinder.

Prob. 12–186

12–187. The cord is attached to the pin at *C* and passes over the two pulleys at *A* and *D*. The pulley at *A* is attached to the smooth collar that travels along the vertical rod. Determine the velocity and acceleration of the end of the cord at *B* if at the instant $s_A = 4$ ft the collar is moving upwards at 5 ft/s, which is decreasing at 2 ft/s².

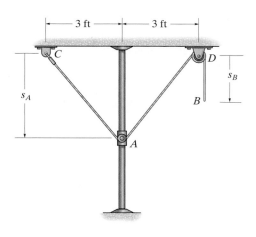

Prob. 12–187

*12–188. The 16-ft-long cord is attached to the pin at C and passes over the two pulleys at A and D. The pulley at A is attached to the smooth collar that travels along the vertical rod. When $s_B = 6$ ft, the end of the cord at B is pulled downwards with a velocity of 4 ft/s and is given an acceleration of 3 ft/s². Determine the velocity and acceleration of the collar at this instant.

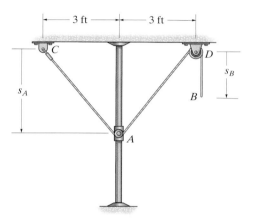

Prob. 12–188

12–189. The crate C is being lifted by moving the roller at A downward with a constant speed of $v_A = 2$ m/s along the guide. Determine the velocity and acceleration of the crate at the instant $s = 1$ m. When the roller is at B, the crate rests on the ground. Neglect the size of the pulley in the calculation. *Hint:* Relate the coordinates x_C and x_A using the problem geometry, then take the first and second time derivatives.

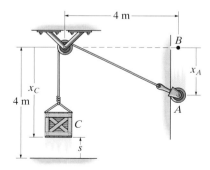

Prob. 12–189

12–190. The girl at C stands near the edge of the pier and pulls in the rope *horizontally* at a constant speed of 6 ft/s. Determine how fast the boat approaches the pier at the instant the rope length AB is 50 ft.

Prob. 12–190

12–191. The man pulls the boy up to the tree limb C by walking backward. If he starts from rest when $x_A = 0$ and moves backward with a constant acceleration $a_A = 0.2$ m/s², determine the speed of the boy at the instant $y_B = 4$ m. Neglect the size of the limb. When $x_A = 0$, $y_B = 8$ m, so that A and B are coincident, i.e., the rope is 16 m long.

Prob. 12–191

*12–192. Collars A and B are connected to the cord that passes over the small pulley at C. When A is located at D, B is 24 ft to the left of D. If A moves at a constant speed of 2 ft/s to the right, determine the speed of B when A is 4 ft to the right of D.

Prob. 12–192

12–193. If block B is moving down with a velocity v_B and has an acceleration a_B, determine the velocity and acceleration of block A in terms of the parameters shown.

Prob. 12–193

12–194. Vertical motion of the load is produced by movement of the piston at A on the boom. Determine the distance the piston or pulley at C must move to the left in order to lift the load 2 ft. The cable is attached at B, passes over the pulley at C, then D, E, F, and again around E, and is attached at G.

Prob. 12–194

12–195. The motion of the collar at A is controlled by a motor at B such that when the collar is at $s_A = 3$ ft it is moving upwards at 2 ft/s and slowing down at 1 ft/s^2. Determine the velocity and acceleration of the cable as it is drawn into the motor B at this instant.

Prob. 12–195

*12–196. The roller at A is moving upward with a velocity of $v_A = 3$ ft/s and has an acceleration of $a_A = 4$ ft/s^2 when $s_A = 4$ ft. Determine the velocity and acceleration of block B at this instant.

Prob. 12–196

12–197. Two planes, A and B, are flying at the same altitude. If their velocities are $v_A = 600$ km/h and $v_B = 500$ km/h such that the angle between their straight-line courses is $\theta = 75°$, determine the velocity of plane B with respect to plane A.

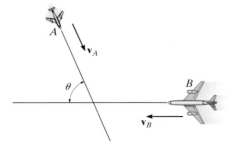

Prob. 12–197

12–198. At the instant shown, cars A and B are traveling at speeds of 30 mi/h and 20 mi/h, respectively. If B is increasing its speed by 1200 mi/h^2, while A maintains a constant speed, determine the velocity and acceleration of B with respect to A.

12–199. At the instant shown, cars A and B are traveling at speeds of 30 m/h and 20 mi/h, respectively. If A is increasing its speed at 400 mi/h^2 whereas the speed of B is decreasing at 800 mi/h^2, determine the velocity and acceleration of B with respect to A.

Probs. 12–198/199

*12–200. Two boats leave the shore at the same time and travel in the directions shown. If $v_A = 20$ ft/s and $v_B = 15$ ft/s, determine the speed of boat A with respect to boat B. How long after leaving the shore will the boats be 800 ft apart?

Prob. 12–200

12–201. At the instant shown, the car at A is traveling at 10 m/s around the curve while increasing its speed at 5 m/s^2. The car at B is traveling at 18.5 m/s along the straightaway and increasing its speed at 2 m/s^2. Determine the relative velocity and relative acceleration of A with respect to B at this instant.

Prob. 12–201

12–202. An aircraft carrier is traveling forward with a velocity of 50 km/h. At the instant shown, the plane at A has just taken off and has attained a forward horizontal air speed of 200 km/h, measured from still water. If the plane at B is traveling along the runway of the carrier at 175 km/h in the direction shown, determine the velocity of A with respect to B.

Prob. 12–202

12–203. Cars A and B are traveling around the circular race track. At the instant shown, A has a speed of 90 ft/s and is increasing its speed at the rate of 15 ft/s², whereas B has a speed of 105 ft/s and is decreasing its speed at 25 ft/s². Determine the relative velocity and relative acceleration of car A with respect to car B at this instant.

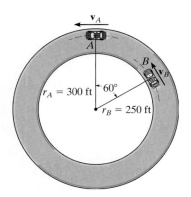

Prob. 12–203

*12–204. The airplane has a speed relative to the wind of 100 mi/h. If the wind relative to the ground is 10 mi/h, determine the angle θ at which the plane must be directed in order to travel in the direction of the runway. Also, what is its speed relative to the runway.

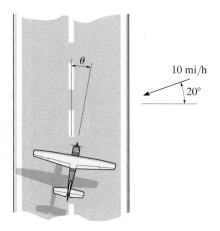

Prob. 12–204

12–205. At the instant shown car A is traveling with a velocity of 30 m/s and has an acceleration of 2 m/s² along the highway. At the same instant B is traveling on the trumpet interchange curve with a speed of 15 m/s, which is decreasing at 0.8 m/s². Determine the relative velocity and relative acceleration of B with respect to A at this instant.

Prob. 12–205

12–206. The boy A is moving in a straight line away from the building at a constant speed of 4 ft/s. The boy C throws the ball B horizontally when A is at $d = 10$ ft. At what speed must C throw the ball so that A can catch it? Also determine the relative speed of the ball with respect to boy A at the instant the catch is made.

Prob. 12–206

12–207. The boy A is moving in a straight line away from the building at a constant speed of 4 ft/s. At what horizontal distance d must he be from C in order to make the catch if the ball is thrown with a horizontal velocity of $v_C = 10$ ft/s? Also determine the relative speed of the ball with respect to the boy A at the instant the catch is made.

*12–208. At a given instant, two particles A and B are moving with a speed of 8 m/s along the paths shown. If B is decelerating at 6 m/s^2 and the speed of A is increasing at 5 m/s^2, determine the acceleration of A with respect to B at this instant.

Prob. 12–207

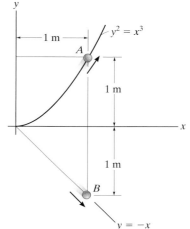

Prob. 12–208

Design Project

12–1D. Design of a Marble-Sorting Device

Marbles roll off the production chute at 0.5 ft/s. Determine the range for the angle $0 \le \theta \le 30°$ for a selected position s for the placement of the hopper relative to the end of the chute. Submit a drawing of the device that shows the path the marbles take.

Fig. 12–1D

Chapter Review

Rectilinear Kinematics

Rectilinear kinematics refers to motion along a straight line. A position coordinate s specifies the location of the particle on the line, and the displacement Δs is the change in this position.
(*See page 5.*)

The average velocity is a vector quantity, defined as the displacement divided by the time interval.

$$v_{avg} = \frac{-\Delta s}{\Delta t}$$

The average speed is a scalar, and is the total distance traveled divided by the time of travel.
(*See page 6.*)

$$(v_{sp})_{avg} = \frac{s_T}{\Delta t}$$

The time, position, velocity, and acceleration are related by three differential equations.

$$v = \frac{ds}{dt}, \qquad a = \frac{dv}{dt}, \qquad a\, ds = v\, dv$$

If the acceleration is known to be constant, then the differential equations relating time, position, velocity, and acceleration can be integrated.
(*See page 8.*)

$$s = s_0 + v_0 t + \tfrac{1}{2} a_c t^2$$

$$v^2 = v_0^2 + 2a_c(s - s_0)$$

$$v = v_0 + a_c t$$

Graphical Solutions

If the motion is erratic, then it can be described by a graph. If one of these graphs is given, then the others can be established using the differential relations between $a, v, s,$ and t.
(*See page 18.*)

$$v = ds/dt, \; a = dv/dt,$$
$$a\, ds = v\, dv.$$

Curvilinear Motion, x, y, z

Curvilinear motion along the path can be resolved into rectilinear motion along the x, y, z axes. The equation of the path is used to relate the motion along each axis. (*See pages 33 and 34.*)

$$v_x = \dot{x} \qquad a_x = \dot{v}_x$$

$$v_y = \dot{y} \qquad a_y = \dot{v}_y$$

$$v_z = \dot{z} \qquad a_z = \dot{v}_z$$

Projectile Motion

Free-flight motion of a projectile follows a parabolic path. It has a constant velocity in the horizontal direction, and a constant acceleration of $g = 9.81$ m/s^2 or 32.2 ft/s^2 in the vertical direction. Any two of the three equations for constant acceleration apply in the vertical direction, and in the horizontal direction only one equation applies.
(*See pages 38 and 39.*)

$$(+\uparrow) \qquad v_y = (v_0)_y + a_c t$$

$$(+\uparrow) \qquad y = y_0 + (v_0)_y t + \tfrac{1}{2} a_c t^2$$

$$(+\uparrow) \qquad v_y^2 = (v_0)_y^2 + 2 a_c (y - y_0)$$

$$(\xrightarrow{+}) \qquad x = x_0 + (v_0)_x t + \tfrac{1}{2} a_c t^2$$

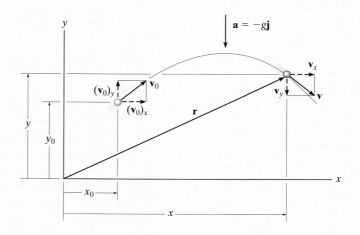

Curvilinear Motion *n*, *t*

If normal and tangential axes are used for the analysis, then **v** is always in the positive *t* direction.

The acceleration has two components. The tangential component, **a**$_t$, accounts for the change in the magnitude of the velocity; a slowing down is in the negative *t* direction, and a speeding up is in the positive *t* direction. The normal component **a**$_n$ accounts for the change in the direction of the velocity. This component is always in the positive *n* direction.
(*See pages 49–51.*)

$$a_t = \dot{v} \quad \text{or} \quad a_t \, ds = v \, dv$$

$$a_n = \frac{v^2}{\rho}$$

Curvilinear Motion, *r*, *θ*, *z*

If the path of motion is expressed in polar coordinates, then the velocity and acceleration components can be related to the time derivatives of *r* and *θ*.

$$v_r = \dot{r}$$
$$v_\theta = r\dot{\theta}$$

To apply the time-derivative equations, it is necessary to determine $r, \dot{r}, \ddot{r}, \theta, \dot{\theta}$ at the instant considered.

$$a_r = \ddot{r} - r\dot{\theta}^2$$
$$a_\theta = r\ddot{\theta} + 2\dot{r}\dot{\theta}$$

If the path $r = f(\theta)$ is given, then the chain rule of calculus must be used to obtain the time derivatives. Once the data are substituted into the equations, then the algebraic sign of the results will indicate the direction of the components of **v** or **a** along each axis.
(*See pages 62–64.*)

Velocity

Acceleration

Absolute Dependent Motion of Two Particles

The dependent motion of blocks that are suspended from pulleys and cables can be related by the geometry of the system. This is done by first establishing position coordinates, measured from a fixed origin to each block so that they are directed along the line of motion of the blocks.

Using geometry and/or trigonometry, the coordinates are then related to the cable length in order to formulate a position coordinate equation.

$$2s_B + h + s_A = l$$

The first time derivative of this equation gives a relationship between the velocities of the blocks, and a second time derivative gives the relationship between their accelerations.
(*See page 77.*)

$$2v_B = -v_A$$
$$2a_B = -a_A$$

Relative-Motion Analysis Using Translating Axes

If two particles A and B undergo independent motions, then these motions can be related to their relative motion using a *translating set of axes* attached to one of the particles (A).

For planar motion, each vector equations produces two scalar equations, one in the x, and the other in the y direction. For solution, the vectors can be expressed in Cartesian form, or the x and y scalar components can be written directly.
(*See pages 83 and 84.*)

$$\mathbf{r}_B = \mathbf{r}_A + \mathbf{r}_{B/A}$$
$$\mathbf{v}_B = \mathbf{v}_A + \mathbf{v}_{B/A}$$
$$\mathbf{a}_B = \mathbf{a}_A + \mathbf{a}_{B/A}$$

The design of conveyors for a bottling plant requires knowledge of the forces that act on them and the ability to predict the motion of the bottles they transport.

13

Kinetics of a Particle: Force and Acceleration

CHAPTER OBJECTIVES

- To state Newton's Laws of Motion and Gravitational Attraction and to define mass and weight.

- To analyze the accelerated motion of a particle using the equation of motion with different coordinate systems.

- To investigate central-force motion and apply it to problems in space mechanics.

13.1 Newton's Laws of Motion

Many of the earlier notions about dynamics were dispelled after 1590 when Galileo performed experiments to study the motions of pendulums and falling bodies. The conclusions drawn from these experiments gave some insight as to the effects of forces acting on bodies in motion. The general laws of motion of a body subjected to forces were not known, however, until 1687, when Isaac Newton first presented three basic laws governing the motion of a particle. In a slightly reworded form, Newton's three laws of motion are

First Law: *A particle originally at rest, or moving in a straight line with a constant velocity, will remain in this state provided the particle is not subjected to an unbalanced force.*

Second Law: *A particle acted upon by an unbalanced force* **F** *experiences an acceleration* **a** *that has the same direction as the force and a magnitude that is directly proportional to the force.**

Third Law: *The mutual forces of action and reaction between two particles are equal, opposite, and collinear.*

---

*Stated another way, the unbalanced force acting on the particle is proportional to the time rate of change of the particle's linear momentum. See footnote † on next page.

The first and third laws were used extensively in developing the concepts of statics. Although these laws are also considered in dynamics, it is Newton's second law of motion that forms the basis for most of this study, since this law relates the accelerated motion of a particle to the forces that act on it.

Measurements of force and acceleration can be recorded in a laboratory so that in accordance with the second law, if a known unbalanced force **F** is applied to a particle, the acceleration **a** of the particle may be measured. Since the force and acceleration are directly proportional, the constant of proportionality, m, may be determined from the ratio $m = F/a$.* The positive scalar m is called the *mass* of the particle. Being constant during any acceleration, m provides a quantitative measure of the resistance of the particle to a change in its velocity.

If the mass of the particle is m, Newton's second law of motion may be written in mathematical form as

$$\mathbf{F} = m\mathbf{a}$$

This equation, which is referred to as the *equation of motion*, is one of the most important formulations in mechanics.* As previously stated, its validity is based solely on *experimental evidence*. In 1905, however, Albert Einstein developed the theory of relativity and placed limitations on the use of Newton's second law for describing general particle motion. Through experiments it was proven that *time* is not an absolute quantity as assumed by Newton; as a result, the equation of motion fails to predict the exact behavior of a particle, especially when the particle's speed approaches the speed of light (0.3 Gm/s). Developments of the theory of quantum mechanics by Erwin Schrödinger and others indicate further that conclusions drawn from using this equation are also invalid when particles are the size of an atom and move close to one another. For the most part, however, these requirements regarding particle speed and size are not encountered in engineering problems, so their effects will not be considered in this book.

---

*Recall that the units of force in the SI system and mass in the FPS system are derived from this equation, where $N = kg \cdot m/s^2$ and slug $= lb \cdot s^2/ft$ (see Sec. 1.3 of *Statics*). If, however, the units of force, mass, length, and time were *all* selected arbitrarily, then it is necessary to write $F = kma$, where k (a dimensionless constant) would have to be determined experimentally in order to preserve the equality.

*Since m is constant, we can also write $\mathbf{F} = d(m\mathbf{v})/dt$, where $m\mathbf{v}$ is the particle's linear momentum.

Newton's Law of Gravitational Attraction.   Shortly after formulating his three laws of motion, Newton postulated a law governing the mutual attraction between any two particles. In mathematical form this law can be expressed as

$$F = G\frac{m_1 m_2}{r^2}$$ (13–1)

where

F = force of attraction between the two particles

G = universal constant of gravitation; according to experimental evidence $G = 66.73(10^{-12})$ m^3/(kg·s^2)

m_1, m_2 = mass of each of the two particles

r = distance between the centers of the two particles

Any two particles or bodies have a mutually attractive gravitational force acting between them. In the case of a particle located at or near the surface of the earth, however, the only gravitational force having any sizable magnitude is that between the earth and the particle. This force is termed the "weight" and, for our purpose, it will be the only gravitational force considered.

Mass and Weight.   *Mass* is a property of matter by which we can compare the response of one body with that of another. As indicated above, this property manifests itself as a gravitational attraction between two bodies and provides a quantitative measure of the resistance of matter to a change in velocity. It is an *absolute* quantity since the measurement of mass can be made at any location. The weight of a body, however, is *not absolute* since it is measured in a gravitational field, and hence its magnitude depends on where the measurement is made. From Eq. 13–1, we can develop a general expression for finding the weight W of a particle having a mass $m_1 = m$. Let m_2 be the mass of the earth and r the distance between the earth's center and the particle. Then, if $g = Gm_2/r^2$, we have

$$W = mg$$

By comparison with $F = ma$, we term g the acceleration due to gravity. For most engineering calculations g is measured at a point on the surface of the earth at sea level, and at a latitude of 45°, which is considered the "standard location."

The mass and weight of a body are measured differently in the SI and FPS systems of units, and the method of defining these units should be thoroughly understood.

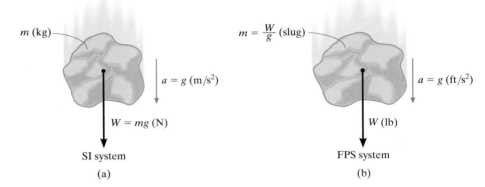

Fig. 13–1

SI System of Units.   In the SI system the mass of the body is specified in kilograms, and the weight must be calculated using the equation of motion, $F = ma$. Hence, if a body has a mass m (kg) and is located at a point where the acceleration due to gravity is g (m/s^2), then the weight is expressed in *newtons* as $W = mg$ (N), Fig. 13–1a. In particular, if the body is located at the "standard location," the acceleration due to gravity is $g = 9.80665$ m/s^2. For calculations, the value $g = 9.81$ m/s^2 will be used, so that

$$W = mg \ (\text{N}) \qquad (g = 9.81 \text{ m/s}^2) \tag{13–2}$$

Therefore, a body of mass 1 kg has a weight of 9.81 N; a 2-kg body weighs 19.62 N; and so on.

FPS System of Units.   In the FPS system the weight of the body is specified in pounds, and the mass must be calculated from $F = ma$. Hence, if a body has a weight W (lb) and is located at a point where the acceleration due to gravity is g (ft/s^2) then the mass is expressed in *slugs* as $m = W/g$ (slug), Fig. 13–1b. Since the acceleration of gravity at the standard location is approximately 32.2 ft/s^2 ($= 9.81$ m/s^2), the mass of the body measured in slugs is

$$m = \frac{W}{g} \ (\text{slug}) \quad (g = 32.2 \text{ ft/s}^2) \tag{13–3}$$

Therefore, a body weighing 32.2 lb has a mass of 1 slug; a 64.4-lb body has a mass of 2 slugs; and so on.

13.2   The Equation of Motion

When more than one force acts on a particle, the resultant force is determined by a vector summation of all the forces; i.e., $\mathbf{F}_R = \Sigma\mathbf{F}$. For this more general case, the equation of motion may be written as

$$\boxed{\Sigma\mathbf{F} = m\mathbf{a}} \qquad (13\text{–}4)$$

To illustrate application of this equation, consider the particle P shown in Fig. 13–2a, which has a mass m and is subjected to the action of two forces, $\mathbf{F}_1$ and $\mathbf{F}_2$. We can graphically account for the magnitude and direction of each force acting on the particle by drawing the particle's *free-body diagram*, Fig. 13–2b. Since the *resultant* of these forces *produces* the vector $m\mathbf{a}$, its magnitude and direction can be represented graphically on the *kinetic diagram*, shown in Fig. 13–2c.* The equal sign written between the diagrams symbolizes the *graphical* equivalency between the free-body diagram and the kinetic diagram; i.e., $\Sigma\mathbf{F} = m\mathbf{a}$.† In particular, note that if $\mathbf{F}_R = \Sigma\mathbf{F} = \mathbf{0}$, then the acceleration is also zero, so that the particle will either remain at *rest* or move along a straight-line path with *constant velocity*. Such are the conditions of *static equilibrium*, Newton's first law of motion.

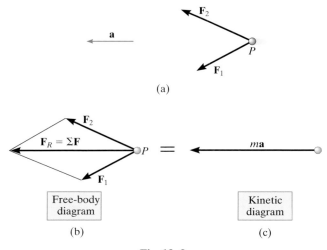

(a)

Free-body diagram

(b)

Kinetic diagram

(c)

Fig. 13–2

*Recall the free-body diagram considers the particle to be free of its surroundings and shows all the forces acting on the particle. The kinetic diagram pertains to the particle's motion as caused by the forces.

†The equation of motion can also be rewritten in the form $\Sigma\mathbf{F} - m\mathbf{a} = \mathbf{0}$. The vector $-m\mathbf{a}$ is referred to as the *inertia force vector*. If it is treated in the same way as a "force vector," then the state of "equilibrium" created is referred to as *dynamic equilibrium*. This method for application is often referred to as the *D'Alembert principle*, named after the French mathematician Jean le Rond d'Alembert.

Inertial Frame of Reference.   Whenever the equation of motion is applied, it is required that measurements of the acceleration be made from a *Newtonian* or *inertial frame of reference. Such a coordinate system does not rotate and is either fixed or translates in a given direction with a constant velocity (zero acceleration).* This definition ensures that the particle's *acceleration* measured by observers in two different inertial frames of reference will always be the *same*. For example, consider the particle P moving with an absolute acceleration $\mathbf{a}_P$ along a straight path as shown in Fig. 13–3. If the observer is *fixed* in the inertial x, y frame of reference, this acceleration, $\mathbf{a}_P$, will be measured by the observer regardless of the direction and magnitude of the velocity $\mathbf{v}_O$ of the frame of reference. On the other hand, if the observer is *fixed* in the noninertial x', y' frame of reference, Fig. 13–3, the observer will not measure the particle's acceleration as $\mathbf{a}_P$. Instead, if the frame is *accelerating* at $\mathbf{a}_{O'}$ the particle will appear to have an acceleration of $\mathbf{a}_{P/O'} = \mathbf{a}_P - \mathbf{a}_{O'}$. Also, if the frame is *rotating*, as indicated by the curl, then the particle will appear to move along a *curved path*, in which case it will appear to have other components of acceleration (see Sec. 16.8). In any case, the measured acceleration from this observer cannot be used in Newton's law of motion to determine the forces acting on the particle.

When studying the motions of rockets and satellites, it is justifiable to consider the inertial reference frame as fixed to the stars, whereas dynamics problems concerned with motions on or near the surface of the earth may be solved by using an inertial frame which is assumed fixed to the earth. Even though the earth both rotates about its own axis and revolves about the sun, the accelerations created by these rotations are relatively small and can be neglected in most computations.

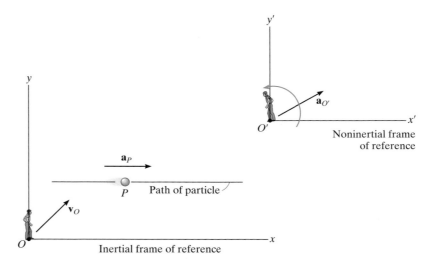

Fig. 13–3

We are all familiar with the sensation one feels when sitting in a car that is subjected to a forward acceleration. Often people think this is caused by a "force" which acts on them and tends to push them back in their seats; however, this is not the case. Instead, this sensation occurs due to their inertia or the resistance of their mass to a change in velocity.

Consider the passenger who is strapped to the seat of a rocket sled. Provided the sled is at rest or is moving with constant velocity, then no force is exerted on his back as shown on his free-body diagram.

At rest or constant velocity

When the thrust of the rocket engine causes the sled to accelerate, then the seat upon which he is sitting exerts a force **F** on him which pushes him forward with the sled. In the photo, notice that the inertia of his head resists this change in motion (acceleration), and so his head moves back against the seat and his face, which is nonrigid, tends to distort.

Acceleration

Upon deceleration the force of the seatbelt **F'** tends to pull his body to a stop, but his head leaves contact with the back of the seat and his face distorts forward, again due to his inertia or tendency to continue to move forward. No force is pulling him forward, although this is the sensation he receives.

Deceleration

13.3 Equation of Motion for a System of Particles

The equation of motion will now be extended to include a system of n particles isolated within an enclosed region in space, as shown in Fig. 13–4a. In particular, there is no restriction in the way the particles are connected, so the following analysis applies equally well to the motion of a solid, liquid, or gas system. At the instant considered, the arbitrary ith particle, having a mass m_i, is subjected to a system of internal forces and a resultant external force. The *resultant internal force*, represented symbolically as $\mathbf{f}_i$, is determined from the forces which the other particles exert on the ith particle. Usually these forces are developed by direct contact, although the summation extends over all n particles within the dashed boundary. The *resultant external force* $\mathbf{F}_i$ represents, for example, the effect of gravitational, electrical, magnetic, or contact forces between the ith particle and adjacent bodies or particles *not* included within the system.

The free-body and kinetic diagrams for the ith particle are shown in Fig. 13–4b. Applying the equation of motion yields

$$\Sigma\mathbf{F} = m\mathbf{a}; \qquad\qquad \mathbf{F}_i + \mathbf{f}_i = m_i\mathbf{a}_i$$

When the equation of motion is applied to each of the other particles of the system, similar equations will result. If all these equations are added together *vectorially*, we obtain

$$\Sigma\mathbf{F}_i + \Sigma\mathbf{f}_i = \Sigma m_i\mathbf{a}_i$$

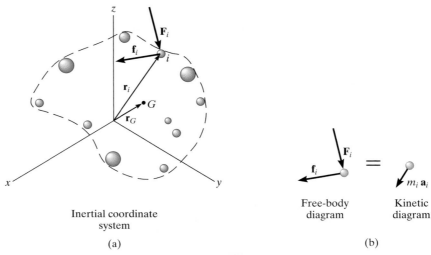

Inertial coordinate
system

(a)

Free-body
diagram

Kinetic
diagram

(b)

Fig. 13–4

The summation of the internal forces, if carried out, will equal zero, since internal forces between particles all occur in equal but opposite collinear pairs. Consequently, only the sum of the external forces will remain, and therefore the equation of motion, written for the system of particles, becomes

$$\Sigma \mathbf{F}_i = \Sigma m_i \mathbf{a}_i \qquad (13\text{–}5)$$

If $\mathbf{r}_G$ is a position vector which locates the *center of mass* G of the particles, Fig. 13–4*a*, then by definition of the center of mass, $m\mathbf{r}_G = \Sigma m_i \mathbf{r}_i$, where $m = \Sigma m_i$ is the total mass of all the particles. Differentiating this equation twice with respect to time, assuming that no mass is entering or leaving the system, yields

$$m\mathbf{a}_G = \Sigma m_i \mathbf{a}_i$$

Substituting this result into Eq. 13–5, we obtain

$$\boxed{\Sigma \mathbf{F} = m\mathbf{a}_G} \qquad (13\text{–}6)$$

Hence, the sum of the external forces acting on the system of particles is equal to the total mass of the particles times the acceleration of its center of mass G. Since in reality all particles must have a finite size to possess mass, Eq. 13–6 justifies application of the equation of motion to a *body* that is represented as a single particle.

Important Points

- The equation of motion is based on experimental evidence and is valid only when applied from an inertial frame of reference.

- The equation of motion states that the *unbalanced force* on a particle causes it to accelerate.

- An inertial frame of reference does not rotate, rather it has axes that either translate with constant velocity or are at rest.

- Mass is a property of matter that provides a quantitative measure of its resistance to a change in velocity. It is an absolute quantity.

- Weight is a force that is caused by the earth's gravitation. It is not absolute; rather it depends on the altitude of the mass from the earth's surface.

13.4   Equations of Motion: Rectangular Coordinates

When a particle is moving relative to an inertial x, y, z frame of reference, the forces acting on the particle, as well as its acceleration, may be expressed in terms of their $\mathbf{i}, \mathbf{j}, \mathbf{k}$ components, Fig. 13–5. Applying the equation of motion, we have

$$\Sigma \mathbf{F} = m\mathbf{a} \qquad \Sigma F_x \mathbf{i} + \Sigma F_y \mathbf{j} + \Sigma F_z \mathbf{k} = m(a_x \mathbf{i} + a_y \mathbf{j} + a_z \mathbf{k})$$

For this equation to be satisfied, the respective $\mathbf{i}, \mathbf{j}, \mathbf{k}$ components on the left side must equal the corresponding components on the right side. Consequently, we may write the following three scalar equations:

$$\begin{array}{l} \Sigma F_x = ma_x \\ \Sigma F_y = ma_y \\ \Sigma F_z = ma_z \end{array} \qquad (13\text{–}7)$$

In particular, if the particle is constrained to move only in the x–y plane, then the first two of these equations are used to specify the motion.

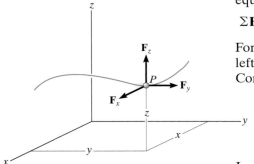

Fig. 13–5

Procedure for Analysis

The equations of motion are used to solve problems which require a relationship between the forces acting on a particle and the accelerated motion they cause.

Free-Body Diagram.

- Select the inertial coordinate system. Most often, rectangular or x, y, z coordinates are chosen to analyze problems for which the particle has *rectilinear motion*.

- Once the coordinates are established, draw the particle's free-body diagram. Drawing this diagram is *very important* since it provides a graphical representation that accounts for *all the forces* ($\Sigma \mathbf{F}$) which act on the particle, and thereby makes it possible to resolve these forces into their x, y, z components.

- The direction and sense of the particle's acceleration $\mathbf{a}$ should also be established. If the senses of its components are unknown, for mathematical convenience assume that they are in the *same direction* as the *positive* inertial coordinate axes.

- The acceleration may be represented as the $m\mathbf{a}$ vector on the kinetic diagram.*

- Identify the unknowns in the problem.

*It is a convention in this text always to use the kinetic diagram as a graphical aid when developing the proofs and theory. The particle's acceleration or its components will be shown as blue colored vectors near the free-body diagram in the examples.

Equations of Motion.

- If the forces can be resolved directly from the free-body diagram, apply the equations of motion in their scalar component form.

- If the geometry of the problem appears complicated, which often occurs in three dimensions, Cartesian vector analysis can be used for the solution.

- *Friction*. If a moving particle contacts a rough surface, it may be necessary to use the *frictional equation*, which relates the coefficient of kinetic friction μ_k to the magnitudes of the frictional and normal forces $\mathbf{F}_f$ and $\mathbf{N}$ acting at the surfaces of contact, i.e., $F_f = \mu_k N$. Remember that $\mathbf{F}_f$ always acts on the free-body diagram such that it opposes the motion of the particle relative to the surface it contacts. If the particle is on the verge of relative motion then the coefficient of static friction should be used.

- *Spring*. If the particle is connected to an *elastic spring* having negligible mass, the spring force F_s can be related to the deformation of the spring by the equation $F_s = ks$. Here k is the spring's stiffness measured as a force per unit length, and s is the stretch or compression defined as the difference between the deformed length l and the undeformed length l_0, i.e., $s = l - l_0$.

Kinematics.

- If the velocity or position of the particle is to be found, it will be necessary to apply the proper kinematic equations once the particle's acceleration is determined from $\Sigma \mathbf{F} = m\mathbf{a}$.

- If *acceleration is a function of time*, use $a = dv/dt$ and $v = ds/dt$ which, when integrated, yield the particle's velocity and position.

- If *acceleration is a function of displacement*, integrate $a\,ds = v\,dv$ to obtain the velocity as a function of position.

- If *acceleration is constant*, use $v = v_0 + a_c t$, $s = s_0 + v_0 t + \frac{1}{2}a_c t^2$, $v^2 = v_0^2 + 2a_c(s - s_0)$ to determine the velocity or position of the particle.

- If the problem involves the dependent motion of several particles, use the method outlined in Sec. 12.9 to relate their accelerations.

- In all cases, make sure the positive inertial coordinate directions used for writing the kinematic equations are the same as those used for writing the equations of motion; otherwise, simultaneous solution of the equations will result in errors.

- If the solution for an unknown vector component yields a negative scalar, it indicates that the component acts in the direction opposite to that which was assumed.

EXAMPLE 13.1

(a)

(b)

Fig. 13–6

The 50-kg crate shown in Fig. 13–6a rests on a horizontal plane for which the coefficient of kinetic friction is $\mu_k = 0.3$. If the crate is subjected to a 400-N towing force as shown, determine the velocity of the crate in 3 s starting from rest.

SOLUTION
Using the equations of motion, we can relate the crate's acceleration to the force causing the motion. The crate's velocity can then be determined using kinematics.

Free-Body Diagram. The weight of the crate is $W = mg = 50 \text{ kg} (9.81 \text{ m/s}^2) = 490.5$ N. As shown in Fig. 13–6b, the frictional force has a magnitude $F = \mu_k N_C$ and acts to the left, since it opposes the motion of the crate. The acceleration **a** is assumed to act horizontally, in the positive x direction. There are two unknowns, namely N_C and a.

Equations of Motion. Using the data shown on the free-body diagram, we have

$$\xrightarrow{+} \Sigma F_x = ma_x; \qquad 400 \cos 30° - 0.3 N_C = 50a \qquad (1)$$

$$+\uparrow \Sigma F_y = ma_y; \qquad N_C - 490.5 + 400 \sin 30° = 0 \qquad (2)$$

Solving Eq. 2 for N_C, substituting the result into Eq. 1, and solving for a yields

$$N_C = 290.5 \text{ N}$$
$$a = 5.19 \text{ m/s}^2$$

Kinematics. Note that the acceleration is *constant*, since the applied force **P** is constant. Since the initial velocity is zero, the velocity of the crate in 3 s is

$$(\xrightarrow{+}) \qquad v = v_0 + a_c t = 0 + 5.19(3)$$
$$= 15.6 \text{ m/s} \rightarrow \qquad \qquad Ans.$$

NOTE: We can also use the alternative procedure of drawing the crate's free-body *and* kinetic diagrams, Fig. 13–6c, prior to applying the equations of motion.

EXAMPLE 13.2

A 10-kg projectile is fired vertically upward from the ground, with an initial velocity of 50 m/s, Fig. 13–7a. Determine the maximum height to which it will travel if (a) atmospheric resistance is neglected; and (b) atmospheric resistance is measured as $F_D = (0.01v^2)$ N, where v is the speed at any instant, measured in m/s.

SOLUTION

In both cases the known force on the projectile can be related to its acceleration using the equation of motion. Kinematics can then be used to relate the projectile's acceleration to its position.

Part (a) Free-Body Diagram. As shown in Fig. 13–7b, the projectile's weight is $W = mg = 10(9.81) = 98.1$ N. We will assume the unknown acceleration **a** acts upward in the *positive z* direction.

Equation of Motion.

$$+\uparrow \Sigma F_z = ma_z; \qquad -98.1 = 10a, \qquad a = -9.81 \text{ m/s}^2$$

The result indicates that the projectile, like every object having free-flight motion near the earth's surface, is subjected to a *constant* downward acceleration of 9.81 m/s².

Kinematics. Initially, $z_0 = 0$ and $v_0 = 50$ m/s, and at the maximum height $z = h$, $v = 0$. Since the acceleration is *constant*, then

$$(+\uparrow) \qquad v^2 = v_0^2 + 2a_c(z - z_0)$$
$$0 = (50)^2 + 2(-9.81)(h - 0)$$
$$h = 127 \text{ m} \qquad \qquad Ans.$$

Part (b) Free-Body Diagram. Since the force $F_D = (0.01v^2)$ N tends to retard the upward motion of the projectile, it acts downward as shown on the free-body diagram, Fig. 13–7c.

Equation of Motion.

$$+\uparrow \Sigma F_z = ma_z; \qquad -0.01v^2 - 98.1 = 10a, \qquad a = -0.001v^2 - 9.81$$

Kinematics. Here the acceleration is *not constant* since F_D depends on the velocity. Since $a = f(v)$, we can relate a to position using

$$(+\uparrow) \; a\,dz = v\,dv; \qquad (-0.001v^2 - 9.81)\,dz = v\,dv$$

Separating the variables and integrating, realizing that initially $z_0 = 0$, $v_0 = 50$ m/s (positive upward), and at $z = h$, $v = 0$, we have

$$\int_0^h dz = -\int_{50}^0 \frac{v\,dv}{0.001v^2 + 9.81} = -500 \ln(v^2 + 9810) \Big|_{50}^0$$
$$h = 114 \text{ m} \qquad \qquad Ans.$$

NOTE: The answer indicates a lower elevation than that obtained in part (a) due to atmospheric resistance.

(a)

(b)

(c)

Fig. 13–7

EXAMPLE 13.3

The baggage truck A shown in the photo has a weight of 900 lb and tows a 550-lb cart B and a 325-lb cart C. For a short time the driving frictional force developed at the wheels of the truck is $F_A = (40t)$ lb, where t is in seconds. If the truck starts from rest, determine its speed in 2 seconds. Also, what is the horizontal force acting on the coupling between the truck and cart B at this instant? Neglect the size of the truck and carts.

(a)

SOLUTION

Free-Body Diagram.   As shown in Fig. 13–8a, it is the frictional driving force that gives both the truck and carts an acceleration. Here we have considered all three vehicles.

Equation of Motion.   Only motion in the horizontal direction has to be considered.

$$\xleftarrow{+} \Sigma F_x = ma_x; \qquad 40t = \left(\frac{900 + 550 + 325}{32.2}\right)a$$
$$a = 0.7256t$$

Kinematics.   Since the acceleration is a function of time, the velocity of the truck is obtained using $a = dv/dt$ with the initial condition that $v_0 = 0$ at $t = 0$. We have

$$\int_0^v dv = \int_0^2 0.7256t \, dt; \qquad v = 0.3628t^2 \Big|_0^2 = 1.45 \text{ ft/s} \qquad \textit{Ans.}$$

Free-Body Diagram.   In order to determine the force between the truck and cart B, we can consider a free-body diagram of the truck so that we can "expose" the coupling force $\mathbf{T}$ as external to the free-body diagram, Fig. 13–8b.

Equation of Motion.   When $t = 2$ s, then

$$\xleftarrow{+} \Sigma F_x = ma_x; \qquad 40(2) - T = \left(\frac{900}{32.2}\right)[0.7256(2)]$$
$$T = 39.4 \text{ lb} \qquad \textit{Ans.}$$

NOTE: Try and obtain this same result by considering a free-body diagram of carts B and C.

(b)

Fig. 13–8

EXAMPLE 13.4

A smooth 2-kg collar C, shown in Fig. 13–9a, is attached to a spring having a stiffness $k = 3$ N/m and an unstretched length of 0.75 m. If the collar is released from rest at A, determine its acceleration and the normal force of the rod on the collar at the instant $y = 1$ m.

SOLUTION

Free-Body Diagram. The free-body diagram of the collar when it is located at the arbitrary position y is shown in Fig. 13–9b. Note that the weight is $W = 2(9.81) = 19.62$ N. Furthermore, the collar is *assumed* to be accelerating so that "**a**" acts downward in the *positive y* direction. There are four unknowns, namely, N_C, F_s, a, and θ.

(a)

Equations of Motion.

$$\xrightarrow{+} \Sigma F_x = ma_x; \qquad -N_C + F_s \cos\theta = 0 \qquad (1)$$

$$+\downarrow \Sigma F_y = ma_y; \qquad 19.62 - F_s \sin\theta = 2a \qquad (2)$$

From Eq. 2 it is seen that the acceleration depends on the magnitude and direction of the spring force. Solution for N_C and a is possible once F_s and θ are known.

The magnitude of the spring force is a function of the stretch s of the spring; i.e., $F_s = ks$. Here the unstretched length is $AB = 0.75$ m, Fig. 13–9a; therefore, $s = CB - AB = \sqrt{y^2 + (0.75)^2} - 0.75$. Since $k = 3$ N/m, then

(b)

Fig. 13–9

$$F_s = ks = 3\left(\sqrt{y^2 + (0.75)^2} - 0.75\right) \qquad (3)$$

From Fig. 13–9a, the angle θ is related to y by trigonometry.

$$\tan\theta = \frac{y}{0.75} \qquad (4)$$

Substituting $y = 1$ m into Eqs. 3 and 4 yields $F_s = 1.50$ N and $\theta = 53.1°$. Substituting these results into Eqs. 1 and 2, we obtain

$$N_C = 0.900 \text{ N} \qquad\qquad Ans.$$
$$a = 9.21 \text{ m/s}^2 \downarrow \qquad\qquad Ans.$$

NOTE: This is not a case of constant acceleration, since the spring force changes both its magnitude and direction as the collar moves downward.

EXAMPLE 13.5

Datum

s_A

C

s_B

A

B

(a)

T T

$2T$

(b)

$2T$

$\mathbf{a}_A$

s_A 981 N

(c)

T

$\mathbf{a}_B$

196.2 N

s_B

(d)

Fig. 13–10

The 100-kg block A shown in Fig. 13–10a is released from rest. If the masses of the pulleys and the cord are neglected, determine the speed of the 20-kg block B in 2 s.

SOLUTION

Free-Body Diagrams.   Since the mass of the pulleys is *neglected*, then for pulley C, $ma = 0$ and we can apply $\Sigma F_y = 0$ as shown in Fig. 13–10b. The free-body diagrams for blocks A and B are shown in Fig. 13–10c and d, respectively. One can see that for A to remain static requires $T = 490.5$ N, whereas for B to remain static requires $T = 196.2$ N. Hence A will move down while B moves up. Here we will assume both blocks accelerate downward, in the direction of $+s_A$ and $+s_B$. The three unknowns are T, a_A, and a_B.

Equations of Motion.   Block A (Fig. 13–10c):

$$+\downarrow \Sigma F_y = ma_y; \qquad 981 - 2T = 100a_A \qquad (1)$$

Block B (Fig. 13–10d):

$$+\downarrow \Sigma F_y = ma_y; \qquad 196.2 - T = 20a_B \qquad (2)$$

Kinematics.   The necessary third equation is obtained by relating a_A to a_B using a dependent motion analysis. Using the technique developed in Sec. 12.9, the coordinates s_A and s_B measure the positions of A and B from the fixed datum, Fig. 13–10a. It is seen that

$$2s_A + s_B = l$$

where l is constant and represents the total vertical length of cord. Differentiating this expression twice with respect to time yields

$$2a_A = -a_B \qquad (3)$$

Notice that in writing Eqs. 1 to 3, the *positive direction was always assumed downward*. It is very important to be *consistent* in this assumption since we are seeking a simultaneous solution of equations. The solution yields

$$T = 327.0 \text{ N}$$
$$a_A = 3.27 \text{ m/s}^2$$
$$a_B = -6.54 \text{ m/s}^2$$

Hence when block A accelerates *downward*, block B accelerates *upward*. Since a_B is constant, the velocity of block B in 2 s is thus

$$(+\downarrow) \qquad\qquad v = v_0 + a_B t$$
$$= 0 + (-6.54)(2)$$
$$= -13.1 \text{ m/s} \qquad\qquad Ans.$$

The negative sign indicates that block B is moving upward.

PROBLEMS

13–1. Determine the gravitational attraction between two spheres which are just touching each other. Each sphere has a mass of 10 kg and a radius of 200 mm.

13–2. By using an inclined plane to retard the motion of a falling object, and thus make the observations more accurate, Galileo was able to determine experimentally that the distance through which an object moves in free fall is proportional to the square of the time for travel. Show that this is the case, i.e., $s \propto t^2$, by determining the time t_B, t_C, and t_D needed for a block of mass m to slide from rest at A to points B, C, and D, respectively. Neglect the effects of friction.

Probs. 13–1/2

13–3. The 300-kg bar B, originally at rest, is being towed over a series of small rollers. Determine the force in the cable when $t = 5$ s, if the motor M is drawing in the cable for a short time at a rate of $v = (0.4t^2)$ m/s, where t is in seconds $(0 \le t \le 6 \text{ s})$. How far does the bar move in 5 s? Neglect the mass of the cable, pulley, and the rollers.

Prob. 13–3

*13–4. A crate having a mass of 60 kg falls horizontally off the back of a truck which is traveling at 80 km/h. Determine the coefficient of kinetic friction between the road and the crate if the crate slides 45 m on the ground with no tumbling along the road before coming to rest. Assume that the initial speed of the crate along the road is 80 km/h.

Prob. 13–4

13–5. The crane lifts the 700-kg bin with an initial acceleration of 3 m/s². Determine the force in each of the supporting cables due to this motion.

Prob. 13–5

13–6. The baggage truck A has a mass of 800 kg and is used to pull the two cars, each with mass 300 kg. If the tractive force F on the truck is F = 480 N, determine the initial acceleration of the truck. What is the acceleration of the truck if the coupling at C suddenly fails? The car wheels are free to roll. Neglect the mass of the wheels.

Prob. 13–6

13–7. The 500-kg fuel assembly for a nuclear reactor is being lifted out from the core of the nuclear reactor using the pulley system shown. It is hoisted upward with a constant acceleration such that $s = 0$ and $v = 0$ when $t = 0$, and $s = 2.5$ m when $t = 1.5$ s. Determine the tension in the cable at A during the motion.

Prob. 13–7

*__13–8.__ The 200-kg crate is suspended from the cable of a crane. Determine the force in the cable when $t = 2$ s if the crate is moving upward with (a) a constant velocity of 2 m/s, and (b) a speed of $v = (0.2t^2 + 2)$ m/s, where t is in seconds.

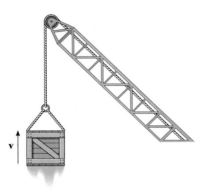

Prob. 13–8

13–9. The elevator E has a mass of 500 kg, and the counterweight at A has a mass of 150 kg. If the motor supplies a constant force of 5 kN on the cable at B, determine the speed of the elevator when $t = 3$ s, starting from rest. Neglect the mass of the pulleys and cable.

13–10. The elevator E has a mass of 500 kg and the counterweight at A has a mass of 150 kg. If the elevator attains a speed of 10 m/s after it rises 40 m, determine the constant force developed in the cable at B. Neglect the mass of the pulleys and cable.

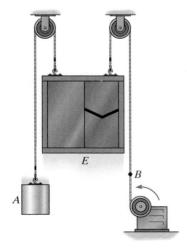

Probs. 13–9/10

13–11. The water-park ride consists of an 800-lb sled which slides from rest down the incline and then into the pool. If the frictional resistance on the incline is $F_r = 30$ lb, and in the pool for a short distance $F_r = 80$ lb, determine how fast the sled is traveling when $s = 5$ ft.

100 ft

100 ft

s

Prob. 13–11

*13–12.** A car of mass m is traveling at a slow velocity v_0. If it is subjected to the drag resistance of the wind, which is proportional to its velocity, i.e., $F_D = kv$, determine the distance and the time the car will travel before its velocity becomes $0.5v_0$. Assume no other frictional forces act on the car.

v

F_D

Prob. 13–12

13–13. Determine the normal force the 10-kg crate A exerts on the smooth cart if the cart is given an acceleration of $a = 2$ m/s² down the plane. Also, what is the acceleration of the crate? Set $\theta = 30°$.

A

a

θ

Prob. 13–13

13–14. Each of the two blocks has a mass m. The coefficient of kinetic friction at all surfaces of contact is μ. If a horizontal force **P** moves the bottom block, determine the acceleration of the bottom block in each case.

B

P   A

(a)

B

P   A

Prob. 13–14   (b)

13–15. The driver attempts to tow the crate using a rope that has a tensile strength of 200 lb. If the crate is originally at rest and has a weight of 500 lb, determine the greatest acceleration it can have if the coefficient of static friction between the crate and the road is $\mu_s = 0.4$, and the coefficient of kinetic friction is $\mu_k = 0.3$.

30°

Prob. 13–15

*13–16.** The 3.5-Mg engine is suspended from a 500-kg spreader beam and hoisted by a crane which gives it an acceleration of 4 m/s² when it has a velocity of 2 m/s. Determine the force in chains AC and AD during the lift.

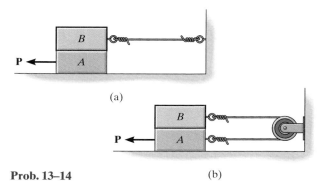

C

60°   60°

A   B

D   E

Prob. 13–16

13–17.   The bullet of mass m is given a velocity due to gas pressure caused by the burning of powder within the chamber of the gun. Assuming this pressure creates a force of $F = F_0 \sin(\pi t/t_0)$ on the bullet, determine the velocity of the bullet at any instant it is in the barrel. What is the bullet's maximum velocity? Also, determine the position of the bullet in the barrel as a function of time.

13–19.   A 40-lb suitcase slides from rest 20 ft down the smooth ramp. Determine the point where it strikes the ground at C. How long does it take to go from A to C?

*13–20.   Solve Prob. 13–19 if the suitcase has an initial velocity down the ramp of $v_A = 10$ ft/s and the coefficient of kinetic friction along AB is $\mu_k = 0.2$.

Prob. 13–17

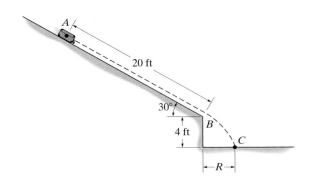

Probs. 13–19/20

13–18.   The 400-lb cylinder at A is hoisted using the motor and the pulley system shown. If the speed of point B on the cable is increased at a constant rate from zero to $v_B = 10$ ft/s in $t = 5$ s, determine the tension in the cable at B to cause the motion.

13–21.   The winding drum D is drawing in the cable at an accelerated rate of 5 m/s². Determine the cable tension if the suspended crate has a mass of 800 kg.

Prob. 13–18

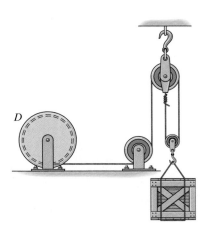

Prob. 13–21

13–22. At a given instant the 5-lb weight A is moving downward with a speed of 4 ft/s. Determine its speed 2 s later. Block B has a weight of 6 lb, and the coefficient of kinetic friction between it and the horizontal plane is $\mu_k = 0.3$. Neglect the mass of the pulleys and cord.

*13–24. At a given instant the 10-lb block A is moving downward with a speed of 6 ft/s. Determine its speed 2 s later. Block B has a weight of 4 lb, and the coefficient of kinetic friction between it and the horizontal plane is $\mu_k = 0.2$. Neglect the mass of the pulleys and cord.

Prob. 13–22

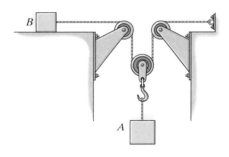

Prob. 13–24

13–25. A freight elevator, including its load, has a mass of 500 kg. It is prevented from rotating due to the track and wheels mounted along its sides. If the motor M develops a constant tension $T = 1.50$ kN in its attached cable, determine the velocity of the elevator when it has moved upward 3 m starting from rest. Neglect the mass of the pulleys and cables.

13–23. A force $F = 15$ lb is applied to the cord. Determine how high the 30-lb block A rises in 2 s starting from rest. Neglect the weight of the pulleys and cord.

Prob. 13–23

Prob. 13–25

13–26.   At the instant shown the 100-lb block A is moving down the plane at 5 ft/s while being attached to the 50-lb block B. If the coefficient of kinetic friction is $\mu_k = 0.2$, determine the acceleration of A and the distance A slides before it stops. Neglect the mass of the pulleys and cables.

*13–28.   The 400-kg mine car is hoisted up the incline using the cable and motor M. For a short time, the force in the cable is $F = (3200t^2)$ N, where t is in seconds. If the car has an initial velocity $v_1 = 2$ m/s when $t = 0$, determine its velocity when $t = 2$ s.

13–29.   The 400-kg mine car is hoisted up the incline using the cable and motor M. For a short time, the force in the cable is $F = (3200t^2)$ N, where t is in seconds. If the car has an initial velocity $v_1 = 2$ m/s at $s = 0$ and $t = 0$, determine the distance it moves up the plane when $t = 2$ s.

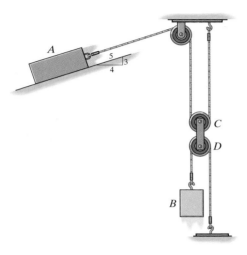

Prob. 13–26

Probs. 13–28/29

13–27.   The safe S has a weight of 200 lb and is supported by the rope and pulley arrangement shown. If the end of the rope is given to a boy B of weight 90 lb, determine his acceleration if in the confusion he doesn't let go of the rope. Neglect the mass of the pulleys and rope.

■13–30.   The tanker has a weight of $800(10^6)$ lb and is traveling forward at $v_0 = 3$ ft/s in still water when the engines are shut off. If the drag resistance of the water is proportional to the speed of the tanker at any instant and can be approximated by $F_D = (400(10^3)v)$ lb, where v is in ft/s, determine the time needed for the tanker's speed to become 1.5 ft/s. Given the initial velocity of $v_0 = 3$ ft/s through what distance must the tanker travel before it stops?

Prob. 13–27

Prob. 13–30

13–31. The spring mechanism is used as a shock absorber for railroad cars. Determine the maximum compression of spring *HI* if the fixed bumper *R* of a 5-Mg railroad car, rolling freely at 2 m/s, strikes the plate *P*. Bar *AB* slides along the guide paths *CE* and *DF*. The ends of all springs are attached to their respective members and are originally unstretched.

Prob. 13–31

13–34. The boy having a weight of 80 lb hangs uniformly from the bar. Determine the force in each of his arms in $t = 2$ s if the bar is moving upward with (a) a constant velocity of 3 ft/s, and (b) a speed of $v = (4t^2)$ ft/s, where t is in seconds.

Prob. 13–34

***13–32.** The 2-kg collar *C* is free to slide along the smooth shaft *AB*. Determine the acceleration of collar *C* if (a) the shaft is fixed from moving, (b) collar *A*, which is fixed to shaft *AB*, moves downward at constant velocity along the vertical rod, and (c) collar *A* is subjected to a downward acceleration of 2 m/s². In all cases, the collar moves in the plane.

13–33. The 2-kg collar *C* is free to slide along the smooth shaft *AB*. Determine the acceleration of collar *C* if collar *A* is subjected to an upward acceleration of 4 m/s².

Probs. 13–32/33

13–35. The 10-kg block *A* rests on the 50-kg plate *B* in the position shown. Neglecting the mass of the rope and pulley, and using the coefficients of kinetic friction indicated, determine the time needed for block *A* to slide 0.5 m *on the plate* when the system is released from rest.

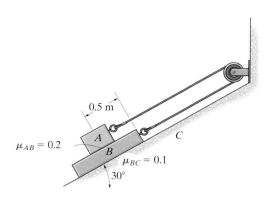

Prob. 13–35

*13–36. Determine the acceleration of block A when the system is released. The coefficient of kinetic friction and the weight of each block are indicated. Neglect the mass of the pulleys and cord.

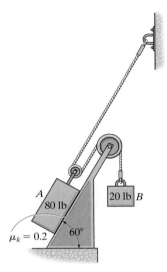

Prob. 13–36

13–38. An electron of mass m is discharged with an initial horizontal velocity of $\mathbf{v}_0$. If it is subjected to two fields of force for which $F_x = F_0$ and $F_y = 0.3F_0$, where F_0 is constant, determine the equation of the path, and the speed of the electron at any time t.

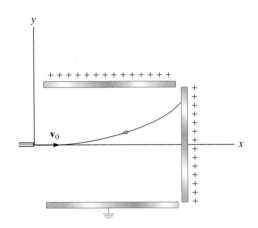

Prob. 13–38

13–37. The conveyor belt is moving at 4 m/s. If the coefficient of static friction between the conveyor and the 10-kg package B is $\mu_s = 0.2$, determine the shortest time the belt can stop so that the package does not slide on the belt.

Prob. 13–37

13–39. The conveyor belt delivers each 12-kg crate to the ramp at A such that the crate's speed is $v_A = 2.5$ m/s, directed down *along* the ramp. If the coefficient of kinetic friction between each crate and the ramp is $\mu_k = 0.3$, determine the speed at which each crate slides off the ramp at B. Assume that no tipping occurs. Take $\theta = 30°$.

Prob. 13–39

*13–40. A parachutist having a mass m opens his parachute from an at-rest position at a very high altitude. If the atmospheric drag resistance is $F_D = kv^2$, where k is a constant, determine his velocity when he has fallen for a time t. What is his velocity when he lands on the ground? This velocity is referred to as the *terminal velocity*, which is found by letting the time of fall $t \to \infty$.

Prob. 13–40

13–41. Block B rests on a smooth surface. If the coefficients of static and kinetic friction between A and B are $\mu_s = 0.4$ and $\mu_k = 0.3$, respectively, determine the acceleration of each block if someone pushes horizontally on block A with a force of (a) $F = 6$ lb, and (b) $F = 50$ lb.

Prob. 13–41

13–42. Blocks A and B each have a mass m. Determine the largest horizontal force P which can be applied to B so that A will not move relative to B. All surfaces are smooth.

13–43. Blocks A and B each have a mass m. Determine the largest horizontal force P which can be applied to B so that A will not slip up B. The coefficient of static friction between A and B is μ_s. Neglect any friction between B and C.

Probs. 13–42/43

*13–44. Each of the three plates has a mass of 10 kg. If the coefficients of static and kinetic friction at each surface of contact are $\mu_s = 0.3$ and $\mu_k = 0.2$, respectively, determine the acceleration of each plate when the three horizontal forces are applied.

Prob. 13–44

13–45. Crate B has a mass m and is released from rest when it is on top of cart A, which has a mass $3m$. Determine the tension in cord CD needed to hold the cart from moving while B is sliding down A. Neglect friction.

Prob. 13–45

■13–46.   The tractor is used to lift the 150-kg load B with the 24-m-long rope, boom, and pulley system. If the tractor is traveling to the right at a constant speed of 4 m/s, determine the tension in the rope when $s_A = 5$ m. When $s_A = 0$, $s_B = 0$.

13–47.   The tractor is used to lift the 150-kg load B with the 24-m-long rope, boom, and pulley system. If the tractor is traveling to the right with an acceleration of 3 m/s² and has a velocity of 4 m/s at the instant $s_A = 5$ m, determine the tension in the rope at this instant. When $s_A = 0$, $s_B = 0$.

13–49.   Block A has a mass m_A and is attached to a spring having a stiffness k and unstretched length l_0. If another block B, having a mass m_B, is pressed against A so that the spring deforms a distance d, determine the distance both blocks slide on the smooth surface before they begin to separate. What is their velocity at this instant?

13–50.   Block A has a mass m_A and is attached to a spring having a stiffness k and unstretched length l_0. If another block B, having a mass m_B, is pressed against A so that the spring deforms a distance d, show that for separation to occur it is necessary that $d > 2\mu_k g(m_A + m_B)/k$, where μ_k is the coefficient of kinetic friction between the blocks and the ground. Also, what is the distance the blocks slide on the surface before they separate?

12 m

s_B

B

A

s_A

Probs. 13–46/47

k

A B

Probs. 13–49/50

*13–48.   Block B has a mass m and is hoisted using the cord and pulley system shown. Determine the magnitude of force $\mathbf{F}$ as a function of the block's vertical position y so that when $\mathbf{F}$ is applied the block rises with a constant acceleration $\mathbf{a}_B$. Neglect the mass of the cord and pulleys.

13–51.   The block A has a mass m_A and rests on the pan B, which has a mass m_B. Both are supported by a spring having a stiffness k that is attached to the bottom of the pan and to the ground. Determine the distance d the pan should be pushed down from the equilibrium position and then released from rest so that separation of the block will take place from the surface of the pan at the instant the spring becomes unstretched.

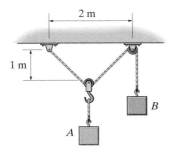

2 m

1 m

B

A

Prob. 13–48

A

B

y d

k

Prob. 13–51

13.5 Equations of Motion: Normal and Tangential Coordinates

When a particle moves over a curved path which is known, the equation of motion for the particle may be written in the tangential, normal, and binormal directions. We have

$$\Sigma \mathbf{F} = m\mathbf{a}$$
$$\Sigma F_t \mathbf{u}_t + \Sigma F_n \mathbf{u}_n + \Sigma F_b \mathbf{u}_b = m\mathbf{a}_t + m\mathbf{a}_n$$

Here ΣF_t, ΣF_n, ΣF_b represent the sums of all the force components acting on the particle in the tangential, normal, and binormal directions, respectively, Fig. 13–11. Note that there is no motion of the particle in the binormal direction, since the particle is constrained to move along the path. The above equation is satisfied provided

$$\boxed{\begin{aligned} \Sigma F_t &= ma_t \\ \Sigma F_n &= ma_n \\ \Sigma F_b &= 0 \end{aligned}}$$

(13–8)

Recall that $a_t \, (= dv/dt)$ represents the time rate of change in the magnitude of velocity. Consequently, if $\Sigma \mathbf{F}_t$ acts in the direction of motion, the particle's speed will increase, whereas if it acts in the opposite direction, the particle will slow down. Likewise, $a_n \, (= v^2/\rho)$ represents the time rate of change in the velocity's direction. Since this vector *always* acts in the positive n direction, i.e., toward the path's center of curvature, then $\Sigma \mathbf{F}_n$, which causes $\mathbf{a}_n$, also acts in this direction. For example, when the particle is constrained to travel in a circular path with a constant speed, there is a normal force exerted on the particle by the constraint in order to change the direction of the particle's velocity ($\mathbf{a}_n$). Since this force is always directed toward the center of the path, it is often referred to as the *centripetal force*.

The centrifuge is used to subject a passenger to very large normal accelerations caused by high rotations. Realize that these accelerations are caused by the unbalanced normal force exerted on the passenger by the seat of the centrifuge.

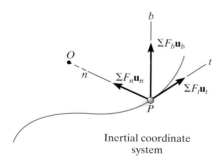

Inertial coordinate system

Fig. 13–11

Procedure for Analysis

When a problem involves the motion of a particle along a *known curved path*, normal and tangential coordinates should be considered for the analysis since the acceleration components can be readily formulated. The method for applying the equations of motion, which relate the forces to the acceleration, has been outlined in the procedure given in Sec. 13.4. Specifically, for t, n, b coordinates it may be stated as follows:

Free-Body Diagram.

- Establish the inertial t, n, b coordinate system at the particle and draw the particle's free-body diagram.
- The particle's normal acceleration $\mathbf{a}_n$ *always* acts in the positive n direction.
- If the tangential acceleration $\mathbf{a}_t$ is unknown, assume it acts in the positive t direction.
- Identify the unknowns in the problem.

Equations of Motion.

- Apply the equations of motion, Eqs. 13–8.

Kinematics.

- Formulate the tangential and normal components of acceleration; i.e., $a_t = dv/dt$ or $a_t = v\,dv/ds$ and $a_n = v^2/\rho$.
- If the path is defined as $y = f(x)$, the radius of curvature at the point where the particle is located can be obtained from $\rho = [1 + (dy/dx)^2]^{3/2}/|d^2y/dx^2|$.

EXAMPLE 13.6

Determine the banking angle θ for the race track so that the wheels of the racing cars shown in Fig. 13–12a will not have to depend upon friction to prevent any car from sliding up or down the track. Assume the cars have negligible size, a mass m, and travel around the curve of radius ρ with a speed v.

(a)

SOLUTION

Before looking at the following solution, give some thought as to why it should be solved using t, n, b coordinates.

Free-Body Diagram.   As shown in Fig. 13–12b, and as stated in the problem, no frictional force acts on the car. Here $\mathbf{N}_C$ represents the *resultant* of the ground on all four wheels. Since a_n can be calculated, the unknowns are N_C and θ.

Equations of Motion.   Using the n, b axes shown,

(b)

Fig. 13–12

$$\xrightarrow{+} \Sigma F_n = ma_n; \qquad N_C \sin\theta = m\frac{v^2}{\rho} \qquad\qquad (1)$$

$$+\uparrow \Sigma F_b = 0; \qquad N_C \cos\theta - mg = 0 \qquad\qquad (2)$$

Eliminating N_C and m from these equations by dividing Eq. 1 by Eq. 2, we obtain

$$\tan\theta = \frac{v^2}{g\rho}$$

$$\theta = \tan^{-1}\left(\frac{v^2}{g\rho}\right) \qquad\qquad Ans.$$

NOTE: The result is independent of the mass of the car. Also, a force summation in the tangential direction is of no consequence to the solution. If it were considered, then $a_t = dv/dt = 0$, since the car moves with *constant speed*. A further analysis of this problem is discussed in Prob. 21–48.

EXAMPLE 13.7

The 3-kg disk D is attached to the end of a cord as shown in Fig. 13–13a. The other end of the cord is attached to a ball-and-socket joint located at the center of a platform. If the platform is rotating rapidly, and the disk is placed on it and released from rest as shown, determine the time it takes for the disk to reach a speed great enough to break the cord. The maximum tension the cord can sustain is 100 N, and the coefficient of kinetic friction between the disk and the platform is $\mu_k = 0.1$.

(a)

SOLUTION

Free-Body Diagram. The frictional force has a magnitude $F = \mu_k N_D = 0.1 N_D$ and a sense of direction that opposes the *relative motion* of the disk with respect to the platform. It is this force that gives the disk a tangential component of acceleration causing v to increase, thereby causing T to increase until it reaches 100 N. The weight of the disk is $W = 3(9.81) = 29.43$ N. Since a_n can be related to v, the unknowns are N_D, a_t, and v.

Equations of Motion.

(b)

Fig. 13–13

$$\Sigma F_n = m a_n; \qquad T = 3\left(\frac{v^2}{1}\right) \qquad (1)$$

$$\Sigma F_t = m a_t; \qquad 0.1 N_D = 3 a_t \qquad (2)$$

$$\Sigma F_b = 0; \qquad N_D - 29.43 = 0 \qquad (3)$$

Setting $T = 100$ N, Eq. 1 can be solved for the critical speed v_{cr} of the disk needed to break the cord. Solving all the equations, we obtain

$$N_D = 29.43 \text{ N}$$
$$a_t = 0.981 \text{ m/s}^2$$
$$v_{cr} = 5.77 \text{ m/s}$$

Kinematics. Since a_t is *constant*, the time needed to break the cord is

$$v_{cr} = v_0 + a_t t$$
$$5.77 = 0 + (0.981)t$$
$$t = 5.89 \text{ s} \qquad \qquad Ans.$$

EXAMPLE 13.8

Design of the ski jump shown in the photo requires knowing the type of forces that will be exerted on the skier and his approximate trajectory. If in this case the jump can be approximated by the parabola shown in Fig. 13–14a, determine the normal force on the 150-lb skier the instant he arrives at the end of the jump, point A, where his velocity is 65 ft/s. Also, what is his acceleration at this point?

SOLUTION

Why consider using n, t coordinates to solve this problem?

Free-Body Diagram.   The free-body diagram for the skier when he is at A is shown in Fig. 13–14b. Since the path is *curved*, there are two components of acceleration, $\mathbf{a}_n$ and $\mathbf{a}_t$. Since a_n can be calculated, the unknowns are a_t and N_A.

(a)

Equations of Motion.

$$+\uparrow \Sigma F_n = ma_n; \quad N_A - 150 = \frac{150}{32.2}\left(\frac{(65)^2}{\rho}\right) \tag{1}$$

$$\xleftrightarrow{} \Sigma F_t = ma_t; \qquad 0 = \frac{150}{32.2}a_t \tag{2}$$

The radius of curvature ρ for the path must be determined at point $A(0, -200 \text{ ft})$. Here $y = \frac{1}{200}x^2 - 200$, $dy/dx = \frac{1}{100}x$, $d^2y/dx^2 = \frac{1}{100}$, so that at $x = 0$,

$$\rho = \frac{[1 + (dy/dx)^2]^{3/2}}{|d^2y/dx^2|}\bigg|_{x=0} = \frac{[1 + (0)^2]^{3/2}}{\left|\frac{1}{100}\right|} = 100 \text{ ft}$$

Substituting into Eq. 1 and solving for N_A, we have

$$N_A = 347 \text{ lb} \qquad\qquad Ans.$$

Kinematics.   From Eq. 2,

$$a_t = 0$$

Thus,

$$a_n = \frac{v^2}{\rho} = \frac{(65)^2}{100} = 42.2 \text{ ft/s}^2$$

$$a_A = a_n = 42.2 \text{ ft/s}^2 \uparrow \qquad\qquad Ans.$$

(b)

Fig. 13–14

NOTE: Apply the equation of motion in the y direction and show that when the skier is in mid air the acceleration is 32.2 ft/s².

EXAMPLE 13.9

$v_0 = 1$ m/s

$ds = 0.5\, d\theta$

θ

$d\theta$ $r = 0.5$ m

(a)

N_B θ $\mathbf{a}_t$

$\mathbf{a}_n$

19.62 N

n t

(b)

Fig. 13–15

Packages, each having a mass of 2 kg, are delivered from a conveyor to a smooth circular ramp with a velocity of $v_0 = 1$ m/s as shown in Fig. 13–15a. If the effective radius of the ramp is 0.5 m, determine the angle $\theta = \theta_{max}$ at which each package begins to leave the surface.

SOLUTION

Free-Body Diagram.   The free-body diagram for a package, when it is located at the *general position* θ, is shown in Fig. 13–15b. The package must have a tangential acceleration $\mathbf{a}_t$, since its *speed* is always *increasing* as it slides downward. The weight is $W = 2(9.81) = 19.62$ N. Specify the three unknowns.

Equations of Motion.

$$+\swarrow \Sigma F_n = ma_n; \qquad -N_B + 19.62 \cos \theta = 2\frac{v^2}{0.5} \qquad (1)$$

$$+\searrow \Sigma F_t = ma_t; \qquad 19.62 \sin \theta = 2a_t \qquad (2)$$

At the instant $\theta = \theta_{max}$, the package leaves the surface of the ramp so that $N_B = 0$. Therefore, there are three unknowns, v, a_t, and θ.

Kinematics.   The third equation for the solution is obtained by noting that the magnitude of tangential acceleration a_t may be related to the speed of the package v and the angle θ. Since $a_t\, ds = v\, dv$ and $ds = r\, d\theta = 0.5\, d\theta$, Fig. 13–15a, we have

$$a_t = \frac{v\, dv}{0.5\, d\theta} \qquad (3)$$

To solve, substitute Eq. 3 into Eq. 2 and separate the variables. This gives

$$v\, dv = 4.905 \sin \theta\, d\theta$$

Integrate both sides, realizing that when $\theta = 0°$, $v_0 = 1$ m/s.

$$\int_1^v v\, dv = 4.905 \int_{0°}^{\theta} \sin \theta\, d\theta$$

$$\left.\frac{v^2}{2}\right|_1^v = -4.905 \cos \theta \bigg|_{0°}^{\theta}; \quad v^2 = 9.81(1 - \cos \theta) + 1$$

Substituting into Eq. 1 with $N_B = 0$ and solving for $\cos \theta_{max}$ yields

$$19.62 \cos \theta_{max} = \frac{2}{0.5}[9.81(1 - \cos \theta_{max}) + 1]$$

$$\cos \theta_{max} = \frac{43.24}{58.86}$$

$$\theta_{max} = 42.7° \qquad\qquad Ans.$$

NOTE: The speed of the package is increasing because its tangential acceleration is increasing with θ, Eq. 2.

PROBLEMS

*13–52. Determine the mass of the sun, knowing that the distance from the earth to the sun is $149.6(10^6)$ km. *Hint:* Use Eq. 13–1 to represent the force of gravity acting on the earth.

13–53. The 1.40-Mg helicopter is traveling at a constant speed of 40 m/s along the horizontal curved path while banking at $\theta = 30°$. Determine the force acting normal to the blade, i.e., in the y' direction, and the radius of curvature of the path.

13–54. The 1.40-Mg helicopter is traveling at a constant speed of 33 m/s along the horizontal curved path having a radius of curvature of $\rho = 300$ m. Determine the force the blade exerts on the frame and the bank angle θ.

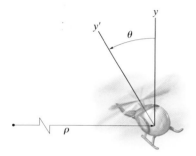

Probs. 13–53/54

13–55. The plane is traveling at a constant speed of 800 ft/s along the curve $y = 20(10^{-6})x^2 + 5000$, where x and y are in feet. If the pilot has a weight of 180 lb, determine the normal and tangential components of the force the seat exerts on the pilot when the plane is at its lowest point.

*13–56. The jet plane is traveling at a constant speed of 1000 ft/s along the curve $y = 20(10^{-6})x^2 + 5000$, where x and y are in feet. If the pilot has a weight of 180 lb, determine the normal and tangential components of the force the seat exerts on the pilot when $y = 10\,000$ ft.

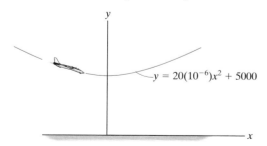

Probs. 13–55/56

13–57. The 600-kg wrecking ball is suspended from the crane by a cable having a negligible mass. If the ball has a speed $v = 8$ m/s at the instant it is at its lowest point, $\theta = 0°$, determine the tension in the cable at this instant. Also, determine the angle θ to which the ball swings before it stops.

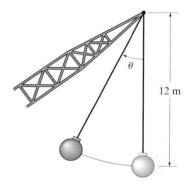

Prob. 13–57

13–58. Prove that if the block is released from rest at point B of a smooth path of *arbitrary shape*, the speed it attains when it reaches point A is equal to the speed it attains when it falls freely through a distance h; i.e., $v = \sqrt{2gh}$.

Prob. 13–58

13–59. The sled and rider have a total mass of 80 kg and start from rest at $A(10\text{ m}, 0)$. If the sled descends the smooth slope, which may be approximated by a parabola, determine the normal force that the ground exerts on the sled at the instant it arrives at point B. Neglect the size of the sled and rider. *Hint:* Use the result of Prob. 13–58.

*13–60. The sled and rider have a total mass of 80 kg and start from rest at $A(10\text{ m}, 0)$. If the sled descends the smooth slope, which may be approximated by a parabola, determine the normal force that the ground exerts on the sled at the instant it arrives at point C. Neglect the size of the sled and rider. *Hint:* Use the result of Prob. 13–58.

Probs. 13–59/60

13–61. At the instant $\theta = 60°$, the boy's center of mass G has a downward speed $v_G = 15$ ft/s. Determine the rate of increase in his speed and the tension in each of the two supporting cords of the swing at this instant. The boy has a weight of 60 lb. Neglect his size and the mass of the seat and cords.

13–62. At the instant $\theta = 60°$, the boy's center of mass G is momentarily at rest. Determine his speed and the tension in each of the two supporting cords of the swing when $\theta = 90°$. The boy has a weight of 60 lb. Neglect his size and the mass of the seat and cords.

Probs. 13–61/62

13–63. If the crest of the hill has a radius of curvature $\rho = 200$ ft, determine the maximum constant speed at which the car can travel over it without leaving the surface of the road. Neglect the size of the car in the calculation. The car has a weight of 3500 lb.

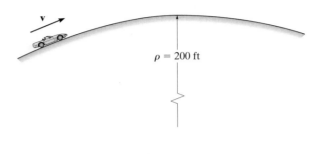

Prob. 13–63

*13–64. The airplane, traveling at a constant speed of 50 m/s, is executing a horizontal turn. If the plane is banked at $\theta = 15°$, when the pilot experiences only a normal force on the seat of the plane, determine the radius of curvature ρ of the turn. Also, what is the normal force of the seat on the pilot if he has a mass of 70 kg.

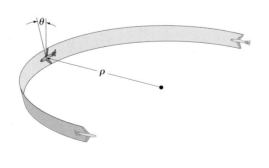

Prob. 13–64

13–65. The 150-lb man lies against the cushion for which the coefficient of static friction is $\mu_s = 0.5$. Determine the resultant normal and frictional forces the cushion exerts on him if, due to rotation about the z axis, he has a constant speed $v = 20$ ft/s. Neglect the size of the man. Take $\theta = 60°$.

13–66. The 150-lb man lies against the cushion for which the coefficient of static friction is $\mu_s = 0.5$. If he rotates about the z axis with a constant speed $v = 30$ ft/s, determine the smallest angle θ of the cushion at which he will begin to slip off.

*13–68.** The 200-kg snowmobile with passenger is traveling down the hill at a constant speed of 6 m/s. Determine the resultant normal force and the resultant frictional force exerted on the tracks at the instant it reaches point A. Neglect the size of the snowmobile.

13–69. The 200-kg snowmobile with passenger is traveling down the hill such that when it is at point A, it is traveling at 4 m/s and increasing its speed at 2 m/s². Determine the resultant normal force and the resultant frictional force exerted on the tracks at this instant. Neglect the size of the snowmobile.

Probs. 13–65/66

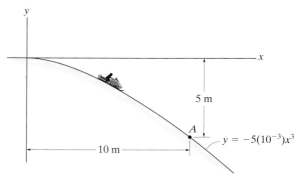

Probs. 13–68/69

13–67. Determine the constant speed of the passengers on the amusement-park ride if it is observed that the supporting cables are directed at $\theta = 30°$ from the vertical. Each chair including its passenger has a mass of 80 kg. Also, what are the components of force in the n, t, and b directions which the chair exerts on a 50-kg passenger during the motion?

13–70. A collar having a mass of 0.75 kg and negligible size slides over the surface of a horizontal circular rod for which the coefficient of kinetic friction is $\mu_k = 0.3$. If the collar is given a speed of 4 m/s and then released at $\theta = 0°$, determine how far, s, it slides on the rod before coming to rest.

Prob. 13–67

Prob. 13–70

13–71. The roller coaster car and passenger have a total weight of 600 lb and starting from rest at A travel down the track that has the shape shown. Determine the normal force of the tracks on the car when the car is at point B, where it has a velocity of 15 ft/s. Neglect friction and the size of the car and passenger.

$y = 40 \cos \left(\frac{\pi}{40}x\right)$

40 ft

40 ft

20 ft

Prob. 13–71

13–73. The 5-kg pendulum bob B is released from rest when $\theta = 0°$. Determine the initial tension in the cord and also at the instant the bob reaches point $D, \theta = 45°$. Neglect the size of the bob.

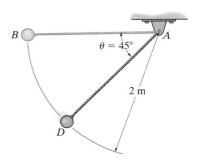

$\theta = 45°$

2 m

Prob. 13–73

*13–72.** The smooth block B, having a mass of 0.2 kg, is attached to the vertex A of the right circular cone using a light cord. The cone is rotating at a constant angular rate about the z axis such that the block attains a speed of 0.5 m/s. At this speed, determine the tension in the cord and the reaction which the cone exerts on the block. Neglect the size of the block.

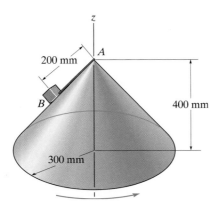

z

200 mm

A

B

400 mm

300 mm

Prob. 13–72

13–74. A ball having a mass of 2 kg and negligible size moves within a smooth vertical circular slot. If it is released from rest when $\theta = 10°$, determine the force of the slot on the ball when the ball arrives at points A and B.

10°

θ

$r = 0.8$ m

A

10°

B

Prob. 13–74

13–75. The rotational speed of the disk is controlled by a 30-g smooth contact arm AB which is spring-mounted on the disk. When the disk is *at rest*, the center of mass G of the arm is located 150 mm from the center O, and the preset compression in the spring is 20 mm. If the initial gap between B and the contact at C is 10 mm, determine the (controlling) speed v_G of the arm's mass center, G, which will close the gap. The disk rotates in the horizontal plane. The spring has a stiffness of $k = 50$ N/m, and its ends are attached to the contact arm at D and to the disk at E.

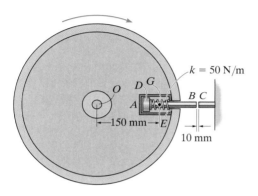

Prob. 13–75

13–77. The 35-kg box has a speed of 2 m/s when it is at A on the smooth ramp. If the surface is in the shape of a parabola, determine the normal force on the box at the instant $x = 3$ m. Also, what is the rate of increase in its speed at this instant?

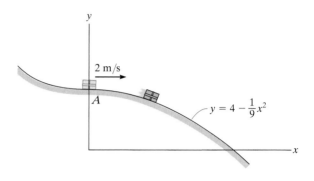

Prob. 13–77

*13–76. The 2-kg spool S fits loosely on the inclined rod for which the coefficient of static friction is $\mu_s = 0.2$. If the spool is located 0.25 m from A, determine the minimum constant speed the spool can have so that it does not slip down the rod.

Prob. 13–76

13–78. The man has a mass of 80 kg and sits 3 m from the center of the rotating platform. Due to the rotation his speed is increased from rest by $\dot{v} = 0.4$ m/s^2. If the coefficient of static friction between his clothes and the platform is $\mu_s = 0.3$, determine the time required to cause him to slip.

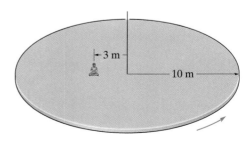

Prob. 13–78

13–79. The collar A, having a mass of 0.75 kg, is attached to a spring having a stiffness of $k = 200$ N/m. When rod BC rotates about the vertical axis, the collar slides outward along the smooth rod DE. If the spring is unstretched when $s = 0$, determine the constant speed of the collar in order that $s = 100$ mm. Also, what is the normal force of the rod on the collar? Neglect the size of the collar.

Prob. 13–79

*13–80. The block has a weight of 2 lb and it is free to move along the smooth slot in the rotating disk. The spring has a stiffness of 2.5 lb/ft and an unstretched length of 1.25 ft. Determine the force of the spring on the block and the tangential component of force which the slot exerts on the side of the block, when the block is at rest with respect to the disk and is traveling with a constant speed of 12 ft/s.

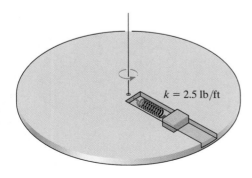

Prob. 13–80

13–81. If the bicycle and rider have a total weight of 180 lb, determine the resultant normal force acting on the bicycle when it is at point A while it is freely coasting at $v_A = 6$ ft/s. Also, compute the increase in the bicyclist's speed at this point. Neglect the resistance due to the wind and the size of the bicycle and rider.

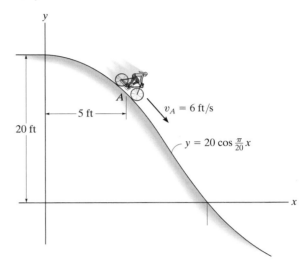

Prob. 13–81

13–82. The 5-lb packages ride on the surface of the conveyor belt. If the belt starts from rest and increases to a constant speed of 12 ft/s in 2 s, determine the maximum angle θ so that none of the packages slip on the inclined surface AB of the belt. The coefficient of static friction between the belt and a package is $\mu_s = 0.3$. At what angle ϕ do the packages first begin to slip off the surface of the belt after the belt is moving at its constant speed of 12 ft/s? Neglect the size of the packages.

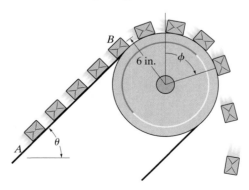

Prob. 13–82

13.6 Equations of Motion: Cylindrical Coordinates

When all the forces acting on a particle are resolved into cylindrical components, i.e., along the unit-vector directions $\mathbf{u}_r$, $\mathbf{u}_\theta$, $\mathbf{u}_z$, Fig. 13–16, the equation of motion may be expressed as

$$\Sigma \mathbf{F} = m\mathbf{a}$$

$$\Sigma F_r\mathbf{u}_r + \Sigma F_\theta\mathbf{u}_\theta + \Sigma F_z\mathbf{u}_z = ma_r\mathbf{u}_r + ma_\theta\mathbf{u}_\theta + ma_z\mathbf{u}_z$$

To satisfy this equation, the respective $\mathbf{u}_r$, $\mathbf{u}_\theta$, $\mathbf{u}_z$ components on the left side must equal the corresponding components on the right side. Consequently, we may write the following three scalar equations of motion:

$$\boxed{\begin{aligned} \Sigma F_r &= ma_r \\ \Sigma F_\theta &= ma_\theta \\ \Sigma F_z &= ma_z \end{aligned}} \tag{13–9}$$

If the particle is constrained to move only in the $r-\theta$ plane, then only the first two of Eqs. 13–9 are used to specify the motion.

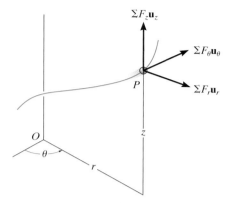

Inertial coordinate system

Fig. 13–16

Tangential and Normal Forces. The most straightforward type of problem involving cylindrical coordinates requires the determination of the resultant force components ΣF_r, ΣF_θ, ΣF_z which cause a particle to move with a *known* acceleration. If, however, the particle's accelerated motion is not completely specified at the given instant, then some information regarding the directions or magnitudes of the forces acting on the particle must be known or computed in order to solve Eqs. 13–9. For example, the force $\mathbf{P}$ causes the particle in Fig. 13–17a to move along a path $r = f(\theta)$. The *normal force* $\mathbf{N}$ which the path exerts on the particle is always *perpendicular to the tangent of the path*, whereas the frictional force $\mathbf{F}$ always acts along the tangent in the opposite direction of motion. The *directions* of $\mathbf{N}$ and $\mathbf{F}$ can be specified relative to the radial coordinate by using the angle ψ (psi), Fig. 13–17b, which is defined between the *extended* radial line and the tangent to the curve.

As the car of weight W descends the spiral track, the resultant normal force which the track exerts on the car can be represented by its three cylindrical components. $-\mathbf{N}_r$ creates a radial acceleration, $-\mathbf{a}_r$, $\mathbf{N}_\theta$ creates a transverse acceleration $\mathbf{a}_\theta$, and the difference, $\mathbf{W} - \mathbf{N}_z$ creates an azimuthal acceleration $-\mathbf{a}_z$.

Fig. 13–17

(a)   (b)

(c)

Fig. 13–17 (cont.)

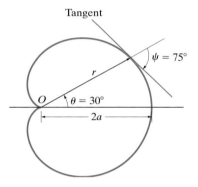

Fig. 13–18

This angle can be obtained by noting that when the particle is displaced a distance ds along the path, Fig. 13–17c, the component of displacement in the radial direction is dr and the component of displacement in the transverse direction is $r\, d\theta$. Since these two components are mutually perpendicular, the angle ψ can be determined from $\tan \psi = r\, d\theta / dr$, or

$$\tan \psi = \frac{r}{dr/d\theta} \qquad (13\text{--}10)$$

If ψ is calculated as a positive quantity, it is measured from the *extended radial line* to the tangent in a counterclockwise sense or in the positive direction of θ. If it is negative, it is measured in the opposite direction to positive θ. For example, consider the cardioid $r = a(1 + \cos \theta)$, shown in Fig. 13–18. Because $dr/d\theta = -a \sin \theta$, then when $\theta = 30°$, $\tan \psi = a(1 + \cos 30°)/(-a \sin 30°) = -3.732$, or $\psi = -75°$, measured clockwise, as shown in the figure.

Procedure for Analysis

Cylindrical or polar coordinates are a suitable choice for the analysis of a problem for which data regarding the angular motion of the radial line r are given, or in cases where the path can be conveniently expressed in terms of these coordinates. Once these coordinates have been established, the equations of motion can be applied in order to relate the forces acting on the particle to its acceleration components. The method for doing this has been outlined in the procedure for analysis given in Sec. 13.4. The following is a summary of this procedure.

Free-Body Diagram.
- Establish the r, θ, z inertial coordinate system and draw the particle's free-body diagram.
- Assume that $\mathbf{a}_r$, $\mathbf{a}_\theta$, $\mathbf{a}_z$ act in the *positive directions* of r, θ, z if they are unknown.
- Identify all the unknowns in the problem.

Equations of Motion.
- Apply the equations of motion, Eqs. 13–9.

Kinematics.
- Use the methods of Sec. 12.8 to determine r and the time derivatives $\dot{r}$, $\ddot{r}$, $\dot{\theta}$, $\ddot{\theta}$, $\ddot{z}$, and then evaluate the acceleration components $a_r = \ddot{r} - r\dot{\theta}^2$, $a_\theta = r\ddot{\theta} + 2\dot{r}\dot{\theta}$, $a_z = \ddot{z}$.
- If any of the acceleration components is computed as a negative quantity, it indicates that it acts in its negative coordinate direction.
- When taking the time derivatives of $r = f(\theta)$, it is very important to use the chain rule of calculus.

EXAMPLE 13.10

The 2-lb block in Fig. 13–19a moves on a smooth horizontal track, such that its path is specified in polar coordinates by the parametric equations $r = (10t^2)$ ft and $\theta = (0.5t)$ rad, where t is in seconds. Determine the magnitude of the tangential force $\mathbf{F}$ causing the motion at the instant $t = 1$ s.

(a)

SOLUTION

Free-Body Diagram. As shown on the block's free-body diagram, Fig. 13–19b, the normal force of the track on the block, $\mathbf{N}$, and the tangential force $\mathbf{F}$ are located at an angle ψ from the r and θ axes. This angle can be obtained from Eq. 13–10. To do so, we must first express the path as $r = f(\theta)$ by eliminating the parameter t between r and θ. This yields $r = 40\theta^2$. Also, when $t = 1$ s, $\theta = 0.5(1\text{ s}) = 0.5$ rad. Thus,

$$\tan\psi = \frac{r}{dr/d\theta} = \frac{40\theta^2}{40(2\theta)}\Big|_{\theta=0.5\text{ rad}} = 0.25$$

$$\psi = 14.04°$$

Because ψ is a positive quantity, it is measured counterclockwise from the r axis to the tangent (the same direction as θ) as shown in Fig. 13–19b. There are presently four unknowns: F, N, a_r and a_θ.

Equations of Motion.

$$+\downarrow\Sigma F_r = ma_r; \qquad F\cos14.04° - N\sin14.04° = \frac{2}{32.2}a_r \qquad (1)$$

$$\uparrow+\Sigma F_\theta = ma_\theta; \qquad F\sin14.04° + N\cos14.04° = \frac{2}{32.2}a_\theta \qquad (2)$$

Kinematics. Since the motion is specified, the coordinates and the required time derivatives can be calculated and evaluated at $t = 1$ s.

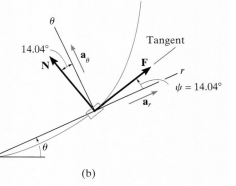

(b)

Fig. 13–19

$$r = 10t^2\Big|_{t=1\text{ s}} = 10\text{ ft} \qquad \theta = 0.5t\Big|_{t=1\text{ s}} = 0.5\text{ rad}$$

$$\dot{r} = 20t\Big|_{t=1\text{ s}} = 20\text{ ft/s} \quad \dot{\theta} = 0.5\text{ rad/s}$$

$$\ddot{r} = 20\text{ ft/s}^2 \qquad \ddot{\theta} = 0$$

$$a_r = \ddot{r} - r\dot{\theta}^2 = 20 - 10(0.5)^2 = 17.5\text{ ft/s}^2$$

$$a_\theta = r\ddot{\theta} + 2\dot{r}\dot{\theta} = 10(0) + 2(20)(0.5) = 20\text{ ft/s}^2$$

Substituting into Eqs. 1 and 2 and solving, we get

$$F = 1.36\text{ lb} \qquad\qquad Ans.$$

$$N = 0.942\text{ lb}$$

NOTE: The tangential axis is in the direction of $\mathbf{F}$, and the normal axis is in the direction of $\mathbf{N}$.

EXAMPLE 13.11

0.4 m

$\dot{\theta} = 0.5$ rad/s

(a)

19.62 N

$\mathbf{F}_P$

N_C

$\mathbf{a}_\theta$

$\mathbf{a}_r$

(b)

Fig. 13–20

The smooth 2-kg cylinder C in Fig. 13–20a has a peg P through its center which passes through the slot in arm OA. If the arm rotates in the *vertical plane* at a constant rate $\dot{\theta} = 0.5$ rad/s, determine the force that the arm exerts on the peg at the instant $\theta = 60°$.

SOLUTION

Why is it a good idea to use polar coordinates to solve this problem?

Free-Body Diagram. The free-body diagram for the cylinder is shown in Fig. 13–20b. The force on the peg, $\mathbf{F}_P$, acts perpendicular to the slot in the arm. As usual, $\mathbf{a}_r$ and $\mathbf{a}_\theta$ are assumed to act in the directions of *positive r* and θ, respectively. Identify the four unknowns.

Equations of Motion. Using the data in Fig. 13–20b, we have

$$+\searrow \Sigma F_r = ma_r; \qquad 19.62 \sin \theta - N_C \sin \theta = 2a_r \qquad (1)$$
$$+\nearrow \Sigma F_\theta = ma_\theta; \qquad 19.62 \cos \theta + F_P - N_C \cos \theta = 2a_\theta \qquad (2)$$

Kinematics. From Fig. 13–20a, r can be related to θ by the equation

$$r = \frac{0.4}{\sin \theta} = 0.4 \csc \theta$$

Since $d(\csc \theta) = -(\csc \theta \cot \theta)\, d\theta$ and $d(\cot \theta) = -(\csc^2 \theta)\, d\theta$, then r and the necessary time derivatives become

$$\dot{\theta} = 0.5 \qquad r = 0.4 \csc \theta$$
$$\ddot{\theta} = 0 \qquad \dot{r} = -0.4(\csc \theta \cot \theta)\dot{\theta}$$
$$= -0.2 \csc \theta \cot \theta$$
$$\ddot{r} = -0.2(-\csc \theta \cot \theta)(\dot{\theta}) \cot \theta - 0.2 \csc \theta(-\csc^2 \theta)\dot{\theta}$$
$$= 0.1 \csc \theta (\cot^2 \theta + \csc^2 \theta)$$

Evaluating these formulas at $\theta = 60°$, we get

$$\dot{\theta} = 0.5 \qquad r = 0.462$$
$$\ddot{\theta} = 0 \qquad \dot{r} = -0.133$$
$$\ddot{r} = 0.192$$
$$a_r = \ddot{r} - r\dot{\theta}^2 = 0.192 - 0.462(0.5)^2 = 0.0770$$
$$a_\theta = r\ddot{\theta} + 2\dot{r}\dot{\theta} = 0 + 2(-0.133)(0.5) = -0.133$$

Substituting these results into Eqs. 1 and 2 with $\theta = 60°$ and solving yields

$$N_C = 19.4 \text{ N} \qquad\qquad F_P = -0.356 \text{ N} \qquad\qquad Ans.$$

The negative sign indicates that $\mathbf{F}_P$ acts opposite to the direction shown in Fig. 13–20b.

EXAMPLE 13.12

A can C, having a mass of 0.5 kg, moves along a grooved horizontal slot shown in Fig. 13–21a. The slot is in the form of a spiral, which is defined by the equation $r = (0.1\theta)$ m, where θ is in radians. If the arm OA is rotating at a constant rate $\dot\theta = 4$ rad/s in the horizontal plane, determine the force it exerts on the can at the instant $\theta = \pi$ rad. Neglect friction and the size of the can.

(a)

SOLUTION

Free-Body Diagram.   The driving force $\mathbf{F}_C$ acts perpendicular to the arm OA, whereas the normal force of the wall of the slot on the can, $\mathbf{N}_C$, acts perpendicular to the tangent to the curve at $\theta = \pi$ rad, Fig. 13–21b. As usual, $\mathbf{a}_r$ and $\mathbf{a}_\theta$ are assumed to act in the *positive directions* of r and θ, respectively. Since the path is specified, the angle ψ which the extended radial line r makes with the tangent, Fig. 13–21c, can be determined from Eq. 13–10. We have $r = 0.1\theta$, so that $dr/d\theta = 0.1$, and therefore

$$\tan\psi = \frac{r}{dr/d\theta} = \frac{0.1\theta}{0.1} = \theta$$

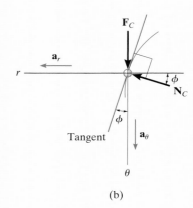

(b)

When $\theta = \pi$, $\psi = \tan^{-1}\pi = 72.3°$, so that $\phi = 90° - \psi = 17.7°$, as shown in Fig. 13–21c. Identify the four unknowns in Fig. 13–21b.

Equations of Motion.   Using $\phi = 17.7°$ and the data shown in Fig. 13–21b, we have

$$\xleftarrow{+} \Sigma F_r = ma_r; \qquad N_C \cos 17.7° = 0.5a_r \qquad (1)$$
$$+\downarrow \Sigma F_\theta = ma_\theta; \qquad F_C - N_C \sin 17.7° = 0.5a_\theta \qquad (2)$$

Kinematics.   The time derivatives of r and θ are

$$\dot\theta = 4 \text{ rad/s} \qquad r = 0.1\theta$$
$$\ddot\theta = 0 \qquad \dot r = 0.1\dot\theta = 0.1(4) = 0.4 \text{ m/s}$$
$$\ddot r = 0.1\ddot\theta = 0$$

At the instant $\theta = \pi$ rad,

$$a_r = \ddot r - r\dot\theta^2 = 0 - 0.1(\pi)(4)^2 = -5.03 \text{ m/s}^2$$
$$a_\theta = r\ddot\theta + 2\dot r\dot\theta = 0 + 2(0.4)(4) = 3.20 \text{ m/s}^2$$

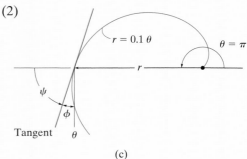

(c)

Fig. 13–21

Substituting these results into Eqs. 1 and 2 and solving yields

$$N_C = -2.64 \text{ N}$$
$$F_C = 0.800 \text{ N} \qquad\qquad \textit{Ans.}$$

What does the negative sign for N_C indicate?

PROBLEMS

13–83. A particle, having a mass of 1.5 kg, moves along a path defined by the equations $r = (4 + 3t)$ m, $\theta = (t^2 + 2)$ rad, and $z = (6 - t^3)$ m, where t is in seconds. Determine the r, θ, and z components of force which the path exerts on the particle when $t = 2$ s.

*13–84. The path of motion of a 5-lb particle in the horizontal plane is described in terms of polar coordinates as $r = (2t + 1)$ ft and $\theta = (0.5t^2 - t)$ rad, where t is in seconds. Determine the magnitude of the unbalanced force acting on the particle when $t = 2$ s.

13–85. The spring-held follower AB has a weight of 0.75 lb and moves back and forth as its end rolls on the contoured surface of the cam, where $r = 0.2$ ft and $z = (0.1 \sin \theta)$ ft. If the cam is rotating at a constant rate of 6 rad/s, determine the force at the end A of the follower when $\theta = 90°$. In this position the spring is compressed 0.4 ft. Neglect friction at the bearing C.

13–86. The spring-held follower AB has a weight of 0.75 lb and moves back and forth as its end rolls on the contoured surface of the cam, where $r = 0.2$ ft and $z = (0.1 \sin 2\theta)$ ft. If the cam is rotating at a constant rate of 6 rad/s, determine the maximum and minimum force the follower exerts on the cam if the spring is compressed 0.2 ft when $\theta = 45°$.

13–87. The 4-kg spool slides along the rotating rod. At the instant shown, the angular rate of rotation of the rod is $\dot{\theta} = 6$ rad/s, which is increasing at $\ddot{\theta} = 2$ rad/s². At this same instant, the spool is moving outward along the rod at 3 m/s, which is increasing at 1 m/s² when $r = 0.5$ m. Determine the radial frictional force and the normal force of the rod on the spool at this instant.

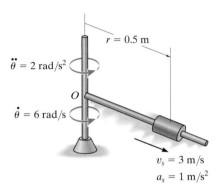

Prob. 13–87

*13–88. The boy of mass 40 kg is sliding down the spiral slide at a constant speed such that his position, measured from the top of the chute, has components $r = 1.5$ m, $\theta = (0.7t)$ rad, and $z = (-0.5t)$ m, where t is in seconds. Determine the components of force $\mathbf{F}_r$, $\mathbf{F}_\theta$, and $\mathbf{F}_z$ which the slide exerts on him at the instant $t = 2$ s. Neglect the size of the boy.

Prob. 13–88

Probs. 13–85/86

13–89. The girl has a mass of 50 kg. She is seated on the horse of the merry-go-round which undergoes constant rotational motion $\dot{\theta} = 1.5$ rad/s. If the path of the horse is defined by $r = 4$ m, $z = (0.5 \sin \theta)$ m, determine the maximum and minimum force F_z the horse exerts on her during the motion.

13–91. The particle has a mass of 0.5 kg and is confined to move along the smooth horizontal slot due to the rotation of the arm OA. Determine the force of the rod on the particle and the normal force of the slot on the particle when $\theta = 30°$. The rod is rotating with a constant angular velocity $\dot{\theta} = 2$ rad/s. Assume the particle contacts only one side of the slot at any instant.

***13–92.** Solve Problem 13–91 if the arm has an angular acceleration of $\ddot{\theta} = 3$ rad/s² and $\dot{\theta} = 2$ rad/s at this instant. Assume the particle contacts only one side of the slot at any instant.

Probs. 13–91/92

Prob. 13–89

13–90. The 0.5-lb particle is guided along the circular path using the slotted arm guide. If the arm has an angular velocity $\dot{\theta} = 4$ rad/s and an angular acceleration $\ddot{\theta} = 8$ rad/s² at the instant $\theta = 30°$, determine the force of the guide on the particle. Motion occurs in the *horizontal plane*.

13–93. A smooth can C, having a mass of 3 kg, is lifted from a feed at A to a ramp at B by a rotating rod. If the rod maintains a constant angular velocity of $\dot{\theta} = 0.5$ rad/s, determine the force which the rod exerts on the can at the instant $\theta = 30°$. Neglect the effects of friction in the calculation and the size of the can so that $r = (1.2 \cos \theta)$ m. The ramp from A to B is circular, having a radius of 600 mm.

Prob. 13–90

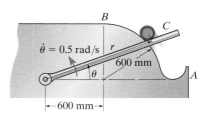

Prob. 13–93

13–94. The 2-lb collar slides along the smooth *horizontal* spiral rod, $r = (2\theta)$ ft, where θ is in radians. If its angular rate of rotation is constant and equals $\dot{\theta} = 4$ rad/s, determine the tangential force P needed to cause the motion and the normal force that the rod exerts on the spool at the instant $\theta = 90°$.

13–95. Solve Prob. 13–94 if the spiral rod is *vertical*.

Probs. 13–94/95

13–97. The smooth particle has a mass of 80 g. It is attached to an elastic cord extending from O to P and due to the slotted arm guide moves along the *horizontal* circular path $r = (0.8 \sin \theta)$ m. If the cord has a stiffness $k = 30$ N/m and an unstretched length of 0.25 m, determine the force of the guide on the particle when $\theta = 60°$. The guide has a constant angular velocity $\dot{\theta} = 5$ rad/s.

13–98. Solve Prob. 13–97 if $\ddot{\theta} = 2$ rad/s^2 when $\dot{\theta} = 5$ rad/s and $\theta = 60°$.

Probs. 13–97/98

13–96. The forked rod is used to move the smooth 2-lb particle around the horizontal path in the shape of a limaçon, $r = (2 + \cos \theta)$ ft. If $\theta = (0.5t^2)$ rad, where t is in seconds, determine the force which the rod exerts on the particle at the instant $t = 1$ s. The fork and path contact the particle on only one side.

Prob. 13–96

13–99. Determine the normal and frictional driving forces that the partial spiral track exerts on the 200-kg motorcycle at the instant $\theta = \frac{5}{3}\pi$ rad, $\dot{\theta} = 0.4$ rad/s, and $\ddot{\theta} = 0.8$ rad/s^2. Neglect the size of the motorcycle.

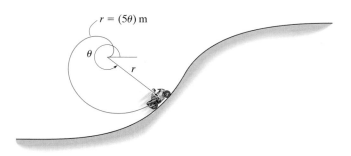

Prob. 13–99

*13–100. Using a forked rod, a smooth cylinder C having a mass of 0.5 kg is forced to move along the *vertical slotted* path $r = (0.5\theta)$ m, where θ is in radians. If the angular position of the arm is $\theta = (0.5t^2)$ rad, where t is in seconds, determine the force of the rod on the cylinder and the normal force of the slot on the cylinder at the instant $t = 2$ s. The cylinder is in contact with only *one edge* of the rod and slot at any instant.

Prob. 13–100

13–101. The ball has a mass of 2 kg and a negligible size. It is originally traveling around the horizontal circular path of radius $r_0 = 0.5$ m such that the angular rate of rotation is $\dot{\theta}_0 = 1$ rad/s. If the attached cord ABC is drawn down through the hole at a constant speed of 0.2 m/s, determine the tension the cord exerts on the ball at the instant $r = 0.25$ m. Also, compute the angular velocity of the ball at this instant. Neglect the effects of friction between the ball and horizontal plane. *Hint:* First show that the equation of motion in the θ direction yields $a_\theta = r\ddot{\theta} + 2\dot{r}\dot{\theta} = (1/r)(d(r^2\dot{\theta})/dt) = 0$. When integrated, $r^2\dot{\theta} = c$, where the constant c is determined from the problem data.

Prob. 13–101

13–102. The smooth surface of the vertical cam is defined in part by the curve $r = (0.2 \cos\theta + 0.3)$ m. If the forked rod is rotating with a constant angular velocity of $\dot{\theta} = 4$ rad/s, determine the force the cam and the rod exert on the 2-kg roller when $\theta = 30°$. The attached spring has a stiffness $k = 30$ N/m and an unstretched length of 0.1 m.

Prob. 13–102

13–103. The collar has a mass of 2 kg and travels along the smooth horizontal rod defined by the equiangular spiral $r = (e^\theta)$ m, where θ is in radians. Determine the tangential force F and the normal force N acting on the collar when $\theta = 90°$, if the force F maintains a constant angular motion $\dot{\theta} = 2$ rad/s.

Prob. 13–103

*13–104. The smooth surface of the vertical cam is defined in part by the curve $r = (0.2 \cos \theta + 0.3)$ m. The forked rod is rotating with an angular acceleration of $\ddot{\theta} = 2$ rad/s², and when $\theta = 45°$ the angular velocity is $\dot{\theta} = 6$ rad/s. Determine the force the cam and the rod exert on the 2-kg roller at this instant. The attached spring has a stiffness $k = 100$ N/m and an unstretched length of 0.1 m.

Prob. 13–104

13–106. Using air pressure, the 0.5-kg ball is forced to move through the tube lying in the *horizontal plane* and having the shape of a logarithmic spiral. If the tangential force exerted on the ball due to the air is 6 N, determine the rate of increase in the ball's speed at the instant $\theta = \pi/2$. What direction does it act in?

Prob. 13–106

13–105. The pilot of an airplane executes a vertical loop which in part follows the path of a "four-leaved rose," $r = (-600 \cos 2\theta)$ ft, where θ is in radians. If his speed at A is a constant $v_P = 80$ ft/s, determine the vertical reaction the seat of the plane exerts on the pilot when the plane is at A. He weighs 130 lb. *Hint:* To determine the time derivatives necessary to compute the acceleration components a_r and a_θ, take the first and second time derivatives of $r = 400(1 + \cos \theta)$. Then, for further information, use Eq. 12–26 to determine $\dot{\theta}$. Also, take the time derivative of Eq. 12–26, noting that $\dot{v}_C = 0$, to determine $\ddot{\theta}$.

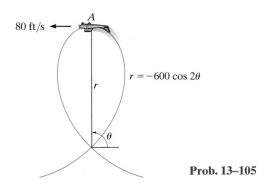

Prob. 13–105

13–107. Solve Prob. 13–106 if the tube lies in a *vertical plane*.

Prob. 13–107

*13–108. The arm is rotating at a rate of $\dot{\theta} = 5$ rad/s when $\ddot{\theta} = 2$ rad/s^2 and $\theta = 90°$. Determine the normal force it must exert on the 0.5-kg particle if the particle is confined to move along the slotted path defined by the *horizontal* hyperbolic spiral $r\theta = 0.2$ m.

Prob. 13–108

13–109. The collar, which has a weight of 3 lb, slides along the smooth rod lying in the *horizontal plane* and having the shape of a parabola $r = 4/(1 - \cos \theta)$, where θ is in radians and r is in feet. If the collar's angular rate is constant and equals $\dot{\theta} = 4$ rad/s, determine the tangential retarding force P needed to cause the motion and the normal force that the collar exerts on the rod at the instant $\theta = 90°$.

Prob. 13–109

13–110. The tube rotates in the horizontal plane at a constant rate of $\dot{\theta} = 4$ rad/s. If a 0.2-kg ball B starts at the origin O with an initial radial velocity of $\dot{r} = 1.5$ m/s and moves outward through the tube, determine the radial and transverse components of the ball's velocity at the instant it leaves the outer end at C, $r = 0.5$ m. *Hint:* Show that the equation of motion in the r direction is $\ddot{r} - 16r = 0$. The solution is of the form $r = Ae^{-4t} + Be^{4t}$. Evaluate the integration constants A and B, and determine the time t when $r = 0.5$ m. Proceed to obtain v_r and v_θ.

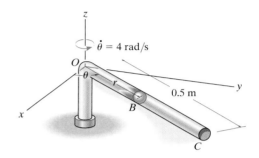

Prob. 13–110

13–111. A 0.2-kg spool slides down along a smooth rod. If the rod has a constant angular rate of rotation $\dot{\theta} = 2$ rad/s in the vertical plane, show that the equations of motion for the spool are $\ddot{r} - 4r - 9.81 \sin \theta = 0$ and $0.8\dot{r} + N_s - 1.962 \cos \theta = 0$, where N_s is the magnitude of the normal force of the rod on the spool. Using the methods of differential equations, it can be shown that the solution of the first of these equations is $r = C_1 e^{-2t} + C_2 e^{2t} - (9.81/8) \sin 2t$. If r, $\dot{r}$, and θ are zero when $t = 0$, evaluate the constants C_1 and C_2 to determine r at the instant $\theta = \pi/4$ rad.

Prob. 13–111

*13.7 Central-Force Motion and Space Mechanics

(a)

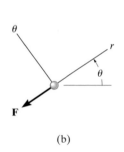

(b)

Fig. 13–22

If a particle is moving only under the influence of a force having a line of action which is always directed toward a fixed point, the motion is called *central-force motion*. This type of motion is commonly caused by electrostatic and gravitational forces.

In order to determine the motion, we will consider the particle P shown in Fig. 13–22a, which has a mass m and is acted upon only by the central force **F**. The free-body diagram for the particle is shown in Fig. 13–22b. Using polar coordinates (r, θ), the equations of motion, Eqs. 13–9, become

$$-F = m\left[\frac{d^2r}{dt^2} - r\left(\frac{d\theta}{dt}\right)^2\right]$$

$$0 = m\left(r\frac{d^2\theta}{dt^2} + 2\frac{dr}{dt}\frac{d\theta}{dt}\right) \qquad (13\text{–}11)$$

The second of these equations may be written in the form

$$\frac{1}{r}\left[\frac{d}{dt}\left(r^2\frac{d\theta}{dt}\right)\right] = 0$$

so that integrating yields

$$r^2\frac{d\theta}{dt} = h \qquad (13\text{–}12)$$

Here h is a constant of integration. From Fig. 13–22a notice that the shaded area described by the radius r, as r moves through an angle $d\theta$, is $dA = \frac{1}{2}r^2\,d\theta$. If the *areal velocity* is defined as

$$\frac{dA}{dt} = \frac{1}{2}r^2\frac{d\theta}{dt} = \frac{h}{2} \qquad (13\text{–}13)$$

then, it is seen that the areal velocity for a particle subjected to central-force motion is *constant*. In other words, the particle will sweep out equal segments of area per unit of time as it travels along the path. To obtain the *path of motion*, $r = f(\theta)$, the independent variable t must be eliminated from Eqs. 13–11. Using the chain rule of calculus and Eq. 13–12, the time derivatives of Eqs. 13–11 may be replaced by

$$\frac{dr}{dt} = \frac{dr}{d\theta}\frac{d\theta}{dt} = \frac{h}{r^2}\frac{dr}{d\theta}$$

$$\frac{d^2r}{dt^2} = \frac{d}{dt}\left(\frac{h}{r^2}\frac{dr}{d\theta}\right) = \frac{d}{d\theta}\left(\frac{h}{r^2}\frac{dr}{d\theta}\right)\frac{d\theta}{dt} = \left[\frac{d}{d\theta}\left(\frac{h}{r^2}\frac{dr}{d\theta}\right)\right]\frac{h}{r^2}$$

Substituting a new dependent variable (xi) $\xi = 1/r$ into the second equation, we have

$$\frac{d^2r}{dt^2} = -h^2\xi^2\frac{d^2\xi}{d\theta^2}$$

Also, the square of Eq. 13–12 becomes

$$\left(\frac{d\theta}{dt}\right)^2 = h^2\xi^4$$

Substituting these last two equations into the first of Eqs. 13–11 yields

$$-h^2\xi^2\frac{d^2\xi}{d\theta^2} - h^2\xi^3 = -\frac{F}{m}$$

or

$$\frac{d^2\xi}{d\theta^2} + \xi = \frac{F}{mh^2\xi^2} \qquad (13\text{–}14)$$

This satellite is subjected to a central force and its orbital motion can be closely predicted using the equations developed in this section.

This differential equation defines the path over which the particle travels when it is subjected to the central force **F**. *

For application, the force of gravitational attraction will be considered. Some common examples of central-force systems which depend on gravitation include the motion of the moon and artificial satellites about the earth, and the motion of the planets about the sun. As a typical problem in space mechanics, consider the trajectory of a space satellite or space vehicle launched into free-flight orbit with an initial velocity $\mathbf{v}_0$, Fig. 13–23. It will be assumed that this velocity is initially *parallel* to the tangent at the surface of the earth, as shown in the figure.† Just after the satellite is released into free flight, the only force acting on it is the gravitational force of the earth. (Gravitational attractions involving other bodies such as the moon or sun will be neglected, since for orbits close to the earth their effect is small in comparison with the earth's gravitation.) According to Newton's law of gravitation, force **F** will always act between the mass centers of the earth and the satellite, Fig. 13–23. From Eq. 13–1, this force of attraction has a magnitude of

$$F = G\frac{M_em}{r^2}$$

where M_e and m represent the mass of the earth and the satellite, respectively, G is the gravitational constant, and r is the distance between

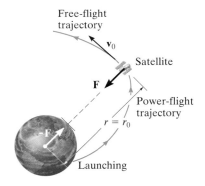

Free-flight trajectory

$\mathbf{v}_0$

Satellite

F

Power-flight trajectory

$r = r_0$

Launching

Fig. 13–23

*In the derivation, **F** is considered positive when it is directed toward point O. If **F** is oppositely directed, the right side of Eq. 13–14 should be negative.

†The case where $\mathbf{v}_0$ acts at some initial angle θ to the tangent is best described using the conservation of angular momentum (see Prob. 15–103).

the mass centers. To obtain the orbital path, we set $\xi = 1/r$ in the foregoing equation and substitute the result into Eq. 13–14. We obtain

$$\frac{d^2\xi}{d\theta^2} + \xi = \frac{GM_e}{h^2} \qquad (13\text{–}15)$$

This second-order ordinary differential equation has constant coefficients and is nonhomogeneous. The solution is represented as the sum of the complementary and particular solutions. The complementary solution is obtained when the term on the right is equal to zero. It is

$$\xi_c = C \cos(\theta - \phi)$$

where C and ϕ are constants of integration. The particular solution is

$$\xi_p = \frac{GM_e}{h^2}$$

Thus, the complete solution to Eq. 13–15 is

$$\xi = \xi_c + \xi_p$$
$$= \frac{1}{r} = C \cos(\theta - \phi) + \frac{GM_e}{h^2} \qquad (13\text{–}16)$$

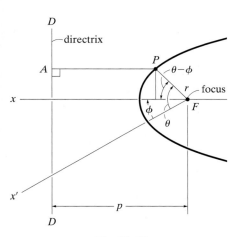

Fig. 13–24

The validity of this result may be checked by substitution into Eq. 13–15.

Equation 13–16 represents the *free-flight trajectory* of the satellite. It is the equation of a conic section expressed in terms of polar coordinates. As shown in Fig. 13–24, a *conic section* is defined as the locus of point P, which moves in a plane in such a way that the ratio of its distance from a fixed point F to its distance from a fixed line is constant. The fixed point is called the *focus*, and the fixed line DD is called the *directrix*. The constant ratio is called the *eccentricity* of the conic section and is denoted by e. Thus,

$$e = \frac{FP}{PA}$$

which may be written in the form

$$FP = r = e(PA) = e[p - r\cos(\theta - \phi)]$$

or

$$\frac{1}{r} = \frac{1}{p}\cos(\theta - \phi) + \frac{1}{ep}$$

Comparing this equation with Eq. 13–16, it is seen that the eccentricity of the conic section for the trajectory is

$$\boxed{e = \frac{Ch^2}{GM_e}} \qquad (13\text{–}17)$$

and the fixed distance from the focus to the directrix is

$$p = \frac{1}{C} \tag{13-18}$$

Provided the polar angle θ is measured from the x axis (an axis of symmetry since it is perpendicular to the directrix), the angle ϕ is zero, Fig. 13–24, and therefore Eq. 13–16 reduces to

$$\frac{1}{r} = C \cos \theta + \frac{GM_e}{h^2} \tag{13-19}$$

The constants h and C are determined from the data obtained for the position and velocity of the satellite at the end of the *power-flight trajectory*. For example, if the initial height or distance to the space vehicle is r_0 (measured from the center of the earth) and its initial speed is v_0 at the beginning of its free flight, Fig. 13–25, then the constant h may be obtained from Eq. 13–12. When $\theta = \phi = 0°$, the velocity $\mathbf{v}_0$ has no radial component; therefore, from Eq. 12–25, $v_0 = r_0(d\theta/dt)$, so that

$$h = r_0^2 \frac{d\theta}{dt}$$

or

$$\boxed{h = r_0 v_0} \tag{13-20}$$

To determine C, use Eq. 13–19 with $\theta = 0°, r = r_0$, and substitute Eq. 13–20 for h:

$$\boxed{C = \frac{1}{r_0}\left(1 - \frac{GM_e}{r_0 v_0^2}\right)} \tag{13-21}$$

The equation for the free-flight trajectory therefore becomes

$$\boxed{\frac{1}{r} = \frac{1}{r_0}\left(1 - \frac{GM_e}{r_0 v_0^2}\right)\cos \theta + \frac{GM_e}{r_0^2 v_0^2}} \tag{13-22}$$

The type of path taken by the satellite is determined from the value of the eccentricity of the conic section as given by Eq. 13–17. If

$$\boxed{\begin{aligned}
&e = 0 \quad \text{free-flight trajectory is a circle} \\
&e = 1 \quad \text{free-flight trajectory is a parabola} \\
&e < 1 \quad \text{free-flight trajectory is an ellipse} \\
&e > 1 \quad \text{free-flight trajectory is a hyperbola}
\end{aligned}} \tag{13-23}$$

Each of these trajectories is shown in Fig. 13–25. From the curves it is seen that when the satellite follows a parabolic path, it is "on the border" of never returning to its initial starting point. The initial launch velocity, $\mathbf{v}_0$, required for the satellite to follow a parabolic path is called the *escape velocity*. The speed, v_e, can be determined by using the second of Eqs. 13–23 with Eqs. 13–17, 13–20, and 13–21. It is left as an exercise to show that

$$v_e = \sqrt{\frac{2GM_e}{r_0}} \tag{13–24}$$

The speed v_c required to launch a satellite into a *circular orbit* can be found using the first of Eqs. 13–23. Since e is related to h and C, Eq. 13–17, C must be zero to satisfy this equation (from Eq. 13–20, h cannot be zero); and therefore, using Eq. 13–21, we have

$$v_c = \sqrt{\frac{GM_e}{r_0}} \tag{13–25}$$

Provided r_0 represents a minimum height for launching, in which frictional resistance from the atmosphere is neglected, speeds at launch which are less than v_c will cause the satellite to reenter the earth's atmosphere and either burn up or crash, Fig. 13–25.

Fig. 13–25

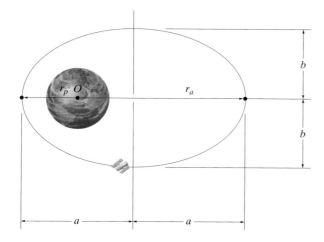

Fig. 13–26

All the trajectories attained by planets and most satellites are elliptical, Fig. 13–26. For a satellite's orbit about the earth, the *minimum distance* from the orbit to the center of the earth O (which is located at one of the foci of the ellipse) is r_p and can be found using Eq. 13–22 with $\theta = 0°$. Therefore;

$$r_p = r_0 \tag{13–26}$$

This minimum distance is called the *perigee* of the orbit. The *apogee* or maximum distance r_a can be found using Eq. 13–22 with $\theta = 180°$.* Thus,

$$r_a = \frac{r_0}{(2GM_e/r_0v_0^2) - 1} \tag{13–27}$$

With reference to Fig. 13–26, the semimajor axis a of the ellipse is

$$a = \frac{r_p + r_a}{2} \tag{13–28}$$

Using analytical geometry, it can be shown that the minor axis b is determined from the equation

$$b = \sqrt{r_p r_a} \tag{13–29}$$

---

*Actually, the terminology perigee and apogee pertains only to orbits about the *earth.* If any other heavenly body is located at the focus of an elliptical orbit, the minimum and maximum distances are referred to respectively as the *periapsis* and *apoapsis* of the orbit.

Furthermore, by direct integration, the area of an ellipse is

$$A = \pi ab = \frac{\pi}{2}(r_p + r_a)\sqrt{r_p r_a} \qquad (13\text{--}30)$$

The areal velocity has been defined by Eq. 13–13, $dA/dt = h/2$. Integrating yields $A = hT/2$, where T is the *period* of time required to make one orbital revolution. From Eq. 13–30, the period is

$$T = \frac{\pi}{h}(r_p + r_a)\sqrt{r_p r_a} \qquad (13\text{--}31)$$

In addition to predicting the orbital trajectory of earth satellites, the theory developed in this section is valid, to a surprisingly close approximation, at predicting the actual motion of the planets traveling around the sun. In this case the mass of the sun, M_s, should be substituted for M_e when the appropriate formulas are used.

The fact that the planets do indeed follow elliptic orbits about the sun was discovered by the German astronomer Johannes Kepler in the early seventeenth century. His discovery was made *before* Newton had developed the laws of motion and the law of gravitation, and so at the time it provided important proof as to the validity of these laws. Kepler's laws, developed after 20 years of planetary observation, are summarized as follows:

1. Every planet moves in its orbit such that the line joining it to the sun sweeps over equal areas in equal intervals of time, whatever the line's length.

2. The orbit of every planet is an ellipse with the sun placed at one of its foci.

3. The square of the period of any planet is directly proportional to the cube of the minor axis of its orbit.

A mathematical statement of the first and second laws is given by Eqs. 13–13 and 13–22, respectively. The third law can be shown from Eq. 13–31 using Eqs. 13–19, 13–28, and 13–29.

EXAMPLE 13.13

A satellite is launched 600 km from the surface of the earth, with an initial velocity of 30 Mm/h acting parallel to the tangent at the surface of the earth, Fig. 13–27. Assuming that the radius of the earth is 6378 km and that its mass is $5.976(10^{24})$ kg, determine (a) the eccentricity of the orbital path, and (b) the velocity of the satellite at apogee.

SOLUTION

Part (a). The eccentricity of the orbit is obtained using Eq. 13–17. The constants h and C are first determined from Eqs. 13–20 and 13–21. Since

$$r_p = r_0 = 6378 \text{ km} + 600 \text{ km} = 6.978(10^6) \text{ m}$$

$$v_0 = 30 \text{ Mm/h} = 8333.3 \text{ m/s}$$

then

$$h = r_p v_0 = 6.978(10^6)(8333.3) = 58.15(10^9) \text{ m}^2/\text{s}$$

$$C = \frac{1}{r_p}\left(1 - \frac{GM_e}{r_p v_0^2}\right)$$

$$= \frac{1}{6.978(10^6)}\left\{1 - \frac{66.73(10^{-12})[5.976(10^{24})]}{6.978(10^6)(8333.3)^2}\right\} = 25.4(10^{-9}) \text{ m}^{-1}$$

Hence,

$$e = \frac{Ch^2}{GM_e} = \frac{2.54(10^{-8})[58.15(10^9)]^2}{66.73(10^{-12})[5.976(10^{24})]} = 0.215 < 1 \qquad Ans.$$

From Eq. 13–23, observe that the orbit is an *ellipse*.

Part (b). If the satellite were launched at the apogee A shown in Fig. 13–27, with a velocity $\mathbf{v}_A$, the same orbit would be maintained provided that

$$h = r_p v_0 = r_a v_A = 58.15(10^9) \text{ m}^2/\text{s}$$

Using Eq. 13–27, we have

$$r_a = \frac{r_p}{\dfrac{2GM_e}{r_p v_0^2} - 1} = \frac{6.978(10^6)}{\dfrac{2[66.73(10^{-12})][5.976(10^{24})]}{6.978(10^6)(8333.3)^2} - 1} = 10.804(10^6)$$

Thus,

$$v_A = \frac{58.15(10^9)}{10.804(10^6)} = 5382.2 \text{ m/s} = 19.4 \text{ Mm/h} \qquad Ans.$$

NOTE: The farther the satellite is from the earth, the slower it moves, which is to be expected since h is constant.

Fig. 13–27

PROBLEMS

In the following problems, except where otherwise indicated, assume that the radius of the earth is 6378 km, the earth's mass is $5.976(10^{24})$ kg, the mass of the sun is $1.99(10^{30})$ kg, and the gravitational constant is $G = 66.73(10^{-12})$ $m^3/(kg \cdot s^2)$.

*13–112. The rocket is in circular orbit about the earth at an altitude of $h = 4$ Mm. Determine the minimum increment in speed it must have in order to escape the earth's gravitational field.

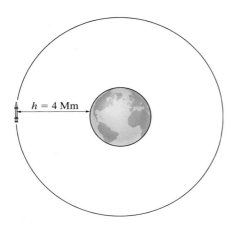

Prob. 13–112

13–113. Prove Kepler's third law of motion. *Hint:* Use Eqs. 13–19, 13–28, 13–29, and 13–31.

13–114. A satellite is to be placed into an elliptical orbit about the earth such that at the perigee of its orbit it has an *altitude* of 800 km, and at apogee its *altitude* is 2400 km. Determine its required launch velocity tangent to the earth's surface at perigee and the period of its orbit.

Prob. 13–114

13–115. The rocket is traveling in free flight along an elliptical trajectory $A'A$. The planet has a mass 0.60 times that of the earth's. If the rocket has an apoapsis and periapsis as shown in the figure, determine the speed of the rocket when it is at point A.

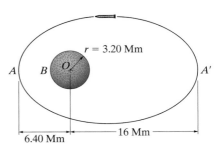

Prob. 13–115

*13–116. An elliptical path of a satellite has an eccentricity $e = 0.130$. If it has a speed of 15 Mm/h when it is at perigee, P, determine its speed when it arrives at apogee, A. Also, how far is it from the earth's surface when it is at A?

Prob. 13–116

13–117. A communications satellite is to be placed into an equatorial circular orbit around the earth so that it always remains directly over a point on the earth's surface. If this requires the period to be 24 hours (approximately), determine the radius of the orbit and the satellite's velocity.

13–118. The rocket is traveling in free flight along an elliptical trajectory $A'A$. The planet has no atmosphere, and its mass is 0.60 times that of the earth's. If the rocket has the apogee and perigee shown, determine the rocket's velocity when it is at point A. Take $G = 34.4(10^{-9})(\text{lb} \cdot \text{ft}^2)/\text{slug}^2$, $M_e = 409(10^{21})$ slug, 1 mi = 5280 ft.

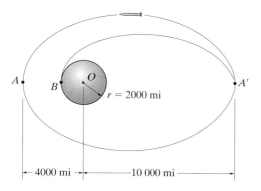

Prob. 13–118

13–119. If the rocket is to land on the surface of the planet, determine the required free-flight speed it must have at A' so that the landing occurs at B. How long does it take for the rocket to land, in going from A' to B? The planet has no atmosphere, and its mass is 0.6 times that of the earth's. Take $G = 34.4(10^{-9})(\text{lb} \cdot \text{ft}^2)/\text{slug}^2$, $M_e = 409(10^{21})$ slug, 1 mi = 5280 ft.

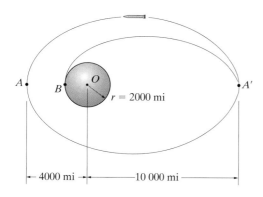

Prob. 13–119

*13–120.** The speed of a satellite launched into a circular orbit about the earth is given by Eq. 13–25. Determine the speed of a satellite launched parallel to the surface of the earth so that it travels in a circular orbit 800 km from the earth's surface.

13–121. The rocket is traveling in free flight along an elliptical trajectory $A'A$. The planet has no atmosphere, and its mass is 0.70 times that of the earth's. If the rocket has an apoapsis and periapsis as shown in the figure, determine the speed of the rocket when it is at point A.

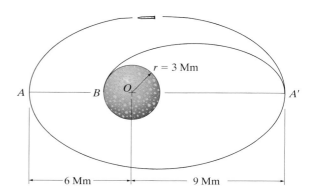

Prob. 13–121

13–122. If the rocket in Prob. 13–121 is to land on the surface of the planet, determine the required free-flight speed it must have at A' so that it strikes the planet at B. How long does it take for the rocket to land, going from A' to B along an elliptical path?

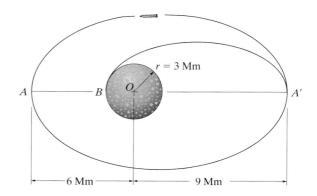

Prob. 13–122

13–123. A satellite S travels in a circular orbit around the earth. A rocket is located at the apogee of its elliptical orbit for which $e = 0.58$. Determine the sudden change in speed that must occur at A so that the rocket can enter the satellite's orbit while in free flight along the blue elliptical trajectory. When it arrives at B, determine the sudden adjustment in speed that must be given to the rocket in order to maintain the circular orbit.

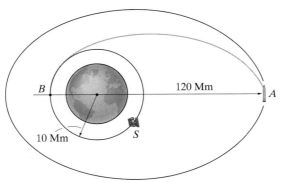

Prob. 13–123

*13–124.** An asteroid is in an elliptical orbit about the sun such that its perihelion distance is $9.30(10^9)$ km. If the eccentricity of the orbit is $e = 0.073$, determine the aphelion distance of the orbit.

13–125. A satellite is in an elliptical orbit around the earth such that $e = 0.156$. If its perigee is 5 Mm, determine its velocity at this point and also the distance OB when it is at point B, located $135°$ away as shown.

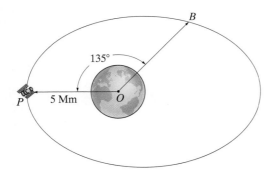

Prob. 13–125

13–126. The rocket is traveling in a free-flight elliptical orbit about the earth such that $e = 0.76$ and its perigee is 9 Mm as shown. Determine its speed when it is at point B. Also determine the sudden decrease in speed the rocket must experience at A in order to travel in a circular orbit about the earth.

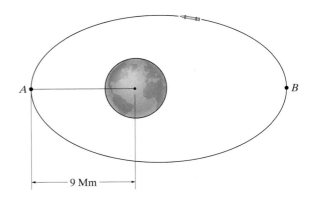

Prob. 13–126

13–127. A rocket is in free-flight elliptical orbit around the planet Venus. Knowing that the periapsis and apoapsis of the orbit are 8 Mm and 26 Mm, respectively, determine (a) the speed of the rocket at point A', (b) the required speed it must attain at A just after braking so that it undergoes an 8-Mm free-flight circular orbit around Venus, and (c) the periods of both the circular and elliptical orbits. The mass of Venus is 0.816 times the mass of the earth.

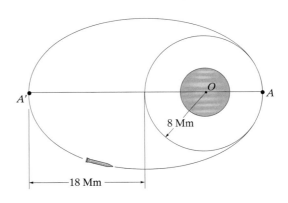

Prob. 13–127

Design Project

13–1D.   Design of a Ramp Catapult

The block B has a mass of 20 kg and is to be catapulted from the table. Design a catapulting mechanism that can be attached to the table and to the container of the block, using cables and pulleys. Neglect the mass of the container, assume the operator can exert a constant tension of 120 N on a single cable during operation, and that the maximum movement of his arm is 0.5 m. The coefficient of kinetic friction between the table and container is $\mu_k = 0.2$. Submit a drawing of your design, and calculate the maximum range R to where the block will strike the ground. Compare your value with that of others in the class.

Fig. 13–1D

Chapter Review

Kinetics

Kinetics is the study of the relationship between forces and the acceleration they cause. This relationship is based on Newton's second law of motion, expressed mathematically as $\Sigma \mathbf{F} = m\mathbf{a}$. (*See page 102.*)

The mass m is the proportionality constant between the resultant force $\Sigma \mathbf{F}$ acting on the particle and the acceleration $\mathbf{a}$ caused by this resultant. The mass represents the quantity of matter contained within the particle. It measures the resistance to a change in its motion.

Before applying the equation of motion, it is important to first draw the particle's *free-body diagram* in order to account for all of the forces that act on the particle. Graphically, this diagram is equal to the *kinetic diagram*, which shows the result of the forces, that is, the $m\mathbf{a}$ vector. (*See page 105.*)

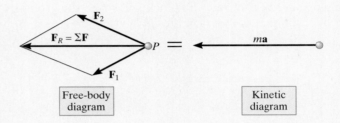

Free-body diagram

Kinetic diagram

Inertial Coordinate Systems

When applying the equation of motion, it is important to make measurements of the acceleration from an inertial coordinate system. This system has axes that do not rotate but are either fixed or translate with a constant velocity. Various types of inertial coordinate systems can be used to apply $\Sigma \mathbf{F} = m\mathbf{a}$ in component form. (*See page 106.*)

Rectangular x, y, z axes are used to describe rectilinear motion along each of the axes.

Normal and tangential n, t axes are often used when the path is known. Recall that $\mathbf{a}_n$ is always directed in the $+n$ direction. It indicates the change in the velocity direction. Also recall that $\mathbf{a}_t$ is tangent to the path. It indicates the change in the velocity magnitude.

Inertial frame of reference

$$\Sigma F_x = ma_x, \ \Sigma F_y = ma_y, \ \Sigma F_z = ma_z$$

$$\Sigma F_t = ma_t, \ \Sigma F_n = ma_n, \ \Sigma F_b = 0$$

$$a_t = dv/dt \text{ or } a_t = v \, dv/ds, \text{ and } a_n = v^2/\rho$$

$$e = \frac{[1 + (dy/dx)^2]^{3/2}}{|d^2y/dx^2|}$$

Cylindrical coordinates are useful when angular motion of the radial coordinate r is specified or when the path can conveniently be described with these coordinates. For some problems, the direction of the forces on the free-body diagram will require finding the angle ψ between the extended radial coordinate and the tangent to the curve.
(*See pages 106, 123, and 135.*)

$$\Sigma F_r = m(\ddot{r} - r\dot{\theta}^2), \quad \Sigma F_\theta = m(r\ddot{\theta} + 2\dot{r}\dot{\theta}), \quad \Sigma F_z = m\ddot{z}$$

$$\tan \psi = \frac{r}{dr/d\theta}$$

Central-Force Motion

When a single force acts upon a particle, such as during the free-flight trajectory of a satellite in a gravitational field, then the motion is referred to as central-force motion. The orbit depends upon the eccentricity e; and as a result, the trajectory can either be circular, parabolic, elliptical, or hyperbolic.
(*See page 147.*)

In order to properly design the loop of this roller coaster it is necessary to ensure that the cars have enough energy to be able to make the loop without leaving the tracks.

14 Kinetics of a Particle: Work and Energy

CHAPTER OBJECTIVES

- To develop the principle of work and energy and apply it to solve problems that involve force, velocity, and displacement.

- To study problems that involve power and efficiency.

- To introduce the concept of a conservative force and apply the theorem of conservation of energy to solve kinetic problems.

14.1 The Work of a Force

In this chapter, we will analyze motion using the concepts of work and energy. We will start by defining the work of a force. Specifically, a force $\mathbf{F}$ will do *work* on a particle only when the particle undergoes a *displacement in the direction of the force*. For example, consider the force $\mathbf{F}$ acting on the particle in Fig. 14–1. If the particle moves along the path s from position $\mathbf{r}$ to a new position $\mathbf{r}'$, the displacement is then $d\mathbf{r} = \mathbf{r}' - \mathbf{r}$. The magnitude of $d\mathbf{r}$ is represented by ds, the differential segment along the path. If the angle between the tails of $d\mathbf{r}$ and $\mathbf{F}$ is θ, Fig. 14–1, then the work dU which is done by $\mathbf{F}$ is a *scalar quantity*, defined by

$$dU = F \, ds \cos \theta$$

Fig. 14–1

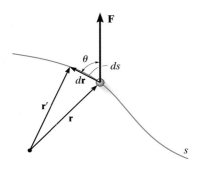

Fig. 14–1

By definition of the dot product (see Eq. C–14) this equation may also be written as

$$dU = \mathbf{F} \cdot d\mathbf{r}$$

This result may be interpreted in one of two ways: either as the product of F and the component of displacement in the direction of the force, i.e., $ds \cos \theta$, or as the product of ds and the component of force in the direction of displacement, i.e., $F \cos \theta$. Note that if $0° \le \theta < 90°$, then the force component and the displacement have the *same sense* so that the work is *positive;* whereas if $90° < \theta \le 180°$, these vectors have an *opposite sense*, and therefore the work is *negative*. Also, $dU = 0$ if the force is *perpendicular* to displacement, since $\cos 90° = 0$, or if the force is applied at a *fixed point*, in which case the displacement is zero.

The basic unit for work in the SI system is called a joule (J). This unit combines the units of force and displacement. Specifically, 1 *joule* of work is done when a force of 1 newton moves 1 meter along its line of action (1 J = 1 N · m). The moment of a force has this same combination of units (N · m); however, the concepts of moment and work are in no way related. A moment is a vector quantity, whereas work is a scalar. In the FPS system work is generally defined by writing the units as ft · lb, which is distinguished from the units for a moment, written as lb · ft.

Work of a Variable Force.   If the particle undergoes a finite displacement along its path from $\mathbf{r}_1$ to $\mathbf{r}_2$ or s_1 to s_2, Fig. 14–2a, the work is determined by integration. If $\mathbf{F}$ is expressed as a function of position, $F = F(s)$, we have

$$U_{1-2} = \int_{\mathbf{r}_1}^{\mathbf{r}_2} \mathbf{F} \cdot d\mathbf{r} = \int_{s_1}^{s_2} F \cos \theta \, ds \qquad (14–1)$$

If the working component of the force, $F \cos \theta$, is plotted versus s, Fig. 14–2b, the integral in this equation can be interpreted as the *area under the curve* from position s_1 to position s_2.

Fig. 14–2

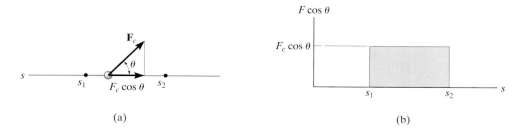

Fig. 14–3

Work of a Constant Force Moving Along a Straight Line.
If the force $\mathbf{F}_c$ has a constant magnitude and acts at a constant angle θ from its straight-line path, Fig. 14–3a, then the component of $\mathbf{F}_c$ in the direction of displacement is $F_c \cos \theta$. The work done by $\mathbf{F}_c$ when the particle is displaced from s_1 to s_2 is determined from Eq. 14–1, in which case

$$U_{1-2} = F_c \cos \theta \int_{s_1}^{s_2} ds$$

or

$$\boxed{U_{1-2} = F_c \cos \theta (s_2 - s_1)} \tag{14–2}$$

Here the work of $\mathbf{F}_c$ represents the *area of the rectangle* in Fig. 14–3b.

Work of a Weight.
Consider a particle of weight $\mathbf{W}$, which moves up along the path s shown in Fig. 14–4 from position s_1 to position s_2. At an intermediate point, the displacement $d\mathbf{r} = dx\mathbf{i} + dy\mathbf{j} + dz\mathbf{k}$. Since $\mathbf{W} = -W\mathbf{j}$, applying Eq. 14–1 yields

$$U_{1-2} = \int \mathbf{F} \cdot d\mathbf{r} = \int_{\mathbf{r}_1}^{\mathbf{r}_2} (-W\mathbf{j}) \cdot (dx\mathbf{i} + dy\mathbf{j} + dz\mathbf{k})$$

$$= \int_{y_1}^{y_2} -W \, dy = -W(y_2 - y_1)$$

or

$$\boxed{U_{1-2} = -W \, \Delta y} \tag{14–3}$$

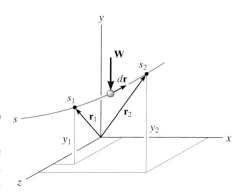

Thus, the work is independent of the path and is equal to the magnitude of the particle's weight times its vertical displacement. In the case shown in Fig. 14–4 the work is *negative*, since W is downward and Δy is upward. Note, however, that if the particle is displaced *downward* $(-\Delta y)$, the work of the weight is *positive*. Why?

Fig. 14–4

Unstretched
position, $s = 0$

(a)

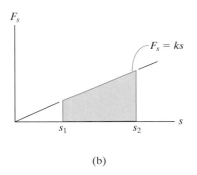

(b)

Unstretched
position, $s = 0$

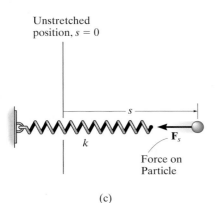

(c)

Fig. 14–5

Work of a Spring Force. The magnitude of force developed in a linear elastic spring when the spring is displaced a distance s from its unstretched position is $F_s = ks$, where k is the spring stiffness. If the spring is elongated or compressed from a position s_1 to a further position s_2, Fig. 14–5a, the work done *on the spring* by $\mathbf{F}_s$ is *positive*, since in each case the force and displacement are in the *same direction*. We require

$$U_{1-2} = \int_{s_1}^{s_2} F_s \, ds = \int_{s_1}^{s_2} ks \, ds$$

$$= \tfrac{1}{2}ks_2^2 - \tfrac{1}{2}ks_1^2$$

This equation represents the trapezoidal area under the line $F_s = ks$, Fig. 14–5b.

If a particle (or body) is attached to a spring, then the force $\mathbf{F}_s$ exerted on the particle is *opposite* to that exerted on the spring, Fig. 14–5c. Consequently, the force will do *negative work* on the particle when the particle is moving so as to further elongate (or compress) the spring. Hence, the above equation becomes

$$U_{1-2} = -\left(\tfrac{1}{2}ks_2^2 - \tfrac{1}{2}ks_1^2\right) \qquad (14\text{–}4)$$

When this equation is used, a mistake in sign can be eliminated if one simply notes the direction of the spring force acting on the particle and compares it with the direction of displacement of the particle—if both are in the *same direction, positive work* results; if they are *opposite* to one another, the *work is negative*.

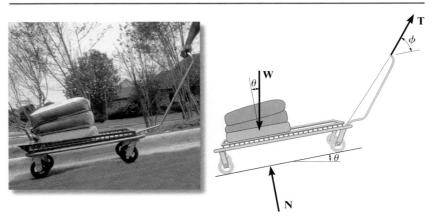

The forces acting on the cart as it is pulled a distance s up the incline, are shown on its free-body diagram. The constant towing force $\mathbf{T}$ does positive work of $U_T = (T \cos \phi)s$, the weight does negative work of $U_W = -(W \sin \theta)s$, and the normal force $\mathbf{N}$ does no work since there is no displacement of this force along its line of action.

EXAMPLE 14.1

The 10-kg block shown in Fig. 14–6a rests on the smooth incline. If the spring is originally stretched 0.5 m, determine the total work done by all the forces acting on the block when a horizontal force $P = 400$ N pushes the block up the plane $s = 2$ m.

(a)

SOLUTION

First the free-body diagram of the block is drawn in order to account for all the forces that act on the block, Fig. 14–6b.

Horizontal Force P.   Since this force is *constant*, the work is determined using Eq. 14–2. The result can be calculated as the force times the component of displacement in the direction of the force; i.e.,

$$U_P = 400 \text{ N } (2 \text{ m} \cos 30°) = 692.8 \text{ J}$$

or the displacement times the component of force in the direction of displacement, i.e.,

$$U_P = 400 \text{ N} \cos 30°(2 \text{ m}) = 692.8 \text{ J}$$

(b)

Spring Force F_s.   In the initial position the spring is stretched $s_1 = 0.5$ m and in the final position it is stretched $s_2 = 0.5 + 2 = 2.5$ m. We require the work to be negative since the force and displacement are in opposite directions. The work of $\mathbf{F}_s$ is thus

Fig. 14–6

$$U_s = -\left[\tfrac{1}{2}(30 \text{ N/M})(2.5 \text{ m})^2 - \tfrac{1}{2}(30 \text{ N/m})(0.5 \text{ m})^2\right] = -90 \text{ J}$$

Weight W.   Since the weight acts in the opposite direction to its vertical displacement, the work is negative; i.e.,

$$U_W = -98.1 \text{ N } (2 \text{ m} \sin 30°) = -98.1 \text{ J}$$

Note that it is also possible to consider the component of weight in the direction of displacement; i.e.,

$$U_W = -(98.1 \sin 30° \text{ N}) \, 2 \text{ m} = -98.1 \text{ J}$$

Normal Force N_B.   This force does *no work* since it is *always* perpendicular to the displacement.

Total Work.   The work of all the forces when the block is displaced 2 m is thus

$$U_T = 692.8 - 90 - 98.1 = 505 \text{ J} \qquad\qquad Ans.$$

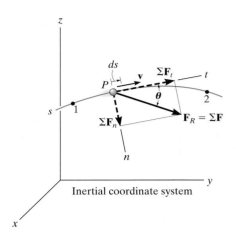

Fig. 14–7

14.2   Principle of Work and Energy

Consider a particle P in Fig. 14–7, which at the instant considered is located on the path as measured from an inertial coordinate system. If the particle has a mass m and is subjected to a system of external forces represented by the resultant $\mathbf{F}_R = \Sigma\mathbf{F}$, then the equation of motion for the particle in the tangential direction is $\Sigma F_t = ma_t$. Applying the kinematic equation $a_t = v\,dv/ds$ and integrating both sides, assuming initially that the particle has a position $s = s_1$ and a speed $v = v_1$, and later at $s = s_2$, $v = v_2$, yields

$$\Sigma \int_{s_1}^{s_2} F_t\,ds = \int_{v_1}^{v_2} mv\,dv$$

$$\Sigma \int_{s_1}^{s_2} F_t\,ds = \tfrac{1}{2}mv_2^2 - \tfrac{1}{2}mv_1^2 \qquad (14\text{–}5)$$

From Fig. 14–7, $\Sigma F_t = \Sigma F \cos\theta$ and since work is defined from Eq. 14–1, the final result may be written as

$$\Sigma U_{1-2} = \tfrac{1}{2}mv_2^2 - \tfrac{1}{2}mv_1^2 \qquad (14\text{–}6)$$

This equation represents the *principle of work and energy* for the particle. The term on the left is the sum of the work done by *all* the forces acting on the particle as the particle moves from point 1 to point 2. The two terms on the right side, which are of the form $T = \tfrac{1}{2}mv^2$, define the particle's final and initial *kinetic energy*, respectively. These terms are always *positive* scalars. Furthermore, Eq. 14–6 must be dimensionally homogeneous so that the kinetic energy has the same units as work, e.g., joules (J) or ft · lb.

When Eq. 14–6 is applied, it is often symbolized in the form

$$\boxed{T_1 + \Sigma U_{1-2} = T_2} \qquad (14\text{–}7)$$

which states that the particle's initial kinetic energy plus the work done by all the forces acting on the particle as it moves from its initial to its final position is equal to the particle's final kinetic energy.

As noted from the derivation, the principle of work and energy represents an integrated form of $\Sigma F_t = ma_t$, obtained by using the kinematic equation $a_t = v\,dv/ds$. As a result, this principle will provide a convenient *substitution* for $\Sigma F_t = ma_t$ when solving those types of kinetic problems which involve force, velocity, and displacement, since these variables are involved in the terms of Eq. 14–7. For example, if a particle's initial speed is known and the work of all the forces acting on the particle can be determined, then Eq. 14–7 provides a *direct means* of obtaining the final speed v_2 of the particle after it undergoes a specified

If an oncoming car strikes these crash barrels, the car's kinetic energy will be transformed into work, which causes the barrels, and to some extent the car, to be deformed. By knowing the amount of energy absorbed by each barrel it is possible to design a crash cushion such as this.

displacement. If instead v_2 is determined by means of the equation of motion, a two-step process is necessary; i.e., apply $\Sigma F_t = ma_t$ to obtain a_t, then integrate $a_t = v\, dv/ds$ to obtain v_2.

Note that the principle of work and energy cannot be used, for example, to determine forces directed *normal* to the path of motion, since these forces do no work on the particle. Instead $\Sigma F_n = ma_n$ must be applied. For curved paths, however, the magnitude of the normal force is a function of speed. Hence, it may be easier to obtain this speed using the principle of work and energy, and then substitute this quantity into the equation of motion $\Sigma F_n = mv^2/\rho$ to obtain the normal force.

Procedure for Analysis

The principle of work and energy is used to solve kinetic problems that involve *velocity, force*, and *displacement*, since these terms are involved in the equation. For application it is suggested that the following procedure be used.

Work (Free-Body Diagram).

- Establish the inertial coordinate system and draw a free-body diagram of the particle in order to account for all the forces that do work on the particle as it moves along its path.

Principle of Work and Energy.

- Apply the principle of work and energy, $T_1 + \Sigma U_{1-2} = T_2$.
- The kinetic energy at the initial and final points is always positive, since it involves the speed squared $\left(T = \frac{1}{2}mv^2\right)$.
- A force does work when it moves through a displacement in the direction of the force.
- Work is *positive* when the force component is in the *same direction* as its displacement, otherwise it is negative.
- Forces that are functions of displacement must be integrated to obtain the work. Graphically, the work is equal to the area under the force-displacement curve.
- The work of a weight is the product of the weight magnitude and the vertical displacement, $U_W = \pm Wy$. It is positive when the weight moves downwards.
- The work of a spring is of the form $U_s = \frac{1}{2}ks^2$, where k is the spring stiffness and s is the stretch or compression of the spring.

Numerical application of this procedure is illustrated in the examples following Sec. 14.3.

14.3 Principle of Work and Energy for a System of Particles

Inertial coordinate system

Fig. 14–8

The principle of work and energy can be extended to include a system of n particles isolated within an enclosed region of space as shown in Fig. 14–8. Here the arbitrary ith particle, having a mass m_i, is subjected to a resultant external force $\mathbf{F}_i$ and a resultant internal force $\mathbf{f}_i$ which each of the other particles exerts on the ith particle. Using Eq. 14–5, which applies in the tangential direction, the principle of work and energy written for the ith particle is thus

$$\tfrac{1}{2} m_i v_{i1}^2 + \int_{s_{i1}}^{s_{i2}} (F_i)_t \, ds + \int_{s_{i1}}^{s_{i2}} (f_i)_t \, ds = \tfrac{1}{2} m_i v_{i2}^2.$$

Similar equations result if the principle of work and energy is applied to each of the other particles of the system. Since both work and kinetic energy are scalars, the results may be added together algebraically, so that

$$\Sigma \tfrac{1}{2} m_i v_{i1}^2 + \Sigma \int_{s_{i1}}^{s_{i2}} (F_i)_t \, ds + \Sigma \int_{s_{i1}}^{s_{i2}} (f_i)_t \, ds = \Sigma \tfrac{1}{2} m_i v_{i2}^2$$

We can write this equation symbolically as

$$\Sigma T_1 + \Sigma U_{1-2} = \Sigma T_2 \qquad (14-8)$$

This equation states that the system's initial kinetic energy (ΣT_1) plus the work done by all the external and internal forces acting on the particles of the system (ΣU_{1-2}) is equal to the system's final kinetic energy (ΣT_2). To maintain this balance of energy, strict accountability of the work done by all the forces must be made. In this regard, note that although the internal forces on adjacent particles occur in equal but opposite collinear pairs, the total work done by each of these forces will, in general, *not cancel* out since the paths over which corresponding particles travel will be *different*. There are, however, two important exceptions to this rule which often occur in practice. If the particles are contained within the boundary of a *translating rigid body*, the internal forces all undergo the same displacement, and therefore the internal work will be zero. Also, particles connected by inextensible cables make up a system that has internal forces which are displaced by an equal amount. In this case, adjacent particles exert equal but opposite internal forces that have components which undergo the same displacement, and therefore the work of these forces cancels. On the other hand, note that if the body is assumed to be *nonrigid*, the particles of the body are displaced along *different paths*, and some of the energy due to force interactions would be given off and lost as heat or stored in the body if permanent deformations occur. We will discuss these effects briefly at the end of this section and in Sec. 15.4. Throughout this text, however, the principle of work and energy will be applied to the solution of problems only when direct accountability of such energy losses does not have to be made.

The procedure for analysis outlined in Sec. 14.2 provides a method for applying Eq. 14–8; however, only one equation applies for the entire system. If the particles are connected by cords, other equations can generally be obtained by using the kinematic principles outlined in Sec. 12.9 in order to relate the particle's speeds. See Example 14.6.

Work of Friction Caused by Sliding.

A special class of problems will now be investigated which requires a careful application of Eq. 14–8. These problems all involve cases where a body is sliding over the surface of another body in the presence of friction. Consider, for example, a block which is translating a distance s over a rough surface as shown in Fig. 14–9a. If the applied force **P** just balances the *resultant* frictional force $\mu_k N$, Fig. 14–9b, then due to equilibrium a constant velocity **v** is maintained, and one would expect Eq. 14–8 to be applied as follows:

$$\tfrac{1}{2}mv^2 + Ps - \mu_k Ns = \tfrac{1}{2}mv^2$$

Indeed this equation is satisfied if $P = \mu_k N$; however, as one realizes from experience, the sliding motion will *generate heat*, a form of energy which seems not to be accounted for in the work-energy equation. In order to explain this paradox and thereby more closely represent the nature of friction, we should actually model the block so that the surfaces of contact are *deformable* (nonrigid).* Recall that the rough portions at the bottom of the block act as "teeth," and when the block slides these teeth *deform slightly* and either break off or vibrate due to interlocking effects and pull away from "teeth" at the contacting surface, Fig. 14–9c. As a result, frictional forces that act on the block at these points are displaced slightly, due to the localized deformations, and then they are replaced by other frictional forces as other points of contact are made. At any instant, the *resultant* **F** of all these frictional forces remains essentially constant, i.e., $\mu_k N$; however, due to the many *localized deformations*, the actual displacement s' of $\mu_k N$ is *not* the same displacement s as the applied force **P**. Instead, s' will be *less* than s $(s' < s)$, and therefore the *external work* done by the resultant frictional force will be $\mu_k Ns'$ and not $\mu_k Ns$. The remaining amount of work, $\mu_k N(s - s')$, manifests itself as an increase in *internal energy*, which in fact causes the block's temperature to rise.

In summary then, Eq. 14–8 can be applied to problems involving sliding friction; however, it should be fully realized that the work of the resultant frictional force is not represented by $\mu_k Ns$; instead, this term represents *both* the external work of friction $(\mu_k Ns')$ and internal work $[\mu_k N(s - s')]$ which is converted into various forms of internal energy, such as heat.*†

(a)

(b)

(c)

Fig. 14–9

---

*See Chapter 8 of *Engineering Mechanics: Statics.*
†See B. A. Sherwood and W. H. Bernard, "Work and Heat Transfer in the Presence of Sliding Friction," *Am. J. Phys.* 52, 1001 (1984).

EXAMPLE 14.2

(a)

(b)

Fig. 14–10

The 3500-lb automobile shown in Fig. 14–10a is traveling down the 10° inclined road at a speed of 20 ft/s. If the driver jams on the brakes, causing his wheels to lock, determine how far s the tires skid on the road. The coefficient of kinetic friction between the wheels and the road is $\mu_k = 0.5$.

SOLUTION

This problem can be solved using the principle of work and energy, since it involves force, velocity, and displacement.

Work (Free-Body Diagram). As shown in Fig. 14–10b, the normal force $\mathbf{N}_A$ does no work since it never undergoes displacement along its line of action. The weight, 3500 lb, is displaced $s \sin 10°$ and does positive work. Why? The frictional force $\mathbf{F}_A$ does both external and internal work when it is *thought* to undergo a displacement s. This work is negative since it is in the opposite direction to displacement. Applying the equation of equilibrium normal to the road, we have

$$+\nwarrow \Sigma F_n = 0; \qquad N_A - 3500 \cos 10° \text{ lb} = 0 \qquad N_A = 3446.8 \text{ lb}$$

Thus,

$$F_A = 0.5 N_A = 1723.4 \text{ lb}$$

Principle of Work and Energy.

$$T_1 + \Sigma U_{1-2} = T_2$$

$$\frac{1}{2}\left(\frac{3500 \text{ lb}}{32.2 \text{ ft/s}^2}\right)(20 \text{ ft/s})^2 + \{3500 \text{ lb}(s \sin 10°) - (1723.4 \text{ lb})s\} = 0$$

Solving for s yields

$$s = 19.5 \text{ ft} \qquad\qquad Ans.$$

NOTE: If this problem is solved by using the equation of motion, *two steps* are involved. First, from the free-body diagram, Fig. 14–10b, the equation of motion is applied along the incline. This yields

$$+\swarrow \Sigma F_s = ma_s; \quad 3500 \sin 10° \text{ lb} - 1723.4 \text{ lb} = \frac{3500 \text{ lb}}{32.2 \text{ ft/s}^2}a$$

$$a = -10.3 \text{ ft/s}^2$$

Then, using the integrated form of $a\, ds = v\, dv$ (kinematics), since a is constant, we have

$$\left(+\swarrow\right) \quad v^2 = v_0^2 + 2a_c(s - s_0);$$

$$(0)^2 = (20 \text{ ft/s})^2 + 2(-10.3 \text{ ft/s}^2)(s - 0)$$

$$s = 19.5 \text{ ft} \qquad\qquad Ans.$$

EXAMPLE 14.3

For a short time the crane in Fig. 14–11a lifts the 2.50-Mg beam with a force of $F = (28 + 3s^2)$ kN. Determine the speed of the beam when it has risen $s = 3$ m. Also, how much time does it take to attain this height starting from rest?

SOLUTION

We can solve part of this problem using the principle of work and energy since it involves force, velocity, and displacement. Kinematics must be used to determine the time.

Work (Free-Body Diagram). As shown on the free-body diagram, Fig. 14–11b, the towing force $\mathbf{F}$ does positive work, which must be determined by integration since this force is a variable. Also, the weight is constant and will do negative work since the displacement is upwards.

Principles of Work and Energy.

$$T_1 + \Sigma U_{1-2} = T_2$$

$$0 + \int_0^s (28 + 3s^2)(10^3)\, ds - (2.50)(10^3)(9.81)s = \tfrac{1}{2}(2.50)(10^3)v^2$$

$$28(10^3)s + (10^3)s^3 - 24.525(10^3)s = 1.25(10^3)v^2$$

$$v = (2.78s + 0.8s^3)^{\frac{1}{2}} \qquad\qquad (1)$$

When $s = 3$ m,

$$v = 5.47 \text{ m/s} \qquad\qquad\qquad Ans.$$

Kinematics. Since we were able to express the velocity as a function of displacement, the time can be determined using $v = ds/dt$. In this case,

$$(2.78s + 0.8s^3)^{\frac{1}{2}} = \frac{ds}{dt}$$

$$t = \int_0^3 \frac{ds}{(2.78s + 0.8s^3)^{\frac{1}{2}}}$$

The integration can be performed numerically using a pocket calculator. The result is

$$t = 1.79 \text{ s} \qquad\qquad\qquad Ans.$$

NOTE: The acceleration of the beam can be determined by integrating Eq. (1) using $v\, dv = a\, ds$, or more directly, by applying the equation of motion, $\Sigma F = ma$.

(a)

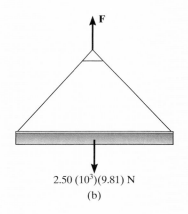

$\mathbf{F}$

2.50 $(10^3)(9.81)$ N

(b)

Fig. 14–11

EXAMPLE 14.4

The platform P, shown in Fig. 14–12a, has negligible mass and is tied down so that the 0.4-m-long cords keep a 1-m-long spring compressed 0.6 m when *nothing* is on the platform. If a 2-kg block is placed on the platform and released from rest after the platform is pushed down 0.1 m, Fig. 14–12b, determine the maximum height h the block rises in the air, measured from the ground.

(a)

$k = 200 \text{ N/m}$

(b)

Fig. 14–12

SOLUTION

Work (Free-Body Diagram).   Since the block is released from rest and later reaches its maximum height, the initial and final velocities are zero. The free-body diagram of the block when it is still in contact with the platform is shown in Fig. 14–12c. Note that the weight does negative work and the spring force does positive work. Why? In particular, the *initial compression* in the spring is $s_1 = 0.6 \text{ m} + 0.1 \text{ m} = 0.7 \text{ m}$. Due to the cords, the spring's *final compression* is $s_2 = 0.6 \text{ m}$ (after the block leaves the platform). The bottom of the block rises from a height of $(0.4 \text{ m} - 0.1 \text{ m}) = 0.3 \text{ m}$ to a final height h.

Principle of Work and Energy.

$$T_1 + \Sigma U_{1-2} = T_2$$

$$\tfrac{1}{2}mv_1^2 + \left\{ -\left(\tfrac{1}{2}ks_2^2 - \tfrac{1}{2}ks_1^2\right) - W\,\Delta y \right\} = \tfrac{1}{2}mv_2^2$$

19.62 N

F_s

(c)

Note that here $s_1 = 0.7 \text{ m} > s_2 = 0.6 \text{ m}$ and so the work of the spring as determined from Eq. 14–4 will indeed be positive once the calculation is made. Thus,

$$0 + \left\{ -\left[\tfrac{1}{2}(200 \text{ N/m})(0.6 \text{ m})^2 - \tfrac{1}{2}(200 \text{ N/m})(0.7 \text{ m})^2\right] \right.$$

$$\left. - (19.62 \text{ N})[h - (0.3 \text{ m})] \right\} = 0$$

Solving yields

$$h = 0.963 \text{ m} \qquad\qquad Ans.$$

EXAMPLE 14.5

Packages having a mass of 2 kg are delivered from a conveyor to a smooth circular ramp with a velocity of $v_0 = 1$ m/s as shown in Fig. 14–13a. If the radius of the ramp is 0.5 m, determine the angle $\theta = \theta_{max}$ at which each package begins to leave the surface.

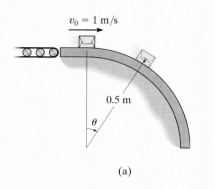

(a)

SOLUTION

Work (Free-Body Diagram). The free-body diagram of the block is shown at the intermediate location θ. The weight $W = 2(9.81) = 19.62$ N does positive work during the displacement. If a package is assumed to leave the surface when $\theta = \theta_{max}$, then the weight moves through a vertical displacement of $[0.5 - 0.5 \cos \theta_{max}]$ m, as shown in the figure.

Principle of Work and Energy.

$$T_1 + \Sigma U_{1-2} = T_2$$
$$\tfrac{1}{2}(2 \text{ kg})(1 \text{ m/s})^2 + \{19.62 \text{ N}(0.5 - 0.5 \cos \theta_{max}) \text{ m}\} = \tfrac{1}{2}(2 \text{ kg})v_2^2$$
$$v_2^2 = 9.81(1 - \cos \theta_{max}) + 1 \qquad (1)$$

(b)

Fig. 14–13

Equation of Motion. There are two unknowns in Eq. 1, θ_{max} and v_2. A second equation relating these two variables may be obtained by applying the equation of motion in the *normal direction* to the forces on the free-body diagram. (The principle of work and energy has replaced application of $\Sigma F_t = ma_t$ as noted in the derivation.) Thus,

$$+\swarrow \Sigma F_n = ma_n; \qquad -N_B + 19.62 \text{ N} \cos \theta = (2 \text{ kg})\left(\frac{v^2}{0.5 \text{ m}}\right)$$

When the package leaves the ramp at $\theta = \theta_{max}$, $N_B = 0$ and $v = v_2$; hence, this equation becomes

$$\cos \theta_{max} = \frac{v_2^2}{4.905} \qquad (2)$$

Eliminating the unknown v_2^2 between Eqs. 1 and 2 gives

$$4.905 \cos \theta_{max} = 9.81(1 - \cos \theta_{max}) + 1$$

Solving, we have

$$\cos \theta_{max} = 0.735$$
$$\theta_{max} = 42.7° \qquad\qquad Ans.$$

NOTE: This problem has also been solved in Example 13.9 using the equation of motion. If the two methods of solution are compared, it will be apparent that a work-energy approach yields a more direct solution.

EXAMPLE 14.6

s_B

B

100 kg

s_A

A 10 kg

Datum

(a)

$\mathbf{T}$ $\mathbf{R}_1$ $\mathbf{R}_2$

B

981 N

A

98.1 N

(b)

Fig. 14–14

The blocks A and B shown in Fig. 14–14a have a mass of 10 kg and 100 kg, respectively. Determine the distance B travels from the point where it is released from rest to the point where its speed becomes 2 m/s.

SOLUTION

This problem may be solved by considering the blocks separately and applying the principle of work and energy to each block. However, the work of the (unknown) cable tension can be eliminated from the analysis by considering blocks A and B together as a *system*. The solution will require simultaneous solution of the equations of work and energy *and* kinematics. To be consistent with our sign convention, we will assume both blocks move in the positive *downward* direction.

Work (Free-Body Diagram).   As shown on the free-body diagram of the system, Fig. 14–14b, the cable force $\mathbf{T}$ and reactions $\mathbf{R}_1$ and $\mathbf{R}_2$ do *no work*, since these forces represent the reactions at the supports and consequently do not move while the blocks are being displaced. The weights both do positive work since, as stated above, they are both assumed to move downward.

Principle of Work and Energy.   Realizing the blocks are released from rest, we have

$$\Sigma T_1 + \Sigma U_{1-2} = \Sigma T_2$$

$$\left\{\tfrac{1}{2}m_A(v_A)_1^2 + \tfrac{1}{2}m_B(v_B)_1^2\right\} + \left\{W_A\,\Delta s_A + W_B\,\Delta s_B\right\} =$$
$$\left\{\tfrac{1}{2}m_A(v_A)_2^2 + \tfrac{1}{2}m_B(v_B)_2^2\right\}$$

$$\{0 + 0\} + \{98.1\text{ N }(\Delta s_A) + 981\text{ N }(\Delta s_B)\} =$$
$$\left\{\tfrac{1}{2}(10\text{ kg})(v_A)_2^2 + \tfrac{1}{2}(100\text{ kg})(2\text{ m/s})^2\right\} \qquad (1)$$

Kinematics.   Using the methods of kinematics discussed in Sec. 12.9, it may be seen from Fig. 14–14a that at any given instant the total length l of all the vertical segments of cable may be expressed in terms of the position coordinates s_A and s_B as

$$s_A + 4s_B = l$$

Hence, a change in position yields the displacement equation

$$\Delta s_A + 4\,\Delta s_B = 0$$
$$\Delta s_A = -4\,\Delta s_B \qquad (2)$$

As required, both of these displacements are positive downward. Taking the time derivative yields

$$v_A = -4v_B = -4(2\text{ m/s}) = -8\text{ m/s}$$

Retaining the negative sign in Eq. 2 and substituting into Eq. 1 yields

$$\Delta s_B = 0.883\text{ m }\downarrow \qquad\qquad Ans.$$

PROBLEMS

14–1. A woman having a mass of 70 kg stands in an elevator which has a downward acceleration of 4 m/s² starting from rest. Determine the work done by her weight and the work of the normal force which the floor exerts on her when the elevator descends 6 m. Explain why the work of these forces is different.

14–2. The 20-lb crate has a velocity of $v_A = 12$ ft/s when it is at A. Determine its velocity after it slides $s = 6$ ft down the plane. The coefficient of kinetic friction between the crate and the plane is $\mu_k = 0.2$.

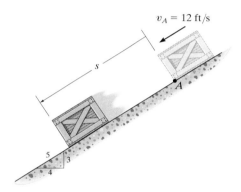

Prob. 14–2

14–3. The 20-kg crate is subjected to a force having a constant direction and a magnitude $F = 100$ N, where s is measured in meters. When $s = 4$ m, the crate is moving to the right with a speed of 8 m/s. Determine its speed when $s = 25$ m. The coefficient of kinetic friction between the crate and the ground is $\mu_k = 0.25$.

Prob. 14–3

*14–4. The "air spring" A is used to protect the support structure B and prevent damage to the conveyor-belt tensioning weight C in the event of a belt failure D. The force developed by the spring as a function of its deflection is shown by the graph. If the weight is 50 lb and is suspended a height $d = 1.5$ ft above the top of the spring, determine the maximum deformation of the spring in the event the conveyor belt fails. Neglect the mass of the pulley and belt.

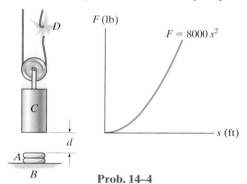

Prob. 14–4

14–5. A car is equipped with a bumper B designed to absorb collisions. The bumper is mounted to the car using pieces of flexible tubing T. Upon collision with a rigid barrier at A, a constant horizontal force $\mathbf{F}$ is developed which causes a car deceleration of $3g = 29.43$ m/s² (the highest safe deceleration for a passenger without a seatbelt). If the car and passenger have a total mass of 1.5 Mg and the car is initially coasting with a speed of 1.5 m/s, determine the magnitude of $\mathbf{F}$ needed to stop the car and the deformation x of the bumper tubing.

Prob. 14–5

14–6. The 100-kg crate is subjected to forces $F_1 = 800$ N and $F_2 = 1.5$ kN, as shown. If it is originally at rest, determine the distance it slides in order to attain a speed of $v = 6$ m/s. The coefficient of kinetic friction between the crate and the surface is $\mu_k = 0.2$.

Prob. 14–6

14–7. Design considerations for the bumper B on the 5-Mg train car require use of a nonlinear spring having the load-deflection characteristics shown in the graph. Select the proper value of k so that the maximum deflection of the spring is limited to 0.2 m when the car, traveling at 4 m/s, strikes the rigid stop. Neglect the mass of the car wheels.

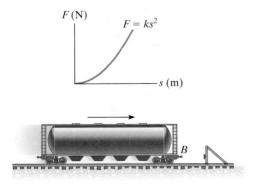

Prob. 14–7

*14–8. Determine the required height h of the roller coaster so that when it is essentially at rest at the crest of the hill it will reach a speed of 100 km/h when it comes to the bottom. Also, what should be the minimum radius of curvature ρ for the track at B so that the passengers do not experience a normal force greater than $4mg = (39.24m)$ N? Neglect the size of the car and passenger.

Prob. 14–8

14–9. When the driver applies the brakes of a light truck traveling 40 km/h, it skids 3 m before stopping. How far will the truck skid if it is traveling 80 km/h when the brakes are applied?

Prob. 14–9

14–10. The 0.5-kg ball of negligible size is fired up the vertical circular track using the spring plunger. The plunger keeps the spring compressed 0.08 m when $s = 0$. Determine how far s it must be pulled back and released so that the ball will begin to leave the track when $\theta = 135°$.

Prob. 14–10

■**14–11.** The force **F**, acting in a constant direction on the 20-kg block, has a magnitude which varies with the position s of the block. Determine how far the block slides before its velocity becomes 5 m/s. When $s = 0$ the block is moving to the right at 2 m/s. The coefficient of kinetic friction between the block and surface is $\mu_k = 0.3$.

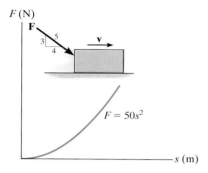

F (N)

$F = 50s^2$

s (m)

Prob. 14–11

*__14–12.__ The force **F**, acting in a constant direction on the 20-kg block, has a magnitude which varies with position s of the block. Determine the speed of the block after it slides 3 m. When $s = 0$ the block is moving to the right at 2 m/s. The coefficient of kinetic friction between the block and surface is $\mu_k = 0.3$.

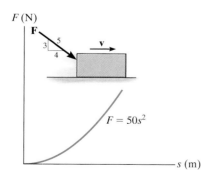

F (N)

$F = 50s^2$

s (m)

Prob. 14–12

14–13. As indicated by the derivation, the principle of work and energy is valid for observers in *any* inertial reference frame. Show that this is so, by considering the 10-kg block which rests on the smooth surface and is subjected to a horizontal force of 6 N. If observer A is in a *fixed* frame x, determine the final speed of the block if it has an initial speed of 5 m/s and travels 10 m, both directed to the right and measured from the fixed frame. Compare the result with that obtained by an observer B, attached to the x' axis and moving at a constant velocity of 2 m/s relative to A. *Hint:* The distance the block travels will first have to be computed for observer B before applying the principle of work and energy.

A

x

B

x'

2 m/s

5 m/s

6 N

10 m

Prob. 14–13

14–14. Determine the velocity of the 60-lb block A if the two blocks are released from rest and the 40-lb block B moves 2 ft up the incline. The coefficient of kinetic friction between both blocks and the inclined planes is $\mu_k = 0.10$.

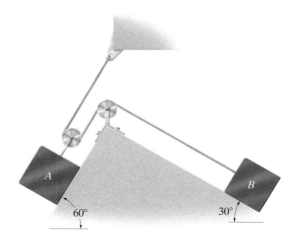

A

B

60°

30°

Prob. 14–14

14–15. Block A has a weight of 60 lb and block B has a weight of 10 lb. Determine the speed of block A after it moves 5 ft down the plane, starting from rest. Neglect friction and the mass of the cord and pulleys.

Prob. 14–15

*14–16. The 3-lb block A rests on a surface for which the coefficient of kinetic friction is $\mu_k = 0.3$. Determine the distance the 8-lb cylinder B must descend so that A has a speed of $v_A = 5$ ft/s starting from rest.

Prob. 14–16

14–17. The 100-lb block slides down the inclined plane for which the coefficient of kinetic friction is $\mu_k = 0.25$. If it is moving at 10 ft/s when it reaches point A, determine the maximum deformation of the spring needed to momentarily arrest the motion.

Prob. 14–17

14–18. The collar has a mass of 20 kg and rests on the smooth rod. Two springs are attached to it and the ends of the rod as shown. Each spring has an uncompressed length of 1 m. If the collar is displaced $s = 0.5$ m and released from rest, determine its velocity at the instant it returns to the point $s = 0$.

Prob. 14–18

14–19. The 2-kg block is subjected to a force having a constant direction and a magnitude $F = (300/(1 + s))$ N, where s is in meters. When $s = 4$ m, the block is moving to the left with a speed of 8 m/s. Determine its speed when $s = 12$ m. The coefficient of kinetic friction between the block and the ground is $\mu_k = 0.25$.

Prob. 14–19

*14–20. The motion of a truck is arrested using a bed of loose stones AB and a set of crash barrels BC. If experiments show that the stones provide a rolling resistance of 160 lb per wheel and the crash barrels provide a resistance as shown in the graph, determine the distance x the 4500-lb truck penetrates the barrels if the truck is coasting at 60 ft/s when it approaches A. Take $s = 50$ ft and neglect the size of the truck.

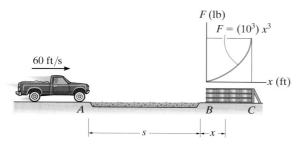

Prob. 14–20

14–22. The collar has a mass of 30 kg and is supported on the rod having a coefficient of kinetic friction $\mu_k = 0.4$. The attached spring has an unstretched length of 0.2 m and a stiffness $k = 50$ N/m. Determine the speed of the collar after the applied force $F = 200$ N causes it to be displaced $s = 1.5$ m from point A. When $s = 0$ the collar is held at rest.

Prob. 14–22

14–21. The crash cushion for a highway barrier consists of a nest of barrels filled with an impact-absorbing material. The barrier stopping force is measured versus the vehicle penetration into the barrier. Determine the distance a car having a weight of 4000 lb will penetrate the barrier if it is originally traveling at 55 ft/s when it strikes the first barrel.

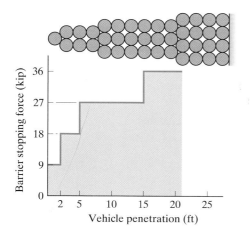

Prob. 14–21

14–23. The 5-lb block is released from rest at A and slides down the smooth circular surface AB. It then continues to slide along the horizontal rough surface until it strikes the spring. Determine how far it compresses the spring before stopping.

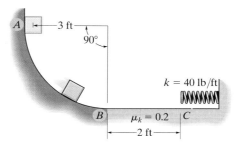

Prob. 14–23

*14–24. The block has a mass of 0.8 kg and moves within the smooth vertical slot. If it starts from rest when the *attached* spring is in the unstretched position at A, determine the *constant* vertical force F which must be applied to the cord so that the block attains a speed $v_B = 2.5$ m/s when it reaches B; $s_B = 0.15$ m. Neglect the size and mass of the pulley. *Hint:* The work of F can be determined by finding the difference Δl in cord lengths AC and BC and using $U_F = F \, \Delta l$.

Prob. 14–24

14–25. The collar has a mass of 5 kg and is moving at 8 m/s when $x = 0$ and a force of $F = 60$ N is applied to it. The direction θ of this force varies such that $\theta = 10x$, where x is in meters and θ is clockwise, measured in degrees. Determine the speed of the collar when $x = 3$ m. The coefficient of kinetic friction between the collar and the rod is $\mu_k = 0.3$.

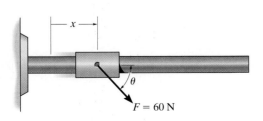

Prob. 14–25

14–26. Cylinder A has a weight of 60 lb and B has a weight of 10 lb. Determine the distance A must descend from rest before it obtains a speed of 8 ft/s. Also, what is the tension in the cord supporting A? Neglect the mass of the cord and pulleys.

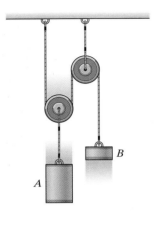

Prob. 14–26

14–27. The conveyor belt delivers each 12-kg crate to the ramp at A such that the crate's velocity is $v_A = 2.5$ m/s, directed down *along* the ramp. If the coefficient of kinetic friction between each crate and the ramp is $\mu_k = 0.3$, determine the speed at which each crate slides off the ramp at B. Assume that no tipping occurs.

Prob. 14–27

*14–28. When the 150-lb skier is at point *A* he has a speed of 5 ft/s. Determine his speed when he reaches point *B* on the smooth slope. For this distance the slope follows the cosine curve shown. Also, what is the normal force on his skis at *B* and his rate of increase in speed? Neglect friction and air resistance.

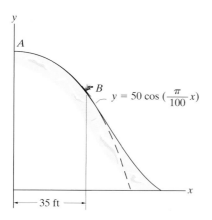

$$y = 50 \cos\left(\frac{\pi}{100}x\right)$$

35 ft

Prob. 14–28

14–30. The catapulting mechanism is used to propel the 10-kg slider *A* to the right along the smooth track. The propelling action is obtained by drawing the pulley attached to rod *BC* rapidly to the left by means of a piston *P*. If the piston applies a constant force $F = 20$ kN to rod *BC* such that it moves it 0.2 m, determine the speed attained by the slider if it was originally at rest. Neglect the mass of the pulleys, cable, piston, and rod *BC*.

Prob. 14–30

14–29. When the 12-lb block *A* is released from rest it lifts the two 15-lb weights *B* and *C*. Determine the maximum distance *A* will fall before its motion is momentarily stopped. Neglect the weight of the cord and the size of the pulleys.

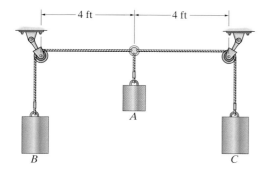

Prob. 14–29

14–31. The collar has a mass of 20 kg and slides along the smooth rod. Two springs are attached to it and the ends of the rod as shown. If each spring has an uncompressed length of 1 m and the collar has a speed of 2 m/s when $s = 0$, determine the maximum compression of each spring due to the back-and-forth (oscillating) motion of the collar.

Prob. 14–31

*14–32. The cyclist travels to point A, pedaling until he reaches a speed $v_A = 8$ m/s. He then coasts freely up the curved surface. Determine the normal force he exerts on the surface when he reaches point B. The total mass of the bike and man is 75 kg. Neglect friction, the mass of the wheels, and the size of the bicycle.

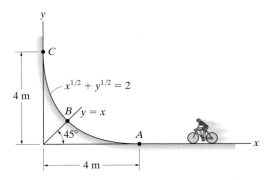

Prob. 14–32

14–33. The cyclist travels to point A, pedaling until he reaches a speed $v_A = 4$ m/s. He then coasts freely up the curved surface. Determine how high he reaches up the surface before he comes to a stop. Also, what are the resultant normal force on the surface at this point and his acceleration? The total mass of the bike and man is 75 kg. Neglect friction, the mass of the wheels, and the size of the bicycle.

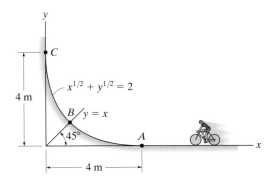

Prob. 14–33

14–34. The 10-lb block is pressed against the spring so as to compress it 2 ft when it is at A. If the plane is smooth, determine the distance d, measured from the wall, to where the block strikes the ground. Neglect the size of the block.

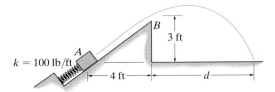

Prob. 14–34

14–35. The man at the window A wishes to throw the 30-kg sack on the ground. To do this he allows it to swing from rest at B to point C, when he releases the cord at $\theta = 30°$. Determine the speed at which it strikes the ground and the distance R.

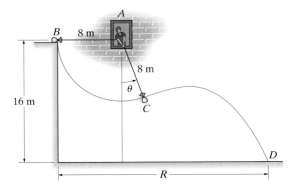

Prob. 14–35

*14–36. A 2-lb block rests on the smooth semicylindrical surface. An elastic cord having a stiffness $k = 2$ lb/ft is attached to the block at B and to the base of the semicylinder at point C. If the block is released from rest at A $(\theta = 0°)$, determine the unstretched length of the cord so the block begins to leave the semicylinder at the instant $\theta = 45°$. Neglect the size of the block.

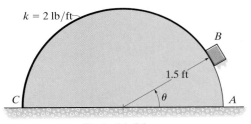

Prob. 14–36

14–37. A rocket of mass m is fired vertically from the surface of the earth, i.e., at $r = r_1$. Assuming no mass is lost as it travels upward, determine the work it must do against gravity to reach a distance r_2. The force of gravity is $F = GM_e m/r^2$ (Eq. 13–1), where M_e is the mass of the earth and r the distance between the rocket and the center of the earth.

Prob. 14–37

14–38. The spring has a stiffness $k = 50$ lb/ft and an *unstretched length* of 2 ft. As shown, it is confined by the plate and wall using cables so that its length is 1.5 ft. A 4-lb block is given a speed v_A when it is at A, and it slides down the incline having a coefficient of kinetic friction $\mu_k = 0.2$. If it strikes the plate and pushes it forward 0.25 ft before stopping, determine its speed at A. Neglect the mass of the plate and spring.

Prob. 14–38

14–39. The "flying car" is a ride at an amusement park which consists of a car having wheels that roll along a track mounted inside a rotating drum. By design the car cannot fall off the track, however motion of the car is developed by applying the car's brake, thereby gripping the car to the track and allowing it to move with a constant speed of the track, $v_t = 3$ m/s. If the rider applies the brake when going from B to A and then releases it at the top of the drum, A, so that the car coasts freely down along the track to B ($\theta = \pi$ rad), determine the speed of the car at B and the normal reaction which the drum exerts on the car at B. Neglect friction during the motion from A to B. The rider and car have a total mass of 250 kg and the center of mass of the car and rider moves along a circular path having a radius of 8 m.

Prob. 14–39

***14–40.** The skier starts from rest at A and travels down the ramp. If friction and air resistance can be neglected, determine his speed v_B when he reaches B. Also, find the distance s to where he strikes the ground at C, if he makes the jump traveling horizontally at B. Neglect the skier's size. He has a mass of 70 kg.

Prob. 14–40

14.4   Power and Efficiency

Power.   *Power* is defined as the amount of work performed per unit of time. Hence, the *power* generated by a machine or engine that performs an amount of work dU within the time interval dt is

$$P = \frac{dU}{dt} \qquad (14\text{–}9)$$

Provided the work dU is expressed by $dU = \mathbf{F} \cdot d\mathbf{r}$, then it is also possible to write

$$P = \frac{dU}{dt} = \frac{\mathbf{F} \cdot d\mathbf{r}}{dt} = \mathbf{F} \cdot \frac{d\mathbf{r}}{dt}$$

or

$$P = \mathbf{F} \cdot \mathbf{v} \qquad (14\text{–}10)$$

Hence, power is a *scalar*, where in the formulation $\mathbf{v}$ represents the velocity of the point which is acted upon by the force $\mathbf{F}$.

The basic units of power used in the SI and FPS systems are the watt (W) and horsepower (hp), respectively. These units are defined as

$$1\ \text{W} = 1\ \text{J/s} = 1\ \text{N} \cdot \text{m/s}$$
$$1\ \text{hp} = 550\ \text{ft} \cdot \text{lb/s}$$

The power output of this locomotive comes from the driving frictional force $\mathbf{F}$ developed at its wheels. It is this force that overcomes the frictional resistance of the cars in tow and is able to lift the weight of the train up the grade.

For conversion between the two systems of units, 1 hp = 746 W.

The term "power" provides a useful basis for determining the type of motor or machine which is required to do a certain amount of work in a given time. For example, two pumps may each be able to empty a reservoir if given enough time; however, the pump having the larger power will complete the job sooner.

Efficiency.   The *mechanical efficiency* of a machine is defined as the ratio of the output of useful power produced by the machine to the input of power supplied to the machine. Hence,

$$\epsilon = \frac{\text{power output}}{\text{power input}} \qquad (14\text{–}11)$$

If energy applied to the machine occurs during the *same time interval* at which it is removed, then the efficiency may also be expressed in terms of the ratio of output energy to input energy; i.e.

$$\epsilon = \frac{\text{energy output}}{\text{energy input}} \qquad (14\text{--}12)$$

Since machines consist of a series of moving parts, frictional forces will always be developed within the machine, and as a result, extra energy or power is needed to overcome these forces. Consequently, *the efficiency of a machine is always less than 1.*

The power requirements of this elevator depend upon the vertical force **F** that acts on the elevator and causes it to move upwards. If the velocity of the elevator is **v**, then the power output is $P = \mathbf{F} \cdot \mathbf{v}$.

Procedure for Analysis

The power supplied to a body can be computed using the following procedure.

- First determine the external force **F** acting on the body which causes the motion. This force is usually developed by a machine or engine placed either within or external to the body.

- If the body is accelerating, it may be necessary to draw its free-body diagram and apply the equation of motion ($\Sigma \mathbf{F} = m\mathbf{a}$) to determine **F**.

- Once **F** and the velocity **v** of the point where **F** is applied have been found, the power is determined by multiplying the force magnitude by the component of velocity acting in the direction of **F**, (i.e., $P = \mathbf{F} \cdot \mathbf{v} = Fv \cos \theta$).

- In some problems the power may be found by calculating the work done by **F** per unit of time ($P_{\text{avg}} = \Delta U / \Delta t$, or $P = dU/dt$).

EXAMPLE 14.7

Datum

$-s_P-$

Datum — P 〇 M

s_C

C

(a)

2T

$\mathbf{a}_C$

75 lb▼

y

(b)

Fig. 14–15

The motor M of the hoist shown in Fig. 14–15a operates with an efficiency of $\epsilon = 0.85$. Determine the power that must be supplied to the motor to lift the 75-lb crate C at the instant point P on the cable has an acceleration of 4 ft/s^2 and a velocity of 2 ft/s. Neglect the mass of the pulley and cable.

SOLUTION

In order to compute the power output of the motor, it is first necessary to determine the tension in the cable since this force is developed by the motor.

From the free-body diagram, Fig. 14–15b, we have

$$+\downarrow \quad \Sigma F_y = ma_y; \quad -2T + 75 \text{ lb} = \frac{75 \text{ lb}}{32.2 \text{ ft/s}^2} a_c \qquad (1)$$

The acceleration of the crate can be obtained by using kinematics to relate it to the known acceleration of point P, Fig. 14–15a. Using the methods of Sec. 12.9, the coordinates s_C and s_P in Fig. 14–15a can be related to a constant portion of cable length l which is changing in the vertical and horizontal directions. We have $2s_C + s_P = l$. Taking the second time derivative of this equation yields

$$2a_C = -a_P \qquad (2)$$

Since $a_P = +4 \text{ ft/s}^2$, then $a_C = (-4 \text{ ft/s}^2)/2 = -2 \text{ ft/s}^2$. What does the negative sign indicate? Substituting this result into Eq. 1 and *retaining* the negative sign since the acceleration in *both* Eqs. 1 and 2 is considered positive downward, we have

$$-2T + 75 \text{ lb} = \frac{75 \text{ lb}}{32.2 \text{ ft/s}^2}(-2 \text{ ft/s}^2)$$

$$T = 39.8 \text{ lb}$$

The power output, measured in units of horsepower, required to draw the cable in at a rate of 2 ft/s is therefore

$$P = \mathbf{T} \cdot \mathbf{v} = (39.8 \text{ lb})(2 \text{ ft/s})[1 \text{ hp}/(550 \text{ ft} \cdot \text{lb/s})]$$

$$= 0.145 \text{ hp}$$

This *power output* requires that the motor provide a *power input* of

$$\text{power input} = \frac{1}{\epsilon}(\text{power output})$$

$$= \frac{1}{0.85}(0.145 \text{ hp}) = 0.170 \text{ hp} \qquad \textit{Ans.}$$

NOTE: Since the velocity of the crate is constantly changing, the power requirement is *instantaneous*.

EXAMPLE 14.8

(a)

(b)

Fig. 14–16

The sports car shown in Fig. 14–16a has a mass of 2 Mg and is traveling at a speed of 25 m/s, when the brakes to all the wheels are applied. If the coefficient of kinetic friction is $\mu_k = 0.35$, determine the power developed by the friction force when the car skids. Then find the car's speed after it has slid 10 m.

SOLUTION

As shown on the free-body diagram, Fig. 14–16b, the normal force $\mathbf{N}_C$ and frictional force $\mathbf{F}_C$ represent the *resultant forces* of all four wheels.

Applying the equation of equilibrium in the y direction to determine N_C, we have

$$+\uparrow \Sigma F_y = 0; \qquad N_C = 19.62 \text{ kN}$$

The kinetic frictional force is therefore

$$F_C = 0.35(19.62 \text{ kN}) = 6.867 \text{ kN}$$

The velocity of the car can be determined when $s = 10$ m by applying the principle of work and energy. Why?

$$T_1 + \Sigma U_{1-2} = T_2$$

$$\tfrac{1}{2}(2000 \text{ kg})(25 \text{ m/s})^2 - 6.867(10^3) \text{ N} (10 \text{ m}) = \tfrac{1}{2}(2000 \text{ kg})v^2$$

$$v = 23.59 \text{ m/s}$$

The power of the frictional force at this instant is therefore

$$P = |\mathbf{F}_C \cdot \mathbf{v}| = 6.867(10^3) \text{ N} (25 \text{ m/s}) = 172 \text{ kW} \qquad \textit{Ans.}$$

PROBLEMS

14–41. A spring having a stiffness of 5 kN/m is compressed 400 mm. The stored energy in the spring is used to drive a machine which requires 90 W of power. Determine how long the spring can supply energy at the required rate.

14–42. Determine the power input for a motor necessary to lift 300 lb at a constant rate of 5 ft/s. The efficiency of the motor is $\epsilon = 0.65$.

14–43. An electrically powered train car draws 30 kW of power. If the car weighs 40 000 lb and starts from rest, determine the maximum speed it attains in 30 s. The mechanical efficiency is $\epsilon = 0.8$.

*14–44. A truck has a weight of 25 000 lb and an engine which transmits a power of 350 hp to *all* the wheels. Assuming that the wheels do not slip on the ground, determine the angle θ of the largest incline the truck can climb at a constant speed of $v = 50$ ft/s.

Prob. 14–44

14–45. An automobile having a mass of 2 Mg travels up a 7° slope at a constant speed of $v = 100$ km/h. If mechanical friction and wind resistance are neglected, determine the power developed by the engine if the automobile has an efficiency $\epsilon = 0.65$.

Prob. 14–45

14–46. The escalator steps move with a constant speed of 0.6 m/s. If the steps are 125 mm high and 250 mm in length, determine the power of a motor needed to lift an average mass of 150 kg per step. There are 32 steps.

14–47. If the escalator in Prob. 14–47 is *not moving*, determine the constant speed at which a man having a mass of 80 kg must walk up the steps to generate 100 W of power—the same amount that is needed to power a standard light bulb.

Probs. 14–46/47

*14–48. An electric streetcar has a weight of 15 000 lb and accelerates along a horizontal straight road from rest such that the power is always 100 hp. Determine how far it must travel to reach a speed of 40 ft/s.

14–49. The 50-lb crate is given a speed of 10 ft/s in $t = 4$ s starting from rest. If the acceleration is constant, determine the power that must be supplied to the motor when $t = 2$ s. The motor has an efficiency $\epsilon = 0.76$. Neglect the mass of the pulley and cable.

Prob. 14–49

14–50. A car has a mass m and accelerates along a horizontal straight road from rest such that the power is always a constant amount P. Determine how far it must travel to reach a speed of v.

14–51. To dramatize the loss of energy in an automobile, consider a car having a weight of 5 000 lb that is traveling at 35 mi/h. If the car is brought to a stop, determine how long a 100-W light bulb must burn to expend the same amount of energy. (1 mi = 5280 ft.)

*****14–52.** Determine the power output of the draw-works motor M necessary to lift the 600-lb drill pipe upward with a constant speed of 4 ft/s. The cable is tied to the top of the oil rig, wraps around the lower pulley, then around the top pulley, and then to the motor.

14–53. The 500-kg elevator starts from rest and travels upward with a constant acceleration $a_c = 2 \text{ m/s}^2$. Determine the power output of the motor M when $t = 3$ s. Neglect the mass of the pulleys and cable.

Prob. 14–53

14–54. The crate has a mass of 150 kg and rests on a surface for which the coefficients of static and kinetic friction are $\mu_s = 0.3$ and $\mu_k = 0.2$, respectively. If the motor M supplies a cable force of $F = (8t^2 + 20)$ N, where t is in seconds, determine the power output developed by the motor when $t = 5$ s.

Prob. 14–52

Prob. 14–54

14–55. The elevator E and its freight have a total mass of 400 kg. Hoisting is provided by the motor M and the 60-kg block C. If the motor has an efficiency of $\epsilon = 0.6$, determine the power that must be supplied to the motor when the elevator is hoisted upward at a constant speed of $v_E = 4$ m/s.

Prob. 14–55

14–57. The block has a mass of 150 kg and rests on a surface for which the coefficients of static and kinetic friction are $\mu_s = 0.5$ and $\mu_k = 0.4$, respectively. If a force $F = (60t^2)$ N, where t is in seconds, is applied to the cable, determine the power developed by the force when $t = 5$ s. *Hint:* First determine the time needed for the force to cause motion.

Prob. 14–57

****14–56.** The 50-kg crate is hoisted up the 30° incline by the pulley system and motor M. If the crate starts from rest and by constant acceleration attains a speed of 4 m/s after traveling 8 m along the plane, determine the power that must be supplied to the motor at this instant. Neglect friction along the plane. The motor has an efficiency of $\epsilon = 0.74$.

Prob. 14–56

14–58. The 50-lb load is hoisted by the pulley system and motor M. If the crate starts from rest and by constant acceleration attains a speed of 15 ft/s after rising $s = 6$ ft, determine the power that must be supplied to the motor at this instant. The motor has an efficiency of $\epsilon = 0.76$. Neglect the mass of the pulleys and cable.

Prob. 14–58

14–59. The 50-lb load is hoisted by the pulley system and motor M. If the motor exerts a constant force of 30 lb on the cable, determine the power that must be supplied to the motor if the load has been hoisted $s = 10$ ft starting from rest. The motor has an efficiency of $\epsilon = 0.76$.

Prob. 14–59

*14–60. The 10-lb collar starts from rest at A and is lifted by applying a constant vertical force of $F = 25$ lb to the cord. If the rod is smooth, determine the power developed by the force at the instant $\theta = 60°$.

Prob. 14–60

14–61. The 10-lb collar starts from rest at A and is lifted with a constant speed of 2 ft/s along the smooth rod. Determine the power developed by the force $\mathbf{F}$ at the instant shown.

Prob. 14–61

14–62. An athlete pushes against an exercise machine with a force that varies with time as shown in the first graph. Also, the velocity of the athlete's arm acting in the same direction as the force varies with time as shown in the second graph. Determine the power applied as a function of time and the work done in $t = 0.3$ s.

14–63. An athlete pushes against an exercise machine with a force that varies with time as shown in the first graph. Also, the velocity of the athlete's arm acting in the same direction as the force varies with time as shown in the second graph. Determine the maximum power developed during the 0.3-second time period.

Probs. 14–62/63

14.5   Conservative Forces and Potential Energy

Conservative Force.   When the work done by a force in moving a particle from one point to another is *independent of the path* followed by the particle, then this force is called a *conservative force*. The weight of a particle and the force of an elastic spring are two examples of conservative forces often encountered in mechanics. The work done by the weight of a particle is *independent of the path* since it depends only on the particle's *vertical displacement*. The work done by a spring force *acting on a particle* is *independent of the path* of the particle, since it depends only on the extension or compression *s* of the spring.

In contrast to a conservative force, consider the force of friction exerted *on a sliding object* by a fixed surface. The work done by the frictional force *depends on the path*—the longer the path, the greater the work. Consequently, *frictional forces are nonconservative*. The work is dissipated from the body in the form of heat.

Potential Energy.   Energy may be defined as the capacity for doing work. When energy comes from the *motion* of the particle, it is referred to as *kinetic energy*. When it comes from the *position* of the particle, measured from a fixed datum or reference plane, it is called potential energy. Thus, *potential energy* is a measure of the amount of work a conservative force will do when it moves from a given position to the datum. In mechanics, the potential energy due to gravity (weight) or an elastic spring is important.

Gravitational Potential Energy.   If a particle is located a distance *y above* an arbitrarily selected datum, as shown in Fig. 14–17, the particle's weight **W** has positive *gravitational potential energy*, V_g, since **W** has the capacity of doing positive work when the particle is moved back down to the datum. Likewise, if the particle is located a distance *y below* the datum, V_g is negative since the weight does negative work when the particle is moved back up to the datum. At the datum $V_g = 0$.

In general, if *y* is *positive upward*, the gravitational potential energy of the particle of weight *W* is*

$$V_g = W y \qquad\qquad (14\text{–}13)$$

*Here the weight is assumed to be *constant*. This assumption is suitable for small differences in elevation Δy. If the elevation change is significant, however, a variation of weight with elevation must be taken into account (see Prob. 14–37).

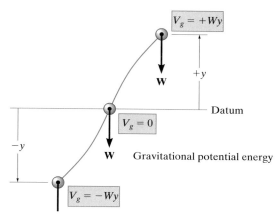

$V_g = +Wy$

W

$+y$

Datum

$V_g = 0$

$-y$

W   Gravitational potential energy

$V_g = -Wy$

Fig. 14–17

Elastic Potential Energy.   When an elastic spring is elongated or compressed a distance s from its unstretched position, the elastic potential energy V_e due to the spring's configuration can be expressed as

$$V_e = +\tfrac{1}{2}ks^2 \qquad (14\text{–}14)$$

Here V_e is *always positive* since, in the deformed position, the force of the spring has the *capacity* or "potential" for always doing positive work on the particle when the spring is returned to its unstretched position, Fig. 14–18.

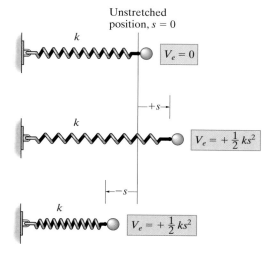

Unstretched position, $s = 0$

k

$V_e = 0$

$+s$

k

$V_e = +\tfrac{1}{2}ks^2$

$-s$

k

$V_e = +\tfrac{1}{2}ks^2$

Elastic potential energy

Fig. 14–18

Potential Function.   In the general case, if a particle is subjected to both gravitational and elastic forces, the particle's potential energy can be expressed as a *potential function*, which is the algebraic sum

$$V = V_g + V_e \qquad (14\text{--}15)$$

Measurement of V depends on the location of the particle with respect to a selected datum in accordance with Eqs. 14–13 and 14–14.

If the particle is located at an arbitrary point (x, y, z) in space, this potential function is then $V = V(x, y, z)$. The work done by a conservative force in moving the particle from point (x_1, y_1, z_1) to point (x_2, y_2, z_2) is measured by the *difference* of this function, i.e.,

$$U_{1\text{--}2} = V_1 - V_2 \qquad (14\text{--}16)$$

For example, the potential function for a particle of weight W suspended from a spring can be expressed in terms of its position, s, measured from a datum located at the unstretched length of the spring, Fig. 14–19. We have

$$V = V_g + V_e$$
$$= -Ws + \tfrac{1}{2}ks^2$$

If the particle moves from s_1 to a lower position s_2, then applying Eq. 14–16 it can be seen that the work of $\mathbf{W}$ and $\mathbf{F}_s$ is

$$U_{1\text{--}2} = V_1 - V_2 = \left(-Ws_1 + \tfrac{1}{2}ks_1^2\right) - \left(-Ws_2 + \tfrac{1}{2}ks_2^2\right)$$
$$= W(s_2 - s_1) - \left(\tfrac{1}{2}ks_2^2 - \tfrac{1}{2}ks_1^2\right)$$

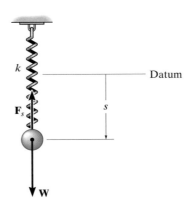

Fig. 14–19

When the displacement along the path is infinitesimal, i.e., from point (x, y, z) to $(x + dx, y + dy, z + dz)$, Eq. 14–16 becomes

$$dU = V(x, y, z) - V(x + dx, y + dy, z + dz)$$
$$= -dV(x, y, z) \tag{14–17}$$

Provided both the force and displacement are defined using rectangular coordinates, then the work can also be expressed as

$$dU = \mathbf{F} \cdot d\mathbf{r} = (F_x \mathbf{i} + F_y \mathbf{j} + F_z \mathbf{k}) \cdot (dx\mathbf{i} + dy\mathbf{j} + dz\mathbf{k})$$
$$= F_x \, dx + F_y \, dy + F_z \, dz$$

Substituting this result into Eq. 14–17 and expressing the differential $dV(x, y, z)$ in terms of its partial derivatives yields

$$F_x \, dx + F_y \, dy + F_z \, dz = -\left(\frac{\partial V}{\partial x} dx + \frac{\partial V}{\partial y} dy + \frac{\partial V}{\partial z} dz \right)$$

Since changes in x, y, and z are all independent of one another, this equation is satisfied provided

$$F_x = -\frac{\partial V}{\partial x}, \qquad F_y = -\frac{\partial V}{\partial y}, \qquad F_z = -\frac{\partial V}{\partial z} \tag{14–18}$$

Thus,

$$\mathbf{F} = -\frac{\partial V}{\partial x}\mathbf{i} - \frac{\partial V}{\partial y}\mathbf{j} - \frac{\partial V}{\partial z}\mathbf{k}$$
$$= -\left(\frac{\partial}{\partial x}\mathbf{i} + \frac{\partial}{\partial y}\mathbf{j} + \frac{\partial}{\partial z}\mathbf{k} \right)V$$

or

$$\mathbf{F} = -\nabla V \tag{14–19}$$

where ∇ (del) represents the vector operator $\nabla = (\partial/\partial x)\mathbf{i} + (\partial/\partial y)\mathbf{j} + (\partial/\partial z)\mathbf{k}$.

Equation 14–19 relates a force $\mathbf{F}$ to its potential function V and thereby provides a mathematical criterion for proving that $\mathbf{F}$ is conservative. For example, the gravitational potential function for a weight located a distance y above a datum is $V_g = Wy$. To prove that $\mathbf{W}$ is conservative, it is necessary to show that it satisfies Eq. 14–19 (or Eq. 14–18), in which case

$$F_y = -\frac{\partial V}{\partial y}; \qquad F = -\frac{\partial}{\partial y}(Wy) = -W$$

The negative sign indicates that $\mathbf{W}$ acts downward, opposite to positive y, which is upward.

14.6   Conservation of Energy

When a particle is acted upon by a system of *both* conservative and nonconservative forces, the portion of the work done by the *conservative forces* can be written in terms of the difference in their potential energies using Eq. 14–16, i.e., $(\Sigma U_{1-2})_{\text{cons.}} = V_1 - V_2$. As a result, the principle of work and energy can be written as

$$T_1 + V_1 + (\Sigma U_{1-2})_{\text{noncons.}} = T_2 + V_2 \qquad (14\text{--}20)$$

Here $(\Sigma U_{1-2})_{\text{noncons.}}$ represents the work of the nonconservative forces acting on the particle. If *only conservative forces* are applied to the body, this term is zero and then we have

$$\boxed{T_1 + V_1 = T_2 + V_2} \qquad (14\text{--}21)$$

This equation is referred to as the *conservation of mechanical energy* or simply the *conservation of energy*. It states that during the motion the sum of the particle's kinetic and potential energies remains *constant*. For this to occur, kinetic energy must be transformed into potential energy, and vice versa. For example, if a ball of weight **W** is dropped from a height h above the ground (datum), Fig. 14–20, the potential energy of the ball is maximum before it is dropped, at which time its kinetic energy is zero. The total mechanical energy of the ball in its initial position is thus

$$E = T_1 + V_1 = 0 + Wh = Wh$$

When the ball has fallen a distance $h/2$, its speed can be determined by using $v^2 = v_0^2 + 2a_c(y - y_0)$, which yields $v = \sqrt{2g(h/2)} = \sqrt{gh}$. The energy of the ball at the mid-height position is therefore

$$E = T_2 + V_2 = \frac{1}{2}\frac{W}{g}\left(\sqrt{gh}\right)^2 + W\frac{h}{2} = Wh$$

Just before the ball strikes the ground, its potential energy is zero and its speed is $v = \sqrt{2gh}$. Here, again, the total energy of the ball is

$$E = T_3 + V_3 = \frac{1}{2}\frac{W}{g}\left(\sqrt{2gh}\right)^2 + 0 = Wh$$

Note that when the ball comes in contact with the ground, it deforms somewhat, and provided the ground is hard enough, the ball will rebound off the surface, reaching a new height h', which will be less than the height h from which it was first released. Neglecting air friction, the difference in height accounts for an energy loss, $E_l = W(h - h')$, which occurs during the collision. Portions of this loss produce noise, localized deformation of the ball and ground, and heat.

The weight of the sacks resting on this platform causes potential energy to be stored in the supporting springs. As each sack is removed, the platform will *rise* slightly since some of the potential energy within the springs will be transferred into an increase in gravitational potential energy of the remaining sacks. Such a device is useful for removing the sacks without having to bend over to pick them up as they are unloaded.

Potential Energy (max)
Kinetic Energy (zero)

Potential Energy and
Kinetic Energy

Potential Energy (zero)
Kinetic Energy (max)

Datum

h

$\dfrac{h}{2}$

Fig. 14–20

System of Particles.   If a system of particles is *subjected only to conservative forces*, then an equation similar to Eq. 14–21 can be written for the particles. Applying the ideas of the preceding discussion, Eq. 14–8 ($\Sigma T_1 + \Sigma U_{1-2} = \Sigma T_2$) becomes

$$\Sigma T_1 + \Sigma V_1 = \Sigma T_2 + \Sigma V_2 \qquad (14\text{–}22)$$

Here, the sum of the system's initial kinetic and potential energies is equal to the sum of the system's final kinetic and potential energies. In other words, $\Sigma T + \Sigma V = $ const.

It is important to remember that only problems involving conservative force systems (weights and springs) may be solved by using the conservation of energy theorem. As stated previously, friction or other drag-resistant forces, which depend upon velocity or acceleration, are nonconservative. A portion of the work done by such forces is transformed into thermal energy, and consequently this energy dissipates into the surroundings and may not be recovered.

Procedure for Analysis

The conservation of energy equation is used to solve problems involving *velocity, displacement,* and *conservative force systems.* It is generally *easier to apply* than the principle of work and energy because the energy equation just requires specifying the particle's kinetic and potential energies at only *two points* along the path, rather than determining the work when the particle moves through a *displacement.* For application it is suggested that the following procedure be used.

Potential Energy.

- Draw two diagrams showing the particle located at its initial and final points along the path.
- If the particle is subjected to a vertical displacement, establish the fixed horizontal datum from which to measure the particle's gravitational potential energy V_g.
- Data pertaining to the elevation y of the particle from the datum and the extension or compression s of any connecting springs can be determined from the geometry associated with the two diagrams.
- Recall $V_g = Wy$, where y is positive upward from the datum and negative downward from the datum; also for a spring, $V_e = \frac{1}{2}ks^2$, which is *always positive.*

Conservation of Energy.

- Apply the equation $T_1 + V_1 = T_2 + V_2$.
- When determining the kinetic energy, $T = \frac{1}{2}mv^2$, the particle's speed v must be measured from an inertial reference frame.

EXAMPLE 14.9

The gantry structure in the photo is used to test the response of an airplane during a crash. As shown in Fig. 14–21a, the plane, having a mass of 8 Mg, is hoisted back until $\theta = 60°$, and then the pull-back cable AC is released when the plane is at rest. Determine the speed of the plane just before crashing into the ground, $\theta = 15°$. Also, what is the maximum tension developed in the supporting cable during the motion? Neglect the effect of lift caused by the wings during the motion and the size of the airplane.

Datum

20 m

θ

A

B

C

(a)

SOLUTION

Since the force of the cable does *no work* on the plane, it must be obtained using the equation of motion. First, however, we must determine the plane's speed at B.

Potential Energy.   For convenience, the datum has been established at the top of the gantry.

Conservation of Energy.

$$T_A + V_A = T_B + V_B$$

$$0 - 8000 \text{ kg } (9.81 \text{ m/s}^2)(20 \cos 60° \text{ m}) =$$

$$\tfrac{1}{2}(8000 \text{ kg})v_B^2 - 8000 \text{ kg } (9.81 \text{ m/s}^2)(20 \cos 15° \text{ m})$$

$$v_B = 13.5 \text{ m/s} \qquad \qquad Ans.$$

Equation of Motion.   Using the data tabulated on the free-body diagram when the plane is at B, Fig. 14–21b, we have

$$+\nwarrow \quad \Sigma F_n = ma_n;$$

$$T - 8000(9.81) \text{ N } \cos 15° = (8000 \text{ kg})\frac{(13.5 \text{ m/s})^2}{20 \text{ m}}$$

$$T = 149 \text{ kN} \qquad \qquad Ans.$$

T

15°

8000(9.81) N

(b)

Fig. 14–21

EXAMPLE 14.10

The ram R shown in Fig. 14–22a has a mass of 100 kg and is released from rest 0.75 m from the top of a spring, A, that has a stiffness $k_A = 12$ kN/m. If a second spring B, having a stiffness $k_B = 15$ kN/m, is "nested" in A, determine the maximum displacement of A needed to stop the downward motion of the ram. The unstretched length of each spring is indicated in the figure. Neglect the mass of the springs.

SOLUTION

Potential Energy. We will *assume* that the ram compresses *both* springs at the instant it comes to rest. The datum is located through the center of gravity of the ram at its initial position, Fig. 14–22b. When the kinetic energy is reduced to zero $(v_2 = 0)$ A is compressed a distance s_A and B compresses $s_B = s_A - 0.1$ m.

Conservation of Energy.

$$T_1 + V_1 = T_2 + V_2$$

$$0 + 0 = 0 + \left\{ \tfrac{1}{2}k_A s_A^2 + \tfrac{1}{2}k_B(s_A - 0.1)^2 - Wh \right\}$$

$$0 + 0 = 0 + \left\{ \tfrac{1}{2}(12\ 000\ \text{N/m})s_A^2 + \tfrac{1}{2}(15\ 000\ \text{N/m})(s_A - 0.1\ \text{m})^2 \right.$$
$$\left. - 981\ \text{N}\ (0.75\ \text{m} + s_A) \right\}$$

Rearranging the terms,

$$13\ 500s_A^2 - 2481s_A - 660.75 = 0$$

Using the quadratic formula and solving for the positive root, we have

$$s_A = 0.331\ \text{m} \qquad\qquad Ans.$$

Since $s_B = 0.331$ m $- 0.1$ m $= 0.231$ m, which is positive, the assumption that *both* springs are compressed by the ram is correct.

NOTE: The second root, $s_A = -0.148$ m, does not represent the physical situation. Since positive s is measured downward, the negative sign indicates that spring A would have to be "extended" by an amount of 0.148 m to stop the ram.

$k_A = 12$ kN/m

A

B

$k_B = 15$ kN/m

0.75 m

0.4 m

0.3 m

(a)

981 N

Datum

981 N

0.75 m

s_A

s_A

$s_B = s_A - 0.1$ m

(b)

Fig. 14–22

EXAMPLE 14.11

(a)

Datum

$y = 1$ m

$\sqrt{(1\text{ m})^2 + (0.75\text{ m})^2} = 1.25$ m

$s_{CB} = 1.25$ m $- 0.75$ m $= 0.5$ m

(b)

Fig. 14–23

A smooth 2-kg collar C, shown in Fig. 14–23a, fits loosely on the vertical shaft. If the spring is unstretched when the collar is in the position A, determine the speed at which the collar is moving when $y = 1$ m, if (a) it is released from rest at A, and (b) it is released at A with an *upward* velocity $v_A = 2$ m/s.

SOLUTION

Part (a) Potential Energy. For convenience, the datum is established through AB, Fig. 14–23b. When the collar is at C, the gravitational potential energy is $-(mg)y$, since the collar is *below* the datum, and the elastic potential energy is $\frac{1}{2}ks_{CB}^2$. Here $s_{CB} = 0.5$ m, which represents the *stretch* in the spring as shown in the figure.

Conservation of Energy.

$$T_A + V_A = T_C + V_C$$
$$0 + 0 = \tfrac{1}{2}mv_C^2 + \left\{\tfrac{1}{2}ks_{CB}^2 - mgy\right\}$$
$$0 + 0 = \left\{\tfrac{1}{2}(2\text{ kg})v_C^2\right\} + \left\{\tfrac{1}{2}(3\text{ N/m})(0.5\text{ m})^2 - 2(9.81)\text{ N }(1\text{ m})\right\}$$
$$v_C = 4.39\text{ m/s} \downarrow \qquad\qquad\qquad Ans.$$

This problem can also be solved by using the equation of motion or the principle of work and energy. Note that in *both* of these methods the variation of the magnitude and direction of the spring force must be taken into account (see Example 13.4). Here, however, the above method of solution is clearly advantageous since the calculations depend *only* on data calculated at the initial and final points of the path.

Part (b) Conservation of Energy. If $v_A = 2$ m/s, using the data in Fig. 14–23b, we have

$$T_A + V_A = T_C + V_C$$
$$\tfrac{1}{2}mv_A^2 + 0 = \tfrac{1}{2}mv_C^2 + \left\{\tfrac{1}{2}ks_{CB}^2 - mgy\right\}$$
$$\tfrac{1}{2}(2\text{ kg})(2\text{ m/s})^2 + 0 = \tfrac{1}{2}(2\text{ kg})v_C^2 + \left\{\tfrac{1}{2}(3\text{ N/m})(0.5\text{ m})^2\right.$$
$$\left. - 2(9.81)\text{ N }(1\text{ m})\right\}$$
$$v_C = 4.82\text{ m/s} \downarrow \qquad\qquad\qquad Ans.$$

NOTE: The kinetic energy of the collar depends only on the *magnitude* of velocity, and therefore it is immaterial if the collar is moving up or down at 2 m/s when released at A.

PROBLEMS

*14–64. Solve Prob. 14–8 using the conservation of energy equation.

14–65. Solve Prob. 14–15 using the conservation of energy equation.

14–66. Solve Prob. 14–18 using the conservation of energy equation.

14–67. Solve Prob. 14–31 using the conservation of energy equation.

*14–68. Solve Prob. 14–36 using the conservation of energy equation.

14–69. Two equal-length springs are "nested" together in order to form a shock absorber. If it is designed to arrest the motion of a 2-kg mass that is dropped $s = 0.5$ m above the top of the springs from an at-rest position, and the maximum compression of the springs is to be 0.2 m, determine the required stiffness of the inner spring, k_B, if the outer spring has a stiffness $k_A = 400$ N/m.

14–70. Determine the smallest amount the spring at B must be compressed against the 0.5-lb block so that when it is released from B it slides along the smooth surface and reaches point A.

14–71. If the spring is compressed 3 in. against the 0.5-lb block and it is released from rest, determine the normal force of the smooth surface on the block when it reaches point $x = 0.5$ ft.

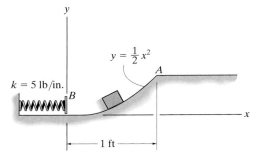

Probs. 14–70/71

*14–72. The girl has a mass of 40 kg and center of mass at G. If she is swinging to a maximum height defined by $\theta = 60°$, determine the force developed along each of the four supporting posts such as AB at the instant $\theta = 0°$. The swing is centrally located between the posts.

Prob. 14–69

Prob. 14–72

14–73. Each of the two elastic rubber bands of the slingshot has an unstretched length of 200 mm. If they are pulled back to the position shown and released from rest, determine the speed of the 25-g pellet just after the rubber bands become unstretched. Neglect the mass of the rubber bands and the change in elevation of the pellet while it is constrained by the rubber bands. Each rubber band has a stiffness of $k = 50$ N/m.

14–74. Each of the two elastic rubber bands of the slingshot has an unstretched length of 200 mm. If they are pulled back to the position shown and released from rest, determine the maximum height the 25-g pellet will reach if it is fired vertically upward. Neglect the mass of the rubber bands and the change in elevation of the pellet while it is constrained by the rubber bands. Each rubber band has a stiffness $k = 50$ N/m.

Probs. 14–73/74

14–75. The bob of the pendulum has a mass of 0.2 kg and is released from rest when it is in the horizontal position shown. Determine its speed and the tension in the cord at the instant the bob passes through its lowest position.

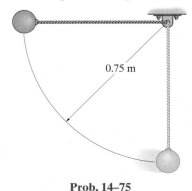

Prob. 14–75

*14–76. The 5-lb collar is released from rest at A and travels along the smooth guide. Determine the speed of the collar just before it strikes the stop at B. The spring has an unstretched length of 12 in.

14–77. The 5-lb collar is released from rest at A and travels along the smooth guide. Determine its speed when its center reaches point C and the normal force it exerts on the rod at this point. The spring has an unstretched length of 12 in., and point C is located just before the end of the curved portion of the rod.

Probs. 14–76/77

14–78. The firing mechanism of a pinball machine consists of a plunger P having a mass of 0.25 kg and a spring of stiffness $k = 300$ N/m. When $s = 0$, the spring is compressed 50 mm. If the arm is pulled back such that $s = 100$ mm and released, determine the speed of the 0.3-kg pinball B *just before* the plunger strikes the stop, i.e., $s = 0$. Assume all surfaces of contact to be smooth. The ball moves in the horizontal plane. Neglect friction, the mass of the spring, and the rolling motion of the ball.

Prob. 14–78

14–79. The roller-coaster car has a mass of 800 kg, including its passenger, and starts from the top of the hill A with a speed $v_A = 3$ m/s. Determine the minimum height h of the hill crest so that the car travels around both inside loops without leaving the track. Neglect friction, the mass of the wheels, and the size of the car. What is the normal reaction on the car when the car is at B and when it is at C?

*14–80. The roller-coaster car has a mass of 800 kg, including its passenger. If it is released from rest at the top of the hill A, determine the minimum height h of the hill crest so that the car travels around both inside loops without leaving the track. Neglect friction, the mass of the wheels, and the size of the car. What is the normal reaction on the car when the car is at B and when it is at C?

Probs. 14–79/80

14–81. The 0.75-kg bob of a pendulum is fired from rest at position A by a spring which has a stiffness of $k = 6$ kN/m and is compressed 125 mm. Determine the speed of the bob and the tension in the cord when the bob is at positions B and C. Point B is located on the path where the radius of curvature is still 0.6 m, i.e., just before the cord becomes horizontal.

Prob. 14–81

14–82. The spring has a stiffness $k = 3$ lb/ft and an unstretched length of 2 ft. If it is attached to the 5-lb smooth collar and the collar is released from rest at A, determine the speed of the collar just before it strikes the end of the rod at B. Neglect the size of the collar.

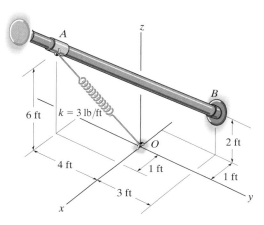

Prob. 14–82

14–83. Just for fun, two 150-lb engineering students A and B intend to jump off the bridge from rest using an elastic cord (bungee cord) having a stiffness $k = 80$ lb/ft. They wish to just reach the surface of the river, when A, attached to the cord, lets go of B at the instant they touch the water. Determine the proper unstretched length of the cord to do the stunt, and calculate the maximum acceleration of student A and the maximum height he reaches above the water after the rebound. From your results, comment on the feasibility of doing this stunt.

Prob. 14–83

*14–84. Two equal-length springs having a stiffness $k_A = 300$ N/m and $k_B = 200$ N/m are "nested" together in order to form a shock absorber. If a 2-kg block is dropped from an at-rest position 0.6 m above the top of the springs, determine their deformation when the block momentarily stops.

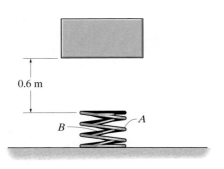

0.6 m

B — A

Prob. 14–84

14–85. The 0.75-kg bob of a pendulum is fired from rest at position A. If the spring is compressed 50 mm and released, determine (a) its stiffness k so that the speed of the bob is zero when it reaches point B, where the radius of curvature is still 0.6 m, and (b) the stiffness k so that when the bob reaches point C the tension in the cord is zero.

C

1.2 m

0.6 m

B

0.6 m

A   k

Prob. 14–85

14–86. The roller-coaster car has a speed of 15 ft/s when it is at the crest of a vertical parabolic track. Determine the car's velocity and the normal force it exerts on the track when it reaches point B. Neglect friction and the mass of the wheels. The total weight of the car and the passengers is 350 lb.

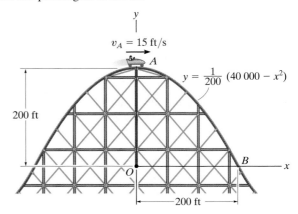

$v_A = 15$ ft/s

A

$y = \frac{1}{200}(40\,000 - x^2)$

200 ft

B x

O

200 ft

Prob. 14–86

14–87. The Raptor is an outside loop roller coaster in which riders are belted into seats resembling ski-lift chairs. Determine the minimum speed v_0 at which the cars should coast down from the top of the hill, so that passengers can just make the loop without leaving contact with their seats. Neglect friction, the size of the car and passenger, and assume each passenger and car has a mass m.

v_0

h

ρ

Prob. 14–87

*14–88. The Raptor is an outside loop roller coaster in which riders are belted into seats resembling ski-lift chairs. If the cars travel at $v_0 = 4$ m/s when they are at the top of the hill, determine their speed when they are at the top of the loop and the reaction of the 70-kg passenger on his seat at this instant. The car has a mass of 50 kg. Take $h = 12$ m, $\rho = 5$ m. Neglect friction and the size of the car and passenger.

Prob. 14–88

14–90. The ball has a weight of 15 lb and is fixed to a rod having a negligible mass. If it is released from rest when $\theta = 0°$, determine the angle θ at which the compressive force in the rod becomes zero.

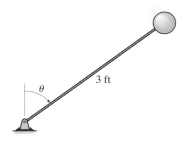

Prob. 14–90

14–89. A block having a mass of 20 kg is attached to four springs. If each spring has a stiffness of $k = 2$ kN/m and an unstretched length of 150 mm, determine the *maximum* downward vertical displacement s_{max} of the block if it is released from rest when $s = 0$.

Prob. 14–89

14–91. The ride at an amusement park consists of a gondola which is lifted to a height of 120 ft at A. If it is released from rest and falls along the parabolic track, determine the speed at the instant $y = 20$ ft. Also determine the normal reaction of the tracks on the gondola at this instant. The gondola and passenger have a total weight of 500 lb. Neglect the effects of friction.

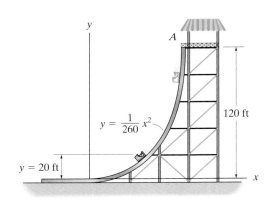

Prob. 14–91

*14–92. The 2-lb collar has a speed of 5 ft/s at A. The attached spring has an unstretched length of 2 ft and a stiffness of $k = 10$ lb/ft. If the collar moves over the smooth rod, determine its speed when it reaches point B, the normal force of the rod on the collar, and the rate of decrease in its speed.

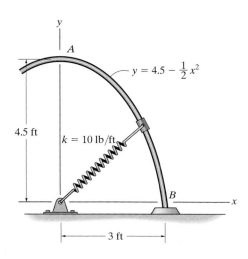

Prob. 14–92

14–93. The 20-lb collar is constrained to move on the smooth rod. It is attached to the three springs which are unstretched when $s = 0$. If the collar is displaced $s = 0.5$ ft and released from rest, determine its speed when $s = 0$.

Prob. 14–93

14–94. A tank car is stopped by two spring bumpers A and B, having a stiffness of $k_A = 15(10^3)$ lb/ft and $k_B = 20(10^3)$ lb/ft, respectively. Bumper A is attached to the car, whereas bumper B is attached to the wall. If the car has a weight of $25(10^3)$ lb and is freely coasting at 3 ft/s, determine the maximum deflection of each spring at the instant the bumpers stop the car.

Prob. 14–94

14–95. If the mass of the earth is M_e, show that the gravitational potential energy of a body of mass m located a distance r from the center of the earth is $V_g = -GM_em/r$. Recall that the gravitational force acting between the earth and the body is $F = G(M_em/r^2)$, Eq. 13–1. For the calculation, locate the datum at $r \to \infty$. Also, prove that F is a conservative force.

*14–96. The double-spring bumper is used to stop the 1500-lb steel billet in the rolling mill. Determine the maximum deflection of the plate A caused by the billet if it strikes the plate with a speed of 8 ft/s. Neglect the mass of the springs, rollers and the plates A and B. Take $k_1 = 3000$ lb/ft, $k_2 = 45000$ lb/ft.

Prob. 14–96

Design Projects

14–1D. Design of a Car Bumper

The body of an automobile is to be protected by a spring-loaded bumper, which is attached to the automobile's frame. Design the bumper so that it will stop a 3500-lb car traveling freely at 5 mi/h and not deform the springs more than 3 in. Submit a sketch of your design showing the placement of the springs and their stiffness. Plot the load-deflection diagram for the bumper during a direct collision with a rigid wall, and also plot the deceleration of the car as a function of the springs' displacement.

5 mi/h

Fig. 14–1D

14–2D. Design of an Elevator Hoist

It is required that an elevator and its contents, having a maximum weight of 500 lb, be lifted $y = 20$ ft, starting from rest and then stopping after 6 seconds. A single motor and cable-winding drum can be mounted anywhere and used for the operation. During any lift or descent the acceleration should not exceed 10 ft/s^2. Design a cable-and-pulley system for the elevator, and estimate the material cost if the cable is \$1.30/ft and pulleys are \$3.50 each. Submit a drawing of your design, and include plots of the power output required of the motor and the elevator's speed versus the height y traveled.

A B

30 ft

20 ft

7 ft

y

Fig. 14–2D

Chapter Review

Work of a Force

A force does work when it has a displacement along its line of action. If the force varies with the displacement, then $U = \int F \, ds$.

Graphically, this represents the area under the $F{-}s$ diagram.

If the force is constant, then for a displacement Δs in the direction of the force, $U = F_c \, \Delta s$. A typical example of this case is work of a weight, $U = W \, \Delta y$. Here, Δy is the vertical displacement.

A spring force, $F = ks$, depends upon the elongation or compression s of the spring. (*See pages 165–168.*)

$$U = \tfrac{1}{2}ks_2^2 - \tfrac{1}{2}ks_1^2$$

The Principle of Work and Energy

If the equation of motion in the tangential direction, $\Sigma F_t = ma_t$, is combined with the kinematic equation, $a_t \, ds = v \, dv$, we obtain the principle of work and energy, where the initial kinetic energy of the particle $\left(T_1 = \tfrac{1}{2}mv_1{}^2\right)$, plus the work done by all the forces that act on the particle as it moves from its initial position to its final position (ΣU_{1-2}), is equal to the final kinetic energy of the particle $\left(T_2 = \tfrac{1}{2}mv_2{}^2\right)$.

$$T_1 + \Sigma U_{1-2} = T_2$$

The principle of work and energy is useful for solving problems that involve force, velocity, and displacement. For application, the free-body diagram of the particle should be drawn in order to identify the forces that do work.
(*See pages 170–173.*)

Power and Efficiency

Power is the time rate of doing work. It is defined by $P = dU/dt$, or $P = \mathbf{F} \cdot \mathbf{v}$. For application, the force $\mathbf{F}$ creating the power and its velocity $\mathbf{v}$ must be specified.

Efficiency represents the ratio of power output to power input. Due to frictional losses, it is always less than one.
(*See pages 188 and 189.*)

$$\epsilon = \frac{\text{power output}}{\text{power input}}$$

Conservation of Energy

A conservative force does work that is independent of its path. Two examples are the weight of a particle and the spring force.

Friction is a nonconservative force since the work depends upon the length of the path. The longer the path, the more work done.

The work done by a conservative force depends upon its position relative to a datum. When this work is referenced from a datum, it is called potential energy. For weight, it is $V_g = Wy$, and for a spring it is $V_e = \frac{1}{2}kx^2$.

Mechanical energy consists of kinetic energy T and gravitational and elastic potential energies V. According to the conservation of energy, this sum is constant and has the same value at any two positions on the path. If only gravitational and spring forces cause motion of the particle, then the conservation-of-energy equation can be used to solve problems involving displacement and velocity.
(*See pages 196–201.*)

$$T_1 + V_1 = T_2 + V_2$$

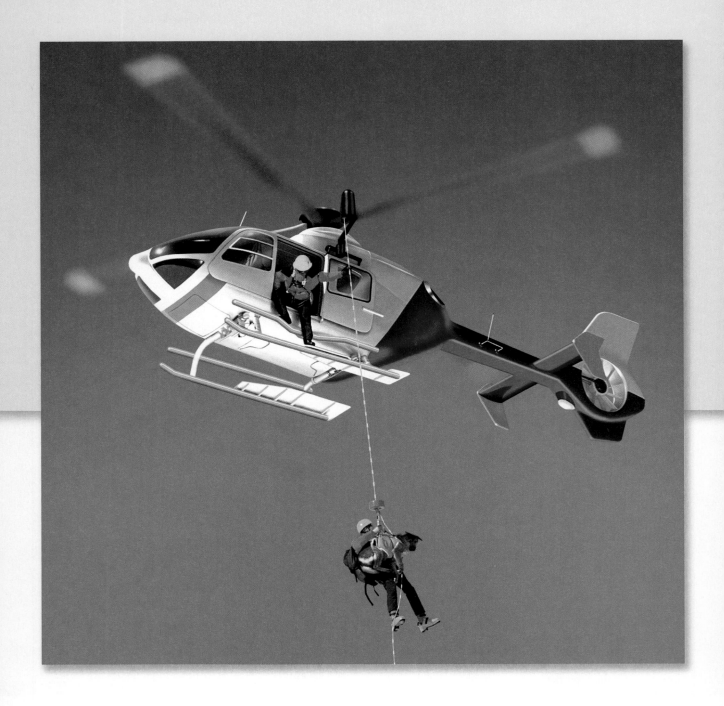

The lift caused by this rotating helicopter blade can be calculated using a steady-flow analysis that is based on the principles of impulse and momentum.

15

Kinetics of a Particle: Impulse and Momentum

CHAPTER OBJECTIVES

- To develop the principle of linear impulse and momentum for a particle.
- To study the conservation of linear momentum for particles.
- To analyze the mechanics of impact.
- To introduce the concept of angular impulse and momentum.
- To solve problems involving steady fluid streams and propulsion with variable mass.

15.1 Principle of Linear Impulse and Momentum

In this section we will integrate the equation of motion with respect to time and thereby obtain the principle of impulse and momentum. The resulting equation will be useful for solving problems involving force, velocity, and time.

The equation of motion for a particle of mass m can be written as

$$\Sigma \mathbf{F} = m\mathbf{a} = m\frac{d\mathbf{v}}{dt} \qquad (15-1)$$

where $\mathbf{a}$ and $\mathbf{v}$ are both measured from an inertial frame of reference. Rearranging the terms and integrating between the limits $\mathbf{v} = \mathbf{v}_1$ at $t = t_1$ and $\mathbf{v} = \mathbf{v}_2$ at $t = t_2$, we have

$$\Sigma \int_{t_1}^{t_2} \mathbf{F}\, dt = m \int_{\mathbf{v}_1}^{\mathbf{v}_2} d\mathbf{v}$$

The impulse tool is used to remove the dent in the automobile fender. To do so its end is first screwed into a hole drilled in the fender, then the weight is gripped and jerked upwards, striking the stop ring. The impulse developed is transferred along the shaft of the tool and pulls suddenly on the dent.

or

$$\Sigma \int_{t_1}^{t_2} \mathbf{F}\, dt = m\mathbf{v}_2 - m\mathbf{v}_1 \qquad (15\text{–}2)$$

This equation is referred to as the *principle of linear impulse and momentum*. From the derivation it can be seen that it is simply a time integration of the equation of motion. It provides a *direct means* of obtaining the particle's final velocity $\mathbf{v}_2$ after a specified time period when the particle's initial velocity is known and the forces acting on the particle are either constant or can be expressed as functions of time. By comparison, if $\mathbf{v}_2$ was determined using the equation of motion, a two-step process would be necessary; i.e., apply $\Sigma \mathbf{F} = m\mathbf{a}$ to obtain $\mathbf{a}$, then integrate $\mathbf{a} = d\mathbf{v}/dt$ to obtain $\mathbf{v}_2$.

Linear Momentum.   Each of the two vectors of the form $\mathbf{L} = m\mathbf{v}$ in Eq. 15–2 is referred to as the particle's linear momentum. Since m is a positive scalar, the linear-momentum vector has the same direction as $\mathbf{v}$, and its magnitude mv has units of mass–velocity, e.g., kg · m/s, or slug · ft/s.

Linear Impulse.   The integral $\mathbf{I} = \int \mathbf{F}\, dt$ in Eq. 15–2 is referred to as the *linear impulse.* This term is a vector quantity which measures the effect of a force during the time the force acts. Since time is a positive scalar, the impulse acts in the same direction as the force, and its magnitude has units of force–time, e.g., N · s or lb · s.* If the force is expressed as a function of time, the impulse may be determined by direct evaluation of the integral. In particular, the magnitude of the impulse $\mathbf{I} = \int_{t_1}^{t_2} \mathbf{F}\, dt$ can be represented experimentally by the shaded area under the curve of force versus time, Fig. 15–1. When the force is constant in both magnitude and direction, the resulting impulse becomes $\mathbf{I} = \int_{t_1}^{t_2} \mathbf{F}_c\, dt = \mathbf{F}_c(t_2 - t_1)$, which represents the shaded rectangular area shown in Fig. 15–2.

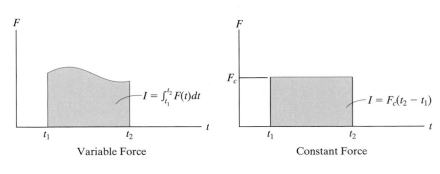

Variable Force

Constant Force

Fig. 15–1   **Fig. 15–2**

*Although the units for impulse and momentum are defined differently, it can be shown that Eq. 15–2 is dimensionally homogeneous.

Principle of Linear Impulse and Momentum. For problem solving, Eq. 15–2 will be rewritten in the form

$$mv_1 + \Sigma \int_{t_1}^{t_2} \mathbf{F}\, dt = mv_2 \qquad (15\text{–}3)$$

which states that the initial momentum of the particle at t_1 plus the sum of all the impulses applied to the particle from t_1 to t_2 is equivalent to the final momentum of the particle at t_2. These three terms are illustrated graphically on the *impulse and momentum diagrams* shown in Fig. 15–3. The two *momentum diagrams* are simply outlined shapes of the particle which indicate the direction and magnitude of the particle's initial and final momenta, mv_1 and mv_2, respectively, Fig. 15–3. Similar to the free-body diagram, the *impulse diagram* is an outlined shape of the particle showing all the impulses that act on the particle when it is located at some intermediate point along its path. In general, whenever the magnitude or direction of a force *varies* with time, the impulse is represented on the impulse diagram as $\int_{t_1}^{t_2}\mathbf{F}\, dt$. If the force is *constant*, the impulse applied to the particle is $\mathbf{F}_c(t_2 - t_1)$, and it acts in the same direction as $\mathbf{F}_c$.

Scalar Equations. If each of the vectors in Eq. 15–3 is resolved into its x, y, z components, we can write symbolically the following three scalar equations:

Fig. 15–3

$$m(v_x)_1 + \Sigma \int_{t_1}^{t_2} F_x\, dt = m(v_x)_2$$

$$m(v_y)_1 + \Sigma \int_{t_1}^{t_2} F_y\, dt = m(v_y)_2 \qquad (15\text{–}4)$$

$$m(v_z)_1 + \Sigma \int_{t_1}^{t_2} F_z\, dt = m(v_z)_2$$

These equations represent the principle of linear impulse and momentum for the particle in the x, y, z directions, respectively.

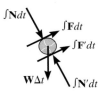

As the wheels of the pitching machine rotate, they apply frictional impulses to the ball, thereby giving it a linear momentum. These impulses are shown on the impulse diagram. Here both the frictional and normal impulses vary with time. By comparison, the weight impulse is constant and is very small since the time Δt the ball is in contact with the wheels is very small.

Procedure for Analysis

The principle of linear impulse and momentum is used to solve problems involving *force*, *time*, and *velocity*, since these terms are involved in the formulation. For application it is suggested that the following procedure be used.*

Free-Body Diagram

- Establish the *x, y, z* inertial frame of reference and draw the particle's free-body diagram in order to account for all the forces that produce impulses on the particle.

- The direction and sense of the particle's initial and final velocities should be established.

- If a vector is unknown, assume that the sense of its components is in the direction of the positive inertial coordinate(s).

- As an alternative procedure, draw the impulse and momentum diagrams for the particle as discussed in reference to Fig. 15–3.

Principle of Impulse and Momentum

- In accordance with the established coordinate system apply the principle of linear impulse and momentum, $m\mathbf{v}_1 + \Sigma \int_{t_1}^{t_2} \mathbf{F}\, dt = m\mathbf{v}_2$. If motion occurs in the *x–y* plane, the two scalar component equations can be formulated by either resolving the vector components of **F** from the free-body diagram, or by using the data on the impulse and momentum diagrams.

- Realize that every force acting on the particle's free-body diagram will create an impulse, even though some of these forces will do no work.

- Forces that are functions of time must be integrated to obtain the impulse. Graphically, the impulse is equal to the area under the force–time curve.

- If the problem involves the dependent motion of several particles, use the method outlined in Sec. 12.9 to relate their velocities. Make sure the positive coordinate directions used for writing these kinematic equations are the *same* as those used for writing the equations of impulse and momentum.

*This procedure will be followed when developing the proofs and theory in the text.

EXAMPLE 15.1

The 100-kg stone shown in Fig. 15–4a is originally at rest on the smooth horizontal surface. If a towing force of 200 N, acting at an angle of 45°, is applied to the stone for 10 s, determine the final velocity and the normal force which the surface exerts on the stone during the time interval.

(a)

SOLUTION

This problem can be solved using the principle of impulse and momentum since it involves force, velocity, and time.

Free-Body Diagram.   See Fig. 15–4b. Since all the forces acting are *constant*, the impulses are simply the product of the force magnitude and 10 s $[\mathbf{I} = \mathbf{F}_c(t_2 - t_1)]$. Note the alternative procedure of drawing the stone's impulse and momentum diagrams, Fig. 15–4c.

Principle of Impulse and Momentum.   Resolving the vectors in Fig. 15–4b along the x, y axes and applying Eqs. 15–4 yields

(b)

$$(\xrightarrow{+}) \qquad m(v_x)_1 + \Sigma \int_{t_1}^{t_2} F_x \, dt = m(v_x)_2$$

$$0 + 200 \text{ N}(10 \text{ s}) \cos 45° = (100 \text{ kg})v_2$$

$$v_2 = 14.1 \text{ m/s} \qquad\qquad Ans.$$

$$(+\uparrow) \qquad m(v_y)_1 + \Sigma \int_{t_1}^{t_2} F_y \, dt = m(v_y)_2$$

$$0 + N_C(10 \text{ s}) - 981 \text{ N}(10 \text{ s}) + 200 \text{ N}(10 \text{ s}) \sin 45° = 0$$

$$N_C = 840 \text{ N} \qquad\qquad Ans.$$

NOTE: Since no motion occurs in the y direction, direct application of the equilibrium equation $\Sigma F_y = 0$ gives the same result for N_C.

(c)

Fig. 15–4

EXAMPLE 15.2

$v_1 = 3$ ft/s

P   30°

(a)

50 lb

y   30°

v

x

$F = 0.3N_C$

$P = 20t$ N_C

(b)

Fig. 15–5

The 50-lb crate shown in Fig. 15–5a is acted upon by a force having a variable magnitude $P = (20t)$ lb, where t is in seconds. Determine the crate's velocity 2 s after **P** has been applied. The initial velocity is $v_1 = 3$ ft/s down the plane, and the coefficient of kinetic friction between the crate and the plane is $\mu_k = 0.3$.

SOLUTION

Free-Body Diagram.   See Fig. 15–5b. Since the magnitude of force $P = 20t$ varies with time, the impulse it creates must be determined by integrating over the 2-s time interval. The weight, normal force, and frictional force (which acts opposite to the direction of motion) are all *constant*, so that the impulse created by each of these forces is simply the magnitude of the force times 2 s.

Principle of Impulse and Momentum.   Applying Eqs. 15–4 in the x direction, we have

$$(+\swarrow) \qquad m(v_x)_1 + \Sigma \int_{t_1}^{t_2} F_x \, dt = m(v_x)_2$$

$$\frac{50 \text{ lb}}{32.2 \text{ ft/s}^2}(3 \text{ ft/s}) + \int_0^2 20t \, dt - 0.3N_C(2 \text{ s}) + (50 \text{ lb})(2 \text{ s}) \sin 30° = \frac{50 \text{ lb}}{32.2 \text{ ft/s}^2}v_2$$

$$4.66 + 40 - 0.6N_C + 50 = 1.55v_2$$

The equation of equilibrium can be applied in the y direction. Why?

$$+\nwarrow\Sigma F_y = 0; \qquad N_C - 50 \cos 30° \text{ lb} = 0$$

Solving,

$$N_C = 43.3 \text{ lb}$$

$$v_2 = 44.2 \text{ ft/s} \swarrow \qquad\qquad Ans.$$

NOTE: We can also solve this problem using the equation of motion. From Fig. 15–5b,

$$+\swarrow\Sigma F_x = ma_x; \quad 20t - 0.3(43.3) + 50 \sin 30° = \frac{50}{32.2}a$$

$$a = 12.88t + 7.734$$

Using kinematics

$$+\swarrow dv = a \, dt; \qquad \int_3^v dv = \int_0^2 (12.88t + 7.734) \, dt$$

$$v = 44.2 \text{ ft/s} \qquad\qquad Ans.$$

By comparison, application of the principle of impulse and momentum eliminates the need for using kinematics ($a = dv/dt$) and thereby yields an easier method for solution.

EXAMPLE 15.3

Blocks A and B shown in Fig. 15–6a have a mass of 3 kg and 5 kg, respectively. If the system is released from rest, determine the velocity of block B in 6 s. Neglect the mass of the pulleys and cord.

Datum

(a)

SOLUTION

Free-Body Diagram. See Fig. 15–6b. Since the weight of each block is constant, the cord tensions will also be constant. Furthermore, since the mass of pulley D is neglected, the cord tension $T_A = 2T_B$. Note that the blocks are both assumed to be traveling downward in the positive coordinate directions, s_A and s_B.

Principle of Impulse and Momentum.

Block A:

$$(+\downarrow) \qquad m(v_A)_1 + \Sigma \int_{t_1}^{t_2} F_y\, dt = m(v_A)_2$$

$$0 - 2T_B(6\text{ s}) + 3(9.81)\text{ N}(6\text{ s}) = (3\text{ kg})(v_A)_2 \qquad (1)$$

Block B:

$$(+\downarrow) \qquad m(v_B)_1 + \Sigma \int_{t_1}^{t_2} F_y\, dt = m(v_B)_2$$

$$0 + 5(9.81)\text{ N}(6\text{ s}) - T_B(6\text{ s}) = (5\text{ kg})(v_B)_2 \qquad (2)$$

Kinematics. Since the blocks are subjected to dependent motion, the velocity of A may be related to that of B by using the kinematic analysis discussed in Sec. 12.9. A horizontal datum is established through the fixed point at C, Fig. 15–6a, and the position coordinates, s_A and s_B, are related to the constant total length l of the vertical segments of the cord by the equation

$$2s_A + s_B = l$$

Taking the time derivative yields

$$2v_A = -v_B \qquad (3)$$

As indicated by the negative sign, when B moves downward A moves upward. Substituting this result into Eq. 1 and solving Eqs. 1 and 2 yields

$$(v_B)_2 = 35.8\text{ m/s} \downarrow$$

$$T_B = 19.2\text{ N} \qquad\qquad Ans.$$

(b)

NOTE: Realize that the *positive* (downward) direction for $\mathbf{v}_A$ and $\mathbf{v}_B$ is *consistent* in Figs. 15–6a and 15–6b and in Eqs. 1 to 3. This is important since we are seeking a simultaneous solution of equations.

Fig. 15–6

15.2   Principle of Linear Impulse and Momentum for a System of Particles

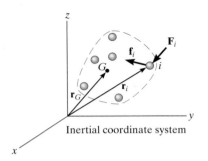

z

$\mathbf{F}_i$

$\mathbf{f}_i$

G

i

$\mathbf{r}_G$

$\mathbf{r}_i$

y

Inertial coordinate system

x

Fig. 15–7

The principle of linear impulse and momentum for a system of particles moving relative to an inertial reference, Fig. 15–7, is obtained from the equation of motion applied to all the particles in the system, i.e.,

$$\Sigma \mathbf{F}_i = \Sigma m_i \frac{d\mathbf{v}_i}{dt} \qquad (15\text{–}5)$$

The term on the left side represents only the sum of the *external forces* acting on the system of particles. Recall that the internal forces $\mathbf{f}_i$ acting between particles do not appear with this summation, since by Newton's third law they occur in equal but opposite collinear pairs and therefore cancel out. Multiplying both sides of Eq. 15–5 by dt and integrating between the limits $t = t_1$, $\mathbf{v}_i = (\mathbf{v}_i)_1$ and $t = t_2$, $\mathbf{v}_i = (\mathbf{v}_i)_2$ yields

$$\Sigma m_i(\mathbf{v}_i)_1 + \Sigma \int_{t_1}^{t_2} \mathbf{F}_i \, dt = \Sigma m_i(\mathbf{v}_i)_2 \qquad (15\text{–}6)$$

This equation states that the initial linear momenta of the system plus the impulses of all the *external forces* acting on the system from t_1 to t_2 are equal to the system's final linear momenta.

Since the location of the mass center G of the system is determined from $m\mathbf{r}_G = \Sigma m_i \mathbf{r}_i$, where $m = \Sigma m_i$ is the total mass of all the particles, Fig. 15–7, then taking the time derivatives, we have

$$m\mathbf{v}_G = \Sigma m_i \mathbf{v}_i$$

which states that the total linear momentum of the system of particles is equivalent to the linear momentum of a "fictitious" aggregate particle of mass $m = \Sigma m_i$ moving with the velocity of the mass center of the system. Substituting into Eq. 15–6 yields

$$m(\mathbf{v}_G)_1 + \Sigma \int_{t_1}^{t_2} \mathbf{F}_i \, dt = m(\mathbf{v}_G)_2 \qquad (15\text{–}7)$$

Here the initial linear momentum of the aggregate particle plus the external impulses acting on the system of particles from t_1 to t_2 is equal to the aggregate particle's final linear momentum. Since in reality all particles must have nonzero size to possess mass, the above equation justifies application of the principle of linear impulse and momentum to a rigid body represented as a single particle.

PROBLEMS

15–1. A 20-lb block slides down a 30° inclined plane with an initial velocity of 2 ft/s. Determine the velocity of the block in 3 s if the coefficient of kinetic friction between the block and the plane is $\mu_k = 0.25$.

15–2. A 2-lb ball is thrown in the direction shown with an initial speed $v_A = 18$ ft/s. Determine the time needed for it to reach its highest point B and the speed at which it is traveling at B. Use the principle of impulse and momentum for the solution.

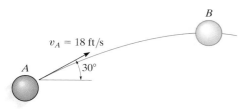

Prob. 15–2

15–3. A 5-lb block is given an initial velocity of 10 ft/s up a 45° smooth slope. Determine the time it will take to travel up the slope before it stops.

*15–4. The baseball has a horizontal speed of 35 m/s when it is struck by the bat B. If it then travels away at an angle of 60° from the horizontal and reaches a maximum height of 50 m, measured from the height of the bat, determine the magnitude of the net impulse of the bat on the ball. The ball has a mass of 400 g. Neglect the weight of the ball during the time the bat strikes the ball.

Prob. 15–4

15–5. The choice of a seating material for moving vehicles depends upon its ability to resist shock and vibration. From the data shown in the graphs, determine the impulses created by a falling weight onto a sample of urethane foam and CONFOR foam.

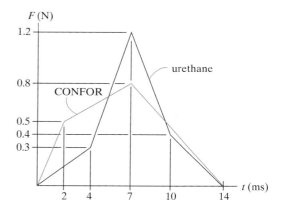

Prob. 15–5

15–6. A man hits the 50-g golf ball such that it leaves the tee at an angle of 40° with the horizontal and strikes the ground at the same elevation a distance of 20 m away. Determine the impulse of the club C on the ball. Neglect the impulse caused by the ball's weight while the club is striking the ball.

Prob. 15–6

15–7. A solid-fueled rocket can be made using a fuel grain with either a hole (a), or starred cavity (b), in the cross section. From experiment the engine thrust-time curves (*T* vs. *t*) for the same amount of propellant using these geometries are shown. Determine the total impulse in both cases.

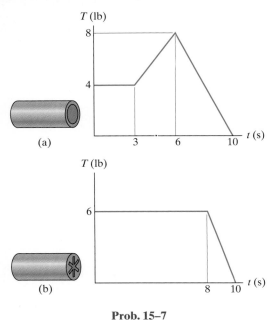

(a)

(b)

Prob. 15–7

15–9. The jet plane has a mass of 250 Mg and a horizontal velocity of 100 m/s when $t = 0$. If *both* engines provide a horizontal thrust which varies as shown in the graph, determine the plane's velocity in $t = 15$ s. Neglect air resistance and the loss of fuel during the motion.

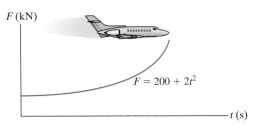

$F = 200 + 2t^2$

Prob. 15–9

***15–8.** During operation the breaker hammer develops on the concrete surface a force which is indicated in the graph. To achieve this the 2-lb spike *S* is fired from rest into the surface at 200 ft/s. Determine the speed of the spike just after rebounding.

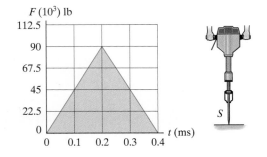

Prob. 15–8

15–10. A man kicks the 200-g ball such that it leaves the ground at an angle of 30° with the horizontal and strikes the ground at the same elevation a distance of 15 m away. Determine the impulse of his foot *F* on the ball. Neglect the impulse caused by the ball's weight while its being kicked.

Prob. 15–10

15–11. The particle P is acted upon by its weight of 3 lb and forces $\mathbf{F}_1$ and $\mathbf{F}_2$, where t is in seconds. If the particle orginally has a velocity of $\mathbf{v}_1 = \{3\mathbf{i} + 1\mathbf{j} + 6\mathbf{k}\}$ ft/s, determine its speed after 2 s.

15–13. From experiments, the time variation of the vertical force on a runner's foot as he strikes and pushes off the ground is shown in the graph. These results are reported for a 1-lb *static* load, i.e., in terms of unit weight. If a runner weighs 175 lb, determine the approximate vertical impulse he exerts on the ground if the impulse occurs in 210 ms.

Prob. 15–11

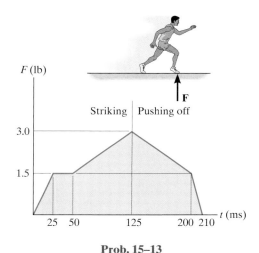

Prob. 15–13

*15–12. The twitch in a muscle of the arm develops a force which can be measured as a function of time as shown in the graph. If the effective contraction of the muscle lasts for a time t_0, determine the impulse developed by the muscle.

15–14. As indicated by the derivation, the principle of impulse and momentum is valid for observers in *any* inertial reference frame. Show that this is so, by considering the 10-kg block which rests on the smooth surface and is subjected to a horizontal force of 6 N. If observer A is in a *fixed* frame x, determine the final speed of the block in 4 s if it has an initial speed of 5 m/s measured from the fixed frame. Compare the result with that obtained by an observer B, attached to the x' axis that moves at a constant velocity of 2 m/s relative to A.

Prob. 15–12

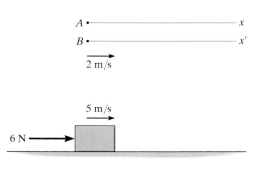

Prob. 15–14

15–15. The 50-lb cabinet is subjected to the force $F = 20(t + 1)$ lb where t is in seconds. If the cabinet is initially moving down the plane with a velocity of 10 ft/s, determine how long it will take before the cabinet comes to a stop. **F** always acts parallel to the plane. Neglect the size of the rollers.

Prob. 15–15

15–17. When the 0.4-lb ball is fired, it leaves the ground at an angle of 40° from the horizontal and strikes the ground at the same elevation a distance of 130 ft away. Determine the impulse given to the ball.

Prob. 15–17

*15–16. If it takes 35 s for the 50-Mg tugboat to increase its speed uniformly to 25 km/h, starting from rest, determine the force of the rope on the tugboat. The propeller provides the propulsion force **F** which gives the tugboat forward motion, whereas the barge moves freely. Also, determine F acting on the tugboat. The barge has a mass of 75 Mg.

Prob. 15–16

15–18. The uniform beam has a weight of 5000 lb. Determine the average tension in each of the two cables AB and AC if the beam is given an upward speed of 8 ft/s in 1.5 s starting from rest. Neglect the mass of the cables.

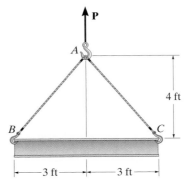

Prob. 15–18

15–19. The 5-kg block is moving downward at $v_1 = 2$ m/s when it is 8 m from the sandy surface. Determine the impulse of the sand on the block necessary to stop its motion. Neglect the distance the block dents into the sand and assume the block does not rebound. Neglect the weight of the block during the impact with the sand.

*15–20. The 5-kg block is falling downward at $v_1 = 2$ m/s when it is 8 m from the sandy surface. Determine the average impulsive force acting on the block by the sand if the motion of the block is stopped in 0.9 s once the block strikes the sand. Neglect the distance the block dents into the sand and assume the block does not rebound. Neglect the weight of the block during the impact with the sand.

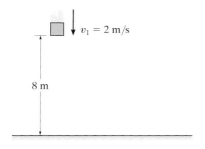

Probs. 15–19/20

15–21. A 50-kg crate rests against a stop block s, which prevents the crate from moving down the plane. If the coefficients of static and kinetic friction between the plane and the crate are $\mu_s = 0.3$ and $\mu_k = 0.2$, respectively, determine the time needed for the force $\mathbf{F}$ to give the crate a speed of 2 m/s up the plane. The force always acts parallel to the plane and has a magnitude of $F = (300t)$ N, where t is in seconds. *Hint:* First determine the time needed to overcome static friction and start the crate moving.

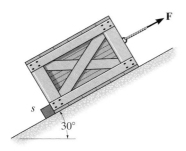

Prob. 15–21

15–22. The 2-lb block has an initial velocity of $v_1 = 10$ ft/s in the direction shown. If a force of $\mathbf{F} = \{0.5\mathbf{i} + 0.2\mathbf{j}\}$ lb acts on the block for $t = 5$ s, determine the final speed of the block. Neglect friction.

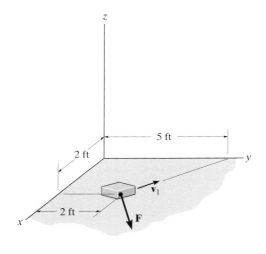

Prob. 15–22

15–23. The tennis ball has a horizontal speed of 15 m/s when it is struck by the racket. If it then travels away at an angle of 25° from the horizontal and reaches a maximum altitude of 10 m, measured from the height of the racket, determine the magnitude of the net impulse of the racket on the ball. The ball has a mass of 180 g. Neglect the weight of the ball during the time the racket strikes the ball.

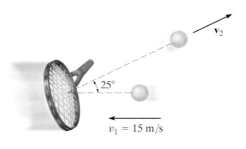

Prob. 15–23

*15–24.   The 40-kg slider block is moving to the right with a speed of 1.5 m/s when it is acted upon by the forces $\mathbf{F}_1$ and $\mathbf{F}_2$. If these loadings vary in the manner shown on the graph, determine the speed of the block at $t = 6$ s. Neglect friction and the mass of the pulleys and cords.

15–26.   The 5-kg package is released from rest at A. It slides down the smooth plane which is inclined at $30°$, onto the rough surface having a coefficient of kinetic friction of $\mu_k = 0.2$. Determine the total time of travel before the package stops sliding. Neglect the size of the package.

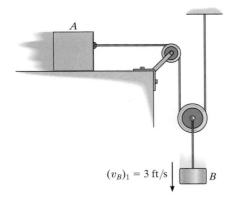

F (N) graph:

Prob. 15–24

Prob. 15–26

15–25.   Determine the velocities of blocks A and B 2 s after they are released from rest. Neglect the mass of the pulleys and cables.

15–27.   Block A weighs 10 lb and block B weighs 3 lb. If B is moving downward with a velocity $(v_B)_1 = 3$ ft/s at $t = 0$, determine the velocity of A when $t = 1$ s. Assume that the horizontal plane is smooth. Neglect the mass of the pulleys and cords.

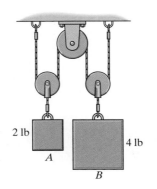

Prob. 15–25

Prob. 15–27

*15–28. Block A weighs 10 lb and block B weighs 3 lb. If B is moving downward with a velocity $(v_B)_1 = 3$ ft/s at $t = 0$, determine the velocity of A when $t = 1$ s. The coefficient of kinetic friction between the horizontal plane and block A is $\mu_A = 0.15$.

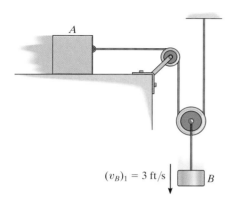

$(v_B)_1 = 3$ ft/s

Prob. 15–28

15–29. A jet plane having a mass of 7 Mg takes off from an aircraft carrier such that the engine thrust varies as shown by the graph. If the carrier is traveling forward with a speed of 40 km/h, determine the plane's airspeed after 5 s.

Prob. 15–29

15–30. The motor pulls on the cable at A with a force $F = (30 + t^2)$ lb, where t is in seconds. If the 17-lb crate is originally at rest at $t = 0$, determine its speed in $t = 4$ s. Neglect the mass of the cable and pulleys. *Hint:* First find the time needed to begin lifting the crate.

Prob. 15–30

15–31. The log has a mass of 500 kg and rests on the ground for which the coefficients of static and kinetic friction are $\mu_s = 0.5$ and $\mu_k = 0.4$, respectively. The winch delivers a horizontal towing force T to its cable at A which varies as shown in the graph. Determine the speed of the log when $t = 5$ s. Originally the tension in the cable is zero. *Hint:* First determine the force needed to begin moving the log.

Prob. 15–31

15.3 Conservation of Linear Momentum for a System of Particles

When the sum of the *external impulses* acting on a system of particles is *zero*, Eq. 15–6 reduces to a simplified form, namely,

$$\Sigma m_i(\mathbf{v}_i)_1 = \Sigma m_i(\mathbf{v}_i)_2 \qquad (15\text{–}8)$$

This equation is referred to as the *conservation of linear momentum*. It states that the total linear momentum for a system of particles remains constant during the time period t_1 to t_2. Substituting $m\mathbf{v}_G = \Sigma m_i\mathbf{v}_i$ into Eq. 15–8, we can also write

$$(\mathbf{v}_G)_1 = (\mathbf{v}_G)_2 \qquad (15\text{–}9)$$

which indicates that the velocity $\mathbf{v}_G$ of the mass center for the system of particles does not change when no external impulses are applied to the system.

The conservation of linear momentum is often applied when particles collide or interact. For application, a careful study of the free-body diagram for the *entire* system of particles should be made in order to identify the forces which create either external or internal impulses and thereby determine in what direction(s) linear momentum is conserved. As stated earlier, the *internal impulses* for the system will always cancel out, since they occur in equal but opposite collinear pairs. If the time period over which the motion is studied is *very short*, some of the external impulses may also be neglected or considered approximately equal to zero. The forces causing these negligible impulses are called *nonimpulsive forces*. By comparison, forces which are very large and act for a very short period of time produce a significant change in momentum and are called *impulsive forces*. They, of course, cannot be neglected in the impulse–momentum analysis.

Impulsive forces normally occur due to an explosion or the striking of one body against another, whereas nonimpulsive forces may include the weight of a body, the force imparted by a slightly deformed spring having a relatively small stiffness, or for that matter, any force that is very small compared to other larger (impulsive) forces. When making this distinction between impulsive and nonimpulsive forces, it is important to realize that this only applies during the time t_1 to t_2. To illustrate, consider the effect of striking a tennis ball with a racket as shown in the photo. During the *very short* time of interaction, the force of the racket on the ball is impulsive since it changes the ball's momentum drastically. By comparison, the ball's weight will have a negligible effect on the

The hammer in the top photo applies an impulsive force to the stake. During this extremely short time of contact the weight of the stake can be considered nonimpulsive, and provided the stake is driven into soft ground, the impulse of the ground acting on the stake can also be considered nonimpulsive. By contrast, if the stake is used in a concrete chipper to break concrete, then two impulsive forces act on the stake: one at its top due to the chipper and the other on its bottom due to the rigidity of the concrete.

change in momentum, and therefore it is nonimpulsive. Consequently, it can be neglected from an impulse–momentum analysis during this time. If an impulse–momentum analysis is considered during the much longer time of flight after the racket–ball interaction, then the impulse of the ball's weight is important since it, along with air resistance, causes the change in the momentum of the ball.

Procedure for Analysis

Generally, the principle of linear impulse and momentum or the conservation of linear momentum is applied to a *system of particles* in order to determine the final velocities of the particles *just after* the time period considered. By applying these equations to the entire system, the internal impulses acting within the system, which may be unknown, are *eliminated* from the analysis. For application it is suggested that the following procedure be used.

Free-Body Diagram.

- Establish the x, y, z inertial frame of reference and draw the free-body diagram for each particle of the system in order to identify the internal and external forces.

- The conservation of linear momentum applies to the system in a given direction when no external forces or if nonimpulsive forces act on the system in that direction.

- Establish the direction and sense of the particles' initial and final velocities. If the sense is unknown, assume it is along a positive inertial coordinate axis.

- As an alternative procedure, draw the impulse and momentum diagrams for each particle of the system.

Momentum Equations.

- Apply the principle of linear impulse and momentum or the conservation of linear momentum in the appropriate directions.

- If it is necessary to determine the *internal impulse* $\int F \, dt$ acting on only one particle of a system, then the particle must be *isolated* (free-body diagram), and the principle of linear impulse and momentum must be applied *to the particle*.

- After the impulse is calculated, and provided the time Δt for which the impulse acts is known, then the *average impulsive force* F_{avg} can be determined from $F_{avg} = \int F \, dt / \Delta t$.

EXAMPLE 15.4

The 15-Mg boxcar A is coasting at 1.5 m/s on the horizontal track when it encounters a 12-Mg tank car B coasting at 0.75 m/s toward it as shown in Fig. 15–8a. If the cars meet and couple together, determine (a) the speed of both cars just after the coupling, and (b) the average force between them if the coupling takes place in 0.8 s.

(a)

SOLUTION

Part (a) Free-Body Diagram.* Here we have considered *both* cars as a single system, Fig. 15–8b. By inspection, momentum is conserved in the x direction since the coupling force **F** is *internal* to the system and will therefore cancel out. It is assumed both cars, when coupled, move at v_2 in the positive x direction.

Conservation of Linear Momentum.

$(\xrightarrow{+})$ $\qquad m_A(v_A)_1 + m_B(v_B)_1 = (m_A + m_B)v_2$

$\qquad (15\,000\text{ kg})(1.5\text{ m/s}) - 12\,000\text{ kg}(0.75\text{ m/s}) = (27\,000\text{ kg})v_2$

$\qquad\qquad\qquad v_2 = 0.5\text{ m/s} \rightarrow$ $\qquad\qquad\qquad$ *Ans.*

Part (b). The average (impulsive) coupling force, $\mathbf{F}_{avg}$, can be determined by applying the principle of linear momentum to *either one* of the cars.

Free-Body Diagram. As shown in Fig. 15–8c, by isolating the boxcar the coupling force is *external* to the car.

Principle of Impulse and Momentum. Since $\int F\, dt = F_{avg}\,\Delta t = F_{avg}(0.8)$, we have

$(\xrightarrow{+})$ $\qquad\qquad m_A(v_A)_1 + \Sigma \int F\, dt = m_A v_2$

$\qquad (15\,000\text{ kg})(1.5\text{ m/s}) - F_{avg}(0.8\text{ s}) = (15\,000\text{ kg})(0.5\text{ m/s})$

$\qquad\qquad\qquad F_{avg} = 18.8\text{ kN}$ $\qquad\qquad\qquad$ *Ans.*

NOTE: Solution was possible here since the boxcar's final velocity was obtained in Part (a). Try solving for F_{avg} by applying the principle of impulse and momentum to the tank car.

*Only horizontal forces are shown on the free-body diagram.

(b)

(c)

Fig. 15–8

EXAMPLE 15.5

The 1200-lb cannon shown in Fig. 15–9a fires an 8-lb projectile with a muzzle velocity of 1500 ft/s relative to the ground. If firing takes place in 0.03 s, determine (a) the recoil velocity of the cannon just after firing, and (b) the average impulsive force acting on the projectile. The cannon support is fixed to the ground, and the horizontal recoil of the cannon is absorbed by two springs.

(a)

SOLUTION

Part (a) Free-Body Diagram.* As shown in Fig. 15–9b, we have considered the projectile and cannon as a single system, since the impulsive forces, **F**, between the cannon and projectile are *internal* to the system and will therefore cancel from the analysis. Furthermore, during the time $\Delta t = 0.03$ s, the two recoil springs which are attached to the support each exert a *nonimpulsive force* $\mathbf{F}_s$ on the cannon. This is because Δt is very short, so that during this time the cannon only moves through a very small distance s. Consequently, $F_s = ks \approx 0$, where k is the spring's stiffness. Hence it may be concluded that momentum for the system is conserved in the *horizontal direction*.

(b)

Conservation of Linear Momentum.

$$(\overset{+}{\rightarrow})\quad m_c(v_c)_1 + m_p(v_p)_1 = -m_c(v_c)_2 + m_p(v_p)_2$$

$$0 + 0 = -\frac{1200\ \text{lb}}{32.2\ \text{ft/s}^2}(v_c)_2 + \frac{8\ \text{lb}}{32.2\ \text{ft/s}^2}(1500\ \text{ft/s})$$

$$(v_c)_2 = 10\ \text{ft/s} \leftarrow \qquad\qquad Ans.$$

Part (b). The average impulsive force exerted by the cannon on the projectile can be determined by applying the principle of linear impulse and momentum to the projectile (or to the cannon). Why?

Principle of Impulse and Momentum. Using the data in Fig. 15–9c, noting that $\int F\, dt = F_{\text{avg}}\, \Delta t = F_{\text{avg}}(0.03)$, we have

$$(\overset{+}{\rightarrow})\qquad m(v_p)_1 + \Sigma \int F\, dt = m(v_p)_2$$

$$0 + F_{\text{avg}}(0.03\ \text{s}) = \frac{8\ \text{lb}}{32.2\ \text{ft/s}^2}(1500\ \text{ft/s})$$

$$F_{\text{avg}} = 12.4(10^3)\ \text{lb} = 12.4\ \text{kip} \qquad\qquad Ans.$$

(c)

Fig. 15–9

NOTE: If the cannon is firmly fixed to its support (no springs), the reactive force of the support on the cannon must be considered as an external impulse to the system, since the support would allow no movement of the cannon.

---

*Only horizontal forces are shown on the free-body diagram.

234 CHAPTER 15 KINETICS OF A PARTICLE: IMPULSE AND MOMENTUM

EXAMPLE 15.6

The 350-Mg tugboat T shown in Fig. 15–10a is used to pull the 50-Mg barge B with a rope R. If the barge is initially at rest and the tugboat is coasting freely with a velocity of $(v_T)_1 = 3$ m/s while the rope is *slack*, determine the velocity of the tugboat *directly after* the rope becomes taut. Assume the rope does not stretch. Neglect the frictional effects of the water.

(a)   (b)

SOLUTION

Free-Body Diagram.* As shown in Fig. 15–10b, we have considered the entire system (tugboat and barge). Hence, the impulsive force created between the tugboat and the barge is *internal* to the system, and therefore momentum of the system is conserved during the instant of towing.

The alternative procedure of drawing the system's impulse and momentum diagrams is shown in Fig. 15–10c.

Conservation of Momentum. Noting that $(v_B)_2 = (v_T)_2$, we have

$$(\xrightarrow{+}) \qquad m_T(v_T)_1 + m_B(v_B)_1 = m_T(v_T)_2 + m_B(v_B)_2$$

$$350(10^3) \text{ kg}(3 \text{ m/s}) + 0 = 350(10^3) \text{ kg}(v_T)_2 + 50(10^3) \text{ kg}(v_T)_2$$

Solving,

$$(v_T)_2 = 2.62 \text{ m/s} \leftarrow \qquad \qquad Ans.$$

NOTE: This value represents the tugboat's velocity *just after* the towing impulse. Use this result and show that the towing *impulse* is 131 kN · s.

(c)

Fig. 15–10

*Only horizontal forces are shown on the free-body diagram.

EXAMPLE 15.7

An 800-kg rigid pile P shown in Fig. 15–11a is driven into the ground using a 300-kg hammer H. The hammer falls from rest at a height $y_0 = 0.5$ m and strikes the top of the pile. Determine the impulse which the hammer imparts on the pile if the pile is surrounded entirely by loose sand so that after striking, the hammer does *not* rebound off the pile.

SOLUTION

Conservation of Energy.   The velocity at which the hammer strikes the pile can be determined using the conservation of energy equation applied to the hammer. With the datum at the top of the pile, Fig. 15–11a, we have

$$T_0 + V_0 = T_1 + V_1$$

$$\tfrac{1}{2}m_H(v_H)_0^2 + W_H y_0 = \tfrac{1}{2}m_H(v_H)_1^2 + W_H y_1$$

$$0 + 300(9.81)\ \text{N}(0.5\ \text{m}) = \frac{1}{2}(300\ \text{kg})(v_H)_1^2 + 0$$

$$(v_H)_1 = 3.13\ \text{m/s}$$

(a)

Free-Body Diagram.   From the physical aspects of the problem, the free-body diagram of the hammer and pile, Fig. 15–11b, indicates that during the *short time* occurring *just before* to *just after* the *collision*, the weights of the hammer and pile and the resistance force $\mathbf{F}_s$ of the sand are all *nonimpulsive*. The impulsive force $\mathbf{R}$ is internal to the system and therefore cancels. Consequently, momentum is conserved in the vertical direction during this short time.

Conservation of Momentum.   Since the hammer does not rebound off the pile just after collision, then $(v_H)_2 = (v_P)_2 = v_2$.

$$(+\downarrow)\qquad m_H(v_H)_1 + m_P(v_P)_1 = m_H v_2 + m_P v_2$$

$$(300\ \text{kg})(3.13\ \text{m/s}) + 0 = (300\ \text{kg})v_2 + (800\ \text{kg})v_2$$

$$v_2 = 0.854\ \text{m/s}$$

(b)

Principle of Impulse and Momentum.   The impulse which the pile imparts to the hammer can now be determined since $\mathbf{v}_2$ is known. From the free-body diagram for the hammer, Fig. 15–11c, we have

$$(+\downarrow)\qquad m_H(v_H)_1 + \Sigma \int_{t_1}^{t_2} F_y\, dt = m_H v_2$$

$$(300\ \text{kg})(3.13\ \text{m/s}) - \int R\, dt = (300\ \text{kg})(0.854\ \text{m/s})$$

$$\int R\, dt = 683\ \text{N} \cdot \text{s} \qquad\qquad Ans.$$

(c)

NOTE: The equal but opposite impulse acts on the pile. Try finding this impulse by applying the principle of impulse and momentum to the pile.

Fig. 15–11

EXAMPLE 15.8

(a)

(b)

Initial position

Final position

(c)

Fig. 15–12

A boy having a mass of 40 kg stands on the back of a 15-kg toboggan which is originally at rest, Fig. 15–12a. If he walks to the front B and stops, determine the distance the toboggan moves. Neglect friction between the bottom of the toboggan and the ground (ice).

SOLUTION

Free-Body Diagram.   The unknown frictional force of the boy's shoes on the bottom of the toboggan can be *excluded* from the analysis if the toboggan and boy on it are considered as a single system. In this way the frictional force **F** becomes *internal* and the conservation of momentum applies, Fig. 15–12b.

Conservation of Momentum.   Since both the initial and final momenta of the system are zero (because the initial and final velocities are zero), the system's momentum must also be zero when the boy is at some intermediate point between A and B. Thus

$$(\xrightarrow{+}) \qquad\qquad -m_b v_b + m_t v_t = 0 \qquad\qquad (1)$$

Here the two unknowns v_b and v_t represent the velocities of the boy moving to the left and the toboggan moving to the right. Both are measured from a *fixed inertial reference* on the ground.

At any instant the *position* of point A on the toboggan and the *position* of the boy must be determined by integration. Since $v = ds/dt$, then $-m_b ds_b + m_t ds_t = 0$. Assuming the initial position of point A to be at the origin, Fig. 15–12c, then at the final position we have $-m_b s_b + m_t s_t = 0$. Since $s_b + s_t = 2$ m, or $s_b = (2 - s_t)$ then

$$-m_b(2 - s_t) + m_t s_t = 0 \qquad\qquad (2)$$

$$s_t = \frac{2m_b}{m_b + m_t} = \frac{2(40)}{40 + 15} = 1.45 \text{ m} \qquad\qquad Ans.$$

NOTE: The problem may also be solved by considering the relative motion of the boy with respect to the toboggan, $\mathbf{v}_{b/t}$. This velocity is related to the velocities of the boy and toboggan by the equation $\mathbf{v}_b = \mathbf{v}_t + \mathbf{v}_{b/t}$, Eq. 12–34. Since positive motion is assumed to be to the right in Eq. 1, $\mathbf{v}_b$ and $\mathbf{v}_{b/t}$ are negative, because the boy's motion is to the left. Hence, in scalar form, $-v_b = v_t - v_{b/t}$, and Eq. 1 then becomes $m_b(v_t - v_{b/t}) + m_t v_t = 0$.
Integrating gives

$$m_b(s_t - s_{b/t}) + m_t s_t = 0$$

Realizing $s_{b/t} = 2$ m, we obtain Eq. 2.

PROBLEMS

*15–32.   A railroad car having a mass of 15 Mg is coasting at 1.5 m/s on a horizontal track. At the same time another car having a mass of 12 Mg is coasting at 0.75 m/s in the opposite direction. If the cars meet and couple together, determine the speed of both cars just after the coupling. Find the difference between the total kinetic energy before and after coupling has occurred, and explain qualitatively what happened to this energy.

15–33.   The car A has a weight of 4500 lb and is traveling to the right at 3 ft/s. Meanwhile a 3000-lb car B is traveling at 6 ft/s to the left. If the cars crash head-on and become entangled, determine their common velocity just after the collision. Assume that the brakes are not applied during collision.

Prob. 15–33

15–34.   The bus B has a weight of 15000 lb and is traveling to the right at 5 ft/s. Meanwhile a 3000-lb car A is traveling at 4 ft/s to the left. If the vehicles crash head-on and become entangled, determine their common velocity just after the collision. Assume that the vehicles are free to roll during collision.

Prob. 15–34

15–35.   The cart has a mass of 3 kg and rolls freely down the slope. When it reaches the bottom, a spring loaded gun fires a 0.5-kg ball out the back with a horizontal velocity of $v_{b/c} = 0.6$ m/s, measured relative to the cart. Determine the final velocity of the cart.

Prob. 15–35

*15–36.   Two men A and B, each having a weight of 160 lb, stand on the 200-lb cart. Each runs with a speed of 3 ft/s measured relative to the cart. Determine the final speed of the cart if (a) A runs and jumps off, then B runs and jumps off the same end, and (b) both run at the same time and jump off at the same time. Neglect the mass of the wheels and assume the jumps are made horizontally.

Prob. 15–36

15–37. A 40-lb box slides from rest down the smooth ramp onto the surface of a 20-lb cart. Determine the speed of the box at the instant it stops sliding on the cart. If someone ties the cart to the ramp at B, determine the horizontal impulse the box will exert at C in order to stop its motion. Neglect friction and the size of the box.

15–39. The barge B weighs 30 000 lb and supports an automobile weighing 3000 lb. If the barge is not tied to the pier P and someone drives the automobile to the other side of the barge for unloading, determine how far the barge moves away from the pier. Neglect the resistance of the water.

Prob. 15–39

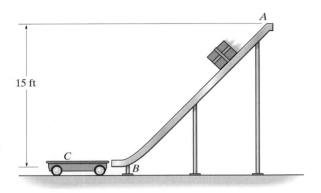

Prob. 15–37

15–38. A 100-lb boy walks forward over the surface of the 60-lb cart with a constant speed of 3 ft/s relative to the cart. Determine the cart's speed and its displacement at the moment he is about to step off. Neglect the mass of the wheels and assume the cart and boy are originally at rest.

*15–40. A 0.03-lb bullet traveling at 1300 ft/s strikes the 10-lb wooden block and exits the other side at 50 ft/s as shown. Determine the speed of the block just after the bullet exits the block, and also determine how far the block slides before it stops. The coefficient of kinetic friction between the block and the surface is $\mu_k = 0.5$.

15–41. A 0.03-lb bullet traveling at 1300 ft/s strikes the 10-lb wooden block and exits the other side at 50 ft/s as shown. Determine the speed of the block just after the bullet exits the block. Also, determine the average normal force on the block if the bullet passes through it in 1 ms, and the time the block slides before it stops. The coefficient of kinetic friction between the block and the surface is $\mu_k = 0.5$.

Prob. 15–38

Probs. 15–40/41

15–42. The man M weighs 150 lb and jumps onto the boat B which has a weight of 200 lb. If he has a horizontal component of velocity *relative to the boat* of 3 ft/s, just before he enters the boat, and the boat is traveling $v_B = 2$ ft/s away from the pier when he makes the jump, determine the resulting velocity of the man and boat.

15–43. The man M weighs 150 lb and jumps onto the boat B which is originally at rest. If he has a horizontal component of velocity of 3 ft/s just before he enters the boat, determine the weight of the boat if it has a velocity of 2 ft/s once the man enters it.

15–45. The 10-lb projectile is fired from ground level with an initial velocity of $v_A = 80$ ft/s in the direction shown. When it reaches its highest point B it explodes into two 5-lb fragments. If one fragment travels vertically upward at 12 ft/s, determine the distance between the fragments after they strike the ground. Neglect the size of the gun.

Probs. 15–42/43

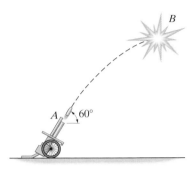

Prob. 15–45

*15–44.** A boy A having a weight of 80 lb and a girl B having a weight of 65 lb stand motionless at the ends of the toboggan, which has a weight of 20 lb. If A walks to B and stops, and both walk back together to the original position of A, determine the final position of the toboggan just after the motion stops. Neglect friction.

15–46. Two boxes A and B, each having a weight of 160 lb, sit on the 500-lb conveyor which is free to roll on the ground. If the belt starts from rest and begins to run with a speed of 3 ft/s, determine the final speed of the conveyor if (a) the boxes are not stacked and A falls off then B falls off, and (b) A is stacked on top of B and both fall off together.

Prob. 15–44

Prob. 15–46

15–47.   The winch on the back of the jeep A is turned on and pulls in the tow rope at $v_{rel} = 2$ m/s. If both the 1.25-Mg car B and the 2.5-Mg jeep A are free to roll, determine their velocities at the instant they meet. If the rope is 5 m long, how long will this take?

15–49.   A tugboat T having a mass of 19 Mg is tied to a barge B having a mass of 75 Mg. If the rope is "elastic" such that it has a stiffness $k = 600$ kN/m, determine the maximum stretch in the rope during the initial towing. Originally both the tugboat and barge are moving in the same direction with speeds $(v_T)_1 = 15$ km/h and $(v_B)_1 = 10$ km/h, respectively. Neglect the resistance of the water.

Prob. 15–47

Prob. 15–49

*15–48.   The 10-kg block is held at rest on the smooth inclined plane by the stop block at A. If the 10-g bullet is traveling at 300 m/s when it becomes embedded in the 10-kg block, determine the distance the block will slide up along the plane before momentarily stopping.

15–50.   The free-rolling ramp has a weight of 120 lb. The crate, whose weight is 80 lb, slides from rest at A, 15 ft down the ramp to B. Determine the ramp's speed when the crate reaches B. Assume that the ramp is smooth, and neglect the mass of the wheels.

Prob. 15–48

Prob. 15–50

15–51. The free-rolling ramp has a weight of 120 lb. If the 80-lb crate is released from rest at A, determine the distance the ramp moves when the crate slides 15 ft down the ramp and reaches the bottom B.

Prob. 15–51

15–53. The free-rolling ramp has a mass of 40 kg. A 10-kg crate is released from rest at A and slides down 3.5 m to point B. If the surface of the ramp is smooth, determine the ramp's speed when the crate reaches B. Also, what is the velocity of the crate?

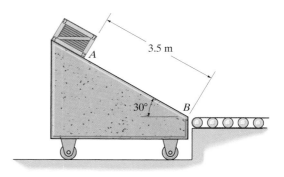

Prob. 15–53

*****15–52.** The boy B jumps off the canoe at A with a velocity of 5 m/s relative to the canoe as shown. If he lands in the second canoe C, determine the final speed of both canoes after the motion. Each canoe has a mass of 40 kg. The boy's mass is 30 kg, and the girl D has a mass of 25 kg. Both canoes are originally at rest.

Prob. 15–52

15–54. Blocks A and B have masses of 40 kg and 60 kg, respectively. They are placed on a smooth surface and the spring connected between them is stretched 2 m. If they are released from rest, determine the speeds of both blocks the instant the spring becomes unstretched.

Prob. 15–54

15.4 Impact

Central impact

(a)

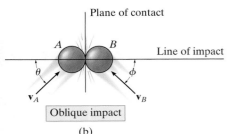

Oblique impact

(b)

Fig. 15–13

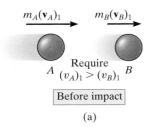

Require
$(v_A)_1 > (v_B)_1$

Before impact

(a)

Impact occurs when two bodies collide with each other during a very *short* period of time, causing relatively large (impulsive) forces to be exerted between the bodies. The striking of a hammer on a nail, or a golf club on a ball, are common examples of impact loadings.

In general, there are two types of impact. *Central impact* occurs when the direction of motion of the mass centers of the two colliding particles is along a line passing through the mass centers of the particles. This line is called the *line of impact*, Fig. 15–13a. When the motion of one or both of the particles is at an angle with the line of impact, Fig. 15–13b, the impact is said to be *oblique impact*.

Central Impact. To illustrate the method for analyzing the mechanics of impact, consider the case involving the central impact of two *smooth* particles A and B shown in Fig. 15–14.

- The particles have the initial momenta shown in Fig. 15–14a. Provided $(v_A)_1 > (v_B)_1$, collision will eventually occur.

- During the collision the particles must be thought of as *deformable* or nonrigid. The particles will undergo a *period of deformation* such that they exert an equal but opposite deformation impulse $\int P \, dt$ on each other, Fig. 15–14b.

- Only at the instant of *maximum deformation* will both particles move with a common velocity **v**, since their relative motion is zero, Fig. 15–14c.

- Afterward a *period of restitution* occurs, in which case the particles will either return to their original shape or remain permanently deformed. The equal but opposite *restitution impulse* $\int R \, dt$ pushes the particles apart from one another, Fig. 15–14d. In reality, the physical properties of any two bodies are such that the deformation impulse is *always greater* than that of restitution, i.e., $\int P \, dt > \int R \, dt$.

- Just after separation the particles will have the final momenta shown in Fig. 15–14e, where $(v_B)_2 > (v_A)_2$.

Deformation impulse
(b)

Maximum deformation
(c)

Restitution impulse
(d)

After impact
(e)

Fig. 15–14

In most problems the initial velocities of the particles will be *known*, and it will be necessary to determine their final velocities $(v_A)_2$ and $(v_B)_2$. In this regard, *momentum* for the *system of particles* is *conserved* since during collision the internal impulses of deformation and restitution *cancel*. Hence, referring to Fig. 15–14a and Fig. 15–14e we require

$(\xrightarrow{+})$ $m_A(v_A)_1 + m_B(v_B)_1 = m_A(v_A)_2 + m_B(v_B)_2$ (15–10)

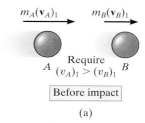

Require $(v_A)_1 > (v_B)_1$

Before impact

(a)

In order to obtain a second equation necessary to solve for $(v_A)_2$ and $(v_B)_2$, we must apply the principle of impulse and momentum to *each* particle. For example, during the deformation phase for particle A, Figs. 15–14a, 15–14b, and 15–14c, we have

$(\xrightarrow{+})$ $m_A(v_A)_1 - \displaystyle\int P \, dt = m_A v$

For the restitution phase, Figs. 15–14c, 15–14d, and 15–14e,

$(\xrightarrow{+})$ $m_A v - \displaystyle\int R \, dt = m_A(v_A)_2$

Effect of A on B   Effect of B on A

Deformation impulse

(b)

The ratio of the restitution impulse to the deformation impulse is called the *coefficient of restitution*, *e*. From the above equations, this value for particle A is

$$ e = \frac{\displaystyle\int R \, dt}{\displaystyle\int P \, dt} = \frac{v - (v_A)_2}{(v_A)_1 - v} $$

Maximum deformation

(c)

In a similar manner, we can establish *e* by considering particle B, Fig. 15–14. This yields

$$ e = \frac{\displaystyle\int R \, dt}{\displaystyle\int P \, dt} = \frac{(v_B)_2 - v}{v - (v_B)_1} $$

Effect of A on B   Effect of B on A

Restitution impulse

(d)

If the unknown *v* is eliminated from the above two equations, the coefficient of restitution can be expressed in terms of the particles' initial and final velocities as

$(\xrightarrow{+})$ $$ \boxed{ e = \frac{(v_B)_2 - (v_A)_2}{(v_A)_1 - (v_B)_1} } $$ (15–11)

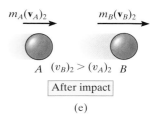

After impact

$(v_B)_2 > (v_A)_2$

(e)

Fig. 15–14

The quality of a manufactured tennis ball is measured by the height of its bounce, which can be related to its coefficient of restitution. Using the mechanics of oblique impact, engineers can design a separation device to remove substandard tennis balls from a production line.

Provided a value for e is specified, Eqs. 15–10 and 15–11 may be solved simultaneously to obtain $(v_A)_2$ and $(v_B)_2$. In doing so, however, it is important to carefully establish a sign convention for defining the positive direction for both $\mathbf{v}_A$ and $\mathbf{v}_B$ and then use it *consistently* when writing *both* equations. As noted from the application shown, and indicated symbolically by the arrow in parentheses, we have defined the positive direction to the right when referring to the motions of both A and B. Consequently, if a negative value results from the solution of either $(v_A)_2$ or $(v_B)_2$, it indicates motion is to the left.

Coefficient of Restitution.

From Figs. 15–14a and 15–14e, it is seen that Eq. 15–11 states that e is equal to the ratio of the relative velocity of the particles' separation *just after impact*, $(v_B)_2 - (v_A)_2$, to the relative velocity of the particles' approach *just before impact*, $(v_A)_1 - (v_B)_1$. By measuring these relative velocities experimentally, it has been found that e varies appreciably with impact velocity as well as with the size and shape of the colliding bodies. For these reasons the coefficient of restitution is reliable only when used with data which closely approximate the conditions which were known to exist when measurements of it were made. In general e has a value between zero and one, and one should be aware of the physical meaning of these two limits.

Elastic Impact ($e = 1$):   If the collision between the two particles is *perfectly elastic*, the deformation impulse $\left(\int \mathbf{P} \, dt \right)$ is equal and opposite to the restitution impulse $\left(\int \mathbf{R} \, dt \right)$. Although in reality this can never be achieved, $e = 1$ for an elastic collision.

Plastic Impact ($e = 0$):   The impact is said to be *inelastic or plastic* when $e = 0$. In this case there is no restitution impulse given to the particles $\left(\int \mathbf{R} \, dt = \mathbf{0} \right)$, so that after collision both particles couple or stick *together* and move with a common velocity.

From the above derivation it should be evident that the principle of work and energy cannot be used for the analysis of impact problems since it is not possible to know how the *internal forces* of deformation and restitution vary or displace during the collision. By knowing the particle's velocities before and after collision, however, the energy loss during collision can be calculated on the basis of the difference in the particle's kinetic energy. This energy loss, $\Sigma U_{1-2} = \Sigma T_2 - \Sigma T_1$, occurs because some of the initial kinetic energy of the particle is transformed into thermal energy as well as creating sound and localized deformation of the material when the collision occurs. In particular, if the impact is *perfectly elastic*, no energy is lost in the collision; whereas if the collision is *plastic*, the energy lost during collision is a maximum.

Procedure for Analysis (Central Impact)

In most cases the *final velocities* of two smooth particles are to be determined *just after* they are subjected to direct central impact. Provided the coefficient of restitution, the mass of each particle, and each particle's initial velocity *just before* impact are known, the solution to the problem can be obtained using the following two equations:

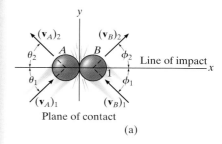

- The conservation of momentum applies to the system of particles, $\Sigma m v_1 = \Sigma m v_2$.
- The coefficient of restitution, $e = [(v_B)_2 - (v_A)_2]/[(v_A)_1 - (v_B)_1]$, relates the relative velocities of the particles along the line of impact, just before and just after collision.

When applying these two equations, the sense of an unknown velocity can be assumed. If the solution yields a negative magnitude, the velocity acts in the opposite sense.

Oblique Impact.   When oblique impact occurs between two smooth particles, the particles move away from each other with velocities having unknown directions as well as unknown magnitudes. Provided the initial velocities are known, four unknowns are present in the problem. As shown in Fig. 15–15a, these unknowns may be represented either as $(v_A)_2, (v_B)_2, \theta_2$, and ϕ_2, or as the x and y components of the final velocities.

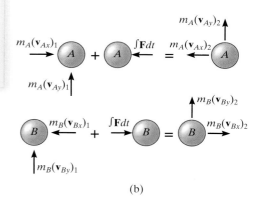

Fig. 15–15

Procedure for Analysis (Oblique Impact)

If the y axis is established within the plane of contact and the x axis along the line of impact, the impulsive forces of deformation and restitution act *only in the x direction*, Fig. 15–15b. Resolving the velocity or momentum vectors into components along the x and y axes, Fig. 15–15b, it is possible to write four independent scalar equations in order to determine $(v_{Ax})_2, (v_{Ay})_2, (v_{Bx})_2$, and $(v_{By})_2$.

- Momentum of the system is conserved *along the line of impact*, x axis, so that $\Sigma m(v_x)_1 = \Sigma m(v_x)_2$.
- The coefficient of restitution, $e = [(v_{Bx})_2 - (v_{Ax})_2]/[(v_{Ax})_1 - (v_{Bx})_1]$, relates the relative-velocity *components* of the particles *along the line of impact* (x axis).
- Momentum of particle A is conserved along the y axis, perpendicular to the line of impact, since no impulse acts on particle A in this direction.
- Momentum of particle B is conserved along the y axis, perpendicular to the line of impact, since no impulse acts on particle B in this direction.

Application of these four equations is illustrated in Example 15.11.

EXAMPLE 15.9

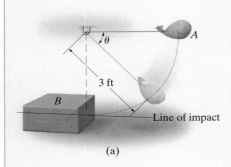

3 ft

B

——Line of impact

(a)

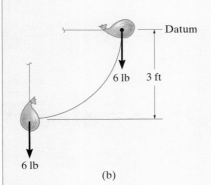

Datum

6 lb   3 ft

6 lb

(b)

$(v_B)_1 = 0$

B A

$(v_A)_1$

Just before impact

B A

$(v_B)_2$ $(v_A)_2$

Just after impact

(c)

Fig. 15–16

The bag A, having a weight of 6 lb, is released from rest at the position $\theta = 0°$, as shown in Fig. 15–16a. After falling to $\theta = 90°$, it strikes an 18-lb box B. If the coefficient of restitution between the bag and box is $e = 0.5$, determine the velocities of the bag and box just after impact and the loss of energy during collision.

SOLUTION

This problem involves central impact. Why? Before analyzing the mechanics of the impact, however, it is first necessary to obtain the velocity of the bag *just before* it strikes the box.

Conservation of Energy.   With the datum at $\theta = 0°$, Fig. 15–16b, we have

$$T_0 + V_0 = T_1 + V_1$$

$$0 + 0 = \frac{1}{2}\left(\frac{6\ \text{lb}}{32.2\ \text{ft/s}^2}\right)(v_A)_1^2 - 6\ \text{lb}(3\ \text{ft}); \qquad (v_A)_1 = 13.9\ \text{ft/s}$$

Conservation of Momentum.   After impact we will assume A and B travel to the left. Applying the conservation of momentum to the system, we have

$$(\xleftarrow{+})\qquad m_B(v_B)_1 + m_A(v_A)_1 = m_B(v_B)_2 + m_A(v_A)_2$$

$$0 + \frac{6\ \text{lb}}{32.2\ \text{ft/s}^2}(13.9\ \text{ft/s}) = \frac{18\ \text{lb}}{32.2\ \text{ft/s}^2}(v_B)_2 + \frac{6\ \text{lb}}{32.2\ \text{ft/s}^2}(v_A)_2$$

$$(v_A)_2 = 13.9 - 3(v_B)_2 \tag{1}$$

Coefficient of Restitution.   Realizing that for separation to occur after collision $(v_B)_2 > (v_A)_2$, Fig. 15–16c, we have

$$(\xleftarrow{+})\qquad e = \frac{(v_B)_2 - (v_A)_2}{(v_A)_1 - (v_B)_1}; \qquad 0.5 = \frac{(v_B)_2 - (v_A)_2}{13.9\ \text{ft/s} - 0}$$

$$(v_A)_2 = (v_B)_2 - 6.95 \tag{2}$$

Solving Eqs. 1 and 2 simultaneously yields

$$(v_A)_2 = -1.74\ \text{ft/s} = 1.74\ \text{ft/s} \rightarrow \quad \text{and} \quad (v_B)_2 = 5.21\ \text{ft/s} \leftarrow \quad Ans.$$

Loss of Energy.   Applying the principle of work and energy to the bag and box just before and just after collision, we have

$$\Sigma U_{1-2} = T_2 - T_1;$$

$$\Sigma U_{1-2} = \left[\frac{1}{2}\left(\frac{18\ \text{lb}}{32.2\ \text{ft/s}^2}\right)(5.21\ \text{ft/s})^2 + \frac{1}{2}\left(\frac{6\ \text{lb}}{32.2\ \text{ft/s}^2}\right)(1.74\ \text{ft/s})^2\right] -$$

$$\left[\frac{1}{2}\left(\frac{6\ \text{lb}}{32.2\ \text{ft/s}^2}\right)(13.9\ \text{ft/s})^2\right]$$

$$\Sigma U_{1-2} = -10.1\ \text{ft} \cdot \text{lb} \qquad Ans.$$

NOTE: The energy loss occurs due to deformation during the collision.

EXAMPLE 15.10

The ball B shown in Fig. 15–17a has a mass of 1.5 kg and is suspended from the ceiling by a 1-m-long elastic cord. If the cord is *stretched* downward 0.25 m and the ball is released from rest, determine how far the cord stretches after the ball rebounds from the ceiling. The stiffness of the cord is $k = 800$ N/m, and the coefficient of restitution between the ball and ceiling is $e = 0.8$. The ball makes a central impact with the ceiling.

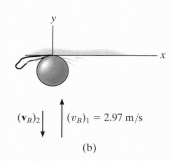

(a)

SOLUTION

First we must obtain the velocity of the ball *just before* it strikes the ceiling using energy methods, then consider the impulse and momentum between the ball and ceiling, and finally again use energy methods to determine the stretch in the cord.

Conservation of Energy. With the datum located as shown in Fig. 15–17a, realizing that initially $y = y_0 = (1 + 0.25)$ m $= 1.25$ m, we have

$$T_0 + V_0 = T_1 + V_1$$

$$\tfrac{1}{2}m(v_B)_0^2 - W_B y_0 + \tfrac{1}{2}ks^2 = \tfrac{1}{2}m(v_B)_1^2 + 0$$

$$0 - 1.5(9.81)\ \text{N}(1.25\ \text{m}) + \tfrac{1}{2}(800\ \text{N/m})(0.25\ \text{m})^2 = \tfrac{1}{2}(1.5\ \text{kg})(v_B)_1^2$$

$$(v_B)_1 = 2.97\ \text{m/s} \uparrow$$

The interaction of the ball with the ceiling will now be considered using the principles of impact.* Since an unknown portion of the mass of the ceiling is involved in the impact, the conservation of momentum for the ball–ceiling system will not be written. The "velocity" of this portion of ceiling is zero since it (or the earth) are assumed to remain at rest *both* before and after impact.

Coefficient of Restitution.   Fig. 15–17b.

$$(+\uparrow)\quad e = \frac{(v_B)_2 - (v_A)_2}{(v_A)_1 - (v_B)_1};\qquad 0.8 = \frac{(v_B)_2 - 0}{0 - 2.97\ \text{m/s}}$$

$$(v_B)_2 = -2.37\ \text{m/s} = 2.37\ \text{m/s} \downarrow$$

(b)

Conservation of Energy.   The maximum stretch s_3 in the cord may be determined by again applying the conservation of energy equation to the ball just after collision. Assuming that $y = y_3 = (1 + s_3)$ m, Fig. 15–17c, then

$$T_2 + V_2 = T_3 + V_3$$

$$\tfrac{1}{2}m(v_B)_2^2 + 0 = \tfrac{1}{2}m(v_B)_3^2 - W_B y_3 + \tfrac{1}{2}ks_3^2$$

$$\tfrac{1}{2}(1.5\ \text{kg})(2.37\ \text{m/s})^2 = 0 - 9.81(1.5)\ \text{N}(1\ \text{m} + s_3) + \tfrac{1}{2}(800\ \text{N/m})s_3^2$$

$$400s_3^2 - 14.72s_3 - 18.94 = 0$$

Solving this quadratic equation for the positive root yields

$$s_3 = 0.237\ \text{m} = 237\ \text{mm} \qquad\qquad Ans.$$

(c)

Fig. 15–17

*The weight of the ball is considered a nonimpulsive force.

EXAMPLE 15.11

(a)

(b)

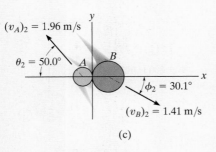

(c)

Fig. 15–18

Two smooth disks A and B, having a mass of 1 kg and 2 kg, respectively, collide with the velocities shown in Fig. 15–18a. If the coefficient of restitution for the disks is $e = 0.75$, determine the x and y components of the final velocity of each disk just after collision.

SOLUTION

The problem involves *oblique impact*. Why? In order to seek a solution, we have established the x and y axes along the line of impact and the plane of contact, respectively, Fig. 15–18a.

Resolving each of the initial velocities into x and y components, we have

$$(v_{Ax})_1 = 3\cos 30° = 2.60 \text{ m/s} \qquad (v_{Ay})_1 = 3\sin 30° = 1.50 \text{ m/s}$$

$$(v_{Bx})_1 = -1\cos 45° = -0.707 \text{ m/s} \quad (v_{By})_1 = -1\sin 45° = -0.707 \text{ m/s}$$

The four unknown velocity components after collision are assumed to act in the positive directions, Fig. 15–18b. Since the impact occurs only in the x direction (line of impact), the conservation of momentum for *both* disks can be applied in this direction. Why?

Conservation of "x" Momentum.   In reference to the momentum diagrams, we have

$$(\xrightarrow{+}) \qquad m_A(v_{Ax})_1 + m_B(v_{Bx})_1 = m_A(v_{Ax})_2 + m_B(v_{Bx})_2$$

$$1 \text{ kg}(2.60 \text{ m/s}) + 2 \text{ kg}(-0.707 \text{ m/s}) = 1 \text{ kg}(v_{Ax})_2 + 2 \text{ kg}(v_{Bx})_2$$

$$(v_{Ax})_2 + 2(v_{Bx})_2 = 1.18 \qquad (1)$$

Coefficient of Restitution (x).   Both disks are *assumed* to have components of velocity in the $+x$ direction after collision, Fig. 15–18b.

$$(\xrightarrow{+}) \qquad e = \frac{(v_{Bx})_2 - (v_{Ax})_2}{(v_{Ax})_1 - (v_{Bx})_1}; \quad 0.75 = \frac{(v_{Bx})_2 - (v_{Ax})_2}{2.60 \text{ m/s} - (-0.707 \text{ m/s})}$$

$$(v_{Bx})_2 - (v_{Ax})_2 = 2.48 \qquad (2)$$

Solving Eqs. 1 and 2 for $(v_{Ax})_2$ and $(v_{Bx})_2$ yields

$$(v_{Ax})_2 = -1.26 \text{ m/s} = 1.26 \text{ m/s} \leftarrow \qquad (v_{Bx})_2 = 1.22 \text{ m/s} \rightarrow \quad Ans.$$

Conservation of "y" Momentum.   The momentum of *each disk* is *conserved* in the y direction (plane of contact), since the disks are smooth and therefore *no* external impulse acts in this direction. From Fig. 15–18b,

$$(+\uparrow) \; m_A(v_{Ay})_1 = m_A(v_{Ay})_2; \quad (v_{Ay})_2 = 1.50 \text{ m/s} \uparrow \qquad Ans.$$

$$(+\uparrow) \; m_B(v_{By})_1 = m_B(v_{By})_2; \; (v_{By})_2 = -0.707 \text{ m/s} = 0.707 \text{ m/s} \downarrow \; Ans.$$

NOTE: Show that when the velocity components are summed, one obtains the results shown in Fig. 15–18c.

PROBLEMS

15–55. Block A has a mass of 3 kg and is sliding on a rough horizontal surface with a velocity $(v_A)_1 = 2$ m/s when it makes a direct collision with block B, which has a mass of 2 kg and is originally at rest. If the collision is perfectly elastic $(e = 1)$, determine the velocity of each block just after collision and the distance between the blocks when they stop sliding. The coefficient of kinetic friction between the blocks and the plane is $\mu_k = 0.3$.

Prob. 15–55

***15–56.** Disks A and B have a mass of 2 kg and 4 kg, respectively. If they have the velocities shown, and $e = 0.4$, determine their velocities just after direct central impact.

Prob. 15–56

15–57. The three balls each weigh 0.5 lb and have a coefficient of restitution of $e = 0.85$. If ball A is released from rest and strikes ball B and then ball B strikes ball C, determine the velocity of each ball after the second collision has occurred. The balls slide without friction.

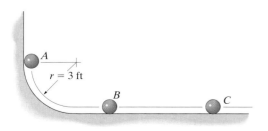

Prob. 15–57

15–58. The 1-lb ball A is thrown so that when it strikes the 10-lb block B it is traveling horizontally at 20 ft/s. If the coefficient of restitution between A and B is $e = 0.6$, and the coefficient of kinetic friction between the plane and the block is $\mu_k = 0.4$, determine the time before block B stops sliding.

15–59. The 1-lb ball A is thrown so that when it strikes the 10-lb block B it is traveling horizontally at 20 ft/s. If the coefficient of restitution between A and B is $e = 0.6$, and the coefficient of kinetic friction between the plane and the block is $\mu_k = 0.4$, determine the distance block B slides on the plane before stopping.

Probs. 15–58/59

***15–60.** The 1-lb ball A is thrown so that when it strikes the 10-lb block B it is traveling horizontally at 20 ft/s. Determine the average normal force exerted between A and B if the impact occurs in 0.02 s. The coefficient of restitution between A and B is $e = 0.6$.

Prob. 15–60

15–61.   The man A has a weight of 175 lb and jumps from rest $h = 8$ ft onto a platform P that has a weight of 60 lb. The platform is mounted on a spring, which has a stiffness $k = 200$ lb/ft. Determine (a) the velocities of A and P just after impact and (b) the maximum compression imparted to the spring by the impact. Assume the coefficient of restitution between the man and the platform is $e = 0.6$, and the man holds himself rigid during the motion.

15–62.   The man A has a weight of 100 lb and jumps from rest onto the platform P that has a weight of 60 lb. The platform is mounted on a spring, which has a stiffness $k = 200$ lb/ft. If the coefficient of restitution between the man and the platform is $e = 0.6$, and the man holds himself rigid during the motion, determine the required height h of the jump if the maximum compression of the spring becomes 2 ft.

Probs. 15–61/62

15–63.   The 10-lb collar B is at rest, and when it is in the position shown the spring is unstretched. If another 1-lb collar A strikes it so that B slides 4 ft on the smooth rod before momentarily stopping, determine the velocity of A just after impact, and the average force exerted between A and B during the impact if the impact occurs in 0.002 s. The coefficient of restitution between A and B is $e = 0.5$.

Prob. 15–63

*15–64.**   If the girl throws the ball with a horizontal velocity of 8 ft/s, determine the distance d so that the ball bounces once on the smooth surface and then lands in the cup at C. Take $e = 0.8$.

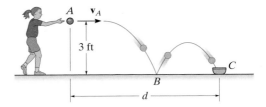

Prob. 15–64

15–65.   The 1 lb ball is dropped from rest and falls a distance of 4 ft before striking the smooth plane at A. If $e = 0.8$, determine the distance d to where it again strikes the plane at B.

15–66.   The 1 lb ball is dropped from rest and falls a distance of 4 ft before striking the smooth plane at A. If it rebounds and in $t = 0.5$ s again strikes the plane at B, determine the coefficient of restitution e between the ball and the plane. Also, what is the distance d?

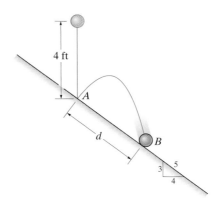

Probs. 15–65/66

15–67. The 2-kg ball is thrown at the suspended 20-kg block with a velocity of 4 m/s. If the coefficient of restitution between the ball and the block is $e = 0.8$, determine the maximum height h to which the block will swing before it momentarily stops.

Prob. 15–67

15–69. A ball is thrown onto a rough floor at an angle θ. If it rebounds at an angle ϕ and the coefficient of kinetic friction is μ, determine the coefficient of restitution e. Neglect the size of the ball. *Hint:* Show that during impact, the average impulses in the x and y directions are related by $I_x = \mu I_y$. Since the time of impact is the same, $F_x \Delta t = \mu F_y \Delta t$ or $F_x = \mu F_y$.

15–70. A ball is thrown onto a rough floor at an angle of $\theta = 45°$. If it rebounds at the same angle $\phi = 45°$, determine the coefficient of kinetic friction between the floor and the ball. The coefficient of restitution is $e = 0.6$. *Hint:* Show that during impact, the average impulses in the x and y directions are related by $I_x = \mu I_y$. Since the time of impact is the same, $F_x \Delta t = \mu F_y \Delta t$ or $F_x = \mu F_y$.

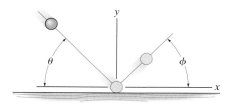

Probs. 15–69/70

*15–68. The 2-kg ball is thrown at the suspended 20-kg block with a velocity of 4 m/s. If the time of impact between the ball and the block is 0.005 s, determine the average normal force exerted on the block during this time. Take $e = 0.8$.

Prob. 15–68

15–71. The 0.2-lb ball bearing travels over the edge A with a velocity of $v_A = 3$ ft/s. Determine the speed at which it rebounds from the smooth inclined plane at B. Take $e = 0.8$.

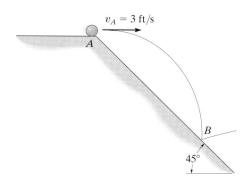

Prob. 15–71

*15–72. The drop hammer H has a weight of 900 lb and falls from rest $h = 3$ ft onto a forged anvil plate P that has a weight of 500 lb. The plate is mounted on a set of springs that have a combined stiffness of $k_T = 500$ lb/ft. Determine (a) the velocity of P and H just after collision and (b) the maximum compression in the springs caused by the impact. The coefficient of restitution between the hammer and the plate is $e = 0.6$. Neglect friction along the vertical guideposts A and B.

Prob. 15–72

15–73. It was observed that a tennis ball when served horizontally 7.5 ft above the ground strikes the smooth ground at B 20 ft away. Determine the initial velocity $\mathbf{v}_A$ of the ball and the velocity $\mathbf{v}_B$ (and θ) of the ball just after it strikes the court at B. Take $e = 0.7$.

15–74. The tennis ball is struck with a horizontal velocity $\mathbf{v}_A$, strikes the smooth ground at B, and bounces upward at $\theta = 30°$. Determine the initial velocity $\mathbf{v}_A$, the final velocity $\mathbf{v}_B$, and the coefficient of restitution between the ball and the ground.

Probs. 15–73/74

15–75. The ping-pong ball has a mass of 2 g. If it is struck with the velocity shown, determine how high h it rises above the end of the smooth table after the rebound. Take $e = 0.8$.

Prob. 15–75

*15–76. The 5-lb box B is dropped from rest 5 ft from the top of the 10-lb plate P, which is supported by the spring having a stiffness of $k = 30$ lb/ft. If $e = 0.6$ between the box and plate, determine the maximum compression imparted to the spring. Neglect the mass of the spring.

Prob. 15–76

15–77. A pitching machine throws the 0.5-kg ball towards the wall with an initial velocity $v_A = 10$ m/s as shown. Determine (a) the velocity at which it strikes the wall at B, (b) the velocity at which it rebounds from the wall if $e = 0.5$, and (c) the distance s from the wall to where it strikes the ground at C.

Prob. 15–77

15–79. The 200-g billiard ball is moving with a speed of 2.5 m/s when it strikes the side of the pool table at A. If the coefficient of restitution between the ball and the side of the table is $e = 0.6$, determine the speed of the ball just after striking the table twice, i.e., at A, then at B. Neglect the size of the ball.

Prob. 15–79

15–78. The 20-lb box slides on the surface for which $\mu_k = 0.3$. The box has a velocity $v = 15$ ft/s when it is 2 ft from the plate. If it strikes the smooth plate, which has a weight of 10 lb and is held in position by an unstretched spring of stiffness $k = 400$ lb/ft, determine the maximum compression imparted to the spring. Take $e = 0.8$ between the box and the plate. Assume that the plate slides smoothly.

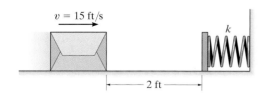

Prob. 15–78

*15–80. The three balls each have the same mass m. If A is released from rest at θ, determine the angle ϕ to which C rises after collision. The coefficient of restitution between each ball is e.

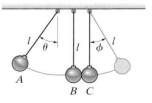

Prob. 15–80

15–81. Two smooth billiard balls A and B each have a mass of 200 g. If A strikes B with a velocity $(v_A)_1 = 1.5$ m/s as shown, determine their final velocities just after collision. Ball B is originally at rest and the coefficient of restitution is $e = 0.85$. Neglect the size of each ball.

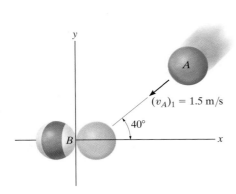

Prob. 15–81

15–82. The two hockey pucks A and B each have a mass of 250 g. If they collide at O and are deflected along the colored paths, determine their speeds just after impact. Assume that the icy surface over which they slide is smooth. *Hint:* Since the y' axis is *not* along the line of impact, apply the conservation of momentum along the x' and y' axes.

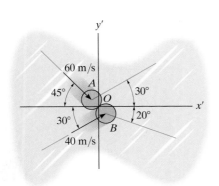

Prob. 15–82

15–83. Two smooth coins A and B, each having the same mass, slide on a smooth surface with the motion shown. Determine the speed of each coin after collision if they move off along the blue paths. *Hint:* Since the line of impact has not been defined, apply the conservation of momentum along the x and y axes, respectively.

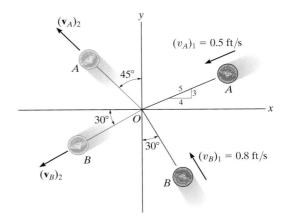

Prob. 15–83

*15–84. The two disks A and B have a mass of 3 kg and 5 kg, respectively. If they collide with the initial velocities shown, determine their velocities just after impact. The coefficient of restitution is $e = 0.65$.

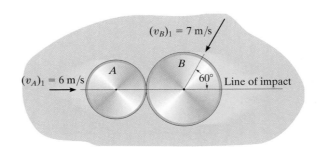

Prob. 15–84

15–85. Two smooth disks A and B each have a mass of 0.5 kg. If both disks are moving with the velocities shown when they collide, determine their final velocities just after collision. The coefficient of restitution is $e = 0.75$.

15–86. Two smooth disks A and B each have a mass of 0.5 kg. If both disks are moving with the velocities shown when they collide, determine the coefficient of restitution between the disks if after collision B travels along a line, $30°$ counterclockwise from the y axis.

15–88. The "stone" A used in the sport of curling slides over the ice track and strikes another "stone" B as shown. If each "stone" is smooth and has a weight of 47 lb, and the coefficient of restitution between the "stones" is $e = 0.8$, determine their speeds just after collision. Initially A has a velocity of 8 ft/s and B is at rest. Neglect friction.

Prob. 15–88

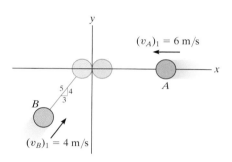

Probs. 15–85/86

15–87. Two smooth disks A and B have the initial velocities shown just before they collide at O. If they have masses $m_A = 8$ kg and $m_B = 6$ kg, determine their speeds just after impact. The coefficient of restitution is $e = 0.5$.

15–89. The two billiard balls A and B are originally in contact with one another when a third ball C strikes each of them at the same time as shown. If ball C remains at rest after the collision, determine the coefficient of restitution. All the balls have the same mass. Neglect the size of each ball.

Prob. 15–87

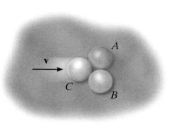

Prob. 15–89

15.5   Angular Momentum

The *angular momentum* of a particle about point O is defined as the "moment" of the particle's linear momentum about O. Since this concept is analogous to finding the moment of a force about a point, the angular momentum, $\mathbf{H}_O$, is sometimes referred to as the *moment of momentum*.

Scalar Formulation.   If a particle is moving along a curve lying in the x–y plane, Fig. 15–19, the angular momentum at any instant can be determined about point O (actually the z axis) by using a scalar formulation. The *magnitude* of $\mathbf{H}_O$ is

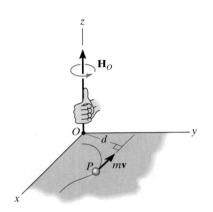

Fig. 15–19

$$(H_O)_z = (d)(mv) \qquad (15\text{--}12)$$

Here d is the moment arm or perpendicular distance from O to the line of action of $m\mathbf{v}$. Common units for $(H_O)_z$ are $\text{kg} \cdot \text{m}^2/\text{s}$ or $\text{slug} \cdot \text{ft}^2/\text{s}$. The *direction* of $\mathbf{H}_O$ is defined by the right-hand rule. As shown, the curl of the fingers of the right hand indicates the sense of rotation of $m\mathbf{v}$ about O, so that in this case the thumb (or $\mathbf{H}_O$) is directed perpendicular to the x–y plane along the $+z$ axis.

Vector Formulation.   If the particle is moving along a space curve, Fig. 15–20, the vector cross product can be used to determine the *angular momentum* about O. In this case

$$\mathbf{H}_O = \mathbf{r} \times m\mathbf{v} \qquad (15\text{--}13)$$

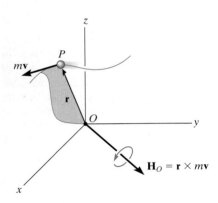

Fig. 15–20

Here $\mathbf{r}$ denotes a position vector drawn from point O to the particle P. As shown in the figure, $\mathbf{H}_O$ is *perpendicular* to the shaded plane containing $\mathbf{r}$ and $m\mathbf{v}$.

In order to evaluate the cross product, $\mathbf{r}$ and $m\mathbf{v}$ should be expressed in terms of their Cartesian components, so that the angular momentum is determined by evaluating the determinant:

$$\mathbf{H}_O = \begin{vmatrix} \mathbf{i} & \mathbf{j} & \mathbf{k} \\ r_x & r_y & r_z \\ mv_x & mv_y & mv_z \end{vmatrix} \qquad (15\text{--}14)$$

15.6   Relation Between Moment of a Force and Angular Momentum

The moments about point O of all the forces acting on the particle in Fig. 15–21a may be related to the particle's angular momentum by using the equation of motion. If the mass of the particle is constant, we may write

$$\Sigma \mathbf{F} = m\dot{\mathbf{v}}$$

The moments of the forces about point O can be obtained by performing a cross-product multiplication of each side of this equation by the position vector $\mathbf{r}$, which is measured in the x, y, z inertial frame of reference. We have

$$\Sigma \mathbf{M}_O = \mathbf{r} \times \Sigma \mathbf{F} = \mathbf{r} \times m\dot{\mathbf{v}}$$

From Appendix C, the derivative of $\mathbf{r} \times m\mathbf{v}$ can be written as

$$\dot{\mathbf{H}}_O = \frac{d}{dt}(\mathbf{r} \times m\mathbf{v}) = \dot{\mathbf{r}} \times m\mathbf{v} + \mathbf{r} \times m\dot{\mathbf{v}}$$

The first term on the right side, $\dot{\mathbf{r}} \times m\mathbf{v} = m(\dot{\mathbf{r}} \times \dot{\mathbf{r}}) = \mathbf{0}$, since the cross product of a vector with itself is zero. Hence, the above equation becomes

$$\boxed{\Sigma \mathbf{M}_O = \dot{\mathbf{H}}_O} \qquad (15\text{–}15)$$

This equation states that *the resultant moment about point O of all the forces acting on the particle is equal to the time rate of change of the particle's angular momentum about point O.* This result is similar to Eq. 15–1, i.e.,

$$\boxed{\Sigma \mathbf{F} = \dot{\mathbf{L}}} \qquad (15\text{–}16)$$

Here $\mathbf{L} = m\mathbf{v}$, so *the resultant force acting on the particle is equal to the time rate of change of the particle's linear momentum.*

From the derivations, it is seen that Eqs. 15–15 and 15–16 are actually another way of stating Newton's second law of motion. In other sections of this book it will be shown that these equations have many practical applications when extended and applied to the solution of problems involving either a system of particles or a rigid body.

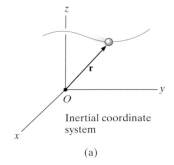

Inertial coordinate system

(a)

Fig. 15–21

System of Particles. An equation having the same form as Eq. 15–15 may be derived for the system of particles shown in Fig. 15–21b. The forces acting on the arbitrary ith particle of the system consist of a resultant *external force* $\mathbf{F}_i$ and a resultant *internal force* $\mathbf{f}_i$. Expressing the moments of these forces about point O, using the form of Eq. 15–15, we have

$$(\mathbf{r}_i \times \mathbf{F}_i) + (\mathbf{r}_i \times \mathbf{f}_i) = (\dot{\mathbf{H}}_i)_O$$

Here $\mathbf{r}_i$ is the position vector drawn from the origin O of an inertial frame of reference to the ith particle, and $(\dot{\mathbf{H}}_i)_O$ is the time rate of change in the angular momentum of the ith particle about O. Similar equations can be written for each of the other particles of the system. When the results are summed vectorially, the result is

$$\Sigma(\mathbf{r}_i \times \mathbf{F}_i) + \Sigma(\mathbf{r}_i \times \mathbf{f}_i) = \Sigma(\dot{\mathbf{H}}_i)_O$$

The second term is zero since the internal forces occur in equal but opposite collinear pairs, and hence the moment of each pair about point O is zero. Dropping the index notation, the above equation can be written in a simplified form as

$$\Sigma\mathbf{M}_O = \dot{\mathbf{H}}_O \tag{15–17}$$

which states that *the sum of the moments about point O of all the external forces acting on a system of particles is equal to the time rate of change of the total angular momentum of the system about point O.* Although O has been chosen here as the origin of coordinates, it actually can represent any *fixed point* in the inertial frame of reference.

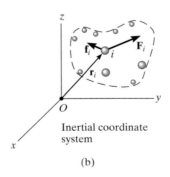

Inertial coordinate system

(b)

Fig. 15–21 (cont.)

EXAMPLE 15.12

The box shown in Fig. 15–22a has a mass m and is traveling down the smooth circular ramp such that when it is at the angle θ it has a speed v. Determine its angular momentum about point O at this instant and the rate of increase in its speed, i.e., a_t.

(a)   (b)

Fig. 15–22

SOLUTION

Since $\mathbf{v}$ is tangent to the path, applying Eq. 15–12 the angular momentum is

$$H_O = rmv \quad \text{↓} \qquad\qquad Ans.$$

The rate of increase in its speed (dv/dt) can be found by applying Eq. 15–15. From the free-body diagram of the block, Fig. 15–22b, it is seen that only the weight $W = mg$ contributes a moment about point O. We have

$$\text{↱}+\Sigma M_O = \dot{H}_O; \qquad mg(r\sin\theta) = \frac{d}{dt}(rmv)$$

Since r and m are constant,

$$mgr\sin\theta = rm\frac{dv}{dt}$$

$$\frac{dv}{dt} = g\sin\theta \qquad\qquad Ans.$$

NOTE: This same result can, of course, be obtained from the equation of motion applied in the tangential direction, Fig. 15–22b, i.e.,

$$+\swarrow\Sigma F_t = ma_t; \qquad mg\sin\theta = m\left(\frac{dv}{dt}\right)$$

$$\frac{dv}{dt} = g\sin\theta \qquad\qquad Ans.$$

15.7   Angular Impulse and Momentum Principles

Principle of Angular Impulse and Momentum.

If Eq. 15–15 is rewritten in the form $\Sigma \mathbf{M}_O \, dt = d\mathbf{H}_O$ and integrated, we have, assuming that at time $t = t_1$, $\mathbf{H}_O = (\mathbf{H}_O)_1$ and at time $t = t_2$, $\mathbf{H}_O = (\mathbf{H}_O)_2$,

$$\Sigma \int_{t_1}^{t_2} \mathbf{M}_O \, dt = (\mathbf{H}_O)_2 - (\mathbf{H}_O)_1$$

or

$$(\mathbf{H}_O)_1 + \Sigma \int_{t_1}^{t_2} \mathbf{M}_O \, dt = (\mathbf{H}_O)_2 \qquad (15\text{–}18)$$

This equation is referred to as the *principle of angular impulse and momentum.* The initial and final angular momenta $(\mathbf{H}_O)_1$ and $(\mathbf{H}_O)_2$ are defined as the moment of the linear momentum of the particle $(\mathbf{H}_O = \mathbf{r} \times m\mathbf{v})$ at the instants t_1 and t_2, respectively. The second term on the left side, $\Sigma \int \mathbf{M}_O \, dt$, is called the *angular impulse.* It is determined by integrating, with respect to time, the moments of all the forces acting on the particle over the time period t_1 to t_2. Since the moment of a force about point O is $\mathbf{M}_O = \mathbf{r} \times \mathbf{F}$, the angular impulse may be expressed in vector form as

$$\text{angular impulse} = \int_{t_1}^{t_2} \mathbf{M}_O \, dt = \int_{t_1}^{t_2} (\mathbf{r} \times \mathbf{F}) \, dt \qquad (15\text{–}19)$$

Here $\mathbf{r}$ is a position vector which extends from point O to any point on the line of action of $\mathbf{F}$.

In a similar manner, using Eq. 15–18, the principle of angular impulse and momentum for a system of particles may be written as

$$\Sigma (\mathbf{H}_O)_1 + \Sigma \int_{t_1}^{t_2} \mathbf{M}_O \, dt = \Sigma (\mathbf{H}_O)_2 \qquad (15\text{–}20)$$

Here the first and third terms represent the angular momenta of the system of particles $[\Sigma \mathbf{H}_O = \Sigma (\mathbf{r}_i \times m\mathbf{v}_i)]$ at the instants t_1 and t_2. The second term is the sum of the angular impulses given to all the particles from t_1 to t_2. Recall that these impulses are created only by the moments of the external forces acting on the system where, for the *i*th particle, $\mathbf{M}_O = \mathbf{r}_i \times \mathbf{F}_i$.

Vector Formulation.   Using impulse and momentum principles, it is therefore possible to write two equations which define the particle's motion, namely, Eqs. 15–3 and 15–18, restated as

$$m\mathbf{v}_1 + \Sigma \int_{t_1}^{t_2} \mathbf{F}\, dt = m\mathbf{v}_2$$

$$(\mathbf{H}_O)_1 + \Sigma \int_{t_1}^{t_2} \mathbf{M}_O\, dt = (\mathbf{H}_O)_2 \qquad (15\text{–}21)$$

Scalar Formulation.   In general, the above equations may be expressed in *x*, *y*, *z* component form, yielding a total of six independent scalar equations. If the particle is confined to move in the *x–y* plane, three independent scalar equations may be written to express the motion, namely,

$$m(v_x)_1 + \Sigma \int_{t_1}^{t_2} F_x\, dt = m(v_x)_2$$

$$m(v_y)_1 + \Sigma \int_{t_1}^{t_2} F_y\, dt = m(v_y)_2 \qquad (15\text{–}22)$$

$$(H_O)_1 + \Sigma \int_{t_1}^{t_2} M_O\, dt = (H_O)_2$$

The first two of these equations represent the principle of linear impulse and momentum in the *x* and *y* directions, which has been discussed in Sec. 15.1, and the third equation represents the principle of angular impulse and momentum about the *z* axis.

Fig. 15–23

Provided air resistance is neglected, the passengers on this amusement-park ride are subjected to a conservation of angular momentum about the axis of rotation. As shown on the free-body diagram, the line of action of the normal force **N** of the seat on the passenger passes through the axis, and the passenger's weight **W** is parallel to it. No angular impulse acts around the z axis.

Conservation of Angular Momentum. When the angular impulses acting on a particle are all zero during the time t_1 to t_2, Eq. 15–18 reduces to the following simplified form:

$$(\mathbf{H}_O)_1 = (\mathbf{H}_O)_2 \qquad (15\text{–}23)$$

This equation is known as the *conservation of angular momentum*. It states that from t_1 to t_2 the particle's angular momentum remains constant. Obviously, if no external impulse is applied to the particle, both linear and angular momentum will be conserved. In some cases, however, the particle's angular momentum will be conserved and linear momentum may not. An example of this occurs when the particle is subjected *only* to a *central force* (see Sec. 13.7). As shown in Fig. 15–23, the impulsive central force **F** is always directed toward point O as the particle moves along the path. Hence, the angular impulse (moment) created by **F** about the z axis passing through point O is always zero, and therefore angular momentum of the particle is conserved about this axis.

From Eq. 15–20, we can also write the conservation of angular momentum for a system of particles, namely,

$$\Sigma(\mathbf{H}_O)_1 = \Sigma(\mathbf{H}_O)_2 \qquad (15\text{–}24)$$

In this case the summation must include the angular momenta of all particles in the system.

Procedure for Analysis

When applying the principles of angular impulse and momentum, or the conservation of angular momentum, it is suggested that the following procedure be used.

Free-Body Diagram.

- Draw the particle's free-body diagram in order to determine any axis about which angular momentum may be conserved. For this to occur, the moments of the forces (or impulses) must be parallel or pass through the axis so as to create zero moment throughout the time period t_1 to t_2.
- The direction and sense of the particle's initial and final velocities should also be established.
- An alternative procedure would be to draw the impulse and momentum diagrams for the particle.

Momentum Equations.

- Apply the principle of angular impulse and momentum, $(\mathbf{H}_O)_1 + \Sigma \int_{t_1}^{t_2} \mathbf{M}_O \, dt = (\mathbf{H}_O)_2$, or if appropriate, the conservation of angular momentum, $(\mathbf{H}_O)_1 = (\mathbf{H}_O)_2$.

EXAMPLE 15.13

The 5-kg block of negligible size rests on the smooth horizontal plane, Fig. 15–24a. It is attached at A to a slender rod of negligible mass. The rod is attached to a ball-and-socket joint at B. If a moment $M = (3t)$ N·m, where t is in seconds, is applied to the rod and a horizontal force $P = 10$ N is applied to the block, determine the speed of the block in 4 s starting from rest.

(a)   (b)

Fig. 15–24

SOLUTION

Free-Body Diagram.   If we consider the system of both the rod and block, Fig. 15–24b, then the resultant force reaction $\mathbf{F}_B$ at the ball-and-socket can be eliminated from the analysis by applying the principle of angular impulse and momentum about the z axis. If this is done, the angular impulses created by the weight and normal reaction $\mathbf{N}_A$ are also eliminated, since they act parallel to the z axis and therefore create zero moment about this axis.

Principle of Angular Impulse and Momentum.

$$(H_z)_1 + \Sigma \int_{t_1}^{t_2} M_z \, dt = (H_z)_2$$

$$(H_z)_1 + \int_{t_1}^{t_2} M \, dt + r_{BA}P(\Delta t) = (H_z)_2$$

$$0 + \int_0^4 3t \, dt + (0.4 \text{ m})(10 \text{ N})(4 \text{ s}) = 5 \text{ kg } (v_A)_2 \, (0.4 \text{ m})$$

$$24 + 16 = 2(v_A)_2$$

$$(v_A)_2 = 20 \text{ m/s} \qquad\qquad Ans.$$

NOTE: The position of the block can be obtained using this result and the principle of work and energy. Only the force P does work.

EXAMPLE 15.14

(a)

(b)

Fig. 15–25

The 0.8-lb ball B, shown in Fig. 15–25a, is attached to a cord which passes through a hole at A in a smooth table. When the ball is $r_1 = 1.75$ ft from the hole, it is rotating around in a circle such that its speed is $v_1 = 4$ ft/s. By applying a force $\mathbf{F}$ the cord is pulled downward through the hole with a constant speed $v_c = 6$ ft/s. Determine (a) the speed of the ball at the instant it is $r_2 = 0.6$ ft from the hole, and (b) the amount of work done by $\mathbf{F}$ in shortening the radial distance from r_1 to r_2. Neglect the size of the ball.

SOLUTION

Part (a) Free-Body Diagram.   As the ball moves from r_1 to r_2, Fig. 15–25b, the cord force $\mathbf{F}$ on the ball always passes through the z axis, and the weight and $\mathbf{N}_B$ are parallel to it. Hence the moments, or angular impulses created by these forces, are all *zero* about this axis. Therefore, the conservation of angular momentum applies about the z axis.

Conservation of Angular Momentum.   The ball's velocity $\mathbf{v}_2$ is resolved into two components. The radial component, 6 ft/s, is known; however, it produces zero angular momentum about the z axis. Thus,

$$\mathbf{H}_1 = \mathbf{H}_2$$

$$r_1 m_B v_1 = r_2 m_B v_2'$$

$$1.75 \text{ ft} \left(\frac{0.8 \text{ lb}}{32.2 \text{ ft/s}^2} \right) 4 = 0.6 \text{ ft} \left(\frac{0.8 \text{ lb}}{32.2 \text{ ft/s}^2} \right) v_2'$$

$$v_2' = 11.67 \text{ ft/s}$$

The speed of the ball is thus

$$v_2 = \sqrt{(11.67)^2 + (6)^2}$$

$$= 13.1 \text{ ft/s}$$

Part (b).   The only force that does work on the ball is $\mathbf{F}$. (The normal force and weight do not move vertically.) The initial and final kinetic energies of the ball can be determined so that from the principle of work and energy we have

$$T_1 + \Sigma U_{1-2} = T_2$$

$$\frac{1}{2} \left(\frac{0.8 \text{ lb}}{32.2 \text{ ft/s}^2} \right) (4 \text{ ft/s})^2 + U_F = \frac{1}{2} \left(\frac{0.8 \text{ lb}}{32.2 \text{ ft/s}^2} \right) (13.1 \text{ ft/s})^2$$

$$U_F = 1.94 \text{ ft} \cdot \text{lb} \qquad\qquad Ans.$$

EXAMPLE 15.15

The 2-kg disk shown in Fig. 15–26a rests on a smooth horizontal surface and is attached to an elastic cord that has a stiffness $k_c = 20$ N/m and is initially unstretched. If the disk is given a velocity $(v_D)_1 = 1.5$ m/s, perpendicular to the cord, determine the rate at which the cord is being stretched and the speed of the disk at the instant the cord is stretched 0.2 m.

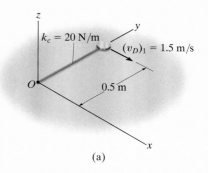

(a)

SOLUTION

Free-Body Diagram.   After the disk has been launched, it slides along the path shown in Fig. 15–26b. By inspection, angular momentum about point O (or the z axis) is *conserved*, since none of the forces produce an angular impulse about this axis. Also, when the distance is 0.7 m, only the component $(v_D')_2$ produces angular momentum of the disk about O.

Conservation of Angular Momentum.   The component $(\mathbf{v}_D')_2$ can be obtained by applying the conservation of angular momentum about O (the z axis), i.e.,

$$(\mathbf{H}_O)_1 = (\mathbf{H}_O)_2$$

$$r_1 m_D (v_D)_1 = r_2 m_D (v_D')_2$$

$(\nearrow +)$ $0.5 \text{ m } (2 \text{ kg})(1.5 \text{ m/s}) = 0.7 \text{ m } (2 \text{ kg})(v_D')_2$

$$(v_D')_2 = 1.07 \text{ m/s}$$

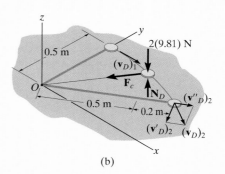

(b)

Fig. 15–26

Conservation of Energy.   The speed of the disk may be obtained by applying the conservation of energy equation at the point where the disk was launched and at the point where the cord is stretched 0.2 m.

$$T_1 + V_1 = T_2 + V_2$$

$$\tfrac{1}{2}(2 \text{ kg})(1.5 \text{ m/s})^2 + 0 = \tfrac{1}{2}(2 \text{ kg})(v_D)_2^2 + \tfrac{1}{2}(20 \text{ N/m})(0.2 \text{ m})^2$$

Thus,

$$(v_D)_2 = 1.36 \text{ m/s} \qquad Ans.$$

Having determined $(v_D)_2$ and its component $(v_D')_2$, the rate of stretch of the cord $(v_D'')_2$ is determined from the Pythagorean theorem,

$$(v_D'')_2 = \sqrt{(v_D)_2^2 - (v_D')_2^2}$$
$$= \sqrt{(1.36)^2 - (1.07)^2}$$
$$= 0.838 \text{ m/s} \qquad Ans.$$

PROBLEMS

15–90. Determine the angular momentum of the 2-lb particle A about point O. Use a Cartesian vector solution.

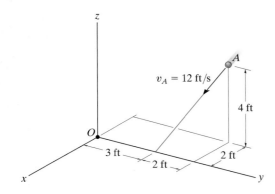

Prob. 15–90

15–91. Determine the angular momentum $\mathbf{H}_O$ of the particle about point O.

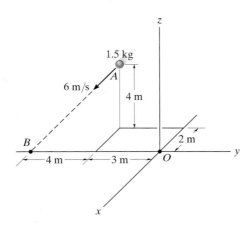

Prob. 15–91

*15–92. Determine the angular momentum $\mathbf{H}_O$ of each of the particles about point O.

15–93. Determine the angular momentum $\mathbf{H}_P$ of each of the particles about point P.

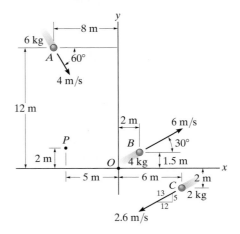

Probs. 15–92/93

15–94. Determine the angular momentum $\mathbf{H}_O$ of the particle about point O.

15–95. Determine the angular momentum $\mathbf{H}_P$ of the particle about point P.

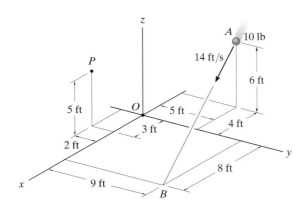

Probs. 15–94/95

*15–96.   Determine the total angular momentum $\mathbf{H}_O$ for the system of three particles about point O. All the particles are moving in the x–y plane.

Prob. 15–96

15–97.   Determine the angular momentum $\mathbf{H}_O$ of each of the two particles about point O. Use a scalar solution.

15–98.   Determine the angular momentum $\mathbf{H}_P$ of each of the two particles about point P. Use a scalar solution.

Probs. 15–97/98

15–99.   The ball B has a mass of 10 kg and is attached to the end of a rod whose mass may be neglected. If the rod is subjected to a torque $M = (3t^2 + 5t + 2)\,\text{N} \cdot \text{m}$, where t is in seconds, determine the speed of the ball when $t = 2$ s. The ball has a speed $v = 2$ m/s when $t = 0$.

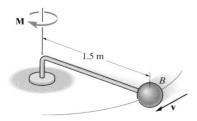

Prob. 15–99

*15–100.   The two blocks A and B each have a mass of 400 g. The blocks are fixed to the horizontal rods, and their initial velocity is 2 m/s in the direction shown. If a couple moment of $M = 0.6\,\text{N} \cdot \text{m}$ is applied about CD of the frame, determine the speed of the blocks when $t = 3$ s. The mass of the frame is negligible, and it is free to rotate about CD. Neglect the size of the blocks.

Prob. 15–100

15–101.   The small cylinder C has a mass of 10 kg and is attached to the end of a rod whose mass may be neglected. If the frame is subjected to a couple $M = (8t^2 + 5)\,\text{N} \cdot \text{m}$, where t is in seconds, and the cylinder is subjected to a force of 60 N, which is always directed as shown, determine the speed of the cylinder when $t = 2$ s. The cylinder has a speed $v_0 = 2$ m/s when $t = 0$.

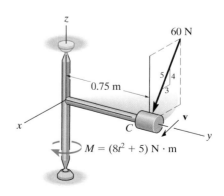

Prob. 15–101

15–102. A box having a weight of 8 lb is moving around in a circle of radius $r_A = 2$ ft with a speed of $(v_A)_1 = 5$ ft/s while connected to the end of a rope. If the rope is pulled inward with a constant speed of $v_r = 4$ ft/s, determine the speed of the box at the instant $r_B = 1$ ft. How much work is done after pulling in the rope from A to B? Neglect friction and the size of the box.

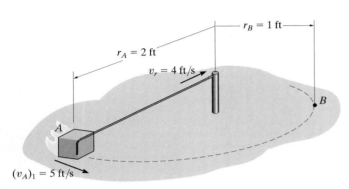

Prob. 15–102

15–103. An earth satellite of mass 700 kg is launched into a free-flight trajectory about the earth with an initial speed of $v_A = 10$ km/s when the distance from the center of the earth is $r_A = 15$ Mm. If the launch angle at this position is $\phi_A = 70°$, determine the speed v_B of the satellite and its closest distance r_B from the center of the earth. The earth has a mass $M_e = 5.976(10^{24})$ kg. *Hint:* Under these conditions, the satellite is subjected only to the earth's gravitational force, $F = GM_e m_s/r^2$, Eq. 13–1. For part of the solution, use the conservation of energy.

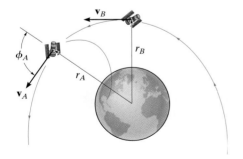

Prob. 15–103

*__15–104.__ The ball B has a weight of 5 lb and is originally rotating in a circle. As shown, the cord AB has a length of 3 ft and passes through the hole A, which is 2 ft above the plane of motion. If 1.5 ft of cord is pulled through the hole, determine the speed of the ball when it moves in a circular path at C.

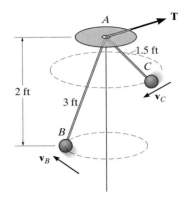

Prob. 15–104

15–105. The 10-lb block rests on a surface for which $\mu_k = 0.5$. It is acted upon by a radial force of 2 lb and a horizontal force of 7 lb, always directed at 30° from the tangent to the path as shown. If the block is initially moving in a circular path with a speed $v_1 = 2$ ft/s at the instant the forces are applied, determine the time required before the tension in cord AB becomes 20 lb. Neglect the size of the block for the calculation.

15–106. The 10-lb block is originally at rest on the smooth surface. It is acted upon by a radial force of 2 lb and a horizontal force of 7 lb, always directed at 30° from the tangent to the path as shown. Determine the time required to break the cord, which requires a tension $T = 30$ lb. What is the speed of the block when this occurs? Neglect the size of the block for the calculation.

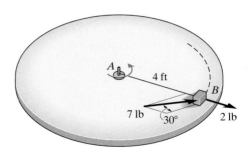

Probs. 15–105/106

15–107. The 800-lb roller-coaster car starts from rest on the track having the shape of a cylindrical helix. If the helix descends 8 ft for every one revolution, determine the time required for the car to attain a speed of 60 ft/s. Neglect friction and the size of the car.

Prob. 15–107

15–109. A small particle having a mass m is placed inside the semicircular tube. The particle is placed at the position shown and released. Apply the principle of angular momentum about point O $(\Sigma M_O = \dot{H}_O)$, and show that the motion of the particle is governed by the differential equation $\ddot{\theta} + (g/R) \sin \theta = 0$.

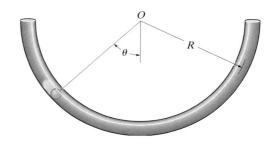

Prob. 15–109

*15–108.** A child having a mass of 50 kg holds her legs up as shown as she swings downward from rest at $\theta_1 = 30°$. Her center of mass is located at point G_1. When she is at the bottom position $\theta = 0°$, she *suddenly* lets her legs come down, shifting her center of mass to position G_2. Determine her speed in the upswing due to this sudden movement and the angle θ_2 to which she swings before momentarily coming to rest. Treat the child's body as a particle.

Prob. 15–108

15–110. A toboggan and rider, having a total mass of 150 kg, enter horizontally tangent to a 90° circular curve with a velocity of $v_A = 70$ km/h. If the track is flat and banked at an angle of 60°, determine the speed v_B and the angle θ of "descent," measured from the horizontal in a vertical x–z plane, at which the toboggan exists at B. Neglect friction in the calculation.

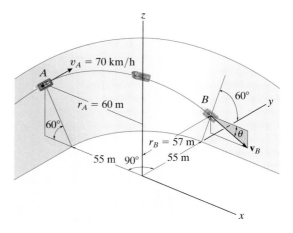

Prob. 15–110

*15.8   Steady Fluid Streams

Knowledge of the forces developed by steadily moving fluid streams is of importance in the design and analysis of turbines, pumps, blades, and fans. To illustrate how the principle of impulse and momentum may be used to determine these forces, consider the diversion of a steady stream of fluid (liquid or gas) by a fixed pipe, Fig. 15–27a. The fluid enters the pipe with a velocity $\mathbf{v}_A$ and exits with a velocity $\mathbf{v}_B$. The impulse and momentum diagrams for the fluid stream are shown in Fig. 15–27b. The force $\Sigma\mathbf{F}$, shown on the impulse diagram, represents the resultant of all the external forces acting on the fluid stream. It is this loading which gives the fluid stream an impulse whereby the original momentum of the fluid is changed in both its magnitude and direction.

Since the flow is steady, $\Sigma\mathbf{F}$ will be *constant* during the time interval dt. During this time the fluid stream is in motion, and as a result a small amount of fluid, having a mass dm, is about to enter the pipe with a velocity $\mathbf{v}_A$ at time t. If this element of mass and the mass of fluid in the pipe are considered as a "closed system," then at time $t + dt$ a corresponding element of mass dm must leave the pipe with a velocity $\mathbf{v}_B$. Also, the fluid stream *within* the pipe section has a mass m and an *average velocity* $\mathbf{v}$ which is constant during the time interval dt. Applying the principle of linear impulse and momentum to the fluid stream, we have

$$dm\,\mathbf{v}_A + m\mathbf{v} + \Sigma\mathbf{F}\,dt = dm\,\mathbf{v}_B + m\mathbf{v}$$

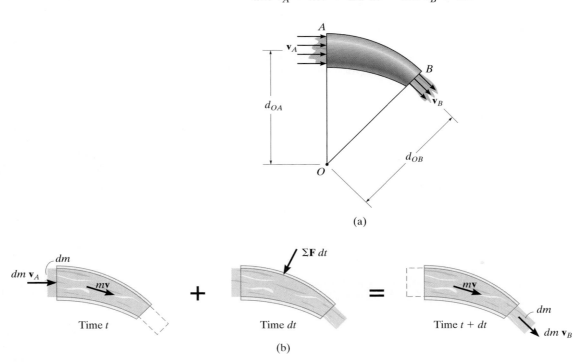

(a)

(b)

Fig. 15–27

Force Resultant.

Solving for the resultant force yields

$$\Sigma \mathbf{F} = \frac{dm}{dt}(\mathbf{v}_B - \mathbf{v}_A) \qquad (15\text{--}25)$$

Provided the motion of the fluid can be represented in the x–y plane, it is usually convenient to express this vector equation in the form of two scalar component equations, i.e.,

$$\Sigma F_x = \frac{dm}{dt}(v_{Bx} - v_{Ax})$$
$$\Sigma F_y = \frac{dm}{dt}(v_{By} - v_{Ay}) \qquad (15\text{--}26)$$

The term dm/dt is called the *mass flow* and indicates the constant amount of fluid which flows either into or out of the pipe per unit of time. If the cross-sectional areas and densities of the fluid at the entrance A and exit B are A_A, ρ_A and A_B, ρ_B, respectively, Fig. 15–27c, then *continuity of mass* requires that $dm = \rho\, dV = \rho_A(ds_A A_A) = \rho_B(ds_B A_B)$. Hence, during the time dt, since $v_A = ds_A/dt$ and $v_B = ds_B/dt$, we have

$$\frac{dm}{dt} = \rho_A v_A A_A = \rho_B v_B A_B = \rho_A Q_A = \rho_B Q_B \qquad (15\text{--}27)$$

Here $Q = vA$ is the volumetric *flow rate*, which measures the volume of fluid flowing per unit of time.

(c)

Fig. 15–27 (cont'd)

The conveyor belt must supply frictional forces to the gravel that falls upon it in order to change the momentum of the gravel stream, so that it begins to travel along the belt.

The air on one side of this fan is essentially at rest, and as it passes through the blades its momentum is increased. To change the momentum of the air flow in this manner, the blades must exert a horizontal thrust on the air stream. As the blades turn faster, the equal but opposite thrust of the air on the blades could overcome the rolling resistance of the wheels on the ground and begin to move the frame of the fan.

Moment Resultant.   In some cases it is necessary to obtain the support reactions on the fluid-carrying device. If Eq. 15–25 does not provide enough information to do this, the principle of angular impulse and momentum must be used. The formulation of this principle applied to fluid streams can be obtained from Eq. 15–17, $\Sigma \mathbf{M}_O = \dot{\mathbf{H}}_O$, which states that the moment of all the external forces acting on the system about point O is equal to the time rate of change of angular momentum about O. In the case of the pipe shown in Fig. 15–27a, the flow is steady in the x–y plane; hence we have

$(\curvearrowright +)$

$$\Sigma M_O = \frac{dm}{dt}(d_{OB}v_B - d_{OA}v_A) \qquad (15\text{–}28)$$

where the moment arms d_{OB} and d_{OA} are directed from O to the *geometric center* or *centroid* of the openings at A and B.

Procedure for Analysis

Problems involving steady flow can be solved using the following procedure.

Kinematic Diagram.

- If the device is *moving*, a *kinematic diagram* may be helpful for determining the entrance and exit velocities of the fluid flowing onto the device, since a *relative-motion analysis* of velocity will be involved.
- The measurement of velocities v_A and v_B must be made by an observer fixed in an inertial frame of reference.
- Once the velocity of the fluid flowing onto the device is determined, the mass flow is calculated using Eq. 15–27.

Free-Body Diagram.

- Draw a free-body diagram of the device which is directing the fluid in order to establish the forces $\Sigma \mathbf{F}$ that act on it. These external forces will include the support reactions, the weight of the device and the fluid contained within it, and the static pressure forces of the fluid at the entrance and exit sections of the device.*

Equations of Steady Flow.

- Apply the equations of steady flow, Eqs. 15–26 and 15–28, using the appropriate components of velocity and force shown on the kinematic and free-body diagrams.

*In the SI system, pressure is measured using the pascal (Pa), where 1 Pa = 1 N/m.

EXAMPLE 15.16

Determine the components of reaction which the fixed pipe joint at A exerts on the elbow in Fig. 15–28a, if water flowing through the pipe is subjected to a static gauge pressure of 100 kPa at A. The discharge at B is $Q_B = 0.2 \text{ m}^3/\text{s}$. Water has a density $\rho_w = 1000 \text{ kg/m}^3$, and the water-filled elbow has a mass of 20 kg and center of mass at G.

(a)

SOLUTION

Using a fixed inertial coordinate system, the velocity of flow at A and B and the mass flow rate can be obtained from Eq. 15–27. Since the density of water is constant, $Q_B = Q_A = Q$. Hence,

$$\frac{dm}{dt} = \rho_w Q = (1000 \text{ kg/m}^3)(0.2 \text{ m}^3/\text{s}) = 200 \text{ kg/s}$$

$$v_B = \frac{Q}{A_B} = \frac{0.2 \text{ m}^3/\text{s}}{\pi(0.05 \text{ m})^2} = 25.46 \text{ m/s} \downarrow$$

$$v_A = \frac{Q}{A_A} = \frac{0.2 \text{ m}^3/\text{s}}{\pi(0.1 \text{ m})^2} = 6.37 \text{ m/s} \rightarrow$$

Free-Body Diagram. As shown on the free-body diagram, Fig. 15–28b, the *fixed* connection at A exerts a resultant couple moment M_O and force components $\mathbf{F}_x$ and $\mathbf{F}_y$ on the elbow. Due to the static pressure of water in the pipe, the pressure force acting on the fluid at A is $F_A = p_A A_A$. Since 1 kPa = 1000 N/m^2,

$$F_A = p_A A_A = [100(10^3) \text{ N/m}^2][\pi(0.1 \text{ m})^2] = 3141.6 \text{ N}$$

There is no static pressure acting at B, since the water is discharged at atmospheric pressure; i.e., the pressure measured by a gauge at B is equal to zero, $p_B = 0$.

(b)

Fig. 15–28

Equations of Steady Flow.

$$\xrightarrow{+} \Sigma F_x = \frac{dm}{dt}(v_{Bx} - v_{Ax}); \quad -F_x + 3141.6 \text{ N} = 200 \text{ kg/s} \, (0 - 6.37 \text{ m/s})$$

$$F_x = 4.41 \text{ kN} \qquad\qquad\qquad\qquad Ans.$$

$$+\uparrow \Sigma F_y = \frac{dm}{dt}(v_{By} - v_{Ay}); \quad -F_y - 20(9.81) \text{ N} = 200 \text{ kg/s} \, (-25.46 \text{ m/s} - 0)$$

$$F_y = 4.90 \text{ kN} \qquad\qquad\qquad\qquad Ans.$$

If moments are summed about point O, Fig. 15–28b, then $\mathbf{F}_x$, $\mathbf{F}_y$, and the static pressure $\mathbf{F}_A$ are eliminated, as well as the moment of momentum of the water entering at A, Fig. 15–28a. Hence,

$$\zeta + \Sigma M_O = \frac{dm}{dt}(d_{OB}v_B - d_{OA}v_A)$$

$$M_O + 20(9.81) \text{ N} (0.125 \text{ m}) = 200 \text{ kg/s} \, [(0.3 \text{ m})(25.46 \text{ m/s}) - 0]$$

$$M_O = 1.50 \text{ kN} \cdot \text{m} \qquad\qquad\qquad\qquad Ans.$$

EXAMPLE 15.17

2 in.

$v_{bl} = 5$ ft/s

$v_w = 25$ ft/s

(a)

A 2-in.-diameter water jet having a velocity of 25 ft/s impinges upon a single moving blade, Fig. 15–29a. If the blade is moving at 5 ft/s away from the jet, determine the horizontal and vertical components of force which the blade is exerting on the water. What power does the water generate on the blade? Water has a specific weight of $\gamma_w = 62.4$ lb/ft³.

SOLUTION

Kinematic Diagram. From a fixed inertial coordinate system, Fig. 15–29b, the rate at which water enters the blade is

$$\mathbf{v}_A = \{25\mathbf{i}\} \text{ ft/s}$$

The *relative-flow velocity* of the water onto the blade is $\mathbf{v}_{w/bl} = \mathbf{v}_w - \mathbf{v}_{bl} = 25\mathbf{i} - 5\mathbf{i} = \{20\mathbf{i}\}$ ft/s. Since the blade is moving with a velocity of $\mathbf{v}_{bl} = \{5\mathbf{i}\}$ ft/s, the velocity of flow at B measured from x, y is the vector sum, shown in Fig. 15–29b. Here,

$$\mathbf{v}_B = \mathbf{v}_{bl} + \mathbf{v}_{w/bl}$$

$$= \{5\mathbf{i} + 20\mathbf{j}\} \text{ ft/s}$$

Thus, the mass flow of water *onto* the blade that undergoes a momentum change is

$$\frac{dm}{dt} = \rho_w(v_{w/bl})A_A = \frac{62.4}{32.2}(20)\left[\pi\left(\frac{1}{12}\right)^2\right] = 0.846 \text{ slug/s}$$

(b)

Free-Body Diagram. The free-body diagram of a section of water acting on the blade is shown in Fig. 15–29c. The weight of the water will be neglected in the calculation, since this force will be small compared to the reactive components $\mathbf{F}_x$ and $\mathbf{F}_y$.

Equations of Steady Flow.

$$\Sigma\mathbf{F} = \frac{dm}{dt}(\mathbf{v}_B - \mathbf{v}_A)$$

$$-F_x\mathbf{i} + F_y\mathbf{j} = 0.846(5\mathbf{i} + 20\mathbf{j} - 25\mathbf{i})$$

Equating the respective $\mathbf{i}$ and $\mathbf{j}$ components gives

$$F_x = 0.846(20) = 16.9 \text{ lb} \leftarrow \qquad\qquad Ans.$$

$$F_y = 0.846(20) = 16.9 \text{ lb} \uparrow \qquad\qquad Ans.$$

The water exerts equal but opposite forces on the blade.

Since the water force which causes the blade to move forward horizontally with a velocity of 5 ft/s is $F_x = 16.9$ lb, then from Eq. 14–10 the power is

(c)

Fig. 15–29

$$P = \mathbf{F} \cdot \mathbf{v}; \qquad P = \frac{16.9 \text{ lb } (5 \text{ ft/s})}{550 \text{ hp/(ft} \cdot \text{lb/s})} = 0.154 \text{ hp} \qquad\qquad Ans.$$

*15.9 Propulsion with Variable Mass

In the previous section we considered the case in which a *constant* amount of mass dm enters and leaves a "*closed system*." There are, however, two other important cases involving mass flow, which are represented by a system that is either gaining or losing mass. In this section we will discuss each of these cases separately.

A System That Loses Mass.   Consider a device such as a rocket which at an instant of time has a mass m and is moving forward with a velocity $\mathbf{v}$, Fig. 15–30*a*. At this same instant the device is expelling an amount of mass m_e with a mass flow velocity $\mathbf{v}_e$. For the analysis, the "closed system" includes *both the mass m of the device and the expelled mass m_e.* The impulse and momentum diagrams for the system are shown in Fig. 15–30*b*. During the time dt, the velocity of the device is increased from $\mathbf{v}$ to $\mathbf{v} + d\mathbf{v}$ since an amount of mass dm_e has been ejected and thereby gained in the exhaust. This increase in forward velocity, however, does not change the velocity $\mathbf{v}_e$ of the expelled mass, since this mass moves at a constant speed once it has been ejected. The impulses are created by $\Sigma\mathbf{F}_s$, which represents the resultant of all the external forces that *act on the system* in the direction of motion. This force resultant *does not include* the force which causes the device to move forward, since this force (called a *thrust*) is *internal to the system*; that is, the thrust acts with equal magnitude but opposite direction on the mass m of the device and the expelled exhaust mass m_e.* Applying the principle of impulse and momentum to the system, Fig. 15–30*b*, we have

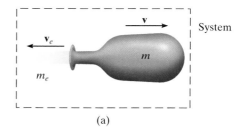

(a)

$$(\xrightarrow{+}) \quad mv - m_e v_e + \Sigma F_s\, dt = (m - dm_e)(v + dv) - (m_e + dm_e)v_e$$

or

$$\Sigma F_s\, dt = -v\, dm_e + m\, dv - dm_e\, dv - v_e\, dm_e$$

(b)

Fig. 15–30

*$\Sigma\mathbf{F}_s$ represents the external resultant force *acting on the system*, which is different from $\Sigma\mathbf{F}$, the resultant force acting only on the device.

Fig. 15–31

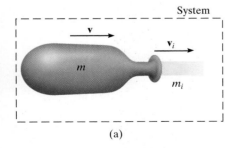

(a)

Fig. 15–32

Without loss of accuracy, the third term on the right side may be neglected since it is a "second-order" differential. Dividing by dt gives

$$\Sigma F_s = m\frac{dv}{dt} - (v + v_e)\frac{dm_e}{dt}$$

The relative velocity of the device as seen by an observer moving with the particles of the ejected mass is $v_{D/e} = (v + v_e)$, and so the final result can be written as

$$\Sigma F_s = m\frac{dv}{dt} - v_{D/e}\frac{dm_e}{dt} \qquad (15\text{–}29)$$

Here the term dm_e/dt represents the rate at which mass is being ejected.

To illustrate an application of Eq. 15–29, consider the rocket shown in Fig. 15–31, which has a weight $\mathbf{W}$ and is moving upward against an atmospheric drag force $\mathbf{F}_D$. The system to be considered consists of the mass of the rocket and the mass of ejected gas m_e. Applying Eq. 15–29 to this system gives

$$(+\uparrow) \qquad -F_D - W = \frac{W}{g}\frac{dv}{dt} - v_{D/e}\frac{dm_e}{dt}$$

The last term of this equation represents the *thrust* $\mathbf{T}$ which the engine exhaust exerts on the rocket, Fig. 15–31. Recognizing that $dv/dt = a$, we may therefore write

$$(+\uparrow) \qquad T - F_D - W = \frac{W}{g}a$$

If a free-body diagram of the rocket is drawn, it becomes obvious that this equation represents an application of $\Sigma F = ma$ for the rocket.

A System That Gains Mass. A device such as a scoop or a shovel may gain mass as it moves forward. For example, the device shown in Fig. 15–32a has a mass m and is moving forward with a velocity $\mathbf{v}$. At this instant, the device is collecting a particle stream of mass m_i. The flow velocity $\mathbf{v}_i$ of this injected mass is constant and independent of the velocity $\mathbf{v}$ such that $v > v_i$. The system to be considered includes both the mass of the device and the mass of the injected particles. The impulse and momentum diagrams for this system are shown in Fig. 15–32b. Along with an increase in mass dm_i gained by the device, there is an assumed increase in velocity $d\mathbf{v}$ during the time interval dt. This increase is caused by the impulse created by $\Sigma\mathbf{F}_s$, the resultant of all the external forces *acting on the system* in the direction of motion. The force summation does not

Time t

$(m + dm_i)(v + dv)$

$(m_i - dm_i)v_i$

Time $t + dt$

(b)

include the retarding force of the injected mass acting on the device. Why? Applying the principle of impulse and momentum to the system, we have

$$(\xrightarrow{+})\quad mv + m_iv_i + \Sigma F_s\,dt = (m + dm_i)(v + dv) + (m_i - dm_i)v_i$$

Using the same procedure as in the previous case, we may write this equation as

$$\Sigma F_s = m\frac{dv}{dt} + (v - v_i)\frac{dm_i}{dt}$$

Since the relative velocity of the device as seen by an observer moving with the particles of the injected mass is $v_{D/i} = (v - v_i)$, the final result can be written as

$$\Sigma F_s = m\frac{dv}{dt} + v_{D/i}\frac{dm_i}{dt} \qquad (15\text{-}30)$$

where dm_i/dt is the rate of mass injected into the device. The last term in this equation represents the magnitude of force $\mathbf{R}$, which the injected mass *exerts on the device*. Since $dv/dt = a$, Eq. 15–30 becomes

$$\Sigma F_s - R = ma$$

This is the application of $\Sigma \mathbf{F} = m\mathbf{a}$, Fig. 15–32c.

As in the case of steady flow, problems which are solved using Eqs. 15–29 and 15–30 should be accompanied by the necessary free-body diagram. With this diagram one can then determine ΣF_s *for the system* and isolate the force exerted on the device by the particle stream.

(c)

The scraper box behind this tractor represents a device that gains mass. If the tractor maintains a constant velocity v, then $dv/dt = 0$ and, because the soil is originally at rest, $v_{D/i} = v$. By Eq. 15–30, the horizontal towing force on the scraper box is then $T = 0 + v(dm/dt)$, where dm/dt is the rate of soil accumulation in the box.

EXAMPLE 15.18

The initial combined mass of a rocket and its fuel is m_0. A total mass m_f of fuel is consumed at a constant rate of $dm_e/dt = c$ and expelled at a constant speed of u relative to the rocket. Determine the maximum velocity of the rocket, i.e., at the instant the fuel runs out. Neglect the change in the rocket's weight with altitude and the drag resistance of the air. The rocket is fired vertically from rest.

SOLUTION

Since the rocket is losing mass as it moves upward, Eq. 15–29 can be used for the solution. The only *external force* acting on the *system* consisting of the rocket and a portion of the expelled mass is the weight **W**, Fig. 15–33. Hence,

$$+\uparrow \Sigma F_s = m\frac{dv}{dt} - v_{D/e}\frac{dm_e}{dt}; \qquad -W = m\frac{dv}{dt} - uc \qquad (1)$$

The rocket's velocity is obtained by integrating this equation.

At any given instant t during the flight, the mass of the rocket can be expressed as $m = m_0 - (dm_e/dt)t = m_0 - ct$. Since $W = mg$, Eq. 1 becomes

$$-(m_0 - ct)g = (m_0 - ct)\frac{dv}{dt} - uc$$

Separating the variables and integrating, realizing that $v = 0$ at $t = 0$, we have

$$\int_0^v dv = \int_0^t \left(\frac{uc}{m_0 - ct} - g\right)dt$$

$$v = -u\ln(m_0 - ct) - gt\Big|_0^t = u\ln\left(\frac{m_0}{m_0 - ct}\right) - gt \qquad (2)$$

Note that lift off requires the first term on the right to be greater than the second during the initial phase of motion. The time t' needed to consume all the fuel is given by

$$m_f = \left(\frac{dm_e}{dt}\right)t' = ct'$$

Hence,

$$t' = m_f/c$$

System

W

T

Fig. 15–33

Substituting into Eq. 2 yields

$$v_{\text{max}} = u\ln\left(\frac{m_0}{m_0 - m_f}\right) - \frac{gm_f}{c} \qquad \qquad Ans.$$

EXAMPLE 15.19

A chain of length l, Fig. 15–34a, has a mass m. Determine the magnitude of force $\mathbf{F}$ required to (a) raise the chain with a constant speed v_c, starting from rest when $y = 0$; and (b) lower the chain with a constant speed v_c, starting from rest when $y = l$.

SOLUTION

Part (a).   As the chain is raised, all the suspended links are given a sudden impulse downward by each added link which is lifted off the ground. Thus, the *suspended portion* of the chain may be considered as a device which is *gaining mass*. The system to be considered is the length of chain y which is suspended by $\mathbf{F}$ at any instant, including the next link which is about to be added but is still at rest, Fig. 15–34b. The forces acting on this system *exclude* the internal forces $\mathbf{P}$ and $-\mathbf{P}$, which act between the added link and the suspended portion of the chain. Hence, $\Sigma F_s = F - mg(y/l)$.

To apply Eq. 15–30, it is also necessary to find the rate at which mass is being added to the system. The velocity $\mathbf{v}_c$ of the chain is equivalent to $\mathbf{v}_{D/i}$. Why? Since v_c is constant, $dv_c/dt = 0$ and $dy/dt = v_c$. Integrating, using the initial condition that $y = 0$ at $t = 0$, gives $y = v_c t$. Thus, the mass of the system at any instant is $m_s = m(y/l) = m(v_c t/l)$, and therefore the *rate* at which mass is *added* to the suspended chain is

$$\frac{dm_i}{dt} = m\left(\frac{v_c}{l}\right)$$

Applying Eq. 15–30 to the system, using this data, we have

$$+\uparrow \Sigma F_s = m\frac{dv_c}{dt} + v_{D/i}\frac{dm_i}{dt}$$

$$F - mg\left(\frac{y}{l}\right) = 0 + v_c m\left(\frac{v_c}{l}\right)$$

Hence,

$$F = (m/l)(gy + v_c^2) \qquad \textit{Ans.}$$

Part (b).   When the chain is being lowered, the links which are expelled (given zero velocity) *do not* impart an impulse to the *remaining* suspended links. Why? Thus, the system in Part (a) cannot be considered. Instead, the equation of motion will be used to obtain the solution. At time t the portion of chain still off the floor is y. The free-body diagram for a suspended portion of the chain is shown in Fig. 15–34c. Thus,

$$+\uparrow \Sigma F = ma; \qquad F - mg\left(\frac{y}{l}\right) = 0$$

$$F = mg\left(\frac{y}{l}\right) \qquad \textit{Ans.}$$

(a)

(b)

(c)

Fig. 15–34

PROBLEMS

15–111. Water is discharged at 16 m/s against the fixed cone diffuser. If the opening diameter of the nozzle is 40 mm, determine the horizontal force exerted by the water on the diffuser. $\rho_w = 1$ Mg/m³.

Prob. 15–111

*15–112. A jet of water having a cross-sectional area of 4 in² strikes the fixed blade with a speed of 25 ft/s. Determine the horizontal and vertical components of force which the blade exerts on the water. $\gamma_w = 62.4$ lb/ft³.

Prob. 15–112

15–113. Water is flowing from the 150-mm-diameter fire hydrant with a velocity $v_B = 15$ m/s. Determine the horizontal and vertical components of force and the moment developed at the base joint A, if the static (gauge) pressure at A is 50 kPa. The diameter of the fire hydrant at A is 200 mm. $\rho_w = 1$ Mg/m³.

Prob. 15–113

15–114. The chute is used to divert the flow of water $Q = 0.6$ m³/s. If the water has a cross-sectional area of 0.05 m² determine the force components at the pin A and roller B necessary for equilibrium. Neglect both the weight of the chute and the weight of the water on the chute. $\rho_w = 1$ Mg/m³.

Prob. 15–114

15–115. The fan draws air through a vent with a speed of 12 ft/s. If the cross-sectional area of the vent is 2 ft², determine the horizontal thrust on the blade. The specific weight of the air is $\gamma_a = 0.076$ lb/ft³.

Prob. 15–115

*15–116. The buckets on the *Pelton wheel* are subjected to a 2-in-diameter jet of water, which has a velocity of 150 ft/s. If each bucket is traveling at 95 ft/s when the water strikes it, determine the power developed by the wheel. $\gamma_w = 62.4$ lb/ft^3.

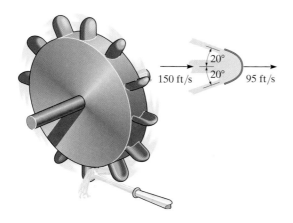

Prob. 15–116

15–117. The 200-kg boat is powered by a fan F which develops a slipstream having a diameter of 0.75 m. If the fan ejects air with a speed of 14 m/s, measured relative to the boat, determine the initial acceleration of the boat if it is initially at rest. Assume that air has a constant density $\rho_a = 1.22$ kg/m^3 and that the entering air is essentially at rest. Neglect the drag resistance of the water.

Prob. 15–117

15–118. The rocket car has a mass of 3 Mg (empty) and carries 150 kg of fuel. If the fuel is consumed at a constant rate of 4 kg/s and ejected from the car with a relative velocity of 250 m/s, determine the maximum speed attained by the car starting from rest. The drag resistance due to the atmosphere is $F_D = (60v^2)$ N, where v is the speed measured in m/s.

Prob. 15–118

15–119. A power lawn mower hovers very close over the ground. This is done by drawing air in at a speed of 6 m/s through an intake unit A, which has a cross-sectional area of $A_A = 0.25$ m^2, and then discharging it at the ground, B, where the cross-sectional area is $A_B = 0.35$ m^2. If air at A is subjected only to atmospheric pressure, determine the air pressure which the lawn mower exerts on the ground when the weight of the mower is freely supported and no load is placed on the handle. The mower has a mass of 15 kg with center of mass at G. Assume that air has a constant density of $\rho_a = 1.22$ kg/m^3.

Prob. 15–119

*15–120. The elbow for a 5-in-diameter buried pipe is subjected to a static pressure of 10 lb/in². The speed of the water passing through it is $v = 8$ ft/s. Assuming the pipe connections at A and B do not offer any vertical force resistance on the elbow, determine the resultant vertical force **F** that the soil must then exert on the elbow in order to hold it in equilibrium. Neglect the weight of the elbow and the water within it. $\gamma_w = 62.4$ lb/ft³.

Prob. 15–120

15–121. The car is used to scoop up water that is lying in a trough at the tracks. Determine the force needed to pull the car forward at constant velocity **v** for each of the three cases. The scoop has a cross-sectional area A and the density of water is ρ_w.

Prob. 15–121

15–122. A rocket has an empty weight of 500 lb and carries 300 lb of fuel. If the fuel is burned at the rate of 15 lb/s and ejected with a relative velocity of 4400 ft/s, determine the maximum speed attained by the rocket starting from rest. Neglect the effect of gravitation on the rocket.

15–123. The boat has a mass of 180 kg and is traveling forward on a river with a constant velocity of 70 km/h, measured *relative* to the river. The river is flowing in the opposite direction at 5 km/h. If a tube is placed in the water, as shown, and it collects 40 kg of water in the boat in 80 s, determine the horizontal thrust T on the tube that is required to overcome the resistance to the water collection. $\rho_w = 1$ Mg/m³.

Prob. 15–123

*15–124. The second stage of a two-stage rocket weighs 2000 lb (empty) and is launched from the first stage with a velocity of 3000 mi/h. The fuel in the second stage weighs 1000 lb. If it is consumed at the rate of 50 lb/s and ejected with a relative velocity of 8000 ft/s, determine the acceleration of the second stage just after the engine is fired. What is the rocket's acceleration just before all the fuel is consumed? Neglect the effect of gravitation.

15–125. The earthmover initially carries 10 m³ of sand having a density of 1520 kg/m³. The sand is unloaded horizontally through a 2.5-m² dumping port P at a rate of 900 kg/s measured relative to the port. If the earthmover maintains a constant resultant tractive force $F = 4$ kN at its front wheels to provide forward motion, determine its acceleration when half the sand is dumped. When empty, the earthmover has a mass of 30 Mg. Neglect any resistance to forward motion and the mass of the wheels. The rear wheels are free to roll.

Prob. 15–125

15–126. The earthmover initially carries 10 m^3 of sand having a density of 1520 kg/m^3. The sand is unloaded horizontally through a 2.5-m^2 dumping port P at a rate of 900 kg/s measured relative to the port. Determine the resultant tractive force $\mathbf{F}$ at its front wheels if the acceleration of the earthmover is 0.1 m/s^2 when half the sand is dumped. When empty, the earthmover has a mass of 30 Mg. Neglect any resistance to forward motion and the mass of the wheels. The rear wheels are free to roll.

Prob. 15–126

15–127. If the chain is lowered at a constant speed determine the normal reaction exerted on the floor as a function of time. The chain has a weight of 5 lb/ft and a total length of 20 ft.

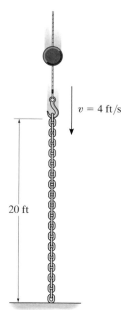

$v = 4 \text{ ft/s}$

20 ft

Prob. 15–127

***15–128.** The rocket has a mass of 65 Mg including the fuel. Determine the constant rate at which the fuel must be burned so that its thrust gives the rocket a speed of 200 ft/s in 10 s starting from rest. The fuel is expelled from the rocket at a relative speed of 3000 ft/s. Neglect the effects of air resistance and assume that g is constant.

$\mathbf{v}$

Prob. 15–128

15–129. The rocket has an initial mass m_0, including the fuel. For practical reasons desired for the crew, it is required that it maintain a constant upward acceleration a_0. If the fuel is expelled from the rocket at a relative speed $v_{e/r}$ determine the rate at which the fuel should be consumed to maintain the motion. Neglect air resistance, and assume that the gravitational acceleration is constant.

$\mathbf{a}_0$

Prob. 15–129

15–130. The 12-Mg jet airplane has a constant speed of 950 km/h when it is flying along a horizontal straight line. Air enters the intake scoops S at the rate of 50 m³/s. If the engine burns fuel at the rate of 0.4 kg/s and the gas (air and fuel) is exhausted relative to the plane with a speed of 450 m/s, determine the resultant drag force exerted on the plane by air resistance. Assume that air has a constant density of 1.22 kg/m³. *Hint:* Since mass both enters and exits the plane, Eqs. 15–29 and 15–30 must be combined to yield

$$\Sigma F_s = m\frac{dv}{dt} - v_{D/e}\frac{dm_e}{dt} + v_{D/i}\frac{dm_i}{dt}$$

Prob. 15–130

15–131. The jet is traveling at a speed of 500 mi/h, 30° with the horizontal. If the fuel is being spent at 3 lb/s, and the engine takes in air at 400 lb/s, whereas the exhaust gas (air and fuel) has a relative speed of 32 800 ft/s, determine the acceleration of the plane at this instant. The drag resistance of the air is $F_D = (0.7v^2)$ lb, where the speed is measured in ft/s. The jet has a weight of 15 000 lb. *Hint:* See Prob. 15–130.

Prob. 15–131

*15–132. The rope has a mass m' per unit length. If the end length $y = h$ is draped off the edge of the table, and released, determine the velocity of its end A for any position y, as the rope uncoils and begins to fall.

Prob. 15–132

15–133. The car has a mass m_0 and is used to tow the smooth chain having a total length l and a mass per unit of length m'. If the chain is originally piled up, determine the tractive force F that must be supplied by the rear wheels of the car, necessary to maintain a constant speed v while the chain is being drawn out.

Prob. 15–133

15–134. Determine the magnitude of force **F** as a function of time, which must be applied to the end of the cord at A to raise the hook H with a constant speed $v = 0.4$ m/s. Initially the chain is at rest on the ground. Neglect the mass of the cord and the hook. The chain has a mass of 2 kg/m.

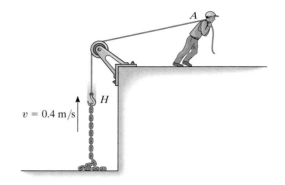

Prob. 15–134

Design Project

15–1D. Design of a Cranberry Selector

The quality of a cranberry depends upon its firmness, which in turn is related to its bounce. Through experiment, it is found that berries that bounce to a height of $2.5 \le h' \le 3.25$ ft, when released from rest at a height of $h = 4$ ft, are appropriate for processing. Using this information, determine the berry's range of the allowable coefficient of restitution, and then design a manner in which good and bad berries can be separated. Submit a drawing of your design, and show calculations as to how the selection and collection of berries is made from your established geometry.

Fig. 15–1D

Chapter Review

Impulse

An impulse that acts on a particle is defined as the product of force and time. Graphically this represents the area under the F–t diagram. If the force is constant, then the impulse becomes $\mathbf{I} = \mathbf{F}_c(t_2 - t_1)$. (*See page 216.*)

Variable Force

Principle of Impulse and Momentum

When the equation of motion, $\Sigma \mathbf{F} = m\mathbf{a}$, and the kinematic equation, $a = dv/dt$, are combined, we obtain the principle of impulse and momentum, where the initial momentum of the particle, $m\mathbf{v}_1$, plus all of the impulses that are applied to the particle during the time t_1 to t_2, $\Sigma \int \mathbf{F}\, dt$, equals the final momentum $m\mathbf{v}_2$ of the particle. This is a vector equation that can be resolved into components and is used to solve problems that involve force, velocity, and time. For application, the free-body diagram should be drawn in order to account for all the impulses that act on the particle. (*See page 217.*)

$$m\mathbf{v}_1 + \Sigma \int_{t_1}^{t_2} \mathbf{F}\, dt = m\mathbf{v}_2$$

Conservation of Linear Momentum

If the principle of impulse and momentum is applied to a system of particles, then the collisions between the particles produce internal impulses that are equal, opposite, and collinear, and therefore cancel from the equation. Furthermore, if an external impulse is small, that is, the force is small and the time is short, then the impulse can be classified as nonimpulsive and can be neglected. Consequently, momentum for the system of particles is conserved.

The conservation-of-momentum equation is useful for finding the final velocity of a particle when internal impulses are exerted between two particles. If the internal impulse is to be determined, then one of the particles is isolated and the principle of impulse and momentum is applied to this particle. (*See page 230.*)

$$\Sigma m_i(\mathbf{v}_i)_1 = \Sigma m_i(\mathbf{v}_i)_2$$

Impact

When two particles A and B have a direct impact, the internal impulse between them is equal, opposite, and collinear. Consequently the conservation of momentum for this system applies along the line of impact.

$$m_A(v_A)_1 + m_B(v_B)_1 = m_A(v_A)_2 + m_B(v_B)_2$$

If the final velocities are unknown, a second equation is needed for solution. We must use the coefficient of restitution, e. This experimentally determined coefficient depends upon the physical properties of the colliding particles. It can be expressed as the ratio of the relative velocity after collision to the relative velocity before collision.

$$e = \frac{(v_B)_2 - (v_A)_2}{(v_A)_1 - (v_B)_1}$$

If the collision is elastic, no energy is lost and $e = 1$. For a plastic collision $e = 0$.

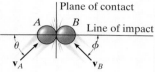

If the impact is oblique, then conservation of momentum for the system and the coefficient-of-restitution equation apply along the line of impact. Also, conservation of momentum for each particle applies perpendicular to this line, because no impulse acts on the particles in this direction. (*See pages 242–244.*)

Principle of Angular Impulse and Momentum

The moment of the linear momentum about an axis (z) is called the angular momentum.

$$(H_O)_z = (d)(mv)$$

The principle of angular impulse and momentum is often used to eliminate unknown impulses by summing the moments about an axis through which the lines of action of these impulses produce no moment. For this reason, a free-body diagram should accompany the solution. (*See pages 256 and 260.*)

$$(\mathbf{H}_O)_1 + \Sigma \int_{t_1}^{t_2} \mathbf{M}_O \, dt = (\mathbf{H}_O)_2$$

Steady Fluid Streams

Impulse-and-momentum methods are often used to determine the forces that a device exerts on the mass flow of a fluid—liquid or gas. To do so, a free-body diagram of the fluid mass in contact with the device is drawn in order to identify these forces. Also, the velocity of the fluid as it flows into and out of the device is calculated. The equations of steady flow involve summing the forces and the moments to determine these reactions. (*See pages 270–272.*)

$$\Sigma \mathbf{F} = \frac{dm}{dt}(\mathbf{v}_B - \mathbf{v}_A)$$

$$\Sigma M_O = \frac{dm}{dt}(d_{OB}v_B - d_{OA}v_A)$$

Propulsion with Variable Mass

Some devices, such as a rocket, lose mass as they are propelled forward. Others gain mass, such as a shovel. We can account for this mass loss or gain by applying the principle of impulse and momentum to the device. From this equation, the force exerted on the device by the mass flow can then be determined. (*See pages 275–277.*)

$$\Sigma F_s = m\frac{dv}{dt} - v_{D/e}\frac{dm_e}{dt}$$

Loses Mass

$$\Sigma F_s = m\frac{dv}{dt} + v_{D/i}\frac{dm_i}{dt}$$

Gains Mass

1

Kinematics and Kinetics of a Particle

The topics and problems presented in Chapters 12 through 15 have all been *categorized* in order to provide a *clear focus* for learning the various problem-solving principles involved. In engineering practice, however, it is most important to be able to *identify* an appropriate method for the solution of a particular problem. In this regard, one must fully understand the limitations and use of the equations of dynamics, and be able to recognize which equations and principles to use for the problem's solution. For these reasons, we will now summarize the equations and principles of particle dynamics and provide the opportunity for applying them to a variety of problems.

Kinematics.   Problems in kinematics require a study only of the geometry of motion only, and do not account for the forces causing the motion. When the equations of kinematics are applied, one should clearly establish a fixed origin and select an appropriate coordinate system used to define the position of the particle. Once the positive direction of each coordinate axis is established, then the directions of the components of position, velocity, and acceleration can be determined from the algebraic sign of their numerical quantities.

Rectilinear Motion.

Variable Acceleration.   If a mathematical (or graphical) relationship is established between *any two* of the *four* variables s, v, a, and t, then a third variable can be determined by solving one of the following equations which relates all three variables.

$$v = \frac{ds}{dt} \quad a = \frac{dv}{dt} \quad a\,ds = v\,dv$$

Constant Acceleration.   Be *absolutely* certain that the acceleration is constant when using the following equations:

$$s = s_0 + v_0 t + \tfrac{1}{2} a_c t^2 \quad v = v_0 + a_c t \quad v^2 = v_0^2 + 2a_c(s - s_0)$$

Curvilinear Motion.

x, y, z Coordinates.   These coordinates are often used when the motion can be resolved into horizontal and vertical components. They are also useful for studying projectile motion since the acceleration of the projectile is *always* downward.

$$\begin{aligned} v_x &= \dot{x} & a_x &= \dot{v}_x \\ v_y &= \dot{y} & a_y &= \dot{v}_y \\ v_z &= \dot{z} & a_z &= \dot{v}_z \end{aligned}$$

n, t, b Coordinates.   These coordinates are particularly advantageous for studying the particle's *acceleration* along a known path. This is because the t and n components of **a** represent the separate changes in the magnitude and direction of the velocity, respectively, and these components can be readily formulated.

$$v = \dot{s}$$

$$a_t = \dot{v} = v\frac{dv}{ds}$$

$$a_n = \frac{v^2}{\rho}$$

where

$$\rho = \left| \frac{[1 + (dy/dx)^2]^{3/2}}{d^2 y/dx^2} \right|$$

when the path $y = f(x)$ is given.

r, θ, z Coordinates.   These coordinates are used when data regarding the angular motion of the radial coordinate r is given to describe the particle's motion. Also, some paths of motion can conveniently be described using these coordinates.

$$\begin{aligned} v_r &= \dot{r} & a_r &= \ddot{r} - r\dot{\theta}^2 \\ v_\theta &= r\dot{\theta} & a_\theta &= r\ddot{\theta} + 2\dot{r}\dot{\theta} \\ v_z &= \dot{z} & a_z &= \ddot{z} \end{aligned}$$

Relative Motion. If the origin of a *translating* coordinate system is established at particle A, then for particle B,

$$\mathbf{r}_B = \mathbf{r}_A + \mathbf{r}_{B/A}$$
$$\mathbf{v}_B = \mathbf{v}_A + \mathbf{v}_{B/A}$$
$$\mathbf{a}_B = \mathbf{a}_A + \mathbf{a}_{B/A}$$

Here the relative motion is measured by an observer fixed in the translating coordinate system.

Kinetics. Problems in kinetics involve the analysis of forces which cause the motion. When applying the equations of kinetics, it is absolutely necessary that measurements of the motion be made from an *inertial coordinate system,* i.e., one that does not rotate and is either fixed or translates with constant velocity. If a problem requires *simultaneous solution* of the equations of kinetics and kinematics, then it is important that the coordinate systems selected for writing each of the equations define the *positive directions* of the axes in the *same* manner.

Equations of Motion. These equations are used to solve for the particle's acceleration or the forces causing the motion. If they are used to determine a particle's position, velocity, or time of motion, then kinematics will also have to be considered in the solution. Before applying the equations of motion, always draw a free-body diagram to identify all the forces acting on the particle. Also, establish the direction of the particle's acceleration or its components. (A kinetic diagram may accompany the solution in order to graphically account for the $m\mathbf{a}$ vector.)

$$\Sigma F_x = ma_x \qquad \Sigma F_n = ma_n \qquad \Sigma F_r = ma_r$$
$$\Sigma F_y = ma_y \qquad \Sigma F_t = ma_t \qquad \Sigma F_\theta = ma_\theta$$
$$\Sigma F_z = ma_z \qquad \Sigma F_b = 0 \qquad \Sigma F_z = ma_z$$

Work and Energy. The equation of work and energy represents an integrated form of the tangential equation of motion, $\Sigma F_t = ma_t$, combined with kinematics ($a_t \, ds = v \, dv$). *It is used to solve problems involving force, velocity, and displacement.* Before applying this equation, *always draw a free-body diagram* in order to identify the forces which do work on the particle.

$$T_1 + \Sigma U_{1-2} = T_2$$

where

$$T = \tfrac{1}{2}mv^2 \qquad \text{(kinetic energy)}$$

$$U_F = \int_{s_1}^{s_2} F \cos\theta \, ds \qquad \text{(work of a variable force)}$$

$$U_{F_c} = F_c \cos\theta(s_2 - s_1) \qquad \text{(work of a constant force)}$$

$$U_W = -W \, \Delta y \qquad \text{(work of a weight)}$$

$$U_s = -\left(\tfrac{1}{2}ks_2^2 - \tfrac{1}{2}ks_1^2\right) \qquad \text{(work of an elastic spring)}$$

If the forces acting on the particle are *conservative forces*, i.e., those that *do not* cause a dissipation of energy, such as friction, then apply the conservation of energy equation. This equation is easier to use than the equation of work and energy since it applies only a *two points* on the path and *does not* require calculation of the work done by a force as the particle moves along the path.

$$T_1 + V_1 = T_2 + V_2$$

where

$$V_g = Wy \qquad \text{(gravitational potential energy)}$$
$$V_e = \tfrac{1}{2}ks^2 \qquad \text{(elastic potential energy)}$$

If the *power* developed by a force is to be determined, use

$$P = \frac{dU}{dt} = \mathbf{F} \cdot \mathbf{v}$$

where **v** is the velocity of a particle acted upon by the force **F**.

Impulse and Momentum.   The equation of *linear impulse and momentum* is an integrated form of the equation of motion, $\Sigma\mathbf{F} = m\mathbf{a}$, combined with kinematics ($\mathbf{a} = d\mathbf{v}/dt$). *It is used to solve problems involving force, velocity, and time.* Before applying this equation, one should *always draw the free-body diagram*, in order to identify all the forces that cause impulses on the particle. From the diagram the impulsive and nonimpulsive forces should be identified. Recall that the nonimpulsive forces can be neglected in the analysis during the time of impact. Also, establish the direction of the particle's velocity just before and just after the impulses are applied. As an alternative procedure, impulse and momentum diagrams may accompany the solution in order to graphically account for the terms in the equation.

$$m\mathbf{v}_1 + \Sigma \int_{t_1}^{t_2} \mathbf{F} \, dt = m\mathbf{v}_2$$

If several particles are involved in the problem, consider applying the *conservation of momentum* to the system in order to eliminate the internal impulses from the analysis. This can be done in a specified direction, provided no external impulses act on the particles in that direction.

$$\Sigma m\mathbf{v}_1 = \Sigma m\mathbf{v}_2$$

If the problem involves impact and the coefficient of restitution e is given, then apply the following equation.

$$e = \frac{(v_B)_2 - (v_A)_2}{(v_A)_1 - (v_B)_1} \qquad \text{(along line of impact)}$$

Remember that during impact the principle of work and energy cannot be used, since the particles deform and therefore the work due to the internal forces will be unknown. The principle of work and energy can be used, however, to determine the energy loss during the collision once the particle's initial and final velocities are determined.

The *principle of angular impulse and momentum and the conservation of angular momentum* may be applied about an axis in order to *eliminate* some of the unknown impulses acting on the particle during the time period when its motion is studied. Investigation of the particle's free-body diagram (or the impulse diagram) will aid in choosing the axis for application.

$$(\mathbf{H}_O)_1 + \Sigma \int_{t_1}^{t_2} \mathbf{M}_O \, dt = (\mathbf{H}_O)_2$$

$$(\mathbf{H}_O)_1 = (\mathbf{H}_O)_2$$

The following problems provide an opportunity for applying the above concepts. They are presented in *random order* so that practice may be gained in identifying the various types of problems and developing the skills necessary for their solution.

REVIEW PROBLEMS

R1–1. The block has a mass of 5 kg and is released from rest at A. It slides down the smooth plane onto the rough surface having a coefficient of kinetic friction of $\mu_k = 0.2$. Determine the total time of travel before the block stops sliding. Also, how far s does the block slide before stopping? Neglect the size of the block.

R1–2. Cartons having a mass of 5 kg are required to move along the assembly line at a constant speed of 8 m/s. Determine the smallest radius of curvature, ρ, for the conveyor so the cartons do not slip. The coefficients of static and kinetic friction between a carton and the conveyor are $\mu_s = 0.7$ and $\mu_k = 0.5$, respectively.

Prob. R1–1

Prob. R1–2

R1–3. A small metal particle passes downward through a fluid medium while being subjected to the attraction of a magnetic field such that its position is observed to be $s = (15t^3 - 3t)$ mm, where t is in seconds. Determine (a) the particle's displacement from $t = 2$ s to $t = 4$ s, and (b) the velocity and acceleration of the particle when $t = 5$ s.

*R1–4. The device shown is designed to produce the experience of weightlessness in a passenger when he reaches point A, $\theta = 90°$, along the path. If the passenger has a mass of 75 kg, determine the minimum speed he should have when he reaches A so that he does not exert a normal reaction on the seat. The chair is pin connected to the frame BC so that he is always seated in an upright position. During the motion his speed remains constant.

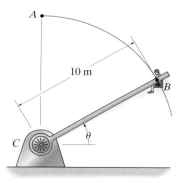

Prob. R1–4

R1–5. The passenger has a mass of 75 kg and always sits in an upright position on the chair. At the instant $\theta = 30°$, he has a speed of 5 m/s and an increase of speed of 2 m/s². Determine the horizontal and vertical forces that the chair exerts on him in order to produce this motion.

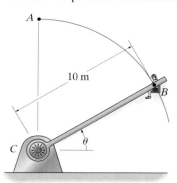

Prob. R1–5

R1–6. The 150-lb man lies against the cushion for which the coefficient of static friction is $\mu_s = 0.5$. Determine the resultant normal and frictional forces the cushion exerts on him if, due to rotation about the z axis, he has a constant speed $v = 20$ ft/s. Neglect the size of the man. Take $\theta = 60°$.

R1–7. The 150-lb man lies against the cushion for which the coefficient of static friction is $\mu_s = 0.5$. If he rotates about the z axis with a constant speed $v = 30$ ft/s, determine the smallest angle θ of the cushion at which he will begin to slip up the cushion.

Probs. R1–6/7

*R1–8. The baggage truck A has a mass of 800 kg and is used to pull each of the 300-kg cars. Determine the tension in the couplings at B and C if the tractive force $\mathbf{F}$ on the truck is $F = 480$ N. What is the speed of the truck when $t = 2$ s, starting from the rest? The car wheels are free to roll. Neglect the mass of the wheels.

R1–9. The baggage truck A has a mass of 800 kg and is used to pull each of the 300-kg cars. If the tractive force $\mathbf{F}$ on the truck is $F = 480$ N, determine the initial acceleration of the truck. What is the acceleration of the truck if the coupling at C suddenly fails? The car wheels are free to roll. Neglect the mass of the wheels.

Probs. R1–8/9

294   REVIEW 1   KINEMATICS AND KINETICS OF A PARTICLE

R1–10. A car is traveling at 80 ft/s when the brakes are suddenly applied, causing a constant deceleration of 10 ft/s². Determine the time required to stop the car and the distance traveled before stopping.

R1–11. Determine the speed of block B if the end of the cable at C is pulled downward with a speed of 10 ft/s. What is the relative velocity of the block with respect to C?

Prob. R1–11

***R1–12.** A hammer head H having a weight of 0.25 lb is moving vertically downward at 40 ft/s when it strikes the head of a nail of negligible mass and drives it into a block of wood. Find the impulse on the nail if it is assumed that the grip at A is loose, the handle has a negligible mass, and he hammer stays in contact with the nail while it comes to rest. Neglect the impulse caused by the weight of the hammer head during contact with the nail.

Prob. R1–12

R1–13. A 4-lb ball B is traveling around in a circle of radius $r_1 = 3$ ft with a speed $(v_B)_1 = 6$ ft/s. If the attached cord is pulled down through the hole with a constant speed $v_r = 2$ ft/s, determine the ball's speed at the instant $r_2 = 2$ ft. How much work has to be done to pull down the cord? Neglect friction and the size of the ball.

R1–14. A 4-lb ball B is traveling around in a circle of radius $r_1 = 3$ ft with a speed $(v_B)_1 = 6$ ft/s. If the attached cord is pulled down through the hole with a constant speed $v_r = 2$ ft/s, determine how much time is required for the ball to reach a speed of 12 ft/s. How far r_2 is the ball from the hole when this occurs? Neglect friction and the size of the ball.

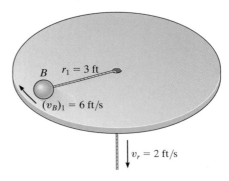

Probs. R1–13/14

R1–15. The block has a mass of 50 kg and rests on the surface of the cart having a mass of 75 kg. If the spring which is attached to the cart and not the block is compressed 0.2 m and the system is released from rest, determine the speed of the block after the spring becomes undeformed. Neglect the mass of the cart's wheels and the spring in the calculation. Also neglect friction. Take $k = 300$ N/m.

***R1–16.** The block has a mass of 50 kg and rests on the surface of the cart having a mass of 75 kg. If the spring which is attached to the cart and not the block is compressed 0.2 m and the system is released from rest, determine the speed of the block with respect to the cart after the spring becomes undeformed. Neglect the mass of the cart's wheels and the spring in the calculation. Also neglect friction. Take $k = 300$ N/m.

Probs. R1–15/16

R1–17. A ball is launched from point A at an angle of 30°. Determine the maximum and minimum speed v_A it can have so that it lands in the container.

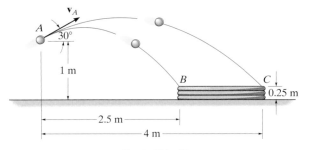

Prob. R1–17

R1–18. At the instant shown, cars A and B are traveling at speeds of 55 mi/h and 40 mi/h, respectively. If B is increasing its speed by 1200 mi/h², while A maintains a constant speed, determine the velocity and acceleration of B with respect to A. Car B moves along a curve having a radius of curvature of 0.5 mi.

R1–19. At the instant shown, cars A and B are traveling at speeds of 55 mi/h and 40 mi/h, respectively. If B is decreasing its speed at 1500 mi/h² while A is increasing its speed at 800 mi/h², determine the acceleration of B with respect to A. Car B moves along a curve having a radius of curvature of 0.75 mi.

Probs. R1–18/19

*R1–20. Four inelastic cables C are attached to a plate P and hold the 1-ft-long spring 0.25 ft in compression when *no weight* is on the plate. There is also an undeformed spring nested within this compressed spring. If the block, having a weight of 10 lb, is moving downward at $v = 4$ ft/s, when it is 2 ft above the plate, determine the maximum compression in each spring after it strikes the plate. Neglect the mass of the plate and springs and any energy lost in the collision.

R1–21. Four inelastic cables C are attached to a plate P and hold the 1-ft-long spring 0.25 ft in compression when *no weight* is on the plate. There is also a 0.5-ft long undeformed spring nested within this compressed spring. Determine the speed v of the 10-lb block when it is 2 ft above the plate, so that after it strikes the plate, it compresses the nested spring, having a stiffness of 50 lb/in., an amount of 0.20 ft. Neglect the mass of the plate and springs and any energy lost in the collision.

Probs. R1–20/21

R1–22. The 2-kg spool S fits loosely on the rotating inclined rod for which the coefficient of static friction is $\mu_s = 0.2$. If the spool is located 0.25 m from A, determine the minimum constant speed the spool can have so that it does not slip down the rod.

R1–23. The 2-kg spool S fits loosely on the inclined rod for which the coefficient of static friction is $\mu_s = 0.2$. If the spool is located 0.25 m from A, determine the maximum constant speed the spool can have so that it does not slip up the rod.

Probs. R1–22/23

*R1–24.   The winding drum D is drawing in the cable at an accelerated rate of 5 m/s^2. Determine the cable tension if the suspended crate has a mass of 800 kg.

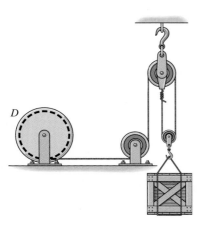

Prob. R1–24

R1–25.   If the hoist H is moving upward at 6 ft/s, determine the speed at which the motor M must draw in the supporting cable.

R1–26.   Starting from rest, the motor M can draw in the cable of the hoist H such that a point P on the cable has an acceleration of $a_{P/H} = (0.4t) \text{ ft/s}^2$, measured with respect to the hoist. Determine the velocity of the hoist when $t = 3$ s.

Probs. R1–25/26

R1–27.   The winch delivers a horizontal towing force F to its cable at A which varies as shown in the graph. Determine the speed of the 70-kg block B when $t = 18$ s. Originally the block is moving upward at $v_1 = 3 \text{ m/s}$.

*R1–28.   The winch delivers a horizontal towing force F to its cable at A which varies as shown in the graph. Determine the speed of the 40-kg block B when $t = 24$ s. Originally the block is resting on the ground.

Probs. R1–27/28

R1–29.   The motor pulls on the cable at A with a force $F = (30 + t^2)$ lb, where t is in seconds. If the 34-lb crate is originally at rest on the ground when $t = 0$, determine its speed when $t = 4$ s. Neglect the mass of the cable and pulleys. *Hint:* First find the time needed to begin lifting the crate.

Prob. R1–29

R1–30. The motor pulls on the cable at A with a force $F = (e^{2t})$ lb, where t is in seconds. If the 34-lb crate is originally at rest on the ground when $t = 0$, determine the crate's velocity when $t = 2$ s. Neglect the mass of the cable and pulleys. *Hint:* First find the time needed to begin lifting the crate.

Prob. R1–30

R1–31. The collar has a mass of 2 kg and travels along the smooth *horizontal* rod defined by the equiangular spiral $r = (e^{\theta})$ m, where θ is in radians. Determine the tangential force F and the normal force N acting on the collar when $\theta = 45°$, if the force F maintains a constant angular motion $\dot{\theta} = 2$ rad/s.

*R1–32. The collar has a mass of 2 kg and travels along the smooth *horizontal* rod defined by the equiangular spiral $r = (e^{\theta})$ m, where θ is in radians. Determine the tangential force F and the normal force N acting on the collar when $\theta = 90°$, if the force F maintains a constant angular motion $\dot{\theta} = 2$ rad/s.

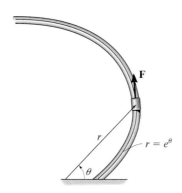

Probs. R1–31/32

R1–33. The acceleration of a particle along a straight line is defined by $a = (2t - 9)$ m/s^2, where t is in seconds. When $t = 0$, $s = 1$ m and $v = 10$ m/s. When $t = 9$ s, determine (a) the particle's position, (b) the total distance traveled, and (c) the velocity. Assume the positive direction is to the right.

R1–34. The 400-kg mine car is hoisted up the incline using the cable and motor M. For a short time, the force in the cable is $F = (3200t^2)$ N, where t is in seconds. If the car has an initial velocity $v_1 = 2$ m/s when $t = 0$, determine its velocity when $t = 2$ s.

R1–35. The 400-kg mine car is hoisted up the incline using the cable and motor M. For a short time, the force in the cable is $F = (3200t^2)$ N, where t is in seconds. If the car has an initial velocity $v_1 = 2$ m/s at $s = 0$ and $t = 0$, determine the distance it moves up the plane when $t = 2$ s.

Probs. R1–34/35

*R1–36. The rocket sled has a mass of 4 Mg and travels from rest along the smooth horizontal track such that it maintains a constant power output of 450 kW. Neglect the loss of fuel mass and air resistance, and determine how far it must travel to reach a speed of $v = 60$ m/s.

Prob. R1–36

R1–37. The collar has a mass of 20 kg and is supported on the smooth rod. The attached springs are undeformed when $d = 0.5$ m. Determine the speed of the collar after the applied force $F = 100$ N causes it to be displaced so that $d = 0.3$ m. When $d = 0.5$ m the collar is at rest.

R1–38. The collar has a mass of 20 kg and is supported on the smooth rod. The attached springs are both compressed 0.4 m when $d = 0.5$ m. Determine the speed of the collar after the applied force $F = 100$ N causes it to be displaced so that $d = 0.3$ m. When $d = 0.5$ m the collar is at rest.

Probs. R1–37/38

R1–39. The assembly consists of two blocks A and B which have masses of 20 kg and 30 kg, respectively. Determine the speed of each block when B descends 1.5 m. The blocks are released from rest. Neglect the mass of the pulleys and cords.

*R1–40. The assembly consists of two blocks A and B, which have masses of 20 kg and 30 kg, respectively. Determine the distance B must descend in order for A to achieve a speed of 3 m/s starting from rest.

Probs. R1–39/40

R1–41. Block A, having a mass m, is released from rest, falls a distance h and strikes the plate B having a mass $2m$. If the coefficient of restitution between A and B is e, determine the velocity of the plate just after collision. The spring has a stiffness k.

R1–42. Block A, having a mass of 2 kg, is released from rest, falls a distance $h = 0.5$ m, and strikes the plate B having a mass of 3 kg. If the coefficient of restitution between A and B is $e = 0.6$, determine the velocity of the block just after collision. The spring has a stiffness $k = 30$ N/m.

Probs. R1–41/42

R1–43. The cylindrical plug has a weight of 2 lb and it is free to move within the confines of the smooth pipe. The spring has a stiffness $k = 14$ lb/ft and when no motion occurs the distance $d = 0.5$ ft. Determine the force of the spring on the plug when the plug is at rest with respect to the pipe. The plug is traveling with a constant speed of 15 ft/s, which is caused by the rotation of the pipe about the vertical axis. Neglect the size of the plug.

Prob. R1–43

*R1–44. A 20-g bullet is fired horizontally into the 300-g block which rests on the smooth surface. After the bullet becomes embedded into the block, the block moves to the right 0.3 m before momentarily coming to rest. Determine the speed $(v_B)_1$ of the bullet. The spring has a stiffness $k = 200$ N/m and is originally unstretched.

R1–45. The 20-g bullet is fired horizontally at $(v_B)_1 = 1200$ m/s into the 300-g block which rests on the smooth surface. Determine the distance the block moves to the right before momentarily coming to rest. The spring has a stiffness $k = 200$ N/m and is originally unstretched.

Probs. R1–44/45

R1–46. The collar has a weight of 8 lb. If it is pushed down so as to compress the spring 2 ft and then released from rest $(h = 0)$, determine its speed when it is displaced $h = 4.5$ ft. The spring is not attached to the collar. Neglect friction.

R1–47. The collar has a weight of 8 lb. If it is released from rest at a height of $h = 2$ ft from the top of the uncompressed spring, determine the speed of the collar after it falls and compresses the spring 0.3 ft.

Probs. R1–46/47

*R1–48. The position of particles A and B are $\mathbf{r}_A = \{3t\mathbf{i} + 9t(2 - t)\mathbf{j}\}$ m and $\mathbf{r}_B = \{3(t^2 - 2t + 2)\mathbf{i} + 3(t - 2)\mathbf{j}\}$ m, respectively, where t is in seconds. Determine the point where the particles collide and their speeds just before the collision. How long does it take before the collision occurs?

R1–49. Determine the speed of the automobile if it has the acceleration shown and is traveling on a road which has a radius of curvature of $\rho = 50$ m. Also, what is the automobile's rate of increase in speed?

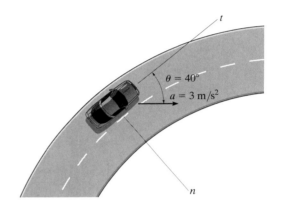

Prob. R1–49

R1–50. If block A of the pulley system is moving downward at 6 ft/s while block C is moving down at 18 ft/s, determine the relative velocity of block B with respect to C.

Prob. R1–50

The wind turbine rotates about a fixed axis with variable angular motion.

16 Planar Kinematics of a Rigid Body

CHAPTER OBJECTIVES

- To classify the various types of rigid-body planar motion.
- To investigate rigid-body translation and show how to analyze motion about a fixed axis.
- To study planar motion using an absolute motion analysis.
- To provide a relative motion analysis of velocity and acceleration using a translating frame of reference.
- To show how to find the instantaneous center of zero velocity and determine the velocity of a point on a body using this method.
- To provide a relative-motion analysis of velocity and acceleration using a rotating frame of reference.

16.1 Rigid-Body Motion

In this chapter, the planar kinematics of a rigid body will be discussed. This study is important for the design of gears, cams, and mechanisms used for many mechanical operations. Once the kinematics of a rigid body is thoroughly understood, then we can apply the equations of motion, which relate the forces on the body to the body's motion.

When all the particles of a rigid body move along paths which are equidistant from a fixed plane, the body is said to undergo *planar motion*. There are three types of rigid body planar motion: in order of increasing complexity, they are

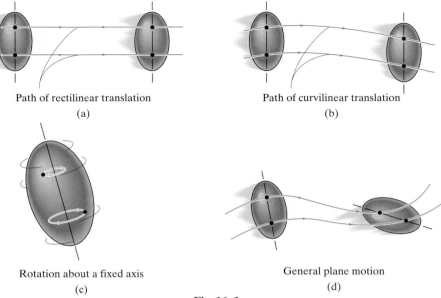

Path of rectilinear translation
(a)

Path of curvilinear translation
(b)

Rotation about a fixed axis
(c)

General plane motion
(d)

Fig. 16–1

1. *Translation.* This type of motion occurs if every line segment on the body remains parallel to its original direction during the motion. When the paths of motion for any two particles of the body are along equidistant straight lines the motion is called *rectilinear translation*, Fig. 16–1a. If the paths of motion are along curved lines which are equidistant the motion is called *curvilinear translation*, Fig. 16–1b.

2. *Rotation about a fixed axis.* When a rigid body rotates about a fixed axis, all the particles of the body, except those which lie on the axis of rotation, move along circular paths, Fig. 16–1c.

3. *General plane motion.* When a body is subjected to general plane motion, it undergoes a combination of translation *and* rotation, Fig. 16–1d. The translation occurs within a reference plane, and the rotation occurs about an axis perpendicular to the reference plane.

In the following sections we will consider each of these motions in detail. Examples of bodies undergoing these motions are shown in Fig. 16–2.

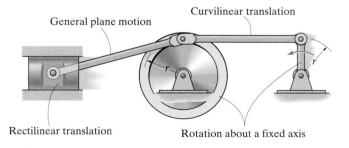

General plane motion

Curvilinear translation

Rectilinear translation

Rotation about a fixed axis

Fig. 16–2

16.2 Translation

Consider a rigid body which is subjected to either rectilinear or curvilinear translation in the x–y plane, Fig. 16–3.

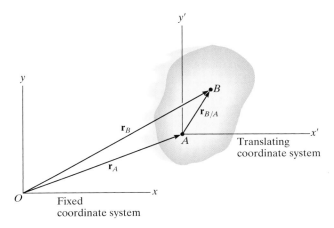

Translating
coordinate system

Fixed
coordinate system

Fig. 16–3

Position. The locations of points A and B in the body are defined from the fixed x, y reference frame by using *position vectors* $\mathbf{r}_A$ and $\mathbf{r}_B$. The translating x',y' coordinate system is *fixed in the body* and has its origin located at A, hereafter referred to as the *base point*. The position of B with respect to A is denoted by the *relative-position vector* $\mathbf{r}_{B/A}$ ("$\mathbf{r}$ of B with respect to A"). By vector addition,

$$\mathbf{r}_B = \mathbf{r}_A + \mathbf{r}_{B/A}$$

Velocity. A relationship between the instantaneous velocities of A and B is obtained by taking the time derivative of the position equation, which yields $\mathbf{v}_B = \mathbf{v}_A + d\mathbf{r}_{B/A}/dt$. Here $\mathbf{v}_A$ and $\mathbf{v}_B$ denote *absolute velocities* since these vectors are measured from the x, y axes. The term $d\mathbf{r}_{B/A}/dt = \mathbf{0}$, since the *magnitude* of $\mathbf{r}_{B/A}$ is *constant* by definition of a rigid body, and because the body is translating the *direction* of $\mathbf{r}_{B/A}$ is *constant*. Therefore,

$$\mathbf{v}_B = \mathbf{v}_A$$

Acceleration. Taking the time derivative of the velocity equation yields a similar relationship between the instantaneous accelerations of A and B:

$$\mathbf{a}_B = \mathbf{a}_A$$

The above two equations indicate that *all points in a rigid body subjected to either rectilinear or curvilinear translation move with the same velocity and acceleration*. As a result, the kinematics of particle motion, discussed in Chapter 12, may also be used to specify the kinematics of points located in a translating rigid body.

Passengers on this amusement ride are subjected to curvilinear translation since the vehicle moves in a circular path yet it always remains in an upright position.

16.3   Rotation about a Fixed Axis

When a body is rotating about a fixed axis, any point P located in the body travels along a *circular path*. To study this motion it is first necessary to discuss the angular motion of the body about the axis.

Angular Motion.   A point is without dimension, and so it has no angular motion. *Only lines or bodies undergo angular motion.* For example, consider the body shown in Fig. 16–4a and the angular motion of a radial line r located within the shaded plane and directed from point O on the axis of rotation to point P.

Angular Position.   At the instant shown, the *angular position* of r is defined by the angle θ, measured between a *fixed* reference line and r.

Angular Displacement.   The change in the angular position, which can be measured as a differential $d\theta$, is called the *angular displacement*.* This vector has a *magnitude* of $d\theta$, measured in degrees, radians, or revolutions, where 1 rev = 2π rad. Since motion is about a *fixed axis*, the direction of $d\theta$ is *always* along the axis. Specifically, the *direction* is determined by the right-hand rule; that is, the fingers of the right hand are curled with the sense of rotation, so that in this case the thumb, or $d\theta$, points upward, Fig. 16–4a. In two dimensions, as shown by the top view of the shaded plane, Fig. 16–4b, both θ and $d\theta$ are directed counterclockwise, and so the thumb points outward from the page.

Angular Velocity.   The time rate of change in the angular position is called the *angular velocity* $\boldsymbol{\omega}$ (omega). Since $d\theta$ occurs during an instant of time dt, then,

$$(\curvearrowright+) \qquad \boxed{\omega = \frac{d\theta}{dt}} \qquad (16\text{–}1)$$

This vector has a *magnitude* which is often measured in rad/s. It is expressed here in scalar form since its *direction* is always along the axis of rotation, i.e., in the same direction as $d\theta$, Fig. 16–4a. When indicating the angular motion in the shaded plane, Fig. 16–4b, we can refer to the sense of rotation as clockwise or counterclockwise. Here we have *arbitrarily* chosen counterclockwise rotations as *positive* and indicated this by the curl shown in parentheses next to Eq. 16–1. Realize, however, that the directional sense of $\boldsymbol{\omega}$ is actually outward from the page.

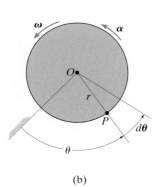

(a)

(b)

Fig. 16–4

---

*It is shown in Sec. 20.1 that finite rotations or finite angular displacements are *not* vector quantities, although differential rotations $d\theta$ are vectors.

Angular Acceleration. The *angular acceleration* $\boldsymbol{\alpha}$ (alpha) measures the time rate of change of the angular velocity. The *magnitude* of this vector may be written as

$(\downarrow+)$

$$\alpha = \frac{d\omega}{dt}$$

(16–2)

Using Eq. 16–1, it is possible to express α as

$(\downarrow+)$

$$\alpha = \frac{d^2\theta}{dt^2}$$

(16–3)

The line of action of $\boldsymbol{\alpha}$ is the same as that for $\boldsymbol{\omega}$, Fig. 16–4a; however, its sense of *direction* depends on whether $\boldsymbol{\omega}$ is increasing or decreasing. In particular, if $\boldsymbol{\omega}$ is decreasing, then $\boldsymbol{\alpha}$ is called an *angular deceleration* and it therefore has a sense of direction which is opposite to $\boldsymbol{\omega}$.

By eliminating dt from Eqs. 16–1 and 16–2, we obtain a differential relation between the angular acceleration, angular velocity, and angular displacement, namely,

$(\downarrow+)$

$$\alpha \, d\theta = \omega \, d\omega$$

(16–4)

The similarity between the differential relations for angular motion and those developed for rectilinear motion of a particle ($v = ds/dt$, $a = dv/dt$, and $a \, ds = v \, dv$) should be apparent.

Constant Angular Acceleration.   If the angular acceleration of the body is constant, $\boldsymbol{\alpha} = \boldsymbol{\alpha}_c$, then Eqs. 16–1, 16–2, and 16–4, when integrated, yield a set of formulas which relate the body's angular velocity, angular position, and time. These equations are similar to Eqs. 12–4 to 12–6 used for rectilinear motion. The results are

$(\downarrow+)$

$(\downarrow+)$

$(\downarrow+)$

$$\omega = \omega_0 + \alpha_c t$$

(16–5)

$$\theta = \theta_0 + \omega_0 t + \tfrac{1}{2}\alpha_c t^2$$

(16–6)

$$\omega^2 = \omega_0^2 + 2\alpha_c(\theta - \theta_0)$$

(16–7)

Constant Angular Acceleration

Here θ_0 and ω_0 are the initial values of the body's angular position and angular velocity, respectively.

(c)

Fig. 16–4 (cont.)

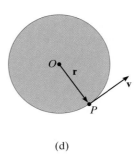

(d)

Motion of Point P.
As the rigid body in Fig. 16–4c rotates, point P travels along a *circular path* of radius r and center at point O. This path is contained within the shaded plane shown in top view, Fig. 16–4d.

Position. The position of P is defined by the position vector $\mathbf{r}$, which extends from O to P.

Velocity. The velocity of P has a magnitude which can be found from its polar coordinate components $v_r = \dot{r}$ and $v_\theta = r\dot{\theta}$, Eqs. 12–25. Since r is constant, the radial component $v_r = \dot{r} = 0$, and so $v = v_\theta = r\dot{\theta}$. Because $\omega = \dot{\theta}$, Eq. 16–1, the velocity is

$$v = \omega r \qquad (16\text{–}8)$$

As shown in Figs. 16–4c and 16–4d, the *direction* of $\mathbf{v}$ is *tangent* to the circular path.

Both the magnitude and direction of $\mathbf{v}$ can also be accounted for by using the cross product of $\boldsymbol{\omega}$ and $\mathbf{r}_P$ (see Appendix C). Here, $\mathbf{r}_P$ is directed from *any point* on the axis of rotation to point P, Fig. 16–4c. We have

$$\mathbf{v} = \boldsymbol{\omega} \times \mathbf{r}_P \qquad (16\text{–}9)$$

The order of the vectors in this formulation is important, since the cross product is not commutative, i.e., $\boldsymbol{\omega} \times \mathbf{r}_P \neq \mathbf{r}_P \times \boldsymbol{\omega}$. In this regard, notice in Fig. 16–4c how the correct direction of $\mathbf{v}$ is established by the right-hand rule. The fingers of the right hand are curled from $\boldsymbol{\omega}$ toward $\mathbf{r}_P$ ($\boldsymbol{\omega}$ "cross" $\mathbf{r}_P$). The thumb indicates the correct direction of $\mathbf{v}$, which is tangent to the path in the direction of motion. From Eq. C–8, the magnitude of $\mathbf{v}$ in Eq. 16–9 is $v = \omega r_P \sin \phi$, and since $r = r_P \sin \phi$, Fig. 16–4c, then $v = \omega r$, which agrees with Eq. 16–8. As a special case, the position vector $\mathbf{r}$ can be chosen for $\mathbf{r}_P$. Here $\mathbf{r}$ lies in the plane of motion and again the velocity of point P is

$$\mathbf{v} = \boldsymbol{\omega} \times \mathbf{r} \qquad (16\text{–}10)$$

Acceleration. The acceleration of P can be expressed in terms of its normal and tangential components.* Since $a_t = dv/dt$ and $a_n = v^2/\rho$, where $\rho = r$, $v = \omega r$, and $\alpha = d\omega/dt$, we have

$$\boxed{a_t = \alpha r} \tag{16–11}$$

$$\boxed{a_n = \omega^2 r} \tag{16–12}$$

The *tangential component of acceleration*, Figs. 16–4e and 16–4f, represents the time rate of change in the velocity's magnitude. If the speed of P is increasing, then $\mathbf{a}_t$ acts in the same direction as $\mathbf{v}$; if the speed is decreasing, $\mathbf{a}_t$ acts in the opposite direction of $\mathbf{v}$; and finally, if the speed is constant, $\mathbf{a}_t$ is zero.

The *normal component of acceleration* represents the time rate of change in the velocity's direction. The *direction* of $\mathbf{a}_n$ is always toward O, the center of the circular path, Figs. 16–4e and 16–4f.

Like the velocity, the acceleration of point P may be expressed in terms of the vector cross product. Taking the time derivative of Eq. 16–9 we have

$$\mathbf{a} = \frac{d\mathbf{v}}{dt} = \frac{d\boldsymbol{\omega}}{dt} \times \mathbf{r}_P + \boldsymbol{\omega} \times \frac{d\mathbf{r}_P}{dt}$$

Recalling that $\boldsymbol{\alpha} = d\boldsymbol{\omega}/dt$, and using Eq. 16–9 $(d\mathbf{r}_P/dt = \mathbf{v} = \boldsymbol{\omega} \times \mathbf{r}_P)$, yields

$$\mathbf{a} = \boldsymbol{\alpha} \times \mathbf{r}_P + \boldsymbol{\omega} \times (\boldsymbol{\omega} \times \mathbf{r}_P) \tag{16–13}$$

From the definition of the cross product, the first term on the right has a magnitude $a_t = \alpha r_P \sin \phi = \alpha r$, and by the right-hand rule, $\boldsymbol{\alpha} \times \mathbf{r}_P$ is in the direction of $\mathbf{a}_t$, Fig. 16–4e. Likewise, the second term has a magnitude $a_n = \omega^2 r_P \sin \phi = \omega^2 r$, and applying the right-hand rule twice, first to determine the result $\mathbf{v}_P = \boldsymbol{\omega} \times \mathbf{r}_P$ then $\boldsymbol{\omega} \times \mathbf{v}_P$, it can be seen that this result is in the same direction as $\mathbf{a}_n$, shown in Fig. 16–4e. Noting that this is also the *same* direction as $-\mathbf{r}$, which lies in the plane of motion, we can express $\mathbf{a}_n$ in a much simpler form as $\mathbf{a}_n = -\omega^2 \mathbf{r}$. Hence, Eq. 16–12 can be identified by its two components as

$$\boxed{\begin{aligned} \mathbf{a} &= \mathbf{a}_t + \mathbf{a}_n \\ &= \boldsymbol{\alpha} \times \mathbf{r} - \omega^2 \mathbf{r} \end{aligned}} \tag{16–14}$$

Since $\mathbf{a}_t$ and $\mathbf{a}_n$ are perpendicular to one another, if needed the magnitude of acceleration can be determined from the Pythagorean theorem; namely, $a = \sqrt{a_n^2 + a_t^2}$, Fig. 16–4f.

---

*Polar coordinates can also be used. Since $a_r = \ddot{r} - r\dot{\theta}^2$ and $a_\theta = r\ddot{\theta} + 2\dot{r}\dot{\theta}$, substituting $\dot{r} = \ddot{r} = 0$, $\dot{\theta} = \omega$, $\ddot{\theta} = \alpha$, we obtain Eqs. 16–11 and 16–12.

(e)

(f)

Fig. 16–4 (cont.)

The many gears used in the operation of a crane all rotate about fixed axes. Engineers must be able to relate their angular motions in order to properly design this gear system.

Important Points

- A body can undergo two types of translation. During rectilinear translation all points follow parallel straight-line paths, and during curvilinear translation the points follow curved paths that are the same shape and are equidistant from one another.

- All the points on a translating body move with the same velocity and acceleration.

- Points located on a body that rotates about a fixed axis follow circular paths.

- The relationship $\alpha \, d\theta = \omega \, d\omega$ is derived from $\alpha = d\omega/dt$ and $\omega = d\theta/dt$ by eliminating dt.

- Once the angular motions ω and α are known, the velocity and acceleration of any point on the body can be determined.

- The velocity always acts tangent to the path of motion.

- The acceleration has two components. The tangential acceleration measures the rate of change in the magnitude of the velocity and can be determined using $a_t = \alpha r$. The normal acceleration measures the rate of change in the direction of the velocity and can be determined from $a_n = \omega^2 r$.

Procedure for Analysis

The velocity and acceleration of a point located on a rigid body that is rotating about a fixed axis can be determined using the following procedure.

Angular Motion.

- Establish the positive sense of direction along the axis of rotation and show it alongside each kinematic equation as it is applied.
- If a relationship is known between any *two* of the four variables α, ω, θ, and t, then a third variable can be obtained by using one of the following kinematic equations which relates all three variables.

$$\omega = \frac{d\theta}{dt} \quad \alpha = \frac{d\omega}{dt} \quad \alpha \, d\theta = \omega \, d\omega$$

- If the body's angular acceleration is *constant*, then the following equations can be used:

$$\omega = \omega_0 + \alpha_c t$$
$$\theta = \theta_0 + \omega_0 t + \tfrac{1}{2}\alpha_c t^2$$
$$\omega^2 = \omega_0^2 + 2\alpha_c(\theta - \theta_0)$$

- Once the solution is obtained, the sense of θ, ω, and α is determined from the algebraic signs of their numerical quantities.

Motion of P.

- In most cases the velocity of P and its two components of acceleration can be determined from the scalar equations

$$v = \omega r$$
$$a_t = \alpha r$$
$$a_n = \omega^2 r$$

- If the geometry of the problem is difficult to visualize, the following vector equations should be used:

$$\mathbf{v} = \boldsymbol{\omega} \times \mathbf{r}_P = \boldsymbol{\omega} \times \mathbf{r}$$
$$\mathbf{a}_t = \boldsymbol{\alpha} \times \mathbf{r}_P = \boldsymbol{\alpha} \times \mathbf{r}$$
$$\mathbf{a}_n = \boldsymbol{\omega} \times (\boldsymbol{\omega} \times \mathbf{r}_P) = -\omega^2 \mathbf{r}$$

Here $\mathbf{r}_P$ is directed from any point on the axis of rotation to point P, whereas $\mathbf{r}$ lies in the plane of motion of P. Either of these vectors, along with $\boldsymbol{\omega}$ and $\boldsymbol{\alpha}$, should be expressed in terms of its $\mathbf{i}$, $\mathbf{j}$, $\mathbf{k}$ components, and, if necessary, the cross products determined by using a determinant expansion (see Eq. C–12).

EXAMPLE 16.1

0.2 m

O

θ

P

a

F

Fig. 16–5

A cord is wrapped around a wheel which is initially at rest as shown in Fig. 16–5. If a force is applied to the cord and gives it an acceleration $a = (4t)$ m/s^2, where t is in seconds, determine as a function of time (a) the angular velocity of the wheel, and (b) the angular position of line OP in radians.

SOLUTION

Part (a). The wheel is subjected to rotation about a fixed axis passing through point O. Thus, point P on the wheel has motion about a circular path, and the acceleration of this point has *both* tangential and normal components. The tangential component is $(a_P)_t = (4t)$ m/s^2, since the cord is wrapped around the wheel and moves *tangent* to it. Hence the angular acceleration of the wheel is

$(\curvearrowright+)$
$$(a_P)_t = \alpha r$$

$$(4t)\text{ m/s}^2 = \alpha(0.2\text{ m})$$

$$\alpha = 20t\text{ rad/s}^2 \downarrow$$

Using this result, the wheel's angular velocity ω can now be determined from $\alpha = d\omega/dt$, since this equation relates α, t, and ω. Integrating, with the initial condition that $\omega = 0$ at $t = 0$, yields

$(\curvearrowright+)$
$$\alpha = \frac{d\omega}{dt} = (20t)\text{ rad/s}^2$$

$$\int_0^\omega d\omega = \int_0^t 20t\, dt$$

$$\omega = 10t^2\text{ rad/s} \downarrow \qquad\qquad Ans.$$

Part (b). Using this result, the angular position θ of OP can be found from $\omega = d\theta/dt$, since this equation relates θ, ω, and t. Integrating, with the initial condition $\theta = 0$ at $t = 0$, we have

$(\curvearrowright+)$
$$\frac{d\theta}{dt} = \omega = (10t^2)\text{ rad/s}$$

$$\int_0^\theta d\theta = \int_0^t 10t^2\, dt$$

$$\theta = 3.33t^3\text{ rad} \qquad\qquad Ans.$$

NOTE: We cannot use the equation of constant angular acceleration, since α is a function of time.

EXAMPLE 16.2

The motor shown in the photo is used to turn a wheel and attached blower contained within the housing. The details of the design are shown in Fig. 16–6a. If the pulley A connected to the motor begins rotating from rest with an angular acceleration of $\alpha_A = 2$ rad/s^2, determine the magnitudes of the velocity and acceleration of point P on the wheel, after the wheel B has turned one revolution. Assume the transmission belt does not slip on the pulley and wheel.

SOLUTION

Angular Motion. First we will convert the one revolution to radians. Since there are 2π rad in one revolution, then

$$\theta_B = 1 \text{ rev}\left(\frac{2\pi \text{ rad}}{1 \text{ rev}}\right) = 6.283 \text{ rad}$$

We can find the angular velocity of pulley A provided we first find its angular displacement. Since the belt does not slip, an equivalent length of belt s must be unraveled from both the pulley and wheel at all times. Thus,

$$s = \theta_A r_A = \theta_B r_B; \qquad \theta_A(0.15 \text{ m}) = 6.283(0.4 \text{ m})$$
$$\theta_A = 16.76 \text{ rad}$$

Since α_A is constant, the angular velocity of pulley A is therefore

$$(\curvearrowright +) \qquad \omega^2 = \omega_0^2 + 2\alpha_c(\theta - \theta_0)$$
$$\omega_A^2 = 0 + 2(2 \text{ rad/s}^2)(16.76 \text{ rad} - 0)$$
$$\omega_A = 8.188 \text{ rad/s}$$

The belt has the same speed and tangential component of acceleration as it passes over the pulley and wheel. Thus,

$$v = \omega_A r_A = \omega_B r_B; \quad 8.188 \text{ rad/s } (0.15 \text{ m}) = \omega_B(0.4 \text{ m})$$
$$\omega_B = 3.070 \text{ rad/s}$$

$$a_t = \alpha_A r_A = \alpha_B r_B; \quad 2 \text{ rad/s}^2 \ (0.15 \text{ m}) = \alpha_B(0.4 \text{ m})$$
$$\alpha_B = 0.750 \text{ rad/s}^2$$

Motion of P. As shown on the kinematic diagram in Fig. 16–6b, we have

$$v_P = \omega_B r_B = 3.070 \text{ rad/s } (0.4 \text{ m}) = 1.23 \text{ m/s} \qquad Ans.$$
$$(a_P)_t = \alpha_B r_B = 0.750 \text{ rad/s}^2 \ (0.4 \text{ m}) = 0.3 \text{ m/s}^2$$
$$(a_P)_n = \omega_B^2 r_B = (3.070 \text{ rad/s})^2(0.4 \text{ m}) = 3.77 \text{ m/s}^2$$

Thus

$$a_P = \sqrt{(0.3)^2 + (3.77)^2} = 3.78 \text{ m/s}^2 \qquad Ans.$$

(a)

(b)

Fig. 16–6

PROBLEMS

16–1. A wheel has an initial clockwise angular velocity of 10 rad/s and a constant angular acceleration of 3 rad/s^2. Determine the number of revolutions it must undergo to acquire a clockwise angular velocity of 15 rad/s. What time is required?

16–2. A flywheel has its angular speed increased uniformly from 15 rad/s to 60 rad/s in 80 s. If the diameter of the wheel is 2 ft, determine the magnitudes of the normal and tangential components of acceleration of a point on the rim of the wheel when $t = 80$ s, and the total distance the point travels during the time period.

16–3. The angular velocity of the disk is defined by $\omega = (5t^2 + 2)$ rad/s, where t is in seconds. Determine the magnitudes of the velocity and acceleration of point A on the disk when $t = 0.5$ s.

Prob. 16–3

*16–4. The figure shows the internal gearing of a "spinner" used for drilling wells. With constant angular acceleration, the motor M rotates the shaft S to 100 rev/min in $t = 2$ s starting from rest. Determine the angular acceleration of the drill-pipe connection D and the number of revolutions it makes during the 2-s start up.

Prob. 16–4

16–5. If gear A starts from rest and has a constant angular acceleration of $\alpha_A = 2$ rad/s^2, determine the time needed for gear B to attain an angular velocity of $\omega_B = 50$ rad/s.

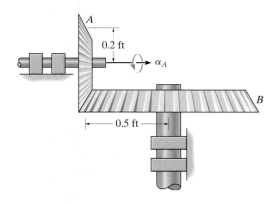

Prob. 16–5

16–6. If the armature A of the electric motor in the drill has a constant angular acceleration of $\alpha_A = 20$ rad/s^2, determine its angular velocity and angular displacement when $t = 3$ s. The motor starts from rest.

Prob. 16–6

16–7. The mechanism for a car window winder is shown in the figure. Here the handle turns the small cog C, which rotates the spur gear S, thereby rotating the fixed-connected lever AB which raises track D in which the window rests. The window is free to slide on the track. If the handle is wound at 0.5 rad/s, determine the speed of points A and E and the speed v_w of the window at the instant $\theta = 30°$.

Prob. 16–7

16–9. The motor M begins rotating at an angular rate of $\omega = 4(1 - e^{-t})$ rad/s, where t is in seconds. If the pulleys and fan have the radii shown, determine the magnitudes of the velocity and acceleration of point P on the fan blade when $t = 0.5$ s. Also, what is the maximum speed of this point?

Prob. 16–9

*16–8. The pinion gear A on the motor shaft is given a constant angular acceleration $\alpha = 3$ rad/s^2. If the gears A and B have the dimensions shown, determine the angular velocity and angular displacement of the output shaft C, when $t = 2$ s starting from rest. The shaft is fixed to B and turns with it.

Prob. 16–8

16–10. The disk is originally rotating at $\omega_0 = 8$ rad/s. If it is subjected to a constant angular acceleration of $\alpha = 6$ rad/s^2, determine the magnitudes of the velocity and the n and t components of acceleration of point A at the instant $t = 0.5$ s.

16–11. The disk is originally rotating at $\omega_0 = 8$ rad/s. If it is subjected to a constant angular acceleration of $\alpha = 6$ rad/s^2, determine the magnitudes of the velocity and the n and t components of acceleration of point B just after the wheel undergoes 2 revolutions.

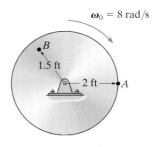

Probs. 16–10/11

*16–12. The anemometer measures the speed of the wind due to the rotation of the three cups. If during a 3-s time period a wind gust causes the cups to have an angular velocity of $\omega = (2t^2 + 3)$ rad/s, where t is in seconds, determine (a) the speed of the cups when $t = 2$ s, (b) the total distance traveled by each cup during the 3-s time period, and (c) the angular acceleration of the cups when $t = 2$ s. Neglect the size of the cups for the calculation.

16–14. The turntable T is driven by the frictional idler wheel A, which simultaneously bears against the inner rim of the turntable and the motor-shaft spindle B. Determine the required diameter d of the spindle if the motor turns it at 25 rad/s and it is required that the turntable rotate at 2 rad/s.

Prob. 16–14

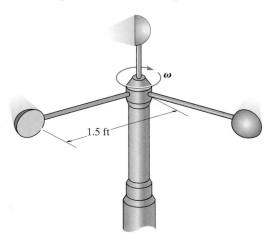

Prob. 16–12

16–13. A motor gives disk A an angular acceleration of $\alpha_A = (0.6t^2 + 0.75)$ rad/s², where t is in seconds. If the initial angular velocity of the disk is $\omega_0 = 6$ rad/s, determine the magnitudes of the velocity and acceleration of block B when $t = 2$ s.

16–15. Gear A is in mesh with gear B as shown. If A starts from rest and has a constant angular acceleration of $\alpha_A = 2$ rad/s², determine the time needed for B to attain an angular velocity of $\omega_B = 50$ rad/s.

Prob. 16–13

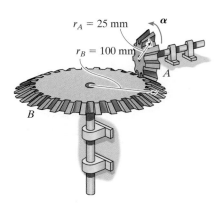

Prob. 16–15

*16–16. The blade on the horizontal-axis windmill is turning with an angular velocity of $\omega_0 = 2$ rad/s. Determine the distance point P on the tip of the blade has traveled if the blade attains an angular velocity of $\omega = 5$ rad/s in 3 s. The angular acceleration is constant. Also, what is the magnitude of the acceleration of this point when $t = 3$ s?

16–17. The blade on the horizontal-axis windmill is turning with an angular velocity of $\omega_0 = 2$ rad/s. If it is given an angular acceleration of $\alpha_c = 0.6$ rad/s^2, determine the angular velocity and the magnitude of acceleration of point P on the tip of the blade when $t = 3$ s.

Probs. 16–16/17

16–18. Starting from rest when $s = 0$, pulley A is given an angular acceleration $\alpha = (6\theta)$ rad/s^2, where θ is in radians. Determine the speed of block B when it has risen $s = 6$ m. The pulley has an inner hub D which is fixed to C and turns with it.

Prob. 16–18

16–19. Starting from rest when $s = 0$, pulley A is given a constant angular acceleration $\alpha_c = 6$ rad/s^2. Determine the speed of block B when it has risen $s = 6$ m. The pulley has an inner hub D which is fixed to C and turns with it.

Prob. 16–19

*16–20. Initially the motor on the circular saw turns its drive shaft at $\omega = (20t^{2/3})$ rad/s, where t is in seconds. If the radii of gears A and B are 0.25 in. and 1 in., respectively, determine the magnitudes of the velocity and acceleration of a tooth C on the saw blade after the drive shaft rotates $\theta = 5$ rad starting from rest.

Prob. 16–20

16–21.   Due to the screw at E, the actuator provides linear motion to the arm at F when the motor turns the gear at A. If the gears have the radii listed in the figure, and the screw at E has a pitch $p = 2$ mm, determine the speed at F when the motor turns A at $\omega_A = 20$ rad/s. *Hint*: The screw pitch indicates the amount of advance of the screw for each full revolution.

$r_A = 10$ mm
$r_B = 50$ mm
$r_C = 15$ mm
$r_D = 60$ mm

Prob. 16–21

16–22.   A motor gives gear A an angular acceleration of $\alpha_A = (0.25\theta^3 + 0.5)$ rad/s^2, where θ is in radians. If this gear is initially turning at $(\omega_A)_0 = 20$ rad/s, determine the angular velocity of gear B after A undergoes an angular displacement of 10 rev.

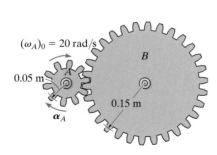

Prob. 16–22

16–23.   A motor gives gear A an angular acceleration of $\alpha_A = (4t^3)$ rad/s^2, where t is in seconds. If this gear is initially turning at $(\omega_A)_0 = 20$ rad/s, determine the angular velocity of gear B when $t = 2$ s.

Prob. 16–23

*16–24.**   For a short time a motor of the random-orbit sander drives the gear A with an angular velocity of $\omega_A = 40(t^3 + 6t)$ rad/s, where t is in seconds. This gear is connected to gear B, which is fixed connected to the shaft CD. The end of this shaft is connected to the eccentric spindle EF and pad P, which causes the pad to orbit around shaft CD at a radius of 15 mm. Determine the magnitudes of the velocity and the tangential and normal components of acceleration of the spindle EF when $t = 2$ s after starting from rest.

Prob. 16–24

16–25. For a short time the motor of the random-orbit sander drives the gear A with an angular velocity of $\omega_A = (5\theta^2)$ rad/s, where θ is in radians. This gear is connected to gear B, which is fixed connected to the shaft CD. The end of this shaft is connected to the eccentric spindle EF and pad P, which causes the pad to orbit around shaft CD at a radius of 15 mm. Determine the magnitudes of the velocity and the tangential and normal components of acceleration of the spindle EF when $\theta = 0.5$ revolutions starting from rest.

Prob. 16–25

16–26. The engine shaft S on the lawnmower rotates at a constant angular rate of 40 rad/s. Determine the magnitudes of the velocity and acceleration of point P on the blade and the distance P travels in 3 seconds. The shaft S is connected to the driver pulley A, and the motion is transmitted to the belt that passes over the idler pulleys at B and C and to the pulley at D. This pulley is connected to the blade and to another belt that drives the other blade.

Prob. 16–26

16–27. The operation of "reverse" for a three-speed automotive transmission is illustrated schematically in the figure. If the crank shaft G is turning with an angular speed of 60 rad/s, determine the angular speed of the drive shaft H. Each of the gears rotates about a fixed axis. Note that gears A and B, C and D, E and F are in mesh. The radii of each of these gears are reported in the figure.

$r_A = 90$ mm
$r_B = r_C = 30$ mm
$r_D = 50$ mm
$r_E = 70$ mm
$r_F = 60$ mm

Prob. 16–27

*16–28. Morse Industrial manufactures the speed reducer shown. If a motor drives the gear shaft S with an angular acceleration of $\alpha = (0.4e^t)$ rad/s^2, where t is in seconds, determine the angular velocity of shaft E when $t = 2$ s after starting from rest. The radius of each gear is listed in the figure. Note that gears B and C are fixed connected to the same shaft.

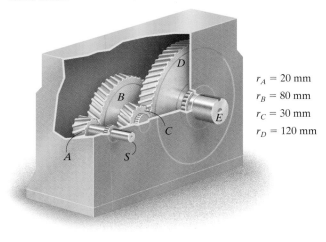

$r_A = 20$ mm
$r_B = 80$ mm
$r_C = 30$ mm
$r_D = 120$ mm

Prob. 16–28

16–29. Morse Industrial manufactures the speed reducer shown. If a motor drives the gear shaft S with an angular acceleration of $\alpha = (4\omega^{-3})$ rad/s², where ω is in rad/s, determine the angular velocity of shaft E when $t = 2$ s after starting from an angular velocity of 1 rad/s when $t = 0$. The radius of each gear is listed in the figure. Note that gears B and C are fixed connected to the same shaft.

$r_A = 20$ mm
$r_B = 80$ mm
$r_C = 30$ mm
$r_D = 120$ mm

Prob. 16–29

16–31. The sphere starts from rest at $\theta = 0°$ and rotates with an angular acceleration of $\alpha = (4\theta)$ rad/s², where θ is in radians. Determine the magnitudes of the velocity and acceleration of point P on the sphere at the instant $\theta = 6$ rad.

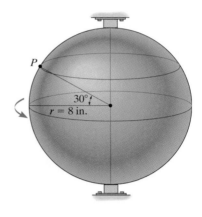

Prob. 16–31

16–30. A tape having a thickness s wraps around the wheel which is turning at a constant rate $\boldsymbol{\omega}$. Assuming the unwrapped portion of tape remains horizontal, determine the acceleration of point P of the unwrapped tape when the radius of the wrapped tape is r. *Hint:* Since $v_P = \omega r$, take the time derivative and note that $dr/dt = \omega(s/2\pi)$.

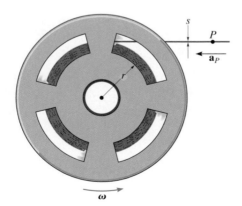

Prob. 16–30

*16–32. The rod assembly is supported by ball-and-socket joints at A and B. At the instant shown it is rotating about the y axis with an angular velocity $\omega = 5$ rad/s and has an angular acceleration $\alpha = 8$ rad/s². Determine the magnitudes of the velocity and acceleration of point C at this instant. Solve the problem using Cartesian vectors and Eqs. 16–9 and 16–13.

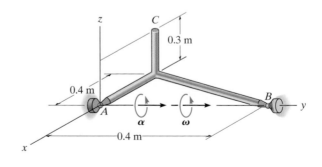

Prob. 16–32

*16.4 Absolute Motion Analysis

A body subjected to *general plane motion* undergoes a *simultaneous* translation and rotation. If the body is represented by a thin slab, the slab translates in the plane and rotates about an axis perpendicular to the plane. The motion can be completely specified by knowing *both* the angular rotation of a line fixed in the body and the motion of a point on the body. One way to define these motions is to use a rectilinear position coordinate *s* to locate the point along its path and an angular position coordinate θ to specify the orientation of the line. The two coordinates are then related using the geometry of the problem. By *direct application* of the time-differential equations $v = ds/dt$, $a = dv/dt$, $\omega = d\theta/dt$, and $\alpha = d\omega/dt$, the *motion* of the point and the *angular motion* of the line can then be related. In some cases, this procedure may also be used to relate the motions of one body to those of a connected body, or to study the motion of a body subjected to rotation about a fixed axis.

The dumping bin on the truck rotates about a fixed axis passing through the pin at *A*. It is operated by the extension of the hydraulic cylinder *BC*. The angular position of the bin can be specified using the angular position coordinate θ, and the position of point *C* on the bin is specified using the rectilinear coordinate *s*. Since *a* and *b* are fixed lengths, then the coordinates can be related by the cosine law, $s = \sqrt{a^2 + b^2 - 2ab\cos\theta}$. Using the chain rule, the time derivative of this equation relates the speed at which the hydraulic cylinder extends, to the angular velocity of the bin, i.e.,

$$v = \frac{1}{2}(a^2 + b^2 - 2ab\cos\theta)^{-\frac{1}{2}}(2ab\sin\theta)\omega.$$

Procedure for Analysis

The velocity and acceleration of a point *P* undergoing rectilinear motion can be related to the angular velocity and angular acceleration of a line contained within a body using the following procedure.

Position Coordinate Equation.

- Locate point *P* using a position coordinate *s*, which is measured from a *fixed origin* and is *directed along the straight-line path of motion* of point *P*.
- Measure from a fixed reference line the angular position θ of a line lying in the body.
- From the dimensions of the body, relate *s* to θ, $s = f(\theta)$, using geometry and/or trigonometry.

Time Derivatives.

- Take the first derivative of $s = f(\theta)$ with respect to time to get a relationship between v and ω.
- Take the second time derivative to get a relationship between a and α.
- In each case the chain rule of calculus must be used when taking the derivatives of the position coordinate equation.

EXAMPLE 16.3

The end of rod R shown in Fig. 16–7 maintains contact with the cam by means of a spring. If the cam rotates about an axis through point O with an angular acceleration α and angular velocity ω, determine the velocity and acceleration of the rod when the cam is in the arbitrary position θ.

Fig. 16–7

SOLUTION

Position Coordinate Equation.   Coordinates θ and x are chosen in order to relate the *rotational motion* of the line segment OA on the cam to the *rectilinear motion* of the rod. These coordinates are measured from the *fixed point O* and may be related to each other using trigonometry. Since $OC = CB = r\cos\theta$, Fig. 16–7, then

$$x = 2r\cos\theta$$

Time Derivatives.   Using the chain rule of calculus, we have

$$\frac{dx}{dt} = -2r(\sin\theta)\frac{d\theta}{dt}$$

$$v = -2r\omega\sin\theta \qquad\qquad Ans.$$

$$\frac{dv}{dt} = -2r\left(\frac{d\omega}{dt}\right)\sin\theta - 2r\omega(\cos\theta)\frac{d\theta}{dt}$$

$$a = -2r(\alpha\sin\theta + \omega^2\cos\theta) \qquad\qquad Ans.$$

NOTE: The negative signs indicate that v and a are opposite to the direction of positive x. This seems reasonable when you visualize the motion.

EXAMPLE 16.4

At a given instant, the cylinder of radius r, shown in Fig. 16–8, has an angular velocity $\boldsymbol{\omega}$ and angular acceleration $\boldsymbol{\alpha}$. Determine the velocity and acceleration of its center G if the cylinder rolls without slipping.

Fig. 16–8

SOLUTION

Position Coordinate Equation.   By inspection, point G moves *horizontally* to the left from G to G' as the cylinder rolls, Fig. 16–8. Consequently its new location G' will be specified by the *horizontal* position coordinate s_G, which is measured from the original position (G) of the cylinder's center. Notice also, that as the cylinder rolls (without slipping), points on its surface contact the ground such that the arc length $A'B$ of contact must be equal to the distance s_G. Consequently, the motion requires the radial line GA to rotate θ to the position $G'A'$. Since the arc $A'B = r\theta$, then G travels a distance

$$s_G = r\theta$$

Time Derivatives.   Taking successive time derivatives of this equation, realizing that r is constant, $\omega = d\theta/dt$, and $\alpha = d\omega/dt$, gives the necessary relationships:

$$s_G = r\theta$$
$$v_G = r\omega \qquad\qquad\qquad Ans.$$
$$a_G = r\alpha \qquad\qquad\qquad Ans.$$

NOTE: Remember that these relationships are valid only if the cylinder (disk, wheel, ball, etc.) rolls *without* slipping.

EXAMPLE 16.5

Fig. 16–9

The large window in Fig. 16–9 is opened using a hydraulic cylinder AB. If the cylinder extends at a constant rate of 0.5 m/s, determine the angular velocity and angular acceleration of the window at the instant $\theta = 30°$.

SOLUTION

Position Coordinate Equation.   The angular motion of the window can be obtained using the coordinate θ, whereas the extension or motion *along the hydraulic cylinder* is defined using a coordinate s, which measures the length from the fixed point A to the moving point B. These coordinates can be related using the law of cosines, namely,

$$s^2 = (2\text{ m})^2 + (1\text{ m})^2 - 2(2\text{ m})(1\text{ m})\cos\theta$$

$$s^2 = 5 - 4\cos\theta \qquad (1)$$

When $\theta = 30°$,

$$s = 1.239\text{ m}$$

Time Derivatives.   Taking the time derivatives of Eq. 1, we have

$$2s\frac{ds}{dt} = 0 - 4(-\sin\theta)\frac{d\theta}{dt}$$

$$s(v_s) = 2(\sin\theta)\omega \qquad (2)$$

Since $v_s = 0.5$ m/s, then at $\theta = 30°$,

$$(1.239\text{ m})(0.5\text{ m/s}) = 2\sin 30°\omega$$

$$\omega = 0.620\text{ rad/s} \qquad\qquad Ans.$$

Taking the time derivative of Eq. 2 yields

$$\frac{ds}{dt}v_s + s\frac{dv_s}{dt} = 2(\cos\theta)\frac{d\theta}{dt}\omega + 2(\sin\theta)\frac{d\omega}{dt}$$

$$v_s^2 + sa_s = 2(\cos\theta)\omega^2 + 2(\sin\theta)\alpha$$

Since $a_s = dv_s/dt = 0$, then

$$(0.5\text{ m/s})^2 + 0 = 2\cos 30°(0.620\text{ rad/s})^2 + 2\sin 30°\alpha$$

$$\alpha = -0.415\text{ rad/s}^2 \qquad\qquad Ans.$$

Because the result is negative, it indicates the window has an angular deceleration.

PROBLEMS

16–33. The bar DC rotates uniformly about the shaft at D with a constant angular velocity $\boldsymbol{\omega}$. Determine the velocity and acceleration of the bar AB, which is confined by the guides to move vertically.

Prob. 16–33

16–34. At the instant shown, $\theta = 60°$, and rod AB is subjected to a deceleration of 16 m/s² when the velocity is 10 m/s. Determine the angular velocity and angular acceleration of link CD at this instant.

Prob. 16–34

16–35. The mechanism is used to convert the constant circular motion ω of rod AB into translating motion of rod CD. Determine the velocity and acceleration of CD for any angle θ of AB.

Prob. 16–35

*16–36. Determine the angular velocity of rod AB when $\theta = 30°$. The shaft and the center of the roller C move forward at a constant rate $v = 5$ m/s.

Prob. 16–36

16–37. Determine the velocity of rod R for any angle θ of the cam C if the cam rotates with a constant angular velocity ω. The pin connection at O does not cause an interference with the motion of A on C.

Prob. 16–37

16–38.   The crankshaft AB is rotating at a constant angular velocity of $\omega = 150$ rad/s. Determine the velocity of the piston P at the instant $\theta = 30°$.

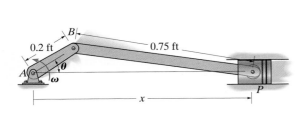

Prob. 16–38

16–39.   At the instant $\theta = 50°$, the slotted guide is moving upward with an acceleration of 3 m/s^2 and a velocity of 2 m/s. Determine the angular acceleration and angular velocity of link AB at this instant. *Note:* The upward motion of the guide is in the negative y direction.

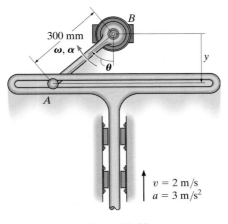

Prob. 16–39

*****16–40.**   Determine the velocity of the rod R for any angle θ of cam C as the cam rotates with a constant angular velocity ω. The pin connection at O does not cause an interference with the motion of plate A on C.

Prob. 16–40

■**16–41.**   The end A of the bar is moving downward along the slotted guide with a constant velocity $\mathbf{v}_A$. Determine the angular velocity ω and angular acceleration $\boldsymbol{\alpha}$ of the bar as a function of its position y.

Prob. 16–41

16–42. The inclined plate moves to the left with a constant velocity **v**. Determine the angular velocity and angular acceleration of the slender rod of length l. The rod pivots about the step at C as it slides on the plate.

Prob. 16–42

*__**16–44.**__ The crate is transported on a platform which rests on rollers, each having a radius r. If the rollers do not slip, determine their angular velocity if the platform moves forward with a velocity **v**.

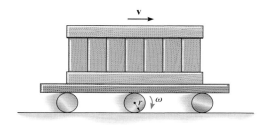

Prob. 16–44

16–43. The bar remains in contact with the floor and with point A. If point B moves to the right with a constant velocity $\mathbf{v}_B$, determine the angular velocity and angular acceleration of the bar as a function of x.

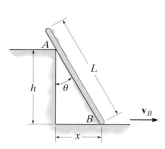

Prob. 16–43

16–45. Bar AB rotates uniformly about the fixed pin A with a constant angular velocity $\boldsymbol{\omega}$. Determine the velocity and acceleration of block C, at the instant $\theta = 60°$.

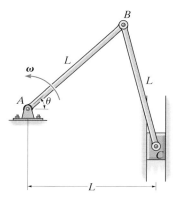

Prob. 16–45

16–46. The bar is confined to move along the vertical and inclined planes. If the velocity of the roller at A is $v_A = 6$ ft/s when $\theta = 45°$, determine the bar's angular velocity and the velocity of roller B at this instant.

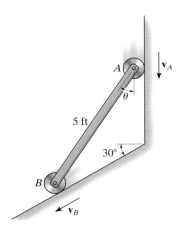

5 ft

30°

Prob. 16–46

16–47. When the bar is at the angle θ, the rod is rotating clockwise at ω and has an angular acceleration of α. Determine the velocity and acceleration of the weight A at this instant. The cord is 20 ft long.

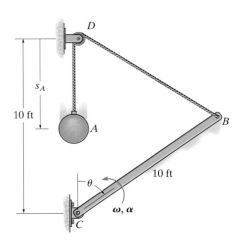

10 ft

10 ft

Prob. 16–47

*16–48. The slotted yoke is pinned at A while end B is used to move the ram R horizontally. If the disk rotates with a constant angular velocity ω, determine the velocity and acceleration of the ram. The crank pin C is fixed to the disk and turns with it.

Prob. 16–48

16–49. The Geneva wheel A provides intermittent rotary motion ω_A for continuous motion $\omega_D = 2$ rad/s of disk D. By choosing $d = 100\sqrt{2}$ mm, the wheel has zero angular velocity at the instant pin B enters or leaves one of the four slots. Determine the magnitude of the angular velocity ω_A of the Geneva wheel at any angle θ for which pin B is in contact with the slot.

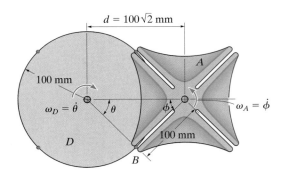

$d = 100\sqrt{2}$ mm

100 mm

$\omega_D = \dot{\theta}$

100 mm

$\omega_A = \dot{\phi}$

Prob. 16–49

16.5 Relative-Motion Analysis: Velocity

The general plane motion of a rigid body can be described as a *combination* of translation and rotation. To view these "component" motions *separately* we will use a *relative-motion analysis* involving two sets of coordinate axes. The x, y coordinate system is fixed and measures the *absolute* position of two points A and B on the body, Fig. 16–10a. The origin of the x', y' coordinate system will be attached to the selected "base point" A, which generally has a *known* motion. The axes of this coordinate system *translate* with respect to the fixed frame but do not rotate with the body.

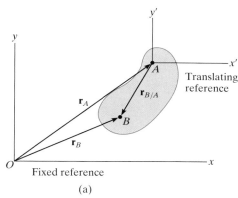

Position. The position vector $\mathbf{r}_A$ in Fig. 16–10a specifies the location of the "base point" A, and the relative-position vector $\mathbf{r}_{B/A}$ locates point B with respect to point A. By vector addition, the *position* of B is then

$$\mathbf{r}_B = \mathbf{r}_A + \mathbf{r}_{B/A}$$

Fig. 16–10

Displacement. During an instant of time dt, points A and B undergo displacements $d\mathbf{r}_A$ and $d\mathbf{r}_B$ as shown in Fig. 16–10b. If we consider the general plane motion by its component parts then the *entire body* first *translates* by an amount $d\mathbf{r}_A$ so that A, the base point, moves to its *final position* and point B moves to B', Fig. 16–10c. The body is then *rotated* about A by an amount $d\theta$ so that B' undergoes a *relative displacement* $d\mathbf{r}_{B/A}$ and thus moves to its final position B. Due to the rotation about A, $dr_{B/A} = r_{B/A}\, d\theta$, and the displacement of B is

$$
\begin{array}{ccc}
d\mathbf{r}_B & = & d\mathbf{r}_A & + & d\mathbf{r}_{B/A} \\
& & & & \text{due to rotation about } A \\
& & \text{due to translation of } A \\
& & \text{due to translation and rotation}
\end{array}
$$

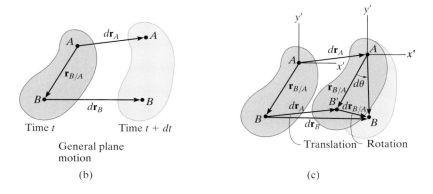

Time t Time $t + dt$

General plane
motion

(b)

Translation   Rotation

(c)

As the slider block A moves horizontally to the left with a velocity $\mathbf{v}_A$, it causes the crank CB to rotate counterclockwise, such that $\mathbf{v}_B$ is directed tangent to its circular path, i.e., upward to the left. The connecting rod AB is subjected to general plane motion, and at the instant shown it has an angular velocity $\boldsymbol{\omega}$.

Velocity.   To determine the relationship between the velocities of points A and B, it is necessary to take the time derivative of the position equation, or simply divide the displacement equation by dt. This yields

$$\frac{d\mathbf{r}_B}{dt} = \frac{d\mathbf{r}_A}{dt} + \frac{d\mathbf{r}_{B/A}}{dt}$$

The terms $d\mathbf{r}_B/dt = \mathbf{v}_B$ and $d\mathbf{r}_A/dt = \mathbf{v}_A$ are measured from the fixed x, y axes and represent the *absolute velocities* of points A and B, respectively. The magnitude of the third term is $r_{B/A}\, d\theta/dt = r_{B/A}\dot{\theta} = r_{B/A}\omega$, where ω is the angular velocity of the body at the instant considered. We will denote this term as the *relative velocity* $\mathbf{v}_{B/A}$, since it represents the velocity of B with respect to A as measured by an observer fixed to the translating x', y' axes. Since the body is rigid, realize that this observer only sees point B move along a *circular arc* that has a radius of curvature $r_{B/A}$. In other words, *the body appears to move as if it were rotating with an angular velocity $\boldsymbol{\omega}$ about the z' axis passing through A.* Consequently, $\mathbf{v}_{B/A}$ has a magnitude of $v_{B/A} = \omega r_{B/A}$ and a *direction* which is perpendicular to $\mathbf{r}_{B/A}$. We therefore have

$$\boxed{\mathbf{v}_B = \mathbf{v}_A + \mathbf{v}_{B/A}} \qquad (16\text{–}15)$$

where

$$\mathbf{v}_B = \text{velocity of point } B$$
$$\mathbf{v}_A = \text{velocity of the base point } A$$
$$\mathbf{v}_{B/A} = \text{relative velocity of ``}B \text{ with respect to } A\text{''}$$

This relative motion is *circular*, the *magnitude* is $v_{B/A} = \omega r_{B/A}$ and the *direction* is perpendicular to $\mathbf{r}_{B/A}$.

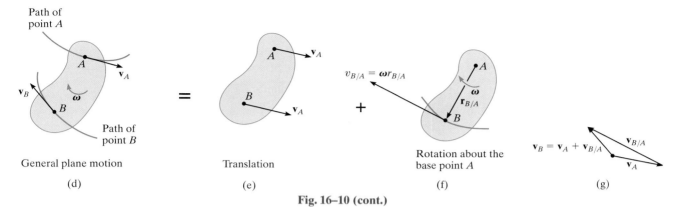

General plane motion
(d)

Translation
(e)

Rotation about the
base point A
(f)

$\mathbf{v}_B = \mathbf{v}_A + \mathbf{v}_{B/A}$
(g)

Fig. 16–10 (cont.)

Each of the three terms in Eq. 16–15 is represented graphically on the *kinematic diagrams* in Figs. 16–10d, 16–10e, and 16–10f. Here it is seen that the velocity of B, Fig. 16–10d, is determined by considering the entire body to translate with a velocity of $\mathbf{v}_A$, Fig. 16–10e, and rotate about A with an angular velocity $\boldsymbol{\omega}$, Fig. 16–10f. Vector addition of these two effects, applied to B, yields $\mathbf{v}_B$, as shown in Fig. 16–10g.

Since the relative velocity $\mathbf{v}_{B/A}$ represents the effect of *circular motion*, about A, this term can be expressed by the cross product $\mathbf{v}_{B/A} = \boldsymbol{\omega} \times \mathbf{r}_{B/A}$, Eq. 16–9. Hence, for application, we can also write Eq. 16–15 as

$$\mathbf{v}_B = \mathbf{v}_A + \boldsymbol{\omega} \times \mathbf{r}_{B/A} \qquad (16\text{–}16)$$

where

$\mathbf{v}_B$ = velocity of B

$\mathbf{v}_A$ = velocity of the base point A

$\boldsymbol{\omega}$ = angular velocity of the body

$\mathbf{r}_{B/A}$ = relative–position vector drawn from A to B

The velocity equation 16–15 or 16–16 may be used in a practical manner to study the general plane motion of a rigid body which is either pin connected to or in contact with other moving bodies. When applying this equation, points A and B should generally be selected as points on the body which are pin-connected to other bodies, or as points in contact with adjacent bodies which have a *known motion*. For example, both points A and B on link AB, Fig. 16–11a, have circular paths of motion since the wheel and link CB move in circular paths. The *directions* of $\mathbf{v}_A$ and $\mathbf{v}_B$ can therefore be established since they are always *tangent* to their paths of motion, Fig. 16–11b. In the case of the wheel in Fig. 16–12, which rolls *without slipping*, point A can be selected at the ground. Here A (momentarily) has zero velocity since the ground does not move. Furthermore, the center of the wheel, B, moves along a horizontal path so that $\mathbf{v}_B$ is horizontal.

(a)

(b)

Fig. 16–11

$v_A = 0$

Fig. 16–12

Procedure for Analysis

The relative velocity equation can be applied either by using Cartesian vector analysis, or by writing the x and y scalar component equations directly. For application, it is suggested that the following procedure be used.

Vector Analysis

Kinematic Diagram.

- Establish the directions of the fixed x, y coordinates and draw a kinematic diagram of the body. Indicate on it the velocities $\mathbf{v}_A$, $\mathbf{v}_B$ of points A and B, the angular velocity $\boldsymbol{\omega}$, and the relative-position vector $\mathbf{r}_{B/A}$.

- If the magnitudes of $\mathbf{v}_A$, $\mathbf{v}_B$, or $\boldsymbol{\omega}$ are unknown, the sense of direction of these vectors may be assumed.

Velocity Equation.

- To apply $\mathbf{v}_B = \mathbf{v}_A + \boldsymbol{\omega} \times \mathbf{r}_{B/A}$, express the vectors in Cartesian vector form and substitute them into the equation. Evaluate the cross product and then equate the respective $\mathbf{i}$ and $\mathbf{j}$ components to obtain two scalar equations.

- If the solution yields a *negative* answer for an *unknown* magnitude, it indicates the sense of direction of the vector is opposite to that shown on the kinematic diagram.

Scalar Analysis

Kinematic Diagram.

- If the velocity equation is to be applied in scalar form, then the magnitude and direction of the relative velocity $\mathbf{v}_{B/A}$ must be established. Draw a kinematic diagram such as shown in Fig. 16–10g, which shows the relative motion. Since the body is considered to be "pinned" momentarily at the base point A, the magnitude is $v_{B/A} = \omega r_{B/A}$. The sense of direction of $\mathbf{v}_{B/A}$ is established from the diagram, such that $\mathbf{v}_{B/A}$ acts perpendicular to $\mathbf{r}_{B/A}$ in accordance with the rotational motion $\boldsymbol{\omega}$ of the body.*

Velocity Equation.

- Write Eq. 16–15 in symbolic form, $\mathbf{v}_B = \mathbf{v}_A + \mathbf{v}_{B/A}$, and underneath each of the terms represent the vectors graphically by showing their magnitudes and directions. The scalar equations are determined from the x and y components of these vectors.

*The notation $\mathbf{v}_B = \mathbf{v}_A + \mathbf{v}_{B/A(\text{pin})}$ may be helpful in recalling that A is "pinned."

EXAMPLE 16.6

The link shown in Fig. 16–13a is guided by two blocks at A and B, which move in the fixed slots. If the velocity of A is 2 m/s downward, determine the velocity of B at the instant $\theta = 45°$.

(a)

SOLUTION (VECTOR ANALYSIS)

Kinematic Diagram.   Since points A and B are restricted to move along the fixed slots and $\mathbf{v}_A$ is directed downward, the velocity $\mathbf{v}_B$ must be directed horizontally to the right, Fig. 16–13b. This motion causes the link to rotate counterclockwise; that is, by the right-hand rule the angular velocity $\boldsymbol{\omega}$ is directed outward, perpendicular to the plane of motion. Knowing the magnitude and direction of $\mathbf{v}_A$ and the lines of action of $\mathbf{v}_B$ and $\boldsymbol{\omega}$, it is possible to apply the velocity equation $\mathbf{v}_B = \mathbf{v}_A + \boldsymbol{\omega} \times \mathbf{r}_{B/A}$ to points A and B in order to solve for the two unknown magnitudes v_B and ω. Since $\mathbf{r}_{B/A}$ is needed, it is also shown in Fig. 16–13b.

(b)

Fig. 16–13

Velocity Equation.   Expressing each of the vectors in Fig. 16–13b in terms of their $\mathbf{i}, \mathbf{j}, \mathbf{k}$ components and applying Eq. 16–16 to A, the base point, and B, we have

$$\mathbf{v}_B = \mathbf{v}_A + \boldsymbol{\omega} \times \mathbf{r}_{B/A}$$
$$v_B \mathbf{i} = -2\mathbf{j} + [\omega \mathbf{k} \times (0.2 \sin 45°\mathbf{i} - 0.2 \cos 45°\mathbf{j})]$$
$$v_B \mathbf{i} = -2\mathbf{j} + 0.2\omega \sin 45°\mathbf{j} + 0.2\omega \cos 45°\mathbf{i}$$

Equating the $\mathbf{i}$ and $\mathbf{j}$ components gives

$$v_B = 0.2\omega \cos 45° \qquad 0 = -2 + 0.2\omega \sin 45°$$

Thus,

$$\omega = 14.1 \text{ rad/s} \curvearrowleft$$
$$v_B = 2 \text{ m/s} \rightarrow \qquad\qquad\qquad Ans.$$

Since both results are *positive*, the *directions* of $\mathbf{v}_B$ and $\boldsymbol{\omega}$ are indeed *correct* as shown in Fig. 16–13b. It should be emphasized that these results are *valid only* at the instant $\theta = 45°$. A recalculation for $\theta = 44°$ yields $v_B = 2.07$ m/s and $\omega = 14.4$ rad/s; whereas when $\theta = 46°$, $v_B = 1.93$ m/s and $\omega = 13.9$ rad/s, etc.

NOTE: Once the velocity of a point (A) on the link and the angular velocity are *known*, the velocity of any other point on the link can be determined. As an exercise, see if you can apply Eq. 16–16 to points A and C or to points B and C and show that when $\theta = 45°$, $v_C = 3.16$ m/s, directed at an angle of 18.4° up from the horizontal.

EXAMPLE 16.7

(a)

(b)

Relative motion
(c)

Fig. 16–14

The cylinder shown in Fig. 16–14a rolls without slipping on the surface of a conveyor belt which is moving at 2 ft/s. Determine the velocity of point A. The cylinder has a clockwise angular velocity $\omega = 15$ rad/s at the instant shown.

SOLUTION I (VECTOR ANALYSIS)

Kinematic Diagram.   Since no slipping occurs, point B on the cylinder has the same velocity as the conveyor, Fig. 16–14b. Also, the angular velocity of the cylinder is known, so we can apply the velocity equation to B, the base point, and A to determine $\mathbf{v}_A$.

Velocity Equation.

$$\mathbf{v}_A = \mathbf{v}_B + \boldsymbol{\omega} \times \mathbf{r}_{A/B}$$

$$(v_A)_x\mathbf{i} + (v_A)_y\mathbf{j} = 2\mathbf{i} + (-15\mathbf{k}) \times (-0.5\mathbf{i} + 0.5\mathbf{j})$$

$$(v_A)_x\mathbf{i} + (v_A)_y\mathbf{j} = 2\mathbf{i} + 7.50\mathbf{j} + 7.50\mathbf{i}$$

so that

$$(v_A)_x = 2 + 7.50 = 9.50 \text{ ft/s} \tag{1}$$

$$(v_A)_y = 7.50 \text{ ft/s} \tag{2}$$

Thus,

$$v_A = \sqrt{(9.50)^2 + (7.50)^2} = 12.1 \text{ ft/s} \qquad Ans.$$

$$\theta = \tan^{-1}\frac{7.50}{9.50} = 38.3° \quad \overset{v_A}{\measuredangle\theta} \qquad Ans.$$

SOLUTION II (SCALAR ANALYSIS)

As an alternative procedure, the scalar components of $\mathbf{v}_A = \mathbf{v}_B + \mathbf{v}_{A/B}$ can be obtained directly. From the kinematic diagram showing the relative "circular" motion $\mathbf{v}_{A/B}$, Fig. 16–14c, we have

$$v_{A/B} = \omega r_{A/B} = (15 \text{ rad/s})\left(\frac{0.5 \text{ ft}}{\cos 45°}\right) = 10.6 \text{ ft/s} \quad \measuredangle 45°$$

Thus,

$$\mathbf{v}_A = \mathbf{v}_B + \mathbf{v}_{A/B}$$

$$\begin{bmatrix} (v_A)_x \\ \rightarrow \end{bmatrix} + \begin{bmatrix} (v_A)_y \\ \uparrow \end{bmatrix} = \begin{bmatrix} 2 \text{ ft/s} \\ \rightarrow \end{bmatrix} + \begin{bmatrix} 106 \text{ ft/s} \\ \measuredangle 45° \end{bmatrix}$$

Equating the x and y components gives the same results as before, namely,

$$(\overset{+}{\rightarrow}) \qquad (v_A)_x = 2 + 10.6 \cos 45° = 9.50 \text{ ft/s}$$

$$(+\uparrow) \qquad (v_A)_y = 0 + 10.6 \sin 45° = 7.50 \text{ ft/s}$$

EXAMPLE 16.8

The collar C in Fig. 16–15a is moving downward with a velocity of 2 m/s. Determine the angular velocities of CB and AB at this instant.

SOLUTION I (VECTOR ANALYSIS)

Kinematic Diagram.   The downward motion of C causes B to move to the right. Also, CB and AB rotate counterclockwise. To solve, we will write the appropriate kinematic equation for each link.

Velocity Equation.

Link CB (general plane motion): See Fig. 16–15b.

$$\mathbf{v}_B = \mathbf{v}_C + \boldsymbol{\omega}_{CB} \times \mathbf{r}_{B/C}$$
$$v_B\mathbf{i} = -2\mathbf{j} + \omega_{CB}\mathbf{k} \times (0.2\mathbf{i} - 0.2\mathbf{j})$$
$$v_B\mathbf{i} = -2\mathbf{j} + 0.2\omega_{CB}\mathbf{j} + 0.2\omega_{CB}\mathbf{i}$$

$$v_B = 0.2\omega_{CB} \qquad (1)$$
$$0 = -2 + 0.2\omega_{CB} \qquad (2)$$
$$\omega_{CB} = 10 \text{ rad/s} \text{ }\curvearrowleft$$
$$v_B = 2 \text{ m/s} \rightarrow \qquad Ans.$$

Link AB (rotation about a fixed axis): See Fig. 16–15c.

$$\mathbf{v}_B = \boldsymbol{\omega}_{AB} \times \mathbf{r}_B$$
$$2\mathbf{i} = \omega_{AB}\mathbf{k} \times (-0.2\mathbf{j})$$
$$2 = 0.2\omega_{AB}$$
$$\omega_{AB} = 10 \text{ rad/s} \text{ }\curvearrowleft \qquad Ans.$$

SOLUTION II (SCALAR ANALYSIS)

The scalar component equations of $\mathbf{v}_B = \mathbf{v}_C + \mathbf{v}_{B/C}$ can be obtained directly. The kinematic diagram in Fig. 16–15d shows the relative "circular" motion $\mathbf{v}_{B/C}$. We have

$$\mathbf{v}_B = \mathbf{v}_C + \mathbf{v}_{B/C}$$

$$\begin{bmatrix} v_B \\ \rightarrow \end{bmatrix} = \begin{bmatrix} 2 \text{ m/s} \\ \downarrow \end{bmatrix} + \begin{bmatrix} \omega_{CB}(0.2\sqrt{2} \text{ m}) \\ \measuredangle 45° \end{bmatrix}$$

Resolving these vectors in the x and y directions yields

$(\xrightarrow{+})$ $\qquad (v_B = 0 + \omega_{CB}(0.2\sqrt{2} \cos 45°)$

$(+\uparrow)$ $\qquad 0 = -2 + \omega_{CB}(0.2\sqrt{2} \sin 45°)$

which is the same as Eqs. 1 and 2.

(a)

(b)

(c)

Relative motion

(d)

Fig. 16–15

EXAMPLE 16.9

(a)

The bar AB of the linkage shown in Fig. 16–16a has a clockwise angular velocity of 30 rad/s when $\theta = 60°$. Determine the angular velocities of member BC and the wheel at this instant.

SOLUTION (VECTOR ANALYSIS)

Kinematic Diagram.   By inspection, the velocities of points B and C are defined by the rotation of link AB and the wheel about their fixed axes. The position vectors and the angular velocity of each member are shown on the kinematic diagram in Fig. 16–16b. To solve, we will write the appropriate kinematic equation for each member.

Velocity Equation.

Link AB (rotation about a fixed axis):

$$\mathbf{v}_B = \boldsymbol{\omega}_{AB} \times \mathbf{r}_B$$
$$= (-30\mathbf{k}) \times (0.2 \cos 60°\mathbf{i} + 0.2 \sin 60°\mathbf{j})$$
$$= \{5.20\mathbf{i} - 3.0\mathbf{j}\} \text{ m/s}$$

Link BC (general plane motion):

$$\mathbf{v}_C = \mathbf{v}_B + \boldsymbol{\omega}_{BC} \times \mathbf{r}_{C/B}$$
$$v_C\mathbf{i} = 5.20\mathbf{i} - 3.0\mathbf{j} + (\omega_{BC}\mathbf{k}) \times (0.2\mathbf{i})$$
$$v_C\mathbf{i} = 5.20\mathbf{i} + (0.2\omega_{BC} - 3.0)\mathbf{j}$$
$$v_C = 5.20 \text{ m/s}$$
$$0 = 0.2\omega_{BC} - 3.0$$
$$\omega_{BC} = 15 \text{ rad/s} \, \text{↰} \qquad\qquad Ans.$$

(b)

Fig. 16–16

Wheel (rotation about a fixed axis):

$$\mathbf{v}_C = \boldsymbol{\omega}_D \times \mathbf{r}_C$$
$$5.20\mathbf{i} = (\omega_D\mathbf{k}) \times (-0.1\mathbf{j})$$
$$5.20 = 0.1\omega_D$$
$$\omega_D = 52 \text{ rad/s} \, \text{↰} \qquad\qquad Ans.$$

NOTE: By inspection, Fig. 16–16a, $v_B = (0.2)(30) = 6$ m/s, ⦨30° and $\mathbf{v}_C$ is directed to the right. As an exercise, use this information and try to obtain ω_{BC} by applying $\mathbf{v}_C = \mathbf{v}_B + \mathbf{v}_{C/B}$ using scalar components.

PROBLEMS

16–50. If h and θ are known, and the speed of A and B is $v_A = v_B = v$, determine the angular velocity $\boldsymbol{\omega}$ of the body and the direction ϕ of $\mathbf{v}_B$.

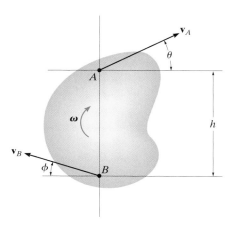

Prob. 16–50

16–51. The wheel is rotating with an angular velocity $\omega = 8$ rad/s. Determine the velocity of the collar A at the instant $\theta = 30°$ and $\phi = 60°$. Also, sketch the location of bar AB when $\theta = 0°$, $30°$, and $60°$ to show its general plane motion.

Prob. 16–51

*16–52. The pinion gear A rolls on the fixed gear rack B with an angular velocity $\omega = 4$ rad/s. Determine the velocity of the gear rack C.

16–53. The pinion gear rolls on the gear racks. If B is moving to the right at 8 ft/s and C is moving to the left at 4 ft/s, determine the angular velocity of the pinion gear and the velocity of its center A.

Probs. 16–52/53

16–54. The shaper mechanism is designed to give a slow cutting stroke and a quick return to a blade attached to the slider at C. Determine the velocity of the slider block C at the instant $\theta = 60°$, if link AB is rotating at 4 rad/s.

16–55. Determine the velocity of the slider block at C at the instant $\theta = 45°$, if link AB is rotating at 4 rad/s.

Probs. 16–54/55

*16–56.   The velocity of the slider block C is 4 ft/s up the inclined groove. Determine the angular velocity of links AB and BC and the velocity of point B at the instant shown.

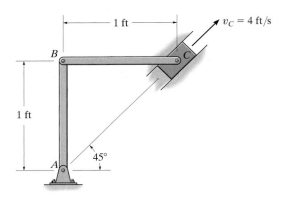

Prob. 16–56

16–57.   Rod AB is rotating with an angular velocity $\omega_{AB} = 5$ rad/s. Determine the velocity of the collar C at the instant $\theta = 60°$ and $\phi = 45°$. Also, sketch the location of bar BC when $\theta = 30°$, $60°$ and $45°$ to show its general plane motion.

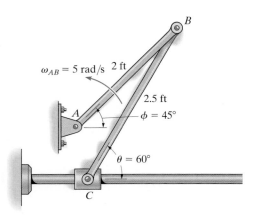

Prob. 16–57

16–58.   If rod CD is rotating with an angular velocity $\omega_{DC} = 8$ rad/s, determine the angular velocities of rods AB and CB at the instant shown.

Prob. 16–58

16–59.   The angular velocity of link AB is $\omega_{AB} = 4$ rad/s. Determine the velocity of the collar at C and the angular velocity of link CB at the instant $\theta = 60°$ and $\phi = 45°$. Link CB is horizontal at this instant. Also, sketch the location of link CB when $\theta = 30°$, $60°$, and $90°$ to show its general plane motion.

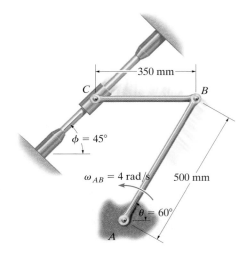

Prob. 16–59

*16–60. The link AB has an angular velocity of 2 rad/s. Determine the velocity of block C at the instant $\theta = 45°$. Also, sketch the location of link BC when $\theta = 60°, 45°$, and $30°$ to show its general plane motion.

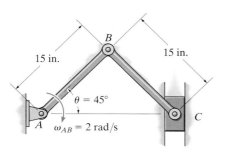

Prob. 16–60

16–63. If, at a given instant, point B has a downward velocity of $v_B = 3$ m/s, determine the velocity of point A at this instant. Notice that for this motion to occur, the wheel must slip at A.

Prob. 16–63

16–61. At the instant shown, the truck is traveling to the right at 3 m/s, while the pipe is rolling counterclockwise at $\omega = 8$ rad/s without slipping at B. Determine the velocity of the pipe's center G.

16–62. At the instant shown, the truck is traveling to the right at 8 m/s. If the spool does not slip at B, determine its angular velocity so that its mass center G appears to an observer on the ground to remain stationary.

Probs. 16–61/62

*16–64. If the link AB is rotating about the pin at A with an angular velocity $\omega_{AB} = 5$ rad/s, determine the velocities of blocks C and E at the instant shown.

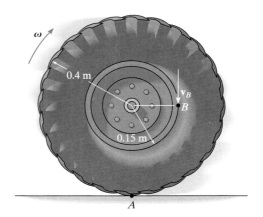

Prob. 16–64

16–65. If disk D has a constant angular velocity $\omega_D = 2$ rad/s, determine the angular velocity of disk A at the instant $\theta = 60°$.

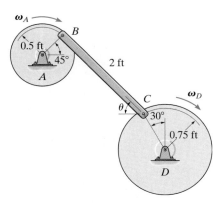

Prob. 16–65

16–66. The planetary gear system is used in an automatic transmission for an automobile. By locking or releasing certain gears, it has the advantage of operating the car at different speeds. Consider the case where the ring gear R is rotating at $\omega_R = 3$ rad/s, and the sun gear S is held fixed, $\omega_S = 0$. Determine the angular velocity of each of the planet gears P and shaft A.

Prob. 16–66

16–67. If bar AB has an angular velocity $\omega_{AB} = 6$ rad/s, determine the velocity of the slider block C at the instant $\theta = 45°$ and $\phi = 30°$. Also, sketch the location of bar BC when $\theta = 30°, 45°,$ and $60°$ to show its general plane motion.

Prob. 16–67

*16–68. If the end of the cord is pulled downward with a speed $v_C = 120$ mm/s, determine the angular velocities of pulleys A and B and the speed of block D. Assume that the cord does not slip on the pulleys.

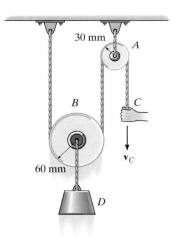

Prob. 16–68

16–69. At the instant shown, the truck is traveling to the right at $8t$ m/s, while the pipe is rolling counterclockwise at $\omega = 2t$ rad/s without slipping at B, where t is in seconds. Determine the velocity of the pipe's center G as a function of time.

16–70. At the instant shown, the truck is traveling to the right at 12 m/s. If the pipe does not slip at B, determine its angular velocity if its mass center G appears to an observer on the ground to be moving to the right at 3 m/s.

Probs. 16–69/70

16–71. The pinion gear A rolls on the fixed gear rack B with an angular velocity $\omega = 4$ rad/s. Determine the velocity of the gear rack C.

*16–72.** Part of an automatic transmission consists of a *fixed* ring gear R, three equal planet gears P, the sun gear S, and the planet carrier C, which is shaded. If the sun gear is rotating at $\omega_S = 6$ rad/s, determine the angular velocity ω_C of the *planet carrier*. Note that C is pin connected to the center of each of the planet gears.

Prob. 16–72

16–73. When the crank on the Chinese windlass is turning, the rope on shaft A unwinds while that on shaft B winds up. Determine the speed of block D if the crank is turning with an angular velocity $\omega = 4$ rad/s. What is the angular velocity of the pulley at C? The rope segments on each side of the pulley are both parallel and vertical, and the rope does not slip on the pulley.

Prob. 16–73

Prob. 16–71

16–74. In an automobile transmission the planet pinions A and B rotate on shafts that are mounted on the planet-pinion carrier CD. As shown, CD is attached to a shaft at E which is aligned with the center of the *fixed* sun-gear S. This shaft is not attached to the sun gear. If CD is rotating at $\omega_{CD} = 8$ rad/s, determine the angular velocity of the ring gear R.

Prob. 16–74

16–75. The cylinder B rolls on the *fixed cylinder A* without slipping. If the connected bar CD is rotating with an angular velocity of $\omega_{CD} = 5$ rad/s, determine the angular velocity of cylinder B.

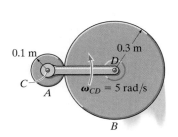

Prob. 16–75

*16–76.** The slider mechanism is used to increase the stroke of travel of one slider with respect to that of another. As shown, when the slider A is moving forward, the attached pinion F rolls on the *fixed* rack D, forcing slider C to move forward. This in turn causes the attached pinion G to roll on the *fixed* rack E, thereby moving slider B. If A has a velocity of $v_A = 4$ ft/s at the instant shown, determine the velocity of B. $r = 0.2$ ft.

Prob. 16–76

16–77. The gauge is used to indicate the safe load acting at the end of the boom, B, when it is in any angular position. It consists of a fixed dial plate D and an indicator arm ACE which is pinned to the plate at C and to a short link EF. If the boom is pin-connected to the trunk frame at G and is rotating downward at $\omega_B = 4$ rad/s, determine the velocity of the dial pointer A at the instant shown, i.e., when EF and AC are in the vertical position.

Prob. 16–77

16.6 Instantaneous Center of Zero Velocity

The velocity of any point B located on a rigid body can be obtained in a very direct way by choosing the base point A to be a point that has *zero velocity* at the instant considered. In this case, $\mathbf{v}_A = \mathbf{0}$, and therefore the velocity equation, $\mathbf{v}_B = \mathbf{v}_A + \boldsymbol{\omega} \times \mathbf{r}_{B/A}$, becomes $\mathbf{v}_B = \boldsymbol{\omega} \times \mathbf{r}_{B/A}$. For a body having general plane motion, point A so chosen is called the *instantaneous center of zero velocity (IC)*, and it lies on the *instantaneous axis of zero velocity*. This axis is always perpendicular to the plane of motion, and the intersection of the axis with this plane defines the location of the *IC*. Since point A is coincident with the *IC*, then $\mathbf{v}_B = \boldsymbol{\omega} \times \mathbf{r}_{B/IC}$ and so point B moves momentarily about the *IC* in a *circular path*; in other words, the body appears to rotate about the instantaneous axis. The *magnitude* of $\mathbf{v}_B$ is simply $v_B = \omega r_{B/IC}$, where ω is the angular velocity of the body. Due to the circular motion, the *direction* of $\mathbf{v}_B$ must always be *perpendicular* to $\mathbf{r}_{B/IC}$.

For example, consider the wheel in Fig. 16–17a. If it rolls *without slipping*, then the point of *contact* with the ground has *zero velocity*. Hence this point represents the *IC* for the wheel, Fig. 16–17b. If it is imagined that the wheel is momentarily pinned at this point, the velocities of points B, C, O, and so on, can be found using $v = \omega r$. Here the radial distances $r_{B/IC}$, $r_{C/IC}$, and $r_{O/IC}$, shown in Fig. 16–17b, must be determined from the geometry of the wheel.

(a)

(b)

Fig. 16–17

The *IC* for this bicycle wheel is at the ground. There the spokes are somewhat visible, whereas at the top of the wheel they become blurred. Note also how points on the side portions of the wheel move as shown by their velocities.

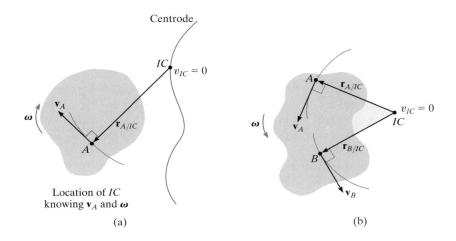

Location of IC
knowing $\mathbf{v}_A$ and $\boldsymbol{\omega}$

(a)

(b)

Fig. 16–18

Location of the IC.

To locate the *IC* we can use the fact that the *velocity* of a point on the body is *always perpendicular* to the *relative-position vector* extending from the *IC* to the point. Several possibilities exist:

- *Given the velocity $\mathbf{v}_A$ of a point A on the body, and the angular velocity $\boldsymbol{\omega}$ of the body*, Fig. 16–18a. In this case, the *IC* is located along the line drawn perpendicular to $\mathbf{v}_A$ at A, such that the distance from A to the *IC* is $r_{A/IC} = v_A/\omega$. Note that the *IC* lies up and to the right of A since $\mathbf{v}_A$ must cause a clockwise angular velocity $\boldsymbol{\omega}$ about the *IC*.

- *Given the lines of action of two nonparallel velocities $\mathbf{v}_A$ and $\mathbf{v}_B$*, Fig. 16–18b. Construct at points A and B line segments that are perpendicular to $\mathbf{v}_A$ and $\mathbf{v}_B$. Extending these perpendiculars to their *point of intersection* as shown locates the *IC* at the instant considered.

- *Given the magnitude and direction of two parallel velocities $\mathbf{v}_A$ and $\mathbf{v}_B$*. Here the location of the *IC* is determined by proportional triangles. Examples are shown in Fig. 16–18c and d. In both cases $r_{A/IC} = v_A/\omega$ and $r_{B/IC} = v_B/\omega$. If d is a known distance between points A and B, then in Fig. 16–18c, $r_{A/IC} + r_{B/IC} = d$ and in Fig. 16–18d, $r_{B/IC} - r_{A/IC} = d$. As a special case, note that if the body is *translating*, $\mathbf{v}_A = \mathbf{v}_B$, then the *IC* would be located at infinity, in which case $r_{A/IC} = r_{B/IC} \to \infty$. This being the case, $\omega = (v_A/r_{A/IC}) = (v_B/r_{B/IC}) \to 0$, as expected.

(c)

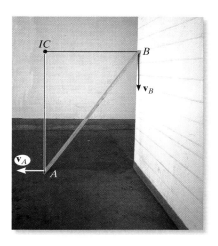

Location of *IC*
knowing $\mathbf{v}_A$ and $\mathbf{v}_B$

(d)

Fig. 16–18 (cont.)

Realize that the point chosen as the instantaneous center of zero velocity for the body *can only be used for an instant of time* since the body changes its position from one instant to the next. The locus of points which define the location of the *IC* during the body's motion is called a *centrode*, Fig. 16–18a, and so each point on the centrode acts as the *IC* for the body only for an instant.

Although the *IC* may be conveniently used to determine the velocity of any point in a body, it generally *does not have zero acceleration* and therefore it *should not* be used for finding the accelerations of points in a body.

As the board slides downward to the left it is subjected to general plane motion. Since the directions of the velocities of its ends *A* and *B* are known, the *IC* is located as shown. At this instant the board will momentarily rotate about this point. Draw the board in several other positions, establish the *IC* for each case and sketch the centrode.

Procedure for Analysis

The velocity of a point on a body which is subjected to general plane motion can be determined with reference to its instantaneous center of zero velocity provided the location of the *IC* is first established using one of the three methods described above.

- As shown on the kinematic diagram in Fig. 16–19, the body is imagined as "extended and pinned" at the *IC* such that, at the instant considered, it rotates about this pin with its angular velocity $\boldsymbol{\omega}$.

- The *magnitude* of velocity for each of the arbitrary points *A*, *B*, and *C* on the body can be determined by using the equation $v = \omega r$, where r is the radial distance from the *IC* to each point.

- The line of action of each velocity vector $\mathbf{v}$ is *perpendicular* to its associated radial line $\mathbf{r}$, and the velocity has a *sense of direction* which tends to move the point in a manner consistent with the angular rotation $\boldsymbol{\omega}$ of the radial line, Fig. 16–19.

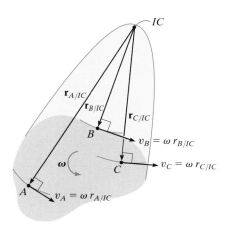

Fig. 16–19

EXAMPLE 16.10

Show how to determine the location of the instantaneous center of zero velocity for (a) member *BC* shown in Fig. 16–20*a*; and (b) the link *CB* shown in Fig. 16–20*b*.

(a)

(b)

Fig. 16–20

(c)

(d)

SOLUTION

Part (a).   As shown in Fig. 16–20*a*, point *B* has a velocity $\mathbf{v}_B$, which is caused by the clockwise rotation of link *AB*. Point *B* moves in a circular path such that $\mathbf{v}_B$ is perpendicular to *AB*, and so it acts at an angle θ from the horizontal as shown in Fig. 16–20*c*. The motion of point *B* causes the piston to move forward *horizontally* with a velocity $\mathbf{v}_C$. When lines are drawn perpendicular to $\mathbf{v}_B$ and $\mathbf{v}_C$, Fig. 16–20*c*, they intersect at the *IC*.

Part (b).   Points *B* and *C* follow circular paths of motion since rods *AB* and *DC* are each subjected to rotation about a fixed axis, Fig. 16–20*b*. Since the velocity is always tangent to the path, at the instant considered, $\mathbf{v}_C$ on rod *DC* and $\mathbf{v}_B$ on rod *AB* are both directed vertically downward, along the axis of link *CB*, Fig. 16–20*d*. Radial lines drawn perpendicular to these two velocities form parallel lines which intersect at "infinity;" i.e., $r_{C/IC} \to \infty$ and $r_{B/IC} \to \infty$. Thus, $\omega_{CB} = (v_C/r_{C/IC}) = \to 0$. As a result, rod *CB* momentarily *translates*. An instant later, however, *CB* will move to a tilted position, causing the instantaneous center to move to some finite location.

EXAMPLE 16.11

Block D shown in Fig. 16–21a moves with a speed of 3 m/s. Determine the angular velocities of links BD and AB, at the instant shown.

(a)

Fig. 16–21

SOLUTION

As D moves to the right, it causes arm AB to rotate clockwise about point A. Hence, $\mathbf{v}_B$ is directed perpendicular to AB. The instantaneous center of zero velocity for BD is located at the intersection of the line segments drawn perpendicular to $\mathbf{v}_B$ and $\mathbf{v}_D$, Fig. 16–21b. From the geometry,

$$r_{B/IC} = 0.4 \tan 45° \text{ m} = 0.4 \text{ m}$$

$$r_{D/IC} = \frac{0.4 \text{ m}}{\cos 45°} = 0.566 \text{ m}$$

(b)

Since the magnitude of $\mathbf{v}_D$ is known, the angular velocity of link BD is

$$\omega_{BD} = \frac{v_D}{r_{D/IC}} = \frac{3 \text{ m/s}}{0.566 \text{ m}} = 5.30 \text{ rad/s} \uparrow \qquad Ans.$$

The velocity of B is therefore

$$v_B = \omega_{BD}(r_{B/IC}) = 5.30 \text{ rad/s} (0.4 \text{ m}) = 2.12 \text{ m/s} \quad \searrow 45°$$

From Fig. 16–21c, the angular velocity of AB is

$$\omega_{AB} = \frac{v_B}{r_{B/A}} = \frac{2.12 \text{ m/s}}{0.4 \text{ m}} = 5.30 \text{ rad/s} \downarrow \qquad Ans.$$

(c)

NOTE: Try and solve this problem by applying $\mathbf{v}_D = \mathbf{v}_B + \mathbf{v}_{B/D}$ to member BD.

EXAMPLE 16.12

The cylinder shown in Fig. 16–22a rolls without slipping between the two moving plates E and D. Determine the angular velocity of the cylinder and the velocity of its center C at the instant shown.

$v_E = 0.25$ m/s

0.125 m

$v_D = 0.4$ m/s

(a)

Fig. 16–22

SOLUTION

Since no slipping occurs, the contact points A and B on the cylinder have the same velocities as the plates E and D, respectively. Furthermore, the velocities $\mathbf{v}_A$ and $\mathbf{v}_B$ are *parallel*, so that by the proportionality of right triangles the IC is located at a point on line AB, Fig. 16–22b. Assuming this point to be a distance x from B, we have

$v_A = 0.25$ m/s

0.25 m

0.125 m

$v_B = 0.4$ m/s B

(b)

$$v_B = \omega x; \qquad\qquad 0.4 \text{ m/s} = \omega x$$

$$v_A = \omega(0.25 \text{ m} - x); \quad 0.25 \text{ m/s} = \omega(0.25 \text{ m} - x)$$

Dividing one equation into the other eliminates ω and yields

$$0.4(0.25 - x) = 0.25x$$

$$x = \frac{0.1}{0.65} = 0.154 \text{ m}$$

Hence, the angular velocity of the cylinder is

$$\omega = \frac{v_B}{x} = \frac{0.4 \text{ m/s}}{0.154 \text{ m}} = 2.60 \text{ rad/s} \,\downarrow \qquad\qquad Ans.$$

The velocity of point C is therefore

$$v_C = \omega r_{C/IC} = 2.60 \text{ rad/s} (0.154 \text{ m} - 0.125 \text{ m})$$

$$= 0.0750 \text{ m/s} \leftarrow \qquad\qquad Ans.$$

PROBLEMS

16–78. Solve Prob. 16–51 using the method of instantaneous center of zero velocity.

16–79. Solve Prob. 16–54 using the method of instantaneous center of zero velocity.

*16–80. Solve Prob. 16–59 using the method of instantaneous center of zero velocity.

16–81. Solve Prob. 16–61 using the method of instantaneous center of zero velocity.

16–82. Solve Prob. 16–62 using the method of instantaneous center of zero velocity.

16–83. Solve Prob. 16–63 using the method of instantaneous center of zero velocity.

*16–84. Solve Prob. 16–65 using the method of instantaneous center of zero velocity.

16–85. The instantaneous center of zero velocity for the body is located at point IC (0.5 m, 2 m). If the body has an angular velocity of 4 rad/s, as shown, determine the velocity of B with respect to A.

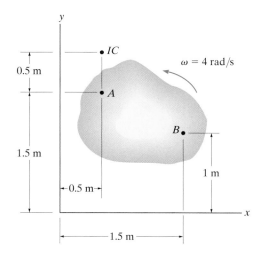

Prob. 16–85

16–86. In each case show graphically how to locate the instantaneous center of zero velocity of link AB. Assume the geometry is known.

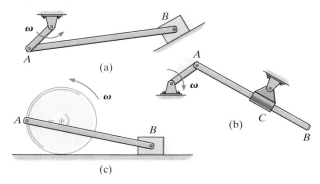

Prob. 16–86

16–87. The disk of radius r is confined to roll without slipping at A and B. If the plates have the velocities shown, determine the angular velocity of the disk.

Prob. 16–87

*16–88. At the instant shown, the disk is rotating at $\omega = 4$ rad/s. Determine the velocities of points A, B, and C.

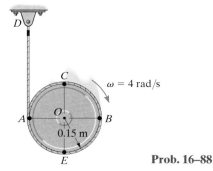

Prob. 16–88

16–89.   The slider block C is moving 4 ft/s up the incline. Determine the angular velocities of links AB and BC and the velocity of point B at the instant shown.

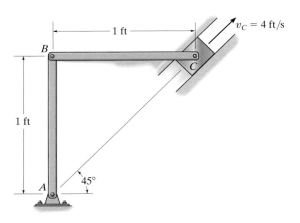

Prob. 16–89

16–90.   Show that if the rim of the wheel and its hub maintain contact with the three tracks as the wheel rolls, it is necessary that slipping occurs at the hub A if no slipping occurs at B. Under these conditions, what is the speed at A if the wheel has an angular velocity ω?

Prob. 16–90

16–91.   The epicyclic gear train is driven by the rotating link DE, which has an angular velocity $\omega_{DE} = 5$ rad/s. If the ring gear F is fixed, determine the angular velocities of gears A, B, and C.

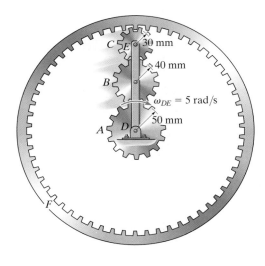

Prob. 16–91

*16–92.   Determine the angular velocity of link AB at the instant shown if block C is moving upward at 12 in./s.

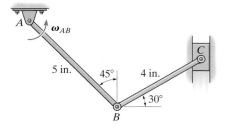

Prob. 16–92

16–93. In an automobile transmission the planet pinions A and B rotate on shafts that are mounted on the planet-pinion carrier CD. As shown, CD is attached to a shaft at E which is aligned with the center of the *fixed* sun gear S. This shaft is not attached to the sun gear. If CD is rotating at $\omega_{CD} = 8$ rad/s, determine the angular velocity of the ring gear R.

Prob. 16–93

16–94. Knowing that the angular velocity of link AB is $\omega_{AB} = 4$ rad/s, determine the velocity of the collar at C and the angular velocity of link CB at the instant shown. Link CB is horizontal at this instant.

16–95. If the collar at C is moving downward to the left at $v_C = 8$ m/s, determine the angular velocity of link AB at the instant shown.

Probs. 16–94/95

*16–96.** Due to slipping, points A and B on the rim of the disk have the velocities shown. Determine the velocities of the center point C and point D at this instant.

16–97. Due to slipping, points A and B on the rim of the disk have the velocities shown. Determine the velocities of the center point C and point E at this instant.

Probs. 16–96/97

16–98. The mechanism used in a marine engine consists of a single crank AB and two connecting rods BC and BD. Determine the velocity of the piston at C the instant the crank is in the position shown and has an angular velocity of 5 rad/s.

16–99. The mechanism used in a marine engine consists of a single crank AB and two connecting rods BC and BD. Determine the velocity of the piston at D the instant the crank is in the position shown and has an angular velocity of 5 rad/s.

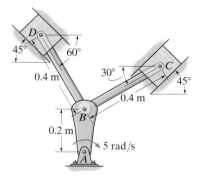

Probs. 16–98/99

*16–100.   The square plate is confined within the slots at A and B. When $\theta = 30°$, point A is moving at $v_A = 8$ m/s. Determine the velocity of point C at this instant.

16–101.   The square plate is confined within the slots at A and B. When $\theta = 30°$, point A is moving at $v_A = 8$ m/s. Determine the velocity of point D at this instant.

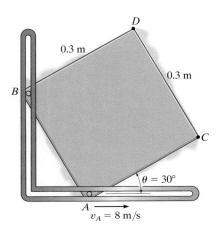

Probs. 16–100/101

16–102.   If the slider block A is moving to the right at $v_A = 8$ ft/s, determine the velocities of blocks B and C at the instant shown.

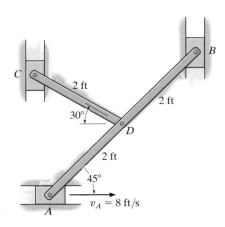

Prob. 16–102

16–103.   The crankshaft AB rotates at $\omega_{AB} = 50$ rad/s about the fixed axis through point A, and the disk at C is held fixed in its support at E. Determine the angular velocity of rod CD at the instant shown.

Prob. 16–103

*16–104.   The mechanism shown is used in a riveting machine. It consists of a driving piston A, three members, and a riveter which is attached to the slider block D. Determine the velocity of D at the instant shown, when the piston at A is traveling at $v_A = 30$ m/s.

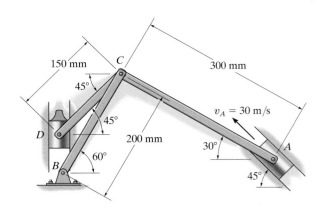

Prob. 16–104

16.7   Relative-Motion Analysis: Acceleration

An equation that relates the accelerations of two points on a rigid body subjected to general plane motion may be determined by differentiating the velocity equation $\mathbf{v}_B = \mathbf{v}_A + \mathbf{v}_{B/A}$ with respect to time. This yields

$$\frac{d\mathbf{v}_B}{dt} = \frac{d\mathbf{v}_A}{dt} + \frac{d\mathbf{v}_{B/A}}{dt}$$

The terms $d\mathbf{v}_B/dt = \mathbf{a}_B$ and $d\mathbf{v}_A/dt = \mathbf{a}_A$ are measured from a set of *fixed* x, y axes and represent the *absolute accelerations* of points B and A. The last term represents the acceleration of B with respect to A as measured by an observer fixed to translating x', y' axes which have their origin at the base point A. In Sec. 16.5 it was shown that to this observer point B appears to move along a *circular arc* that has a radius of curvature $r_{B/A}$. Consequently, $\mathbf{a}_{B/A}$ can be expressed in terms of its tangential and normal components of motion; i.e., $\mathbf{a}_{B/A} = (\mathbf{a}_{B/A})_t + (\mathbf{a}_{B/A})_n$, where $(a_{B/A})_t = \alpha r_{B/A}$ and $(a_{B/A})_n = \omega^2 r_{B/A}$. Hence, the relative-acceleration equation can be written in the form

$$\boxed{\mathbf{a}_B = \mathbf{a}_A + (\mathbf{a}_{B/A})_t + (\mathbf{a}_{B/A})_n} \qquad (16\text{–}17)$$

where

$\quad \mathbf{a}_B =$ acceleration of point B

$\quad \mathbf{a}_A =$ acceleration of point A

$(\mathbf{a}_{B/A})_t =$ relative tangential acceleration component of "B with respect to A." The *magnitude* is $(a_{B/A})_t = \alpha r_{B/A}$, and the *direction* is perpendicular to $\mathbf{r}_{B/A}$.

$(\mathbf{a}_{B/A})_n =$ relative normal acceleration component of "B with respect to A." The *magnitude* is $(a_{B/A})_n = \omega^2 r_{B/A}$, and the *direction* is always from B towards A.

Each of the four terms in Eq. 16–17 is represented graphically on the *kinematic diagrams* shown in Fig. 16–23. Here it is seen that at a given instant the acceleration of B, Fig. 16–23a, is determined by considering the body to translate with an acceleration $\mathbf{a}_A$, Fig. 16–23b, and simultaneously rotate about the base point A with an instantaneous angular velocity $\boldsymbol{\omega}$ and angular acceleration $\boldsymbol{\alpha}$, Fig. 16–23c. Vector addition of these two effects, applied to B, yields $\mathbf{a}_B$, as shown in Fig. 16–23d. It should be noted from Fig. 16–23a that since points A and B move along *curved paths*, the accelerations of these points will have *both tangential and normal components*. (Recall that the acceleration of a point is *tangent to the path* only when the path is *rectilinear* or when it is an inflection point on a curve.)

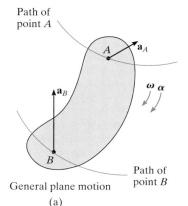

Path of point A

General plane motion

(a)

Translation

(b)

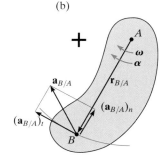

Rotation about the base point A

(c)

$$\mathbf{a}_B = \mathbf{a}_A + (\mathbf{a}_{B/A})_t + (\mathbf{a}_{B/A})_n$$

(d)

Fig. 16–23

(a)

(b)

Fig. 16–24

Since the relative-acceleration components represent the effect of *circular motion* observed from translating axes having their origin at the base point A, these terms can be expressed as $(\mathbf{a}_{B/A})_t = \boldsymbol{\alpha} \times \mathbf{r}_{B/A}$ and $(\mathbf{a}_{B/A})_n = -\omega^2 \mathbf{r}_{B/A}$, Eq. 16–14. Hence, Eq. 16–17 becomes

$$\mathbf{a}_B = \mathbf{a}_A + \boldsymbol{\alpha} \times \mathbf{r}_{B/A} - \omega^2 \mathbf{r}_{B/A} \qquad (16\text{–}18)$$

where

$\mathbf{a}_B =$ acceleration of point B

$\mathbf{a}_A =$ acceleration of the base point A

$\boldsymbol{\alpha} =$ angular acceleration of the body

$\boldsymbol{\omega} =$ angular velocity of the body

$\mathbf{r}_{B/A} =$ relative-position vector drawn from A to B

If Eq. 16–17 or 16–18 is applied in a practical manner to study the accelerated motion of a rigid body which is pin connected to two other bodies, it should be realized that points which are *coincident at the pin* move with the *same acceleration*, since the path of motion over which they travel is the *same*. For example, point B lying on either rod AB or BC of the crank mechanism shown in Fig. 16–24a has the same acceleration, since the rods are pin connected at B. Here the motion of B is along a *curved path*, so that $\mathbf{a}_B$ can be expressed in terms of its tangential and normal components. At the other end of rod BC point C moves along a *straight-lined path*, which is defined by the piston. Hence, $\mathbf{a}_C$ is horizontal, Fig. 16–24b.

If two bodies contact one another *without slipping*, and the *points in contact* move along *different paths*, the *tangential components* of acceleration of the points will be the *same*; however, the *normal components* will *not* be the same. For example, consider the two meshed gears in Fig. 16–25a. Point A is located on gear B and a coincident point A' is located on gear C. Due to the rotational motion, $(\mathbf{a}_A)_t = (\mathbf{a}_{A'})_t$; however, since both points follow different curved paths, $(\mathbf{a}_A)_n \neq (\mathbf{a}_{A'})_n$ and therefore $\mathbf{a}_A \neq \mathbf{a}_{A'}$, Fig. 16–25b.

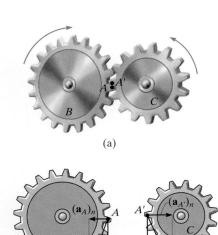

(a)

(b)

Fig. 16–25

Procedure for Analysis

The relative acceleration equation can be applied between any two points A and B on a body either by using a Cartesian vector analysis, or by writing the x and y scalar component equations directly.

Velocity Analysis.

- Determine the angular velocity ω of the body by using a velocity analysis as discussed in Sec. 16.5 or 16.6. Also, determine the velocities $\mathbf{v}_A$ and $\mathbf{v}_B$ of points A and B *if these points move along curved paths.*

Vector Analysis

Kinematic Diagram.

- Establish the directions of the fixed x, y coordinates and draw the kinematic diagram of the body. Indicate on it $\mathbf{a}_A$, $\mathbf{a}_B$, ω, α, and $\mathbf{r}_{B/A}$.

- If points A and B move along *curved paths*, then their accelerations should be indicated in terms of their tangential and normal components, i.e., $\mathbf{a}_A = (\mathbf{a}_A)_t + (\mathbf{a}_A)_n$ and $\mathbf{a}_B = (\mathbf{a}_B)_t + (\mathbf{a}_B)_n$.

Acceleration Equation.

- To apply $\mathbf{a}_B = \mathbf{a}_A + \alpha \times \mathbf{r}_{B/A} - \omega^2 \mathbf{r}_{B/A}$ express the vectors in Cartesian vector form and substitute them into the equation. Evaluate the cross product and then equate the respective $\mathbf{i}$ and $\mathbf{j}$ components to obtain two scalar equations.

- If the solution yields a *negative* answer for an *unknown* magnitude, it indicates that the sense of direction of the vector is opposite to that shown on the kinematic diagram.

Scalar Analysis

Kinematic Diagram.

- If the equation $\mathbf{a}_B = \mathbf{a}_A + (\mathbf{a}_{B/A})_t + (\mathbf{a}_{B/A})_n$ is applied, then the magnitudes and directions of the relative-acceleration components $(\mathbf{a}_{B/A})_t$ and $(\mathbf{a}_{B/A})_n$ must be established. To do this draw a kinematic diagram such as shown in Fig. 16–23c. Since the body is considered to be momentarily "pinned" at the base point A, the *magnitudes* are $(a_{B/A})_t = \alpha r_{B/A}$ and $(a_{B/A})_n = \omega^2 r_{B/A}$. Their *sense of direction* is established from the diagram such that $(\mathbf{a}_{B/A})_t$ acts perpendicular to $\mathbf{r}_{B/A}$, in accordance with the rotational motion α of the body, and $(\mathbf{a}_{B/A})_n$ is directed from B towards A.*

Acceleration Equation.

- Represent the vectors in $\mathbf{a}_B = \mathbf{a}_A + (\mathbf{a}_{B/A})_t + (\mathbf{a}_{B/A})_n$ graphically by showing their magnitudes and directions underneath each term. The scalar equations are determined from the x and y components of these vectors.

*The notation $\mathbf{a}_B = \mathbf{a}_A + (\mathbf{a}_{B/A(\text{pin})})_t + (\mathbf{a}_{B/A(\text{pin})})_n$ may be helpful in recalling that A is assumed to be pinned.

The mechanism for a window is shown. Here CA rotates about a fixed axis through C, and AB undergoes general plane motion. Since point A moves along a curved path it has two components of acceleration, whereas point B moves along a straight track and the direction of its acceleration is specified.

EXAMPLE 16.13

$v_A = 2$ m/s
$a_A = 3$ m/s^2

(a)

(b)

The rod AB shown in Fig. 16–26a is confined to move along the inclined planes at A and B. If point A has an acceleration of 3 m/s^2 and a velocity of 2 m/s, both directed down the plane at the instant the rod becomes horizontal, determine the angular acceleration of the rod at this instant.

SOLUTION I (VECTOR ANALYSIS)

We will apply the acceleration equation to points A and B on the rod. To do so it is first necessary to determine the angular velocity of the rod. Show that it is $\omega = 0.283$ rad/s ↺ using either the velocity equation or the method of instantaneous centers.

Kinematic Diagram.   Since points A and B both move along straight-line paths, they have *no* components of acceleration normal to the paths. There are two unknowns in Fig. 16–26b, namely, a_B and α.

Acceleration Equation.   Applying Eq. 16–18 to points A and B on the rod, and expressing each of the vectors in Cartesian vector form, we have

$$\mathbf{a}_B = \mathbf{a}_A + \boldsymbol{\alpha} \times \mathbf{r}_{B/A} - \omega^2 \mathbf{r}_{B/A}$$

$$a_B \cos 45°\mathbf{i} + a_B \sin 45°\mathbf{j} = 3 \cos 45°\mathbf{i} - 3 \sin 45°\mathbf{j} + (\alpha\mathbf{k}) \times (10\mathbf{i}) - (0.283)^2(10\mathbf{i})$$

Carrying out the cross product and equating the $\mathbf{i}$ and $\mathbf{j}$ components yields

$$a_B \cos 45° = 3 \cos 45° - (0.283)^2(10) \tag{1}$$

$$a_B \sin 45° = -3 \sin 45° + \alpha(10) \tag{2}$$

Solving, we have

$$a_B = 1.87 \text{ m/s}^2 \; \measuredangle 45°$$

$$\alpha = 0.344 \text{ rad/s}^2 \, ↺ \qquad\qquad Ans.$$

SOLUTION II (SCALAR ANALYSIS)

As an alternative procedure, the scalar component equations 1 and 2 can be obtained directly. From the kinematic diagram, showing the relative-acceleration components $(\mathbf{a}_{B/A})_t$ and $(\mathbf{a}_{B/A})_n$, Fig. 16–26c, we have

$$\mathbf{a}_B = \mathbf{a}_A + (\mathbf{a}_{B/A})_t + (\mathbf{a}_{B/A})_n$$

$$\begin{bmatrix} a_B \\ \measuredangle 45° \end{bmatrix} = \begin{bmatrix} 3 \text{ m/s}^2 \\ \measuredangle 45° \end{bmatrix} + \begin{bmatrix} \alpha(10 \text{ m}) \\ \uparrow \end{bmatrix} + \begin{bmatrix} (0.283 \text{ rad/s})^2(10 \text{ m}) \\ \leftarrow \end{bmatrix}$$

Equating the x and y components yields Eqs. 1 and 2, and the solution proceeds as before.

(c)

Fig. 16–26

EXAMPLE 16.14

At a given instant, the cylinder of radius r, shown in Fig. 16–27a, has an angular velocity $\boldsymbol{\omega}$ and angular acceleration $\boldsymbol{\alpha}$. Determine the velocity and acceleration of its center G if it rolls without slipping.

(a)

SOLUTION (VECTOR ANALYSIS)

As the cylinder rolls, point G moves along a straight line, and point A, located on the rim of the cylinder, moves along a curved path called a *cycloid*, Fig. 16–27b. We will apply the velocity and acceleration equations to these two points.

Velocity Analysis.   Since no slipping occurs, at the instant A contacts the ground, $\mathbf{v}_A = \mathbf{0}$. Thus, from the kinematic diagram in Fig. 16–27c we have

$$\mathbf{v}_G = \mathbf{v}_A + \boldsymbol{\omega} \times \mathbf{r}_{G/A}$$

$$v_G\mathbf{i} = \mathbf{0} + (-\omega\mathbf{k}) \times (r\mathbf{j})$$

$$v_G = \omega r \qquad (1) \; Ans.$$

(b)

This same result can also be obtained directly by noting that point A represents the instantaneous center of zero velocity.

Kinematic Diagram.   The acceleration of point G is horizontal since it moves along a *straight-line path. Just before* point A touches the ground, its velocity is directed downward along the y axis, Fig. 16–27b, and just after contact, its velocity is directed *upward*. For this reason, point A begins to accelerate upward when it leaves the ground at A, Fig. 16–27d. The magnitudes of $\mathbf{a}_A$ and $\mathbf{a}_G$ are unknown.

Acceleration Equation.

$$\mathbf{a}_G = \mathbf{a}_A + \boldsymbol{\alpha} \times \mathbf{r}_{G/A} - \omega^2\mathbf{r}_{G/A}$$

$$a_G\mathbf{i} = a_A\mathbf{j} + (-\alpha\mathbf{k}) \times (r\mathbf{j}) - \omega^2(r\mathbf{j})$$

Evaluating the cross product and equating the $\mathbf{i}$ and $\mathbf{j}$ components yields

$$a_G = \alpha r \qquad (2) \; Ans.$$

$$a_A = \omega^2 r \qquad (3)$$

(c)

(d)

Fig. 16–27

NOTE: These important results, that $v_G = \omega r$ and $a_G = \alpha r$, were also obtained in Example 16.4. They apply to any circular object, such as a ball, pulley, disk, etc., that rolls *without* slipping. Also, the fact that $a_A = \omega^2 r$ indicates that the instantaneous center of zero velocity, point A, *is not* a point of zero acceleration.

EXAMPLE 16.15

The ball rolls without slipping and has the angular motion shown in Fig. 16–28a. Determine the accelerations of point B and point A at this instant.

Fig. 16–28

SOLUTION (VECTOR ANALYSIS)

Kinematic Diagram.   Using the results of the previous example, the center of the ball has an acceleration of $a_O = \alpha r = (4 \text{ rad/s}^2)(0.5 \text{ ft}) = 2 \text{ ft/s}^2$. We will apply the acceleration equation to points O and B and points O and A.

Acceleration Equation.

For point B, Fig. 16–28b,

$$\mathbf{a}_B = \mathbf{a}_O + \boldsymbol{\alpha} \times \mathbf{r}_{B/O} - \omega^2 \mathbf{r}_{B/O}$$

$$\mathbf{a}_B = -2\mathbf{i} + (4\mathbf{k}) \times (0.5\mathbf{i}) - (6)^2(0.5\mathbf{i})$$

$$\mathbf{a}_B = \{-20\mathbf{i} + 2\mathbf{j}\} \text{ ft/s}^2 \qquad\qquad Ans.$$

For point A, Fig. 16–28c,

$$\mathbf{a}_A = \mathbf{a}_O + \boldsymbol{\alpha} \times \mathbf{r}_{A/O} - \omega^2 \mathbf{r}_{A/O}$$

$$\mathbf{a}_A = -2\mathbf{i} + (4\mathbf{k}) \times (0.5\mathbf{j}) - (6)^2(0.5\mathbf{j})$$

$$\mathbf{a}_A = \{-4\mathbf{i} - 18\mathbf{j}\} \text{ ft/s}^2 \qquad\qquad Ans.$$

EXAMPLE 16.16

The spool shown in Fig. 16–29a unravels from the cord, such that at the instant shown it has an angular velocity of 3 rad/s and an angular acceleration of 4 rad/s². Determine the acceleration of point B.

SOLUTION I (VECTOR ANALYSIS)

The spool "appears" to be rolling downward without slipping at point A. Therefore, we can use the results of Example 16.14 to determine the acceleration of point G, i.e.,

$$a_G = \alpha r = 4 \text{ rad/s}^2 \ (0.5 \text{ ft}) = 2 \text{ ft/s}^2$$

We will apply the acceleration equation at points G and B.

Kinematic Diagram.   Point B moves along a *curved path* having an *unknown* radius of curvature.* Its acceleration will be represented by its unknown x and y components as shown in Fig. 16–29b.

Acceleration Equation.

$$\mathbf{a}_B = \mathbf{a}_G + \boldsymbol{\alpha} \times \mathbf{r}_{B/G} - \omega^2 \mathbf{r}_{B/G}$$

$$(a_B)_x\mathbf{i} + (a_B)_y\mathbf{j} = -2\mathbf{j} + (-4\mathbf{k}) \times (0.75\mathbf{j}) - (3)^2(0.75\mathbf{j})$$

Equating the **i** and **j** terms, the component equations are

$$(a_B)_x = 4(0.75) = 3 \text{ ft/s}^2 \rightarrow \qquad (1)$$

$$(a_B)_y = -2 - 6.75 = -8.75 \text{ ft/s}^2 = 8.75 \text{ ft/s}^2 \downarrow \qquad (2)$$

The magnitude and direction of $\mathbf{a}_B$ are therefore

$$a_B = \sqrt{(3)^2 + (8.75)^2} = 9.25 \text{ ft/s}^2 \qquad \qquad Ans.$$

$$\theta = \tan^{-1}\frac{8.75}{3} = 71.1° \quad \diagdown\theta \qquad \qquad Ans.$$

SOLUTION II (SCALAR ANALYSIS)

This problem may be solved by writing the scalar component equations directly. The kinematic diagram in Fig. 16–29c shows the relative-acceleration components $(\mathbf{a}_{B/G})_t$ and $(\mathbf{a}_{B/G})_n$. Thus,

$$\mathbf{a}_B = \mathbf{a}_G + (\mathbf{a}_{B/G})_t + (\mathbf{a}_{B/G})_n$$

$$\begin{bmatrix} (a_B)_x \\ \rightarrow \end{bmatrix} + \begin{bmatrix} (a_B)_y \\ \uparrow \end{bmatrix}$$

$$= \begin{bmatrix} 2 \text{ ft/s}^2 \\ \downarrow \end{bmatrix} + \begin{bmatrix} 4 \text{ rad/s}^2 \ (0.75 \text{ ft}) \\ \rightarrow \end{bmatrix} + \begin{bmatrix} (3 \text{ rad/s})^2(0.75 \text{ ft}) \\ \downarrow \end{bmatrix}$$

The x and y components yield Eqs. 1 and 2 above.

*Realize that the path's radius of curvature ρ is *not* equal to the radius of the spool since the spool is *not* rotating about point G. Furthermore, ρ is *not* defined as the distance from A (IC) to B, since the location of the IC depends only on the velocity of a point and *not* the geometry of its path.

(a)

(b)

(c)

Fig. 16–29

EXAMPLE 16.17

The collar C in Fig. 16–30a is moving downward with an acceleration of 1 m/s². At the instant shown, it has a speed of 2 m/s which gives links CB and AB an angular velocity $\omega_{AB} = \omega_{CB} = 10$ rad/s. (See Example 16.8.) Determine the angular accelerations of CB and AB at this instant.

(a)

SOLUTION (VECTOR ANALYSIS)

Kinematic Diagram.   The kinematic diagrams of *both* links AB and CB are shown in Fig. 16–30b. To solve, we will apply the appropriate kinematic equation to each link.

Acceleration Equation.

Link AB (rotation about a fixed axis):

$$\mathbf{a}_B = \boldsymbol{\alpha}_{AB} \times \mathbf{r}_B - \omega_{AB}^2 \mathbf{r}_B$$
$$\mathbf{a}_B = (\alpha_{AB}\mathbf{k}) \times (-0.2\mathbf{j}) - (10)^2(-0.2\mathbf{j})$$
$$\mathbf{a}_B = 0.2\alpha_{AB}\mathbf{i} + 20\mathbf{j}$$

(b)

Fig. 16–30

Note that $\mathbf{a}_B$ has two components since it moves along a *curved path*.

Link BC (general plane motion): Using the result for $\mathbf{a}_B$ and applying Eq. 16–18, we have

$$\mathbf{a}_B = \mathbf{a}_C + \boldsymbol{\alpha}_{CB} \times \mathbf{r}_{B/C} - \omega_{CB}^2 \mathbf{r}_{B/C}$$

$$0.2\alpha_{AB}\mathbf{i} + 20\mathbf{j} = -1\mathbf{j} + (\alpha_{CB}\mathbf{k}) \times (0.2\mathbf{i} - 0.2\mathbf{j}) - (10)^2(0.2\mathbf{i} - 0.2\mathbf{j})$$

$$0.2\alpha_{AB}\mathbf{i} + 20\mathbf{j} = -1\mathbf{j} + 0.2\alpha_{CB}\mathbf{j} + 0.2\alpha_{CB}\mathbf{i} - 20\mathbf{i} + 20\mathbf{j}$$

Thus,

$$0.2\alpha_{AB} = 0.2\alpha_{CB} - 20$$
$$20 = -1 + 0.2\alpha_{CB} + 20$$

Solving,

$$\alpha_{CB} = 5 \text{ rad/s}^2 \;\text{↰} \qquad\qquad \textit{Ans.}$$
$$\alpha_{AB} = -95 \text{ rad/s}^2 = 95 \text{ rad/s}^2 \;\text{↲} \qquad \textit{Ans.}$$

EXAMPLE 16.18

The crankshaft AB of an engine turns with a clockwise angular acceleration of 20 rad/s^2, Fig. 16–31a. Determine the acceleration of the piston at the instant AB is in the position shown. At this instant $\omega_{AB} = 10$ rad/s and $\omega_{BC} = 2.43$ rad/s.

(a)

SOLUTION (VECTOR ANALYSIS)

Kinematic Diagram.   The kinematic diagrams for both AB and BC are shown in Fig. 16–31b. Here $\mathbf{a}_C$ is vertical since C moves along a straight-line path.

Acceleration Equation.   Expressing each of the position vectors in Cartesian vector form

$$\mathbf{r}_B = \{-0.25 \sin 45°\mathbf{i} + 0.25 \cos 45°\mathbf{j}\} \text{ ft} = \{-0.177\mathbf{i} + 0.177\mathbf{j}\} \text{ ft}$$

$$\mathbf{r}_{C/B} = \{0.75 \sin 13.6°\mathbf{i} + 0.75 \cos 13.6°\mathbf{j}\} \text{ ft} = \{0.176\mathbf{i} + 0.729\mathbf{j}\} \text{ ft}$$

Crankshaft AB (rotation about a fixed axis):

$$\mathbf{a}_B = \boldsymbol{\alpha}_{AB} \times \mathbf{r}_B - \omega_{AB}^2 \mathbf{r}_B$$

$$= (-20\mathbf{k}) \times (-0.177\mathbf{i} + 0.177\mathbf{j}) - (10)^2(-0.177\mathbf{i} + 0.177\mathbf{j})$$

$$= \{21.21\mathbf{i} - 14.14\mathbf{j}\} \text{ ft/s}^2$$

Connecting Rod BC (general plane motion): Using the result for $\mathbf{a}_B$ and noting that $\mathbf{a}_C$ is in the vertical direction, we have

$$\mathbf{a}_C = \mathbf{a}_B + \boldsymbol{\alpha}_{BC} \times \mathbf{r}_{C/B} - \omega_{BC}^2 \mathbf{r}_{C/B}$$

$$a_C\mathbf{j} = 21.21\mathbf{i} - 14.14\mathbf{j} + (\alpha_{BC}\mathbf{k}) \times (0.176\mathbf{i} + 0.729\mathbf{j}) - (2.43)^2(0.176\mathbf{i} + 0.729\mathbf{j})$$

$$a_C\mathbf{j} = 21.21\mathbf{i} - 14.14\mathbf{j} + 0.176\alpha_{BC}\mathbf{j} - 0.729\alpha_{BC}\mathbf{i} - 1.04\mathbf{i} - 4.30\mathbf{j}$$

$$0 = 20.17 - 0.729\alpha_{BC}$$

$$a_C = 0.176\alpha_{BC} - 18.45$$

Solving yields

$$\alpha_{BC} = 27.7 \text{ rad/s}^2 \,\curvearrowleft$$

$$a_C = -13.6 \text{ ft/s}^2 \qquad\qquad Ans.$$

NOTE: Since the piston is moving upward, the negative sign for a_C indicates that the piston is decelerating, i.e., $\mathbf{a}_C = \{-13.6\mathbf{j}\}$ ft/s^2. This causes the speed of the piston to decrease until AB becomes vertical, at which time the piston is momentarily at rest.

(b)

Fig. 16–31

PROBLEMS

16–105. At a given instant the bottom A of the ladder has an acceleration $a_A = 4$ ft/s^2 and velocity $v_A = 6$ ft/s, both acting to the left. Determine the acceleration of the top of the ladder, B, and the ladder's angular acceleration at this same instant.

16–106. At a given instant the top B of the ladder has an acceleration $a_B = 2$ ft/s^2 and a velocity of $v_B = 4$ ft/s, both acting downward. Determine the acceleration of the bottom A of the ladder, and the ladder's angular acceleration at this instant.

Probs. 16–105/106

16–107. At a given instant the top end A of the bar has the velocity and acceleration shown. Determine the acceleration of the bottom B and the bar's angular acceleration at this instant.

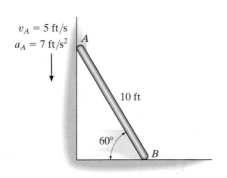

Prob. 16–107

*16–108.** The 10-ft rod slides down the inclined plane, such that when it is at B it has the motion shown. Determine the velocity and acceleration of A at this instant.

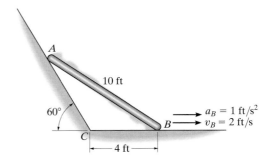

Prob. 16–108

16–109. The wheel is moving to the right such that it has an angular velocity $\omega = 2$ rad/s and angular acceleration $\alpha = 4$ rad/s^2 at the instant shown. If it does not slip at A, determine the acceleration of point B.

Prob. 16–109

16–110. Determine the angular acceleration of link AB at the instant $\theta = 90°$ if the collar C has a velocity of $v_C = 4$ ft/s and deceleration of $a_C = 3$ ft/s^2 as shown.

Prob. 16–110

16–111. The flywheel rotates with an angular velocity $\omega = 2$ rad/s and an angular acceleration $\alpha = 6$ rad/s^2. Determine the angular acceleration of links AB and BC at this instant.

Prob. 16–111

*16–112. At a given instant the wheel is rotating with the angular velocity and angular acceleration shown. Determine the acceleration of block B at this instant.

Prob. 16–112

16–113. The disk is moving to the left such that it has an angular acceleration $\alpha = 8$ rad/s^2 and angular velocity $\omega = 3$ rad/s at the instant shown. If it does not slip at A, determine the acceleration of point B.

16–114. The disk is moving to the left such that it has an angular acceleration $\alpha = 8$ rad/s^2 and angular velocity $\omega = 3$ rad/s at the instant shown. If it does not slip at A, determine the acceleration of point D.

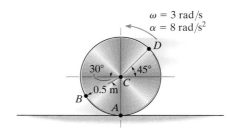

Probs. 16–113/114

16–115. The hoop is cast on the rough surface such that it has an angular velocity $\omega = 4$ rad/s and an angular acceleration $\alpha = 5$ rad/s^2. Also, its center has a velocity $v_O = 5$ m/s and a deceleration $a_O = 2$ m/s^2. Determine the acceleration of point A at this instant.

16–117. The disk rotates with an angular velocity $\omega = 5$ rad/s and an angular acceleration $\alpha = 6$ rad/s^2. Determine the angular acceleration of link CB at this instant.

Prob. 15–115

Prob. 16–117

*16–116. The hoop is cast on the rough surface such that it has an angular velocity $\omega = 4$ rad/s and an angular acceleration $\alpha = 5$ rad/s^2. Also, its center has a velocity of $v_O = 5$ m/s and a deceleration $a_O = 2$ m/s^2. Determine the acceleration of point B at this instant.

16–118. At a given instant the slider block B is moving to the right with the motion shown. Determine the angular acceleration of link AB and the acceleration of point A at this instant.

Prob. 16–116

Prob. 16–118

16–119. The closure is manufactured by the LCN Company and is used to control the restricted motion of a heavy door. If the door to which is it connected has an angular acceleration of 3 rad/s^2, determine the angular acceleration of links BC and CD. Originally the door is not rotating but is hinged at A.

Prob. 16–119

16–121. At the given instant member AB has the angular motions shown. Determine the velocity and acceleration of the slider block C at this instant.

Prob. 16–121

16–120. Rod AB has the angular motion shown. Determine the acceleration of the collar C at this instant.

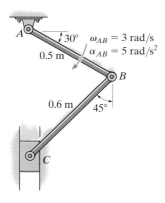

Prob. 16–120

16–122. At a given instant gears A and B have the angular motions shown. Determine the angular acceleration of gear C and the acceleration of its center point D at this instant. Note that the inner hub of gear C is in mesh with gear A and its outer rim is in mesh with gear B.

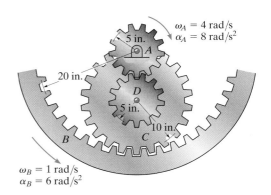

Prob. 16–122

16–123.   The tied crank and gear mechanism gives rocking motion to crank AC, necessary for the operation of a printing press. If link DE has the angular motion shown, determine the respective angular velocities of gear F and crank AC at this instant, and the angular acceleration of crank AC.

Prob. 16–123

16–125.   The wheel rolls without slipping such that at the instant shown it has an angular velocity ω and angular acceleration α. Determine the velocity and acceleration of point B on the rod at this instant.

Prob. 16–125

*16–124.   At a given instant the wheel is rotating with the angular velocity and angular acceleration shown. Determine the acceleration of block B at this instant.

Prob. 16–124

16–126.   The disk rolls without slipping such that it has an angular acceleration of $\alpha = 4 \text{ rad/s}^2$ and angular velocity of $\omega = 2 \text{ rad/s}$ at the instant shown. Determine the accelerations of points A and B on the link and the link's angular acceleration at this instant. Assume point A lies on the periphery of the disk, 150 mm from C.

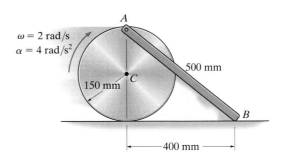

Prob. 16–126

16–127. Determine the angular acceleration of link *AB* if link *CD* has the angular velocity and angular deceleration shown.

Prob. 16–127

***16–128.** The slider block *B* is moving to the right with an acceleration of 2 ft/s². At the instant shown, its velocity is 6 ft/s. Determine the angular acceleration of link *AB* and the acceleration of point *A* at this instant.

Prob. 16–128

16–129. The ends of the bar *AB* are confined to move along the paths shown. At a given instant, *A* has a velocity of $v_A = 4$ ft/s and an acceleration of $a_A = 7$ ft/s². Determine the angular velocity and angular acceleration of *AB* at this instant.

Prob. 16–129

16–130. The mechanism produces intermittent motion of link *AB*. If the sprocket *S* is turning with an angular acceleration $\alpha_S = 2$ rad/s² and has an angular velocity $\omega_S = 6$ rad/s at the instant shown, determine the angular velocity and angular acceleration of link *AB* at this instant. The sprocket *S* is mounted on a shaft which is *separate* from a collinear shaft attached to *AB* at *A*. The pin at *C* is attached to one of the chain links such that it moves vertically downward.

Prob. 16–130

16.8   Relative-Motion Analysis using Rotating Axes

In the previous sections the relative-motion analysis for velocity and acceleration was described using a translating coordinate system. This type of analysis is useful for determining the motion of points on the *same* rigid body, or the motion of points located on several pin-connected rigid bodies. In some problems, however, rigid bodies (mechanisms) are constructed such that *sliding* will occur at their connections. The kinematic analysis for such cases is best performed if the motion is analyzed using a coordinate system which both *translates* and *rotates*. Furthermore, this frame of reference is useful for analyzing the motions of two points on a mechanism which are *not* located in the *same* rigid body and for specifying the kinematics of particle motion when the particle is moving along a rotating path.

In the following analysis two equations are developed which relate the velocity and acceleration of two points, one of which is the origin of a moving frame of reference subjected to both a translation and a rotation in the plane.* Due to the generality in the derivation which follows, these two points may represent either two particles moving independently of one another or two points located on the same (or different) rigid bodies.

Position.   Consider the two points A and B shown in Fig. 16–32a. Their location is specified by the position vectors $\mathbf{r}_A$ and $\mathbf{r}_B$, which are measured from the fixed X, Y, Z coordinate system. As shown in the figure, the "base point" A represents the origin of the x, y, z coordinate system, which is assumed to be both translating and rotating with respect to the X, Y, Z system. The position of B with respect to A is specified by the relative-position vector $\mathbf{r}_{B/A}$. The components of this vector may be expressed either in terms of unit vectors along the X, Y axes, i.e., $\mathbf{I}$ and $\mathbf{J}$, or by unit vectors along the x, y axes, i.e., $\mathbf{i}$ and $\mathbf{j}$. For the development which follows, $\mathbf{r}_{B/A}$ will be measured relative to the moving x, y frame of reference. Thus, if B has coordinates (x_B, y_B), Fig. 16–32a, then

$$\mathbf{r}_{B/A} = x_B\mathbf{i} + y_B\mathbf{j}$$

Using vector addition, the three position vectors in Fig. 16–32a are related by the equation

$$\boxed{\mathbf{r}_B = \mathbf{r}_A + \mathbf{r}_{B/A}} \tag{16–19}$$

At the instant considered, point A has a velocity $\mathbf{v}_A$ and an acceleration $\mathbf{a}_A$, while the angular velocity and angular acceleration of the x, y axes are $\boldsymbol{\Omega}$ (omega) and $\dot{\boldsymbol{\Omega}} = d\boldsymbol{\Omega}/dt$, respectively. All these vectors are measured from the X, Y, Z frame of reference, although they may be

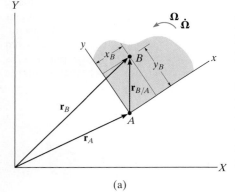

(a)

Fig. 16–32

*The more general, three-dimensional motion of the points is developed in Sec. 20.4.

expressed in terms of either $\mathbf{I}$, $\mathbf{J}$, $\mathbf{K}$ or $\mathbf{i}$, $\mathbf{j}$, $\mathbf{k}$ components. Since planar motion is specified, by the right-hand rule Ω and $\dot{\Omega}$ are always directed *perpendicular* to the reference plane of motion, whereas $\mathbf{v}_A$ and $\mathbf{a}_A$ lie in this plane.

Velocity.   The velocity of point B is determined by taking the time derivative of Eq. 16–19, which yields

$$\mathbf{v}_B = \mathbf{v}_A + \frac{d\mathbf{r}_{B/A}}{dt} \qquad (16\text{–}20)$$

The last term in this equation is evaluated as follows:

$$
\begin{aligned}
\frac{d\mathbf{r}_{B/A}}{dt} &= \frac{d}{dt}(x_B\mathbf{i} + y_B\mathbf{j}) \\[2mm]
&= \frac{dx_B}{dt}\mathbf{i} + x_B\frac{d\mathbf{i}}{dt} + \frac{dy_B}{dt}\mathbf{j} + y_B\frac{d\mathbf{j}}{dt} \\[2mm]
&= \left(\frac{dx_B}{dt}\mathbf{i} + \frac{dy_B}{dt}\mathbf{j}\right) + \left(x_B\frac{d\mathbf{i}}{dt} + y_B\frac{d\mathbf{j}}{dt}\right) \qquad (16\text{–}21)
\end{aligned}
$$

The two terms in the first set of parentheses represent the components of velocity of point B as measured by an observer attached to the moving x, y, z coordinate system. These terms will be denoted by vector $(\mathbf{v}_{B/A})_{xyz}$. In the second set of parentheses the instantaneous time rate of change of the unit vectors $\mathbf{i}$ and $\mathbf{j}$ is measured by an observer located in the fixed X, Y, Z coordinate system. These changes, $d\mathbf{i}$ and $d\mathbf{j}$, are due *only* to the instantaneous *rotation* $d\theta$ of the x, y, z axes, causing $\mathbf{i}$ to become $\mathbf{i}' = \mathbf{i} + d\mathbf{i}$ and $\mathbf{j}$ to become $\mathbf{j}' = \mathbf{j} + d\mathbf{j}$, Fig. 16–32b. As shown, the *magnitudes* of both $d\mathbf{i}$ and $d\mathbf{j}$ equal $1\,(d\theta)$, since $i = i' = j = j' = 1$. The *direction* of $d\mathbf{i}$ is defined by $+\mathbf{j}$, since $d\mathbf{i}$ is tangent to the path described by the arrowhead of $\mathbf{i}$ in the limit as $\Delta t \to dt$. Likewise, $d\mathbf{j}$ acts in the $-\mathbf{i}$ direction, Fig. 16–32b. Hence,

$$\frac{d\mathbf{i}}{dt} = \frac{d\theta}{dt}(\mathbf{j}) = \Omega\mathbf{j} \qquad \frac{d\mathbf{j}}{dt} = \frac{d\theta}{dt}(-\mathbf{i}) = -\Omega\mathbf{i}$$

(b)

Viewing the axes in three dimensions, Fig. 16–32c, and noting that $\Omega = \Omega\mathbf{k}$, we can express the above derivatives in terms of the cross product as

$$\frac{d\mathbf{i}}{dt} = \Omega \times \mathbf{i} \qquad \frac{d\mathbf{j}}{dt} = \Omega \times \mathbf{j} \qquad (16\text{–}22)$$

Substituting these results into Eq. 16–21 and using the distributive property of the vector cross product, we obtain

$$\frac{d\mathbf{r}_{B/A}}{dt} = (\mathbf{v}_{B/A})_{xyz} + \Omega \times (x_B\mathbf{i} + y_B\mathbf{j}) = (\mathbf{v}_{B/A})_{xyz} + \Omega \times \mathbf{r}_{B/A} \qquad (16\text{–}23)$$

(c)

Fig. 16–32 (cont.)

Hence, Eq. 16–20 becomes

$$\mathbf{v}_B = \mathbf{v}_A + \boldsymbol{\Omega} \times \mathbf{r}_{B/A} + (\mathbf{v}_{B/A})_{xyz} \tag{16–24}$$

where

$\mathbf{v}_B =$ velocity of B, measured from the X, Y, Z reference

$\mathbf{v}_A =$ velocity of the origin A of the x, y, z reference, measured from the X, Y, Z reference

$(\mathbf{v}_{B/A})_{xyz} =$ relative velocity of "B with respect to A," as measured by an observer attached to the *rotating* x, y, z reference

$\boldsymbol{\Omega} =$ angular velocity of the x, y, z reference, measured from the X, Y, Z reference

$\mathbf{r}_{B/A} =$ relative position of "B with respect to A"

Comparing Eq. 16–24 with Eq. 16–16 ($\mathbf{v}_B = \mathbf{v}_A + \boldsymbol{\Omega} \times \mathbf{r}_{B/A}$), which is valid for a translating frame of reference, it can be seen that the only difference between the equations is represented by the term $(\mathbf{v}_{B/A})_{xyz}$.

When applying Eq. 16–24 it is often useful to understand what each of the terms represents. In order of appearance, they are as follows:

$\mathbf{v}_B$ $\begin{cases} \text{absolute velocity of } B \\ \\ \end{cases}$ $\left.\begin{array}{l} \text{motion of } B \text{ observed} \\ \text{from the } X, Y, Z \text{ frame} \end{array}\right\}$

(equals)

$\mathbf{v}_A$ $\begin{cases} \text{absolute velocity of origin} \\ \text{of } x, y, z \text{ frame} \end{cases}$

(plus)

$\left.\begin{array}{l} \text{motion of } x, y, z \text{ frame} \\ \text{observed from the} \\ X, Y, Z \text{ frame} \end{array}\right\}$

$\boldsymbol{\Omega} \times \mathbf{r}_{B/A}$ $\begin{cases} \text{angular velocity effect caused} \\ \text{by rotation of } x, y, z \text{ frame} \end{cases}$

(plus)

$(\mathbf{v}_{B/A})_{xyz}$ $\begin{cases} \text{relative velocity of } B \\ \text{with respect to } A \end{cases}$ $\left.\begin{array}{l} \text{motion of } B \text{ observed} \\ \text{from the } x, y, z \text{ frame} \end{array}\right\}$

Acceleration.   The acceleration of B, observed from the X, Y, Z coordinate system, may be expressed in terms of its motion measured with respect to the rotating or moving system of coordinates by taking the time derivative of Eq. 16–24, i.e.,

$$\frac{d\mathbf{v}_B}{dt} = \frac{d\mathbf{v}_A}{dt} + \frac{d\mathbf{\Omega}}{dt} \times \mathbf{r}_{B/A} + \mathbf{\Omega} \times \frac{d\mathbf{r}_{B/A}}{dt} + \frac{d(\mathbf{v}_{B/A})_{xyz}}{dt}$$

$$\mathbf{a}_B = \mathbf{a}_A + \dot{\mathbf{\Omega}} \times \mathbf{r}_{B/A} + \mathbf{\Omega} \times \frac{d\mathbf{r}_{B/A}}{dt} + \frac{d(\mathbf{v}_{B/A})_{xyz}}{dt} \qquad (16\text{–}25)$$

Here $\dot{\mathbf{\Omega}} = d\mathbf{\Omega}/dt$ is the angular acceleration of the x, y, z coordinate system. For planar motion $\mathbf{\Omega}$ is always perpendicular to the plane of motion, and therefore $\dot{\mathbf{\Omega}}$ measures *only the change in magnitude* of $\mathbf{\Omega}$. The derivative $d\mathbf{r}_{B/A}/dt$ in Eq. 16–25 is defined by Eq. 16–23, so that

$$\mathbf{\Omega} \times \frac{d\mathbf{r}_{B/A}}{dt} = \mathbf{\Omega} \times (\mathbf{v}_{B/A})_{xyz} + \mathbf{\Omega} \times (\mathbf{\Omega} \times \mathbf{r}_{B/A}) \qquad (16\text{–}26)$$

Finding the time derivative of $(\mathbf{v}_{B/A})_{xyz} = (v_{B/A})_x \mathbf{i} + (v_{B/A})_y \mathbf{j}$,

$$\frac{d(\mathbf{v}_{B/A})_{xyz}}{dt} = \left[\frac{d(v_{B/A})_x}{dt}\mathbf{i} + \frac{d(v_{B/A})_y}{dt}\mathbf{j}\right] + \left[(v_{B/A})_x \frac{d\mathbf{i}}{dt} + (v_{B/A})_y \frac{d\mathbf{j}}{dt}\right]$$

The two terms in the first set of brackets represent the components of acceleration of point B as measured by an observer attached to the moving coordinate system. These terms will be denoted by $(\mathbf{a}_{B/A})_{xyz}$. The terms in the second set of brackets can be simplified using Eqs. 16–22.

$$\frac{d(\mathbf{v}_{B/A})_{xyz}}{dt} = (\mathbf{a}_{B/A})_{xyz} + \mathbf{\Omega} \times (\mathbf{v}_{B/A})_{xyz}$$

Substituting this and Eq. 16–26 into Eq. 16–25 and rearranging terms,

$$\boxed{\mathbf{a}_B = \mathbf{a}_A + \dot{\mathbf{\Omega}} \times \mathbf{r}_{B/A} + \mathbf{\Omega} \times (\mathbf{\Omega} \times \mathbf{r}_{B/A}) + 2\mathbf{\Omega} \times (\mathbf{v}_{B/A})_{xyz} + (\mathbf{a}_{B/A})_{xyz}}$$

$$(16\text{–}27)$$

where

$\mathbf{a}_B =$ acceleration of B, measured from the X, Y, Z reference

$\mathbf{a}_A =$ acceleration of the origin A of the x, y, z reference, measured from the X, Y, Z reference

$(\mathbf{a}_{B/A})_{xyz}, (\mathbf{v}_{B/A})_{xyz} =$ relative acceleration and relative velocity of "B with respect to A," as measured by an observer attached to the *rotating x, y, z* reference

$\dot{\mathbf{\Omega}}, \mathbf{\Omega} =$ angular acceleration and angular velocity of the x, y, z reference, measured from the X, Y, Z reference

$\mathbf{r}_{B/A} =$ relative position of "B with respect to A"

If Eq. 16–27 is compared with Eq. 16–18, written in the form $\mathbf{a}_B = \mathbf{a}_A + \dot{\boldsymbol{\Omega}} \times \mathbf{r}_{B/A} + \boldsymbol{\Omega} \times (\boldsymbol{\Omega} \times \mathbf{r}_{B/A})$, which is valid for a translating frame of reference, it can be seen that the difference between the equations is represented by the terms $2\boldsymbol{\Omega} \times (\mathbf{v}_{B/A})_{xyz}$ and $(\mathbf{a}_{B/A})_{xyz}$. In particular, $2\boldsymbol{\Omega} \times (\mathbf{v}_{B/A})_{xyz}$ is called the *Coriolis acceleration*, named after the French engineer G. C. Coriolis, who was the first to determine it. This term represents the difference in the acceleration of B as measured from nonrotating and rotating x, y, z axes. As indicated by the vector cross product, the Coriolis acceleration will *always* be perpendicular to both $\boldsymbol{\Omega}$ and $(\mathbf{v}_{B/A})_{xyz}$. It is an important component of the acceleration which must be considered whenever rotating reference frames are used. This often occurs, for example, when studying the accelerations and forces which act on rockets, long-range projectiles, or other bodies having motions whose measurements are significantly affected by the rotation of the earth.

The following interpretation of the terms in Eq. 16–27 may be useful when applying this equation to the solution of problems.

$\mathbf{a}_B$ {absolute acceleration of B } motion of B observed from the X, Y, Z frame

(equals)

$\mathbf{a}_A$ { absolute acceleration of origin of x, y, z frame

(plus)

$\dot{\boldsymbol{\Omega}} \times \mathbf{r}_{B/A}$ { angular acceleration effect caused by rotation of x, y, z frame   } motion of x, y, z frame observed from the X, Y, Z frame

(plus)

$\boldsymbol{\Omega} \times (\boldsymbol{\Omega} \times \mathbf{r}_{B/A})$ { angular velocity effect caused by rotation of x, y, z frame

(plus)

$2\boldsymbol{\Omega} \times (\mathbf{v}_{B/A})_{xyz}$ { combined effect of B moving relative to x, y, z coordinates and rotation of x, y, z frame   } interacting motion

(plus)

$(\mathbf{a}_{B/A})_{xyz}$ { relative acceleration of B with respect to A } motion of B observed from the x, y, z frame

Procedure for Analysis

Equations 16–24 and 16–27 can be applied to the solution of problems involving the planar motion of particles or rigid bodies using the following procedure.

Coordinate Axes.

- Choose an appropriate location for the origin and proper orientation of the axes for both the X, Y, Z and moving x, y, z reference frames.

- Most often solutions are easily obtained if at the instant considered:

  (1) the origins are coincident

  (2) the corresponding axes are collinear

  (3) the corresponding axes are parallel

- The moving frame should be selected fixed to the body or device along which the relative motion occurs.

Kinematic Equations.

- After defining the origin A of the moving reference and specifying the moving point B, Eqs. 16–24 and 16–27 should be written in symbolic form

$$\mathbf{v}_B = \mathbf{v}_A + \boldsymbol{\Omega} \times \mathbf{r}_{B/A} + (\mathbf{v}_{B/A})_{xyz}$$
$$\mathbf{a}_B = \mathbf{a}_A + \dot{\boldsymbol{\Omega}} \times \mathbf{r}_{B/A} + \boldsymbol{\Omega} \times (\boldsymbol{\Omega} \times \mathbf{r}_{B/A}) + 2\boldsymbol{\Omega} \times (\mathbf{v}_{B/A})_{xyz} + (\mathbf{a}_{B/A})_{xyz}$$

- The Cartesian components of all these vectors may be expressed along either the X, Y, Z axes or the x, y, z axes. The choice is arbitrary provided a consistent set of unit vectors is used.

- Motion of the moving reference is expressed by $\mathbf{v}_A$, $\mathbf{a}_A$, $\boldsymbol{\Omega}$, and $\dot{\boldsymbol{\Omega}}$; and motion of B with respect to the moving reference is expressed by $\mathbf{r}_{B/A}$, $(\mathbf{v}_{B/A})_{xyz}$, and $(\mathbf{a}_{B/A})_{xyz}$.

The rotation of the dumping bin of the truck about point C is operated by the extension of the hydraulic cylinder AB. To determine the rotation of the bin due to this extension, we can use the equations of relative motion and fix the x, y axes to the cylinder so that the relative motion of the cylinder's extension occurs along the y axis.

EXAMPLE 16.19

Fig. 16–33

At the instant $\theta = 60°$, the rod in Fig. 16–33 has an angular velocity of 3 rad/s and an angular acceleration of 2 rad/s². At this same instant, the collar C is traveling outward along the rod such that when $x = 0.2$ m the velocity is 2 m/s and the acceleration is 3 m/s², both measured relative to the rod. Determine the Coriolis acceleration and the velocity and acceleration of the collar at this instant.

SOLUTION

Coordinate Axes.   The origin of both coordinate systems is located at point O, Fig. 16–33. Since motion of the collar is reported relative to the rod, the moving x, y, z frame of reference is *attached* to the rod.

Kinematic Equations.

$$\mathbf{v}_C = \mathbf{v}_O + \boldsymbol{\Omega} \times \mathbf{r}_{C/O} + (\mathbf{v}_{C/O})_{xyz} \tag{1}$$

$$\mathbf{a}_C = \mathbf{a}_O + \dot{\boldsymbol{\Omega}} \times \mathbf{r}_{C/O} + \boldsymbol{\Omega} \times (\boldsymbol{\Omega} \times \mathbf{r}_{C/O}) + 2\boldsymbol{\Omega} \times (\mathbf{v}_{C/O})_{xyz} + (\mathbf{a}_{C/O})_{xyz} \tag{2}$$

It will be simpler to express the data in terms of $\mathbf{i}, \mathbf{j}, \mathbf{k}$ component vectors rather than $\mathbf{I}, \mathbf{J}, \mathbf{K}$ components. Hence,

Motion of moving reference	Motion of C with respect to moving reference
$\mathbf{v}_O = \mathbf{0}$	$\mathbf{r}_{C/O} = \{0.2\mathbf{i}\}$ m
$\mathbf{a}_O = \mathbf{0}$	$(\mathbf{v}_{C/O})_{xyz} = \{2\mathbf{i}\}$ m/s
$\boldsymbol{\Omega} = \{-3\mathbf{k}\}$ rad/s	$(\mathbf{a}_{C/O})_{xyz} = \{3\mathbf{i}\}$ m/s²
$\dot{\boldsymbol{\Omega}} = \{-2\mathbf{k}\}$ rad/s²	

From Eq. 2, the Coriolis acceleration is defined as

$$\mathbf{a}_{\text{Cor}} = 2\boldsymbol{\Omega} \times (\mathbf{v}_{C/O})_{xyz} = 2(-3\mathbf{k}) \times (2\mathbf{i}) = \{-12\mathbf{j}\} \text{ m/s}^2 \quad \textit{Ans.}$$

This vector is shown dashed in Fig. 16–33. If desired, it may be resolved into $\mathbf{I}, \mathbf{J}$ components acting along the X and Y axes, respectively.

The velocity and acceleration of the collar are determined by substituting the data into Eqs. 1 and 2 and evaluating the cross products, which yields

$$\mathbf{v}_C = \mathbf{v}_O + \boldsymbol{\Omega} \times \mathbf{r}_{C/O} + (\mathbf{v}_{C/O})_{xyz}$$

$$= \mathbf{0} + (-3\mathbf{k}) \times (0.2\mathbf{i}) + 2\mathbf{i}$$

$$= \{2\mathbf{i} - 0.6\mathbf{j}\} \text{ m/s} \qquad\qquad\qquad\qquad \textit{Ans.}$$

$$\mathbf{a}_C = \mathbf{a}_O + \dot{\boldsymbol{\Omega}} \times \mathbf{r}_{C/O} + \boldsymbol{\Omega} \times (\boldsymbol{\Omega} \times \mathbf{r}_{C/O}) + 2\boldsymbol{\Omega} \times (\mathbf{v}_{C/O})_{xyz} + (\mathbf{a}_{C/O})_{xyz}$$

$$= \mathbf{0} + (-2\mathbf{k}) \times (0.2\mathbf{i}) + (-3\mathbf{k}) \times [(-3\mathbf{k}) \times (0.2\mathbf{i})] + 2(-3\mathbf{k}) \times (2\mathbf{i}) + 3\mathbf{i}$$

$$= \mathbf{0} - 0.4\mathbf{j} - 1.80\mathbf{i} - 12\mathbf{j} + 3\mathbf{i}$$

$$= \{1.20\mathbf{i} - 12.4\mathbf{j}\} \text{ m/s}^2 \qquad\qquad\qquad\qquad \textit{Ans.}$$

EXAMPLE 16.20

The rod AB, shown in Fig. 16–34, rotates clockwise such that it has an angular velocity $\omega_{AB} = 3$ rad/s and angular acceleration $\alpha_{AB} = 4$ rad/s^2 when $\theta = 45°$. Determine the angular motion of rod DE at this instant. The collar at C is pin connected to AB and slides over rod DE.

SOLUTION

Coordinate Axes.   The origin of both the fixed and moving frames of reference is located at D, Fig. 16–34. Furthermore, the x, y, z reference is attached to and rotates with rod DE so that the relative motion of the collar is easy to follow.

Kinematic Equations.

$$\mathbf{v}_C = \mathbf{v}_D + \boldsymbol{\Omega} \times \mathbf{r}_{C/D} + (\mathbf{v}_{C/D})_{xyz} \tag{1}$$

$$\mathbf{a}_C = \mathbf{a}_D + \dot{\boldsymbol{\Omega}} \times \mathbf{r}_{C/D} + \boldsymbol{\Omega} \times (\boldsymbol{\Omega} \times \mathbf{r}_{C/D}) + 2\boldsymbol{\Omega} \times (\mathbf{v}_{C/D})_{xyz} + (\mathbf{a}_{C/D})_{xyz} \tag{2}$$

All vectors will be expressed in terms of $\mathbf{i}, \mathbf{j}, \mathbf{k}$ components.

Motion of moving reference	Motion of C with respect to moving reference
$\mathbf{v}_D = \mathbf{0}$	$\mathbf{r}_{C/D} = \{0.4\mathbf{i}\}$ m
$\mathbf{a}_D = \mathbf{0}$	$(\mathbf{v}_{C/D})_{xyz} = (v_{C/D})_{xyz}\mathbf{i}$
$\boldsymbol{\Omega} = -\omega_{DE}\mathbf{k}$	$(\mathbf{a}_{C/D})_{xyz} = (a_{C/D})_{xyz}\mathbf{i}$
$\dot{\boldsymbol{\Omega}} = -\alpha_{DE}\mathbf{k}$	

Motion of C: Since the collar moves along a *circular path*, its velocity and acceleration can be determined using Eqs. 16–9 and 16–14.

$$\mathbf{v}_C = \boldsymbol{\omega}_{AB} \times \mathbf{r}_{C/A} = (-3\mathbf{k}) \times (0.4\mathbf{i} + 0.4\mathbf{j}) = \{1.2\mathbf{i} - 1.2\mathbf{j}\} \text{ m/s}$$

$$\mathbf{a}_C = \boldsymbol{\alpha}_{AB} \times \mathbf{r}_{C/A} - \omega_{AB}^2 \mathbf{r}_{C/A}$$

$$= (-4\mathbf{k}) \times (0.4\mathbf{i} + 0.4\mathbf{j}) - (3)^2(0.4\mathbf{i} + 0.4\mathbf{j}) = \{-2\mathbf{i} - 5.2\mathbf{j}\} \text{ m/s}^2$$

Fig. 16–34

Substituting the data into Eqs. 1 and 2, we have

$$\mathbf{v}_C = \mathbf{v}_D + \boldsymbol{\Omega} \times \mathbf{r}_{C/D} + (\mathbf{v}_{C/D})_{xyz}$$

$$1.2\mathbf{i} - 1.2\mathbf{j} = \mathbf{0} + (-\omega_{DE}\mathbf{k}) \times (0.4\mathbf{i}) + (v_{C/D})_{xyz}\mathbf{i}$$

$$1.2\mathbf{i} - 1.2\mathbf{j} = \mathbf{0} - 0.4\omega_{DE}\mathbf{j} + (v_{C/D})_{xyz}\mathbf{i}$$

$$(v_{C/D})_{xyz} = 1.2 \text{ m/s}$$

$$\omega_{DE} = 3 \text{ rad/s} \downarrow \qquad\qquad \textit{Ans.}$$

$$\mathbf{a}_C = \mathbf{a}_D + \dot{\boldsymbol{\Omega}} \times \mathbf{r}_{C/D} + \boldsymbol{\Omega} \times (\boldsymbol{\Omega} \times \mathbf{r}_{C/D}) + 2\boldsymbol{\Omega} \times (\mathbf{v}_{C/D})_{xyz} + (\mathbf{a}_{C/D})_{xyz}$$

$$-2\mathbf{i} - 5.2\mathbf{j} = \mathbf{0} + (-\alpha_{DE}\mathbf{k}) \times (0.4\mathbf{i}) + (-3\mathbf{k}) \times [(-3\mathbf{k}) \times (0.4\mathbf{i})]$$

$$+ 2(-3\mathbf{k}) \times (1.2\mathbf{i}) + (a_{C/D})_{xyz}\mathbf{i}$$

$$-2\mathbf{i} - 5.2\mathbf{j} = -0.4\alpha_{DE}\mathbf{j} - 3.6\mathbf{i} - 7.2\mathbf{j} + (a_{C/D})_{xyz}\mathbf{i}$$

$$(a_{C/D})_{xyz} = 1.6 \text{ m/s}^2$$

$$\alpha_{DE} = -5 \text{ rad/s}^2 = 5 \text{ rad/s}^2 \uparrow \qquad\qquad \textit{Ans.}$$

EXAMPLE 16.21

Two planes A and B are flying at the same elevation and have the motions shown in Fig. 16–35. Determine the velocity and acceleration of A as measured by the pilot of B.

SOLUTION

Coordinate Axes.   Since the relative motion of A with respect to the pilot in B is being sought, the x, y, z axes are attached to plane B, Fig. 16–35. At the *instant* considered, the origin B coincides with the origin of the fixed X, Y, Z frame.

Kinematic Equations.

$$\mathbf{v}_A = \mathbf{v}_B + \mathbf{\Omega} \times \mathbf{r}_{A/B} + (\mathbf{v}_{A/B})_{xyz} \tag{1}$$

$$\mathbf{a}_A = \mathbf{a}_B + \dot{\mathbf{\Omega}} \times \mathbf{r}_{A/B} + \mathbf{\Omega} \times (\mathbf{\Omega} \times \mathbf{r}_{A/B}) + 2\mathbf{\Omega} \times (\mathbf{v}_{A/B})_{xyz} + (\mathbf{a}_{A/B})_{xyz} \tag{2}$$

Motion of Moving Reference:

$$\mathbf{v}_B = \{600\mathbf{j}\} \text{ km/h}$$

$$(a_B)_n = \frac{v_B^2}{\rho} = \frac{(600)^2}{400} = 900 \text{ km/h}^2$$

$$\mathbf{a}_B = (\mathbf{a}_B)_n + (\mathbf{a}_B)_t = \{900\mathbf{i} - 100\mathbf{j}\} \text{ km/h}^2$$

$$\Omega = \frac{v_B}{\rho} = \frac{600 \text{ km/h}}{400 \text{ km}} = 1.5 \text{ rad/h} \downarrow \qquad \mathbf{\Omega} = \{-1.5\mathbf{k}\} \text{ rad/h}$$

$$\dot{\Omega} = \frac{(a_B)_t}{\rho} = \frac{100 \text{ km/h}^2}{400 \text{ km}} = 0.25 \text{ rad/h}^2 \uparrow \qquad \dot{\mathbf{\Omega}} = \{0.25\mathbf{k}\} \text{ rad/h}^2$$

Motion of A with Respect to Moving Reference:

$$\mathbf{r}_{A/B} = \{-4\mathbf{i}\} \text{ km} \quad (\mathbf{v}_{A/B})_{xyz} = ? \quad (\mathbf{a}_{A/B})_{xyz} = ?$$

Substituting the data into Eqs. 1 and 2, realizing that $\mathbf{v}_A = \{700\mathbf{j}\}$ km/h and $\mathbf{a}_A = \{50\mathbf{j}\}$ km/h^2, we have

$$\mathbf{v}_A = \mathbf{v}_B + \mathbf{\Omega} \times \mathbf{r}_{A/B} + (\mathbf{v}_{A/B})_{xyz}$$

$$700\mathbf{j} = 600\mathbf{j} + (-1.5\mathbf{k}) \times (-4\mathbf{i}) + (\mathbf{v}_{A/B})_{xyz}$$

$$(\mathbf{v}_{A/B})_{xyz} = \{94\mathbf{j}\} \text{ km/h} \qquad\qquad Ans.$$

$$\mathbf{a}_A = \mathbf{a}_B + \dot{\mathbf{\Omega}} \times \mathbf{r}_{A/B} + \mathbf{\Omega} \times (\mathbf{\Omega} \times \mathbf{r}_{A/B}) + 2\mathbf{\Omega} \times (\mathbf{v}_{A/B})_{xyz} + (\mathbf{a}_{A/B})_{xyz}$$

$$50\mathbf{j} = (900\mathbf{i} - 100\mathbf{j}) + (0.25\mathbf{k}) \times (-4\mathbf{i})$$

$$+ (-1.5\mathbf{k}) \times [(-1.5\mathbf{k}) \times (-4\mathbf{i})] + 2(-1.5\mathbf{k}) \times (94\mathbf{j}) + (\mathbf{a}_{A/B})_{xyz}$$

$$(\mathbf{a}_{A/B})_{xyz} = \{-1191\mathbf{i} + 151\mathbf{j}\} \text{ km/h}^2 \qquad\qquad Ans.$$

Fig. 16–35

NOTE: The solution of this problem should be compared with that of Example 12.26.

PROBLEMS

16–131. Block A, which is attached to a cord, moves along the slot of a horizontal forked rod. At the instant shown, the cord is pulled down through the hole at O with an acceleration of 4 m/s² and its velocity is 2 m/s. Determine the acceleration of the block at this instant. The rod rotates about O with a constant angular velocity $\omega = 4$ rad/s.

Prob. 16–131

16–133. The collar E is attached to, and pivots about, rod AB while it slides on rod CD. If rod AB has an angular velocity of 6 rad/s and an angular acceleration of 1 rad/s², both acting clockwise, determine the angular velocity and the angular acceleration of rod CD at the instant shown.

Prob. 16–133

*16–132. The ball B of negligible size rolls through the tube such that at the instant shown it has a velocity of 5 ft/s and an acceleration of 3 ft/s², measured relative to the tube. If the tube has an angular velocity of $\omega = 3$ rad/s and an angular acceleration of $\alpha = 5$ rad/s² at this same instant, determine the velocity and acceleration of the ball.

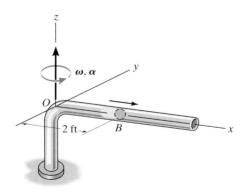

Prob. 16–132

16–134. Block B moves along the slot in the platform with a constant speed of 2 ft/s, measured relative to the platform in the direction shown. If the platform is rotating at a constant rate of $\omega = 5$ rad/s, determine the velocity and acceleration of the block at the instant $\theta = 60°$.

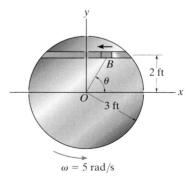

Prob. 16–134

16–135. While the swing bridge is closing with a constant rotation of 0.5 rad/s, a man runs along the roadway at a constant speed of 5 ft/s relative to the roadway. Determine his velocity and acceleration at the instant $d = 15$ ft.

16–137. A girl stands at A on a platform which is rotating with a constant angular velocity $\omega = 0.5$ rad/s. If she walks at a constant speed of $v = 0.75$ m/s measured relative to the platform, determine her acceleration (a) when she reaches point D in going along the path $ADC, d = 1$ m; and (b) when she reaches point B if she follows the path $ABC, r = 3$ m.

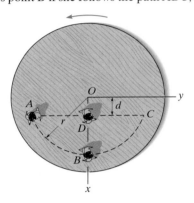

Prob. 16–135

Prob. 16–137

*16–136. While the swing bridge is closing with a constant rotation of 0.5 rad/s, a man runs along the roadway such that when $d = 10$ ft he is running outward from the center at 5 ft/s with an acceleration of 2 ft/s², both measured relative to the roadway. Determine his velocity and acceleration at this instant.

16–138. A girl stands at A on a platform which is rotating with an angular acceleration $\alpha = 0.2$ rad/s² and at the instant shown has an angular velocity $\omega = 0.5$ rad/s. If she walks at a constant speed $v = 0.75$ m/s measured relative to the platform, determine her acceleration (a) when she reaches point D in going along the path $ADC, d = 1$ m; and (b) when she reaches point B if she follows the path $ABC, r = 3$ m.

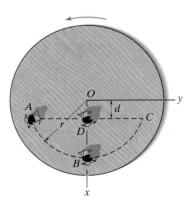

Prob. 16–136

Prob. 16–138

16–139. Rod AB rotates counterclockwise with a constant angular velocity $\omega = 3$ rad/s. Determine the velocity and acceleration of point C located on the double collar when $\theta = 45°$. The collar consists of two pin-connected slider blocks which are constrained to move along the circular path and the rod AB.

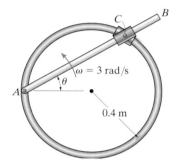

Prob. 16–139

16–141. Block B of the mechanism is confined to move within the slot member CD. If AB is rotating at a constant rate of $\omega_{AB} = 3$ rad/s, determine the angular velocity and angular acceleration of member CD at the instant shown.

Prob. 16–141

*****16–140.** A ride in an amusement park consists of a rotating platform P, having a constant angular velocity $\omega_P = 1.5$ rad/s, and four cars, C, mounted on the platform, which have constant angular velocities $\omega_{C/P} = 2$ rad/s measured relative to the platform. Determine the velocity and acceleration of the passenger at B at the instant shown.

Prob. 16–140

16–142. The "quick-return" mechanism consists of a crank AB, slider block B, and slotted link CD. If the crank has the angular motion shown, determine the angular motion of the slotted link at this instant.

Prob. 16–142

16–143. At a given instant, rod AB has the angular motions shown. Determine the angular velocity and angular acceleration of rod CD at this instant. There is a collar at C.

Prob. 16–143

16–145. The gear has the angular motion shown. Determine the angular velocity and angular acceleration of the slotted link BC at this instant. The peg at A is fixed to the gear.

Prob. 16–145

*16–144. At the instant shown, rod AB has an angular velocity $\omega_{AB} = 3$ rad/s and an angular acceleration $\alpha_{AB} = 5$ rad/s^2. Determine the angular velocity and angular acceleration of rod CD at this instant. The collar at C is pin-connected to CD and slides over AB.

Prob. 16–144

16–146. The disk rotates with the angular motion shown. Determine the angular velocity and angular acceleration of the slotted link AC at this instant. The peg at B is fixed to the disk.

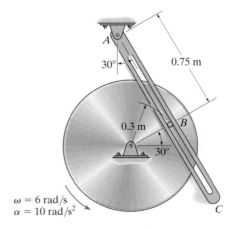

Prob. 16–146

16–147. A ride in an amusement park consists of a rotating arm AB having a constant angular velocity $\omega_{AB} = 2$ rad/s about point A and a car mounted at the end of the arm which has a constant angular velocity $\boldsymbol{\omega}' = \{-0.5\mathbf{k}\}$ rad/s, measured relative to the arm. At the instant shown, determine the velocity and acceleration of the passenger at C.

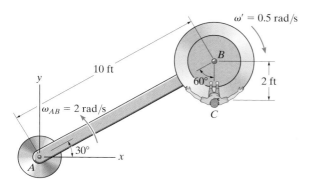

Prob. 16–147

16–149. The cars on the amusement-park ride rotate around the axle at A with a constant angular velocity $\omega_{A/f} = 2$ rad/s, measured relative to the frame AB. At the same time the frame rotates around the main axle support at B with a constant angular velocity $\omega_f = 1$ rad/s. Determine the velocity and acceleration of the passenger at C at the instant shown.

Prob. 16–149

*16–148.** A ride in an amusement park consists of a rotating arm AB that has an angular acceleration of $\alpha_{AB} = 1$ rad/s^2 when $\omega_{AB} = 2$ rad/s at the instant shown. Also at this instant the car mounted at the end of the arm has a relative angular acceleration of $\boldsymbol{\alpha}' = \{-0.6\mathbf{k}\}$ rad/s^2 when $\boldsymbol{\omega}' = \{-0.5\mathbf{k}\}$ rad/s. Determine the velocity and acceleration of the passenger C at this instant.

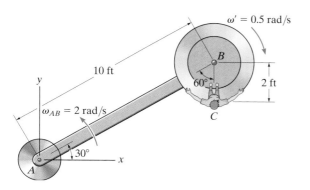

Prob. 16–148

16–150. The block B of the "quick-return" mechanism is confined to move within the slot in member CD. If AB is rotating at a constant rate of $\omega_{AB} = 3$ rad/s, determine the angular velocity and angular acceleration of member CD at the instant shown.

Prob. 16–150

Design Projects

16–1D. Design of a Belt Transmission System

The wheel A is used in a textile mill and must be rotated counterclockwise at 4 rad/s. This can be done using a motor which is mounted on the platform at the location shown. If the shaft B on the motor can rotate clockwise 50 rad/s, design a method for transmitting the rotation from B to A. Use a series of belts and pulleys as a basis for your design. A belt drives the rotation of wheel A by wrapping around its outer surface, and a pulley can be attached to the shaft of the motor as well as anywhere else. Do not let the length of any belt be longer than 6 ft. Submit a drawing of your design and the calculations of the kinematics. Also, determine the total cost of materials if any belt costs \$2.50, and any pulley costs \$$2r$, where r is the radius of the pulley in inches.

Fig. 16–1D

16–2D. Design of an Oscillating Link Mechanism

The operation of a sewing machine requires the 200-mm-long bar to oscillate back-and-forth through an angle of 60° every 0.2 seconds. A motor having a drive shaft which turns at 40 rad/s is available to provide the necessary power. Specify the location of the motor and design a mechanism required to perform the motion. Submit a drawing of your design, showing the placement of the motor, and compute the velocity and acceleration of the end A of the link as a function of its angle of rotation $0° \le \theta \le 60°$.

Fig. 16–2D

16–3D. Design of a Retractable Aircraft Landing-Gear Mechanism

The nose wheel of a small plane is attached to member AB, which is pinned to the aircraft frame at B. Design a mechanism that will allow the wheel to be fully retracted forward; i.e., rotated clockwise 90°, in $t \leq 4$ seconds. Use a hydraulic cylinder which has a closed length of 1.25 ft and, if needed, a fully extended length of 2 ft. Make sure that your design holds the wheel in a stable position when the wheel is on the ground. Show plots of the angular velocity and angular acceleration of AB versus its angular position $0° \leq \theta \leq 90°$.

Fig. 16–3D

16–4D. Design of a Saw Link Mechanism

The saw blade in a lumber mill is required to remain in the horizontal position and undergo a complete back-and-forth motion in 2 seconds. An electric motor, having a shaft rotation of 50 rad/s, is available to power the saw and can be located anywhere. Design a mechanism that will transfer the rotation of the motor's shaft to the saw blade. Submit drawings of your design and calculations of the kinematics of the saw blade. Include a plot of the velocity and acceleration of the saw blade as a function of its horizontal position. Note that to cut through the log the blade must be allowed to move freely downward as well as back and forth.

Fig. 16–4D

Chapter Review

Rigid-Body Planar Motion

A rigid body undergoes three types of planar motion: translation, rotation about a fixed axis, and general plane motion. (*See pages 301 and 302.*)

Path of rectilinear translation

Translation

When a body has rectilinear translation, all the particles of the body travel along straight-line paths. If the paths have the same radius of curvature, then curvilinear translation occurs. Provided we know the motion of one of the particles, then the motion of all of the others is also known. (*See page 303.*)

Path of curvilinear translation

Rotation about a Fixed Axis

For this type of motion, all of the particles move along circular paths. Here, all line segments in the body undergo the same angular displacement, angular velocity, and angular acceleration.

Once the angular motion of the body is known, then the velocity of any particle a distance r from the axis can be obtained.

The acceleration of any particle has two components. The tangential component accounts for the change in the magnitude of the velocity, and the normal component accounts for the change in the velocity's direction. (*See pages 304–308.*)

Rotation about a fixed axis

$$\omega = d\theta/dt \qquad\qquad \omega = \omega_0 + \alpha_c t$$
$$\alpha = d\omega/dt \qquad\qquad \theta = \theta_0 + \omega_0 t + \tfrac{1}{2}\alpha_c t^2$$
$$\alpha\, d\theta = \omega\, d\omega \qquad\qquad \omega^2 = \omega_0^2 + 2\alpha_c(\theta - \theta_0)$$

$$v = \omega r \qquad\qquad a_t = \alpha r$$
$$a_n = \omega^2 r$$

General Plane Motion

When a body undergoes general plane motion, it simultaneously translates and rotates. There are several methods for analyzing this motion.

Absolute Motion Analysis

If the motion of a point on a body or the angular motion of a line is known, then it may be possible to relate this motion to that of another point or line using an absolute motion analysis. To do so, linear position coordinates s or angular position coordinates θ are established (measured from a fixed point or line). These position coordinates are then related using the geometry of the body. The time derivative of this equation gives the relationship between the velocities and/or the angular velocities. A second time derivative relates the accelerations and/or the angular accelerations. (*See page 319.*)

General plane motion

Relative-Motion Analysis

General plane motion can also be analyzed using a relative-motion analysis between two points A and B. This method considers the motion in parts: first a translation of the selected base point A, then a relative "rotation" of the body about point A, which is measured from a translating axis. Since the relative motion is viewed as circular motion about the base point, point B will have a velocity $\mathbf{v}_{B/A}$ that is tangent to the circle. It also has two components of acceleration, $(\mathbf{a}_{B/A})_t$ and $(\mathbf{a}_{B/A})_n$. It is important to also realize that $\mathbf{a}_A$ and $\mathbf{a}_B$ will have two components if these points move along curved paths.
(*See pages 327–329.*)

$$\mathbf{v}_B = \mathbf{v}_A + \boldsymbol{\omega} \times \mathbf{r}_{B/A}$$

$$\mathbf{a}_B = \mathbf{a}_A + \boldsymbol{\alpha} \times \mathbf{r}_{B/A} - \omega^2 \mathbf{r}_{B/A}$$

Instantaneous Center of Zero Velocity

If the base point A is selected as having zero velocity, then the relative velocity equation becomes $\mathbf{v}_B = \boldsymbol{\omega} \times \mathbf{r}_{B/A}$. In this case, motion appears as if the body is rotating about an instantaneous axis.

The instantaneous center of rotation (IC) can be established provided the directions of the velocities of any two points on the body are known. Since a radial line r will always be perpendicular to each velocity, then the IC is at the point of intersection of these two radial lines. Its measured location is determined from the geometry of the body. Once it is established, then the velocity of any point P on the body can be determined from $v = \omega r$, where r extends from the IC to point P.
(*See pages 341–343 and 351–353.*)

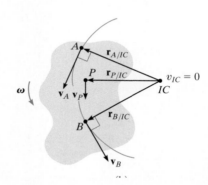

Relative Motion using Rotating Axes

Problems that involve connected members that slide relative to one another or points not located on the same body can be analyzed using a relative-motion analysis referenced from a rotating frame. In particular, the term $2\boldsymbol{\Omega} \times (\mathbf{v}_{B/A})_{xyz}$ is called the Coriolis acceleration.
(*See pages 366–371.*)

$$\mathbf{v}_B = \mathbf{v}_A + \boldsymbol{\Omega} \times \mathbf{r}_{B/A} + (\mathbf{v}_{B/A})_{xyz}$$

$$\mathbf{a}_B = \mathbf{a}_A + \dot{\boldsymbol{\Omega}} \times \mathbf{r}_{B/A} + \boldsymbol{\Omega} \times (\boldsymbol{\Omega} \times \mathbf{r}_{B/A}) + 2\boldsymbol{\Omega} \times (\mathbf{v}_{B/A})_{xyz} + (\mathbf{a}_{B/A})_{xyz}$$

The forces acting on this dragster as it begins to accelerate are quite severe and must be accounted for in the design of its structure.

17 Planar Kinetics of a Rigid Body: Force and Acceleration

CHAPTER OBJECTIVES

- To introduce the methods used to determine the mass moment of inertia of a body.
- To develop the planar kinetic equations of motion for a symmetric rigid body.
- To discuss applications of these equations to bodies undergoing translation, rotation about a fixed axis, and general plane motion.

17.1 Moment of Inertia

Since a body has a definite size and shape, an applied nonconcurrent force system may cause the body to both translate and rotate. The translational aspects of the motion were studied in Chapter 13 and are governed by the equation $\mathbf{F} = m\mathbf{a}$. It will be shown in Sec. 17.2 that the rotational aspects, caused by a moment $\mathbf{M}$, are governed by an equation of the form $\mathbf{M} = I\boldsymbol{\alpha}$. The symbol I in this equation is termed the moment of inertia. By comparison, the *moment of inertia* is a measure of the resistance of a body to *angular acceleration* ($\mathbf{M} = I\boldsymbol{\alpha}$) in the same way that *mass* is a measure of the body's resistance to *acceleration* ($\mathbf{F} = m\mathbf{a}$).

The flywheel on the engine of this tractor has a large moment of inertia about its axis of rotation. Once it is set into motion, it will be difficult to stop, and this in turn will prevent the engine from stalling and instead will allow it to maintain a constant power.

We define the *moment of inertia* as the integral of the "second moment" about an axis of all the elements of mass *dm* which compose the body.* For example, the body's moment of inertia about the *z* axis in Fig. 17–1 is

Fig. 17–1

$$I = \int_m r^2 \, dm \qquad (17\text{–}1)$$

Here the "moment arm" *r* is the perpendicular distance from the *z* axis to the arbitrary element *dm*. Since the formulation involves *r*, the value of *I* is different for each axis about which it is computed. In the study of planar kinetics, the axis which is generally chosen for analysis passes through the body's mass center *G* and is always perpendicular to the plane of motion. The moment of inertia computed about this axis will be denoted as I_G. Since *r* is squared in Eq. 17–1, the mass moment of inertia is always a *positive* quantity. Common units used for its measurement are $kg \cdot m^2$ or $slug \cdot ft^2$.

If the body consists of material having a variable density, $\rho = \rho \, (x, y, z)$, the elemental mass *dm* of the body may be expressed in terms of its density and volume as $dm = \rho \, dV$. Substituting *dm* into Eq. 17–1, the body's moment of inertia is then computed using *volume elements* for integration; i.e.,

$$I = \int_V r^2 \rho \, dV \qquad (17\text{–}2)$$

---

*Another property of the body, which measures the symmetry of the body's mass with respect to a coordinate system, is the product of inertia. This property applies to the three-dimensional motion of a body and will be discussed in Chapter 21.

In the special case of ρ being a *constant*, this term may be factored out of the integral, and the integration is then purely a function of geometry,

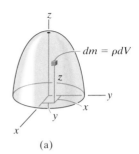

$$I = \rho \int_V r^2 \, dV \qquad (17\text{--}3)$$

When the elemental volume chosen for integration has infinitesimal dimensions in all three directions, e.g., $dV = dx \, dy \, dz$, Fig. 17–2a, the moment of inertia of the body must be determined using "triple integration." The integration process can, however, be simplified to a *single integration* provided the chosen elemental volume has a differential size or thickness in only *one direction*. Shell or disk elements are often used for this purpose.

Procedure for Analysis

To obtain the moment of inertia by integration, we will consider only symmetric bodies having surfaces which are generated by revolving a curve about an axis. An example of such a body which is generated about the z axis is shown in Fig. 17–2a. Two types of differential elements can be chosen.

Shell Element.

- If a *shell element* having a height z, radius $r = y$, and thickness dy is chosen for integration, Fig. 17–2b, then the volume is $dV = (2\pi y)(z) \, dy$.
- This element may be used in Eq. 17–2 or 17–3 for determining the moment of inertia I_z of the body about the z axis, since the *entire element*, due to its "thinness," lies at the *same* perpendicular distance $r = y$ from the z axis (see Example 17.1).

Disk Element.

- If a disk element having a radius y and a thickness dz is chosen for integration, Fig. 17–2c, then the volume is $dV = (\pi y^2) \, dz$.
- This element is *finite* in the radial direction, and consequently its parts *do not* all lie at the *same radial distance r* from the z axis. As a result, Eq. 17–2 or 17–3 *cannot* be used to determine I_z directly. Instead, to perform the integration it is first necessary to determine the moment of inertia *of the element* about the z axis and then integrate this result (see Example 17.2).

Fig. 17–2

EXAMPLE 17.1

Determine the moment of inertia of the cylinder shown in Fig. 17–3a about the z axis. The density of the material, ρ, is constant.

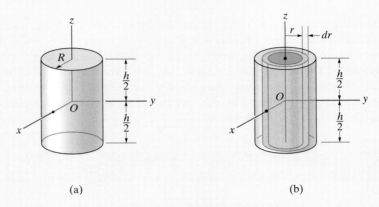

(a)   (b)

Fig. 17–3

SOLUTION

Shell Element. This problem may be solved using the *shell element* in Fig. 17–3b and single integration. The volume of the element is $dV = (2\pi r)(h)\, dr$, so that its mass is $dm = \rho\, dV = \rho(2\pi h r\, dr)$. Since the *entire element* lies at the same distance r from the z axis, the moment of inertia *of the element* is

$$dI_z = r^2\, dm = \rho 2\pi h r^3\, dr$$

Integrating over the entire region of the cylinder yields

$$I_z = \int_m r^2\, dm = \rho 2\pi h \int_0^R r^3\, dr = \frac{\rho\pi}{2} R^4 h$$

The mass of the cylinder is

$$m = \int_m dm = \rho 2\pi h \int_0^R r\, dr = \rho\pi h R^2$$

so that

$$I_z = \frac{1}{2} m R^2 \qquad\qquad Ans.$$

EXAMPLE 17.2

A solid is formed by revolving the shaded area shown in Fig. 17–4a about the y axis. If the density of the material is 5 slug/ft³, determine the moment of inertia about the y axis.

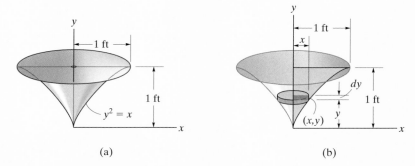

Fig. 17–4

SOLUTION

Disk Element.   The moment of inertia will be computed using a *disk element*, as shown in Fig. 17–4b. Here the element intersects the curve at the arbitrary point (x, y) and has a mass

$$dm = \rho \, dV = \rho(\pi x^2) \, dy$$

Although all portions of the element are *not* located at the same distance from the y axis, it is still possible to determine the moment of inertia dI_y *of the element* about the y axis. In the preceding example it was shown that the moment of inertia of a cylinder about its longitudinal axis is $I = \frac{1}{2}mR^2$, where m and R are the mass and radius of the cylinder. Since the height of the cylinder is not involved in this formula, the cylinder itself can be thought of as a disk. Thus, for the disk element in Fig. 17–4b, we have

$$dI_y = \tfrac{1}{2}(dm)x^2 = \tfrac{1}{2}[\rho(\pi x^2) \, dy]x^2$$

Substituting $x = y^2$, $\rho = 5$ slug/ft³, and integrating with respect to y, from $y = 0$ to $y = 1$ ft, yields the moment of inertia for the entire solid.

$$I_y = \frac{\pi(5)}{2} \int_0^1 x^4 \, dy = \frac{\pi(5)}{2} \int_0^1 y^8 \, dy = 0.873 \text{ slug} \cdot \text{ft}^2 \qquad Ans.$$

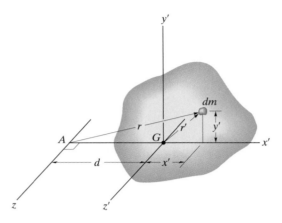

Fig. 17–5

Parallel-Axis Theorem.   If the moment of inertia of the body about an axis passing through the body's mass center is known, then the moment of inertia about any other *parallel axis* may be determined by using the *parallel-axis theorem*. This theorem can be derived by considering the body shown in Fig. 17–5. The z' axis passes through the mass center G, whereas the corresponding *parallel z axis* lies at a constant distance d away. Selecting the differential element of mass dm, which is located at point (x', y'), and using the Pythagorean theorem, $r^2 = (d + x')^2 + y'^2$, we can express the moment of inertia of the body about the z axis as

$$I = \int_m r^2 \, dm = \int_m [(d + x')^2 + y'^2] \, dm$$

$$= \int_m (x'^2 + y'^2) \, dm + 2d \int_m x' \, dm + d^2 \int_m dm$$

Since $r'^2 = x'^2 + y'^2$, the first integral represents I_G. The second integral equals *zero*, since the z' axis passes through the body's mass center, i.e., $\int x' \, dm = \bar{x}' \int dm = 0$ since $\bar{x}' = 0$. Finally, the third integral represents the total mass m of the body. Hence, the moment of inertia about the z axis can be written as

$$\boxed{I = I_G + md^2} \qquad (17\text{--}4)$$

where

I_G = moment of inertia about the z' axis passing through the mass center G

m = mass of the body

d = perpendicular distance between the parallel axes

Radius of Gyration.   Occasionally, the moment of inertia of a body about a specified axis is reported in handbooks using the *radius of gyration, k.* This value has units of length, and when it and the body's mass m are known, the body's moment of inertia is determined from the equation

$$\boxed{I = mk^2 \quad \text{or} \quad k = \sqrt{\dfrac{I}{m}}} \qquad (17\text{--}5)$$

Note the *similarity* between the definition of k in this formula and r in the equation $dI = r^2 \, dm$, which defines the moment of inertia of an elemental mass dm of the body about an axis.

Composite Bodies.   If a body is constructed of a number of simple shapes such as disks, spheres, and rods, the moment of inertia of the body about any axis z can be determined by adding algebraically the moments of inertia of all the composite shapes computed about the z axis. Algebraic addition is necessary since a composite part must be considered as a negative quantity if it has already been counted as a piece of another part—for example, a "hole" subtracted from a solid plate. The parallel-axis theorem is needed for the calculations if the center of mass of each composite part does not lie on the z axis. For the calculation, then, $I = \Sigma(I_G + md^2)$. Here I_G for each of the composite parts is computed by integration or can be determined from a table, such as the one given on the inside back cover of this book.

EXAMPLE 17.3

If the plate shown in Fig. 17–6a has a density of 8000 kg/m³ and a thickness of 10 mm, determine its moment of inertia about an axis directed perpendicular to the page and passing through point O.

Fig. 17–6

SOLUTION

The plate consists of two composite parts, the 250-mm-radius disk *minus* a 125-mm-radius disk, Fig. 17–6b. The moment of inertia about O can be determined by computing the moment of inertia of each of these parts about O and then adding the results *algebraically*. The calculations are performed by using the parallel-axis theorem in conjunction with the data listed in the table on the inside back cover.

Disk.   The moment of inertia of a disk about the centroidal axis perpendicular to the plane of the disk is $I_G = \frac{1}{2}mr^2$. The mass center of the disk is located at a distance of 0.25 m from point O. Thus,

$$m_d = \rho_d V_d = 8000 \text{ kg/m}^3 \, [\pi(0.25 \text{ m})^2(0.01 \text{ m})] = 15.71 \text{ kg}$$

$$(I_d)_O = \frac{1}{2}m_d r_d^2 + m_d d^2$$

$$= \frac{1}{2}(15.71 \text{ kg})(0.25 \text{ m})^2 + (15.71 \text{ kg})(0.25 \text{ m})^2$$

$$= 1.473 \text{ kg} \cdot \text{m}^2$$

Hole.   For the 125-mm-radius disk (hole), we have

$$m_h = \rho_h V_h = 8000 \text{ kg/m}^3 \, [\pi(0.125 \text{ m})^2(0.01 \text{ m})] = 3.93 \text{ kg}$$

$$(I_h)_O = \frac{1}{2}m_h r_h^2 + m_h d^2$$

$$= \frac{1}{2}(3.93 \text{ kg})(0.125 \text{ m})^2 + (3.93 \text{ kg})(0.25 \text{ m})^2$$

$$= 0.276 \text{ kg} \cdot \text{m}^2$$

The moment of inertia of the plate about point O is therefore

$$I_O = (I_d)_O - (I_h)_O$$

$$= 1.473 \text{ kg} \cdot \text{m}^2 - 0.276 \text{ kg} \cdot \text{m}^2$$

$$= 1.20 \text{ kg} \cdot \text{m}^2 \qquad\qquad \textit{Ans.}$$

EXAMPLE 17.4

The pendulum in Fig. 17–7 is suspended from point O and consists of two thin rods, each having a weight of 10 lb. Determine the pendulum's moment of inertia about an axis passing through (a) the pin at O, and (b) the mass center G of the pendulum.

Fig. 17–7

SOLUTION

Part (a).   Using the table on the inside back cover, the moment of inertia of rod OA about an axis perpendicular to the page and passing through the end point O of the rod is $I_O = \frac{1}{3}ml^2$. Hence,

$$(I_{OA})_O = \frac{1}{3}ml^2 = \frac{1}{3}\left(\frac{10\ \text{lb}}{32.2\ \text{ft/s}^2}\right)(2\ \text{ft})^2 = 0.414\ \text{slug} \cdot \text{ft}^2$$

This same value is obtained using $I_G = \frac{1}{12}ml^2$ and the parallel-axis theorem.

$$(I_{OA})_O = \frac{1}{12}ml^2 + md^2 = \frac{1}{12}\left(\frac{10\ \text{lb}}{32.2\ \text{ft/s}^2}\right)(2\ \text{ft})^2 + \left(\frac{10\ \text{lb}}{32.2\ \text{ft/s}^2}\right)(1\ \text{ft})^2$$

$$= 0.414\ \text{slug} \cdot \text{ft}^2$$

For rod BC we have

$$(I_{BC})_O = \frac{1}{12}ml^2 + md^2 = \frac{1}{12}\left(\frac{10\ \text{lb}}{32.2\ \text{ft/s}^2}\right)(2\ \text{ft})^2 + \left(\frac{10\ \text{lb}}{32.2\ \text{ft/s}^2}\right)(2\ \text{ft})^2$$

$$= 1.346\ \text{slug} \cdot \text{ft}^2$$

The moment of inertia of the pendulum about O is therefore

$$I_O = 0.414 + 1.346 = 1.76\ \text{slug} \cdot \text{ft}^2 \qquad\qquad Ans.$$

Part (b).   The mass center G will be located relative to the pin at O. Assuming this distance to be $\bar{y}$, Fig. 17–7, and using the formula for determining the mass center, we have

$$\bar{y} = \frac{\Sigma \tilde{y}m}{\Sigma m} = \frac{1(10/32.2) + 2(10/32.2)}{(10/32.2) + (10/32.2)} = 1.50\ \text{ft}$$

The moment of inertia I_G may be computed in the same manner as I_O, which requires successive applications of the parallel-axis theorem to transfer the moments of inertia of rods OA and BC to G. A more direct solution, however, involves using the result for I_O, i.e.,

$$I_O = I_G + md^2; \quad 1.76\ \text{slug} \cdot \text{ft}^2 = I_G + \left(\frac{20\ \text{lb}}{32.2\ \text{ft/s}^2}\right)(1.50\ \text{ft})^2$$

$$I_G = 0.362\ \text{slug} \cdot \text{ft}^2 \qquad\qquad Ans.$$

PROBLEMS

17–1. The right circular cone is formed by revolving the shaded area around the x axis. Determine the moment of inertia I_x and express the result in terms of the total mass m of the cone. The cone has a constant density ρ.

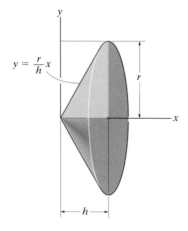

$$y = \frac{r}{h}x$$

Prob. 17–1

17–3. The solid is formed by revolving the shaded area around the y axis. Determine the radius of gyration k_y. The specific weight of the material is $\gamma = 380$ lb/ft^3.

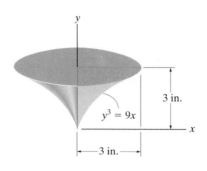

$$y^3 = 9x$$

3 in.

3 in.

Prob. 17–3

17–2. Determine the moment of inertia of the thin ring about the z axis. The ring has a mass m.

R

Prob. 17–2

***17–4.** Determine the moment of inertia I_x of the sphere and express the result in terms of the total mass m of the sphere. The sphere has a constant density ρ.

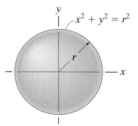

$$x^2 + y^2 = r^2$$

r

Prob. 17–4

17–5. Determine the radius of gyration k_x of the paraboloid. The density of the material is $\rho = 5 \text{ Mg/m}^3$.

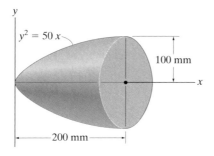

$y^2 = 50x$

100 mm

200 mm

Prob. 17–5

17–7. Determine the radius of gyration k_x of the body. The specific weight of the material is $\gamma = 380 \text{ lb/ft}^3$.

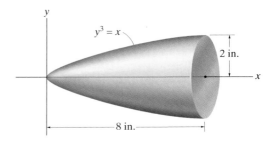

$y^3 = x$

2 in.

8 in.

Prob. 17–7

17–6. Determine the moment of inertia of the semiellipsoid with respect to the x axis and express the result in terms of the mass m of the semiellipsoid. The material has a constant density ρ.

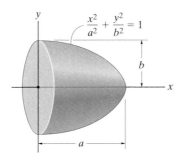

$\dfrac{x^2}{a^2} + \dfrac{y^2}{b^2} = 1$

b

a

Prob. 17–6

***17–8.** Determine the moment of inertia of the ellipsoid with respect to the x axis and express the result in terms of the mass m of the ellipsoid. The material has a constant density ρ.

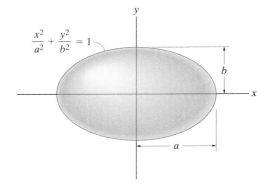

$\dfrac{x^2}{a^2} + \dfrac{y^2}{b^2} = 1$

b

a

Prob. 17–8

17–9. Determine the moment of inertia of the homogeneous pyramid of mass m with respect to the z axis. The density of the material is ρ. *Suggestion:* Use a rectangular plate element having a volume of $dV = (2x)(2y)\,dz$.

17–11. Determine the moment of inertia of the thin plate about an axis perpendicular to the page and passing through the pin at O. The plate has a hole in its center. Its thickness is 50 mm, and the material has a density of $\rho = 50\ \text{kg/m}^3$.

Prob. 17–9

Prob. 17–11

17–10. The concrete shape is formed by rotating the shaded area about the y axis. Determine the moment of inertia I_y. The specific weight of concrete is $\gamma = 150\ \text{lb/ft}^3$.

*17–12. Determine the moment of inertia I_z of the frustum of the cone which has a conical depression. The material has a density of $200\ \text{kg/m}^3$.

Prob. 17–10

Prob. 17–12

17–13. Determine the moment of inertia of the assembly about an axis which is perpendicular to the page and passes through the center of mass G. The material has a specific weight of $\gamma = 90 \text{ lb/ft}^3$.

17–14. Determine the moment of inertia of the assembly about an axis which is perpendicular to the page and passes through point O. The material has a specific weight of $\gamma = 90 \text{ lb/ft}^3$.

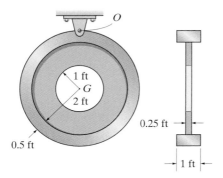

Probs. 17–13/14

17–15. The wheel consists of a thin ring having a mass of 10 kg and four spokes made from slender rods, each having a mass of 2 kg. Determine the wheel's moment of inertia about an axis perpendicular to the page and passing through point A.

Prob. 17–15

*17–16. The slender rods have a weight of 3 lb/ft. Determine the moment of inertia of the assembly about an axis perpendicular to the page and passing through point A.

Prob. 17–16

17–17. Each of the three rods has a mass m. Determine the moment of inertia of the assembly about an axis which is perpendicular to the page and passes through the center point O.

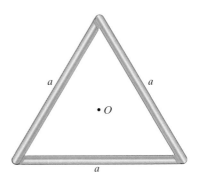

Prob. 17–17

17–18. The slender rods have a weight of 3 lb/ft. Determine the moment of inertia of the assembly about an axis perpendicular to the page and passing through point A.

Prob. 17–18

17–19. The pendulum consists of a plate having a weight of 12 lb and a slender rod having a weight of 4 lb. Determine the radius of gyration of the pendulum about an axis perpendicular to the page and passing through point O.

Prob. 17–19

***17–20.** Determine the moment of inertia of the overhung crank about the x axis. The material is steel having a density of $\rho = 7.85$ Mg/m³.

17–21. Determine the moment of inertia of the overhung crank about the x' axis. The material is steel having a density of $\rho = 7.85$ Mg/m³.

Probs. 17–20/21

17–22. Determine the moment of inertia of the solid steel assembly about the x axis. Steel has a specific weight of $\gamma_{st} = 490$ lb/ft³.

Prob. 17–22

17–23. The pendulum consists of two slender rods AB and OC which have a mass of 3 kg/m. The thin plate has a mass of 12 kg/m². Determine the location $\bar{y}$ of the center of mass G of the pendulum, then calculate the moment of inertia of the pendulum about an axis perpendicular to the page and passing through G.

Prob. 17–23

17.2   Planar Kinetic Equations of Motion

In the following analysis we will limit our study of planar kinetics to rigid bodies which, along with their loadings, are considered to be *symmetrical with respect to a fixed reference plane.* * In this case the path of motion of each particle of the body is a plane curve parallel to a fixed reference plane. Since the motion of the body may be viewed within the reference plane, all the forces (and couple moments) acting on the body can then be projected onto the plane. An example of an arbitrary body of this type is shown in Fig. 17–8a. Here the *inertial frame of reference x, y, z* has its origin *coincident* with the arbitrary point P in the body. By definition, *these axes do not rotate and are either fixed or translate with constant velocity.*

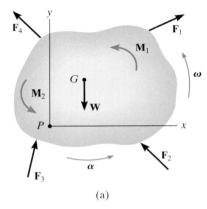

(a)

Fig. 17–8

Equation of Translational Motion.   The external forces shown on the body in Fig. 17–8a represent the effect of gravitational, electrical, magnetic, or contact forces between adjacent bodies. Since this force system has been considered previously in Sec. 13.3 for the analysis of a system of particles, the resulting Eq. 13–6 may be used here, in which case

$$\Sigma \mathbf{F} = m\mathbf{a}_G$$

This equation is referred to as the *translational equation of motion* for the mass center of a rigid body. It states that *the sum of all the external forces acting on the body is equal to the body's mass times the acceleration of its mass center G.*

   For motion of the body in the x–y plane, the translational equation of motion may be written in the form of two independent scalar equations, namely,

$$\Sigma F_x = m(a_G)_x$$
$$\Sigma F_y = m(a_G)_y$$

*By doing this, the rotational equation of motion reduces to a rather simplified form. The more general case of body shape and loading is considered in Chapter 21.

Particle free-body diagram

(b)

Particle kinetic diagram

(c)

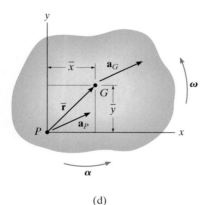

(d)

Fig. 17–8 (cont.)

Equation of Rotational Motion.   We will now determine the effects caused by the moments of the external force system computed about an axis perpendicular to the plane of motion (the z axis) and passing through point P. As shown on the free-body diagram of the ith particle, Fig. 17–8b, $\mathbf{F}_i$ represents the *resultant external force* acting on the particle, and $\mathbf{f}_i$ is the *resultant of the internal forces* caused by interactions with adjacent particles. If the particle has a mass m_i and at the instant considered its acceleration is $\mathbf{a}_i$, then the kinetic diagram is constructed as shown in Fig. 17–8c. If moments of the forces acting on the particle are summed about point P, we require

$$\mathbf{r} \times \mathbf{F}_i + \mathbf{r} \times \mathbf{f}_i = \mathbf{r} \times m_i \mathbf{a}_i$$

or

$$(\mathbf{M}_P)_i = \mathbf{r} \times m_i \mathbf{a}_i$$

The moments about P can be expressed in terms of the acceleration of point P, Fig. 17–8d. If the body has an angular acceleration $\boldsymbol{\alpha}$ and angular velocity $\boldsymbol{\omega}$, then using Eq. 16–18 we have

$$(\mathbf{M}_P)_i = m_i \mathbf{r} \times (\mathbf{a}_P + \boldsymbol{\alpha} \times \mathbf{r} - \omega^2 \mathbf{r})$$
$$= m_i[\mathbf{r} \times \mathbf{a}_P + \mathbf{r} \times (\boldsymbol{\alpha} \times \mathbf{r}) - \omega^2(\mathbf{r} \times \mathbf{r})]$$

The last term is zero, since $\mathbf{r} \times \mathbf{r} = \mathbf{0}$. Expressing the vectors with Cartesian components and carrying out the cross-product operations yields

$$(M_P)_i \mathbf{k} = m_i\{(x\mathbf{i} + y\mathbf{j}) \times [(a_P)_x \mathbf{i} + (a_P)_y \mathbf{j}]$$
$$+ (x\mathbf{i} + y\mathbf{j}) \times [\alpha\mathbf{k} \times (x\mathbf{i} + y\mathbf{j})]\}$$
$$(M_P)_i \mathbf{k} = m_i[-y(a_P)_x + x(a_P)_y + \alpha x^2 + \alpha y^2]\mathbf{k}$$
$$\zeta(M_P)_i = m_i[-y(a_P)_x + x(a_P)_y + \alpha r^2]$$

Letting $m_i \rightarrow dm$ and integrating with respect to the entire mass m of the body, we obtain the resultant moment equation

$$\zeta \Sigma M_P = -\left(\int_m y\, dm\right)(a_P)_x + \left(\int_m x\, dm\right)(a_P)_y + \left(\int_m r^2\, dm\right)\alpha$$

Here ΣM_P represents only the moment of the *external forces* acting on the body about point P. The resultant moment of the internal forces is zero, since for the entire body these forces occur in equal and opposite collinear pairs and thus the moment of each pair of forces about P cancels. The integrals in the first and second terms on the right are used to locate the body's center of mass G with respect to P, since $\bar{y}m = \int y\, dm$ and $\bar{x}m = \int x\, dm$, Fig. 17–8d. Also, the last integral represents the body's moment of inertia computed about the z axis, i.e., $I_P = \int r^2\, dm$. Thus,

$$\zeta \Sigma M_P = -\bar{y}m(a_P)_x + \bar{x}m(a_P)_y + I_P \alpha \qquad (17\text{–}6)$$

It is possible to reduce this equation to a simpler form if point P coincides with the mass center G for the body. If this is the case, then $\bar{x} = \bar{y} = 0$, and therefore*

$$\Sigma M_G = I_G \alpha \qquad (17\text{–}7)$$

This rotational equation of motion states that the sum of the moments of all the external forces computed about the body's mass center G is equal to the product of the moment of inertia of the body about an axis passing through G and the body's angular acceleration.

Equation 17–6 can also be rewritten in terms of the x and y components of $\mathbf{a}_G$ and the body's moment of inertia I_G. If point G is located at point $(\bar{x}, \bar{y})$, Fig. 17–8d, then by the parallel-axis theorem, $I_p = I_G + m(\bar{x}^2 + \bar{y}^2)$. Substituting into Eq. 17–6 and rearranging terms, we get

$$\downarrow \Sigma M_P = \bar{y}m[-(a_P)_x + \bar{y}\alpha] + \bar{x}m[(a_P)_y + \bar{x}\alpha] + I_G\alpha \qquad (17\text{–}8)$$

From the kinematic diagram of Fig. 17–8d, $\mathbf{a}_P$ can be expressed in terms of $\mathbf{a}_G$ as

$$\mathbf{a}_G = \mathbf{a}_P + \boldsymbol{\alpha} \times \bar{\mathbf{r}} - \omega^2 \bar{\mathbf{r}}$$

$$(a_G)_x \mathbf{i} + (a_G)_y \mathbf{j} = (a_P)_x \mathbf{i} + (a_P)_y \mathbf{j} + \alpha \mathbf{k} \times (\bar{x}\mathbf{i} + \bar{y}\mathbf{j}) - \omega^2(\bar{x}\mathbf{i} + \bar{y}\mathbf{j})$$

Carrying out the cross product and equating the respective $\mathbf{i}$ and $\mathbf{j}$ components yields the two scalar equations

$$(a_G)_x = (a_P)_x - \bar{y}\alpha - \bar{x}\omega^2$$
$$(a_G)_y = (a_P)_y + \bar{x}\alpha - \bar{y}\omega^2$$

From these equations, $[-(a_P)_x + \bar{y}\alpha] = [-(a_G)_x - \bar{x}\omega^2]$ and $[(a_P)_y + \bar{x}\alpha] = [(a_G)_y + \bar{y}\omega^2]$. Substituting these results into Eq. 17–8 and simplifying gives

$$\downarrow \Sigma M_P = -\bar{y}m(a_G)_x + \bar{x}m(a_G)_y + I_G\alpha \qquad (17\text{–}9)$$

This important result indicates that when moments of the external forces shown on the free-body diagram are summed about point P, Fig. 17–8e, they are equivalent to the sum of the "kinetic moments" of the components of $m\mathbf{a}_G$ about P plus the "kinetic moment" of $I_G\boldsymbol{\alpha}$, Fig. 17–8f. In other words, when the "kinetic moments," $\Sigma(\mathcal{M}_k)_P$, are computed, Fig. 17–8f, the vectors $m(\mathbf{a}_G)_x$ and $m(\mathbf{a}_G)_y$ are treated as sliding vectors; that is, they can act at *any point along their line of action*. In a similar manner, $I_G\boldsymbol{\alpha}$ can be treated as a *free vector* and can therefore act at *any point*. It is important to keep in mind that $m\mathbf{a}_G$ and $I_G\boldsymbol{\alpha}$ are not the same as a force or a couple moment. Instead, they are caused by the external effects of forces and couple moments acting on the body. With this in mind we can therefore write Eq. 17–9 in a more general form as

$$\Sigma M_P = \Sigma(\mathcal{M}_k)_P \qquad (17\text{–}10)$$

Free-body diagram

(e)

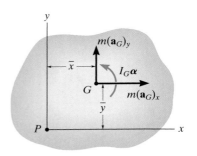

Kinetic diagram

(f)

Fig. 17–8 (cont.)

---

*It also reduces to this same simple form $\Sigma M_P = I_P\alpha$ if point P is a *fixed point* (see Eq. 17–16) or the acceleration of point P is directed along the line PG.

General Application of the Equations of Motion. To summarize this analysis, *three* independent scalar equations may be written to describe the general plane motion of a symmetrical rigid body.

$$\Sigma F_x = m(a_G)_x$$
$$\Sigma F_y = m(a_G)_y$$
$$\Sigma M_G = I_G\alpha \quad \text{or} \quad \Sigma M_P = \Sigma(\mathcal{M}_k)_P \qquad (17\text{–}11)$$

When applying these equations, one should *always* draw a free-body diagram, Fig. 17–8e, in order to account for the terms involved in ΣF_x, ΣF_y, ΣM_G, or ΣM_P. In some problems it may also be helpful to draw the *kinetic diagram* for the body. This diagram graphically accounts for the terms $m(\mathbf{a}_G)_x$, $m(\mathbf{a}_G)_y$, and $I_G\boldsymbol{\alpha}$, and it is especially convenient when used to determine the components of $m\mathbf{a}_G$ and the moment terms in $\Sigma(\mathcal{M}_k)_P$.*

17.3 Equations of Motion: Translation

When a rigid body undergoes a *translation*, Fig. 17–9a, all the particles of the body have the *same acceleration*, so that $\mathbf{a}_G = \mathbf{a}$. Furthermore, $\boldsymbol{\alpha} = \mathbf{0}$, in which case the rotational equation of motion applied at point G reduces to a simplified form, namely, $\Sigma M_G = 0$. Application of this and the translational equations of motion will now be discussed for each of the two types of translation.

Rectilinear Translation. When a body is subjected to *rectilinear translation*, all the particles of the body (slab) travel along parallel straight-line paths. The free-body and kinetic diagrams are shown in Fig. 17–9b. Since $I_G\boldsymbol{\alpha} = \mathbf{0}$, only $m\mathbf{a}_G$ is shown on the kinetic diagram. Hence, the equations of motion which apply in this case become

$$\boxed{\begin{aligned}\Sigma F_x &= m(a_G)_x \\ \Sigma F_y &= m(a_G)_y \\ \Sigma M_G &= 0\end{aligned}} \qquad (17\text{–}12)$$

The last equation requires that the sum of the moments of all the external forces (and couple moments) computed about the body's center of mass be equal to zero. It is possible, of course, to sum moments about other points on or off the body, in which case the moment of $m\mathbf{a}_G$ must

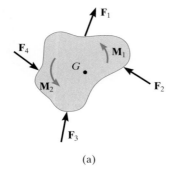

(a)

Fig. 17–9

---

*For this reason, the kinetic diagram will be used in the solution of an example problem whenever $\Sigma M_P = \Sigma(\mathcal{M}_k)_P$ is applied.

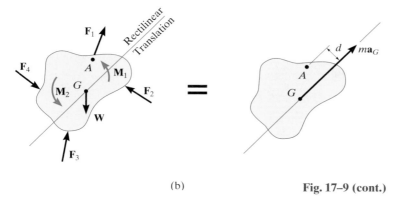

(b)   **Fig. 17–9 (cont.)**

be taken into account. For example, if point A is chosen, which lies at a perpendicular distance d from the line of action of $m\mathbf{a}_G$, the following moment equation applies:

$$\zeta + \Sigma M_A = \Sigma (\mathcal{M}_k)_A; \qquad \Sigma M_A = (ma_G)d$$

Here the sum of moments of the external forces and couple moments about A (ΣM_A, free-body diagram) equals the moment of $m\mathbf{a}_G$ about A ($\Sigma (\mathcal{M}_k)_A$, kinetic diagram).

Curvilinear Translation.

When a rigid body is subjected to *curvilinear translation*, all the particles of the body travel along *parallel curved paths*. For analysis, it is often convenient to use an inertial coordinate system having an origin which is coincident with the body's mass center at the instant considered, and axes which are oriented in the normal and tangential directions to the path of motion, Fig. 17–9c. The three scalar equations of motion are then

$$\boxed{\begin{aligned} \Sigma F_n &= m(a_G)_n \\ \Sigma F_t &= m(a_G)_t \\ \Sigma M_G &= 0 \end{aligned}}$$

(17–13)

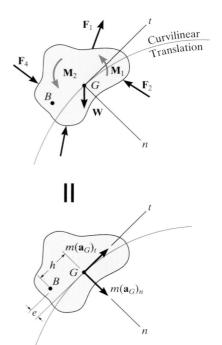

(c)

Here $(a_G)_t$ and $(a_G)_n$ represent, respectively, the magnitudes of the tangential and normal components of acceleration of point G.

If the moment equation $\Sigma M_G = 0$ is replaced by a moment summation about the arbitrary point B, Fig. 17–9c, it is necessary to account for the moments, $\Sigma (\mathcal{M}_k)_B$, of the two components $m(\mathbf{a}_G)_n$ and $m(\mathbf{a}_G)_t$ about this point. From the kinetic diagram, h and e represent the perpendicular distances (or "moment arms") from B to the lines of action of the components. The required moment equation therefore becomes

$$\zeta + \Sigma M_B = \Sigma (\mathcal{M}_k)_B; \qquad \Sigma M_B = e[m(a_G)_t] - h[m(a_G)_n]$$

The free-body and kinetic diagrams for this boat and trailer are drawn first in order to apply the equations of motion. Here the forces on the free-body diagram cause the effect shown on the kinetic diagram. If moments are summed about the mass center, G, then $\Sigma M_G = 0$. However, if moments are summed about point B then $\curvearrowright + \Sigma M_B = m a_G(d)$.

Procedure for Analysis

Kinetic problems involving rigid-body *translation* can be solved using the following procedure.

Free-Body Diagram.

- Establish the x, y or n, t inertial coordinate system and draw the free-body diagram in order to account for all the external forces and couple moments that act on the body.

- The direction and sense of the acceleration of the body's mass center $\mathbf{a}_G$ should be established.

- Identify the unknowns in the problem.

- If it is decided that the rotational equation of motion $\Sigma M_P = \Sigma (\mathcal{M}_k)_P$ is to be used in the solution, then consider drawing the kinetic diagram, since it graphically accounts for the components $m(\mathbf{a}_G)_x$, $m(\mathbf{a}_G)_y$ or $m(\mathbf{a}_G)_t$, $m(\mathbf{a}_G)_n$ and is therefore convenient for "visualizing" the terms needed in the moment sum $\Sigma (\mathcal{M}_k)_P$.

Equations of Motion.

- Apply the three equations of motion in accordance with the established sign convention.

- To simplify the analysis, the moment equation $\Sigma M_G = 0$ can be replaced by the more general equation $\Sigma M_P = \Sigma (\mathcal{M}_k)_P$, where point P is usually located at the intersection of the lines of action of as many unknown forces as possible.

- If the body is in contact with a *rough surface* and slipping occurs, use the frictional equation $F = \mu_k N$. Remember, $\mathbf{F}$ always acts on the body so as to oppose the motion of the body relative to the surface it contacts.

Kinematics.

- Use kinematics to determine the velocity and position of the body.

- For *rectilinear translation* with *variable acceleration*, use
$$a_G = dv_G/dt \quad a_G ds_G = v_G dv_G \quad v_G = ds_G/dt$$

- For *rectilinear translation* with *constant acceleration*, use
$$v_G = (v_G)_0 + a_G t \qquad v_G^2 = (v_G)_0^2 + 2a_G[s_G - (s_G)_0]$$
$$s_G = (s_G)_0 + (v_G)_0 t + \tfrac{1}{2} a_G t^2$$

- For *curvilinear translation*, use
$$(a_G)_n = v_G^2/\rho = \omega^2 \rho$$
$$(a_G)_t = dv_G/dt, \ (a_G)_t ds_G = v_G dv_{G,} \ (a_G)_t = \alpha \rho$$

EXAMPLE 17.5

The car shown in Fig. 17–10a has a mass of 2 Mg and a center of mass at G. Determine the car's acceleration if the "driving" wheels in the back are always slipping, whereas the front wheels freely rotate. Neglect the mass of the wheels. The coefficient of kinetic friction between the wheels and the road is $\mu_k = 0.25$.

(a)

SOLUTION I

Free-Body Diagram. As shown in Fig. 17–10b, the rear-wheel frictional force $\mathbf{F}_B$ pushes the car forward, and since *slipping occurs*, $F_B = 0.25N_B$. The frictional forces acting on the *front wheels* are *zero*, since these wheels have negligible mass.* There are three unknowns in the problem, N_A, N_B, and a_G. Here we will sum moments about the mass center. The car (point G) is assumed to accelerate to the left, i.e., in the negative x direction, Fig. 17–10b.

Equations of Motion.

$$\xrightarrow{+} \Sigma F_x = m(a_G)_x; \qquad -0.25N_B = -(2000 \text{ kg})a_G \qquad (1)$$

$$+\uparrow \Sigma F_y = m(a_G)_y; \qquad N_A + N_B - 2000(9.81) \text{ N} = 0 \qquad (2)$$

$$\zeta + \Sigma M_G = 0; \qquad -N_A(1.25 \text{ m}) - 0.25N_B(0.3 \text{ m}) + N_B(0.75 \text{ m}) = 0 \qquad (3)$$

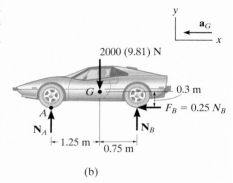

(b)

Solving,

$$a_G = 1.59 \text{ m/s}^2 \leftarrow$$

$$N_A = 6.88 \text{ kN}$$

$$N_B = 12.7 \text{ kN} \qquad\qquad\qquad Ans.$$

SOLUTION II

Free-Body and Kinetic Diagrams. If the "moment" equation is applied about point A, then the unknown N_A will be eliminated from the equation. To "visualize" the moment of $m\mathbf{a}_G$ about A, we will include the kinetic diagram as part of the analysis, Fig. 17–10c.

Equation of Motion. We require

$$\zeta + \Sigma M_A = \Sigma(\mathcal{M}_k)_A; \qquad N_B(2 \text{ m}) - 2000(9.81) \text{ N}(1.25 \text{ m}) =$$

$$(2000 \text{ kg})a_G(0.3 \text{ m})$$

Solving this and Eq. 1 for a_G leads to a simpler solution than that obtained from Eqs. 1 to 3.

(c)

Fig. 17–10

*With negligible wheel mass, $I\alpha = 0$ and the frictional force at A required to turn the wheel is zero. If the wheels' mass were included, then the problem solution for this case would be more involved, since a general-plane-motion analysis of the wheels would have to be considered (see Sec. 17.5).

EXAMPLE 17.6

The motorcycle shown in Fig. 17–11a has a mass of 125 kg and a center of mass at G_1, while the rider has a mass of 75 kg and a center of mass at G_2. Determine the minimum coefficient of static friction between the wheels and the pavement in order for the rider to do a "wheely," i.e., lift the front wheel off the ground as shown in the photo. What acceleration is necessary to do this? Neglect the mass of the wheels and assume that the front wheel is free to roll.

(a)

Fig. 17–11

SOLUTION

Free-Body and Kinetic Diagrams.   In this problem we will consider both the motorcycle and the rider as the "system" to be analyzed. It is possible first to determine the location of the center of mass for this "system" by using the equations $\bar{x} = \Sigma \tilde{x} m / \Sigma m$ and $\bar{y} = \Sigma \tilde{y} m / \Sigma m$. Here, however, we will consider the separate weight and mass of each of its *component parts* as shown on the free-body and kinetic diagrams, Fig. 17–11b. Both parts move with the *same* acceleration and we have assumed that the front wheel is *about* to leave the ground, so that the normal reaction $N_A \approx 0$. The three unknowns in the problem are N_B, F_B, and a_G.

(b)

Equations of Motion.

$$\xrightarrow{+} \Sigma F_x = m(a_G)_x; \qquad F_B = (75 \text{ kg} + 125 \text{ kg})a_G \qquad (1)$$

$$+\uparrow \Sigma F_y = m(a_G)_y; \qquad N_B - 735.75 \text{ N} - 1226.25 \text{ N} = 0 \qquad (2)$$

$$\zeta+\Sigma M_B = \Sigma(\mathcal{M}_k)_B; \quad -(735.75 \text{ N})(0.4 \text{ m}) - (1226.25 \text{ N})(0.8 \text{ m}) =$$
$$-(75 \text{ kg } a_G)(0.9 \text{ m}) - (125 \text{ kg } a_G)(0.6 \text{ m})$$

Solving,

$$a_G = 8.95 \text{ m/s}^2 \rightarrow \qquad \qquad Ans.$$
$$N_B = 1962 \text{ N}$$
$$F_B = 1790 \text{ N}$$

Thus the minimum coefficient of static friction is

$$(\mu_s)_{\min} = \frac{F_B}{N_B} = \frac{1790 \text{ N}}{1962 \text{ N}} = 0.912 \qquad \qquad Ans.$$

EXAMPLE 17.7

A uniform 50-kg crate rests on a horizontal surface for which the coefficient of kinetic friction is $\mu_k = 0.2$. Determine the crate's acceleration if a force of $P = 600$ N is applied to the crate as shown in Fig. 17–12a.

(a)

Fig. 17–12

SOLUTION

Free-Body Diagram.   The force **P** can cause the crate either to slide or to tip over. As shown in Fig. 17–12b, it is assumed that the crate slides, so that $F = \mu_k N_C = 0.2 N_C$. Also, the resultant normal force $\mathbf{N}_C$ acts at O, a distance x (where $0 < x \leq 0.5$ m) from the crate's center line.* The three unknowns are N_C, x, and a_G.

(b)

Equations of Motion.

$$\xrightarrow{+} \Sigma F_x = m(a_G)_x; \qquad 600 \text{ N} - 0.2 N_C = (50 \text{ kg}) a_G \qquad (1)$$

$$+\uparrow \Sigma F_y = m(a_G)_y; \qquad N_C - 490.5 \text{ N} = 0 \qquad (2)$$

$$\zeta + \Sigma M_G = 0; \quad -600 \text{ N}(0.3 \text{ m}) + N_C(x) - 0.2 N_C(0.5 \text{ m}) = 0 \quad (3)$$

Solving, we obtain

$$N_C = 490 \text{ N}$$

$$x = 0.467 \text{ m}$$

$$a_G = 10.0 \text{ m/s}^2 \rightarrow \qquad \qquad Ans.$$

Since $x = 0.467$ m < 0.5 m, indeed the crate slides as originally assumed.

NOTE: If the solution had given a value of $x > 0.5$ m, the problem would have to be reworked with the assumption that tipping occurred. If this were the case, $\mathbf{N}_C$ would act at the *corner point A* and $F \leq 0.2 N_C$.

---

*The line of action of $\mathbf{N}_C$ does not necessarily pass through the mass center G ($x = 0$), since $\mathbf{N}_C$ must counteract the tendency for tipping caused by **P**. See Sec. 8.1 of *Engineering Mechanics: Statics*.

EXAMPLE 17.8

(a)

Fig. 17–13

The 100-kg beam *BD* shown in Fig. 17–13a is supported by two rods having negligible mass. Determine the force created in each rod if at the instant $\theta = 30°$ and $\omega = 6$ rad/s.

SOLUTION

Free-Body Diagram.   The beam moves with *curvilinear translation* since points *B* and *D* and the center of mass *G* all move along circular paths, each path having the same radius of 0.5 m. Using normal and tangential coordinates, the free-body diagram for the beam is shown in Fig. 17–13b. Because of the *translation*, *G* has the *same* motion as the pin at *B*, which is connected to both the rod and the beam. By studying the angular motion of rod *AB*, Fig. 17–13c, note that the tangential component of acceleration acts downward to the left due to the clockwise direction of $\boldsymbol{\alpha}$. Furthermore, the normal component of acceleration is *always* directed toward the center of curvature (toward point *A* for rod *AB*). Since the angular velocity of *AB* is 6 rad/s, then

$$(a_G)_n = \omega^2 r = (6 \text{ rad/s})^2(0.5 \text{ m}) = 18 \text{ m/s}^2$$

The three unknowns are T_B, T_D, and $(a_G)_t$. The directions of $(\mathbf{a}_G)_n$ and $(\mathbf{a}_G)_t$ have been established, and are indicated on the coordinate axes.

(b)

(c)

Equations of Motion.

$$+\nwarrow \Sigma F_n = m(a_G)_n; \quad T_B + T_D - 981 \cos 30° \text{ N} = 100 \text{ kg}(18 \text{ m/s}^2) \quad (1)$$

$$+\swarrow \Sigma F_t = m(a_G)_t; \qquad 981 \sin 30° = 100 \text{ kg}(a_G)_t \qquad\qquad (2)$$

$$\zeta + \Sigma M_G = 0; \quad -(T_B \cos 30°)(0.4 \text{ m}) + (T_D \cos 30°)(0.4 \text{ m}) = 0 \quad (3)$$

Simultaneous solution of these three equations gives

$$T_B = T_D = 1.32 \text{ kN} \quad \nwarrow 30°$$

$$(a_G)_t = 4.90 \text{ m/s}^2 \qquad\qquad\qquad\qquad Ans.$$

NOTE: It is also possible to apply the equations of motion along horizontal and vertical x, y axes, but the solution becomes more involved.

PROBLEMS

*17–24. Determine the greatest possible acceleration of the 975-kg race car so that its front tires do not leave the ground or the tires slip on the track. The coefficients of static and kinetic friction are $\mu_s = 0.8$ and $\mu_k = 0.6$, respectively. Neglect the mass of the tires. The car has rear-wheel drive and the front tires are free to roll.

17–25. Determine the greatest possible acceleration of the 975-kg race car so that its front wheels do not leave the ground or the tires slip on the track. The coefficients of static and kinetic friction are $\mu_s = 0.8$ and $\mu_k = 0.6$, respectively. Neglect the mass of the tires. The car has four-wheel drive.

Probs. 17–24/25

17–26. The 2-lb bottle rests on the check-out conveyor at a grocery store. If the coefficient of static friction is $\mu_s = 0.2$, determine the largest acceleration the conveyor can have without causing the bottle to slip or tip. The center of gravity is at G.

Prob. 17–26

17–27. The assembly has a mass of 8 Mg and is hoisted using the boom and pulley system. If the winch at B draws in the cable with an acceleration of 2 m/s^2, determine the compressive force in the hydraulic cylinder needed to support the boom. The boom has a mass of 2 Mg and mass center at G.

Prob. 17–27

*17–28. The jet aircraft has a total mass of 22 Mg and a center of mass at G. Initially at take-off the engines provide a thrust $2T = 4$ kN and $T' = 1.5$ kN. Determine the acceleration of the plane and the normal reactions on the nose wheel and each of the *two* wing wheels located at B. Neglect the mass of the wheels and, due to low velocity, neglect any lift caused by the wings.

Prob. 17–28

17–29. The lift truck has a mass of 70 kg and mass center at *G*. If it lifts the 120-kg spool with an acceleration of 3 m/s², determine the reactions of each of the four wheels on the ground. The loading is symmetric. Neglect the mass of the movable arm *CD*.

17–30. The lift truck has a mass of 70 kg and mass center at *G*. Determine the largest upward acceleration of the 120-kg spool so that no reaction of the wheels on the ground exceeds 600 N.

Probs. 17–29/30

17–31. The door has a weight of 200 lb and a center of gravity at *G*. Determine how far the door moves in 2 s, starting from rest, if a man pushes on it at *C* with a horizontal force *F* = 30 lb. Also, find the vertical reactions at the rollers *A* and *B*.

*17–32. The door has a weight of 200 lb and a center of gravity at *G*. Determine the constant force *F* that must be applied to the door to push it open 12 ft to the right in 5 s, starting from rest. Also, find the vertical reactions at the rollers *A* and *B*.

Probs. 17–31/32

17–33. The fork lift has a boom with a mass of 800 kg and a mass center at *G*. If the vertical acceleration of the boom is 4 m/s², determine the horizontal and vertical reactions at the pin *A* and on the short link *BC* when the 1.25-Mg load is lifted.

Prob. 17–33

17–34. The pipe has a mass of 800 kg and is being towed behind the truck. If the acceleration of the truck is $a_t = 0.5$ m/s², determine the angle θ and the tension in the cable. The coefficient of kinetic friction between the pipe and the ground is $\mu_k = 0.1$.

17–35. The pipe has a mass of 800 kg and is being towed behind a truck. If the angle $\theta = 30°$, determine the acceleration of the truck and the tension in the cable. The coefficient of kinetic friction between the pipe and the ground is $\mu_k = 0.1$.

Probs. 17–34/35

*17–36. The pipe has a mass of 460 kg and is held in place on the truck bed using the two boards A and B. Determine the greatest acceleration of the truck so that the pipe begins to lose contact at A and the bed of the truck and starts to pivot about B. Assume board B will not slip on the bed of the truck, and the pipe is smooth. Also, what force does board B exert on the pipe during the acceleration?

17–38. The sports car has a mass of 1.5 Mg and a center of mass at G. Determine the shortest time it takes for it to reach a speed of 80 km/h, starting from rest, if the engine only drives the rear wheels, whereas the front wheels are free rolling. The coefficient of static friction between the wheels and the road is $\mu_s = 0.2$. Neglect the mass of the wheels for the calculation. If driving power could be supplied to all four wheels, what would be the shortest time for the car to reach a speed of 80 km/h?

Prob. 17–36

Prob. 17–38

17–37. The drop gate at the end of the trailer has a mass of 1.25 Mg and mass center at G. If it is supported by the cable AB and hinge at C, determine the tension in the cable when the truck begins to accelerate at 5 m/s². Also, what are the horizontal and vertical components of reaction at the hinge C?

17–39. The crate of mass m is supported on a cart of negligible mass. Determine the maximum force P that can be applied a distance d from the cart bottom without causing the crate to tip on the cart.

Prob. 17–37

Prob. 17–39

*17–40. The car accelerates uniformly from rest to 88 ft/s in 15 seconds. If it has a weight of 3800 lb and a center of gravity at G, determine the normal reaction of *each wheel* on the pavement during the motion. Power is developed at the rear wheels, whereas the front wheels are free to roll. Neglect the mass of the wheels and take the coefficients of static and kinetic friction to be $\mu_s = 0.4$ and $\mu_k = 0.2$, respectively.

17–42. The 1.6-Mg car shown has been "raked" by increasing the height of its center of mass to $h = 0.2$ m. This was done by raising the springs on the rear axle. If the coefficient of kinetic friction between the rear wheels and the ground is $\mu_k = 0.3$, show that the car can accelerate slightly faster than its counterpart for which $h = 0$. Neglect the mass of the wheels and driver and assume the front wheels at B are free to roll while the rear wheels slip.

Prob. 17–40

Prob. 17–42

17–41. Block A weighs 50 lb and the platform weighs 10 lb. If $P = 100$ lb, determine the normal force exerted by block A on B. Neglect the weight of the pulleys and bars of the triangular frame.

17–43. The forklift and operator have a combined weight of 10 000 lb and center of mass at G. If the forklift is used to lift the 2000-lb concrete pipe, determine the maximum vertical acceleration it can give to the pipe so that it does not tip forward on its front wheels.

*17–44. The forklift and operator have a combined weight of 10 000 lb and center of mass at G. If the forklift is used to lift the 2000-lb concrete pipe, determine the normal reactions on each of its four wheels if the pipe is given an upward acceleration of 4 ft/s^2.

Prob. 17–41

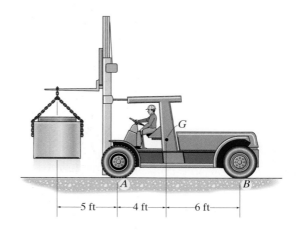

Probs. 17–43/44

17–45. The van has a weight of 4500 lb and center of gravity at G_v. It carries a fixed 800-lb load which has a center of gravity at G_l. If the van is traveling at 40 ft/s, determine the distance it skids before stopping. The brakes cause *all* the wheels to lock or skid. The coefficient of kinetic friction between the wheels and the pavement is $\mu_k = 0.3$. Compare this distance with that of the van being empty. Neglect the mass of the wheels.

Prob. 17–45

17–46. The "muscle car" is designed to do a "wheeley," i.e., to be able to lift its front wheels off the ground in the manner shown when it accelerates. If the 1.35-Mg car has a center of mass at G, determine the minimum torque that must be developed at both rear wheels in order to do this. Also, what is the smallest necessary coefficient of static friction assuming the thick-walled rear wheels do not slip on the pavement? Neglect the mass of the wheels.

Prob. 17–46

17–47. The bicycle and rider have a mass of 80 kg with center of mass located at G. If the coefficient of kinetic friction at the rear tire is $\mu_B = 0.8$, determine the normal reactions at the tires A and B, and the deceleration of the rider, when the rear wheel locks for braking. What is the normal reaction at the rear wheel when the bicycle is traveling at constant velocity and the brakes are not applied? Neglect the mass of the wheels.

*17–48. The bicycle and rider have a mass of 80 kg with center of mass located at G. Determine the minimum coefficient of kinetic friction between the road and the wheels so that the rear wheel B starts to lift off the ground when the rider applies the brakes to the front wheel. Neglect the mass of the wheels.

Probs. 17–47/48

17–49. The dresser has a weight of 80 lb and is pushed along the floor. If the coefficient of static friction at A and B is $\mu_s = 0.3$ and the coefficient of kinetic friction is $\mu_k = 0.2$, determine the smallest horizontal force P needed to cause motion. If this force is increased slightly, determine the acceleration of the dresser. Also, what are the normal reactions at A and B when it begins to move?

17–50. The dresser has a weight of 80 lb and is pushed along the floor. If the coefficient of static friction at A and B is $\mu_s = 0.3$ and the coefficient of kinetic friction is $\mu_k = 0.2$, determine the maximum horizontal force P that can be applied without causing the dresser to tip over.

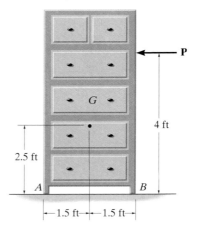

Probs. 17–49/50

17–51. The crate C has a weight of 150 lb and rests on the truck elevator for which the coefficient of static friction is $\mu_s = 0.4$. Determine the largest initial angular acceleration α, starting from rest, which the parallel links AB and DE can have without causing the crate to slip. No tipping occurs.

***17–52.** The two 3-lb rods EF and HI are fixed (welded) to the link AC at E. Determine the normal force N_E, shear force V_E, and moment M_E, which the bar AC exerts on FE at E if at the instant $\theta = 30°$ link AB has an angular velocity $\omega = 5$ rad/s and an angular acceleration $\alpha = 8$ rad/s^2 as shown.

Prob. 17–51

Prob. 17–52

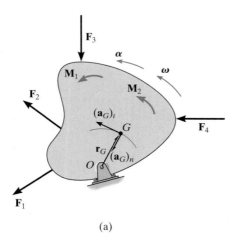

(a)

Fig. 17–14

17.4 Equations of Motion: Rotation about a Fixed Axis

Consider the rigid body (or slab) shown in Fig. 17–14a, which is constrained to rotate in the vertical plane about a fixed axis perpendicular to the page and passing through the pin at O. The angular velocity and angular acceleration are caused by the external force and couple moment system acting on the body. Because the body's center of mass G moves in a *circular path*, the acceleration of this point is represented by its tangential and normal components. The *tangential component of acceleration* has a *magnitude* of $(a_G)_t = \alpha r_G$ and must act in a *direction* which is *consistent* with the body's angular acceleration α. The *magnitude* of the *normal component of acceleration* is $(a_G)_n = \omega^2 r_G$. This component is *always directed* from point G to O, regardless of the direction of ω.

The free-body and kinetic diagrams for the body are shown in Fig. 17–14b. The weight of the body, $W = mg$, and the pin reaction $\mathbf{F}_O$ are included on the free-body diagram since they represent external forces acting on the body. The two components $m(\mathbf{a}_G)_t$ and $m(\mathbf{a}_G)_n$, shown on the kinetic diagram, are associated with the tangential and normal acceleration components of the body's mass center. These vectors act in the same *direction* as the acceleration components and have *magnitudes* of $m(a_G)_t$ and $m(a_G)_n$. The $I_G\boldsymbol{\alpha}$ vector acts in the same *direction* as $\boldsymbol{\alpha}$ and has a *magnitude* of $I_G\alpha$, where I_G is the body's moment of inertia calculated about an axis which is perpendicular to the page and passing through G. From the derivation given in Sec. 17.2, the equations of motion which apply to the body may be written in the form

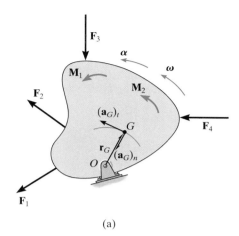

(a)

$$\Sigma F_n = m(a_G)_n = m\omega^2 r_G$$
$$\Sigma F_t = m(a_G)_t = m\alpha r_G \qquad (17\text{--}14)$$
$$\Sigma M_G = I_G\alpha$$

The moment equation may be replaced by a moment summation about any arbitrary point P on or off the body provided one accounts for the moments $\Sigma(\mathcal{M}_k)_P$ produced by $I_G\boldsymbol{\alpha}$, $m(\mathbf{a}_G)_t$, and $m(\mathbf{a}_G)_n$ about the point. Often it is convenient to sum moments about the pin at O in order to eliminate the *unknown* force $\mathbf{F}_O$. From the kinetic diagram, Fig. 17–14b, this requires

$$\zeta + \Sigma M_O = \Sigma(\mathcal{M}_k)_O; \qquad \Sigma M_O = r_G m(a_G)_t + I_G\alpha \qquad (17\text{--}15)$$

Note that the moment of $m(\mathbf{a}_G)_n$ is not included in the summation since the line of action of this vector passes through O. Substituting $(a_G)_t = r_G\alpha$, we may rewrite the above equation as $\zeta + \Sigma M_O = (I_G + mr_G^2)\alpha$. From the parallel-axis theorem, $I_O = I_G + md^2$, and therefore the term in parentheses represents the *moment of inertia of the body about the fixed axis of rotation passing through O.* Consequently, we can write the three equations of motion for the body as

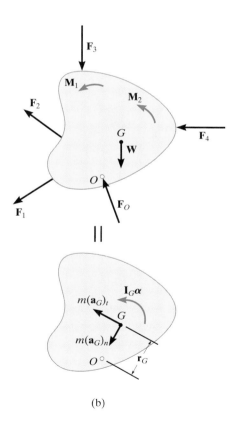

$$\Sigma F_n = m(a_G)_n = m\omega^2 r_G$$
$$\Sigma F_t = m(a_G)_t = m\alpha r_G \qquad (17\text{--}16)$$
$$\Sigma M_O = I_O\alpha$$

When using these equations, remember that "$I_O\alpha$" accounts for the "moment" of *both* $m(\mathbf{a}_G)_t$ and $I_G\boldsymbol{\alpha}$ about point O, Fig. 17–14b. In other words, $\Sigma M_O = \Sigma(\mathcal{M}_k)_O = I_O\alpha$, as indicated by Eqs. 17–15 and 17–16.

*The result $\Sigma M_O = I_O\alpha$ can also be obtained *directly* from Eq. 17–6 by selecting point P to coincide with O, realizing that $(a_P)_x = (a_P)_y = 0$.

(b)

Fig. 17–14 (cont.)

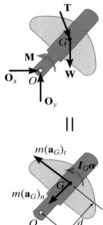

The crank on the oil-pumping rig undergoes rotation about a fixed axis which is caused by a driving torque **M** of the motor. The loadings shown on the free-body diagram cause the effects shown on the kinetic diagram. If moments are summed about the mass center, G, then $\Sigma M_G = I_G \alpha$. However, if moments are summed about point O, noting that $(a_G)_t = \alpha d$, then $\zeta + \Sigma M_O = I_G \alpha + m(a_G)_t d + m(a_G)_n(0) = (I_G + md^2)\alpha = I_O \alpha$.

Procedure for Analysis

Kinetic problems which involve the rotation of a body about a fixed axis can be solved using the following procedure.

Free-Body Diagram.

- Establish the inertial x, y or n, t coordinate system and specify the direction and sense of the accelerations $(\mathbf{a}_G)_n$ and $(\mathbf{a}_G)_t$ and the angular acceleration $\boldsymbol{\alpha}$ of the body. Recall that $(\mathbf{a}_G)_t$ must act in a direction which is in accordance with $\boldsymbol{\alpha}$, whereas $(\mathbf{a}_G)_n$ always acts toward the axis of rotation, point O.

- Draw the free-body diagram to account for all the external forces and couple moments that act on the body.

- Compute the moment of inertia I_G or I_O.

- Identify the unknowns in the problem.

- If it is decided that the rotational equation of motion $\Sigma M_P = \Sigma(\mathcal{M}_k)_P$ is to be used, i.e., P is a point other than G or O, then consider drawing the kinetic diagram in order to help "visualize" the "moments" developed by the components $m(\mathbf{a}_G)_n$, $m(\mathbf{a}_G)_t$, and $I_G \boldsymbol{\alpha}$ when writing the terms for the moment sum $\Sigma(\mathcal{M}_k)_P$.

Equations of Motion.

- Apply the three equations of motion in accordance with the established sign convention.

- If moments are summed about the body's mass center, G, then $\Sigma M_G = I_G \alpha$, since $(m\mathbf{a}_G)_t$ and $(m\mathbf{a}_G)_n$ create no moment about G.

- If moments are summed about the pin support O on the axis of rotation, then $(m\mathbf{a}_G)_n$ creates no moment about G, and it can be shown that $\Sigma M_O = I_O \alpha$.

Kinematics.

- Use kinematics if a complete solution cannot be obtained strictly from the equations of motion.

- If the *angular acceleration is variable*, use

$$\alpha = \frac{d\omega}{dt} \qquad \alpha \, d\theta = \omega \, d\omega \qquad \omega = \frac{d\theta}{dt}$$

- If the *angular acceleration is constant*, use

$$\omega = \omega_0 + \alpha_c t$$
$$\theta = \theta_0 + \omega_0 t + \tfrac{1}{2}\alpha_c t^2$$
$$\omega^2 = \omega_0^2 + 2\alpha_c(\theta - \theta_0)$$

EXAMPLE 17.9

The 30-kg uniform disk shown in Fig. 17–15a is pin supported at its center. If it starts from rest, determine the number of revolutions it must make to attain an angular velocity of 20 rad/s. Also, what are the reactions at the pin? The disk is acted upon by a constant force $F = 10$ N, which is applied to a cord wrapped around its periphery, and a constant couple moment $M = 5$ N·m. Neglect the mass of the cord in the calculation.

(a)

SOLUTION

Free-Body Diagram. Fig. 17–15b. Note that the mass center is not subjected to an acceleration; however, the disk has a clockwise angular acceleration.

The moment of inertia of the disk about the pin is

$$I_O = \tfrac{1}{2}mr^2 = \frac{1}{2}(30 \text{ kg})(0.2 \text{ m})^2 = 0.6 \text{ kg} \cdot \text{m}^2$$

The three unknowns are O_x, O_y, and α.

(b)

Fig 17–15

Equations of Motion.

$$\xrightarrow{+} \Sigma F_x = m(a_G)_x; \qquad O_x = 0 \qquad\qquad Ans.$$
$$+\uparrow \Sigma F_y = m(a_G)_y; \qquad O_y - 294.3 \text{ N} - 10 \text{ N} = 0$$
$$O_y = 304 \text{ N} \qquad\qquad Ans.$$
$$\zeta + \Sigma M_O = I_O \alpha; \qquad -10 \text{ N}(0.2 \text{ m}) - 5 \text{ N} \cdot \text{m} = -(0.6 \text{ kg} \cdot \text{m}^2)\alpha$$
$$\alpha = 11.7 \text{ rad/s}^2 \; \zeta$$

Kinematics. Since α is constant and is clockwise, the number of radians the disk must turn to obtain a clockwise angular velocity of 20 rad/s is

$$\zeta +$$
$$\omega^2 = \omega_0^2 + 2\alpha_c(\theta - \theta_0)$$
$$(-20 \text{ rad/s})^2 = 0 + 2(-11.7 \text{ rad/s}^2)(\theta - 0)$$
$$\theta = -17.1 \text{ rad} = 17.1 \text{ rad} \; \zeta$$

Hence,

$$\theta = 17.1 \text{ rad}\left(\frac{1 \text{ rev}}{2\pi \text{ rad}}\right) = 2.73 \text{ rev} \; \zeta \qquad\qquad Ans.$$

EXAMPLE 17.10

The 20-kg slender rod shown in Fig. 17–16a is rotating in the vertical plane, and at the instant shown it has an angular velocity of $\omega = 5$ rad/s. Determine the rod's angular acceleration and the horizontal and vertical components of reaction at the pin at this instant.

(a)

SOLUTION

Free-Body and Kinetic Diagrams. Fig. 17–16b. As shown on the kinetic diagram, point G moves in a circular path and so has two components of acceleration. It is important that the tangential component $a_t = \alpha r_G$ act downward since it must be in accordance with the angular acceleration $\boldsymbol{\alpha}$ of the rod. The three unknowns are O_n, O_t, and α.

(b)

Fig. 17–16

Equations of Motion.

$$\xleftarrow{+} \Sigma F_n = m\omega^2 r_G; \qquad O_n = (20\ \text{kg})(5\ \text{rad/s})^2(1.5\ \text{m})$$

$$+\downarrow \Sigma F_t = m\alpha r_G; \qquad -O_t + 20(9.81)\ \text{N} = (20\ \text{kg})(\alpha)(1.5\ \text{m})$$

$$\zeta + \Sigma M_G = I_G \alpha; \quad O_t(1.5\ \text{m}) + 60\ \text{N} \cdot \text{m} = \left[\tfrac{1}{12}(20\ \text{kg})(3\ \text{m})^2\right]\alpha$$

Solving

$$O_n = 750\ \text{N} \qquad O_t = 19.0\ \text{N} \qquad \alpha = 5.90\ \text{rad/s}^2 \qquad Ans.$$

A more direct solution to this problem would be to sum moments about point O to eliminate $\mathbf{O}_n$ and $\mathbf{O}_t$ and obtain a *direct solution* for α. Here,

$$\zeta + \Sigma M_O = \Sigma(\mathcal{M}_k)_O; \quad 60\ \text{N} \cdot \text{m} + 20(9.81)\ \text{N}(1.5\ \text{m}) =$$

$$\left[\tfrac{1}{12}(20\ \text{kg})(3\ \text{m})^2\right]\alpha + [20\ \text{kg}(\alpha)(1.5\ \text{m})](1.5\ \text{m})$$

$$\alpha = 5.90\ \text{rad/s}^2 \qquad Ans.$$

Also, since $I_O = \tfrac{1}{3}ml^2$ for a slender rod, we can apply

$$\zeta + \Sigma M_O = I_O \alpha; \quad 60\ \text{N} \cdot \text{m} + 20(9.81)\ \text{N}(1.5\ \text{m}) = \left[\tfrac{1}{3}(20\ \text{kg})(3\ \text{m})^2\right]\alpha$$

$$\alpha = 5.90\ \text{rad/s}^2 \qquad Ans.$$

NOTE: By comparison, the last equation provides the simplest solution for α and *does not* require use of the kinetic diagram.

EXAMPLE 17.11

The drum shown in Fig. 17–17a has a mass of 60 kg and a radius of gyration $k_O = 0.25$ m. A cord of negligible mass is wrapped around the periphery of the drum and attached to a block having a mass of 20 kg. If the block is released, determine the drum's angular acceleration.

(a)

SOLUTION I

Free-Body Diagram.   Here we will consider the drum and block separately, Fig. 17–17b. Assuming the block accelerates *downward* at **a**, it creates a *counterclockwise* angular acceleration α of the drum.
   The moment of inertia of the drum is

$$I_O = mk_O^2 = (60 \text{ kg})(0.25 \text{ m})^2 = 3.75 \text{ kg} \cdot \text{m}^2$$

There are five unknowns, namely O_x, O_y, T, a, and α.

Equations of Motion.   Applying the translational equations of motion $\Sigma F_x = m(a_G)_x$ and $\Sigma F_y = m(a_G)_y$ to the drum is of no consequence to the solution, since these equations involve the unknowns O_x and O_y. Thus, for the drum and block, respectively,

$$\zeta + \Sigma M_O = I_O\alpha; \qquad T(0.4 \text{ m}) = (3.75 \text{ kg} \cdot \text{m}^2)\alpha \qquad (1)$$

$$+\uparrow\Sigma F_y = m(a_G)_y; \qquad -20(9.81) \text{ N} + T = -20a \qquad (2)$$

Kinematics.   Since the point of contact A between the cord and drum has a tangential component of acceleration **a**, Fig. 17–17a, then

$$\zeta + a = \alpha r; \qquad a = \alpha(0.4) \qquad (3)$$

Solving the above equations,

$$T = 106 \text{ N}$$

$$a = 4.52 \text{ m/s}^2$$

$$\alpha = 11.3 \text{ rad/s}^2 \, \zeta \qquad\qquad Ans.$$

(b)

SOLUTION II

Free-Body and Kinetic Diagrams.   The cable tension T can be eliminated from the analysis by considering the drum and block as a *single system*, Fig. 17–17c. The kinetic diagram is shown since moments will be summed about point O.

Equations of Motion.   Using Eq. 3 and applying the moment equation about O to eliminate the unknowns O_x and O_y, we have

$$\zeta + \Sigma M_O = \Sigma(\mathcal{M}_k)_O; \quad 20(9.81) \text{ N}(0.4 \text{ m}) =$$
$$(3.75 \text{ kg} \cdot \text{m}^2)\alpha + [20 \text{ kg}(0.4 \text{ m } \alpha)](0.4 \text{ m})$$
$$\alpha = 11.3 \text{ rad/s}^2 \, \zeta \qquad\qquad Ans.$$

NOTE: If the block were *removed* and a force of 20(9.81) N were applied to the cord, show that $\alpha = 20.9$ rad/s^2 and explain the reason for the difference in the results.

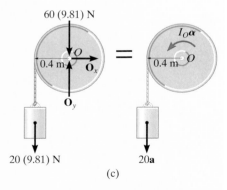

(c)

Fig. 17–17

EXAMPLE 17.12

||

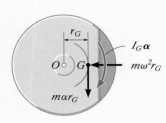

(b)

Fig. 17–18

The unbalanced 50-lb flywheel shown in Fig. 17–18a has a radius of gyration of $k_G = 0.6$ ft about an axis passing through its mass center G. If it has a clockwise angular velocity of 8 rad/s at the instant shown, determine the horizontal and vertical components of reaction at the pin O.

SOLUTION

Free-Body and Kinetic Diagrams.   Since G moves in a circular path, it will have both normal and tangential components of acceleration. Also, since α, which is caused by the flywheel's weight, acts clockwise, the tangential component of acceleration will act downward. Why? The vectors $m(a_G)_t = m\alpha r_G$, $m(a_G)_n = m\omega^2 r_G$, and $I_G\alpha$ are shown on the kinematic diagram in Fig. 17–18b. Here, the moment of inertia of the flywheel about its mass center is determined from the radius of gyration and the flywheel's mass; i.e., $I_G = mk_G^2 = (50 \text{ lb}/32.2 \text{ ft/s}^2)(0.6 \text{ ft})^2 = 0.559 \text{ slug} \cdot \text{ft}^2$.

The three unknowns are O_n, O_t, and α.

Equations of Motion.

$$\xleftrightarrow{} \Sigma F_n = m\omega^2 r_G; \qquad O_n = \left(\frac{50 \text{ lb}}{32.2 \text{ ft/s}^2}\right)(8 \text{ rad/s})^2(0.5 \text{ ft}) \qquad (1)$$

$$+\downarrow \Sigma F_t = m\alpha r_G; \qquad -O_t + 50 \text{ lb} = \left(\frac{50 \text{ lb}}{32.2 \text{ ft/s}^2}\right)(\alpha)(0.5 \text{ ft}) \qquad (2)$$

$$\zeta + \Sigma M_G = I_G\alpha; \qquad 80 \text{ lb} \cdot \text{ft} + O_t(0.5 \text{ ft}) = (0.559 \text{ slug} \cdot \text{ft}^2)\alpha \qquad (3)$$

Solving,

$$\alpha = 111 \text{ rad/s}^2 \qquad O_n = 49.7 \text{ lb} \qquad O_t = -36.1 \text{ lb} \qquad \textit{Ans.}$$

Moments can also be summed about point O in order to eliminate $\mathbf{O}_n$ and $\mathbf{O}_t$ and thereby obtain a *direct solution* for $\boldsymbol{\alpha}$, Fig. 17–18b. This can be done in one of *two* ways, i.e., by using either $\Sigma M_O = \Sigma(\mathcal{M}_k)_O$ or $\Sigma M_O = I_O\alpha$. If the first of these equations is applied, we have

$$\zeta + \Sigma M_O = \Sigma(\mathcal{M}_k)_O; \qquad 80 \text{ lb} \cdot \text{ft} + 50 \text{ lb}(0.5 \text{ ft}) =$$

$$(0.559 \text{ slug} \cdot \text{ft}^2)\alpha + \left[\left(\frac{50 \text{ lb}}{32.2 \text{ ft/s}^2}\right)\alpha(0.5 \text{ ft})\right](0.5 \text{ ft})$$

$$105 = 0.947\alpha \qquad (4)$$

NOTE: If $\Sigma M_O = I_O\alpha$ is applied, then by the parallel-axis theorem the moment of inertia of the flywheel about O is

$$I_O = I_G + mr_G^2 = 0.559 + \left(\frac{50}{32.2}\right)(0.5)^2 = 0.947 \text{ slug} \cdot \text{ft}^2$$

Hence, from the free-body diagram, Fig. 17–18b, we require

$$\zeta + \Sigma M_O = I_O\alpha; \qquad 80 \text{ lb} \cdot \text{ft} + 50 \text{ lb}(0.5 \text{ ft}) = (0.947 \text{ slug} \cdot \text{ft}^2)\alpha$$

which is the same as Eq. 4. Solving for α and substituting into Eq. 2 yields the answer for O_t obtained previously.

EXAMPLE 17.13

The slender rod shown in Fig. 17–19a has a mass m and length l and is released from rest when $\theta = 0°$. Determine the horizontal and vertical components of force which the pin at A exerts on the rod at the instant $\theta = 90°$.

(a)

SOLUTION

Free-Body Diagram.   The free-body diagram for the rod is shown when the rod is in the general position θ, Fig. 17–19b. For convenience, the force components at A are shown acting in the n and t directions. Note that $\boldsymbol{\alpha}$ acts clockwise.

The moment of inertia of the rod about point A is $I_A = \frac{1}{3}ml^2$.

Equations of Motion.   Moments will be summed about A in order to eliminate the reactive forces there.

$$+\nwarrow\Sigma F_n = m\omega^2 r_G; \qquad A_n - mg\sin\theta = m\omega^2(l/2) \qquad (1)$$

$$+\swarrow\Sigma F_t = m\alpha r_G; \qquad A_t + mg\cos\theta = m\alpha(l/2) \qquad (2)$$

$$\curvearrowright+\Sigma M_A = I_A\alpha; \qquad mg\cos\theta(l/2) = \left(\tfrac{1}{3}ml^2\right)\alpha \qquad (3)$$

(b)

Fig. 17–19

Kinematics.   For a given angle θ there are four unknowns in the above three equations: A_n, A_t, ω, and α. As shown by Eq. 3, α is *not constant*; rather, it depends on the position θ of the rod. The necessary fourth equation is obtained using kinematics, where α and ω can be related to θ by the equation

$$(\curvearrowright+) \qquad\qquad \omega\,d\omega = \alpha\,d\theta \qquad (4)$$

Note that the positive clockwise direction for this equation *agrees* with that of Eq. 3. This is important since we are seeking a simultaneous solution.

In order to solve for ω at $\theta = 90°$, eliminate α from Eqs. 3 and 4, which yields

$$\omega\,d\omega = (1.5\,g/l)\cos\theta\,d\theta$$

Since $\omega = 0$ at $\theta = 0°$, we have

$$\int_0^\omega \omega\,d\omega = (1.5\,g/l)\int_{0°}^{90°}\cos\theta\,d\theta$$

$$\omega^2 = 3\,g/l$$

Substituting this value into Eq. 1 with $\theta = 90°$ and solving Eqs. 1 to 3 yields

$$\alpha = 0 \qquad A_t = 0 \qquad A_n = 2.5\,mg \qquad\qquad Ans.$$

NOTE: If $\Sigma M_A = \Sigma(\mathcal{M}_k)_A$ is used, one must account for the moments of $I_G\alpha$ and $m(\mathbf{a}_G)_t$ about A. Here, however, we have used $\Sigma M_A = I_A\alpha$.

PROBLEMS

17–53. The 80-kg disk is supported by a pin at A. If it is released from rest from the position shown, determine the initial horizontal and vertical components of reaction at the pin.

Prob. 17–53

17–54. The 10-kg wheel has a radius of gyration $k_A = 200$ mm. If the wheel is subjected to a moment $M = (5t)$ N $\cdot$ m, where t is in seconds, determine its angular velocity when $t = 3$ s starting from rest. Also, compute the reactions which the fixed pin A exerts on the wheel during the motion.

Prob. 17–54

17–55. The fan blade has a mass of 2 kg and a moment of inertia $I_O = 0.18$ kg $\cdot$ m^2 about an axis passing through its center O. If it is subjected to a moment of $M = 3(1 - e^{-0.2t})$ N $\cdot$ m, where t is in seconds, determine its angular velocity when $t = 4$ s starting from rest.

Prob. 17–55

*17–56. The 10-lb rod is pin connected to its support at A and has an angular velocity $\omega = 4$ rad/s when it is in the horizontal position shown. Determine its angular acceleration and the horizontal and vertical components of reaction which the pin exerts on the rod at this instant.

Prob. 17–56

17–57. The pendulum consists of a 15-lb disk and a 10-lb slender rod. Determine the horizontal and vertical components of reaction that the pin O exerts on the rod just as it passes the horizontal position, at which time its angular velocity is $\omega = 8$ rad/s.

Prob. 17–57

17–58. The pendulum consists of a uniform 5-kg plate and a 2-kg slender rod. Determine the horizontal and vertical components of reaction that the pin O exerts on the rod at the instant $\theta = 30°$, at which time its angular velocity is $\omega = 3$ rad/s.

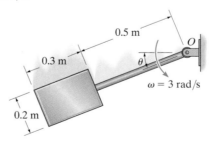

Prob. 17–58

17–59. The 10-lb bar is pinned at its center O and connected to a torsional spring. The spring has a stiffness $k = 5$ lb·ft/rad, so that the torque developed is $M = (5\theta)$ lb·ft, where θ is in radians. If the bar is released from rest when it is vertical at $\theta = 90°$, determine its angular velocity at the instant $\theta = 0°$.

*17–60. The 10-lb bar is pinned at its center O and connected to a torsional spring. The spring has a stiffness $k = 5$ lb·ft/rad, so that the torque developed is $M = (5\theta)$ lb·ft, where θ is in radians. If the bar is released from rest when it is vertical at $\theta = 90°$, determine its angular velocity at the instant $\theta = 45°$.

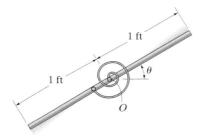

Probs. 17–59/60

17–61. The 20-kg roll of paper has a radius of gyration $k_A = 90$ mm about an axis passing through point A. It is pin-supported at both ends by two brackets AB. If the roll rests against a wall for which the coefficient of kinetic friction is $\mu_k = 0.2$ and a vertical force $F = 30$ N is applied to the end of the paper, determine the angular acceleration of the roll as the paper unrolls.

Prob. 17–61

17–62. The cylinder has a radius r and mass m and rests in the trough for which the coefficient of kinetic friction at A and B is μ_k. If a horizontal force **P** is applied to the cylinder, determine the cylinder's angular acceleration when it begins to spin.

Prob. 17–62

17–63. The uniform slender rod has a mass of 5 kg. If the cord at A is cut, determine the reaction at the pin O, (a) when the rod is still in the horizontal position, and (b) when the rod swings to the vertical position.

Prob. 17–63

*17–64. The bar has a mass m and length l. If it is released from rest from the position $\theta = 30°$, determine its angular acceleration and the horizontal and vertical components of reaction at the pin O.

Prob. 17–64

17–65. The kinetic diagram representing the general rotational motion of a rigid body about a fixed axis at O is shown in the figure. Show that $I_G\alpha$ may be eliminated by moving the vectors $m(\mathbf{a}_G)_t$ and $m(\mathbf{a}_G)_n$ to point P, located a distance $r_{GP} = k_G^2/r_{OG}$ from the center of mass G of the body. Here k_G represents the radius of gyration of the body about G. The point P is called the *center of percussion* of the body.

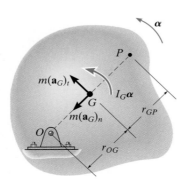

Prob. 17–65

17–66. Determine the position r_P of the center of percussion P of the 10-lb slender bar. (See Prob. 17–65.) What is the horizontal force A_x at the pin when the bar is struck at P with a force of $F = 20$ lb?

Prob. 17–66

17–67. The 4-kg slender rod is supported horizontally by a spring at A and a cord at B. Determine the angular acceleration of the rod and the acceleration of the rod's mass center at the instant the cord at B is cut. *Hint:* The stiffness of the spring is not needed for the calculation.

Prob. 17–67

***17–68.** In order to experimentally determine the moment of inertia I_G of a 4-kg connecting rod, the rod is suspended horizontally at A by a cord and at B by a bearing and piezoelectric sensor, an instrument used for measuring force. Under these equilibrium conditions, the force at B is measured as 14.6 N. If, at the instant the cord is released, the reaction at B is measured as 9.3 N, determine the value of I_G. The support at B does not move when the measurement is taken. For the calculation, the horizontal location of G must be determined.

Prob. 17–68

17–69. The 10-lb disk D is subjected to a counterclockwise moment of $M = (10t)$ lb·ft, where t is in seconds. Determine the angular velocity of the disk 2 s after the moment is applied. Due to the spring the plate P exerts a constant force of 100 lb on the disk. The coefficients of static and kinetic friction between the disk and the plate are $\mu_s = 0.3$ and $\mu_k = 0.2$, respectively. *Hint:* First find the time needed to start the disk rotating.

Prob. 17–69

17–70. The furnace cover has a mass of 20 kg and a radius of gyration $k_G = 0.25$ m about its mass center G. If an operator applies a force $F = 120$ N to the handle in order to open the cover, determine the cover's initial angular acceleration and the horizontal and vertical components of reaction which the pin at A exerts on the cover at the instant the cover begins to open. Neglect the mass of the handle BAC in the calculation.

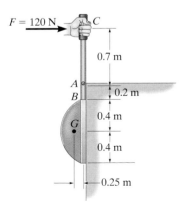

Prob. 17–70

17–71. The variable-resistance motor is often used for appliances, pumps, and blowers. By applying a current through the stator S, an electromagnetic field is created that "pulls in" the nearest rotor poles. The result of this is to create a torque of 4 N·m about the bearing at A. If the rotor is made from iron and has a 3-kg cylindrical core of 50-mm diameter and eight extended slender rods, each having a mass of 1 kg and 100-mm length, determine its angular velocity in 5 seconds starting from rest.

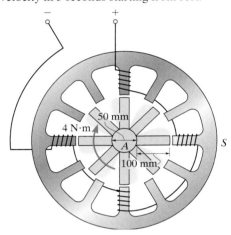

Prob. 17–71

*****17–72.** Determine the angular acceleration of the 25-kg diving board and the horizontal and vertical components of reaction at the pin A the instant the man jumps off. Assume that the board is uniform and rigid, and that at the instant he jumps off the spring is compressed a maximum amount of 200 mm, $\omega = 0$, and the board is horizontal. Take $k = 7$ kN/m.

Prob. 17–72

17–73. The disk has a mass of 20 kg and is originally spinning at the end of the strut with an angular velocity of $\omega = 60$ rad/s. If it is then placed against the wall, for which the coefficient of kinetic friction is $\mu_k = 0.3$, determine the time required for the motion to stop. What is the force in strut BC during this time?

Prob. 17–73

17–74. The relay switch consists of an electromagnet E and a 20-g armature AB (slender bar) which is pinned at A and lies in the vertical plane. When the current is turned off, the armature is held open against the smooth stop at B by the spring CD, which exerts an upward vertical force $F_s = 0.85$ N on the armature at C. When the current is turned on, the electromagnet attracts the armature at E with a vertical force $F = 0.8$ N. Determine the initial angular acceleration of the armature when the contact BF begins to close.

Prob. 17–74

17–75. The two blocks A and B have a mass m_A and m_B, respectively, where $m_B > m_A$. If the pulley can be treated as a disk of mass M, determine the acceleration of block A. Neglect the mass of the cord and any slipping on the pulley.

Prob. 17–75

*17–76. The rod has a length L and mass m. If it is released from rest when $\theta \approx 0°$, determine its angular velocity as a function of θ. Also, express the horizontal and vertical components of reaction at the pin O as a function of θ.

Prob. 17–76

17–77. The two-bar assembly is released from rest in the position shown. Determine the initial bending moment at the fixed joint B. Each bar has a mass m and length l.

Prob. 17–77

17–78. Disk A has a weight of 5 lb and disk B has a weight of 10 lb. If no slipping occurs between them, determine the couple moment **M** which must be applied to disk A to give it an angular acceleration of 4 rad/s².

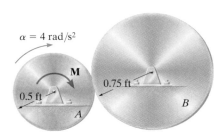

$\alpha = 4 \text{ rad/s}^2$

M

0.75 ft

0.5 ft

A

B

Prob. 17–78

17–79. The wheel has a mass of 25 kg and a radius of gyration $k_B = 0.15$ m. It is originally spinning at $\omega_1 = 40$ rad/s. If it is placed on the ground, for which the coefficient of kinetic friction is $\mu_C = 0.5$, determine the time required for the motion to stop. What are the horizontal and vertical components of reaction which the pin at A exerts on AB during this time? Neglect the mass of AB.

0.4 m

A

0.3 m

0.2 m

B

ω

C

Prob. 17–79

*****17–80.** The cord is wrapped around the inner core of the spool. If a 5-lb block B is suspended from the cord and released from rest, determine the spool's angular velocity when $t = 3$ s. Neglect the mass of the cord. The spool has a weight of 180 lb and the radius of gyration about the axle A is $k_A = 1.25$ ft. Solve the problem in two ways, first by considering the "system" consisting of the block and spool, and then by considering the block and spool separately.

2.75 ft

1.5 ft A

s

B

Prob. 17–80

■17–81. A 40-kg boy sits on top of the large wheel which has a mass of 400 kg and a radius of gyration $k_G = 5.5$ m. If the boy essentially starts from rest at $\theta = 0°$, and the wheel begins to rotate freely, determine the angle at which the boy begins to slip. The coefficient of static friction between the wheel and the boy is $\mu_s = 0.5$. Neglect the size of the boy in the calculation.

8 m

θ

Prob. 17–81

17–82. The "Catherine wheel" is a firework that consists of a coiled tube of powder which is pinned at its center. If the powder burns at a constant rate of 20 g/s such as that the exhaust gases always exert a force having a constant magnitude of 0.3 N, directed tangent to the wheel, determine the angular velocity of the wheel when 75% of the mass is burned off. Initially, the wheel is at rest and has a mass of 100 g and a radius of $r = 75$ mm. For the calculation, consider the wheel to always be a thin disk.

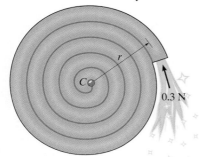

Prob. 17–82

17–83. The bar has a weight per length of w. If it is rotating in the vertical plane at a constant rate ω about point O, determine the internal normal force, shear force, and moment as a function of x and θ.

Prob. 17–83

*17–84. A force $F = 2$ lb is applied perpendicular to the axis of the 5-lb rod and moves from O to A at a constant rate of 4 ft/s. If the rod is at rest when $\theta = 0°$ and $\mathbf{F}$ is at O when $t = 0$, determine the rod's angular velocity at the instant the force is at A. Through what angle has the rod rotated when this occurs? The rod rotates in the *horizontal plane*.

Prob. 17–84

17–85. Block A has a mass m and rests on a surface having a coefficient of kinetic friction μ_k. The cord attached to A passes over a pulley at C and is attached to a block B having a mass $2m$. If B is released, determine the acceleration of A. Assume that the cord does not slip over the pulley. The pulley can be approximated as a thin disk of radius r and mass $\frac{1}{4}m$. Neglect the mass of the cord.

Prob. 17–85

17–86. The slender rod of mass m is released from rest when $\theta = 45°$. At the same instant ball B having the same mass m is released. Will B or the end A of the rod have the greatest speed when they pass the horizontal ($\theta = 0°$)? What is the difference in their speeds?

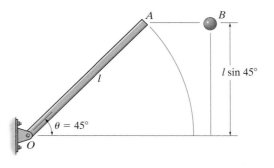

Prob. 17–86

17.5   Equations of Motion: General Plane Motion

The rigid body (or slab) shown in Fig. 17–20a is subjected to general plane motion caused by the externally applied force and couple-moment system. The free-body and kinetic diagrams for the body are shown in Fig. 17–20b. If an x and y inertial coordinate system is chosen as shown, the three equations of motion may be written as

$$\begin{array}{l} \Sigma F_x = m(a_G)_x \\ \Sigma F_y = m(a_G)_y \\ \Sigma M_G = I_G \alpha \end{array} \qquad (17\text{–}17)$$

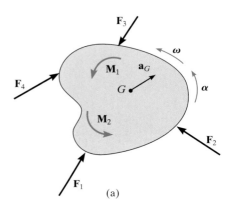

(a)

Fig. 17–20

In some problems it may be convenient to sum moments about some point P other than G. This is usually done in order to eliminate unknown forces from the moment summation. When used in this more general sense, the three equations of motion become

$$\begin{array}{l} \Sigma F_x = m(a_G)_x \\ \Sigma F_y = m(a_G)_y \\ \Sigma M_P = \Sigma(\mathcal{M}_k)_P \end{array} \qquad (17\text{–}18)$$

Here $\Sigma(\mathcal{M}_k)_P$ represents the moment sum of $I_G\alpha$ and $m\mathbf{a}_G$ (or its components) about P as determined by the data on the kinetic diagram.

(b)

(a)

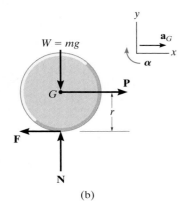

(b)

Fig. 17–21

Frictional Rolling Problems.   There is a class of planar kinetics problems which deserves special mention. These problems involve wheels, cylinders, or bodies of similar shape, which roll on a *rough* plane surface. Because of the applied loadings, it may not be known if the body *rolls without slipping*, or if it *slides as it rolls*. For example, consider the homogeneous disk shown in Fig. 17–21a, which has a mass m and is subjected to a known horizontal force $\mathbf{P}$. The free-body diagram is shown in Fig. 17–21b. Since $\mathbf{a}_G$ is directed to the right and α is clockwise, we have

$$\xrightarrow{+} \Sigma F_x = m(a_G)_x; \qquad\qquad P - F = ma_G \qquad (17\text{–}19)$$

$$+\uparrow \Sigma F_y = m(a_G)_y; \qquad\qquad N - mg = 0 \qquad (17\text{–}20)$$

$$\zeta + \Sigma M_G = I_G\alpha; \qquad\qquad Fr = I_G\alpha \qquad (17\text{–}21)$$

A fourth equation is needed since these *three equations* contain *four unknowns: F, N, α, and a_G.*

No Slipping.   If the frictional force F is great enough to allow the disk to roll without slipping, then a_G can be related to α by the kinematic equation,*

$$(\zeta+) \qquad\qquad\qquad a_G = \alpha r \qquad\qquad (17\text{–}22)$$

The four unknowns are determined by *solving simultaneously* Eqs. 17–19 to 17–22. When the solution is obtained, the assumption of no slipping must be *checked*. Recall that no slipping occurs provided $F \le \mu_s N$, where μ_s is the coefficient of static friction. If the inequality is satisfied, the problem is solved. However, if $F > \mu_s N$, the problem must be *reworked*, since then the disk slips as it rolls.

Slipping.   In the case of slipping, there is relative motion at the points of contact between the disk and the ground, and so α and a_G are *independent of one another* so that Eq. 17–22 does not apply. Instead, the magnitude of the frictional force is related to the magnitude of the normal force using the coefficient of kinetic friction μ_k, i.e.,

$$F = \mu_k N \qquad\qquad (17\text{–}23)$$

In this case Eqs. 17–19 to 17–21 and 17–23 are used for the solution. It is important to keep in mind that whenever Eq. 17–22 or 17–23 is applied, it is necessary to have consistency in the directional sense of the vectors. In the case of Eq. 17–22, $\mathbf{a}_G$ must be directed to the right when $\boldsymbol{\alpha}$ is clockwise, since the rolling motion requires it. And in Eq. 17–23, $\mathbf{F}$ must be directed to the left to prevent the assumed slipping motion to the right, Fig. 17–21b. On the other hand, if these equations are *not used* for the solution, these vectors can have *any* assumed directional sense. Then if the calculated numerical value of these quantities is negative, the vectors act in their opposite sense of direction. Examples 17.15 and 17.16 illustrate these concepts numerically.

*See Example 16.3 or 16.14.

Procedure for Analysis

Kinetic problems involving general plane motion of a rigid body can be solved using the following procedure.

Free-Body Diagram.

- Establish the x, y inertial coordinate system and draw the free-body diagram for the body.

- Specify the direction and sense of the acceleration of the mass center, $\mathbf{a}_G$, and the angular acceleration $\boldsymbol{\alpha}$ of the body.

- Compute the moment of inertia I_G.

- Identify the unknowns in the problem.

- If it is decided that the rotational equation of motion $\Sigma M_P = \Sigma(\mathcal{M}_k)_P$ is to be used, then consider drawing the kinetic diagram in order to help "visualize" the "moments" developed by the components $m(\mathbf{a}_G)_x$, $m(\mathbf{a}_G)_y$, and $I_G\boldsymbol{\alpha}$ when writing the terms in the moment sum $\Sigma(\mathcal{M}_k)_P$.

Equations of Motion.

- Apply the three equations of motion in accordance with the established sign convention.

- When friction is present, there is the possibility for motion with no slipping or tipping. Each possibility for motion should be considered.

Kinematics.

- Use kinematics if a complete solution cannot be obtained strictly from the equations of motion.

- If the body's motion is *constrained* due to its supports, additional equations may be obtained by using $\mathbf{a}_B = \mathbf{a}_A + \mathbf{a}_{B/A}$, which relates the accelerations of any two points A and B on the body.

- When a wheel, disk, cylinder, or ball *rolls without slipping*, then $a_G = \alpha r$.

As the soil compactor moves forward, the roller has general plane motion. The forces shown on the roller's free-body diagram cause the effects shown on the kinetic diagram. If moments are summed about the mass center, G, then $\Sigma M_G = I_G\alpha$. However, if moments are summed about point A then $\zeta + \Sigma M_A = I_G\alpha + (ma_G)d$.

EXAMPLE 17.14

(a)

(b)

(c)

Fig. 17–22

The spool in Fig. 17–22a has a mass of 8 kg and a radius of gyration of $k_G = 0.35$ m. If cords of negligible mass are wrapped around its inner hub and outer rim as shown, determine the spool's angular acceleration.

SOLUTION I

Free-Body Diagram. Fig. 17–22b. The 100-N force causes $\mathbf{a}_G$ to act upward. Also, $\boldsymbol{\alpha}$ acts clockwise, since the spool winds around the cord at A.

There are three unknowns T, a_G, and α. The moment of inertia of the spool about its mass center is

$$I_G = mk_G^2 = 8 \text{ kg}(0.35 \text{ m})^2 = 0.980 \text{ kg} \cdot \text{m}^2$$

Equations of Motion.

$$+\uparrow \Sigma F_y = m(a_G)_y; \qquad T + 100 \text{ N} - 78.48 \text{ N} = (8 \text{ kg})a_G \qquad (1)$$

$$\zeta + \Sigma M_G = I_G \alpha; \quad 100 \text{ N}(0.2 \text{ m}) - T(0.5 \text{ m}) = (0.980 \text{ kg} \cdot \text{m}^2)\alpha \quad (2)$$

Kinematics. A complete solution is obtained if kinematics is used to relate a_G to α. In this case the spool "rolls without slipping" on the cord at A. Hence, we can use the results of Example 16.3 or 16.14, so that

$$(\zeta +)a_G = \alpha r, \qquad a_G = 0.5\alpha \qquad (3)$$

Solving Eqs. 1 to 3, we have

$$\alpha = 10.3 \text{ rad/s}^2 \qquad\qquad Ans.$$
$$a_G = 5.16 \text{ m/s}^2$$
$$T = 19.8 \text{ N}$$

SOLUTION II

Equations of Motion. We can eliminate the unknown T by summing moments about point A. From the free-body and kinetic diagrams Figs. 17–22b and 17–22c, we have

$$\zeta + \Sigma M_A = \Sigma (\mathcal{M}_k)_A; \qquad 100 \text{ N}(0.7 \text{ m}) - 78.48 \text{ N}(0.5 \text{ m})$$
$$= (0.980 \text{ kg} \cdot \text{m}^2)\alpha + [(8 \text{ kg})a_G](0.5 \text{ m})$$

Using Eq. (3),

$$\alpha = 10.3 \text{ rad/s}^2 \qquad\qquad Ans.$$

EXAMPLE 17.15

The 50-lb wheel shown in Fig. 17–23a has a radius of gyration $k_G = 0.70$ ft. If a 35-lb · ft couple moment is applied to the wheel, determine the acceleration of its mass center G. The coefficients of static and kinetic friction between the wheel and the plane at A are $\mu_s = 0.3$ and $\mu_k = 0.25$, respectively.

(a)

SOLUTION

Free-Body Diagram.   By inspection of Fig. 17–23b, it is seen that the couple moment causes the wheel to have a clockwise angular acceleration of $\boldsymbol{\alpha}$. As a result, the acceleration of the mass center, $\mathbf{a}_G$, is directed to the right. The moment of inertia is

$$I_G = mk_G^2 = \frac{50 \text{ lb}}{32.2 \text{ ft/s}^2}(0.70 \text{ ft})^2 = 0.761 \text{ slug} \cdot \text{ft}^2$$

The unknowns are N_A, F_A, a_G, and α.

Equations of Motion.

$$\xrightarrow{+} \Sigma F_x = m(a_G)_x; \qquad F_A = \frac{50 \text{ lb}}{32.2 \text{ ft/s}^2} a_G \qquad (1)$$

$$+\uparrow \Sigma F_y = m(a_G)_y; \qquad N_A - 50 \text{ lb} = 0 \qquad (2)$$

$$\zeta+\Sigma M_G = I_G\alpha; \quad 35 \text{ lb} \cdot \text{ft} - 1.25 \text{ ft}(F_A) = (0.761 \text{ slug} \cdot \text{ft}^2)\alpha \qquad (3)$$

(b)

Fig. 17–23

A fourth equation is needed for a complete solution.

Kinematics (No Slipping).   If this assumption is made, then

$$(\zeta+) \qquad\qquad a_G = (1.25 \text{ ft})\alpha \qquad (4)$$

Solving Eqs. 1 to 4,

$$N_A = 50.0 \text{ lb} \qquad F_A = 21.3 \text{ lb}$$
$$\alpha = 11.0 \text{ rad/s}^2 \qquad a_G = 13.7 \text{ ft/s}^2$$

The original assumption of no slipping requires $F_A \leq \mu_s N_A$. However, since 21.3 lb $> 0.3(50 \text{ lb}) = 15$ lb, the wheel slips as it rolls.

(Slipping).   Equation 4 is not valid, and so $F_A = \mu_k N_A$, or

$$F_A = 0.25N_A \qquad (5)$$

Solving Eqs. 1 to 3 and 5 yields

$$N_A = 50.0 \text{ lb} \qquad F_A = 12.5 \text{ lb}$$
$$\alpha = 25.5 \text{ rad/s}^2$$
$$a_G = 8.05 \text{ ft/s}^2 \rightarrow \qquad\qquad Ans.$$

EXAMPLE 17.16

(a)

(b)

Fig. 17–24

The uniform slender pole shown in Fig. 17–24a has a mass of 100 kg and a moment of inertia $I_G = 75 \text{ kg} \cdot \text{m}^2$. If the coefficients of static and kinetic friction between the end of the pole and the surface are $\mu_s = 0.3$ and $\mu_k = 0.25$, respectively, determine the pole's angular acceleration at the instant the 400-N horizontal force is applied. The pole is originally at rest.

SOLUTION

Free-Body Diagram.   Figure 17–24b. The path of motion of the mass center G will be along an unknown curved path having a radius of curvature ρ, which is initially parallel to the y axis. There is no normal or y component of acceleration since the pole is originally at rest, i.e., $\mathbf{v}_G = \mathbf{0}$, so that $(a_G)_y = v_G^2/\rho = 0$. We will assume the mass center accelerates to the right and that the pole has a clockwise angular acceleration of $\boldsymbol{\alpha}$. The unknowns are N_A, F_A, a_G, and α.

Equations of Motion.

$$\xrightarrow{+} \Sigma F_x = m(a_G)_x; \qquad 400 \text{ N} - F_A = (100 \text{ kg})a_G \qquad (1)$$

$$+\uparrow \Sigma F_y = m(a_G)_y; \qquad N_A - 981 \text{ N} = 0 \qquad (2)$$

$$\zeta + \Sigma M_G = I_G\alpha; \quad F_A(1.5 \text{ m}) - 400 \text{ N}(1 \text{ m}) = (75 \text{ kg} \cdot \text{m}^2)\alpha \qquad (3)$$

A fourth equation is needed for a complete solution.

Kinematics (No Slipping).   In this case point A acts as a "pivot" so that indeed, if α is clockwise, then a_G is directed to the right.

$$\zeta + a_G = \alpha r_{AG}; \qquad\qquad a_G = (1.5 \text{ m})\alpha \qquad (4)$$

Solving Eqs. 1 to 4 yields

$$N_A = 981 \text{ N} \qquad F_A = 300 \text{ N}$$

$$a_G = 1 \text{ m/s}^2 \qquad \alpha = 0.667 \text{ rad/s}^2$$

Testing the original assumption of no slipping requires $F_A \leq \mu_s N_A$. However, 300 N > 0.3(981 N) = 294 N. (Slips at A.)

(Slipping).   For this case Eq. 4 does *not* apply. Instead the frictional equation $F_A = \mu_k N_A$ is used. Hence,

$$F_A = 0.25N_A \qquad (5)$$

Solving Eqs. 1 to 3 and 5 simultaneously yields

$$N_A = 981 \text{ N} \qquad F_A = 245 \text{ N} \qquad a_G = 1.55 \text{ m/s}^2$$

$$\alpha = -0.428 \text{ rad/s}^2 = 0.428 \text{ rad/s}^2 \, \zeta \qquad\qquad\qquad Ans.$$

EXAMPLE 17.17

The 30-kg wheel shown in Fig. 17–25a has a mass center at G and a radius of gyration $k_G = 0.15$ m. If the wheel is originally at rest and released from the position shown, determine its angular acceleration. No slipping occurs.

(a)

SOLUTION

Free-Body and Kinetic Diagrams.   The two unknowns $\mathbf{F}_A$ and $\mathbf{N}_A$ shown on the free-body diagram, Fig. 17–25b, can be eliminated from the analysis by summing moments about point A. The kinetic diagram accompanies the solution in order to illustrate application of $\Sigma(\mathcal{M}_k)_A$. Since point G moves along a curved path, the two components $m(\mathbf{a}_G)_x$ and $m(\mathbf{a}_G)_y$ are shown on the kinetic diagram, Fig. 17–25b.

The moment of inertia is

$$I_G = mk_G^2 = 30(0.15)^2 = 0.675 \text{ kg} \cdot \text{m}^2$$

There are five unknowns, N_A, F_A, $(a_G)_x$, $(a_G)_y$, and α.

Equation of Motion.   Applying the rotational equation of motion about point A, to eliminate N_A, and F_A, we have

$$\zeta + \Sigma M_A = \Sigma(\mathcal{M}_k)_A; \qquad\qquad 30(9.81) \text{ N}(0.1 \text{ m}) =$$
$$(0.675 \text{ kg} \cdot \text{m}^2)\alpha + (30 \text{ kg})(a_G)_x(0.25 \text{ m}) + (30 \text{ kg})(a_G)_y(0.1 \text{ m}) \quad (1)$$

There are three unknowns in this equation: $(a_G)_x$, $(a_G)_y$, and α.

Kinematics.   Using kinematics, $(a_G)_x$, $(a_G)_y$ will be related to α. As shown in Fig. 17–25c, these vectors must have the same directional sense as the corresponding vectors on the kinetic diagram since we are seeking a simultaneous solution with Eq. 1. Since no slipping occurs, $a_O = \alpha r = \alpha(0.25 \text{ m})$, directed to the left, Fig. 17–25c. Also, $\omega = 0$, since the wheel is originally at rest. Applying the acceleration equation to point O (base point) and point G, we have

$$\mathbf{a}_G = \mathbf{a}_O + \boldsymbol{\alpha} \times \mathbf{r}_{G/O} - \omega^2 \mathbf{r}_{G/O}$$
$$-(a_G)_x\mathbf{i} - (a_G)_y\mathbf{j} = -\alpha(0.25)\mathbf{i} + (\alpha\mathbf{k}) \times (-0.1\mathbf{i}) - \mathbf{0}$$

Expanding and equating the respective $\mathbf{i}$ and $\mathbf{j}$ components, we have

$$(a_G)_x = \alpha(0.25) \qquad\qquad (2)$$
$$(a_G)_y = \alpha(0.1) \qquad\qquad (3)$$

Solving Eqs. 1 to 3 yields

$$\alpha = 10.3 \text{ rad/s}^2 \, \zeta \qquad\qquad Ans.$$
$$(a_G)_x = 2.58 \text{ m/s}^2$$
$$(a_G)_y = 1.03 \text{ m/s}^2$$

(b)

(c)

Fig. 17–25

NOTE: As an exercise, show that $F_A = 77.4$ N and $N_A = 263$ N.

PROBLEMS

17–87. If the disk in Fig. 17–21a rolls without slipping, show that when moments are summed about the instantaneous center of zero velocity, IC, it is possible to use the moment equation $\Sigma M_{IC} = I_{IC}\alpha$, where I_{IC} represents the moment of inertia of the disk calculated about the instantaneous axis of zero velocity.

*17–88. The 20-kg punching bag has a radius of gyration about its center of mass G of $k_G = 0.4$ m. If it is initially at rest and is subjected to a horizontal force $F = 30$ N, determine the initial angular acceleration of the bag and the tension in the supporting cable AB.

Prob. 17–88

17–89. The trailer has a mass of 580 kg and a mass center at G, whereas the spool has a mass of 200 kg, mass center at O, and a radius of gyration about an axis passing through O of $k_O = 0.45$ m. If a force of 60 N is applied to the cable, determine the angular acceleration of the spool and the acceleration of the trailer. The wheels have negligible mass and are free to roll.

Prob. 17–89

17–90. The rocket has a weight of 20 000 lb, mass center at G, and radius of gyration about the mass center of $k_G = 21$ ft when it is fired. Each of its two engines provides a thrust $T = 50\,000$ lb. At a given instant, engine A suddenly fails to operate. Determine the angular acceleration of the rocket and the acceleration of its nose B.

Prob. 17–90

17–91. The spool and wire wrapped around its core have a mass of 20 kg and a centroidal radius of gyration $k_G = 250$ mm. If the coefficient of kinetic friction at the ground is $\mu_k = 0.1$, determine the angular acceleration of the spool when the 30-N·m couple is applied.

Prob. 17–91

*17–92. The uniform 50-lb board is suspended from cords at C and D. If these cords are subjected to constant forces of 30 lb and 45 lb, respectively, determine the acceleration of the board's center and the board's angular acceleration. Assume the board is a thin plate. Neglect the mass of the pulleys at E and F.

Prob. 17–92

17–93. The spool has a mass of 500 kg and a radius of gyration $k_G = 1.30$ m. It rests on the surface of a conveyor belt for which the coefficient of static friction is $\mu_s = 0.5$ and the coefficient of kinetic friction is $\mu_k = 0.4$. If the conveyor accelerates at $a_C = 1$ m/s^2, determine the initial tension in the wire and the angular acceleration of the spool. The spool is originally at rest.

17–94. The spool has a mass of 500 kg and a radius of gyration $k_G = 1.30$ m. It rests on the surface of a conveyor belt for which the coefficient of static friction is $\mu_s = 0.5$. Determine the greatest acceleration a_C of the conveyor so that the spool will not slip. Also, what are the initial tension in the wire and the angular acceleration of the spool? The spool is originally at rest.

Probs. 17–93/94

17–95. The wheel has a weight of 30 lb and a radius of gyration of $k_G = 0.6$ ft. If the coefficients of static and kinetic friction between the wheel and the plane are $\mu_s = 0.2$ and $\mu_k = 0.15$, determine the wheel's angular acceleration as it rolls down the incline. Set $\theta = 12°$.

*17–96. The wheel has a weight of 30 lb and a radius of gyration of $k_G = 0.6$ ft. If the coefficients of static and kinetic friction between the wheel and the plane are $\mu_s = 0.2$ and $\mu_k = 0.15$, determine the maximum angle θ of the inclined plane so that the wheel rolls without slipping.

Probs. 17–95/96

17–97. The truck carries the spool which has a weight of 500 lb and a radius of gyration of $k_G = 2$ ft. Determine the angular acceleration of the spool if it is not tied down on the truck and the truck begins to accelerate at 3 ft/s^2. Assume the spool does not slip on the bed of the truck.

17–98. The truck carries the spool which has a weight of 200 lb and a radius of gyration of $k_G = 2$ ft. Determine the angular acceleration of the spool if it is not tied down on the truck and the truck begins to accelerate at 5 ft/s^2. The coefficients of static and kinetic friction between the spool and the truck bed are $\mu_s = 0.15$ and $\mu_k = 0.1$, respectively.

Probs. 17–97/98

17–99.   The spool has a mass of 75 kg and a radius of gyration $k_G = 0.380$ m. It rests on the inclined surface for which the coefficient of kinetic friction is $\mu_k = 0.15$. If the spool is released from rest and slips at A, determine the initial tension in the cord and the angular acceleration of the spool.

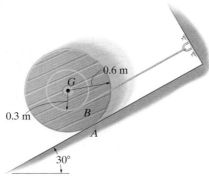

Prob. 17–99

*17–100.   A uniform rod having a weight of 10 lb is pin supported at A from a roller which rides on a horizontal track. If the rod is originally at rest, and a horizontal force of $F = 15$ lb is applied to the roller, determine the acceleration of the roller. Neglect the mass of the roller and its size d in the computations.

17–101.   Solve Prob. 17–100 assuming that the roller at A is replaced by a slider block having a negligible mass. The coefficient of kinetic friction between the block and the track is $\mu_k = 0.2$. Neglect the dimension d and the size of the block in the computations.

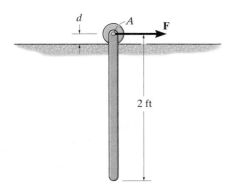

Probs. 17–100/101

17–102.   The lawn roller has a mass of 80 kg and a radius of gyration $k_G = 0.175$ m. If it is pushed forward with a force of 200 N when the handle is at 45°, determine its angular acceleration. The coefficients of static and kinetic friction between the ground and the roller are $\mu_s = 0.12$ and $\mu_k = 0.1$, respectively.

Prob. 17–102

17–103.   The slender 150-lb bar is supported by two cords AB and AC. If cord AC suddenly breaks, determine the initial angular acceleration of the bar and the tension in cord AB.

Prob. 17–103

*17–104. A long strip of paper is wrapped into two rolls, each having a mass of 8 kg. Roll A is pin supported about its center whereas roll B is not centrally supported. If B is brought into contact with A and released from rest, determine the initial tension in the paper between the rolls and the angular acceleration of each roll. For the calculation, assume the rolls to be approximated by cylinders.

Prob. 17–104

17–105. The uniform bar of mass m and length L is balanced in the vertical position when the horizontal force **P** is applied to the roller at A. Determine the bar's initial angular acceleration and the acceleration of its top point B.

Prob. 17–105

17–106. A woman sits in a rigid position in the middle of the swing. The combined weight of the woman and swing is 180 lb and the radius of gyration about the center of mass G is $k_G = 2.5$ ft. If a man pushes on the swing with a horizontal force $F = 20$ lb as shown, determine the initial angular acceleration and the tension in each of the two supporting chains AB. During the motion, assume that the chain segment CAD remains rigid. The swing is originally at rest.

Prob. 17–106

17–107. A girl sits snugly inside a large tire such that both the girl and tire have a total weight of 185 lb, a center of mass at G, and a radius of gyration $k_G = 1.65$ ft about G. If the tire rolls freely down the incline, determine the normal and frictional forces it exerts on the ground when it is in the position shown and has an angular velocity of 6 rad/s. Assume that the tire does not slip as it rolls.

Prob. 17–107

*17–108. The 10-lb hoop or thin ring is given an initial angular velocity of 6 rad/s when it is placed on the surface. If the coefficient of kinetic friction between the hoop and the surface is $\mu_k = 0.3$, determine the distance the hoop moves before it stops slipping.

$\omega_0 = 6$ rad/s

6 in.

O

Prob. 17–108

17–109. The 15-lb circular plate is suspended from a pin at A. If the pin is connected to a track which is given an acceleration $a_A = 3$ ft/s^2, determine the horizontal and vertical components of reaction at A and the acceleration of the plate's mass center G. The plate is originally at rest.

$\mathbf{a}_A$

A

G

2 ft

Prob. 17–109

17–110. Wheel C has a mass of 60 kg and a radius of gyration of 0.4 m, whereas wheel D has a mass of 40 kg and a radius of gyration of 0.35 m. Determine the angular acceleration of each wheel at the instant shown. Neglect the mass of the link and assume that the assembly does not slip on the plane.

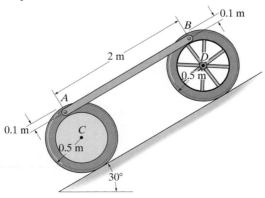

0.1 m

B

2 m

D

0.5 m

A

0.1 m

C

0.5 m

30°

Prob. 17–110

■17–111. The assembly consists of an 8-kg disk and a 10-kg bar which is pin connected to the disk. If the system is released from rest, determine the angular acceleration of the disk. The coefficients of static and kinetic friction between the disk and the inclined plane are $\mu_s = 0.6$ and $\mu_k = 0.4$, respectively. Neglect friction at B.

*17–112. Solve Prob. 17–111 if the bar is removed. The coefficients of static and kinetic friction between the disk and inclined plane are $\mu_s = 0.15$ and $\mu_k = 0.1$, respectively.

B

1 m

A

0.3 m

C

30°

Probs. 17–111/112

17–113. A "lifted" truck can become a road hazard since the bumper is high enough to ride up a standard car in the event the car is rear-ended. As a model of this case consider the truck to have a mass of 2.70 Mg, a mass center G, and a radius of gyration about G of $k_G = 1.45$ m. Determine the horizontal and vertical components of acceleration of the mass center G, and the angular acceleration of the truck, at the moment its front wheels at C have just left the ground and its smooth front bumper begins to ride up the back of the stopped car so that point B has a velocity of $v_B = 8$ m/s at 20° from the horizontal. Assume the wheels are free to roll, and neglect the size of the wheels and the deformation of the material.

G

20°

1.3 m

B

STATE POLICE

A

C

0.4 m

1.6 m — 1.2 m

Prob. 17–113

Design Projects

17–1D. Design of a Dynamometer

In order to test the dynamic strength of cables, an instrument called a *dynamometer* must be used that will measure the tension in a cable when it hoists a very heavy object with accelerated motion. Design such an instrument, based on the use of single or multiple springs so that it can be used on the cable supporting the 300-kg pipe that is given an upward acceleration of 2 m/s^2. Submit a drawing and explain how your dynamometer operates.

Fig. 17–1D

17–2D. Design of a Small Elevator Brake

A small household elevator is operated using a hoist. For safety purposes it is necessary to install a braking mechanism which will automatically engage in case the cable fails during operation. Design the braking mechanism using steel members and springs. The elevator and its contents are assumed to have a mass of 300 kg, and it travels at 2.5 m/s. The maximum allowable deceleration to stop the motion is to be 4 m/s^2. Assume the coefficient of kinetic friction between any steel members and the walls of the elevator shaft is $\mu_k = 0.3$. The gap between the elevator frame and each wall of the shaft is 50 mm. Submit a scale drawing of your design along with a force analysis to show that your design will arrest the motion as required. Discuss the safety and reliability of the mechanism.

Fig. 17–2D

17–3D. Safety Performance of a Bicycle

One of the most common accidents one can have on a bicycle is to flip over the handle bars. Obtain the necessary measurements of a standard-size bicycle and its mass and center of mass. Consider yourself as the rider, with center of mass at your navel. Perform an experiment to determine the coefficient of kinetic friction between the wheels and the pavement. With this data, calculate the possibility of flipping over when (a) only the rear brakes are applied, (b) only the front brakes are applied, and (c) both front and rear brakes are applied simultaneously. What effect does the height of the seat have on these results? Suggest a way to improve the bicycle's design, and write a report on the safety of cycling based on this analysis.

Fig. 17–3D

Chapter Review

Moment of Inertia

The moment of inertia is a measure of the resistance of a body to a change in its angular velocity. It is defined by $I = \int r^2 \, dm$ and will be different for each axis about which it is computed. For a body having axial symmetry, the integration is usually performed using disk or shell elements.

Many bodies are composed of simple shapes. If this is the case, then tabular values of I can be used, such as the ones given on the inside back cover of this book. To obtain the moment of inertia of a composite body about any specified axis, the moment of inertia of each part is determined about the axis and the results are added together. Doing this often requires use of the parallel-axis theorem.

Handbooks may also report values of the radius of gyration k for the body. If the body's mass is known, then the mass moment of inertia is determined from $I = mk^2$.

(*See pages 385–387.*)

$$I = I_G + md^2$$

Planar Equations of Motion

The equations of motion define the translational, and rotational motion of a rigid body. In order to account for all of the terms in these equations, a free-body diagram should always accompany their application, and for some problems, it may also be convenient to draw the kinetic diagram.

(*See pages 399–402.*)

Translation

Here $I_G \alpha = 0$ since $\alpha = 0$.

(*See pages 402–404.*)

$$\Sigma F_x = m(a_G)_x$$

$$\Sigma F_y = m(a_G)_y$$

$$\Sigma M_G = 0$$

Rectilinear translation

$$\Sigma F_n = m(a_G)_n$$

$$\Sigma F_t = m(a_G)_t$$

$$\Sigma M_G = 0$$

Curvilinear translation

Rotation About a Fixed Axis

For fixed axis rotation, the kinetic vector $m(\mathbf{a}_G)_n$ produces no moment about the axis of rotation, and so the rotational equation of motion reduces to a simplified form about point O.

(*See pages 414–416.*)

$$\Sigma F_n = m(a_G)_n = m\omega^2 r_G$$

$$\Sigma F_t = m(a_G)_t = m\alpha r_G$$

$$\Sigma M_G = I_G \alpha$$

or

$$\Sigma F_n = m(a_G)_n = m\omega^2 r_G$$

$$\Sigma F_t = m(a_G)_t = m\alpha r_G$$

$$\Sigma M_O = I_O \alpha$$

General Plane Motion

If the body is constrained by its supports, then additional equations of kinematics can be obtained by using $\mathbf{a}_G = \mathbf{a}_A + \mathbf{a}_{G/A}$ to relate the accelerations of any two G to another point A on the body.

(*See pages 429–431.*)

$$\Sigma F_x = m(a_G)_x$$

$$\Sigma F_y = m(a_G)_y$$

$$\Sigma M_G = I_G \alpha$$

or

$$\Sigma F_x = m(a_G)_x$$

$$\Sigma F_y = m(a_G)_y$$

$$\Sigma M_P = \Sigma (\mathcal{M}_k)_P$$

The principle of work and energy plays an important role in the motion of the draw works used to lift pipe on this drilling rig.

18 Planar Kinetics of a Rigid Body: Work and Energy

CHAPTER OBJECTIVES

- To develop formulations for the kinetic energy of a body, and define the various ways a force and couple do work.

- To apply the principle of work and energy to solve rigid-body planar kinetic problems that involve force, velocity, and displacement.

- To show how the conservation of energy can be used to solve rigid-body planar kinetic problems.

18.1 Kinetic Energy

In this chapter we will apply work and energy methods to problems involving force, velocity, and displacement related to the planar motion of a rigid body. But first it will be necessary to develop a means of obtaining the body's kinetic energy when the body is subjected to translation, rotation about a fixed axis, or general plane motion.

To do this we will consider the rigid body shown in Fig. 18–1, which is represented here by a *slab* moving in the inertial x–y reference plane. An arbitrary ith particle of the body, having a mass dm, is located at r from the arbitrary point P. If at the *instant* shown the particle has a velocity $\mathbf{v}_i$, then the particle's kinetic energy is $T_i = \frac{1}{2} dm\, v_i^2$.

Fig. 18–1

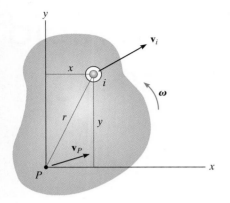

Fig. 18–1

The kinetic energy of the entire body is determined by writing similar expressions for each particle of the body and integrating the results, i.e.,

$$T = \frac{1}{2} \int_m dm \, v_i^2$$

This equation may also be expressed in terms of the velocity of point P. If the body has an angular velocity $\boldsymbol{\omega}$, then from Fig. 18–1 we have

$$\mathbf{v}_i = \mathbf{v}_P + \mathbf{v}_{i/P}$$
$$= (v_P)_x \mathbf{i} + (v_P)_y \mathbf{j} + \omega \mathbf{k} \times (x\mathbf{i} + y\mathbf{j})$$
$$= [(v_P)_x - \omega y]\mathbf{i} + [(v_P)_y + \omega x]\mathbf{j}$$

The square of the magnitude of $\mathbf{v}_i$ is thus

$$\mathbf{v}_i \cdot \mathbf{v}_i = v_i^2 = [(v_P)_x - \omega y]^2 + [(v_P)_y + \omega x]^2$$
$$= (v_P)_x^2 - 2(v_P)_x \omega y + \omega^2 y^2 + (v_P)_y^2 + 2(v_P)_y \omega x + \omega^2 x^2$$
$$= v_P^2 - 2(v_P)_x \omega y + 2(v_P)_y \omega x + \omega^2 r^2$$

Substituting this into the equation of kinetic energy yields

$$T = \frac{1}{2}\left(\int_m dm \right)v_P^2 - (v_P)_x \omega \left(\int_m y \, dm \right) + (v_P)_y \omega \left(\int_m x \, dm \right) + \frac{1}{2}\omega^2 \left(\int_m r^2 \, dm \right)$$

The first integral on the right represents the entire mass m of the body. Since $\bar{y}m = \int y \, dm$ and $\bar{x}m = \int x \, dm$, the second and third integrals locate the body's center of mass G with respect to P. The last integral represents the body's moment of inertia I_P, computed about the z axis passing through point P. Thus,

$$T = \tfrac{1}{2}mv_P^2 - (v_P)_x \omega \bar{y}m + (v_P)_y \omega \bar{x}m + \tfrac{1}{2}I_P\omega^2 \qquad (18\text{–}1)$$

As a special case, if point P coincides with the mass center G for the body, then $\bar{y} = \bar{x} = 0$, and therefore

$$T = \tfrac{1}{2}mv_G^2 + \tfrac{1}{2}I_G\omega^2 \qquad (18\text{–}2)$$

Here I_G is the moment of inertia for the body about an axis which is perpendicular to the plane of motion and passes through the mass center. Both terms on the right side are *always positive*, since the velocities are squared. Furthermore, it may be verified that these terms have units of length times force, common units being $m \cdot N$ or $ft \cdot lb$. Recall, however, that in the SI system the unit of energy is the joule (J), where $1 \, J = 1 \, m \cdot N$.

Translation. When a rigid body of mass m is subjected to either rectilinear or curvilinear *translation*, the kinetic energy due to rotation is zero, since $\boldsymbol{\omega} = \mathbf{0}$. From Eq. 18–2, the kinetic energy of the body is therefore

$$T = \tfrac{1}{2}mv_G^2 \tag{18–3}$$

where v_G is the magnitude of the translational velocity $\mathbf{v}$ at the instant considered, Fig. 18–2.

Rotation About a Fixed Axis. When a rigid body is *rotating about a fixed axis* passing through point O, Fig. 18–3, the body has both *translational* and *rotational* kinetic energy as defined by Eq. 18–2, i.e.,

$$T = \tfrac{1}{2}mv_G^2 + \tfrac{1}{2}I_G\omega^2 \tag{18–4}$$

The body's kinetic energy may also be formulated by noting that $v_G = r_G\omega$, in which case $T = \tfrac{1}{2}(I_G + mr_G^2)\omega^2$. By the parallel-axis theorem, the terms inside the parentheses represent the moment of inertia I_O of the body about an axis perpendicular to the plane of motion and passing through point O. Hence,*

$$T = \tfrac{1}{2}I_O\omega^2 \tag{18–5}$$

From the derivation, this equation will give the same result as Eq. 18–4, since it accounts for *both* the translational and rotational kinetic energies of the body.

General Plane Motion. When a rigid body is subjected to general plane motion, Fig. 18–4, it has an angular velocity $\boldsymbol{\omega}$ and its mass center has a velocity $\mathbf{v}_G$. Hence, the kinetic energy is defined by Eq. 18–2, i.e.,

$$T = \tfrac{1}{2}mv_G^2 + \tfrac{1}{2}I_G\omega^2 \tag{18–6}$$

Here it is seen that the total kinetic energy of the body consists of the *scalar* sum of the body's *translational* kinetic energy, $\tfrac{1}{2}mv_G^2$, and *rotational* kinetic energy about its mass center, $\tfrac{1}{2}I_G\omega^2$.

Because energy is a scalar quantity, the total kinetic energy for a system of *connected* rigid bodies is the sum of the kinetic energies of all its moving parts. Depending on the type of motion, the kinetic energy of *each body* is found by applying Eq. 18–2 or the alternative forms mentioned above.

---

*The similarity between this derivation and that of $\Sigma M_O = I_O\alpha$, Eq. 17–16, should be noted. Also note that the same result can be obtained directly from Eq. 18–1 by selecting point P at O, realizing that $\mathbf{v}_O = \mathbf{0}$.

Translation

Fig. 18–2

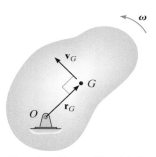

Rotation About a Fixed Axis

Fig. 18–3

General Plane Motion

Fig. 18–4

The total kinetic energy of this soil compactor consists of the kinetic energy of the body or frame of the machine due to its translation, and the translational and rotational kinetic energies of the roller and the wheels due to their general plane motion. Here we exclude the additional kinetic energy developed by the moving parts of the engine and drive train.

EXAMPLE 18.1

(a)

(b)

Fig. 18–5

The system of three elements shown in Fig. 18–5a consists of a 6-kg block B, a 10-kg disk D, and a 12-kg cylinder C. If no slipping occurs, determine the total kinetic energy of the system at the instant shown.

SOLUTION

In order to compute the kinetic energy of the disk and cylinder, it is first necessary to determine ω_D, ω_C, and v_G, Fig. 18–5a. From the *kinematics* of the disk,

$$v_B = r_D\omega_D; \qquad 0.8 \text{ m/s} = (0.1 \text{ m})\omega_D \qquad \omega_D = 8 \text{ rad/s}$$

Since the cylinder rolls without slipping, the instantaneous center of zero velocity is at the point of contact with the ground, Fig. 18–5b. Hence,

$$v_E = r_{E/IC}\omega_C; \qquad 0.8 \text{ m/s} = (0.2 \text{ m})\omega_C \qquad \omega_C = 4 \text{ rad/s}$$

$$v_G = r_{G/IC}\omega_C; \qquad v_G = (0.1 \text{ m})(4 \text{ rad/s}) = 0.4 \text{ m/s}$$

Block.

$$T_B = \tfrac{1}{2}m_B v_B^2 = \tfrac{1}{2}(6 \text{ kg})(0.8 \text{ m/s})^2 = 1.92 \text{ J}$$

Disk.

$$T_D = \tfrac{1}{2}I_D\omega_D^2 = \tfrac{1}{2}\left(\tfrac{1}{2}m_D r_D^2\right)\omega_D^2$$

$$= \tfrac{1}{2}\left[\tfrac{1}{2}(10 \text{ kg})(0.1 \text{ m})^2\right](8 \text{ rad/s})^2 = 1.60 \text{ J}$$

Cylinder.

$$T_C = \tfrac{1}{2}mv_G^2 + \tfrac{1}{2}I_G\omega_C^2 = \tfrac{1}{2}mv_G^2 + \tfrac{1}{2}\left(\tfrac{1}{2}m_C r_C^2\right)\omega_C^2$$

$$= \tfrac{1}{2}(12 \text{ kg})(0.4 \text{ m/s})^2 + \tfrac{1}{2}\left[\tfrac{1}{2}(12 \text{ kg})(0.1 \text{ m})^2\right](4 \text{ rad/s})^2 = 1.44 \text{ J}$$

The total kinetic energy of the system is therefore

$$T = T_B + T_D + T_C$$

$$= 1.92 \text{ J} + 1.60 \text{ J} + 1.44 \text{ J} = 4.96 \text{ J} \qquad \textit{Ans.}$$

18.2   The Work of a Force

Several types of forces are often encountered in planar kinetics problems involving a rigid body. The work of each of these forces has been presented in Sec. 14.1 and is listed below as a summary.

Work of a Variable Force.   If an external force **F** acts on a rigid body, the work done by the force when it moves along the path s, Fig. 18–6, is defined as

$$U_F = \mathbf{F} \cdot \mathbf{r} = \int_s F \cos \theta \, ds \qquad (18\text{–}7)$$

Here θ is the angle between the "tails" of the force vector and the differential displacement. In general, the integration must account for the variation of the force's direction and magnitude.

Fig. 18–6

Work of a Constant Force.   If an external force $\mathbf{F}_c$ acts on a rigid body, Fig. 18–7, and maintains a constant magnitude F_c and constant direction θ, while the body undergoes a translation s, Eq. 18–7 can be integrated so that the work becomes

$$U_{F_c} = (F_c \cos \theta)s \qquad (18\text{–}8)$$

Here $F_c \cos \theta$ represents the magnitude of the component of force in the direction of displacement.

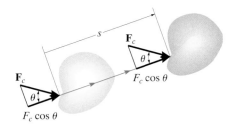

Fig. 18–7

Work of a Weight.

The weight of a body does work only when the body's center of mass G undergoes a *vertical displacement* Δy. If this displacement is *upward*, Fig. 18–8, the work is negative, since the weight and displacement are in opposite directions.

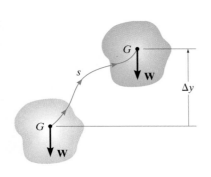

$$U_W = -W\,\Delta y \qquad (18\text{–}9)$$

Likewise, if the displacement is *downward* $(-\Delta y)$ the work becomes *positive*. In both cases the elevation change is considered to be small so that $\mathbf{W}$, which is caused by gravitation, is constant.

Fig. 18–8

Work of a Spring Force.

If a linear elastic spring is attached to a body, the spring force $F_s = ks$ *acting on the body* does work when the spring either stretches or compresses from s_1 to a *further* position s_2. In both cases the work will be *negative* since the *displacement of the body* is in the opposite direction to the force, Fig. 18–9. The work done is

Unstretched position of spring, $s = 0$

$$U_s = -\left(\tfrac{1}{2}ks_2^2 - \tfrac{1}{2}ks_1^2\right) \qquad (18\text{–}10)$$

where $|s_2| > |s_1|$.

Fig. 18–9

Forces That Do No Work.

There are some external forces that do no work when the body is displaced. These forces can act either at *fixed points* on the body, or they can have a direction *perpendicular to their displacement*. Examples include the reactions at a pin support about which a body rotates, the normal reaction acting on a body that moves along a fixed surface, and the weight of a body when the center of gravity of the body moves in a *horizontal plane*, Fig. 18–10. A rolling resistance force $\mathbf{F}_r$ acting on a round body as it *rolls without slipping* over a rough surface also does no work, Fig. 18–10.* This is because, during any *instant of time* dt, $\mathbf{F}_r$ acts at a point on the body which has *zero velocity* (instantaneous center, *IC*), and so the work done by the force on the point is zero. In other words, the point is not displaced in the direction of the force during this instant. Since $\mathbf{F}_r$ contacts successive points for only an instant, the work of $\mathbf{F}_r$ will be zero.

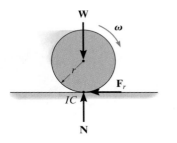

Fig. 18–10

---

*The work done by the frictional force *when the body slips* is discussed in Sec. 14.3.

18.3   The Work of a Couple

When a body subjected to a couple undergoes general plane motion, the two couple forces do work *only* when the body undergoes a *rotation*. To show this, consider the body in Fig. 18–11a, which is subjected to a couple moment $M = Fr$. Any general differential displacement of the body can be considered as a translation plus rotation. When the body *translates*, such that the *component of displacement* along the line of action of the forces is ds_t, Fig. 18–11b, clearly the "positive" work of one force *cancels* the "negative" work of the other. If the body undergoes a differential rotation $d\theta$ about an axis which is perpendicular to the plane of the couple and intersects the plane at point O, Fig. 18–11c, then each force undergoes a displacement $ds_\theta = (r/2)\,d\theta$ in the direction of the force. Hence, the total work done is

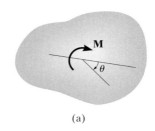

(a)

$$dU_M = F\left(\frac{r}{2}d\theta\right) + F\left(\frac{r}{2}d\theta\right) = (Fr)\,d\theta$$
$$= M\,d\theta$$

Translation
(b)

Here the line of action of $d\theta$ is parallel to the line of action of M. This is *always the case for general plane motion*, since **M** and $d\boldsymbol{\theta}$ are perpendicular to the plane of motion. Furthermore, the resultant work is *positive* when **M** and $d\boldsymbol{\theta}$ have the *same sense of direction* and *negative* if these vectors have an *opposite sense of direction*.

When the body rotates in the plane through a finite angle θ measured in radians, from θ_1 to θ_2, the work of a couple is

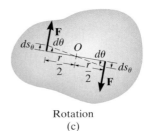

Rotation
(c)

$$U_M = \int_{\theta_1}^{\theta_2} M\,d\theta \qquad (18\text{–}11)$$

Fig. 18–11

If the couple moment **M** has a *constant magnitude*, then

$$U_M = M(\theta_2 - \theta_1) \qquad (18\text{–}12)$$

Here the work is *positive* provided **M** and $(\boldsymbol{\theta}_2 - \boldsymbol{\theta}_1)$ are in the same direction.

EXAMPLE 18.2

0.75 m

$M = 50$ N · m

$k = 30$ N/m

θ

B

$P = 80$ N

2 m

1 m

(a)

$\mathbf{A}_y$

$\mathbf{A}_x$

θ

50 N · m

$\mathbf{F}_s$

$P = 80$ N

1.5 m

98.1 N

0.5 m

1 m

(b)

Fig. 18–12

The bar shown in Fig. 18–12a has a mass of 10 kg and is subjected to a couple moment of $M = 50$ N · m and a force of $P = 80$ N, which is always applied perpendicular to the end of the bar. Also, the spring has an unstretched length of 0.5 m and remains in the vertical position due to the roller guide at B. Determine the total work done by all the forces acting on the bar when it has rotated downward from $\theta = 0°$ to $\theta = 90°$.

SOLUTION

First the free-body diagram of the bar is drawn in order to account for all the forces that act on it, Fig. 18–12b.

Weight W. Since the weight $10(9.81)$ N = 98.1 N is displaced downward 1.5 m, the work is

$$U_W = 98.1 \text{ N}(1.5 \text{ m}) = 147.2 \text{ J}$$

Why is the work positive?

Couple Moment M. The couple moment rotates through an angle of $\theta = \pi/2$ rad. Hence,

$$U_M = 50 \text{ N} \cdot \text{m}(\pi/2) = 78.5 \text{ J}$$

Spring Force F_s. When $\theta = 0°$ the spring is stretched $(0.75 \text{ m} - 0.5) = 0.25$ m, and when $\theta = 90°$, the stretch is $(2 \text{ m} + 0.75 \text{ m}) - 0.5 \text{ m} = 2.25$ m. Thus,

$$U_s = -\left[\tfrac{1}{2}(30 \text{ N/m})(2.25 \text{ m})^2 - \tfrac{1}{2}(30 \text{ N/m})(0.25 \text{ m})^2\right] = -75.0 \text{ J}$$

By inspection the spring does negative work on the bar since F_s acts in the opposite direction to displacement. This checks with the result.

Force P. As the bar moves downward, the force is displaced through a distance of $(\pi/2)(3 \text{ m}) = 4.712$ m. The work is positive. Why?

$$U_P = 80 \text{ N}(4.712 \text{ m}) = 377.0 \text{ J}$$

Pin Reactions. Forces $\mathbf{A}_x$ and $\mathbf{A}_y$ do no work since they are not displaced.

Total Work. The work of all the forces when the bar is displaced is thus

$$U = 147.2 \text{ J} + 78.5 \text{ J} - 75.0 \text{ J} + 377.0 \text{ J} = 528 \text{ J} \qquad Ans.$$

18.4   Principle of Work and Energy

By applying the principle of work and energy developed in Sec. 14.2 to each of the particles of a rigid body and adding the results algebraically, since energy is a scalar, the principle of work and energy for a rigid body becomes

$$T_1 + \Sigma U_{1-2} = T_2 \qquad (18\text{-}13)$$

This equation states that the body's initial translational *and* rotational kinetic energy, plus the work done by all the external forces and couple moments acting on the body as the body moves from its initial to its final position, is equal to the body's final translational *and* rotational kinetic energy. Note that the work of the body's *internal forces* does not have to be considered since the body is rigid. These forces occur in equal but opposite collinear pairs, so that when the body moves, the work of one force cancels that of its counterpart. Furthermore, since the body is rigid, *no relative movement* between these forces occurs, so that no internal work is done.

When several rigid bodies are pin connected, connected by inextensible cables, or in mesh with one another, Eq. 18–13 may be applied to the entire system of connected bodies. In all these cases the internal forces, which hold the various members together, do no work and hence are eliminated from the analysis.

The work of the torque or moment developed by the driving gears on the two motors is transformed into kinetic energy of rotation of the drum of the mixer.

Procedure for Analysis

The principle of work and energy is used to solve kinetic problems that involve *velocity, force,* and *displacement,* since these terms are involved in the formulation. For application, it is suggested that the following procedure be used.

Kinetic Energy (Kinematic Diagrams).

- The kinetic energy of a body is made up of two parts. Kinetic energy of translation is referenced to the velocity of the mass center, $T = \frac{1}{2}mv_G^2$, and kinetic energy of rotation is determined from knowing the moment of inertia about the mass center, $T = \frac{1}{2}I_G\omega^2$. In the special case of rotation about a fixed axis, these two kinetic energies are combined and can be expressed as $T = \frac{1}{2}I_O\omega^2$, where I_O is the moment of inertia about the axis of rotation.

- *Kinematic diagrams* for velocity may be useful for determining v_G and ω or for establishing a *relationship* between v_G and ω.*

Work (Free-Body Diagram).

- Draw a free-body diagram of the body when it is located at an intermediate point along the path in order to account for all the forces and couple moments which do work on the body as it moves along the path.

- A force does work when it moves through a displacement in the direction of the force.

- Forces that are functions of displacement must be integrated to obtain the work. Graphically, the work is equal to the area under the force–displacement curve.

- The work of a weight is the product of its magnitude and the vertical displacement, $U_W = Wy$. It is positive when the weight moves downwards.

- The work of a spring is of the form $U_s = \frac{1}{2}ks^2$, where k is the spring stiffness and s is the stretch or compression of the spring.

- The work of a couple is the product of the couple moment and the angle in radians through which it rotates.

- Since *algebraic addition* of the work terms is required, it is important that the proper sign of each term be specified. Specifically, work is *positive* when the force (couple moment) is in the *same direction* as its displacement (rotation); otherwise, it is negative.

Principle of Work and Energy.

- Apply the principle of work and energy, $T_1 + \Sigma U_{1-2} = T_2$. Since this is a scalar equation, it can be used to solve for only one unknown when it is applied to a single rigid body.

*A brief review of Secs. 16.5 to 16.7 may prove helpful when solving problems, since computations for kinetic energy require a kinematic analysis of velocity.

EXAMPLE 18.3

The 30-kg disk shown in Fig. 18–13a is pin supported at its center. Determine the number of revolutions it must make to attain an angular velocity of 20 rad/s starting from rest. It is acted upon by a constant force $F = 10$ N, which is applied to a cord wrapped around its periphery, and a constant couple moment $M = 5$ N·m. Neglect the mass of the cord in the calculation.

$M = 5$ N·m

0.2 m

O

$F = 10$ N

(a)

SOLUTION

Kinetic Energy. Since the disk rotates about a fixed axis, the kinetic energy can be computed using $T = \frac{1}{2}I_O\omega^2$, where the moment of inertia is $I_O = \frac{1}{2}mr^2$. Initially, the disk is at rest, so that

$$T_1 = 0$$

$$T_2 = \tfrac{1}{2}I_O\omega_2^2 = \tfrac{1}{2}\left[\tfrac{1}{2}(30 \text{ kg})(0.2 \text{ m})^2\right](20 \text{ rad/s})^2 = 120 \text{ J}$$

Work (Free-Body Diagram). As shown in Fig. 18–13b, the pin reactions $\mathbf{O}_x$ and $\mathbf{O}_y$ and the weight (294.3 N) do no work, since they are not displaced. The *couple moment*, having a constant magnitude, does positive work $U_M = M\theta$ as the disk *rotates* through a clockwise angle of θ rad, and the *constant force* $\mathbf{F}$ does positive work $U_{F_c} = Fs$ as the cord *moves* downward $s = \theta r = \theta(0.2 \text{ m})$.

Principle of Work and Energy.

$$\{T_1\} + \{\Sigma U_{1-2}\} = \{T_2\}$$

$$\{T_1\} + \{M\theta + Fs\} = \{T_2\}$$

$$\{0\} + \{(5 \text{ N}\cdot\text{m})\theta + (10 \text{ N})\theta(0.2 \text{ m})\} = \{120 \text{ J}\}$$

$$\theta = 17.1 \text{ rad} = 17.1 \text{ rad}\left(\frac{1 \text{ rev}}{2\pi \text{ rad}}\right) = 2.73 \text{ rev} \qquad Ans.$$

NOTE: This problem has also been solved in Example 17.9. Compare the two methods of solution and note that since force, velocity, and displacement θ are involved, a work–energy approach yields a more direct solution.

294.3 N

$M = 5$ N·m

O

0.2 m

$\mathbf{O}_x$

$\mathbf{O}_y$

$F = 10$ N

(b)

Fig. 18–13

EXAMPLE 18.4

The 700-kg pipe is equally suspended from the two tines of the fork lift shown in the photo. It is undergoing a swinging motion such that when $\theta = 30°$ it is momentarily at rest. Determine the normal and frictional forces acting on each tine which are needed to support the pipe at the instant $\theta = 0°$. Measurements of the pipe and the suspender are shown in Fig. 18–14a. Neglect the mass of the suspender and the thickness of the pipe.

(a)

Fig. 18–14

SOLUTION

We must use the equations of motion to find the forces on the tines since these forces do no work. Before doing this, however, we will apply the principle of work and energy to determine the angular velocity of the pipe when $\theta = 0°$.

Kinetic Energy (Kinematic Diagram).   Since the pipe is originally at rest, then

$$T_1 = 0$$

The final kinetic energy may be computed with reference to either the fixed point O or the center of mass G. For the calculation we will consider the pipe to be a thin ring so that $I_G = mr^2$. If point G is considered, we have

$$T_2 = \tfrac{1}{2}m(v_G)_2^2 + \tfrac{1}{2}I_G\omega_2^2$$
$$= \tfrac{1}{2}(700 \text{ kg})[(0.4 \text{ m})\omega_2]^2 + \tfrac{1}{2}[700 \text{ kg}(0.15 \text{ m})^2]\omega_2^2$$
$$= 63.875\omega_2^2$$

If point O is considered then the parallel-axis theorem must be used to determine I_O. Hence,

$$T_2 = \tfrac{1}{2}I_O\omega_2^2 = \tfrac{1}{2}[700 \text{ kg}(0.15 \text{ m})^2 + 700 \text{ kg}(0.4 \text{ m})^2]\omega_2^2$$
$$= 63.875\omega_2^2$$

Work (Free-Body Diagram). Fig. 18–14b. The normal and frictional forces on the tines do no work since they do not move as the pipe swings. The weight, centered at G, does positive work since the weight moves downward through a vertical distance $\Delta y = 0.4 \text{ m} - 0.4 \cos 30° \text{ m} = 0.05359 \text{ m}$.

Principle of Work and Energy.

$$\{T_1\} + \{\Sigma U_{1-2}\} = \{T_2\}$$

$$\{0\} + \{700(9.81) \text{ N}(0.05359 \text{ m})\} = \{63.875\omega_2^2\}$$

$$\omega_2 = 2.40 \text{ rad/s}$$

(b)

Equations of Motion. Referring to the free-body and kinetic diagrams shown in Fig. 18–14c, and using the result for ω_2, we have

$$\xleftarrow{+} \Sigma F_t = m(a_G)_t; \quad F_T = 700(a_G)_t$$

$$+\uparrow \Sigma F_n = m(a_G)_n; \quad N_T - 700(9.81) \text{ N} = 700 \text{ kg}(2.40 \text{ rad/s})^2(0.4 \text{ m})$$

$$\curvearrowright +\Sigma M_O = I_O\alpha; \quad 0 = [700 \text{ kg}(0.15 \text{ m})^2 + 700 \text{ kg}(0.4 \text{ m})^2]\alpha$$

Since $(a_G)_t = 0.4\alpha$, then

$$\alpha = 0, \quad (a_G)_t = 0$$

$$F_T = 0$$

$$N_T = 8.48 \text{ kN}$$

There are two tines used to support the load, therefore

$$F'_T = 0 \qquad\qquad Ans.$$

$$N'_T = \frac{8.48 \text{ kN}}{2} = 4.24 \text{ kN} \qquad\qquad Ans.$$

NOTE: Due to the swinging motion the tines are subjected to a *greater* normal force than would be the case if the load was static. In that case $N'_T = 700(9.81) \text{ N}/2 = 3.43 \text{ kN}$.

(c)

EXAMPLE 18.5

(a)

(b)

(c)

Fig. 18–15

The wheel shown in Fig. 18–15a weighs 40 lb and has a radius of gyration $k_G = 0.6$ ft about its mass center G. If it is subjected to a clockwise couple moment of 15 lb·ft and rolls from rest without slipping, determine its angular velocity after its center G moves 0.5 ft. The spring has a stiffness $k = 10$ lb/ft and is initially unstretched when the couple moment is applied.

SOLUTION

Kinetic Energy (Kinematic Diagram).   Since the wheel is initially at rest,

$$T_1 = 0$$

The kinematic diagram of the wheel when it is in the final position is shown in Fig. 18–15b. Hence, the final kinetic energy is

$$T_2 = \tfrac{1}{2}m(v_G)_2^2 + \tfrac{1}{2}I_G\omega_2^2$$

$$= \frac{1}{2}\left(\frac{40\ \text{lb}}{32.2\ \text{ft/s}^2}\right)(v_G)_2^2 + \frac{1}{2}\left[\frac{40\ \text{lb}}{32.2\ \text{ft/s}^2}(0.6\ \text{ft})^2\right]\omega_2^2$$

The velocity of the mass center can be related to the angular velocity from the instantaneous center of zero velocity (IC), i.e., $(v_G)_2 = 0.8\omega_2$. Substituting into the above equation and simplifying, we have

$$T_2 = 0.621\omega_2^2$$

Work (Free-Body Diagram).   As shown in Fig. 18–15c, only the spring force $\mathbf{F}_s$ and the couple moment do work. The normal force does not move along its line of action and the frictional force does *no work*, since the wheel does not slip as it rolls.

The work of $\mathbf{F}_s$ may be computed using $U_s = -\tfrac{1}{2}ks^2$. Here the work is negative since $\mathbf{F}_s$ is in the opposite direction to displacement. Since the wheel does not slip when the center G moves 0.5 ft, then the wheel rotates $\theta = s_G/r_{G/IC} = 0.5\ \text{ft}/0.8\ \text{ft} = 0.625$ rad, Fig. 18–15b. Hence, the spring stretches $s_A = \theta r_{A/IC} = 0.625\ \text{rad}(1.6\ \text{ft}) = 1$ ft.

Principle of Work and Energy.

$$\{T_1\} + \{\Sigma U_{1-2}\} = \{T_2\}$$

$$\{T_1\} + \{M\theta - \tfrac{1}{2}ks^2\} = \{T_2\}$$

$$\{0\} + \left\{15\ \text{lb}\cdot\text{ft}(0.625\ \text{rad}) - \frac{1}{2}(10\ \text{lb/ft})(1\ \text{ft})^2\right\} = \{0.621\omega_2^2\ \text{ft}\cdot\text{lb}\}$$

$$\omega_2 = 2.65\ \text{rad/s} \downarrow \qquad\qquad\qquad \textit{Ans.}$$

EXAMPLE 18.6

The 10-kg rod shown in Fig. 18–16a is constrained so that its ends move along the grooved slots. The rod is initially at rest when $\theta = 0°$. If the slider block at B is acted upon by a horizontal force $P = 50$ N, determine the angular velocity of the rod at the instant $\theta = 45°$. Neglect friction and the mass of blocks A and B.

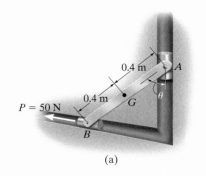

(a)

SOLUTION

Why can the principle of work and energy be used to solve this problem?

Kinetic Energy (Kinematic Diagrams).   Two kinematic diagrams of the rod, when it is, in the initial position 1 and final position 2, are shown in Fig. 18–16b. When the rod is in position 1, $T_1 = 0$ since $(\mathbf{v}_G)_1 = \boldsymbol{\omega}_1 = \mathbf{0}$. In position 2 the angular velocity is $\boldsymbol{\omega}_2$ and the velocity of the mass center is $(\mathbf{v}_G)_2$. Hence, the kinetic energy is

$$T_2 = \tfrac{1}{2}m(v_G)_2^2 + \tfrac{1}{2}I_G\omega_2^2$$
$$= \tfrac{1}{2}(10 \text{ kg})(v_G)_2^2 + \tfrac{1}{2}\left[\tfrac{1}{12}(10 \text{ kg})(0.8 \text{ m})^2\right]\omega_2^2$$
$$= 5(v_G)_2^2 + 0.267(\omega_2)^2$$

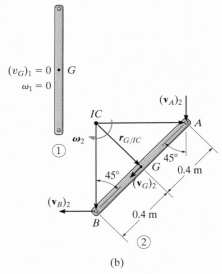

(b)

The two unknowns $(v_G)_2$ and ω_2 may be related from the instantaneous center of zero velocity for the rod. Fig. 18–16b. It is seen that as A moves downward with a velocity $(\mathbf{v}_A)_2$, B moves horizontally to the left with a velocity $(\mathbf{v}_B)_2$. Knowing these directions, the IC is determined as shown in the figure. Hence,

$$(v_G)_2 = r_{G/IC}\omega_2 = (0.4 \tan 45° \text{ m})\omega_2$$
$$= 0.4\omega_2$$

Therefore,
$$T_2 = 0.8\omega_2^2 + 0.267\omega_2^2 = 1.067\omega_2^2$$

Work (Free-Body Diagram).   Fig. 18–16c. The normal forces $\mathbf{N}_A$ and $\mathbf{N}_B$ do no work as the rod is displaced. Why? The 98.1-N weight is displaced a vertical distance of $\Delta y = (0.4 - 0.4 \cos 45°)$ m; whereas the 50-N force moves a horizontal distance of $s = (0.8 \sin 45°)$ m. Both of these forces do positive work. Why?

Principle of Work and Energy.
$$\{T_1\} + \{\Sigma U_{1-2}\} = \{T_2\}$$
$$\{T_1\} + \{W\,\Delta y + Ps\} = \{T_2\}$$
$$\{0\} + \{98.1 \text{ N}(0.4 \text{ m} - 0.4 \cos 45° \text{ m}) + 50 \text{ N}(0.8 \sin 45° \text{ m})\}$$
$$= \{1.067\omega_2^2 \text{ J}\}$$

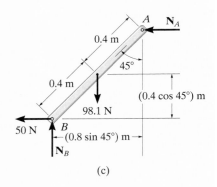

(c)

Solving for ω_2 gives
$$\omega_2 = 6.11 \text{ rad/s } \downarrow \qquad\qquad Ans.$$

Fig. 18–16

PROBLEMS

18–1. At a given instant the body of mass m has an angular velocity ω and its mass center has a velocity $\mathbf{v}_G$. Show that its kinetic energy can be represented as $T = \frac{1}{2}I_{IC}\omega^2$, where I_{IC} is the moment of inertia of the body computed about the instantaneous axis of zero velocity, located a distance $r_{G/IC}$ from the mass center as shown.

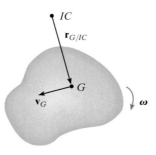

Prob. 18–1

18–2. The wheel is made from a 5-kg thin ring and two 2-kg slender rods. If the torsional spring attached to the wheel's center has a stiffness $k = 2\,\text{N} \cdot \text{m/rad}$, so that the torque on the center of the wheel is $M = (2\theta)\,\text{N} \cdot \text{m}$, where θ is in radians, determine the maximum angular velocity of the wheel if it is rotated two revolutions and then released from rest.

Prob. 18–2

18–3. At the instant shown, the 30-lb disk has a counterclockwise angular velocity of 5 rad/s when its center has a velocity of 20 ft/s. Determine the kinetic energy of the disk at this instant.

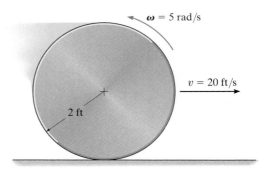

Prob. 18–3

*18–4. The uniform rectangular plate weighs 30 lb. If the plate is pinned at A and has an angular velocity of 3 rad/s, determine the kinetic energy of the plate.

Prob. 18–4

18–5. At the instant shown, link AB has an angular velocity $\omega_{AB} = 2$ rad/s. If each link is considered as a uniform slender bar with a weight of 0.5 lb/in., determine the total kinetic energy of the system.

Prob. 18–5

18–6. Determine the kinetic energy of the system of three links. Links AB and CD each weigh 10 lb, and link BC weighs 20 lb.

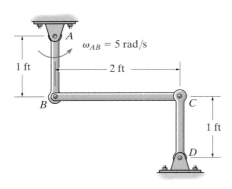

Prob. 18–6

18–7. The mechanism consists of two rods, AB and BC, which weigh 10 lb and 20 lb, respectively, and a 4-lb block at C. Determine the kinetic energy of the system at the instant shown, when the block is moving at 3 ft/s.

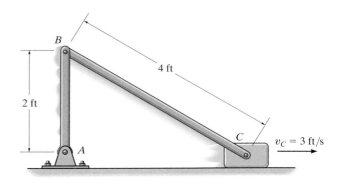

Prob. 18–7

*18–8. Solve Prob. 17–59 using the principle of work and energy.

18–9. A force of $P = 20$ N is applied to the cable, which causes the 175-kg reel to turn since it is resting on the two rollers A and B of the dispenser. Determine the angular velocity of the reel after it has made two revolutions starting from rest. Neglect the mass of the rollers and the mass of the cable. The radius of gyration of the reel about its center axis is $k_G = 0.42$ m.

Prob. 18–9

18–10.   The rotary screen S is used to wash limestone. When empty it has a mass of 800 kg and a radius of gyration of $k_G = 1.75$ m. Rotation is achieved by applying a torque of $M = 280$ N $\cdot$ m about the drive wheel A. If no slipping occurs at A and the supporting wheel at B is free to roll, determine the angular velocity of the screen after it has rotated 5 revolutions. Neglect the mass of A and B.

Prob. 18–10

18–12.   The soap-box car has a weight of 110 lb, including the passenger but *excluding* its four wheels. Each wheel has a weight of 5 lb, radius of 0.5 ft, and a radius of gyration $k = 0.3$ ft, computed about an axis passing through the wheel's axle. Determine the car's speed after it has traveled 100 ft starting from rest. The wheels roll without slipping. Neglect air resistance.

Prob. 18–12

18–11.   A yo-yo has a weight of 0.3 lb and a radius of gyration $k_O = 0.06$ ft. If it is released from rest, determine how far it must descend in order to attain an angular velocity $\omega = 70$ rad/s. Neglect the mass of the string and assume that the string is wound around the central peg such that the mean radius at which it unravels is $r = 0.02$ ft.

Prob. 18–11

18–13.   The pendulum of the Charpy impact machine has a mass of 50 kg and a radius of gyration of $k_A = 1.75$ m. If it is released from rest when $\theta = 0°$, determine its angular velocity just before it strikes the specimen S, $\theta = 90°$.

Prob. 18–13

18–14. The 10-kg pulley has a radius of gyration about O of $k_O = 0.21$ m. If a motor M supplies a force to the cable of $P = 800(3 - 2e^{-x})$ N, where x is the amount of cable wound up in meters, determine the speed of the 50-kg crate when it has been hoisted 2 m starting from rest. Neglect the mass of the cable and assume the cable does not slip on the pulley.

Prob. 18–14

18–15. The uniform pipe has a mass of 16 Mg and radius of gyration about the z axis of $k_G = 2.7$ m. If the worker pushes on it with a horizontal force of 50 N, applied perpendicular to the pipe, determine the pipe's angular velocity when it has rotated 90° about the z axis, starting from rest. Assume the pipe does not swing.

Prob. 18–15

*18–16. The 4-kg slender rod is subjected to the force and couple moment. When it is in the position shown it has an angular velocity $\omega_1 = 6$ rad/s. Determine its angular velocity at the instant it has rotated downward 90°. The force is always applied perpendicular to the axis of the rod. Motion occurs in the vertical plane.

Prob. 18–16

18–17. The 4-kg slender rod is subjected to the force and couple moment. When the rod is in the position shown it has a angular velocity $\omega_1 = 6$ rad/s. Determine its angular velocity at the instant it has rotated 360°. The force is always applied perpendicular to the axis of the rod and motion occurs in the vertical plane.

Prob. 18–17

18–18. The elevator car E has a mass of 1.80 Mg and the counterweight C has a mass of 2.30 Mg. If a motor turns the driving sheave A with a constant torque of $M = 100$ N · m, determine the speed of the elevator when it has ascended 10 m starting from rest. Each sheave A and B has a mass of 150 kg and a radius of gyration of $k = 0.2$ m about its mass center or pinned axis. Neglect the mass of the cable and assume the cable does not slip on the sheaves.

18–19. The elevator car E has a mass of 1.80 Mg and the counterweight C has a mass of 2.30 Mg. If a motor turns the driving sheave A with a torque of $M = (0.06\theta^2 + 7.5)$ N · m, where θ is in radians, determine the speed of the elevator when it has ascended 12 m starting from rest. Each sheave A and B has a mass of 150 kg and a radius of gyration of $k = 0.2$ m about its mass center or pinned axis. Neglect the mass of the cable and assume the cable does not slip on the sheaves.

***18–20.** The wheel has a mass of 100 kg and a radius of gyration $k_O = 0.2$ m. A motor supplies a torque $M = (40\theta + 900)$ N · m, where θ is in radians, about the drive shaft at O. Determine the speed of the loading car, which has a mass of 300 kg, after it travels $s = 4$ m. Initially the car is at rest when $s = 0$ and $\theta = 0°$. Neglect the mass of the attached cable and the mass of the car's wheels.

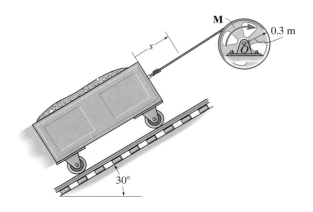

Prob. 18–20

18–21. The gear has a weight of 15 lb and a radius of gyration $k_G = 0.375$ ft. If the spring is unstretched when the torque $M = 6$ lb · ft is applied, determine the gear's angular velocity after its mass center G has moved to the left 2 ft.

Probs. 18–18/19

Prob. 18–21

18–22. The 20-kg disk is originally at rest, and the spring holds it in equilibrium. A couple moment of $M = 30$ N·m is then applied to the disk as shown. Determine its angular velocity at the instant its mass center G has moved 0.8 m down along the inclined plane. The disk rolls without slipping.

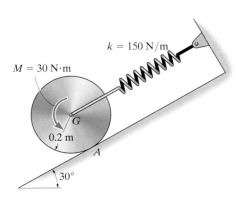

Prob. 18–22

18–23. The 20-kg disk is originally at rest, and the spring holds it in equilibrium. A couple moment of $M = 30$ N·m is then applied to the disk as shown. Determine how far the center of mass of the disk travels down along the incline, measured from the equilibrium position, before it stops. The disk rolls without slipping.

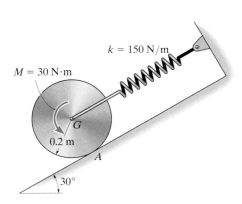

Prob. 18–23

*18–24. The linkage consists of two 8-lb rods AB and CD and a 10-lb bar AD. When $\theta = 0°$, rod AB is rotating with an angular velocity $\omega_{AB} = 2$ rad/s. If rod CD is subjected to a couple moment $M = 15$ lb·ft and bar AD is subjected to a horizontal force $P = 20$ lb as shown, determine ω_{AB} at the instant $\theta = 90°$.

18–25. The linkage consists of two 8-lb rods AB and CD and a 10-lb bar AD. When $\theta = 0°$, rod AB is rotating with an angular velocity $\omega_{AB} = 2$ rad/s. If rod CD is subjected to a couple moment $M = 15$ lb·ft and bar AD is subjected to a horizontal force $P = 20$ lb as shown, determine ω_{AB} at the instant $\theta = 45°$.

Probs. 18–24/25

18–26. The spool has a weight of 500 lb and a radius of gyration of $k_G = 1.75$ ft. A horizontal force of $P = 15$ lb is applied to a cable wrapped around its inner core. If the spool is originally at rest, determine its angular velocity after the mass center G has moved 6 ft to the left. The spool rolls without slipping. Neglect the mass of the cable.

Prob. 18–26

18–27. The double pulley consists of two parts that are attached to one another. It has a weight of 50 lb and a centroidal radius of gyration of $k_O = 0.6$ ft and is turning with an angular velocity of 20 rad/s clockwise. Determine the kinetic energy of the system. Assume that neither cable slips on the pulley.

$\omega = 20$ rad/s

0.5ft   1 ft

O

B 30 lb

A 20 lb

Prob. 18–27

18–29. The 1500-lb cement bucket is hoisted using a motor that supplies a torque of $M = 2000$ lb·ft to the axle of the wheel. If the wheel has a weight of 115 lb and a radius of gyration about O of $k_O = 0.95$ ft, determine the speed of the bucket when it has been hoisted 10 ft starting from rest.

1.25 ft

O

$M = 2000$ lb · ft

Prob. 18–29

*18–28. The system consists of a 20-lb disk A, a 4-lb slender rod BC, and a 1-lb smooth collar C. If the disk rolls without slipping, determine the velocity of the collar at the instant the rod becomes horizontal, i.e., $\theta = 0°$. The system is released from rest when $\theta = 45°$.

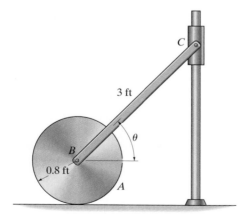

C

3 ft

θ

B

0.8 ft

A

Prob. 18–28

18–30. The assembly consists of two 15-lb slender rods and a 20-lb disk. If the spring is unstretched when $\theta = 45°$ and the assembly is released from rest at this position, determine the angular velocity of rod AB at the instant $\theta = 0°$. The disk rolls without slipping.

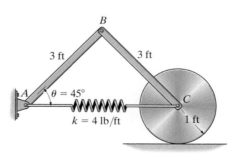

B

3 ft   3 ft

A $\theta = 45°$ C

$k = 4$ lb/ft   1 ft

Prob. 18–30

18–31. The uniform door has a mass of 20 kg and can be treated as a thin plate having the dimensions shown. If it is connected to a torsional spring at A, which has a stiffness of $k = 80$ N · m/rad, determine the required initial twist of the spring in radians so that the door has an angular velocity of 12 rad/s when it closes at $\theta = 0°$ after being opened at $\theta = 90°$ and released from rest. *Hint:* For a torsional spring $M = k\theta$, when k is the stiffness and θ is the angle of twist.

Prob. 18–31

*18–32.** The uniform slender bar has a mass m and a length L is subjected to a uniform distributed load w_0 which is always erected perpendicular to the axis of the bar. If it is released from n from the position shown, determine its angular velocity at the instant it has rotated 90°. Solve the problem for rotation in (a) the horizontal plane, and (b) the vertical plane.

Prob. 18–32

18–33. A ball of mass m and radius r is cast onto the horizontal surface such that it rolls without slipping. Determine the minimum speed v_G of its mass center G so that it rolls completely around the loop of radius $R + r$ without leaving the track.

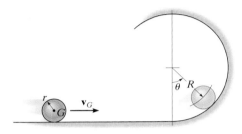

Prob. 18–33

18–34. The beam has a weight of 1500 lb and is being raised to a vertical position by pulling very slowly on its bottom end A. If the cord fails when $\theta = 60°$ and the beam is essentially at rest, determine the speed of A at the instant cord BC becomes vertical. Neglect friction and the mass of the cords, and treat the beam as a slender rod.

Prob. 18–34

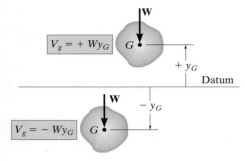

Gravitational potential energy

Fig. 18–17

18.5   Conservation of Energy

When a force system acting on a rigid body consists only of *conservative forces*, the conservation of energy theorem may be used to solve a problem that otherwise would be solved using the principle of work and energy. This theorem is often easier to apply since the work of a conservative force is *independent of the path* and depends only on the initial and final positions of the body. It was shown in Sec. 14.5 that the work of a conservative force may be expressed as the difference in the body's potential energy measured from an arbitrarily selected reference or datum.

Gravitational Potential Energy.   Since the total weight of a body can be considered concentrated at its center of gravity, the *gravitational potential energy* of the body is determined by knowing the height of the body's center of gravity above or below a horizontal datum. Measuring y_G as *positive upward*, the gravitational potential energy of the body is thus

$$V_g = W y_G \qquad (18\text{--}14)$$

Here the potential energy is *positive* when y_G is positive, since the weight has the ability to do *positive work* when the body is moved back to the datum, Fig. 18–17. Likewise, if the body is located *below* the datum $(-y_G)$, the gravitational potential energy is *negative*, since the weight does *negative work* when the body is returned to the datum.

Elastic Potential Energy.   The force developed by an elastic spring is also a conservative force. The *elastic potential energy* which a spring imparts to an attached body when the spring is elongated or compressed from an initial undeformed position $(s = 0)$ to a final position s, Fig. 18–18, is

$$V_e = +\tfrac{1}{2} k s^2 \qquad (18\text{--}15)$$

Unstretched position of spring, $s = 0$

Elastic potential energy

Fig. 18–18

In the deformed position, the spring force acting *on the body* always has the capacity for doing positive work when the spring is returned back to its original undeformed position (see Sec. 14.5).

Conservation of Energy.   In general, if a body is subjected to both gravitational and elastic forces, the total *potential energy* is expressed as a potential function V represented as the algebraic sum

$$V = V_g + V_e \qquad (18\text{-}16)$$

Here measurement of V depends on the location of the body with respect to the selected datum.

Realizing that the work of conservative forces can be written as a difference in their potential energies, i.e., $(\Sigma U_{1\text{-}2})_{\text{cons}} = V_1 - V_2$, Eq. 14–16, we can rewrite the principle of work and energy for a rigid body as

$$T_1 + V_1 + (\Sigma U_{1\text{-}2})_{\text{noncons}} = T_2 + V_2 \qquad (18\text{-}17)$$

Here $(\Sigma U_{1\text{-}2})_{\text{noncons}}$ represents the work of the nonconservative forces such as friction. If this term is zero, then

$$T_1 + V_1 = T_2 + V_2 \qquad (18\text{-}18)$$

This equation is referred to as the conservation of mechanical energy. It states that the *sum* of the potential and kinetic energies of the body remains *constant* when the body moves from one position to another. It also applies to a system of smooth, pin-connected rigid bodies, bodies connected by inextensible cords, and bodies in mesh with other bodies. In all these cases the forces acting at the points of contact are *eliminated* from the analysis, since they occur in equal but opposite collinear pairs and each pair of forces moves through an equal distance when the system undergoes a displacement.

It is important to remember that only problems involving conservative force systems may be solved by using Eq. 18–18. As stated in Sec. 14.5, friction or other drag-resistant forces, which depend on velocity or acceleration, are nonconservative. The work of such forces is transformed into thermal energy used to heat up the surfaces of contact, and consequently this energy is dissipated into the surroundings and may not be recovered. Therefore, problems involving frictional forces can be solved by using either the principle of work and energy written in the form of Eq. 18–17, if it applies, or the equations of motion.

The torsional springs located at the top of the garage door wind up as the door is lowered. When the door is raised, the potential energy stored in the springs is then transferred into gravitational potential energy of the door's weight, thereby making it easy to open.

Procedure for Analysis

The conservation of energy equation is used to solve problems involving *velocity, displacement*, and *conservative force systems*. For application it is suggested that the following procedure be used.

Potential Energy.

- Draw two diagrams showing the body located at its initial and final positions along the path.

- If the center of gravity, G, is subjected to a *vertical displacement*, establish a fixed horizontal datum from which to measure the body's gravitational potential energy V_g.

- Data pertaining to the elevation y_G of the body's center of gravity from the datum and the extension or compression of any connecting springs can be determined from the problem geometry and listed on the two diagrams.

- Recall that the potential energy $V = V_g + V_e$. Here $V_g = W y_G$, which can be positive or negative, and $V_e = \frac{1}{2}ks^2$, which is always positive.

Kinetic Energy.

- The kinetic energy of the body consists of two parts, namely translational kinetic energy, $T = \frac{1}{2}mv_G^2$, and rotational kinetic energy, $T = \frac{1}{2}I_G\omega^2$.

- Kinematic diagrams for velocity may be useful for determining v_G and ω for establishing a *relationship* between these quantities.

Conservation of Energy.

- Apply the conservation of energy equation $T_1 + V_1 = T_2 + V_2$.

EXAMPLE 18.7

The 10-kg rod AB shown in Fig. 18–19a is confined so that its ends move in the horizontal and vertical slots. The spring has a stiffness of $k = 800$ N/m and is unstretched when $\theta = 0°$. Determine the angular velocity of AB when $\theta = 0°$, if the rod is released from rest when $\theta = 30°$. Neglect the mass of the slider blocks.

(a)

SOLUTION

Potential Energy.   The two diagrams of the rod, when it is located at its initial and final positions, are shown in Fig. 18–19b. The datum, used to measure the gravitational potential energy, is placed in line with the rod when $\theta = 0°$.

When the rod is in position 1, the center of gravity G is located *below the datum* so that the gravitational potential energy is *negative*. Furthermore, (positive) elastic potential energy is stored in the spring, since it is stretched a distance of $s_1 = (0.4 \sin 30°)$ m. Thus,

$$V_1 = -W y_1 + \tfrac{1}{2} k s_1^2$$
$$= -98.1 \text{ N}(0.2 \sin 30° \text{ m}) + \tfrac{1}{2}(800 \text{ N/m})(0.4 \sin 30° \text{ m})^2 = 6.19 \text{ J}$$

When the rod is in position 2, the potential energy of the rod is zero, since the spring is unstretched, $s_2 = 0$, and the center of gravity G is located at the datum. Thus,

$$V_2 = 0$$

Kinetic Energy.   The rod is released from rest from position 1, thus $(\mathbf{v}_G)_1 = \mathbf{0}$ and $\boldsymbol{\omega}_1 = \mathbf{0}$, and

$$T_1 = 0$$

In position 2, the angular velocity is $\boldsymbol{\omega}_2$ and the rod's mass center has a velocity of $(\mathbf{v}_G)_2$. Thus,

$$T_2 = \tfrac{1}{2} m (v_G)_2^2 + \tfrac{1}{2} I_G \omega_2^2$$
$$= \tfrac{1}{2}(10 \text{ kg})(v_G)_2^2 + \tfrac{1}{2}\left[\tfrac{1}{12}(10 \text{ kg})(0.4 \text{ m})^2\right]\omega_2^2$$

(b)

Using *kinematics*, $(\mathbf{v}_G)_2$ can be related to $\boldsymbol{\omega}_2$ as shown in Fig. 18–19c. At the instant considered, the instantaneous center of zero velocity (IC) for the rod is at point A; hence, $(v_G)_2 = (r_{G/IC})\omega_2 = (0.2)\omega_2$. Substituting into the above expression and simplifying, we get

$$T_2 = 0.267\omega_2^2$$

(c)

Conservation of Energy.

$$\{T_1\} + \{V_1\} = \{T_2\} + \{V_2\}$$
$$\{0\} + \{6.19\} = \{0.267\omega_2^2\} + \{0\}$$
$$\omega_2 = 4.82 \text{ rad/s} \,\ ⤸ \qquad \qquad \textit{Ans.}$$

Fig. 18–19

EXAMPLE 18.8

$k = 2$ lb/ft

4 ft

G

0.75 ft

3 ft

(a)

$s_2 = 3$ ft

$s_1 = 4$ ft

30 lb

30 lb

②   ①

(b)

ω_2

$(v_G)_2$

0.75 ft

IC

(c)

Fig. 18–20

The disk shown in Fig. 18–20a has a weight of 30 lb and a radius of gyration of $k_G = 0.6$ ft, and it is attached to a spring which has a stiffness $k = 2$ lb/ft and an unstretched length of 1 ft. If the disk is released from rest in the position shown and rolls without slipping, determine its angular velocity at the instant G moves 3 ft to the left.

SOLUTION

Potential Energy. Two diagrams of the disk, when it is located in its initial and final positions, are shown in Fig. 18–20b. A gravitational datum is not needed here since the weight is not displaced vertically. From the problem geometry the spring is stretched $s_1 = \left(\sqrt{3^2 + 4^2} - 1\right) = 4$ ft and $s_2 = (4 - 1) = 3$ ft in the initial and final positions, respectively. Hence,

$$V_1 = \tfrac{1}{2}ks_1^2 = \tfrac{1}{2}(2 \text{ lb/ft})(4 \text{ ft})^2 = 16 \text{ J}$$

$$V_2 = \tfrac{1}{2}ks_2^2 = \tfrac{1}{2}(2 \text{ lb/ft})(3 \text{ ft})^2 = 9 \text{ J}$$

Kinetic Energy. The disk is released from rest so that $(\mathbf{v}_G)_1 = \mathbf{0}$, $\boldsymbol{\omega}_1 = \mathbf{0}$, and

$$T_1 = 0$$

In the final position,

$$T_2 = \frac{1}{2}m(v_G)_2^2 + \frac{1}{2}I_G\omega_2^2$$

$$= \frac{1}{2}\left(\frac{30 \text{ lb}}{32.2 \text{ ft/s}^2}\right)(v_G)_2^2 + \frac{1}{2}\left[\left(\frac{30 \text{ lb}}{32.2 \text{ ft/s}^2}\right)(0.6 \text{ ft})^2\right]\omega_2^2$$

Since the disk rolls without slipping, $(\mathbf{v}_G)_2$ can be related to $\boldsymbol{\omega}_2$ from the instantaneous center of zero velocity, Fig. 18–20c, i.e., $(v_G)_2 = (0.75 \text{ ft})\omega_2$. Substituting and simplifying yields

$$T_2 = 0.430\omega_2^2$$

Conservation of Energy.

$$\{T_1\} + \{V_1\} = \{T_2\} + \{V_2\}$$

$$\{0\} + \{16\} = \{0.430\omega_2^2\} + \{9\}$$

$$\omega_2 = 4.04 \text{ rad/s} \,\gamma \qquad\qquad Ans.$$

NOTE: If the principle of work and energy were used to solve this problem, then the work of the spring would have to be determined by considering both the change in magnitude and direction of the spring force.

EXAMPLE 18.9

The 10-kg homogeneous disk shown in Fig. 18–21*a* is attached to a uniform 5-kg rod *AB*. If the assembly is released from rest when $\theta = 60°$, determine the angular velocity of the rod when $\theta = 0°$. Assume that the disk rolls without slipping. Neglect friction along the guide and the mass of the collar at *B*.

(a)

SOLUTION

Potential Energy.   Two diagrams for the rod and disk, when they are located at their initial and final positions, are shown in Fig. 18–21*b*. For convenience the datum passes through point *A*.

When the system is in position 1, the rod's weight has positive potential energy. Thus,

$$V_1 = W_R y_1 = 49.05 \text{ N}(0.3 \sin 60° \text{ m}) = 12.74 \text{ J}$$

When the system is in position 2, both the weight of the rod and the weight of the disk have zero potential energy. Why? Thus,

$$V_2 = 0$$

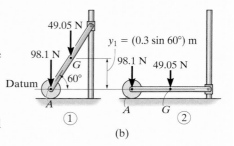

(b)

Kinetic Energy.   Since the entire system is at rest at the initial position,

$$T_1 = 0$$

In the final position the rod has an angular velocity $(\boldsymbol{\omega}_R)_2$ and its mass center has a velocity $(\mathbf{v}_G)_2$, Fig. 18–21*c*. Since the rod is *fully extended* in this position, the disk is momentarily at rest, so $(\boldsymbol{\omega}_D)_2 = \mathbf{0}$ and $(\mathbf{v}_A)_2 = \mathbf{0}$. For the rod $(\mathbf{v}_G)_2$ can be related to $(\boldsymbol{\omega}_R)_2$ from the instantaneous center of zero velocity, which is located at point *A*, Fig. 18–21*c*. Hence, $(v_G)_2 = r_{G/IC}(\omega_R)_2$ or $(v_G)_2 = 0.3(\omega_R)_2$. Thus,

$$T_2 = \frac{1}{2}m_R(v_G)_2^2 + \frac{1}{2}I_G(\omega_R)_2^2 + \frac{1}{2}m_D(v_A)_2^2 + \frac{1}{2}I_A(\omega_D)_2^2$$

$$= \frac{1}{2}(5 \text{ kg})[0.3 \text{ m}(\omega_R)_2]^2 + \frac{1}{2}\left[\frac{1}{12}(5 \text{ kg})(0.6 \text{ m})^2\right](\omega_R)_2^2 + 0 + 0$$

$$= 0.3(\omega_R)_2^2$$

(c)

Fig. 18–21

Conservation of Energy.

$$\{T_1\} + \{V_1\} = \{T_2\} + \{V_2\}$$

$$\{0\} + \{12.74\} = \{0.3(\omega_R)_2^2\} + \{0\}$$

$$(\omega_R)_2 = 6.52 \text{ rad/s} \, \downarrow \qquad\qquad Ans.$$

PROBLEMS

18–35. Solve Prob. 18–13 using the conservation of energy equation.

*18–36. Solve Prob. 18–12 using the conservation of energy equation.

18–37. Solve Prob. 18–30 using the conservation of energy equation.

18–38. Solve Prob. 18–11 using the conservation of energy equation.

18–39. Solve Prob. 18–34 using the conservation of energy equation.

*18–40. Solve Prob. 18–28 using the conservation of energy equation.

18–41. The spool has a mass of 50 kg and a radius of gyration $k_O = 0.280$ m. If the 20-kg block A is released from rest, determine the distance the block must fall in order for the spool to have an angular velocity $\omega = 5$ rad/s. Also, what is the tension in the cord while the block is in motion? Neglect the mass of the cord.

18–42. When the slender 10-kg bar AB is horizontal it is at rest and the spring is unstretched. Determine the stiffness k of the spring so that the motion of the bar is momentarily stopped when it has rotated downward 90°.

Prob. 18–42

18–43. The 15-lb disk is rotating about pin A in the vertical plane with an angular velocity $\omega = 2$ rad/s when $\theta = 0°$. Determine its angular velocity at the instant $\theta = 90°$. Also, compute the horizontal and vertical components of reaction at A at this instant.

Prob. 18–41

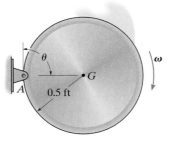

Prob. 18–43

*18–44. The door is made from one piece, whose ends move along the horizontal and vertical tracks. If the door is in the open position, $\theta = 0°$, and then released, determine the speed at which its end A strikes the stop at C. Assume the door is a 180-lb thin plate having a width of 10 ft.

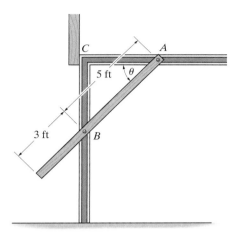

Prob. 18–44

18–45. The overhead door BC is pushed slightly from its open position and then rotates downward about the pin at A. Determine its angular velocity just before its end B strikes the floor. Assume the door is a thin plate having a mass of 180 kg and length of 6 m. Neglect the mass of the supporting frame AB and AC.

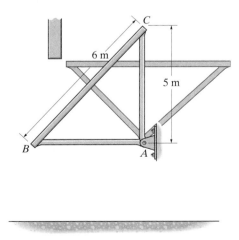

Prob. 18–45

18–46. The 80-lb cylinder is attached to the 10-lb slender rod which is pinned at point A. At the instant $\theta = 30°$ the rod has an angular velocity of $\omega_0 = 1$ rad/s as shown. Determine the angle θ to which the rod swings before it momentarily stops.

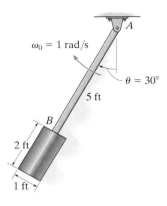

Prob. 18–46

18–47. The compound disk pulley consists of a hub and attached outer rim. If it has a mass of 3 kg and a radius of gyration $k_G = 45$ mm, determine the speed of block A after A descends 0.2 m from rest. Blocks A and B each have a mass of 2 kg. Neglect the mass of the cords.

Prob. 18–47

*18–48.   The 15-kg semicircular segment is released from rest in the position shown. Determine the velocity of point A when it has rotated counterclockwise 90°. Assume that the segment rolls without slipping on the surface. The moment of inertia about its mass center is $I_G = 0.25$ kg · m².

Prob. 18–48

18–50.   The assembly consists of a 3-kg pulley A and 10-kg pulley B. If a 2-kg block is suspended from the cord, determine the block's speed after it descends 0.5 m starting from rest. Neglect the mass of the cord and treat the pulleys as thin disks. No slipping occurs.

Prob. 18–50

18–49.   The uniform 150-lb stone (rectangular block) is being turned over on its side by pulling the vertical cable *slowly* upward until the stone begins to tip. If it then falls freely ($\mathbf{T} = \mathbf{0}$) from an essentially balanced at-rest position, determine the speed at which the corner A strikes the pad at B. The stone does not slip at its corner C as it falls.

Prob. 18–49

18–51.   A uniform ladder having a weight of 30 lb is released from rest when it is in the vertical position. If it is allowed to fall freely, determine the angle θ at which the bottom end A starts to lift off the ground. For the calculation, assume the ladder to be a slender rod and neglect friction at A.

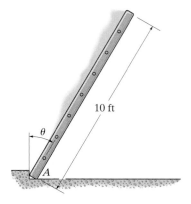

Prob. 18–51

*18–52. The 25-lb slender rod AB is attached to a spring BC which, has an unstretched length of 4 ft. If the rod is released from rest when $\theta = 30°$, determine its angular velocity at the instant $\theta = 90°$.

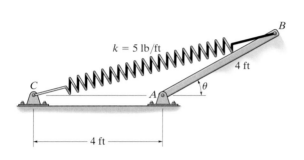

Prob. 18–52

18–53. The 25-lb slender rod AB is attached to a spring BC which has an unstretched length of 4 ft. If the rod is released from rest when $\theta = 30°$, determine the angular velocity of the rod the instant the spring becomes unstretched.

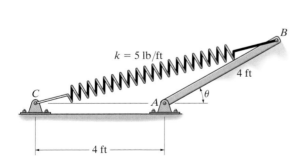

Prob. 18–53

18–54. A chain that has a negligible mass is draped over the sprocket which has a mass of 2 kg and a radius of gyration of $k_O = 50$ mm. If the 4-kg block A is released from rest in the position $s = 1$ m, determine the angular velocity of the sprocket at the instant $s = 2$ m.

Prob. 18–54

18–55. Solve Prob. 18–54 if the chain has a mass of 0.8 kg/m. For the calculation neglect the portion of the chain that wraps over the sprocket.

Prob. 18–55

18–56. Pulley *A* has a weight of 30 lb and a centroidal radius of gyration $k_B = 0.6$ ft. Determine the speed of the 20-lb crate *C* at the instant $s = 10$ ft. Initially, the crate is released from rest when $s = 5$ ft. The pulley at *P* "rolls" downward on the cord without slipping. For the calculation, neglect the mass of this pulley and the cord as it unwinds from the inner and outer hubs of pulley *A*.

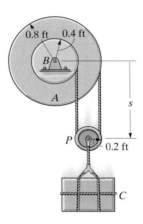

Prob. 18–56

18–57. The assembly consists of two 8-lb bars which are pinconnected to the two 10-lb disks. If the bars are released from rest when $\theta = 60°$, determine their angular velocities at the instant $\theta = 0°$. Assume the disks roll without slipping.

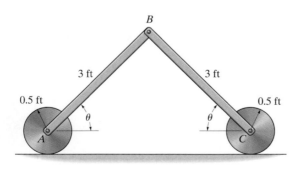

Prob. 18–57

18–58. The assembly consists of two 8-lb bars which are pinconnected to the two 10-lb disks. If the bars are released from rest when $\theta = 60°$, determine their angular velocities at the instant $\theta = 30°$. Assume the disks roll without slipping.

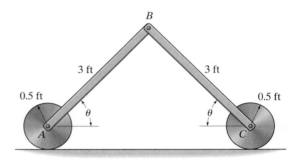

Prob. 18–58

18–59. The end *A* of the garage door *AB* travels along the horizontal track, and the end of member *BC* is attached to a spring at *C*. If the spring is originally unstretched, determine the stiffness *k* so that when the door falls downward from rest in the position shown, it will have zero angular velocity the moment it closes, i.e., when it and *BC* become vertical. Neglect the mass of member *BC* and assume the door is a thin plate having a weight of 200 lb and a width and height of 12 ft. There is a similar connection and spring on the other side of the door.

Prob. 18–59

Chapter Review

Kinetic Energy

The kinetic energy of a rigid body that undergoes planar motion can be referenced to its mass center. It includes a scalar sum of its translational and rotational kinetic energies.
(*See pages 445–448.*)

Translation

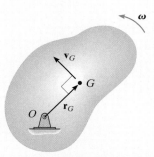

Rotation About a Fixed Axis

General Plane Motion

Translation

$$T = \tfrac{1}{2}mv_G^2$$

Rotation About a Fixed Axis

$$T = \tfrac{1}{2}mv_G^2 + \tfrac{1}{2}I_G\omega^2 \qquad \text{or} \qquad T = \tfrac{1}{2}I_O\omega^2$$

General Plane Motion

$$T = \tfrac{1}{2}mv_G^2 + \tfrac{1}{2}I_G\omega^2$$

Work of a Force and a Couple Moment

A force does work when it undergoes a displacement ds in the direction of the force. In particular, the frictional and normal forces that act on a cylinder or sphere that rolls *without slipping* will do no work, since the normal force does not undergo a displacement and the frictional force acts on successive points on the surface of the body. (*See pages 449–451.*)

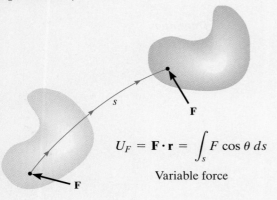

$$U_F = \mathbf{F} \cdot \mathbf{r} = \int_s F \cos \theta \, ds$$

Variable force

$$U_W = -W \, \Delta y$$

Weight

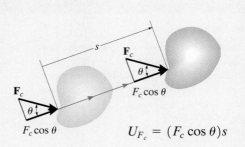

$$U_{F_c} = (F_c \cos \theta)s$$

Constant force

$$U_s = -\left(\tfrac{1}{2}ks_2^2 - \tfrac{1}{2}ks_1^2\right)$$

Spring force

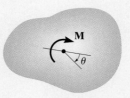

$$U_M = \int_{\theta_1}^{\theta_2} M \, d\theta$$

$$U_M = M(\theta_2 - \theta_1)$$

Constant magnitude

Principle of Work and Energy

Problems that involve velocity, force, and displacement can be solved using the principle of work and energy. The kinetic energy is the sum of both its rotational and translational parts. For application, a free-body diagram should be drawn in order to account for the work of all of the forces and couple moments that act on the body as it moves along the path.
(*See page 453.*)

$$T_1 = \Sigma U_{1-2} = T_2$$

Conservation of Energy

If a rigid body is subjected only to conservative forces, then the conservation-of-energy equation can be used to solve the problem. This equation requires that the sum of the potential and kinetic energies of the body remain the same at any two points along the path.

$$T_1 + V_1 = T_2 + V_2$$

The potential energy is the sum of its gravitational and elastic potential energies. The gravitational potential energy will be positive if the body's center of mass is located above a datum. If it is below the datum, then it will be negative. The elastic potential energy is always positive, regardless if the spring is stretched or compressed.
(*See pages 468–470.*)

$$V_g = +W y_G$$

$$V_g = -W y_G$$

Gravitational potential energy

$$V_e = \tfrac{1}{2} k s^2$$

Elastic potential energy

$$V = V_g + V_e$$

The launching of this weather satellite requires application of impulse and momentum principles to accurately predict its orbital angular motion and proper orientation.

19

Planar Kinetics of a Rigid Body: Impulse and Momentum

CHAPTER OBJECTIVES

- To develop formulations for the linear and angular momentum of a body.
- To apply the principles of linear and angular impulse and momentum to solve rigid-body planar kinetic problems that involve force, velocity, and time.
- To discuss application of the conservation of momentum.
- To analyze the mechanics of eccentric impact.

19.1 Linear and Angular Momentum

In this chapter we will use the principles of linear and angular impulse and momentum to solve problems involving force, velocity, and time as related to the planar motion of a rigid body. Before doing this, we will first formalize the methods for obtaining a body's linear and angular momentum when the body is subjected to translation, rotation about a fixed axis, and general plane motion. Here we will assume the body is symmetric with respect to an inertial x–y reference plane.

Particle momentum
diagram

(a)

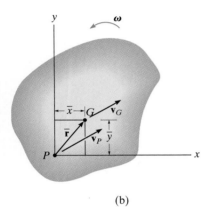

(b)

Fig. 19–1

Linear Momentum.
The linear momentum of a rigid body is determined by summing vectorially the linear momenta of all the particles of the body, i.e., $\mathbf{L} = \Sigma m_i \mathbf{v}_i$. Since $\Sigma m_i \mathbf{v}_i = m\mathbf{v}_G$ (see Sec. 15.2) we can also write

$$\mathbf{L} = m\mathbf{v}_G \tag{19–1}$$

This equation states that the body's linear momentum is a vector quantity having a *magnitude* mv_G, which is commonly measured in units of kg · m/s or slug · ft/s and a *direction* defined by v_G the velocity of the body's mass center.

Angular Momentum.
Consider the body in Fig. 19–1a, which is subjected to general plane motion. At the instant shown, the arbitrary point P has a velocity $\mathbf{v}_P$, and the body has an angular velocity $\boldsymbol{\omega}$. If the velocity of the ith particle of the body is to be determined, Fig. 19–1a, then

$$\mathbf{v}_i = \mathbf{v}_P + \mathbf{v}_{i/P} = \mathbf{v}_P + \boldsymbol{\omega} \times \mathbf{r}$$

The angular momentum of particle i about point P is equal to the "moment" of the particle's linear momentum about P, Fig. 19–1a. Thus,

$$(\mathbf{H}_P)_i = \mathbf{r} \times m_i \mathbf{v}_i$$

Expressing $\mathbf{v}_i$ in terms of $\mathbf{v}_P$ and using Cartesian vectors, we have

$$(H_P)_i \mathbf{k} = m_i(x\mathbf{i} + y\mathbf{j}) \times [(v_P)_x\mathbf{i} + (v_P)_y\mathbf{j} + \omega\mathbf{k} \times (x\mathbf{i} + y\mathbf{j})]$$
$$(H_P)_i = -m_i y(v_P)_x + m_i x(v_P)_y + m_i \omega r^2$$

Letting $m_i \rightarrow dm$ and integrating over the entire mass m of the body, we obtain

$$H_P = -\left(\int_m y\, dm \right)(v_P)_x + \left(\int_m x\, dm \right)(v_P)_y + \left(\int_m r^2\, dm \right)\omega$$

Here H_P represents the angular momentum of the body about an axis (the z axis) perpendicular to the plane of motion and passing through point P. Since $\overline{y}m = \int y\, dm$ and $\overline{x}m = \int x\, dm$, the integrals for the first and second terms on the right are used to locate the body's center of mass G with respect to P, Fig. 19–1b. Also, the last integral represents the body's moment of inertia computed about the z axis, i.e., $I_P = \int r^2\, dm$. Thus,

$$H_P = -\overline{y}m(v_P)_x + \overline{x}m(v_P)_y + I_P\omega \tag{19–2}$$

This equation reduces to a simpler form if point P coincides with the mass center G for the body,* in which case $\overline{x} = \overline{y} = 0$. Hence,

*It also reduces to the same simple form, $H_P = I_P\omega$, if point P is a *fixed point* (see Eq. 19–9) or the velocity of point P is directed along the line PG.

$$H_G = I_G\omega \qquad\qquad (19\text{--}3)$$

This equation states that the angular momentum of the body computed about G is equal to the product of the moment of inertia of the body about an axis passing through G and the body's angular velocity. It is important to realize that $\mathbf{H}_G$ is a vector quantity having a *magnitude* $I_G\omega$, which is commonly measured in units of $kg \cdot m^2/s$ or $slug \cdot ft^2/s$, and a *direction* defined by $\boldsymbol{\omega}$, which is always perpendicular to the plane of motion.

Equation 19–2 can also be rewritten in terms of the x and y components of the velocity of the body's mass center, $(v_G)_x$ and $(v_G)_y$, and the body's moment of inertia I_G. Since point G is located at coordinates $(\bar{x}, \bar{y})$, then by the parallel-axis theorem, $I_P = I_G + m(\bar{x}^2 + \bar{y}^2)$. Substituting into Eq. 19–2 and rearranging terms, we have

$$H_P = \bar{y}m[-(v_P)_x + \bar{y}\omega] + \bar{x}m[(v_P)_y + \bar{x}\omega] + I_G\omega \qquad (19\text{--}4)$$

From the kinematic diagram of Fig. 19–1b, $\mathbf{v}_G$ can be expressed in terms of $\mathbf{v}_P$ as

$$\mathbf{v}_G = \mathbf{v}_P + \boldsymbol{\omega} \times \bar{\mathbf{r}}$$
$$(v_G)_x\mathbf{i} + (v_G)_y\mathbf{j} = (v_P)_x\mathbf{i} + (v_P)_y\mathbf{j} + \omega\mathbf{k} \times (\bar{x}\mathbf{i} + \bar{y}\mathbf{j})$$

Carrying out the cross product and equating the respective $\mathbf{i}$ and $\mathbf{j}$ components yields the two scalar equations

$$(v_G)_x = (v_P)_x - \bar{y}\omega$$
$$(v_G)_y = (v_P)_y + \bar{x}\omega$$

Substituting these results into Eq. 19–4 yields

$$H_P = -\bar{y}m(v_G)_x + \bar{x}m(v_G)_y + I_G\omega \qquad (19\text{--}5)$$

With reference to Fig. 19–1c, *this result indicates that when the angular momentum of the body is computed about point P, it is equivalent to the moment of the linear momentum* $m\mathbf{v}_G$ *or its components* $m(\mathbf{v}_G)_x$ *and* $m(\mathbf{v}_G)_y$ *about P plus the angular momentum* $I_G\boldsymbol{\omega}$. Note that since $\boldsymbol{\omega}$ is a free vector, $\mathbf{H}_G$ *can act at any point on the body* provided it preserves its same magnitude and direction. Furthermore, since angular momentum is equal to the "moment" of the linear momentum, the *line of action of* $\mathbf{L}$ *must pass through the body's mass center G* in order to preserve the correct magnitude of $\mathbf{H}_P$ when "moments" are computed about P, Fig. 19–1c. Using these results, we will now consider three types of motion.

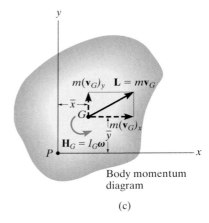

Body momentum diagram

(c)

Fig. 19–1 (cont.)

Translation.   When a rigid body of mass m is subjected to either rectilinear or curvilinear *translation*, Fig. 19–2a, its mass center has a velocity of $\mathbf{v}_G = \mathbf{v}$ and $\boldsymbol{\omega} = \mathbf{0}$ for the body. Hence, the linear momentum and the angular momentum computed about G become

$$
\begin{array}{l}
L = mv_G \\
H_G = 0
\end{array}
\tag{19–6}
$$

If the angular momentum is computed about any other point A on or off the body, Fig. 19–2a, the "moment" of the linear momentum $\mathbf{L}$ must be computed about the point. Since d is the "moment arm" as shown in the figure, then in accordance with Eq. 19–5, $H_A = (d)(mv_G)\,\gtrdot$.

Rotation About a Fixed Axis.   When a rigid body is *rotating about a fixed axis* passing through point O, Fig. 19–2b, the linear momentum and the angular momentum computed about G are

$$
\begin{array}{l}
L = mv_G \\
H_G = I_G\omega
\end{array}
\tag{19–7}
$$

It is sometimes convenient to compute the angular momentum of the body about point O. In this case it is necessary to account for the "moment" of *both* $\mathbf{L}$ and $\mathbf{H}_G$ about O. Noting that $\mathbf{L}$ (or $\mathbf{v}_G$) is always *perpendicular to* $\mathbf{r}_G$, we have

$$
(\zeta+)\ H_O = I_G\omega + r_G(mv_G)
\tag{19–8}
$$

This equation may be *simplified* by first substituting $v_G = r_G\omega$, in which case $H_O = (I_G + mr_G^2)\omega$, and, by the parallel-axis theorem, noting that the terms inside the parentheses represent the moment of inertia I_O of the body about an axis perpendicular to the plane of motion and passing through point O. Hence,*

$$
H_O = I_O\,\boldsymbol{\omega}
\tag{19–9}
$$

For the calculation, then, either Eq. 19–8 or 19–9 can be used.

---

*The similarity between this derivation and that of Eq. 17–16 ($\Sigma M_O = I_O\alpha$) and Eq. 18–5 $\left(T = \frac{1}{2}I_O\omega^2\right)$ should be noted. Also note that the same result can be obtained from Eq. 19–2 by selecting point P at O, realizing that $(v_O)_x = (v_O)_y = 0$.

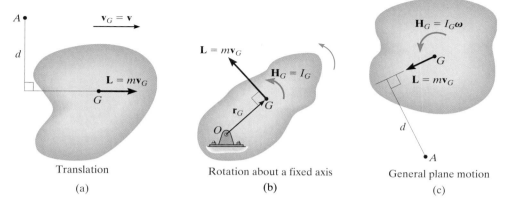

Fig. 19–2

General Plane Motion.   When a rigid body is subjected to general plane motion, Fig. 19–2c, the linear momentum and the angular momentum computed about G become

$$\boxed{\begin{aligned} L &= mv_G \\ H_G &= I_G \boldsymbol{\omega} \end{aligned}}$$

$$(19\text{--}10)$$

If the angular momentum is computed about a point A located either on or off the body, Fig. 19–2c, it is necessary to find the moments of *both* $\mathbf{L}$ and $\mathbf{H}_G$ about this point. In this case,

$$\zeta + \qquad H_A = I_G \omega + (d)(mv_G)$$

Here d is the moment arm, as shown in the figure.

As the pendulum swings downward, its angular momentum about point O can be determined by computing the moment of $I_G \boldsymbol{\omega}$ and mv_G about O. This is $H_O = I_G \omega + (mv_G)d$. Since $v_G = \omega d$, then $H_O = I_G \omega + m(\omega d)d = (I_G + md^2)\omega = I_O \omega$.

EXAMPLE 19.1

At a given instant the 10-kg disk and 5-kg bar have the motions shown in Fig. 19–3a. Determine their angular momenta about point G and about point B for the disk and about G and about the IC for the bar at this instant.

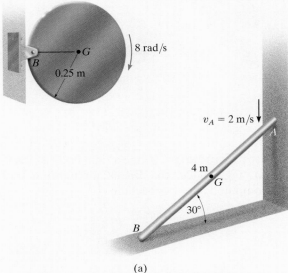

(a)

SOLUTION

Disk.   Since the disk is *rotating about a fixed axis* (through point B), then $v_G = (8 \text{ rad/s})(0.25 \text{ m}) = 2 \text{ m/s}$, Fig. 19–3b. Hence

$$(\curvearrowright)+H_G = I_G\omega = \left[\tfrac{1}{2}(10 \text{ kg})(0.25 \text{ m})^2\right](8 \text{ rad/s}) = 2.50 \text{ kg} \cdot \text{m}^2/\text{s} \; Ans.$$

$$(\curvearrowright)+H_B = I_G\omega + (mv_G)r_G = 2.50 \text{ kg} \cdot \text{m}^2/\text{s} + (10 \text{ kg})(2 \text{ m/s})(0.25 \text{ m})$$

$$= 7.50 \text{ kg} \cdot \text{m}^2/\text{s} \downarrow \qquad\qquad\qquad\qquad Ans.$$

Also, from the table on the inside back cover, $I_B = (3/2)mr^2$, so that

$$(\curvearrowright)+H_B = I_B\omega = \left[\tfrac{3}{2}(10 \text{ kg})(0.25 \text{ m})^2\right](8 \text{ rad/s}) = 7.50 \text{ kg} \cdot \text{m}^2/\text{s} \downarrow \; Ans.$$

Bar.   The bar undergoes *general plane motion*. The IC is established in Fig. 19–3c, so that $\omega = (2 \text{ m/s})/(3.464 \text{ m}) = 0.5774 \text{ rad/s}$ and $v_G = (0.5774 \text{ rad/s})(2 \text{ m}) = 1.155 \text{ m/s}$. Thus,

$$(\curvearrowright)+H_G = I_G\omega = \left[\tfrac{1}{12}(5 \text{ kg})(4 \text{ m})^2\right](0.5774 \text{ rad/s}) = 3.85 \text{ kg} \cdot \text{m}^2/\text{s} \downarrow Ans.$$

Moments of $I_G\omega$ and mv_G about the IC yield

$$(\curvearrowright)+H_{IC} = I_G\omega + d(mv_G) = 3.85 \text{ kg} \cdot \text{m}^2/\text{s} + (2 \text{ m})(5 \text{ kg})(1.155 \text{ m/s})$$

$$= 15.4 \text{ kg} \cdot \text{m}^2/\text{s} \downarrow \qquad\qquad\qquad\qquad Ans.$$

(b)

(c)

Fig. 19–3

19.2 Principle of Impulse and Momentum

Like the case for particle motion, the principle of impulse and momentum for a rigid body is developed by *combining* the equation of motion with kinematics. The resulting equation will allow a *direct solution to problems involving force, velocity, and time.*

Principle of Linear Impulse and Momentum.

The equation of translational motion for a rigid body can be written as $\Sigma\mathbf{F} = m\mathbf{a}_G = m(d\mathbf{v}_G/dt)$. Since the mass of the body is constant,

$$\Sigma\mathbf{F} = \frac{d}{dt}(m\mathbf{v}_G)$$

Multiplying both sides by dt and integrating from $t = t_1$, $\mathbf{v}_G = (\mathbf{v}_G)_1$ to $t = t_2$, $\mathbf{v}_G = (\mathbf{v}_G)_2$ yields

$$\Sigma\int_{t_1}^{t_2}\mathbf{F}\,dt = m(\mathbf{v}_G)_2 - m(\mathbf{v}_G)_1 \qquad (19\text{--}11)$$

This equation is referred to as the *principle of linear impulse and momentum*. It states that the sum of all the impulses created by the *external force system* which acts on the body during the time interval t_1 to t_2 is equal to the change in the linear momentum of the body during the time interval, Fig. 19–4.

Principle of Angular Impulse and Momentum.

If the body has *general plane motion* we can write $\Sigma M_G = I_G\alpha = I_G(d\omega/dt)$. Since the moment of inertia is constant,

$$\Sigma M_G = \frac{d}{dt}(I_G\omega)$$

Multiplying both sides by dt and integrating from $t = t_1$, $\omega = \omega_1$ to $t = t_2$, $\omega = \omega_2$ gives

$$\Sigma\int_{t_1}^{t_2}M_G\,dt = I_G\omega_2 - I_G\omega_1 \qquad (19\text{--}12)$$

In a similar manner, for *rotation about a fixed axis* passing through point O, Eq. 17–16 ($\Sigma M_O = I_O\alpha$) when integrated becomes

$$\Sigma\int_{t_1}^{t_2}M_O\,dt = I_O\omega_2 - I_O\omega_1 \qquad (19\text{--}13)$$

Equations 19–12 and 19–13 are referred to as the *principle of angular impulse and momentum*. Both equations state that the sum of the angular impulses acting on the body during the time interval t_1 to t_2 is equal to the change in the body's angular momentum during this time interval. The angular impulse is determined by integrating the moments about point G or O of all the external forces and couple moments applied to the body.

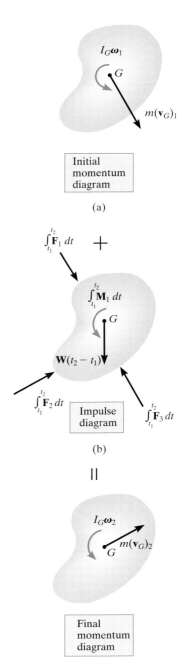

Initial momentum diagram

(a)

Impulse diagram

(b)

Final momentum diagram

(c)

Fig. 19–4

Initial
momentum
diagram

(a)

Impulse
diagram

(b)

||

Final
momentum
diagram

(c)

Fig. 19–4 (repeated)

To summarize these concepts, if motion is occurring in the x–y plane, using impulse and momentum principles the following *three scalar equations* may be written which describe the *planar motion* of the body:

$$m(v_{Gx})_1 + \Sigma \int_{t_1}^{t_2} F_x \, dt = m(v_{Gx})_2$$

$$m(v_{Gy})_1 + \Sigma \int_{t_1}^{t_2} F_y \, dt = m(v_{Gy})_2 \qquad (19\text{–}14)$$

$$I_G\omega_1 + \Sigma \int_{t_1}^{t_2} M_G \, dt = I_G\omega_2$$

The first two of these equations represent the principle of linear impulse and momentum in the x–y plane, Eq. 19–11, and the third equation represents the principle of angular impulse and momentum about the z axis, which passes through the body's mass center G, Eq. 19–12.

The terms in Eqs. 19–14 can be shown graphically by drawing a set of impulse and momentum diagrams for the body, Fig. 19–4. Note that the linear momenta $m\mathbf{v}_G$ are applied at the body's mass center, Figs. 19–4a and 19–4c; whereas the angular momenta $I_G \boldsymbol{\omega}$ are free vectors, and therefore, like a couple moment, they may be applied at any point on the body. When the impulse diagram is constructed, Fig. 19–4b, the forces $\mathbf{F}$ and moment $\mathbf{M}$ vary with time, and are indicated by the integrals. However, if $\mathbf{F}$ and $\mathbf{M}$ are *constant* from t_1 to t_2, integration of the impulses yields $\mathbf{F}(t_2 - t_1)$ and $\mathbf{M}(t_2 - t_1)$, respectively. Such is the case for the body's weight $\mathbf{W}$, Fig. 19–4b.

Equations 19–14 may also be applied to an entire system of connected bodies rather than to each body separately. This eliminates the need to include reactive impulses which occur at the connections since they are *internal* to the system. The resultant equations may be written in symbolic form as

$$\left(\Sigma \begin{array}{c} \text{syst. linear} \\ \text{momentum} \end{array}\right)_{x1} + \left(\Sigma \begin{array}{c} \text{syst. linear} \\ \text{impulse} \end{array}\right)_{x(1-2)} = \left(\Sigma \begin{array}{c} \text{syst. linear} \\ \text{momentum} \end{array}\right)_{x2}$$

$$\left(\Sigma \begin{array}{c} \text{syst. linear} \\ \text{momentum} \end{array}\right)_{y1} + \left(\Sigma \begin{array}{c} \text{syst. linear} \\ \text{impulse} \end{array}\right)_{y(1-2)} = \left(\Sigma \begin{array}{c} \text{syst. linear} \\ \text{momentum} \end{array}\right)_{y2}$$

$$\left(\Sigma \begin{array}{c} \text{syst. angular} \\ \text{momentum} \end{array}\right)_{O1} + \left(\Sigma \begin{array}{c} \text{syst. angular} \\ \text{impulse} \end{array}\right)_{O(1-2)} = \left(\Sigma \begin{array}{c} \text{syst. angular} \\ \text{momentum} \end{array}\right)_{O2}$$

$$(19\text{–}15)$$

As indicated by the third equation, the system's angular momentum and angular impulse must be computed with respect to the *same fixed reference point O* for all the bodies of the system.

Procedure For Analysis

Impulse and momentum principles are used to solve kinetic problems that involve *velocity, force*, and *time* since these terms are involved in the formulation.

Free-Body Diagram.

- Establish the *x, y, z* inertial frame of reference and draw the free-body diagram in order to account for all the forces and couple moments that produce impulses on the body.
- The direction and sense of the initial and final velocity of the body's mass center, $\mathbf{v}_G$, and the body's angular velocity $\boldsymbol{\omega}$ should be established. If any of these motions is unknown, assume that the sense of its components is in the direction of the positive inertial coordinates.
- Compute the moment of inertia I_G or I_O.
- As an alternative procedure, draw the impulse and momentum diagrams for the body or system of bodies. Each of these diagrams represents an outlined shape of the body which graphically accounts for the data required for each of the three terms in Eqs. 19–14 or 19–15, Fig. 19–4. These diagrams are particularly helpful in order to visualize the "moment" terms used in the principle of angular impulse and momentum, if application is about a point other than the body's mass center G or a fixed point O.

Principle of Impulse and Momentum.

- Apply the three scalar equations of impulse and momentum.
- The angular momentum of a rigid body rotating about a fixed axis is the moment of $m\mathbf{v}_G$ plus $I_G\boldsymbol{\omega}$ about the axis. This is equal to $H_O = I_O\boldsymbol{\omega}$, where I_O is the moment of inertia of the body about the axis.
- All the forces acting on the body's free-body diagram will create an impulse; however, some of these forces will do no work.
- Forces that are functions of time must be integrated to obtain the impulse. Graphically, the impulse is equal to the area under the force–time curve.
- The principle of angular impulse and momentum is often used to eliminate unknown impulsive forces that are parallel or pass through a common axis, since the moment of these forces is zero about this axis.

Kinematics.

- If more than three equations are needed for a complete solution, it may be possible to relate the velocity of the body's mass center to the body's angular velocity using *kinematics*. If the motion appears to be complicated, kinematic (velocity) diagrams may be helpful in obtaining the necessary relation.

EXAMPLE 19.2

0.75 ft

M = 4 lb · ft

F = 10 lb

(a)

The 20-lb disk shown in Fig. 19–5a is assumed to be uniform and is pin supported at its center. If it is acted upon by a constant couple moment of 4 lb·ft and a force of 10 lb which is applied to a cord wrapped around its periphery, determine the angular velocity of the disk two seconds after starting from rest. Also, what are the force components of reaction at the pin?

SOLUTION

Since angular velocity, force, and time are involved in the problems we will apply the principles of impulse and momentum to the solution.

Free-Body Diagram. Fig. 19–5b. The disk's mass center does not move; however, the loading causes the disk to rotate clockwise.

The moment of inertia of the disk about its fixed axis of rotation is

$$I_A = \frac{1}{2}mr^2 = \frac{1}{2}\left(\frac{20 \text{ lb}}{32.2 \text{ ft/s}^2}\right)(0.75 \text{ ft})^2 = 0.175 \text{ slug} \cdot \text{ft}^2$$

y

ω

x

4 lb · ft

20 lb

A_x A   0.75 ft

A_y

10 lb

(b)

Fig. 19–5

Principle of Impulse and Momentum.

$(\xrightarrow{+})$
$$m(v_{Ax})_1 + \Sigma \int_{t_1}^{t_2} F_x \, dt = m(v_{Ax})_2$$

$$0 + A_x(2 \text{ s}) = 0$$

$(+\uparrow)$
$$m(v_{Ay})_1 + \Sigma \int_{t_1}^{t_2} F_y \, dt = m(v_{Ay})_2$$

$$0 + A_y(2 \text{ s}) - 20 \text{ lb}(2 \text{ s}) - 10 \text{ lb}(2 \text{ s}) = 0$$

$(\curvearrowright+)$
$$I_A\omega_1 + \Sigma \int_{t_1}^{t_2} M_A \, dt = I_A\omega_2$$

$$0 + 4 \text{ lb} \cdot \text{ft}(2 \text{ s}) + [10 \text{ lb}(2 \text{ s})](0.75 \text{ ft}) = 0.175\omega_2$$

Solving these equations yields

$$A_x = 0 \qquad\qquad Ans.$$
$$A_y = 30 \text{ lb} \qquad\qquad Ans.$$
$$\omega_2 = 132 \text{ rad/s} \, \curvearrowright \qquad\qquad Ans.$$

EXAMPLE 19.3

The 100-kg spool shown in Fig. 19–6a has a radius of gyration $k_G = 0.35$ m. A cable is wrapped around the central hub of the spool, and a horizontal force having a variable magnitude of $P = (t + 10)$ N is applied, where t is in seconds. If the spool is initially at rest, determine its angular velocity in 5 s. Assume that the spool rolls without slipping at A.

$P = (t + 10)$ N

(a)

SOLUTION

Free-Body Diagram. By inspection of the free-body diagram, Fig. 19–6b, the *variable* force P will cause the friction force F_A to be variable, and thus the impulses created by both P and F_A must be determined by integration. The force **P** causes the mass center to have a velocity $\mathbf{v}_G$ to the right, and the spool has a clockwise angular velocity $\boldsymbol{\omega}$.

The moment of inertia of the spool about its mass center is

$$I_G = mk_G^2 = 100 \text{ kg}(0.35 \text{ m})^2 = 12.25 \text{ kg} \cdot \text{m}^2$$

Principle of Impulse and Momentum.

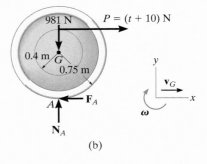

$P = (t + 10)$ N

(b)

$(\xrightarrow{+})$
$$m(v_G)_1 + \Sigma \int F_x \, dt = m(v_G)_2$$

$$0 + \int_0^{5 \text{ s}} (t + 10) \text{ N } dt - \int F_A \, dt = 100 \text{ kg}(v_G)_2$$

$$62.5 - \int F_A \, dt = 100(v_G)_2 \qquad (1)$$

$(\curvearrowright+)$
$$I_G\omega_1 + \Sigma \int M_G \, dt = I_G\omega_2$$

$$0 + \left[\int_0^{5 \text{ s}} (t + 10) \text{ N } dt\right](0.4 \text{ m}) + \left(\int F_A \, dt\right)(0.75 \text{ m}) = (12.25 \text{ kg} \cdot \text{m}^2)\omega_2$$

$$25 + \left(\int F_A \, dt\right)(0.75) = 12.25\omega_2 \qquad (2)$$

Kinematics. Since the spool does not slip, the instantaneous center of zero velocity is at point A, Fig. 19–6c. Hence, the velocity of G can be expressed in terms of the spool's angular velocity as $(v_G)_2 = (0.75 \text{ m})\omega_2$. Substituting this into Eq. 1 and eliminating the unknown impulse $\int F_A \, dt$ between Eqs. 1 and 2, we obtain

$$\omega_2 = 1.05 \text{ rad/s} \downarrow \qquad\qquad Ans.$$

NOTE: A more direct solution can be obtained by applying the principle of angular impulse and momentum about point A. As an exercise, do this and show that one obtains the same result.

(c)

Fig. 19–6

EXAMPLE 19.4

The block shown in Fig. 19–7a has a mass of 6 kg. It is attached to a cord which is wrapped around the periphery of a 20-kg disk that has a moment of inertia $I_A = 0.40 \text{ kg} \cdot \text{m}^2$. If the block is initially moving downward with a speed of 2 m/s, determine its speed in 3 s. Neglect the mass of the cord in the calculation.

0.2 m

A

B

$v_B = 2$ m/s

(a)

SOLUTION I

Free-Body Diagram.   The free-body diagrams of the block and disk are shown in Fig. 19–7b. All the forces are *constant* since the weight of the block causes the motion. The downward motion of the block, $\mathbf{v}_B$, causes $\boldsymbol{\omega}$ of the disk to be clockwise.

Principle of Impulse and Momentum.   We can eliminate $\mathbf{A}_x$ and $\mathbf{A}_y$ from the analysis by applying the principle of angular impulse and momentum about point A. Hence

Disk

196.2 N

y

v_B

$\boldsymbol{\omega}$

x

A

0.2 m $\mathbf{A}_x$

$\mathbf{A}_y$

$\mathbf{T}$

$\mathbf{T}$

58.86 N

(b)

Fig. 19–7

$(\curvearrowright +)$
$$I_A \omega_1 + \Sigma \int M_A \, dt = I_A \omega_2$$

$$0.40 \text{ kg} \cdot \text{m}^2(\omega_1) + T(3 \text{ s})(0.2 \text{ m}) = (0.4 \text{ kg} \cdot \text{m}^2)\omega_2$$

Block

$(+\uparrow)$
$$m_B(v_B)_1 + \Sigma \int F_y \, dt = m_B(v_B)_2$$

$$-6 \text{ kg}(2 \text{ m/s}) + T(3 \text{ s}) - 58.86 \text{ N}(3 \text{ s}) = -6 \text{ kg}(v_B)_2$$

Kinematics.   Since $\omega = v_B/r$, then $\omega_1 = (2 \text{ m/s})/(0.2 \text{ m}) = 10 \text{ rad/s}$ and $\omega_2 = (v_B)_2/0.2 \text{ m} = 5(v_B)_2$. Substituting and solving the equations simultaneously for $(v_B)_2$ yields

$$(v_B)_2 = 13.0 \text{ m/s} \downarrow \qquad\qquad\qquad Ans.$$

SOLUTION II

Impulse and Momentum Diagrams.   We can obtain $(v_B)_2$ *directly* by considering the *system* consisting of the block, the cord, and the disk. The impulse and momentum diagrams have been drawn to clarify application of the principle of angular impulse and momentum about point A, Fig. 19–7c.

Principle of Angular Impulse and Momentum.   Realizing that $\omega_1 = 10$ rad/s and $\omega_2 = 5(v_B)_2$, we have

$$(\curvearrowright +)\left(\sum \begin{array}{c}\text{syst. angular} \\ \text{momentum}\end{array} \right)_{A1} + \left(\sum \begin{array}{c}\text{syst. angular} \\ \text{impulse}\end{array} \right)_{A(1-2)} = \left(\sum \begin{array}{c}\text{syst. angular} \\ \text{momentum}\end{array} \right)_{A2}$$

$$6 \text{ kg}(2 \text{ m/s})(0.2 \text{ m}) + 0.4 \text{ kg} \cdot \text{m}^2(10 \text{ rad/s}) + 58.86 \text{ N}(3 \text{ s})(0.2 \text{ m})$$
$$= 6 \text{ kg}(v_B)_2(0.2 \text{ m}) + 0.40 \text{ kg} \cdot \text{m}^2[5(v_B)_2(0.2 \text{ m})]$$
$$(v_B)_2 = 13.0 \text{ m/s} \downarrow \qquad\qquad Ans.$$

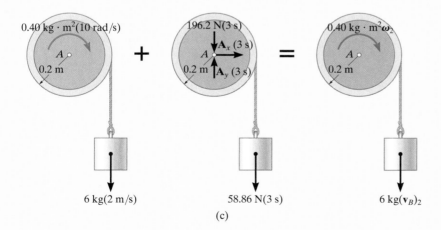

(c)

Fig. 19–7 (cont.)

EXAMPLE 19.5

(a)

(b)

Fig. 19–8

The Charpy impact test is used in materials testing to determine the energy absorption characteristics of a material during impact. The test is performed using the pendulum shown in Fig. 19–8a, which has a mass m, mass center at G, and a radius of gyration k_G about G. Determine the distance r_P from the pin at A to the point P where the impact with the specimen S should occur so that the horizontal force at the pin is essentially zero during the impact. For the calculation, assume the specimen absorbs all the pendulum's kinetic energy gained during the time it falls and thereby stops the pendulum from swinging when $\theta = 0°$.

SOLUTION

Free-Body Diagram.   As shown on the free-body diagram, Fig. 19–8b, the conditions of the problem require the horizontal impulse at A to be zero. Just before impact, the pendulum has a clockwise angular velocity ω_1, and the mass center of the pendulum is moving to the left at $(v_G)_1 = \bar{r}\omega_1$.

Principle of Impulse and Momentum.   We will apply the principle of angular impulse and momentum about point A. Thus,

$$(\circlearrowright+) \qquad\qquad I_A\omega_1 + \Sigma M_A\, dt = I_A\omega_2$$

$$I_A\omega_1 - \left(\int F\, dt \right)r_P = 0$$

$$(\xrightarrow{+}) \qquad\qquad m(v_G)_1 + \Sigma F\, dt = m(v_G)_2$$

$$-m(\bar{r}\omega_1) + \int F\, dt = 0$$

Eliminating the impulse $\int F\, dt$ and substituting $I_A = mk_G^2 + m\bar{r}^2$ yields

$$[mk_G^2 + m\bar{r}^2]\omega_1 - m(\bar{r}\omega_1)r_P = 0$$

Factoring out $m\omega_1$ and solving for r_P, we obtain

$$r_P = \bar{r} + \frac{k_G^2}{\bar{r}} \qquad\qquad Ans.$$

NOTE: The point P, so defined, is called the *center of percussion*. By placing the striking point at P, the force developed at the pin will be minimized. Many sports rackets, clubs, etc. are designed so that collision with the object being struck occurs at the center of percussion. As a consequence, no "sting" or little sensation occurs in the hand of the player. (Also see Probs. 17–65 and 19–1.)

PROBLEMS

19–1. The rigid body (slab) has a mass m and is rotating with an angular velocity $\boldsymbol{\omega}$ about an axis passing through the fixed point O. Show that the momenta of all the particles composing the body can be represented by a single vector having a magnitude mv_G and acting through point P, called the *center of percussion*, which lies at a distance $r_{P/G} = k_G^2/r_{G/O}$ from the mass center G. Here k_G is the radius of gyration of the body, computed about an axis perpendicular to the plane of motion and passing through G.

Prob. 19–1

19–2. At a given instant, the body has a linear momentum $\mathbf{L} = m\mathbf{v}_G$ and an angular momentum $\mathbf{H}_G = I_G\boldsymbol{\omega}$ computed about its mass center. Show that the angular momentum of the body computed about the instantaneous center of zero velocity IC can be expressed as $\mathbf{H}_{IC} = I_{IC}\boldsymbol{\omega}$, where I_{IC} represents the body's moment of inertia computed about the instantaneous axis of zero velocity. As shown, the IC is located at a distance $r_{G/IC}$ away from the mass center G.

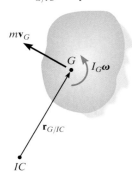

Prob. 19–2

19–3. Show that if a slab is rotating about a fixed axis perpendicular to the slab and passing through its mass center G, the angular momentum is the same when computed about any other point P on the slab.

Prob. 19–3

*19–4. Gear A rotates along the inside of the circular gear rack R. If A has a weight of 4 lb and a radius of gyration of $k_B = 0.5$ ft, determine its angular momentum about point C when $\omega_{CB} = 30$ rad/s and (a) $\omega_R = 0$, (b) $\omega_R = 20$ rad/s.

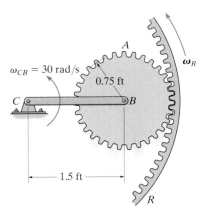

Prob. 19–4

19–5. Solve Prob. 17–55 using the principle of impulse and momentum.

19–6. Solve Prob. 17–54 using the principle of impulse and momentum.

19–7. Solve Prob. 17–69 using the principle of impulse and momentum.

***19–8.** Solve Prob. 17–80 using the principle of impulse and momentum.

19–9. Solve Prob. 17–73 using the principle of impulse and momentum.

19–10. A flywheel has a mass of 60 kg and a radius of gyration of $k_G = 150$ mm about an axis of rotation passing through its mass center. If a motor supplies a clockwise torque having a magnitude of $M = (5t)$ N·m, where t is in seconds, determine the flywheel's angular velocity in $t = 3$ s. Initially the flywheel is rotating clockwise at $\omega_1 = 2$ rad/s.

19–11. A wire of negligible mass is wrapped around the outer surface of the 2-kg disk. If the disk is released from rest, determine its angular velocity in 3 s.

***19–12.** The spool has a mass of 30 kg and a radius of gyration $k_O = 0.25$ m. Block A has a mass of 25 kg, and block B has a mass of 10 kg. If they are released from rest, determine the time required for block A to attain a speed of 2 m/s. Neglect the mass of the ropes.

Prob. 19–12

19–13. The man pulls the rope off the reel with a constant force of 8 lb in the direction shown. If the reel has a weight of 250 lb and radius of gyration $k_G = 0.8$ ft about the trunnion (pin) at A, determine the angular velocity of the reel in 3 s starting from rest. Neglect friction and the weight of rope that is removed.

Prob. 19–11

Prob. 19–13

19–14. Angular motion is transmitted from a driver wheel A to the driven wheel B by friction between the wheels at C. If A always rotates at a constant rate of 16 rad/s, and the coefficient of kinetic friction between the wheels is $\mu_k = 0.2$, determine the time required for B to reach a constant angular velocity once the wheels make contact with a normal force of 50 N. What is the final angular velocity of wheel B? Wheel B has a mass of 90 kg and a radius of gyration about its axis of rotation of $k_G = 120$ mm.

Prob. 19–14

19–15. The 4-kg slender rod rests on a smooth floor. If it is kicked so as to receive a horizontal impulse $I = 8 \text{ N} \cdot \text{s}$ at point A as shown, determine its angular velocity and the speed of its mass center.

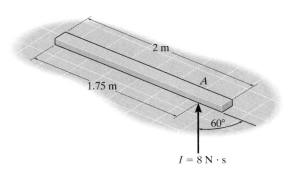

Prob. 19–15

*19–16. A cord of negligible mass is wrapped around the outer surface of the 50-lb cylinder and its end is subjected to a constant horizontal force of $P = 2$ lb. If the cylinder rolls without slipping at A, determine its angular velocity in 4 s starting from rest. Neglect the thickness of the cord.

Prob. 19–16

19–17. The drum has a mass of 70 kg, a radius of 300 mm, and radius of gyration $k_O = 125$ mm. If the coefficients of static and kinetic friction at A are $\mu_s = 0.4$ and $\mu_k = 0.3$, respectively, determine the drum's angular velocity 2 s after it is released from rest. Take $\theta = 30°$.

Prob. 19–17

19–18. The double pulley consists of two wheels which are attached to one another and turn at the same rate. The pulley has a mass of 15 kg and a radius of gyration $k_O = 110$ mm. If the block at A has a mass of 40 kg, determine the speed of the block in 3 s after a constant force $F = 2$ kN is applied to the rope wrapped around the inner hub of the pulley. The block is originally at rest. Neglect the mass of the rope.

*19–20. The two gears A and B have weights and radii of gyration of $W_A = 15$ lb, $k_A = 0.5$ ft and $W_B = 10$ lb, $k_B = 0.35$ ft, respectively. If a motor transmits a couple moment to gear B of $M = 2(1 - e^{-0.5t})$ lb · ft, where t is in seconds, determine the angular velocity of gear A in $t = 5$ s, starting from rest.

Prob. 19–18

Prob. 19–20

19–19. The spool has a weight of 30 lb and a radius of gyration $k_O = 0.45$ ft. A cord is wrapped around its inner hub and the end subjected to a horizontal force $P = 5$ lb. Determine the spool's angular velocity in 4 s starting from rest. Assume the spool rolls without slipping.

19–21. Spool B is at rest and spool A is rotating at 6 rad/s when the slack in the cord connecting them is taken up. Determine the angular velocity of each spool immediately after the cord is jerked tight by the spinning of spool A. The weights and radii of gyration of A and B are $W_A = 30$ lb, $k_A = 0.8$ ft and $W_B = 15$ lb, $k_B = 0.6$ ft, respectively.

Prob. 19–19

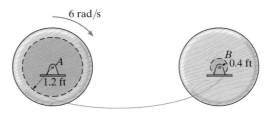

Prob. 19–21

19–22. A 4-kg disk A is mounted on arm BC, which has a negligible mass. If a torque of $M = (5e^{0.5t})$ N·m, where t is in seconds, is applied to the arm at C, determine the angular velocity of BC in 2 s starting from rest. Solve the problem assuming that (a) the disk is set in a smooth bearing at B so that it rotates with curvilinear translation, (b) the disk is fixed to the shaft BC, and (c) the disk is given an initial freely spinning angular velocity of $\boldsymbol{\omega}_D = \{-80\mathbf{k}\}$ rad/s prior to application of the torque.

*19–24. If the hoop has a weight W and radius r and is thrown onto a *rough surface* with a velocity $\mathbf{v}_G$ parallel to the surface, determine the amount of backspin, $\boldsymbol{\omega}_0$, it must be given so that it stops spinning at the same instant that its forward velocity is zero. It is not necessary to know the coefficient of kinetic friction at A for the calculation.

Prob. 19–22

Prob. 19–24

19–23. The inner hub of the wheel rests on the horizontal track. If it does not slip at A, determine the speed of the 10-lb block in 2 s after the block is released from rest. The wheel has a weight of 30 lb and a radius of gyration $k_G = 1.30$ ft. Neglect the mass of the pulley and cord.

19–25. The 10-lb rectangular plate is at rest on a smooth *horizontal* floor. If it is given the horizontal impulses shown, determine its angular velocity and the velocity of the mass center.

Prob. 19–23

Prob. 19–25

19–26. The ball of mass m and radius r rolls along an inclined plane for which the coefficient of static friction is μ. If the ball is released from rest, determine the maximum angle θ for the incline so that it rolls without slipping at A.

*19–28. The slender rod has a mass m and is suspended at its end A by a cord. If the rod receives a horizontal blow giving it an impulse $\mathbf{I}$ at its bottom B, determine the location y of the point P about which the rod appears to rotate during the impact.

Prob. 19–26

Prob. 19–28

19–27. The spool has a weight of 75 lb and a radius of gyration $k_O = 1.20$ ft. If the block B weighs 60 lb, and a force $P = 25$ lb is applied to the cord, determine the speed of the block in 5 s starting from rest. Neglect the mass of the cord.

19–29. A thin rod having a mass of 4 kg is balanced vertically as shown. Determine the height h at which it can be struck with a horizontal force $\mathbf{F}$ and not slip on the floor. This requires that the frictional force at A be essentially zero.

Prob. 19–27

Prob. 19–29

19–30. The square plate has a mass m and is suspended at its corner A by a cord. If it receives a horizontal impulse $\mathbf{I}$ at corner B, determine the location y of the point P about which the plate appears to rotate during the impact.

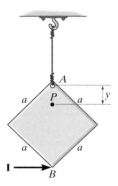

Prob. 19–30

19–31. Determine the height h of the bumper of the pool table, so that when the pool ball of mass m strikes it, no frictional force will be developed between the ball and the table at A. Assume the bumper exerts only a horizontal force on the ball.

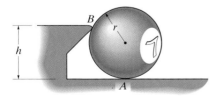

Prob. 19–31

*19–32. The double pulley consists of two wheels which are attached to one another and turn at the same rate. The pulley has a mass of 15 kg and a radius of gyration of $k_O = 110$ mm. If the block at A has a mass of 40 kg and the container at B has a mass of 85 kg, including its contents, determine the speed of the container when $t = 3$ s after it is released from rest.

Prob. 19–32

19–33. The crate has a mass m_c. Determine the constant speed v_0 it acquires as it moves down the conveyor. The rollers each have a radius of r, mass m, and are spaced d apart. Note that friction causes each roller to rotate when the crate comes in contact with it.

Prob. 19–33

19.3 Conservation of Momentum

Conservation of Linear Momentum.   If the sum of all the *linear impulses* acting on a system of connected rigid bodies is *zero*, the linear momentum of the system is constant, or conserved. Consequently, the first two of Eqs. 19–15 reduce to the form

$$\left(\sum \begin{array}{c} \text{syst. linear} \\ \text{momentum} \end{array} \right)_1 = \left(\sum \begin{array}{c} \text{syst. linear} \\ \text{momentum} \end{array} \right)_2 \qquad (19\text{–}16)$$

This equation is referred to as the *conservation of linear momentum*.

Without inducing appreciable errors in the calculations, it may be possible to apply Eq. 19–16 in a specified direction for which the linear impulses are small or *nonimpulsive*. Specifically, nonimpulsive forces occur when small forces act over very short periods of time. Typical examples include the force of a slightly deformed spring, the initial contact with soft ground, and in some cases the weight of the body.

Conservation of Angular Momentum.   The angular momentum of a system of connected rigid bodies is conserved about the system's center of mass G, or a fixed point O, when the sum of all the angular impulses created by the external forces acting on the system is zero or appreciably small (nonimpulsive) when computed about these points. The third of Eqs. 19–15 then becomes

$$\left(\sum \begin{array}{c} \text{syst. angular} \\ \text{momentum} \end{array} \right)_{O1} = \left(\sum \begin{array}{c} \text{syst. angular} \\ \text{momentum} \end{array} \right)_{O2} \qquad (19\text{–}17)$$

This equation is referred to as the *conservation of angular momentum*. In the case of a single rigid body, Eq. 19–17 applied to point G becomes $(I_G\omega)_1 = (I_G\omega)_2$. To illustrate an application of this equation, consider a swimmer who executes a somersault after jumping off a diving board. By tucking his arms and legs in close to his chest, he *decreases* his body's moment of inertia and thus *increases* his angular velocity ($I_G\omega$ must be constant). If he straightens out just before entering the water, his body's moment of inertia is *increased*, and his angular velocity *decreases*. Since the weight of his body creates a linear impulse during the time of motion, this example also illustrates how the angular momentum of a body can be conserved and yet the linear momentum is *not*. Such cases occur whenever the external forces creating the linear impulse pass through either the center of mass of the body or a fixed axis of rotation.

Provided the initial linear or angular velocity of the body is known, the conservation of linear or angular momentum is used to determine the respective final linear or angular velocity of the body *just after* the time period considered. Furthermore, by applying these equations to a *system* of bodies, the internal impulses acting within the system, which may be unknown, are eliminated from the analysis, since they occur in equal but opposite collinear pairs. If it is necessary to determine an *internal impulsive force* acting on only one body of a system of connected bodies, the body must be *isolated* (free-body diagram) and the principle of linear or angular impulse and momentum applied *to the body*. After the impulse $\int F\, dt$ is calculated; then, provided the time Δt for which the impulse acts is known, the *average impulsive force* F_{avg} can be determined from $F_{\text{avg}} = \left(\int F\, dt \right)/\Delta t$.

Procedure for Analysis

The conservation of linear or angular momentum should be applied using the following procedure.

Free-Body Diagram.

- Establish the x, y inertial frame of reference and draw the free-body diagram for the body or system of bodies during the time of impact. From this diagram classify each of the applied forces as being either "impulsive" or "nonimpulsive."

- By inspection of the free-body diagram, the *conservation of linear momentum* applies in a given direction when *no* external impulsive forces act on the body or system in that direction; whereas the *conservation of angular momentum* applies about a fixed point O or at the mass center G of a body or system of bodies when all the external impulsive forces acting on the body or system create zero moment (or zero angular impulse) about O or G.

- As an alternative procedure, draw the impulse and momentum diagrams for the body or system of bodies. These diagrams are particularly helpful in order to visualize the "moment" terms used in the conservation of angular momentum equation, when it has been decided that angular momenta are to be computed about a point other than the body's mass center G.

Conservation of Momentum.

- Apply the conservation of linear or angular momentum in the appropriate directions.

Kinematics.

- If the motion appears to be complicated, kinematic (velocity) diagrams may be helpful in obtaining the necessary kinematic relations.

EXAMPLE 19.6

(a)

$r' = (0.2 - 0.03)$ m

$+$

$\parallel$

(b)

(c)

Fig. 19–9

The 10-kg wheel shown in Fig. 19–9a has a moment of inertia $I_G = 0.156$ kg $\cdot$ m^2. Assuming that the wheel does not slip or rebound, determine the minimum velocity $\mathbf{v}_G$ it must have to just roll over the obstruction at A.

SOLUTION

Impulse and Momentum Diagrams. Since no slipping or rebounding occurs, the wheel essentially *pivots* about point A during contact. This condition is shown in Fig. 19–9b, which indicates, respectively, the momentum of the wheel *just before impact*, the impulses given to the wheel *during impact*, and the momentum of the wheel *just after impact*. Only two impulses (forces) act on the wheel. By comparison, the force at A is much greater than that of the weight, and since the time of impact is very short, the weight can be considered nonimpulsive. The impulsive force $\mathbf{F}$ at A has both an unknown magnitude and an unknown direction θ. To eliminate this force from the analysis, note that angular momentum about A is essentially *conserved* since $(98.1 \, \Delta t)d \approx 0$.

Conservation of Angular Momentum. With reference to Fig. 19–9b,

$$(\curvearrowleft +) \qquad\qquad (H_A)_1 = (H_A)_2$$

$$r'm(v_G)_1 + I_G\omega_1 = rm(v_G)_2 + I_G\omega_2$$

$$(0.2 \text{ m} - 0.03 \text{ m})(10 \text{ kg})(v_G)_1 + (0.156 \text{ kg} \cdot \text{m}^2)(\omega_1) =$$

$$(0.2 \text{ m})(10 \text{ kg})(v_G)_2 + (0.156 \text{ kg} \cdot \text{m}^2)(\omega_2)$$

Kinematics. Since no slipping occurs, in general $\omega = v_G/r = v_G/0.2$ m $= 5v_G$. Substituting this into the above equation and simplifying yields

$$(v_G)_2 = 0.892(v_G)_1 \qquad\qquad (1)$$

Conservation of Energy.* In order to roll over the obstruction, the wheel must pass position 3 shown in Fig. 19–9c. Hence, if $(v_G)_2$ [or $(v_G)_1$] is to be a minimum, it is necessary that the kinetic energy of the wheel at position 2 be equal to the potential energy at position 3. Constructing the datum through the center of gravity, as shown in the figure, and applying the conservation of energy equation, we have

$$\{T_2\} + \{V_2\} = \{T_3\} + \{V_3\}$$

$$\left\{\tfrac{1}{2}(10 \text{ kg})(v_G)_2^2 + \tfrac{1}{2}(0.156 \text{ kg} \cdot \text{m}^2)\omega_2^2\right\} + \{0\} =$$

$$\{0\} + \{(98.1 \text{ N})(0.03 \text{ m})\}$$

Substituting $\omega_2 = 5(v_G)_2$ and Eq. 1 into this equation, and solving,

$$(v_G)_1 = 0.729 \text{ m/s} \rightarrow \qquad\qquad Ans.$$

*This principle *does not* apply during *impact*, since energy is *lost* during the collision; however, just after impact, position in Fig. 19–9c, it can be used.

EXAMPLE 19.7

The 5-kg slender rod shown in Fig. 19–10a is pinned at O and is initially at rest. If a 4-g bullet is fired into the rod with a velocity of 400 m/s, as shown in the figure, determine the angular velocity of the rod just after the bullet becomes embedded in it.

(a)

SOLUTION

Impulse and Momentum Diagrams. The impulse which the bullet exerts on the rod can be eliminated from the analysis, and the angular velocity of the rod just after impact can be determined by considering the bullet and rod as a single system. To clarify the principles involved, the impulse and momentum diagrams are shown in Fig. 19–10b. The momentum diagrams are drawn *just before and just after impact*. During impact, the bullet and rod exchange equal but *opposite internal impulses* at A. As shown on the impulse diagram, the impulses that are external to the system are due to the reactions at O and the weights of the bullet and rod. Since the time of impact, Δt, is very short, the rod moves only a slight amount, and so the "moments" of the weight impulses about point O are essentially zero. Therefore angular momentum is conserved about this point.

(b)

Conservation of Angular Momentum. From Fig. 19–10b, we have

$$(\downarrow+) \qquad\qquad \Sigma(H_O)_1 = \Sigma(H_O)_2$$

$$m_B(v_B)_1 \cos 30°(0.75 \text{ m}) = m_B(v_B)_2(0.75 \text{ m}) + m_R(v_G)_2(0.5 \text{ m}) + I_G\omega_2$$

$$(0.004 \text{ kg})(400 \cos 30° \text{ m/s})(0.75 \text{ m}) =$$

$$(0.004 \text{ kg})(v_B)_2(0.75 \text{ m}) + (5 \text{ kg})(v_G)_2(0.5 \text{ m}) + \left[\tfrac{1}{12}(5 \text{ kg})(1 \text{ m})^2\right]\omega_2$$

or

$$1.039 = 0.003(v_B)_2 + 2.50(v_G)_2 + 0.417\omega_2 \qquad\qquad (1)$$

Kinematics. Since the rod is pinned at O, from Fig. 19–10c we have

$$(v_G)_2 = (0.5 \text{ m})\omega_2 \qquad (v_B)_2 = (0.75 \text{ m})\omega_2$$

Substituting into Eq. 1 and solving yields

$$\omega_2 = 0.623 \text{ rad/s} \curvearrowleft \qquad\qquad Ans.$$

(c)

Fig. 19–10

19.4   Eccentric Impact

The concepts involving central and oblique impact of particles were presented in Sec. 15.4. We will now expand this treatment and discuss the eccentric impact of two bodies. *Eccentric impact* occurs when the line connecting the *mass centers* of the two bodies *does not* coincide with the line of impact.* This type of impact often occurs when one or both of the bodies are constrained to rotate about a fixed axis. Consider, for example, the collision at C between the two bodies A and B, shown in Fig. 19–11a. It is assumed that just before collision B is rotating counterclockwise with an angular velocity $(\omega_B)_1$, and the velocity of the contact point C located on A is $(\mathbf{u}_A)_1$. Kinematic diagrams for both bodies just before collision are shown in Fig. 19–11b. Provided the bodies are smooth, the *impulsive forces* they exert on each other *are directed along the line of impact.* Hence, the component of velocity of point C on body B, which is directed along the line of impact, is $(v_B)_1 = (\omega_B)_1 r$, Fig. 19–11b. Likewise, on body A the component of velocity $(\mathbf{u}_A)_1$ along the line of impact is $(\mathbf{v}_A)_1$. In order for a collision to occur, $(v_A)_1 > (v_B)_1$.

During the impact an equal but opposite impulsive force **P** is exerted between the bodies which *deforms* their shapes at the point of contact. The resulting impulse is shown on the impulse diagrams for both bodies, Fig. 19–11c. Note that the impulsive force created at point C on the rotating body creates impulsive pin reactions at O. On these diagrams it is assumed that the impact creates forces which are much larger than the nonimpulsive weights of the bodies, which are not shown. When the deformation at point C is a maximum, C on both the bodies moves with a common velocity **v** along the line of impact, Fig. 19–11d. A period of *restitution* then occurs in which the bodies tend to regain their original shapes. The restitution phase creates an equal but opposite impulsive force **R** acting between the bodies as shown on the impulse diagram, Fig. 19–11e. After restitution the bodies move apart such that point C on body B has a velocity $(\mathbf{v}_B)_2$ and point C on body A has a velocity $(\mathbf{u}_A)_2$, Fig. 19–11f, where $(v_B)_2 > (v_A)_2$.

In general, a problem involving the impact of two bodies requires determining the *two unknowns* $(v_A)_2$ and $(v_B)_2$, assuming $(v_A)_1$ and $(v_B)_1$ are known (or can be determined using kinematics, energy methods, the equations of motion, etc.). To solve such problems, two equations must be written. The *first equation* generally involves application of *the conservation of angular momentum to the two bodies*. In the case of both bodies A and B, we can state that angular momentum is conserved about point O since the impulses at C are internal to the system and the impulses at O create zero moment (or zero angular impulse) about point O. The *second equation* is obtained using the

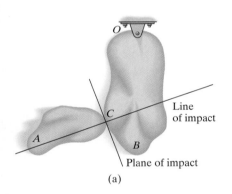

An example of eccentric impact occurring between this bowling ball and pin.

Line of impact

Plane of impact

(a)

Fig. 19–11

---

*When these lines coincide, central impact occurs and the problem can be analyzed as discussed in Sec. 15.4.

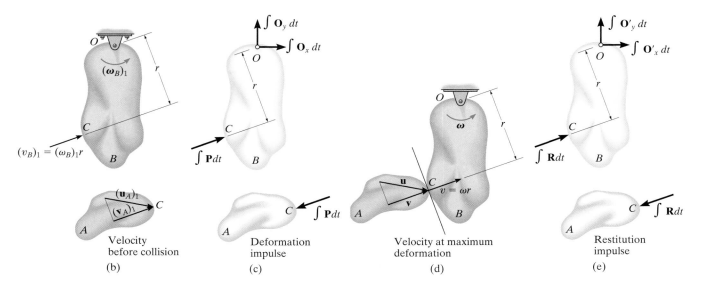

Velocity
before collision
(b)

Deformation
impulse
(c)

Velocity at maximum
deformation
(d)

Restitution
impulse
(e)

definition of the *coefficient of restitution, e*, which is a ratio of the restitution impulse to the deformation impulse. To establish a useful form of this equation we must first apply the principle of angular impulse and momentum about point O to bodies B and A separately. Combining the results, we then obtain the necessary equation. Proceeding in this manner, the principle of impulse and momentum applied to body B from the time just before the collision to the instant of maximum deformation, Figs. 19–11b, 19–11c, and 19–11d, becomes

$(\zeta+)$
$$I_O(\omega_B)_1 + r\int P\,dt = I_O\omega \qquad (19\text{–}18)$$

Here I_O is the moment of inertia of body B about point O. Similarly, applying the principle of angular impulse and momentum from the instant of maximum deformation to the time just after the impact, Figs. 19–11d, 19–11e, and 19–11f, yields

$(\zeta+)$
$$I_O\omega + r\int R\,dt = I_O(\omega_B)_2 \qquad (19\text{–}19)$$

Solving Eqs. 19–18 and 19–19 for $\int P\,dt$ and $\int R\,dt$, respectively, and formulating e, we have

$$e = \frac{\displaystyle\int R\,dt}{\displaystyle\int P\,dt} = \frac{r(\omega_B)_2 - r\omega}{r\omega - r(\omega_B)_1} = \frac{(v_B)_2 - v}{v - (v_B)_1}$$

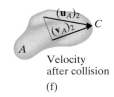

Velocity
after collision
(f)

Fig. 19–11 (cont.)

In the same manner, we may write an equation which relates the magnitudes of velocity $(v_A)_1$ and $(v_A)_2$ of body A. The result is

$$e = \frac{v - (v_A)_2}{(v_A)_1 - v}$$

Combining the above equations by eliminating the common velocity v yields the desired result, i.e.,

$(+\nearrow)$

$$e = \frac{(v_B)_2 - (v_A)_2}{(v_A)_1 - (v_B)_1} \qquad (19\text{--}20)$$

This equation is identical to Eq. 15–11, which was derived for the central impact between two particles. Equation 19–20 states that the coefficient of restitution is equal to the ratio of the relative velocity of *separation* of the points of contact (C) *just after impact* to the relative velocity at which the points *approach* one another *just* before impact. In deriving this equation, we assumed that the points of contact for both bodies move up and to the right *both* before and after impact. If motion of any one of the contacting points occurs down and to the left, the velocity of this point is considered a negative quantity in Eq. 19–20.

As stated previously, when Eq. 19–20 is used in conjunction with the conservation of angular momentum for the bodies, it provides a useful means of obtaining the velocities of two colliding bodies just after collision.

During impact the columns of many highway signs are intended to break out of their supports and easily collapse at their joints. This is shown by the slotted connections at their base and the breaks at the column's midsection. The mechanics of eccentric impact is used in the design of these structures.

EXAMPLE 19.8

The 10-lb slender rod is suspended from the pin at A, Fig. 19–12a. If a 2-lb ball B is thrown at the rod and strikes its center with a horizontal velocity of 30 ft/s, determine the angular velocity of the rod just after impact. The coefficient of restitution is $e = 0.4$.

(a)

SOLUTION

Conservation of Angular Momentum.   Consider the ball and rod as a system, Fig. 19–12b. Angular momentum is conserved about point A since the impulsive force between the rod and ball is *internal*. Also, the *weights* of the ball and rod are *nonimpulsive*. Noting the directions of the velocities of the ball and rod just after impact as shown on the kinematic diagram, Fig. 19–12c, we require

$(\downarrow+)$ $(H_A)_1 = (H_A)_2$

$$m_B(v_B)_1(1.5 \text{ ft}) = m_B(v_B)_2(1.5 \text{ ft}) + m_R(v_G)_2(1.5 \text{ ft}) + I_G\omega_2$$

$$\left(\frac{2 \text{ lb}}{32.2 \text{ ft/s}^2}\right)(30 \text{ ft/s})(1.5 \text{ ft}) = \left(\frac{2 \text{ lb}}{32.2 \text{ ft/s}^2}\right)(v_B)_2(1.5 \text{ ft}) +$$

$$\left(\frac{10 \text{ lb}}{32.2 \text{ ft/s}^2}\right)(v_G)_2(1.5 \text{ ft}) + \left[\frac{1}{12}\left(\frac{10 \text{ lb}}{32.2 \text{ ft/s}^2}\right)(3 \text{ ft})^2\right]\omega_2$$

Since $(v_G)_2 = 1.5\omega_2$ then

$$2.795 = 0.09317(v_B)_2 + 0.9317\omega_2 \qquad (1)$$

Coefficient of Restitution.   With reference to Fig. 19–12c, we have

$(\xrightarrow{+})$ $e = \dfrac{(v_G)_2 - (v_B)_2}{(v_B)_1 - (v_G)_1}$ $0.4 = \dfrac{(1.5 \text{ ft})\omega_2 - (v_B)_2}{30 \text{ ft/s} - 0}$

$$12.0 = 1.5\omega_2 - (v_B)_2$$

Solving,

$$(v_B)_2 = -6.52 \text{ ft/s} = 6.52 \text{ ft/s} \leftarrow$$
$$\omega_2 = 3.65 \text{ rad/s} \, \curvearrowleft \qquad\qquad Ans.$$

(b)

(c)

Fig. 19–12

PROBLEMS

19–34. Two wheels A and B have masses m_A and m_B, and radii of gyration about their central vertical axes of k_A and k_B, respectively. If they are freely rotating in the same direction at ω_A and ω_B about the same vertical axis, determine their common angular velocity after they are brought into contact and slipping between them stops.

19–35. The Hubble Space Telescope is powered by two solar panels as shown. The body of the telescope has a mass of 11 Mg and radii of gyration $k_x = 1.64$ m and $k_y = 3.85$ m, whereas the solar panels can be considered as thin plates, each having a mass of 54 kg. Due to an internal drive, the panels are given an angular velocity of $\{0.6\mathbf{j}\}$ rad/s, measured relative to the telescope. Determine the angular velocity of the telescope due to the rotation of the panels. Prior to rotating the panels, the telescope was originally traveling at $\mathbf{v}_G = \{-400\mathbf{i} + 250\mathbf{j} + 175\mathbf{k}\}$ m/s. Neglect its orbital rotation.

*19–36. The platform swing consists of a 200-lb flat plate suspended by four rods of negligible weight. When the swing is at rest, the 150-lb man jumps off the platform when his center of gravity G is 10 ft from the pin at A. This is done with a horizontal velocity of 5 ft/s, measured relative to the swing at the level of G. Determine the angular velocity he imparts to the swing just after jumping off.

Prob. 19–36

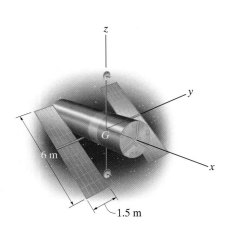

Prob. 19–35

19–37. Each of the two slender rods and the disk have the same mass m. Also, the length of each rod is equal to the diameter d of the disk. If the assembly is rotating with an angular velocity ω_1 when the rods are directed outward, determine the angular velocity of the assembly if by internal means the rods are brought to an upright vertical position.

Prob. 19–37

19–38. The rod has a length L and mass m. A smooth collar having a negligible size and one-fourth the mass of the rod is placed on the rod at its midpoint. If the rod is freely rotating at ω about its end and the collar is released, determine the rod's angular velocity just before the collar flies off the rod. Also, what is the speed of the collar as it leaves the rod?

Prob. 19–38

19–39. A man has a moment of inertia I_z about the z axis. He is originally at rest and standing on a small platform which can turn freely. If he is handed a wheel which is rotating at ω and has a moment of inertia I about its spinning axis, determine his angular velocity if (a) he holds the wheel upright as shown, (b) turns the wheel out, $\theta = 90°$, and (c) turns the wheel downward, $\theta = 180°$.

Prob. 19–39

*19–40. The space satellite has a mass of 125 kg and a moment of inertia $I_z = 0.940 \text{ kg} \cdot \text{m}^2$, excluding the four solar panels A, B, C, and D. Each solar panel has a mass of 20 kg and can be approximated as a thin plate. If the satellite is originally spinning about the z axis at a constant rate $\omega_z = 0.5 \text{ rad/s}$ when $\theta = 90°$, determine the rate of spin if all the panels are raised and reach the upward position, $\theta = 0°$, at the same instant

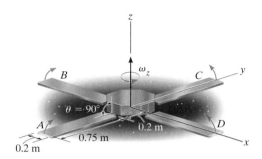

Prob. 19–40

19–41. The 2-kg rod ACB supports the two 4-kg disks at its ends. If both disks are given a clockwise angular velocity $(\omega_A)_1 = (\omega_B)_1 = 5 \text{ rad/s}$ while the rod is held stationary and then released, determine the angular velocity of the rod after both disks have stopped spinning relative to the rod due to frictional resistance at the pins A and B. Motion is in the *horizontal plane*. Neglect friction at pin C.

Prob. 19–41

19–42.   Disk A has a weight of 20 lb. An inextensible cable is attached to the 10-lb weight and wrapped around the disk. The weight is dropped 2 ft before the slack is taken up. If the impact is perfectly elastic, i.e., $e = 1$, determine the angular velocity of the disk just after impact.

0.5 ft

A

10 lb

Prob. 19–42

19–43.   A thin disk of mass m has an angular velocity ω_1 while rotating on a smooth surface. Determine its new angular velocity just after the hook at its edge strikes the peg P and the disk starts to rotate about P without rebounding.

P

r

ω_1

Prob. 19–43

***19–44.**   The pendulum consists of a 5-lb slender rod AB and a 10-lb wooden block. A projectile weighing 0.2 lb is fired into the center of the block with a velocity of 1000 ft/s. If the pendulum is initially at rest, and the projectile embeds itself into the block, determine the angular velocity of the pendulum just after the impact.

A

2 ft

B

$v = 1000$ ft/s

1 ft

1 ft

Prob. 19–44

19–45.   The pendulum consists of a slender 2-kg rod AB and 5-kg disk. It is released from rest without rotating. When it falls 0.3 m, the end A strikes the hook S, which provides a permanent connection. Determine the angular velocity of the pendulum after it has rotated 90°. Treat the pendulum's weight during impact as a nonimpulsive force.

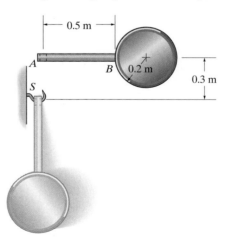

0.5 m

A

B 0.2 m

0.3 m

S

Prob. 19–45

19–46. A horizontal circular platform has a weight of 300 lb and a radius of gyration $k_z = 8$ ft about the z axis passing through its center O. The platform is free to rotate about the z axis and is initially at rest. A man having a weight of 150 lb begins to run along the edge in a circular path of radius 10 ft. If he has a speed of 4 ft/s and maintains this speed relative to the platform, determine the angular velocity of the platform. Neglect friction.

*19–48.** Two children A and B, each having a mass of 30 kg, sit at the edge of the merry-go-round which is rotating at $\omega = 2$ rad/s. Excluding the children, the merry-go-round has a mass of 180 kg and a radius of gyration $k_z = 0.6$ m. Determine the angular velocity of the merry-go-round if A jumps off horizontally in the $-n$ direction with a speed of 2 m/s, measured with respect to the merry-go-round. What is the merry-go-round's angular velocity if B then jumps off horizontally in the $+t$ direction with a speed of 2 m/s, measured with respect to the merry-go-round? Neglect friction and the size of each child.

Prob. 19–46

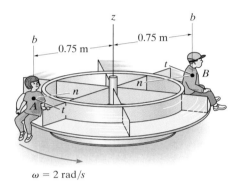

$\omega = 2$ rad/s

Prob. 19–48

19–47. The square plate has a weight W and is rotating on the smooth surface with a constant angular velocity ω_0. Determine the new angular velocity of the plate just after its corner strikes the peg P and the plate starts to rotate about P without rebounding.

19–49. A 7-g bullet having a velocity of 800 m/s is fired into the edge of the 5-kg disk as shown. Determine the angular velocity of the disk just after the bullet becomes embedded in it. Also, calculate how far θ the disk will swing until it stops. The disk is originally at rest.

Prob. 19–47

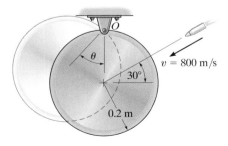

Prob. 19–49

19–50.   The two disks each weigh 10 lb. If they are released from rest when $\theta = 30°$, determine θ after they collide and rebound from each other. The coefficient of restitution is $e = 0.75$. When $\theta = 0°$, the disks hang so that they just touch one another.

*19–52.   The pendulum consists of a 10-lb solid ball and 4-lb rod. If it is released from rest when $\theta_1 = 0°$, determine the angle θ_2 after the ball strikes the wall, rebounds, and the pendulum swings up to the point of momentary rest. Take $e = 0.6$.

Prob. 19–52

Prob. 19–50

19–51.   The 15-lb rod AB is released from rest in the vertical position. If the coefficient of restitution between the floor and the cushion at B is $e = 0.7$, determine how high the end of the rod rebounds after impact with the floor.

19–53.   The plank has a weight of 30 lb, center of gravity at G, and it rests on the two sawhorses at A and B. If the end D is raised 2 ft above the top of the sawhorses and is released from rest, determine how high end C will rise from the top of the sawhorses after the plank falls so that it rotates clockwise about A, strikes and pivots on the sawhorses at B, and rotates clockwise off the sawhorse at A.

Prob. 19–51

Prob. 19–53

19–54. Tests of impact on the fixed crash dummy are conducted using the 300-lb ram that is released from rest at $\theta = 30°$, and allowed to fall and strike the dummy at $\theta = 90°$. If the coefficient of restitution between the dummy and the ram is $e = 0.4$, determine the angle θ to which the ram will rebound before momentarily coming to rest.

*19–56. A solid ball with a mass m is thrown on the ground such that at the instant of contact it has an angular velocity $\boldsymbol{\omega}_1$ and velocity components $(\mathbf{v}_G)_{x1}$ and $(\mathbf{v}_G)_{y1}$ as shown. If the ground is rough so no slipping occurs, determine the components of the velocity of its mass center just after impact. The coefficient of restitution is e.

Prob. 19–56

Prob. 19–54

19–55. The solid ball of mass m is dropped with a velocity $\mathbf{v}_1$ onto the edge of the rough step. If it rebounds horizontally off the step with a velocity $\mathbf{v}_2$, determine the angle θ at which contact occurs. Assume no slipping when the ball strikes the step. The coefficient of restitution is e.

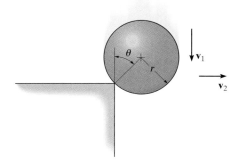

Prob. 19–55

Chapter Review

Linear and Angular Momentum

The linear and angular momentum of a rigid body can be referenced to its mass center G.

If the angular momentum is to be determined about an axis other than the one passing through the mass center, then the angular momentum is determined by summing vector $\mathbf{H}_G$ and the moment of vector $\mathbf{L}$ about this axis.

(*See pages 483–487.*)

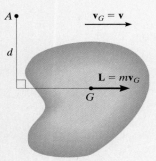

Translation

$$L = mv_G$$

$$H_G = 0$$

$$H_A = (mv_G)d$$

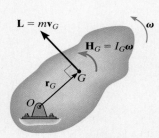

Rotation about a fixed axis

$$L = mv_G$$

$$H_G = I_G\omega$$

$$H_O = I_O\omega$$

General plane motion

$$L = mv_G$$

$$H_G = I_G\omega$$

$$H_A = I_G\omega + (mv_G)d$$

Principle of Impulse and Momentum

The principles of linear and angular impulse and momentum are used to solve problems that involve force, velocity, and time. Before applying these equations, it is important to establish the x, y, z inertial coordinate system. The free-body diagram for the body should also be drawn in order to account for all of the forces and couple moments that produce impulses on the body.

(*See pages 489–490.*)

$$m(v_{Gx})_1 + \Sigma \int_{t_1}^{t_2} F_x \, dt = m(v_{Gx})_2$$

$$m(v_{Gy})_1 + \Sigma \int_{t_1}^{t_2} F_y \, dt = m(v_{Gy})_2$$

$$I_G\omega_1 + \Sigma \int_{t_1}^{t_2} M_G \, dt = I_G\omega_2$$

Conservation of Momentum

Provided the sum of the linear impulses acting on a system of connected rigid bodies is zero in a particular direction, then the linear momentum for the system is conserved in this direction. Conservation of angular momentum occurs if the impulses pass through an axis or are parallel to it. Momentum is also conserved if the external forces are small and thereby create nonimpulsive forces on the system. A free-body diagram should accompany any application in order to classify the forces as impulsive or nonimpulsive and to determine an axis about which the angular momentum may be conserved.

(*See pages 504–505.*)

$$\left(\Sigma \begin{array}{l} \text{syst. linear} \\ \text{momentum} \end{array}\right)_1 = \left(\Sigma \begin{array}{l} \text{syst. linear} \\ \text{momentum} \end{array}\right)_2$$

$$\left(\Sigma \begin{array}{l} \text{syst. angular} \\ \text{momentum} \end{array}\right)_{O1} = \left(\Sigma \begin{array}{l} \text{syst. angular} \\ \text{momentum} \end{array}\right)_{O2}$$

Eccentric Impact

If the line of impact does not coincide with the line connecting the mass centers of two colliding bodies, then eccentric impact will occur. If the motion of the bodies just after the impact is to be determined, then it is necessary to consider a conservation of momentum equation for the system and use the coefficient of restitution equation.

(*See pages 508–510.*)

$$e = \frac{(v_B)_2 - (v_A)_2}{(v_A)_1 - (v_B)_1}$$

2 Planar Kinematics and Kinetics of a Rigid Body

Having presented the various topics in planar kinematics and kinetics in Chapters 16 through 19, we will now summarize these principles and provide an opportunity for applying them to the solution of various types of problems.

Kinematics. Here we are interested in studying the geometry of motion, without concern for the forces which cause the motion. Before solving a planar kinematics problem, it is *first* necessary to *classify the motion* as being either rectilinear or curvilinear translation, rotation about a fixed axis, or general plane motion. In particular, problems involving general plane motion can be solved either with reference to a fixed axis (absolute motion analysis) or using translating or rotating frames of reference (relative motion analysis). The choice generally depends upon the type of constraints and the problem's geometry. In all cases, application of the necessary equations may be clarified by drawing a kinematic diagram. Remember that the *velocity* of a point is always *tangent* to its path of motion, and the *acceleration* of a point can have *components* in the *n–t* directions when the path is *curved*.

Translation. When the body moves with rectilinear or curvilinear translation, *all* the points on the body have the *same motion*.

$$\mathbf{v}_B = \mathbf{v}_A \qquad \mathbf{a}_B = \mathbf{a}_A$$

Rotation About a Fixed Axis.
Angular Motion.
Variable Angular Acceleration.   Provided a mathematical relationship is given between *any two* of the four variables θ, ω, α, and t, then a *third* variable can be determined by solving one of the following equations which relate all three variables.

$$\omega = \frac{d\theta}{dt} \qquad \alpha = \frac{d\omega}{dt} \qquad \alpha \, d\theta = \omega \, d\omega$$

Constant Angular Acceleration.   The following equations apply when it is *absolutely certain* that the angular acceleration is constant.

$$\theta = \theta_0 + \omega_0 t + \tfrac{1}{2}\alpha_c t^2 \qquad \omega = \omega_0 + \alpha_c t \qquad \omega^2 = \omega_0^2 + 2\alpha_c(\theta - \theta_0)$$

Motion of Point P.   Once ω and α have been determined, then the circular motion of point P can be specified using the following scalar or vector equations.

$$v = \omega r \qquad\qquad \mathbf{v} = \boldsymbol{\omega} \times \mathbf{r}$$
$$a_t = \alpha r \quad a_n = \omega^2 r \qquad \mathbf{a} = \boldsymbol{\alpha} \times \mathbf{r} + \boldsymbol{\omega} \times (\boldsymbol{\omega} \times \mathbf{r})$$

General Plane Motion—Relative-Motion Analysis.   Recall that when *translating axes* are placed at the "base point" A, the *relative motion* of point B with respect to A is simply *circular motion of B about A*. The following equations apply to two points A and B located on the *same* rigid body.

$$\mathbf{v}_B = \mathbf{v}_A + \mathbf{v}_{B/A} = \mathbf{v}_A + (\boldsymbol{\omega} \times \mathbf{r}_{B/A})$$
$$\mathbf{a}_B = \mathbf{a}_A + \mathbf{a}_{B/A} = \mathbf{a}_A + \boldsymbol{\alpha} \times \mathbf{r}_{B/A} + \boldsymbol{\omega} \times (\boldsymbol{\omega} \times \mathbf{r}_{B/A})$$

Rotating and translating axes are often used to analyze the motion of rigid bodies which are connected together by collars or slider blocks.

$$\mathbf{v}_B = \mathbf{v}_A + \boldsymbol{\Omega} \times \mathbf{r}_{B/A} + (\mathbf{v}_{B/A})_{xyz}$$
$$\mathbf{a}_B = \mathbf{a}_A + \dot{\boldsymbol{\Omega}} \times \mathbf{r}_{B/A} + \boldsymbol{\Omega} \times (\boldsymbol{\Omega} \times \mathbf{r}_{B/A}) + 2\boldsymbol{\Omega} \times (\mathbf{v}_{B/A})_{xyz} + (\mathbf{a}_{B/A})_{xyz}$$

Kinetics.   To analyze the forces which cause the motion we must use the principles of kinetics. When applying the necessary equations, it is important to first establish the inertial coordinate system and define the positive directions of the axes. The *directions* should be the *same* as those selected when writing any equations of kinematics provided *simultaneous solution* of equations becomes necessary.

Equations of Motion.   These equations are used to determine accelerated motions or forces causing the motion. If used to determine position, velocity, or time of motion, then kinematics will have to be considered for part of the solution. Before applying the equations of motion, *always draw a free-body diagram* in order to identify all the

forces acting on the body. Also, establish the directions of the acceleration of the mass center and the angular acceleration of the body. (A kinetic diagram may also be drawn in order to represent $m\mathbf{a}_G$ and $I_G\boldsymbol{\alpha}$ graphically. This diagram is particularly convenient for resolving $m\mathbf{a}_G$ into components and for identifying the terms in the moment sum $\Sigma(\mathcal{M}_k)_P$.)

The three equations of motion are

$$\Sigma F_x = m(a_G)_x$$
$$\Sigma F_y = m(a_G)_y$$
$$\Sigma M_G = I_G\alpha \quad \text{or} \quad \Sigma M_P = \Sigma(\mathcal{M}_k)_P$$

In particular, if the body is *rotating about a fixed axis*, moments may also be summed about point O on the axis, in which case

$$\Sigma M_O = \Sigma(\mathcal{M}_k)_O = I_O\alpha$$

Work and Energy.   *The equation of work and energy is used to solve problems involving force, velocity, and displacement.* Before applying this equation, *always draw a free-body diagram* of the body in order to identify the forces which do work. Recall that the kinetic energy of the body is due to translational motion of the mass center, $\mathbf{v}_G$, *and* rotational motion of the body, $\boldsymbol{\omega}$.

$$T_1 + \Sigma U_{1-2} = T_2$$

where

$$T = \tfrac{1}{2}mv_G^2 + \tfrac{1}{2}I_G\omega^2$$

$$U_F = \int F \cos\theta \, ds \qquad \text{(variable force)}$$

$$U_{F_c} = F_c \cos\theta(s_2 - s_1) \qquad \text{(constant force)}$$

$$U_W = -W\,\Delta y \qquad \text{(weight)}$$

$$U_s = -\left(\tfrac{1}{2}ks_2^2 - \tfrac{1}{2}ks_1^2\right) \qquad \text{(spring)}$$

$$U_M = M\theta \qquad \text{(constant couple moment)}$$

If the forces acting on the body are *conservative forces*, then apply the *conservation of energy equation.* This equation is easier to use than the equation of work and energy, since it applies only at *two points* on the path and *does not* require calculation of the work done by a force as the body moves along the path.

$$T_1 + V_1 = T_2 + V_2$$

where

$$V_g = Wy \text{ (gravitational potential energy)}$$
$$V_e = \tfrac{1}{2}ks^2 \text{ (elastic potential energy)}$$

Impulse and Momentum.   *The principles of linear and angular impulse and momentum are used to solve problems involving force, velocity, and time.* Before applying the equations, *draw a free-body diagram* in order to identify all the forces which cause linear and angular impulses on the body. Also, establish the directions of the velocity of the mass center and the angular velocity of the body just before and just after the impulses are applied. (As an alternative procedure, the impulse and momentum diagrams may accompany the solution in order to graphically account for the terms in the equations. These diagrams are particularly advantageous when computing the angular impulses and angular momenta about a point other than the body's mass center.)

$$m(\mathbf{v}_G)_1 + \Sigma \int \mathbf{F}\, dt = m(\mathbf{v}_G)_2$$

$$(\mathbf{H}_G)_1 + \Sigma \int \mathbf{M}_G\, dt = (\mathbf{H}_G)_2$$

or

$$(\mathbf{H}_O)_1 + \Sigma \int \mathbf{M}_O\, dt = (\mathbf{H}_O)_2$$

Conservation of Momentum.   If nonimpulsive forces or no impulsive forces act on the body in a particular direction, or if the motions of several bodies are involved in the problem, then consider applying the conservation of linear or angular momentum for the solution. Investigation of the free-body diagram (or the impulse diagram) will aid in determining the directions for which the impulsive forces are zero, or axes about which the impulsive forces cause zero angular momentum. For these cases,

$$m(\mathbf{v}_G)_1 = m(\mathbf{v}_G)_2$$

$$(\mathbf{H}_O)_1 = (\mathbf{H}_O)_2$$

The problems that follow involve application of all the above concepts. They are presented in *random order* so that practice may be gained at identifying the various types of problems and developing the skills necessary for their solution.

REVIEW PROBLEMS

R2–1. An automobile transmission consists of the planetary gear system shown. If the ring gear R is held fixed so that $\omega_R = 0$, and the shaft s, which is fixed to the sun gear S, is rotating at 20 rad/s, determine the angular velocity of each planet gear P and the angular velocity of the connecting rack D, which is free to turn about the center shaft s.

R2–2. An automobile transmission consists of the planetary gear system shown. If the ring gear R is rotating at $\omega_R = 2$ rad/s, and the shaft s, which is fixed to the sun gear S, is rotating at 20 rad/s, determine the angular velocity of each planet gear P and the angular velocity of the connecting rack D, which is free to turn about the center shaft s.

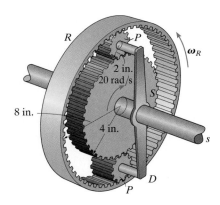

Prob. R2–2

R2–3. The 6-lb slender rod AB is released from rest when it is in the *horizontal position* so that it begins to rotate clockwise. A 1-lb ball is thrown at the rod with a velocity $v = 50$ ft/s. The ball strikes the rod at C at the instant the rod is in the vertical position as shown. Determine the angular velocity of the rod just after the impact. Take $e = 0.7$ and $d = 2$ ft.

*R2–4. The 6-lb slender rod AB is originally at rest, suspended in the vertical position. A 1-lb ball is thrown at the rod with a velocity $v = 50$ ft/s and strikes the rod at C. Determine the angular velocity of the rod just after the impact. Take $e = 0.7$ and $d = 2$ ft.

R2–5. The 6-lb slender rod is originally at rest, suspended in the vertical position. Determine the distance d where the 1-lb ball, traveling at $v = 50$ ft/s, should strike the rod so that it does not create a horizontal impulse at A. What is the rod's angular velocity just after the impact? Take $e = 0.5$.

Probs. R2–3/4/5

R2–6. At a given instant, the wheel is rotating with the angular motions shown. Determine the acceleration of the collar at A at this instant.

Prob. R2–6

R2–7. The transmission gear fixed onto the frame of an electric train turns at a constant rate of $\omega_t = 30$ rad/s. This gear is in mesh with the gear that is fixed to the axle of the engine. Determine the velocity of the train, assuming the wheels do not slip on the track.

Prob. R2–7

R2–9. The gear rack has a mass of 6 kg, and the gears each have a mass of 4 kg and a radius of gyration of $k = 30$ mm. If the rack is originally moving downward at 2 m/s, when $s = 0$, determine the speed of the rack when $s = 600$ mm. The gears are free to turn about their centers, A and B.

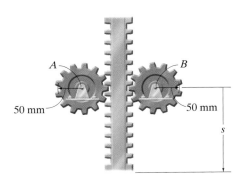

Prob. R2–9

*R2–8. The 50-kg cylinder has an angular velocity of 30 rad/s when it is brought into contact with the surface at C. If the coefficient of kinetic friction is $\mu_k = 0.2$, determine how long it takes for the cylinder to stop spinning. What force is developed in the link AB during this time? The axis of the cylinder is connected to *two* symmetrical links. (Only AB is shown.) For the computation, neglect the weight of the links.

Prob. R2–8

R2–10. The gear has a mass of 2 kg and a radius of gyration $k_A = 0.15$ m. The connecting link (slender rod) and slider block at B have a mass of 4 kg and 1 kg, respectively. If the gear has an angular velocity $\omega = 8$ rad/s at the instant $\theta = 45°$, determine the gear's angular velocity when $\theta = 0°$.

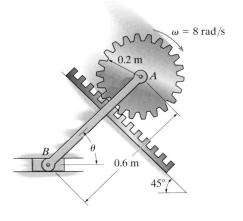

Prob. R2–10

R2–11. The rotation of link *AB* creates an oscillating movement of gear *F*. If *AB* has an angular velocity of $\omega_{AB} = 6$ rad/s, determine the angular velocity of gear *F* at the instant shown. Gear *E* is rigidly attached to arm *CD* and pinned at *D* to a fixed point.

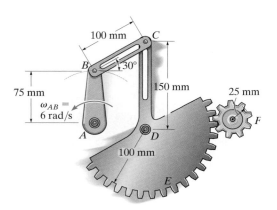

Prob. R2–11

*R2–12. The revolving door consists of four doors which are attached to an axle *AB*. Each door can be assumed to be a 50-lb thin plate. Friction at the axle contributes a moment of 2 lb · ft which resists the rotation of the doors. If a woman passes through one door by always pushing with a force *P* = 15 lb perpendicular to the plane of the door as shown, determine the door's angular velocity after it has rotated 90°. The doors are originally at rest.

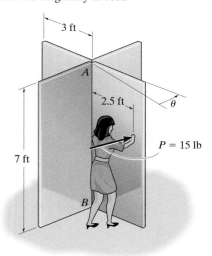

Prob. R2–12

R2–13. The 10-lb cylinder rests on the 20-lb dolly. If the system is released from rest, determine the angular velocity of the cylinder in 2 s. The cylinder does not slip on the dolly. Neglect the mass of the wheels on the dolly.

R2–14. Solve Prob. R2–13 if the coefficients of static and kinetic friction between the cylinder and the dolly are $\mu_s = 0.3$ and $\mu_k = 0.2$, respectively.

Probs. R2–13/14

R2–15. Gears *C* and *H* each have a weight of 0.4 lb and a radius of gyration about their mass center of $(k_H)_B = (k_C)_A = 2$ in. The uniform link *AB* has a weight of 0.2 lb and a radius of gyration of $(k_{AB})_A = 3$ in., whereas link *DE* has a weight of 0.15 lb and a radius of gyration of $(k_{DE})_B = 4.5$ in. If a couple moment of $M = 3$ lb · ft is applied to link *AB* and the assembly is originally at rest, determine the angular velocity of link *DE* when link *AB* has rotated 360°. Gear *C* is fixed from rotating and motion occurs in the horizontal plane. Also, gear *H* and link *DE* rotate together about the same shaft at *B*.

Prob. R2–15

*R2–16. The inner hub of the roller bearing is rotating with an angular velocity of $\omega_i = 6$ rad/s, while the outer hub is rotating in the opposite direction at $\omega_o = 4$ rad/s. Determine the angular velocity of each of the rollers if they roll on the hubs without slipping.

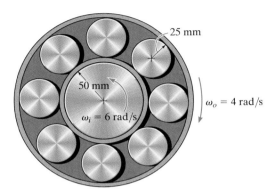

Prob. R2–16

R2–17. The hoop (thin ring) has a mass of 5 kg and is released down the inclined plane such that it has a backspin $\omega = 8$ rad/s and its center has a velocity $v_G = 3$ m/s as shown. If the coefficient of kinetic friction between the hoop and the plane is $\mu_k = 0.6$, determine how long the hoop rolls before it stops slipping.

R2–18. The hoop (thin ring) has a mass of 5 kg and is released down the inclined plane such that it has a backspin $\omega = 8$ rad/s and its center has a velocity $v_G = 3$ m/s as shown. If the coefficient of kinetic friction between the hoop and the plane is $\mu_k = 0.6$, determine the hoop's angular velocity in 1 s.

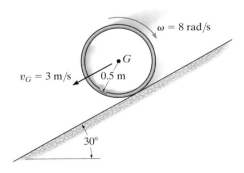

Probs. R2–17/18

R2–19. Determine the angular velocity of rod CD at the instant $\theta = 30°$. Rod AB moves to the left at a constant rate $v_{AB} = 5$ m/s.

*R2–20. Determine the angular acceleration of rod CD at the instant $\theta = 30°$. Rod AB has zero velocity, i.e., $v_{AB} = 0$, and an acceleration of $a_{AB} = 2$ m/s^2 to the right when $\theta = 30°$.

Probs. R2–19/20

R2–21. The slender 15-kg bar is initially at rest and standing in the vertical position when the bottom end A is displaced slightly to the right. If the track in which it moves is smooth, determine the speed at which end A strikes the corner D. The bar is constrained to move in the vertical plane. Neglect the mass of the cord BC.

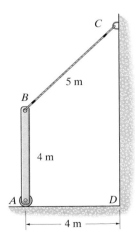

Prob. R2–21

R2–22. The four-wheeler has a weight of 335 lb and center of gravity at G_1, whereas the rider has a weight of 150 lb and center of gravity at G_2. If the engine can develop enough torque to cause the wheels to slip, determine the largest coefficient of static friction between the rear wheels and the ground so that the vehicle will accelerate without tipping over. What is this maximum acceleration? In order to *increase* the acceleration, should the rider crouch down or sit up straight from the position shown? Explain. The front wheels are free to roll. Neglect the mass of the wheels in the calculation.

Prob. R2–22

R2–23. The four-wheeler has a weight of 335 lb and center of gravity at G_1, whereas the rider has a weight of 150 lb and center of gravity at G_2. If the coefficient of static friction between the rear wheels and the ground is $\mu_s = 0.3$, determine the greatest acceleration the vehicle can have. The front wheels are free to roll. Neglect the mass of the wheels in the calculation.

Prob. R2–23

*R2–24. The pavement roller is traveling down the incline at $v_1 = 5$ ft/s when the motor is disengaged. Determine the speed of the roller when it has traveled 20 ft down the plane. The body of the roller, excluding the rollers, has a weight of 8000 lb and a center of gravity at G. Each of the two rear rollers weighs 400 lb and has a radius of gyration of $k_A = 3.3$ ft. The front roller has a weight of 800 lb and a radius of gyration of $k_B = 1.8$ ft. The rollers do not slip as they rotate.

Prob. R2–24

R2–25. If the rod starts from rest in the position shown and a motor drives it for a short time with an angular acceleration of $\alpha = (1.5e^t)$ rad/s², where t is in seconds, determine the magnitude of the angular velocity and the angular displacement of the rod when $t = 3$ s. Locate the point on the rod which has the greatest velocity and acceleration, and compute the magnitudes of the velocity and acceleration of this point when $t = 3$ s. The rod is defined by $z = 0.25 \sin(\pi y)$, where the argument for the sine is given in radians and y is in meters.

Prob. R2–25

R2–26. The disk has a mass M and a radius R. If a block of mass m is attached to the cord, determine the angular acceleration of the disk when the block is released from rest. Also, what is the distance the block falls from rest in the time t?

Prob. R2–26

R2–27. The tub of the mixer has a weight of 70 lb and a radius of gyration $k_G = 1.3$ ft about its center of gravity. If a constant torque $M = 60$ lb · ft is applied to the dumping wheel, determine the angular velocity of the tub when it has rotated $\theta = 90°$. Originally the tub is at rest when $\theta = 0°$.

*R2–28. Solve Prob. R2–27 if the applied torque is $M = (50\theta)$ lb · ft, where θ is in radians.

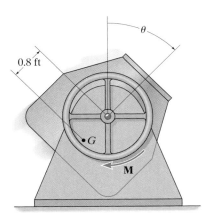

Probs. R2–27/28

R2–29. The spool has a weight of 30 lb and a radius of gyration $k_O = 0.45$ ft. A cord is wrapped around the spool's inner hub and its end subjected to a horizontal force $P = 5$ lb. Determine the spool's angular velocity in 4 s starting from rest. Assume the spool rolls without slipping.

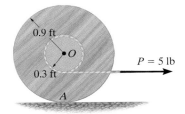

Prob. R2–29

R2–30. The slender rod of length L and mass m is released from rest when $\theta = 0°$. Determine, as a function of θ, the normal and frictional forces which are exerted on the ledge at A as the rod falls downward. At what angle θ does it begin to slip if the coefficient of static friction at A is μ_s?

Prob. R2–30

R2–31. A sphere and cylinder are released from rest on the ramp at $t = 0$. If each has a mass m and a radius r, determine their angular velocities at time t. Assume no slipping occurs.

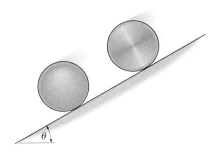

Prob. R2–31

*R2–32. At a given instant, link AB has an angular acceleration $\alpha_{AB} = 12 \text{ rad/s}^2$ and an angular velocity $\omega_{AB} = 4 \text{ rad/s}$. Determine the angular velocity and angular acceleration of link CD at this instant.

R2–33. At a given instant, link CD has an angular acceleration $\alpha_{CD} = 5 \text{ rad/s}^2$ and angular velocity $\omega_{CD} = 2 \text{ rad/s}$. Determine the angular velocity and angular acceleration of link AB at this instant.

R2–35. The bar is confined to move along the vertical and inclined planes. If the velocity of the roller at A is $v_A = 6 \text{ ft/s}$ when $\theta = 45°$, determine the bar's angular velocity and the velocity of B at this instant.

*R2–36. The bar is confined to move along the vertical and inclined planes. If the roller at A has a constant velocity of $v_A = 6 \text{ ft/s}$, determine the bar's angular acceleration and the acceleration of B when $\theta = 45°$.

Probs. R2–32/33

Probs. R2–35/36

R2–34. The spool and the wire wrapped around its core have a mass of 50 kg and a centroidal radius of gyration of $k_G = 235 \text{ mm}$. If the coefficient of kinetic friction at the surface is $\mu_k = 0.15$, determine the angular acceleration of the spool after it is released from rest.

R2–37. The 15-lb cylinder is initially at rest on a 5-lb plate. If a couple moment of $M = 40 \text{ lb} \cdot \text{ft}$ is applied to the cylinder, determine the angular acceleration of the cylinder and the time needed for the end B of the plate to travel 3 ft and strike the wall. Assume the cylinder does not slip on the plate, and neglect the mass of the rollers under the plate.

Prob. R2–34

Prob. R2–37

R2–38. Each gear has a mass of 2 kg and a radius of gyration about its pinned mass centers A and B of $k_g = 40$ mm. Each link has a mass of 2 kg and a radius of gyration about its pinned ends A and B of $k_l = 50$ mm. If originally the spring is unstretched when the couple moment $M = 20$ N·m is applied to link AC, determine the angular velocities of the links at the instant link AC rotates $\theta = 45°$. Each gear and link is connected together and rotates about the fixed pins A and B.

Prob. R2–38

R2–39. The 5-lb rod AB supports the 3-lb disk at its end. If the disk is given an angular velocity $\omega_D = 8$ rad/s while the rod is held stationary and then released, determine the angular velocity of the rod after the disk has stopped spinning relative to the rod due to frictional resistance at the bearing A. Motion is in the *horizontal plane*. Neglect friction at the fixed bearing B.

Prob. R2–39

*R2–40. If link AB is rotating at $\omega_{AB} = 3$ rad/s, determine the angular velocity of link CD at the instant shown.

R2–41. If link CD is rotating at $\omega_{CD} = 5$ rad/s, determine the angular velocity of link AB at the instant shown.

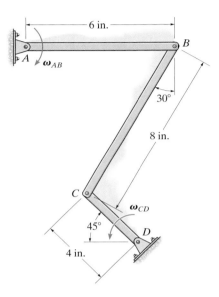

Probs. R2–40/41

R2–42. The 15-kg disk is pinned at O and is initially at rest. If a 10-g bullet is fired into the disk with a velocity of 200 m/s, as shown, determine the maximum angle θ to which the disk swings. The bullet becomes embedded in the disk.

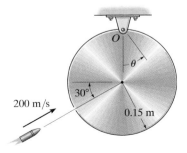

Prob. R2–42

R2–43. The disk is rotating at a constant rate of 4 rad/s as it falls freely, its center G having an acceleration of 32.2 ft/s². Determine the accelerations of points A and B on the rim of the disk at the instant shown.

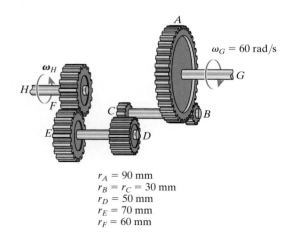

Prob. R2–43

R2–45. Shown is the internal gearing of a "spinner" used for drilling wells. With constant angular acceleration, the motor M rotates the shaft S to 100 rev/min in $t = 2$ s starting from rest. Determine the angular acceleration of the drill-pipe connection D and the number of revolutions it makes during the 2-s startup.

Prob. R2–45

*R2–44. The operation of "reverse" for a three-speed automotive transmission is illustrated schematically in the figure. If the crankshaft G is turning with an angular speed of 60 rad/s, determine the angular speed of the drive shaft H. Each of the gears rotates about a fixed axis. Note that gears A and B, C and D, E and F are in mesh. The radius of each of these gears is reported in the figure.

$r_A = 90$ mm
$r_B = r_C = 30$ mm
$r_D = 50$ mm
$r_E = 70$ mm
$r_F = 60$ mm

Prob. R2–44

R2–46. The plate weighs 40 lb and is supported by a roller at A. If a horizontal force of $F = 70$ lb is suddenly applied to the roller, determine the acceleration of the center of the roller at the instant the force is applied. The plate has a moment of inertia about its center of mass of $I_G = 0.414$ slug · ft². Neglect the weight and the size d of the roller.

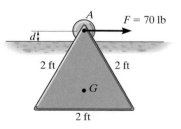

Prob. R2–46

R2–47. The 15-kg cylinder is rotating with an angular velocity of $\omega = 40$ rad/s. If a force $F = 6$ N is applied to link AB, as shown, determine the time needed to stop the rotation. The coefficient of kinetic friction between AB and the cylinder is $\mu_k = 0.4$.

R2–49. The wheel has a mass of 25 kg and a radius of gyration $k_B = 0.15$ m. It is originally spinning at $\omega_1 = 40$ rad/s. If it is placed on the ground, for which the coefficient of kinetic friction is $\mu_k = 0.5$, determine the time required for the motion to stop. What are the horizontal and vertical components of reaction which the pin at A exerts on AB during this time? Neglect the mass of AB.

Prob. R2–47

Prob. R2–49

*****R2–48.** If link AB is rotating at $\omega_{AB} = 6$ rad/s, determine the angular velocities of links BC and CD at the instant shown.

R2–50. At the *start* of take-off, the propeller on the 2-Mg plane exerts a horizontal thrust of 600 N on the plane. Determine the plane's acceleration and the vertical reactions at the nose wheel A and each of the *two* wing wheels B. Neglect the lifting force of the wings since the plane is originally at rest. The mass center is at G.

Prob. R2–48

Prob. R2–50

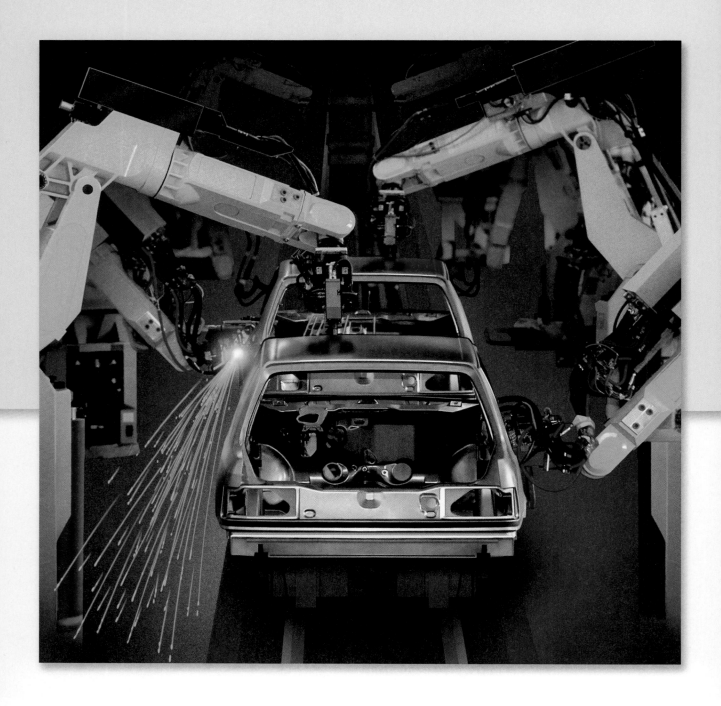

The three-dimensional motion of these industrial robots used in the manufacturing of automobiles must be accurately specified.

20 Three-Dimensional Kinematics of a Rigid Body

CHAPTER OBJECTIVES

- To analyze the kinematics of a body subjected to rotation about a fixed axis and general plane motion.
- To provide a relative-motion analysis of a rigid body using translating and rotating axes.

20.1 Rotation About a Fixed Point

When a rigid body rotates about a fixed point, the distance r from the point to a particle P located on the body is the *same* for *any position* of the body. Thus, the path of motion for the particle lies on the *surface of a sphere* having a radius r and centered at the fixed point. Since motion along this path occurs only from a series of rotations made during a finite time interval, we will first develop a familiarity with some of the properties of rotational displacements.

The boom can rotate up and down, and because it is hinged at a point on the vertical axis about which it turns, it is subjected to rotation about a fixed point.

Euler's Theorem.   Euler's theorem states that two "component" rotations about different axes passing through a point are equivalent to a single resultant rotation about an axis passing through the point. If more than two rotations are applied, they can be combined into pairs, and each pair can be further reduced to combine into one rotation.

Finite Rotations.   If component rotations used in Euler's theorem are *finite*, it is important that the *order* in which they are applied be maintained. This is because finite rotations do *not* obey the law of vector addition, and hence they cannot be classified as vector quantities. To show this, consider the two finite rotations $\boldsymbol{\theta}_1 + \boldsymbol{\theta}_2$ applied to the block in Fig. 20–1a. Each rotation has a magnitude of 90° and a direction defined by the right-hand rule, as indicated by the arrow. The resultant orientation of the block is shown at the right. When these two rotations are applied in the order $\boldsymbol{\theta}_2 + \boldsymbol{\theta}_1$, as shown in Fig. 20–1b, the resultant position of the block is *not* the same as it is in Fig. 20–1a. Consequently, *finite rotations* do not obey the commutative law of addition ($\boldsymbol{\theta}_1 + \boldsymbol{\theta}_2 \neq \boldsymbol{\theta}_2 + \boldsymbol{\theta}_1$), and therefore *they cannot be classified as vectors*. If smaller, yet finite, rotations had been used to illustrate this point, e.g., 10° instead of 90°, the *resultant* orientation of the block after each combination of rotations would also be different; however, in this case, the difference is only a small amount.

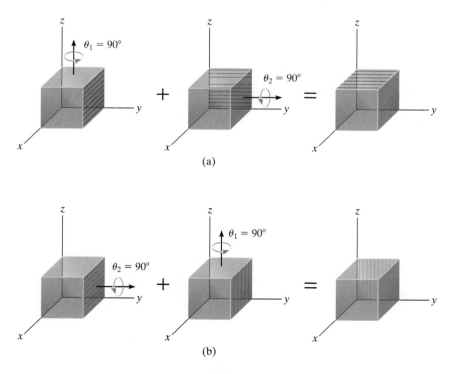

Fig. 20–1

Infinitesimal Rotations.   When defining the angular motions of a body subjected to three-dimensional motion, only rotations which are *infinitesimally small* will be considered. *Such rotations may be classified as vectors, since they can be added vectorially in any manner.* To show this, let us for purposes of simplicity consider the rigid body itself to be a sphere which is allowed to rotate about its central fixed point O, Fig. 20–2a. If we impose two infinitesimal rotations $d\boldsymbol{\theta}_1 + d\boldsymbol{\theta}_2$ on the body, it is seen that point P moves along the path $d\boldsymbol{\theta}_1 \times \mathbf{r} + d\boldsymbol{\theta}_2 \times \mathbf{r}$ and ends up at P'. Had the two successive rotations occurred in the order $d\boldsymbol{\theta}_2 + d\boldsymbol{\theta}_1$, then the resultant displacements of P would have been $d\boldsymbol{\theta}_2 \times \mathbf{r} + d\boldsymbol{\theta}_1 \times \mathbf{r}$. Since the vector cross product obeys the distributive law, by comparison $(d\boldsymbol{\theta}_1 + d\boldsymbol{\theta}_2) \times \mathbf{r} = (d\boldsymbol{\theta}_2 + d\boldsymbol{\theta}_1) \times \mathbf{r}$. Here infinitesimal rotations $d\boldsymbol{\theta}$ are vectors, since these quantities have both a magnitude and direction for which the order of (vector) addition is not important, i.e., $d\boldsymbol{\theta}_1 + d\boldsymbol{\theta}_2 = d\boldsymbol{\theta}_2 + d\boldsymbol{\theta}_1$. Furthermore, as shown in Fig. 20–2a, the two "component" rotations $d\boldsymbol{\theta}_1$ and $d\boldsymbol{\theta}_2$ are equivalent to a single resultant rotation $d\boldsymbol{\theta} = d\boldsymbol{\theta}_1 + d\boldsymbol{\theta}_2$, a consequence of Euler's theorem.

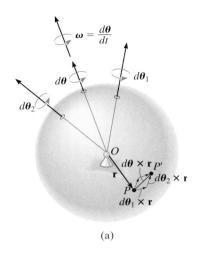

(a)

Angular Velocity.   If the body is subjected to an angular rotation $d\boldsymbol{\theta}$ about a fixed point, the angular velocity of the body is defined by the time derivative,

$$\boldsymbol{\omega} = \dot{\boldsymbol{\theta}} \tag{20–1}$$

The line specifying the direction of $\boldsymbol{\omega}$, which is collinear with $d\boldsymbol{\theta}$, is referred to as the *instantaneous axis of rotation*, Fig. 20–2b. In general, this axis changes direction during each instant of time. Since $d\boldsymbol{\theta}$ is a vector quantity, so too is $\boldsymbol{\omega}$, and it follows from vector addition that if the body is subjected to two component angular motions, $\boldsymbol{\omega}_1 = \dot{\boldsymbol{\theta}}_1$ and $\boldsymbol{\omega}_2 = \dot{\boldsymbol{\theta}}_2$, the resultant angular velocity is $\boldsymbol{\omega} = \boldsymbol{\omega}_1 + \boldsymbol{\omega}_2$.

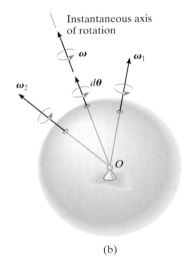

(b)

Fig. 20–2

Angular Acceleration.   The body's angular acceleration is determined from the time derivative of the angular velocity, i.e.,

$$\boldsymbol{\alpha} = \dot{\boldsymbol{\omega}} \tag{20–2}$$

For motion about a fixed point, $\boldsymbol{\alpha}$ must account for a change in *both* the magnitude and direction of $\boldsymbol{\omega}$, so that, in general, $\boldsymbol{\alpha}$ is not directed along the instantaneous axis of rotation, Fig. 20–3.

As the direction of the instantaneous axis of rotation (or the line of action of $\boldsymbol{\omega}$) changes in space, the locus of points defined by the axis generates a fixed *space cone*, Fig. 20–4. If the change in this axis is viewed with respect to the rotating body, the locus of the axis generates a *body cone*.

Fig. 20–3

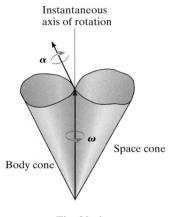

Fig. 20–4

At any given instant, these cones are tangent along the instantaneous axis of rotation, and when the body is in motion, the body cone appears to roll either on the inside or the outside surface of the fixed space cone. Provided the paths defined by the open ends of the cones are described by the head of the $\boldsymbol{\omega}$ vector, $\boldsymbol{\alpha}$ must act tangent to these paths at any given instant, since the time rate of change of $\boldsymbol{\omega}$ is equal to $\boldsymbol{\alpha}$. Fig. 20–4.

Velocity.   Once $\boldsymbol{\omega}$ is specified, the velocity of any point P on a body rotating about a fixed point can be determined using the same methods as for a body rotating about a fixed axis (Sec. 16.3). Hence, by the cross product,

$$\mathbf{v} = \boldsymbol{\omega} \times \mathbf{r} \qquad (20\text{–}3)$$

Here $\mathbf{r}$ defines the position of P measured from the fixed point O, Fig. 20–3.

Acceleration.   If $\boldsymbol{\omega}$ and $\boldsymbol{\alpha}$ are known at a given instant, the acceleration of any point P on the body can be obtained by time differentiation of Eq. 20–3, which yields

$$\mathbf{a} = \boldsymbol{\alpha} \times \mathbf{r} + \boldsymbol{\omega} \times (\boldsymbol{\omega} \times \mathbf{r}) \qquad (20\text{–}4)$$

The form of this equation is the same as that developed in Sec. 16.3, which defines the acceleration of a point located on a body subjected to rotation about a fixed axis.

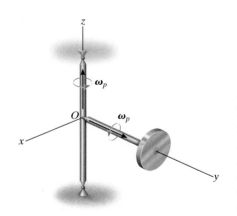

Fig. 20–5

*20.2   The Time Derivative of a Vector Measured from Either a Fixed or Translating-Rotating System

In many types of problems involving the motion of a body about a fixed point, the angular velocity $\boldsymbol{\omega}$ is specified in terms of its component angular motions. For example, the disk in Fig. 20–5 spins about the horizontal y axis at $\boldsymbol{\omega}_s$ while it rotates or precesses about the vertical z axis at $\boldsymbol{\omega}_p$. Therefore, its resultant angular velocity is $\boldsymbol{\omega} = \boldsymbol{\omega}_s + \boldsymbol{\omega}_p$. If the angular acceleration $\boldsymbol{\alpha} = \dot{\boldsymbol{\omega}}$ of such a body is to be determined, it is sometimes easier to compute the time derivative of $\boldsymbol{\omega}$ by using a coordinate system that has a *rotation* defined by one or more of the components of $\boldsymbol{\omega}$.* For this reason, and for other uses later, an equation will now be derived, which relates the time derivative of any vector $\mathbf{A}$ defined from a translating-rotating reference to its time derivative defined from a fixed reference.

---

*In the case of the spinning disk, Fig. 20–5, the x, y, z axes may be given an angular velocity of $\boldsymbol{\omega}_p$.

Consider the x, y, z axes of the moving frame of reference to be rotating about an axis with an angular velocity $\boldsymbol{\Omega}$, which is measured from the fixed X, Y, Z axes, Fig. 20–6a. In the following discussion, it will be convenient to express vector $\mathbf{A}$ in terms of its $\mathbf{i}, \mathbf{j}, \mathbf{k}$ components, which define the directions of the moving axes. Hence,

$$\mathbf{A} = A_x\mathbf{i} + A_y\mathbf{j} + A_z\mathbf{k}$$

In general, the time derivative of $\mathbf{A}$ must account for the change in both the vector's magnitude and direction. However, if this derivative is taken *with respect to the moving frame of reference*, only a change in the magnitudes of the components of $\mathbf{A}$ must be accounted for, since the directions of the components do not change with respect to the moving reference. Hence,

$$(\dot{\mathbf{A}})_{xyz} = \dot{A}_x\mathbf{i} + \dot{A}_y\mathbf{j} + \dot{A}_z\mathbf{k} \qquad (20\text{–}5)$$

When the time derivative of $\mathbf{A}$ is taken *with respect to the fixed frame of reference*, the *directions* of $\mathbf{i}$, $\mathbf{j}$, and $\mathbf{k}$ change only on account of the *rotation* $\boldsymbol{\Omega}$ of the axes and not their translation. Hence, in general,

$$\dot{\mathbf{A}} = \dot{A}_x\mathbf{i} + \dot{A}_y\mathbf{j} + \dot{A}_z\mathbf{k} + A_x\dot{\mathbf{i}} + A_y\dot{\mathbf{j}} + A_z\dot{\mathbf{k}}$$

The time derivatives of the unit vectors will now be considered. For example, $\dot{\mathbf{i}} = d\mathbf{i}/dt$ represents only a change in the *direction* of $\mathbf{i}$ with respect to time, since $\mathbf{i}$ has a fixed magnitude of 1 unit. As shown in Fig. 20–6b, the change, $d\mathbf{i}$, is *tangent to the path* described by the arrowhead of $\mathbf{i}$ as $\mathbf{i}$ moves due to the rotation $\boldsymbol{\Omega}$. Accounting for both the magnitude and direction of $d\mathbf{i}$, we can therefore define $\dot{\mathbf{i}}$ using the cross product, $\dot{\mathbf{i}} = \boldsymbol{\Omega} \times \mathbf{i}$. In general,

$$\dot{\mathbf{i}} = \boldsymbol{\Omega} \times \mathbf{i} \qquad \dot{\mathbf{j}} = \boldsymbol{\Omega} \times \mathbf{j} \qquad \dot{\mathbf{k}} = \boldsymbol{\Omega} \times \mathbf{k}$$

These formulations were also developed in Sec. 16.8, regarding planar motion of the axes. Substituting the results into the above equation and using Eq. 20–5 yields

$$\boxed{\dot{\mathbf{A}} = (\dot{\mathbf{A}})_{xyz} + \boldsymbol{\Omega} \times \mathbf{A}} \qquad (20\text{–}6)$$

This result is important, and will be used throughout Sec. 20.4 and Chapter 21. It states that the time derivative of *any vector* $\mathbf{A}$ as observed from the fixed X, Y, Z frame of reference is equal to the time rate of change of $\mathbf{A}$ as observed from the x, y, z translating-rotating frame of reference, Eq. 20–5, plus $\boldsymbol{\Omega} \times \mathbf{A}$, the change of $\mathbf{A}$ caused by the rotation of the x, y, z frame. As a result, Eq. 20–6 should always be used whenever $\boldsymbol{\Omega}$ produces a change in the direction of $\mathbf{A}$ as seen from the X, Y, Z reference. If this change does not occur, i.e., $\boldsymbol{\Omega} = \mathbf{0}$, then $\dot{\mathbf{A}} = (\dot{\mathbf{A}})_{xyz}$, and so the time rate of change of $\mathbf{A}$ as observed from both coordinate systems will be the *same*.

(a)

(b)

Fig. 20–6

EXAMPLE 20.1

The disk shown in Fig. 20–7a is spinning about its horizontal axis with a constant angular velocity $\omega_s = 3$ rad/s, while the horizontal platform on which the disk is mounted is rotating about the vertical axis at a constant rate $\omega_p = 1$ rad/s. Determine the angular acceleration of the disk and the velocity and acceleration of point A on the disk when it is in the position shown.

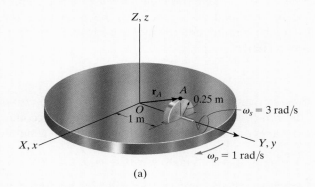

(a)

Fig. 20–7

SOLUTION
Point O represents a fixed point of rotation for the disk if one considers a hypothetical extension of the disk to this point. To determine the velocity and acceleration of point A, it is first necessary to determine the resultant angular velocity $\boldsymbol{\omega}$ and angular acceleration $\boldsymbol{\alpha}$ of the disk, since these vectors are used in Eqs. 20–3 and 20–4.

Angular Velocity.   The angular velocity, which is measured from X, Y, Z, is simply the vector addition of the two component motions. Thus,

$$\boldsymbol{\omega} = \boldsymbol{\omega}_s + \boldsymbol{\omega}_p = \{3\mathbf{j} - 1\mathbf{k}\} \text{ rad/s}$$

At first glance, it may not appear that the disk is actually rotating with this angular velocity, since it is generally more difficult to imagine the resultant of angular motions in comparison with linear motions. To further understand the angular motion, consider the disk as being replaced by a cone (a body cone), which is rolling over the stationary space cone, Fig. 20–7b. The instantaneous axis of rotation is along the line of contact of the cones. This axis defines the direction of the resultant $\boldsymbol{\omega}$, which has components $\boldsymbol{\omega}_s$ and $\boldsymbol{\omega}_p$.

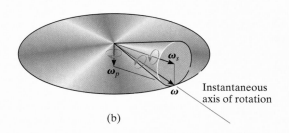

(b)

Angular Acceleration.   Since the magnitude of $\boldsymbol{\omega}$ is constant, only a change in its direction, as seen from a fixed reference, creates the angular acceleration $\boldsymbol{\alpha}$ of the disk. One way to obtain $\boldsymbol{\alpha}$ is to compute the time derivative of *each of the two components* of $\boldsymbol{\omega}$ using Eq. 20–6. At the instant shown in Fig. 20–7a, imagine the fixed X, Y, Z and a rotating x, y, z frame to be coincident. If the rotating x, y, z frame is chosen to have an angular velocity of $\boldsymbol{\Omega} = \boldsymbol{\omega}_p = \{-1\mathbf{k}\}$ rad/s, then $\boldsymbol{\omega}_s$ will *always* be directed along the y (not Y) axis, and the time rate of change of $\boldsymbol{\omega}_s$ as seen from x, y, z is *zero*; i.e., $(\dot{\boldsymbol{\omega}}_s)_{xyz} = \mathbf{0}$ (the magnitude and direction of $\boldsymbol{\omega}_s$ is constant). Thus, by Eq. 20–6,

$$\dot{\boldsymbol{\omega}}_s = (\dot{\boldsymbol{\omega}}_s)_{xyz} + \boldsymbol{\omega}_p \times \boldsymbol{\omega}_s = \mathbf{0} + (-1\mathbf{k}) \times (3\mathbf{j}) = \{3\mathbf{i}\} \text{ rad/s}^2$$

By the same choice of axes rotation, $\boldsymbol{\Omega} = \boldsymbol{\omega}_p$, or even with $\boldsymbol{\Omega} = \mathbf{0}$, the time derivative $(\dot{\boldsymbol{\omega}}_p)_{xyz} = \mathbf{0}$, since $\boldsymbol{\omega}_p$ has a constant magnitude and direction. Hence,

$$\dot{\boldsymbol{\omega}}_p = (\dot{\boldsymbol{\omega}}_p)_{xyz} + \boldsymbol{\omega}_p \times \boldsymbol{\omega}_p = \mathbf{0} + \mathbf{0} = \mathbf{0}$$

The angular acceleration of the disk is therefore

$$\boldsymbol{\alpha} = \dot{\boldsymbol{\omega}} = \dot{\boldsymbol{\omega}}_s + \dot{\boldsymbol{\omega}}_p = \{3\mathbf{i}\} \text{ rad/s}^2 \qquad \textit{Ans.}$$

Velocity and Acceleration.   Since $\boldsymbol{\omega}$ and $\boldsymbol{\alpha}$ have been determined, the velocity and acceleration of point A can be computed using Eqs. 20–3 and 20–4. Realizing that $\mathbf{r}_A = \{1\mathbf{j} + 0.25\mathbf{k}\}$ m, Fig. 20–7a, we have

$$\mathbf{v}_A = \boldsymbol{\omega} \times \mathbf{r}_A = (3\mathbf{j} - 1\mathbf{k}) \times (1\mathbf{j} + 0.25\mathbf{k}) = \{1.75\mathbf{i}\} \text{ m/s} \qquad \textit{Ans.}$$

$$\mathbf{a}_A = \boldsymbol{\alpha} \times \mathbf{r}_A + \boldsymbol{\omega} \times (\boldsymbol{\omega} \times \mathbf{r}_A)$$

$$= (3\mathbf{i}) \times (1\mathbf{j} + 0.25\mathbf{k}) + (3\mathbf{j} - 1\mathbf{k}) \times [(3\mathbf{j} - 1\mathbf{k}) \times (1\mathbf{j} + 0.25\mathbf{k})]$$

$$= \{-2.50\mathbf{j} - 2.25\mathbf{k}\} \text{ m/s}^2 \qquad \textit{Ans.}$$

EXAMPLE 20.2

At the instant $\theta = 60°$, the gyrotop in Fig. 20–8 has three components of angular motion directed as shown and having magnitudes defined as:

Spin: $\omega_s = 10$ rad/s, increasing at the rate of 6 rad/s²

Nutation: $\omega_n = 3$ rad/s, increasing at the rate of 2 rad/s²

Precession: $\omega_p = 5$ rad/s, increasing at the rate of 4 rad/s²

Determine the angular velocity and angular acceleration of the top.

SOLUTION

Angular Velocity. The top is rotating about the fixed point O. If the fixed and rotating frames are coincident at the instant shown, then the angular velocity can be expressed in terms of $\mathbf{i}, \mathbf{j}, \mathbf{k}$ components, appropriate to the x, y, z frame; i.e.,

$$\boldsymbol{\omega} = -\omega_n\mathbf{i} + \omega_s \sin\theta\mathbf{j} + (\omega_p + \omega_s \cos\theta)\mathbf{k}$$

$$= -3\mathbf{i} + 10\sin 60°\mathbf{j} + (5 + 10\cos 60°)\mathbf{k}$$

$$= \{-3\mathbf{i} + 8.66\mathbf{j} + 10\mathbf{k}\}\text{ rad/s} \qquad\qquad Ans.$$

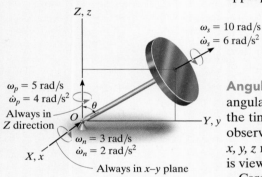

$\omega_s = 10$ rad/s
$\dot\omega_s = 6$ rad/s²

$\omega_p = 5$ rad/s
$\dot\omega_p = 4$ rad/s²
Always in
Z direction

$\omega_n = 3$ rad/s
$\dot\omega_n = 2$ rad/s²

Always in x–y plane

Fig. 20–8

Angular Acceleration. As in the solution of Example 20.1, the angular acceleration $\boldsymbol{\alpha}$ will be determined by investigating separately the time rate of change of *each of the angular velocity components* as observed from the fixed X, Y, Z reference. We will choose an $\boldsymbol{\Omega}$ for the x, y, z reference so that the component of $\boldsymbol{\omega}$ which is being considered is viewed as having a *constant direction* when observed from x, y, z.

Careful examination of the motion of the top reveals that $\boldsymbol{\omega}_s$ has a *constant direction* relative to x, y, z if these axes rotate at $\boldsymbol{\Omega} = \boldsymbol{\omega}_n + \boldsymbol{\omega}_p$. Thus,

$$\dot{\boldsymbol{\omega}}_s = (\dot{\boldsymbol{\omega}}_s)_{xyz} + (\boldsymbol{\omega}_n + \boldsymbol{\omega}_p) \times \boldsymbol{\omega}_s$$

$$= (6\sin 60°\mathbf{j} + 6\cos 60°\mathbf{k}) + (-3\mathbf{i} + 5\mathbf{k}) \times (10\sin 60°\mathbf{j} + 10\cos 60°\mathbf{k})$$

$$= \{-43.30\mathbf{i} + 20.20\mathbf{j} - 22.98\mathbf{k}\}\text{ rad/s}^2$$

Since $\boldsymbol{\omega}_n$ *always* lies in the fixed X–Y plane, this vector has a *constant direction* if the motion is viewed from axes x, y, z having a rotation of $\boldsymbol{\Omega} = \boldsymbol{\omega}_p$ (not $\boldsymbol{\Omega} = \boldsymbol{\omega}_s + \boldsymbol{\omega}_p$). Thus,

$$\dot{\boldsymbol{\omega}}_n = (\dot{\boldsymbol{\omega}}_n)_{xyz} + \boldsymbol{\omega}_p \times \boldsymbol{\omega}_n = -2\mathbf{i} + (5\mathbf{k}) \times (-3\mathbf{i}) = \{-2\mathbf{i} - 15\mathbf{j}\}\text{ rad/s}^2$$

Finally, the component $\boldsymbol{\omega}_p$ is *always directed* along the Z axis so that here it is not necessary to think of x, y, z as rotating, i.e., $\boldsymbol{\Omega} = \mathbf{0}$. Expressing the data in terms of the $\mathbf{i}, \mathbf{j}, \mathbf{k}$ components, we therefore have

$$\dot{\boldsymbol{\omega}}_p = (\dot{\boldsymbol{\omega}}_p)_{xyz} + \mathbf{0} \times \boldsymbol{\omega}_p = \{4\mathbf{k}\}\text{ rad/s}^2$$

Thus, the angular acceleration of the top is

$$\boldsymbol{\alpha} = \dot{\boldsymbol{\omega}}_s + \dot{\boldsymbol{\omega}}_n + \dot{\boldsymbol{\omega}}_p = \{-45.3\mathbf{i} + 5.20\mathbf{j} - 19.0\mathbf{k}\}\text{ rad/s}^2 \quad Ans.$$

20.3   General Motion

Shown in Fig. 20–9 is a rigid body subjected to general motion in three dimensions for which the angular velocity is $\boldsymbol{\omega}$ and the angular acceleration is $\boldsymbol{\alpha}$. If point A has a known motion of $\mathbf{v}_A$ and $\mathbf{a}_A$, the motion of any other point B may be determined by using a relative-motion analysis. In this section a *translating coordinate system* will be used to define the relative motion, and in the next section a reference that is both rotating and translating will be considered.

If the origin of the translating coordinate system x, y, z $(\boldsymbol{\Omega} = \mathbf{0})$ is located at the "base point" A, then, at the instant shown, the motion of the body may be regarded as the sum of an instantaneous translation of the body having a motion of $\mathbf{v}_A$ and $\mathbf{a}_A$ and a rotation of the body about an instantaneous axis passing through the base point. Since the body is rigid, the motion of point B measured by an observer located at A is the same as *motion of the body about a fixed point*. This relative motion occurs about the instantaneous axis of rotation and is defined by $\mathbf{v}_{B/A} = \boldsymbol{\omega} \times \mathbf{r}_{B/A}$, Eq. 20–3, and $\mathbf{a}_{B/A} = \boldsymbol{\alpha} \times \mathbf{r}_{B/A} + \boldsymbol{\omega} \times (\boldsymbol{\omega} \times \mathbf{r}_{B/A})$, Eq. 20–4. For translating axes, the relative motions are related to absolute motions by $\mathbf{v}_B = \mathbf{v}_A + \mathbf{v}_{B/A}$ and $\mathbf{a}_B = \mathbf{a}_A + \mathbf{a}_{B/A}$, Eqs. 16–15 and 16–17, so that the absolute velocity and acceleration of point B can be determined from the equations

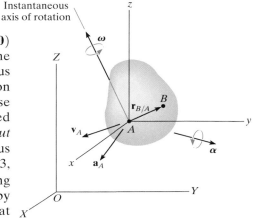

Fig. 20–9

$$\boxed{\mathbf{v}_B = \mathbf{v}_A + \boldsymbol{\omega} \times \mathbf{r}_{B/A}} \qquad (20\text{–}7)$$

and

$$\boxed{\mathbf{a}_B = \mathbf{a}_A + \boldsymbol{\alpha} \times \mathbf{r}_{B/A} + \boldsymbol{\omega} \times (\boldsymbol{\omega} \times \mathbf{r}_{B/A})} \qquad (20\text{–}8)$$

These two equations are identical to those describing the general plane motion of a rigid body, Eqs. 16–16 and 16–18. However, difficulty in application arises for three-dimensional motion, because $\boldsymbol{\alpha}$ measures the change in *both* the magnitude and direction of $\boldsymbol{\omega}$. (Recall that, for general plane motion, $\boldsymbol{\alpha}$ and $\boldsymbol{\omega}$ are always parallel or perpendicular to the plane of motion, and therefore $\boldsymbol{\alpha}$ measures only a change in the magnitude of $\boldsymbol{\omega}$.) In some problems, the constraints or connections of a body will require that the directions of the angular motions or displacement paths of points on the body be defined. As illustrated in the following example, this information is useful for obtaining some of the terms in the above equations.

EXAMPLE 20.3

(a)

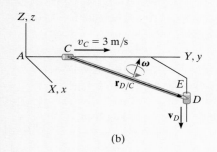

(b)

Fig. 20–10

One end of the rigid bar CD shown in Fig. 20–10a slides along the horizontal member AB, and the other end slides along the vertical member EF. If the collar at C is moving towards B at a speed of 3 m/s, determine the velocity of the collar at D and the angular velocity of the bar at the instant shown. The bar is connected to the collars at its end points by ball-and-socket joints.

SOLUTION

Bar CD is subjected to general motion. Why? The velocity of point D on the bar may be related to the velocity of point C by the equation

$$\mathbf{v}_D = \mathbf{v}_C + \boldsymbol{\omega} \times \mathbf{r}_{D/C}$$

The fixed and translating frames of reference are assumed to coincide at the instant considered, Fig. 20–10b. We have

$$\mathbf{v}_D = -v_D\mathbf{k} \qquad \mathbf{v}_C = \{3\mathbf{j}\} \text{ m/s}$$

$$\mathbf{r}_{D/C} = \{1\mathbf{i} + 2\mathbf{j} - 0.5\mathbf{k}\} \text{ m} \qquad \boldsymbol{\omega} = \omega_x\mathbf{i} + \omega_y\mathbf{j} + \omega_z\mathbf{k}$$

Substituting these quantities into the above equation gives

$$-v_D\mathbf{k} = 3\mathbf{j} + \begin{vmatrix} \mathbf{i} & \mathbf{j} & \mathbf{k} \\ \omega_x & \omega_y & \omega_z \\ 1 & 2 & -0.5 \end{vmatrix}$$

Expanding and equating the respective $\mathbf{i}, \mathbf{j}, \mathbf{k}$ components yields

$$-0.5\omega_y - 2\omega_z = 0 \tag{1}$$
$$0.5\omega_x + 1\omega_z + 3 = 0 \tag{2}$$
$$2\omega_x - 1\omega_y + v_D = 0 \tag{3}$$

These equations contain four unknowns.* A fourth equation can be written if the direction of $\boldsymbol{\omega}$ is specified. In particular, any component of $\boldsymbol{\omega}$ acting along the bar's axis has no effect on moving the collars. This is because the bar is *free to rotate* about its axis. Therefore, if $\boldsymbol{\omega}$ is specified as acting *perpendicular* to the axis of the bar, then $\boldsymbol{\omega}$ must have a unique magnitude to satisfy the above equations. Perpendicularity is guaranteed provided the dot product of $\boldsymbol{\omega}$ and $\mathbf{r}_{D/C}$ is zero (see Eq. C–14 of Appendix C). Hence,

$$\boldsymbol{\omega} \cdot \mathbf{r}_{D/C} = (\omega_x\mathbf{i} + \omega_y\mathbf{j} + \omega_z\mathbf{k}) \cdot (1\mathbf{i} + 2\mathbf{j} - 0.5\mathbf{k}) = 0$$

$$1\omega_x + 2\omega_y - 0.5\omega_z = 0 \tag{4}$$

Solving Eqs. 1 through 4 simultaneously yields

$$\omega_x = -4.86 \text{ rad/s} \quad \omega_y = 2.29 \text{ rad/s} \quad \omega_z = -0.571 \text{ rad/s} \quad Ans.$$

$$v_D = 12.0 \text{ m/s} \downarrow \qquad\qquad Ans.$$

---

*Although this is the case, the magnitude of $\mathbf{v}_D$ can be obtained. For example, solve Eqs. 1 and 2 for ω_y and ω_x in terms of ω_z and substitute into Eq. 3. It will be noted that ω_z will *cancel* out, which will allow a solution for v_D.

PROBLEMS

20–1.   The ladder of the fire truck rotates around the z axis with an angular velocity $\omega_1 = 0.15$ rad/s, which is increasing at 0.8 rad/s^2. At the same instant it is rotating upward at a constant rate $\omega_2 = 0.6$ rad/s. Determine the velocity and acceleration of point A located at the top of the ladder at this instant.

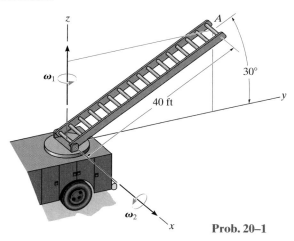

Prob. 20–1

20–2.   The ladder of the fire truck rotates around the z axis with an angular velocity of $\omega_1 = 0.15$ rad/s, which is increasing at 0.2 rad/s^2. At the same instant it is rotating upwards at $\omega_2 = 0.6$ rad/s while increasing at 0.4 rad/s^2. Determine the velocity and acceleration of point A located at the top of the ladder at this instant.

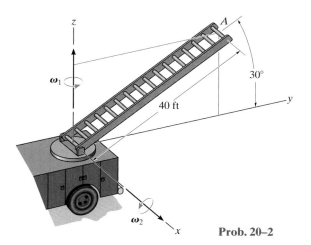

Prob. 20–2

20–3.   The antenna is following the motion of a jet plane. At the instant $\theta = 25°$ and $\phi = 75°$, the constant angular rates of change are $\dot{\theta} = 0.4$ rad/s and $\dot{\phi} = 0.6$ rad/s. Determine the velocity and acceleration of the signal horn A at this instant. The distance OA is 0.8 m.

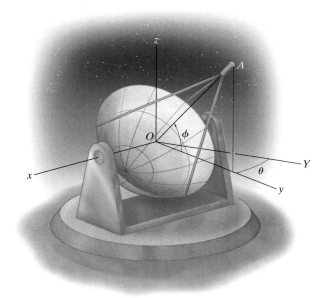

Prob. 20–3

*20–4.   The propeller of an airplane is rotating at a constant speed $\omega_s \mathbf{i}$, while the plane is undergoing a turn at a constant rate ω_t. Determine the angular acceleration of the propeller if (a) the turn is horizontal, i.e., $\omega_t \mathbf{k}$, and (b) the turn is vertical, downward, i.e., $\omega_t \mathbf{j}$.

Prob. 20–4

20–5. Gear A is fixed while gear B is free to rotate on the shaft S. If the shaft is turning about the z axis at $\omega_z = 5$ rad/s, while increasing at 2 rad/s², determine the velocity and acceleration of point P at the instant shown. The face of gear B lies in a vertical plane.

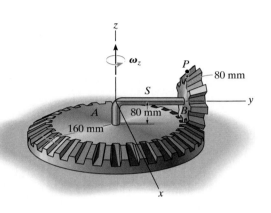

Prob. 20–5

20–6. The conical spool rolls on the plane without slipping. If the axle has an angular velocity of $\omega_1 = 3$ rad/s and an angular acceleration of $\alpha_1 = 2$ rad/s² at the instant shown, determine the angular velocity and angular acceleration of the spool at this instant.

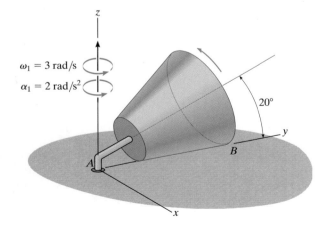

Prob. 20–6

20–7. At a given instant, the antenna has an angular motion $\omega_1 = 3$ rad/s and $\dot{\omega}_1 = 2$ rad/s² about the z axis. At this same instant $\theta = 30°$, the angular motion about the x axis is $\omega_2 = 1.5$ rad/s, and $\dot{\omega}_2 = 4$ rad/s². Determine the velocity and acceleration of the signal horn A at this instant. The distance from O to A is $d = 3$ ft.

Prob. 20–7

*20–8.** The cone rolls without slipping such that at the instant shown $\omega_z = 4$ rad/s and $\dot{\omega}_z = 3$ rad/s². Determine the velocity and acceleration of point A at this instant.

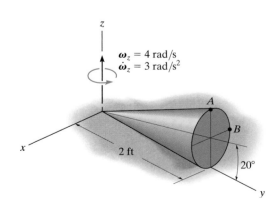

Prob. 20–8

20–9. The cone rolls without slipping such that at the instant shown $\omega_z = 4$ rad/s and $\dot{\omega}_z = 3$ rad/s². Determine the velocity and acceleration of point B at this instant.

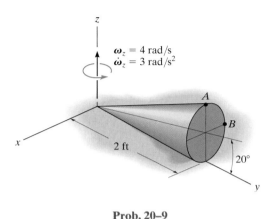

$\omega_z = 4$ rad/s
$\dot{\omega}_z = 3$ rad/s²

2 ft

20°

Prob. 20–9

20–11. Gear A is fixed to the crankshaft S, while gear C is fixed and B are free to rotate. The crankshaft is turning at 80 rad/s about its axis. Determine the magnitudes of the angular velocity of the propeller and the angular acceleration of gear B.

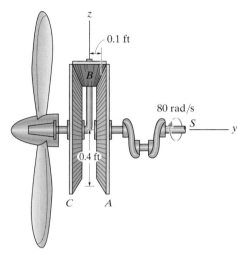

0.1 ft

80 rad/s

0.4 ft

Prob. 20–11

20–10. If the plate gears A and B are rotating with the angular velocities shown, determine the angular velocity of gear C about the shaft DE. What is the angular velocity of DE about the y axis?

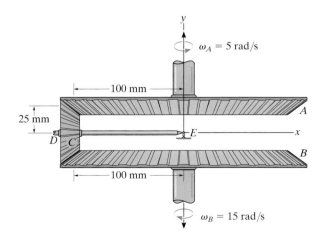

$\omega_A = 5$ rad/s

100 mm

25 mm

100 mm

$\omega_B = 15$ rad/s

Prob. 20–10

*20–12. The right circular cone rotates about the z axis at a constant rate of $\omega_1 = 4$ rad/s without slipping on the horizontal plane. Determine the magnitudes of the velocity and acceleration of points B and C.

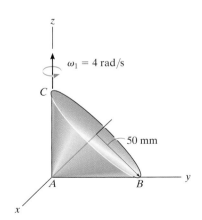

$\omega_1 = 4$ rad/s

50 mm

Prob. 20–12

20–13.   Shaft BD is connected to a ball-and-socket joint at B, and a beveled gear A is attached to its other end. The gear is in mesh with a fixed gear C. If the shaft and gear A are *spinning* with a constant angular velocity $\omega_1 = 8$ rad/s, determine the angular velocity and angular acceleration of gear A.

20–15.   The truncated double cone rotates about the z axis at $\omega_z = 0.4$ rad/s without slipping on the horizontal plane. If at this same instant ω_z is increasing at $\dot{\omega}_z = 0.5$ rad/s², determine the velocity and acceleration of point A on the cone.

Prob. 20–13

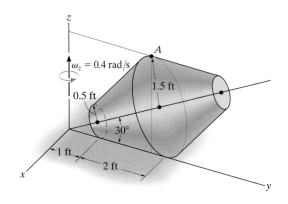

Prob. 20–15

20–14.   The truncated double cone rotates about the z axis at a constant rate $\omega_z = 0.4$ rad/s without slipping on the horizontal plane. Determine the velocity and acceleration of point A on the cone.

*20–16.   The bevel gear A rolls on the fixed gear B. If at the instant shown the shaft to which A is attached is rotating at 2 rad/s and has an angular acceleration of 4 rad/s², determine the angular velocity and angular acceleration of gear A.

Prob. 20–14

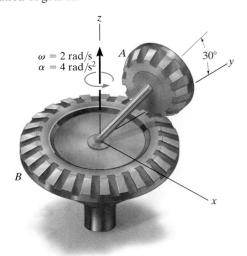

Prob. 20–16

20–17.   The differential of an automobile allows the two rear wheels to rotate at different speeds when the automobile travels along a curve. For operation, the rear axles are attached to the wheels at one end and have beveled gears A and B on their other ends. The differential case D is placed over the left axle but can rotate about C independent of the axle. The case supports a pinion gear E on a shaft, which meshes with gears A and B. Finally, a ring gear G is *fixed* to the differential case so that the case rotates with the ring gear when the latter is driven by the drive pinion H. This gear, like the differential case, is free to rotate about the left wheel axle. If the drive pinion is turning at $\omega_H = 100 \text{ rad/s}$ and the pinion gear E is spinning about its shaft at $\omega_E = 30 \text{ rad/s}$, determine the angular velocity, ω_A and ω_B, of each axle.

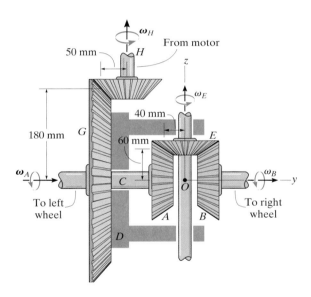

Prob. 20–17

20–18.   Rod AB is attached to the rotating arm using ball and-socket joints. If AC is rotating with a constant angular velocity of 8 rad/s about the pin at C, determine the angular velocity of link BD at the instant shown.

20–19.   Rod AB is attached to the rotating arm using ball and-socket joints. If AC is rotating about point C with an angular velocity of 8 rad/s and has an angular acceleration of 6 rad/s^2 at the instant shown, determine the angular velocity and angular acceleration of link BD at this instant.

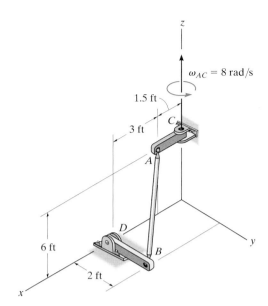

Probs. 20–18/19

*20–20.**   If the rod is attached with ball-and-socket joints to smooth collars A and B at its end points, determine the speed of B at the instant shown if A is moving downward at a constant speed of $v_A = 8 \text{ ft/s}$. Also, determine the angular velocity of the rod if it is directed perpendicular to the axis of the rod.

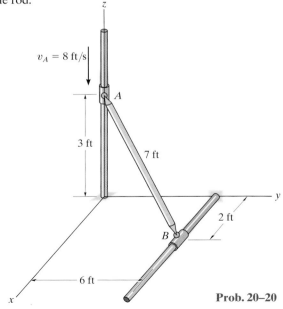

Prob. 20–20

20–21. If the collar at A is moving downward with an acceleration $\mathbf{a}_A = \{-5\mathbf{k}\}$ ft/s^2, at the instant its speed is $v_A = 8$ ft/s, determine the acceleration of the collar at B at this instant.

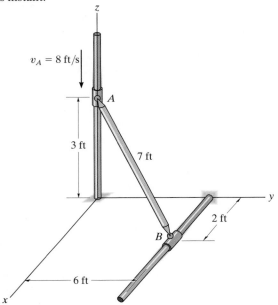

Prob. 20–21

20–22. Rod AB is attached to a disk and a collar by ball and-socket joints. If the disk is rotating at a constant angular velocity $\boldsymbol{\omega} = \{2\mathbf{i}\}$ rad/s, determine the velocity and acceleration of the collar at A at the instant shown. Assume the angular velocity is directed perpendicular to the rod.

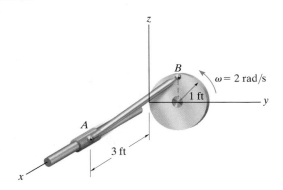

Prob. 20–22

20–23. Rod AB is attached to a disk and a collar by ball and-socket joints. If the disk is rotating with an angular acceleration $\alpha = \{4\mathbf{i}\}$ rad/s^2, and at the instant shown has an angular velocity $\boldsymbol{\omega} = \{2\mathbf{i}\}$ rad/s, determine the velocity and acceleration of the collar at A at the instant shown.

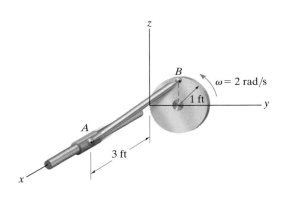

Prob. 20–23

***20–24.** Disk A is rotating at a constant angular velocity of 10 rad/s. If rod BC is joined to the disk and a collar by ball-and-socket joints, determine the velocity of collar B at the instant shown. Also, what is the rod's angular velocity $\boldsymbol{\omega}_{BC}$ if it is directed perpendicular to the axis or the rod?.

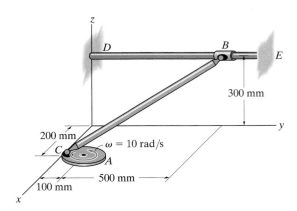

Prob. 20–24

20–25. Solve Prob. 20–24 if the connection at B consists of a pin as shown in the figure below, rather than a ball-and-socket joint. *Hint:* The constraint allows rotation of the rod both along the bar (**j** direction) and along the axis of the pin (**n** direction). Since there is no rotational component in the **u** direction, i.e., perpendicular to **n** and **j** where $\mathbf{u} = \mathbf{j} \times \mathbf{n}$, an additional equation for solution can be obtained from $\boldsymbol{\omega} \cdot \mathbf{u} = 0$. The vector **n** is in the same direction as $\mathbf{r}_{B/C} \times \mathbf{r}_{D/C}$.

Prob. 20–25

20–26. The rod AB is attached to collars at its ends by ball-and-socket joints. If collar A has a speed $v_A = 20$ ft/s, determine the speed of collar B at the instant shown. Assume the angular velocity of the rod is directed perpendicular to the rod.

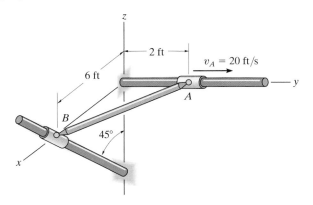

Prob. 20–26

20–27. The rod is attached to smooth collars A and B at its ends using ball-and-socket joints. Determine the speed of B at the instant shown if A is moving at $v_A = 6$ m/s. Also, determine the angular velocity of the rod if it is directed perpendicular to the axis of the rod.

*20–28. If the collar at A in Prob. 20–27 has a deceleration of $\mathbf{a}_A = \{-5\mathbf{i}\}$ m/s^2, at the instant shown, determine the acceleration of collar B at this instant.

Probs. 20–27/28

20–29. Rod AB is attached to collars at its ends by using ball-and-socket joints. If collar A moves along the fixed rod at $v_A = 8$ ft/s, determine the angular velocity of the rod and the velocity of collar B at the instant shown. Assume that the rod's angular velocity is directed perpendicular to the axis of the rod.

20–30. Rod AB is attached to collars at its ends by using ball-and-socket joints. If collar A moves along the fixed rod with a velocity $v_A = 8$ ft/s and has an acceleration $a_A = 4$ ft/s^2 at the instant shown, determine the angular acceleration of the rod and the acceleration of collar B at this instant. Assume that the rod's angular velocity and angular acceleration are directed perpendicular to the axis of the rod.

Probs. 20–29/30

*20.4   Relative-Motion Analysis Using Translating and Rotating Axes

The most general way to analyze the three-dimensional motion of a rigid body requires the use of x, y, z axes that both translate and rotate relative to a second frame X, Y, Z. This analysis also provides a means for determining the motions of two points A and B located on separate members of a mechanism, and for determining the relative motion of one particle with respect to another when one or both particles are moving along *rotating paths*.

As shown in Fig. 20–11, the locations of points A and B are specified relative to the X, Y, Z frame of reference by position vectors $\mathbf{r}_A$ and $\mathbf{r}_B$. The base point A represents the origin of the x, y, z coordinate system, which is translating and rotating with respect to X, Y, Z. At the instant considered, the velocity and acceleration of point A are $\mathbf{v}_A$ and $\mathbf{a}_A$, respectively, and the angular velocity and angular acceleration of the x, y, z axes are $\boldsymbol{\Omega}$ and $\dot{\boldsymbol{\Omega}} = d\boldsymbol{\Omega}/dt$, respectively. All these vectors are *measured* with respect to the X, Y, Z frame of reference, although they may be expressed in Cartesian component form along either set of axes.

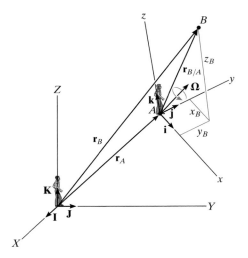

Fig. 20–11

Position.   If the position of "B with respect to A" is specified by the *relative-position vector* $\mathbf{r}_{B/A}$, Fig. 20–11, then, by vector addition,

$$\boxed{\mathbf{r}_B = \mathbf{r}_A + \mathbf{r}_{B/A}} \qquad (20\text{–}9)$$

where

$\mathbf{r}_B$ = position of B

$\mathbf{r}_A$ = position of the origin A

$\mathbf{r}_{B/A}$ = relative positionof "B with respect to A"

Velocity.   The velocity of point B measured from X, Y, Z is determined by taking the time derivative of Eq. 20–9, which yields

$$\dot{\mathbf{r}}_B = \dot{\mathbf{r}}_A + \dot{\mathbf{r}}_{B/A}$$

The first two terms represent $\mathbf{v}_B$ and $\mathbf{v}_A$. The last term is evaluated by applying Eq. 20–6, since $\mathbf{r}_{B/A}$ is measured between two points in a rotating reference. Hence,

$$\dot{\mathbf{r}}_{B/A} = (\dot{\mathbf{r}}_{B/A})_{xyz} + \mathbf{\Omega} \times \mathbf{r}_{B/A} = (\mathbf{v}_{B/A})_{xyz} + \mathbf{\Omega} \times \mathbf{r}_{B/A} \quad (20\text{–}10)$$

Here $(\mathbf{v}_{B/A})_{xyz}$ is the relative velocity of B with respect to A measured from x, y, z. Thus,

$$\boxed{\mathbf{v}_B = \mathbf{v}_A + \mathbf{\Omega} \times \mathbf{r}_{B/A} + (\mathbf{v}_{B/A})_{xyz}} \qquad (20\text{–}11)$$

where

$\mathbf{v}_B$ = velocity of B

$\mathbf{v}_A$ = velocity of the origin A of the x, y, z frame of reference

$(\mathbf{v}_{B/A})_{xyz}$ = relative velocity of "B with respect to A" as measured by an observer attached to the rotating x, y, z frame of reference

$\mathbf{\Omega}$ = angular velocity of the x, y, z frame of reference

$\mathbf{r}_{B/A}$ = relative position of "B with respect to A"

Acceleration.   The acceleration of point B measured from X, Y, Z is determined by taking the time derivative of Eq. 20–11, which yields

$$\dot{\mathbf{v}}_B = \dot{\mathbf{v}}_A + \dot{\boldsymbol{\Omega}} \times \mathbf{r}_{B/A} + \boldsymbol{\Omega} \times \dot{\mathbf{r}}_{B/A} + \frac{d}{dt}(\mathbf{v}_{B/A})_{xyz}$$

The time derivatives defined in the first and second terms represent $\mathbf{a}_B$ and $\mathbf{a}_A$ respectively. The fourth term is evaluated using Eq. 20–10, and the last term is evaluated by applying Eq. 20–6, which yields

$$\frac{d}{dt}(\mathbf{v}_{B/A})_{xyz} = (\dot{\mathbf{v}}_{B/A})_{xyz} + \boldsymbol{\Omega} \times (\mathbf{v}_{B/A})_{xyz}$$

$$= (\mathbf{a}_{B/A})_{xyz} + \boldsymbol{\Omega} \times (\mathbf{v}_{B/A})_{xyz}$$

Here $(\mathbf{a}_{B/A})_{xyz}$ is the relative acceleration of B with respect to A measured from x, y, z. Substituting this result and Eq. 20–10 into the above equation and simplifying, we have

$$\mathbf{a}_B = \mathbf{a}_A + \dot{\boldsymbol{\Omega}} \times \mathbf{r}_{B/A} + \boldsymbol{\Omega} \times (\boldsymbol{\Omega} \times \mathbf{r}_{B/A}) + 2\boldsymbol{\Omega} \times (\mathbf{v}_{B/A})_{xyz} + (\mathbf{a}_{B/A})_{xyz}$$

$$(20\text{–}12)$$

where

$$\mathbf{a}_B = \text{acceleration of } B$$

$$\mathbf{a}_A = \text{acceleration of the origin } A \text{ of the } x, y, z \text{ frame of reference}$$

$$(\mathbf{a}_{B/A})_{xyz}, (\mathbf{v}_{B/A})_{xyz} = \text{relative acceleration and relative velocity of } ``B \text{ with respect to } A" \text{ as measured by an observer attached to the rotating } x, y, z \text{ frame of reference}$$

$$\dot{\boldsymbol{\Omega}}, \boldsymbol{\Omega} = \text{angular acceleration and angular velocity of the } x, y, z \text{ frame of reference}$$

$$\mathbf{r}_{B/A} = \text{relative position of } ``B \text{ with respect to } A"$$

Complicated spatial motion of the concrete bucket B occurs due to the rotation of the boom about the Z axis, motion of the carriage A along the boom, and extension and swinging of the cable AB. A translating-rotating x, y, z coordinate system can be established on the carriage, and a relative-motion analysis can then be applied to study this motion.

Equations 20–11 and 20–12 are identical to those used in Sec. 16.8 for analyzing relative plane motion.* In that case, however, application is simplified since $\boldsymbol{\Omega}$ and $\dot{\boldsymbol{\Omega}}$ have a *constant direction* which is always perpendicular to the plane of motion. For three-dimensional motion, $\dot{\boldsymbol{\Omega}}$ must be computed by using Eq. 20–6, since $\dot{\boldsymbol{\Omega}}$ depends on the change in *both* the magnitude and direction of $\boldsymbol{\Omega}$.

*Refer to Sec. 16.8 for an interpretation of the terms

Procedure for Analysis

Three-dimensional motion of particles or rigid bodies can be analyzed with Eqs. 20–11 and 20–12 by using the following procedure.

Coordinate Axes.

- Select the location and orientation of the X, Y, Z and x, y, z coordinate axes. Most often solutions are easily obtained if at the instant considered:

    (1) the origins are *coincident*

    (2) the axes are collinear

    (3) the axes are parallel

- If several components of angular velocity are involved in a problem, the calculations will be reduced if the x, y, z axes are selected such that only one component of angular velocity is observed in this frame $(\mathbf{\Omega}_{xyz})$ and the frame rotates with $\mathbf{\Omega}$ defined by the other components of angular velocity.

Kinematic Equations.

- After the origin of the moving reference, A, is defined and the moving point B is specified, Eqs. 20–11 and 20–12 should be written in symbolic form as

$$\mathbf{v}_B = \mathbf{v}_A + \mathbf{\Omega} \times \mathbf{r}_{B/A} + (\mathbf{v}_{B/A})_{xyz}$$

$$\mathbf{a}_B = \mathbf{a}_A + \dot{\mathbf{\Omega}} \times \mathbf{r}_{B/A} + \mathbf{\Omega} \times (\mathbf{\Omega} \times \mathbf{r}_{B/A}) + 2\mathbf{\Omega} \times (\mathbf{v}_{B/A})_{xyz} + (\mathbf{a}_{B/A})_{xyz}$$

- If $\mathbf{r}_A$ and $\mathbf{\Omega}$ appear to *change direction* when observed from the fixed X, Y, Z reference use a set of primed reference axes, x', y', z' having $\mathbf{\Omega}' = \mathbf{\Omega}$ and Eq. 20–6 to determine $\dot{\mathbf{\Omega}}$ and the motion $\mathbf{v}_A$ and $\mathbf{a}_A$ of the origin of the moving x, y, z axes.

- If $(\mathbf{r}_{B/A})_{xyz}$ and $\mathbf{\Omega}_{xyz}$ appear to change direction as observed from x, y, z, then use a set of primed reference axes x', y', z' having $\mathbf{\Omega}' = \mathbf{\Omega}_{xyz}$ and Eq. 20–6 to determine $\dot{\mathbf{\Omega}}_{xyz}$ and the relative motion $(\mathbf{v}_{B/A})_{xyz}$ and $(\mathbf{a}_{B/A})_{xyz}$.

- After the final forms of $\dot{\mathbf{\Omega}}, \mathbf{v}_A, \mathbf{a}_A, \dot{\mathbf{\Omega}}_{xyz}, (\mathbf{v}_{B/A})_{xyz}$, and $(\mathbf{a}_{B/A})_{xyz}$ are obtained, numerical problem data may be substituted and the kinematic terms evaluated. The components of all these vectors may be selected either along the X, Y, Z axes or along x, y, z. The choice is arbitrary, provided a consistent set of unit vectors is used.

EXAMPLE 20.4

A motor and attached rod AB have the angular motions shown in Fig. 20–12. A collar C on the rod is located 0.25 m from A and is moving downward along the rod with a velocity of 3 m/s and an acceleration of 2 m/s^2. Determine the velocity and acceleration of C at this instant.

SOLUTION

Coordinate Axes.

The origin of the fixed X, Y, Z reference is chosen at the center of the platform, and the origin of the moving, x, y, z frame at point A, Fig. 20–12. Since the collar is subjected to two components of angular motion, $\boldsymbol{\omega}_p$ and $\boldsymbol{\omega}_M$, it will be viewed as having an angular velocity of $\boldsymbol{\Omega}_{xyz} = \boldsymbol{\omega}_M$ in x, y, z. The x, y, z axes will be attached to the platform so that $\boldsymbol{\Omega} = \boldsymbol{\omega}_p$.

Fig. 20–12

Kinematic Equations.   Equations 20–11 and 20–12, applied to points C and A, become

$$\mathbf{v}_C = \mathbf{v}_A + \mathbf{\Omega} \times \mathbf{r}_{C/A} + (\mathbf{v}_{C/A})_{xyz}$$

$$\mathbf{a}_C = \mathbf{a}_A + \dot{\mathbf{\Omega}} \times \mathbf{r}_{C/A} + \mathbf{\Omega} \times (\mathbf{\Omega} \times \mathbf{r}_{C/A}) + 2\mathbf{\Omega} \times (\mathbf{v}_{C/A})_{xyz} + (\mathbf{a}_{C/A})_{xyz}$$

Motion of A.   Here $\mathbf{r}_A$ changes direction relative to XYZ. To find the time derivatives of $\mathbf{r}_A$ we will use a set of x', y', z' axes coincident with the X, Y, Z axes that rotate at $\mathbf{\Omega}' = \mathbf{\Omega} = \boldsymbol{\omega}_p$. Thus,

$$\mathbf{\Omega} = \boldsymbol{\omega}_p = \{5\mathbf{k}\}\ \text{rad/s} \quad (\mathbf{\Omega}\ \text{does not change direction relative to } XYZ.)$$

$$\dot{\mathbf{\Omega}} = \dot{\boldsymbol{\omega}}_p = \{2\mathbf{k}\}\ \text{rad/s}^2$$

$$\mathbf{r}_A = \{2\mathbf{i}\}\ \text{m}$$

$$\mathbf{v}_A = \dot{\mathbf{r}}_A = (\dot{\mathbf{r}}_A)_{x'y'z'} + \boldsymbol{\omega}_p \times \mathbf{r}_A = 0 + 5\mathbf{k} \times 2\mathbf{i} = \{10\mathbf{j}\}\ \text{m/s}$$

$$\mathbf{a}_A = \ddot{\mathbf{r}}_A = [(\ddot{\mathbf{r}}_A)_{x'y'z'} + \boldsymbol{\omega}_p \times (\dot{\mathbf{r}}_A)_{x'y'z'}] + \dot{\boldsymbol{\omega}}_p \times \mathbf{r}_A + \boldsymbol{\omega}_p \times \dot{\mathbf{r}}_A$$

$$= [0 + 0] + 2\mathbf{k} \times 2\mathbf{i} + 5\mathbf{k} \times 10\mathbf{j} = \{-50\mathbf{i} + 4\mathbf{j}\}\ \text{m/s}^2$$

Motion of C with Respect to A.   Here $(\mathbf{r}_{C/A})_{xyz}$ changes direction relative to $x\,y\,z$. To find the time derivatives of $(\mathbf{r}_{C/A})_{xyz}$ use a set of x'', y'', z'' axes that rotate at $\mathbf{\Omega}'' = \mathbf{\Omega}_{xyz} = \boldsymbol{\omega}_M$. Thus,

$$\mathbf{\Omega}_{xyz} = \boldsymbol{\omega}_M = \{3\mathbf{i}\}\ \text{rad/s} \quad (\mathbf{\Omega}_{xyz}\ \text{does not change direction relative to } xyz.)$$

$$\dot{\mathbf{\Omega}}_{xyz} = \dot{\boldsymbol{\omega}}_M = \{1\mathbf{i}\}\ \text{rad/s}^2$$

$$(\mathbf{r}_{C/A})_{xyz} = \{-0.25\mathbf{k}\}\ \text{m}$$

$$(\mathbf{v}_{C/A})_{xyz} = (\dot{\mathbf{r}}_{C/A})_{xyz} = (\dot{\mathbf{r}}_{C/A})_{xyz} + \boldsymbol{\omega}_M \times (\mathbf{r}_{C/A})_{xyz}$$

$$= -3\mathbf{k} + [3\mathbf{i} \times (-0.25\mathbf{k})] = \{0.75\mathbf{j} - 3\mathbf{k}\}\ \text{m/s}$$

$$(\mathbf{a}_{C/A})_{xyz} = (\ddot{\mathbf{r}}_{C/A})_{xyz} = [(\ddot{\mathbf{r}}_{C/A})_{xyz} + \boldsymbol{\omega}_M \times (\dot{\mathbf{r}}_{C/A})_{xyz}] + \dot{\boldsymbol{\omega}}_M \times (\mathbf{r}_{C/A})_{xyz} + \boldsymbol{\omega}_M \times (\mathbf{r}_{C/A})_{xyz}$$

$$= [-2\mathbf{k} + 3\mathbf{i} \times (-3\mathbf{k})] + (1\mathbf{i}) \times (-0.25\mathbf{k}) + (3\mathbf{i}) \times (0.75\mathbf{j} - 3\mathbf{k})$$

$$= \{18.25\mathbf{j} + 0.25\mathbf{k}\}\ \text{m/s}^2$$

Motion of C.

$$\mathbf{v}_C = \mathbf{v}_A + \mathbf{\Omega} \times \mathbf{r}_{C/A} + (\mathbf{v}_{C/A})_{xyz}$$

$$= 10\mathbf{j} + [5\mathbf{k} \times (-0.25\mathbf{k})] + (0.75\mathbf{j} - 3\mathbf{k})$$

$$= \{10.8\mathbf{j} - 3\mathbf{k}\}\ \text{m/s} \qquad\qquad\qquad Ans.$$

$$\mathbf{a}_C = \mathbf{a}_A + \dot{\mathbf{\Omega}} \times \mathbf{r}_{C/A} + \mathbf{\Omega} \times (\mathbf{\Omega} \times \mathbf{r}_{C/A}) + 2\mathbf{\Omega} \times (\mathbf{v}_{C/A})_{xyz} + (\mathbf{a}_{C/A})_{xyz}$$

$$= (-50\mathbf{i} + 4\mathbf{j}) + [2\mathbf{k} \times (-0.25\mathbf{k})] + 5\mathbf{k} \times [5\mathbf{k} \times (-0.25\mathbf{k})]$$

$$+ 2[5\mathbf{k} \times (0.75\mathbf{j} - 3\mathbf{k})] + (18.25\mathbf{j} + 0.25\mathbf{k})$$

$$= \{-57.5\mathbf{i} + 22.2\mathbf{j} + 0.25\mathbf{k}\}\ \text{m/s}^2 \qquad\qquad Ans.$$

EXAMPLE 20.5

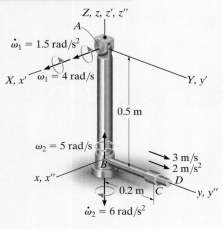

$\dot{\omega}_1 = 1.5\ \text{rad/s}^2$

$\omega_1 = 4\ \text{rad/s}$

Z, z, z', z''

A

X, x'

Y, y'

0.5 m

$\omega_2 = 5\ \text{rad/s}$

x, x''

B

3 m/s
2 m/s²

D

0.2 m C

y, y''

$\dot{\omega}_2 = 6\ \text{rad/s}^2$

Fig. 20–13

The pendulum shown in Fig. 20–13 consists of two rods; AB is pin supported at A and swings only in the Y–Z plane, whereas a bearing at B allows the attached rod BD to spin about rod AB. At a given instant, the rods have the angular motions shown. Also, a collar C, located 0.2 m from B, has a velocity of 3 m/s and an acceleration of 2 m/s² along the rod. Determine the velocity and acceleration of the collar at this instant.

SOLUTION I

Coordinate Axes.   The origin of the fixed X, Y, Z frame will be chosen at A. Motion of the collar is conveniently observed from B, so the origin of the x, y, z frame is located at this point. We will choose $\Omega = \omega_1$ and $\Omega_{xyz} = \omega_2$.

Kinematic Equations.

$$\mathbf{v}_C = \mathbf{v}_B + \Omega \times \mathbf{r}_{C/B} + (\mathbf{v}_{C/B})_{xyz}$$

$$\mathbf{a}_C = \mathbf{a}_B + \dot{\Omega} \times \mathbf{r}_{C/B} + \Omega \times (\Omega \times \mathbf{r}_{C/B}) + 2\Omega \times (\mathbf{v}_{C/B})_{xyz} + (\mathbf{a}_{C/B})_{xyz}$$

Motion of B. To find the time derivatives of $\mathbf{r}_B$ let the x', y', z' axes rotate with $\Omega = \omega_1$. Then

$$\Omega = \omega_1 = \{4\mathbf{i}\}\ \text{rad/s} \qquad \dot{\Omega} = \dot{\omega}_1 = \{1.5\mathbf{i}\}\ \text{rad/s}^2$$

$$\mathbf{r}_B = \{-0.5\mathbf{k}\}\ \text{m}$$

$$\mathbf{v}_B = \dot{\mathbf{r}}_B = (\dot{\mathbf{r}}_B)_{x'y'z'} + \omega_1 \times \mathbf{r}_B = 0 + 4\mathbf{i} \times (-0.5\mathbf{k}) = \{2\mathbf{j}\}\ \text{m/s}$$

$$\mathbf{a}_B = \ddot{\mathbf{r}}_B = [(\ddot{\mathbf{r}}_B)_{x'y'z'} + \omega_1 \times (\dot{\mathbf{r}}_B)_{x'y'z'}] + \dot{\omega}_1 \times \mathbf{r}_B + \omega_1 \times \dot{\mathbf{r}}_B$$

$$= [0 + 0] + 1.5\mathbf{i} \times (-0.5\mathbf{k}) + 4\mathbf{i} \times 2\mathbf{j} = \{0.75\mathbf{j} + 8\mathbf{k}\}\ \text{m/s}^2$$

Motion of C with Respect to B. To find the time derivatives of $(\mathbf{r}_{C/B})_{xyz}$, let the x'', y'', z'' axes rotate with $\Omega_{xyz} = \omega_2$. Then

$$\Omega_{xyz} = \omega_2 = \{5\mathbf{k}\}\ \text{rad/s} \qquad \dot{\Omega}_{xyx} = \dot{\omega}_2 = \{-6\mathbf{k}\}\ \text{rad/s}^2$$

$$(\mathbf{r}_{C/B})_{xyz} = \{0.2\mathbf{j}\}\ \text{m}$$

$$(\mathbf{v}_{C/B})_{xyz} = (\dot{\mathbf{r}}_{C/B})_{xyz} = (\dot{\mathbf{r}}_{C/B})_{x''y''z''} + \omega_2 \times (\mathbf{r}_{C/B})_{xyz} = 3\mathbf{j} + 5\mathbf{k} \times 0.2\mathbf{j} = \{-1\mathbf{i} + 3\mathbf{j}\}\ \text{m/s}$$

$$(\mathbf{a}_{C/B})_{xyz} = (\ddot{\mathbf{r}}_{C/B})_{xyz} = [(\ddot{\mathbf{r}}_{C/B})_{x''y''z''} + \omega_2 \times (\dot{\mathbf{r}}_{C/B})_{x''y''z''}] + \dot{\omega}_2 \times (\mathbf{r}_{C/B})_{xyz} + \omega_2 \times (\dot{\mathbf{r}}_{C/B})_{xyz}$$

$$= (2\mathbf{j} + 5\mathbf{k} \times 3\mathbf{j}) + (-6\mathbf{k} \times 0.2\mathbf{j}) + [5\mathbf{k} \times (-1\mathbf{i} + 3\mathbf{j})]$$

$$= \{-28.8\mathbf{i} - 3\mathbf{j}\}\ \text{m/s}^2$$

Motion of C.

$$\mathbf{v}_C = \mathbf{v}_B + \Omega \times \mathbf{r}_{C/B} + (\mathbf{v}_{C/B})_{xyz} = 2\mathbf{j} + 4\mathbf{i} \times 0.2\mathbf{j} + (-1\mathbf{i} + 3\mathbf{j})$$

$$= \{-1\mathbf{i} + 5\mathbf{j} + 0.8\mathbf{k}\}\ \text{m/s} \qquad\qquad Ans.$$

$$\mathbf{a}_C = \mathbf{a}_B + \dot{\Omega} \times \mathbf{r}_{C/B} + \Omega \times (\Omega \times \mathbf{r}_{C/B}) + 2\Omega \times (\mathbf{v}_{C/B})_{xyz} + (\mathbf{a}_{C/B})_{xyz}$$

$$= (0.75\mathbf{j} + 8\mathbf{k}) + (1.5\mathbf{i} \times 0.2\mathbf{j}) + [4\mathbf{i} \times (4\mathbf{i} \times 0.2\mathbf{j})]$$

$$+ 2[4\mathbf{i} \times (-1\mathbf{i} + 3\mathbf{j})] + (-28.8\mathbf{i} - 3\mathbf{j})$$

$$= \{-28.8\mathbf{i} - 5.45\mathbf{j} + 32.3\mathbf{k}\}\ \text{m/s}^2 \qquad\qquad Ans.$$

SOLUTION II

Coordinate Axes.   Here we will let the x, y, z axes rotate at

$$\Omega = \omega_1 + \omega_2 = \{4\mathbf{i} + 5\mathbf{k}\} \text{ rad/s}$$

Then $\dot{\Omega}_{xyz} = \mathbf{0}$.

Motion of B.   From the constraints of the problem ω_1 does not change direction relative to X, Y, Z; however, the direction of ω_2 is changed by ω_1. Thus, to obtain $\dot{\Omega}$ consider x', y', z' axes coincident with the X, Y, Z axes at A, such that $\dot{\Omega} = \omega_1$. Then taking the derivative of its components,

$$\dot{\Omega} = \dot{\omega}_1 + \dot{\omega}_2 = [(\dot{\omega}_1)_{x'y'z'} + \omega_1 \times \omega_1] + [(\dot{\omega}_2)_{x'y'z'} + \omega_1 \times \omega_2]$$

$$= [1.5\mathbf{i} + \mathbf{0}] + [-6\mathbf{k} + 4\mathbf{i} \times 5\mathbf{k}] = \{1.5\mathbf{i} - 20\mathbf{j} - 6\mathbf{k}\} \text{ rad/s}^2$$

Also, ω_1 changes the direction of $\mathbf{r}_B$ so that the time derivatives of $\mathbf{r}_B$ can be computed using the primed axes defined above. Hence,

$$\mathbf{v}_B = \dot{\mathbf{r}}_B = (\dot{\mathbf{r}}_B)_{x'y'z'} + \omega_1 \times \mathbf{r}_B$$

$$= \mathbf{0} + 4\mathbf{i} \times (-0.5\mathbf{k}) = \{2\mathbf{j}\} \text{ m/s}$$

$$\mathbf{a}_B = \ddot{\mathbf{r}}_B = [(\ddot{\mathbf{r}}_B)_{x'y'z'} + \omega_1 \times (\dot{\mathbf{r}}_B)_{x'y'z'}] + \dot{\omega}_1 \times \mathbf{r}_B + \omega_1 \times \dot{\mathbf{r}}_B$$

$$= [\mathbf{0} + \mathbf{0}] + 1.5\mathbf{i} \times (-0.5\mathbf{k}) + 4\mathbf{i} \times 2\mathbf{j} = \{0.75\mathbf{j} + 8\mathbf{k}\} \text{ m/s}^2$$

Motion of C with Respect to B.

$$\Omega_{xyz} = \mathbf{0}$$

$$\dot{\Omega}_{xyz} = \mathbf{0}$$

$$(\mathbf{r}_{C/B})_{xyz} = \{0.2\mathbf{j}\} \text{ m}$$

$$(\mathbf{v}_{C/B})_{xyz} = \{3\mathbf{j}\} \text{ m/s}$$

$$(\mathbf{a}_{C/B})_{xyz} = \{2\mathbf{j}\} \text{ m/s}^2$$

Motion of C.

$$\mathbf{v}_C = \mathbf{v}_B + \Omega \times \mathbf{r}_{C/B} + (\mathbf{v}_{C/B})_{xyz}$$

$$= 2\mathbf{j} + [(4\mathbf{i} + 5\mathbf{k}) \times (0.2\mathbf{j})] + 3\mathbf{j}$$

$$= \{-1\mathbf{i} + 5\mathbf{j} + 0.8\mathbf{k}\} \text{ m/s} \qquad\qquad Ans.$$

$$\mathbf{a}_C = \mathbf{a}_B + \dot{\Omega} \times \mathbf{r}_{C/B} + \Omega \times (\Omega \times \mathbf{r}_{C/B}) + 2\Omega \times (\mathbf{v}_{C/B})_{xyz} + (\mathbf{a}_{C/B})_{xyz}$$

$$= (0.75\mathbf{j} + 8\mathbf{k}) + [(1.5\mathbf{i} - 20\mathbf{j} - 6\mathbf{k}) \times (0.2\mathbf{j})]$$

$$+ (4\mathbf{i} + 5\mathbf{k}) \times [(4\mathbf{i} + 5\mathbf{k}) \times 0.2\mathbf{j}] + 2[(4\mathbf{i} + 5\mathbf{k}) \times 3\mathbf{j}] + 2\mathbf{j}$$

$$= \{-28.8\mathbf{i} - 5.45\mathbf{j} + 32.3\mathbf{k}\} \text{ m/s}^2 \qquad\qquad Ans.$$

PROBLEMS

20–31. Solve Example 20.5 such that the x, y, z axes move with curvilinear translation, $\Omega = 0$, in which case the collar appears to have both an angular velocity $\Omega_{xyz} = \omega_1 + \omega_2$ and radial motion.

*20–32. Solve Example 20.5 by fixing x, y, z axes to rod BD so that $\Omega = \omega_1 + \omega_2$. In this case the collar appears only to move radially outward along BD; hence $\Omega_{xyz} = 0$.

20–33. At a given instant, rod BD is rotating about the y axis with an angular velocity $\omega_{BD} = 2$ rad/s and an angular acceleration $\dot{\omega}_{BD} = 5$ rad/s^2. Also, when $\theta = 60°$ link AC is rotating downward such that $\dot{\theta} = 2$ rad/s and $\ddot{\theta} = 8$ rad/s^2. Determine the velocity and acceleration of point A on the link at this instant.

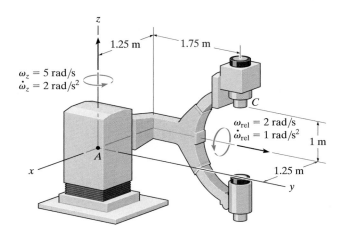

Prob. 20–34

20–35. At the instant shown, the frame of the brush cutter is traveling forward in the x direction with a constant velocity of 1 m/s, and the cab is rotating about the vertical axis with a constant angular velocity of $\omega_1 = 0.5$ rad/s. At the same instant the boom AB has an angular velocity of $\dot{\theta} = 0.8$ rad/s in the direction shown. Determine the velocity and acceleration of point B at the connection to the mower at this instant.

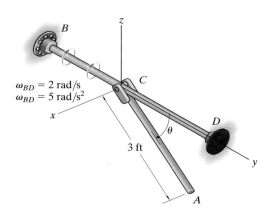

Prob. 20–33

20–34. During the instant shown the frame of the X-ray camera is rotating about the vertical axis at $\omega_z = 5$ rad/s and $\dot{\omega}_z = 2$ rad/s^2. Relative to the frame the arm is rotating at $\omega_{rel} = 2$ rad/s and $\dot{\omega}_{rel} = 1$ rad/s^2. Determine the velocity and acceleration of the center of the camera C at this instant.

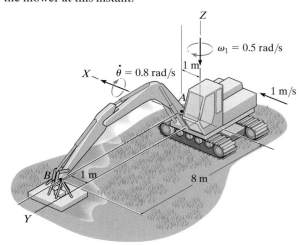

Prob. 20–35

*20–36. At the instant shown, the frame of the brush cutter is traveling forward in the x direction with a constant velocity of 1 m/s, and the cab is rotating about the vertical axis with an angular velocity of $\omega_1 = 0.5$ rad/s, which is increasing at $\dot{\omega}_1 = 0.4$ rad/s². At the same instant the boom AB has an angular velocity of $\dot{\theta} = 0.8$ rad/s, which is increasing at $\ddot{\theta} = 0.9$ rad/s². Determine the velocity and acceleration of point B at the connection to the mower at this instant.

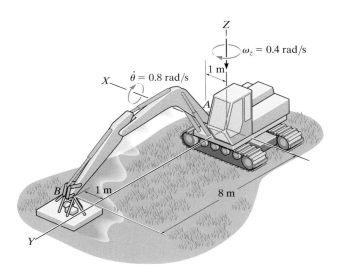

Prob. 20–36

20–37. At the instant shown, rod BD is rotating about the vertical axis with an angular velocity $\omega_{BD} = 7$ rad/s and an angular acceleration $\alpha_{BD} = 4$ rad/s². Also, $\theta = 60°$ and link AC is rotating downward such that $\dot{\theta} = 2$ rad/s and $\ddot{\theta} = 3$ rad/s². Determine the velocity and acceleration of point A on the link at this instant.

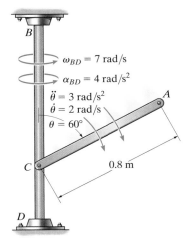

Prob. 20–37

20–38. The boom AB of the locomotive crane is rotating about the Z axis with an angular velocity $\omega_1 = 0.5$ rad/s, which is increasing at $\dot{\omega}_1 = 3$ rad/s². At this same instant, $\theta = 30°$ and the boom is rotating upward at a constant rate of $\dot{\theta} = 3$ rad/s. Determine the velocity and acceleration of the tip B of the boom at this instant.

Prob. 20–38

20–39. The locomotive crane is traveling to the right at 2 m/s and has an acceleration of 1.5 m/s², while the boom is rotating about the Z axis with an angular velocity $\omega_1 = 0.5$ rad/s, which is increasing at $\dot{\omega}_1 = 3$ rad/s². At this same instant, $\theta = 30°$ and the boom is rotating upward at a constant rate $\dot{\theta} = 3$ rad/s. Determine the velocity and acceleration of the tip B of the boom at this instant.

Prob. 20–39

*20–40. At a given instant, the rod has the angular motions shown, while the collar C is moving down *relative* to the rod with a velocity of 6 ft/s and an acceleration of 2 ft/s². Determine the collar's velocity and acceleration at this instant.

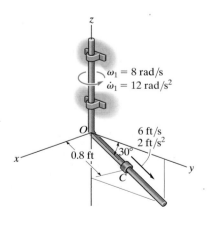

Prob. 20–40

20–41. At the instant shown, the arm OA of the conveyor belt is rotating about the z axis with a constant angular velocity $\omega_1 = 6$ rad/s, while at the same instant the arm is rotating upward at a constant rate $\omega_2 = 4$ rad/s. If the conveyor is running at a constant rate $\dot{r} = 5$ ft/s, determine the velocity and acceleration of the package P at the instant shown. Neglect the size of the package.

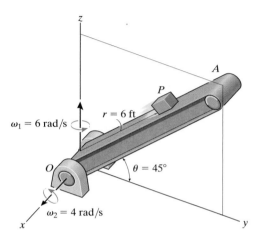

Prob. 20–41

20–42. At the instant shown, the arm OA of the conveyor belt is rotating about the z axis with a constant angular velocity $\omega_1 = 6$ rad/s, while at the same instant the arm is rotating upward at a constant rate $\omega_2 = 4$ rad/s. If the conveyor is running at a rate $\dot{r} = 5$ ft/s, which is increasing at $\ddot{r} = 8$ ft/s², determine the velocity and acceleration of the package P at the instant shown. Neglect the size of the package.

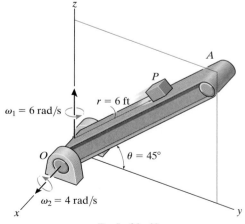

Prob. 20–42

20–43. At the given instant, the rod is spinning about the z axis with an angular velocity $\omega_1 = 3$ rad/s and angular acceleration $\dot{\omega}_1 = 4$ rad/s². At this same instant, the disk is spinning at $\omega_2 = 2$ rad/s when $\dot{\omega}_2 = 1$ rad/s², both measured *relative* to the rod. Determine the velocity and acceleration of point P on the disk at this instant.

20–45. At the instant shown, the base of the robotic arm is turning about the z axis with an angular velocity of $\omega_1 = 4$ rad/s, which is increasing at $\dot{\omega}_1 = 3$ rad/s². Also, the boom segment BC is rotating at a constant rate of $\omega_{BC} = 8$ rad/s. Determine the velocity and acceleration of the part C held in its grip at this instant.

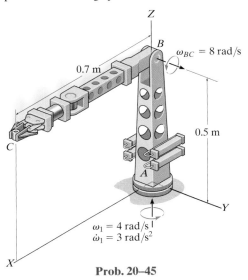

Prob. 20–43

Prob. 20–45

*20–44. At a given instant, the crane is moving along the track with a velocity $v_{CD} = 8$ m/s and acceleration of 9 m/s². Simultaneously, it has the angular motions shown. If the trolley T is moving outwards along the boom AB with a relative speed of 3 m/s and relative acceleration of 5 m/s², determine the velocity and acceleration of the trolley.

20–46. At the instant shown, the base of the robotic arm is turning about the z axis with an angular velocity of $\omega_1 = 4$ rad/s, which is increasing at $\dot{\omega}_1 = 3$ rad/s². Also, the boom segment BC is rotating at $\omega_{BC} = 8$ rad/s, which is increasing at $\dot{\omega}_{BC} = 2$ rad/s². Determine the velocity and acceleration of the part C held in its grip at this instant.

Prob. 20–44

Prob. 20–46

20–47.   The load is being lifted upward at a constant rate of 9 m/s relative to the crane boom AB. At the instant shown, the boom is rotating about the vertical axis at a constant rate of $\omega_1 = 4$ rad/s, and the trolley T is moving outward along the boom at a constant rate of $v_t = 2$ m/s. Furthermore, at this same instant the retractable arm supporting the load is vertical and is swinging in the y–z plane at an angular rate of 5 rad/s, with an increase in the rate of swing of 7 rad/s². Determine the velocity and acceleration of the center G of the load at this instant, i.e., when $s = 4$ m and $h = 3$ m.

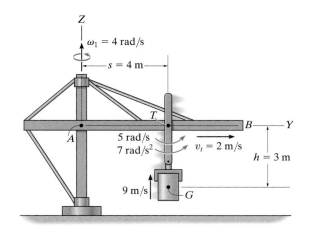

Prob. 20–47

Chapter Review

Rotation About a Fixed Point

When a body rotates about a fixed point O, then points on the body follow a path that lies on the surface of a sphere. Infinitesimal rotations are vector quantities, whereas finite rotations are not.

Since the angular acceleration is a time rate of change in the angular velocity, then we must account for both the magnitude and directional changes of $\boldsymbol{\omega}$ when finding its derivative. To do this, the angular velocity is often specified in terms of its component motions, such that some of these components will remain constant relative to rotating x, y, z axes. If this is the case, then the time derivative relative to the fixed axis can be determined using $\dot{\mathbf{A}} = (\dot{\mathbf{A}})_{xyz} + \boldsymbol{\Omega} \times \mathbf{A}$.

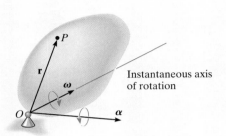

Once $\boldsymbol{\omega}$ and $\boldsymbol{\alpha}$ are known, the velocity and acceleration of any point P in the body can then be determined.
(*See pages 535–538.*)

$$\mathbf{v} = \boldsymbol{\omega} \times \mathbf{r}$$

$$\mathbf{a} = \boldsymbol{\alpha} \times \mathbf{r} + \boldsymbol{\omega} \times (\boldsymbol{\omega} \times \mathbf{r})$$

Relative Motion Analysis Using Translating and Rotating Axes

The motion of two points A and B on a body, a series of connected bodies, or located on two different paths, can be related using a relative motion analysis with rotating and translating axes at A.

When applying these equations, it is important to account for both the magnitude and directional changes of $\mathbf{r}_A$, $(\mathbf{r}_{B/A})_{xyz}$, $\boldsymbol{\Omega}$, and $\boldsymbol{\Omega}_{xyz}$ when taking their time derivatives to find $\mathbf{v}_A$, $\mathbf{a}_A$, $(\mathbf{v}_{B/A})_{xyz}$, $(\mathbf{a}_{B/A})_{xyz}$, and $\dot{\boldsymbol{\Omega}}$, and $\dot{\boldsymbol{\Omega}}_{xyz}$. To do this properly, one must use Eq. 20–6.
(*See pages 538–539.*)

$$\mathbf{v}_B = \mathbf{v}_A + \boldsymbol{\Omega} \times \mathbf{r}_{B/A} + (\mathbf{v}_{B/A})_{xyz}$$

$$\mathbf{a}_B = \mathbf{a}_A + \dot{\boldsymbol{\Omega}} \times \mathbf{r}_{B/A} + \boldsymbol{\Omega} \times (\boldsymbol{\Omega} \times \mathbf{r}_{B/A}) + \\ 2\boldsymbol{\Omega} \times (\mathbf{v}_{B/A})_{xyz} + (\mathbf{a}_{B/A})_{xyz}$$

General Motion

If the body undergoes general motion, then the motion of a point B in the body can be related to the motion of another point A using a relative motion analysis, along with translating axes at A.
(*See page 543.*)

$$\mathbf{v}_B = \mathbf{v}_A + \boldsymbol{\omega} \times \mathbf{r}_{B/A}$$

$$\mathbf{a}_B = \mathbf{a}_A + \boldsymbol{\alpha} \times \mathbf{r}_{B/A} + \boldsymbol{\omega} \times (\boldsymbol{\omega} \times \mathbf{r}_{B/A})$$

The design of amusement-park rides requires a force analysis that depends on their three-dimensional motion.

21 Three-Dimensional Kinetics of a Rigid Body

CHAPTER OBJECTIVES

- To introduce the methods for finding the moments of inertia and products of inertia of a body about various axes.

- To show how to apply the principles of work and energy and linear and angular momentum to a rigid body having three-dimensional motion.

- To develop and apply the equations of motion in three dimensions.

- To study the motion of a gyroscope and torque-free motion.

*21.1   Moments and Products of Inertia

When studying the planar kinetics of a body, it was necessary to introduce the moment of inertia I_G, which was computed about an axis perpendicular to the plane of motion and passing through the body's mass center G. For the kinetic analysis of three-dimensional motion it will sometimes be necessary to calculate six inertial quantities. These terms, called the moments and products of inertia, describe in a particular way the distribution of mass for a body relative to a given coordinate system that has a specified orientation and point of origin.

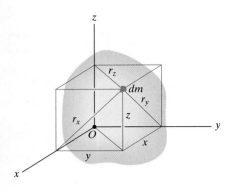

Fig. 21–1

Moment of Inertia.   Consider the rigid body shown in Fig. 21–1. The *moment of inertia* for a differential element dm of the body about any one of the three coordinate axes is defined as the product of the mass of the element and the square of the shortest distance from the axis to the element. For example, as noted in the figure, $r_x = \sqrt{y^2 + z^2}$, so that the mass moment of inertia of dm about the x axis is

$$dI_{xx} = r_x^2\, dm = (y^2 + z^2)\, dm$$

The moment of inertia I_{xx} for the body is determined by integrating this expression over the entire mass of the body. Hence, for each of the axes, we may write

$$
\begin{aligned}
I_{xx} &= \int_m r_x^2\, dm = \int_m (y^2 + z^2)\, dm \\[4pt]
I_{yy} &= \int_m r_y^2\, dm = \int_m (x^2 + z^2)\, dm \\[4pt]
I_{zz} &= \int_m r_z^2\, dm = \int_m (x^2 + y^2)\, dm
\end{aligned}
\tag{21–1}
$$

Here it is seen that the moment of inertia is *always a positive quantity*, since it is the summation of the product of the mass dm, which is always positive, and the distances squared.

Product of Inertia.   The *product of inertia* for a differential element dm is defined with respect to a set of *two orthogonal planes* as the product of the mass of the element and the perpendicular (or shortest) distances from the planes to the element. For example, this distance is x to the y–z plane and it is y to the x–z plane, Fig. 21–1. The product of inertia dI_{xy} for the element dm is therefore

$$dI_{xy} = xy\, dm$$

Note also that $dI_{yx} = dI_{xy}$. By integrating over the entire mass, the product of inertia of the body for each combination of planes may be expressed as

$$
\begin{aligned}
I_{xy} &= I_{yx} = \int_m xy\, dm \\[4pt]
I_{yz} &= I_{zy} = \int_m yz\, dm \\[4pt]
I_{xz} &= I_{zx} = \int_m xz\, dm
\end{aligned}
\tag{21–2}
$$

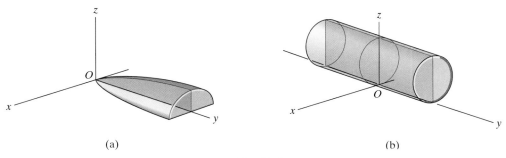

(a)   (b)

Fig. 21–2

Unlike the moment of inertia, which is always positive, the product of inertia may be positive, negative, or zero. The result depends on the signs of the two defining coordinates, which vary independently from one another. In particular, if either one or both of the orthogonal planes are *planes of symmetry* for the mass, the *product of inertia* with respect to these planes will be *zero*. In such cases, elements of mass will occur in *pairs* located on each side of the plane of symmetry. On one side of the plane the product of inertia for the element will be positive, while on the other side the product of inertia for the corresponding element will be negative, the sum therefore yielding zero. Examples of this are shown in Fig. 21–2. In the first case, Fig. 21–2a, the y–z plane is a plane of symmetry, and hence $I_{xy} = I_{xz} = 0$. Calculation of I_{yz} will yield a *positive* result, since all elements of mass are located using only positive y and z coordinates. For the cylinder, with the coordinate axes located as shown in Fig. 21–2b, the x–z and y–z planes are both planes of symmetry. Thus, $I_{xy} = I_{yz} = I_{zx} = 0$.

Parallel-Axis and Parallel-Plane Theorems.   The techniques of integration used to determine the moment of inertia of a body were described in Sec. 17.1. Also discussed were methods to determine the moment of inertia of a composite body, i.e., a body that is composed of simpler segments, as tabulated on the inside back cover. In both of these cases the *parallel-axis theorem* is often used for the calculations. This theorem, which was developed in Sec. 17.1, allows us to transfer the moment of inertia of a body from an axis passing through its mass center G to a parallel axis passing through some other point. If G has coordinates x_G, y_G, z_G defined from the x, y, z axes, Fig. 21–3, then the parallel-axis equations used to calculate the moments of inertia about the x, y, z axes are

$$
\begin{aligned}
I_{xx} &= (I_{x'x'})_G + m(y_G^2 + z_G^2) \\
I_{yy} &= (I_{y'y'})_G + m(x_G^2 + z_G^2) \\
I_{zz} &= (I_{z'z'})_G + m(x_G^2 + y_G^2)
\end{aligned}
\qquad (21\text{–}3)
$$

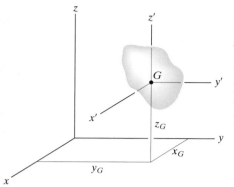

Fig. 21–3

The products of inertia of a composite body are computed in the same manner as the body's moments of inertia. Here, however, the *parallel-plane theorem* is important. This theorem is used to transfer the products of inertia of the body from a set of three orthogonal planes passing through the body's mass center to a corresponding set of three parallel planes passing through some other point O. Defining the perpendicular distances between the planes as x_G, y_G and z_G, Fig. 21–3, the parallel-plane equations can be written as

$$
\begin{aligned}
I_{xy} &= (I_{x'y'})_G + m x_G y_G \\
I_{yz} &= (I_{y'z'})_G + m y_G z_G \\
I_{zx} &= (I_{z'x'})_G + m z_G x_G
\end{aligned}
\qquad (21\text{–}4)
$$

The derivation of these formulas is similar to that given for the parallel-axis equation, Sec. 17.1.

Inertia Tensor.   The inertial properties of a body are therefore completely characterized by nine terms, six of which are independent of one another. This set of terms is defined using Eqs. 21–1 and 21–2 and can be written as

$$
\begin{pmatrix}
I_{xx} & -I_{xy} & -I_{xz} \\
-I_{yx} & I_{yy} & -I_{yz} \\
-I_{zx} & -I_{zy} & I_{zz}
\end{pmatrix}
$$

This array is called an *inertia tensor*. It has a unique set of values for a body when it is computed for each location of the origin O and orientation of the coordinate axes.

In general, for point O we can specify a unique axes inclination for which the products of inertia for the body are zero when computed with respect to these axes. When this is done, the inertia tensor is said to be "diagonalized" and may be written in the simplified form

$$
\begin{pmatrix}
I_x & 0 & 0 \\
0 & I_y & 0 \\
0 & 0 & I_z
\end{pmatrix}
$$

The dynamics of the space shuttle while it orbits the earth can be predicted only if its moments and products of inertia are known.

Here $I_x = I_{xx}$, $I_y = I_{yy}$, and $I_z = I_{zz}$ are termed the *principal moments of inertia* for the body, which are computed from the *principal axes of inertia*. Of these three principal moments of inertia, one will be a maximum and another a minimum of the body's moment of inertia.

Although mathematical determination of the directions of principal axes of inertia will not be discussed here (see Prob. 21–21), there are many cases, however, in which the principal axes may be determined by inspection. From the previous discussion it was noted that if the coordinate axes are oriented such that *two* of the three orthogonal planes containing the axes are planes of *symmetry* for the body, then all the products of inertia for the body are zero with respect to the coordinate planes, and hence the coordinate axes are principal axes of inertia. For example, the x, y, z axes shown in Fig. 21–2b represent the principal axes of inertia for the cylinder at point O.

Moment of Inertia About an Arbitrary Axis.

Consider the body shown in Fig. 21–4, where the nine elements of the inertia tensor have been computed for the x, y, z axes having an origin at O. Here we wish to determine the moment of inertia of the body about the Oa axis, for which the direction is defined by the unit vector $\mathbf{u}_a$. By definition $I_{Oa} = \int b^2 \, dm$, where b is the *perpendicular distance* from dm to Oa. If the position of dm is located using $\mathbf{r}$, then $b = r \sin \theta$, which represents the *magnitude* of the cross product $\mathbf{u}_a \times \mathbf{r}$. Hence, the moment of inertia can be expressed as

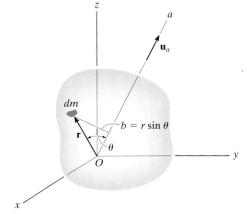

$$I_{Oa} = \int_m |(\mathbf{u}_a \times \mathbf{r})|^2 \, dm = \int_m (\mathbf{u}_a \times \mathbf{r}) \cdot (\mathbf{u}_a \times \mathbf{r}) \, dm$$

Provided $\mathbf{u}_a = u_x \mathbf{i} + u_y \mathbf{j} + u_z \mathbf{k}$ and $\mathbf{r} = x\mathbf{i} + y\mathbf{j} + z\mathbf{k}$, so that $\mathbf{u}_a \times \mathbf{r} = (u_y z - u_z y)\mathbf{i} + (u_z x - u_x z)\mathbf{j} + (u_x y - u_y x)\mathbf{k}$, then, after substituting and performing the dot-product operation, we can write the moment of inertia as

$$I_{Oa} = \int_m [(u_y z - u_z y)^2 + (u_z x - u_x z)^2 + (u_x y - u_y x)^2] \, dm$$

$$= u_x^2 \int_m (y^2 + z^2) \, dm + u_y^2 \int_m (z^2 + x^2) \, dm + u_z^2 \int_m (x^2 + y^2) \, dm$$

$$- 2u_x u_y \int_m xy \, dm - 2u_y u_z \int_m yz \, dm - 2u_z u_x \int_m zx \, dm$$

Recognizing the integrals to be the moments and products of inertia of the body, Eqs. 21–1 and 21–2, we have

$$\boxed{I_{Oa} = I_{xx} u_x^2 + I_{yy} u_y^2 + I_{zz} u_z^2 - 2I_{xy} u_x u_y - 2I_{yz} u_y u_z - 2I_{zx} u_z u_x} \quad (21\text{–}5)$$

Thus, if the inertia tensor is specified for the x, y, z axes, the moment of inertia of the body about the inclined Oa axis can be found by using Eq. 21–5. For the calculation, the direction cosines u_x, u_y, u_z of the axes must be determined. These terms specify the cosines of the coordinate direction angles α, β, γ made between the positive Oa axis and the positive x, y, z axes, respectively (see Appendix C).

Fig. 21–4

EXAMPLE 21.1

(a)

Determine the moment of inertia of the bent rod shown in Fig. 21–5a about the Aa axis. The mass of each of the three segments is shown in the figure.

SOLUTION

Before applying Eq. 21–5, it is first necessary to determine the moments and products of inertia of the rod about the x, y, z axes. This is done using the formula for the moment of inertia of a slender rod, $I = \frac{1}{12}ml^2$, and the parallel-axis and parallel-plane theorems, Eqs. 21–3 and 21–4. Dividing the rod into three parts and locating the mass center of each segment, Fig. 21–5b, we have

$$I_{xx} = \left[\frac{1}{12}(2)(0.2)^2 + 2(0.1)^2\right] + [0 + 2(0.2)^2]$$
$$+ \left[\frac{1}{12}(4)(0.4)^2 + 4((0.2)^2 + (0.2)^2)\right] = 0.480 \text{ kg} \cdot \text{m}^2$$

$$I_{yy} = \left[\frac{1}{12}(2)(0.2)^2 + 2(0.1)^2\right] + \left[\frac{1}{12}(2)(0.2)^2 + 2((-0.1)^2 + (0.2)^2)\right]$$
$$+ [0 + 4((-0.2)^2 + (0.2)^2)] = 0.453 \text{ kg} \cdot \text{m}^2$$

$$I_{zz} = [0 + 0] + \left[\frac{1}{12}(2)(0.2)^2 + 2(0.1)^2\right] + \left[\frac{1}{12}(4)(0.4)^2 + \right.$$
$$\left. 4((-0.2)^2 + (0.2)^2)\right] = 0.400 \text{ kg} \cdot \text{m}^2$$

$$I_{xy} = [0 + 0] + [0 + 0] + [0 + 4(-0.2)(0.2)] = -0.160 \text{ kg} \cdot \text{m}^2$$

$$I_{yz} = [0 + 0] + [0 + 0] + [0 + 4(0.2)(0.2)] = 0.160 \text{ kg} \cdot \text{m}^2$$

$$I_{zx} = [0 + 0] + [0 + 2(0.2)(-0.1)] +$$
$$[0 + 4(0.2)(-0.2)] = -0.200 \text{ kg} \cdot \text{m}^2$$

The Aa axis is defined by the unit vector

$$\mathbf{u}_{Aa} = \frac{\mathbf{r}_D}{r_D} = \frac{-0.2\mathbf{i} + 0.4\mathbf{j} + 0.2\mathbf{k}}{\sqrt{(-0.2)^2 + (0.4)^2 + (0.2)^2}} = -0.408\mathbf{i} + 0.816\mathbf{j} + 0.408\mathbf{k}$$

(b)

Fig. 21–5

Thus,

$$u_x = -0.408 \qquad u_y = 0.816 \qquad u_z = 0.408$$

Substituting these results into Eq. 21–5 yields

$$I_{Aa} = I_{xx}u_x^2 + I_{yy}u_y^2 + I_{zz}u_z^2 - 2I_{xy}u_xu_y - 2I_{yz}u_yu_z - 2I_{zx}u_zu_x$$
$$= 0.480(-0.408)^2 + (0.453)(0.816)^2 + 0.400(0.408)^2$$
$$- 2(-0.160)(-0.408)(0.816) - 2(0.160)(0.816)(0.408)$$
$$- 2(-0.200)(0.408)(-0.408)$$
$$= 0.169 \text{ kg} \cdot \text{m}^2 \qquad \qquad \textit{Ans.}$$

PROBLEMS

21–1. Show that the sum of the moments of inertia of a body, $I_{xx} + I_{yy} + I_{zz}$, is independent of the orientation of the x, y, z axes and thus depends only on the location of the origin.

21–2. Determine the moment of inertia of the cylinder with respect to the a–a axis of the cylinder. The cylinder has a mass m.

Prob. 21–2

21–3. Determine the moments of inertia I_x and I_y of the paraboloid of revolution. The mass of the paraboloid is 20 slug.

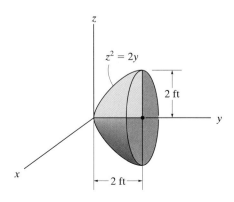

Prob. 21–3

*21–4. Determine the product of inertia I_{xy} of the object formed by revolving the shaded area about the line $x = 5$ ft. Express the result in terms of the density of the material, ρ.

21–5. Determine the moment of inertia I_y of the object formed by revolving the shaded area about the line $x = 5$ ft. Express the result in terms of the density of the material, ρ.

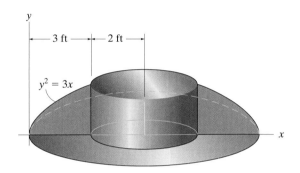

Probs. 21–4/5

21–6. Determine by direct integration the product of inertia I_{yz} for the homogeneous prism. The density of the material is ρ. Express the result in terms of the mass m of the prism.

21–7. Determine by direct integration the product of inertia I_{xy} for the homogeneous prism. The density of the material is ρ. Express the result in terms of the mass m of the prism.

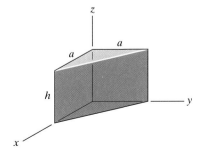

Probs. 21–6/7

*21–8. Determine the radii of gyration k_x and k_y for the solid formed by revolving the shaded area about the y axis. The density of the material is ρ.

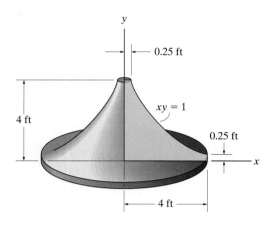

Prob. 21–8

21–10. Determine the elements of the inertia tensor for the cube with respect to the x, y, z coordinate system. The mass of the cube is m.

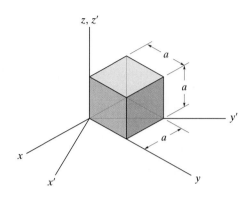

Prob. 21–10

21–9. Determine the mass moment of inertia of the homogeneous block with respect to its centroidal x' axis. The mass of the block is m.

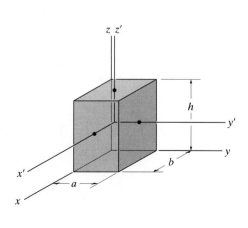

Prob. 21–9

21–11. Compute the moment of inertia of the rod-and-thin-ring assembly about the z axis. The rods and ring have a mass of 2 kg/m.

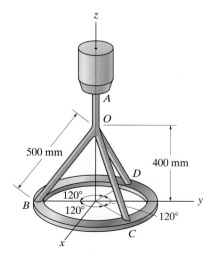

Prob. 21–11

*21–12. Determine the moment of inertia of the cone about the z' axis. The weight of the cone is 15 lb, the *height* is $h = 1.5$ ft, and the radius is $r = 0.5$ ft.

Prob. 21–12

21–14. The assembly consists of two square plates A and B which have a mass of 3 kg each and a rectangular plate C which has a mass of 4.5 kg. Determine the moments of inertia I_x, I_y and I_z.

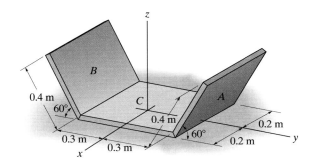

Prob. 21–14

21–13. The bent rod has a weight of 1.5 lb/ft. Locate the center of gravity $G(\bar{x}, \bar{y})$ and determine the principal moments of inertia $I_{x'}$, $I_{y'}$, and $I_{z'}$ of the rod with respect to the x', y', z' axes.

21–15. Determine the moment of inertia I_x of the composite plate assembly. The plates have a specific weight of 6 lb/ft^2.

*21–16. Determine the product of inertia I_{yz} of the composite plate assembly. The plates have a weight of 6 lb/ft^2.

Prob. 21–13

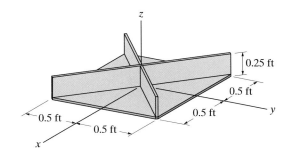

Probs. 21–15/16

21–17.   Determine the moment of inertia of the composite body about the *aa* axis. The cylinder weighs 20 lb, and each hemisphere weighs 10 lb.

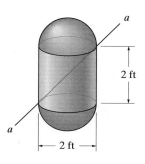

Prob. 21–17

21–18.   Determine the moment of inertia about the *z* axis of the assembly which consists of the 1.5-kg rod *CD* and the 7-kg disk.

Prob. 21–18

21–19.   The assembly consists of a 15-lb plate *A*, 40-lb plate *B*, and four 7-lb rods. Determine the moments of inertia of the assembly with respect to the principal *x*, *y*, *z* axes.

Prob. 21–19

***21–20.**   The thin plate has a weight of 5 lb and each of the four rods weighs 3 lb. Determine the moment of inertia of the assembly about the *z* axis.

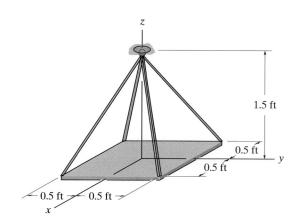

Prob. 21–20

*21.2   Angular Momentum

In this section we will develop the necessary equations used to determine the angular momentum of a rigid body about an arbitrary point. These equations will provide a means for developing both the principle of impulse and momentum and the equations of rotational motion for a rigid body.

Consider the rigid body in Fig. 21–6, which has a mass m and center of mass at G. The X, Y, Z coordinate system represents an inertial frame of reference, and hence, its axes are fixed or translate with a constant velocity. The angular momentum as measured from this reference will be computed relative to the arbitrary point A. The position vectors $\mathbf{r}_A$ and $\boldsymbol{\rho}_A$ are drawn from the origin of coordinates to point A and from A to the ith particle of the body. If the particle's mass is m_i, the angular momentum about point A is

$$(\mathbf{H}_A)_i = \boldsymbol{\rho}_A \times m_i \mathbf{v}_i$$

where $\mathbf{v}_i$ represents the particle's velocity measured from the X, Y, Z coordinate system. If the body has an angular velocity $\boldsymbol{\omega}$ at the instant considered, $\mathbf{v}_i$ may be related to the velocity of A by applying Eq. 20–7, i.e.,

$$\mathbf{v}_i = \mathbf{v}_A + \boldsymbol{\omega} \times \boldsymbol{\rho}_A$$

Thus,

$$(\mathbf{H}_A)_i = \boldsymbol{\rho}_A \times m_i(\mathbf{v}_A + \boldsymbol{\omega} \times \boldsymbol{\rho}_A)$$
$$= (\boldsymbol{\rho}_A m_i) \times \mathbf{v}_A + \boldsymbol{\rho}_A \times (\boldsymbol{\omega} \times \boldsymbol{\rho}_A) m_i$$

Summing all the particles of the body requires an integration, and since $m_i \rightarrow dm$, we have

$$\mathbf{H}_A = \left(\int_m \boldsymbol{\rho}_A \, dm \right) \times \mathbf{v}_A + \int_m \boldsymbol{\rho}_A \times (\boldsymbol{\omega} \times \boldsymbol{\rho}_A) \, dm \qquad (21\text{–}6)$$

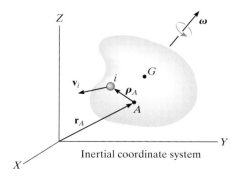

Inertial coordinate system

Fig. 21–6

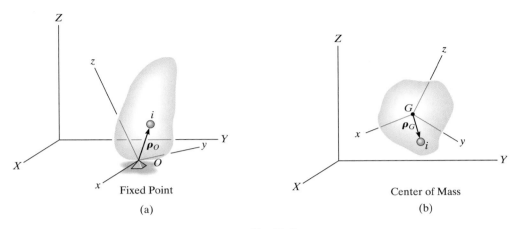

Fig. 21–7

Fixed Point O.

If A becomes a *fixed point* O in the body, Fig. 21–7a, then $\mathbf{v}_A = \mathbf{0}$ and Eq. 21–6 reduces to

$$\mathbf{H}_O = \int_m \boldsymbol{\rho}_O \times (\boldsymbol{\omega} \times \boldsymbol{\rho}_O)\, dm \qquad (21-7)$$

Center of Mass G.

If A is located at the *center of mass* G of the body, Fig. 21–7b, then $\int_m \boldsymbol{\rho}_A\, dm = \mathbf{0}$ and

$$\mathbf{H}_G = \int_m \boldsymbol{\rho}_G \times (\boldsymbol{\omega} \times \boldsymbol{\rho}_G)\, dm \qquad (21-8)$$

Arbitrary Point A.

In general, A may be some point other than O or G, Fig. 21–7c, in which case Eq. 21–6 may nevertheless be simplified to the following form (see Prob. 21–22).

$$\mathbf{H}_A = \boldsymbol{\rho}_{G/A} \times m\mathbf{v}_G + \mathbf{H}_G \qquad (21-9)$$

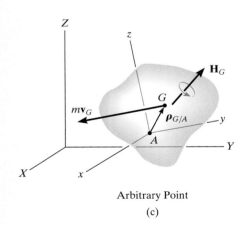

Arbitrary Point

(c)

Fig. 21–7

Here the angular momentum consists of two parts—the moment of the linear momentum $m\mathbf{v}_G$ of the body about point A added (vectorially) to the angular momentum $\mathbf{H}_G$. Equation 21–9 may also be used for computing the angular momentum of the body about a fixed point O; the results, of course, will be the same as those computed using the more convenient Eq. 21–7.

Rectangular Components of H.

To make practical use of Eqs. 21–7 through 21–9, the angular momentum must be expressed in terms of its scalar components. For this purpose, it is convenient to choose a second set of x, y, z axes having an arbitrary orientation relative to the X, Y, Z axes, Fig. 21–7, and for a general formulation, note that Eqs. 21–7 and 21–8 are both of the form

$$\mathbf{H} = \int_m \boldsymbol{\rho} \times (\boldsymbol{\omega} \times \boldsymbol{\rho}) \, dm$$

Expressing $\mathbf{H}$, $\boldsymbol{\rho}$, and $\boldsymbol{\omega}$ in terms of x, y, and z components, we have

$$H_x\mathbf{i} + H_y\mathbf{j} + H_z\mathbf{k} = \int_m (x\mathbf{i} + y\mathbf{j} + z\mathbf{k}) \times [(\omega_x\mathbf{i} + \omega_y\mathbf{j} + \omega_z\mathbf{k})$$
$$\times (x\mathbf{i} + y\mathbf{j} + z\mathbf{k})] \, dm$$

Expanding the cross products and combining terms yields

$$H_x\mathbf{i} + H_y\mathbf{j} + H_z\mathbf{k} = \left[\omega_x \int_m (y^2 + z^2)\, dm - \omega_y \int_m xy \, dm - \omega_z \int_m xz \, dm \right]\mathbf{i}$$
$$+ \left[-\omega_x \int_m xy \, dm + \omega_y \int_m (x^2 + z^2)\, dm - \omega_z \int_m yz \, dm \right]\mathbf{j}$$
$$+ \left[-\omega_x \int_m zx \, dm - \omega_y \int_m yz \, dm + \omega_z \int_m (x^2 + y^2)\, dm \right]\mathbf{k}$$

Equating the respective $\mathbf{i}$, $\mathbf{j}$, $\mathbf{k}$ components and recognizing that the integrals represent the moments and products of inertia, we obtain

$$\boxed{\begin{aligned} H_x &= I_{xx}\omega_x - I_{xy}\omega_y - I_{xz}\omega_z \\ H_y &= -I_{yx}\omega_x + I_{yy}\omega_y - I_{yz}\omega_z \\ H_z &= -I_{zx}\omega_x - I_{zy}\omega_y + I_{zz}\omega_z \end{aligned}} \qquad (21\text{--}10)$$

These three equations represent the scalar form of the $\mathbf{i}$, $\mathbf{j}$, $\mathbf{k}$ components of $\mathbf{H}_O$ or $\mathbf{H}_G$ (given in vector form by Eqs. 21–7 and 21–8). The angular momentum of the body about the arbitrary point A, other than the fixed point O or the center of mass G, may also be expressed in scalar form. Here it is necessary to use Eq. 21–9 to represent $\boldsymbol{\rho}_{G/A}$ and $\mathbf{v}_G$ as Cartesian vectors, carry out the cross-product operation, and substitute the components, Eqs. 21–10, for $\mathbf{H}_G$.

Equations 21–10 may be simplified further if the x, y, z coordinate axes are oriented such that they become *principal axes of inertia* for the body at the point. When these axes are used, the products of inertia $I_{xy} = I_{yz} = I_{zx} = 0$, and if the principal moments of inertia about the x, y, z axes are represented as $I_x = I_{xx}$, $I_y = I_{yy}$, and $I_z = I_{zz}$, the three components of angular momentum become

$$\boxed{H_x = I_x\omega_x \qquad H_y = I_y\omega_y \qquad H_z = I_z\omega_z} \qquad (21\text{--}11)$$

The motion of the astronaut is controlled by use of small directional jets attached to his or her space suit. The impulses these jets provide must be carefully specified in order to prevent tumbling and loss of orientation.

Principle of Impulse and Momentum. Now that the formulation of the angular momentum for a body has been developed, the *principle of impulse and momentum*, as discussed in Sec. 19.2, may be used to solve kinetic problems which involve *force, velocity, and time*. For this case, the following two vector equations are available:

$$m(\mathbf{v}_G)_1 + \Sigma \int_{t_1}^{t_2} \mathbf{F}\, dt = m(\mathbf{v}_G)_2 \qquad (21\text{–}12)$$

$$(\mathbf{H}_O)_1 + \Sigma \int_{t_1}^{t_2} \mathbf{M}_O\, dt = (\mathbf{H}_O)_2 \qquad (21\text{–}13)$$

In three dimensions each vector term can be represented by three scalar components, and therefore a total of *six scalar equations* can be written. Three equations relate the linear impulse and momentum in the *x, y, z* directions, and the other three equations relate the body's angular impulse and momentum about the *x, y, z* axes. Before applying Eqs. 21–12 and 21–13 to the solution of problems, the material in Secs. 19.2 and 19.3 should be reviewed.

*21.3   Kinetic Energy

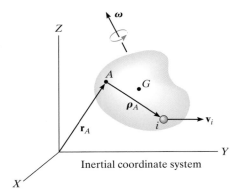

Inertial coordinate system

Fig. 21–8

In order to apply the principle of work and energy to the solution of problems involving general rigid body motion, it is first necessary to formulate expressions for the kinetic energy of the body. In this regard, consider the rigid body shown in Fig. 21–8, which has a mass m and center of mass at G. The kinetic energy of the *i*th particle of the body having a mass m_i and velocity $\mathbf{v}_i$, measured relative to the inertial X, Y, Z frame of reference, is

$$T_i = \tfrac{1}{2} m_i v_i^2 = \tfrac{1}{2} m_i (\mathbf{v}_i \cdot \mathbf{v}_i)$$

Provided the velocity of an arbitrary point A in the body is known, $\mathbf{v}_i$ may be related to $\mathbf{v}_A$ by the equation $\mathbf{v}_i = \mathbf{v}_A + \boldsymbol{\omega} \times \boldsymbol{\rho}_A$, where $\boldsymbol{\omega}$ is the angular velocity of the body, measured from the X, Y, Z coordinate system, and $\boldsymbol{\rho}_A$ is a position vector drawn from A to i. Using this expression for $\mathbf{v}_i$, the kinetic energy for the particle may be written as

$$T_i = \tfrac{1}{2} m_i (\mathbf{v}_A + \boldsymbol{\omega} \times \boldsymbol{\rho}_A) \cdot (\mathbf{v}_A + \boldsymbol{\omega} \times \boldsymbol{\rho}_A)$$
$$= \tfrac{1}{2} (\mathbf{v}_A \cdot \mathbf{v}_A) m_i + \mathbf{v}_A \cdot (\boldsymbol{\omega} \times \boldsymbol{\rho}_A) m_i + \tfrac{1}{2} (\boldsymbol{\omega} \times \boldsymbol{\rho}_A) \cdot (\boldsymbol{\omega} \times \boldsymbol{\rho}_A) m_i$$

The kinetic energy for the entire body is obtained by summing the kinetic energies of all the particles of the body. This requires an integration, and since $m_i \rightarrow dm$, we get

$$T = \tfrac{1}{2} m (\mathbf{v}_A \cdot \mathbf{v}_A) + \mathbf{v}_A \cdot \left(\boldsymbol{\omega} \times \int_m \boldsymbol{\rho}_A\, dm \right) + \tfrac{1}{2} \int_m (\boldsymbol{\omega} \times \boldsymbol{\rho}_A) \cdot (\boldsymbol{\omega} \times \boldsymbol{\rho}_A)\, dm$$

The last term on the right may be rewritten using the vector identity $\mathbf{a} \times \mathbf{b} \cdot \mathbf{c} = \mathbf{a} \cdot \mathbf{b} \times \mathbf{c}$, where $\mathbf{a} = \boldsymbol{\omega}$, $\mathbf{b} = \boldsymbol{\rho}_A$, and $\mathbf{c} = \boldsymbol{\omega} \times \boldsymbol{\rho}_A$. The final result is

$$T = \tfrac{1}{2}m(\mathbf{v}_A \cdot \mathbf{v}_A) + \mathbf{v}_A \cdot \left(\boldsymbol{\omega} \times \int_m \boldsymbol{\rho}_A \, dm\right)$$

$$+ \tfrac{1}{2}\boldsymbol{\omega} \cdot \int_m \boldsymbol{\rho}_A \times (\boldsymbol{\omega} \times \boldsymbol{\rho}_A) \, dm \qquad (21\text{--}14)$$

This equation is rarely used because of the computations involving the integrals. Simplification occurs, however, if the reference point A is either a fixed point O or the center of mass G.

Fixed Point O.   If A is a *fixed point* O in the body, Fig. 21–7a, then $\mathbf{v}_A = \mathbf{0}$, and using Eq. 21–7, we can express Eq. 21–14 as

$$T = \tfrac{1}{2}\boldsymbol{\omega} \cdot \mathbf{H}_O$$

If the x, y, z axes represent the principal axes of inertia for the body, then $\boldsymbol{\omega} = \omega_x \mathbf{i} + \omega_y \mathbf{j} + \omega_z \mathbf{k}$ and $\mathbf{H}_O = I_x \omega_x \mathbf{i} + I_y \omega_y \mathbf{j} + I_z \omega_z \mathbf{k}$. Substituting into the above equation and performing the dot-product operations yields

$$T = \tfrac{1}{2}I_x \omega_x^2 + \tfrac{1}{2}I_y \omega_y^2 + \tfrac{1}{2}I_z \omega_z^2 \qquad (21\text{--}15)$$

Center of Mass G.   If A is located at the *center of mass* G of the body, Fig. 21–7b, then $\int \boldsymbol{\rho}_A \, dm = \mathbf{0}$ and, using Eq. 21–8, we can write Eq. 21–14 as

$$T = \tfrac{1}{2}mv_G^2 + \tfrac{1}{2}\boldsymbol{\omega} \cdot \mathbf{H}_G$$

In a manner similar to that for a fixed point, the last term on the right side may be represented in scalar form, in which case

$$T = \tfrac{1}{2}mv_G^2 + \tfrac{1}{2}I_x \omega_x^2 + \tfrac{1}{2}I_y \omega_y^2 + \tfrac{1}{2}I_z \omega_z^2 \qquad (21\text{--}16)$$

Here it is seen that the kinetic energy consists of two parts; namely, the translational kinetic energy of the mass center, $\tfrac{1}{2}mv_G^2$, and the body's rotational kinetic energy.

Principle of Work and Energy.   Having formulated the kinetic energy for a body, the *principle of work and energy* may be applied to solve kinetics problems which involve *force, velocity, and displacement*. For this case only one scalar equation can be written for each body, namely,

$$T_1 + \Sigma U_{1\text{--}2} = T_2 \qquad (21\text{--}17)$$

Before applying this equation, the material in Chapter 18 should be reviewed.

EXAMPLE 21.2

(a)

The rod in Fig. 21–9a has a weight of 1.5 lb/ft. Determine its angular velocity just after the end A falls onto the hook at E. The hook provides a permanent connection for the rod due to the spring-lock mechanism S. Just before striking the hook the rod is falling downward with a speed $(v_G)_1 = 10$ ft/s.

SOLUTION

The principle of impulse and momentum will be used since impact occurs.

Impulse and Momentum Diagrams.   Fig. 21–9b. During the short time Δt, the impulsive force $\mathbf{F}$ acting at A changes the momentum of the rod. (The impulse created by the rod's weight $\mathbf{W}$ during this time is small compared to $\int \mathbf{F}\,dt$, so that it is neglected, i.e., the weight is a nonimpulsive force.) Hence, the angular momentum of the rod is *conserved* about point A since the moment of $\int \mathbf{F}\,dt$ about A is zero.

Conservation of Angular Momentum.   Equation 21–9 must be used for computing the angular momentum of the rod, since A does not become a *fixed point* until *after* the impulsive interaction with the hook. Thus, with reference to Fig. 21–9b, $(\mathbf{H}_A)_1 = (\mathbf{H}_A)_2$, or

$$\mathbf{r}_{G/A} \times m(\mathbf{v}_G)_1 = \mathbf{r}_{G/A} \times m(\mathbf{v}_G)_2 + (\mathbf{H}_G)_2 \tag{1}$$

From Fig. 21–9a, $\mathbf{r}_{G/A} = \{-0.667\mathbf{i} + 0.5\mathbf{j}\}$ ft. Furthermore, the primed axes are principal axes of inertia for the rod because $I_{x'y'} = I_{x'z'} = I_{z'y'} = 0$. Hence, from Eqs. 21–11, $(\mathbf{H}_G)_2 = I_{x'}\omega_x\mathbf{i} + I_{y'}\omega_y\mathbf{j} + I_{z'}\omega_z\mathbf{k}$. The principal moments of inertia are $I_{x'} = 0.0272$ slug·ft², $I_{y'} = 0.0155$ slug·ft², $I_{z'} = 0.0427$ slug·ft² (see Prob. 21–13). Substituting into Eq. 1, we have

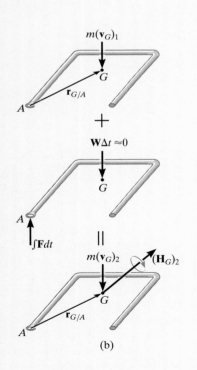

$$(-0.667\mathbf{i} + 0.5\mathbf{j}) \times \left[\left(\frac{4.5}{32.2}\right)(-10\mathbf{k})\right] = (-0.667\mathbf{i} + 0.5\mathbf{j}) \times \left[\left(\frac{4.5}{32.2}\right)(-v_G)_2\mathbf{k}\right]$$
$$+\ 0.0272\omega_x\mathbf{i} + 0.0155\omega_y\mathbf{j} + 0.0427\omega_z\mathbf{k}$$

Expanding and equating the respective $\mathbf{i}$, $\mathbf{j}$, and $\mathbf{k}$ components yields

$$-0.699 = -0.0699(v_G)_2 + 0.0272\omega_x \tag{2}$$
$$-0.932 = -0.0932(v_G)_2 + 0.0155\omega_y \tag{3}$$
$$0 = 0.0427\omega_z \tag{4}$$

Kinematics.   There are four unknowns in the above equations; however, another equation may be obtained by relating $\boldsymbol{\omega}$ to $(\mathbf{v}_G)_2$ using *kinematics*. Since $\omega_z = 0$ (Eq. 4) and after impact the rod rotates about the fixed point A, Eq. 20–3 may be applied, in which case $(\mathbf{v}_G)_2 = \boldsymbol{\omega} \times \mathbf{r}_{G/A}$, or

$$-(v_G)_2\mathbf{k} = (\omega_x\mathbf{i} + \omega_y\mathbf{j}) \times (-0.667\mathbf{i} + 0.5\mathbf{j})$$
$$-(v_G)_2 = 0.5\omega_x + 0.667\omega_y \tag{5}$$

Solving Eqs. 2 through 5 simultaneously yields

$$(\mathbf{v}_G)_2 = \{-8.41\mathbf{k}\} \text{ ft/s} \qquad \boldsymbol{\omega} = \{-4.09\mathbf{i} - 9.55\mathbf{j}\} \text{ rad/s} \qquad Ans.$$

Fig. 21–9

EXAMPLE 21.3

A 5-N·m torque is applied to the vertical shaft CD shown in Fig. 21–10a, which allows the 10-kg gear A to turn freely about CE. Assuming that gear A starts from rest, determine the angular velocity of CD after it has turned two revolutions. Neglect the mass of shaft CD and axle CE and assume that gear A can be approximated by a thin disk. Gear B is fixed.

(a)

SOLUTION

The principle of work and energy may be used for the solution. Why?

Work.   If shaft CD, the axle CE, and gear A are considered as a system of connected bodies, only the applied torque $\mathbf{M}$ does work. For two revolutions of CD, this work is $\Sigma U_{1-2} = (5\text{ N}\cdot\text{m})(4\pi\text{ rad}) = 62.83$ J.

Kinetic Energy.   Since the gear is initially at rest, its initial kinetic energy is zero. A kinematic diagram for the gear is shown in Fig. 21–10b. If the angular velocity of CD is taken as $\boldsymbol{\omega}_{CD}$, then the angular velocity of gear A is $\boldsymbol{\omega}_A = \boldsymbol{\omega}_{CD} + \boldsymbol{\omega}_{CE}$. The gear may be imagined as a portion of a massless extended body which is rotating about the *fixed point C*. The instantaneous axis of rotation for this body is along line CH, because both points C and H on the body (gear) have zero velocity and must therefore lie on this axis. This requires that the components $\boldsymbol{\omega}_{CD}$ and $\boldsymbol{\omega}_{CE}$ be related by the equation $\omega_{CD}/0.1\text{ m} = \omega_{CE}/0.3\text{ m}$ or $\omega_{CE} = 3\omega_{CD}$. Thus,

(b)

Fig. 21–10

$$\boldsymbol{\omega}_A = -\omega_{CE}\mathbf{i} + \omega_{CD}\mathbf{k} = -3\omega_{CD}\mathbf{i} + \omega_{CD}\mathbf{k} \qquad (1)$$

The x, y, z axes in Fig. 21–10a represent *principal axes of inertia* at C for the gear. Since point C is a fixed point of rotation, Eq. 21–15 may be applied to determine the kinetic energy, i.e.,

$$T = \tfrac{1}{2}I_x\omega_x^2 + \tfrac{1}{2}I_y\omega_y^2 + \tfrac{1}{2}I_z\omega_z^2 \qquad (2)$$

Using the parallel-axis theorem, the moments of inertia of the gear about point C are as follows:

$$I_x = \tfrac{1}{2}(10\text{ kg})(0.1\text{ m})^2 = 0.05\text{ kg}\cdot\text{m}^2$$

$$I_y = I_z = \tfrac{1}{4}(10\text{ kg})(0.1\text{ m})^2 + 10\text{ kg}(0.3\text{ m})^2 = 0.925\text{ kg}\cdot\text{m}^2$$

Since $\omega_x = -3\omega_{CD}$, $\omega_y = 0$, $\omega_z = \omega_{CD}$, Eq. 2 becomes

$$T_A = \tfrac{1}{2}(0.05)(-3\omega_{CD})^2 + 0 + \tfrac{1}{2}(0.925)(\omega_{CD})^2 = 0.6875\omega_{CD}^2$$

Principle of Work and Energy.   Applying the principle of work and energy, we obtain

$$T_1 + \Sigma U_{1-2} = T_2$$

$$0 + 62.83 = 0.6875\omega_{CD}^2$$

$$\omega_{CD} = 9.56\text{ rad/s} \qquad\qquad Ans.$$

PROBLEMS

21–21. If a body contains *no planes of symmetry*, the principal moments of inertia can be determined mathematically. To show how this is done, consider the rigid body which is spinning with an angular velocity $\boldsymbol{\omega}$, directed along one of its principal axes of inertia. If the principal moment of inertia about this axis is I, the angular momentum can be expressed as $\mathbf{H} = I\boldsymbol{\omega} = I\omega_x\mathbf{i} + I\omega_y\mathbf{j} + I\omega_z\mathbf{k}$. The components of $\mathbf{H}$ may also be expressed by Eqs. 21–10, where the inertia tensor is assumed to be known. Equate the $\mathbf{i}, \mathbf{j},$ and $\mathbf{k}$ components of both expressions for $\mathbf{H}$ and consider $\omega_x, \omega_y,$ and ω_z to be unknown. The solution of these three equations is obtained provided the determinant of the coefficients is zero. Show that this determinant, when expanded, yields the cubic equation

$$I^3 - (I_{xx} + I_{yy} + I_{zz})I^2 + (I_{xx}I_{yy} + I_{yy}I_{zz} + I_{zz}I_{xx}$$
$$- I_{xy}^2 - I_{yz}^2 - I_{zx}^2)I - (I_{xx}I_{yy}I_{zz} - 2I_{xy}I_{yz}I_{zx}$$
$$- I_{xx}I_{yz}^2 - I_{yy}I_{zx}^2 - I_{zz}I_{xy}^2) = 0$$

The three positive roots of I, obtained from the solution of this equation, represent the principal moments of inertia I_x, I_y, and I_z.

21–22. Show that if the angular momentum of a body is determined with respect to an arbitrary point A, then $\mathbf{H}_A$ can be expressed by Eq. 21–9. This requires substituting $\boldsymbol{\rho}_A = \boldsymbol{\rho}_G + \boldsymbol{\rho}_{G/A}$ into Eq. 21–6 and expanding, noting that $\int \boldsymbol{\rho}_G \, dm = \mathbf{0}$ by definition of the mass center and $\mathbf{v}_G = \mathbf{v}_A + \boldsymbol{\omega} \times \boldsymbol{\rho}_{G/A}$.

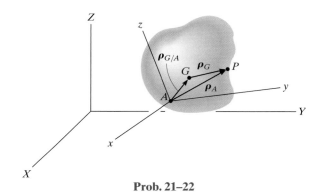

Prob. 21–22

21–23. The 5-kg thin plate is suspended at O using a ball-and-socket joint. It is rotating with a constant angular velocity $\boldsymbol{\omega} = \{2\mathbf{k}\}$ rad/s when the corner A strikes the hook at S, which provides a permanent connection. Determine the angular velocity of the plate immediately after impact.

Prob. 21–23

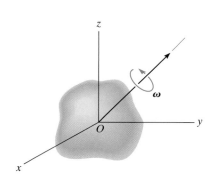

Prob. 21–21

*21–24. Rod AB has a weight of 6 lb and is attached to two smooth collars at its end points by ball-and-socket joints. If collar A is moving downward at a speed of 8 ft/s, determine the kinetic energy of the rod at the instant shown. Assume that at this instant the angular velocity of the rod is directed perpendicular to the rod's axis.

21–25. At the instant shown the collar at A on the 6-lb rod AB has a velocity of $v_A = 8$ ft/s. Determine the kinetic energy of the rod after the collar has descended 3 ft. Neglect friction and the thickness of the rod. Neglect the mass of the collar and the collar is attached to the rod using ball-and-socket joints.

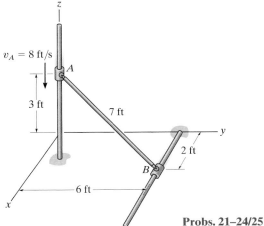

Probs. 21–24/25

21–26. The 4-lb rod AB is attached to the 1-lb collar at A and a 2-lb link BC using ball-and-socket joints. If the rod is released from rest in the position shown, determine the angular velocity of the link after it has rotated 180°.

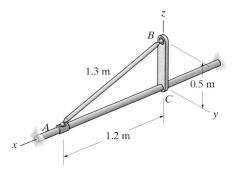

Prob. 21–26

21–27. The rod weighs 3 lb/ft and is suspended from parallel cords at A and B. If the rod has an angular velocity of 2 rad/s about the z axis at the instant shown, determine how high the center of the rod rises at the instant the rod momentarily stops swinging.

Prob. 21–27

*21–28. The assembly consists of a 4-kg rod AB which is connected to link OA and the collar at B by ball-and-socket joints. When $\theta = 0°$, $y = 600$ mm, the system is at rest, the spring is unstretched, and a couple moment $M = 7$ N·m is applied to the link at O. Determine the angular velocity of the link at the instant $\theta = 90°$. Neglect the mass of the link.

21–29. The assembly consists of a 4-kg rod AB which is connected to link OA and the collar at B by ball-and-socket joints. When $\theta = 0°$, $y = 600$ mm, the system is at rest, the spring is unstretched, and a couple moment $M = (4\theta + 2)$ N·m, where θ is in radians, is applied to the link at O. Determine the angular velocity of the link at the instant $\theta = 90°$. Neglect the mass of the link.

Probs. 21–28/29

21–30. The circular plate has a weight of 19 lb and a diameter of 1.5 ft. If it is released from rest and falls horizontally 2.5 ft onto the hook at S, which provides a permanent connection, determine the velocity of the mass center of the plate just after the connection with the hook is made.

Prob. 21–30

21–31. A thin plate, having a mass of 4 kg, is suspended from one of its corners by a ball-and-socket joint O. If a stone strikes the plate perpendicular to its surface at an adjacent corner A with an impulse of $\mathbf{I}_s = \{-60\mathbf{i}\}$ N·s, determine the instantaneous axis of rotation for the plate and the impulse created at O.

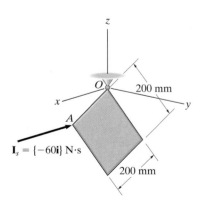

Prob. 21–31

*21–32. Rod AB has a weight of 6 lb and is attached to two smooth collars at its ends by ball-and-socket joints. If collar A is moving downward with a speed of 8 ft/s when $z = 3$ ft, determine the speed of A at the instant $z = 0$. The spring has an unstretched length of 2 ft. Neglect the mass of the collars. Assume the angular velocity of rod AB is perpendicular to its axis.

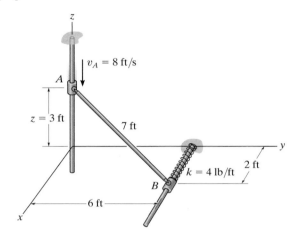

Prob. 21–32

21–33. The circular disk has a weight of 15 lb and is mounted on the shaft AB at an angle of 45° with the horizontal. Determine the angular velocity of the shaft when $t = 3$ s if a constant torque $M = 2$ lb · ft is applied to the shaft. The shaft is originally spinning at $\omega_1 = 8$ rad/s when the torque is applied.

21–34. The circular disk has a weight of 15 lb and is mounted on the shaft AB at an angle of 45° with the horizontal. Determine the angular velocity of the shaft when $t = 2$ s if a torque $M = (4e^{0.1t})$ lb · ft, where t is in seconds, is applied to the shaft. The shaft is originally spinning at $\omega_1 = 8$ rad/s when the torque is applied.

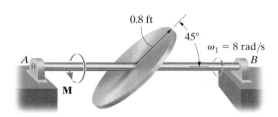

Probs. 21–33/34

21–35. The 15-kg rectangular plate is free to rotate about the y axis because of the bearing supports at A and B. When the plate is balanced in the vertical plane, a 3-g bullet is fired into it, perpendicular to its surface, with a velocity $\mathbf{v} = \{-2000\mathbf{i}\}$ m/s. Compute the angular velocity of the plate at the instant it has rotated $180°$. If the bullet strikes corner D with the same velocity $\mathbf{v}$, instead of at C, does the angular velocity remain the same? Why or why not?

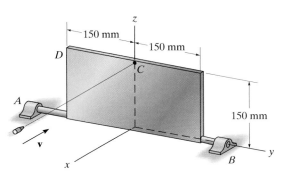

Prob. 21–35

21–37. The 15-lb plate is subjected to a force $F = 8$ lb which is always directed perpendicular to the face of the plate. If the plate is originally at rest, determine its angular velocity after it has rotated one revolution ($360°$). The plate is supported by ball-and-socket joints at A and B.

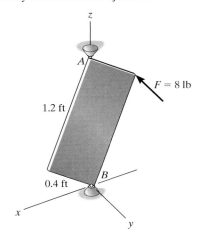

Prob. 21–37

***21–36.** The rod assembly has a mass of 2.5 kg/m and is rotating with a constant angular velocity of $\boldsymbol{\omega} = \{2\mathbf{k}\}$ rad/s when the looped end at C encounters a hook at S, which provides a permanent connection. Determine the angular velocity of the assembly immediately after impact.

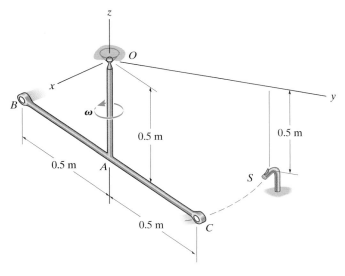

Prob. 21–36

21–38. The space capsule has a mass of 3.5 Mg and the radii of gyration are $k_x = k_z = 0.8$ m and $k_y = 0.5$ m. If it is traveling with a velocity $\mathbf{v}_G = \{600\mathbf{j}\}$ m/s, compute its angular velocity just after it is struck by a meteoroid having a mass of 0.60 kg and a velocity $\mathbf{V}_m = \{-200\mathbf{i} - 400\mathbf{j} + 200\mathbf{k}\}$ m/s. Assume that the meteoroid embeds itself into the capsule at point A and that the capsule initially has no angular velocity.

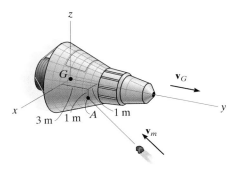

Prob. 21–38

*21.4   Equations of Motion

Having become familiar with the techniques used to describe both the inertial properties and the angular momentum of a body, we can now write the equations which describe the motion of the body in their most useful forms.

Equations of Translational Motion.   The *translational motion* of a body is defined in terms of the acceleration of the body's mass center, which is measured from an inertial X, Y, Z reference. The equation of translational motion for the body can be written in vector form as

$$\Sigma \mathbf{F} = m\mathbf{a}_G \qquad (21\text{--}18)$$

or by the three scalar equations

$$
\begin{aligned}
\Sigma F_x &= m(a_G)_x \\
\Sigma F_y &= m(a_G)_y \\
\Sigma F_z &= m(a_G)_z
\end{aligned}
\qquad (21\text{--}19)
$$

Here, $\Sigma \mathbf{F} = \Sigma F_x \mathbf{i} + \Sigma F_y \mathbf{j} + \Sigma F_z \mathbf{k}$ represents the sum of all the external forces acting on the body.

Equations of Rotational Motion.   In Sec. 15.6, we developed Eq. 15–17, namely,

$$\Sigma \mathbf{M}_O = \dot{\mathbf{H}}_O \qquad (21\text{--}20)$$

which states that the sum of the moments about a fixed point O of all the external forces acting on a system of particles (contained in a rigid body) is equal to the time rate of change of the total angular momentum of the body about point O. When moments of the external forces acting on the particles are summed about the system's *mass center* G, one again obtains the same simple form of Eq. 21–20, relating the moment summation $\Sigma \mathbf{M}_G$ to the angular momentum $\mathbf{H}_G$. To show this, consider the system of particles in Fig. 21–11, where X, Y, Z represents an inertial frame of reference and the x, y, z axes, with origin at G, *translate* with respect to this frame. In general, G is *accelerating*, so by definition the translating frame is *not* an inertial reference. The angular momentum of the ith particle with respect to this frame is, however,

$$(\mathbf{H}_i)_G = \mathbf{r}_{i/G} \times m_i \mathbf{v}_{i/G}$$

where $\mathbf{r}_{i/G}$ and $\mathbf{v}_{i/G}$ represent the relative position and relative velocity of the ith particle with respect to G. Taking the time derivative we have

$$(\dot{\mathbf{H}}_i)_G = \dot{\mathbf{r}}_{i/G} \times m_i \mathbf{v}_{i/G} + \mathbf{r}_{i/G} \times m_i \dot{\mathbf{v}}_{i/G}$$

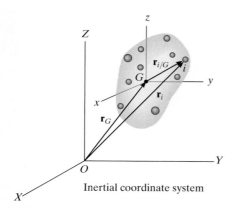

Inertial coordinate system

Fig. 21–11

By definition, $\mathbf{v}_{i/G} = \dot{\mathbf{r}}_{i/G}$. Thus, the first term on the right side is zero since the cross product of equal vectors is zero. Also, $\mathbf{a}_{i/G} = \dot{\mathbf{v}}_{i/G}$, so that

$$(\dot{\mathbf{H}}_i)_G = (\mathbf{r}_{i/G} \times m_i \mathbf{a}_{i/G})$$

Similar expressions can be written for the other particles of the body. When the results are summed, we get

$$\dot{\mathbf{H}}_G = \Sigma(\mathbf{r}_{i/G} \times m_i \mathbf{a}_{i/G})$$

Here $\dot{\mathbf{H}}_G$ is the time rate of change of the total angular momentum of the body computed relative to point G.

The relative acceleration for the ith particle is defined by the equation $\mathbf{a}_{i/G} = \mathbf{a}_i - \mathbf{a}_G$, where $\mathbf{a}_i$ and $\mathbf{a}_G$ represent, respectively, the accelerations of the ith particle and point G measured with respect to the *inertial frame of reference*. Substituting and expanding, using the distributive property of the vector cross product, yields

$$\dot{\mathbf{H}}_G = \Sigma(\mathbf{r}_{i/G} \times m_i \mathbf{a}_i) - (\Sigma m_i \mathbf{r}_{i/G}) \times \mathbf{a}_G$$

By definition of the mass center, the sum $(\Sigma m_i \mathbf{r}_{i/G}) = (\Sigma m_i)\bar{\mathbf{r}}$ is equal to zero, since the position vector $\bar{\mathbf{r}}$ relative to G is zero. Hence, the last term in the above equation is zero. Using the equation of motion, the product $m_i \mathbf{a}_i$ may be replaced by the resultant *external force* $\mathbf{F}_i$ acting on the ith particle. Denoting $\Sigma \mathbf{M}_G = (\Sigma \mathbf{r}_{i/G} \times \mathbf{F}_i)$, the final result may be written as

$$\Sigma \mathbf{M}_G = \dot{\mathbf{H}}_G \qquad (21\text{--}21)$$

The rotational equation of motion for the body will now be developed from either Eq. 21–20 or 21–21. In this regard, the scalar components of the angular momentum $\mathbf{H}_O$ or $\mathbf{H}_G$ are defined by Eqs. 21–10 or, if principal axes of inertia are used either at point O or G, by Eqs. 21–11. If these components are computed about x, y, z axes that are *rotating* with an angular velocity $\mathbf{\Omega}$, which may be *different* from the body's angular velocity $\boldsymbol{\omega}$, then the time derivative $\dot{\mathbf{H}} = d\mathbf{H}/dt$, as used in Eqs. 21–20 and 21–21, must account for the rotation of the x, y, z axes as measured from the inertial X, Y, Z axes. Hence, the time derivative of $\mathbf{H}$ must be determined from Eq. 20-6, in which case Eqs. 21–20 and 21–21 become

$$\Sigma \mathbf{M}_O = (\dot{\mathbf{H}}_O)_{xyz} + \mathbf{\Omega} \times \mathbf{H}_O$$
$$\Sigma \mathbf{M}_G = (\dot{\mathbf{H}}_G)_{xyz} + \mathbf{\Omega} \times \mathbf{H}_G \qquad (21\text{--}22)$$

Here $(\dot{\mathbf{H}})_{xyz}$ is the time rate of change of $\mathbf{H}$ measured from the x, y, z reference.

There are three ways in which one can define the motion of the x, y, z axes. Obviously, motion of this reference should be chosen to yield the simplest set of moment equations for the solution of a particular problem.

x, y, z Axes Having Motion $\Omega = 0$. If the body has general motion, the x, y, z axes may be chosen with origin at G, such that the axes only *translate* relative to the inertial X, Y, Z frame of reference. Doing this simplifies Eq. 21–22, since $\Omega = 0$. However, the body may have a rotation $\boldsymbol{\omega}$ about these axes, and therefore the moments and products of inertia of the body would have to be expressed as *functions of time*. In most cases this would be a difficult task, so that such a choice of axes has restricted value.

x, y, z Axes Having Motion $\Omega = \omega$. The x, y, z axes may be chosen such that they are *fixed in and move with the body*. The moments and products of inertia of the body relative to these axes will be *constant* during the motion. Since $\Omega = \omega$, Eqs. 21–22 become

$$\Sigma \mathbf{M}_O = (\dot{\mathbf{H}}_O)_{xyz} + \boldsymbol{\omega} \times \mathbf{H}_O$$
$$\Sigma \mathbf{M}_G = (\dot{\mathbf{H}}_G)_{xyz} + \boldsymbol{\omega} \times \mathbf{H}_G \tag{21–23}$$

We may express each of these vector equations as three scalar equations using Eqs. 21–10. Neglecting the subscripts O and G yields

$$\Sigma M_x = I_{xx}\dot{\omega}_x - (I_{yy} - I_{zz})\omega_y\omega_z - I_{xy}(\dot{\omega}_y - \omega_z\omega_x)$$
$$\qquad - I_{yz}(\omega_y^2 - \omega_z^2) - I_{zx}(\dot{\omega}_z + \omega_x\omega_y)$$

$$\Sigma M_y = I_{yy}\dot{\omega}_y - (I_{zz} - I_{xx})\omega_z\omega_x - I_{yz}(\dot{\omega}_z - \omega_x\omega_y)$$
$$\qquad - I_{zx}(\omega_z^2 - \omega_x^2) - I_{xy}(\dot{\omega}_x + \omega_y\omega_z) \tag{21–24}$$

$$\Sigma M_z = I_{zz}\dot{\omega}_z - (I_{xx} - I_{yy})\omega_x\omega_y - I_{zx}(\dot{\omega}_x - \omega_y\omega_z)$$
$$\qquad - I_{xy}(\omega_x^2 - \omega_y^2) - I_{yz}(\dot{\omega}_y + \omega_z\omega_x)$$

Notice that for a rigid body symmetric with respect to the x–y reference plane, and undergoing general plane motion in this plane, $I_{xz} = I_{yz} = 0$, and $\omega_x = \omega_y = d\omega_x/dt = d\omega_y/dt = 0$. Equations 21–24 reduce to the form $\Sigma M_x = \Sigma M_y = 0$, and $\Sigma M_z = I_{zz}\alpha_z$ (where $\alpha_z = \dot{\omega}_z$), which is essentially the third of Eqs. 17–16 or 17–17 depending on the choice of point O or G for summing moments.

If the x, y, and z axes are chosen as *principal axes of inertia*, the products of inertia are zero, $I_{xx} = I_x$, etc., and Eqs. 21–24 reduce to the form

$$\boxed{\begin{aligned} \Sigma M_x &= I_x\dot{\omega}_x - (I_y - I_z)\omega_y\omega_z \\ \Sigma M_y &= I_y\dot{\omega}_y - (I_z - I_x)\omega_z\omega_x \\ \Sigma M_z &= I_z\dot{\omega}_z - (I_x - I_y)\omega_x\omega_y \end{aligned}} \tag{21–25}$$

This set of equations is known historically as the *Euler equations of motion*, named after the Swiss mathematician Leonhard Euler, who first developed them. They apply *only* for moments summed about either point O or G.

When applying these equations it should be realized that $\dot{\omega}_x$, $\dot{\omega}_y$, $\dot{\omega}_z$ represent the time derivatives of the magnitudes of the x, y, z components of $\boldsymbol{\omega}$ as observed from x, y, z. Since the x, y, z axes are rotating at $\boldsymbol{\Omega} = \boldsymbol{\omega}$, then, from Eq. 20–6, it should be noted that $\dot{\boldsymbol{\omega}} = (\dot{\boldsymbol{\omega}})_{xyz} + \boldsymbol{\omega} \times \boldsymbol{\omega}$. Since $\boldsymbol{\omega} \times \boldsymbol{\omega} = \mathbf{0}$, $\dot{\boldsymbol{\omega}} = (\dot{\boldsymbol{\omega}})_{xyz}$. This important result indicates that the required time derivative of $\boldsymbol{\omega}$ can be obtained either by first finding the components of $\boldsymbol{\omega}$ along the x, y, z axes when these axes are oriented in a general position *and then* taking the time derivative of the magnitudes of these components, i.e., $(\dot{\boldsymbol{\omega}})_{xyz}$, or by finding the time derivative of $\boldsymbol{\omega}$ with respect to the X, Y, Z axes, i.e., $\dot{\boldsymbol{\omega}}$, and then determining the components $\dot{\omega}_x$, $\dot{\omega}_y$, $\dot{\omega}_z$. In practice, it is generally easier to compute $\dot{\omega}_x$, $\dot{\omega}_y$, $\dot{\omega}_z$ on the basis of finding $\dot{\boldsymbol{\omega}}$. See Example 21.5.

x, y, z Axes Having Motion $\Omega \neq \omega$.

To simplify the calculations for the time derivative of $\boldsymbol{\omega}$, it is often convenient to choose the x, y, z axes having an angular velocity $\boldsymbol{\Omega}$ which is different from the angular velocity $\boldsymbol{\omega}$ of the body. This is particularly suitable for the analysis of spinning tops and gyroscopes which are *symmetrical* about their spinning axes.* When this is the case, the moments and products of inertia remain constant during the motion.

Equations 21–22 are applicable for such a set of chosen axes. Each of these two vector equations may be reduced to a set of three scalar equations which are derived in a manner similar to Eqs. 21–25,† i.e.,

$$\boxed{\begin{aligned}
\Sigma M_x &= I_x \dot{\omega}_x - I_y \Omega_z \omega_y + I_z \Omega_y \omega_z \\
\Sigma M_y &= I_y \dot{\omega}_y - I_z \Omega_x \omega_z + I_x \Omega_z \omega_x \\
\Sigma M_z &= I_z \dot{\omega}_z - I_x \Omega_y \omega_x + I_y \Omega_x \omega_y
\end{aligned}} \qquad (21\text{–}26)$$

Here Ω_x, Ω_y, Ω_z represent the x, y, z components of $\boldsymbol{\Omega}$, measured from the inertial frame of reference, and $\dot{\omega}_x$, $\dot{\omega}_y$, $\dot{\omega}_z$ must be determined relative to the x, y, z axes that have the rotation $\boldsymbol{\Omega}$. See Example 21.6.

Any one of these sets of moment equations, Eqs. 21–24, 21–25, or 21–26, represents a series of three first-order nonlinear differential equations. These equations are "coupled," since the angular-velocity components are present in all the terms. Success in determining the solution for a particular problem therefore depends upon what is unknown in these equations. Difficulty certainly arises when one attempts to solve for the unknown components of $\boldsymbol{\omega}$, given the external moments as functions of time. Further complications can arise if the moment equations are coupled to the three scalar equations of translational motion, Eqs. 21–19. This can happen because of the existence of kinematic constraints which relate the rotation of the body to the translation of its mass center, as in the case of a hoop

*A detailed discussion of such devices is given in Sec. 21.5.
†See Prob. 21–41.

which rolls without slipping. Problems that require the simultaneous solution of differential equations are generally solved using numerical methods with the aid of a computer. In many engineering problems, however, we are given information about the motion of the body and are required to determine the applied moments acting on the body. Most of these problems have direct solutions, so that there is no need to resort to computer techniques.

Procedure for Analysis

Problems involving the three-dimensional motion of a rigid body can be solved using the following procedure.

Free-Body Diagram.

- Draw a *free-body diagram* of the body at the instant considered and specify the x, y, z coordinate system. The origin of this reference must be located either at the body's mass center G, or at point O, considered fixed in an inertial reference frame and located either in the body or on a massless extension of the body.

- Unknown reactive forces can be shown having a positive sense of direction.

- Depending on the nature of the problem, decide what type of rotational motion Ω the x, y, z coordinate system should have, i.e., $\Omega = 0$, $\Omega = \omega$, or $\Omega \neq \omega$. When choosing, one should keep in mind that the moment equations are simplified when the axes move in such a manner that they represent principal axes of inertia for the body at all times.

- Compute the necessary moments and products of inertia for the body relative to the x, y, z axes.

Kinematics.

- Determine the x, y, z components of the body's angular velocity and compute the time derivatives of ω.

- Note that if $\Omega = \omega$, then $\dot{\omega} = (\dot{\omega})_{xyz}$, and we can either find the components of ω along the x, y, z axes when the axes are oriented in a general position, and then take the time derivative of the magnitudes of these components, $(\dot{\omega})_{xyz}$; or we can find the time derivative of ω with respect to the X, Y, Z axes, $\dot{\omega}$, and then determine the components $\dot{\omega}_x$, $\dot{\omega}_y$, $\dot{\omega}_z$.

Equations of Motion.

- Apply either the two vector equations 21–18 and 21–22, or the six scalar component equations appropriate for the x, y, z coordinate axes chosen for the problem.

EXAMPLE 21.4

The gear shown in Fig. 21–12a has a mass of 10 kg and is mounted at an angle of $10°$ with a rotating shaft having negligible mass. If $I_z = 0.1 \text{ kg} \cdot \text{m}^2$, $I_x = I_y = 0.05 \text{ kg} \cdot \text{m}^2$, and the shaft is rotating with a constant angular velocity of $\omega = 30 \text{ rad/s}$, determine the reactions that the bearing supports A and B exert on the shaft at the instant shown.

(a)

SOLUTION

Free-Body Diagram. Fig. 21–12b. The origin of the x, y, z coordinate system is located at the gear's center of mass G, which is also a fixed point. The axes are fixed in and rotate with the gear, since these axes will then always represent the principal axes of inertia for the gear. Hence $\boldsymbol{\Omega} = \boldsymbol{\omega}$.

Kinematics. As shown in Fig. 21–12c, the angular velocity $\boldsymbol{\omega}$ of the gear is constant in magnitude and is always directed along the axis of the shaft AB. Since this vector is measured from the X, Y, Z inertial frame of reference, for any position of the x, y, z axes,

$$\omega_x = 0 \qquad \omega_y = -30 \sin 10° \qquad \omega_z = 30 \cos 10°$$

These components remain constant for any general orientation of the x, y, z axes, and so $\dot{\omega}_x = \dot{\omega}_y = \dot{\omega}_z = 0$. Also note that since $\boldsymbol{\Omega} = \boldsymbol{\omega}$, then $\dot{\boldsymbol{\omega}} = (\dot{\boldsymbol{\omega}})_{xyz}$ and we can find these time derivatives relative to the X, Y, Z axes. In this regard $\boldsymbol{\omega}$ has a constant magnitude and direction $(+Z)$ and so $\dot{\boldsymbol{\omega}} = \mathbf{0}$, and so $\dot{\omega}_x = \dot{\omega}_y = \dot{\omega}_z = 0$.
Furthermore, since G is a fixed point, $(a_G)_x = (a_G)_y = (a_G)_z = 0$.

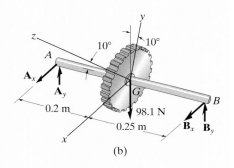

(b)

Equations of Motion. Applying Eqs. 21–25 ($\boldsymbol{\Omega} = \boldsymbol{\omega}$) yields

$$\Sigma M_x = I_x \dot{\omega}_x - (I_y - I_z)\omega_y\omega_z$$

$$-(A_Y)(0.2) + (B_Y)(0.25) = 0 - (0.05 - 0.1)(-30 \sin 10°)(30 \cos 10°)$$

$$-0.2A_Y + 0.25B_Y = -7.70 \qquad (1)$$

$$\Sigma M_y = I_y \dot{\omega}_y - (I_z - I_x)\omega_z\omega_x$$

$$A_X(0.2) \cos 10° - B_X(0.25) \cos 10° = 0 + 0$$

$$A_X = 1.25B_X \qquad (2)$$

$$\Sigma M_z = I_z \dot{\omega}_z - (I_x - I_y)\omega_x\omega_y$$

$$A_X(0.2) \sin 10° - B_X(0.25) \sin 10° = 0 + 0$$

$$A_X = 1.25B_X$$

Applying Eqs. 21–19, we have

$$\Sigma F_X = m(a_G)_X; \qquad\qquad A_X + B_X = 0 \qquad (3)$$

$$\Sigma F_Y = m(a_G)_Y; \qquad A_Y + B_Y - 98.1 = 0 \qquad (4)$$

$$\Sigma F_Z = m(a_G)_Z; \qquad\qquad\qquad 0 = 0$$

Solving Eqs. 1 through 4 simultaneously gives

$$A_X = B_X = 0 \qquad A_Y = 71.6 \text{ N} \qquad B_Y = 26.5 \text{ N} \qquad\qquad Ans.$$

(c)

Fig. 21–12

EXAMPLE 21.5

The airplane shown in Fig. 21–13a is in the process of making a steady *horizontal* turn at the rate of ω_p. During this motion, the airplane's propeller is spinning at the rate of ω_s. If the propeller has two blades, determine the moments which the propeller shaft exerts on the propeller at the instant the blades are in the vertical position. For simplicity, assume the blades to be a uniform slender bar having a moment of inertia I about an axis perpendicular to the blades and passing through their center, and having zero moment of inertia about a longitudinal axis.

(a)

SOLUTION

Free-Body Diagram.   Fig. 21–13b. The effect of the connecting shaft on the propeller is indicated by the resultants $\mathbf{F}_R$ and $\mathbf{M}_R$. (The propeller's weight is assumed to be negligible.) The x, y, z axes will be taken fixed to the propeller, since these axes always represent the principal axes of inertia for the propeller. Thus, $\mathbf{\Omega} = \boldsymbol{\omega}$. The moments of inertia I_x and I_y are equal ($I_x = I_y = I$) and $I_z = 0$.

Kinematics.   The angular velocity of the propeller observed from the X, Y, Z axes, coincident with the x, y, z axes, Fig. 21–13c, is $\boldsymbol{\omega} = \boldsymbol{\omega}_s + \boldsymbol{\omega}_p = \omega_s \mathbf{i} + \omega_p \mathbf{k}$, so that the x, y, z components of $\boldsymbol{\omega}$ are

$$\omega_x = \omega_s \qquad \omega_y = 0 \qquad \omega_z = \omega_p$$

Since $\mathbf{\Omega} = \boldsymbol{\omega}$, then $\dot{\boldsymbol{\omega}} = (\dot{\boldsymbol{\omega}})_{xyz}$. The time derivative of $\dot{\boldsymbol{\omega}}$ can be determined with respect to the fixed X, Y, Z axes. To do this, Eq. 20–6 must be used since $\boldsymbol{\omega}$ is changing direction relative to X, Y, Z. Since $\boldsymbol{\omega} = \boldsymbol{\omega}_s + \boldsymbol{\omega}_p$, then $\dot{\boldsymbol{\omega}} = \dot{\boldsymbol{\omega}}_s + \dot{\boldsymbol{\omega}}_p$. The time rate of change of each of these components relative to the X, Y, Z axes can be obtained by using a third coordinate system x', y', z', which has an angular velocity $\mathbf{\Omega}' = \boldsymbol{\omega}_p$ and is coincident with the X, Y, Z axes at the instant shown. Thus

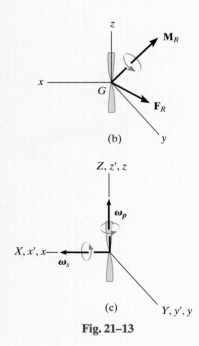

(b)

(c)

Fig. 21–13

$$\dot{\boldsymbol{\omega}} = (\dot{\boldsymbol{\omega}})_{x'y'z'} + \boldsymbol{\omega}_p \times \boldsymbol{\omega}$$

$$= (\dot{\boldsymbol{\omega}}_s)_{x'y'z'} + (\dot{\boldsymbol{\omega}}_p)_{x'y'z'} + \boldsymbol{\omega}_p \times (\boldsymbol{\omega}_s + \boldsymbol{\omega}_p)$$

$$= \mathbf{0} + \mathbf{0} + \boldsymbol{\omega}_p \times \boldsymbol{\omega}_s + \boldsymbol{\omega}_p \times \boldsymbol{\omega}_p$$

$$= \mathbf{0} + \mathbf{0} + \omega_p \mathbf{k} \times \omega_s \mathbf{i} + \mathbf{0} = \omega_p \omega_s \mathbf{j}$$

Since the X, Y, Z axes are coincident with the x, y, z axes at the instant shown, the components of $\dot{\boldsymbol{\omega}}$ along x, y, z are therefore

$$\dot{\omega}_x = 0 \qquad \dot{\omega}_y = \omega_p \omega_s \qquad \dot{\omega}_z = 0$$

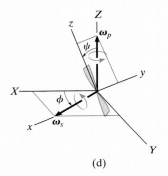

(d)

These same results can also be determined by direct calculation of $(\dot{\boldsymbol{\omega}})_{xyz}$; however, this will involve a bit more work. To do this, it will be necessary to view the propeller (or the x, y, z axes) in some *general position* such as shown in Fig. 21–13d. Here the plane has turned through an angle ϕ (phi) and the propeller has turned through an angle ψ (psi) relative to the plane. Notice that $\boldsymbol{\omega}_p$ is always directed along the fixed Z axis and $\boldsymbol{\omega}_s$ follows the x axis. Thus the general components of $\boldsymbol{\omega}$ are

$$\omega_x = \omega_s \qquad \omega_y = \omega_p \sin \psi \qquad \omega_z = \omega_p \cos \psi$$

Since ω_s and ω_p are constant, the time derivatives of these components become

$$\dot{\omega}_x = 0 \qquad \dot{\omega}_y = \omega_p \cos \psi \, \dot{\psi} \qquad \dot{\omega}_z = -\omega_p \sin \psi \, \dot{\psi}$$

But $\phi = \psi = 0°$ and $\dot{\psi} = \omega_s$ at the instant considered. Thus,

$$\omega_x = \omega_s \qquad \omega_y = 0 \qquad \omega_z = \omega_p$$

$$\dot{\omega}_x = 0 \qquad \dot{\omega}_y = \omega_p \omega_s \qquad \dot{\omega}_z = 0$$

which are the same results as those obtained previously.

Equations of Motion. Using Eqs. 21–25, we have

$$\Sigma M_x = I_x \dot{\omega}_x - (I_y - I_z)\omega_y \omega_z = I(0) - (I - 0)(0)\omega_p$$

$$M_x = 0 \qquad\qquad\qquad Ans.$$

$$\Sigma M_y = I_y \dot{\omega}_y - (I_z - I_x)\omega_z \omega_x = I(\omega_p \omega_s) - (0 - I)\omega_p \omega_s$$

$$M_y = 2I\omega_p \omega_s \qquad\qquad Ans.$$

$$\Sigma M_z = I_z \dot{\omega}_z - (I_x - I_y)\omega_x \omega_y = 0(0) - (I - I)\omega_s(0)$$

$$M_z = 0 \qquad\qquad\qquad Ans.$$

EXAMPLE 21.6

(a)

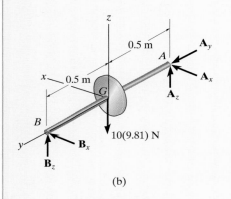

(b)

Fig. 21–14

The 10-kg flywheel (or thin disk) shown in Fig. 21–14a rotates (spins) about the shaft at a constant angular velocity of $\omega_s = 6$ rad/s. At the same time, the shaft is rotating (precessing) about the bearing at A with an angular velocity of $\omega_p = 3$ rad/s. If A is a thrust bearing and B is a journal bearing, determine the components of force reaction at each of these supports due to the motion.

SOLUTION I

Free-Body Diagram. Fig. 21–14b. The origin of the x, y, z coordinate system is located at the center of mass G of the flywheel. Here we will let these coordinates have an angular velocity of $\boldsymbol{\Omega} = \boldsymbol{\omega}_p = \{3\mathbf{k}\}$ rad/s. Although the wheel spins relative to these axes, the moments of inertia *remain constant*,* i.e.,

$$I_x = I_z = \tfrac{1}{4}(10 \text{ kg})(0.2 \text{ m})^2 = 0.1 \text{ kg} \cdot \text{m}^2$$

$$I_y = \tfrac{1}{2}(10 \text{ kg})(0.2 \text{ m})^2 = 0.2 \text{ kg} \cdot \text{m}^2$$

Kinematics. From the coincident inertial X, Y, Z frame of reference, Fig. 21–41c, the flywheel has an angular velocity of $\boldsymbol{\omega} = \{6\mathbf{j} + 3\mathbf{k}\}$ rad/s, so that

$$\omega_x = 0, \quad \omega_y = 6 \text{ rad/s}, \quad \omega_z = 3 \text{ rad/s}$$

The time derivative of $\boldsymbol{\omega}$ must be determined relative to the x, y, z axes. In this case both $\boldsymbol{\omega}_p$ and $\boldsymbol{\omega}_s$ do not change and so

$$\dot{\omega}_x = 0, \quad \dot{\omega}_y = 0, \quad \dot{\omega}_z = 0$$

Equations of Motion. Applying Eqs. 21–26 ($\boldsymbol{\Omega} \neq \boldsymbol{\omega}$) yields

$$\Sigma M_x = I_x \dot{\omega}_x - I_y \Omega_z \omega_y + I_z \Omega_y \omega_z$$

$$-A_z(0.5) + B_z(0.5) = 0 - (0.2)(3)(6) + 0 = -3.6$$

$$\Sigma M_y = I_y \dot{\omega}_y - I_z \Omega_x \omega_z + I_x \Omega_z \omega_x$$

$$0 = 0 - 0 + 0$$

$$\Sigma M_z = I_z \dot{\omega}_z - I_x \Omega_y \omega_x + I_y \Omega_x \omega_y$$

$$A_x(0.5) - B_x(0.5) = 0 - 0 + 0$$

---

*This would not be true for the propeller in Example 21.5.

Applying Eqs. 21–19, we have

$$\Sigma F_X = m(a_G)_X; \qquad A_x + B_x = 0$$
$$\Sigma F_Y = m(a_G)_Y; \qquad A_y = -10(0.5)(3)^2$$
$$\Sigma F_Z = m(a_G)_Z; \qquad A_z + B_z - 10(9.81) = 0$$

Solving these equations, we obtain

$$A_x = 0 \quad A_y = -45.0 \text{ N} \quad A_z = 52.6 \text{ N} \qquad Ans.$$
$$B_x = 0 \qquad\qquad\qquad B_z = 45.4 \text{ N} \qquad Ans.$$

NOTE: If the precession ω_p had not occurred, the reactions at A and B would be equal to 49.05 N. In this case, however, the difference in the reactions is caused by the "gyroscopic moment" created whenever a spinning body precesses about another axis. We will study this effect in detail in the next section.

SOLUTION II

This example can also be solved using Euler's equations of motion, Eqs. 21–25. In this case $\Omega = \omega = \{6\mathbf{j} + 3\mathbf{k}\}$ rad/s, and the time derivative $(\dot{\omega})_{xyz}$ can be conveniently obtained with reference to the fixed X, Y, Z axes since $\dot{\omega} = (\dot{\omega})_{xyz}$. This calculation can be performed by choosing x', y', z' axes to have an angular velocity of $\Omega' = \omega_p$, Fig. 21–14c, so that

$\omega_p = 3$ rad/s

Z, z, z'

X, x, x'

$\omega_s = 6$ rad/s

Y, y, y'

(c)

$$\dot{\omega} = (\dot{\omega})_{x'y'z'} + \omega_p \times \omega = 0 + 3\mathbf{k} \times (6\mathbf{j} + 3\mathbf{k}) = \{-18\mathbf{i}\} \text{ rad/s}^2$$
$$\dot{\omega}_x = -18 \text{ rad/s} \quad \dot{\omega}_y = 0 \quad \dot{\omega}_z = 0$$

The moment equations then become

$$\Sigma M_x = I_x \dot{\omega}_x - (I_y - I_z)\omega_y \omega_z$$
$$-A_x(0.5) + B_z(0.5) = 0.1(-18) - (0.2 - 0.1)(6)(3) = -3.6$$
$$\Sigma M_y = I_y \dot{\omega}_y - (I_z - I_x)\omega_z \omega_x$$
$$0 = 0 - 0$$
$$\Sigma M_z = I_z \dot{\omega}_z - (I_x - I_y)\omega_x \omega_y$$
$$A_x(0.5) - B_x(0.5) = 0 - 0$$

The solution then proceeds as before.

PROBLEMS

21–39. Derive the scalar form of the rotational equation of motion along the x axis when $\mathbf{\Omega} \neq \boldsymbol{\omega}$ and the moments and products of inertia of the body are *not constant* with respect to time.

*21–40. Derive the scalar form of the rotational equation of motion along the x axis when $\mathbf{\Omega} \neq \boldsymbol{\omega}$ and the moments and products of inertia of the body are *constant* with respect to time.

21–41. Derive the Euler equations of motion for $\mathbf{\Omega} \neq \boldsymbol{\omega}$, i.e., Eqs. 21–26.

21–42. The 40-kg flywheel (disk) is mounted 20 mm off its true center at G. If the shaft is rotating at a constant speed $\omega = 8$ rad/s, determine the maximum reactions exerted on the journal bearings at A and B.

21–43. The 40-kg flywheel (disk) is mounted 20 mm off its true center at G. If the shaft is rotating at a constant speed $\omega = 8$ rad/s, determine the minimum reactions exerted on the journal bearings at A and B during the motion.

21–45. The 4-lb bar rests along the smooth corners of an open box. At the instant shown, the box has a velocity $\mathbf{v} = \{3\mathbf{j}\}$ ft/s and an acceleration $\mathbf{a} = \{-6\mathbf{j}\}$ ft/s². Determine the x, y, z components of force which the corners exert on the bar.

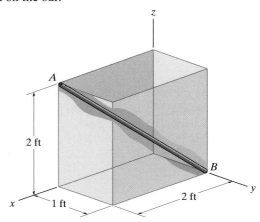

Probs. 21–44/45

21–46. The conical pendulum consists of a bar of mass m and length L that is supported by the pin at its end A. If the pin is subjected to a rotation $\boldsymbol{\omega}$, determine the angle θ that the bar makes with the vertical as it rotates. Also, determine the components of reaction at the pin.

Prob. 21–46

Probs. 21–42/43

*21–44. The 4-lb bar rests along the smooth corners of an open box. At the instant shown, the box has a velocity $\mathbf{v} = \{5\mathbf{k}\}$ ft/s and an acceleration $\mathbf{a} = \{2\mathbf{k}\}$ ft/s². Determine the x, y, z components of force which the corners exert on the bar.

21–47. The 20-lb plate is mounted on the shaft AB so that the plane of the plate makes an angle $\theta = 30°$ with the vertical. If the shaft is turning in the direction shown with an angular velocity of 25 rad/s, determine the vertical reactions at the bearing supports A and B when the plate is in the position shown.

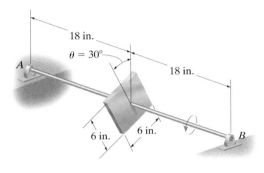

Prob. 21–47

*21–48. The car is traveling around the curved road of radius ρ such that its mass center has a constant speed v_G. Write the equations of rotational motion with respect to the x, y, z axes. Assume that the car's six moments and products of inertia with respect to these axes are known.

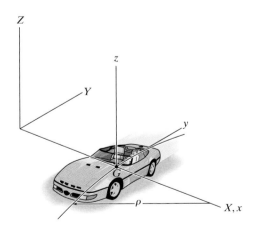

Prob. 21–48

21–49. The rod assembly is supported by journal bearings at A and B, which develops only x and y force reactions on the shaft. If the shaft AB is rotating in the direction shown at $\boldsymbol{\omega} = \{-5\mathbf{j}\}$ rad/s, determine the reactions at the bearings when the assembly is in the position shown. Also, what is the shaft's angular acceleration? The mass of each rod is 1.5 kg/m.

Prob. 21–49

21–50. The rod assembly is supported by journal bearings at A and B, which develops only x and z force reactions on the shaft. If the shaft AB is subjected to a couple moment $\mathbf{M} = \{8\mathbf{j}\}$ N · m, and at the instant shown the shaft has an angular velocity of $\boldsymbol{\omega} = \{-5\mathbf{j}\}$ rad/s, determine the reactions at the bearings of the assembly at this instant. Also, what is the shaft's angular acceleration? The mass of each rod is 1.5 kg/m.

Prob. 21–50

21–51. The rod assembly has a weight of 5 lb/ft. It is supported at B by a smooth journal bearing, which develops x and y force reactions, and at A by a smooth thrust bearing, which develops x, y, and z force reactions. If a 50-lb·ft torque is applied along rod AB, determine the components of reaction at the bearings when the assembly has an angular velocity $\omega = 10$ rad/s at the instant shown.

Prob. 21–51

*21–52. The rod AB supports the 10-lb sphere. If the rod is pinned at A to the vertical shaft which is rotating at a constant rate $\omega = \{7\mathbf{k}\}$ rad/s, determine the angle θ of the rod during the motion. Neglect the mass of the rod in the calculation.

21–53. The rod AB supports the 10-lb sphere. If the rod is pinned at A to the vertical shaft which is rotating with an angular acceleration $\alpha = \{2\mathbf{k}\}$ rad/s^2, and at the instant shown the shaft has an angular velocity $\omega = \{7\mathbf{k}\}$ rad/s, determine the angle θ of the rod at this instant. Neglect the mass of the rod in the calculation.

Probs. 21–52/53

21–54. The *thin rod* has a mass of 0.8 kg and a total length of 150 mm. It is rotating about its midpoint at a constant rate while the table to which its axle A is fastened is rotating at 2 rad/s. Determine the x, y, z moment components which the axle exerts on the rod when the rod is in any position θ.

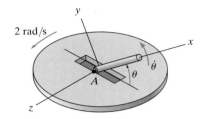

Prob. 21–54

21–55. The cylinder has a mass of 30 kg and is mounted on an axle that is supported by bearings at A and B. If the axle is turning at $\omega = \{-40\mathbf{j}\}$ rad/s, determine the vertical components of force acting at the bearings at this instant.

Prob. 21–55

*21–56. The cylinder has a mass of 30 kg and is mounted on an axle which is supported by bearings at A and B. If the axle is subjected to a couple moment $\mathbf{M} = \{-30\mathbf{j}\}$ N·m, and at the instant shown has an angular velocity $\omega = \{-40\mathbf{j}\}$ rad/s, determine the vertical components of force acting at the bearings at this instant.

Prob. 21–56

21–57. The uniform hatch door, having a mass of 15 kg and a mass center at G, is supported in the horizontal plane by bearings at A and B. If a vertical force $F = 300$ N is applied to the door as shown, determine the components of reaction at the bearings and the angular acceleration of the door. The bearing at A will resist a component of force in the y direction, whereas the bearing at B will not. For the calculation, assume the door to be a thin plate and neglect the size of each bearing. The door is originally at rest.

Prob. 21–57

21–58. The man sits on a swivel chair which is rotating with a constant angular velocity of 3 rad/s. He holds the uniform 5-lb rod AB horizontal. He suddenly gives it an angular acceleration of 2 rad/s^2, measured relative to him, as shown. Determine the required force and moment components at the grip, A, necessary to do this. Establish axes at the rod's center of mass G, with $+z$ upward, and $+y$ directed along the axis of the rod towards A.

Prob. 21–58

21–59. Four spheres are connected to shaft AB. If $m_C = 1$ kg and $m_E = 2$ kg, determine the mass of D and F and the angles of the rods, θ_D and θ_F, so that the shaft is dynamically balanced, that is, so that the bearings at A and B exert only vertical reactions on the shaft as it rotates. Neglect the mass of the rods.

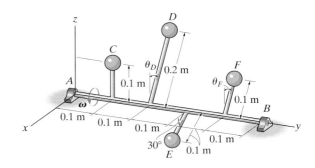

Prob. 21–59

*21–60. The bent uniform rod ACD has a weight of 5 lb/ft and is supported at A by a pin and at B by a cord. If the vertical shaft rotates with a constant angular velocity $\omega = 20$ rad/s, determine the x, y, z components of force and moment developed at A and the tension in the cord.

Prob. 21–60

(a)

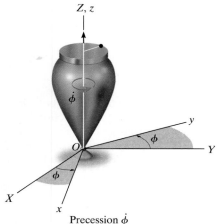

Precession $\dot{\phi}$

(b)

*21.5   Gyroscopic Motion

In this section the equations defining the motion of a body (top) which is symmetrical with respect to an axis and moving about a fixed point lying on the axis will be developed. These equations will then be applied to study the motion of a particularly interesting device, the gyroscope.

The body's motion will be analyzed using *Euler angles* ϕ, θ, ψ (phi, theta, psi). To illustrate how these angles define the position of a body, consider the top shown in Fig. 21–15a. The top is attached to point O and has an orientation relative to the fixed X, Y, Z axes at some instant of time as shown in Fig. 21–15d. To define this final position, a second set of x, y, z axes is needed. For purposes of discussion, assume that this reference is fixed in the top. Starting with the X, Y, Z and x, y, z axes in coincidence, Fig. 21–15a, the final position of the top is determined using the following three steps:

1. Rotate the top about the Z (or z) axis through an angle ϕ ($0 \le \phi < 2\pi$), Fig. 21–15b.

2. Rotate the top about the x axis through an angle θ ($0 \le \theta \le \pi$), Fig. 21–15c.

3. Rotate the top about the z axis through an angle ψ ($0 \le \psi < 2\pi$) to obtain the final position, Fig. 20-15d.

The sequence of these three angles, ϕ, θ, then ψ, must be maintained, since finite rotations are *not vectors* (see Fig. 20–1). Although this is the case, the differential rotations $d\phi$, $d\theta$, and $d\psi$ are vectors, and thus the angular velocity $\boldsymbol{\omega}$ of the top can be expressed in terms of the time derivatives of the Euler angles. The angular-velocity components $\dot{\phi}$, $\dot{\theta}$, and $\dot{\psi}$ are known as the *precession, nutation,* and *spin,* respectively.

Nutation $\dot{\theta}$

(c)

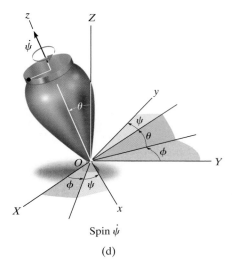

Spin $\dot{\psi}$

(d)

Fig. 21–15

Their positive directions are shown in Fig. 21–16. It is seen that these vectors are not all perpendicular to one another; however, $\boldsymbol{\omega}$ of the top can still be expressed in terms of these three components.

In our case the body (top) is symmetric with respect to the z or spin axis. If we consider the top to be oriented so that at the instant considered the spin angle $\psi = 0$ and *the x, y, z axes follow the motion of the body only in nutation and precession*, i.e., $\boldsymbol{\Omega} = \boldsymbol{\omega}_p + \boldsymbol{\omega}_n$, then the nutation and spin are always directed along the x and z axes, respectively, Fig. 21–16. Hence, the angular velocity of the body is specified only in terms of the Euler angle θ, i.e.,

$$\boldsymbol{\omega} = \omega_x \mathbf{i} + \omega_y \mathbf{j} + \omega_z \mathbf{k}$$
$$= \dot{\theta}\mathbf{i} + (\dot{\phi} \sin \theta)\mathbf{j} + (\dot{\phi} \cos \theta + \dot{\psi})\mathbf{k} \qquad (21\text{–}27)$$

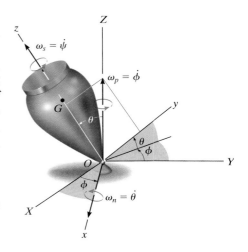

Fig. 21–16

Since motion of the axes is not affected by the spin component,

$$\boldsymbol{\Omega} = \Omega_x \mathbf{i} + \Omega_y \mathbf{j} + \Omega_z \mathbf{k}$$
$$= \dot{\theta}\mathbf{i} + (\dot{\phi} \sin \theta)\mathbf{j} + (\dot{\phi} \cos \theta)\mathbf{k} \qquad (21\text{–}28)$$

The x, y, z axes in Fig. 21–16 represent *principal axes of inertia* of the body for *any* spin of the body about these axes. Hence, the moments of inertia are constant and will be represented as $I_{xx} = I_{yy} = I$ and $I_{zz} = I_z$. Since $\boldsymbol{\Omega} \neq \boldsymbol{\omega}$, Eqs. 21–26 are used to establish the rotational equations of motion. Substituting into these equations the respective angular-velocity components defined by Eqs. 21–27 and 21–28, their corresponding time derivatives, and the moment of inertia components, yields

$$\Sigma M_x = I(\ddot{\theta} - \dot{\phi}^2 \sin \theta \cos \theta) + I_z \dot{\phi} \sin \theta (\dot{\phi} \cos \theta + \dot{\psi})$$
$$\Sigma M_y = I(\ddot{\phi} \sin \theta + 2\dot{\phi}\dot{\theta} \cos \theta) - I_z \dot{\theta}(\dot{\phi} \cos \theta + \dot{\psi}) \qquad (21\text{–}29)$$
$$\Sigma M_z = I_z(\ddot{\psi} + \ddot{\phi} \cos \theta - \dot{\phi}\dot{\theta} \sin \theta)$$

Each moment summation applies only at the fixed point O or the center of mass G of the body. Since the equations represent a coupled set of nonlinear second-order differential equations, in general a closed-form solution may not be obtained. Instead, the Euler angles ϕ, θ, and ψ may be obtained graphically as functions of time using numerical analysis and computer techniques.

A special case, however, does exist for which simplification of Eqs. 21–29 is possible. Commonly referred to as *steady precession*, it occurs when the nutation angle θ, precession $\dot{\phi}$, and spin $\dot{\psi}$ all remain *constant*. Equations 21–29 then reduce to the form

$$\boxed{\Sigma M_x = -I\dot{\phi}^2 \sin \theta \cos \theta + I_z \dot{\phi} \sin \theta (\dot{\phi} \cos \theta + \dot{\psi})} \qquad (21\text{–}30)$$

$$\Sigma M_y = 0$$
$$\Sigma M_z = 0$$

Equation 21–30 may be further simplified by noting that, from Eq. 21–27, $\omega_z = \dot{\phi} \cos \theta + \dot{\psi}$, so that

$$\Sigma M_x = -I\dot{\phi}^2 \sin \theta \cos \theta + I_z \dot{\phi} \sin \theta \omega_z$$

or

$$\Sigma M_x = \dot{\phi} \sin \theta (I_z \omega_z - I\dot{\phi} \cos \theta) \qquad (21\text{--}31)$$

It is interesting to note what effects the spin $\dot{\psi}$ has on the moment about the x axis. In this regard consider the spinning rotor shown in Fig. 21–17. Here $\theta = 90°$, in which case Eq. 21–30 reduces to the form

$$\Sigma M_x = I_z \dot{\phi} \dot{\psi}$$

or

$$\Sigma M_x = I_z \Omega_y \omega_z \qquad (21\text{--}32)$$

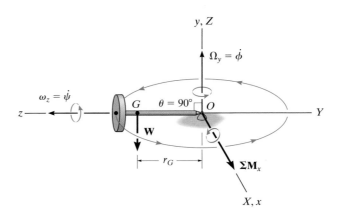

Fig. 21–17

From the figure it is seen that vectors $\Sigma \mathbf{M}_x, \mathbf{\Omega}_y$, and $\boldsymbol{\omega}_z$ all act along their respective *positive axes* and therefore are mutually perpendicular. Instinctively, one would expect the rotor to fall down under the influence of gravity! However, this is not the case at all, provided the product $I_z \Omega_y \omega_z$ is correctly chosen to counterbalance the moment $\Sigma M_x = W r_G$ of the rotor's weight about O. This unusual phenomenon of rigid-body motion is often referred to as the *gyroscopic effect*.

Perhaps a more intriguing demonstration of the gyroscopic effect comes from studying the action of a *gyroscope*, frequently referred to as a *gyro*. A gyro is a rotor which spins at a very high rate about its axis of symmetry. This rate of spin is considerably greater than its precessional rate of rotation about the vertical axis. Hence, for all practical purposes, the angular momentum of the gyro can be assumed directed among its axis of spin. Thus, for the gyro rotor shown in Fig. 21–18, $\omega_z \gg \Omega_y$, and the magnitude of the angular momentum about point O, as determined from Eqs. 21–11, reduces to the form $H_O = I_z \omega_z$. Since both the magnitude and direction of $\mathbf{H}_O$ are constant as observed from x, y, z, direct application of Eq. 21–22 yields

Fig. 21–18

$$\boxed{\Sigma \mathbf{M}_x = \Omega_y \times \mathbf{H}_O} \tag{21–33}$$

Using the right-hand rule applied to the cross product, it is seen that Ω_y always swings $\mathbf{H}_O$ (or $\boldsymbol{\omega}_z$) toward the sense of $\Sigma \mathbf{M}_x$. In effect, the *change in direction* of the gyro's angular momentum, $d\mathbf{H}_O$, is equivalent to the angular impulse caused by the gyro's weight about O, i.e., $d\mathbf{H}_O = \Sigma \mathbf{M}_x \, dt$, Eq. 21–20. Also, since $H_O = I_z \omega_z$ and $\Sigma M_x, \Omega_y$, and H_O are mutually perpendicular, Eq. 21–33 reduces to Eq. 21–32.

When a gyro is mounted in gimbal rings, Fig. 21–19, it becomes *free* of external moments applied to its base. Thus, in theory, its angular momentum $\mathbf{H}$ will never precess but, instead, maintain its same fixed orientation along the axis of spin when the base is rotated. This type of gyroscope is called a *free gyro* and is useful as a gyrocompass when the spin axis of the gyro is directed north. In reality, the gimbal mechanism is never completely free of friction, so such a device is useful only for the local navigation of ships and aircraft. The gyroscopic effect is also useful as a means of stabilizing both the rolling motion of ships at sea and the trajectories of missiles and projectiles. Furthermore, this effect is of significant importance in the design of shafts and bearings for rotors which are subjected to forced precessions.

The spinning of the gyro within the frame of this toy gyroscope produces angular momentum $\mathbf{H}_O$, which is changing direction as the frame precesses $\boldsymbol{\omega}_p$ about the vertical axis. The gyroscope will not fall down since the moment of its weight $\mathbf{W}$ about the support is balanced by the change in the direction of $\mathbf{H}_O$.

Fig. 21–19

EXAMPLE 21.7

$\omega_s = 100$ rad/s

G 60°

$\omega_p = \dot{\phi}$

O

50 mm

(a)

Fig. 21–20

The top shown in Fig. 21–20a has a mass of 0.5 kg and is precessing about the vertical axis at a constant angle of $\theta = 60°$. If it spins with an angular velocity $\omega_s = 100$ rad/s, determine the precessional velocity ω_p. Assume that the axial and transverse moments of inertia of the top are $0.45(10^{-3})$ kg·m^2 and $1.20(10^{-3})$ kg·m^2, respectively, measured with respect to the fixed point O.

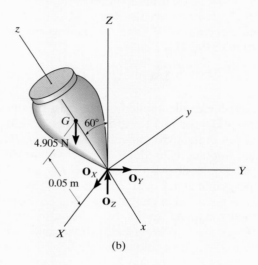

(b)

SOLUTION

Equation 21–30 will be used for the solution since the motion is *steady precession*. As shown on the free-body diagram, Fig. 21–20b, the coordinate axes are established in the usual manner, that is, with the positive z axis in the direction of spin, the positive Z axis in the direction of precession, and the positive x axis in the direction of the moment $\Sigma \mathbf{M}_x$ (refer to Fig. 21–16). Thus,

$$\Sigma M_x = -I\dot{\phi}^2 \sin\theta \cos\theta + I_z\dot{\phi}\sin\theta(\dot{\phi}\cos\theta + \dot{\psi})$$

$$4.905 \text{ N}(0.05 \text{ m})\sin 60° = -[1.20(10^{-3}) \text{ kg·m}^2 \, \dot{\phi}^2]\sin 60° \cos 60°$$

$$+ [0.45(10^{-3}) \text{ kg·m}^2]\dot{\phi}\sin 60°(\dot{\phi}\cos 60° + 100 \text{ rad/s})$$

or

$$\dot{\phi}^2 - 120.0\dot{\phi} + 654.0 = 0 \qquad (1)$$

Solving this quadratic equation for the precession gives

$$\dot{\phi} = 114 \text{ rad/s} \qquad \text{(high precession)} \qquad \qquad Ans.$$

and

$$\dot{\phi} = 5.72 \text{ rad/s} \qquad \text{(low precession)} \qquad \qquad Ans.$$

NOTE: In reality, low precession of the top would generally be observed, since high precession would require a larger kinetic energy.

EXAMPLE 21.8

The 1-kg disk shown in Fig. 21–21a is spinning about its axis with a constant angular velocity $\omega_D = 70$ rad/s. The block at B has a mass of 2 kg, and by adjusting its position s one can change the precession of the disk about its supporting pivot at O. Determine the position s that will enable the disk to have a constant precessional velocity $\omega_p = 0.5$ rad/s about the pivot. Neglect the weight of the shaft.

(a)

(b)

(c)   **Fig. 21–21**

SOLUTION

The free-body diagram of the disk is shown in Fig. 21–21b, where $\mathbf{F}$ represents the force reaction of the shaft on the disk. The origin for both the x, y, z and X, Y, Z coordinate systems is located at point O, which represents a *fixed point* for the disk. (Although point O does not lie on the disk, imagine a massless extension of the disk to this point.) In the conventional sense, the Z axis is chosen along the axis of precession, and the z axis is along the axis of spin, so that $\theta = 90°$. Since the precession is *steady*, Eq. 21–31 may be used for the solution. This equation reduces to

$$\Sigma M_x = \dot{\phi} I_z \omega_z$$

which is the same as Eq. 21–32. Substituting the required data gives

$$9.81 \text{ N}(0.2 \text{ m}) - F(0.2 \text{ m}) = 0.5 \text{ rad/s}\left[\tfrac{1}{2}(1 \text{ kg})(0.05 \text{ m})^2\right](-70 \text{ rad/s})$$

$$F = 10.0 \text{ N}$$

As shown on the free-body diagram of the shaft and block B, Fig. 21–21c, summing moments about the x axis requires

$$(19.62 \text{ N})s = (10.0 \text{ N})(0.2 \text{ m})$$

$$= 0.102 \text{ m} = 102 \text{ mm} \qquad Ans.$$

*21.6 Torque-Free Motion

When the only external force acting on a body is caused by gravitation, the general motion of the body is referred to as *torque-free motion*. This type of motion is characteristic of planets, artificial satellites, and projectiles—provided the effects of air friction are neglected.

In order to describe the characteristics of this motion, the distribution of the body's mass will be assumed *axisymmetric*. The satellite shown in Fig. 21–22 is an example of such a body, where the z axis represents an axis of symmetry. The origin of the x, y, z coordinates is located at the mass center G, such that $I_{zz} = I_z$ and $I_{xx} = I_{yy} = I$ for the body. Since gravitation is the only external force present, the summation of moments about the mass center is zero. From Eq. 21–21, this requires the angular momentum of the body to be constant, i.e.,

$$\mathbf{H}_G = \text{constant}$$

At the instant considered, it will be assumed that the inertial frame of reference is oriented so that the positive Z axis is directed along $\mathbf{H}_G$ and the y axis lies in the plane formed by the z and Z axes, Fig. 21–22. The Euler angle formed between Z and z is θ, and therefore, with this choice of axes the angular momentum may be expressed as

$$\mathbf{H}_G = H_G \sin \theta \mathbf{j} + H_G \cos \theta \mathbf{k}$$

Furthermore, using Eqs. 21–11, we have

$$\mathbf{H}_G = I\omega_x \mathbf{i} + I\omega_y \mathbf{j} + I_z \omega_z \mathbf{k}$$

where ω_x, ω_y, ω_z represent the x, y, z components of the body's angular velocity. Equating the respective $\mathbf{i}$, $\mathbf{j}$, and $\mathbf{k}$ components of the above two equations yields

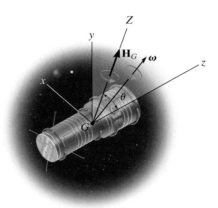

Fig. 21–22

$$\omega_x = 0 \qquad \omega_y = \frac{H_G \sin \theta}{I} \qquad \omega_z = \frac{H_G \cos \theta}{I_z} \qquad (21\text{--}34)$$

or

$$\boldsymbol{\omega} = \frac{H_G \sin \theta}{I}\mathbf{j} + \frac{H_G \cos \theta}{I_z}\mathbf{k} \qquad (21\text{--}35)$$

In a similar manner, equating the respective $\mathbf{i}$, $\mathbf{j}$, $\mathbf{k}$ components of Eq. 21–27 to those of Eq. 21–34, we obtain

$$\dot{\theta} = 0$$

$$\dot{\phi} \sin \theta = \frac{H_G \sin \theta}{I}$$

$$\dot{\phi} \cos \theta + \dot{\psi} = \frac{H_G \cos \theta}{I_z}$$

Solving, we get

$$\theta = \text{constant}$$
$$\dot{\phi} = \frac{H_G}{I} \qquad\qquad (21\text{--}36)$$
$$\dot{\psi} = \frac{I - I_z}{I I_z} H_G \cos \theta$$

Thus, for torque-free motion of an axisymmetrical body, the angle θ formed between the angular-momentum vector and the spin of the body remains constant. Furthermore, the angular momentum $\mathbf{H}_G$, precession $\dot{\phi}$, and spin $\dot{\psi}$ for the body remain constant at all times during the motion. Eliminating H_G from the second and third of Eqs. 21–36 yields the following relationship between the spin and precession:

$$\dot{\psi} = \frac{I - I_z}{I_z}\dot{\phi}\cos \theta \qquad (21\text{--}37)$$

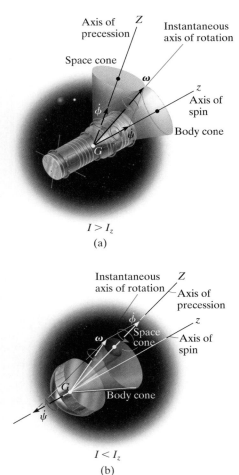

$I > I_z$

(a)

$I < I_z$

(b)

Fig. 21–23

As shown in Fig. 21–23a, the body precesses about the Z axis, which is fixed in direction, while it spins about the z axis. These two components of angular motion may be studied by using a simple cone model, introduced in Sec. 20.1. The *space cone* defining the precession is fixed from rotating, since the precession has a fixed direction, while the *body cone* rotates around the space cone's outer surface without slipping. Try to imagine this motion. The interior angle of each cone is chosen such that the resultant angular velocity of the body is directed along the line of contact of the two cones. This line of contact represents the instantaneous axis of rotation for the body cone, and hence the angular velocity of both the body cone and the body must be directed along this line. Since the spin is a function of the moments of inertia I and I_z of the body, Eq. 21–36, the cone model in Fig. 21–23a is satisfactory for describing the motion, provided $I > I_z$. Torque-free motion which meets these requirements is called *regular precession*. If $I < I_z$, the spin is negative and the precession positive. This motion is represented by the satellite motion shown in Fig. 21–23b $(I < I_z)$. The cone model may again be used to represent the motion; however, to preserve the correct vector addition of spin and precession to obtain the angular velocity $\boldsymbol{\omega}$, the inside surface of the body cone must roll on the outside surface of the (fixed) space cone. This motion is referred to as *retrograde precession*.

Satellites are often given a spin before they are launched. If their angular momentum is not collinear with the axis of spin, they will exhibit precession. In the photo on the left, regular precession will occur since $I > I_z$, and in the photo on the right, retrograde precession will occur since $I < I_z$.

EXAMPLE 21.9

The motion of a football is observed using a slow-motion projector. From the film, the spin of the football is seen to be directed 30° from the horizontal, as shown in Fig. 21–24a. Also, the football is precessing about the vertical axis at a rate $\dot{\phi} = 3$ rad/s. If the ratio of the axial to transverse moments of inertia of the football is $\frac{1}{3}$, measured with respect to the center of mass, determine the magnitude of the football's spin and its angular velocity. Neglect the effect of air resistance.

(a)   (b)

Fig. 21–24

SOLUTION

Since the weight of the football is the only force acting, the motion is torque-free. In the conventional sense, if the z axis is established along the axis of spin and the Z axis along the precession axis, as shown in Fig. 21–24b, then the angle $\theta = 60°$. Applying Eq. 21–37, the spin is

$$\dot{\psi} = \frac{I - I_z}{I_z}\dot{\phi}\cos\theta = \frac{I - \frac{1}{3}I}{\frac{1}{3}I}(3)\cos 60°$$

$$= 3 \text{ rad/s} \qquad\qquad Ans.$$

Using Eqs. 21–34, where $H_G = \dot{\phi}I$ (Eq. 21–36), we have

$$\omega_x = 0$$

$$\omega_y = \frac{H_G \sin\theta}{I} = \frac{3I \sin 60°}{I} = 2.60 \text{ rad/s}$$

$$\omega_z = \frac{H_G \cos\theta}{I_z} = \frac{3I \cos 60°}{\frac{1}{3}I} = 4.50 \text{ rad/s}$$

Thus,

$$\omega = \sqrt{(\omega_x)^2 + (\omega_y)^2 + (\omega_z)^2}$$

$$= \sqrt{(0)^2 + (2.60)^2 + (4.50)^2}$$

$$= 5.20 \text{ rad/s} \qquad\qquad Ans.$$

PROBLEMS

21–61. Show that the angular velocity of a body, in terms of Euler angles ϕ, θ, and ψ, may be expressed as $\omega = (\dot{\phi} \sin \theta \sin \psi + \dot{\theta} \cos \psi)\mathbf{i} + (\dot{\phi} \sin \theta \cos \psi - \dot{\theta} \sin \psi)\mathbf{j} + (\dot{\phi} \cos \theta + \dot{\psi})\mathbf{k}$, where $\mathbf{i}$, $\mathbf{j}$, and $\mathbf{k}$ are directed along the x, y, z axes as shown in Fig. 21–15d.

21–62. A thin rod is initially coincident with the Z axis when it is given three rotations defined by the Euler angles $\phi = 30°$, $\theta = 45°$, and $\psi = 60°$. If these rotations are given in the order stated, determine the coordinate direction angles α, β, γ of the axis of the rod with respect to the X, Y, and Z axes. Are these directions the same for any order of the rotations? Why?

21–63. The turbine on a ship has a mass of 400 kg and is mounted on bearings A and B as shown. Its center of mass is at G, its radius of gyration is $k_z = 0.3$ m, and $k_x = k_y = 0.5$ m. If it is spinning at 200 rad/s, determine the vertical reactions at the bearings when the ship undergoes each of the following motions: (a) rolling, $\omega_1 = 0.2$ rad/s, (b) turning, $\omega_2 = 0.8$ rad/s, (c) pitching, $\omega_3 = 1.4$ rad/s.

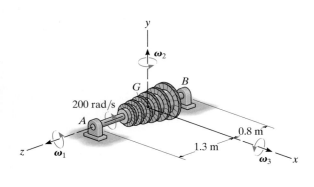

Prob. 21–63

*21–64. An airplane descends at a steep angle and then levels off horizontally to land. If the propeller is turning clockwise when observed from the rear of the plane, determine the direction in which the plane tends to turn as caused by the gyroscopic effect as it levels off.

21–65. The propeller on a single-engine airplane has a mass of 15 kg and a centroidal radius of gyration of 0.3 m computed about the axis of spin. When viewed from the front of the airplane, the propeller is turning clockwise at 350 rad/s about the spin axis. If the airplane enters a vertical curve having a radius of 80 m and is traveling at 200 km/h, determine the gyroscopic bending moment which the propeller exerts on the bearings of the engine when the airplane is in its lowest position.

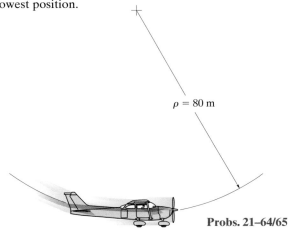

$\rho = 80$ m

Probs. 21–64/65

21–66. The rotor assembly on the engine of a jet airplane consists of the turbine, drive shaft, and compressor. The total mass is 700 kg, the radius of gyration about the shaft axis is $k_{AB} = 0.35$ m, and the mass center is at G. If the rotor has an angular velocity $\omega_{AB} = 1000$ rad/s, and the plane is pulling out of a vertical curve while traveling at 250 m/s, determine the components of reaction at the bearings A and B due to the gyroscopic effect.

$\rho = 1.30$ km

250 m/s

$\omega_{AB} = 1000$ rad/s

0.8 m   0.4 m

Prob. 21–66

21–67. The motor weighs 50 lb and has a radius of gyration of 0.2 ft about the z axis. The shaft of the motor is supported by bearings at A and B, and is turning at a constant rate of $\boldsymbol{\omega}_s = \{100\mathbf{k}\}$ rad/s, while the frame has an angular velocity of $\boldsymbol{\omega}_y = \{2\mathbf{j}\}$ rad/s. Determine the moment which the bearing forces at A and B exert on the shaft due to this motion.

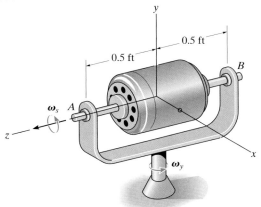

Prob. 21–67

21–69. A wheel of mass m and radius r rolls with constant spin $\boldsymbol{\omega}$ about a circular path having a radius a. If the angle of inclination is θ, determine the rate of precession. Treat the wheel as a thin ring. No slipping occurs.

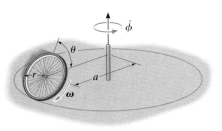

Prob. 21–69

*21–68. The conical top has a mass of 0.8 kg, and the moments of inertia are $I_x = I_y = 3.5(10^{-3})$ kg·m² and $I_z = 0.8(10^{-3})$ kg·m². If it spins freely in the ball-and-socket joint at A with an angular velocity $\omega_s = 750$ rad/s, compute the precession of the top about the axis of the shaft AB.

Prob. 21–68

21–70. The top consists of a thin disk that has a weight of 8 lb and a radius of 0.3 ft. The rod has a negligible mass and a length of 0.5 ft. If the top is spinning with an angular velocity $\omega_s = 300$ rad/s, determine the steady-state precessional angular velocity ω_p of the rod when $\theta = 40°$.

21–71. Solve Prob. 21–70 when $\theta = 90°$.

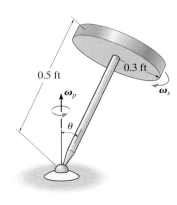

Probs. 21–70/71

*21–72.   The top has a mass of 3 lb and can be considered as a solid cone. If it is observed to precess about the vertical axis at a constant rate of 5 rad/s, determine its spin.

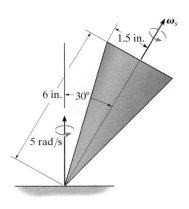

Prob. 21–72

21–74.   The car is traveling at $v_C = 100$ km/h around the horizontal curve having a radius of 80 m. If each wheel has a mass of 16 kg, a radius of gyration $k_G = 300$ mm about its spinning axis, and a radius of 400 mm, determine the difference between the normal forces of the rear wheels, caused by the gyroscopic effect. The distance between the wheels is 1.30 m.

Prob. 21–74

21–73.   The toy gyroscope consists of a rotor R which is attached to the frame of negligible mass. If it is observed that the frame is precessing about the pivot point O at $\omega_p = 2$ rad/s, determine the angular velocity ω_R of the rotor. The stem OA moves in the horizontal plane. The rotor has a mass of 200 g and a radius of gyration $k_{OA} = 20$ mm about OA.

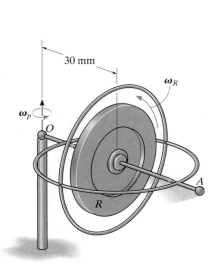

Prob. 21–73

21–75.   The projectile shown is subjected to torque-free motion. The transverse and axial moments of inertia are I and I_z, respectively. If θ represents the angle between the precessional axis Z and the axis of symmetry z, and β is the angle between the angular velocity $\boldsymbol{\omega}$ and the z axis, show that β and θ are related by the equation $\tan \theta = (I/I_z) \tan \beta$.

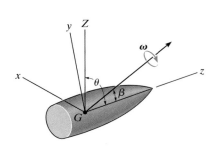

Prob. 21–75

*21–76. While the rocket is in free flight, it has a spin of 3 rad/s and precesses about an axis measured 10° from the axis of spin. If the ratio of the axial to transverse moments of inertia of the rocket is 1/15, computed about axes which pass through the mass center G, determine the angle which the resultant angular velocity makes with the spin axis. Construct the body and space cones used to describe the motion. Is the precession regular or retrograde?

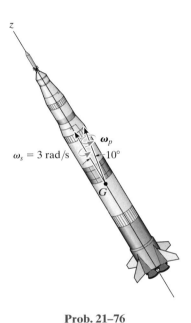

Prob. 21–76

21–77. The projectile has a mass of 0.9 kg and axial and transverse radii of gyration of $k_z = 20$ mm and $k_t = 25$ mm, respectively. If it is spinning at $\omega_s = 6$ rad/s when it leaves the barrel of a gun, determine its angular momentum. Precession occurs about the Z axis.

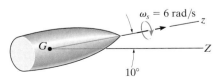

Prob. 21–77

21–78. The satellite has a mass of 1.8 Mg, and about axes passing through the mass center G the axial and transverse radii of gyration are $k_z = 0.8$ m and $k_t = 1.2$ m, respectively. If it is spinning at $\omega_s = 6$ rad/s when it is launched, determine its angular momentum. Precession occurs about the Z axis.

Prob. 21–78

21–79. The 4-kg disk is thrown with a spin $\omega_z = 6$ rad/s. If the angle θ is measured as 160°, determine the precession about the Z axis.

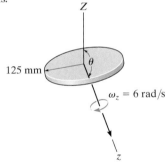

Prob. 21–79

*21–80. The radius of gyration about an axis passing through the axis of symmetry of the 1.6-Mg space capsule is $k_z = 1.2$ m, and about any transverse axis passing through the center of mass G, $k_t = 1.8$ m. If the capsule has a known steady-state precession of two revolutions per hour about the Z axis, determine the rate of spin about the z axis.

Prob. 21–80

Chapter Review

Moments and Products of Inertia

A body has six components of inertia for any specified x, y, z axes. Three of these are moments of inertia about each of the axes, I_x, I_y, I_z, and three are products of inertia, each defined from two orthogonal planes, I_{xy}, I_{yz}, I_{xz}. If either one or both of the planes are planes of symmetry, then the product of inertia with respect to these planes will be zero.

The moments and products of inertia can be determined by direct integration or by using tabulated values. If these quantities are to be determined with respect to axes or planes that do not pass through the mass center, then parallel-axis and parallel-plane theorems must be used. Provided the six components of inertia are known, then the moment of inertia about any axis can be determined using the inertia transformation equation.
(*See pages 567–571.*)

$$I_{xx} = \int_m r_x^2\, dm = \int_m (y^2 + z^2)\, dm \qquad I_{xy} = I_{yx} = \int_m xy\, dm$$

$$I_{yy} = \int_m r_y^2\, dm = \int_m (x^2 + z^2)\, dm \qquad I_{yz} = I_{zy} = \int_m yz\, dm$$

$$I_{zz} = \int_m r_z^2\, dm = \int_m (x^2 + y^2)\, dm \qquad I_{xz} = I_{zx} = \int_m xz\, dm$$

$$I_{Oa} = I_{xx}u_x^2 + I_{yy}u_y^2 + I_{zz}u_z^2 - 2I_{xy}u_xu_y - 2I_{yz}u_yu_z - 2I_{zx}u_zu_x$$

Principal Moments of Inertia

At any point on or off the body, the x, y, z axes can be oriented so that the products of inertia will be zero. The resulting moments of inertia are called the principal moments of inertia, one of which will be a maximum and the other a minimum moment of inertia for the body.
(*See page 570.*)

$$\begin{pmatrix} I_x & 0 & 0 \\ 0 & I_y & 0 \\ 0 & 0 & I_z \end{pmatrix}$$

Principle of Impulse and Momentum

The angular momentum for a body can be determined about any arbitrary point A.

Once the linear and angular momentum for the body have been formulated, then the principle of impulse and momentum can be used to solve problems that involve force, velocity, and time.
(*See pages 577–580.*)

$$m(\mathbf{v}_G)_1 + \Sigma \int_{t_1}^{t_2} \mathbf{F}\, dt = m(\mathbf{v}_G)_2 \qquad (\mathbf{H}_O)_1 + \Sigma \int_{t_1}^{t_2} \mathbf{M}_O\, dt = (\mathbf{H}_O)_2$$

$$\mathbf{H}_A = \boldsymbol{\rho}_{G/A} \times m\mathbf{v}_G + \mathbf{H}_G$$

Arbitrary Point

$$\mathbf{H}_O = \int_m \boldsymbol{\rho}_O \times (\boldsymbol{\omega} \times \boldsymbol{\rho}_O)\, dm \qquad \text{where}$$

Fixed Print O

$$\mathbf{H}_G = \int_m \boldsymbol{\rho}_G \times (\boldsymbol{\omega} \times \boldsymbol{\rho}_G)\, dm$$

Center of Mass

$$H_x = I_{xx}\omega_x - I_{xy}\omega_y - I_{xz}\omega_z$$

$$H_y = -I_{yx}\omega_x + I_{yy}\omega_y - I_{yz}\omega_z$$

$$H_z = -I_{zx}\omega_x - I_{zy}\omega_y + I_{zz}\omega_z$$

Principle of Work and Energy

The kinetic energy for a body is usually determined relative to a fixed point or the body's mass center.

These formulations can be used with the principle of work and energy to solve problems that involve force, velocity, and displacement. (*See pages 581–582.*)

$$T = \tfrac{1}{2}I_x\omega_x^2 + \tfrac{1}{2}I_y\omega_y^2 + \tfrac{1}{2}I_z\omega_z^2 \qquad T = \tfrac{1}{2}mv_G^2 + \tfrac{1}{2}I_x\omega_x^2 + \tfrac{1}{2}I_y\omega_y^2 + \tfrac{1}{2}I_z\omega_z^2$$

$$\text{Fixed Point} \qquad\qquad\qquad \text{Center of Mass}$$

$$T_1 + \Sigma U_{1-2} = T_2$$

Equations of Motion

There are three scalar equations of translational motion for a rigid body that moves in three dimensions.

$$\Sigma F_x = m(a_G)_x$$
$$\Sigma F_y = m(a_G)_y$$
$$\Sigma F_z = m(a_G)_z$$

The three scalar equations of rotational motion depend upon the location of the x, y, z reference. Most often, these axes are oriented so that they are principal axes of inertia. If the axes are fixed in and move with the rotation $\boldsymbol{\omega}$ of the body, then the equations are referred to as the Euler equations of motion.

$$\Sigma M_x = I_x\dot{\omega}_x - (I_y - I_z)\omega_y\omega_z$$
$$\Sigma M_y = I_y\dot{\omega}_y - (I_z - I_x)\omega_z\omega_x$$
$$\Sigma M_z = I_z\dot{\omega}_z - (I_x - I_y)\omega_x\omega_y$$
$$\boldsymbol{\Omega} = \boldsymbol{\omega}$$

A free-body diagram should always accompany the application of the equations of motion. (*See pages 588–592.*)

$$\Sigma M_x = I_x\dot{\omega}_x - I_y\Omega_z\omega_y + I_z\Omega_y\omega_z$$
$$\Sigma M_y = I_y\dot{\omega}_y - I_z\Omega_x\omega_z + I_x\Omega_z\omega_x$$
$$\Sigma M_z = I_z\dot{\omega}_z - I_x\Omega_y\omega_x + I_y\Omega_x\omega_y$$
$$\boldsymbol{\Omega} \neq \boldsymbol{\omega}$$

Gyroscopic Motion

The angular motion of a gyroscope is best described using the changes in motion of the three Euler angles. These angular velocity components are the precession $\dot{\phi}$, the nutation $\dot{\theta}$, and the spin $\dot{\psi}$.

If $\ddot{\psi} = 0$ and $\dot{\phi}$ and θ are constant, then the motion is referred to as steady precession.

It is the spin of a gyro rotor that is responsible for holding the rotor from falling downward, and instead causing it to precess about a vertical axis. This phenomenon is called the gyroscopic effect. (*See pages 602–605.*)

$$\Sigma M_x = -I\dot{\phi}^2 \sin\theta \cos\theta + I_z\dot{\phi}\sin\theta(\dot{\phi}\cos\theta + \dot{\psi})$$

$$\Sigma M_y = 0, \ \Sigma M_z = 0$$

Torque-Free Motion

A body that is only subjected to a gravitational force will have no moments on it about its mass center, and so the motion is described as torque-free motion. The angular momentum for the body will remain constant. This causes the body to have both a spin and a precession. The behavior depends upon the size of the moment of inertia of a symmetric body about the spin axis, I_z, versus that about a perpendicular axis, I. (*See pages 608–610.*)

$$\theta = \text{constant}$$

$$\dot{\phi} = \frac{H_G}{I}$$

$$\dot{\psi} = \frac{I - I_z}{II_z}H_G\cos\theta$$

Spring suspensions can induce vibrations in moving vehicles, such as this railroad car. In order to predict the behavior we must use a vibrational analysis.

22

Vibrations

CHAPTER OBJECTIVES

- To discuss undamped one-degree-of-freedom vibration of a rigid body using the equation of motion and energy methods.

- To study the analysis of undamped forced vibration and viscous damped forced vibration.

- To introduce the concept of electrical circuit analogs to study vibrational motion.

*22.1   Undamped Free Vibration

A *vibration* is the periodic motion of a body or system of connected bodies displaced from a position of equilibrium. In general, there are two types of vibration, free and forced. *Free vibration* occurs when the motion is maintained by gravitational or elastic restoring forces, such as the swinging motion of a pendulum or the vibration of an elastic rod. *Forced vibration* is caused by an external periodic or intermittent force applied to the system. Both of these types of vibration may be either damped or undamped. *Undamped* vibrations can continue indefinitely because frictional effects are neglected in the analysis. Since in reality both internal and external frictional forces are present, the motion of all vibrating bodies is actually *damped*.

Equilibrium
position

(a)

(b)

Fig. 22–1

The simplest type of vibrating motion is undamped free vibration, represented by the model shown in Fig. 22–1a. The block has a mass m and is attached to a spring having a stiffness k. Vibrating motion occurs when the block is released from a displaced position x so that the spring pulls on the block. The block will attain a velocity such that it will proceed to move out of equilibrium when $x = 0$, and provided the supporting surface is smooth, oscillation will continue indefinitely.

The time-dependent path of motion of the block may be determined by applying the equation of motion to the block when it is in the displaced position x. The free-body diagram is shown in Fig. 22–1b. The elastic restoring force $F = kx$ is always directed toward the equilibrium position, whereas the acceleration $\mathbf{a}$ is assumed to act in the direction of *positive displacement*. Noting that $a = d^2x/dt^2 = \ddot{x}$, we have

$$\xrightarrow{+} \Sigma F_x = ma_x; \qquad\qquad -kx = m\ddot{x}$$

Note that the acceleration is proportional to the block's displacement. Motion described in this manner is called *simple harmonic motion*. Rearranging the terms into a "standard form" gives

$$\ddot{x} + \omega_n^2 x = 0 \qquad\qquad (22\text{–}1)$$

The constant ω_n is called the *circular frequency* or *natural frequency*, expressed in 20 rad/s, and in this case

$$\boxed{\omega_n = \sqrt{\frac{k}{m}}} \qquad\qquad (22\text{–}2)$$

Equilibrium
position

(a)

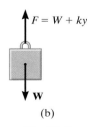

(b)

Fig. 22–2

Equation 22–1 may also be obtained by considering the block to be suspended and measuring the displacement y from the block's *equilibrium position*, Fig. 22–2a. When the block is in equilibrium, the spring exerts an upward force of $F = W = mg$ on the block. Hence, when the block is displaced a distance y downward from this position, the magnitude of the spring force is $F = W + ky$, Fig. 22–2b. Applying the equation of motion gives

$$+\downarrow \Sigma F_y = ma_y; \qquad\qquad -W - ky + W = m\ddot{y}$$

or

$$\ddot{y} + \omega_n^2 y = 0$$

which is the same form as Eq. 22–1, where ω_n is defined by Eq. 22–2.

Equation 22–1 is a homogeneous, second-order, linear, differential equation with constant coefficients. It can be shown, using the methods of differential equations, that the general solution is

$$x = A \sin \omega_n t + B \cos \omega_n t \qquad (22\text{–}3)$$

where A and B represent two constants of integration. The block's velocity and acceleration are determined by taking successive time derivatives, which yields

$$v = \dot{x} = A\omega_n \cos \omega_n t - B\omega_n \sin \omega_n t \qquad (22\text{–}4)$$

$$a = \ddot{x} = -A\omega_n^2 \sin \omega_n t - B\omega_n^2 \cos \omega_n t \qquad (22\text{–}5)$$

When Eqs. 22–3 and 22–5 are substituted into Eq. 22–1, the differential equation is indeed satisfied, showing that Eq. 22–3 is the solution to Eq. 22–1.

The integration constants A and B in Eq. 22–3 are generally determined from the initial conditions of the problem. For example, suppose that the block in Fig. 22–1a has been displaced a distance x_1 to the right from its equilibrium position and given an initial (positive) velocity $\mathbf{v}_1$ directed to the right. Substituting $x = x_1$ at $t = 0$ into Eq. 22–3 yields $B = x_1$. Since $v = v_1$ at $t = 0$, using Eq. 22–4 we obtain $A = v_1/\omega_n$. If these values are substituted into Eq. 22–3, the equation describing the motion becomes

$$x = \frac{v_1}{\omega_n} \sin \omega_n t + x_1 \cos \omega_n t \qquad (22\text{–}6)$$

Equation 22–3 may also be expressed in terms of simple sinusoidal motion. Let

$$A = C \cos \phi \qquad (22\text{–}7)$$

and

$$B = C \sin \phi \qquad (22\text{–}8)$$

where C and ϕ are new constants to be determined in place of A and B. Substituting into Eq. 22–3 yields

$$x = C \cos \phi \sin \omega_n t + C \sin \phi \cos \omega_n t$$

Since $\sin(\theta + \phi) = \sin \theta \cos \phi + \cos \theta \sin \phi$, then

$$x = C \sin(\omega_n t + \phi) \qquad (22\text{–}9)$$

If this equation is plotted on an x-versus-$\omega_n t$ axis, the graph shown in Fig. 22–3 is obtained. The maximum displacement of the block from its

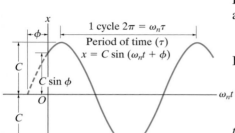

Fig. 22–3

equilibrium position is defined as the *amplitude* of vibration. From either the figure or Eq. 22–9 the amplitude is C. The angle ϕ is called the *phase angle* since it represents the amount by which the curve is displaced from the origin when $t = 0$. The constants C and ϕ are related to A and B by Eqs. 22–7 and 22–8. Squaring and adding these two equations, the amplitude becomes

$$C = \sqrt{A^2 + B^2} \qquad (22\text{–}10)$$

If Eq. 22–8 is divided by Eq. 22–7, the phase angle is

$$\phi = \tan^{-1}\frac{B}{A} \qquad (22\text{–}11)$$

Note that the sine curve, Eq. 22–9, completes one *cycle* in time $t = \tau$ (tau) when $\omega_n \tau = 2\pi$, or

$$\boxed{\tau = \frac{2\pi}{\omega_n}} \qquad (22\text{–}12)$$

This length of time is called a *period*, Fig. 22–3. Using Eq. 22–2, the period may also be represented as

$$\tau = 2\pi\sqrt{\frac{m}{k}} \qquad (22\text{–}13)$$

The *frequency f* is defined as the number of cycles completed per unit of time, which is the reciprocal of the period:

$$\boxed{f = \frac{1}{\tau} = \frac{\omega_n}{2\pi}} \qquad (22\text{–}14)$$

or

$$f = \frac{1}{2\pi}\sqrt{\frac{k}{m}} \qquad (22\text{–}15)$$

The frequency is expressed in cycles/s. This ratio of units is called a *hertz* (Hz), where 1 Hz = 1 cycle/s = 2π rad/s.

When a body or system of connected bodies is given an initial displacement from its equilibrium position and released, it will vibrate with the *natural frequency*, ω_n. Provided the body has a single degree of freedom, that is, it requires only one coordinate to specify completely the position of the system at any time, then the vibrating motion of the body will have the same characteristics as the simple harmonic motion of the block and spring just presented. Consequently, the body's motion is described by a differential equation of the same "standard form" as Eq. 22–1, i.e.,

$$\boxed{\ddot{x} + \omega_n^2 x = 0} \qquad (22\text{–}16)$$

Hence, if the natural frequency ω_n of the body is known, the period of vibration τ, natural frequency f, and other vibrating characteristics of the body can be established using Eqs. 22–3 through 22–15.

Important Points

- Free vibration occurs when the motion is maintained by gravitational or elastic restoring forces.
- The amplitude is the maximum displacement of the body.
- The period is the time required to complete one cycle.
- The frequency is the number of cycles completed per unit of time, where 1 Hz = 1 cycle/s.
- Only one position-coordinate system is needed to describe the location of a one-degree-of-freedom system.

Procedure for Analysis

As in the case of the block and spring, the natural frequency ω_n of a rigid body or system of connected rigid bodies having a single degree of freedom can be determined using the following procedure:

Free-Body Diagram.

- Draw the free-body diagram of the body when the body is displaced by a *small amount* from its equilibrium position.
- Locate the body with respect to its equilibrium position by using an appropriate *inertial coordinate q*. The acceleration of the body's mass center $\mathbf{a}_G$ or the body's angular acceleration $\boldsymbol{\alpha}$ should have a sense which is in the *positive direction* of the position coordinate.
- If the rotational equation of motion $\Sigma M_P = \Sigma(\mathcal{M}_k)_P$ is to be used, then it may be beneficial to also draw the kinetic diagram since it graphically accounts for the components $m(\mathbf{a}_G)_x$, $m(\mathbf{a}_G)_y$, and $I_G\boldsymbol{\alpha}$, and thereby makes it convenient for visualizing the terms needed in the moment sum $\Sigma(\mathcal{M}_k)_P$.

Equation of Motion.

- Apply the equation of motion to relate the elastic or gravitational *restoring* forces and couple moments acting on the body to the body's accelerated motion.

Kinematics.

- Using kinematics, express the body's accelerated motion in terms of the second time derivative of the position coordinate, $\ddot{q}$.
- Substitute the result into the equation of motion and determine ω_n by rearranging the terms so that the resulting equation is of the "standard form," $\ddot{q} + \omega_n^2 q = 0$.

EXAMPLE 22.1

(a)

$W = mg$

(b)

Fig. 22–4

Determine the period of vibration for the simple pendulum shown in Fig. 22–4a. The bob has a mass m and is attached to a cord of length l. Neglect the size of the bob.

SOLUTION

Free-Body Diagram.   Motion of the system will be related to the position coordinate $(q =) \theta$, Fig. 22–4b. When the bob is displaced by an angle θ, the *restoring force* acting on the bob is created by the weight component $mg \sin \theta$. Furthermore, $\mathbf{a}_t$ acts in the direction of *increasing s* (or θ).

Equation of Motion.   Applying the equation of motion in the *tangential direction*, since it involves the restoring force, yields

$$+\nearrow \Sigma F_t = ma_t; \qquad -mg \sin \theta = ma_t \qquad (1)$$

Kinematics. $a_t = d^2 s/dt^2 = \ddot{s}$. Furthermore, s may be related to θ by the equation $s = l\theta$, so that $a_t = l\ddot{\theta}$. Hence, Eq. 1 reduces to

$$\ddot{\theta} + \frac{g}{l} \sin \theta = 0 \qquad (2)$$

The solution of this equation involves the use of an elliptic integral. For *small displacements*, however, $\sin \theta \approx \theta$, in which case

$$\ddot{\theta} + \frac{g}{l} \theta = 0 \qquad (3)$$

Comparing this equation with Eq. 22–16 ($\ddot{x} + \omega_n^2 x = 0$), it is seen that $\omega_n = \sqrt{g/l}$. From Eq. 22–12, the period of time required for the bob to make one complete swing is therefore

$$\tau = \frac{2\pi}{\omega_n} = 2\pi \sqrt{\frac{l}{g}} \qquad \qquad Ans.$$

This interesting result, originally discovered by Galileo Galilei through experiment, indicates that the period depends only on the length of the cord and not on the mass of the pendulum bob or the angle θ.

NOTE: The solution of Eq. 3 is given by Eq. 22–3, where $\omega_n = \sqrt{g/l}$ and θ is substituted for x. Like the block and spring, the constants A and B in this problem may be determined if, for example, one knows the displacement and velocity of the bob at a given instant.

EXAMPLE 22.2

The 10-kg rectangular plate shown in Fig. 22–5a is suspended at its center from a rod having a torsional stiffness $k = 1.5 \text{ N} \cdot \text{m/rad}$. Determine the natural period of vibration of the plate when it is given a small angular displacement θ in the plane of the plate.

(a)

SOLUTION

Free-Body Diagram.   Fig. 22–5b. Since the plate is displaced in its own plane, the torsional *restoring* moment created by the rod is $M = k\theta$. This moment acts in the direction opposite to the angular displacement θ. The angular acceleration $\ddot{\theta}$ acts in the direction of *positive* θ.

Equation of Motion.

$$\Sigma M_O = I_O \alpha; \qquad -k\theta = I_O \ddot{\theta}$$

or

$$\ddot{\theta} + \frac{k}{I_O}\theta = 0$$

Since this equation is in the "standard form," the natural frequency is $\omega_n = \sqrt{k/I_O}$.

From the table on the inside back cover, the moment of inertia of the plate about an axis coincident with the rod is $I_O = \frac{1}{12}m(a^2 + b^2)$. Hence,

$$I_O = \frac{1}{12}(10 \text{ kg})[(0.2 \text{ m})^2 + (0.3 \text{ m})^2] = 0.108 \text{ kg} \cdot \text{m}^2$$

The natural period of vibration is therefore,

$$\tau = \frac{2\pi}{\omega_n} = 2\pi\sqrt{\frac{I_O}{k}} = 2\pi\sqrt{\frac{0.108}{1.5}} = 1.69 \text{ s} \qquad \textit{Ans.}$$

(b)

Fig. 22–5

EXAMPLE 22.3

(a)

(b)

(c)

Fig. 22–6

The bent rod shown in Fig. 22–6a has a negligible mass and supports a 5-kg collar at its end. Determine the natural period of vibration for the system.

SOLUTION

Free-Body and Kinetic Diagrams. Fig. 22–6b. Here the rod is displaced by a small amount θ from the equilibrium position. Since the spring is subjected to an initial compression of x_{st} for equilibrium, then when the displacement $x > x_{st}$ the spring exerts a force of $F_s = kx - kx_{st}$ on the rod. To obtain the "standard form," Eq. 22–16, $5\mathbf{a}_y$ acts *upward*, which is in accordance with positive θ displacement.

Equation of Motion. Moments will be summed about point B to eliminate the unknown reaction at this point. Since θ is small,

$$\zeta + \Sigma M_B = \Sigma(\mathcal{M}_k)_B;$$

$$kx(0.1\text{ m}) - kx_{st}(0.1\text{ m}) + 49.05\text{ N}(0.2\text{ m}) = -(5\text{ kg})a_y(0.2\text{ m})$$

The second term on the left side, $-kx_{st}(0.1\text{ m})$, represents the moment created by the spring force which is necessary to hold the collar in *equilibrium*, i.e., at $x = 0$. Since this moment is equal and opposite to the moment 49.05(0.2) created by the weight of the collar, these two terms cancel in the above equation, so that

$$kx(0.1) = -5a_y(0.2) \qquad (1)$$

Kinematics. The positions of the spring and the collar may be related to the angle θ, Fig. 22–6c. Since θ is small, $x = (0.1\text{ m})\theta$ and $y = (0.2\text{ m})\theta$. Therefore, $a_y = \ddot{y} = 0.2\ddot{\theta}$. Substituting into Eq. 1 yields

$$400(0.1\theta)\,0.1 = -5(0.2\ddot{\theta})0.2$$

Rewriting this equation in the "standard form" gives

$$\ddot{\theta} + 20\theta = 0$$

Compared with $\ddot{x} + \omega_n^2 x = 0$ (Eq. 22–16), we have

$$\omega_n^2 = 20 \qquad \omega_n = 4.47\text{ rad/s}$$

The natural period of vibration is therefore

$$\tau = \frac{2\pi}{\omega_n} = \frac{2\pi}{4.47} = 1.40\text{ s} \qquad\qquad Ans.$$

EXAMPLE 22.4

A 10-lb block is suspended from a cord that passes over a 15-lb disk, as shown in Fig. 22–7a. The spring has a stiffness $k = 200$ lb/ft. Determine the natural period of vibration for the system.

(a)   (b)

SOLUTION

Free-Body and Kinetic Diagrams. Fig. 22–7b. The *system* consists of the disk, which undergoes a rotation defined by the angle θ, and the block, which translates by an amount s. The vector $I_O\ddot{\theta}$ acts in the direction of *positive* θ, and consequently $m_B\mathbf{a}_s$ acts downward in the direction of *positive s*.

Equation of Motion. Summing moments about point O to eliminate the reactions $\mathbf{O}_x$ and $\mathbf{O}_y$, realizing that $I_O = \frac{1}{2}mr^2$, yields

$$\zeta+\Sigma M_O = \Sigma(\mathcal{M}_k)_O;$$

$$10\text{ lb}(0.75\text{ ft}) - F_s(0.75\text{ ft})$$

$$= \frac{1}{2}\left(\frac{15\text{ lb}}{32.2\text{ ft/s}^2}\right)(0.75\text{ ft})^2\ddot{\theta} + \left(\frac{10\text{ lb}}{32.2\text{ ft/s}^2}\right)a_s(0.75\text{ ft}) \qquad (1)$$

Kinematics. As shown on the kinematic diagram in Fig. 22–7c, a small positive displacement θ of the disk causes the block to lower by an amount $s = 0.75\theta$; hence, $a_s = \ddot{s} = 0.75\ddot{\theta}$. When $\theta = 0°$, the spring force required for *equilibrium* of the disk is 10 lb, acting to the right. For position θ, the spring force is $F_s = (200\text{ lb/ft})(0.75\theta\text{ ft}) + 10\text{ lb}$. Substituting these results into Eq. 1 and simplifying yields

$$\ddot{\theta} + 368\theta = 0$$

Hence,

$$\omega_n^2 = 368 \qquad \omega_n = 19.2\text{ rad/s}$$

Therefore, the natural period of vibration is

$$T = \frac{2\pi}{\omega_n} = \frac{2\pi}{19.2} = 0.328\text{ s} \qquad\qquad Ans.$$

(c)

Fig. 22–7

PROBLEMS

22–1. When a 20-lb weight is suspended from a spring, the spring is stretched a distance of 4 in. Determine the natural frequency and the period of vibration for a 10-lb weight attached to the same spring.

22–2. A spring has a stiffness of 600 N/m. If a 4-kg block is attached to the spring, pushed 50 mm above its equilibrium position, and released from rest, determine the equation which describes the block's motion. Assume that positive displacement is measured downward.

22–3. When a 3-kg block is suspended from a spring, the spring is stretched a distance of 60 mm. Determine the natural frequency and the period of vibration for a 0.2-kg block attached to the same spring.

*22–4. An 8-kg block is suspended from a spring having a stiffness $k = 80$ N/m. If the block is given an upward velocity of 0.4 m/s when it is 90 mm above its equilibrium position, determine the equation which describes the motion and the maximum upward displacement of the block measured from the equilibrium position. Assume that positive displacement is measured downward.

22–5. A 2-lb weight is suspended from a spring having a stiffness $k = 2$ lb/in. If the weight is pushed 1 in. upward from its equilibrium position and then released from rest, determine the equation which describes the motion. What is the amplitude and the natural frequency of the vibration?

22–6. A 6-lb weight is suspended from a spring having a stiffness $k = 3$ lb/in. If the weight is given an upward velocity of 20 ft/s when it is 2 in. above its equilibrium position, determine the equation which describes the motion and the maximum upward displacement of the weight, measured from the equilibrium position. Assume positive displacement is downward.

■22–7. A spring is stretched 175 mm by an 8-kg block. If the block is displaced 100 mm downward from its equilibrium position and given a downward velocity of 1.50 m/s, determine the differential equation which describes the motion. Assume that positive displacement is measured downward. Use the Runge–Kutta method to determine the position of the block, measured from its unstretched position, when $t = 0.22$ s. (See Appendix B.) Use a time increment of $\Delta t = 0.02$ s.

*22–8. If the block in Prob. 22–7 is given an upward velocity of 4 m/s when it is displaced downward a distance of 60 mm from its equilibrium position, determine the equation which describes the motion. What is the amplitude of the motion? Assume that positive displacement is measured downward.

22–9. Determine the frequency of vibration for the block. The springs are originally compressed Δ.

Prob. 22–9

22–10. A pendulum has a 0.4-m-long cord and is given a tangential velocity of 0.2 m/s toward the vertical from a position $\theta = 0.3$ rad. Determine the equation which describes the angular motion.

22–11. A platform, having an unknown mass, is supported by *four* springs, each having the same stiffness k. When nothing is on the platform, the period of vertical vibration is measured as 2.35 s; whereas if a 3-kg block is supported on the platform, the period of vertical vibration is 5.23 s. Determine the mass of a block placed on the (empty) platform which causes the platform to vibrate vertically with a period of 5.62 s. What is the stiffness k of each of the springs?

Prob. 22–11

*22–12. If the lower end of the 30-kg slender rod is displaced a small amount and released from rest, determine the natural frequency of vibration. Each spring has a stiffness of $k = 500$ N/m and is unstretched when the rod is hanging vertically.

Prob. 22–12

22–13. The body of arbitrary shape has a mass m, mass center at G, and a radius of gyration about G of k_G. If it is displaced a slight amount θ from its equilibrium position and released, determine the natural period of vibration.

Prob. 22–13

22–14. Determine to the nearest degree the maximum angular displacement of the bob if it is initially displaced $\theta = 0.2$ rad from the vertical and given a tangential velocity of 0.4 m/s away from the vertical.

Prob. 22–14

22–15. The semicircular disk weighs 20 lb. Determine the natural period of vibration if it is displaced a small amount and released.

Prob. 22–15

*22–16. The square plate has a mass m and is suspended at its corner by the pin O. Determine the natural period of vibration if it is displaced a small amount and released.

Prob. 22–16

22–17. The disk has a weight of 10 lb and rolls without slipping on the horizontal surface as it oscillates about its equilibrium position. If the disk is displaced, by rolling it counterclockwise 0.4 rad, determine the equation which describes its oscillatory motion when it is released.

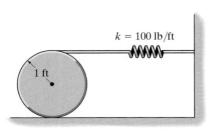

$k = 100$ lb/ft

1 ft

Prob. 22–17

22–18. The pointer on a metronome supports a 0.4-lb slider A, which is positioned at a fixed distance from the pivot O of the pointer. When the pointer is displaced, a torsional spring at O exerts a restoring torque on the pointer having a magnitude $M = (1.2\theta)$ lb · ft, where θ represents the angle of displacement from the vertical, measured in radians. Determine the natural period of vibration when the pointer is displaced a small amount θ and released. Neglect the mass of the pointer.

A

0.25 ft

O

$k = 1.2$ lb · ft/rad

Prob. 22–18

22–19. The block has a mass m and is supported by a rigid bar of negligible mass. If the spring has a stiffness k, determine the natural period of vibration for the block.

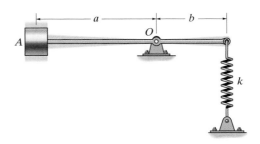

a

b

O

A

k

Prob. 22–19

*22–20. The disk, having a weight of 15 lb, is pinned at its center O and supports the block A that has a weight of 3 lb. If the belt which passes over the disk is not allowed to slip at its contacting surface, determine the natural period of vibration of the system.

0.75 ft

O

$k = 80$ lb/ft

A

Prob. 22–20

22–21. While standing in an elevator, the man holds a pendulum which consists of an 18-in. cord and a 0.5-lb bob. If the elevator is descending with an acceleration $a = 4$ ft/s², determine the natural period of vibration for small amplitudes of swing.

$a = 4$ ft/s²

Prob. 22–21

22–22. The 50-lb spool is attached to two springs. If the spool is displaced a small amount and released, determine the natural period of vibration. The radius of gyration of the spool is $k_G = 1.5$ ft. The spool rolls without slipping.

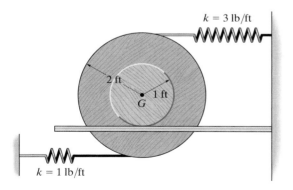

$k = 3$ lb/ft

2 ft

1 ft

G

$k = 1$ lb/ft

Prob. 22–22

22–23. Determine the natural frequency for small oscillations of the 10-lb sphere when the rod is displaced a slight distance and released. Neglect the size of the sphere and the mass of the rod. The spring has an unstretched length of 1 ft.

$k = 5$ lb/ft

1 ft

θ

1 ft

1 ft

Prob. 22–23

*22–24. The bar has a length l and mass m. It is supported at its ends by rollers of negligible mass. If it is given a small displacement and released, determine the natural frequency of vibration.

R

A

B

l

Prob. 22–24

22–25. The 25-lb weight is fixed to the end of the rod assembly. If both springs are unstretched when the assembly is in the position shown, determine the natural period of vibration for the weight when it is displaced slightly and released. Neglect the size of the block and the mass of the rods.

6 in. 6 in.

$k = 2$ lb/in.

$k = 2$ lb/in.

12 in.

Prob. 22–25

*22.2 Energy Methods

The simple harmonic motion of a body, discussed in the previous section, is due only to gravitational and elastic restoring forces acting on the body. Since these types of forces are *conservative*, it is also possible to use the conservation of energy equation to obtain the body's natural frequency or period of vibration. To show how to do this, consider again the block and spring in Fig. 22–8. When the block is displaced an arbitrary amount x from the equilibrium position, the kinetic energy is $T = \frac{1}{2}mv^2 = \frac{1}{2}m\dot{x}^2$ and the potential energy is $V = \frac{1}{2}kx^2$. Since energy is conserved, it is necessary that

$$T + V = \text{constant}$$

$$\tfrac{1}{2}m\dot{x}^2 + \tfrac{1}{2}kx^2 = \text{constant} \qquad (22\text{--}17)$$

The differential equation describing the *accelerated motion* of the block can be obtained by *differentiating* this equation with respect to time; i.e.,

$$m\dot{x}\ddot{x} + kx\dot{x} = 0$$

$$\dot{x}(m\ddot{x} + kx) = 0$$

Since the velocity $\dot{x}$ is not *always* zero in a vibrating system,

$$\ddot{x} + \omega_n^{\,2}x = 0 \qquad \omega_n = \sqrt{k/m}$$

which is the same as Eq. 22–1.

If the energy equation is written for a *system of connected bodies*, the natural frequency or the equation of motion can also be determined by time differentiation. It is *not necessary* to dismember the system to account for reactive and connective forces because they do no work.

Equilibrium
position

Fig. 22–8

The suspension of a railroad car consists of a set of springs which are mounted between the frame of the car and the wheel truck. This will give the car a natural frequency of vibration which can be determined.

Procedure for Analysis

The circular or natural frequency ω_n of a body or system of connected bodies can be determined by applying the conservation of energy equation using the following procedure.

Energy Equation.

- Draw the body when it is displaced by a *small amount* from its equilibrium position and define the location of the body from its equilibrium position by an appropriate position coordinate q.

- Formulate the equation of energy for the body, $T + V = $ constant, in terms of the position coordinate.

- In general, the kinetic energy must account for both the body's translational and rotational motion, $T = \frac{1}{2}mv_G^2 + \frac{1}{2}I_G\omega_n^2$, Eq. 18–2.

- The potential energy is the sum of the gravitational and elastic potential energies of the body, $V = V_g + V_e$, Eq. 18–16. In particular, V_g should be measured from a datum for which $q = 0$ (equilibrium position).

Time Derivative.

- Take the time derivative of the energy equation using the chain rule of calculus and factor out the common terms. The resultant differential equation represents the equation of motion for the system. The value of ω_n is obtained after rearranging the terms in the "standard form," $\ddot{q} + \omega_n^2 q = 0$.

EXAMPLE 22.5

(a)

(b)

Fig. 22–9

The thin hoop shown in Fig. 22–9a is supported by a peg at O. Determine the natural period of oscillation for small amplitudes of swing. The hoop has a mass m.

SOLUTION

Energy Equation.   A diagram of the hoop when it is displaced a small amount $(q=)\theta$ from the equilibrium position is shown in Fig. 22–9b. Using the table on the inside back cover and the parallel-axis theorem to determine I_O, we can express the kinetic energy as

$$T = \tfrac{1}{2}I_O\omega_n{}^2 = \tfrac{1}{2}[mr^2 + mr^2]\dot{\theta}^2 = mr^2\dot{\theta}^2$$

If a horizontal datum is placed through the center of gravity of the hoop when $\theta = 0$, then the center of gravity moves upward $r(1 - \cos\theta)$ in the displaced position. For *small angles*, $\cos\theta$ may be replaced by the first two terms of its power series expansion, $\cos\theta = 1 - \theta^2/2 + \cdots$. Therefore, the potential energy is

$$V = mgr\left[1 - \left(1 - \frac{\theta^2}{2}\right)\right] = mgr\frac{\theta^2}{2}$$

The total energy in the system is

$$T + V = mr^2\dot{\theta}^2 + mgr\frac{\theta^2}{2}$$

Time Derivative.

$$mr^2 2\dot{\theta}\ddot{\theta} + mgr\theta\dot{\theta} = 0$$
$$mr\dot{\theta}(2r\ddot{\theta} + g\theta) = 0$$

Since $\dot{\theta}$ is not always equal to zero, from the terms in parentheses,

$$\ddot{\theta} + \frac{g}{2r}\theta = 0$$

Hence,

$$\omega_n = \sqrt{\frac{g}{2r}}$$

so that

$$\tau = \frac{2\pi}{\omega_n} = 2\pi\sqrt{\frac{2r}{g}} \qquad\qquad Ans.$$

EXAMPLE 22.6

A 10-kg block is suspended from a cord wrapped around a 5-kg disk, as shown in Fig. 22–10a. If the spring has a stiffness $k = 200$ N/m, determine the natural period of vibration for the system.

(a)

SOLUTION

Energy Equation.   A diagram of the block and disk when they are displaced by respective amounts s and θ from the equilibrium position is shown in Fig. 22–10b. Since $s = (0.15\text{ m})\theta$, the kinetic energy of the system is

$$T = \tfrac{1}{2}m_b v_b^2 + \tfrac{1}{2}I_O \omega_d^2$$

$$= \tfrac{1}{2}(10\text{ kg})[(0.15\text{ m})\dot{\theta}]^2 + \tfrac{1}{2}\big[\tfrac{1}{2}(5\text{ kg})(0.15\text{ m})^2\big](\dot{\theta})^2$$

$$= 0.141(\dot{\theta})^2$$

Establishing the datum at the equilibrium position of the block and realizing that the spring stretches s_{st} for equilibrium, we can write the potential energy as

(b)

Fig. 22–10

$$V = \tfrac{1}{2}k(s_{st} + s)^2 - Ws$$

$$= \tfrac{1}{2}(200\text{ N/m})[s_{st} + (0.15\text{ m})\theta]^2 - 98.1\text{ N}[(0.15\text{ m})\theta]$$

The total energy for the system is, therefore,

$$T + V = 0.141(\dot{\theta})^2 + 100(s_{st} + 0.15\theta)^2 - 14.72\theta$$

Time Derivative.

$$0.281(\dot{\theta})\ddot{\theta} + 200(s_{st} + 0.15\theta)0.15\dot{\theta} - 14.72\dot{\theta} = 0$$

Since $s_{st} = 98.1/200 = 0.4905$ m, the above equation reduces to the "standard form"

$$\ddot{\theta} + 16\theta = 0$$

so that

$$\omega_n = \sqrt{16} = 4\text{ rad/s}$$

Thus,

$$\tau = \frac{2\pi}{\omega_n} = \frac{2\pi}{4} = 1.57\text{ s} \qquad\qquad Ans.$$

PROBLEMS

22–26. Solve Prob. 22–13 using energy methods.

22–27. Solve Prob. 22–15 using energy methods.

*22–28. Solve Prob. 22–16 using energy methods.

22–29. Solve Prob. 22–20 using energy methods.

22–30. The uniform rod of mass m is supported by a pin at A and a spring at B. If the end B is given a small downward displacement and released, determine the natural period of vibration.

Prob. 22–30

22–31. Determine the differential equation of motion of the 3-kg block when it is displaced slightly and released. The surface is smooth and the springs are originally unstretched.

Prob. 22–31

*22–32. Determine the natural period of vibration of the 10-lb semicircular disk.

Prob. 22–32

22–33. The 7-kg disk is pin-connected at its midpoint. Determine the natural period of vibration of the disk if the springs have sufficient tension in them to prevent the cord from slipping on the disk as it oscillates. *Hint:* Assume that the initial stretch in each spring is δ_O. This term will cancel out after taking the time derivative of the energy equation.

Prob. 22–33

22–34. The 5-lb sphere is attached to a rod of negligible mass and rests in the horizontal position. Determine the natural frequency of vibration. Neglect the size of the sphere.

Prob. 22–34

22–35. The bar has a mass of 8 kg and is suspended from two springs such that when it is in equilibrium, the springs make an angle of 45° with the horizontal as shown. Determine the natural period of vibration if the bar is pulled down a short distance and released. Each spring has a stiffness of $k = 40$ N/m.

Prob. 22–35

*22–36. Determine the natural period of vibration of the 3-kg sphere. Neglect the mass of the rod and the size of the sphere.

Prob. 22–36

22–37. The slender rod has a weight of 4 lb/ft. If it is supported in the horizontal plane by a ball-and-socket joint at A and a cable at B, determine the natural frequency of vibration when the end B is given a small horizontal displacement and then released.

Prob. 22–37

22–38. Determine the natural frequency of vibration of the 20-lb disk. Assume the disk does not slip on the inclined surface.

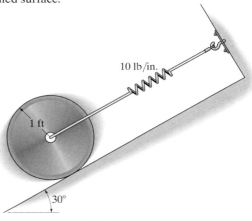

Prob. 22–38

22–39. If the disk has a mass of 8 kg, determine the natural frequency of vibration. The springs are originally unstretched.

Prob. 22–39

*22–40. Determine the differential equation of motion of the 3-kg spool. Assume that it does not slip at the surface of contact as it oscillates. The radius of gyration of the spool about its center of mass is $k_G = 125$ mm.

Prob. 22–40

*22.3   Undamped Forced Vibration

Undamped forced vibration is considered to be one of the most important types of vibrating motion in engineering work. The principles which describe the nature of this motion may be used to analyze the forces which cause vibration in many types of machines and structures.

Equilibrium position

(a)

Periodic Force.

The block and spring shown in Fig. 22–11a provide a convenient model which represents the vibrational characteristics of a system subjected to a periodic force $F = F_0 \sin \omega_0 t$. This force has an amplitude of F_0 and a *forcing frequency* ω_0. The free-body diagram for the block when it is displaced a distance x is shown in Fig. 22–11b. Applying the equation of motion yields

$$\xrightarrow{+} \Sigma F_x = ma_x; \qquad F_0 \sin \omega_0 t - kx = m\ddot{x}$$

or

$$\ddot{x} + \frac{k}{m}x = \frac{F_0}{m}\sin \omega_0 t \tag{22-18}$$

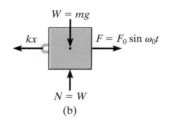

(b)

Fig. 22–11

This equation is referred to as a nonhomogeneous second-order differential equation. The general solution consists of a complementary solution, x_c, *plus* a particular solution, x_p.

The *complementary solution* is determined by setting the term on the right side of Eq. 22–18 equal to zero and solving the resulting homogeneous equation, which is equivalent to Eq. 22–1. The solution is defined by Eq. 22–3, i.e.,

$$x_c = A \sin \omega_n t + B \cos \omega_n t \tag{22-19}$$

where ω_n is the natural frequency, $\omega_n = \sqrt{k/m}$, Eq. 22–2.

Since the motion is periodic, the *particular solution* of Eq. 22–18 may be determined by assuming a solution of the form

$$x_p = C \sin \omega_0 t \tag{22-20}$$

where C is a constant. Taking the second time derivative and substituting into Eq. 22–18 yields

$$-C\omega_0^2 \sin \omega_0 t + \frac{k}{m}(C \sin \omega_0 t) = \frac{F_0}{m}\sin \omega_0 t$$

Factoring out $\sin \omega_0 t$ and solving for C gives

$$C = \frac{F_0/m}{(k/m) - \omega_0^2} = \frac{F_0/k}{1 - (\omega_0/\omega_n)^2} \tag{22-21}$$

Substituting into Eq. 22–20, we obtain the particular solution

$$\boxed{x_p = \frac{F_0/k}{1 - (\omega_0/\omega_n)^2}\sin \omega_0 t} \tag{22-22}$$

Shaker tables provide forced vibration and are used to separate out granular materials.

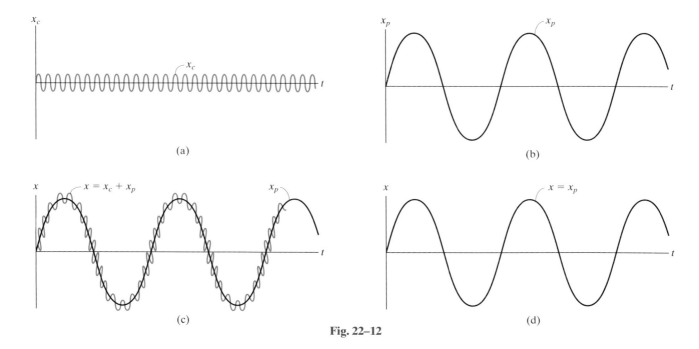

x_c

x_c

t

(a)

x_p

x_p

t

(b)

x

$x = x_c + x_p$

x_p

t

(c)

x

$x = x_p$

t

(d)

Fig. 22–12

The *general solution* is therefore

$$x = x_c + x_p = A \sin \omega_n + B \cos \omega_n + \frac{F_0/k}{1 - (\omega_0/\omega_n)^2} \sin \omega_0 t \quad (22\text{–}23)$$

Here x describes two types of vibrating motion of the block. The *complementary solution* x_c defines the *free vibration*, which depends on the circular frequency $\omega_n = \sqrt{k/m}$ and the constants A and B, Fig. 22–12a. Specific values for A and B are obtained by evaluating Eq. 22–23 at a given instant when the displacement and velocity are known. The *particular solution* x_p describes the *forced vibration* of the block caused by the applied force $F = F_0 \sin \omega_0 t$, Fig. 22–12b. The resultant vibration x is shown in Fig. 22–12c. Since all vibrating systems are subject to *friction*, the free vibration, x_c, will in time dampen out. For this reason the free vibration is referred to as *transient*, and the forced vibration is called *steady-state*, since it is the only vibration that remains, Fig. 22–12d.

From Eq. 22–21 it is seen that the *amplitude* of forced vibration depends on the *frequency ratio* ω_0/ω_n. If the *magnification factor* MF is defined as the ratio of the amplitude of steady-state vibration, $(x_p)_{\max}$, to the static deflection F_0/k, which would be produced by the amplitude of the periodic force F_0, then, from Eq. 22–22,

The soil compactor operates by forced vibration developed by an internal motor. It is important that the forcing frequency not be close to the natural frequency of vibration, which is determined when the motor is turned off; otherwise resonance will occur and the machine will become uncontrollable.

MF

Fig. 22–13

(a)

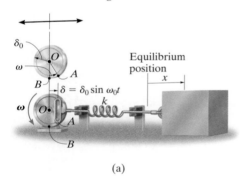

(b)

Fig. 22–14

$$MF = \frac{(x_p)_{max}}{F_0/k} = \frac{1}{1 - (\omega_0/\omega_n)^2} \qquad (22–24)$$

This equation is graphed in Fig. 22–13. Note that for $\omega_0 \approx 0$, the MF ≈ 1. In this case, because of the very low frequency $\omega_0 \ll \omega_n$, the magnitude of the force $\mathbf{F}$ changes slowly and so the vibration of the block will be in phase with the applied force $\mathbf{F}$. If the force or displacement is applied with a frequency close to the natural frequency of the system, i.e., $\omega_0/\omega_n \approx 1$, the amplitude of vibration of the block becomes extremely large. This occurs because the force $\mathbf{F}$ is applied to the block so that it always follows the motion of the block. This condition is called *resonance*, and in practice, resonating vibrations can cause tremendous stress and rapid failure of parts.* When the cyclic force $F_0 \sin \omega t$ is applied at high frequencies ($\omega_0 > \omega_n$), the value of the MF becomes negative, indicating that the motion of the block is out of phase with the force. Under these conditions, as the block is displaced to the right, the force acts to the left, and vice versa. For extremely high frequencies ($\omega_0 \gg \omega_n$) the block remains almost stationary, and hence the MF is approximately zero.

Periodic Support Displacement.   Forced vibrations can also arise from the periodic excitation of the support of a system. The model shown in Fig. 22–14a represents the periodic vibration of a block which is caused by harmonic movement $\delta = \delta_0 \sin \omega_0 t$ of the support. The free-body diagram for the block in this case is shown in Fig. 22–14b. The coordinate x is measured from the point of zero displacement of the support, i.e., when the radial line OA coincides with OB, Fig. 22–14a. Therefore, general displacement of the spring is $(x - \delta_0 \sin \omega_0 t)$. Applying the equation of motion yields

$$\xrightarrow{+} F_x = ma_x; \qquad -k(x - \delta_0 \sin \omega_0 t) = m\ddot{x}$$

or

$$\ddot{x} + \frac{k}{m}x = \frac{k\delta_0}{m}\sin \omega_0 t \qquad (22–25)$$

By comparison, this equation is identical to the form of Eq. 22–18, *provided F_0 is replaced by $k\delta_0$*. If this substitution is made into the solutions defined by Eqs. 22–21 to 22–23, the results are appropriate for describing the motion of the block when subjected to the support displacement $\delta = \delta_0 \sin \omega_0 t$.

*A swing has a natural period of vibration, as determined in Example 22.1. If someone pushes on the swing only when it reaches its highest point, neglecting drag or wind resistance, resonance will occur since the natural and forcing frequencies will be equal.

EXAMPLE 22.7

The instrument shown in Fig. 22–15 is rigidly attached to a platform P, which in turn is supported by *four* springs, each having a stiffness $k = 800$ N/m. Initially the platform is at rest when the floor is subjected to a displacement $\delta = 10 \sin(8t)$ mm, where t is in seconds. If the instrument is constrained to move vertically and the total mass of the instrument and platform is 20 kg, determine the vertical displacement y of the platform as a function of time, measured from the equilibrium position. What floor vibration is required to cause resonance?

SOLUTION
Since the induced vibration is caused by the displacement of the supports, the motion is described by Eq. 22–23, with F_0 replaced by $k\delta_0$, i.e.,

$$y = A \sin \omega_n t + B \cos \omega_n t + \frac{\delta_0}{1 - (\omega_0/\omega_n)^2} \sin \omega_0 t \qquad (1)$$

Here $\delta = \delta_0 \sin \omega_0 t = 10 \sin(8t)$ mm, so that

$$\delta_0 = 10 \text{ mm} \qquad \omega_0 = 8 \text{ rad/s}$$

$$\omega_n = \sqrt{\frac{k}{m}} = \sqrt{\frac{4(800 \text{ N/m})}{20 \text{ kg}}} = 12.6 \text{ rad/s}$$

From Eq. 22–22, with $k\delta_0$ replacing F_0, the amplitude of vibration caused by the floor displacement is

$$(y_{\omega_n})_{\max} = \frac{\delta_0}{1 - (\omega_0/\omega_n)^2} = \frac{10}{1 - [(8 \text{ rad/s})/(12.6 \text{ rad/s})]^2} = 16.7 \text{ mm}$$
$$(2)$$

Hence, Eq. 1 and its time derivative become

$$y = A \sin(12.6t) + B \cos(12.6t) + 16.7 \sin(8t)$$

$$\dot{y} = A(12.6) \cos(12.6t) - B(12.6) \sin(12.6t) + 133.3 \cos(8t)$$

The constants A and B are evaluated from these equations. Since $y = 0$ and $\dot{y} = 0$ at $t = 0$, then

$$0 = 0 + B + 0 \qquad\qquad B = 0$$
$$0 = A(12.6) - 0 + 133.3 \qquad A = -10.5$$

The vibrating motion is therefore described by the equation

$$y = -10.5 \sin(12.6t) + 16.7 \sin(8t) \qquad\qquad Ans.$$

Resonance will occur when the amplitude of vibration caused by the floor displacement approaches infinity. From Eq. 2, this requires

$$\omega_0 = \omega_n = 12.6 \text{ rad/s} \qquad\qquad Ans.$$

Fig. 22–15

*22.4   Viscous Damped Free Vibration

The vibration analysis considered thus far has not included the effects of friction or damping in the system, and as a result, the solutions obtained are only in close agreement with the actual motion. Since all vibrations die out in time, the presence of damping forces should be included in the analysis.

In many cases damping is attributed to the resistance created by the substance, such as water, oil, or air, in which the system vibrates. Provided the body moves slowly through this substance, the resistance to motion is directly proportional to the body's speed. The type of force developed under these conditions is called a *viscous damping force*. The magnitude of this force is expressed by an equation of the form

$$F = c\dot{x} \tag{22–26}$$

where the constant c is called the *coefficient of viscous damping* and has units of $N \cdot s/m$ or $lb \cdot s/ft$.

The vibrating motion of a body or system having viscous damping may be characterized by the block and spring shown in Fig. 22–16a. The effect of damping is provided by the *dashpot* connected to the block on the right side. Damping occurs when the piston P moves to the right or left within the enclosed cylinder. The cylinder contains a fluid, and the motion of the piston is retarded since the fluid must flow around or through a small hole in the piston. The dashpot is assumed to have a coefficient of viscous damping c.

If the block is displaced a distance x from its equilibrium position, the resulting free-body diagram is shown in Fig. 22–16b. Both the spring force kx and the damping force $c\dot{x}$ oppose the forward motion of the block, so that applying the equation of motion yields

$$\xrightarrow{+} \Sigma F_x = ma_x; \qquad -kx - c\dot{x} = m\ddot{x}$$

or

$$m\ddot{x} + c\dot{x} + kx = 0 \tag{22–27}$$

This linear, second-order, homogeneous, differential equation has solutions of the form

$$x = e^{\lambda t}$$

where e is the base of the natural logarithm and λ (lambda) is a constant. The value of λ may be obtained by substituting this solution into Eq. 22–27, which yields

$$m\lambda^2 e^{\lambda t} + c\lambda e^{\lambda t} + ke^{\lambda t} = 0$$

or

$$e^{\lambda t}(m\lambda^2 + c\lambda + k) = 0$$

Equilibrium position

x

k

P

c

(a)

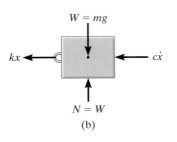

$W = mg$

kx

$c\dot{x}$

$N = W$

(b)

Fig. 22–16

Since $e^{\lambda t}$ is never zero, a solution is possible provided

$$m\lambda^2 + c\lambda + k = 0$$

Hence, by the quadratic formula, the two values of λ are

$$\lambda_1 = -\frac{c}{2m} + \sqrt{\left(\frac{c}{2m}\right)^2 - \frac{k}{m}}$$

$$\lambda_2 = -\frac{c}{2m} - \sqrt{\left(\frac{c}{2m}\right)^2 - \frac{k}{m}}$$

$$(22\text{–}28)$$

The general solution of Eq. 22–27 is therefore a linear combination of exponentials which involves both of these roots. There are three possible combinations of λ_1 and λ_2 which must be considered. Before discussing these combinations, however, we will first define the *critical damping coefficient* c_c as the value of c which makes the radical in Eqs. 22–28 equal to zero; i.e.,

$$\left(\frac{c_c}{2m}\right)^2 - \frac{k}{m} = 0$$

or

$$\boxed{c_c = 2m\sqrt{\frac{k}{m}} = 2m\omega_n} \qquad (22\text{–}29)$$

Here the value of ω_n is the natural frequency $\omega_n = \sqrt{k/m}$, Eq. 22–2.

Overdamped System.
When $c > c_c$, the roots λ_1 and λ_2 are both real. The general solution of Eq. 22–27 may then be written as

$$x = Ae^{\lambda_1 t} + Be^{\lambda_2 t} \qquad (22\text{–}30)$$

Motion corresponding to this solution is *nonvibrating*. The effect of damping is so strong that when the block is displaced and released, it simply creeps back to its original position without oscillating. The system is said to be *overdamped*.

Critically Damped System.
If $c = c_c$, then $\lambda_1 = \lambda_2 = -c_c/2m = -\omega_n$. This situation is known as *critical damping*, since it represents a condition where c has the smallest value necessary to cause the system to be nonvibrating. Using the methods of differential equations, it may be shown that the solution to Eq. 22–27 for critical damping is

$$x = (A + Bt)e^{-\omega_n t} \qquad (22\text{–}31)$$

Underdamped System.

Most often $c < c_c$, in which case the system is referred to as *underdamped*. In this case the roots λ_1 and λ_2 are complex numbers, and it may be shown that the general solution of Eq. 22–27 can be written as

$$x = D[e^{-(c/2m)t} \sin(\omega_d t + \phi)] \qquad (22\text{--}32)$$

where D and ϕ are constants generally determined from the initial conditions of the problem. The constant ω_d is called the *damped natural frequency* of the system. It has a value of

$$\omega_d = \sqrt{\frac{k}{m} - \left(\frac{c}{2m}\right)^2} = \omega_n \sqrt{1 - \left(\frac{c}{c_c}\right)^2} \qquad (22\text{--}33)$$

where the ratio c/c_c is called the *damping factor*.

The graph of Eq. 22–32 is shown in Fig. 22–17. The initial limit of motion, D, diminishes with each cycle of vibration, since motion is confined within the bounds of the exponential curve. Using the damped natural frequency ω_d, the period of damped vibration may be written as

$$\tau_d = \frac{2\pi}{\omega_d} \qquad (22\text{--}34)$$

Since $\omega_d < \omega_n$, Eq. 22–33, the period of damped vibration, τ_d, will be greater than that of free vibration, $\tau = 2\pi/\omega_n$.

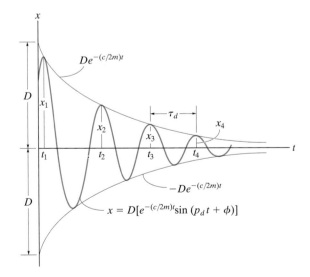

Fig. 22–17

*22.5 Viscous Damped Forced Vibration

The most general case of single-degree-of-freedom vibrating motion occurs when the system includes the effects of forced motion and induced damping. The analysis of this particular type of vibration is of practical value when applied to systems having significant damping characteristics.

If a dashpot is attached to the block and spring shown in Fig. 22–11a, the differential equation which describes the motion becomes

$$m\ddot{x} + c\dot{x} + kx = F_0 \sin \omega_0 t \tag{22–35}$$

A similar equation may be written for a block and spring having a periodic support displacement, Fig. 22–14a, which includes the effects of damping. In that case, however, F_0 is replaced by $k\delta_0$. Since Eq. 22–35 is nonhomogeneous, the general solution is the sum of a complementary solution, x_c, and a particular solution, x_p. The complementary solution is determined by setting the right side of Eq. 22–35 equal to zero and solving the homogeneous equation, which is equivalent to Eq. 22–27. The solution is therefore given by Eq. 22–30, 22–31, or 22–32, depending on the values of λ_1 and λ_2. Because all systems contain friction, however, this solution will dampen out with time. Only the particular solution, which describes the *steady-state vibration* of the system, will remain. Since the applied forcing function is harmonic, the steady-state motion will also be harmonic. Consequently, the particular solution will be of the form

$$x_p = A' \sin \omega_0 t + B' \cos \omega_0 t \tag{22–36}$$

The constants A' and B' are determined by taking the necessary time derivatives and substituting them into Eq. 22–35, which after simplification yields

$$(-A'm\omega_0^2 - cB'\omega_0 + kA') \sin \omega_0 t +$$
$$(-B'm\omega_0^2 + cA'\omega_0 + kB') \cos \omega_0 t = F_0 \sin \omega_0 t$$

Since this equation holds for all time, the constant coefficients of $\sin \omega_0 t$ and $\cos \omega_0 t$ may be equated; i.e.,

$$-A'm\omega_0^2 - cB'\omega_0 + kA' = F_0$$
$$-B'm\omega_0^2 + cA'\omega_0 + kB' = 0$$

Solving for A' and B', realizing that $\omega_n^2 = k/m$, yields

$$A' = \frac{(F_0/m)(\omega_n^2 - \omega_0^2)}{(\omega_n^2 - \omega_0^2)^2 + (c\omega_0/m)^2}$$

$$B' = \frac{-F_0(c\omega_0/m^2)}{(\omega_n^2 - \omega_0^2)^2 + (c\omega_0/m)^2}$$

(22–37)

It is also possible to express Eq. 22–36 in a form similar to Eq. 22–9,

$$x_p = C' \sin(\omega_0 t - \phi')$$

(22–38)

in which case the constants C' and ϕ' are

$$C' = \frac{F_0/k}{\sqrt{[1 - (\omega_0/\omega_n)^2]^2 + [2(c/c_c)(\omega_0/\omega_n)]^2}}$$

(22–39)

$$\phi' = \tan^{-1}\left[\frac{2(c/c_c)(\omega_0/\omega_n)}{1 - (\omega_0/\omega_n)^2}\right]$$

The angle ϕ' represents the phase difference between the applied force and the resulting steady-state vibration of the damped system.

The *magnification factor* MF has been defined in Sec. 22.3 as the ratio of the amplitude of deflection caused by the forced vibration to the deflection caused by a static force $\mathbf{F}_0$. From Eq. 22–38, the forced vibration has an amplitude of C'; thus,

$$\text{MF} = \frac{C'}{F_0/k} = \frac{1}{\sqrt{[1 - (\omega_0/\omega_n)^2]^2 + [2(c/c_c)(\omega_0/\omega_n)]^2}}$$

(22–40)

The MF is plotted in Fig. 22–18 versus the frequency ratio ω_0/ω_n for various values of the damping factor c/c_c. It can be seen from this graph that the magnification of the amplitude increases as the damping factor decreases. Resonance obviously occurs only when the damping factor is zero and the frequency ratio equals 1.

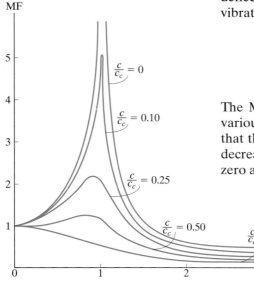

Fig. 22–18

EXAMPLE 22.8

The 30-kg electric motor shown in Fig. 22–19 is supported by *four* springs, each spring having a stiffness of 200 N/m. If the rotor R is unbalanced such that its effect is equivalent to a 4-kg mass located 60 mm from the axis of rotation, determine the amplitude of vibration when the rotor is turning at $\omega_0 = 10$ rad/s. The damping factor is $c/c_c = 0.15$.

Fig. 22–19

SOLUTION

The periodic force which causes the motor to vibrate is the centrifugal force due to the unbalanced rotor. This force has a constant magnitude of

$$F_0 = ma_n = mr\omega_0^2 = 4 \text{ kg}(0.06 \text{ m})(10 \text{ rad/s})^2 = 24 \text{ N}$$

Since $F = F_0 \sin \omega_0 t$, where $\omega_0 = 10$ rad/s, then

$$F = 24 \sin 10t$$

The stiffness of the entire system of four springs is $k = 4(200 \text{ N/m}) = 800 \text{ N/m}$. Therefore, the natural frequency of vibration is

$$\omega_n = \sqrt{\frac{k}{m}} = \sqrt{\frac{800 \text{ N/m}}{30 \text{ kg}}} = 5.16 \text{ rad/s}$$

Since the damping factor is known, the steady-state amplitude may be determined from the first of Eqs. 22–39, i.e.,

$$C' = \frac{F_0/k}{\sqrt{[1 - (\omega_0/\omega_n)^2]^2 + [2(c/c_c)(\omega_0/\omega_n)]^2}}$$

$$= \frac{24/800}{\sqrt{[1 - (10/5.16)^2]^2 + [2(0.15)(10/5.16)]^2}}$$

$$= 0.0107 \text{ m} = 10.7 \text{ mm} \hspace{3cm} Ans.$$

(a)

*22.6 Electrical Circuit Analogs

The characteristics of a vibrating mechanical system may be represented by an electric circuit. Consider the circuit shown in Fig. 22–20a, which consists of an inductor L, a resistor R, and a capacitor C. When a voltage $E(t)$ is applied, it causes a current of magnitude i to flow through the circuit. As the current flows past the inductor the voltage drop is $L(di/dt)$, when it flows across the resistor the drop is Ri, and when it arrives at the capacitor the drop is $(1/C)\int i\, dt$. Since current cannot flow past a capacitor, it is only possible to measure the charge q acting on the capacitor. The charge may, however, be related to the current by the equation $i = dq/dt$. Thus, the voltage drops which occur across the inductor, resistor, and capacitor may be written as $L\, d^2q/dt^2$, $R\, dq/dt$, and q/C, respectively. According to Kirchhoff's voltage law, the applied voltage balances the sum of the voltage drops around the circuit. Therefore,

(b)

Fig. 22–20

$$L\frac{d^2q}{dt^2} + R\frac{dq}{dt} + \frac{1}{C}q = E(t) \qquad (22\text{–}41)$$

Consider now the model of a single-degree-of-freedom mechanical system, Fig. 22–20b, which is subjected to both a general forcing function $F(t)$ and damping. The equation of motion for this system was established in the previous section and can be written as

$$m\frac{d^2x}{dt^2} + c\frac{dx}{dt} + kx = F(t) \qquad (22\text{–}42)$$

By comparison, it is seen that Eqs. 22–41 and 22–42 have the same form, and hence mathematically the problem of analyzing an electric circuit is the same as that of analyzing a vibrating mechanical system. The analogs between the two equations are given in Table 22–1.

This analogy has important application to experimental work, for it is much easier to simulate the vibration of a complex mechanical system using an electric circuit, which can be constructed on an analog computer, than to make an equivalent mechanical spring-and-dashpot model.

TABLE 22–1 Electrical–Mechanical Analogs			
Electrical		Mechanical	
Electric charge	q	Displacement	x
Electric current	i	Velocity	dx/dt
Voltage	$E(t)$	Applied force	$F(t)$
Inductance	L	Mass	m
Resistance	R	Viscous damping coefficient	c
Reciprocal of capacitance	$1/C$	Spring stiffness	k

PROBLEMS

22–41. Use a block-and-spring model like that shown in Fig. 22–14a but suspended from a vertical position and subjected to a periodic support displacement of $\delta = \delta_0 \cos \omega_0 t$, determine the equation of motion for the system, and obtain its general solution. Define the displacement y measured from the static equilibrium position of the block when $t = 0$.

22–42. The 20-lb block is attached to a spring having a stiffness of 20 lb/ft. A force $F = (6 \cos 2t)$ lb, where t is in seconds, is applied to the block. Determine the maximum speed of the block after frictional forces cause the free vibrations to dampen out.

*22–44. If the block is subjected to the impressed force $F = F_0 \cos \omega_0 t$, show that the differential equation of motion is $\ddot{y} + (k/m)y = (F_0/m) \cos \omega_0 t$, where y is measured from the equilibrium position of the block. What is the general solution of this equation?

Prob. 22–44

Prob. 22–42

22–45. The light elastic rod supports a 4-kg sphere. When an 18-N vertical force is applied to the sphere, the rod deflects 14 mm. If the wall oscillates with harmonic frequency of 2 Hz and has an amplitude of 15 mm, determine the amplitude of vibration for the sphere.

22–43. A 4-lb weight is attached to a spring having a stiffness $k = 10$ lb/ft. The weight is drawn downward a distance of 4 in. and released from rest. If the support moves with a vertical displacement $\delta = (0.5 \sin 4t)$ in., where t is in seconds, determine the equation which describes the position of the weight as a function of time.

Prob. 22–45

22–46. Use a block-and-spring model like that shown in Fig. 22–14a, but suspended from a vertical position and subjected to a periodic support displacement $\delta = \delta_0 \sin \omega_0 t$, determine the equation of motion for the system, and obtain its general solution. Define the displacement y measured from the static equilibrium position of the block when $t = 0$.

22–47. A 5-kg block is suspended from a spring having a stiffness of 300 N/m. If the block is acted upon by a vertical force $F = (7 \sin 8t)$ N, where t is in seconds, determine the equation which describes the motion of the block when it is pulled down 100 mm from the equilibrium position and released from rest at $t = 0$. Assume that positive displacement is downward.

$k = 300$ N/m

$F = 7 \sin 8t$

Prob. 22–47

*22–48.** The 4-kg circular disk is attached to three springs, each spring having a stiffness $k = 180$ N/m. If the disk is immersed in a fluid and given a downward velocity of 0.3 m/s at the equilibrium position, determine the equation which describes the motion. Assume that positive displacement is measured downward, and that fluid resistance acting on the disk furnishes a damping force having a magnitude $F = (60|v|)$ N, where v is in m/s.

120° 120°
120°

Prob. 22–48

22–49. The instrument is centered uniformly on a platform P, which in turn is supported by *four* springs, each spring having a stiffness $k = 130$ N/m. If the floor is subjected to a vibration $\omega_0 = 7$ Hz, having a vertical displacement amplitude $\delta_0 = 0.17$ ft, determine the vertical displacement amplitude of the platform and instrument. The instrument and the platform have a total weight of 18 lb.

k k

Prob. 22–49

22–50. The 450-kg trailer is pulled with a constant speed over the surface of a bumpy road, which may be approximated by a cosine curve having an amplitude of 50 mm and wave length of 4 m. If the two springs s which support the trailer each have a stiffness of 800 N/m, determine the speed v which will cause the greatest vibration (resonance) of the trailer. Neglect the weight of the wheels.

v

s 100 mm

2 m   2 m

Prob. 22–50

22–51. Determine the amplitude of vibration of the trailer in Prob. 22–50 if the speed $v = 15$ km/h.

Prob. 22–51

22–53. What will be the amplitude of steady-state vibration of the motor in Prob. 22–52 if the angular velocity of the flywheel is 20 rad/s?

22–54. Determine the angular velocity of the flywheel in Prob. 22–52 which will produce an amplitude of vibration of 0.25 in.

Probs. 22–53/54

*22–52. The electric motor turns an eccentric flywheel which is equivalent to an unbalanced 0.25-lb weight located 10 in. from the axis of rotation. If the static deflection of the beam is 1 in. because of the weight of the motor, determine the angular velocity of the flywheel at which resonance will occur. The motor weights 150 lb. Neglect the mass of the beam.

Prob. 22–52

22–55. The engine is mounted on a foundation block which is spring-supported. Describe the steady-state vibration of the system if the block and engine have a total weight of 1500 lb and the engine, when running, creates an impressed force $F = (50 \sin 2t)$ lb, where t is in seconds. Assume that the system vibrates only in the vertical direction, with the positive displacement measured downward, and that the total stiffness of the springs can be represented as $k = 2000$ lb/ft.

Prob. 22–55

*22–56. Determine the rotational speed ω of the engine in Prob. 22–55 which will cause resonance.

Prob. 22–56

22–57. The block, having a weight of 12 lb, is immersed in a liquid such that the damping force acting on the block has a magnitude of $F = (0.7|v|)$ lb, where v is in ft/s. If the block is pulled down 0.62 ft and released from rest, determine the position of the block as a function of time. The spring has a stiffness of $k = 53$ lb/ft. Assume that positive displacement is downward.

Prob. 22–57

22–58. A 7-lb block is suspended from a spring having a stiffness of $k = 75$ lb/ft. The support to which the spring is attached is given simple harmonic motion which may be expressed as $\delta = (0.15 \sin 2t)$ ft, where t is in seconds. If the damping factor is $c/c_c = 0.8$, determine the phase angle ϕ of forced vibration.

22–59. Determine the magnification factor of the block, spring, and dashpot combination in Prob. 22–58.

*22–60. The bar has a weight of 6 lb. If the stiffness of the spring is $k = 8$ lb/ft and the dashpot has a damping coefficient $c = 60$ lb $\cdot$ s/ft, determine the differential equation which describes the motion in terms of the angle θ of the bar's rotation. Also, what should be the damping coefficient of the dashpot if the bar is to be critically damped?

Prob. 22–60

22–61. A block having a mass of 7 kg is suspended from a spring that has a stiffness $k = 600$ N/m. If the block is given an upward velocity of 0.6 m/s from its equilibrium position at $t = 0$, determine its position as a function of time. Assume that positive displacement of the block is downward and that motion takes place in a medium which furnishes a damping force $F = (50|v|)$ N, where v is in m/s.

22–62. The damping factor, c/c_c, may be determined experimentally by measuring the successive amplitudes of vibrating motion of a system. If two of these maximum displacements can be approximated by x_1 and x_2, as shown in Fig. 22–17, show that the ratio $\ln x_1/x_2 = 2\pi(c/c_c)/\sqrt{1 - (c/c_e)^2}$. The quantity $\ln x_1/x_2$ is called the *logarithmic decrement*.

22–63. Determine the differential equation of motion for the damped vibratory system shown. What type of motion occurs?

Prob. 22–63

*22–64. The 20-kg block is subjected to the action of the harmonic force $F = (90 \cos 6t)$ N, where t is in seconds. Write the equation which describes the steady-state motion.

22–66. The 10-kg block-spring-damper system is continually damped. If the block is displaced to $x = 50$ mm and released from rest, determine the time required for it to return to the position $x = 2$ mm.

Prob. 22–64

Prob. 22–66

22–65. Draw the electrical circuit that is equivalent to the mechanical system shown. What is the differential equation which describes the charge q in the circuit?

Prob. 22–65

Chapter Review

Undamped Free Vibration

A body has free vibration when gravitational or elastic restoring forces cause the motion. This motion is undamped when friction forces are neglected. The periodic motion of an undamped, freely vibrating body can be studied by displacing the body from the equilibrium position and then applying the equation of motion along the path.

For a one-degree-of-freedom system, the resulting differential equation can be written in terms of its circular or natural frequency ω_n.

(*See pages 619–623.*)

$$\ddot{x} + \omega_n^2 x = 0 \qquad \tau = \frac{2\pi}{\omega_n} \qquad f = \frac{1}{\tau} = \frac{\omega_n}{2\pi}$$

Energy Methods

Provided the restoring forces acting on the body are gravitational and elastic, then conservation of energy can also be used to determine its simple harmonic motion. To do this, the body is displaced a small amount from its equilibrium position, and an expression for its kinetic and potential energy is written. The time derivative of this equation can then be rearranged in the standard form $\ddot{x} + \omega_n^2 x = 0$. Knowing ω_n, the period and frequency of the motion can then be obtained.

(*See page 632–633.*)

Undamped Forced Vibration

When the equation of motion is applied to a body, which is subjected to a periodic force, or a periodic support displacement having a frequency ω_0, then the displacement consists of a complementary solution and a particular solution. The complementary solution is caused by the free vibration and can be neglected. The particular solution is caused by the forced vibration.

Resonance will occur if the natural frequency of vibration ω_n is equal to the forcing frequency ω_0. This should be avoided, since the motion will tend to become unbounded.

(*See pages 638–640.*)

$$x_p = \frac{F_0/k}{1 - (\omega_0/\omega_n)^2} \sin \omega_0 t$$

Viscous Damped Free Vibration

A viscous damping force is caused by fluid drag on the system as it vibrates. If the motion is slow, this drag force will be proportional to the velocity, that is, $F = c\dot{x}$. Here c is the coefficient of viscous damping. By comparing its value to the critical damping coefficient $c_c = 2m\omega_n$, we can specify the type of vibration that occurs. If $c > c_c$, it is an overdamped system; if $c = c_c$, it is a critically damped system; if $c < c_c$, it is an underdamped system.

(*See pages 642–644.*)

Viscous Damped Forced Vibration

The most general type of vibration for a one-degree-of-freedom system occurs when the system is damped and subjected to periodic forced motion. The solution provides insight as to how the damping factor, c/c_c, and the frequency ratio, ω_0/ω_n, influence the vibration.

Resonance is avoided provided $c/c_c \neq 0$ and $\omega_0/\omega_n \neq 1$.

(*See page 645–646.*)

Electrical Circuit Analogs

The vibrating motion of a complex system can be studied by modeling it as an electrical circuit. This is possible since the differential equations that govern the behavior of each system are the same.

(*See page 648.*)

Mathematical Expressions

Quadratic Formula

If $ax^2 + bx + c = 0$, then $x = \dfrac{-b \pm \sqrt{b^2 - 4ac}}{2a}$

Hyperbolic Functions

$\sinh x = \dfrac{e^x - e^{-x}}{2}$, $\cosh x = \dfrac{e^x + e^{-x}}{2}$, $\tanh x = \dfrac{\sinh x}{\cosh x}$

Trigonometric Identities

$\sin \theta = \dfrac{A}{C}$, $\csc \theta = \dfrac{C}{A}$

$\cos \theta = \dfrac{B}{C}$, $\sec \theta = \dfrac{C}{B}$

$\tan \theta = \dfrac{A}{B}$, $\cot \theta = \dfrac{B}{A}$

$\sin^2 \theta + \cos^2 \theta = 1$

$\sin(\theta \pm \phi) = \sin \theta \cos \phi \pm \cos \theta \sin \phi$

$\sin 2\theta = 2 \sin \theta \cos \theta$

$\cos(\theta \pm \phi) = \cos \theta \cos \phi \mp \sin \theta \sin \phi$

$\cos 2\theta = \cos^2 \theta - \sin^2 \theta$

$\cos \theta = \pm\sqrt{\dfrac{1 + \cos 2\theta}{2}}$, $\sin \theta = \pm\sqrt{\dfrac{1 - \cos 2\theta}{2}}$

$\tan \theta = \dfrac{\sin \theta}{\cos \theta}$

$1 + \tan^2 \theta = \sec^2 \theta \qquad 1 + \cot^2 \theta = \csc^2 \theta$

Power-Series Expansions

$\sin x = x - \dfrac{x^3}{3!} + \cdots \qquad \sinh x = x + \dfrac{x^3}{3!} + \cdots$

$\cos x = 1 - \dfrac{x^2}{2!} + \cdots \qquad \cosh x = 1 + \dfrac{x^2}{2!} + \cdots$

Derivatives

$\dfrac{d}{dx}(u^n) = nu^{n-1}\dfrac{du}{dx}$

$\dfrac{d}{dx}(uv) = u\dfrac{dv}{dx} + v\dfrac{du}{dx}$

$\dfrac{d}{dx}\left(\dfrac{u}{v}\right) = \dfrac{v\dfrac{du}{dx} - u\dfrac{dv}{dx}}{v^2}$

$\dfrac{d}{dx}(\cot u) = -\csc^2 u\dfrac{du}{dx}$

$\dfrac{d}{dx}(\sec u) = \tan u \sec u\dfrac{du}{dx}$

$\dfrac{d}{dx}(\csc u) = -\csc u \cot u\dfrac{du}{dx}$

$\dfrac{d}{dx}(\sin u) = \cos u\dfrac{du}{dx}$

$\dfrac{d}{dx}(\cos u) = -\sin u\dfrac{du}{dx}$

$\dfrac{d}{dx}(\tan u) = \sec^2 u\dfrac{du}{dx}$

$\dfrac{d}{dx}(\sinh u) = \cosh u\dfrac{du}{dx}$

$\dfrac{d}{dx}(\cosh u) = \sinh u\dfrac{du}{dx}$

Integrals

$$\int x^n \, dx = \frac{x^{n+1}}{n+1} + C, \; n \neq -1$$

$$\int \frac{dx}{a+bx} = \frac{1}{b}\ln(a+bx) + C$$

$$\int \frac{dx}{a+bx^2} = \frac{1}{2\sqrt{-ba}}\ln\left[\frac{a+x\sqrt{-ab}}{a-x\sqrt{-ab}}\right] + C, \; ab < 0$$

$$\int \frac{x \, dx}{a+bx^2} = \frac{1}{2b}\ln(bx^2+a) + C,$$

$$\int \frac{x^2 \, dx}{a+bx^2} = \frac{x}{b} - \frac{a}{b\sqrt{ab}}\tan^{-1}\frac{x\sqrt{ab}}{a} + C, \; ab > 0$$

$$\int \frac{dx}{a^2-x^2} = \frac{1}{2a}\ln\left[\frac{a+x}{a-x}\right] + C, \; a^2 > x^2$$

$$\int \sqrt{a+bx} \, dx = \frac{2}{3b}\sqrt{(a+bx)^3} + C$$

$$\int x\sqrt{a+bx} \, dx = \frac{-2(2a-3bx)\sqrt{(a+bx)^3}}{15b^2} + C$$

$$\int x^2\sqrt{a+bx} \, dx = \frac{2(8a^2-12abx+15b^2x^2)\sqrt{(a+bx)^3}}{105b^3} + C$$

$$\int \sqrt{a^2-x^2} \, dx = \frac{1}{2}\left[x\sqrt{a^2-x^2} + a^2\sin^{-1}\frac{x}{a}\right] + C, \; a > 0$$

$$\int x\sqrt{a^2-x^2} \, dx = -\frac{1}{3}\sqrt{(a^2-x^2)^3} + C$$

$$\int x^2\sqrt{a^2-x^2} \, dx = -\frac{x}{4}\sqrt{(a^2-x^2)^3}$$

$$+ \frac{a^2}{8}\left(x\sqrt{a^2-x^2} + a^2\sin^{-1}\frac{x}{a}\right) + C, \; a > 0$$

$$\int \sqrt{x^2 \pm a^2} \, dx = \frac{1}{2}\left[x\sqrt{x^2 \pm a^2} \pm a^2\ln\left(x + \sqrt{x^2 \pm a^2}\right)\right] + C$$

$$\int x\sqrt{x^2 \pm a^2} \, dx = \frac{1}{3}\sqrt{(x^2 \pm a^2)^3} + C$$

$$\int x^2\sqrt{x^2 \pm a^2} \, dx = \frac{x}{4}\sqrt{(x^2 \pm a^2)^3} \mp \frac{a^2}{8}x\sqrt{x^2 \pm a^2}$$

$$- \frac{a^4}{8}\ln\left(x + \sqrt{x^2 \pm a^2}\right) + C$$

$$\int \frac{dx}{\sqrt{a+bx}} = \frac{2\sqrt{a+bx}}{b} + C$$

$$\int \frac{x \, dx}{\sqrt{x^2 \pm a^2}} = \sqrt{x^2 \pm a^2} + C$$

$$\int \frac{dx}{\sqrt{a+bx+cx^2}} = \frac{1}{\sqrt{c}}\ln\left[\sqrt{a+bx+cx^2}\right.$$

$$\left. + x\sqrt{c} + \frac{b}{2\sqrt{c}}\right] + C, \; c > 0$$

$$= \frac{1}{\sqrt{-c}}\sin^{-1}\left(\frac{-2cx-b}{\sqrt{b^2-4ac}}\right) + C, \; c < 0$$

$$\int \sin x \, dx = -\cos x + C$$

$$\int \cos x \, dx = \sin x + C$$

$$\int x\cos(ax) \, dx = \frac{1}{a^2}\cos(ax) + \frac{x}{a}\sin(ax) + C$$

$$\int x^2\cos(ax) \, dx = \frac{2x}{a^2}\cos(ax)$$

$$+ \frac{a^2x^2-2}{a^3}\sin(ax) + C$$

$$\int e^{ax} \, dx = \frac{1}{a}e^{ax} + C$$

$$\int xe^{ax} \, dx = \frac{e^{ax}}{a^2}(ax-1) + C$$

$$\int \sinh x \, dx = \cosh x + C$$

$$\int \cosh x \, dx = \sinh x + C$$

B

Numerical and Computer Analysis

Occasionally the application of the laws of mechanics will lead to a system of equations for which a closed-form solution is difficult or impossible to obtain. When confronted with this situation, engineers will often use a numerical method which in most cases can be programmed on a microcomputer or "programmable" pocket calculator. Here we will briefly present a computer program for solving a set of linear algebraic equations, and three numerical methods which can be used to solve an algebraic or transcendental equation, evaluate a definite integral, and solve an ordinary differential equation. Application of each method will be explained by example, and an associated computer program written in Microsoft BASIC, which is designed to run on most personal computers, is provided.* A text on numerical analysis should be consulted for further discussion regarding a check of the accuracy of each method and the inherent errors that can develop from the methods.

Linear Algebraic Equations

Application of the equations of static equilibrium or the equations of motion sometimes requires solving a set of linear algebraic equations. The computer program listed in Fig. B–1 can be used for this purpose. It is based on the method of a Gaussian elimination and can solve at

---

*Similar types of programs can be written or purchased for programmable pocket calculators.

```
1 PRINT"Linear system of equations" : PRINT        20 PRINT "Unknowns"                       39 NEXT I
2 DIM A (10, 11)                                   21 FOR I = 1 TO N                        40 FOR I = M + 1 TO N
3 INPUT" Input number of equations : ", N          22 PRINT " X (" ; I ; ") = " ; A (I, N + 1)   41 FC = A (I, M) / A (M, M)
4 PRINT                                            23 NEXT I                                42 FOR J = M + 1 TO N + 1
5 PRINT"A   coefficients"                          24 END                                   43 A (I, J) = A (I , J) – FC * A (M, J)
6 FOR I = 1 TO N                                   25 REM Subroutine Guassian               44 NEXT J
7 FOR J = 1 TO N                                   26 FOR M = 1 TO N                         45 NEXT I
8 PRINT "A(" ; I ; "," ; J ;                       27 NP = M                                 46 NEXT M
9 INPUT" ) =" , A ( I, J )                         28 BG = ABS (A( M, M ))                   47 A (N, N + 1) = A (N, N + 1) / A (N, N)
10 NEXT J                                          29 FOR I = M TO N                         48 FOR I = N – 1 TO 1 STEP – 1
11 NEXT I                                          30 IF ABS (A ( I, M)) < = BG  THEN 33     49 SM = 0
12 PRINT                                           31 BG = ABS (A (I, M))                    50 FOR J = I + 1 TO N
13 PRINT " B   coefficients "                      32 NP = I                                 51 SM = SM + A (I, J) * A (J, N + 1)
14 FOR I = 1 TO N                                  33 NEXT I                                 52 NEXT J
15 PRINT " B ( " ; I ;                             34 IF NP = M THEN 40                      53 A (I, N + 1) = (A ( I, N + 1) – SM) / A (I, I)
16 INPUT ") = " , A (I, N + 1)                     35 FOR I = M TO N + 1                     54 NEXT I
17 NEXT I                                          36 TE = A (M, I)                          55 RETURN
18 GOSUB 25                                        37 A (M, I) = A (NP, I)
19 PRINT                                           38 A (NP, I) = TE
```

Fig. B–1

most 10 equations with 10 unknowns. To do so, the equations should first be written in the following general format:

$$A_{11}x_1 + A_{12}x_2 + \cdots + A_{1n}x_n = B_1$$
$$A_{21}x_1 + A_{22}x_2 + \cdots + A_{2n}x_n = B_2$$
$$\vdots$$
$$A_{n1}x_1 + A_{n2}x_2 + \cdots + A_{nn}x_n = B_n$$

The "A" and "B" coefficients are "called" for when running the program. The output presents the unknowns $x_1, \ldots, x_n$.

EXAMPLE B.1

Solve the two equations

$$3x_1 + x_2 = 4$$
$$2x_1 - x_2 = 10$$

SOLUTION
When the program begins to run, it first calls for the number of equations (2); then the A coefficients in the sequence $A_{11} = 3$, $A_{12} = 1$, $A_{21} = 2$, $A_{22} = -1$; and finally the B coefficients $B_1 = 4$, $B_2 = 10$. The output appears as

Unknowns

$X(1) = 2.8$ *Ans.*

$X(2) = -4.4$ *Ans.*

Simpson's Rule

Simpson's rule is a numerical method that can be used to determine the area under a curve given as a graph or as an explicit function $y = f(x)$. Likewise, it can be used to find the value of a definite integral which involves the function $y = f(x)$. To do so, the area must be subdivided into an *even number* of strips or intervals having a width h. The curve between three consecutive ordinates is approximated by a parabola, and the entire area or definite integral is then determined from the formula

$$\int_{x_0}^{x_n} f(x)\, dx \approx \frac{h}{3}[y_0 + 4(y_1 + y_3 + \cdots + y_{n-1})$$
$$+ 2(y_2 + y_4 + \cdots + y_{n-2}) + y_n] \qquad \text{(B–1)}$$

The computer program for this equation is given in Fig. B–2. For its use, we must first specify the function (on line 6 of the program). The upper and lower limits of the integral and the number of intervals are called for when the program is executed. The value of the integral is then given as the output.

```
1  PRINT" Simpson's rule" : PRINT
2  PRINT" To execute this program : " : PRINT
3  PRINT"      1– Modify right–hand side of the equation given below,
4  PRINT"           then press RETURN key"
5  PRINT"      2– Type RUN 6" : PRINT : EDIT 6
6  DEF FNF (X) = LOG (X)
7  PRINT : INPUT" Enter Lower Limit = ",  A
8  INPUT"    Enter Upper Limit = ", B
9  INPUT"    Enter Number (even) of Intervals = ", N%
10  H= (B–A) / N%;  AR=FNF (A) : X=A+H
11  FOR J%=2  TO  N%
12  K=2* (2–J%+2*INT (J%/2))
13  AR=AR+K*FNF (X)
14  X=X+H: NEXT J%
15  AR=H* (AR+FNF (B)) / 3
16  PRINT"    Integral = ", AR
17  END
```

Fig. B–2

EXAMPLE B.2

Evaluate the definite integral

$$\int_2^5 \ln x \, dx$$

SOLUTION

The interval $x_0 = 2$ to $x_6 = 5$ will be divided into six equal parts ($n = 6$), each having a width $h = (5 - 2)/6 = 0.5$. We then compute $y = f(x) = \ln x$ at each point of subdivision.

n	x_n	y_n
0	2	0.693
1	2.5	0.916
2	3	1.099
3	3.5	1.253
4	4	1.386
5	4.5	1.504
6	5	1.609

Thus, Eq. B–1 becomes

$$\int_2^5 \ln x \, dx \approx \frac{0.5}{3} \, [0.693 + 4(0.916 + 1.253 + 1.504) + 2(1.099 + 1.386) + 1.609]$$

$$\approx 3.66 \qquad\qquad Ans.$$

This answer is equivalent to the exact answer to three significant figures. Obviously, accuracy to a greater number of significant figures can be improved by selecting a smaller interval h (or larger n).

Using the computer program, we first specify the function $\ln x$, line 6 in Fig. B–2. During execution, the program input requires the upper and lower limits 2 and 5 and the number of intervals $n = 6$. The output appears as

$$\text{Integral} = 3.66082 \qquad\qquad Ans.$$

The Secant Method

The secant method is used to find the real roots of an algebraic or transcendental equation $f(x) = 0$. The method derives its name from the fact that the formula used is established from the slope of the secant line to the graph $y = f(x)$. This slope is $[f(x_n) - f(x_{n-1})]/(x_n - x_{n-1})$, and the secant formula is

$$x_{n+1} = x_n - f(x_n)\left[\frac{x_n - x_{n-1}}{f(x_n) - f(x_{n-1})}\right] \qquad (B-2)$$

For application it is necessary to provide two initial guesses, x_0 and x_1, and thereby evaluate x_2 from Eq. B–2 $(n = 1)$. One then proceeds to reapply Eq. B–2 with x_1 and the calculated value of x_2 and obtain x_3 $(n = 2)$, etc., until the value $x_{n+1} \approx x_n$. One can see this will occur if x_n is approaching the root of the function $f(x) = 0$, since the correction term on the right of Eq. B–2 will tend toward zero. In particular, the larger the slope, the smaller the correction to x_n, and the faster the root will be found. On the other hand, if the slope is very small in the neighborhood of the root, the method leads to large corrections for x_n, and convergence to the root is slow and may even lead to a failure to find it. In such cases other numerical techniques must be used for solution.
A computer program based on Eq. B–2 is listed in Fig. B–3. We must first specify the function on line 7 of the program. When the program is executed, two initial guesses, x_0 and x_1, must be entered in order to approximate the solution. The output specifies the value of the root. If it cannot be determined, this is so stated.

```
1  PRINT" Secant method" : PRINT
2  PRINT" To execute this program : " : PRINT
3  PRINT"        1)  Modify right hand side of the equation given below."
4  PRINT"             then press RETURN key."
5  PRINT"        2)  Type RUN 7"
6  PRINT : EDIT 7
7  DEF FNF (X) = .5 * SIN (X) –2 * COS (X) + 1.3
8  INPUT"Enter point #1 =", X
9  INPUT"Enter point #1 =", X1
10  IF X = X1 THEN 14
11  EP = .00001 : TL = 2E – 20
12  FP = (FNF (X1) – FNF(X)) / (X1 – X)
13  IF ABS (FP) > TL THEN 15
14  PRINT"Root can not be found." : END
15  DX = FNF (X1)/FP
16  IF ABS (DX) >EP THEN 19
17  PRINT "Root = " ; X1; "    Function evaluated at this root = " ; FNF (X1)
18  END
19  X = X1 : X1 = X1 – DX
20  GOTO 12
```

Fig. B–3

EXAMPLE B.3

Determine the root of the equation

$$f(x) = 0.5 \sin x - 2 \cos x + 1.30 = 0$$

SOLUTION

Guesses of the initial roots will be $x_0 = 45°$ and $x_1 = 30°$. Applying Eq. B–2,

$$x_2 = 30° - (-0.1821)\frac{(30° - 45°)}{(-0.1821 - 0.2393)} = 36.48°$$

Using this value in Eq. B–2, along with $x_1 = 30°$, we have

$$x_3 = 36.48° - (-0.0108)\frac{36.48° - 30°}{(-0.0108 + 0.1821)} = 36.89°$$

Repeating the process with this value and $x_2 = 36.48°$ yields

$$x_4 = 36.89° - (0.0005)\left[\frac{36.89° - 36.48°}{(0.0005 + 0.0108)}\right] = 36.87°$$

Thus $x = 36.9°$ is appropriate to three significant figures.

 If the problem is solved using the computer program, first we specify the function, line 7 in Fig. B–3. During execution, the first and second guesses must be entered in radians. Choosing these to be 0.8 rad and 0.5 rad, the result appears as

Root = 0.6435022

Function evaluated at this root = 1.66893E-06

This result converted from radians to degrees is therefore

$$x = 36.9° \qquad\qquad Ans.$$

Runge-Kutta Method

The Runge-Kutta method is used to solve an ordinary differential equation. It consists of applying a set of formulas which are used to find specific values of y for corresponding incremental values h in x. The formulas given in general form are as follows:

First-Order Equation.

To integrate $\dot{x} = f(t, x)$ step by step, use

$$x_{i+1} = x_i + \frac{1}{6}(k_1 + 2k_2 + 2k_3 + k_4) \tag{B–3}$$

where

$$k_1 = hf(t_i, x_i)$$

$$k_2 = hf\left(t_i + \frac{h}{2}, x_i + \frac{k_1}{2}\right)$$

$$k_3 = hf\left(t_i + \frac{h}{2}, x_i + \frac{k_2}{2}\right) \tag{B–4}$$

$$k_4 = hf(t_i + h, x_i + k_3)$$

Second-Order Equation.

To integrate $\ddot{x} = f(t, x, \dot{x})$, use

$$x_{i+1} = x_i + h\left[\dot{x}_i + \frac{1}{6}(k_1 + k_2 + k_3)\right]$$

$$\dot{x}_{i+1} = \dot{x}_i + \frac{1}{6}(k_1 + 2k_2 + 2k_3 + k_4) \tag{B–5}$$

where

$$k_1 = hf(t_i, x_i, \dot{x}_i)$$

$$k_2 = hf\left(t_i + \frac{h}{2}, x_i + \frac{h}{2}\dot{x}_i, \dot{x}_i + \frac{k_1}{2}\right)$$

$$k_3 = hf\left(t_i + \frac{h}{2}, x_i + \frac{h}{2}\dot{x}_i + \frac{h}{4}k_1, \dot{x}_i + \frac{k_2}{2}\right) \tag{B–6}$$

$$k_4 = hf\left(t_i + h, x_i + h\dot{x}_i + \frac{h}{2}k_2, \dot{x}_i + k_3\right)$$

To apply these equations, one starts with initial values $t_i = t_0$, $x_i = x_0$ and $\dot{x}_i = \dot{x}_0$ (for the second-order equation). Choosing an increment h for t_0, the four constants k are computed, and these results are substituted into Eq. B–3 or B–5 in order to compute $x_{i+1} = x_1$, $\dot{x}_{i+1} = \dot{x}_1$, corresponding to $t_{i+1} = t_1 = t_0 + h$. Repeating this process using t_1, x_1, $\dot{x}_1$ and h, values for x_2, $\dot{x}_2$ and $t_2 = t_1 + h$ are then computed, etc.

Computer programs which solve first- and second-order differential equations by this method are listed in Figs. B–4 and B–5, respectively. In order to use these programs, the operator specifies the function $\dot{x} = f(t, x)$ or $\ddot{x} = f(t, x, \dot{x})$ (line 7), the initial values t_0, x_0, $\dot{x}_0$ (for second-order equation), the final time t_n, and the step size h. The output gives the values of t, x, and $\dot{x}$ for each time increment until t_n is reached.

```
 1  PRINT" Runge–Kutta  Method for 1–st order Differential Equation " : PRINT
 2  PRINT" To  execute  this  program  : " : PRINT
 3  PRINT"     1)  Modify right hand side of the equation given below , "
 4  PRINT"         then Press RETURN key"
 5  PRINT"     2) Type   RUN 7"
 6  PRINT : EDIT 7
 7  DEF FNF (T , X) = 5*T+X
 8  CLS : PRINT" Initial  Conditions" : PRINT
 9  INPUT" Input   t   = " , T
10  INPUT"        x   = " , X
11  INPUT" Final   t   = " , T1
12  INPUT " step  size  = " , H : PRINT
13  PRINT"          t               x"
14  IF  T > = T1 + H  THEN  23
15  PRINT  USING" # # # # # # . # # # # #" ; T ; X
16  K1 = H*FNF (T, X)
17  K2 = H*FNF (T+. 5*H, X+. 5*K1)
18  K3 = H*FNF (T+. 5*H, X+. 5*K2)
19  K4 = H*FNF (T+H, X+K3)
20  T = T+H
21  X = X + (K1 + K2 + K2 + K3 + K3 + K4) / 6
22  GOTO 14
23  END
```

Fig. B–4

```
 1  PRINT" Runge–Kutta Method for 2–nd order Differential Equation" : PRINT
 2  PRINT"  To execute this program : " : PRINT
 3  PRINT"      1)   Modify right hand side of the equation given below, "
 4  PRINT"           then Press RETURN key"
 5  PRINT"      2)   Type RUN 7"
 6  PRINT : EDIT 7
 7  DEF  FNF (T, X, XD) =
 8  INPUT" Input    t    = " , T
 9  INPUT"         x    = " , X
10  INPUT"       dx/dt    = " , XD
11  INPUT" Final    t    = " , T1
12  INPUT" step size = " , H : PRINT
13  PRINT"         t             x           dx/dt"
14  IF  T> = T1 + H THEN 24
15  PRINT  USING"######, #####" ; T ; X ; XD
16  K1=H*FNF (T, X, XD)
17  K2=H*FNF (T + .5*H, X+ .5*H*XD, XD + .5*K1)
18  K3=H*FNF (T + .5*H, X+ (.5*H)*(XD + .5*K1), XD + .5*K2)
19  K4=H*FNF (T+H, X+H*XD + .5*H*K2, XD + K3)
20  T=T+H
21  X=X+H*XD+H* (K1 + K2 + K3) /6
22  XD=XD+ (K1 + K2 + K2 + K3 + K3 + K4)/6
23  GOTO 14
24  END
```

Fig. B–5

EXAMPLE B.4

Solve the differential equations $\dot{x} = 5t + x$. Obtain the results for two steps using time increments of $h = 0.02$ s. At $t_0 = 0$, $x_0 = 0$.

SOLUTION

This is a first-order equation, so Eqs. B–3 and B–4 apply. Thus, for $t_0 = 0$, $x_0 = 0$, $h = 0.02$, we have

$$k_1 = 0.02(0 + 0) = 0$$

$$k_2 = 0.02[5(0.01) + 0] = 0.001$$

$$k_3 = 0.02[5(0.01) + 0.0005] = 0.00101$$

$$k_4 = 0.02[5(0.02) + 0.00101] = 0.00202$$

$$x_1 = 0 + \tfrac{1}{6}[0 + 2(0.001) + 2(0.00101) + 0.00202] = 0.00101$$

Using the values $t_1 = 0 + 0.02 = 0.02$ and $x_1 = 0.00101$ with $h = 0.02$, the value for x_2 is now computed from Eqs. B–3 and B–4.

$$k_1 = 0.02[5(0.02) + 0.00101] = 0.00202$$

$$k_2 = 0.02[5(0.03) + 0.00202] = 0.00304$$

$$k_3 = 0.02[5(0.03) + 0.00253] = 0.00305$$

$$k_4 = 0.02[5(0.04) + 0.00406] = 0.00408$$

$$x_2 = 0.00101 + \tfrac{1}{6}[0.00202 + 2(0.00304) + 2(0.00305) + 0.00408]$$

$$= 0.00405 \qquad\qquad\qquad\qquad\qquad Ans.$$

To solve this problem using the computer program in Fig. B–4, the function is first entered on line 7, then the data $t_0 = 0$, $x_0 = 0$, $t_n = 0.04$, and $h = 0.02$ is specified. The results appear as

t	x	
0.00000	0.00000	
0.02000	0.00101	Ans.
0.04000	0.00405	

Vector Analysis

The following discussion provides a brief review of vector analysis. A more detailed treatment of these topics is given in *Engineering Mechanics: Statics*.

Vector. A vector, **A**, is a quantity which has magnitude and direction, and adds according to the parallelogram law. As shown in Fig. C–1, **A** = **B** + **C**, where **A** is the *resultant vector* and **B** and **C** are *component vectors*.

Unit Vector. A unit vector, $\mathbf{u}_A$, has a magnitude of one "dimensionless" unit and acts in the same direction as **A**. It is determined by dividing **A** by its magnitude A, i.e,

$$\mathbf{u}_A = \frac{\mathbf{A}}{A} \qquad\qquad (C\text{–}1)$$

Fig. C–1

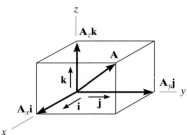

Fig. C–2

Cartesian Vector Notation.

The directions of the positive x, y, z axes are defined by the Cartesian unit vectors $\mathbf{i}, \mathbf{j}, \mathbf{k}$, respectively.

As shown in Fig. C–2, vector $\mathbf{A}$ is formulated by the addition of its x, y, z components as

$$\mathbf{A} = A_x\mathbf{i} + A_y\mathbf{j} + A_z\mathbf{k} \tag{C–2}$$

The *magnitude* of $\mathbf{A}$ is determined from

$$A = \sqrt{A_x^2 + A_y^2 + A_z^2} \tag{C–3}$$

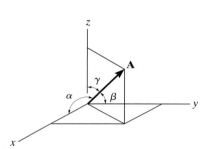

Fig. C–3

The *direction* of $\mathbf{A}$ is defined in terms of its *coordinate direction angles*, α, β, γ, measured from the *tail* of $\mathbf{A}$ to the *positive x, y, z* axes, Fig. C–3. These angles are determined from the *direction cosines* which represent the $\mathbf{i}, \mathbf{j}, \mathbf{k}$ components of the unit vector $\mathbf{u}_A$; i.e., from Eqs. C–1 and C–2

$$\mathbf{u}_A = \frac{A_x}{A}\mathbf{i} + \frac{A_y}{A}\mathbf{j} + \frac{A_z}{A}\mathbf{k} \tag{C–4}$$

so that the direction cosines are

$$\cos \alpha = \frac{A_x}{A} \qquad \cos \beta = \frac{A_y}{A} \qquad \cos \gamma = \frac{A_z}{A} \tag{C–5}$$

Hence, $\mathbf{u}_A = \cos \alpha\mathbf{i} + \cos \beta\mathbf{j} + \cos \gamma\mathbf{k}$, and using Eq. C–3, it is seen that

$$\cos^2 \alpha + \cos^2 \beta + \cos^2 \gamma = 1 \tag{C–6}$$

The Cross Product.

The cross product of two vectors $\mathbf{A}$ and $\mathbf{B}$, which yields the resultant vector $\mathbf{C}$, is written as

$$\mathbf{C} = \mathbf{A} \times \mathbf{B} \tag{C–7}$$

and reads $\mathbf{C}$ equals $\mathbf{A}$ "cross" $\mathbf{B}$. The *magnitude* of $\mathbf{C}$ is

$$C = AB \sin \theta \tag{C–8}$$

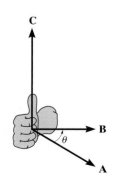

Fig. C–4

where θ is the angle made between the *tails* of $\mathbf{A}$ and $\mathbf{B}$ ($0° \leq \theta \leq 180°$). The *direction* of $\mathbf{C}$ is determined by the right-hand rule, whereby the fingers of the right hand are curled *from $\mathbf{A}$ to $\mathbf{B}$* and the thumb points in the direction of $\mathbf{C}$, Fig. C–4. This vector is perpendicular to the plane containing vectors $\mathbf{A}$ and $\mathbf{B}$.

The vector cross product is *not* commutative, i.e., $\mathbf{A} \times \mathbf{B} \neq \mathbf{B} \times \mathbf{A}$. Rather,

$$\mathbf{A} \times \mathbf{B} = -\mathbf{B} \times \mathbf{A} \qquad (C–9)$$

The distributive law is valid; i.e.,

$$\mathbf{A} \times (\mathbf{B} + \mathbf{D}) = \mathbf{A} \times \mathbf{B} + \mathbf{A} \times \mathbf{D} \qquad (C–10)$$

And the cross product may be multiplied by a scalar m in any manner; i.e.,

$$m(\mathbf{A} \times \mathbf{B}) = (m\mathbf{A}) \times \mathbf{B} = \mathbf{A} \times (m\mathbf{B}) = (\mathbf{A} \times \mathbf{B})m \quad (C–11)$$

Equation C–7 can be used to find the cross product of any pair of Cartesian unit vectors. For example, to find $\mathbf{i} \times \mathbf{j}$, the magnitude is $(i)(j) \sin 90° = (1)(1)(1) = 1$, and its direction $+\mathbf{k}$ is determined from the right-hand rule, applied to $\mathbf{i} \times \mathbf{j}$, Fig. C–2. A simple scheme shown in Fig. C–5 may be helpful in obtaining this and other results when the need arises. If the circle is constructed as shown, then "crossing" two of the unit vectors in a *counterclockwise* fashion around the circle yields a *positive* third unit vector, e.g., $\mathbf{k} \times \mathbf{i} = \mathbf{j}$. Moving *clockwise*, a *negative* unit vector is obtained, e.g., $\mathbf{i} \times \mathbf{k} = -\mathbf{j}$.

If $\mathbf{A}$ and $\mathbf{B}$ are expressed in Cartesian component form, then the cross product, Eq. C–7, may be evaluated by expanding the determinant

Fig. C–5

$$\mathbf{C} = \mathbf{A} \times \mathbf{B} = \begin{vmatrix} \mathbf{i} & \mathbf{j} & \mathbf{k} \\ A_x & A_y & A_z \\ B_x & B_y & B_z \end{vmatrix} \qquad (C–12)$$

which yields

$$\mathbf{C} = (A_y B_z - A_z B_y)\mathbf{i} - (A_x B_z - A_z B_x)\mathbf{j} + (A_x B_y - A_y B_x)\mathbf{k}$$

Recall that the cross product is used in statics to define the moment of a force $\mathbf{F}$ about point O, in which case

$$\mathbf{M}_O = \mathbf{r} \times \mathbf{F} \qquad (C–13)$$

where $\mathbf{r}$ is a position vector directed from point O to *any point* on the line of action of $\mathbf{F}$.

The Dot Product. The dot product of two vectors **A** and **B**, which yields a scalar, is defined as

$$\mathbf{A} \cdot \mathbf{B} = AB \cos \theta \qquad \text{(C–14)}$$

and reads **A** "dot" **B**. The angle θ is formed between the *tails* of **A** and **B** ($0° \leq \theta \leq 180°$).

The dot product is commutative; i.e.,

$$\mathbf{A} \cdot \mathbf{B} = \mathbf{B} \cdot \mathbf{A} \qquad \text{(C–15)}$$

The distributive law is valid; i.e.,

$$\mathbf{A} \cdot (\mathbf{B} + \mathbf{D}) = \mathbf{A} \cdot \mathbf{B} + \mathbf{A} \cdot \mathbf{D} \qquad \text{(C–16)}$$

And scalar multiplication can be performed in any manner, i.e.,

$$m(\mathbf{A} \cdot \mathbf{B}) = (m\mathbf{A}) \cdot \mathbf{B} = \mathbf{A} \cdot (m\mathbf{B}) = (\mathbf{A} \cdot \mathbf{B})m \qquad \text{(C–17)}$$

Using Eq. C–14, the dot product between any two Cartesian vectors can be determined. For example, $\mathbf{i} \cdot \mathbf{i} = (1)(1) \cos 0° = 1$ and $\mathbf{i} \cdot \mathbf{j} = (1)(1) \cos 90° = 0$.

If **A** and **B** are expressed in Cartesian component form, then the dot product, Eq. C–14, can be determined from

$$\mathbf{A} \cdot \mathbf{B} = A_x B_x + A_y B_y + A_z B_z \qquad \text{(C–18)}$$

The dot product may be used to determine the *angle θ formed between two vectors*. From Eq. C–14,

$$\theta = \cos^{-1}\left(\frac{\mathbf{A} \cdot \mathbf{B}}{AB}\right) \qquad \text{(C–19)}$$

It is also possible to find the *component of a vector in a given direction* using the dot product. For example, the magnitude of the component (or projection) of vector **A** in the direction of **B**, Fig. C–6, is defined by $A \cos \theta$. From Eq. C–14, this magnitude is

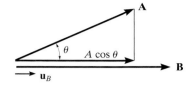

Fig. C–6

$$A \cos \theta = \mathbf{A} \cdot \frac{\mathbf{B}}{B} = \mathbf{A} \cdot \mathbf{u}_B \qquad \text{(C–20)}$$

where $\mathbf{u}_B$ represents a unit vector acting in the direction of **B**, Fig. C–6.

Differentiation and Integration of Vector Functions. The rules for differentiation and integration of the sums and products of scalar functions also apply to vector functions. Consider, for example, the two vector functions $\mathbf{A}(s)$ and $\mathbf{B}(s)$. Provided these functions are smooth and continuous for all s, then

$$\frac{d}{ds}(\mathbf{A} + \mathbf{B}) = \frac{d\mathbf{A}}{ds} + \frac{d\mathbf{B}}{ds} \qquad \text{(C–21)}$$

$$\int (\mathbf{A} + \mathbf{B})\, ds = \int \mathbf{A}\, ds + \int \mathbf{B}\, ds \qquad \text{(C–22)}$$

For the cross product,

$$\frac{d}{ds}(\mathbf{A} \times \mathbf{B}) = \left(\frac{d\mathbf{A}}{ds} \times \mathbf{B} \right) + \left(\mathbf{A} \times \frac{d\mathbf{B}}{ds} \right) \qquad \text{(C–23)}$$

Similarly, for the dot product,

$$\frac{d}{ds}(\mathbf{A} \cdot \mathbf{B}) = \frac{d\mathbf{A}}{ds} \cdot \mathbf{B} + \mathbf{A} \cdot \frac{d\mathbf{B}}{ds} \qquad \text{(C–24)}$$

Review for the Fundamentals of Engineering Examination

The Fundamentals of Engineering (FE) exam is given semiannually by the National Council of Engineering Examiners (NCEE) and is one of the requirements for obtaining a Professional Engineering License. A portion of this exam contains problems in dynamics, and this appendix provides a review of the subject matter most often asked on this exam. Before solving any of the problems, you should review the sections indicated in each chapter in order to become familiar with the boldfaced definitions and the procedures used to solve the various types of problems. Also, review the example problems in these sections.

The following problems are arranged in the same sequence as the topics in each chapter. Besides helping as preparation for the FE exam, these problems also provide additional examples for general practice of the subject matter. Partial solutions and answers to *all the problems* are given at the back of this appendix.

Chapter 12—Review Sections 12.1, 12.4–12.6, 12.8–12.9

D–1. The position of a particle is $s = (0.5t^3 + 4t)$ ft, where t is in seconds. Determine the velocity and the acceleration of the particle when $t = 3$ s.

D–2. After traveling a distance of 100 m, a particle reaches a velocity of 30 m/s, starting from rest. Determine its constant acceleration.

D–3. A particle moves in a straight line such that $s = (12t^3 + 2t^2 + 3t)$ m, where t is in seconds. Determine the velocity and acceleration of the particle when $t = 2$ s.

D–4. A particle moves along a straight line such that $a = (4t^2 - 2)$ m/s^2, where t is in seconds. When $t = 0$, the particle is located 2 m to the left of the origin, and when $t = 2$ s, it is 20 m to the left of the origin. Determine the position of the particle when $t = 4$ s.

D–5. Determine the speed at which the basketball at A must be thrown at the angle of 30° so that it makes it to the basket at B.

Prob. D–5

D–6. A particle moves with curvilinear motion in the x–y plane such that the y component of motion is described by the equation $y = (7t^3)$ m, where t is in seconds. If the particle starts from rest at the origin when $t = 0$, and maintains a *constant* acceleration in the x direction of 12 m/s^2, determine the particle's speed when $t = 2$ s.

D–7. Water is sprayed at an angle of 90° from the slope at 20 m/s. Determine the range R.

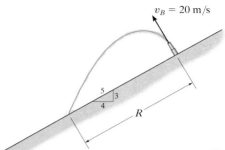

Prob. D–7

D–8. An automobile is traveling with a *constant speed* along a horizontal circular curve that has a radius of $\rho = 250$ m. If the magnitude of acceleration is $a = 1.5$ m/s^2, determine the speed at which the automobile is traveling.

D–9. A boat is traveling along a circular path having a radius of 30 m. Determine the magnitude of the boat's acceleration if at a given instant the boat's speed is $v = 6$ m/s and the rate of increase in speed is $\dot{v} = 2$ m/s^2.

D–10. A train travels along a horizontal circular curve that has a radius of 600 m. If the speed of the train is uniformly increased from 40 km/h to 60 km/h in 5 s, determine the magnitude of the acceleration at the instant the speed of the train is 50 km/h.

D–11. At a given instant, the automobile has a speed of 25 m/s and an acceleration of 3 m/s^2 acting in the direction shown. Determine the radius of curvature of the path and the rate of increase of the automobile's speed.

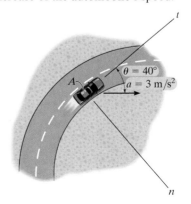

Prob. D–11

D–12. At the instant shown, cars A and B are traveling at the speeds shown. If B is accelerating at 1200 km/h² while A maintains a constant speed, determine the velocity and acceleration of A with respect to B.

Prob. D–12

D–13. Determine the speed of point P on the cable in order to lift the platform at 2 m/s.

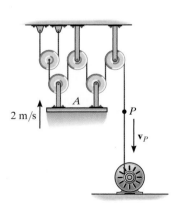

Prob. D–13

Chapter 13—Review Sections 13.1–13.5

D–14. The effective weight of a man in an elevator varies between 130 lb and 170 lb while he is riding in the elevator. When the elevator is at rest the man weighs 153 lb. Determine how fast the elevator car can accelerate, going up and going down.

D–15. Neglecting friction and the mass of the pulley and cord, determine the acceleration at which the 4-kg block B will descend. What is the tension in the cord? Block A has a mass of 2 kg.

Prob. D–15

D–16. The blocks are suspended over a pulley by a rope. Neglecting the mass of the rope and the pulley, determine the acceleration of both blocks and the tension in the rope.

Prob. D–16

D–17. Block B rests upon a smooth surface. If the coefficients of static and kinetic friction between A and B are $\mu_s = 0.4$ and $\mu_k = 0.3$, respectively, determine the acceleration of each block if $P = 6$ lb.

D–19. Determine the maximum speed that the jeep can travel over the crest of the hill and not lose contact with the road.

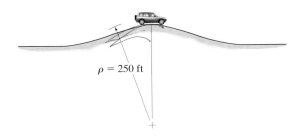

Prob. D–17

Prob. D–19

D–18. The block rests at a distance of 2 m from the center of the platform. If the coefficient of static friction between the block and the platform is $\mu_s = 0.3$, determine the maximum speed which the block can attain before it begins to slip. Assume the angular motion of the disk is slowly increasing.

D–20. A pilot weighs 150 lb and is traveling at a constant speed of 120 ft/s. Determine the normal force he exerts on the seat of the plane when he is upside down at A. The loop has a radius of curvature of 400 ft.

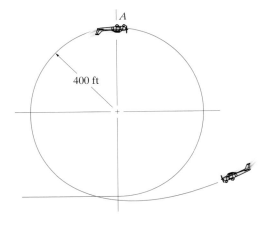

Prob. D–18

Prob. D–20

D–21. The sports car is traveling along a 30° banked road having a radius of curvature of $\rho = 500$ ft. If the coefficient of static friction between the tires and the road is $\mu_s = 0.2$, determine the maximum safe speed for travel so no slipping occurs. Neglect the size of the car.

Prob. D–21

D–22. The 5-lb pendulum bob B is released from rest when $\theta = 0°$. Determine the tension in string BC immediately after it is released and when the pendulum reaches point D, where $\theta = 90°$.

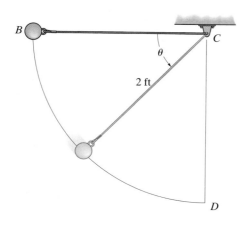

Prob. D–22

Chapter 14—Review All Sections

D–23. A 15 000-lb freight car is pulled along a horizontal track. If the car starts from rest and attains a velocity of 40 ft/s after traveling a distance of 300 ft, determine the total work done on the car by the towing force in this distance if the rolling frictional force between the car and track is 80 lb.

D–24. The 20-lb block resting on the 30° inclined plane is acted upon by a 40-lb force. If the block's initial velocity is 5 ft/s down the plane, determine its velocity after it has traveled 10 ft down the plane. The coefficient of kinetic friction between the block and the plane is $\mu_k = 0.2$.

Prob. D–24

D–25. The 3-kg block is subjected to the action of the two forces shown. If the block starts from rest, determine the distance it has moved when it attains a velocity of 10 m/s. The coefficient of kinetic friction between the block and the surface is $\mu_k = 0.2$.

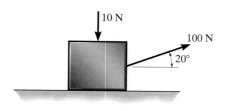

Prob. D–25

D–26. The 6-lb ball is to be fired from rest using a spring having a stiffness of $k = 40$ lb/ft. Determine how far the spring must be compressed so that when the ball reaches a height of 8 ft it has a velocity of 6 ft/s.

Prob. D–26

D–28. The 5-lb collar is released from rest at A and travels along the frictionless guide. Determine the speed of the collar when it strikes the stop B. The spring has an unstretched length of 0.5 ft.

Prob. D–28

D–27. The 2-kg pendulum bob is released from rest when it is at A. Determine the speed of the bob and the tension in the cord when the bob passes through its lowest position, B.

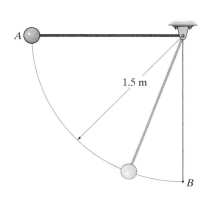

Prob. D–27

D–29. The 2-kg collar is given a downward velocity of 4 m/s when it is at A. If the spring has an unstretched length of 1 m and a stiffness of $k = 30$ N/m, determine the velocity of the collar at $s = 1$ m.

Prob. D–29

Chapter 15—Review Sections 15.1–15.4

D–30. A 30-ton engine exerts a constant horizontal force of $40(10^3)$ lb on a train having three cars that have a total weight of 250 tons. If the rolling resistance is 10 lb per ton for both the engine and cars, determine how long it takes to increase the speed of the train from 20 ft/s to 30 ft/s. What is the driving force exerted by the engine wheels on the tracks?

D–31. A 5-kg block is moving up a 30° inclined plane with an initial velocity of 3 m/s. If the coefficient of kinetic friction between the block and the plane is $\mu_k = 0.3$, determine how long a 100-N horizontal force must act on the block in order to increase the velocity of the block to 10 m/s up the plane.

D–32. The 10-lb block A attains a velocity of 1 ft/s in 5 seconds, starting from rest. Determine the tension in the cord and the coefficient of kinetic friction between block A and the horizontal plane. Neglect the weight of the pulley. Block B has a weight of 8 lb.

Prob. D–32

D–33. Determine the velocity of each block 10 seconds after the blocks are released from rest. Neglect the mass of the pulleys.

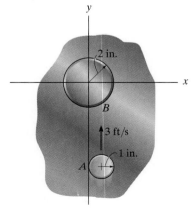

Prob. D–33

D–34. The two blocks have a coefficient of restitution of $e = 0.5$. If the surface is smooth, determine the velocity of each block after impact.

Prob. D–34

D–35. A 6-kg disk A has an initial velocity of $(v_A)_1 = 20$ m/s and strikes head-on disk B that has a mass of 24 kg and is originally at rest. If the collision is perfectly elastic, determine the speed of each disk after the collision and the impulse which disk A imparts to disk B.

D–36. Blocks A and B weigh 5 lb and 10 lb, respectively. After striking block B, A slides 2 in. to the right, and B slides 3 in. to the right. If the coefficient of kinetic friction between the blocks and the surface is $\mu_k = 0.2$, determine the coefficient of restitution between the blocks. Block B is originally at rest.

Prob. D–36

D–37. Disk A weighs 2 lb and is sliding on the smooth horizontal plane with a velocity of 3 ft/s. Disk B weighs 11 lb and is initially at rest. If after the impact A has a velocity of 1 ft/s, directed along the positive x axis, determine the speed of disk B after impact.

Prob. D–37

Chapter 16—Review Sections 16.3, 16.5–16.7

D–38. If gear A is rotating clockwise with an angular velocity of $\omega_A = 3$ rad/s, determine the angular velocities of gears B and C. Gear B is one unit, having radii of 2 in. and 5 in.

Prob. D–38

D–39. The spin drier of a washing machine has a constant angular acceleration of 2 rev/s^2, starting from rest. Determine how many turns it makes in 10 seconds and its angular velocity when $t = 5$ s.

D–40. Starting from rest, point P on the cord has a constant acceleration of 20 ft/s^2. Determine the angular acceleration and angular velocity of the disk after it has completed 10 revolutions. How many revolutions will the disk turn after it has completed 10 revolutions and P continues to move downward for 4 seconds longer?

Prob. D–40

D–41. The center of the wheel has a velocity of 3 m/s. At the same time, it is slipping and has a clockwise angular velocity of $\omega = 2$ rad/s. Determine the velocity of point A at the instant shown.

Prob. D–41

D–42. A cord is wrapped around the inner core of the gear and it is pulled with a constant velocity of 2 ft/s. Determine the velocity of the center of the gear, C.

Prob. D–42

D–43.   The center of the wheel is moving to the right with a speed of 2 m/s. If no slipping occurs at the ground, A, determine the velocity of point B at the instant shown.

Prob. D–43

D–44.   If the velocity of the slider block at B is 2 ft/s to the left, compute the velocity of the block at A and the angular velocity of the rod at the instant shown.

Prob. D–44

D–45.   Determine the angular velocity of link AB at the instant shown.

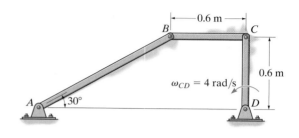

Prob. D–45

D–46.   When the slider block C is in the position shown, the link AB has a clockwise angular velocity of 2 rad/s. Determine the velocity of block C at this instant.

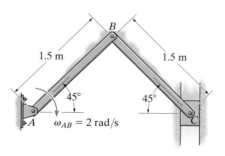

Prob. D–46

D–47.   The center of the pulley is being lifted vertically with an acceleration of 3 m/s², and at the instant shown its velocity is 2 m/s. Determine the accelerations of points A and B. Assume that the rope does not slip on the pulley's surface.

Prob. D–47

Chapter 17—Review All Sections

D–49.   The 3500-lb car has a center of mass located at G. Determine the normal reactions of both front and both rear wheels on the road and the acceleration of the car if it is rolling freely down the incline. Neglect the weight of the wheels.

Prob. D–49

D–48.   At a given instant, the slider block A has the velocity and deceleration shown. Determine the acceleration of block B and the angular acceleration of the link at this instant.

D–50.   The 20-lb link AB is pinned to a moving frame at A and held in a vertical position by means of a string BC which can support a maximum tension of 10 lb. Determine the maximum acceleration of the link without breaking the string. What are the corresponding components of reaction at the pin A?

Prob. D–48

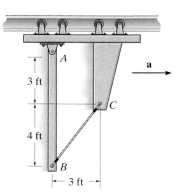

Prob. D–50

D–51. The 50-lb triangular plate is released from rest. Determine the initial angular acceleration of the plate and the horizontal and vertical components of reaction at B. The moment of inertia of the plate about the pinned axis B is $I_B = 2.30$ slug · ft^2.

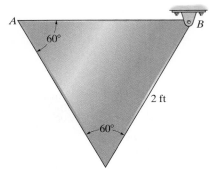

Prob. D–51

D–52. The 20-kg slender rod is pinned at O. Determine the vertical reaction at O just after the cable is cut.

Prob. D–52

D–53. The 20-kg wheel has a radius of gyration of $k_G = 0.8$ m. Determine the angular acceleration of the wheel if no slipping occurs.

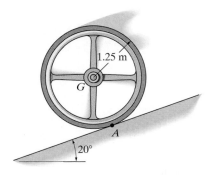

Prob. D–53

D–54. The 15-kg wheel has a wire wrapped around its inner hub and is released from rest on the inclined plane, for which the coefficient of kinetic friction is $\mu_k = 0.1$. If the centroidal radius of gyration of the wheel is $k_G = 0.8$ m, determine the angular acceleration of the wheel.

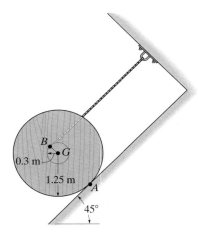

Prob. D–54

D–55. The 2-kg gear is at rest on the surface of a gear rack. If the rack is suddenly given an acceleration of 5 m/s^2, determine the initial angular acceleration of the gear. The radius of gyration of the gear is $k_G = 0.3$ m.

Prob. D–55

SOLUTIONS AND ANSWERS

D–1. $v = \dfrac{ds}{dt} = 1.5t^2 + 4|_{t=3} = 17.5 \text{ ft/s}$ *Ans.*

$a = \dfrac{dv}{dt} = 3t|_{t=3} = 9 \text{ ft/s}^2$ *Ans.*

D–2. $(30)^2 = (0)^2 + 2a(100 - 0)$

$a = 4.5 \text{ m/s}^2$ *Ans.*

D–3. $v = \dfrac{ds}{dt} = 36t^2 + 4t + 3|_{t=2} = 155 \text{ m/s}$ *Ans.*

$a = \dfrac{dv}{dt} = 72t + 4|_{t=2} = 148 \text{ m/s}^2$ *Ans.*

D–4. $v = \int (4t^2 - 2)\, dt$

$v = \dfrac{4}{3}t^3 - 2t + C_1$

$s = \int \left(\dfrac{4}{3}t^3 - 2t + C_1 \right) dt$

$s = \dfrac{1}{3}t^4 - t^2 + C_1 t + C_2$

$t = 0, \ s = -2, \ C_2 = -2$

$t = 2, \ s = -20, \ C_1 = -9.67$

$t = 4, \ s = 28.7 \text{ m}$ *Ans.*

D–5. $\overset{+}{\rightarrow} s = s_0 + v_0 t$

$10 = 0 + v_A \cos 30°t$

$+\uparrow s = s_0 + v_0 t + \dfrac{1}{2}a_c t^2$

$3 = 1.5 + v_A \sin 30°t + \dfrac{1}{2}(-9.81)t^2$

$t = 0.9335, \ v_A = 12.4 \text{ m/s}$ *Ans.*

D–6. $v = v_0 + a_c t$

$v_x = 0 + 12(2) = 24 \text{ m/s}$

$v_y = \dfrac{dy}{dt} = 21t^2|_{t=2} = 84 \text{ m/s}$

$v = \sqrt{(24)^2 + (84)^2} = 87.4 \text{ m/s}$ *Ans.*

D–7. $(\overset{+}{\leftarrow})\, s = s_0 + v_0 t$

$R\left(\dfrac{4}{5}\right) = 0 + 20\left(\dfrac{3}{5}\right)t$

$(+\uparrow)\, s = s_0 + v_0 t + \dfrac{1}{2}a_c t^2$

$-R\left(\dfrac{3}{5}\right) = 0 + 20\left(\dfrac{4}{5}\right)t + \dfrac{1}{2}(-9.81)t^2$

$t = 5.10 \text{ s}$

$R = 76.5 \text{ m}$ *Ans.*

D–8. $a_t = 0$

$a_n = a = 1.5 = \dfrac{v^2}{250}; \ v = 19.4 \text{ m/s}$ *Ans.*

D–9. $a_t = 2 \text{ m/s}^2$

$a_n = \dfrac{v^2}{\rho} = \dfrac{(6)^2}{30} = 1.20 \text{ m/s}^2$

$a = \sqrt{(2)^2 + (1.20)^2} = 2.33 \text{ m/s}^2$ *Ans.*

D–10. $a_t = \dfrac{\Delta v}{\Delta t} = \dfrac{60 - 40}{[(5 - 0)/3600]} = 14\,400 \text{ km/h}^2$

$a_n = \dfrac{v^2}{\rho} = \dfrac{(50)^2}{0.6} = 4167 \text{ km/h}^2$

$a = \sqrt{(14.4)^2 + (4.167)^2}\, 10^3 = 15.0(10^3) \text{ km/h}^2$ *Ans.*

D–11. $a_t = 3 \cos 40° = 2.30 \text{ m/s}^2$ *Ans.*

$a_n = \dfrac{v^2}{\rho}; \ 3 \sin 40° = \dfrac{(25)^2}{\rho}, \ \rho = 324 \text{ m}$ *Ans.*

D–12. $\mathbf{v}_A = \mathbf{v}_B + \mathbf{v}_{A/B}$

$-20 \cos 45°\mathbf{i} + 20 \sin 45°\mathbf{j} = 65\mathbf{i} + \mathbf{v}_{A/B}$

$\mathbf{v}_{A/B} = -79.14\mathbf{i} + 14.14\mathbf{j}$

$v_{A/B} = \sqrt{(-79.14)^2 + (14.14)^2} = 80.4 \text{ km/h}$ *Ans.*

$\mathbf{a}_A = \mathbf{a}_B + \mathbf{a}_{A/B}$

$\dfrac{(20)^2}{0.1}\cos 45°\mathbf{i} + \dfrac{(20)^2}{0.1}\sin 45°\mathbf{j} = 1200\mathbf{i} + \mathbf{a}_{A/B}$

$\mathbf{a}_{A/B} = 1628\mathbf{i} + 2828\mathbf{j}$

$a_{A/B} = \sqrt{(1628)^2 + (2828)^2} = 3.26(10^3) \text{ km/h}^2$ *Ans.*

D–13. $4s_A + s_P = l$

$v_P = -4v_A = -4(-2) = 8 \text{ m/s}$ *Ans.*

D–14. $+\uparrow \Sigma F_y = ma_y; \ 170 - 153 = \dfrac{153}{32.2}a$

$a = 3.58 \text{ ft/s}^2 \uparrow$ *Ans.*

$+\downarrow \Sigma F_y = ma_y; \ 153 - 130 = \dfrac{153}{32.2}a'$

$a' = 4.84 \text{ ft/s}^2 \downarrow$ *Ans.*

D–15.   Block B:

$+\downarrow \Sigma F_y = ma_y$; $4(9.81) - T = 4a$

Block A:

$\xleftrightarrow{} \Sigma F_x = ma_x$; $T = 2a$

$T = 13.1$ N, $a = 6.54$ m/s^2 *Ans.*

D–16.   Block A:

$+\downarrow \Sigma F_y = ma_y$; $15(9.81) - T = -15a$

Block B:

$+\downarrow \Sigma F_y = ma_y$; $25(9.81) - T = 25a$

$a = 2.45$ m/s^2, $T = 184$ N   *Ans.*

D–17.   Blocks A and B:

$\xrightarrow{} \Sigma F_x = ma_x$; $6 = \dfrac{70}{32.2}a$; $a = 2.76$ ft/s^2

Check if slipping occurs between A and B.

$\xrightarrow{} \Sigma F_x = ma_x$; $6 - F = \dfrac{20}{32.2}(2.76)$;

$F = 4.29$ lb $< 0.4(20) = 8$ lb

$a_A = a_B = 2.76$ ft/s^2 *Ans.*

D–18. $\Sigma F_n = m\dfrac{v^2}{\rho}$; $(0.3)m(9.81) = m\dfrac{v^2}{2}$

$v = 2.43$ m/s   *Ans.*

D–19. $+\downarrow \Sigma F_n = ma_n$; $m(32.2) = m\left(\dfrac{v^2}{250}\right)$

$v = 89.7$ ft/s   *Ans.*

D–20. $+\downarrow \Sigma F_n = ma_n$; $150 + N_p = \dfrac{150}{32.2}\left(\dfrac{(120)^2}{400}\right)$

$N_p = 17.7$ lb   *Ans.*

D–21. $\xleftrightarrow{} \Sigma F_n = ma_n$; $N_c \sin 30° + 0.2N_c \cos 30° = m\dfrac{v^2}{500}$

$+\uparrow \Sigma F_b = 0$;

$N_c \cos 30° - 0.2N_c \sin 30° - m(32.2) = 0$

$v = 119$ ft/s   *Ans.*

D–22.   At B:

$\xrightarrow{} \Sigma F_n = ma_n$, $T = \left(\dfrac{5}{32.2}\right)\left(\dfrac{(0)^2}{2}\right) = 0$ *Ans.*

In the general position,

$+\nearrow \Sigma F_n = ma_n$; $T - 5 \sin \theta = \left(\dfrac{5}{32.2}\right)\left(\dfrac{v^2}{2}\right)$

$\searrow + \Sigma F_t = ma_t$; $5 \cos \theta = \dfrac{5}{32.2}\left(\dfrac{v\, dv}{2\, d\theta}\right)$

$\displaystyle\int_0^{90°} 64.4 \cos \theta\, d\theta = \int_0^v v\, dv$

$v = 11.3$ ft/s

When $\theta = 90°$; $T = 15$ lb   *Ans.*

D–23. $T_1 + \Sigma U_{1-2} = T_2$

$0 + U_{1-2} - 80(300) = \dfrac{1}{2}\left(\dfrac{15000}{32.2}\right)(40)^2$

$U_{1-2} = 397(10^3)$ ft · lb   *Ans.*

D–24. $T_1 + \Sigma U_{1-2} = T_2$

$\dfrac{1}{2}\left(\dfrac{20}{32.2}\right)(5)^2 + 40(10) - $

$(0.2)(20 \cos 30°)(10) + 20(10 \sin 30°) = \dfrac{1}{2}\left(\dfrac{20}{32.2}\right)v^2$

$v = 39.0$ ft/s   *Ans.*

D–25. $+\uparrow \Sigma F_y = ma_y$; $N_b + 100 \sin 20° - 10 - 3(9.81) = 0$

$N_b = 5.23$ N

$T_1 + \Sigma U_{1-2} = T_2$

$0 + (100 \cos 20°)d - 0.2(5.23)d = \dfrac{1}{2}(3)(10)^2$

$d = 1.61$ m   *Ans.*

D–26. $T_1 + V_1 = T_2 + V_2$

$0 + \dfrac{1}{2}(40)(x)^2 = \dfrac{1}{2}\left(\dfrac{6}{32.2}\right)(6)^2 + 6(8)$

$x = 1.60$ ft   *Ans.*

D–27. $T_A + V_A = T_B + V_B$

$0 + 2(9.81)(1.5) = \dfrac{1}{2}(2)(v_B)^2 + 0$

$v_B = 5.42$ m/s   *Ans.*

$+\uparrow \Sigma F_n = ma_n$; $T - 2(9.81) = 2\left(\dfrac{(5.42)^2}{1.5}\right)$

$T = 58.9$ N   *Ans.*

D–28. $T_A + V_A = T_B + V_B$

$0 + \dfrac{1}{2}(4)(2.5 - 0.5)^2 + 5(2.5)$

$= \dfrac{1}{2}\left(\dfrac{5}{32.2}\right)v_B^2 + \dfrac{1}{2}(4)(1 - 0.5)^2$

$v_B = 16.0$ ft/s   *Ans.*

D–29. $T_1 + V_1 = T_2 + V_2$

$\dfrac{1}{2}(2)(4)^2 + \dfrac{1}{2}(30)(2 - 1)^2$

$= \dfrac{1}{2}(2)(v)^2 - 2(9.81)(1) + \dfrac{1}{2}(30)(\sqrt{5} - 1)^2$

$v = 5.26$ m/s   *Ans.*

D–30. Three cars:

$$\xrightarrow{+} mv_1 + \Sigma \int F \, dt = mv_2$$

$$\frac{250(2000)}{32.2}(20) + 40(10^3)(t) - 10(250)t$$

$$= \frac{250(2000)}{32.2}(30)$$

$$t = 4.14 \text{ s} \qquad\qquad Ans.$$

Engine:

$$\xrightarrow{+} mv_1 + \Sigma \int F \, dt = mv_2$$

$$\frac{30(2000)}{32.2}(20) + F(4.14) - 10(30)(4.14)$$

$$= \frac{30(2000)}{32.2}(30)$$

$$F = 4800 \text{ lb} \qquad\qquad Ans.$$

D–31. $+\nwarrow \Sigma F_y = 0;$

$$N_b - 5(9.81) \cos 30° - 100 \sin 30° = 0$$

$$N_b = 92.48 \text{ N}$$

$$+\nearrow mv_1 + \Sigma \int F \, dt = mv_2$$

$$5(3) + (100 \cos 30°)t - 5(9.81) \sin 30°(t) -$$

$$0.3(92.48)t = 5(10)$$

$$t = 1.02 \text{ s} \qquad\qquad Ans.$$

D–32. Block B:

$$(+\downarrow) \, mv_1 + \int F \, dt = mv_2$$

$$0 + 8(5) - T(5) = \frac{8}{32.2}(1)$$

$$T = 7.95 \text{ lb} \qquad\qquad Ans.$$

Block A:

$$(\xrightarrow{+}) \, mv_1 + \int F \, dt = mv_2$$

$$0 + 7.95(5) - \mu_k(10)(5) = \frac{10}{32.2}(1)$$

$$\mu_k = 0.789 \qquad\qquad Ans.$$

D–33. $2s_A + s_B = l$

$$2v_A = -v_B$$

$$+\downarrow m(v_A)_1 + \int F \, dt = m(v_A)_2$$

$$0 + 10(9.81)(10) - 2T(10) = 10(v_A)_2$$

$$+\downarrow m(v_B)_1 + \int F \, dt = m(v_B)_2$$

$$0 + 50(9.81)(10) - T(10) = 50(v_B)_2$$

$$T = 70.1 \text{ N}$$

$$(v_A)_2 = -42.0 \text{ m/s} = 42.0 \text{ m/s} \uparrow \qquad Ans.$$

$$(v_B)_2 = 84.1 \text{ m/s} \downarrow \qquad\qquad Ans.$$

D–34. $\xrightarrow{+} \Sigma mv_1 = \Sigma mv_2; \; 5(2) - 2(5) = 5(v_A)_2 + 2(v_B)_2$

$$\xrightarrow{+} e = \frac{(v_B)_2 - (v_A)_2}{(v_A)_1 - (v_B)_1}; \; 0.5 = \frac{(v_B)_2 - (v_A)_2}{2 - (-5)}$$

$$(v_A)_2 = -1 \text{ m/s} = 1 \text{ m/s} \leftarrow \qquad Ans.$$

$$(v_B)_2 = 2.5 \text{ m/s} \rightarrow \qquad\qquad Ans.$$

D–35. $\xrightarrow{+} \Sigma mv_1 = \Sigma mv_2; \; 6(20) + 0 = 6(v_A)_2 + 24(v_B)_2$

$$\xrightarrow{+} e = \frac{(v_B)_2 - (v_A)_2}{(v_A)_1 - (v_B)_1}; \; 1 = \frac{(v_B)_2 - (v_A)_2}{20 - 0}$$

$$(v_A)_2 = -12 \text{ m/s} = 12 \text{ m/s} \leftarrow \qquad Ans.$$

$$(v_B)_2 = 8 \text{ m/s} \rightarrow \qquad\qquad Ans.$$

Disk A:

$$mv_1 + \int F \, dt = mv_2;$$

$$6(20) - \int F \, dt = -6(12)$$

$$\int F \, dt = 192 \text{ N} \cdot \text{s} \qquad\qquad Ans.$$

D–36. After collision: $T_1 + \Sigma U_{1-2} = T_2$

$$\frac{1}{2}\left(\frac{5}{32.2}\right)(v_A)_2^2 - 0.2(5)\left(\frac{2}{12}\right) = 0$$

$$(v_A)_2 = 1.465 \text{ ft/s}$$

$$\frac{1}{2}\left(\frac{10}{32.2}\right)(v_B)_2^2 - 0.2(10)\left(\frac{3}{12}\right) = 0$$

$$(v_B)_2 = 1.794 \text{ ft/s}$$

$$\Sigma mv_1 = \Sigma mv_2$$

$$\frac{5}{32.2}(v_A)_1 + 0 = \frac{5}{32.2}(1.465) + \frac{10}{32.2}(1.794)$$

$$(v_A)_1 = 5.054$$

$$e = \frac{(v_B)_2 - (v_A)_2}{(v_A)_1 - (v_B)_1} = \frac{1.794 - 1.465}{5.054 - 0} = 0.0652 \; Ans.$$

D–37. $\Sigma m(v_x)_1 = \Sigma m(v_x)_2$

$$0 + 0 = \frac{2}{32.2}(1) + \frac{11}{32.2}(v_{Bx})_2$$

$$(v_{Bx})_2 = -0.1818 \text{ ft/s}$$

$$\Sigma m(v_y)_1 = \Sigma m(v_y)_2$$

$$\frac{2}{32.2}(3) + 0 = 0 + \frac{11}{32.2}(v_{By})_2$$

$$(v_{By})_2 = 0.545 \text{ ft/s}$$

$$(v_B)_2 = \sqrt{(-0.1818)^2 + (0.545)^2} = 0.575 \text{ ft/s} \; Ans.$$

D–38. $\omega_B(5) = 3(4)$

$$\omega_B = 2.40 \text{ rad/s} \curvearrowright \qquad\qquad Ans.$$

$$\omega_C(3) = 2.40(2)$$

$$\omega_C = 1.60 \text{ rad/s} \curvearrowleft \qquad\qquad Ans.$$

D–39. $\theta = \theta_0 + \omega_0 t + \frac{1}{2}\alpha_c t^2$

$\theta = 0 + 0 + \frac{1}{2}(2)(10)^2 = 100 \text{ rev}$ *Ans.*

$\omega = \omega_0 + \alpha_c t$

$\omega = 0 + 2(5) = 10 \text{ rev/s}$ *Ans.*

D–40. $\alpha = \dfrac{a_t}{r} = \dfrac{20}{2} = 10 \text{ rad/s}^2$ *Ans.*

$10 \text{ rev} = 20\pi \text{ rad}$

$\omega^2 = \omega_0^2 + 2\,\alpha_c(\theta - \theta_0)$

$\omega^2 = 0 + 2(10)(20\pi - 0)$

$\omega = 35.4 \text{ rad/s}$ *Ans*

$\theta = \theta_0 + \omega_0 t + \frac{1}{2}\alpha_c t^2$

$\theta = 0 + 35.4(4) + \frac{1}{2}(10)(4)^2 = 222 \text{ rad}$

$\theta = \dfrac{222}{2\pi} = 35.3 \text{ rev.}$ *Ans.*

D–41. $\mathbf{v}_A = \mathbf{v}_C + \boldsymbol{\omega} \times \mathbf{r}_{A/C}$

$v_A \mathbf{i} = 3\mathbf{i} + (-2\mathbf{k}) \times (-0.5\mathbf{j})$

$v_A = 2 \text{ m/s} \rightarrow$ *Ans.*

D–42. $\mathbf{v}_A = \mathbf{v}_B + \boldsymbol{\omega} \times \mathbf{r}_{A/B}$

$2\mathbf{i} = 0 + (-\omega\mathbf{k}) \times (1.5\mathbf{j})$

$\omega = 1.33 \text{ rad/s} \downarrow$

$\mathbf{v}_C = \mathbf{v}_B + \boldsymbol{\omega} \times \mathbf{r}_{C/B}$

$v_C \mathbf{i} = 0 + (-1.33\mathbf{k}) \times (1\mathbf{j})$

$v_C = 1.33 \text{ ft/s} \rightarrow$ *Ans.*

D–43. $\mathbf{v}_O = \mathbf{v}_A + \boldsymbol{\omega} \times \mathbf{r}_{O/A}$

$2\mathbf{i} = 0 + (-\omega\mathbf{k}) \times 1.5\mathbf{j}$

$\omega = 1.33 \text{ rad/s} \downarrow$

$\mathbf{v}_B = \mathbf{v}_O + \boldsymbol{\omega} \times \mathbf{r}_{B/O}$

$\mathbf{v}_B = 2\mathbf{i} + (-1.33\mathbf{k}) \times (1.5\cos 30°\mathbf{i} + 1.5\sin 30°\mathbf{j})$

$\mathbf{v}_B = \{3\mathbf{i} - 1.73\mathbf{j}\} \text{ m/s}$ *Ans.*

D–44. $\mathbf{v}_A = \mathbf{v}_B + \boldsymbol{\omega} \times \mathbf{r}_{A/B}$

$-v_A\mathbf{j} = -2\mathbf{i} + \omega\mathbf{k} \times (-3\mathbf{i} - 4\mathbf{j})$

$v_A = 1.5 \text{ ft/s} \downarrow$ *Ans.*

$\omega = 0.5 \text{ rad/s} \nwarrow$ *Ans.*

D–45. $v_C = 4(0.6) = 2.4 \text{ m/s} \leftarrow$

$\mathbf{v}_B = \mathbf{v}_C + \boldsymbol{\omega} \times \mathbf{r}_{B/C}$

$-v_B \sin 30°\mathbf{i} + v_B \cos 30°\mathbf{j}$

$\quad = -2.4\mathbf{i} + (\omega_{BC}\mathbf{k}) \times (-0.6\mathbf{i})$

$v_B = 4.80 \text{ m/s} \nwarrow \;\; \omega_{BC} = 6.93 \text{ rad/s} \downarrow$

$\omega_{AB} = \dfrac{4.80}{0.6/\sin 30°} = 4 \text{ rad/s} \nwarrow$ *Ans.*

D–46. $v_B = 2(1.5) = 3 \text{ m/s} \searrow$

$\mathbf{v}_C = \mathbf{v}_B + \boldsymbol{\omega} \times \mathbf{r}_{C/B}$

$-v_C\mathbf{j} = 3\cos 45°\mathbf{i} - 3\sin 45°\mathbf{j}$

$\quad + (\omega\mathbf{k}) \times (1.5\cos 45°\mathbf{i} - 1.5\sin 45°\mathbf{j})$

$v_C = 4.24 \text{ m/s} \downarrow$ *Ans.*

$\omega = 2 \text{ rad/s} \downarrow$

D–47. $\omega = \dfrac{v_B}{r_{B/C}} = \dfrac{2}{0.4} = 5 \text{ rad/s} \downarrow$

$\alpha = \dfrac{a_B}{r_{B/C}} = \dfrac{3}{0.4} = 7.5 \text{ rad/s}^2 \downarrow$

$\mathbf{a}_A = \mathbf{a}_B + \boldsymbol{\alpha} \times \mathbf{r}_{A/B} - \omega^2 \mathbf{r}_{A/B}$

$\mathbf{a}_A = 3\mathbf{j} + (-7.5\mathbf{k}) \times (-0.4\mathbf{j}) - (5)^2(-0.4\mathbf{j})$

$\mathbf{a}_A = \{-3\mathbf{i} + 13\mathbf{j}\} \text{ m/s}^2$ *Ans.*

$a_B = 3 \text{ m/s}^2$ *Ans.*

D–48. $\mathbf{v}_B = \mathbf{v}_A + \boldsymbol{\omega} \times \mathbf{r}_{B/A}$

$-v_B\mathbf{j} = -6\mathbf{i} + \omega\mathbf{k} \times (3\cos 45°\mathbf{i} + 3\sin 45°\mathbf{j})$

$\omega = 2.828 \text{ rad/s} \downarrow, \;\; v_B = 6 \text{ ft/s} \downarrow$

$\mathbf{a}_B = \mathbf{a}_A + \boldsymbol{\alpha} \times \mathbf{r}_{B/A} - \omega^2 \mathbf{r}_{B/A}$

$-a_B\mathbf{j} = 2\mathbf{i} + (\alpha\mathbf{k}) \times (3\cos 45°\mathbf{i} + 3\sin 45°\mathbf{j})$

$\qquad\qquad - (2.828)^2(3\cos 45°\mathbf{i} + 3\sin 45°\mathbf{j})$

$a_B = 31.9 \text{ ft/s}^2 \downarrow$ *Ans.*

$\alpha = 7.06 \text{ rad/s}^2 \downarrow$ *Ans.*

D–49. $+\swarrow\Sigma F_x = m(a_G)_x;\ 3500\sin 20° = \dfrac{3500}{32.2}a_G$

$a_G = 11.0\ \text{ft/s}^2$ *Ans.*

$+\nwarrow\Sigma F_y = m(a_G)_y;\ N_A + N_B - 3500\cos 20° = 0$

$\llcorner+\Sigma M_G = 0;\ N_B(4) - N_A(3) = 0$

$N_A = 1879\ \text{lb}$ *Ans.*

$N_B = 1410\ \text{lb}$ *Ans.*

D–50. $\llcorner+\Sigma M_A = \Sigma(M_k)_A;\ 10\left(\dfrac{3}{5}\right)(7) = \dfrac{20}{32.2}a(3.5)$

$a = 19.32\ \text{ft/s}^2$ *Ans.*

$\xrightarrow{+}\Sigma F_x = m(a_G)_x;\ A_x + 10\left(\dfrac{3}{5}\right) = \dfrac{20}{32.2}(19.32)$

$A_x = 6\ \text{lb}$ *Ans.*

$+\uparrow\Sigma F_y = m(a_G)_y;\ A_y - 20 + 10\left(\dfrac{4}{5}\right) = 0$

$A_y = 12\ \text{lb}$ *Ans.*

D–51. $\llcorner+\Sigma M_B = I_B\alpha;\ 50(1) = 2.30\alpha$

$\alpha = 21.7\ \text{rad/s}^2$ *Ans.*

$a_G = \left(\dfrac{1}{\cos 30°}\right)(21.7) = 25.1\ \text{m/s}^2$

$\xrightarrow{+}\Sigma F_x = m(a_G)_x;$

$B_x = \left(\dfrac{50}{32.2}\right)25.1\sin 30° = 19.5\ \text{lb}$ *Ans.*

$+\uparrow\Sigma F_y = m(a_G)_y;$

$B_y - 50 = -\left(\dfrac{50}{32.2}\right)25.1\cos 30°$

$B_y = 16.2\ \text{lb}$ *Ans.*

D–52. $\curvearrowright+\Sigma M_O = I_O\alpha;$

$20(9.81)(0.5) = \left[\dfrac{1}{12}(20)(3)^2 + 20(0.5)^2\right]\alpha$

$\alpha = 4.90\ \text{rad/s}^2$

$+\uparrow\Sigma F_y = m(a_G)_y;$

$O_y - 20(9.81) = -20[4.90\ (0.5)]$

$O_y = 147\ \text{N}$ *Ans.*

D–53. No slipping, so that

$a_G = 1.25\alpha$

$\llcorner+\Sigma M_A = \Sigma(M_k)_A;\ 20(9.81)\sin 20°(1.25)$

$= 20(0.8)^2\alpha + (20a_G)(1.25)$

$\alpha = 1.90\ \text{rad/s}^2$ *Ans.*

$a_G = 2.38\ \text{m/s}^2$

D–54. $+\nwarrow\Sigma F_y = 0;\ N_A - 15(9.81)\cos 45° = 0;$

$N_A = 104.1\ \text{N}$

$a_G = \alpha(0.3)$

$\llcorner+\Sigma M_B = \Sigma(M_k)_B;\ (0.1)(104.1)(1.55) -$

$15(9.81)\sin 45°(0.3) = -15(0.8)^2\alpha - 15(a_G)(0.3)$

$a_G = 0.413\ \text{m/s}^2$

$\alpha = 1.38\ \text{rad/s}^2$ *Ans.*

D–55. $\xrightarrow{+}\Sigma F_x = m(a_G)_x;\ F_A = 2a_G$

$\llcorner+\Sigma M_G = I_G\alpha;\ F_A(0.4) = 2(0.3)^2\alpha$

$\mathbf{a}_A = \mathbf{a}_G + \boldsymbol{\alpha}\times\mathbf{r}_{A/G} - \omega^2 r_{A/G}$

$5\mathbf{i} = a_G\mathbf{i} + (\alpha\mathbf{k})\times(-0.4\mathbf{j}) - \mathbf{0}$

$5 = a_G + 0.4\alpha$

$F_A = 3.60\ \text{N}$

$a_G = 1.80\ \text{m/s}^2$

$\alpha = 8\ \text{rad/s}^2$ *Ans.*

Answers to Selected Problems

Chapter 12

12–1. $v = 59.5$ ft/s; $t = 1.29$ s

12–2. $a_c = 6.40$ ft/s^2; $t = 12.5$ s

12–3. $v_2 = 59.5$ ft/s; $t = 1.29$ s

12–5. $t = 30$ s; $s = 792$ m

12–6. $s = 123$ ft; $a = 2.99$ ft/s^2

12–7. $a_{max} = 42$ ft/s^2; $v_{max} = 135$ ft/s

12–9. $v_{avg} = 4$ m/s; $(v_{sp})_{avg} = 6$ m/s; $a|_{t=4} = 18$ m/s^2

12–10. $v = (20 - 2s)$ m/s; $s = 10$ m

12–11. $v = 32$ m/s; $s = 67$ m; $d = 66$ m

12–13. $a = -12$ m/s^2; $s = 0$; $s_T = 8$ m; $(v_{sp})_{avg} = 2.67$ m/s

12–14. $v_{avg} = 0$; $(v_{sp})_{avg} = 3$ m/s; $a|_{t=6} = 2$ m/s^2

12–15. $v|_{t=2} = 0.25$ m/s; $s|_{t=2} = 3.5$ m

12–17. $d_A = 41.0$ ft; $d_B = 200$ ft; $\Delta s_{AB} = 152$ ft

12–18. $t = 21.9$ s

12–19. $\Delta s = 48.3$ ft

12–21. $s|_{t=3} = 54$ m

12–22. $v = 322$ m/s; $t = 19.3$ s

12–23. $t = 5.62$ s

12–25. $t = 0.372$ s; $v|_{t=0.372} = 26.6$ m/s

12–26. $v_B = 31.4$ ft/s

12–27. $s|_{t=6} = 41.6$ m; $v|_{t=6} = 10.4$ m/s

12–29. $v = -(2.667s^3 + 7333.3)^{\frac{1}{2}}$ m/s

12–30. $t = 32$ s

12–31. $t = 48.3$ s

12–33. (a) $v = 45.5$ m/s; (b) $v = 100$ m/s

12–34. $v = 11.2$ km/s

12–35. $v = 3.02$ km/s $\downarrow$

12–37. $t = 70.4$ s; $s = 12\ 943$ ft; $v = 238$ ft/s; $a = 3$ ft/s^2

12–38. $t = 7.48$ s

12–39. $t = 10$ s; $v = -55$ ft/s; $a = -6$ ft/s^2

12–41. $v_{max} = 0.894$ m/s; $t' = 0.447$ s

12–42. $t = 0.2$ s; $a = 100$ m/s^2; $s = 1$ m

12–43. $t = 11.25$ s

12–45. $t = 40$ s; $a = 8$ m/s^2; $s = 2800$ m

12–46. $s = 144$ m

12–47. $s = 3600$ ft; $(v_{sp})_{avg} = 30$ ft/s

12–49. $a_{max} = 8$ ft/s^2

12–50. $(v_{sp})_{avg} = 37.8$ ft/s; $s_T = 1.13(10^3)$ ft

12–51. At $s = 75$ ft, $v = 27.4$ ft/s;
At $s = 125$ ft, $v = 31.62$ ft/s;
At $s = 125$ ft, $v = 37.4$ ft/s

12–53. $\Delta s = 87.5$ m; Car A is ahead of car B.

12–54. $t = 60$ s; $v = 540$ m/s; $s = 10\ 125$ m

12–55. $t = 13.1$ s; $s|_{t=13.1} = 523$ ft

12–57. $v_{max} = 120$ m/s; $t = 41.9$ s

12–58. $t = 9.88$ s

12–59. At $s = 50$ m, $a = 0.32$ m/s^2;
At $s = 150$ m, $a = -0.32$ m/s^2

12–61. At $s = 400$ m, $v = 34.6$ m/s

12–62. At $s = 100$ m, $a = 16$ m/s^2;
At $s = 150$ m, $a = 4.5$ m/s^2

12–63. For $s = 40$ ft, $v = 12.7$ ft/s;
For $s = 90$ ft, $v = 22.8$ ft/s;
For $s = 200$ ft, $v = 36.1$ ft/s;
When $s = 40$ ft, $v = 12.7$ ft/s;
When $s = 90$ ft, $v = 22.8$ ft/s;
When $s = 200$ ft, $v = 36.1$ ft/s

12–65. At $s = 100$ m, $a = 4.48$ m/s^2;
At $s = 200$ m, $a = 7.04$ m/s^2

12–66. (4 ft, 2 ft, 6 ft)

12–67. $a = 80.2$ m/s^2, (42.7 m, 16.0 m, 14.0 m)

12–69. $v = 9.68$ m/s; $a = 16.8$ m/s^2

12–70. $v_{avg} = \{10\mathbf{i} - 10\mathbf{j}\}$ m/s

12–71. $v_{sp} = 4.28$ m/s

12–73. $\mathbf{a}_{AB} = \{0.404\mathbf{i} + 7.07\mathbf{j}\}$ m/s^2;
$\mathbf{a}_{AC} = \{2.50\mathbf{i}\}$ m/s^2

12–74. $v_x = 0.398$ ft/s; $v_y = 3.98$ ft/s

12–75. $a_x = \dfrac{c}{2k}(y + ct^2)$; $a_y = c$

690

12–77. $d = 204$ m; $v = 41.8$ m/s;
$a = 4.66$ m/s^2

12–78. $v_0 = 15.3$ ft/s; $v_B = 81.7$ ft/s

12–79. $v = 201$ m/s; $a = 405$ m/s^2

12–81. $\theta = 45°$

12–82. $t = 3.55$ s; $v = 32.0$ m/s

12–83. $h = 12.9$ ft

12–85. $\theta = 76.0°$; $v_A = 49.8$ ft/s;
$h = 39.7$ ft

12–86. $s = 4.52$ ft

12–87. $v_A = 36.7$ ft/s; $h = 11.5$ ft

12–89. $R = \dfrac{v_0^2}{g}\sin 2\theta$; $t = \dfrac{2v_0}{g}\sin\theta$;
$h = \dfrac{v_0^2}{2g}\sin^2\theta$

12–90. $v_A = 89.7$ ft/s

12–91. $R = 66.3$ ft

12–93. $d = 25.4$ ft

12–94. $v = 81.9$ ft/s

12–95. $v_{min} = 0.838$ m/s; $v_{max} = 1.76$ m/s

12–97. The two roots are $\theta_D = 14.7°$;
$\theta_c = 75.3°$; $\Delta t = t_c - t_D = 1.45$ s

12–98. $x = 5.15$ ft; $y = 1.33$ ft

12–99. $R = \left(\dfrac{v_0^2\sin 2\theta}{2g}\right) + \dfrac{v_0}{2g}\sqrt{v_0^2\sin^2 2\theta + 8gh\cos^2\theta}$

12–101. $a = 21.6$ ft/s^2; $\Delta s = 40.5$ ft

12–102. $a_t = 35.0$ ft/s^2; $\rho = 2.64(10^3)$ ft

12–103. $a_p = 180$ ft/s^2; $a_{p'} = 72$ ft/s^2

12–105. $v = 1.96$ m/s; $a = 0.930$ m/s^2

12–106. $a = 0.921$ m/s^2

12–107. $a = 4.22$ m/s^2

12–109. $x = 0$, $y = -4$ m; $(a)_{max} = 50$ m/s^2

12–110. $v = 19.9$ ft/s; $a = 16.0$ ft/s^2

12–111. $a_t = 3.62$ m/s^2; $\rho = 29.6$ m

12–113. $v = 97.2$ ft/s; $a = 42.6$ ft/s^2

12–114. $v = 115$ ft/s; $a = 58.4$ ft/s^2

12–115. $v = 4.58$ m/s; $a = 0.653$ m/s^2

12–117. $a_{max} = 10.4$ m/s^2

12–118. $a_{min} = 3.09$ m/s^2

12–119. $v_B = 3.68$ m/s; $a_B = 4.98$ m/s^2

12–121. $v = 1.5$ m/s; $a = 0.117$ m/s^2

12–122. $y = -0.0766x^2$ (Parabola);
$v = 8.37$ m/s; $a_n = 9.38$ m/s^2; $a_t = 2.88$ m/s^2

12–123. $v = 43.0$ m/s; $a = 6.52$ m/s^2

12–125. (a) $s_A = 1.40$ m; $s_B = 3$ m;
(b) $\mathbf{r}_A = \{1.38\mathbf{i} + 0.195\mathbf{j}\}$ m;
$\mathbf{r}_B = \{-2.82\mathbf{i} + 0.873\mathbf{j}\}$ m;
(c) $\Delta r = 4.26$ m

12–126. $t = 14.3$ s; $a_B = 0.45$ m/s^2

12–127. $t = 1.21$ s

12–129. $v_t = 2$ m/s; $v_n = 0$; $a_t = 0$;
$a_n = \dfrac{4\sin x}{(1 + \cos^2 x)^{3/2}}$

12–130. $\mathbf{u}_b = 0.581\mathbf{i} + 0.161\mathbf{j} + 0.798\mathbf{k}$

12–131. $d = 17.0$ m; $a_A = 190$ m/s^2;
$a_B = 12.8$ m/s^2

12–133. At $s = 10$ m, $v = 4.58$ m/s;
$a = 0.653$ m/s^2

12–134. $a = 35.0$ m/s^2; $s = 67.1$ ft

12–135. $\alpha = 52.5°$; $\beta = 142°$; $\gamma = 85.1°$ or
$\alpha = 128°$; $\beta = 37.9°$; $\gamma = 94.9°$

12–137. $v_r = -1.66$ m/s; $v_\theta = -2.07$ m/s;
$a_r = -4.20$ m/s^2; $a_\theta = 2.97$ m/s^2

12–138. $v_r = 0.955$ in./s; $v_\theta = 21$ in./s;
$a_r = -63$ in./s^2; $a_\theta = 5.73$ in./s^2

12–139. $v_r = 0.955$ in./s; $v_\theta = 21$ in./s;
$a_r = -62.4$ in./s^2; $a_\theta = 19.7$ in./s^2

12–141. $v_r = -2.33$ m/s; $v_\theta = 7.91$ m/s;
$a_r = -158$ m/s^2; $a_\theta = -18.6$ m/s^2

12–142. $a = 2.32$ m/s^2

12–143. $v_r = 16.0$ ft/s; $v_\theta = 1.94$ ft/s;
$a_r = 7.76$ ft/s^2; $a_\theta = 1.94$ ft/s^2

12–145. $a_r = -6.67$ m/s^2; $a_\theta = 3$ m/s^2

12–146. $a = 48.3$ in./s^2

12–147. $v_r = 1.20$ m/s; $v_\theta = 1.26$ m/s;
$a_r = -3.77$ m/s^2; $a_\theta = 7.20$ m/s^2

12–149. $v_r = 1.20$ m/s; $v_\theta = 1.50$ m/s;
$a_r = -4.50$ m/s^2; $a_\theta = 7.20$ m/s^2

12–150. $v = 12$ ft/s; $a = 0.646$ ft/s^2

12–151. $v_r = -16.9$ m/s; $v_\theta = 4.87$ m/s;
$a_r = -89.4$ m/s^2; $a_\theta = -53.7$ m/s^2

12–153. $\dot{\theta} = 1.30$ rad/s; $a = 2.55$ m/s^2

12–154. $\dot{\theta} = 0.6$ rad/s

12–155. $v_r = -2.80$ m/s; $v_\theta = 19.8$ m/s

12–157. $v = 5.95$ ft/s; $a = 3.44$ ft/s^2

12–158. $v = 24.1$ ft/s; $a = 8.17$ ft/s^2

12–159. $v = 1.32$ m/s

12–161. $\dot{\theta} = 0.0178$ rad/s

12–162. $\ddot{\theta} = 0.00404$ rad/s^2

12–163. $v = 86.6$ ft/s; $a = 266$ ft/s^2

12–165. $v = 0.242$ m/s; $a = 0.169$ m/s^2;
$(v_A)_x = -0.162$ m/s;
$(v_A)_y = 0.180$ m/s

12–166. $a = 6.54$ m/s^2

12–167. $\dot{\theta} = 0.75$ rad/s

12–169. $v = 7.83$ m/s; $a = 2.27$ m/s^2

12–170. $\mathbf{v} = \{-116\,\mathbf{u}_r - 163\,\mathbf{u}_z\}$ mm/s;
$\mathbf{a} = \{-5.81\mathbf{u}_r - 8.14\mathbf{u}_z\}$ mm/s^2

12–171. $v_r = 0$; $v_\theta = 12$ ft/s; $a_r = -216$ ft/s^2; $a_\theta = 0$

12–173. $v_B = 0.5$ m/s ↑

12–174. $v_A = 12$ ft/s →

12–175. $t = 160$ s

12–177. $v_P = 12$ ft/s

12–178. $\Delta s_B = 2$ ft ↑

12–179. $v_D = 1.5$ ft/s ↑

12–181. $v_B = 1$ ft/s ↑

12–182. $v_{B/C} = 30$ ft/s ↑

12–183. (a) $t = 1.22$ s;
(b) $v_{A/B} = 2.90$ m/s ↑

12–185. $v_B = v_C = 2$ m/s ↑

12–186. $v_C = 0.667$ m/s ↑

12–187. $v_B = 8$ ft/s ↑; $a_B = 6.80$ ft/s^2 ↑

12–189. $v_C = 1.2$ m/s ↑; $a_C = 0.512$ m/s^2 ↑

12–190. $v_B = 6.08$ ft/s ←

12–191. $v_B = 1.41$ m/s ↑

12–193. $v_A = -v_B\left(1 + \left(\dfrac{h}{s_A}\right)^2\right)^{1/2}$;

$a_A = -a_B\left(1 + \left(\dfrac{h}{s_B}\right)^2\right)^{1/2} + \dfrac{v_A v_B h^2}{s_A^3}\left(1 + \left(\dfrac{h}{s_A}\right)^2\right)^{-1/2}$

12–194. $\Delta s_C = 2$ ft

12–195. $v_B = 1.20$ ft/s ↓; $a_B = 1.11$ ft/s^2 ↑

12–197. $v_{A/B} = 875$ km/h; $\theta = 41.5°$ ↗

12–198. $v_{B/A} = 26.5$ mi/h; $\theta = 40.9°$ ∠θ;
$a_{B/A} = 1.79(10^3)$ mi/h^2; $\theta = 72.0°$ ∠θ

12–199. $a_{B/A} = 1955$ mi/h^2; $\theta = 0.767°$ ↘θ

12–201. $v_{A/B} = 13.4$ m/s; $\theta = 31.7°$ θ↙;
$a_{A/B} = 4.32$ m/s^2; $\theta = 79.0°$ ↘θ

12–202. $v_{A/B} = 49.1$ km/h; $\theta = 67.2°$ θ↗

12–203. $v_{A/B} = 98.4$ ft/s; $\theta = 67.6°$ θ↗;
$a_{A/B} = 19.8$ ft/s^2; $\theta = 57.4°$ ∠θ

12–205. $v_{B/A} = 26.0$ m/s; $\theta = 30°$ θ↘;
$a_{B/A} = 1.98$ m/s^2; $\phi = 35.2°$ θ↗

12–206. $v_C = 13.0$ ft/s; $v_{B/A} = 37.0$ ft/s

12–207. $d = 11.1$ ft; $v_{B/A} = 36.4$ ft/s

Chapter 13

13–1. $F = 41.7$ nN

13–2. $t_B = 1.09$ s; $t_C = 1.54$ s;
$t_D = 2.32$ s

13–3. $T = 1.20$ kN; $s = 16.7$ m

13–5. $T = 5.60$ kN

13–6. $a = 0.343$ m/s^2; $a = 0.436$ m/s^2

13–7. $T = 3.01$ kN

13–9. $v_E = 7.23$ m/s

13–10. $F = 4.25$ kN

13–11. $v_2 = 77.9$ ft/s

13–13. $N = 88.1$ N; $a = 1$ m/s^2

13–14. (a) $a_A = \dfrac{P}{m} - 3\,\mu g$;

(b) $a_A = \dfrac{P}{2m} - 2\,\mu g$

13–15. $a = 3.43$ ft/s^2

13–17. $v_{\max} = \dfrac{2F_0 t_0}{\pi m}$; $v = \left(\dfrac{F_0 t_0}{\pi m}\right)\left(1 - \cos\left(\dfrac{\pi t}{t_0}\right)\right)$;

$s = \left(\dfrac{F_0 t_0}{\pi m}\right)\left(t - \dfrac{t_0}{\pi}\sin\left(\dfrac{\pi t}{t_0}\right)\right)$

13–18. $T = 206$ lb

13–19. $R = 5.30$ ft; Total time $= 1.82$ s

13–21. $T = 4.92$ kN

13–22. $v_A = 44.6$ ft/s

13–23. $s = 64.4$ ft

13–25. $v = 3.62$ m/s

13–26. $a_A = 1.29$ ft/s^2; $s = 9.70$ ft

13–27. $a_B = 2.30$ ft/s^2 ↑

13–29. $s = 5.43$ m

13–30. $t = 43.1$ s; $s_{max} = 186$ ft

13–31. $x = 250$ mm

13–33. $a_C = 7.49$ m/s^2; $\theta = 22.8°$ $\theta \angle$

13–34. (a) $T = 40$ lb; (b) $T = 59.9$ lb

13–35. $t = 0.519$ s

13–37. $t = 2.04$ s

13–38. $v = \dfrac{1}{m}\sqrt{1.09F_0^2 t^2 + 2F_0 t m v_0 + m^2 v_0^2}$;

$x = \dfrac{y}{0.3} + v_0\left(\sqrt{\dfrac{2m}{0.3F_0}}\right)y^{\frac{1}{2}}$

13–39. $v_B = 4.52$ m/s

13–41. (a) $a_A = a_B = 3.86$ ft/s^2;

(b) $a_A = 70.8$ ft/s^2; $a_B = 6.44$ ft/s^2

13–42. $P = 2mg \tan\theta$

13–43. $P = 2mg\left(\dfrac{\sin\theta + \mu_s\cos\theta}{\cos\theta - \mu_s\sin\theta}\right)$

13–45. $T = \dfrac{mg}{2}\sin 2\theta$

13–46. $T = 1.63$ kN

13–47. $T = 1.80$ kN

13–49. $N = 0$ when $x = d$;

$v = \sqrt{\dfrac{kd^2}{(m_A + m_B)}}$

13–50. $N = 0$, then $x = d$ for separation.

13–51. $d = \dfrac{(m_A + m_B)g}{k}$

13–53. $N = 15.9$ N, $\rho = 282$ m

13–54. $F_R = 14.6$ kN; $\theta = 20.3°$

13–55. $F_n = 323$ lb; $F_t = 0$

13–57. $T = 9.09$ kN; $\theta = 43.3°$

13–59. $N_b = 1.57$ kN

13–61. $a_t = 16.1$ ft/s^2; $T = 46.9$ lb

13–62. $T = 38.0$ lb; $v = 9.29$ ft/s

13–63. $v = 80.2$ ft/s

13–65. $N = 277$ lb; $F = 13.4$ lb

13–66. $\theta = 47.5°$

13–67. $v = 6.30$ m/s; $F_n = 283$ N; $F_t = 0$; $F_b = 490$ N

13–69. $F = 1.23$ kN; $N_s = 924$ N

13–70. $s = 0.581$ m

13–71. $N_t = 182$ lb

13–73. $T = 0$; $T = 104$ N

13–74. At A: $N_s = 38.6$ N

At B: $N_s = 96.6$ N

13–75. $v_G = 2.83$ m/s

13–77. $N = 180$ N; $a_t = 5.44$ m/s^2

13–78. $t = 7.39$ s

13–79. $v = 1.63$ m/s; $N = 7.36$ N

13–81. $a_t = 29.4$ ft/s^2; $N = 69.0$ lb

13–82. $\theta = 15.0°$; $\phi = 12.6°$

13–83. $F_r = -240$ N; $F_\theta = 66$ N;

$F_z = -3.28$ N

13–85. $F_A = 4.46$ lb

13–86. $(F_A)_{max} = 2.06$ lb; $(F_A)_{min} = 0.335$ lb

13–87. $F_{frict} = 68$ N; $N_C = 153$ N

13–89. $(F_z)_{max} = 547$ N; $(F_z)_{min} = 434$ N;

13–90. $F = 0.143$ lb

13–91. $N = 5.79$ N; $F = 1.78$ N

13–93. $F = 17.0$ N

13–94. $P = 1.68$ lb; $N_s = 4.77$ lb

13–95. $P = 2.75$ lb; $N_s = 3.08$ lb

13–97. $F = 7.67$ N

13–98. $F = 7.82$ N

13–99. $F = 5.07$ kN; $N = 2.74$ kN

13–101. $\dot\theta = 4.00$ rad/s; $T = 8$ N

13–102. $N = 0.328$ N; $F = 10.5$ N

13–103. $N_C = 54.4$ N; $F = 54.4$ N

13–105. $N = 85.3$ lb

13–106. $a_t = 12$ m/s^2; $5.71°$ $\theta \searrow$

13–107. $a_t = 11.0$ m/s^2; $5.71°$ $\theta \searrow$

13–109. $P = 12.6$ lb; $N = 4.22$ lb

13–110. $v_r = 2.50$ m/s; $v_\theta = 2$ m/s

13–111. $r = 0.198$ m

13–114. $v_0 = 7.82$ km/s; $T = 1.97$ h

13–115. $v_p = 7.31$ km/s

13–117. $r = 42.2$ Mm; $v = 3.07$ km/s

13–118. $v_0 = 23.9(10^3)$ ft/s

13–119. $v_{A'} = 7.30(10^3)$ ft/s; $T = 1.69$ h

13–121. $v_p = 7.47$ km/s

13–122. $v_a = 3.94$ km/s; $t = 46.1$ min

13–123. $\Delta v = 466$ m/s; $\Delta v_c = 2.27$ km/s

13–125. $v_0 = 9.60$ km/s; $r = 6.50$ Mm

13–126. $v_a = 1.20$ km/s; $\Delta v = 2.17$ km/s

13–127. (a) $v_A = 2.43$ km/s;

(b) $v_A'' = 6.38$ km/s;

(c) $T_c = 2.19$ h; $T_e = 6.78$ h

Chapter 14

14–1. $U_W = 4.12$ kJ; $U_{Np} = -2.44$ kJ; The difference accounts for a change in kinetic energy.

14–2. $v_2 = 17.7$ ft/s

14–3. $v = 7.07$ m/s

14–5. $F_{avg} = 44.1$ kN; $x = 38.2$ mm

14–6. $s = 0.933$ m

14–7. $k = 15.0$ MN/m^2

14–9. $d = 12$ m

14–10. $s = 179$ mm

14–11. $s = 3.41$ m

14–13. At A: $v_2 = 6.08$ m/s; At B: $v_2 = 4.08$ m/s

14–14. $v_A = 0.771$ ft/s

14–15. $v_A = 7.18$ ft/s

14–17. $d = 2.56$ ft

14–18. $v_C = 1.37$ m/s

14–19. $v_2 = 15.4$ m/s

14–21. $s = 9.29$ ft

14–22. $v = 1.67$ m/s

14–23. $s = 0.782$ ft

14–25. $v = 6.89$ m/s

14–26. $s = 2.48$ ft; $T_A = 36.0$ lb

14–27. $v_B = 4.52$ m/s

14–29. $y = 3.81$ ft

14–30. $v_A = 28.3$ m/s

14–31. $s = 0.730$ m

14–33. $y = 0.815$ m; $N_b = 568$ N; $a = a_t = 6.23$ m/s^2

14–34. $d = 36.2$ ft

14–35. $v_D = 17.7$ m/s; $R = 33.0$ m

14–37. $U_{1-2} = GM_e m\left(\dfrac{1}{r_1} - \dfrac{1}{r_2}\right)$

14–38. $v_A = 5.80$ ft/s

14–39. $v_B = 18.0$ m/s; $N_B = 12.5$ kN

14–41. $t = 4.44$ s

14–42. $P_{in} = 4.20$ hp

14–43. $v = 29.2$ ft/s

14–45. $P_{in} = 102$ kW

14–46. $P = 12.6$ kW

14–47. $v = 0.285$ m/s

14–49. $P_{in} = 0.644$ hp

14–50. $s = \dfrac{mv^3}{3P}$

14–51. $t = 46.2$ min

14–53. $P_O = 35.4$ kW

14–54. $P_O = 1.12$ kW

14–55. $P_i = 22.2$ kW

14–57. $P = 58.1$ kW

14–58. $P_i = 2.84$ hp

14–59. $P_i = 1.63$ hp

14–61. $P = 0.0364$ hp

14–62. At $0 \le t \le 0.2$, $P = 53.3t$ kW; At $0.2 \le t \le 0.3$, $P = (160t - 533t^2)$ kW; $U = 1.69$ kJ

14–63. $P_{max} = 10.7$ kW

14–65. $v_A = 7.18$ ft/s

14–66. $v = 1.37$ m/s

14–67. $s = 0.730$ m

14–69. $k_B = 287$ N/m

14–70. $x = 1.10$ in.

14–71. $N = 3.04$ lb

14–73. $v = 2.86$ m/s

14–74. $h = 416$ mm

14–75. $v_2 = 3.84$ m/s; $T = 5.89$ N

14–77. $v = 12.6$ ft/s; $N_C = 18.9$ lb

14–78. $v_2 = 3.30$ m/s

14–79. $h = 24.5$ m; $N_B = 0$; $N_C = 16.8$ kN

14–81. $v_B = 10.6$ m/s; $T_B = 142$ N; $v_C = 9.47$ m/s; $T_C = 48.7$ N

14–82. $v_B = 27.2$ ft/s

14–83. $l = 90$ ft; $h = 219$ ft; $a = 483$ ft/s^2

14–85. (a) $k = 3.53$ kN/m;

(b) $k = 14.1$ kN/m

14–86. $v_B = 114$ ft/s; $N_B = 29.1$ lb

14–87. $v_0 = \sqrt{g(5\rho - 2h)}$

14–89. $s_{max} = 49.0$ mm

14–90. $\theta = 48.2°$

14–91. $v = 80.2$ ft/s; $N_G = 952$ lb

14–93. $v = 4.49$ ft/s

14–94. $x_A = 0.516$ ft; $x_B = 0.387$ ft

Chapter 15

15–1. $v = 29.4$ ft/s

15–2. $t = 0.280$ s; $v_B = 15.6$ ft/s

15–3. $t = 0.439$ s

15–5. $I_C = 6.55$ N · ms; $I_U = 6.05$ N · ms

15–6. $\int F \, dt = 0.706$ N · s; $40° \angle \theta$

15–7. (a) $I = 46$ lb · s; (b) $I = 54$ lb · s

15–9. $v = 121$ m/s

15–10. $I = 2.61$ N · s; $30° \angle \theta$

15–11. $v = 150$ ft/s

15–13. $\int F \, dt = 70.2$ lb · s

15–14. At A: $v = 7.40$ m/s;

At B: $v = 5.40$ m/s

15–15. $t = 2.95$ s

15–17. $0 + \Sigma \int F \, dt = 0.810$ lb · s

15–18. $F_{avg} = 3.64$ kip

15–19. $I = \int F \, dt = 63.4$ N · s

15–21. $t = 1.93$ s

15–22. $v = 42.4$ ft/s

15–23. $I = 8.49$ N · s; $17.3° \angle \theta$

15–25. $v_B = 21.5$ ft/s $\downarrow$; $v_A = 21.5$ ft/s $\uparrow$

15–26. $t = 5.47$ s

15–27. $(v_A)_2 = 10.5$ ft/s $\rightarrow$

15–29. $v_2 = 16.1$ m/s

15–30. $v_2 = 10.1$ ft/s

15–31. $v_2 = 7.65$ m/s

15–33. $v_2 = 0.600$ ft/s $\leftarrow$

15–34. $v = 3.5$ ft/s $\rightarrow$

15–35. $v_c = 5.04$ m/s $\leftarrow$

15–37. $v_2 = 20.7$ ft/s; $\int F \, dt = 38.6$ lb · s

15–38. $v_C = 1.88$ ft/s $\leftarrow$; $s_C = 3.75$ ft

15–39. $s_b = 18.2$ ft

15–41. $v_B = 3.48$ ft/s; $N = 504$ lb; $t = 0.216$ s

15–42. $(v_B)_2 = 3.29$ ft/s

15–43. $W_B = 75$ lb

15–45. $R = 145$ ft

15–46. (a) 1.17 ft/s; (b) 1.17 ft/s

15–47. $v_B = 1.33$ m/s; $v_A = 0.667$ m/s;

$t = 2.5$ s

15–49. $x = 0.221$ m

15–50. $v_r = 8.93$ ft/s

15–51. $s_r = 4.8$ ft

15–53. $v_r = 1.10$ m/s; $v_c = 5.43$ m/s, $\theta = 35.8°$ $\angle\theta$

15–54. $v_A = 3.29$ m/s; $v_B = 2.19$ m/s

15–55. $(v_A)_2 = 0.400$ m/s; $(v_B)_2 = 2.40$ m/s;

$d = 0.951$ m

15–57. $(v_A)_2 = 1.04$ ft/s; $(v_B)_3 = 0.964$ ft/s;

$(v_C)_3 = 11.9$ ft/s

15–58. $t = 0.226$ s

15–59. $d = 0.329$ ft

15–61. $v_{p2} = 27.0$ ft/s; $v_{m2} = 13.4$ ft/s; $x = 2.61$ ft

15–62. $h = 4.82$ ft

15–63. $(v_A)_2 = 42.8$ ft/s $\leftarrow$; $F = 2.49$ kip

15–65. $d = 13.8$ ft

15–66. $e = 0.502$; $d = 7.23$ ft

15–67. $h = 21.8$ mm

15–69. $e = \dfrac{\sin \phi}{\sin \theta} \left(\dfrac{\cos \theta - \mu \sin \theta}{\mu \sin \phi + \cos \phi} \right)$

15–70. $\mu = 0.25$

15–71. $(v_B)_2 = 6.59$ ft/s

15–73. $v_A = 29.3$ ft/s; $v_{B2} = 33.1$ ft/s; $\theta = 27.7°$ $\angle\theta$

15–74. $v_A = 29.3$ ft/s; $v_{B_2} = 33.8$ ft/s; $e = 0.770$

15–75. $h = 0.390$ m

15–77. (a) $(v_B)_1 = 8.81$ m/s; $\theta_1 = 10.5°$ $\angle\theta$

(b) $(v_B)_2 = 4.62$ m/s; $\theta_2 = 20.3°$ $\theta\searrow$

(c) $s = 3.96$ m

15–78. $s = 0.456$ ft

15–79. $(v_B)_3 = 1.50$ m/s

15–81. $(v_B)_2 = 1.06$ m/s $\leftarrow$; $(v_A)_2 = 0.968$ m/s;

$(\theta_A)_2 = 5.11°$ $\angle$

15–82. $(v_A)_2 = 6.90$ m/s; $(v_B)_2 = 75.7$ m/s

15–83. $(v_B)_2 = 0.298$ ft/s; $(v_A)_2 = 0.766$ ft/s

15–85. $v_A = 1.35$ m/s →; $v_B = 5.89$ m/s;
$\theta = 32.9°$ 🡖

15–86. $e = 0.0113$

15–87. $v_B = 3.50$ m/s; $v_A = 6.47$ m/s

15–89. $e = \dfrac{2}{3}$

15–90. $(\mathbf{H}_A)_0 = \{-1.83\mathbf{i} - 0.913\mathbf{k}\}$ slug · ft²/s

15–91. $\mathbf{H}_o = \{42\mathbf{i} + 21\mathbf{k}\}$ kg · m²/s

15–93. $(H_A)_P = -57.6$ kg · m²/s;
$(H_B)_P = 94.4$ kg · m²/s;
$(H_C)_P = -41.2$ kg · m²/s

15–94. $\mathbf{H}_O = \{-16.8\mathbf{i} + 14.9\mathbf{j} - 23.6\mathbf{k}\}$ slug · ft²/s

15–95. $\mathbf{H}_P = \{-14.3\mathbf{i} - 9.32\mathbf{j} - 34.8\mathbf{k}\}$ slug · ft²/s

15–97. $(H_A)_O = 72.0$ kg · m²/s ↓;
$(H_B)_O = 59.5$ kg · m²/s ↓

15–98. $(H_A)_P = 66.0$ kg · m²/s ↓;
$(H_B)_P = 73.9$ kg · m²/s ↓

15–99. $v = 3.47$ m/s

15–101. $v = 13.4$ m/s

15–102. $v_B = 10.8$ ft/s;
$U_{AB} = 11.3$ ft · lb

15–103. $v_B = 10.2$ km/s; $r_B = 13.8$ Mm

15–105. $t = 3.41$ s

15–106. $t = 0.910$ s

15–107. $t = 11.9$ s

15–110. $v_B = 21.9$ m/s; $\theta = 20.9°$

15–111. $F_x = 11.0$ N

15–113. $A_x = 3.98$ kN; $A_y = 3.81$ kN;
$M = 1.99$ kN · m

15–114. $A_y = 7.20$ kN; $B_x = 4.97$ kN;
$A_x = 2.23$ kN

15–115. $T = 0.680$ lb

15–117. $a = 0.528$ m/s²

15–118. $v = 3.93$ m/s

15–119. $p = 452$ Pa

15–121. $F = v^2 \rho A$

15–122. $v_{max} = 2068$ ft/s

15–123. $T = 9.72$ N

15–125. $a = 112$ mm/s²

15–126. $F = 3.55$ kN

15–127. $R = (20t + 2.48)$ lb

15–129. $\dfrac{dm}{dt} = m_0\left(\dfrac{(a_0 + g)}{v_{e/r}}\right)e^{[(a_0+g)/v_{e/r}]t}$

15–130. $F_D = 11.5$ kN

15–131. $a = 37.5$ ft/s²

15–133. $F = m'v^2$

15–134. $F = (7.85t + 0.320)$ N

Review 1

R1–1. $t = 5.47$ s; $s = 15.0$ m

R1–2. $\rho = 9.32$ m

R1–3. (a) $\Delta s = 834$ mm; (b) $v = 1.12$ m/s;
$a = 0.450$ m/s²

R1–5. $F_x = 237$ N ←; $F_y = 772$ N ↑

R1–6. $N = 277$ lb; $F = 13.4$ lb

R1–7. $\theta = 47.5°$

R1–9. $a = 0.343$ m/s²; and $a_2 = 0.436$ m/s²

R1–10. $t = 8$ s; $s = 320$ ft

R1–11. $v_B = 3.33$ ft/s ↑; $v_{B/C} = 13.3$ ft/s ↑

R1–13. $v_2 = 9.22$ ft/s; $\Sigma U_{1-2} = 3.04$ ft · lb

R1–14. $r_2 = 1.52$ ft; $t = 0.739$ s

R1–15. $v_b = 0.379$ m/s →

R1–17. $(v_A)_{min} = 4.32$ m/s; $(v_A)_{max} = 5.85$ m/s

R1–18. $v_{B/A} = 28.5$ mi/h; $\theta = 44.5°$;
$a_{B/A} = 3.42(10^3)$ mi/h²; $\theta = 80.6°$ 🡖θ

R1–19. $a_{B/A} = 3.35(10^3)$ mi/h²; $\theta = 19.1°$

R1–21. $v = 20.4$ ft/s

R1–22. $v = 0.969$ m/s

R1–23. $v = 1.48$ m/s

R1–25. $v_{P/H} = 18$ ft/s

R1–26. $v_H = 0.6$ ft/s ↑

R1–27. $v = 21.8$ m/s

R1–29. $v_2 = 10.1$ ft/s

R1–30. $v_2 = 2.13$ ft/s

R1–31. $N_C = 24.8$ N; $F = 24.8$ N

R1–33. (a) $s = -30.5$ m;
(b) $s_{tot} = 56.0$ m;
(c) $v|_{t=9} = 10$ m/s

R1–34. $v = 14.1$ m/s

R1–35. $s = 5.43$ m

R1–37. $v_C = 2.36$ m/s

R1–38. $v_C = 2.34$ m/s

R1–39. $v_A = 1.54$ m/s; $v_B = 4.62$ m/s

R1–41. $(v_B)_2 = \dfrac{1}{3}\sqrt{2gh}(1 + e)$

R1–42. $(v_A)_3 = 0.125$ m/s

R1–43. $F_s = 4.90$ lb

R1–45. $x = 3$ m

R1–46. $v_2 = 13.9$ ft/s

R1–47. $v_2 = 11.7$ ft/s

R1–49. $v = 9.82$ m/s; $a_t = 2.30$ m/s²

R1–50. $v_{B/C} = 39$ ft/s ↑

Chapter 16

16–1. $\theta = 3.32$ rev.; $t = 1.67$ s

16–2. $a_t = 0.562$ ft/s²; $a_n = 3600$ ft/s²; $s = 3000$ ft

16–3. $v_A = 2.60$ m/s; $a_A = 9.35$ m/s²

16–5. $t = 62.5$ s

16–6. $\omega = 60$ rad/s; $\theta = 90$ rad

16–7. $v_A = v_E = 40$ mm/s; $v_w = 34.6$ mm/s

16–9. $v_P = 6.30$ in./s; $a_P = 10.0$ in./s²;
$\quad$ $v_P = 16.0$ in./s

16–10. $v_A = 22$ ft/s; $(a_A)_t = 12.0$ ft/s²;
$\quad$ $(a_A)_n = 242$ ft/s²

16–11. $v_B = 22.0$ ft/s; $(a_B)_t = 9.00$ ft/s²;
$\quad$ $(a_B)_n = 322$ ft/s²

16–13. $v_B = 1.37$ m/s; $a_B = 0.472$ m/s²

16–14. $d = 1.44$ in.

16–15. $t = 100$ s

16–17. $\omega = 3.80$ rad/s; $a_P = 217$ ft/s²

16–18. $v_B = 14.7$ m/s

16–19. $v = 1.34$ m/s

16–21. $v_F = 0.318$ mm/s

16–22. $\omega_B = 465$ rad/s

16–23. $\omega_B = 12$ rad/s

16–25. $v_E = 0.185$ m/s; $(a_E)_t = 5.81$ m/s²;
$\quad$ $(a_E)_n = 2.28$ m/s²

16–26. $s_P = 36$ m; $v_P = 12$ m/s; $a_P = 720$ m/s²

16–27. $\omega_H = 126$ rad/s

16–29. $\omega_E = 0.150$ rad/s

16–30. $a = \dfrac{s}{2\pi}\omega^2$

16–31. $v = 6.93$ ft/s; $a = 84.3$ ft/s²

16–33. $v_{AB} = \omega l \cos\theta$; $a_{AB} = -\omega^2 l \sin\theta$

16–34. $\omega = -19.2$ rad/s; $\alpha = -183$ rad/s²

16–35. $v_{CD} = -\omega l \sin\theta$; $a_{CD} = -\omega^2 l \cos\theta$

16–37. $v = -\left(\dfrac{r_1^2\omega \sin 2\theta}{2\sqrt{r_1^2\cos^2\theta + r_2^2 + 2r_1 r_2}} + r_1\omega \sin\theta\right)$

16–38. $v_P = 18.5$ ft/s ←

16–39. $\omega = 8.70$ rad/s; $\alpha = -50.5$ rad/s²

16–41. $\omega = \dfrac{rv_A}{y\sqrt{y^2 - r^2}}$; $\alpha = \dfrac{rv_A^2(2y^2 - r^2)}{y^2(y^2 - r^2)^{\frac{3}{2}}}$

16–42. $\omega = \dfrac{-v(\sin\phi)}{l\cos(\phi - \theta)}$; $\alpha = \dfrac{-v^2\sin^2\phi\sin(\phi - \theta)}{l^2\cos^3(\phi - \theta)}$

16–43. $\omega = \dfrac{h}{h^2 + x^2}v_B$; $\alpha = \dfrac{-2xh}{(h^2 + x^2)^2}v_B^2$

16–45. $v_C = L\omega$↑; $a_C = 0.577L\omega^2$ ↑

16–46. $\omega = 1.08$ rad/s; $v_B = 4.39$ ft/s

16–47. $v_A = \dfrac{-7.071\omega \sin\theta}{(1 - \cos\theta)^{\frac{1}{2}}}$;

$\quad a_A = 7.07\left[\dfrac{(\omega \sin\theta)^2}{2(1 - \cos\theta)^{\frac{3}{2}}} - \dfrac{(\alpha \sin\theta + \omega^2 \cos\theta)}{(1 - \cos\theta)^{\frac{1}{2}}}\right]$

16–49. $\omega_A = 2\left(\dfrac{\sqrt{2}\cos\theta - 1}{3 - 2\sqrt{2}\cos\theta}\right)$

16–50. $\phi = \theta$; $\omega = \dfrac{2v}{h}\cos\theta$

16–51. $v_A = 2.40$ m/s →

16–53. $\omega = 20$ rad/s; $v_A = 2$ ft/s →

16–54. $v_C = 1.64$ m/s

16–55. $v_C = 1.70$ m/s

16–57. $v_C = 5.18$ ft/s →

16–58. $\omega_{BC} = 1.07$ rad/s; $\omega_{AB} = 9.62$ rad/s

16–59. $v_C = 2.45$ m/s ↙; $\omega_{CB} = 7.81$ rad/s ↑

16–61. $v_G = 9$ m/s ←

16–62. $\omega = 5.33$ rad/s

16–63. $v_A = 8$ m/s ←

16–65. $\omega_A = 0$

16–66. $\omega_p = 6$ rad/s ↑; $\omega_A = 2$ rad/s ↑

16–67. $v_C = 1.34$ m/s ←

16–69. $v_G = 5t$ m/s →

16–70. $\omega = 6$ rad/s ↑

16–71. $v_C = 2.40$ ft/s ↑

16–73. $\omega = 4$ rad/s ↑; $v_D = 100$ mm/s ↓

16–74. $\omega_R = 11.4$ rad/s

16–75. $\omega_B = 6.67$ rad/s ↑

16–77. $v_A = 1.36$ m/s →

16–78. $v_A = 2.40$ m/s →

16–79. $v_C = 1.64$ m/s ←

16–81. $v_G = 9$ m/s ←

16–82. $\omega = 5.33$ rad/s

16–83. $v_A = 8$ m/s ←

16–85. $\theta = 63.4°$ ⤢; $v_{B/A} = 4.47$ m/s

16–87. $\omega = 1.5\dfrac{v}{r}$

16–89. $\omega_{BC} = 2.83$ rad/s; $v_B = 2.83$ ft/s; $\omega_{AB} = 2.83$ rad/s

16–90. $v_A = \omega(r_2 - r_1)$ →

16–91. $\omega_C = 26.7$ rad/s; $\omega_B = 28.75$ rad/s; $\omega_A = 14.0$ rad/s

16–93. $\omega_R = 11.4$ rad/s ↓

16–94. $\omega_{CB} = 7.81$ rad/s; $v_C = 2.45$ m/s

16–95. $\omega_{AB} = 13.1$ rad/s

16–97. $v_C = 2.50$ ft/s; $v_E = 7.91$ ft/s

16–98. $v_C = 0.897$ m/s ↗

16–99. $v_D = 0.518$ m/s ↘

16–101. $v_D = 5.72$ m/s; $\theta = 36.2°$ ↘

16–102. $v_B = 8$ ft/s ↑; $v_C = 2.93$ ft/s ↓

16–103. $\omega_{CD} = 57.7$ rad/s ↑

16–105. $\alpha = 1.47$ rad/s²; $a_B = 24.9$ ft/s² ↓

16–106. $\alpha = 0.0962$ rad/s² ↓; $a_A = 0.385$ ft/s² →

16–107. $\alpha = 0.332$ rad/s²; $a_B = 7.88$ ft/s² ←

16–109. $a_B = 13.9$ ft/s²; $\theta = 8.80°$ ⤢θ

16–110. $\alpha_{AB} = 36.2$ rad/s² ↓

16–111. $\alpha_{AB} = 0.75$ rad/s² ↑; $\alpha_{BC} = 3.94$ rad/s² ↑

16–113. $a_B = 2.25$ m/s²; $\theta = 32.6°$ ↘

16–114. $\theta = 2.02°$ ↗; $a_D = 10.0$ m/s²

16–115. $a_A = 5.94$ m/s²; $\theta = 53.9°$ ↗

16–117. $\alpha_{BC} = 5.21$ rad/s² ↑

16–118. $\alpha_{AB} = 3$ rad/s² ↓; $\theta = 59.7°$ ↘; $a_A = 13.9$ ft/s²

16–119. $\alpha_{CB} = 9.67$ rad/s² ↓; $\alpha_{DC} = 3$ rad/s² ↓

16–121. $v_C = 9.38$ in./s; $a_C = 54.7$ in./s²

16–122. $\alpha = 10.67$ rad/s² ↑; $a_D = 14.1$ in./s²

16–123. $\omega_{AC} = 0$; $\omega_F = 10.7$ rad/s; $\alpha_{AC} = 28.7$ rad/s² ↓

16–125. $v_B = 1.58\omega a$; $a_B = 1.58\alpha a - 1.77\omega^2 a$

16–126. $a_A = 1.34$ m/s²; $\theta = 26.6°$ ↘; $a_B = 1.65$ m/s²; $\alpha_{AB} = 1.5$ rad/s²

16–127. $\alpha_{AB} = 36$ rad/s² ↓

16–129. $\omega = 4.73$ rad/s ↑; $\alpha = 131$ rad/s² ↓

16–130. $\omega_{AB} = 7.17$ rad/s ↓; $\alpha_{AB} = 23.1$ rad/s² ↑

16–131. $\mathbf{a}_A = \{-5.60\mathbf{i} - 16\mathbf{j}\}$ m/s²

16–133. $\omega_{CD} = 0$; $\alpha_{CD} = 36$ rad/s²

16–134. $\mathbf{v}_B = \{-12.0\mathbf{i} + 5.77\mathbf{j}\}$ ft/s; $\mathbf{a}_B = \{-28.9\mathbf{i} - 70.0\mathbf{j}\}$ ft/s²

16–135. $\mathbf{v}_m = \{7.5\mathbf{i} - 5\mathbf{j}\}$ ft/s; $\mathbf{a}_m = \{5\mathbf{i} + 3.75\mathbf{j}\}$ ft/s²

16–137. (a) $\mathbf{a}_B = \{-1\mathbf{i}\}$ m/s²; (b) $\mathbf{a}_B = \{-1.69\mathbf{i}\}$ m/s²

16–138. (a) $\mathbf{a}_B = \{-1\mathbf{i} + 0.2\mathbf{j}\}$ m/s²; (b) $\mathbf{a}_B = \{-1.69\mathbf{i} + 0.6\mathbf{j}\}$ m/s²

16–139. $v_C = 2.40$ m/s; $a_C = \{-14.4\mathbf{j}\}$ m/s²

16–141. $\omega_{CD} = 0.750$ rad/s; $\alpha_{CD} = 1.95$ rad/s²

16–142. $\omega_{CD} = 0.866$ rad/s ↑; $\alpha_{CD} = 3.23$ rad/s² ↑

16–143. $\omega_{CD} = 10$ rad/s ↓; $\alpha_{CD} = 24$ rad/s² ↓

16–145. $\omega_{BC} = 0.720$ rad/s ↑; $\alpha_{BC} = 2.02$ rad/s² ↑

16–146. $\omega_{AC} = 0$; $\alpha_{AC} = 14.4$ rad/s² ↓

16–147. $\mathbf{v}_C = \{-7.00\mathbf{i} + 17.3\mathbf{j}\}$ ft/s; $\mathbf{a}_C = \{-34.6\mathbf{i} - 15.5\mathbf{j}\}$ ft/s²

16–149. $\mathbf{v}_C = \{-7.5\mathbf{i} - 37.0\mathbf{j}\}$ ft/s; $\mathbf{a}_C = \{85.0\mathbf{i} - 7.5\mathbf{j}\}$ ft/s²

16–150. $\omega_{CD} = 0.375$ rad/s ↓; $\alpha_{CD} = 2.44$ rad/s² ↑

Chapter 17

17–1. $I_x = \dfrac{3}{10}mr^2$

17–2. $I_z = mR^2$

17–3. $k_y = 1.56$ in.

17–5. $k_x = 57.7$ mm

17–6. $I_x = \dfrac{2}{5}mb^2$

17–7. $k_x = 1.20$ in.

17–9. $I_z = \dfrac{m}{10}a^2$

17–10. $I_y = 2.25$ slug · ft²

17–11. $I_O = 6.23$ kg · m²

17–13. $I_G = 118$ slug · ft²

17–14. $I_O = 282$ slug $\cdot$ ft^2

17–15. $I_A = 7.67$ kg $\cdot$ m^2

17–17. $I_O = \dfrac{1}{2}ma^2$

17–18. $I_A = 1.58$ slug $\cdot$ ft^2

17–19. $k_O = 3.15$ ft

17–21. $I_{x'} = 7.20(10^{-3})$ kg $\cdot$ m^2

17–22. $I_x = 5.64$ slug $\cdot$ ft^2

17–23. $\bar{y} = 0.888$ m; $I_G = 5.61$ kg $\cdot$ m^2

17–25. $a_G = 6.78$ m/s^2

17–26. $a_G = 6.04$ ft/s^2

17–27. $F_{CD} = 289$ kN

17–29. $N_A = 568$ N; $N_B = 544$ N

17–30. $a = 3.96$ m/s^2

17–31. $N_B = 95.0$ lb; $N_A = 105$ lb; $s = 9.66$ ft

17–33. $F_{CB} = 41.9$ kN; $A_x = 41.9$ kN;
$A_y = 28.3$ kN

17–34. $T = 1.52$ kN; $\theta = 18.6°$

17–35. $T = 2.38$ kN; $a = 1.33$ m/s^2

17–37. $T = 15.7$ kN; $C_x = 8.92$ kN;
$C_y = 16.3$ kN

17–38. For rear-wheel drive: $t = 17.5$ s;
For 4-wheel drive: $t = 11.3$ s

17–39. $P_{\max} = \dfrac{mgb}{2\left(d - \frac{h}{2}\right)}$

17–41. $R = 33.3$ lb

17–42. $h = 0.2$ m: $a_G = 1.41$ m/s^2;
$h = 0$: $a_G = 1.38$ m/s^2

17–43. $a = 96.6$ ft/s^2

17–45. $s = 82.8$ ft

17–46. $M = 6.74$ kN $\cdot$ m; $\mu_{\min} = 1.64$

17–47. Deceleration: $a_G = 2.26$ m/s^2;
$N_B = 226$ N; $N_A = 559$ N
Equilibrium: $N_A = 330$ N; $N_B = 454$ N

17–49. $a_G = 3.22$ ft/s^2; $N_B = 14.7$ lb;
$N_A = 65.3$ lb

17–50. $P = 53.3$ lb

17–51. $\alpha = 9.67$ rad/s^2

17–53. $A_x = 0$; $A_y = 262$ N

17–54. $\omega = 56.2$ rad/s; $A_x = 0$;
$A_y = 98.1$ N

17–55. $\omega = 20.8$ rad/s

17–57. $O_x = 142$ lb; $O_y = 4.29$ lb

17–58. $O_x = 54.8$ N; $O_y = 41.2$ N

17–59. $\omega = 10.9$ rad/s

17–61. $\alpha = 7.28$ rad/s^2

17–62. $\alpha = -\dfrac{2\,\mu_k}{mr}\left(\dfrac{mg - \mu_k P}{\cos\theta(1 + \mu_k{}^2)}\right) + \dfrac{2P}{mr}$

17–63. (a) $F_O = 28.0$ N; (b) $F_O = 91.1$ N

17–66. $r_P = 2.67$ ft; $A_x = 0$

17–67. $\alpha = 14.7$ rad/s^2; $a_G = 4.90$ m/s^2

17–69. $\omega = 96.6$ rad/s

17–70. $\alpha = 3.60$ rad/s^2; $A_x = 163$ N;
$A_y = 214$ N

17–71. $\omega = 380$ rad/s

17–73. $F_{CB} = 193$ N; $t = 3.11$

17–74. $\alpha = 891$ rad/s^2

17–75. $a = \dfrac{g(M_B - M_A)}{\frac{1}{2}M + M_B + M_A}$

17–77. $M = 0.3gml$

17–78. $M = 0.233$ lb $\cdot$ ft

17–79. $A_x = 89.2$ N; $A_y = 66.9$ N; $t = 1.26$ s

17–81. $\theta = 29.8°$; Slipping occurs first.

17–82. $\omega = 800$ rad/s

17–83. $N = wx\left[\dfrac{\omega^2}{g}\left(L - \dfrac{x}{2}\right) + \cos\theta\right]$; $V = wx\sin\theta$;
$M = \dfrac{1}{2}wx^2\sin\theta$

17–85. $a = \dfrac{8}{25}(2 - \mu_k)g$

17–86. Point A has the greatest speed; $\Delta v = 0.267\sqrt{gl}$

17–89. $a = 0.0769$ m/s^2; $\alpha = 0.741$ rad/s^2

17–90. $\alpha = 0.274$ rad/s^2; $a_B = 49.0$ ft/s^2;
$\theta = 80.3°$ $\theta\searrow$

17–91. $\alpha = 8.89$ rad/s^2

17–93. $\alpha = 1.25$ rad/s; $T = 2.32$ kN

17–94. $T = 3.13$ kN; $\alpha = 1.68$ rad/s;
$a_C = 1.35$ m/s^2

17–95. $\alpha = 4.35$ rad/s^2

17–97. $\alpha = 0.692$ rad/s^2

17–98. $\alpha = 1.15$ rad/s^2

17–99. $\alpha = 4.65$ rad/s^2; $T = 359$ N

17–101. $a_A = 167$ ft/s^2

17–102. $\alpha = 5.01$ rad/s^2

17–103. $\alpha = 4.18$ rad/s^2; $T_{AB} = 43.3$ lb

17–105. $\alpha = \dfrac{6P}{mL}$; $a_B = \dfrac{2P}{m}$

17–106. $T = 90$ lb; $\alpha = 0.229$ rad/s^2

17–107. $F_T = 40.2$ lb; $N_T = 329$ lb

17–109. $A_y = 15.0$ lb; $A_x = 0.466$ lb; $a_G = 1.00$ ft/s^2

17–110. $\alpha = 6.21$ rad/s^2

17–111. $\alpha = 13.4$ rad/s^2

17–113. $(a_G)_x = 1.82$ m/s^2; $(a_G)_y = 1.69$ m/s^2; $\alpha = 0.283$ rad/s^2

Chapter 18

18–2. $\omega = 14.1$ rad/s

18–3. $T = 210$ ft·lb

18–5. $T = 0.0188$ ft·lb

18–6. $T_T = 10.4$ ft·lb

18–7. $T_T = 3.82$ ft·lb

18–9. $\omega = 2.02$ rad/s

18–10. $\omega = 6.92$ rad/s

18–11. $s = 0.304$ ft

18–13. $\omega_2 = 2.83$ rad/s

18–14. $v = 9.42$ m/s

18–15. $\omega = 0.0636$ rad/s

18–17. $\omega = 11.2$ rad/s

18–18. $v = 4.97$ m/s

18–19. $v = 5.34$ m/s

18–21. $\omega = 7.08$ rad/s

18–22. $\omega = 11.0$ rad/s

18–23. $s = 2.00$ m

18–25. $\omega_{AB} = 5.92$ rad/s

18–26. $\omega = 1.32$ rad/s

18–27. $T = 283$ ft·lb

18–29. $v = 6.41$ ft/s

18–30. $\omega_{AB} = 4.28$ rad/s

18–31. $\theta_0 = 1.66$ rad

18–33. $v_G = 3\sqrt{\dfrac{3}{7}gR}$

18–34. $v_A = 14.2$ ft/s

18–35. $\omega = 2.83$ rad/s

18–37. $\omega_{AB} = 4.28$ rad/s

18–38. $s = 0.304$ ft

18–39. $v_G = 14.2$ ft/s

18–41. $s = 0.301$ m; $T = 163$ N

18–42. $k = 42.8$ N/m

18–43. $\omega = 9.48$ rad/s; $A_x = 20.9$ lb; $A_y = 5.00$ lb

18–45. $\omega = 2.03$ rad/s

18–46. $\theta = 39.3°$

18–47. $v_A = 1.40$ m/s

18–49. $v_A = 11.9$ ft/s

18–50. $v_C = 1.52$ m/s

18–51. $\theta = 70.5°$

18–53. $\omega = 2.82$ rad/s

18–54. $\omega = 41.8$ rad/s

18–55. $\omega = 39.3$ rad/s

18–57. $\omega = 5.28$ rad/s

18–58. $\omega_{AB} = 2.21$ rad/s

18–59. $k = 100$ lb/ft

Chapter 19

19–5. $\omega = 20.8$ rad/s

19–6. $\omega = 56.2$ rad/s; $A_x = 0$; $A_y = 98.1$ N

19–7. $\omega = 96.6$ rad/s

19–9. $T_{BC} = 193$ N; $t = 3.11$ s

19–10. $\omega = 18.7$ rad/s

19–11. $\omega_2 = 245$ rad/s

19–13. $\omega = 6.04$ rad/s

19–14. $\omega_B = 3.56$ rad/s; $t = 5.12$ s

19–15. $v_G = 2$ m/s; $\omega = 3.90$ rad/s

19–17. $\omega = 27.9$ rad/s

19–18. $v_A = 24.1$ m/s

19–19. $\omega = 12.7$ rad/s

19–21. $\omega_A = 1.70$ rad/s; $\omega_B = 5.10$ rad/s

19–22. (a) $\omega_{BC} = 68.7$ rad/s; (b) $\omega_{BC} = 66.8$ rad/s; (c) $\omega_{BC} = 68.7$ rad/s

19–23. $v_B = 34.0$ ft/s

19–25. $\omega = 119$ rad/s; $v_G = 64.5$ ft/s; $\theta = 22.5°$ ∡θ

19–26. $\theta = \tan^{-1}(3.5\,\mu)$

19–27. $v_B = 4.26$ ft/s

19–29. $h = 0.533$ m

19–30. $y = \dfrac{\sqrt{2}}{3} a$

19–31. $h = \dfrac{7}{5} r$

19–33. $v_0 = \sqrt{(2gd \sin \theta)\left(\dfrac{m_c}{m}\right)}$

19–34. $\omega = \dfrac{m_A k_A^2 \omega_A + m_B k_B^2 \omega_B}{m_A k_A^2 + m_B k_B^2}$

19–35. $\omega_T = 1.19(10^{-3})$ rad/s

19–37. $\omega' = \dfrac{11}{3} \omega_1$

19–38. $\omega' = \dfrac{19}{28} \omega$; $v'' = 0.985\omega L$

19–39. (a) $\omega_M = 0$; (b) $\omega_M = \dfrac{I}{I_z} \omega$; (c) $\omega_M = \dfrac{2I}{I_z} \omega$

19–41. $\omega = 0.0906$ rad/s

19–42. $\omega = 22.7$ rad/s

19–43. $\omega_2 = \dfrac{1}{3} \omega_1$

19–45. $\omega = 6.45$ rad/s

19–46. $\omega = 0.175$ rad/s

19–47. $\omega = \dfrac{1}{4} \omega_0$

19–49. $\omega = 3.23$ rad/s; $\theta = 32.8°$

19–50. $\theta = 22.4°$

19–51. $h_B = 0.980$ ft

19–53. $h_C = 0.500$ ft

19–54. $\theta = 66.9°$

19–55. $\theta = \tan^{-1}\left(\sqrt{\dfrac{7}{5} e}\right)$

Review 2

R2–1. $\omega_P = 20$ rad/s; $\omega_D = 6.67$ rad/s

R2–2. $\omega_P = 24$ rad/s; $\omega_D = 5.33$ rad/s

R2–3. $\omega_2 = 3.82$ rad/s

R2–5. $d = 2$ ft; $\omega_2 = 6.82$ rad/s

R2–6. $a_A = 12.5$ m/s^2 ←

R2–7. $v_A = 45$ ft/s →

R2–9. $v_2 = 3.46$ m/s

R2–10. $\omega_2 = 13.3$ rad/s

R2–11. $\omega_F = 12$ rad/s ↓

R2–13. $\omega = 0$

R2–14. $\omega = 0$; No friction force develops.

R2–15. $\omega_{DE} = 132$ rad/s

R2–17. $t = 1.32$ s

R2–18. $\omega = 2.19$ rad/s ↖

R2–19. $\omega_{CD} = 4.17$ rad/s

R2–21. $v_A = 4.00$ m/s

R2–22. $a = 53.5$ ft/s^2; $\mu_s = 1.66$

R2–23. $a = 4.94$ ft/s^2

R2–25. 28.6 rad/s; 24.1 rad; 7.16 m/s; 205 m/s^2

R2–26. $\alpha = \dfrac{2mg}{R(M + 2m)}$; $h = \dfrac{mg}{M + 2m} t^2$

R2–27. $\omega = 3.89$ rad/s

R2–29. $\omega_2 = 12.7$ rad/s

R2–30. $F = 2.5mg \sin \theta$; $N = 0.25mg \cos \theta$; $\theta = \tan^{-1}(0.1 \mu_s)$

R2–31. $(\omega_S)_2 = \dfrac{5g \sin \theta}{7r} t$; $(\omega_C)_2 = \dfrac{2g \sin \theta}{3r} t$

R2–33. $\omega_{AB} = 1.47$ rad/s ↖; $\alpha_{AB} = 4.93$ rad/s^2 ↙

R2–34. $\alpha = 2.66$ rad/s^2 ↙

R2–35. $\omega = 1.08$ rad/s; $v_B = 4.39$ ft/s

R2–37. $\alpha = 73.3$ rad/s^2; $t = 0.296$ s

R2–38. $\omega = 30.7$ rad/s

R2–39. $\omega = 0.0708$ rad/s

R2–41. $\omega_{AB} = 3.72$ rad/s

R2–42. $\theta = 4.45°$

R2–43. $a_A = 56.2$ ft/s^2 ↓; $a_B = 40.2$ ft/s^2; $\theta = 53.3°$ ↗

R2–45. $\alpha_D = 2.09$ rad/s^2; $\theta_D = 0.667$ rev

R2–46. $a_A = 282$ ft/s^2

R2–47. $t = 10.4$ s

R2–49. $A_x = 89.2$ N; $A_y = 66.9$ N; $t = 1.26$ s

R2–50. $a_G = 0.3$ m/s^2; $N_B = 7.09$ kN; $N_A = 5.43$ kN

Chapter 20

20–1. $v_A = \{-5.20\mathbf{i} - 12\mathbf{j} + 20.8\mathbf{k}\}$ ft/s; $a_A = \{-24.1\mathbf{i} - 13.3\mathbf{j} - 7.20\mathbf{k}\}$ ft/s^2

20–2. $v_A = \{-5.20\mathbf{i} - 12\mathbf{j} + 20.8\mathbf{k}\}$ ft/s; $a_A = \{-3.33\mathbf{i} - 21.3\mathbf{j} + 6.66\mathbf{k}\}$ ft/s^2

20–3. $\mathbf{v}_A = \{0.0828\mathbf{i} - 0.464\mathbf{j} + 0.124\mathbf{k}\}$ m/s;
$\mathbf{a}_A = \{-0.371\mathbf{i} - 0.108\mathbf{j} - 0.278\mathbf{k}\}$ m/s^2

20–5. $\mathbf{v}_P = \{-0.800\mathbf{i} + 0.400\mathbf{j} + 0.800\mathbf{k}\}$ m/s;
$\mathbf{a}_P = \{-10.3\mathbf{i} - 3.84\mathbf{j} + 0.320\mathbf{k}\}$ m/s^2

20–6. $\boldsymbol{\omega} = \{-8.24\mathbf{j}\}$ rad/s;
$\boldsymbol{\alpha} = \{24.7\mathbf{i} - 5.49\mathbf{j}\}$ rad/s^2

20–7. $\mathbf{v}_A = \{-7.79\mathbf{i} - 2.25\mathbf{j} + 3.90\mathbf{k}\}$ ft/s;
$\mathbf{a}_A = \{8.30\mathbf{i} - 35.2\mathbf{j} + 7.02\mathbf{k}\}$ ft/s^2

20–9. $\mathbf{v}_B = \{-7.06\mathbf{i} - 7.52\mathbf{k}\}$ ft/s;
$\mathbf{a}_B = \{77.3\mathbf{i} - 28.3\mathbf{j} - 5.64\mathbf{k}\}$ ft/s^2

20–10. $\omega_s = 40$ rad/s; (Angular velocity of C about DE); $\omega_p = 5$ rad/s (Angular velocity of DE about y axis)

20–11. $\boldsymbol{\omega}_P = \{-40\mathbf{j}\}$ rad/s; $\boldsymbol{\alpha} = \{-6400\mathbf{i}\}$ rad/s^2

20–13. $\boldsymbol{\omega} = \{4.35\mathbf{i} + 12.7\mathbf{j}\}$ rad/s;
$\boldsymbol{\alpha} = \{-26.1\mathbf{k}\}$ rad/s^2

20–14. $\mathbf{v}_A = \{-1.80\mathbf{i}\}$ ft/s;
$\mathbf{a}_A = (-0.720\mathbf{j} - 0.831\mathbf{k})$ ft/s^2

20–15. $\mathbf{v}_A = \{-1.80\mathbf{i}\}$ ft/s;
$\mathbf{a}_A = \{-0.750\mathbf{i} - 0.720\mathbf{j} - 0.831\mathbf{k}\}$ ft/s^2

20–17. $\omega_A = 47.8$ rad/s; $\omega_B = 7.78$ rad/s

20–18. $\boldsymbol{\omega}_{BD} = \{-2.00\mathbf{i}\}$ rad/s

20–19. $\boldsymbol{\omega}_{BD} = \{-2.00\mathbf{i}\}$ rad/s; $\boldsymbol{\alpha}_{BD} = \{31.6\mathbf{i}\}$ rad/s^2

20–21. $\mathbf{a}_B = \{-96.5\mathbf{i}\}$ ft/s^2

20–22. $\mathbf{v}_A = \{0.667\mathbf{i}\}$ ft/s; $\mathbf{a}_A = \{-0.148\mathbf{i}\}$ ft/s^2

20–23. $\mathbf{v}_A = \{0.66667\mathbf{i} + 2\mathbf{j}\}$ ft/s; $\mathbf{a}_A = \{1.19\mathbf{i}\}$ ft/s^2

20–25. $\boldsymbol{\omega}_{BC} = \{0.769\mathbf{i} - 2.31\mathbf{j} + 0.513\mathbf{k}\}$ rad/s;
$\mathbf{v}_B = \{-0.333\mathbf{j}\}$ m/s

20–26. $v_B = 9.43$ ft/s

20–27. $\boldsymbol{\omega} = \{1.33\mathbf{i} + 2.67\mathbf{j} + 3.33\mathbf{k}\}$ rad/s;
$\mathbf{v}_B = \{3.00\mathbf{j}\}$ m/s

20–29. $\boldsymbol{\omega} = \{-0.440\mathbf{i} + 0.293\mathbf{j} - 0.238\mathbf{k})$ rad/s;
$\mathbf{v}_B = \{-2.82\mathbf{i} + 3.76\mathbf{j}\}$ ft/s

20–30. $\boldsymbol{\alpha} = \{0.413\mathbf{i} + 0.622\mathbf{j} - 0.412(10^{-3})\mathbf{k}\}$ rad/s^2;
$\mathbf{a}_B = \{-5.98\mathbf{i} + 7.98\mathbf{j}\}$ ft/s^2

20–31. $\mathbf{v}_C = \{-1.00\mathbf{i} + 5.00\mathbf{j} + 0.800\mathbf{k}\}$ m/s;
$\mathbf{a}_C = \{-28.8\mathbf{i} - 5.45\mathbf{j} + 32.3\mathbf{k}\}$ m/s^2

20–33. $\mathbf{v}_A = \{5.20\mathbf{i} - 5.20\mathbf{j} - 3.00\mathbf{k}\}$ ft/s;
$\mathbf{a}_A = \{25\mathbf{i} - 26.8\mathbf{j} + 8.78\mathbf{k}\}$ ft/s^2

20–34. $\mathbf{v}_C = \{-6.75\mathbf{i} - 6.25\mathbf{j}\}$ m/s;
$\mathbf{a}_C = \{28.75\mathbf{i} - 26.25\mathbf{j} - 4\mathbf{k}\}$ m/s^2

20–35. $\mathbf{v}_B = \{5\mathbf{i} - 0.5\mathbf{j} + 6.4\mathbf{k}\}$ m/s;
$\mathbf{a}_B = \{-0.25\mathbf{i} - 7.12\mathbf{j}\}$ m/s^2

20–37. $\mathbf{v}_A = \{-4.85\mathbf{i} + 0.800\mathbf{j} - 1.39\mathbf{k}\}$ m/s;
$\mathbf{a}_A = \{-14.0\mathbf{i} - 35.5\mathbf{j} - 3.68\mathbf{k}\}$ m/s^2

20–38. $\mathbf{v}_B = \{-10.2\mathbf{i} - 30\mathbf{j} + 52.0\mathbf{k}\}$ m/s;
$\mathbf{a}_B = \{-31.0\mathbf{i} - 161\mathbf{j} - 90\mathbf{k}\}$ m/s^2

20–39. $\mathbf{v}_B = \{-10.2\mathbf{i} - 28\mathbf{j} + 52.0\mathbf{k}\}$ m/s;
$\mathbf{a}_B = \{-33.0\mathbf{i} - 159\mathbf{j} - 90\mathbf{k}\}$ m/s^2

20–41. $\mathbf{v}_P = \{-25.5\mathbf{i} - 13.4\mathbf{j} + 20.5\mathbf{k}\}$ ft/s;
$\mathbf{a}_P = \{161\mathbf{i} - 249\mathbf{j} - 39.6\mathbf{k}\}$ ft/s^2

20–42. $\mathbf{v}_P = \{-25.5\mathbf{i} - 13.4\mathbf{j} + 20.5\mathbf{k}\}$ ft/s;
$\mathbf{a}_P = \{161\mathbf{i} - 243\mathbf{j} - 33.9\mathbf{k}\}$ ft/s^2

20–43. $\mathbf{v}_P = \{-10.5\mathbf{i} + 9.00\mathbf{j} - 1.00\mathbf{k}\}$ ft/s;
$\mathbf{a}_P = \{-41.0\mathbf{i} - 17.5\mathbf{j} - 0.500\mathbf{k}\}$ ft/s^2

20–45. $\mathbf{v}_C = \{2.80\mathbf{j} - 5.60\mathbf{k}\}$ m/s;
$\mathbf{a}_C = \{-56\mathbf{i} + 2.1\mathbf{j}\}$ m/s^2

20–46. $\mathbf{v}_C = \{2.80\mathbf{j} - 5.60\mathbf{k}\}$ m/s;
$\mathbf{a}_C = \{-56\mathbf{i} + 2.1\mathbf{j} - 1.40\mathbf{k}\}$ m/s^2

20–47. $\mathbf{v}_G = \{-16\mathbf{i} + 17\mathbf{j} + 9\mathbf{k}\}$ m/s;
$\mathbf{a}_G = \{-136\mathbf{i} - 133\mathbf{j} + 75\mathbf{k}\}$ m/s^2

Chapter 21

21–2. $I_{aa} = \dfrac{m}{12}(3a^2 + 4h^2)$

21–3. $I_x = 53.3$ slug $\cdot$ ft^2; $I_y = 26.7$ slug $\cdot$ ft^2

21–5. $I_y = 4.48(10^3)\rho$

21–6. $I_{yz} = \dfrac{m}{6}ah$

21–7. $I_{xy} = \dfrac{m}{12}a^2$

21–9. $I_{x'} = \dfrac{m}{12}(a^2 + h^2)$

21–10. $\begin{pmatrix} \frac{2}{3}ma^2 & \frac{1}{4}ma^2 & \frac{1}{4}ma^2 \\ \frac{1}{4}ma^2 & \frac{2}{3}ma^2 & -\frac{1}{4}ma^2 \\ \frac{1}{4}ma^2 & -\frac{1}{4}ma^2 & \frac{2}{3}ma^2 \end{pmatrix}$

21–11. $I_z = 0.429$ kg $\cdot$ m^2

21–13. $\bar{y} = 0.5$ ft; $\bar{x} = -0.667$ ft;
$I_{x'} = 0.0272$ slug $\cdot$ ft^2; $I_{y'} = 0.0155$ slug $\cdot$ ft^2;
$I_{z'} = 0.0427$ slug $\cdot$ ft^2

21–14. $I_x = 1.36 \text{ kg} \cdot \text{m}^2$; $I_y = 0.380 \text{ kg} \cdot \text{m}^2$;
$I_z = 1.26 \text{ kg} \cdot \text{m}^2$

21–15. $I_x = 0.0292 \text{ slug} \cdot \text{ft}^2$

21–17. $I_{aa} = 1.13 \text{ slug} \cdot \text{ft}^2$

21–18. $I_z = 0.0915 \text{ kg} \cdot \text{m}^2$

21–19. $I_z = 16.3 \text{ slug} \cdot \text{ft}^2$; $I_x = 20.2 \text{ slug} \cdot \text{ft}^2$;
$I_y = 20.2 \text{ slug} \cdot \text{ft}^2$

21–23. $\boldsymbol{\omega} = \{-0.750\mathbf{j} + 1.00\mathbf{k}\} \text{ rad/s}$

21–25. $T_2 = 15.5 \text{ ft} \cdot \text{lb}$

21–26. $\omega_x = 19.7 \text{ rad/s}$

21–27. $h = 2.24 \text{ in.}$

21–29. $\omega_{OA} = 17.3 \text{ rad/s}$

21–30. $\mathbf{v}_G = \{-10.2\mathbf{k}\} \text{ ft/s}$

21–31. $\mathbf{u}_{IA} = 0.141\mathbf{j} - 0.990\mathbf{k}$; $\int \mathbf{F}_O \, dt = \{8.57\mathbf{i}\} \text{ N} \cdot \text{s}$

21–33. $\omega_2 = 61.7 \text{ rad/s}$

21–34. $\omega_2 = 87.2 \text{ rad/s}$

21–35. $\omega_{AB} = 21.4 \text{ rad/s}$

21–37. $\omega = 58.4 \text{ rad/s}$

21–38. $\boldsymbol{\omega} = \{-0.107\mathbf{i} - 0.107\mathbf{k}\} \text{ rad/s}$

21–39. $\Sigma M_x = \dfrac{d}{dt}(I_x\omega_x - I_{xy}\omega_y - I_{xz}\omega_z)$
$\qquad - \Omega_z(I_y\omega_y - I_{yz}\omega_z - I_{yx}\omega_x)$
$\qquad + \Omega_y(I_z\omega_z - I_{zx}\omega_x - I_{zy}\omega_y)$

21–41. $\Sigma M_x = I_x\dot{\omega}_x - I_y\Omega_z\omega_y + I_z\Omega_y\omega_z$

21–42. $F_A = 277 \text{ N}$; $F_B = 166 \text{ N}$

21–43. $F_A = 213 \text{ N}$; $F_B = 128 \text{ N}$

21–45. $B_z = 4 \text{ lb}$; $A_x = -2.00 \text{ lb}$; $A_y = 0.627 \text{ lb}$;
$B_x = 2.00 \text{ lb}$; $B_y = -1.37 \text{ lb}$

21–46. $\theta = \cos^{-1}\left(\dfrac{3g}{2L\omega^2}\right)$

21–47. $F_A = 8.83 \text{ lb}$; $F_B = 11.2 \text{ lb}$

21–49. $\dot{\omega}_y = 25.9 \text{ rad/s}^2$; $B_x = -0.0791 \text{ N}$;
$B_z = 12.3 \text{ N}$; $A_x = -1.17 \text{ N}$; $A_z = 12.3 \text{ N}$

21–50. $\dot{\omega}_y = 25.9 \text{ rad/s}^2$; $B_x = -0.0791 \text{ N}$;
$B_z = 12.3 \text{ N}$; $A_x = -1.17 \text{ N}$; $A_z = 12.3 \text{ N}$

21–51. $B_x = -12.5 \text{ lb}$; $B_y = -46.4 \text{ lb}$; $A_x = -15.6 \text{ lb}$;
$A_y = -46.8 \text{ lb}$; $A_z = 50 \text{ lb}$

21–53. $\theta = 70.8°$

21–54. $\Sigma M_x = 0$; $\Sigma M_y = (-0.036 \sin \theta) \text{ N} \cdot \text{m}$;
$\Sigma M_z = (0.003 \sin 2\theta) \text{ N} \cdot \text{m}$

21–55. $A_x = B_x = 0$; $A_y = -1.09 \text{ kN}$;
$B_y = 1.38 \text{ kN}$

21–57. $\dot{\omega}_y = -102 \text{ rad/s}^2$; $A_x = B_x = 0$;
$A_y = 0$; $A_z = 297 \text{ N}$; $B_z = -143 \text{ N}$

21–58. $A_x = 0$; $A_y = 4.89 \text{ lb}$; $A_z = 5.47 \text{ lb}$;
$M_x = -8.43 \text{ lb} \cdot \text{ft}$; $M_y = 0$; $M_z = 0$

21–59. $\theta_D = 139°$; $m_D = 0.661 \text{ kg}$;
$\theta_F = 40.9°$; $m_F = 1.32 \text{ kg}$

21–62. $\alpha = 69.3°$; $\beta = 127.8°$; $\gamma = 45°$; No

21–63. (a) $A_y = 1.49 \text{ kN}$; $B_y = 2.43 \text{ kN}$
(b) $A_y = -1.24 \text{ kN}$; $B_y = 5.17 \text{ kN}$
(c) $A_y = 1.49 \text{ kN}$; $B_y = 2.43 \text{ kN}$

21–65. $M_x = 328 \text{ N} \cdot \text{m}$

21–66. $A_y = 13.7 \text{ kN}$; $B_y = -13.7 \text{ kN}$

21–67. $M_x = 12.4 \text{ lb} \cdot \text{ft}$; $M_y = 0$; $M_z = 0$

21–69. $\dot{\phi} = \left(\dfrac{2g \text{ ctn } \theta}{a + r \cos \theta}\right)^{1/2}$

21–70. $\omega_p = 1.21 \text{ rad/s}$ (Low precession);
$\omega_p = 76.3 \text{ rad/s}$ (High precession)

21–71. $\omega_p = 1.19 \text{ rad/s}$

21–73. $\omega_R = 368 \text{ rad/s}$

21–74. $\Delta F = 53.4 \text{ N}$

21–77. $H_G = 6.09(10^{-3}) \text{ kg} \cdot \text{m}^2/\text{s}$

21–78. $H_G = 12.5 \text{ Mg} \cdot \text{m}^2/\text{s}$

21–79. $\phi = 12.8 \text{ rad/s}$; This is a case of retrograde precession since $I_z > I$.

Chapter 22

22–1. $\tau = 0.452 \text{ s}$; $f = 2.21 \text{ Hz}$

22–2. $x = -0.05 \cos(12.2t) \text{ m}$

22–3. $f = 7.88 \text{ Hz}$; $\tau = 0.127 \text{ s}$

22–5. $f = 3.13 \text{ Hz}$; $C = 1 \text{ in.}$;
$y = (0.0833 \cos 19.7t) \text{ ft}$

22–6. $y = [-1.44 \sin(13.9t) - 0.167 \cos(13.9t)] \text{ ft}$;
$C = 1.45 \text{ ft}$

22–7. $\ddot{y} = -56.1y$; $y = 192 \text{ mm}$

22–9. $f = \dfrac{1}{\pi}\sqrt{\dfrac{k}{m}}$

22–10. $\theta = -0.101 \sin(4.95t) + 0.3 \cos(4.95t)$

22–11. $k = 1.36 \text{ N/m}$; $m_B = 3.58 \text{ kg}$; $m_p = 0.759 \text{ kg}$

22–13. $\tau = 2\pi\sqrt{\dfrac{k_G^2 + d^2}{gd}}$

22–14. $C = 16°$

22–15. $\tau = 1.18$ s

22–17. $\theta = 0.4\cos(29.3t)$

22–18. $\tau = 0.167$ s

22–19. $\tau = 2\pi\left(\dfrac{a}{b}\right)\sqrt{\dfrac{m}{k}}$

22–21. $\tau = 1.45$ s

22–22. $\tau = 2.67$ s

22–23. $f = 0.597$ Hz

22–25. $\tau = 0.910$ s

22–26. $\tau = 2\pi\sqrt{\dfrac{(k_G^2 + d^2)}{gd}}$

22–27. $\tau = 1.18$ s

22–29. $\tau = 0.401$ s

22–30. $\tau = 2\pi\sqrt{\dfrac{m}{3k}}$

22–31. $\ddot{x} + 333x = 0$

22–33. $\tau = 0.339$ s

22–34. $f = 0.851$ Hz

22–35. $\tau = 2.81$ s

22–37. $f = 1.28$ Hz

22–38. $f = 1.81$ Hz

22–39. $f = 2.25$ Hz

22–41. $y = A\sin\omega_n t + B\cos\omega_n t + \dfrac{\frac{k\delta_0}{m}}{\left(\frac{k}{m} - \omega_0^2\right)}\cos\omega_0 t$

22–42. $v_{max} = 0.685$ ft/s

22–43. $y = \{-0.0232\sin 8.97t + 0.333\cos 8.97t + 0.0520\sin 4t\}$ ft

22–45. $(x_p)_{max} = 29.5$ mm

22–46. $\ddot{y} + \omega_n^2 y = \dfrac{k\delta_0}{m}\sin\omega_0 t$;

$y = A\sin\omega_n t + B\cos\omega_n t + \dfrac{\delta_0}{1 - \left(\frac{\omega_0}{\omega_n}\right)^2}\sin\omega_0 t$

22–47. $y = (361\sin 7.75t + 100\cos 7.75t - 350\sin 8t)$ mm

22–49. $(x_p)_{max} = 1.89$ in.

22–50. $v_R = 1.20$ m/s

22–51. $(x_p)_{max} = 4.53$ mm

22–53. $C = 0.490$ in.

22–54. $\omega = 19.0$ rad/s

22–55. $x_p = (0.0276\sin 2t)$ ft

22–57. $y = 0.622[e^{-0.939t}\sin(11.9t + 1.49)]$

22–58. $\phi' = 9.89°$

22–59. MF $= 0.997$

22–61. $y = \{-0.0702[e^{-3.57t}\sin(8.54t)]\}$ m

22–63. $\ddot{y} + 16\dot{y} + 4y = 0$; Since $c > c_c$, the system will not vibrate. Therefore, it is **overdamped**.

22–65. $L\ddot{q} + R\dot{q} + \left(\dfrac{2}{C}\right)q = 0$

22–66. $t = 3.99$ s

Index

A

Absolute acceleration, 84, 351
Absolute dependent–motion analysis of particles, 77–82, 99
Absolute motion analysis, 319–322, 382
Absolute velocity, 83
Acceleration, 7–8, 32, 34, 50–51, 52, 64, 66, 84, 289, 305–307, 351–359, 369–370, 537–538, 554
    absolute, 84, 351
    angular, 64, 305, 537–358
    average, 7
    centripetal, 51
    constant, 8, 289
    constant angular, 305
    Coriolis, 370
    curved path, 351, 353
    curvilinear motion, 32, 34, 50–51, 52, 64
    cylindrical components, 64, 66
    deceleration, 7
    instantaneous, 7
    kinematic diagrams, 351, 353
    normal, 50–51, 52, 307
    particles, 7–8, 32, 34, 50–51, 52, 64, 66, 84
    procedure for analysis of, 353
    rectilinear motion, 7–8, 289
    rectilinear path, 351
    relative, 84, 351–352
    relative–acceleration equations, 351–352, 554
    relative–motion analysis, 351–359, 369–370, 554
    review of, 289
    rotating axis, 369–370
    rotational motion, 305–307, 537–538
    scalar analysis, 353
    tangential, 50–51, 52, 307
    three–dimensional motion, 537–538, 554
    variable, 289
    vector analysis, 353
Amplitude, 622
Analysis, 77–82, 83–87, 99, 290, 319–322, 327–334, 351–359, 366–374, 382–383, 552–559, 565, 658–666, 667–670
    absolute dependent–motion of particles, 77–82, 99
    absolute motion, 319–322, 382
    numerical and computer analysis, 658–666
    relative–motion, 83–87, 99, 327–334, 351–359, 366–374, 383, 552–559, 565
    scalar, 330, 353
    vector, 330, 353, 667–670
Angular acceleration/ deceleration, 64, 305, 357–358
Angular displacement, 304
Angular impulse, 260–265, 287, 489–490, 491, 518
    angular momentum and, principle of, 260–265, 489–490, 491, 518
    defined, 260
    free–body diagram, 262, 491
    kinematics for, 491
    principles of, 260–265, 287
    procedure for analysis of, 262, 491
    resultant equations, 490
    scalar formulation, 261, 490
    vector formulation, 261
Angular momentum, 256, 257–260, 260–265, 287, 483–488, 489–496, 504–507, 518–519, 577–580
    angular impulse and, principle of, 260–265, 489–490, 491, 518
    arbitrary point, 578
    center of mass, 578
    conservation of, 262, 504–507, 519
    defined, 256
    fixed point, 578
    free–body diagram, 262, 491, 505
    general plane motion, 487
    kinematics for, 491, 505
    moment of a force, relation between, 257–260
    particles, 256, 257–260, 260–265, 287
    principle axes of inertia, 579
    principle of impulse and momentum, 580
    principles of, 260–265, 287
    procedure for analysis of, 262, 491, 505
    rectangular components, 578–579
    resultant equations, 490
    right–hand rule, 256
    rigid–body planar motion, 483–488, 489–496, 504–507, 518–519
    rotation about a fixed axis, 486
    scalar formulation, 256, 490
    system of particles, 258
    three–dimensional motion, 577–580
    translation, 486
    vector formulation, 256
Angular motion, 304–305, 309
Angular position, 304
Angular velocity, 63, 304, 537
Apogee, defined, 155
Arbitrary axes, moment of inertia about a, 517
Arbitrary point, angular momentum of, 578
Areal velocity, 150
Average acceleration, 7
Average impulse force, 231